Energy Storage and Conversion Systems

Energy Storage and Conversion Systems

Editors

Alon Kuperman
Alessandro Lampasi

MDPI

Basel • Beijing • Wuhan • Barcelona • Belgrade • Novi Sad • Cluj • Manchester

Editors

Alon Kuperman
Ben-Gurion University of the Negev
Beer-Sheva, Israel

Alessandro Lampasi
ENEA
Frascati, Italy

Editorial Office
MDPI
St. Alban-Anlage 66
4052 Basel, Switzerland

This is a reprint of articles from the Topic published online in the open access journals *Energies* (ISSN 1996-1073), *Electronics* (ISSN 2079-9292), and *Processes* (ISSN 2227-9717) (available at: https://www.mdpi.com/topics/energy).

For citation purposes, cite each article independently as indicated on the article page online and as indicated below:

Lastname, A.A.; Lastname, B.B. Article Title. *Journal Name* **Year**, *Volume Number*, Page Range.

ISBN 978-3-0365-9322-7 (Hbk)
ISBN 978-3-0365-9323-4 (PDF)
doi.org/10.3390/books978-3-0365-9323-4

Contents

About the Editors ix

Preface xi

Hongzhe Zhang, Fang Ye, Hang Guo and Xiaoke Yan
Sodium-Potassium Alloy Heat Pipe under Geyser Boiling Experimental Study: Heat Transfer Analysis
Reprinted from: *Energies* **2021**, *14*, 7582, doi:10.3390/en14227582 1

Wei Wu, Yusong Gao and Zhijian Jin
Magnetic Field Saturation of Non-Insulation High-Temperature Superconducting Coils during Overcurrent
Reprinted from: *Electronics* **2021**, *10*, 2789, doi:10.3390/electronics10222789 17

Stephan Renninger, Paul Rößner, Jan Stein, Maike Lambarth and Kai Peter Birke
Towards High Efficiency CO_2 Utilization by Glow Discharge Plasma
Reprinted from: *Processes* **2021**, *9*, 2063, doi:10.3390/pr9112063 33

Ahmed M. Fares, Matias Kippke, Mohamed Rashed, Christian Klumpner and Serhiy Bozhko
Development of a Smart Supercapacitor Energy Storage System for Aircraft Electric Power Systems
Reprinted from: *Energies* **2021**, *14*, 8056, doi:10.3390/en14238056 39

Amirhossein Banaei and Amir Zanj
A Review on the Challenges of Using Zeolite 13X as Heat Storage Systems for the Residential Sector
Reprinted from: *Energies* **2021**, *14*, 8062, doi:10.3390/en14238062 53

Watcharakorn Pinthurat and Branislav Hredzak
Distributed Control Strategy of Single-Phase Battery Systems for Compensation of Unbalanced Active Powers in a Three-Phase Four-Wire Microgrid
Reprinted from: *Energies* **2021**, *14*, 8287, doi:10.3390/en14248287 67

Shuyan Zhang, Kaoshe Zhang, Gang Zhang, Tuo Xie, Jiaxing Wen, Chao Feng and Weihong Ben
The Bi-Level Optimization Model Research for Energy-Intensive Load and Energy Storage System Considering Congested Wind Power Consumption
Reprinted from: *Processes* **2022**, *10*, 51, doi:10.3390/pr10010051 85

Miftah Altwieb, Rakesh Mishra, Aliyu M. Aliyu and Krzysztof J. Kubiak
Heat Transfer Enhancement by Perforated and Louvred Fin Heat Exchangers
Reprinted from: *Energies* **2022**, *15*, 400, doi:10.3390/en15020400 111

Jorge Olmedo-González, Guadalupe Ramos-Sánchez, Erika Paola Garduño-Ruiz and Rosa de Guadalupe González-Huerta
Analysis of Stand-Alone Photovoltaic—Marine Current Hybrid System and the Influence on Daily and Seasonal Energy Storage
Reprinted from: *Energies* **2022**, *15*, 468, doi:10.3390/en15020468 127

Vaclav Novotny, Vit Basta, Petr Smola and Jan Spale
Review of Carnot Battery Technology Commercial Development
Reprinted from: *Energies* **2022**, *15*, 647, doi:10.3390/en15020647 149

Ruben Rafael Boros and István Bodnár
Grid and PV Fed Uninterruptible Induction Motor Drive Implementation and Measurements †
Reprinted from: *Energies* **2022**, *15*, 708, doi:10.3390/en15030708 . . . . **183**

Junfang Li, Yue Xing and Donghui Zhang
Planning Method and Principles of the Cloud Energy Storage Applied in the Power Grid Based on Charging and Discharging Load Model for Distributed Energy Storage Devices
Reprinted from: *Processes* **2022**, *10*, 194, doi:10.3390/pr10020194 . . . . **201**

Rekabra Youssef, Md Sazzad Hosen, Jiacheng He, Mohammed AL-Saadi, Joeri Van Mierlo and Maitane Berecibar
Thermal Performance Improvement for Different Strategies of Battery Thermal Management Systems Combined with Jute—A Comparison Study †
Reprinted from: *Energies* **2022**, *15*, 873, doi:10.3390/en15030873 . . . . **225**

Bizhong Xia, Bo Ye and Jianwen Cao
Polarization Voltage Characterization of Lithium-Ion Batteries Based on a Lumped Diffusion Model and Joint Parameter Estimation Algorithm
Reprinted from: *Energies* **2022**, *15*, 1150, doi:10.3390/en15031150 . . . . **241**

Kangqi Tian, Li Song, Yongyan Chen, Xiaofeng Jiao, Rui Feng and Rui Tian
Stress Coupling Analysis and Failure Damage Evaluation of Wind Turbine Blades during Strong Winds
Reprinted from: *Energies* **2022**, *15*, 1339, doi:10.3390/en15041339 . . . . **263**

Xinsheng Zhang, Fei Xiao, Ruitian Wang, Wei Kang and Beichao Yang
Modeling and Design of High-Power Enhanced Leakage-Inductance-Integrated Medium-Frequency Transformers for DAB Converters
Reprinted from: *Energies* **2022**, *15*, 1361, doi:10.3390/en15041361 . . . . **283**

Hongyuan Yuan, Youjun Han, Yu Zhou, Zongke Chen, Juan Du and Hailong Pei
State of Charge Dual Estimation of a Li-ion Battery Based on Variable Forgetting Factor Recursive Least Square and Multi-Innovation Unscented Kalman Filter Algorithm
Reprinted from: *Energies* **2022**, *12*, 1529, doi:10.3390/en15041529 . . . . **307**

Yang Guo and Ziguang Lu
A Robust Algorithm for State-of-Charge Estimation under Model Uncertainty and Voltage Sensor Bias
Reprinted from: *Energies* **2022**, *15*, 1537, doi:10.3390/en15041537 . . . . **329**

Lawrie Swinfen-Styles, Seamus D. Garvey, Donald Giddings, Bruno Cárdenas and James P. Rouse
Analysis of a Wind-Driven Air Compression System Utilising Underwater Compressed Air Energy Storage
Reprinted from: *Energies* **2022**, *15*, 2142, doi:10.3390/en15062142 . . . . **347**

Sandro Hiller, Christian Hartmann, Babette Hebenstreit and Stefan Arzbacher
Solidified-Air Energy Storage: Conceptualization and Thermodynamic Analysis
Reprinted from: *Energies* **2022**, *15*, 2159, doi:10.3390/en15062159 . . . . **375**

Ralf Peters, Nils Wegener, Remzi Can Samsun, Felix Schorn, Julia Riese, Marcus Grünewald and Detlef Stolten
A Techno-Economic Assessment of Fischer–Tropsch Fuels Based on Syngas from Co-Electrolysis
Reprinted from: *Processes* **2022**, *10*, 699, doi:10.3390/pr10040699 . . . . **389**

Phan Anh Duong, Borim Ryu, Chongmin Kim, Jinuk Lee and Hokeun Kang
Energy and Exergy Analysis of an Ammonia Fuel Cell Integrated System for Marine Vessels
Reprinted from: *Energies* **2022**, *15*, 3331, doi:10.3390/en15093331 . . . . **431**

Nimat Shamim, Vilayanur V. Viswanathan, Edwin C. Thomsen, Guosheng Li, David M. Reed and Vincent L. Sprenkle
Valve Regulated Lead Acid Battery Evaluation under Peak Shaving and Frequency Regulation Duty Cycles
Reprinted from: *Energies* **2022**, *15*, 3389, doi:10.3390/en15093389 . . . . **453**

Chenliang Li, Ruizhi Li and Yingke Zhou
Preparation of a Honeycomb-like FeNi(OH/P) Nanosheet Array as a High-Performance Cathode for Hybrid Supercapacitors
Reprinted from: *Energies* **2022**, *15*, 3877, doi:10.3390/en15113877 . . . . **473**

Xiaoyu Li, Tengyuan Wang, Jiaxu Li, Yong Tian and Jindong Tian
Energy Consumption Estimation for Electric Buses Based on a Physical and Data-Driven Fusion Model
Reprinted from: *Energies* **2022**, *15*, 4160, doi:10.3390/en15114160 . . . . **489**

Paolo Maria Congedo, Cristina Baglivo, Simone Panico, Domenico Mazzeo and Nicoletta Matera
Optimization of Micro-CAES and TES Systems for Trigeneration
Reprinted from: *Energies* **2022**, *15*, 6232, doi:10.3390/en15176232 . . . . **507**

Anti Kur, Jo Darkwa, John Calautit, Rabah Boukhanouf and Mark Worall
Solid–Gas Thermochemical Energy Storage Materials and Reactors for Low to High-Temperature Applications: A Concise Review
Reprinted from: *Energies* **2023**, *16*, 756, doi:10.3390/en16020756 . . . . **521**

Wenwen Ye, Dourna Jamshideasli and Jay M. Khodadadi
Improved Performance of Latent Heat Energy Storage Systems in Response to Utilization of High Thermal Conductivity Fins
Reprinted from: *Energies* **2023**, *16*, 1277, doi:10.3390/en16031277 . . . . **557**

About the Editors

Alon Kuperman

Alon Kuperman earned his Ph.D. degree in electrical and computer engineering from the Ben-Gurion University of the Negev, Beersheba, Israel, in 2006. He served as a Marie Curie Training Site Member at Imperial College London, London, U.K, from 2003 to 2006 and later as an Honorary Research Fellow at the University of Liverpool, Liverpool, U.K., from 2008 to 2009. He is currently affiliated with the School of Electrical and Computer Engineering, Ben-Gurion University, Israel, where he heads the Power and Energy Systems track and directs the Applied Energy Laboratory. His research interests include all aspects of energy conversion and applied control.

Alessandro Lampasi

Alessandro Lampasi earned his M.Sc. degree in electronic engineering and a Ph.D. degree in electrical engineering from the University of Rome Sapienza, Rome, Italy, in 2002 and 2006, respectively. He is currently affiliated with the Italian National Agency for New Technologies, Energy and Sustainable Economic Development (ENEA), Frascati, Italy, and the consortium for the construction of the Divertor Tokamak Test (DTT) facility, where he is in charge of several international projects concerning nuclear fusion, power systems and energy storage. His current research interests include modeling and measurement techniques in the fields of nuclear fusion, power electronics, energy storage and applied electromagnetics.

Preface

Energy storage and conversion are crucial research and industry topics, especially within the context of sustainable development. Scientific and technological progress in these fields have the potential to improve energy use efficiency across traditional, renewable and unconventional sources.

Energy storage technologies, including batteries, fuel cells, supercapacitors (ultracapacitors) and superconducting magnetic energy storage (SMES), combined with decreasing costs are creating new scenarios and opportunities in the development and market of energy generation, grids, industrial plants, complex systems and consumer electronics.

This reprint collects some of the latest developments and applications in these interdisciplinary fields, offering a common framework for authors from different research areas.

Alon Kuperman and Alessandro Lampasi
Editors

energies

Article

Sodium-Potassium Alloy Heat Pipe under Geyser Boiling Experimental Study: Heat Transfer Analysis

Hongzhe Zhang [1], Fang Ye [1,*], Hang Guo [1] and Xiaoke Yan [2]

[1] MOE Key Laboratory of Enhanced Heat Transfer and Energy Conservation, Beijing Key Laboratory of Heat Transfer and Energy Conversion, College of Energy and Power Engineering, Beijing University of Technology, Beijing 100124, China; ZhangHongZhe@emails.bjut.edu.cn (H.Z.); hangguo@sohu.com (H.G.)
[2] National Institute of Metrology, Beijing 100013, China; yanxk@nim.ac.cn
* Correspondence: yefang@bjut.edu.cn; Tel.: +86-10-67396661 (ext. 8002)

Abstract: In the geyser boiling mode, the working fluid state is divided into a boiling process and a quiet process, and the sodium-potassium (Na-K) alloy heat pipe can discontinuously transfer heat at each boiling. The overheating of the liquid working fluid at the bottom causes short-term boiling and forms slug bubble, the strong condensing ability quickly conducts heat from the evaporator section. And geyser boiling can occur before the working fluid forms continuous flow, so it transfers more heat at lower temperatures than natural convection cooling. In this study, the heat transfer process of a Na-K alloy heat pipe with forced convection cooling under different heating power was experimental studied. The geyser boiling mode can make the Na-K alloy heat pipe work below 650 °C and reduce the start-up time. In the process of geyser boiling, the heat transfer quantity was increased by the boiling frequency and the amount of vapor produced in a single boiling. The boiling temperature had no obvious change with the increased of heating power, and the condenser section temperature increased with the heating power.

Keywords: sodium-potassium alloy; geyser boiling; forced convection cooling; start-up performance; heat transfer performance

Citation: Zhang, H.; Ye, F.; Guo, H.; Yan, X. Sodium-Potassium Alloy Heat Pipe under Geyser Boiling Experimental Study: Heat Transfer Analysis. *Energies* **2021**, *14*, 7582. https://doi.org/10.3390/en14227582

Academic Editors: Alon Kuperman and Alessandro Lampasi

Received: 11 October 2021
Accepted: 10 November 2021
Published: 12 November 2021

1. Introduction

Heat pipes rely on the phase transformation of the working fluid to transfer heat by using latent heat. Therefore, heat pipes have high heat transfer performance [1], isothermal performance [2] and thermal response speed [3]. The working temperature of high temperature heat pipes is higher than 750 K. Sodium, potassium and other alkali metals are often used as working fluid for high temperature heat pipe because of their high boiling temperature, low saturation pressure, high latent heat of vaporization and high heat transfer performance. Alkali metal heat pipes have been used in many high temperature applications such as high temperature isothermal heater [4], thermometric calibration system [5,6], solar energy utilization system [7,8], residual heat removal system [9–12], high temperature thermal management [13–15] and so on. The alloy of suitable ratio of sodium and potassium is liquid at room temperature, so the sodium-potassium (Na-K) alloy heat pipe has great advantages in cold-start and working fluid filling. Na-K alloy was the first used as coolant in nuclear industry, Anderson et al. [16] proposed to replace pure metals such as sodium and potassium as high temperature heat pipe working fluid. The thermal properties and flow characteristics of Na-K (78% by mass of potassium) alloy were studied by Serizawa et al. [17] and Timothy et al. [18]. The results showed that the heat transfer performance of the alloy is better than sodium and potassium. The start-up performance of liquid metal heat pipes is very different from the low temperature heat pipes, the effect of compressibility must be considered at start-up stage because of the low pressure and low density of vapor. The continuous flow can be formed only after the working fluid reaches a certain temperature, then the vapor density and pressure increase significantly and the heat pipe can transfer heat effectively. Similar to other alkali metal heat pipes, the heat transfer limits such as sonic limit and viscosity limit also appear in the start-up process of Na-K alloy heat pipes. Na-K alloy heat pipes also have several advantages, different mass ratios of sodium and potassium can be selected according to the requirements of working temperature. Due to the melting point of the Na-K alloy is below room temperature when the mass fraction of potassium is between 46% and 89%. Therefore, there is no frozen limit at room temperature and it is beneficial to the manufacture of heat pipe such as the high temperature oscillating heat pipe [19], and there is no solidification limit, the working fluid will not solidify into solid state in the condenser section, avoid that the working fluid cannot return and the evaporator section dry burning.

At present, Na-K alloy heat pipes have been applied in many fields such as molten salt residual heat utilization system [20], high temperature cooling system [21] and so on. Working conditions have a great influence on the start-up heat transfer performance of Na-K alloy heat pipe. When the length of condenser section is too long, the working fluid is not easy to fill the whole heat pipe [22,23]. When the length of the condenser section is too small, the condenser amount of the working medium is small, which increases the gas pressure [23,24]. The inclination angle of heat pipe will affect the effective heating area of evaporator and the driving force of reflux liquid, and the most suitable inclination angle of Na-K alloy heat pipe is 55° [25]. The Na-K alloy heat pipe has been proved to be an effective high thermal conductivity element, the heat transfer ability and service temperature of Na-K alloy heat pipe are between those of sodium heat pipe and potassium heat pipe in full start state. Because the melting point of Na-K alloy is lower than room temperature, it is safer in working fluid filling and cold start process, so the Na-K alloy heat pipe is a kind of safer heat pipe which can replace sodium heat pipe. And there is no solidification phenomenon in geyser boiling. The geyser boiling is a kind of repeated process in which the working fluid suddenly boils and returns to calm.

In this article, Section 2 Literature review presents related works on the geyser boiling. Section 3 Experimental methods presents the equipment, experimental system and experimental method. Section 4 Results and discussions presents working process of Na-K alloy heat pipe and effects of heating power on geyser boiling process, heat pipe wall temperature distribution and heat transfer performance. Section 5 Conclusions presents the main conclusions in this study.

2. Literature Review

Noie et al. [26] studied the influence of the inclination angle on the heat transfer coefficient of the condenser under different liquid charge ratios. It was found that geyser boiling occurred when the liquid charge ratios is greater than 30%.

Emani et al. [27] studied the effects of liquid charge ratio, inclination and cooling water mass flow on the geyser boiling. It was found that when the inclination angle was reduced, the geyser period and the temperature fluctuation range were reduced, and the geyser boiling phenomenon disappeared when the inclination angle was less than 15°. The mass flow rate of the cooling water has little effect on the geyser boiling phenomenon.

Lin et al. [28] studied the influence of heating power, condenser temperature, liquid charge ratio and evaporator length on water and ethanol geyser boiling. The correlation equation of heat transfer coefficient in geyser boiling was proposed.

Wang et al. [29] presented a combined CFD/visualization study and used the VOF method to improve the Lee model. The improved model had better predictive performance, and the heat transfer behavior obtained was closer to the actual phenomenon in the visualization experiment.

Some literature studied how to suppress geyser boiling. Casarosa et al. [30] studied the influence of heating power and condenser pressure on geyser boiling. When the condenser pressure remained constant, the boiling frequency increased linearly with the increase of heating power. When the heating power was constant, increasing the condenser pressure reduced the geyser boiling, and even eliminated the geyser boiling.

Kujawska et al. [31] studied the geyser boiling phenomenon of heat pipes using nanofluids. It was found that nanofluids can reduce or even inhibit geyser boiling. The deposition of nanoparticles on the evaporator wall increases the number of nucleation points that can form vapor bubbles and prevents the formation of gas plugs.

Although most literatures hope to eliminate geyser boiling [32,33], some literatures have noticed that geyser boiling can be applied. Kuncoro et al. [34] studied the effects of temperature and pressure on the geyser boiling of water and R113. The study found that the temperature distribution inside the liquid has a great influence on geyser boiling. And in this research, it was found that the start-up time of the heat pipe is reduced after geyser boiling occurs.

Tecchio et al. [35] studied the effects of heat flux and vapor pressure on geyser boiling in loop thermosyphons with liquid charge rates of 0.5 and 0.9. Geyser boiling occurred when the heat flux was higher than 12.5 kW/m^2 and vapor pressures was below 25 kPa, and the dimensionless pressure rates were about 1.0. And it was noted that the heat transfer quantity obviously increased when geyser boiling occurred.

Jia et al. [22] studied the effect of heating temperature on the heat transfer performance of Na-K alloy heat pipe. Under cooling water cooling, geyser boiling phenomenon occurred, and the start-up temperature and working temperature of heat pipe were lower than those under natural convection. Compared with the study of heat transfer performance of Na-K alloy heat pipe under natural convection cooling by Guo et al. [24], geyser boiling can reduce the lowest working temperature of Na-K alloy heat pipe. Under forced convection cooling of cooling water, when the heating temperature reached 600 °C, geyser boiling occurred in the heat pipe and the heat transfer quantity increased. But under the same heating conditions and natural convection cooling, the phase transformation process of working fluid became violent only when the heating temperature reached 725 °C [23,24]. Moreover, geyser boiling can transfer more heat at lower temperatures. When the heating temperature was 650 °C, the heat transfer quantity under natural convection cooling of air was 475 W [24], and the heat transfer quantity under cooling water was 790 W [22].

Therefore, the heat pipe could transfer heat effectively in the geyser boiling state, and the working temperature range of the heat pipe is greatly improved. The influence of heating power on the heat transfer process of Na-K alloy heat pipe under forced convective cooling, along with the variation of wall temperature during the geyser boiling process under forced convective cooling need to be further studied.

In this study, the influence of heating power on heat transfer process of Na-K (77.8% by mass of potassium) alloy heat pipe under forced convective cooling was studied experimentally, and the variation of wall temperature during geyser boiling process under forced convective cooling was analyzed.

3. Experimental Methods

3.1. Experimental Setup and Procedure

The main equipment includes Na-K alloy heat pipe, heating furnace, power controller, cooling water circulator, data collector, computer and so on. Figure 1 shows the experimental system of heat pipe under forced convection cooling.

Figure 2 shows the experimental procedures. During the experiment, the heat pipe was heated from room temperature. The heating condition is that the heating power was constant at 800 W, 1000 W, 1200 W and 1400 W respectively. The cooling condition is that the cooling water flow was constant at 16 mL/s, and the temperature of constant temperature water tank was constant at 20 °C. Each temperature was obtained through the data collector, and the acquisition frequency was 3 s/time. In geyser boiling, the wall temperature of the heat pipe changes periodically and drastically. The start-up process of heat pipe is from the beginning of heating to the first boiling. From the beginning of heating to the first boiling, it is the start-up process of the heat pipe. At least 20 stable and complete boiling fluctuation cycles need to be recorded in each group of experiment. Because the heat pipe always kept geyser boiling, the time-average values of the wall temperature, the temperature of the cooling water jacket, the outlet and inlet temperature of the cooling water in the calculation are taken within 5 boiling after the stable fluctuation.

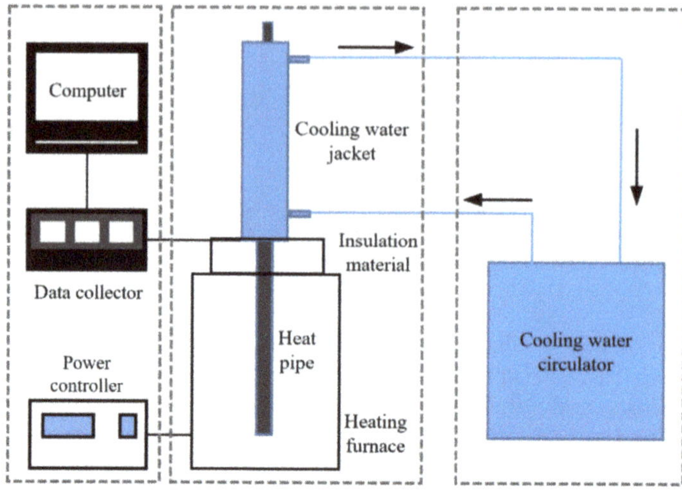

Figure 1. Experimental system of Na-K heat pipe under forced convection cooling.

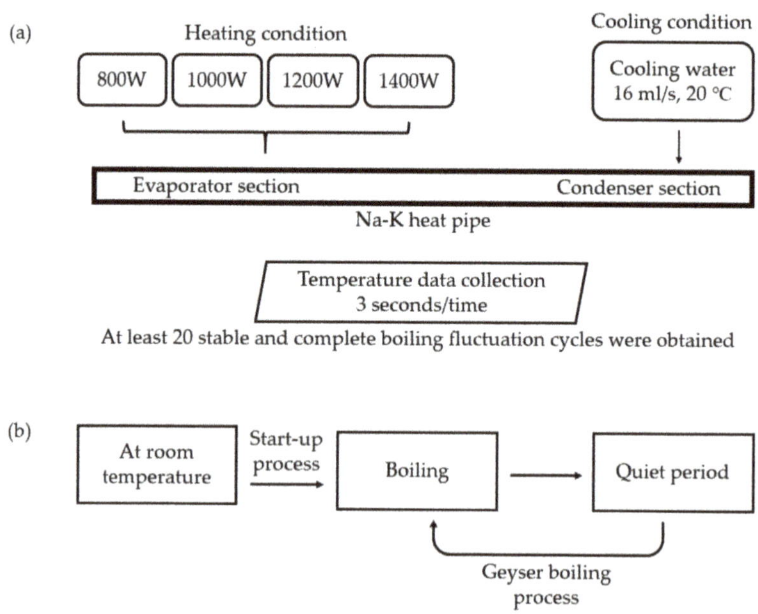

Figure 2. Experimental procedures of Na-K heat pipe of (**a**) Experimental conditions; (**b**) Operation status of Na-K heat pipe.

The temperature of the outer wall of heat pipe, the outlet and inlet of the cooling water, the ambient and the outer wall of the cooling water jacket were measured. Figure 3 shows the distribution of temperature measuring points on the outer wall, these measuring points were numbered as P1 to P15 in turn. The evaporator section was 580 mm long with 5 measuring points, the insulation section was 90 mm long with 1 measuring point, the condenser section was 330 mm long with 9 measuring points due to the large axial temperature variation of the condenser section. The heat pipe was always placed vertically during the experiment.

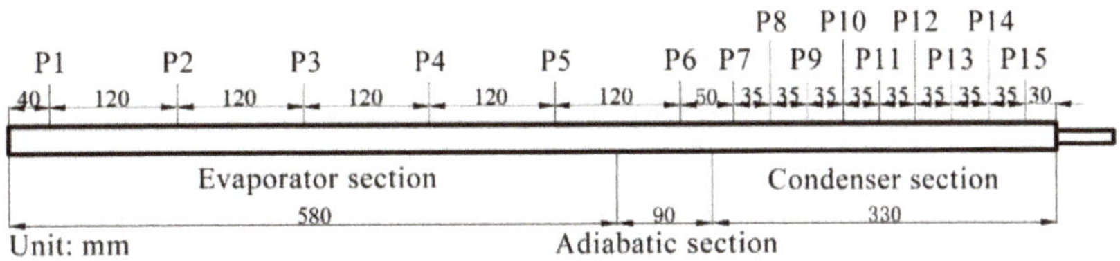

Figure 3. Location of temperature measuring points.

A Na-K alloy gravity heat pipe was used in the experiment. The mass fraction of metal potassium and sodium in the Na-K alloy are 77.8% and 22.2% respectively. Compared with sodium and potassium, the Na-K alloy has low melting point and it is liquid at room temperature. As an alkali metal heat pipe working fluid, Na-K alloy is easy to fill and will not condense in the condenser section. Table 1 presents the thermophysical properties of the Na-K alloy.

Table 1. Thermophysical properties of the Na-K (77.8% by mass of potassium) alloy.

Properties	Values
Melting temperature (0.1 MPa)	−12.6 °C
Boiling temperature (0.1 MPa)	785.1 °C
Vapor pressure (700 °C)	0.489 kPa
Density (700 °C)	731.1 kg/m^3
Specific heat (700 °C)	1.01 kJ/(kg·K)
Viscosity (700 °C)	1.5×10^{-3} Pa·s
Latent heat of vaporization (0.1 MPa)	3453.5 kJ/kg

The shape of heat pipe is cylindrical, the parameters of the Na-K alloy heat pipe is presented in Table 2. In our previous work [24], the Na-K heat pipe had best start-up performance under the length of evaporator, condenser, and adiabatic section in Table 2.

Table 2. Parameters of the Na-K alloy heat pipe.

Parameters	Specifications
Total length	1000 mm
Outer diameter	25 mm
Inner diameter	20 mm
Working fluid weight	70 g (about 78 cm^3, 25.3% of the heat pipe internal volume)
Container material	Inconel 600
Evaporator section length	580 mm
Condenser section length	330 mm
Adiabatic section length	90 mm

The heating system is composed of single-phase full digital thyristor power controller, voltage regulating power supply and heating furnace, which can realize the constant temperature heating and the constant power heating of Na-K alloy heat pipe.

The power controller precisely controls the heating power by adjusting the voltage and current of the circuit through the thyristor controller. In this experiment, the constant power output mode was mainly used. The depth of heating furnace is 1000 mm, the resistance heating wire was used, and the heating temperature range is 20~1000 °C. The maximum temperature difference is less than 20 °C when heating at 800 °C constant temperature. The heating furnace can be heated at 0~7 kW constant power by the power controller and the control accuracy is ±0.01 kW.

The cooling system is composed of annulus type cooling water jacket and cooling water circulator. The cooling water jacket is made by stainless steel, the length of water jacket is 330 mm, which is the same as the length of condenser section. The cooling water in the jacket was upward flow and the gap between the cooling water jacket and the condensation section of the heat pipe was filled with copper powder. The temperature control accuracy of the cooling water circulator is ±0.1 °C and the flow control accuracy is ±0.3 mL/s.

The temperature data were recorded by a data collector with voltage accuracy of 0.004%.

Because the gap between cooling water jacket and heat pipe is very small and the volume of standard thermocouple is too large to be arranged on the condenser section wall. Therefore, the self-made K-type thermocouples which were welded with 0.2 mm diameter nickel-chromium alloy wires and nickel-aluminum alloy wires were used after calibrated by China Institute of Metrology, and the uncertainty of all self-made K-type thermocouples is 1.2 K.

3.2. Data Processing

The calculation methods of the equivalent heat transfer coefficient, the equivalent thermal resistance, and the heat transfer coefficient of evaporator section and condenser section are shown in Equations (1)–(5). The equations were taken from the National standards of China "Testing method for heat transfer performance of heat pipes (GB/T14812-2008)". The heat transfer quantity can be calculated by subtracting the heat leakage of cooling water jacket from the cooling water.

$$Q = Gc_p(T_{j,out} - T_{j,in}) - h_{ja}(T_a - T_j)A_j \tag{1}$$

The h_{ja} is the convective heat transfer coefficient between water jacket out wall and air, it is about 20 W/(m^2·°C). The equivalent heat transfer coefficient can be calculated by considering the heat pipe as a uniform solid,

$$K = \frac{Q}{(T_e - T_c)A_{hp}} \tag{2}$$

The heat leakage of heat pipe insulation section and the heat resistance of heat pipe wall is small, so the insulation section temperature can be used as the temperature of working fluid vapor of heat pipe. The heat transfer coefficient of evaporator section and condenser section can be defined as

$$h_e = \frac{Q}{(T_e - T_a)A_e} \tag{3}$$

$$h_c = \frac{Q}{(T_a - T_c)A_c} \tag{4}$$

The equivalent thermal resistance can be defined as

$$R = \frac{T_e - T_c}{Q} \tag{5}$$

In Equations (2)–(5), the average wall temperature was taken as the average temperature of each section.

$$T_e = \frac{T_1 + T_2 + T_3 + T_4 + T_5}{5} \tag{6}$$

$$T_c = \frac{T_7 + T_8 + \cdots + T_{15}}{9} \tag{7}$$

According to the measurement accuracy of temperature and heating power, the uncertainty of thermal resistance and effective thermal conductivity was calculated. The uncertainty of thermal resistance is defined as [36]

$$\frac{\delta R}{R}=\sqrt{\left(\frac{\delta T}{T}\right)^2+\left(\frac{\delta Q}{Q}\right)^2} \tag{8}$$

Therefor, the maximum relative thermal resistance uncertainty was ±0.41%.

3.3. Experimental Repeatability

Figure 4 shows the reproducibility of the time-average wall temperature. Three experiments under the same conditions were completed in three different days. The heating power was 800 W, the cooling water temperature was 20 °C and the cooling water flow rate was 16 mL/s. From this Figure, the experimental error of time-average wall temperature of heat pipe was less than 2%. Therefore, it could be considered that the experimental results were accurate.

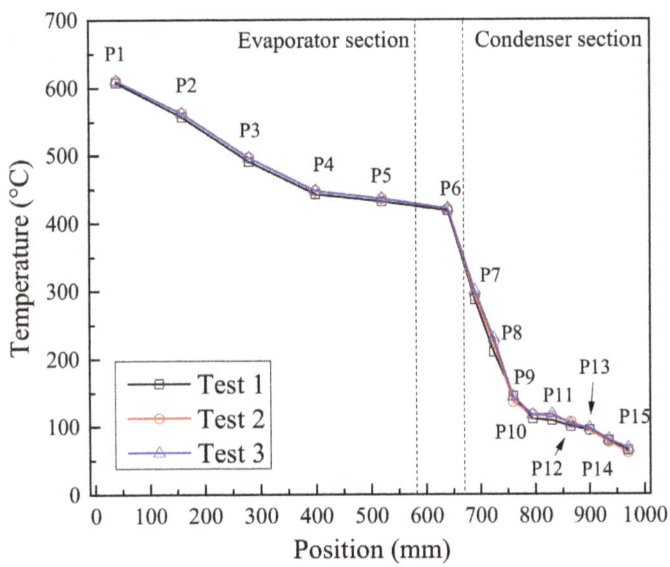

Figure 4. Reproducibility of time-average wall temperature of Na-K alloy heat pipe.

4. Results and Discussions

4.1. Start-Up Process

During the start-up process of Na-K alloy heat pipe, the flow state of working fluid will transform from the rarefied vapor flow to the continuum vapor flow. The Na-K alloy heat pipe can only achieve efficient heat transfer when the continuum vapor flow is established. Knudsen number can be used to determine whether the working fluid forms continuous flow, the Knudsen number can be calculated by the ratio of mean free path of working fluid and diameter of the vapor flow passage.

$$Kn=\frac{\lambda}{D} \tag{9}$$

It is generally believed that continuous flow can be formed only when Knudsen number is less than 0.01 [20,37]. The temperature that can make the working fluid vapor form a continuous flow is named the transition temperature, it can be defined as [20,37]

$$T_{\text{tr}} \geq \frac{\pi M}{2 \times 10^{-4} R_{\text{g}}} \left(\frac{\mu_v}{\rho_v D} \right)^2 \tag{10}$$

The vapor transition temperature is mainly affected by the height of vapor space which is the inner diameter of heat pipe. The transition temperature of Na-K (77.8% by mass of potassium) alloy is about 340.2 °C when the inner diameter of the heat pipe is 20 mm.

Figure 5 shows the temperature variation of each measurement point under different heating power during the start-up process and working process. The sonic limit appeared in the start-up process under all heating power. The sonic limit is very common in the start-up process of alkali metal heat pipes, and it will disappear with the increase of vapor temperature [38]. An intense boiling occurred in the heat pipe after heating for 25 to 40 min under all heating power, the evaporator section temperature decreased sharply while the condenser section temperature increased rapidly. Because a large amount of high temperature vapor flowed to the condenser section, and the sound of liquid working fluid impacting the end cover of condenser section could be heard clearly during the boiling.

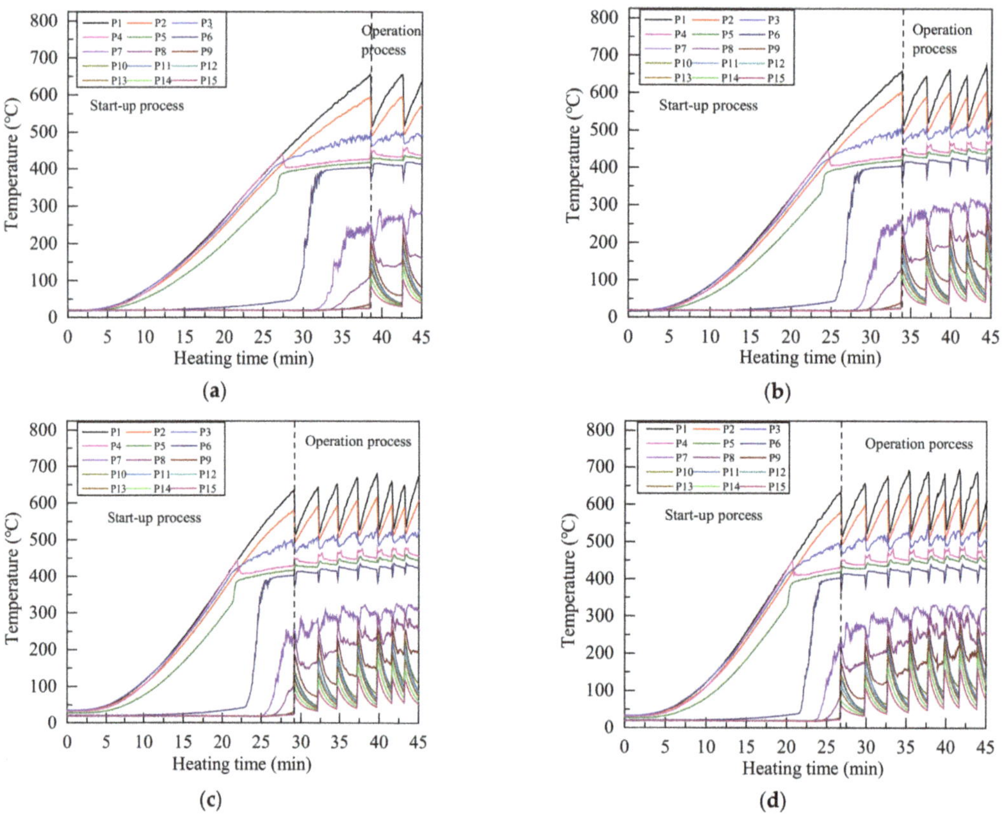

Figure 5. Temperature distribution of Na-K alloy heat pipe under heating power of (**a**) 800 W; (**b**) 1000 W; (**c**) 1200 W; (**d**) 1400 W.

Due to the intense boiling and strong forced convection cooling, the evaporator section temperature decreased greatly after the first boiling, and a large amount of heat was taken away by the cooling water, so the heat pipe can not maintain continuous boiling state. Therefore, the condenser section temperature began to decrease, and the evaporator section temperature raised. About 2 to 3 min later, another intense boiling occurred and this process was repeated, then the geyser boiling occurred in the heat pipe [28,38]. Because the melting point of Na-K alloy is lower than room temperature and there is no solidification limit, the working fluid will not solidify into solid state in the condenser section. Even if the condenser section temperature was reduced to 40 °C during the geyser boiling process, the Na-K alloy heat pipe can work normally.

Since measuring point 1 to measuring point 5 were in the evaporator section, the temperature rises gently in the start-up stage. When the heating time was 20 min to 25 min, the temperature of measuring point 4 decreases slightly, because the liquid level of liquid Na-K alloy is lower than measuring point 4. At this time, the Na-K alloy had begun a large number of phase transformation on the liquid surface, and the temperature change of measuring point 4 in the start-up stage was very small.

The flow state of Na-K alloy transformed from the rarefied vapor flow to the continuum vapor flow when temperature reached the transition temperature, and the flow of Na-K alloy increased greatly, which greatly increased the wall temperature of heat pipe. The temperature of measuring point 6 in the adiabatic section rose sharply at 20 min to 30 min, indicating that the working fluid in the adiabatic section had changed to continuous vapor flow at this time. Since the liquid level of Na-K alloy is lower than measuring point 4, measuring point 5 in the evaporation section also had the process of flow state transformation, and the temperature of measuring point 5 also increased greatly in a short time. The temperature from measuring point 4 to measuring point 6 was maintained at about 400 °C after flow state transformation, which is close to the theoretical transition temperature of Na-K alloy.

In the start-up state, because there were few Na-K alloys entering the condenser section, the temperature of only a few measuring points increased, and most measuring points were at room temperature. Therefore, geyser boiling can occur before continuous flow is formed in the condenser section under forced convection cooling.

4.2. Working Process

In the working state, when the temperature of measuring point 1 at the bottom of the heat pipe reached about 650 °C, a violent boiling occurred at the bottom of the heat pipe to produce Taylor bubble, which is a kind of elastic flow bubble. A large amount of vapor carried liquid Na-K alloy into the condenser section, reduced the temperature of each measuring point in the evaporator section and increased the temperature of each measuring point in the condenser section, and the sound of liquid working fluid hitting the top of the heat pipe could be heard. Then, due to the strong cooling capacity of the cooling water, the heat input in the evaporation section was not enough to maintain the continuous boiling of Na-K alloy, so that the temperature of each measuring point in the condenser section decreased rapidly, and the temperature of each measuring point in the evaporator section rose again. A new boiling occurred when the temperature of measuring point 1 reached about 650 °C again, and then the heat pipe repeated this process to form geyser boiling.

Compared with the temperature fluctuation under different heating power in Figure 5, with the increased of heating power, the temperature distribution at the beginning and the end of each single boiling was very close. Because the working fluid returned to the same state after each boiling, increasing the heating power only increases the boiling frequency.

The time-average wall temperature of each measurement point is the average value of temperature within 5 boiling after the stable fluctuation. Figure 6 shows the time-average wall temperature distribution of heat pipe under different heating power. The average temperature of condenser section was obviously increased with the heating power, and the temperature uniformity was slightly improved. Due to the strong heat transfer ability of forced convection, there was a large temperature gradient along the heat pipe, the vapor temperature decreased rapidly after entered the condenser section. The temperature gradient at 795 mm to 970 mm decreased obviously, indicating that there was less Na-K alloy here and the flow state is rarefied vapor flow.

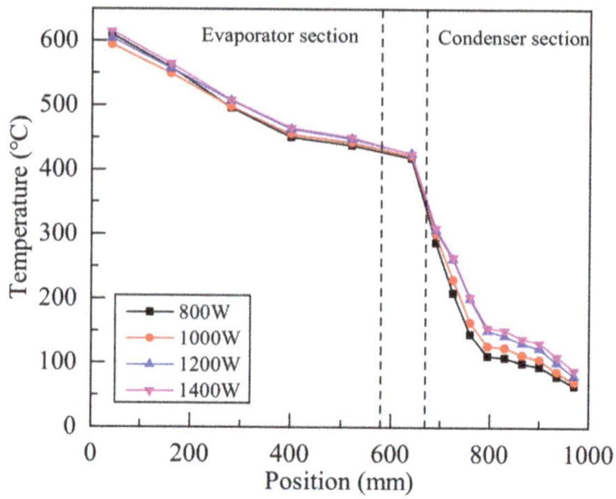

Figure 6. Average temperature distribution of heat pipe under different heating power.

Figure 7 shows the start-up time and average temperature of evaporator section during each boiling under different heating power. The start-up time is the first time of intense boiling of the heat pipe. The average temperature of evaporator section was calculated according to Equation (6) from Figure 6. The start-up time decreased as the heating power increases, indicating that the heating power has greater influence on the temperature of working fluid in the non-boiling state. The next boiling time will be reduced, and it can be seen in Figure 5 that the fluctuation frequency increased at higher heating power. Geyser boiling would not make the maximum temperature of evaporator section continue to increase, the temperature of P6 in the insulation section was maintained at about 400 °C. That meant the starting condition of each boiling and the temperature of vapor was very close to 400 °C during a single boiling, so the average temperature of evaporator section had small differences, and the temperature of the evaporation section is relatively close in Figure 6. It means that Na-K alloy heat pipes can work at lower temperature under geyser boiling. With the increased of heating power, the mass of vapor produced in single boiling increased, and the average temperature of evaporation section increased after a single boiling.

Figure 8 shows the average temperature difference of some measurement point at the condenser section in each single boiling and the average fluctuation cycle of each single boiling. With the increased of heating power, the evaporation of the working fluid increased, and the condenser section temperature increased obviously after the end of single boiling. Since the average temperature of evaporation section under different heating power in Figure 7 has little difference, it shows that the mass flow of working fluid during single boiling increases with the heating power. When the heating power was 800 W, the geyser boiling cycle was obviously longer, so the temperature difference in the condenser section was obvious in Figure 6. The lower heating power made the temperature rise slowly in the evaporator section, it took a longer time to reach the required boiling temperature. As the heating power reached 1400 W, the mass of vapor reaching the condenser section was higher, and it needed more time to cool due to the heat dissipation condition of the condenser section, so the temperature fluctuation frequency was close to the frequency under 1200 W heating.

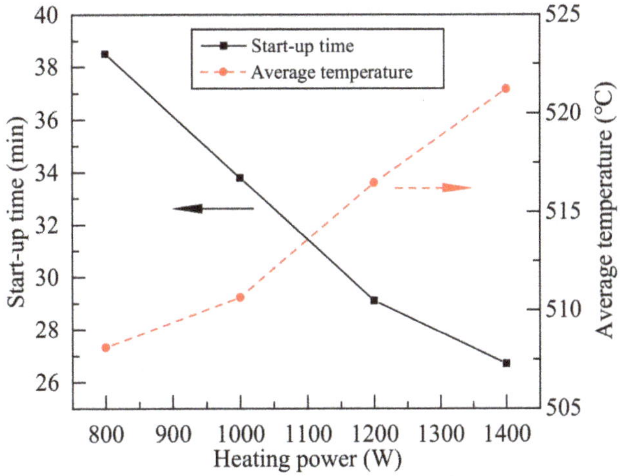

Figure 7. Start-up time and average temperature of evaporator section during the each boiling under different heating power.

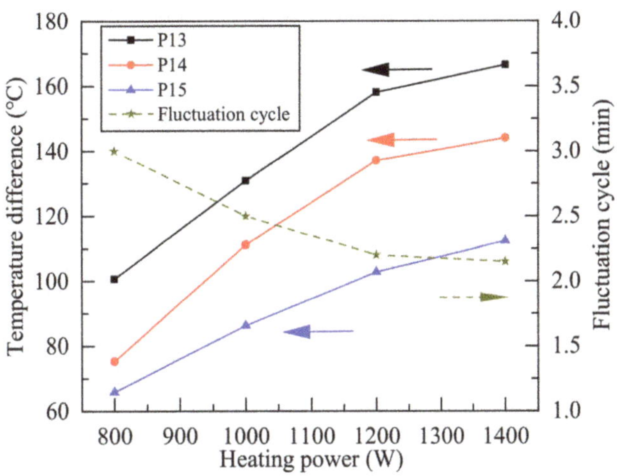

Figure 8. Average temperature difference and fluctuation cycle of condenser section during geyser boiling.

4.3. Heat Transfer Performance

The temperature change in the middle of the condenser section was most obvious in a single boiling. The wall temperature was greatly affected by the cooling water when the vapor just entered the condenser section. Because the vapor temperature was higher before boiling, the wall temperature change was small at the entrance of condenser section. When the vapor arrived the end of the condenser section, most of the heat had been taken away by the cooling water, the cooling water temperature was also increased and the cooling capacity decreased, so the temperature change of the heat pipe was also small at end of the condenser section.

Figures 9 and 10 show the heat transfer quantity, the equivalent heat transfer coefficient, the equivalent thermal resistance, and the surface heat transfer coefficient of evaporator section and condenser section of heat pipe during working process according to Equations (1)–(5). With the increase of heating power, the heat transfer performance of heat pipe was improved. Because the average temperature of the evaporator section changed little, it can be further proved that the mass flow of working fluid increases with the heating power in single boiling. The heat transferred by the heat pipe was only about half of the input of the heating controller, and a lot of heat was lost by the heating furnace. In this study, the minimum thermal resistance was about 0.6 K/W, and equivalent heat transfer coefficient could reach about 5500 W/(m^2·°C).

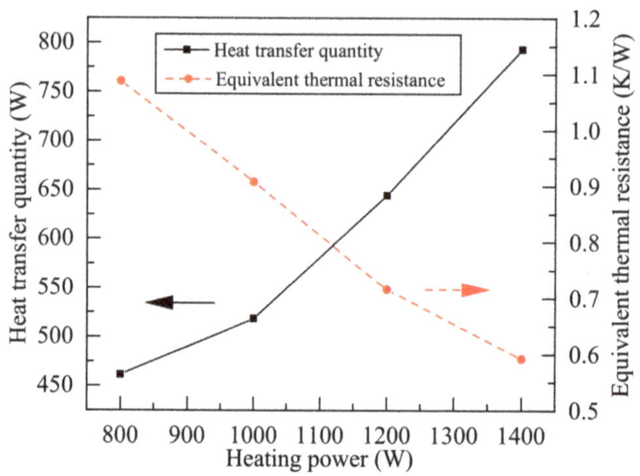

Figure 9. Heat transfer quantity and equivalent thermal resistance of heat pipe under different heating power.

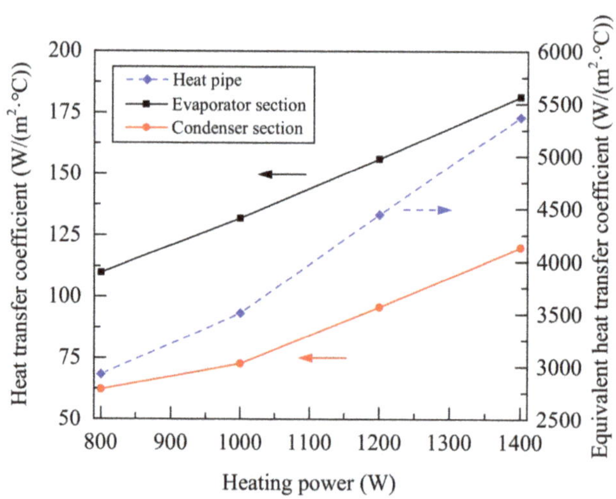

Figure 10. Heat transfer coefficient of heat pipe under different heating power.

Compared with Figures 6, 9 and 10, although there was a large temperature gradient along the heat pipe under 800 W to 1400 W heating, the increased of heating power could improve the evaporation rate and promoted the heat transfer of the evaporator section. Due to a large amount of vapor entered the condenser section, the heat transfer of the condenser section had been promoted and the vapor flow rate was increased, so the heat transfer coefficients increased linearly.

Table 3 presents the operating parameters of Na-K heat pipe under different heating power. From the Table 3, the effective heat transfer of heat pipe increased with the heating power, but the maximum temperature of heat pipe and the average temperature of evaporation section (T_e) change little, while the fluctuation cycle of geyser boiling decreased and the average temperature of condensation section (T_c) increased. Therefor, the geyser boiling mainly increases heat transfer by increasing boiling frequency rather than increasing the temperature of Na-K heat pipe.

Table 3. Operating parameters of Na-K heat pipe under different heating power.

Heating Power	Maximum Temperature of Heat Pipe	Average Temperature of Evaporator Section (T_e)	Fluctuation Cycle	Average Temperature of Condenser Section (T_c)	Effective Heat Transfer Quantity
800 W	696.3 °C	508.1 °C	3.02 min	133.0 °C	461.2 W
1000 W	672.7 °C	510.7 °C	2.51 min	146.6 °C	518.5 W
1200 W	679.5 °C	516.5 °C	2.20 min	166.3 °C	644.2 W
1400 W	684.4 °C	521.2 °C	2.15 min	172.8 °C	793.1 W

Although geyser boiling will cause temperature fluctuation and shell vibration. Geyser boiling can make Na-K alloy heat pipe working at lower temperature. Table 3 presents the maximum temperature of heat pipe under geyser boiling was only about 700 °C when the effective heat transfer quantity of heat pipe reached 800 W. Compared with our previous work of Guo et al. [24], under the natural convection cooling, the effective heat transfer quantity can reach 800 W only when the maximum temperature is above 800 °C. Therefore, geyser boiling reduced the operating temperature of the Na-K alloy heat pipe by 100 °C.

Geyser boiling also can make Na-K alloy heat pipe have higher heat transfer quantity at lower temperature. Figure 6 and Table 3 present that when the maximum average temperature of the evaporator section was 600 °C, the effective heat transfer quantity under geyser boiling reached 800 W. While the Na-K alloy heat pipe has not start at the same temperature under natural convection cooling conditions, and the effective heat transfer was less than 50 W [24,25].

5. Conclusions

Due to geyser boiling can occur before continuous flow is formed in the condenser section under forced convection cooling, the start-up capacity and heat transfer capacity of a Na-K alloy heat pipe under forced convection cooling were experimentally studied at different heating power (800 W, 1000 W, 1200 W and 1400 W). The Na-K alloy heat pipe can work in geyser boiling mode, and transfer a lot of heat quantity at lower temperature than natural convection cooling. The main conclusions were as follows:

(1) Geyser boiling can reduce the minimum operating temperature of Na-K alloy heat pipe, and can have higher heat transfer capacity than natural convection cooling at lower temperature.

(2) In the geyser boiling process, the increase in the average temperature of the condensing section is caused by the increase in the mass flow of the Na-K alloy working fluid. And the small change in the average temperature of the evaporation section is caused by the small change in the boiling temperature.

(3) The increase of heating power leads to the increase of Na-K alloy working fluid mass flow during single boiling, which further leads to a linear increase of heat transfer capacity. But because of the cooling capacity, the boiling frequency does not increase linearly.

Author Contributions: Conceptualization, H.Z. and H.G.; Data curation, H.Z.; Formal analysis, H.Z.; Funding acquisition, F.Y. and X.Y.; Investigation, H.Z.; Methodology, H.Z.; Project administration, F.Y. and X.Y.; Resources, F.Y. and X.Y.; Supervision, H.G. and F.Y.; Visualization, H.Z.; Writing-original draft, H.Z.; Writing-review and editing, H.G. and F.Y. All authors listed have made a substantial, direct, and intellectual contribution to the work. All authors have read and agreed to the published version of the manuscript.

Funding: This work was funded by the National Key R&D Program of China [grant number 2017YFF0205901].

Data Availability Statement: The supporting data will be made available on request.

Conflicts of Interest: The authors declare that they have no conflict of interest.

Nomenclature

A	Surface area, m^2
A_{hp}	Heat pipe axial section area, m^2
c_p	Specific heat of cooling water, J/(kg·K)
D	Height of vapor space, m
G	Mass flow of cooling water, kg/s
h	Heat transfer coefficient, W/(m^2·K)
h_{ja}	Convective heat transfer coefficient between outer wall of cooling water jacket and air, W/(m^2·K)
K	Equivalent heat transfer coefficient of the heat pipe, W/(m^2·K)
M	Molar mass, g/mol;
P	Measuring point
Q	Heat transfer quantity of heat pipe, W
R	Thermal resistance, °C/W
R_g	Universal gas constant, J/(K·mol)
T	Time-averaged temperature, °C
T_{tr}	Vapor transition temperature, K

Greek symbols

ρ	Density, kg/m^3
μ	Viscosity, Pa·s
λ	Length of mean free path of vapor, m

Subscripts

1–15	Measuring point 1 to 15
a	Adiabatic section
c	Condenser section
e	Evaporation section
j	Water jacket
v	Vapor

References

1. Panda, K.K.; Dulera, I.V.; Basak, A. Numerical simulation of high temperature sodium heat pipe for passive heat removal in nuclear reactors. *Nucl. Eng. Des.* **2017**, *323*, 376–385. [CrossRef]
2. Kusuma, M.H.; Putra, N.; Antariksawan, A.R.; Koestoer, R.A.; Widodo, S.; Ismarwanti, S.; Verlambang, B.T. Passive cooling system in a nuclear spent fuel pool using a vertical straight wickless-heat pipe. *Int. J. Therm. Sci.* **2018**, *126*, 162–171. [CrossRef]
3. Yan, X.K.; Duan, Y.N.; Ma, C.F.; Lv, Z.F. Construction of sodium heat-pipe furnaces and the isothermal characteristics of the furnaces. *Int. J. Thermophys.* **2011**, *32*, 494–504. [CrossRef]
4. Astrua, A.M.; Iacomini, L.; Battuello, M. The combined use of a gas-controlled heat pipe and a copper point to improve the calibration of thermocouples up to 1100 °C. *Int. J. Thermophys.* **2008**, *32*, 1838–1847. [CrossRef]
5. Bertiglia, F.; Iacomini, L.; Moro, F.; Merlone, A. Comparison of two potassium-filled gas-controlled heat pipes. *Int. J. Thermophys.* **2015**, *32*, 3393–3403. [CrossRef]
6. Mahboobe, M.; Qiu, S.G.; Saeed, T. Numerical investigation of hydrodynamics and thermal performance of a specially configured heat pipe for high-temperature thermal energy storage systems. *Appl. Therm. Eng.* **2015**, *81*, 325–337.
7. Liao, Z.; Faghri, A. Thermal analysis of a heat pipe solar central receiver for concentrated solar power tower. *Appl. Therm. Eng.* **2016**, *102*, 952–960. [CrossRef]
8. Yuan, Y.; Shan, J.Q.; Zhang, B.; Gou, J.L.; Zhang, B.; Lu, T.Y.; Ge, L.; Yang, Z.J. Study on startup characteristics of heat pipe cooled and AMTEC conversion space reactor system. *Prog. Nucl. Energy* **2016**, *86*, 18–30. [CrossRef]
9. Wang, C.L.; Liu, L.; Liu, M.H.; Zhang, D.L.; Tian, W.X.; Qiu, S.Z.; Su, G.H. Conceptual design and analysis of heat pipe cooled silo cooling system for the transportable fluoride-salt-cooled high-temperature reactor. *Ann. Nucl. Energy* **2017**, *109*, 458–468. [CrossRef]

10. Liu, M.H.; Zhang, D.L.; Wang, C.L.; Qiu, S.Z.; Su, G.H.; Tian, W.X. Experimental study on heat transfer performance between fluoride salt and heat pipes in the new conceptual passive residual heat removal system of molten salt reactor. *Nucl. Eng. Des.* **2018**, *339*, 215–224. [CrossRef]
11. Behi, H.; Kalogiannis, T.; Patil, M.S.; Mierlo, J.V.; Berecibar, M. A new concept of air cooling and heat pipe for electric vehicles in fast discharging. *Energies* **2021**, *14*, 6477. [CrossRef]
12. Xiao, H.; George, F. Design and fabrication of hybrid bi-modal wick structure for heat pipe application. *J. Porous Mat.* **2008**, *15*, 635–642.
13. Behi, H.; Behi, M.; Ghanbarpour, A.; Karimi, D.; Azad, A.; Ghanbarpour, M.; Behnia, M. Enhancement of the thermal energy storage using heat-pipe-assisted phase change material. *Energies* **2021**, *14*, 6176. [CrossRef]
14. Chen, J.; Dong, J.B.; Yao, Y. Experimental study on the starting-up and heat transfer characteristics of a pulsating heat pipe under local low-frequency vibrations. *Energies* **2021**, *14*, 6310. [CrossRef]
15. Anderson, W.G. Sodium-Potassium (NaK) Heat Pipe. In Proceedings of the 29th National Heat Transfer Conference, Atlanta, GA, USA, 8–11 August 1993; Volume 236, pp. 47–53.
16. Serizawa, A.; Ida, T.; Takahashi, O.; Michiyoshi, I. MHD effect on NaK-nitrogen two-phase flow and heat transfer in a vertical round tube. *Int. J. Multiph. Flow* **1990**, *16*, 761–788. [CrossRef]
17. Schriener, T.M.; El-Genk, M.S. Convection heat transfer of NaK-78 liquid metal in a circular tube and a tri-lobe channel. *Int. J. Heat Mass Transf.* **2015**, *86*, 234–243. [CrossRef]
18. Ji, Y.L.; Wu, M.K.; Feng, Y.M.; Yu, C.R.; Chu, L.L.; Chang, C.; Li, Y.T.; Xiao, X.; Ma, H.B. An experimental investigation on the heat transfer performance of a liquid metal high-temperature oscillating heat pipe. *Int. J. Heat Mass Transf.* **2020**, *149*, 119198. [CrossRef]
19. Wang, C.L.; Guo, Z.P.; Zhang, D.L.; Qiu, S.Z.; Tian, W.X.; Wu, Y.W.; Su, G.H. Transient behavior of the sodium-potassium alloy heat pipe in passive residual heat removal system of molten salt reactor. *Prog. Nucl. Energy* **2013**, *68*, 142–152. [CrossRef]
20. Li, T.; Jiang, Y.Y.; Li, Z.G.; Liu, Q.; Tang, D.W. Loop thermosiphon as a feasible cooling method for the stators of gas turbine. *Appl. Therm. Eng.* **2016**, *109*, 449–453. [CrossRef]
21. Jia, X.J.; Guo, H.; Guo, Q.; Yan, X.K.; Ye, F.; Ma, C.F. Effect of heating temperature on start-up and heat transfer performance of Na-K alloy heat pipe. *Acta Energ. Sol. Sin.* **2019**, *40*, 17–23. (In Chinese)
22. Zhang, H.Z.; Jia, X.J.; Guo, H.; Guo, Q.; Yan, X.K.; Ye, F.; Ma, C.F. Effect of cooling water parameters on heat transfer performance of sodium-potassium alloy heat pipe. *J. Chem. Ind. Eng.* **2017**, *S1*, 105–110. (In Chinese)
23. Guo, H.; Guo, Q.; Yan, X.K.; Ye, F.; Ma, C.F. Experimental investigation on heat transfer performance of high-temperature thermosyphon charged with sodium-potassium alloy. *Appl. Therm. Eng.* **2018**, *139*, 402–408. [CrossRef]
24. Guo, Q.; Guo, H.; Yan, X.K.; Ye, F.; Ma, C.F. Influence of inclination angle on the start-up performance of a sodium-potassium alloy heat pipe. *Heat Transf. Eng.* **2017**, *4*, 1–9. [CrossRef]
25. Noie, S.H.; Sarmastiemami, M.R.; Khoshnoodi, M. Effect of inclination angle and filling ratio on thermal performance of a two-phase closed thermosyphon under normal operating conditions. *Heat Transfer. Eng.* **2007**, *28*, 365–371. [CrossRef]
26. Emani, M.R.S.; Noie, S.H.; Khoshnoodi, M.; Mosavian, M.T.H.; Kianifar, A. Investigation of geyser boiling phenomenon in a two-phase closed thermosyphon. *Heat Transfer Eng.* **2009**, *30*, 408–415. [CrossRef]
27. Lin, T.F.; Lin, W.T.; Tsay, Y.L.; Wu, J.C. Experimental investigation of geyser boiling in an annular two phase closed thermosyphon. *Int. J. Heat Mass Transfer* **1995**, *38*, 295–307. [CrossRef]
28. Wang, X.; Wang, Y.; Chen, H.; Zhu, Y.A. Combined CFD/visualization investigation of heat transfer behaviors during geyser boiling in two-phase closed thermosyphon. *Int. J. Heat Mass Transfer* **2018**, *121*, 703–714. [CrossRef]
29. Casarosa, C.; Latrofa, E.; Shelginski, A. The geyser effect in a two phase thermosyphon. *Int. J. Heat Mass Transfer* **1983**, *6*, 933–941. [CrossRef]
30. Kujawska, A.; Zajaczkowski, B.; Wilde, L.M.; Buschmann, M.H. Geyser boiling in a thermosyphon with nanofluids and surfactant solution. *Int. J. Therm. Sci.* **2019**, *139*, 195–216. [CrossRef]
31. Morgan, S.K.; Brady, H.F. Elimination of the geyser effect in missiles. *Adv. Cryog. Eng.* **1962**, *7*, 206–213.
32. Howard, F.S. Geysering inhibitor for vertical cryogenic transfer piping. *Adv. Cryog. Eng.* **1973**, *18*, 162–169.
33. Kuncoro, H.; Rao, Y.F.; Fukuda, K. An experimental study on the mechanism of geysering in a closed two-phase thermosyphon. *Int. J. Multiph. Flow* **1995**, *21*, 1243–1252. [CrossRef]
34. Tecchio, C.; Oliveira, J.L.G.; Paiva, K.V.; Mantelli, M.B.H.; Galdolfi, R.; Ribeiro, L.G.S. Geyser boiling phenomenon in two-phase closed loop-thermosyphons. *Int. J. Heat Mass Transfer* **2017**, *111*, 29–40. [CrossRef]
35. Bao, K.L.; Wang, X.H.; Fang, Y.B.; Jia, X.S.; Han, X.H.; Chen, G.M. Effects of the surfactant solution on the performance of the pulsating heat pipe. *Appl. Therm. Eng.* **2020**, *178*, 115678. [CrossRef]
36. Jang, J.H. Startup characteristics of a potassium heat pipe from the frozen state. *J. Thermophys. Heat Transf.* **1995**, *9*, 117–122. [CrossRef]
37. Wang, C.L.; Liu, X.; Liu, M.H.; Tang, S.M.; Tian, Z.X.; Zhang, D.L.; Tian, W.X.; Qiu, S.Z.; Su, G.H. Experimental study on heat transfer limit of high temperature potassium heat pipe for advanced reactors. *Ann. Nucl. Energy* **2021**, *151*, 107935. [CrossRef]
38. Mahan, J.R.; Felske, J.D. Radiation heat transfer: A statistical approach. *Appl. Mech. Rev.* **2002**, *56*, B15–B16. [CrossRef]

electronics

MDPI

Article

Magnetic Field Saturation of Non-Insulation High-Temperature Superconducting Coils during Overcurrent

Wei Wu *, Yusong Gao and Zhijian Jin

School of Electronic Information and Electrical Engineering, Shanghai Jiao Tong University, Shanghai 200240, China; yusongao@sjtu.edu.cn (Y.G.); zjjin@sjtu.edu.cn (Z.J.)
* Correspondence: wei.wu@sjtu.edu.cn

Abstract: Non-insulation high-temperature superconducting coils provide a much lower risk of burnout in fault/abnormal conditions, such as hot-spot quench and overcurrent. This study employs an equivalent circuit grid model, coupled with magnetic field calculation and the *E–J* power law of superconductors, to deeply and systematically investigate the overcurrent charging process in a double-pancake non-insulation coil. An evident saturation of the magnetic field in the axial direction of the coil was observed and verified by experiments. Experimentally, the entire process, including the behavior of the magnetic field, was consistent with the numerical results. Based on the verified model, two main points were addressed: (1) Transient current distribution inside the coil during overcurrent charging was studied. Potential quenching risks were found to be at the innermost and outermost turn near the electrodes, as well as the pancake-to-pancake connection part. (2) Magnetic field saturation, which is a unique phenomenon in non-insulation superconducting coils during overcurrent charging, was studied in detail and first quantitatively defined by a new concept "converged load factor". Its relationship with turn-to-turn resistivity was revealed.

Keywords: high-temperature superconductor; non-insulation coil; overcurrent; numerical modeling; magnetic field saturation

Citation: Wu, W.; Gao, Y.; Jin, Z. Magnetic Field Saturation of Non-Insulation High-Temperature Superconducting Coils during Overcurrent. *Electronics* **2021**, *10*, 2789. https://doi.org/10.3390/electronics10222789

Academic Editors: Alon Kuperman and Alessandro Lampasi

Received: 14 October 2021
Accepted: 12 November 2021
Published: 14 November 2021

1. Introduction

One of the advantages of non-insulation (NI) high-temperature superconducting (HTS) coils is the capability to operate under the fault (typically overcurrent and hot-spot quench) conditions with a much lower risk of burnout [1]. The high stability against fault conditions [2,3] puts NI coils forward as a promising option for the large-scale application of DC high-temperature superconducting magnets, such as in maglev trains [4], motors [5,6], TOKAMAK systems [7], and high-field NMR [8]. Overcurrent conditions with an operating current higher than the critical current sometimes occur and play an important role in achieving a maximum field and in the elimination of screening currents [7].

Several studies have focused on the performance of NI–HTS coils operating under overcurrent conditions. Experimentally, a saturation of the magnetic field was observed in various overcurrent tests [7,9–11], and burnout [11] of both the innermost and outermost turns close to the electrodes occurred [12]. Similar phenomena were also observed during overcurrent tests of HTS coils with turn-to-turn metal insulation [13,14]. Numerically, the partial element equivalent circuit (PEEC) [15] or equivalent circuit grid (ECG) [16], coupled with a thermal model, which was proposed to numerically analyze the behaviors of NI coils during overcurrent conditions, suggested that a considerable amount of Joule heat is generated near the outer electrode, initiating the quench propagation and ascribing the magnetic field saturation to a local decrease in critical current [17]. Overcurrent test results in [18,19] also demonstrate the capability of NI coils to operate steadily in the saturated phase without leading to thermally induced degradation.

To further investigate the evolution of current distributions, understand the more quantitative characteristics during the transition to the magnetic-field saturation phase, and clarify the potential risks of NI coils during overcurrent excitation, this study adopted an equivalent circuit grid model [15–17] coupled with a magnetic field the *E–J* power law of superconductors, calculated by the method mentioned in [20], to realize a real-time circuit-field simulation. In addition, a double pancake NI test coil was wound and charged with exquisite excitation procedures to validate the model. The responses of both the magnetic field and coil voltage were recorded and compared with the numerical results. The main contributions of this paper are as follows: (1) Detailed current distributions inside the coil during overcurrent charging are presented and discussed. Potential quenching risks were found to be at the innermost and outermost turn near the electrodes, as well as at the pancake-to-pancake connection part. (2) Magnetic field saturation, which is a unique phenomenon in non-insulation superconducting coils during overcurrent charging, was studied in detail and first quantitatively defined by a new concept "converged load factor". Its relationship with turn-to-turn resistivity was revealed.

2. Model Description

2.1. Double-Pancake ECG Model

The ECG model [16] in Figure 1 was adopted to calculate the distributions of both the spiral and radial currents of a NI–DP coil. The number of total turns in one pancake was defined as N_t. Each turn was equally divided into N_e arc elements. The total number of spiral and radial elements for one pancake was defined as $N_i = N_t \times N_e$ and $N_j = (N_t - 1) \times N_e$, respectively. Each spiral element consists of its own inductance, its mutual inductance with the other elements, and spiral resistance. The *k*-th spiral resistance $R_{i,k}$ is composed of the resistance of the HTS layer R_{sc} and that of the other metal layers R_{mt}, as illustrated in Figure 2. $R_{j,k}$ represents the *k*-th radial resistance.

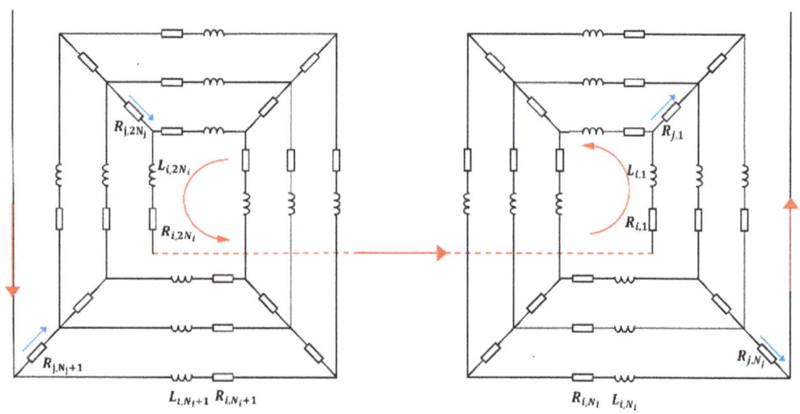

Figure 1. ECG model of a NI–DP coil.

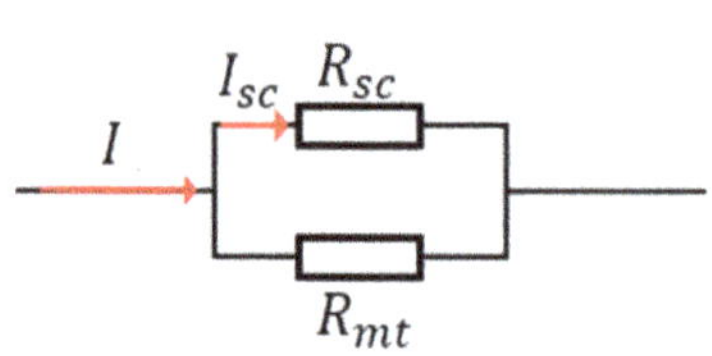

Figure 2. Circuit of a spiral element.

Considering that a DP coil is wound spirally with a single tape, the upper pancake is reflectional rather than translationally symmetric [21] with respect to the lower pancake, which should be considered when numbering spiral elements and calculating the mutual inductance between an arc element of the upper pancake and one of the lower pancakes.

The relationship between the spiral and radial currents of the DP ECG model can be obtained according to Kirchhoff's law at each circuit node. The governing equations are the following Equation (1):

$$\begin{cases} I_k - I_{k+1} - J_k = 0 & ;k \in [1, N_e] \\ I_k - I_{k+1} + J_{k-N_e} - J_k = 0 & ;k \in [N_e + 1, N_j] \\ I_k - I_{k+1} + J_{k-N_e} = 0 & ;k \in [N_j + 1, N_i - 1] \\ I_k + J_{k-N_e} = I_{op} & ;k \in [N_i, N_i + 1] \\ I_k - I_{k-1} + J_{k-N_e} = 0 & ;k \in [N_i + 2, N_i + N_e] \\ I_k - I_{k-1} + J_{k-N_e} - J_{k-2N_e} = 0 & ;k \in [N_i + N_e + 1, N_i + N_j - 1] \\ I_k - I_{k-1} - J_{k-2N_e} = 0 & ;k \in [N_i + N_j, 2N_i] \end{cases} \tag{1}$$

where I_k, J_k, and I_{op} denote the current in the k-th spiral element, radial element, and power supply, respectively.

The governing equations of each circuit loop derived from Kirchhoff's voltage law are the following Equation (2):

$$\begin{cases} \sum\limits_{p=2}^{N_e+1} U_p - J_k R_{j,k} = 0 & ;k = 1 \\ U_k - U_{k+N_e} - J_{k-1} R_{j,k-1} + J_k R_{j,k} = 0 & ;k \in [2, 2N_j - 1] \\ \sum\limits_{p=2N_i-N_e}^{2Ni} U_p - J_k R_{j,k} = 0 & ;k = 2N_j \end{cases} \tag{2}$$

where U_k denotes the voltage drop along the k-th spiral element, consisting of both the inductive and resistive voltages, as shown by the following Equation (3):

$$U_k = \sum_{m=1}^{2N_i} M_{k,m} \frac{dI_m}{dt} + I_k R_{i,k} \tag{3}$$

where $M_{k,m}$ represents the self-inductance of the k-th spiral element if $k = m$ and the mutual inductance between the k-th and m-th spiral elements if $k \neq m$. The self-inductance and mutual inductance are calculated by integrating Neumann's formula [22,23].

Equations (1)–(3) can be expressed in a matrix form (Equation (4)):

$$\begin{cases} A_1 I + A_2 J = b \\ B_1 \frac{dI}{dt} + B_2 I + B_3 J = 0 \end{cases} \tag{4}$$

where $I = [I_1\ I_2 \ldots I_{2Ni}]^T$ and $J = [J_1\ J_2 \ldots J_{2Nj}]^T$.

For the aforementioned ECG model [16], A_1 is always a non-singular square matrix, and consequently, unlike the previously proposed method [16], the radial current vector J is selected as the state variable, and the spiral current vector I can be derived, as shown by Equation (5).

$$I = A_1^{-1}(b - A_2 J) \tag{5}$$

To solve the system of ordinary differential Equation (4), iterative methods including the Runge–Kutta fourth-order method were adopted, and the calculation and post-processing were conducted in MATLAB R2021b. The geometry of the coil in profiles of current distribution [24,25] in the radial direction was enlarged for better illustration.

2.2. Coupling of Magnetic Fields and the DP ECG Model

To calculate the field-dependent critical current effectively, a two-dimensional axisymmetric model mentioned in [20] was used as Equation (6).

$$\begin{aligned} B(r,\phi,z) &= -\mu_0 I_\phi \frac{\partial A(r,\phi,z)}{\partial z}\hat{r} + \mu_0 I_\phi \frac{1}{r}\frac{\partial (rA(r,\phi,z))}{\partial r}\hat{z} \\ &= B_{\text{per}}\hat{r} + B_{\text{par}}\hat{z} \end{aligned} \tag{6}$$

The magnetic vector potentials $A(r,\phi,z)$ can be calculated by integrating the current density multiplied by an integral kernel [20]. Numerically, only two linear transformations are needed to obtain the parallel component B_{par} and perpendicular component B_{per} of the magnetic field by multiplying the current density with two pre-calculated constant matrices. Therefore, the coupling of the magnetic field and the DP ECG model can be performed within several milliseconds. The calculated parallel and perpendicular components of the magnetic field B_{par} and B_{per} are used to calculate the field-dependent critical current by Equation (7) [26,27]:

$$I_c(\mathbf{B}) = I_{c0} \times \left[1 + \frac{\sqrt{(kB_{\text{par}})^2 + B_{\text{per}}^2}}{B_c}\right]^{-\alpha} \tag{7}$$

where I_{c0} = 167 A, k = 0.518, α = 0.74, and B_c = 106 mT. The parameters are obtained by fitting the above elliptical function [26,27] with the measured data of a short sample under an external parallel and perpendicular magnetic field.

The high nonlinearity of the E–J power law causes the resistance of spiral elements to change drastically with respect to the ratio of I/I_c. The following model [28] shown by Figure 2 and Equation (8), consisting of two parallel resistances, is used to calculate the overall resistance of spiral elements based on the rule of mixtures [29]. The n-value was set to be 21.

$$\frac{E_0 l}{R_{\text{mt}}}\left(\frac{I_{\text{sc}}}{I_c}\right)^n + I_{\text{sc}} = I \tag{8}$$

where l denotes the length of a spiral element. Once the field-dependent critical current I_c is obtained from Equation (7), the nonlinear Equation (8) can be solved by iterative methods such as Newton's method to obtain I_{sc} (the current in the HTS layer), and the overall resistance R_i can be calculated by Equation (9).

$$\begin{cases} R_{\text{sc}} = \frac{E_0 l}{I_c}\left(\frac{I_{\text{sc}}}{I_c}\right)^{n-1} \\ R_i = \frac{R_{\text{sc}} R_{\text{mt}}}{R_{\text{sc}} + R_{\text{mt}}} \end{cases} \tag{9}$$

In terms of overcurrent excitation, the presented model focused on the field–circuit coupling and real-time update of the spiral resistance with time-varying conditions of currents and magnetic fields.

The high nonlinearity of the E–J power law also makes the system of the odes in Equation (4) difficult to converge, particularly when local I/I_c is approximately 1. It is recommended that the larger the I/I_c ratio, the smaller the time step that should be used to obtain converged solutions.

3. Experimental Setup and Procedure

3.1. Experimental Setup

The experimental DP coil was wound with a single piece of GdBCO tape, and the specifications of the tape and coil are listed in Table 1.

Table 1. Details of the tape and coil.

Tape Parameters	Value
Manufacturer	Shanghai Superconductor Technology Co., Ltd., Shanghai, China
Tape width/thickness	6 [mm]/185 [μm]
HTS layer width/thickness	4.75 [mm]/1 [μm]
Copper stabilizer thickness	40 [μm]
Hastelloy substrate thickness	50 [μm]
Coil Parameters	**Value**
Total turns	35 × 2
Coil inner radius	5 [mm]
Distance between the upper pancake and the lower pancake	3 [mm]
Total tape length	23.46 [m]
HTS leads length	0.38 [m] × 2
Self-inductance	815.45 [μH]
Characteristic resistance	27.725 [μΩ]
Calculated turn-to-turn contact resistivity (ρ_{ct})	6.47 [μΩ cm^2]
Magnet constant	0.7 [mT/A]

The measured critical current of the coil was 114.6 A at 77 K according to the 1 μV/cm criterion. A Hall sensor was fixed at 1.8 cm above the center of the DP coil along the z-axis, as illustrated in Figure 3.

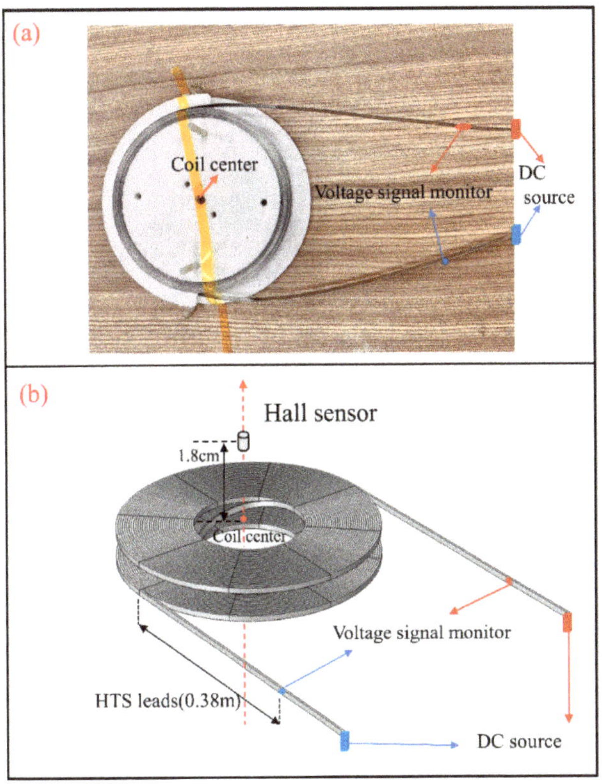

Figure 3. (**a**) Photography and (**b**) schematic diagram of the sample NI–DP coil.

Notably, there was a 0.38 m long tape extending from the outermost turn of each pancake to attach to the DC source, preventing the heat generated in the copper terminals from affecting the coil, and the voltage drop along the HTS leads should be considered, particularly in overcurrent situations.

3.2. Experimental Procedure

The step-rising excitation method was adopted to measure the stabilized magnetic flux density and voltage signal of the DP coil operating at currents of different amplitudes.

First, the DP coil was excited to 10 A, and when the voltage and magnetic field signals were stabilized, the power supply current increased by 10 A at a rate of 0.278 A/s. When the voltage was no longer stabilized at approximately 0 V, which indicated a clear index loss occurring along the spiral HTS layer, the power supply current increased by 5 A instead until it reached 150 A. Thereafter, the power supply current started to decrease at 5705 s from 150 to 0 A at a rate of 1 A/s. When the coil was completely demagnetized, normal charge (50.1 A) and sudden discharge tests (from 6400 to 7200 s) were conducted to confirm whether the coil was degraded under the previous overcurrent test. Finally, after confirming that the coil was normal, overcurrent excitation (120.3 A) and sudden discharge tests (from 7200 to 8000 s) were performed to cross-validate the performance of the coil under overcurrent conditions, with the test performed from 0 to 6400 s. The key time points and corresponding power supply currents are listed in Table 2.

Table 2. Time points and power supply current.

Time [s]	I_{op} [A]	Time [s]	I_{op} [A]
56	10	3504	110.2
380	20.2	3665	115
689	30.3	3904	120.3
968	40.2	4564	125.1
1226	50.1	4908	130.2
1494	60.2	5161	135
1730	70.1	5343	140.1
2020	75.2	5493	145.1
2283	80	5624	150
2515	85.1	5705	150
2751	90.1	6222	0
2978	95.2	6657	50.1
3153	100	7239	0
3346	105.1	7450	120.3

4. Results and Verification of Numerical Model

The experimental data and the numerical results are shown and compared in Figure 4, where B_{exp}, B_{sim}, I_{op}, U_{exp}, and U_{sim} denote the experimental magnetic field flux density, simulated magnetic field flux density, current of power supply, experimental coil voltage, and simulated coil voltage, respectively. U_{sim} denotes the sum of the voltage drops of each spiral element in the DP–ECG model as defined in Equation (10).

$$U_{sim} = \sum_{k=1}^{2N_i} \left(\sum_{m=1}^{2N_i} M_{k,m} \frac{dI_m}{dt} + I_k R_{i,k} \right) \tag{10}$$

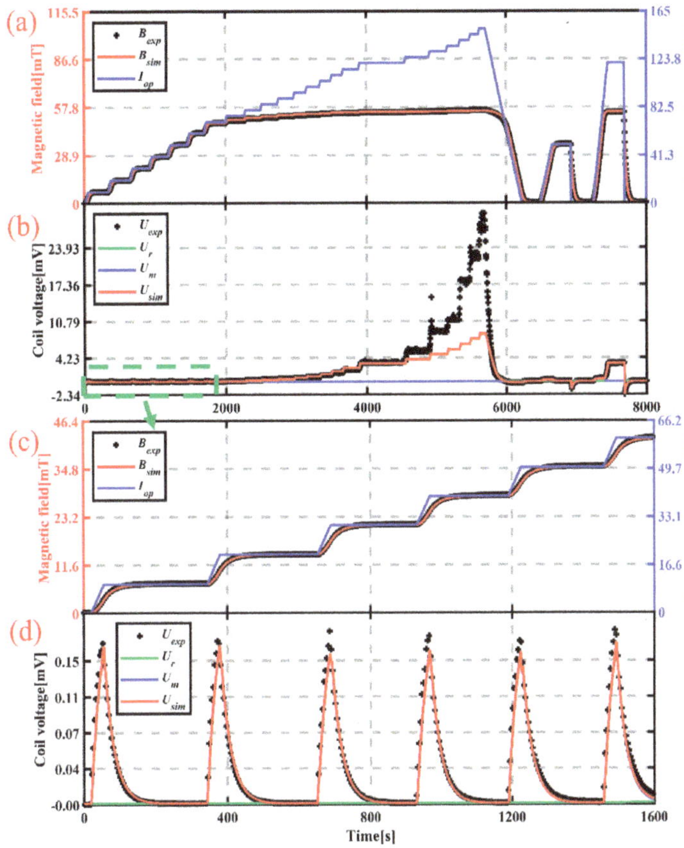

Figure 4. Experimental and numerical results of center magnetic field and terminal voltage (**a**,**b**) from 0 to 8000 s and (**c**,**d**) from 0 to 1600 s.

4.1. Behavior of Magnetic Flux Density during the Overcurrent Charging Process

In terms of the magnetic field, the calculated results are consistent with the experimental results, as shown in Figure 4.

From 0 to 2000 s, the NI–DP coil was charged to 10.0, 20.2, 30.3, 40.2, 50.1, 60.2, and 70.1 A, and at every stage, the DC source was maintained for a few seconds to record the stabilized magnetic field flux density and voltage signals. The ratio of the stabilized center magnetic field flux density to the power supply current remained at approximately 0.7 mT/A, which can be described as the linear increase process [17,19]. The voltage's responses to the increase in the operating current by 10 A at a rate of 0.278 A/s were almost identical, and the terminal voltage dropped to nearly zero when the operating current stabilized at each stage. It is clear that the current only flowed along the spiral HTS layer path and the resistance of the HTS layer was almost null when the steady operating current was far below the critical current.

As shown in Figure 4, starting from 2000 s, the power supply current increased from 70.1 A, with an amplitude of 5 A each time at a rate of 0.278 A/s. The magnetic field flux density no longer corresponded linearly to the stabilized operating current but reached a plateau. The maximum center magnetic field flux density that the NI–DP coil could generate was approximately 55 mT, and the average current flowing along the spiral path was 78.6 A, which was estimated by the magnet constant 0.7 mT/A. The overloaded input current flowed along the radial path [30], which provided no significant contribution to the magnetic field but resulted in a terminal voltage rise and Joule heat.

It was observed that the Joule heat led to no irreversible quenching during the entire test, mainly attributed to the equivalent contact resistivity, which was found to be as low as 6.47 $\mu\Omega{\cdot}cm^2$, and the small amount of generated heat could be dissipated quickly in an LN_2 bath.

The NI–DP coil exhibited great stability during the overcurrent test from 0 to 6400 s. The subsequent normal charge and sudden discharge test from 6400 to 7200 s indicated that the coil was not degraded during the previous overcurrent test, as the DP coil maintained the same capability of generating a saturated magnetic field and the coil voltage dropped to nearly zero with the power supply current stabilized at 50.1 A, which was much lower than the critical current of the coil.

The final test from 7200 to 8000 s exhibited similar performance to the NI–DP coil in an overcurrent state, and the maximum center magnetic field flux density stabilized at approximately 55 mT, which was consistent with the previous test from 0 to 6400 s. The overloaded current caused the terminal voltage to stabilize at 3.3 mV, as the power supply current remained at 120.3 A.

4.2. Voltage Drop in HTS Leads

As shown in Figure 4, the simulated magnetic field exhibited acceptable consistency with experimental results throughout the entire test process, whereas the calculated coil voltage exhibited a clear inconsistency from 4564 to 5800 s, during which the magnetic field remained saturated and relatively stable even as the power supply current increased further, but the coil voltage increased more dramatically, compared with the calculated results.

The main discrepancy in the terminal voltage from 4564 to 5800 s occurred when the power supply current exceeds 120.3 A, while there were no marked fluctuations in the magnetic field. The extra voltage drop could be attributed to the HTS leads extending from the outermost turn of each pancake, as depicted in Figure 3, because the operating current flowing in the HTS leads was as large as the power supply current and had no significant contribution to the magnetic field but was capable of generating a considerable voltage drop along the voltage measuring path.

The model depicted in Figure 2 was used to numerically verify the presented ratiocination. The extra voltage drop along the HTS leads U_l can be estimated using the following Equation (11):

$$U_l(t) = 2\int_0^{l_{ds}} E_0\left(\frac{I_{sc}(t)}{I_c(l)}\right)^n dl \approx 2E_0 \times l_{ds} \times \left(\frac{I_{sc}(t)}{I_{ceq}}\right)^n \quad (11)$$

where l_{ds} denotes the length of a single HTS lead, which is 0.38 m, and E_0 is the critical electric field (1 μV/cm and n = 21).

As the magnetic field remained almost stable from 4564 to 5800 s, it is reasonable to adopt a time-independent and approximated critical current of the entire HTS leads, denoted as I_{ceq}. $I_{sc}(t)$ is the current flowing along the HTS layer of the leads with respect to $I_{op}(t)$, which can be calculated by solving the non-linear Equation (8). Figure 5 shows that with I_{ceq} equal to 107.5 A, the extra voltage can be explained.

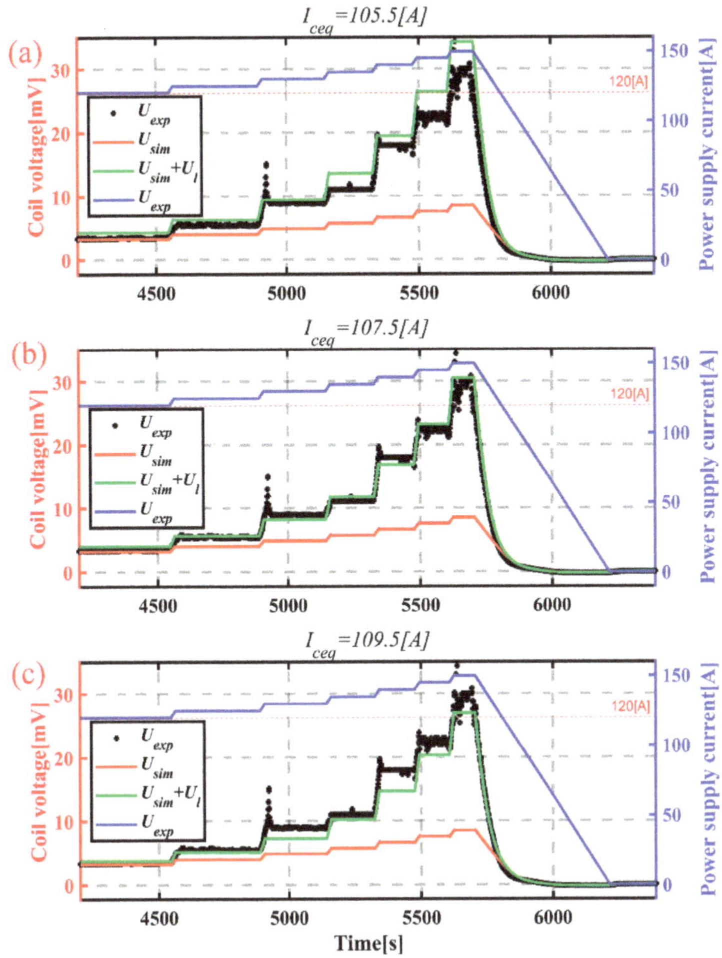

Figure 5. Extra voltage drop U_l along HTS leads with I_{ceq} of (**a**) 105.5 A, (**b**) 107.5 A, and (**c**) 109.5 A.

5. Transient Behavior during Overcurrent Charging

Both the magnetic field and coil voltage results exhibited good consistency with the experimental results from our analysis. Thus, this model can be used to investigate the mechanism of the transition from the normal charging condition to the overcurrent charging condition of the magnetic field and the detailed current distributions inside the coil.

5.1. Analysis of the Transition Process to Saturation of Magnetic Field

The experimental and computed results in Figure 6a from 6400 to 8000 s were selected to compare the difference in the performance of the NI–DP coil between normal excitation and overcurrent excitation. The term U_{sim} is decomposed into the resistive component U_r and inductive component U_m, which are calculated using Equations (12) and (13), respectively.

$$U_r = \sum_{k=1}^{2N_i} I_k R_{i,k} \tag{12}$$

$$U_m = \sum_{k=1}^{2N_i} \left(\sum_{m=1}^{2N_i} M_{k,m} \frac{dI_m}{dt} \right) \tag{13}$$

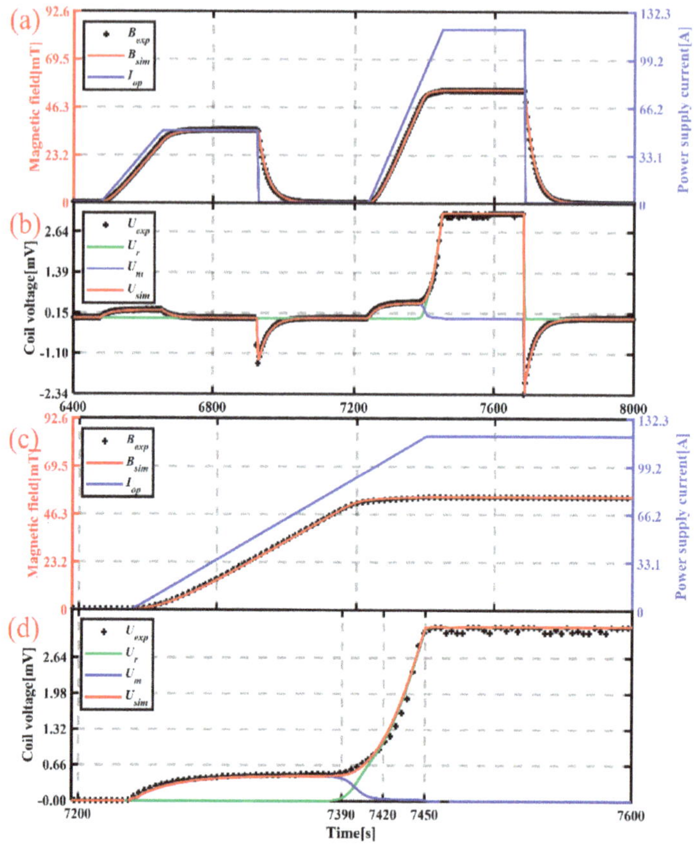

Figure 6. Experimental and numerical results (**a**,**b**) from 6400 to 7600 s and (**c**,**d**) from 7200 to 7600 s.

As shown in Figure 6a, the resistive voltage U_r was nearly zero before the power supply current exceeded 86.14 A at 7390 s for the first time and during the sudden discharge process. At 7390 s, as shown in Figure 6b, U_r started increasing rapidly, and the magnetic field began to saturate, whereas U_m decreased gradually to zero at 7420 s, even with a ramped-up power supply.

Thus, from the perspective of voltage signals, the simulated results suggested that as the spiral current increased, the increased index resistance of the spiral path, in turn, drove the ascending rate of the spiral current down, which was reflected in the decreasing inductive voltage and, consequently, the saturation of the magnetic field (from 7390 to 7420 s).

5.2. Current Distributions Inside the Coil

The profiles of the spiral current, load factor I/I_c, and radial current density at 6800, 7390, and 7600 s are illustrated in Figure 7.

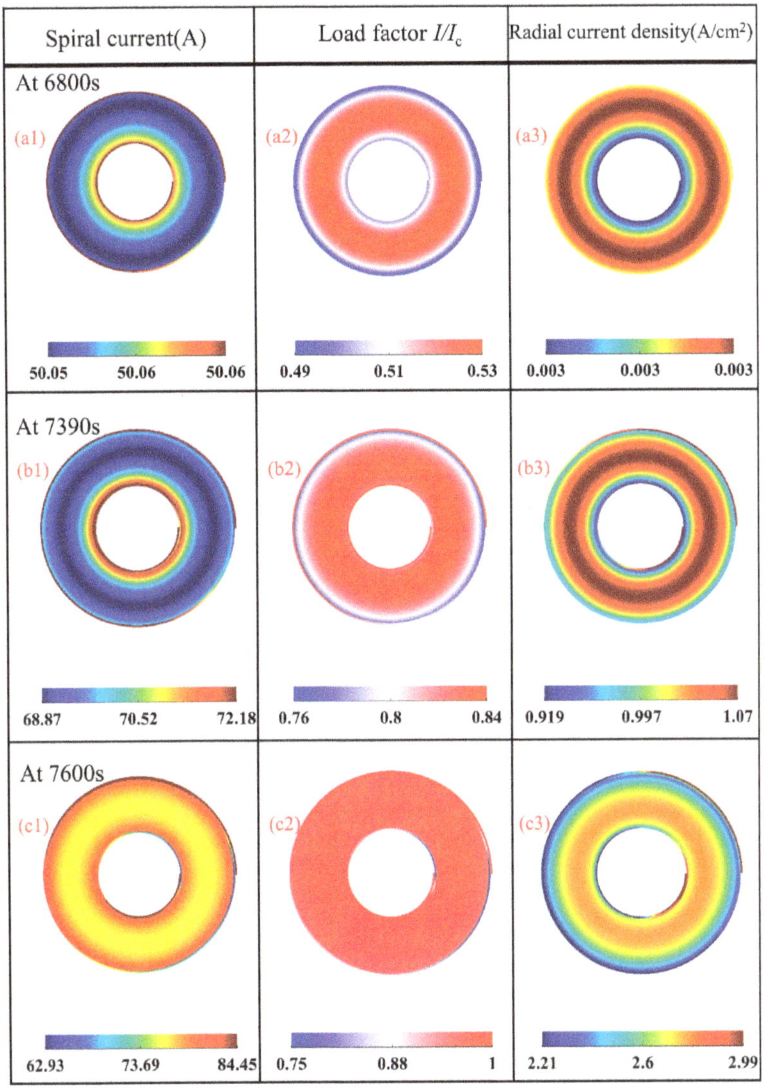

Figure 7. Profiles of current distribution and load factor of the upper pancake at 6800, 7390, and 7600 s. (**a1–c1**) Distribution of spiral current at 6800 s, 7200 s and 7600 s, respectively; (**a2–c2**) Distribution of load factor at 6800 s, 7200 s and 7600 s, respectively; (**a3–c3**) Distribution of radial current density at 6800 s, 7200 s and 7600 s, respectively.

It was observed that the spiral current was almost evenly distributed along the spiral HTS layer path and was close to the power supply current maintained at 50.1 A at 6800 s, while the radial current density was rather small. In contrast, the radial current density was much larger when I_{op} was maintained at 120.3 A at 7600 s, as depicted in Figure 7(a1,a3,c1,c3).

At 7390 s, when the power supply current reached 86.15 A and the magnetic field started to saturate, the overall index resistance was sufficiently large to initiate saturation even when none of the local load factors I/I_c exceeded 0.85, as shown in Figure 7(b2).

As shown in Figure 7(c1,c3), the highest spiral current and radial current density were both distributed in the outermost turn near the electrodes and innermost turn near the port where the spiral current flowed to the other pancake during the overcurrent state. A similar phenomenon was observed in [15] during charging without overcurrent.

The largest local load factors (>1) were located near the current input and output parts of the innermost and outermost turns, as depicted in Figure 7(c2). This caused considerable spiral resistance, impeding the operating current flowing along the spiral path but making it flow along the radial path to the middle turns. This implied that both the innermost and outermost turns of each pancake near the ports where the power supply current was input and output were the most vulnerable areas when NI–DP coils operated under overcurrent conditions with bad cooling conditions or when the power supply current increased further.

This deduction may provide a reasonable explanation for the overcurrent tests of NI coils [11], and even coils with turn-to-turn metal insulation [13], which reported burnout of the innermost and outermost turns near the electrodes—that is, irreversible quenching was initiated at both the innermost and outermost turns close to current terminals, owing to excessive overcurrent excitation, and thereafter propagated to the middle turns. The ratiocination was different from the analysis in [18], which indicates that quenching occurs from the exterior to the interior turns. Therefore, more emphasis should be placed on the protection of turns close to electrodes and joints between pancakes in the application of HTS magnets consisting of multiple DP coils, especially for the magnets in conduction-cooled conditions.

In addition, the critical current of the coil, 114.6 A, determined by the 1 μV/cm criteria, seriously overestimated the maximum spiral operating current because the main voltage drop was located along small pieces of HTS tape in the boundary turns, while most of the middle turns remained in a subcritical state according to the 1 μV/cm criteria. Therefore, it is more reasonable to evaluate the spiral current-carrying capacity from the perspective of the magnetic field [15,19].

5.3. Converged Load Factor for NI Coils

The non-uniform distribution of the magnetic field resulted in a non-uniform local critical current. Therefore, a non-uniform spiral current distribution was observed when the NI–DP coil operated under overcurrent conditions, as depicted in Figure 7(c1). However, most of the local load factors I/I_c inside the coil converged at approximately 0.97, as depicted in Figure 7(c2) The convergence of load factors during overcurrent indicated that most of the spiral elements inside the coil reached their own maximum current-carrying capacity.

Thus, a concept defined as the converged load factor (CLF), which is the maximum load factor in the coil at the saturation state, was proposed to reflect an overall quasi-equilibrium mechanism between the index loss and contact resistivity existing in NI coils when operating at a reasonable overcurrent state, at which the NI coil reached its maximum potential to generate a stable magnetic field without heat accumulation and thermal degradation in an LN_2 bath.

Figure 8 shows the simulated CLFs and saturated magnetic flux density of the NI–DP coils with respect to the contact resistivity. The simulated results demonstrate that for a given NI coil, the CLF can be lower than 1 with low contact resistivity. The high thermal stability of NI coils with low contact resistivity was at the expense of the time-consuming charging delay and also generated a lower saturated magnetic field, which underuses the full potential of HTS tapes [15].

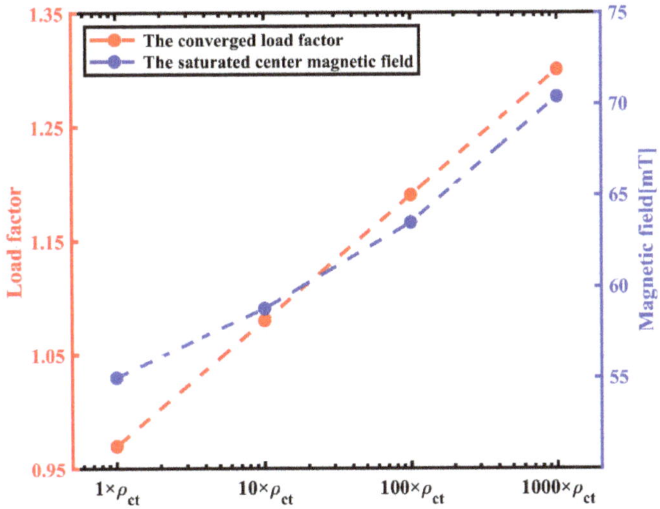

Figure 8. Converged load factors and saturated center magnetic fields vs. contact resistivity.

6. Conclusions

An ECG model coupled with magnetic field calculation and *E–J* power law of superconductors was adopted to further investigate the behavior of NI–DP coils during overcurrent charging. Experimentally, the entire process of magnetic-field saturation was consistent with the simulated results both in the magnetic flux density and voltage. The analysis of coil voltage signals and current distributions revealed the evolution of an overall quasi-equilibrium between the index loss and contact resistivity inside the coil, which drove the NI coil into a saturated state of the magnetic flux density in the coil's axial direction.

The potential risk of irreversible quenching at both the innermost and outermost turns was also found to provide a possible explanation for the near-electrode burnout reports of NI coils during overcurrent excitations. Therefore, it is essential to strengthen the protection of the boundary turns of NI pancakes in HTS magnets, particularly under conduction cooling conditions.

A concept defined as the converged load factor (CLF), which is the maximum load factor in the coil at the saturation state, was proposed to quantitatively characterize the spiral-current-carrying capacity of NI coils during overcurrent conditions. The typical CLF values of NI coils with a turn-to-turn resistivity ranging from 6.47 to 6470 $\mu\Omega\cdot cm^2$, which covers the range of typical NI coils, were calculated to be from 0.95 to 1.30. Although the CLF value was limited to around 1, the thermal stability of NI coils allowed the operation at a load factor evidently higher than insulated coils, which were unstable with the operation current close to their critical current in a real application system. From the viewpoint of achieving a higher magnetic field, turn-to-turn resistivity of NI coils as large as several thousands of $\mu\Omega\cdot cm^2$ of was recommended, however, with compromised thermal stability and suggested to be applied in static conditions with lower risk of quenching.

In the future, the long-term operation of NI coils in overcurrent conditions and the detailed mechanism/transient behavior of the overcurrent during the elimination process of screening current should be studied.

Author Contributions: Conceptualization, W.W.; methodology, Y.G. and W.W.; software, Y.G.; validation, W.W. and Y.G.; formal analysis, W.W. and Y.G.; investigation, W.W. and Y.G.; resources, W.W. and Z.J.; data curation, Y.G.; writing—original draft preparation, Y.G. and W.W.; writing—review and editing, W.W. and Y.G.; visualization, Y.G.; supervision, Z.J.; project administration, W.W.

and Z.J.; funding acquisition, W.W. All authors have read and agreed to the published version of the manuscript.

Funding: This research was funded by National Natural Science Foundation of China [51977130] and the APC was funded by the authors.

Data Availability Statement: The data presented in this study are available on request from the corresponding author. The data are not publicly available due to unfinished study on the data in future work.

Acknowledgments: The authors would like to thank Longbiao Wang from Shanghai Jiao Tong University for his help in the short sample test, and Xin Yu and Chao Zhang from Shanghai Superconductor Technology Co., Ltd. for their help in winding the test NI–DP coil.

Conflicts of Interest: The authors declare no conflict of interest.

References

1. Hahn, S.; Park, D.K.; Bascunan, J.; Iwasa, Y. HTS Pancake Coils Without Turn-to-Turn Insulation. *IEEE Trans. Appl. Supercond.* **2011**, *21*, 1592–1595. [CrossRef]
2. Tran, M.Q.; Elsisi, M.; Mahmoud, K.; Liu, M.K.; Lehtonen, M.; Darwish, M.M.F. Experimental Setup for Online Fault Diagnosis of Induction Machines via Promising IoT and Machine Learning: Towards Industry 4.0 Empowerment. *IEEE Access* **2021**, *9*, 115429–115441. [CrossRef]
3. Elsisi, M.; Tran, M.Q.; Mahmoud, K.; Mansour, D.E.A.; Lehtonen, M.; Darwish, M.M.F. Towards Secured Online Monitoring for Digitalized GIS Against Cyber-Attacks Based on IoT and Machine Learning. *IEEE Access* **2021**, *9*, 78415–78427. [CrossRef]
4. Lu, L.; Wu, W.; Yu, X.; Jin, Z. High-Temperature Superconducting Non-Insulation Closed-Loop Coils for Electro-Dynamic Suspension System. *Electronics* **2021**, *10*, 1980. [CrossRef]
5. Weng, F.J.; Zhang, M.; Lan, T.; Wang, Y.W.; Yuan, W.J. Fully superconducting machine for electric aircraft propulsion: Study of AC loss for HTS stator. *Supercond. Sci. Technol.* **2020**, *33*, 104002. [CrossRef]
6. Ma, G.; Gong, T.; Zhang, H.; Wang, Z.; Li, X.; Yang, C.; Liu, K.; Zhang, W. Experiment and Simulation of REBCO Conductor Coils for an HTS Linear Synchronous Motor. *IEEE Trans. Appl. Supercond.* **2017**, *27*, 5201805. [CrossRef]
7. Brittles, G. Recent progress in HTS magnet development at Tokamak Energy. In Proceedings of the 9th Cryogenic Cluster Day, Rutherford Appleton Laboratory, Oxfordshire, UK, 24 May 2019.
8. Li, Y.; Park, D.; Yan, Y.; Choi, Y.; Lee, J.; Michael, P.C.; Chen, S.; Qu, T.; Bascuñán, J.; Iwasa, Y. Magnetization and screening current in an 800 MHz (18.8 T) REBCO nuclear magnetic resonance insert magnet: Experimental results and numerical analysis. *Supercond. Sci. Technol.* **2019**, *32*, 105007. [CrossRef]
9. Kim, H.-W.; Hur, J.; Kim, S.-W.; Ha, D.-W.; Ko, R.-K.; Hong, J.-P.; Kim, J.H.; Kim, H.M.; Joo, J.-H.; Kim, S.-B.; et al. Electrical Characteristic Analysis According to Contact Resistance Between Turns of HTS Coil. *IEEE Trans. Appl. Supercond.* **2016**, *26*, 4601504. [CrossRef]
10. Kim, H.W.; Jo, Y.S.; Kim, S.W.; Kim, H.M.; Jeong, J.S.; Hong, J.P.; Hur, J. Determining the Operating Current of No-Insulation Field Coils in HTS Generators. *IEEE Trans. Magn.* **2015**, *51*, 9000404. [CrossRef]
11. Yanagisawa, Y.; Sato, K.; Yanagisawa, K.; Nakagome, H.; Jin, X.; Takahashi, M.; Maeda, H. Basic mechanism of self-healing from thermal runaway for uninsulated REBCO pancake coils. *Phys. C Supercond. Its Appl.* **2014**, *499*, 40–44. [CrossRef]
12. Kim, J.M.; Kim, J.C.; Kim, Y.G.; Hong, S.J.; Park, S.J.; Kim, J.H.; Kim, H.M.; Choi, Y.S.; Lee, H.G. Investigation about the effects of metal-clad winding on the electromagnetic characteristics of the GdBCO racetrack coils in a time-varying magnetic field. *Results Phys.* **2018**, *11*, 400–405. [CrossRef]
13. Kim, J.; Yoon, S.; Cheon, K.; Shin, K.H.; Hahn, S.; Kim, D.L.; Lee, S.; Lee, H.; Moon, S.H. Effect of Resistive Metal Cladding of HTS Tape on the Characteristic of No-Insulation Coil. *IEEE Trans. Appl. Supercond.* **2016**, *26*, 4601906. [CrossRef]
14. Wang, R.C.; Ma, G.T.; Zhou, P.B.; Gong, T.Y.; Li, S.L. Thermo-electromagnetic modeling of coated superconductor coils with metal insulation. *Supercond. Sci. Technol.* **2021**, *34*, 115017. [CrossRef]
15. Wang, X.D.; Wang, T.; Nakada, E.; Ishiyama, A.; Itoh, R.; Noguchi, S. Charging Behavior in No-Insulation REBCO Pancake Coils. *IEEE Trans. Appl. Supercond.* **2015**, *25*, 4601805. [CrossRef]
16. Wang, Y.; Song, H.; Xu, D.; Li, Z.Y.; Jin, Z.; Hong, Z. An equivalent circuit grid model for no-insulation HTS pancake coils. *Supercond. Sci. Technol.* **2015**, *28*, 045017. [CrossRef]
17. Wang, T.; Noguchi, S.; Wang, X.; Arakawa, I.; Minami, K.; Monma, K.; Ishiyama, A.; Hahn, S.; Iwasa, Y. Analyses of Transient Behaviors of No-Insulation REBCO Pancake Coils During Sudden Discharging and Overcurrent. *IEEE Trans. Appl. Supercond.* **2015**, *25*, 4603409. [CrossRef]
18. Liu, X.Y.; Xie, Z.T.; Ning, F.P.; Wang, Y.B.; Wang, X.W.; Zhang, G.Q.; Zhao, L.; Xu, Q.J.; Zhu, Z. Preliminary Study on Different Winding Processes of ReBCO Coils. *IEEE Trans. Appl. Supercond.* **2020**, *30*, 4603406. [CrossRef]
19. Liu, X.Y.; Zhu, Z.A.; Ning, F.P.; Xie, Z.T.; Tong, X.Y.; Zhang, G.Q.; Xu, Q.J.; Zhao, L.; Zhao, W.; Hou, Z.L.; et al. The Research on No-Insulation ReBCO Racetrack Coil. *IEEE Trans. Appl. Supercond.* **2019**, *29*, 4602305. [CrossRef]

20. Lai, L.F.; Gu, C. AC loss calculation in REBCO coils or stacks by solving the equation of motion for current using an integration approach. *Supercond. Sci. Technol.* **2021**, *34*, 015003. [CrossRef]
21. Liu, D.H.; Li, D.K.; Zhang, W.W.; Yong, H.D.; Zhou, Y.H. Electromagnetic-thermal-mechanical behaviors of a no-insulation double-pancake coil induced by a quench in the self field and the high field. *Supercond. Sci. Technol.* **2021**, *34*, 025014. [CrossRef]
22. Durmus, F.; Karagol, S. Mutual Inductance Calculation Formula for Planar Square Coils. In Proceedings of the 2018 2nd International Symposium on Multidisciplinary Studies and Innovative Technologies, IEEE, Ankara, Turkey, 19–21 October 2018.
23. Liu, Y.Z.; Ou, J.; Gyuraki, R.; Schreiner, F.; de Sousa, W.T.B.; Noe, M.; Grilli, F. Study of contact resistivity of a no-insulation superconducting coil. *Supercond. Sci. Technol.* **2021**, *34*, 035009. [CrossRef]
24. Sayed, A.M.; Abouelatta, M.A.; Badawi, M.; Mahmoud, K.; Lehtonen, M.; Darwish, M.M.F. Novel accurate modeling of dust loaded wire-duct precipitators using FDM-FMG method on one fine computational domains. *Electr. Power Syst. Res.* **2022**, *203*, 107634. [CrossRef]
25. Abouelatta, M.A.; Ward, S.A.; Sayed, A.M.; Mahmoud, K.; Lehtonen, M.; Darwish, M.M.F. Fast Corona Discharge Assessment Using FDM integrated With Full Multigrid Method in HVDC Transmission Lines Considering Wind Impact. *IEEE Access* **2020**, *8*, 225872–225883. [CrossRef]
26. Grilli, F.; Sirois, F.; Zermeno, V.M.R.; Vojenciak, M. Self-Consistent Modeling of the Ic of HTS Devices: How Accurate do Models Really Need to Be? *IEEE Trans. Appl. Supercond.* **2014**, *24*, 8000508. [CrossRef]
27. Rostila, L.; Lehtonen, J.; Mikkonen, R.; Souc, J.; Seiler, E.; Melisek, T.; Vojenciak, M. How to determine critical current density in YBCO tapes from voltage-current measurements at low magnetic fields. *Supercond. Sci. Technol.* **2007**, *20*, 1097–1100. [CrossRef]
28. Wang, Y.; Chan, W.K.; Schwartz, J. Self-protection mechanisms in no-insulation (RE)Ba2Cu3Oxhigh temperature superconductor pancake coils. *Supercond. Sci. Technol.* **2016**, *29*, 045007. [CrossRef]
29. Chan, W.K.; Masson, P.J.; Luongo, C.; Schwartz, J. Three-Dimensional Micrometer-Scale Modeling of Quenching in High-Aspect-Ratio YBa2Cu3O7-δ Coated Conductor Tapes—Part I: Model Development and Validation. *IEEE Trans. Appl. Supercond.* **2010**, *20*, 2370–2380. [CrossRef]
30. Wang, X.; Hahn, S.; Kim, Y.; Bascunan, J.; Voccio, J.; Lee, H.; Iwasa, Y. Turn-to-turn contact characteristics for an equivalent circuit model of no-insulation ReBCO pancake coil. *Supercond. Sci. Technol.* **2013**, *26*, 035012. [CrossRef]

processes

MDPI

Communication

Towards High Efficiency CO_2 Utilization by Glow Discharge Plasma

Stephan Renninger *, Paul Rößner, Jan Stein, Maike Lambarth and Kai Peter Birke

Electrical Energy Storage Systems, Institute for Photovoltaics, University of Stuttgart, Pfaffenwaldring 47, 70569 Stuttgart, Germany; paul.roessner@ipv.uni-stuttgart.de (P.R.); jan.stein@ipv.uni-stuttgart.de (J.S.); maike.lambarth@ipv.uni-stuttgart.de (M.L.); peter.birke@ipv.uni-stuttgart.de (K.P.B.)
* Correspondence: stephan.renninger@ipv.uni-stuttgart.de; Tel.: +49-711-6856-1508

Abstract: Plasma technology reaches rapidly increasing efficiency in catalytic applications. One such application is the splitting reaction of CO_2 to oxygen and carbon monoxide. This reaction could be a cornerstone of power-to-X processes that utilize electricity to produce value-added compounds such as chemicals and fuels. However, it poses problems in practice due to its highly endothermal nature and challenging selectivity. In this communication a glow discharge plasma reactor is presented that achieves high energy efficiency in the CO_2 splitting reaction. To achieve this, a magnetic field is used to increase the discharge volume. Combined with laminar gas flow, this leads to even energy distribution in the working gas. Thus, the reactor achieves very high energy efficiency of up to 45% while also reaching high CO_2 conversion efficiency. These results are briefly explained and then compared to other plasma technologies. Lastly, cutting edge energy efficiencies of competing technologies such as CO_2 electrolysis are discussed in comparison.

Keywords: plasma catalysis; CO_2 splitting; glow discharge; energy efficiency; CO_2 electrolysis; CO_2 utilization; magnetic field; power-to-X

Citation: Renninger, S.; Rößner, P.; Stein, J.; Lambarth, M.; Birke, K.P. Towards High Efficiency CO_2 Utilization by Glow Discharge Plasma. *Processes* **2021**, 9, 2063. https://doi.org/10.3390/pr9112063

Academic Editors: Alon Kuperman and Alessandro Lampasi

Received: 11 October 2021
Accepted: 15 November 2021
Published: 18 November 2021

1. Introduction

Many fields offer solutions for CO_2 utilization. Among them are thermochemical processes, electrolysis [1] and plasma catalysis; the latter has the smallest technology readiness level (TRL) but also offers a large potential for future improvements [2–4]. Attention is focused mostly on four types of plasma reactors: dielectric barrier discharges (DBDs) [5], gliding arc (GA) [6,7], atmospheric pressure glow discharges (APGD) [8,9] and microwave (MW) plasmas [10,11]. Increasing their energy efficiency and conversion at ambient pressure is the main point of concern. We recently presented reactors using a direct current APGD, which delivered promising results [12]. To further improve the previous design, it was scaled and now uses a laminar gas flow instead of a turbulent one. To be used industrially, a plasma reactor should operate at ambient pressure. However, this makes it hard to maintain a stable discharge. Two major difficulties are the negative differential resistance [13] and glow-to-arc transition [14]. A discharge thus tends to form a narrow, low-resistance arc that can damage the plasma source and is not useful in catalysis. To suppress these effects, current control strategies or vortex gas flow [14] are often used in recent studies to disperse the plasma [8]. The reactor setup presented here uses a magnetic field instead to force the plasma into a large disc-like volume. This approach is viable for various discharge forms, such as gliding arc plasma reactors [15,16]. A laminar gas flow can be then used to introduce energy into the working gas as homogeneously as possible. This communication aims to give an update on the ongoing design process for an improved plasma reactor for the CO_2 splitting reaction.

2. Materials and Methods

The reactor vessel was a glass tube with an inner diameter of 38 mm. Direct current formed a discharge between two copper electrodes. One was an axial rod, the other a ring. CO_2 was introduced into the reactor through an injection plate with 22 axial nozzles in concentric nozzles, arranged in concentric circles. The injector was placed 120 mm above the plane of the electrodes. An axial magnetic field was provided by permanent magnets below the electrode assembly, and the field strength was 30 mT on the central axis. A packed bed made from zirconia balls was placed 2 mm below the discharge plane inside the ring electrode. It serves to suppress thermal currents in the gas and could also help with quenching the hot exhaust gas. Zirconia was chosen because it is chemically inert, non-conductive and can be used to carry catalysts in future experiments. The reactor assembly is shown in Figure 1. The input gas flow V_{in} consisted of pure CO_2. It was measured by an analogue rotameter and adjusted using a needle valve. The assembly was calibrated using a displacement cylinder. The exhaust gas was characterized using non-dispersive infrared sensors (SmartGas Flow Evo; Heilbronn; Germany); measurements at a flow of V_{in} =1.4 SLM were confirmed by a gas chromatographer (Trace 1310 Thermo Scientific; Waltham, MA, USA). The sensors were placed 1 m downstream from the reactor in the exhaust gas pipe. Power was provided to the electrodes by a custom current-limiting driver circuit. It delivers direct current for ignition (up to 25 kV) and is sustaining of the discharge (<2 kV). Mean burn voltage of the discharge and mean current were measured. Mean values are deemed sufficient here, because a large choke inductor of 1.5 H was placed on the output of the driver circuit, leading to low current ripple. Voltage ripple was typically around 15%. The discharge power P_d was calculated from the power supplied to the driver circuit by a lab power supply and the known driver efficiency. To confirm these values, they can also be calculated as the product of burn voltage and current. CO_2 conversion X is calculated using Equation (1), while energy efficiency η is calculated by Equation (2). They use the concentrations of CO and CO_2 in the exhaust gas. ΔH_r = 12.6 J SCC^{-1} (standard cubic centimeter) is the reaction enthalpy of the CO_2 splitting reaction. Measurements of the gas concentrations were taken after a steady state in exhaust gas concentrations occurred.

$$X = \frac{c_{CO,out}}{c_{CO,out} + c_{CO2,out}} \tag{1}$$

$$\eta = \frac{X \, \Delta H_r \, \dot{V}_{in}}{P_{el}} \tag{2}$$

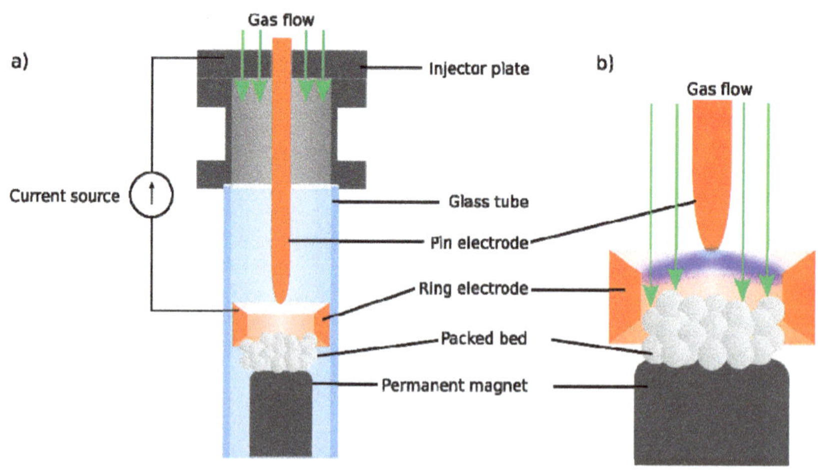

Figure 1. Overview over the plasma reactor assembly (**a**) and schematic view of the discharge (**b**).

3. Results

The achieved CO_2 conversion X and energy efficiency η for different discharge power P_d and gas flow rates V_{in} is shown in Figure 2. The highest conversion X was achieved at a discharge power of P_d = 165 W. The best energy efficiency η that was achieved is 45% at the highest gas flow rate of 1.25 SLM. The effectiveness of the magnetic field could be determined visibly: the discharge rotates quickly, so it gives the appearance of a disk to the naked eye. This leads to very homogeneous energy input into the gas. Quantizing the influence of the magnetic field will be the subject of future experiments.

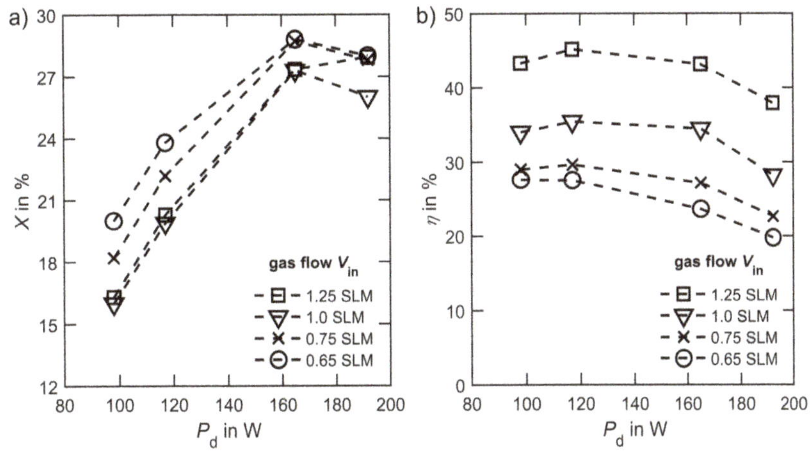

Figure 2. Performance of the reactor at different gas flow rates V_{in} and discharge power P_d. In (**a**), the conversion X is shown while (**b**) displays energy efficiency η.

4. Discussion

4.1. Performance of the Reactor

Conversion depends strongly on discharge power P_d. One reason is that higher power equals higher specific energy input. Additionally, the properties of the plasma, such as temperature, electron density and reduced electric field, can also be expected to change with the discharge power. In this reactor the rotation of the discharge filament accelerates at higher discharge powers. This can also be expected to have a positive influence on the conversion, since the gas will be swept more efficiently by the plasma. However, the conversion does not increase with power indefinitely. At the highest discharge power of P_d = 192W, conversion reduces. One reason could be the heating of the packed bed. This heating reduces the quenching rate, which increases the rate of the recombination reaction, thus again forming CO_2 [11]. Energy efficiency seems to mainly depend on gas flow rate but also decreases at high discharge powers.

4.2. Comparison to Other Technologies

The results achieved compare well to other plasma-based systems, as shown in Figure 3a. They were selected based on performance from a broader range of systems previously reviewed [17], considering more recent work. Gliding arcs provide high efficiency at ambient pressure; a vortex is often used to increase the discharge volume [6,7]. Gliding arcs also obtain good results without vortex flow [18]. Glow discharges can also benefit from vortex gas flow [8]. Increasing their stability is possible by operation in non-self-sustaining mode [9]. DBDs that moderate efficiency and conversion could be boosted by using a burst mode, where high power density is applied intermittently [5]. Microwave plasmas reach the most promising results to date [10]. However, these were obtained at very low pressures, and at ambient pressures even after utilizing precise quenching, efficiency is lower, yet still impressive [11]. The highest efficiencies reported in literature were achieved at a very low pressure by radio frequency excitation at a pressure of just 40 Pa [19]. A common theme in the results seems to be that a homogeneous energy input into the gas results in good performance. Vortex gas flow as used in [6–8] can distribute energy well but is ultimately a chaotic process that will not lead to even energy distribution. In contrast, the combination of laminar flow and a disk-like discharge can distribute energy very evenly.

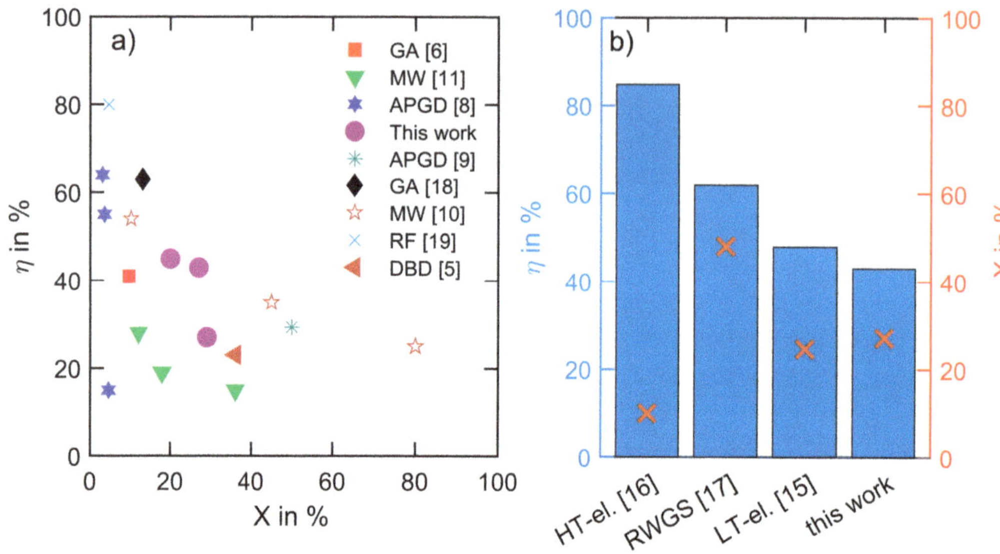

Figure 3. A comparison of the achieved energy efficiency η and conversion X for CO_2 splitting by plasma is shown in (**a**). In (**b**) the results are compared to high temperature electrolysis (HT-el.), low temperature electrolysis (LT-el.) and reverse water gas shift (RWGS).

In the following, a quantitative assessment of different CO_2 conversion technologies regarding energy efficiency η is given. The comparison has no claim for completeness and focuses only on the actual conversion step of CO_2 to CO; influences on a systemic level or scaling effects are not included here. This approach allows a comparison of vastly different technologies but is not intended as a ranking given that each technology has its own ideal configuration in which the full potential can be realized. Figure 3b shows calculated energy efficiencies drawn from recent publications (see Appendix A for the calculation). The technologies included are low temperature, gas-phase CO_2 electrolysis [20], high temperature CO_2 electrolysis in a solid oxide electrolysis cell [21], a thermochemical approach (Reverse Water Gas Shift, RWGS) [22] and the plasma approach reported in this work. The energy efficiencies given in Figure 3b show that each technology has the potential to enable a reasonable application. This is reasoned on the basis that systemic effects of the individual application have the potential to outweigh the differences inherent to the energy efficiency of the CO_2 conversion.

5. Conclusions

The presented glow discharge plasma reactor achieves a competitive CO_2 conversion of 27% and energy efficiency of 42%. This is a respectable performance since the process was running at ambient pressure. We attribute this good performance to the efficient sweeping of the gas by the discharge due to the magnetic field. In general, the energy efficiency of plasma-based systems is gaining ground compared to competing technologies such as electrolysis and thermochemical approaches. Focus thus shifts to scalability, lifespan and, most importantly, integration. After all, none of the presented technologies manage to produce pure product gases; their separation is a major task for which few technologies are available. The integration of electrochemical oxygen pumps or separation membranes into plasma reactor systems will be a future focus. Our results illustrate that plasma technology can play an important role in CO_2 utilization, which is a cornerstone of a fossil-free economy.

Author Contributions: Conceptualization, S.R., J.S. and P.R.; methodology, S.R; investigation, S.R., M.L. and J.S.; resources, K.P.B.; writing—original draft preparation, S.R.; writing—review and editing, P.R. and M.L.; visualization, S.R.; supervision, K.P.B.; project administration, M.L.; funding acquisition, K.P.B. All authors have read and agreed to the published version of the manuscript.

Funding: This work was funded by the Federal Ministry for Economic Affairs and Energy (Germany) in the scope of their initiative "Energy transition in the transport sector" and the associated "PlasmaFuel" project (Funding code: 03EIV161A).

Informed Consent Statement: Not applicable.

Data Availability Statement: The data is not filed into a public repository but will be kindly provided upon request.

Conflicts of Interest: The authors declare no conflict of interest.

Appendix A

The energy efficiency of electrolysis is calculated considering electrical E_{el} and thermal energy E_{th} input following Equation (A1). Thermal energy is calculated by using the heat capacity c_p and temperature difference from ambient ΔT. Electrical energy is calculated using the Faradaic efficiency η_F, cell voltage U_{cell} at a current density of 200 mA cm^{-2}, electron number z and the Faraday constant F.

$$\eta = \frac{E_{use}}{E_{th} + E_{el}} = \frac{\Delta H_r^0}{c_p \Delta T + U_{cell} \eta_{FE}\, zF} \quad \text{(A1)}$$

For the thermochemical approach, energy input is the sum of thermal energy used for gas heating and utilized hydrogen. Hydrogen was weighed as an energy expense of E_{H2} = 350 kJ mol^{-1}.

References

1. Saravanan, A.; Kumar, P.S.; Vo, D.-V.N.; Jeevanantham, S.; Bhuvaneswari, V.; Narayanan, V.A.; Yaashikaa, P.; Swetha, S.; Reshma, B. A comprehensive review on different approaches for CO_2 utilization and conversion pathways. *Chem. Eng. Sci.* **2021**, *236*, 116515. [CrossRef]
2. Bogaerts, A.; Centi, G. Plasma technology for CO_2 conversion: A personal perspective on prospects and gaps. *Front. Energy Res.* **2020**, *8*, 111. [CrossRef]
3. Bogaerts, A.; Tu, X.; Whitehead, J.C.; Centi, G.; Lefferts, L.; Guaitella, O.; Azzolina-Jurry, F.; Kim, H.-H.; Murphy, A.B.; Schneider, W.F. The 2020 plasma catalysis roadmap. *J. Phys. D Appl. Phys.* **2020**, *53.44*, 443001. [CrossRef]
4. Grim, R.G.; Huang, Z.; Guarnieri, M.T.; Ferrell, J.R.; Tao, L.; Schaidle, J.A. Transforming the carbon economy: Challenges and opportunities in the convergence of low-cost electricity and reductive CO_2 utilization. *Energy Environ. Sci.* **2019**, *13*, 472–494. [CrossRef]
5. Ozkan, A.; Dufour, T.; Silva, T.; Britun, N.; Snyders, R.; Reniers, F.; Bogaerts, A. DBD in burst mode: Solution for more efficient CO_2 conversion? *Plasma Sources Sci. Technol.* **2016**, *25*, 055005. [CrossRef]
6. Trenchev, G.; Bogaerts, A. Dual-vortex plasmatron: A novel plasma source for CO_2 conversion. *J. CO_2 Util.* **2020**, *39*, 101152. [CrossRef]
7. Ramakers, M.; Medrano, J.A.; Trenchev, G.; Gallucci, F.; Bogaerts, A. Revealing the arc dynamics in a gliding arc plasmatron: A better insight to improve CO_2 conversion. *Plasma Sources Sci. Technol.* **2017**, *26*, 125002. [CrossRef]
8. Bharathi, R.; Sarathi, R.; Vinu, R. Development of a Swirl-Induced Rotating Glow Discharge Reactor for CO_2 Conversion: Fluid Dynamics and Discharge Dynamics Studies. *Energy Technol.* **2020**, *8.12*, 2000535.
9. Andreev, S.; Zakharov, V.; Ochkin, V.; Savinov, S. Plasma-chemical CO_2 decomposition in a non-self-sustained discharge with a controlled electronic component of plasma. *Spectrochim. Acta Part A Mol. Biomol. Spectrosc.* **2004**, *60*, 3361–3369. [CrossRef]
10. Bongers, W.; Bouwmeester, H.J.; Wolf, B.; Peeters, F.; Welzel, S.; Bekerom, D.V.D.; Harder, N.D.; Goede, A.; Graswinckel, M.; Groen, P.W.; et al. Plasma-driven dissociation of CO_2 for fuel synthesis. *Plasma Process. Polym.* **2016**, *14*, e1600126. [CrossRef]

11. Kim, H.; Song, S.; Tom, C.P.; Xie, F. Carbon dioxide conversion in an atmospheric pressure microwave plasma reactor: Improving efficiencies by enhancing afterglow quenching. *J. CO_2 Util.* **2019**, *37*, 240–247. [CrossRef]
12. Renninger, S.; Lambarth, M.; Birke, K.P. High efficiency CO_2-splitting in atmospheric pressure glow discharge. *J. CO_2 Util.* **2020**, *42*, 101322. [CrossRef]
13. Saifutdinov, A.I.; Timerkaev, B.A. Features of Transient Processes in DC Microdischarges in Molecular Gases: From a Glow Discharge to an Arc Discharge with a Unfree or Free Cathode Regime. *JETP Lett.* **2020**, *112*, 405–412. [CrossRef]
14. Rabinovich, A.; Nirenberg, G.; Kocagoz, S.; Surace, M.; Sales, C.; Fridman, A. Scaling Up of Non-Thermal Gliding Arc Plasma Systems for Industrial Applications. *Plasma Chem. Plasma Process.* **2021**, 1–16. [CrossRef]
15. Li, L.; Zhang, H.; Li, X.; Huang, J.; Kong, X.; Xu, R.; Tu, X. Magnetically enhanced gliding arc discharge for CO_2 activation. *J. CO_2 Util.* **2019**, *35*, 28–37. [CrossRef]
16. Zhang, H.; Li, L.; Li, X.; Wang, W.; Yan, J.; Tu, X. Warm plasma activation of CO_2 in a rotating gliding arc discharge reactor. *J. CO_2 Util.* **2018**, *27*, 472–479. [CrossRef]
17. Snoeckx, R.; Bogaerts, A. Plasma technology—A novel solution for CO_2 conversion? *Chem. Soc. Rev.* **2017**, *46.19*, 5805–5863. [CrossRef] [PubMed]
18. Kim, S.C.; Lim, M.S.; Chun, Y.N. Reduction Characteristics of Carbon Dioxide Using a Plasmatron. *Plasma Chem. Plasma Process.* **2013**, *34*, 125–143. [CrossRef]
19. Spencer, L.F.; Gallimore, A.D. Efficiency of CO_2 Dissociation in a Radio-Frequency Discharge. *Plasma Chem. Plasma Process.* **2010**, *31*, 79–89. [CrossRef]
20. Liu, Z.; Yang, H.; Kutz, R.; Masel, R.I. CO_2 Electrolysis to CO and O_2 at High Selectivity, Stability and Efficiency Using Sustainion Membranes. *J. Electrochem. Soc.* **2018**, *165*, J3371–J3377. [CrossRef]
21. Kaur, G.; Kulkarni, A.P.; Giddey, S. CO_2 reduction in a solid oxide electrolysis cell with a ceramic composite cathode: Effect of load and thermal cycling. *Int. J. Hydrogen Energy* **2018**, *43*, 21769–21776. [CrossRef]
22. Zonetti, C.P.; Letichevsky, S.; Gaspar, A.B.; Sousa-Aguiar, E.F.; Appel, L.G. The NixCe0.75Zr0.25−xO2 solid solution and the RWGS. *Appl. Catal. A Gen.* **2014**, *475*, 48–54. [CrossRef]

energies

MDPI

Article

Development of a Smart Supercapacitor Energy Storage System for Aircraft Electric Power Systems

Ahmed M. Fares [1,2,*], Matias Kippke [1], Mohamed Rashed [1], Christian Klumpner [1] and Serhiy Bozhko [1]

1 Department of Electrical and Electronic Engineering, University of Nottingham, Nottingham NG7 2TU, UK; mkippke@gmail.com (M.K.); eezmr1@exmail.nottingham.ac.uk (M.R.); christian.klumpner@nottingham.ac.uk (C.K.); Serhiy.bozhko@nottingham.ac.uk (S.B.)
2 National Authority for Remote Sensing and Space Sciences, P.O. Box 1564, Cairo 11769, Egypt
* Correspondence: ahmed.fares@nottingham.ac.uk

Abstract: This paper presents the development of a supercapacitor energy storage system (ESS) aimed to minimize weight, which is very important for aerospace applications, whilst integrating smart functionalities like voltage monitoring, equalization, and overvoltage protection for the cells. The methodology for selecting the supercapacitor cells type/size is detailed to achieve the safest and most energy-dense ESS. Additionally, the development of the interface electronics for cells' voltage monitoring and overvoltage protection is presented. The proposed design implements a modular distributed architecture coordinated using communication buses to minimize the wirings and associated complexity and to enable system reconfiguration and expansions, as well as fault diagnoses. Validating the proposed ESS functionalities has been done via experimental testing and the results are presented and discussed.

Keywords: aircraft electrical power systems; more electric aircraft; supercapacitors energy storage

Citation: Fares, A.M.; Kippke, M.; Rashed, M.; Klumpner, C.; Bozhko, S. Development of a Smart Supercapacitor Energy Storage System for Aircraft Electric Power Systems. *Energies* **2021**, *14*, 8056. https://doi.org/10.3390/en14238056

Academic Editors: Alon Kuperman and Alessandro Lampasi

Received: 2 November 2021
Accepted: 29 November 2021
Published: 2 December 2021

1. Introduction

The newly introduced different types of electrically driven systems in future more electric aircraft (MEA) significantly increase power demands, and thus increasing generation capacity requirements for the onboard electric power system (EPS) [1–5]. Some specific loads with high dynamics such as flight actuation systems are characterized by significant power requirement peaks. These types of loads impose significant challenges for EPS rating definition and EPS stability and power quality [1,6]. To minimize these effects, the first option is to sufficiently increase the rated power of the onboard power generation system to handle these power peaks. This means the power generators should be sized based on the peak power rather than the average power, leading to a significant increase of EPS weight and volume. The second option is to keep generators sizing based on the average power consumption whilst implementing an energy storage system (ESS), that should be sized to smooth these high power peaks, i.e., to supply the high dynamic transient loads such that these become invisible for the generators. Implementation of the ESS in this way allows enhancing the design of the MEA EPS by reducing the sizing of the power generators, as well as adding the capability of storing the regenerative energy produced by some of the loads like the electromechanical actuator (EMA) for future re-use instead of dissipating it in braking resistors. In addition, smoothing of the power demand surges seen by generators lead to reduced power cables size following the average power consumption rather than its peak power/current requirement, which results in further reduction of EPS weight.

A wide range of ESSs is available in the market with different characteristics for different applications [7,8]. Aircraft applications impose design constraints for the weight of onboard equipment to minimize fuel consumption. Accordingly, the ESS needs to fulfill its required power and energy ratings at minimum weight. This makes the electrochemical

ESS a better choice for aircraft EPS. There are two representative devices, which are the lithium-ion batteries that are characterized by high specific energy (100–300 Wh/kg), and supercapacitors (SC) that that characterized by high specific power (20 kW/kg) at low specific energy (5–10 Wh/kg) [9]. Accordingly, the ESS selection depends on the specific requirements of the applications. For centralized ESS where a central energy storage unit supplies all the targeted loads, and thus requires large energy capacity ESS, the use of Li-ion batteries is more adequate. In contrast, distributed ESSs where distributed energy storage units are localized at each load, and thus do not require high energy capacity but need significant power capability to cover the short-term peak power demands; therefore, in these, the use of SC is more adequate [10]. Much research has been focused on the utilization of the SC-based ESS for aircraft EPS and its corresponding benefits; however, most of the research is focused on studying the energy management strategies [6,11–13] rather than studying the design optimization for the SC-ESS itself, including its associated management subsystems in terms of ESS weight and reliability, which are very important parameters for aircraft applications.

In the paper, the design and implementation of SC-based ESS and associated management subsystems for SC cells' voltage monitoring, balancing, and overvoltage protection are proposed.

The paper is organized into five sections. In Section 2, the sizing and optimal design of the SC energy storage system are investigated. In Section 3, The design of cells' voltage monitoring and overvoltage protection subsystem is presented. In Section 4, an experimental evaluation of the proposed system is presented, and Section 5 summarizes the conclusions of this work.

2. Sizing of the SC Energy Storage System

The supercapacitor energy storage system (SC-ESS) as shown in Figure 1 should be interfaced to the DC power bus through a DC/DC converter that controls the SC-ESS charge/discharge process. The SC-ESS is constructed based on series/parallel combinations of SC cells to achieve the required voltage level and energy capacity. The SC-ESS presented in this paper is designed to fulfill the energy and power requirements listed in Table 1 based on the power consumption of an EMA load profile in an aircraft. Accordingly, the equivalent capacitance of the SC-ESS can be estimated based on these requirements as [14]:

$$C_{eq} = \frac{2E_{str}}{V_{max}^2 - V_{min}^2} \tag{1}$$

where E_{str} is the targeted energy capacity of the SC-ESS whilst V_{max} and V_{min} are its targeted maximum and minimum operating voltages, respectively. Accordingly, the selected minimum and maximum voltages strongly affect the SC-ESS sizing to achieve the required energy capacity (E_{str}).

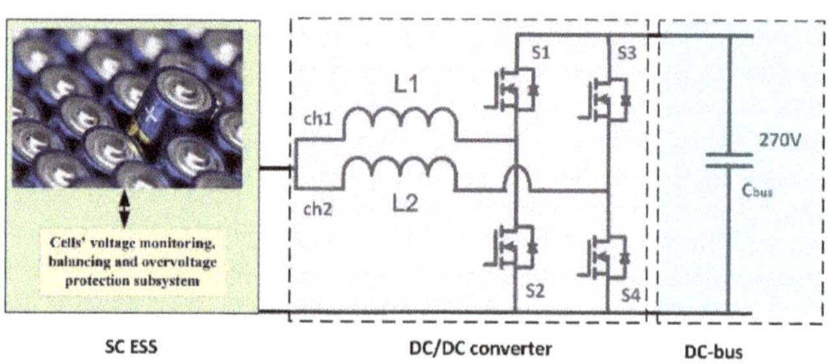

Figure 1. Supercapacitor energy storage system block diagram.

The common design technique is selecting the SC-ESS maximum voltage at a value close to the DC bus voltage whilst keeping its minimum voltage around 50% of that value to limit the current ratings of both the SC-ESS and the interfaced DC/DC converter, as this impacts the size/weight of their magnetics/semiconductors. Accordingly, by setting the minimum and maximum SC-ESS voltages, the capacitance of the single SC cell in the string, which consists of n series-connected cells, can be estimated as follows:

$$C_{cell} = \frac{C_{eq}}{n} \ ; n = \frac{V_{max}}{V_{cell-max}} \tag{2}$$

where $V_{cell\text{-}max}$ is the maximum operating voltage of the single SC cell (2.7 V for most SC cells). This capacitance can be obtained directly using a single cell or by implementing a parallel combination of two or more cells based on the available standard capacitances.

Applying the SC-ESS design requirements listed in Table 1 to Equations (1) and (2) and selecting V_{max} = 260 V and $V_{cell\text{-}max}$ = 2.5 V, yield the required capacity of the single SC cell to be around 150 F. However, changing the targeted SC-ESS operating voltage range can enable utilization of other cells with higher/lower capacities that might have different characteristics and, hence, make a choice based on the best cell. Accordingly, to identify the best SC cell to be implemented to achieve minimum overall weight for the SC-ESS, the characteristics of the different commercially available SC cells in the range of 100–400 F have been evaluated focusing on their specific energy (Wh/kg) as it significantly affects the overall SC-ESS weight. Figure 2 shows the specific energy of the different SC cells within the selected range. As can be seen, the 150 F SC cell that matches the design calculations at V_{max} = 260 V has lower specific energy (4.7 Wh/kg) compared to the 350 F cell produced by Company1, which has 25% more specific energy (5.9 Wh/kg). Accordingly, implementing this cell in the targeted SC-ESS resulted in a 25% reduction in the overall SC cells weight. However, implementing such large capacity cells requires a reduction of V_{max} according to (1) and (2), which adds the advantages of reducing the number of series-connected cells and thus reducing the complexity of cells' voltage monitoring and overvoltage protection subsystem. In the meantime, it also adds the disadvantage of increasing the current rating of the interfaced DC/DC converter, thus increasing its size and weight.

Table 1. SC-ESS Targeted Specifications.

Parameters	Value
Maximum power	2 kW
Bus voltage	270 V
Energy capacity	10.9 Wh

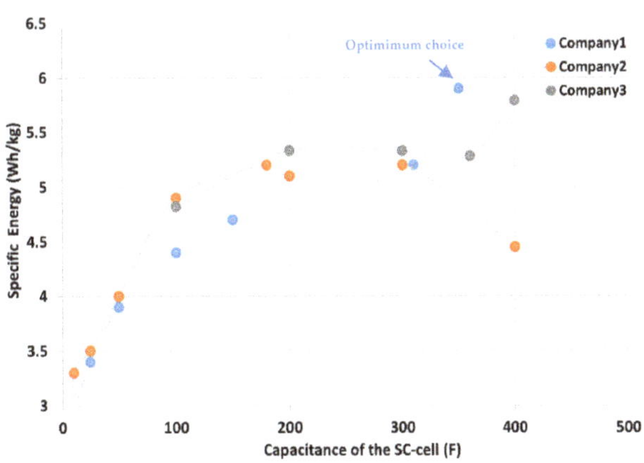

Figure 2. Comparing specific energy vs. nominal cell capacity for a range of SC cells.

The two design options (based on 150 F and 350 F cells) are investigated based on the specifications of the targeted SC-ESS listed in Table 2 and the design parameters for designs are presented. As it can be seen, design B, which utilizes the 350 F SC cell, has reduced weight by nearly 22% compared to design A. However, the maximum current required to deliver the given power (2 kW) at minimum voltage is increased significantly and this may increase the thermal stress of the SC-ESS and increase the weight/size of the interfaced converter. As per the investigations done in [15,16], implementing a two-channel interleaved converter in conjunction with wide band-gap semiconductor switches can minimize the effects of the increased current rating on the converter size/weight. Additionally, the investigations carried out in [15] show minor differences in terms of thermal stresses between the two cells due to the better thermal characteristics and reduced internal resistance of the 350 F SC cell.

Table 2. Comparing Design Options for SC-ESS.

Parameters	Design A	Design B
Utilized SC cell capacity	150 F	350 F
Maximum operating voltage	260 V	120 V
Minimum operating Voltage	120 V	65 V
No. of series-connected cells/string	106	48
The maximum operating current of the SC-ESS	16.7 A	31 A
No. of parallel strings	1	1
Total no. of utilized SC cells	106	48
Total weight of the utilized SC cells	3.4 kg	2.8 kg

3. SC Cells' Voltage Monitoring and Overvoltage Protection Subsystem

Based on the proven weight savings in the targeted SC-ESS according to design B that utilizes 350 F SC cells, this design is considered for the SC-ESS and its associated subsystems are presented in this paper. To facilitate the design, manufacturing, and maintenance of the SC-ESS, a modular distributed architecture has been considered such that the targeted system is based on four series-connected modules, each of these consisting of 12 series-connected SC cells with associated voltage monitoring and overvoltage protection circuits, as presented in the following subsections.

3.1. *SC Cells' Voltage Monitoring Subsystem*

Monitoring the voltages of the SC cells is essential to ensure safe SC-ESS operation, as well as assessing the cells' state of charge (SoC) and state of health (SoH). To monitor all 48 cells with a minimum wirings/complexity, a distributed cells' voltage monitoring subsystem is implemented such that each module has a localized cells' voltage monitoring board and all the boards alongside the modules are sharing the data via a communication channel which selected to be based on serial peripheral interface (SPI) protocol. The module's voltage monitoring board as illustrated in Figure 3 consists of signal conditioning electronics with associated filtration for adaptation of the measured signals to a suitable level, and some application-specific integrated circuits (ASICs) for analog to digital conversion (ADC) and communication protocols.

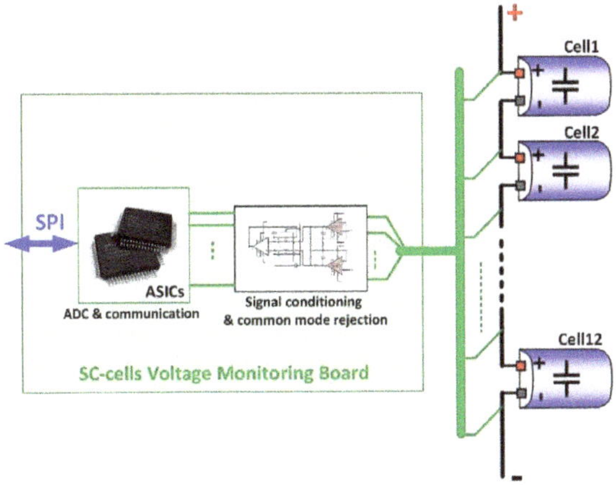

Figure 3. Block diagram representing cells' voltage monitoring within a single SC module as a part of the full voltage monitoring subsystem of the SC-ESS.

3.2. SC Cells' Overvoltage Protection Subsystem

The maximum operating voltage of the SC cells should be strictly limited at the recommended level by the manufacturer as exceeding this limit may result in a catastrophic failure of the cell. Accordingly, an overvoltage protection subsystem is mandatory to ensure SC-ESS safe operation. Overvoltage protection of the cells can be achieved by either preventive or corrective actions. Overvoltage protection subsystems based on preventive action are employing one of the cells' voltages active or passive voltage balancing techniques [17–19] that act during the charging process to prevent the weak cells within the string from reaching the maximum voltage limit whilst other stronger cells are still charging. On the other hand, overvoltage protection subsystems based on corrective action are employing shunt circuits that act only if one or more cells hit the maximum voltage limit during the charging process to completely bypass those cells preventing them from further charging whilst allowing other cells within the string to continue charging. The overvoltage protection (OVP) in the proposed SC-ESS design is built to allow both preventive and corrective actions to ensure maximum reliability for the OVP subsystem. The block diagram shown in Figure 4 represents the different parts of the proposed OVP subsystem.

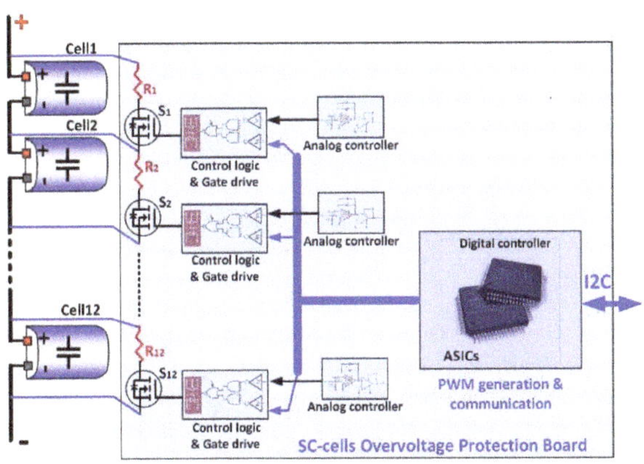

Figure 4. Block diagram represents the design of the SC cells' overvoltage protection within a single module as apart of the full overvoltage protection subsystem of the SC-ESS.

The subsystem consists of a shunt circuit per cell, formed by a semiconductor switch (Si) and a power resistor (Ri), which is controlled by two independent controllers: the first controller is built based on a voltage comparator with associated filtering and signal conditioning electronics. Each SC cell has its controller that works independently from other cells' controllers or system central controller and activates the shunt circuit of the cell automatically once the cell voltage hits the preset limit; hence, it performs the corrective actions of the OVP subsystem. The second controller is built based on a digital platform that implements digital electronics (ASICs) that receives commands via the inter-integrated circuit I2C communication channel from the central controller. Each SC module has its digital controller that controls the shunt circuits of the entire cells based on pulse width modulation (PWM) with different duty cycles defined by the central controller according to the degree of mismatching between measured cells' voltages to perform cells' voltage balancing as a preventive action for overvoltage protection.

The resistance of the shunt resistor should be selected based on the level of degradation/capacity fade of the SC cells to be mitigated. Under the specified constant charging power (2 kW), the current of the SC-ESS decreases as the voltage increases with the increase in the SoC. Accordingly, the value of the current at which the degraded SC cell hits the maximum voltage limit (2.5 V) is changing according to the level of its capacity fading (as a percentage of the initial capacitance), as illustrated in Figure 5. As it can be seen, with no capacity fade in the SC cells, all the cells hit the limit when the voltage of the SC-ESS reaches 120 V (100% SoC), where the current of the SC-ESS is 16.7 A, and hence, no actions are required from the OVP subsystem, since charging is going to be stopped anyway. On the other hand, introducing 10% capacity to one or more cells causes the voltage of these specific cells to hit the limit early whilst the voltage of the SC-ESS is still at around 108 V. Continuing the charging process under this condition forces the OVP subsystem to bypass degraded cells by activating their corresponding shunt circuits. The current of the SC-ESS at this condition is around 18.6 A; hence, the shunt circuit associated with the cell should be able to handle the full value of this current/power. For the proposed SC-ESS, the shunt resistor is selected to be 100 mΩ to allow a shunt current of 25 A at maximum cell voltage limit (2.5 V). This allows mitigating the capacity fade of ≈33%, as illustrated in Figure 5.

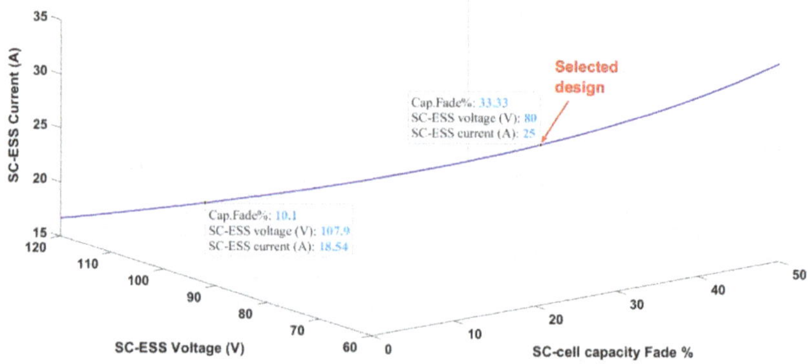

Figure 5. Illustration of how the individual SC cell capacity fades (%) impact the voltages and current levels of SC-ESS at which the corresponding shunt circuits of degraded cells need to be activated, considering 2 kW constant power charging condition.

The preventive action of the OVP subsystem does not affect the sizing of the shunt circuit as the balancing mechanism should operate with a significantly lower power compared to the protective action. Hence, sizing of the shunt circuit based on protective action supports both protective and preventive actions. Accordingly, the balancing act can operate with continuous current (defined by cell voltage and the resistance of the shunt circuit) for fast balancing, or with PWM current for slower operations, which gives more flexibility to design the balancing algorithms.

The block diagram of the developed SC-ESS with all subsystems is shown in Figure 6. The full system consists of four series-connected modules, each with a power board (connecting SC cells) and two detachable boards for voltage monitoring and OVP as can be seen in Figure 7. The aluminum enclosure of the module is used as a heat sink for the shunt resistors for more savings in weight.

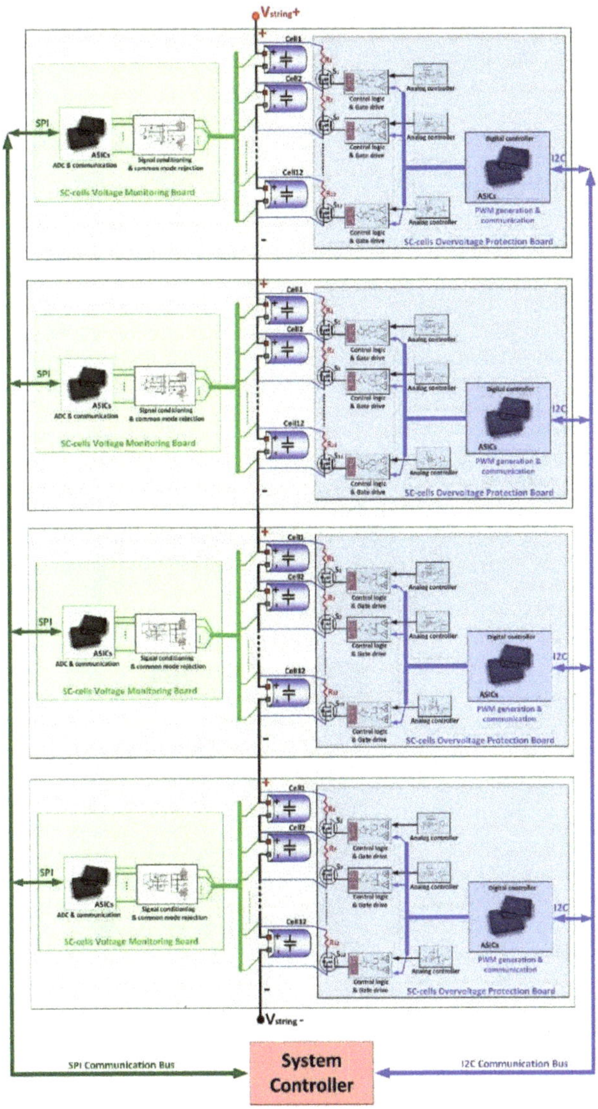

Figure 6. Block diagram for the full architecture of the proposed SC-ESS.

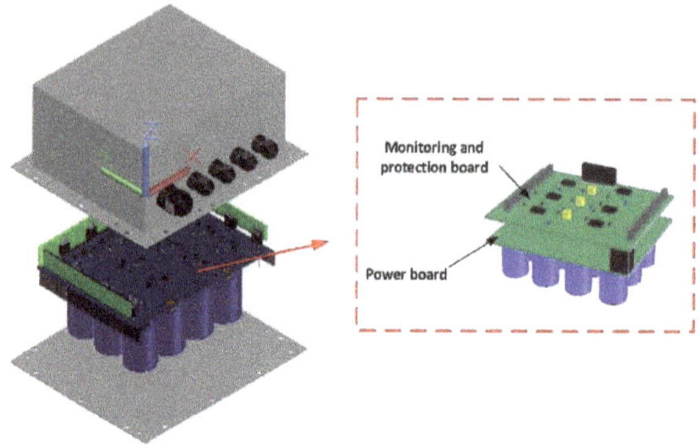

Figure 7. Assembly of one module as a part of the proposed SC-ESS.

4. Experimental Validations

To validate the proposed SC-ESS design, a test setup based on the manufactured four modules has been constructed. The test setup as shown in Figure 8 consists of a power supply that acts as a charger, the electronic load that acts as the system load, and the Arduino Mega 2560 board that acts as the SC-ESS controller. The Arduino board is interfaced with the PC to perform data-logging as well.

The evaluation of the proposed SC-ESS design with associated management electronics is achieved by three tests as detailed below.

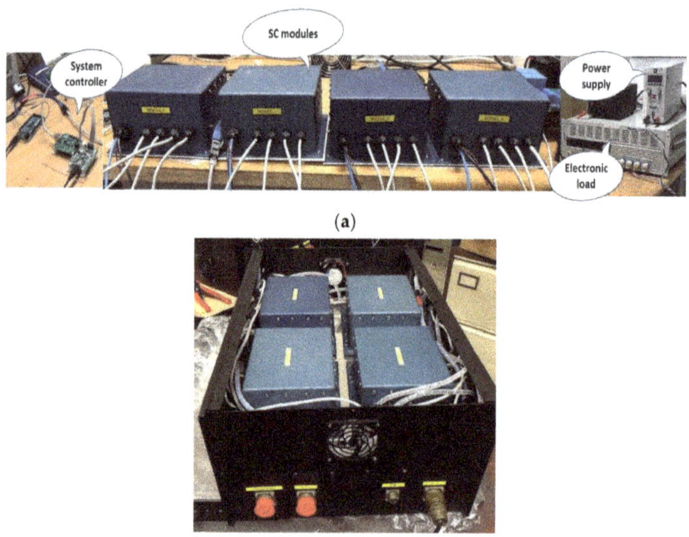

(**a**)

(**b**)

Figure 8. Experimental setup to validate the proposed design for SC-ESS: (**a**) test setup; (**b**) assembled SC-ESS.

The first test is reported in Figure 9, in which the operation and performance of the protective action of the OVP subsystem are evaluated. It is conducted under continuous charge–rest–discharge cycling with 15 A constant current charge and 1.8 kW constant power discharge whilst voltages of the SC cells are measured by the voltage monitoring subsystem and the stack current (SC-ESS current) are measured by a current prob. The test started with forced voltages mismatches between the SC cells of 0.25 V to emulate a capacity mismatch of 12.5%, which caused the cells that were initially at higher voltages to hit the maximum voltage limit (2.5 V) at t = 38 s, and hence the OVP subsystem activated the shunt circuits associated with those cells. Accordingly, their voltages were kept at the limit whilst other cells continued charging until they hit the limit as well (at t = 47 s in this experiment) and the shunt circuits for these cells were also activated to maintain their voltages at the limit until the charging process stopped (at t = 62 s). The overshoots of cell voltage due to the switching of the shunt circuit were maintained at ≤10 mV to ensure the maximum utilization of cells' capacity. Hence, the test confirmed that the OVP subsystem enabled safe operation for SC-ESS that was able to complete the charge cycles and protect the SC cells that were at the risk of overvoltage under the imposed voltages mismatching that emulates SC cell capacity fade due to degradation.

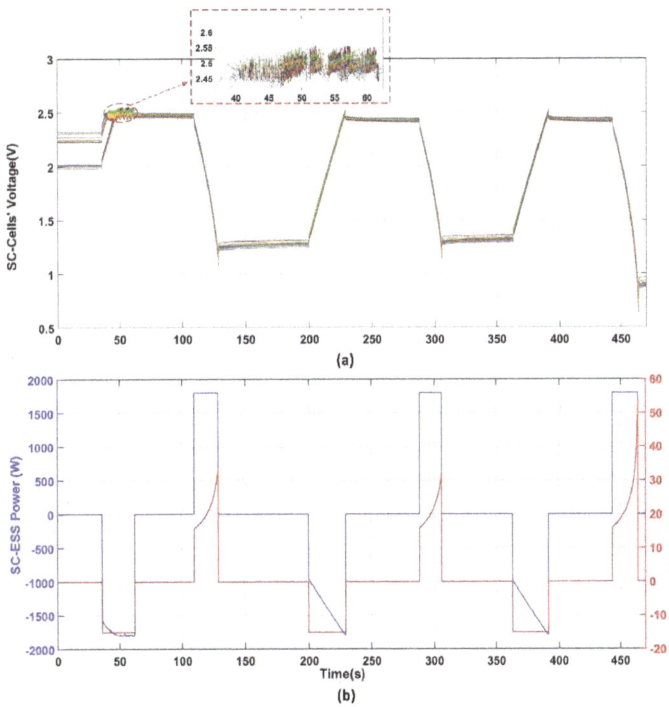

Figure 9. Performing of protective action to protect the SC cells by OVP subsystem during continuous charge/discharge cycling: (**a**) SC cells' voltages; (**b**) charge/discharge power and current.

The second test, which is reported in Figure 10, evaluates the operation of the OVP subsystem under given discharging commands from the central controller to demonstrate its ability to execute these commands when required. These may happen under some conditions like a safe discharge of the ESS. The test starts with SC cells having unequal voltages, and because of applying multiple charging cycles (at t = 49 s and 130 s) with a current of 15 A, the voltage raises to the overvoltage limit; hence, activation of the associated shunt circuits by the OVP preventive action was required at t = 160 s, which continued in operation until the charging cycle stopped at t = 172 s. Accordingly, as the shunt resistors were thermally coupled with the modules' enclosures, the power dissipated in these resistors causes temperatures (temp.) of the SC-modules' enclosures to be increased, as can be seen for modules 1 and 2 recorded temps. (M1 and M2 case temp.) in Figure 10. At t = 360 s, a controlled discharge for the cells of modules 1 and 3 with a duty cycle of 20% followed by faster discharging with larger duty cycles (50% and 75%) based on the received commands from the central controller were reported. Accordingly, the energy stored in these modules dissipated in the shunt circuits, causing temp. increase, as can be seen for M1 case temp. recorded. When the voltage reached 0.4 V, the discharge was stopped. At t = 580 s, the remaining cells at modules 2 and 4 were also commanded to discharge with a 75% duty cycle and this caused a fast discharge to the same minimum cell voltage as the other cells (0.4 V) to achieve a balanced state for all cells while the energy dissipated in these modules also caused its temp. to be increased, as can be seen for M2 case temp. recorded. This test confirmed the capability of the OVP subsystem to perform a discharge cycle for specific SC cells by a specific discharging rate controlled by the duty cycle that is selected by the central controller and commanded via the communication bus.

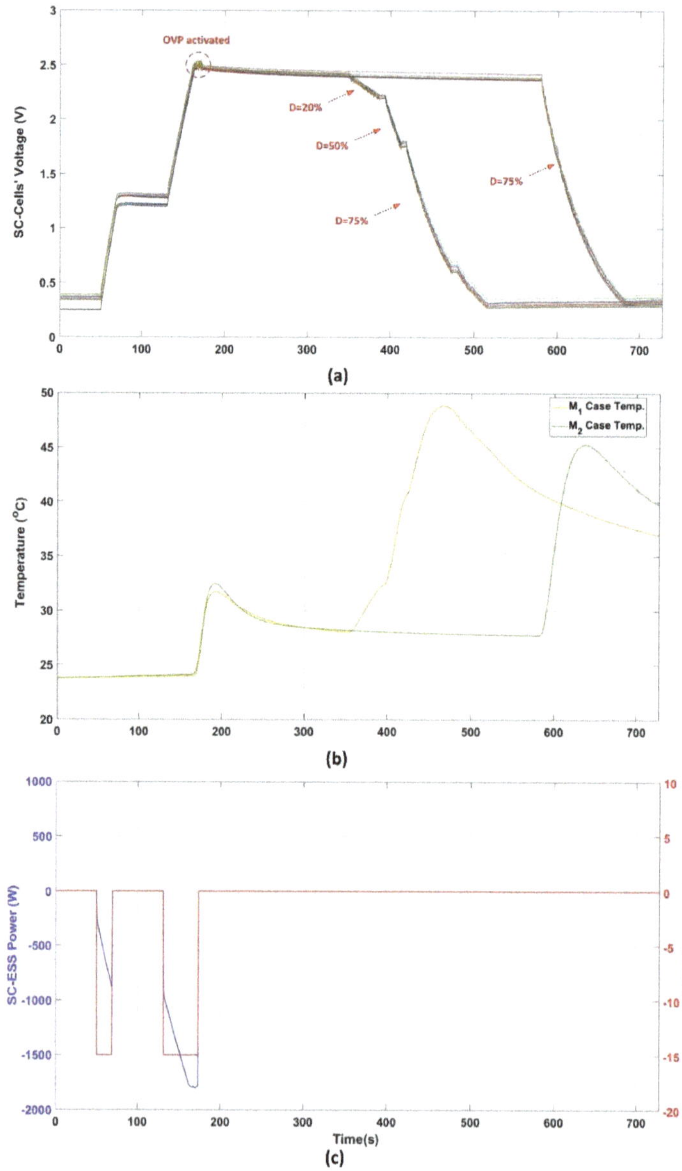

Figure 10. Performing of protective action to protect the SC cells by OVP subsystem during charge/discharge as well as responding to discharge commands from the central controller: (**a**) SC cells' voltages; (**b**) SC-modules 1 and 2 enclosures' temperatures; (**c**) Discharge power and current.

The third test reported in Figure 11 evaluates the capability of the OVP subsystem to perform active dissipative SC cells' voltage balancing as a preventive action to protect the cells from overvoltage. The test started with imposed mismatching between the SC cells, then a discharging process by a constant power of 1.3 kW was introduced at t = 15 until t = 25 s to allow for sufficient charging time for demonstrating the balancing mechanism. After a rest time provided to the system, the charging of the SC-ESS started by a constant current of 5 A at t = 47 s, where the mismatching in the SC cells' voltages was detected by the system controller. Following this, it activates the balancing mechanism in which the SC cells' shunt circuits equalize the cells' voltages in a PWM manner. As can be seen, the balancing mechanism succeeded to achieve balanced cells' voltages in a considerable short time due to the ability of the shunt circuit to handle a high current.

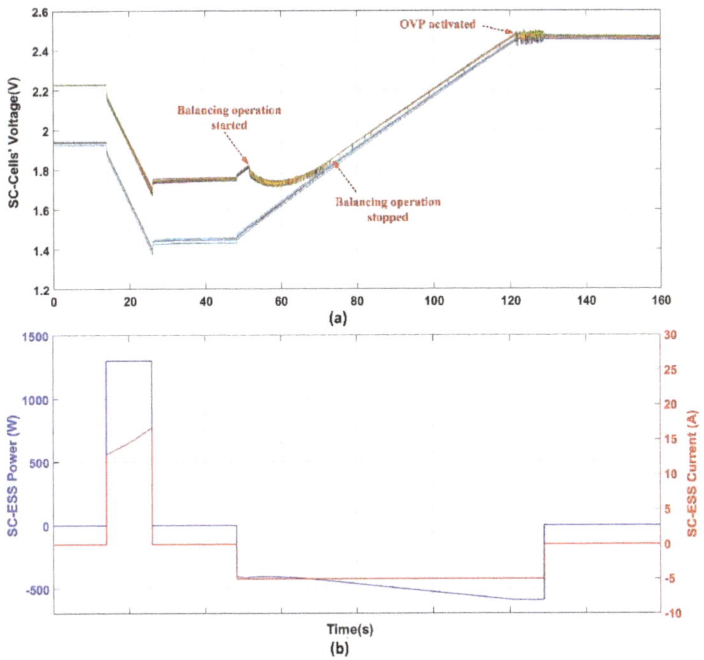

Figure 11. Performing preventive action by OVP subsystem via active dissipative balancing during the continuous charge cycle: (**a**) SC cells' voltages; (**b**) Charge/discharge power and current.

In the three experimental tests, the results confirmed the proposed functionalities of the SC-ESS in terms of cells' voltage monitoring, fast balancing, and overvoltage protection.

5. Conclusions

In this paper, the design and implementation of a smart super-capacitors-based energy storage system are proposed. The SC-ESS sizing was optimized for minimum weight by utilizing the highest energy density cells of the available range and adjusting the operating voltage range, accordingly. In addition, the proposed design minimizes the wirings for cells' voltage monitoring and overvoltage protections by employing a distributed modular architecture coordinated via standard communication buses. SC cells overvoltage protection subsystem in the proposed design implements two independent preventive and protective actions for the highest reliability to ensure safe operation of the system. Finally, the design is validated based on experimental testing of the manufactured modules and the results showed the achievement of the proposed functionalities.

Author Contributions: Conceptualization, A.M.F., M.R. and C.K.; methodology, A.M.F.; software, A.M.F. and M.K.; validation, A.M.F.; formal analysis, A.M.F., M.R. and C.K.; investigation, A.M.F. and C.K.; resources, S.B. and C.K.; data curation, A.M.F.; writing—original draft preparation, A.M.F.; writing—review and editing, A.M.F., C.K. and S.B.; visualization, A.M.F.; supervision, C.K. and S.B.; funding acquisition, S.B. All authors have read and agreed to the published version of the manuscript.

Funding: The project received funding from the Clean Sky 2 Joint Undertaking under the European Union's Horizon 2020 research and innovation program under grant agreement No 755485.

Institutional Review Board Statement: Not applicable.

Informed Consent Statement: Not applicable.

Data Availability Statement: Not applicable.

Conflicts of Interest: The authors declare no conflict of interest. The funders had no role in the design of the study; in the collection, analyses, or interpretation of data; in the writing of the manuscript, or in the decision to publish the results.

References

1. Buticchi, G.; Bozhko, S.; Liserre, M.; Wheeler, P.; Al-Haddad, K. On-Board Microgrids for the More Electric Aircraft—Technology Review. *IEEE Trans. Ind. Electron.* **2019**, *66*, 5588–5599. [CrossRef]
2. Wheeler, P.; Bozhko, S. The More Electric Aircraft: Technology and challenges. *IEEE Electrif. Mag.* **2014**, *2*, 6–12. [CrossRef]
3. Lamantia, A.; Giuliani, F.; Castellazzi, A. Power Scalable Bi-Directional DC-DC Conversion Solutions for Future Aircraft Applications. *Energies* **2020**, *13*, 5470. [CrossRef]
4. Ojeda-Rodríguez, Á.; González-Vizuete, P.; Bernal-Méndez, J.; Martín-Prats, M.A. A Survey on Bidirectional DC/DC Power Converter Topologies for the Future Hybrid and All Electric Aircrafts. *Energies* **2020**, *13*, 4883. [CrossRef]
5. Rigogiannis, N.; Voglitsis, D.; Jappe, T.; Papanikolaou, N. Voltage Transients Mitigation in the DC Distribution Network of More/All Electric Aircrafts. *Energies* **2020**, *13*, 4123. [CrossRef]
6. Todd, R.; Wu, D.; Girio, J.A.d.S.; Poucand, M.; Forsyth, A.J. Supercapacitor-based energy management for future aircraft systems. In Proceedings of the 2010 Twenty-Fifth Annual IEEE Applied Power Electronics Conference and Exposition (APEC), Palm Springs, CA, USA, 21–25 February 2010; pp. 1306–1312.
7. Nadeem, F.; Hussain, S.M.S.; Tiwari, P.K.; Goswami, A.K.; Ustun, T.S. Comparative Review of Energy Storage Systems, Their Roles, and Impacts on Future Power Systems. *IEEE Access* **2019**, *7*, 4555–4585. [CrossRef]
8. Al Shaqsi, A.Z.; Sopian, K.; Al-Hinai, A. Review of energy storage services, applications, limitations, and benefits. *Energy Rep.* **2020**, *6*, 288–306. [CrossRef]
9. Zuo, W.; Li, R.; Zhou, C.; Li, Y.; Xia, J.; Liu, J. Battery-Supercapacitor Hybrid Devices: Recent Progress and Future Prospects. *Adv. Sci.* **2017**, *4*, 1600539. [CrossRef] [PubMed]
10. Mohamed Rashed, C.K. Design and evaluation of an energy storage system for helicopters. In Proceedings of the 9th International Conference on Power Electronics, Machines and Drives, Liverpool, UK, 17–19 April 2018.
11. Chen, J.; Song, Q. A Decentralized Energy Management Strategy for a Fuel Cell/Supercapacitor-Based Auxiliary Power Unit of a More Electric Aircraft. *IEEE Trans. Ind. Electron.* **2019**, *66*, 5736–5747. [CrossRef]
12. Chen, J.; Song, Q.; Yin, S.; Chen, J. On the Decentralized Energy Management Strategy for the All-Electric APU of Future More Electric Aircraft Composed of Multiple Fuel Cells and Supercapacitors. *IEEE Trans. Ind. Electron.* **2020**, *67*, 6183–6194. [CrossRef]
13. Saenger, P.; Devillers, N.; Deschinkel, K.; Péra, M.; Couturier, R.; Gustin, F. Optimization of Electrical Energy Storage System Sizing for an Accurate Energy Management in an Aircraft. *IEEE Trans. Veh. Technol.* **2017**, *66*, 5572–5583. [CrossRef]
14. Navarro, G.; Blanco, M.; Torres, J.; Nájera, J.; Santiago, Á.; Santos-Herran, M.; Ramírez, D.; Lafoz, M. Dimensioning Methodology of an Energy Storage System Based on Supercapacitors for Grid Code Compliance of a Wave Power Plant. *Energies* **2021**, *14*, 985. [CrossRef]
15. Fares, A.; Klumpner, C.; Rashed, M. Design Considerations to Optimise Supercapacitor-based Energy Storage Systems for Aerospace Applications. In Proceedings of the 2018 IEEE International Conference on Electrical Systems for Aircraft, Railway, Ship Propulsion and Road Vehicles & International Transportation Electrification Conference (ESARS-ITEC), Nottingham, UK, 7–9 November 2018; pp. 1–8.
16. Fritz, N.; Rashed, M.; Klumpner, C. Power Density Optimization of a DC/DC Converter for an Aircraft Supercapacitors Energy Storage. In Proceedings of the 2018 IEEE International Conference on Electrical Systems for Aircraft, Railway, Ship Propulsion and Road Vehicles & International Transportation Electrification Conference (ESARS-ITEC), Nottingham, UK, 7–9 November 2018; pp. 1–9.
17. Ibanez, F.M. Analyzing the Need for a Balancing System in Supercapacitor Energy Storage Systems. *IEEE Trans. Power Electron.* **2018**, *33*, 2162–2171. [CrossRef]

18. Guo, M.; Zhang, X.; Li, H.; Liao, H.; Liao, Y.; Meng, Z.; Zhang, H.; Huang, Z. Cooperative Cell Balancing of Supercapacitors with Adaptive Observers. In Proceedings of the 2020 IEEE Energy Conversion Congress and Exposition (ECCE), Detroit, MI, USA, 11–15 October 2020; pp. 588–593.
19. Wang, X.; Cheng, K.W.; Fong, Y.C. Non-Equal Voltage Cell Balancing for Battery and Super-Capacitor Source Package Management System Using Tapped Inductor Techniques. *Energies* **2018**, *11*, 1037. [CrossRef]

energies

MDPI

Review

A Review on the Challenges of Using Zeolite 13X as Heat Storage Systems for the Residential Sector

Amirhossein Banaei * and Amir Zanj

College of Science and Engineering, Flinders University, Adelaide, SA 5042, Australia; amir.zanj@flinders.edu.au
* Correspondence: Amirhossein.banaei@flinders.edu.au

Abstract: In recent years, several attempts have been made to promote renewable energy in the residential sector to help reducing its CO_2 emissions. Among existing approaches utilizing substances capable of directly storing and transporting thermal energy has recently become a point of interest. Zeolite 13X with exceptional capacity to safely store thermal energy for long periods and release heat due to its unique molecular structure is known to be one of the best options serving this purpose. However, the application of this ceramic as a heat storage material in the residential sector is associated with significant challenges dictated by the limitations of the sector, such as space restrictions and affordability. The current review attempts to explore the extent of these challenges, mainly related to design and efficiency from different perspectives. The main aim here is to provide a clear vision for a better understanding of the state of the art of this technology and to help to identify possible solutions fostering the adaptation of this technology to the residential sector.

Keywords: TES system; heat adsorption; thermal storage system efficiency; residential sector heating

Citation: Banaei, A.; Zanj, A. A Review on the Challenges of Using Zeolite 13X as Heat Storage Systems for the Residential Sector. *Energies* 2021, *14*, 8062. https://doi.org/10.3390/en14238062

Academic Editors: Alon Kuperman and Alessandro Lampasi

Received: 27 October 2021
Accepted: 24 November 2021
Published: 2 December 2021

Publisher's Note: MDPI stays neutral with regard to jurisdictional claims in published maps and institutional affiliations.

1. Introduction

One solution to fully exploit renewable energy in residential heating is to develop a technology by which the thermal energy received by the sun could be directly stored during the day and released whenever needed. Doing so would result in a significant reduction in the residential share of fossil fuel consumption and CO_2 emissions while providing residential heating. The key factor to attain such goal lays in finding suitable substances capable of storing high level of thermal energy and efficiently substituting them in a residential thermal system.

In recent years, zeolite 13X has received considerable attention to be used as an efficient heat storage substance in thermal systems. Due to its special molecular structure, which contains well-defined microchannel and cavities, zeolite 13X can store heat by removing humidity and release heat when humidity is introduced to the compound, which gives zeolite 13X a unique heat storage property. The stored heat will be confined as long as no humidity is introduced to the system. This feature provides a simple, safe, and affordable mechanism for portable thermal energy. Despite these unique features, Thermal Energy Storage (TES) systems containing zeolite13X have not yet acquired a significant share in residential heating due to their poor operating performance for the sector.

2. Thermal Energy Storage Systems

To provide a better understanding of TES challenges, in this section, briefly, the underpinning theory, as well their systematic characteristics, are explained. In general, TES systems have been used widely in the industrial sector (with some large-scale residential applications). These systems are divided into three main categories based on their storage methods: Sensible heat, Latent heat, and thermochemical heat, as presented in Figure 1. As can be seen in all of these systems, a high-capacity storage substance is implemented to receive, save, hold, or carry the thermal energy. The performance process of these substances can be characterized in three main steps, charging, storage, and discharging.

In Figure 2, these three processes are presented for the different TES categories. As presented in Figure 2a, Sensible heat storage systems use the simple method of storing heat by discharging excess heat from the material without phase change. In Latent heat storage (Figure 2b), the stored heat is the result of a phase change; thus, they can attain a higher thermal density compared to the Sensible method. In Thermochemical systems (Figure 2c), unlike the two other methods, the stored heat is obtained from a reversible thermochemical reaction.

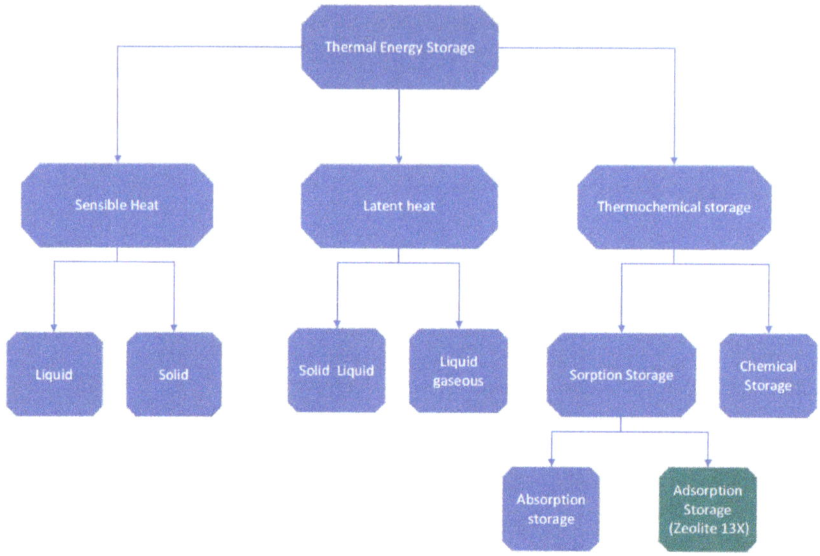

Figure 1. Classification of different thermal storage methods.

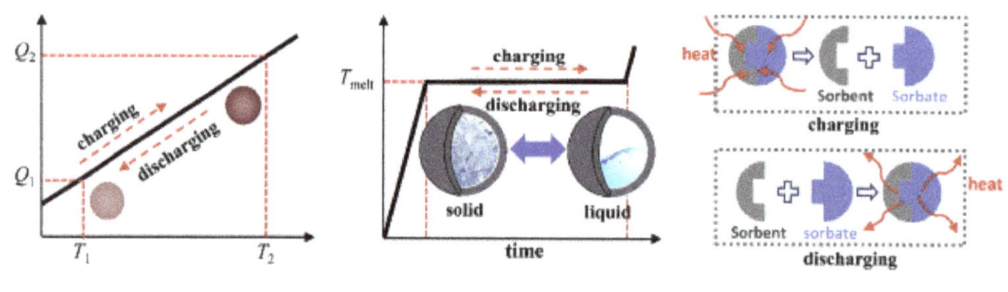

Figure 2. Methods of thermal energy storage (**a**) Sensible heat, (**b**) latent heat, (**c**) thermochemical heat (Sorption heat) [1].

Thermochemical systems are particularly efficient in storing heat for long-term storage applications because heat loss from the system is low.

In addition, Thermochemical storage systems, due to their high-density storage characteristics, provide a spatially efficient and compact storage system compared to the other heat storage. A volumetric comparison between the three main categories is presented in Figure 3; as can be seen, for instance, replacement of latent system by Thermochemical would save up to 10 times the required space, which makes these types of systems advantageous for residential sector application [2].

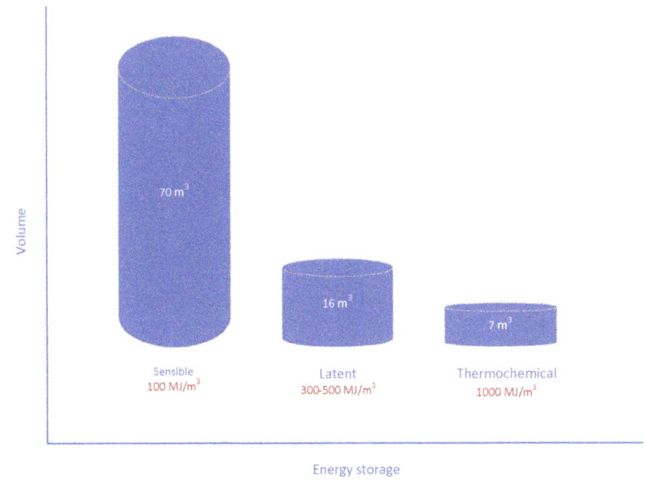

Figure 3. Volume needed to full cover the annual storage need of an energy efficient passive house (6480 MJ) [3].

Given that zeolite 13X stocks heat by adsorption method that belongs to the thermochemical category, in this review, the focus will be on adsorption heat storage.

Adsorption is the movement of atoms or molecules from a bulk phase (which might be solid, liquid, or gas) to a solid or liquid surface. Heat is released as a result of attractive interactions between the surface (adsorbent) and the molecules being adsorbed (adsorbate) [2], as presented in Figure 4.

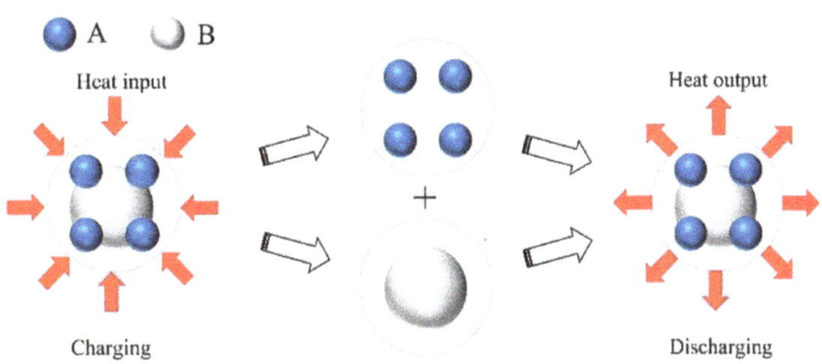

Figure 4. Charging and discharging thermochemical heat by sorption reaction [4].

The attractive potential for this thermochemical reaction can be explained with Lennard-Jones potential (Figure 5) equation as a function of distance:

$$V_{LJ}(r) = 4\varepsilon\left[\left(\frac{\sigma}{r}\right)^{12} - \left(\frac{\sigma}{r}\right)^{6}\right] \quad (1)$$

where σ is the distance at which the potential is zero, and ε is the depth of the potential well. According to Equation (1), higher ε means more heat would be released after adsorption. The extent of this potential varies from 8 to 800 kJ mol^{-1}. The amount of adsorption depends on the inherent properties of the material in reaction, such as specific surface area, and the affinity of the couple adsorbent/adsorbate. The adsorption process also is subject to operating conditions, such as temperature, pressure, and concentrate of the adsorbate. For instance, high specific surface area, high pressure, and low temperature increase adsorption reaction rate.

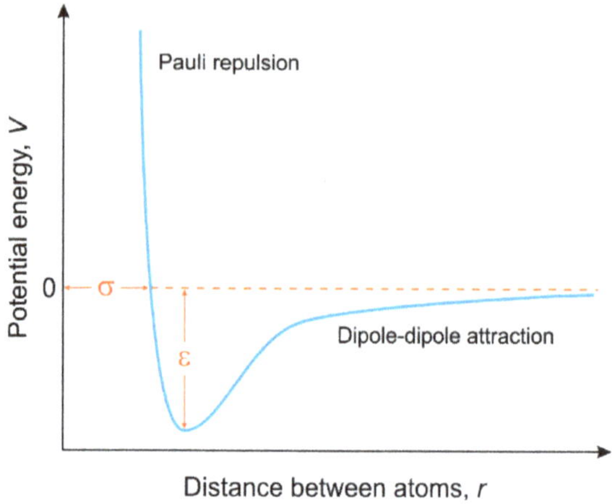

Figure 5. Lennard-Jones potential.

3. Adsorption Heat Materials

Generally, all solids are adsorbents, but only those with a high specific surface area are interesting for sorption applications. In this section, a few adsorbent families are briefly introduced.

In industrial applications, activated carbons are the most often used adsorbents. It is mostly employed in the purification of gases. Non-polar activated carbons are commonly utilized for water treatment. As a result of their poor affinity for water, they are not employed in heat storage applications. Silica gel is a synthetically porous type of sodium silicone dioxide. It may be represented as $SiO_2.nH_2O$ in its chemical compound. Treating synthesis (PH and presence of cations) control carefully permits pores to be controlled. Its large specific surface area (from 600 to 800 $m^2.g^{-1}$) and hydrophilic characteristics make it an excellent desiccant, and it has a very high adsorption capacity at low pressures and temperatures. When silica gel is saturated with water, it loses its capacity to create heat for long periods of time, which is one of its limitations as a heat storage medium.

Zeolites are porous crystalline minerals composed of silicon (Si), aluminum (Al), and oxygen (O) atoms. In tetrahedral configurations, each Si or Al atom is linked to four oxygen atoms (SiO_4 and AlO_4). Each oxygen atom is shared with another tetrahedron that connects Si or Al atoms, as presented in Figure 6. This configuration results in atomic angles remaining the same and, thus, a uniform distribution in pore sizes. On the other hand, since each oxygen atom is shared between two tetrahedral Si or Al atoms, the stoichiometric composition of each tetrahedral unit is SiO_2 or AlO_2 with minimum Si/Al ratio to be 1.0 without any upper limit. Given the fact that rich Aluminium sieves has a high affinity for water (since each Al atom introduces a negative charge in the material and polar molecules, such as water, are sensitive to these charges), it demonstrates a high level of hydrophilic behavior. Conversely, since rich silicon sieves have a hydrophobic behavior, heat induced transition between hydrophilic and hydrophobic behavior (usually occurs at a Si/Al ratio between 8 and 10 [5], plus uniform pore sizes, introduces zeolites as an exceptional adsorbent fit for TES systems.

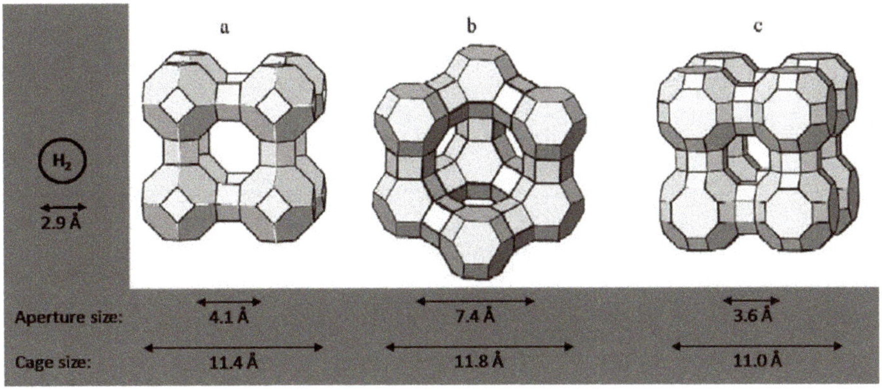

Figure 6. Zeolite framework structures: (**a**) A, (**b**) X and Y, (**c**) Rho. The H_2 molecule is shown to scale for comparison [6].

There are various type of zeolite and their composite used for heat storage [7]. Among the zeolite family, zeolite 13X (Figure 6b) has the largest permeability for water molecules due to its inner hole size, which plays a key role in heat storage efficiency.

In addition, while employing zeolitic composite as heat storage may be more efficient, the use of added materials and the synthesis process raises health concerns that zeolite 13X, in turn, does not.

To make use of this exceptional adsorptive capacity, zeolites must be placed inside reactor beds to undergo the aforesaid three performance processes (charging, storing, and discharging). The efficiency of performance processes is closely related to controlling the reactor bed operational condition, such as temperature, flow rate, and moisture level.

There exist varieties of reactor systems that can serve the purpose, among them open adsorption systems with the fixed bed being the most used type. A schematic of these systems is presented in Figure 7. To charge zeolite in this system, a stream of hot air with minimum 120 °C must be injected into the reactor, where zeolite is placed. The injected hot stream into the bed, while charging zeolites, also extracts and transports its trapped moisture.

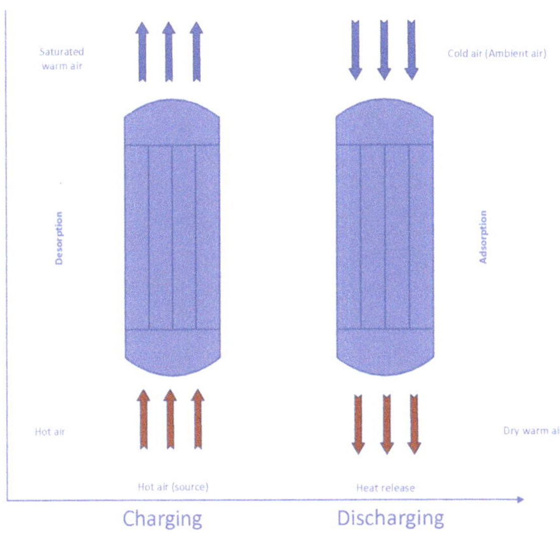

Figure 7. Open adsorption process.

Regarding the storage step, thanks to the unique zeolite structure, the stored heat will remain untouched until moisture re-enters the reactor, which makes this system needless for thermal isolation (a specific advantage compared to other systems). In these systems, heat sorption capacity is defined as:

$$Q\text{ sorbtion} = -V\ |\Delta H|\ \Delta q \tag{2}$$

where V is the bed volume, $|\Delta H|$ is the specific sorption heat, and Δq presents the sorbate uptake capacity between charging and discharging phase [8]. During the discharge (adsorption) process, ambient airflow with precisely controlled humidity enters the reactor. Given that zeolite is hydrophilic at ambient temperature, it absorbs the humidity and releases the stored energy in the form of heat. The outlet air temperature and the discharging duration depend not only on the inherent property of zeolite but also on the effectiveness of reactor system design and its operational conditions. This makes reactors a key player of the heat storage systems, after selecting appropriate materials.

Given that the proper access to zeolite's unique storage capacity and its efficient delivery bonds to reactor parameters, the design and optimization of reactors would then become one of the main challenges in using heat storage systems in the residential sector. Figure 8 presents the reactor most important variables in three main categories: Reactor Structure, Auxiliary Equipment, and Design Parameters, each introducing their challenges.

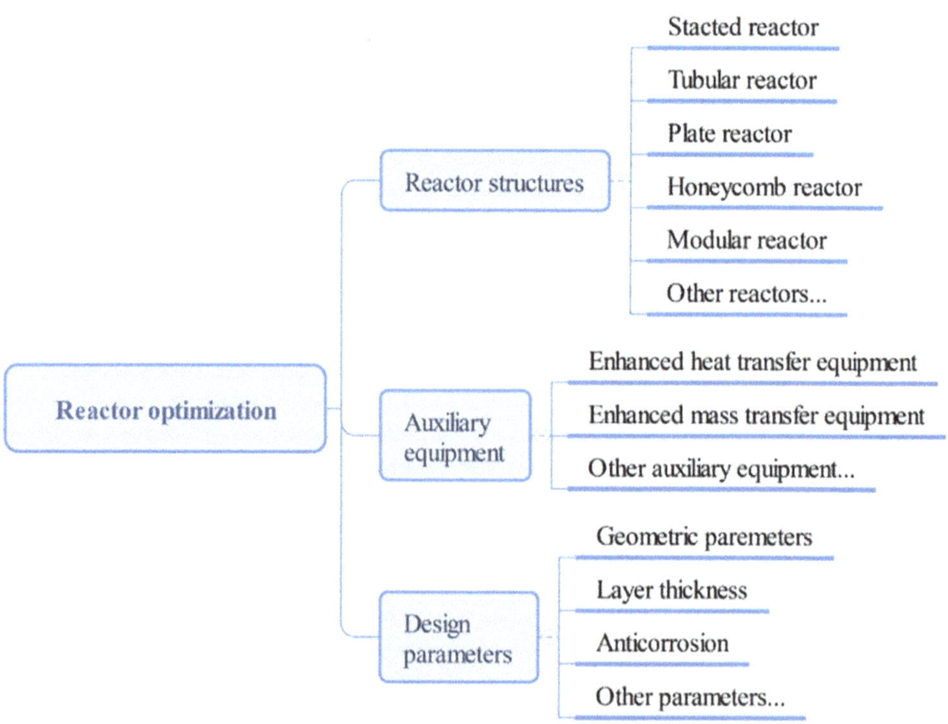

Figure 8. Influence impact on reactor optimization [9].

The main goal in TES systems is to optimize mass/heat transfer and decrease heat loss. To attain this, several configurations have been examined so far, such as staced, tabular, honeycomb, plate, modular, etc., in all these configurations parameters, such as reactor diameter, length, and connectors, have significant effects on the efficiency of heat storage and have been investigated by many researchers. Anderson et al. [10] found that bed length and heat loss have direct relationship, so, by reducing the length, heat loss can be reduced. Lahmidi et al. [11] used a stacted reactor. In this design, to improve mass transfer a nozzle device is added to the system, which increases the interaction area between solid and water vapor conducive. To further improve mass and heat capacity Stitou et al. [12] developed a pilot plan using high thermal conductivity of ENG. Layer thickness is another key parameter for reactor design, directly affecting the hydration time. Van Essen et al. [13] observed that decreasing the thickness speeds up the hydration process. Oktariani et al. developed system for generation steam by using a zeolite 13X-water system [14]. They found that the flow direction of feeding water from the top of the reactor using nozzle configuration could conform a better result than feeding water from the reactor bottom. Considering the aforementioned facts, it seems simpler reactors, such as a high efficiency staced reactor, could be the better option for the residential sector; however, thanks to the advancement of fabrication technology, modification to reactor design is yet to present future enhancements.

Evaluation of an efficient design can be tested using TES systematic performance parameters. In general, performance parameters can be expressed by three terms: thermal power density with unit (kW·m^{-3}), thermal storage density with unit (kWh·m^{-3}), and the Coefficient of Performance (COP) with the first two being proportional, and can be obtained from various design and condition parameters and the third to be defined as (Figure 9):

$$\mathrm{COP} = \frac{Q_H}{W} \tag{3}$$

where W and Q_H are the power required to run the discharging process and the amount of extracted heat, respectively. As can be seen, the provided COP in the article, although present in the effectiveness of the discharge process, leaves the charging process untouched. To cover the whole performance cycle, it might be better to introduce TES cycle thermal efficiency η as:

$$\eta = \frac{Q_H}{W + Q_{in}} \tag{4}$$

where Q_{in} presents the provided heat to the system. It is now vivid that the thermal efficiency of the TES systems indeed depends on a variety of factors, on top of which the effectiveness of the charging process is present. The multivariable nature of TES ongoing performance process opens several challenges for further optimization and TES systems new applications (e.g., residential applications), to be investigated in the next section.

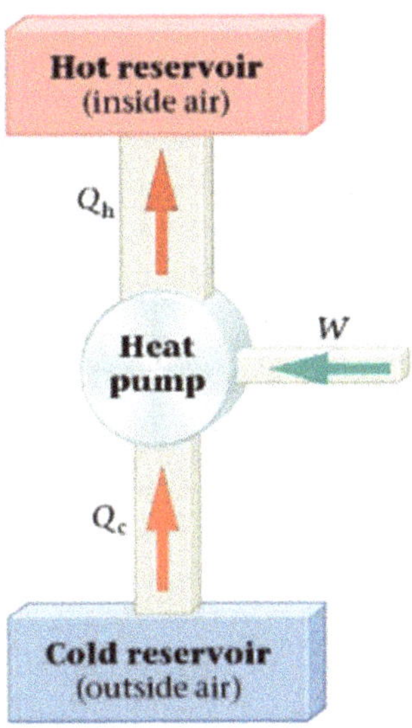

Figure 9. Coefficient of performance [15].

4. Technical Challenge

According to the facts mentioned in the previous section, one can easily conclude that to introduce adsorption systems in the residential sector a compact system must be designed due to space restrictions. However, given that providing the required amount of household power is directly proportional to the amount of zeolite, TES sizing would become a significant limiting factor in the development of an adsorption system for residential application. As a result, many researchers are attempting to improve TES system efficiency to make them capable of storing more energy in less space (highly efficient system). This transition introduces several challenges in executing TES performance processes (charge and discharge), such as efficient charging resources, precise humidity control, reactor design, manufacturing, etc. In this section, the main obstacles in achieving such a compact and efficient system will be discussed.

4.1. Required Temperature Supply

To begin the charging phase in adsorption systems, the inlet air temperature must reach the hydration reaction temperature. Depending on the materials employed, different storage systems require different input temperatures; for zeolite 13X, this temperature must be above 120 °C [16]. As a way to increase efficiency, many researchers investigated the effect of increasing input temperature on system function. Johannes et al., in 2015, used open-source heat storage, including two containers of zeolite 13X. They examined two different temperatures levels for the charging phase and compared their effects on the system charging phase. They discovered increasing temperature from 120 °C to 180 °C can enhance storage density by 40%, while decreasing charging time by 7% (Figure 10).

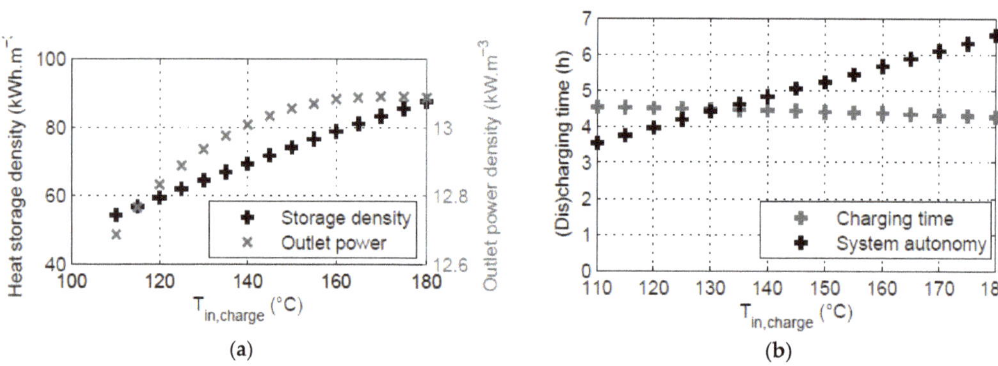

Figure 10. The effect inlet temperature on the heat storage density (**a**) and the charging time (**b**) [17].

Required high temperature flow can be supplied from various resources. Table 1 illustrates some of the prototypes that were built, as well as the temperatures and heat sources that were used. As can be seen in the highlighted column, electric heaters have been employed to deliver the required heat in almost all instances. However, electric heaters have high consumption due to their low COP as a heating system. With a typical heater (consuming between 2 to 4 kWh), achieving higher temperature input flow seems a less than ideal solution to provide Q_{in}, unless a cost-effective power source can be provided. This necessity makes the intake power supply of compact TES systems a significant challenge to be addressed for residential applications.

Table 1. Used temperature for prototypes.

Setup Name	Year	Inlet Temperature [°C]	Heat Source	Outlet Temperature [°C]	Energy Density [kWh/m^3]	Max Power [kW]
Alebeek [16]	2018	180	Electrical heater	13	54	4.4
STAID [18]	2015	120–180	Electrical heater	20	114	2.25
ASIC [19]	2014	230–180	Electrical heater	25	148	1.5
E-HUB/ECN [20]	2013	185	Oil to air exchanger (Electrical heater)	25–60	58	0.4
MONDESTORE [21]	2008	180	Electrical heater	25	57	
MONOSORP [11]	2006	170	Electrical heater	20	120	1.5

4.2. Relative Humidity Control

The level of humidity entering to reaction area in the collector with open systems has the most influence on power density and storage density [22]. Relative humidity has direct effect on the sorbate uptake (as reflected in Equation (2)), which makes it a key parameter for storage density. Most prototypes employ an electrical humidifier, water tank, and sensor to supply and control the humidity of the system's incoming air [23]. Figure 11 shows how lowering the relative humidity in the charging process from 0.5 to 0.1 percent enhances both heat storage density (a) and charging time (b). It also demonstrates that storage density and power density are very sensitive to humidity changes. A change in humidity of 0.4 percent caused a significant power change, as can be seen in Figure 11. These characteristics must, thus, be considered to build an appropriate power system.

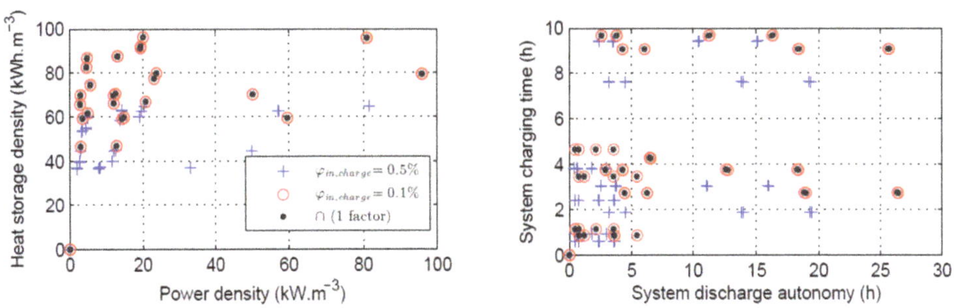

(a) Heat density against power on charge (b) Charging time against autonomy on charge

Figure 11. The influence of inlet relative humidity on the heat storage density and power density (**a**) as well as the charging time (**b**) [18].

For household applications, the inlet relative humidity becomes more important as unstable ambient air is used for the charging and discharging process [24]. Due to the strong dependency of stored and released heat on relative humidity, the use of TES system in the residential sector would introduce a significant challenge that necessitates implementing precise humidity monitoring for closed loop control, especially when variable setpoints are desired to adjust variable system's needs.

4.3. Reduce Reactor Size

Reducing the size of the reactor itself can improve their residential development. Smaller reactors occupy less space and are simpler to integrate with an energy source. However, as mentioned earlier, due to the direct relationship between zeolite volume and system power, constructing a smaller reactor faces numerous configuration and manufacturing limitations. Several investigations have been performed to identify and improve effective size criteria, such as bed length, cross section area, and tank volume by Kuznik et al. [22]

and Michel, Mazet, and Neveu [24]. Gondre et al. [17] found that there exists a linear between the outlet power versus cross section area, heat storage capacity versus storage tank volume, and charging time and autonomy versus bed length, as shown in Figure 12.

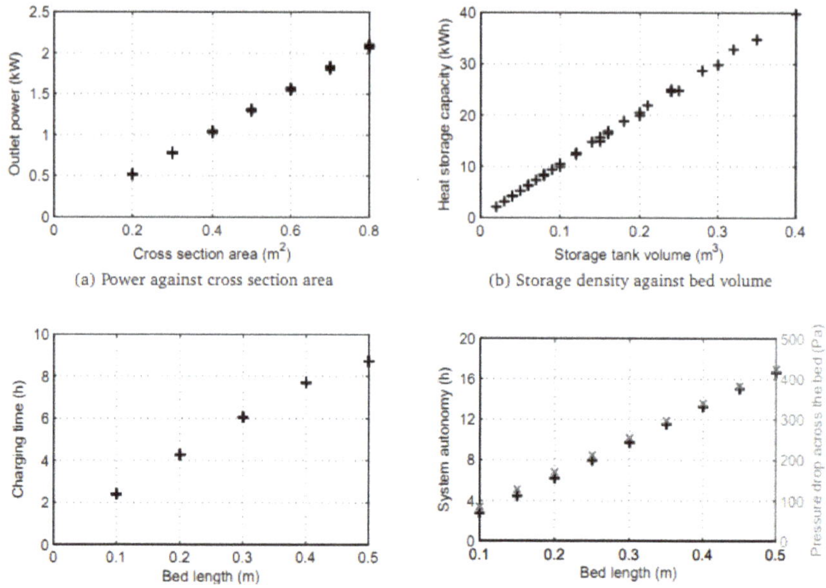

(a) Power against cross section area

(b) Storage density against bed volume

(c) Charging time against bed length

(d) Autonomy against bed length

Figure 12. (**a**) The influence of cross section area on outlet power; (**b**) The impact of storage tank volume on hear storage capacity; (**c**) The influence of bed length against chrging time; (**d**) The impact of bed length on discharging time [17].

Their experiment presented the following significant outcomes, as illustrated in the Figure 12:

- outlet power is directly proportional to cross section area (a)
- heat storage capacity is directly equivalent to storage tank volume (b)
- charging time and discharging time is directly proportional to bed length (c,d)

Reducing the size of the reactor, as predicted, reduces the power of the system. Therefore, ways to optimize the reactor's dimensions should be investigated if residential applications are in the list.

4.4. System Output Power

Another point of interest in TES system optimization is the output power during its discharge process. Many researchers have been able to enhance TES system output power by selectively adjusting the discharge process parameters. Several prototypes in laboratories were designed for this purpose, presented in Figure 13. Among them, Jahannes et al. developed a high power open sorption system (STAID), which contains two reactors of 80 kg of zeolite13X [18]. The discharge parameter of this system was adjusted to generate heat for the residential sector during peak hours. Input ambient air flow temperature was considered to be 20 °C, and outlet temperature was 57 °C. It was observed that their system delivered 6 h of continuous heating during the discharge phase, equal to a maximum 2.25 kw of thermal power output.

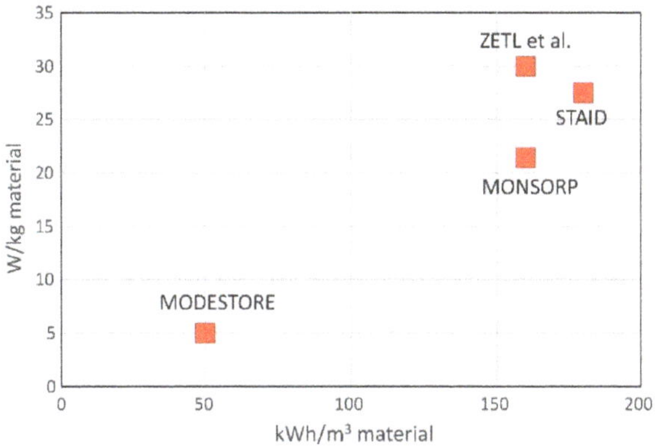

Figure 13. Comparison of mass power obtained from different physical adsorption systems, where STAID is the present work [18].

Zettl et al., of the Austria Solar Innovation Center (ASIC), used different techniques to increase the discharge power. They designed a prototype with a rotating bed that was capable to generate maximum outlet power 1.5 kW only using 50 kg zeolite 4A. The purpose of using rotating bed was to avoid the formation of dead zone to increase the outlet power. The input temperature was 25 °C, and maximum outlet temperature was 60 °C [19].

ADEnergy research Center of the Netherlands developed an open sorption concept using two beds with 150 kg zeolite 13X with the compact bed. This system was designed to supply warm air for the residential sector. The air is humidified with 12 mbar water vapor pressure, and air flow rate is 80 m^3/h. This system generates maximum 0.4 kW, and output temperature is 70 °C [25].

MonoSorp prototype was designed as heat storage system with opened bed for space heating. The input temperature is around 20 °C, and the maximum outlet temperature is approximately 42 °C. They used zeolite as extruded honeycomb structures to avoid pressure loss. This system was able to deliver maximum thermal power of 1.5 kW [13].

Figure 13 collectively present these TES systems specific power output concerning their involved substance mass and volume The research process, as shown in Figure 13, is aimed at improving extractable power based on material mass (W/kg material) and improving heat storage density (kwh/m^3), which, in addition to designing smaller systems, enables these systems to be used for a longer period of time, allowing them to deliver the required heat not for several hours, but for many days.

Although the current attempt has already proven the feasibility of utilizing such systems for residential use, it seems that further improvement of system output power is required market justifications. This would become a significant challenge as power optimization would directly point toward zeolite's physical limitations.

4.5. *Efficiency*

As mentioned before, zeolites have the unique capacity to store heat, and, due to their high structural endurance, they can tolerate numerous thermal cycles. These characteristics make zeolites known for their high thermal efficiency; however, utilizing them in TES systems has yet to reveal anything near to their optimum capacity. In this regard, in the literature, one can find two different efficiencies reported by scientists and engineers which indicates zeolite material efficiency and its system efficiency. Figure 14 shows these efficiencies for TES system containing zeolite in an open bed reactor. In this prototype, developed by Kuznik, the reported material efficiency was about 70%. However, one can clearly see that, when it is placed in the storage systems, this efficiency drops to around 36%.

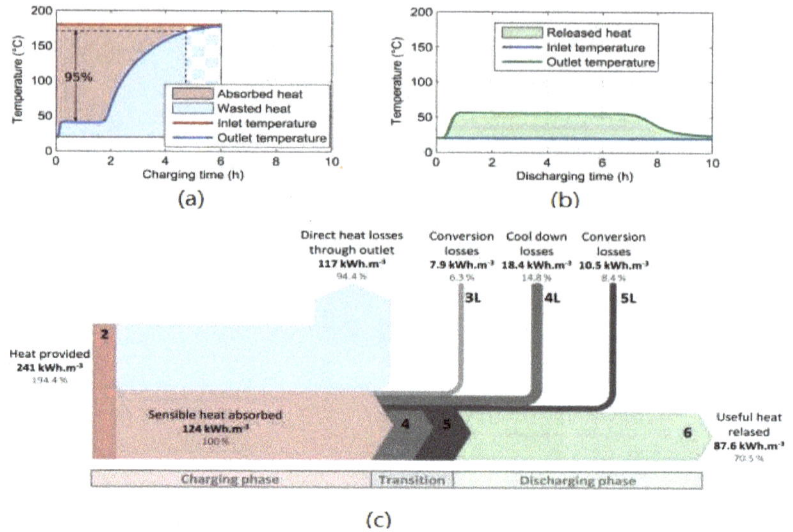

Figure 14. (**a**) The influence of charging time on the temperature profile (**b**) The impact of discharging time on the temperature profile; (**c**) Heat losses in the TES system [22].

As can be seen in Figure 14c, there exist several losses associated with all the three TES main processes which drop the engineering efficiency significantly. Nearly half of the injected heat is directly discharged through the outlet in the charging process.

The type of reactor and the inherent feature of zeolite can be responsible for the large amount of energy wasted during the charging phase. Zeolites, despite their high energy storage capacity, have poor thermal conductivity. Thermal conductivity is very important for increasing the internal temperature of zeolites to a level where the zeolite's internal moisture can release as a gas. Figure 14a shows the consequences of low thermal conductivity, as the reactor outlet temperature begins to increase after two hours of the charging process begins.

During the storing period, there exist charge conversion losses, cool down losses, and, finally, discharge conversion losses for the discharge process. There is no energy loss during the discharging phase, and the outlet temperature remains constant for six hours Figure 14b. All these losses, on the other hand, provide a window of opportunity to be addressed by engineers which, of course, brings many new challenges to the table.

5. Conclusions

Different aspects of utilizing zeolite 13X as a heat storage medium have been briefly discussed, and the main challenges have been summarized. It was found that zeolite 13X has high potential to be used in the residential sector; however, there exist several challenges that should be addressed prior for this to happen. This review demonstrated and discussed the variety of challenges from different perspectives, such as compact system design, charging supply, humidity management, system output power, and system efficiency. It was concluded that, to bring this technology to the residential sector, compact efficient designs are required, including techniques that lead to increasing the contact area while reducing the size of the system, a contradiction to be solved. In addition, it was concluded that, to make system energy intake justified, more affordable energy resources should be identified and implemented. Overall, the current review reveals that, although the implication of the zeolites TES system as a heating system seems feasible, they are, by far, set to become a serious competitor for current heating appliances in the residential market. This outcome, although it may negate the business side of the technology, offers several nitce opportunities for potential experts of the field.

Author Contributions: Conceptualization, A.B. and A.Z.; methodology, A.B.; investigation, A.B.; resources, A.B.; writing—original draft preparation, A.B.; writing—review and editing, A.B. and A.Z.; visualization, A.B.; supervision, A.Z.; project administration, A.B. All authors have read and agreed to the published version of the manuscript.

Funding: The paper has not been supported by any funding resources.

Institutional Review Board Statement: Not applicable.

Informed Consent Statement: Not applicable.

Data Availability Statement: Not applicable.

Conflicts of Interest: The authors declare no conflict of interest. The funders had no role in the design of the study; in the collection, analyses, or interpretation of data; in the writing of the manuscript, or in the decision to publish the results.

References

1. Zhang, Y.; Wang, R. Sorption Thermal Energy Storage: Concept, Process, Applications and Perspectives. *Energy Storage Mater.* **2020**, *27*, 352–369. [CrossRef]
2. Dincer, I.; Ezan, M.A. *Thermal Energy Storage Methods BT—Heat Storage: A Unique Solution for Energy Systems*; Springer International Publishing: Cham, Switzerland, 2018; pp. 57–84. [CrossRef]
3. Tatsidjodoung, P.; Le Pierrès, N.; Luo, L. A Review of Potential Materials for Thermal Energy Storage in Building Applications. *Renew. Sustain. Energy Rev.* **2013**, *18*, 327–349. [CrossRef]
4. Lin, J.; Zhao, Q.; Huang, H.; Wu, Q.; Xiao, Y. Applications of low-temperature thermochemical energy storage systems for salt hydrates based on material classification: A review. *Sol. Energy* **2020**, *214*, 149–178. [CrossRef]
5. Ruthven, D.M. *Principles of Adsorption and Adsorption Processes*; John Wiley & Sons: Hoboken, NJ, USA, 1984.
6. Edge, J. *Hydorgen Adsorption and Dynamics in Clay Minerals*; UCL (University College London): London, UK, 2014; p. 269.
7. Rocky, K.A.; Pal, A.; Rupam, T.H.; Palash, M.L.; Saha, B.B. Recent Advances of Composite Adsorbents for Heat Transformation Applications. *Therm. Sci. Eng. Prog.* **2021**, *23*, 100900. [CrossRef]
8. Bales, C.; Gantenbein, P.; Hauer, A.; Jaehnig, D.; Kerskes, H.; Henning, T.N.; Visscher, K.; Laevemann, E.; Peltzer, M.; Henning, H.-M. *Thermal Energy Storage for Solar and Low Energy Buildings State of the Art*; International Energy Agency: Bern, Switzerland, 2005.
9. Zhao, Q.; Lin, J.; Huang, H.; Wu, Q.; Shen, Y.; Xiao, Y. Optimization of Thermochemical Energy Storage Systems Based on Hydrated Salts: A Review. *Energy Build.* **2021**, *244*, 111035. [CrossRef]
10. Anderson, R.; Shiri, S.; Bindra, H.; Morris, J.F. Experimental Results and Modeling of Energy Storage and Recovery in a Packed Bed of Alumina Particles. *Appl. Energy* **2014**, *119*, 521–529. [CrossRef]
11. Lahmidi, H.; Mauran, S.; Goetz, V. Definition, Test and Simulation of a Thermochemical Storage Process Adapted to Solar Thermal Systems. *Sol. Energy* **2006**, *80*, 883–893. [CrossRef]
12. Stitou, D.; Mazet, N.; Mauran, S. Experimental Investigation of a Solid/Gas Thermochemical Storage Process for Solar Air-Conditioning. *Energy* **2012**, *41*, 261–270. [CrossRef]
13. Bales, C.; Gantenbein, P.; Jaenig, D.; Kerskes, H.; Summer, K.; van Essen, M.; Weber, R. *Laboratory Tests of Chemical Reactions and Prototype Sorption Storage Units. A Report of IEA Solar Heating and Cooling Programme—Task 32: Advanced Storage Concepts for Solar and Low Energy Buildings*; Solar Energy Research Center SERC Högskolan Dalarna: Borlänge, Sweden, 2008; p. 55.
14. Oktariani, E.; Tahara, K.; Nakashima, K.; Noda, A.; Xue, B.; Wijayanta, A.T.; Nakaso, K.; Fukai, J. Experimental Investigation on the Adsorption Process for Steam Generation Using a Zeolite-Water System. *J. Chem. Eng. Jpn.* **2012**, *45*, 355–362. [CrossRef]
15. Tolich, K. *Lecture 11 Second Law of Thermodynamics*; Canvas Learning Management System. University of Washington; Available online: https://canvas.uw.edu/files/40309275/download?download_frd=1&verifier=wfvQGeN0R7J6PC33VKp7O1cbquwVpa6FELtQAxl0 (accessed on 1 December 2021).
16. van Alebeek, R.; Scapino, L.; Beving, M.A.J.M.; Gaeini, M.; Rindt, C.C.M.; Zondag, H.A. Investigation of a Household-Scale Open Sorption Energy Storage System Based on the Zeolite 13X/Water Reacting Pair. *Appl. Therm. Eng.* **2018**, *139*, 325–333. [CrossRef]
17. Gondre, D. Numerical Modeling and Analysis of Heat and Mass Transfers in an Adsorption Heat Storage Tank: Influences of Material Properties, Operating Conditions and System Design on Storage Performances. Ph.D. Thesis, University of Lyon, Lyon, France, 21 March 2016.
18. Johannes, K.; Kuznik, F.; Hubert, J.L.; Durier, F.; Obrecht, C. Design and Characterisation of a High Powered Energy Dense Zeolite Thermal Energy Storage System for Buildings. *Appl. Energy* **2015**, *159*, 80–86. [CrossRef]
19. Zettl, B.; Englmair, G.; Steinmaurer, G. Development of a Revolving Drum Reactor for Open-Sorption Heat Storage Processes. *Appl. Therm. Eng.* **2014**, *70*, 42–49. [CrossRef]
20. Zondag, H.; Kikkert, B.; Smeding, S.; de Boer, R.; Bakker, M. Prototype Thermochemical Heat Storage with Open Reactor System. *Appl. Energy* **2013**, *109*, 360–365. [CrossRef]

21. Bales, C.; Gantenbein, P.; Jaenig, D.; Kerskes, H.; Summer, K.; van Essen, M.; Weber, R. *Laboratory Tests of Chemical Reactions and Prototype Sorption Storage Units. Report B4 of Subtask B*; IEA-SHC: Paris, France, 2008.
22. Kuznik, F.; Gondre, D.; Johannes, K.; Obrecht, C.; David, D. Numerical Modelling and Investigations on a Full-Scale Zeolite 13X Open Heat Storage for Buildings. *Renew. Energy* **2019**, *132*, 761–772. [CrossRef]
23. Donkers, P.A.J.; Pel, L.; Adan, O.C.G. Experimental Studies for the Cyclability of Salt Hydrates for Thermochemical Heat Storage. *J. Energy Storage* **2016**, *5*, 25–32. [CrossRef]
24. Michel, B.; Mazet, N.; Neveu, P. Experimental Investigation of an Innovative Thermochemical Process Operating with a Hydrate Salt and Moist Air for Thermal Storage of Solar Energy: Global Performance. *Appl. Energy* **2014**, *129*, 177–186. [CrossRef]
25. de Boer, R.; Smeding, S.F.; Zondag, H.A.; Krol, G. Development of a Prototype System for Seasonal Solar Heat Storage Using an Open Sorption Process. In Proceedings of the Eurotherm Seminar, Lleida, Spain, 28–30 May 2014; pp. 1–9.

energies

MDPI

Article

Distributed Control Strategy of Single-Phase Battery Systems for Compensation of Unbalanced Active Powers in a Three-Phase Four-Wire Microgrid

Watcharakorn Pinthurat and Branislav Hredzak *

School of Electrical Engineering and Telecommunications, The University of New South Wales, Sydney 2052, Australia; w.pinthurat@student.unsw.edu.au
* Correspondence: b.hredzak@unsw.edu.au

Abstract: Unbalanced active powers can affect power quality and system reliability due to high penetration and uneven allocation of single-phase photovoltaic (PV) rooftop systems and load demands in a three-phase four-wire microgrid. This paper proposes a distributed control strategy to alleviate the unbalanced active powers using distributed single-phase battery storage systems. In order to balance the unbalanced active powers at the point of common coupling (PCC) in a distributed manner, the agents (households' single-phase battery storage systems) must have information on the active powers and phases. Inspired by supervised learning, a clustering approach was developed to use labels in order to match the three-phase active powers at the PCC with the agents' phases. This enables the agent to select the correct active power data from the three-phase active powers. Then, a distributed power balancing control strategy is applied by all agents to compensate the unbalanced active powers. Each agent calculates the average grid power based on information received from its neighbours so that all agents can then cooperatively operate in either charging or discharging modes to achieve the compensation. As an advantage, the proposed distributed control strategy offers the battery owners flexibility to participate in the strategy. Case studies comparing performance of local, centralized, and the proposed distributed strategy on a modified IEEE-13-bus test system with real household PV powers and load demands are provided.

Keywords: multi-agents; single-phase battery storage system; rooftop PV unit; unbalanced active powers; distributed control; current unbalance factor; voltage unbalance factor

Citation: Pinthurat, W.; Hredzak, B. Distributed Control Strategy of Single-Phase Battery Systems for Compensation of Unbalanced Active Powers in a Three-Phase Four-Wire Microgrid. *Energies* **2021**, *14*, 8287. https://doi.org/10.3390/en14248287

Academic Editors: Alon Kuperman and Alessandro Lampasi

Received: 13 November 2021
Accepted: 2 December 2021
Published: 9 December 2021

1. Introduction

In recent years, unbalanced conditions in three-phase four-wire power systems, due to high penetration and uneven allocation of single-phase rooftop PV systems and load demands, have become exacerbated [1]. Specifically, the unbalanced active powers among phases and between single-phase PV systems and local loads can have a negative effect on power quality and system reliability due to large current flowing through neutral wires. The unbalanced conditions in the three-phase four-wire power systems can be described as a condition when the voltages or currents are not equal in the magnitudes and/or the phase angles of voltages and currents are not equal in consecutive sequence of phasors [2]. Since the voltages and/or currents are unbalanced, the active powers among phases in the three-phase system are also unbalanced.

There are several strategies to mitigate the unbalanced conditions. Traditionally, the unbalanced conditions were considered to be static since the penetration and variation of renewable energy sources (RESs) in power systems were low. Traditional compensation strategies, include using static synchronous compensators (STATCOMs) [3], passive devices, such as shunt capacitors [4], equalizing power generation, and load consumption [5], or using STATCOM with delta cascaded H-bridge (CHB) converter [6]. Moreover, some traditional methods to mitigate the neutral current and neutral to ground voltage (NGV) rise by resizing the neutral conductor, improving grounding and installing a passive harmonic filter were introduced in [7–10], respectively. However, owing to high variation in a number of single-phase PV rooftop systems distributed in the three-phase four-wire power systems, balancing between single-phase loads and rooftop PV sources and among the three phases becomes more difficult and, hence, the traditional methods cannot properly manage the unbalanced conditions.

Recently, the unbalanced conditions have been mainly compensated by designing new power electronic converters, managing electric vehicles (EVs) based on arrival and departure times, and employing energy storage systems (ESSs). The control strategies based on these approaches can be broadly divided into three main frameworks: centralized, decentralized, and distributed. A centralized controller requires information from all agents to be sent to a central control unit. The communication system should be fast and reliable, which can be challenging in practical distribution systems. A decentralized controller can make decisions based on only local observations, but the capability of resources may not be fully utilized due to the lack of cooperation between agents. A distributed controller is able to achieve cooperative control using the agent's own (and neighbouring agents') information [11].

1.1. Unbalance Compensation by Power Electronic Converters

As an active compensation strategy, three-phase four-leg topology based PV-VSI inverters with fixed capacity [12] and with dynamic capacity [13] were designed to alleviate the neutral current caused by the unbalanced loads. However, different network parameters, e.g., negative and zero sequence components of line impedance can affect the system stability and additional switching devices are required for the compensation. In [14], a distributed control method of a single-phase H-bridge PV-VSI was proposed for compensating the voltage unbalance factor (VUF) at a critical bus. Steinmetz design was employed for calculating required reactive power injections at different PV and load connections. However, if an upstream network (medium voltage side) was imbalanced and downstream loads were not constant power, using the Steinmetz design to calculate the required reactive powers for compensation may not be applicable. Furthermore, a reactive power control method with a centralized controller was developed for single-phase H-bridge VSI of DGs to compensate zero and negative sequence current components [15]. Power factors of DGs were controlled to obtain required reactive power, and Karush–Kuhn–Tucker (KKT) optimization algorithm using instantaneous power analysis was used to minimize the unbalanced conditions. The required reactive powers were shared among DGs through communication links by considering capacity ratings of the VSIs. However, there was a trade-off between compensation of the negative sequence current component and the zero sequence current component. The authors in [16] developed a current control strategy based asynchronous parallel pattern search (APPS) method for three single-phase full-bridge VSIs to reduce the voltage unbalance. Different indicators of the voltage unbalance were discussed. Moreover, a geometric norm based on a phasor diagram was introduced to examine the unbalanced conditions with different indicators. Then, the geometric norm was used as an objective function for obtaining asymmetrical current references for the three single-phase full bridge VSIs to inject the required current to the grid. However, the new power converters required three single-phase full bridge inverters, twelve switching devices, and three isolated DC voltage sources, thus increased additional cost.

1.2. Unbalance Compensation by Management of EVs/PEVs

Some strategies using EV chargers for the unbalance compensation were proposed. Impact of uncoordinated plug-in EVs was investigated in [17]. Two coordinated control strategies were proposed for PEVs to regulate bus voltages and minimize voltage unbalance. First, a centralized active power charging control method was introduced for charging PEVs using genetic algorithm (GA). Second, a decentralized reactive power discharging control system was developed for the voltage regulation and voltage unbalance compensation.

PEV inverter was locally assigned to discharge the reactive power based on reactive power droop controller at the bus having a poor voltage profile. In [18], two distributed consensus algorithms were proposed for PV units and PEVs to regulate bus voltages and to reduce voltage unbalance. The first algorithm was developed to maximize utilization of ES capacity of PEVs subject to ES constraints by controlling charge and discharge of PEV ESs, while the second algorithm was used to minimize the active power curtailment (APC) of PV units in case of PEVs having insufficient capacity for reduction of voltage rise. Arrival and departure times of PEVs used for verification of the proposed algorithms were selected randomly. In [19], an energy management system (EMS) of PV units and PEVs using decision making optimization based demand response (DR) was introduced to minimize VUF. PEV owners could decide to participate in various charging and discharging options offered by an aggregator. A central communication system was used as all PEVs were required to send all data to the aggregator. Owner preferences offered flexibility, while the combination of PVs and PEVs ensured minimizing VUF. Comprehensive indices represented by current waveforms were derived in [20]. The indices were then used to analyse the unbalanced conditions and harmonic distortion at the fundamental and harmonic currents of three-phase EV chargers during unbalanced charging and grid voltage unbalance. Two different EV technologies were used for the verification.

1.3. Unbalance Compensation by ESSs

ESSs have widely been employed to minimize the unbalanced conditions in three-phase four-wire power systems. Central energy storage, also called community energy storage (CES), with a centralized controller [21], and single-phase distributed energy storage system (DESs) with a distributed cooperative control strategy [22], respectively, were proposed to alleviate the neutral current and NGV rise in a multi-grounded three-phase four-wire distribution system. The current unbalance factor (CUF) was used to quantify the the unbalanced conditions. The authors assumed that all single-phase household installed PV systems were willing to participate in the control strategy. A three-phase damping control strategy for a CES connected at the end of line feeder was proposed in [23] to address over-voltage and voltage unbalance. The control strategy was applied to split DC-bus capacitors of VSI to consume (deliver) asymmetrical (negative and zero sequence components) phase currents from (to) the grid by adjusting the damping conductance. Based on the damping conductance, the VSI was able to operate in resistive mode. Comparing the proposed controller with the positive sequence control method, less current was required. However, the damping capacitance was a function of the VSI capacity; hence, performance of the proposed control strategy may be limited by the inverter size. Furthermore, a fuzzy logic control strategy was proposed in [24] for controlling individual single-phase DES based on the phase voltage deviation and the battery SoCs to mitigate voltage unbalance and voltage rise. Park's transformation was used to obtain the positive, negative, and zero sequence voltage components. It was also used to determine which phase DES should take action for the balancing process.

In addition, a distributed control strategy was proposed for single-phase distributed generators (DGs) to alleviate the unbalanced powers at PCC, but without considering variability and fluctuation of RESs [25]. It can be observed from the literature review that few studies focused on using single-phase distributed ESSs with distributed control strategies to mitigate the unbalanced conditions. The comparison of existing techniques for compensation of unbalanced conditions is summarized in Table 1. As illustrated in the table, the strategy proposed in this paper employs single-phase distributed battery storage systems and a distributed control strategy to minimize VUF and CUF in a modified IEEE-13-bus test system.

Table 1. Comparison of techniques for compensation of unbalanced conditions.

Ref.	Converter	EV	DES	CES	Centralized	Decentralized	Distributed	VUF	CUF	1-ϕ	Test System
[13]	✓					✓			✓		Real-44-bus
[14]	✓						✓	✓		✓	IEEE-13-bus
[15]	✓				✓					✓	IEEE-13-bus
[16]	✓				✓						Simplified
[17]		✓			✓	✓		✓			Real-74-bus
[18]		✓					✓	✓			IEEE European
[19]		✓			✓			✓			IEEE-123-bus
[20]		✓				✓		✓			Simplified
[21]				✓	✓				✓		Australian DS
[22]			✓				✓		✓	✓	Australian DS
[23]				✓		✓		✓			Simplified
[24]			✓			✓		✓			Simplified
Our			✓				✓	✓	✓	✓	IEEE-13-bus

To employ single-phase battery storage systems distributed in a three-phase four-wire microgrid for compensation of unbalanced conditions, the battery storage systems must know information about the active powers and phases at the PCC so that the agents (single-phase battery storage systems) can select the correct phase to perform the balancing. To achieve this task, clustering algorithms based on k-means, hierarchical clustering, a self-organization map, and expectation maximization in [26–29], can be applied. However, the mentioned algorithms have to be operated in a centralized manner, and cannot be directly applied to a multi-agent distributed system with a unidirectional communication graph. Thus, distributed clustering approach-based supervised learning has to be developed in order to match the phase active powers at the PCC with the agent phases using labelled data.

Motivated by the above discussion, this paper presents a distributed control strategy for compensation of unbalanced active powers in a three-phase four-wire microgrid. First, the phase active powers at the PCC are required to be labelled and sent to an agent via a unidirectional communication link. Each agent labels its own data with its own phase. The agent receives information about the labelled active powers at the PCC and agent phase labels from a neighbour agent. Then, it compares its own phase label with the received phase labels and selects the active power, having the same phase by applying the proposed clustering approach. In the next step, the proposed distributed control strategy is applied. Single-phase battery storage systems (agents) will cooperatively charge/discharge their active powers to minimize the unbalanced active powers. A battery owner can choose to not participate in the control strategy, e.g., due to violation of state of charge or power limits. This offers participation flexibility to the battery owner, while the active powers are still being balanced by the other participating owners. It should be noted that once the active powers have been balanced, the voltages also become balanced, as the voltage is also a function of the active power. The main contributions of this paper can be summarized as:

1. A distributed clustering method for labelling was developed. The three-phase active powers at the point of common coupling are labelled with phase labels a, b and c. Then, the values of the labelled active powers are sent to a neighbouring agent in a distributed manner via unidirectional communication network. Subsequently, the agents use the labelled active powers to determine to select its own phase data.
2. Within each agent, the average grid active power is calculated based on information received from the neighbouring agent. Then, using the proposed distributed power balancing control strategy, the battery storage systems cooperatively charge/discharge their active powers to minimize the difference between the average grid power and the phase active power to which they belong. Agents are allowed to disconnect and reconnect without affecting the balancing operation.
3. Modified IEEE-13-bus test system with real household PV powers and load demand over 24 h are used to verify the performance of the proposed strategy.

The rest of the paper is organized as follows. Section 2 describes the test system used to verify the proposed strategy, it introduces indicators to measure the unbalanced conditions, the constraints on battery storage systems, as well as a local active power control method. Centralized and proposed distributed power balancing control strategies are presented in Section 3. Section 4 verifies the performance of the proposed control strategy using real household PV powers and load demands. Finally, Section 5 concludes the paper.

2. Formulation of Unbalanced Active Power Problem

2.1. System Description

For verification, the IEEE-13-bus test system in [14] is modified with single-phase households equipped with rooftop PV systems, battery storage systems, and load demands, as seen in Figure 1. There is a transformer between the buses 650 and 632 to step-down the voltage level from the main grid from 33 kV to 4.16 kV. It is assumed that single-phase households with battery storage systems are connected to bus numbers 646, 611, 652, and 680, whereas households without battery storage systems are connected to other buses. Schematic diagram of a single-phase household is shown in Figure 2. Also, bus numbers 633 and 634 are three-phase balanced buses. There are nine agents ($N = 9$) considered in this paper. Each agent is single-phase and consists of a rooftop PV system, a local load and a battery storage system. The capacity of all battery storage systems is selected as 25 kW· h. Connection of the agents is given in Table 2. Real data of household PV powers and load demands over 24 h, taken from [30] are used for each agent, as shown in Figures 3 and 4 respectively.

Table 2. Locations of distributed single-phase battery systems.

Agent	1	2	3	4	5	6	7	8	9
Bus	646	646	611	652	652	652	680	680	680
Phase	*a*	*b*	*b*	*a*	*c*	*c*	*a*	*b*	*c*

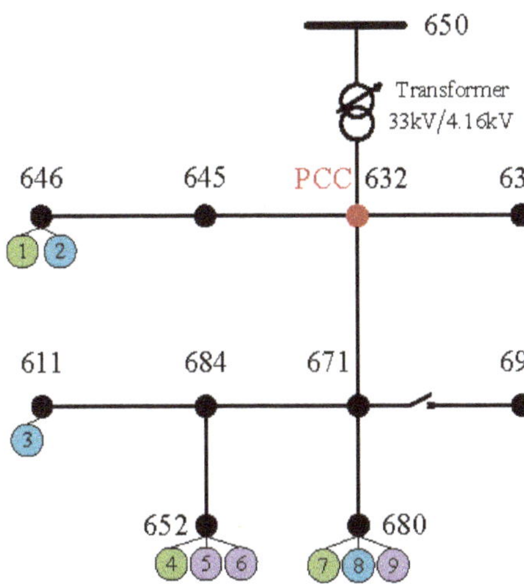

Figure 1. Modified IEEE-13-bus test system.

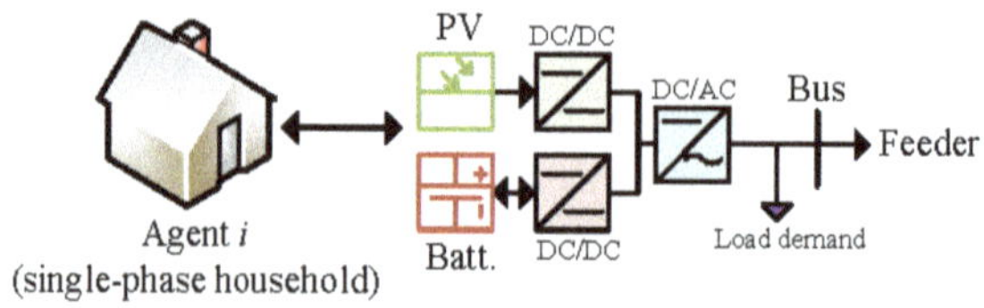

Figure 2. Illustrative example of agent i (single-phase household). During the normal mode of operation, the rooftop PV unit is operated at maximum power and the battery storage system can operate in either the charging or discharging mode to compensate the power mismatch between PV generation and load demand.

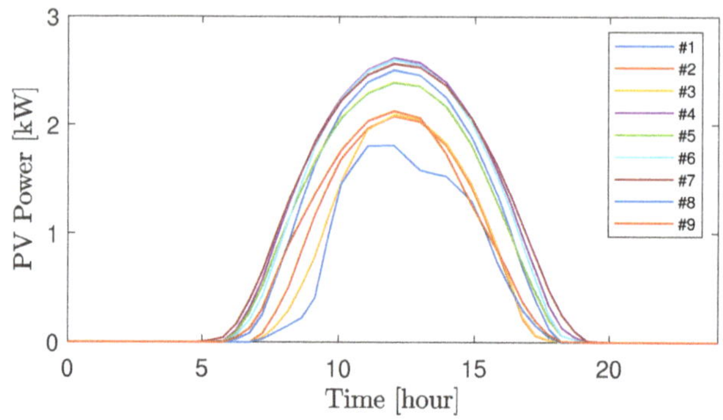

Figure 3. PV powers of all agents over 24 h.

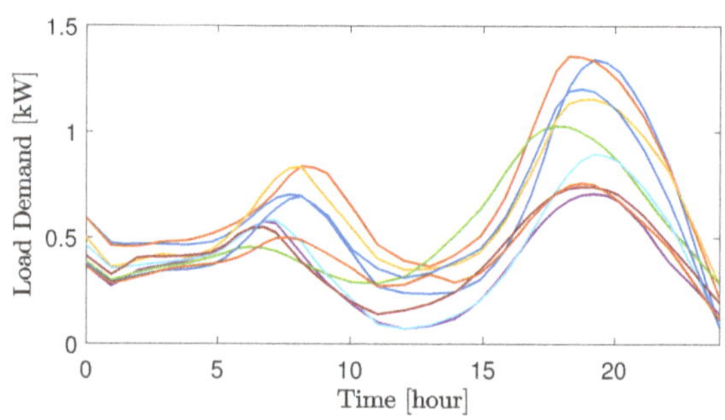

Figure 4. Load demands of all agents over 24 h.

2.2. Indicators for Measurement of Unbalanced Conditions

In three-phase four-wire power networks, the unbalance factor (UF), considering both negative and zero sequence current components is widely employed to measure voltage and current unbalances caused by high variation and uneven allocation of rooftop PV systems and load demands [1], and is defined as,

$$\mathrm{UF}_{\%} = \frac{\sqrt{|G_n|^2 + |G_z|^2}}{|G_p|} \times 100,$$
$$\begin{bmatrix} G_p \\ G_n \\ G_z \end{bmatrix} = \frac{1}{3}\begin{bmatrix} 1 & a & a^2 \\ 1 & a^2 & a \\ 1 & 1 & 1 \end{bmatrix} \times \begin{bmatrix} G_a \\ G_b \\ G_{c}, \end{bmatrix}, \tag{1}$$

where $a = e^{j(\frac{2\pi}{3})}$; G_p, G_n and G_z are positive, negative, and zero sequence components, respectively. The unbalance factor can be represented by the voltage unbalance factor

(VUF) or the current unbalance factor (CUF) if G_* is replaced by V_* or I_*. The standard for the voltage unbalance factor that is widely adopted in the literature is less than 2%.

2.3. Constraints on Battery Storage Systems

In order to achieve good performance of unbalance compensation, all agents are required to participate at all time. However, some agents may have faulty battery systems, or battery constraints have been reached, or the owners may decide not to participate in the control strategy for a period of time. The battery state of charge is one of the constraints that should be considered. The battery state of charge (SoC) can be estimated as,

$$SoC_i(t) = SoC_i(0) - \frac{1}{E_{bi}} \int I_{bi} dt. \tag{2}$$

Differentiating both sides of Equation (2) gives $\dot{SoC_i} = -I_{bi}/E_{bi}$. Let V_{bi} be the output voltage of the i-th battery storage system. Then, the output power of the i-th battery storage system can be obtained as $P_{bi} = V_{bi} \cdot I_{bi}$ and the SoC of the i-th battery storage system can be defined as [31],

$$P_{bi} = -V_{bi} \cdot \dot{SoC_i} \cdot E_{bi}, \tag{3}$$

where E_{bi} is the battery capacity (W·s) and P_{bi} the active power of the i-th battery storage system. As it can be seen from Equation (3), the SoC is only a function of the battery output power if the battery voltage V_{bi} is assumed to be constant. For safe operation, the constraints on battery storage systems can be bounded as $P_{bi} \in [P_{bi}^{min}, P_{bi}^{max}], [SoC^{min}, SoC^{max}]$.

2.4. Local Active Power Control Strategy

Typically, a battery storage system is employed to locally compensate for the power mismatch between load demand and renewable generation (rooftop PV). The power exchanged with the grid by the i-th household at phase ϕ is,

$$P_{gi}^{\phi} = P_{bi}^{\phi} + (P_{pvi}^{\phi} - P_{li}^{\phi}), \tag{4}$$

where P_{gi}^{ϕ} is the power exchanged with the grid; P_{bi}^{ϕ}, P_{pvi}^{ϕ} and P_{li}^{ϕ} are the battery output power, PV output power, and load demand of the i-th household at phase ϕ, respectively.

As it can be seen from Equation (4), under the normal mode of operation, if the i-th household at phase ϕ has a PV source and a battery storage system, and both are properly sized, the unbalanced power caused by the power mismatch between load and PV generation can be locally compensated by the battery storage system. For example, if $P_{pvi}^{\phi} > P_{li}^{\phi}$ or $P_{pvi}^{\phi} < P_{li}^{\phi}$ the battery system will operate either in charging ($P_{bi}^{\phi} < 0$) or discharging ($P_{bi}^{\phi} > 0$) mode, and the power exchanged with the grid by the i-th household will be close to zero, $P_{gi}^{\phi} \approx 0$.

However, not all households have battery storage systems. Some households may have only a PV source or no battery system and no PV source. Moreover, the battery storage systems may not be evenly distributed among phases throughout the power system. As a result, there will be the unbalanced phase powers and reverse power flow at some phases at the PCC even if loads are balanced as $P_g^a \neq P_g^b \neq P_g^c$, and $P_g^{\phi} = \sum_{i=1}^{N_{\phi}} P_{gi}^{\phi}, \phi \in \{a,b,c\}$ is the total power exchanged with the grid. Therefore, a new and effective strategy for compensating the unbalanced active powers at the PCC is necessary.

3. Active Power Balancing Control Strategy

Figure 5 illustrates a conceptual framework of the active power balancing strategy. Distributed single-phase battery storage systems in the microgrid as shown in Figure 1 can be employed to alleviate the unbalanced conditions. Two control strategies for the compensation are given in this section, (i) centralized and (ii) distributed.

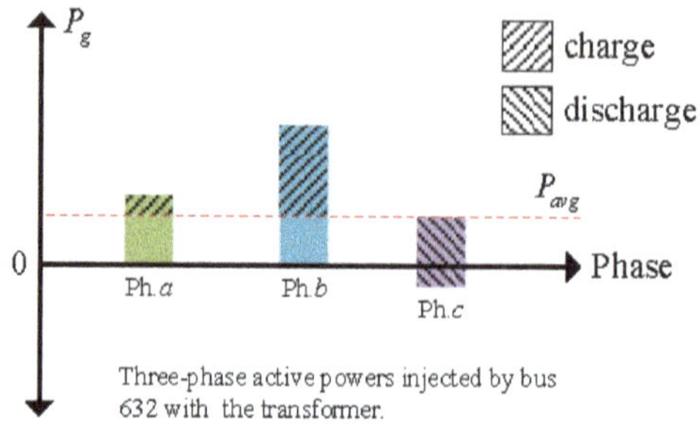

Figure 5. Conceptual framework of the active power balancing control strategy. The active powers in all three phases are balanced by adjusting them to be equal to the average grid power P_{avg} using single-phase battery storage systems distributed in the microgrid.

3.1. Centralized Active Power Balancing Strategy

A centralized controller requires that all agents send information to a central control unit, and a fast bidirectional communication system is required. As mentioned earlier, this control strategy may not be attractive due to lack of cost-effectiveness and risk of a single-point of failure. A schematic diagram of the centralized control approach applied for mitigating the unbalanced active powers is shown in Figure 6. The central control unit located at the PCC measures phases active powers and calculates the required active power for the compensation of each phase. Each single-phase battery storage agents sends its phase connection information to the central control unit. Then, the required active power from each agent at each phase can be obtained as,

$$P_{bci}^{\phi} = \frac{1}{N_{\phi}} \int \alpha_c (P_{avg} - P_g^{\phi}) dt, \phi \in \{a, b, c\}, \tag{5}$$

where P_{bci}^{ϕ} is the required active power from agent i at phase ϕ for the unbalance compensation, N_{ϕ} is the number of agents at each phase that participate in the balancing, α_c is the centralized control gain, $P_{avg} = (P_g^a + P_g^b + P_g^c)/3$ is the average grid power and P_g^{ϕ} is the grid active power at phase ϕ.

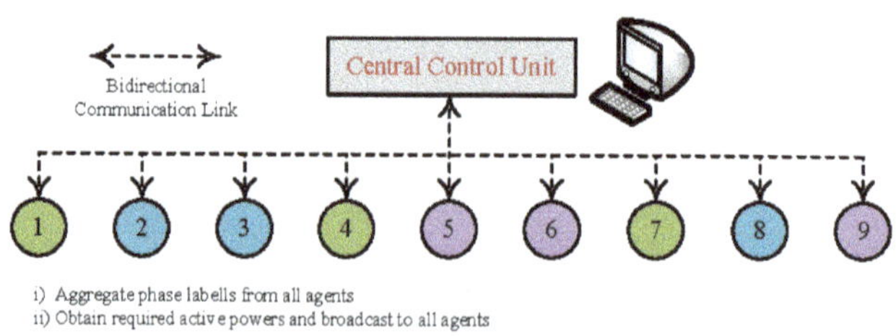

Figure 6. Schematic diagram of the centralized power balancing control strategy. The central control unit aggregates agents of each phase. Then, it calculates and broadcasts to all agents how much active power is required from each of them to compensate the unbalanced active powers. Bidirectional communication links are required.

3.2. Distributed Active Power Balancing Strategy

The proposed distributed active power balancing strategy employs unidirectional communication between neighbouring agents. In this subsection, preliminaries of a communication graph are introduced.

3.2.1. Communication Graph

A sparse graph, $\mathcal{G}(\mathcal{V},\mathcal{E})$ represents distributed communication links among neighbours, where $\mathcal{V} = \{1,\ldots,N\}$ and $\mathcal{E}$ denote the nodes (agents or households) and edges respectively. The node, $\mathcal{E}$, has elements (i,j), in which $(i,j) \in \mathcal{E}$ if node i can communicate with node j via a communication link [32]. The neighbours of the node i are denoted as $\mathcal{N}_i$. Node j is said to be a neighbour of node i if $(i,j) \in \mathcal{E}$. The adjacent matrix of the communication graph is expressed by,

$$\mathcal{A} = [a_{ij}] \in \mathbb{R}^{N\times N}, a_{ij} = \begin{cases} \alpha, (i,j) \in \mathcal{E} \\ 0, \text{otherwise} \end{cases}, \tag{6}$$

where α denotes the coupling gain.

The graph Laplacian matrix is defined as,

$$\mathcal{L} = \mathcal{D} - \mathcal{A}, \tag{7}$$

where $\mathcal{D} = diag\{d_i\}$, and the in-degree of the graph is represented as $d_i = \sum_{j=1}^{N} a_{ij}$.

3.2.2. Distributed Power Balancing Control Strategy

It is critical for each agent to know information about the active powers at the PCC (both magnitudes and phases). Inspired by supervised machine learning that classifies data based on similar labelled features, the following clustering approach is developed. Let $k \in \{1,2,3\}$ be a set of clusters (3-clusters) containing information about the three phases $\{1,2,3\}$ representing phases a, b and c respectively. Moreover, denote $x_i, i \in \{1,\ldots,N\}$ where $x_i \in \{1,2,3\}$ as locally labelled data for each agent. Hence, the agent i selects its own phase data as,

$$P_g^k \in \left\{ k\text{-th phase} \middle| k_i = \underset{k\in\{1,2,3\}}{\arg\min} \parallel x_i - k \parallel \right\}, \tag{8}$$

where k_i represents the selected phase.

Meanwhile, agent calculates the average grid power based on the active powers received from neighbouring agents as,

$$P_{avg} = \frac{1}{3}(P_g^1 + P_g^2 + P_g^3). \tag{9}$$

Then, based on [32], the battery storage system of the agent i will charge/discharge its active power P_{bdi} to make the active power at its phase is equal to the average grid power as,

$$P_{bdi} = \gamma_i \cdot \alpha_d \cdot a_{ij} \int (P_{avg} - P_g^k)dt, k \in \{1,2,3\}, \tag{10}$$

where $\gamma_i \in \{0,1\}$ is the willingness factor for the agent i to participate in the controller ($\gamma_i = 1$ means agent agrees to participate, 0 means otherwise), P_{bdi} is the active power of agent i, α_d is the distributed control gain, P_{avg} is the average grid power in Equation (9) and P_g^k is the grid active power at the k-th phase, where $k \in \{1,2,3\}$. Schematic diagram of the distributed power balancing control strategy is illustrated in Figure 7, while the procedure of the proposed distributed power balancing strategy of agent i is summarized as in Figure 8.

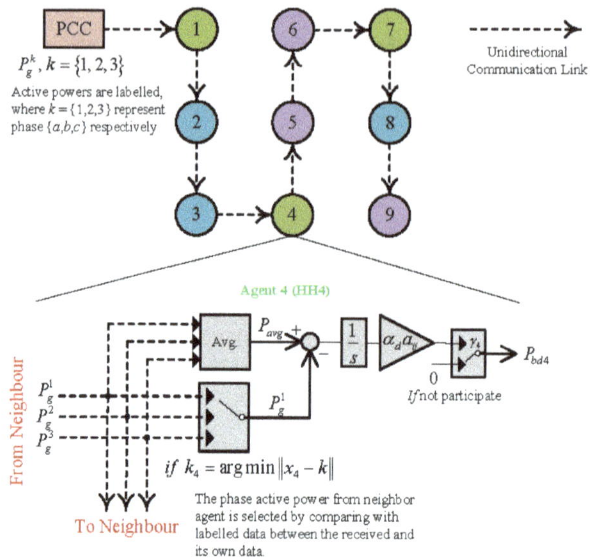

Figure 7. Schematic diagram of the distributed power balancing control strategy. Active powers exchanged by bus 623 with the transformer are first labelled with the value of the active power and the phase. Then, each agent receives the labelled powers, and passes them to the neighbouring agent via unidirectional communication links. Within each agent, the average grid power, P_{avg} is obtained and the agent selects the participating phase by comparing its label with the labelled grid powers. Finally, the agent adjusts its active powers according to the obtained average power, and decides whether to contribute by setting the willingness factor, γ_i.

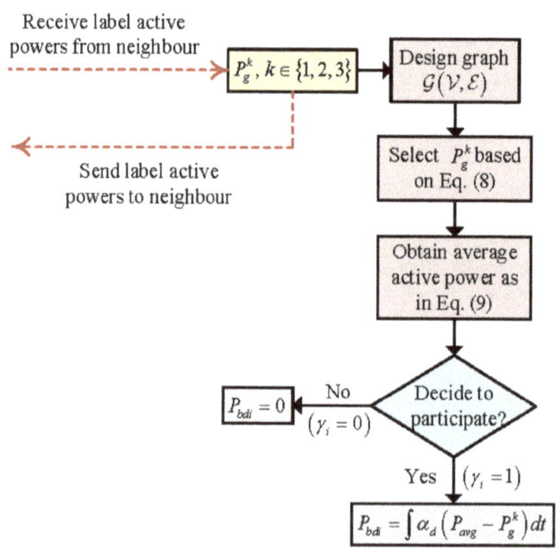

Figure 8. Distributed power balancing control strategy for the *i*-th agent (single-phase household). The distributed communication is represented by the dashed red lines.

4. Verification Results

The performance of the proposed strategy was verified on a modified IEEE-13-bus test system in Figure 1 with real data of rooftop PV units powers and load demands over 24 h, taken from [30]. Parameters are shown in Table 3. The effectiveness of three control strategies is compared. The first strategy is the local active power control strategy in Equation (4). Only the power mismatch between the PV unit and the load demand is locally compensated within each household; hence, it might fail to achieve the unbalance compensation at the PCC even if the loads are balanced. The second strategy is the centralized controller based on Equation (5). This controller can effectively balance the unbalanced active powers, and is given in order to compare its performance with the proposed distributed control strategy. The third strategy is the proposed distributed control strategy based on Equation (10). The agent can only communicate with its neighbours via a unidirectional communication system.

Table 3. Case study parameters.

Parameter	Symbol	Value
Number of agents	N	9
Microgrid voltage level	V_{mg}	4.16 kV-LL (1.0 p.u.)
Battery capacity	E_{bi}	25 kW·h
Coupling gain	α	1
Centralized control gain	α_c	1.5×10^{-3}
Distributed control gain	α_d	1.425×10^{-4}
Willingness factor	γ_i	$\{0,1\}$

4.1. Local Power Balancing Control Strategy

A local active power controller, given in Equation (4), is applied to illustrate that it fails to compensate the unbalanced conditions in the system. The power mismatch of each agent is locally compensated within the household, and no powers are consumed/injected from/to the grid ($P_{gi}^{\phi} \approx 0$). As a consequence, the active powers flowing from bus 632 to the transformer are always negative and the loads are supplied by importing the active powers from the grid. The unbalanced active powers and voltages are shown in Figures 9 and 10, respectively. Batteries active powers of the local control strategy is depicted in Figure 11. These unbalanced conditions are caused by uneven allocation of PV units and loads.

4.2. Centralized Power Balancing Control Strategy

In this control strategy, bidirectional communication links are required between agents and the central control unit. All agents that participate in the controller send information about the phase to the control unit, and then the control unit sends the required active powers back to each agent based on Equation (6). Hence, the unbalanced active powers and voltages at the PCC can be alleviated as it can be seen in Figures 12 and 13, respectively. Batteries active powers of the centralized control strategy can be seen in Figure 14.

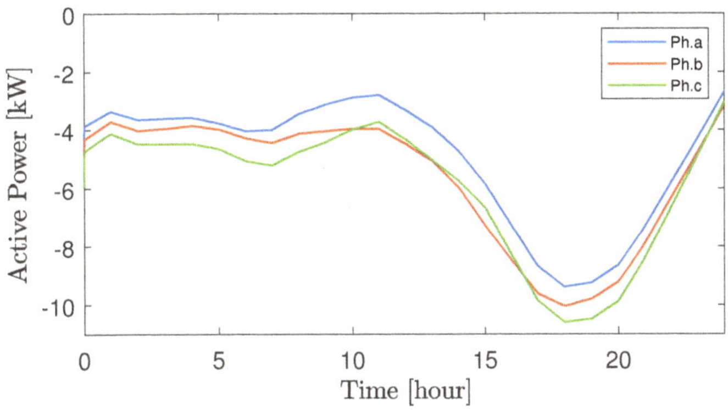

Figure 9. Active powers exchanged by bus 632 at PCC with the transformer using the local active power control strategy.

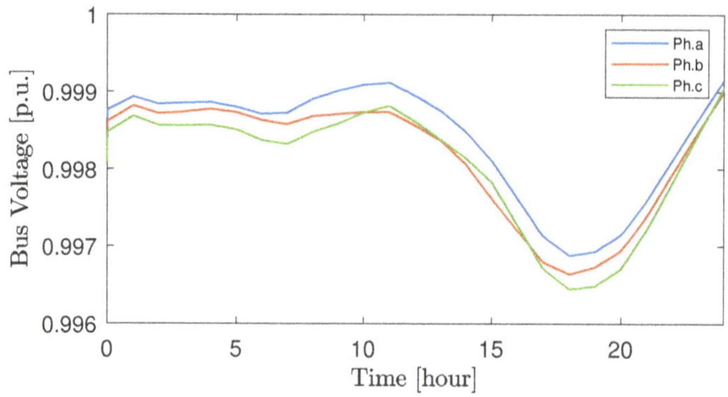

Figure 10. Bus voltages at bus 623 at PCC using the local active power control strategy.

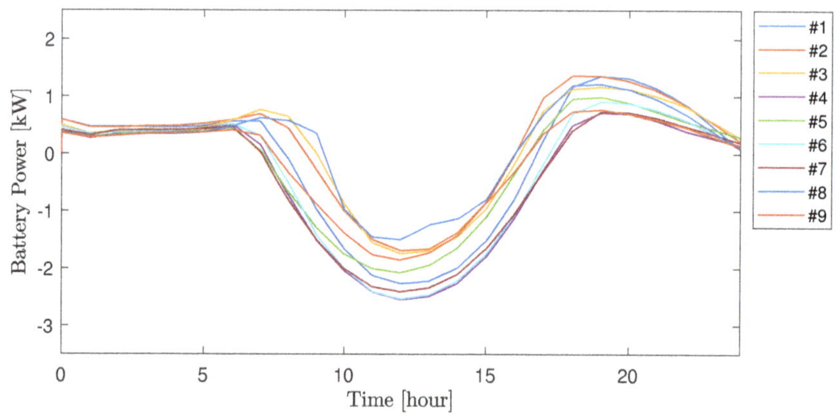

Figure 11. Batteries active powers of all agents during the balancing operation using the local control strategy.

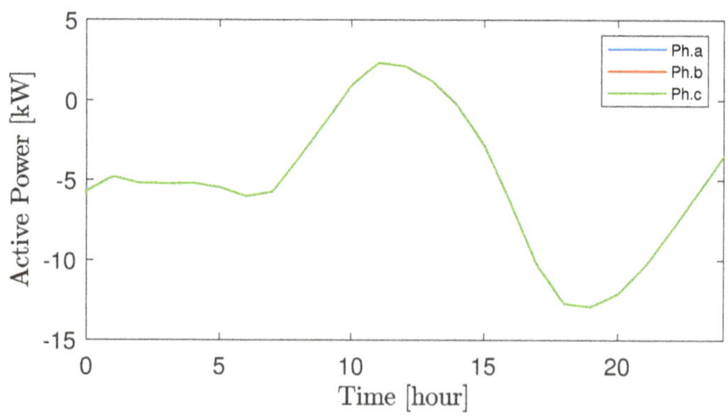

Figure 12. Active powers exchanged by bus 632 at PCC with the transformer using the centralized control strategy.

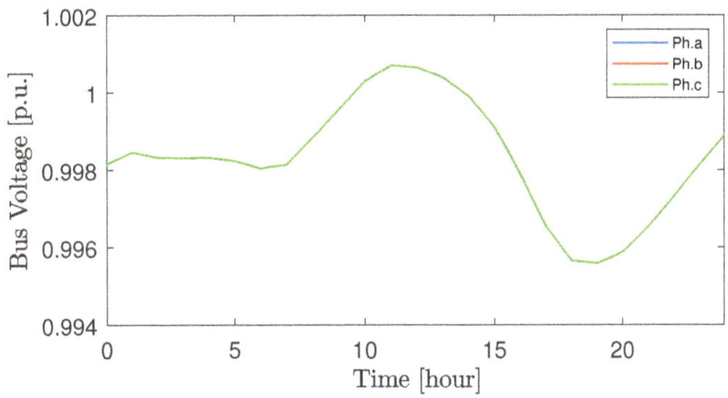

Figure 13. Bus voltages at bus 623 at PCC using the centralized control strategy.

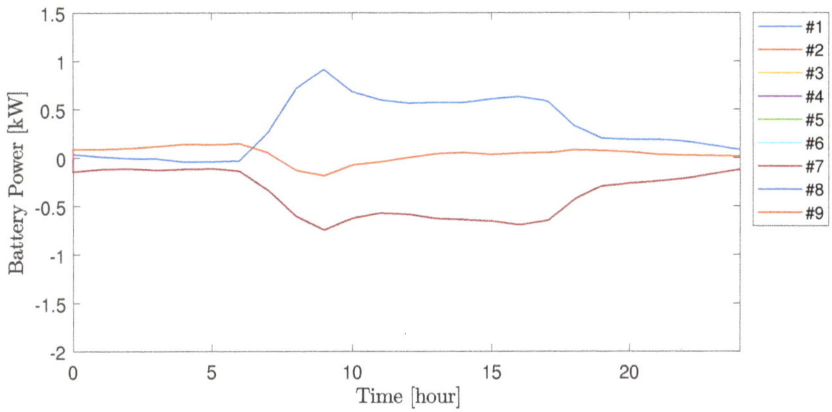

Figure 14. Batteries active powers of all agents during the balancing operation using the centralized control strategy.

4.3. Distributed Power Balancing Control Strategy

This strategy only required unidirectional communication links between agents and labelled data of active powers of each phase at the PCC. As it can be seen in Figures 15 and 16, the active powers of each phase exchanged by bus 623 with the transformer and the phase voltages at the bus are balanced. The results are similar to the results obtained by the centralized control strategy in Figures 12 and 13, respectively. However, in the proposed control scheme, the active power required for the unbalanced compensation is calculated by the agent (single-phase household) as illustrated in Figure 8. The agent can decide whether to participate in the control scheme without affecting the power balancing operation. Batteries active powers during the balancing can be seen in Figure 17.

Furthermore, during the balancing operation, at $t = 12.00$, h agent-1 is not willing to participate in the control strategy ($\gamma_i = 0$), so it contributes zero active power. At $t = 20.00$ h, the agent-1 is reconnected ($\gamma_i = 1$) and starts contributing the same active powers as the other agents at the same phase. During the disconnection and reconnection of the agent-1, the active powers and bus voltages are still maintained/balanced, even though there is some transient response at $t = 12.00$ h and $t = 20.00$ h.

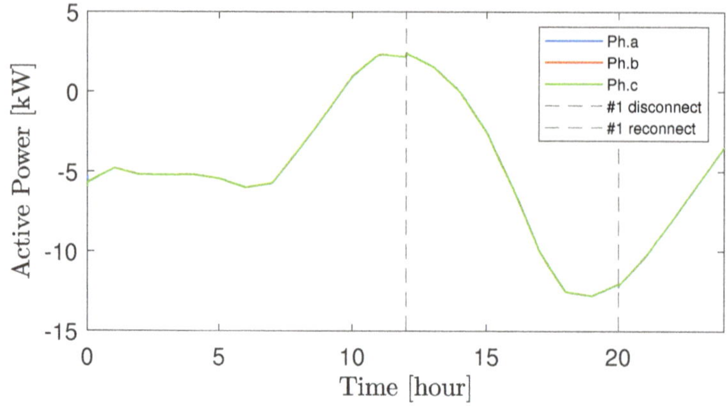

Figure 15. Active powers exchanged by bus 632 at PCC with the transformer using the proposed distributed control strategy.

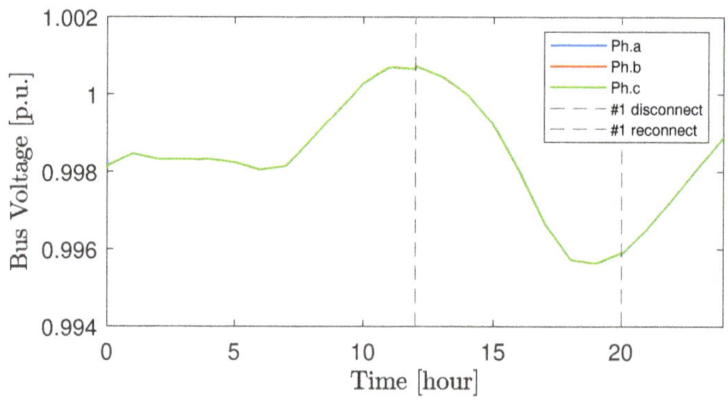

Figure 16. Bus voltages at bus 623 at PCC using the proposed distributed control strategy.

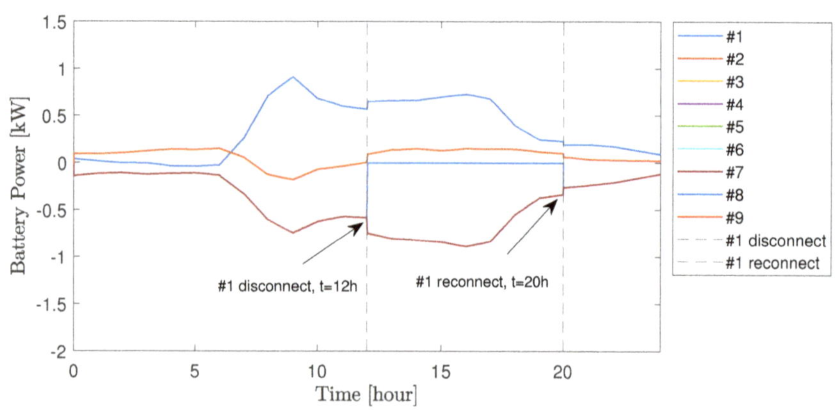

Figure 17. Batteries active powers of all agents during the balancing operation. It can been observed that the battery storage systems belonging to the same phase contributed the same amount of active power. At time $t = 12.00$ h to $t = 20.00$ h, the agent-1 is disconnected and contributes zero active power. After the agent-1 is reconnected it starts sharing the same active power as the other agents at the phase.

4.4. Comparison of Balancing Control Strategies

The unbalance factor in Equation (1) is adopted to quantify the unbalanced conditions. Two widely used unbalance factors are used in this subsection: the current unbalance factor (CUF) and the voltage unbalance factor (VUF). Figure 18 shows the current unbalance factor during operation. As it can be seen, the highest CUF is when the battery storage systems of all agents are switched-off (*batt. off*). The CUF obtained by the local power balancing control strategy denoted as *local* is decreased, but it is still greater than 20%. On the contrary, the CUFs for the *central* and *distributed* control strategies are substantially reduced to less than 0.5%.

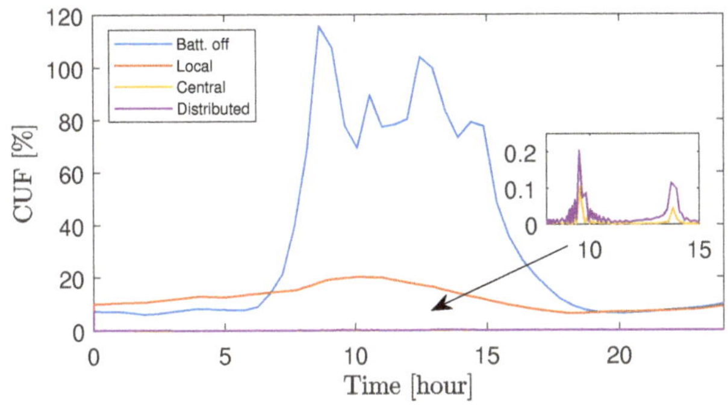

Figure 18. Comparison of the current unbalance factor for different control strategies. *Batt. off, local, central,* and *distributed* correspond to battery switched-off, local power controller, centralized power controller, and distributed power controller, respectively. The inset figure shows the difference between the *central* and *distributed* controllers.

Similarly, the VUF, when all battery storage systems are switched-off (*Batt. off*), is the highest. Meanwhile, the VUF obtained by the *Local* power balancing control strategy is decreased, but it is still higher than the standard (VUF $<$ 2%) [1] during midday as shown in Figure 19. However, the VUFs for the *Central* and *distributed* control strategies are significantly decreased to around 2%.

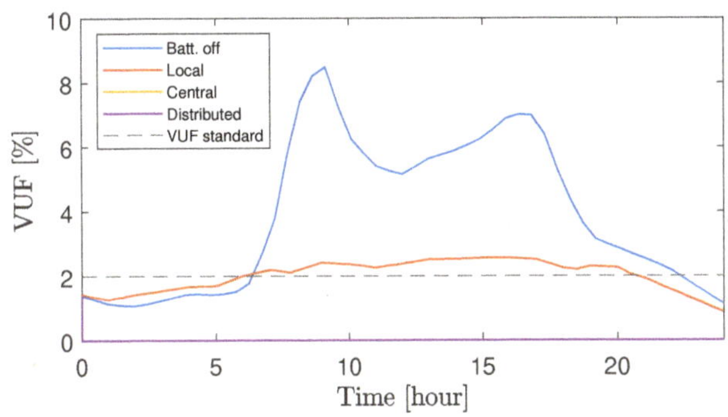

Figure 19. Comparison of the voltage unbalance factor for different control strategies. *Batt. off, local, central,* and *distributed* correspond to battery switched-off, local power controller, centralized power controller, and distributed power controller, respectively.

5. Conclusions

This paper presented a distributed control strategy to alleviate the unbalanced active powers, caused by high penetration and uneven allocation of PV rooftop systems and loads, using single-phase battery storage systems distributed in a three-phase four-wire microgrid. First, the three-phase active powers at the PCC were phase labelled and then the labelled powers were sent to agents in a distributed manner via unidirectional communication links. Within each agent, the labelled power data were compared with the agent data, to select its own phase active power. Furthermore, the average grid power was obtained and then the proposed distributed power balancing control strategy was applied. The battery storage systems were cooperatively operated to minimize the difference between the phase active power and the average power so that the active powers at the PCC become balanced. The battery's owner can decide not to participate in the control strategy for a period of time due to violation of battery constraints or for other reasons. The effectiveness of the proposed strategy was verified on the modified IEEE-13-bus test system with real-time data of rooftop PV systems and load demands.

Further work will focus on developing a method for partitioning a large-scale power system into smaller sub-systems based on geographical location, and applying the proposed distributed power balancing control strategy for compensation of unbalanced active powers in each sub-system.

Author Contributions: Conceptualization, W.P. and B.H.; methodology, W.P. and B.H.; simulation, W.P.; validation,W.P. and B.H.; visualization, W.P.; formal analysis, W.P.; writing—original draft preparation, W.P.; writing—review and editing, W.P. and B.H.; supervision, B.H. All authors have read and agreed to the published version of the manuscript.

Funding: This research received no external funding.

Institutional Review Board Statement: Not applicable.

Informed Consent Statement: Not applicable.

Data Availability Statement: Not applicable.

Acknowledgments: W.P. gratefully acknowledges the scholarship from the Ministry of Higher Education, Science, Research, and Innovation, Thailand.

Conflicts of Interest: The authors declare no conflict of interest.

References

1. Rafi, F.H.M.; Hossain, M.; Rahman, M.S.; Taghizadeh, S. An overview of unbalance compensation techniques using power electronic converters for active distribution systems with renewable generation. *Renew. Sustain. Energy Rev.* **2020**, *125*, 109812. [CrossRef]
2. Von Jouanne, A.; Banerjee, B. Assessment of voltage unbalance. *IEEE Trans. Power Deliv.* **2001**, *16*, 782–790. [CrossRef]
3. Castilla, M.; Miret, J.; Camacho, A.; Matas, J.; García de Vicuña, L. Voltage Support Control Strategies for Static Synchronous Compensators Under Unbalanced Voltage Sags. *IEEE Trans. Ind. Electron.* **2014**, *61*, 808–820. [CrossRef]
4. Yunus, K.; De La Parra, H.Z.; Reza, M. Distribution grid impact of plug-in electric vehicles charging at fast charging stations using stochastic charging model. In Proceedings of the 2011 14th European Conference on Power Electronics and Applications, Birmingham, UK, 30 August–1 September 2011; pp. 1–11.
5. Shahnia, F.; Wolfs, P.J.; Ghosh, A. Voltage unbalance reduction in low voltage feeders by dynamic switching of residential customers among three phases. *IEEE Trans. Smart Grid* **2014**, *5*, 1318–1327. [CrossRef]
6. Wu, P.H.; Chen, H.C.; Chang, Y.T.; Cheng, P.T. Delta-Connected Cascaded H-Bridge Converter Application in Unbalanced Load Compensation. *IEEE Trans. Ind. Appl.* **2017**, *53*, 1254–1262. [CrossRef]
7. Lefcourt, A. *Effects of Electrical Voltage/Current on Farm Animals: How to Detect and Remedy Problems. Agriculture Handbook*; Technical Report; Agricultural Research Service: Albany, CA, USA, 1991.
8. Surbrook, T.C.; Reese, N.D.; Kehrle, A.M. Stray Voltage: Sources and Solutions. *IEEE Trans. Ind. Appl.* **1986**, *IA-22*, 210–215. [CrossRef]
9. Zhu, J.; Chow, M.Y.; Zhang, F. Phase balancing using mixed-integer programming [distribution feeders]. *IEEE Trans. Power Syst.* **1998**, *13*, 1487–1492. [CrossRef]
10. Hsu, Y.Y.; Yi, J.H.; Liu, S.; Chen, Y.; Feng, H.; Lee, Y. Transformer and feeder load balancing using a heuristic search approach. *IEEE Trans. Power Syst.* **1993**, *8*, 184–190. [CrossRef]
11. Zeraati, M.; Golshan, M.E.H.; Guerrero, J.M. Voltage Quality Improvement in Low Voltage Distribution Networks Using Reactive Power Capability of Single-Phase PV Inverters. *IEEE Trans. Smart Grid* **2019**, *10*, 5057–5065. [CrossRef]
12. Rafi, F.H.M.; Hossain, M.J.; Lu, J. Improved Neutral Current Compensation with a Four-Leg PV Smart VSI in a LV Residential Network. *IEEE Trans. Power Deliv.* **2017**, *32*, 2291–2302. [CrossRef]
13. Md Rafi, F.H.; Hossain, M.J.; Town, G.; Lu, J. Smart Voltage-Source Inverters with a Novel Approach to Enhance Neutral-Current Compensation. *IEEE Trans. Ind. Electron.* **2019**, *66*, 3518–3529. [CrossRef]

14. Yao, M.; Hiskens, I.A.; Mathieu, J.L. Mitigating Voltage Unbalance Using Distributed Solar Photovoltaic Inverters. *IEEE Trans. Power Syst.* **2021**, *36*, 2642–2651. [CrossRef]
15. Nejabatkhah, F.; Li, Y.W. Flexible Unbalanced Compensation of Three-Phase Distribution System Using Single-Phase Distributed Generation Inverters. *IEEE Trans. Smart Grid* **2019**, *10*, 1845–1857. [CrossRef]
16. Neukirchner, L.; Görbe, P.; Magyar, A. Voltage unbalance reduction in the domestic distribution area using asymmetric inverters. *J. Clean. Prod.* **2017**, *142*, 1710–1720. [CrossRef]
17. Jabalameli, N.; Su, X.; Ghosh, A. Online centralized charging coordination of PEVs with decentralized var discharging for mitigation of voltage unbalance. *IEEE Power Energy Technol. Syst. J.* **2019**, *6*, 152–161. [CrossRef]
18. Zeraati, M.; Hamedani Golshan, M.E.; Guerrero, J.M. A Consensus-Based Cooperative Control of PEV Battery and PV Active Power Curtailment for Voltage Regulation in Distribution Networks. *IEEE Trans. Smart Grid* **2019**, *10*, 670–680. [CrossRef]
19. Akhavan-Rezai, E.; Shaaban, M.F.; El-Saadany, E.F.; Karray, F. Managing demand for plug-in electric vehicles in unbalanced LV systems with photovoltaics. *IEEE Trans. Ind. Inform.* **2017**, *13*, 1057–1067. [CrossRef]
20. Hernández, J.C.; Langella, R.; Cano, A.; Testa, A. Unbalance characteristics of fundamental and harmonic currents of three-phase electric vehicle battery chargers. *IET Gener. Transm. Distrib.* **2020**, *14*, 6220–6229. [CrossRef]
21. Alam, M.J.E.; Muttaqi, K.M.; Sutanto, D. Community energy storage for neutral voltage rise mitigation in four-wire multigrounded LV feeders with unbalanced solar PV allocation. *IEEE Trans. Smart Grid* **2015**, *6*, 2845–2855. [CrossRef]
22. Alam, M.J.E.; Muttaqi, K.M.; Sutanto, D. Alleviation of neutral-to-ground potential rise under unbalanced allocation of rooftop PV using distributed energy storage. *IEEE Trans. Sustain. Energy* **2015**, *6*, 889–898. [CrossRef]
23. Bozalakov, D.V.; Mnati, M.J.; Laveyne, J.; Van den Bossche, A.; Vandevelde, L. Voltage unbalance and overvoltage mitigation by using the three-phase damping control strategy in battery storage applications. In Proceedings of the 2018 7th International Conference on Renewable Energy Research and Applications (ICRERA), Paris, France, 14–17 October 2018; pp. 753–759.
24. Wong, J.; Lim, Y.S.; Morris, S.; Morris, E.; Chua, K.H. Fuzzy-driven energy storage system for mitigating voltage unbalance factor on distribution network with photovoltaic system. *AIP Conf. Proc.* **2017**, *1828*, 020019.
25. Ferreira, D.M.; Brandao, D.I.; Bergna-Diaz, G.; Tedeschi, E.; Silva, S.M. Distributed Control Strategy for Low-Voltage Three-Phase Four-Wire Microgrids: Consensus Power-Based Control. *IEEE Trans. Smart Grid* **2021**, *12*, 3215–3231. [CrossRef]
26. Yu, S.S.; Chu, S.W.; Wang, C.M.; Chan, Y.K.; Chang, T.C. Two improved k-means algorithms. *Appl. Soft Comput.* **2018**, *68*, 747–755. [CrossRef]
27. Geva, A.B. Hierarchical unsupervised fuzzy clustering. *IEEE Trans. Fuzzy Syst.* **1999**, *7*, 723–733. [CrossRef]
28. Flanagan, J.A. Self-organisation in Kohonen's SOM. *Neural Netw.* **1996**, *9*, 1185–1197. [CrossRef]
29. Aci, M.; Avci, M. K nearest neighbor reinforced expectation maximization method. *Expert Syst. Appl.* **2011**, *38*, 12585–12591. [CrossRef]
30. Sharma, V.; Haque, M.H.; Aziz, S.M. PV generation and load profile data of net zero energy homes in South Australia. *Data Brief* **2019**, *25*, 104235. [CrossRef] [PubMed]
31. Pinthurat, W.; Hredzak, B. Decentralized Frequency Control of Battery Energy Storage Systems Distributed in Isolated Microgrid. *Energies* **2020**, *13*, 3026. [CrossRef]
32. Lewis, F.L.; Zhang, H.; Hengster-Movric, K.; Das, A. *Cooperative Control of Multi-Agent Systems: Optimal and Adaptive Design Approaches*; Springer Science & Business Media: Berlin/Heidelberg, Germany, 2013.

processes

MDPI

Article

The Bi-Level Optimization Model Research for Energy-Intensive Load and Energy Storage System Considering Congested Wind Power Consumption

Shuyan Zhang [1], Kaoshe Zhang [1,2], Gang Zhang [1,2,*], Tuo Xie [1,2], Jiaxing Wen [1], Chao Feng [3] and Weihong Ben [4]

1 School of Electrical Engineering, Xi'an University of Technology, Xi'an 710048, China; 2190420016@stu.xaut.edu.cn (S.Z.); zhangks@xaut.edu.cn (K.Z.); xiet@xaut.edu.cn (T.X.); 2201920079@stu.xaut.edu.cn (J.W.)
2 State Key Laboratory of Eco-Hydraulics in Northwest Arid Region, Xi'an University of Technology, Xi'an 710048, China
3 State Grid Qinghai Electric Power Company, Xining 810001, China; qhfengchao@hotmail.com
4 Haixi Electric Power Supply Company, State Grid Qinghai Electric Power Company, Golmud 817077, China; benweihong@hotmail.com
* Correspondence: zhanggang3463003@xaut.edu.cn; Tel.: +86-1311-049-1071

Abstract: Due to the uncertainty of wind power output, the congestion of wind power has become prominent. Exactly how to improve the capacity of wind power consumption has become a problem that needs to be studied urgently. In this paper, an energy storage system and energy-extensive load with adjustable characteristics are used as an important means of consuming wind power. Firstly, we analyze the reasons for the congestion according to the characteristics of wind power output, and establish a model of the grid's ability to integrate wind power based on the concept of a wind power admissible interval. Secondly, we analyze the energy-extensive load regulation characteristics and establish an energy-extensive load dispatch model. Thirdly, on the basis of considering the energy-extensive load and energy storage system adjustment constraints, a bi-level optimization model is established. The upper level determines the configured capacity of the energy storage system with the goal of minimizing the total economic investment of the energy storage system, and the lower level coordinates the dispatching with the goal of maximizing wind power consumption and minimizing system operating costs. Finally, a certain region is taken as an example to verify the validity of the proposed method.

Keywords: wind power consumption; energy storage system; energy-intensive load; uncertainty of wind power

Citation: Zhang, S.; Zhang, K.; Zhang, G.; Xie, T.; Wen, J.; Feng, C.; Ben, W. The Bi-Level Optimization Model Research for Energy-Intensive Load and Energy Storage System Considering Congested Wind Power Consumption. *Processes* **2022**, *10*, 51. https://doi.org/10.3390/pr10010051

Academic Editors: Alon Kuperman and Alessandro Lampasi

Received: 8 November 2021
Accepted: 23 December 2021
Published: 27 December 2021

1. Introduction

By the end of 2020, Chinese-installed wind power capacity has continued to grow to 281 million kilowatts. However, wind power output is volatile and random [1]. When large-scale wind power is integrated into the grid, the wind power consumption of the wind farm is hindered due to the insufficient peak shaving capacity of the system, which results in a large number of wind abandonment [2]. To improve the consumption level of wind power, the energy storage resources [3] and the load-side resources need to be fully utilized at the same time [4]. In recent years, due to the rapid development of energy storage technology, energy storage devices have gradually been deployed into new energy systems [5]. This strategy can effectively increase the rate of new energy consumption, which has attracted wide attention from many researchers and governments [6]. Besides, enterprises with energy-intensive load are usually built near large-scale wind power bases [7], so the load is highly concentrated and large in capacity, making the control of the load more flexible [8]. Therefore, to alleviate the problem of Chinese wind power consumption, it is feasible to use energy storage systems and load-side to consume congested wind power on-site.

At present, the existing references have researched the load-side participation in wind power consumption. Reference [9] divides the energy-intensive load into interruptible and translatable loads according to the response mode, and comprehensively considers all available factors on the source side, grid side, and load side. On this basis, a source-grid-load comprehensive planning model is established. However, this model does not consider the power consumption characteristics and adjustment methods of an energy-intensive load. Since improper adjustment of the energy-intensive load can cause serious losses, in order to make reasonable use of the adjustable performance of the energy-intensive load, it is necessary to carry out fine modeling of the electrical characteristics of load. Since demand response is playing an increasingly important role in balancing short-term supply and demand, researchers propose three different methods to integrate demand response into a unit combination optimization model that considers operating constraints [10]. However, this model is established when the wind power forecasting is accurate and does not consider the volatility of wind power. In order to alleviate the problems of grid integration and safe operation of the power system caused by the uncertainty of wind power, pumped storage and demand response participate in the process of grid operation as auxiliary services In addition, the Lagrangian relaxation method is proposed to solve the unit combination problem [11]. However, the adjustment cost of an energy-intensive load is not considered in this process, which will lead to excessively high overall operating costs of the system Aiming at the uncertainty of renewable energy output, reference [12] proposes a two-stage robust scheduling model. Due to the high flexibility of demand response, this model can meet electricity demand with minimal energy costs and maximize the use of clean energy potential. However, due to the complexity of the model, it is not suitable for grid dispatch calculation. Reference [13] uses the dynamic adjustment capabilities of hydropower and energy-intensive load to propose an optimal wind power-solar capacity allocation method to reduce the uncertainty of output. However, the risk constraints of energy-intensive load and wind power are not considered. When the discretely adjustable energy-intensive load participates in the consumption of wind power, since it cannot be continuously adjusted in a short time, the adjustment increment of the energy-intensive load does not match the output of wind power, which increases the risk of wind power curtailment or load shedding of the energy-intensive load.

On the other hand, the energy storage system can store the power during the low load period and release it during the peak load period [14]. Joint dispatch with wind power can effectively reduce the wind power curtailment rate [15]. Therefore, there are currently many studies that combine energy storage and wind power into a joint system for optimal dispatch [16]. By analyzing the negative impact of wind speed variability on the large-scale grid integration of wind power, the researcher proposes to use energy storage systems to mitigate it [17]. Based on the reliability analysis under the unit operation and technical constraints, the AC power flow model is used to determine the scale of the energy storage system. However, the operating cost of the energy storage system is not considered, which leads to excessively high system operating costs, and is not conducive to the economic operation of the system. Reference [18] is based on the complementary characteristics of solar and wind energy, and proposes a method to optimize the configuration of renewable energy by using battery energy storage technology so as to make the system more reliable However, this research does not take into account the uncertainty of renewable energy output, which affects the planning and operation of the energy storage system, thereby reducing the applicability and reliability of the results. In view of the fact that wind power cannot be accurately predicted, reference [19] proposes an approach for planning and operating an energy storage system for a wind farm in the electricity market while using electrochemical batteries to compensate for changes in power generation. However, this method does not explain how to determine the capacity of the energy storage system and cannot guarantee that the capacity is the optimal value. The energy storage capacity should be optimally configured to improve the overall investment benefit. Reference [20] proposes a multi-objective optimal scheduling model based on the operating characteristics of the battery energy storage system and the uncertainty of wind power output, which reduces the risk of the integrated power system with wind farms and batteries. Although the scheduling model considers the uncertainty of wind power, it does not quantify the risk of wind power curtailment. Besides, the existing references mostly focus on the exploration of the effect of energy-intensive load or energy storage system alone. Few references analyze the effective coordination between the energy-intensive load and energy storage system.

In view of the above problems, this paper takes the energy-intensive load and energy storage system together as an important means to consume wind power and jointly participate in the optimal dispatch of the power grid. Firstly, the regulation characteristics of the energy-intensive load are analyzed, and the energy-intensive load dispatching model is established. On the basis of fully considering the uncertainty of wind power, the risk constraints of the energy-intensive load and wind power have been established. At the same time, taking into account the adjustment cost and adjustment constraints of the energy-intensive load and energy storage system, a bi-level optimization model considering the congested wind power consumption is established. The upper level determines the configured capacity of the energy storage system with the goal of minimizing the total economic investment of the energy storage system, and the lower level coordinates the dispatching with the goal of maximizing wind power consumption and minimizing system operating costs. The simulation results show that the above method can effectively improve the consumption capacity of wind power and reduce the operating cost of the system.

2. Uncertainty Analysis of Wind Power

Wind power output has strong randomness and volatility. When large-scale wind farms are integrated into the grid, the safe and stable operation of the system will be affected. Therefore, the uncertainty of wind power output needs to be analyzed. This chapter firstly proposes the concept of wind power admissible interval to represent the power grid's ability to integrate wind power. Then, the characteristics of wind power output are analyzed, considering the uncertainty of wind speed changes, and a probability distribution model is usually used to describe it. On this basis, analysis and research are carried out according to the wind curtailment situation outside of the capacity of the grid, and the curtailment risk is characterized by the wind curtailment expected value.

2.1. The Admissible Region of Wind Power

Energy-intensive load and energy storage system are mainly used to consume wind curtailment. Therefore, the acceptance level of power grid to wind power should be calculated to evaluate the wind curtailment situation in the future [21].

The calculation of the wind curtailment index is closely related to the grid's acceptance level to wind power. This paper uses the admissible region of wind power (ARWP) to indicate the acceptance level of wind power in power grid [22]. The acceptance region of the power grid for the output of a wind farm is shown in Figure 1. The blue solid line in the figure is the planned output of the wind farm, and the red dotted line is the admissible wind power output range of the power grid without curtailed wind or reduced load.

According to the concept of ARWP, the wind power output satisfies the following relationship:

$$\begin{cases} w_{i,t}^{l} \leq w_{i,t} \leq w_{i,t}^{u} \\ w_{i,t} = w_{i,t}^{p} + \Delta\hat{w}_{i,t}^{p} \end{cases} \tag{1}$$

where $w_{i,t}^{p}$, $\Delta\hat{w}_{i,t}$, $w_{i,t}$, $w_{i,t}^{u}$, and $w_{i,t}^{l}$ represent planned output, wind power output fluctuation, actual output, the upper boundary before coordinated dispatching, and the lower boundary before coordinated dispatching of the i-th wind farm at time t, respectively.

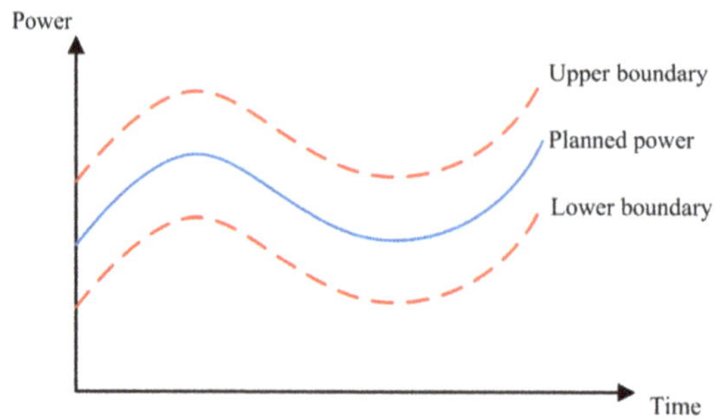

Figure 1. ARWP of a wind farm.

2.2. *Distribution of Wind Power Output*

Wind power output is highly uncertain. In this paper, the uncertainty of wind power output is described as a probability function that obeys a normal distribution near the predicted point [23]. As shown in Figure 2.

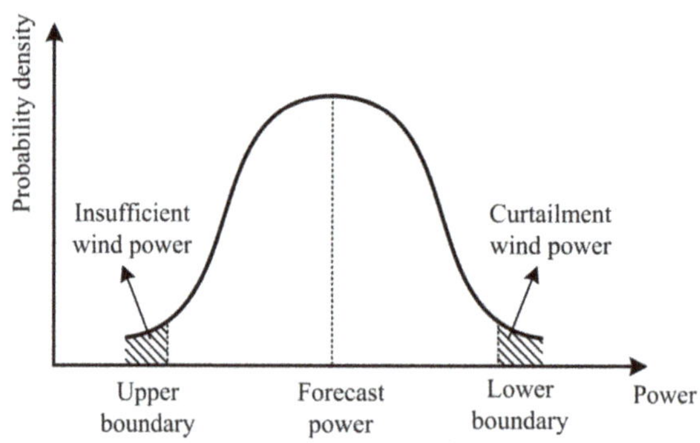

Figure 2. Probability density function of a wind farm.

The distribution of wind farm output is:

$$(w^{p}_{i,t} + \Delta\hat{w}^{p}_{i,t}) \sim N(w^{f}_{i,t}, (\sigma_i + t\Delta\sigma_i)^2) \tag{2}$$

where $N(w^{f}_{i,t}, (\sigma_i + t\Delta\sigma_i)^2)$ represents the normal distribution with expectation $w^{f}_{i,t}$ and variance $(\sigma_i + t\Delta\sigma_i)^2$; $w^{f}_{i,t}$ is the predicted output of wind farm i at time t; σ_i is the initial standard deviation of wind farm i load forecasting; and $\Delta\sigma_i$ is the standard deviation increment of wind farm i load forecasting process with time scale.

2.3. *Risk Analysis of Wind Curtailment*

Due to the randomness and volatility of wind power output, prediction errors are prone to occur when predicting wind power output, which will increase the uncertainty of large-scale wind power integrated into the grid. It will have a great impact on the peak shaving capacity of the power grid, which will lead to the obstruction of wind power consumption and a large amount of wind curtailment [24]. Figure 3 is the schematic diagram of congested wind power consumption.

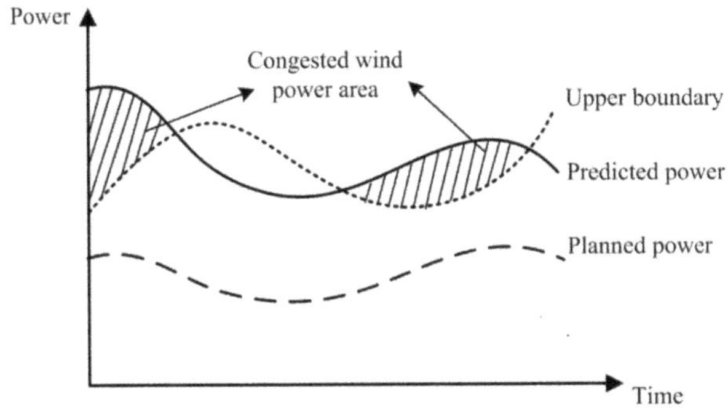

Figure 3. Schematic diagram of congested wind power consumption.

Based on the above analysis, the risk of wind curtailment of wind farm i can be expressed as:

$$C_i^{u,VaR}(w_{i,t}^u) = \rho^u \sum_{t=1}^{T} E_{i,t}^u(w_{i,t}^u) = \rho^u \sum_{t=1}^{T} \int_{w_{i,t}^u}^{w_{i,t}^{\max}} (x - w_{i,t}^u) f_{i,t}(x) dx \quad (3)$$

where ρ^u is the penalty for wind curtailment; $E_{i,t}^u(\cdot)$ is the wind curtailment expectation of wind farm i at time t; T is the time scale of dispatching control; $w_{i,t}^{\max}$ is the upper limit of output of wind farm i at time t, taking the installed capacity of wind farm; $f_{i,t}(x)$ is the probability density function of wind farm i output at time t.

According to Formula (3), $C_i^{u,VaR}$ is a complex nonlinear nonconvex function. It will not only increase the difficulty of finding the global optimal solution, but also increase the computational complexity and time. Therefore, this paper linearizes $E_{i,t}^u(w_{i,t}^u)$ piecewise, and the piecewise linearization models with different values are shown in Formula (4).

$$\begin{cases} E_{i,t}^u(w_{i,t}^u) = \sum\limits_{s=1}^{n} a_{i,t}^{u,s} w_{i,t}^{u,s} + b_{i,t}^{u,s} z_{i,t}^{u,s} \\ 0 \le w_{i,t}^{u,s} \le M z_{i,t}^{u,s} \\ \sum\limits_{s=1}^{n} w_{i,t}^{u,s} = w_{i,t}^u \\ \sum\limits_{s=1}^{n} z_{i,t}^{u,s} = 1 \end{cases} \quad (4)$$

where n is the total number of sections; $w_{i,t}^{u,s}$ and $z_{i,t}^{u,s}$ are respectively the continuous and discrete auxiliary variables of the s-th segment of the upper boundary of the ARWP of wind farm i at time t before the load participates in the coordination; $a_{i,t}^{u,s}$ and $b_{i,t}^{u,s}$ are respectively the slope and intercept of the s-th segment of the ARWP upper boundary of wind farm i at time t, which can be obtained in advance from the distribution of wind farm i. Here, M is a preset large number constant.

From Formula (4), it can be seen that $C_i^{u,VaR}(\cdot)$ can be changed from a complex function to a series of mixed integer linear constraints, which is easy to solve.

3. Model of Energy-Intensive Load Dispatching

The uncertainty of wind power output imposes a burden on the regulation of the power grid. When the regulation capacity of conventional power sources is insufficient, the energy-intensive load can be adjusted to ensure the balance between supply and demand of the power system. The premise for using energy-intensive load to consume congested wind power is to have an accurate understanding of load power characteristics. Therefore, it is necessary to analyze the characteristics of different types of energy-intensive load regulation and establish a mathematical model of energy-intensive load regulation.

On this basis, combined with the wind curtailment situation outside the capacity of the grid analyzed in Section 2.3, the risk constraints related to load consumption increment and wind curtailment volume are constructed. The flow chart of this process is shown in Figure 4.

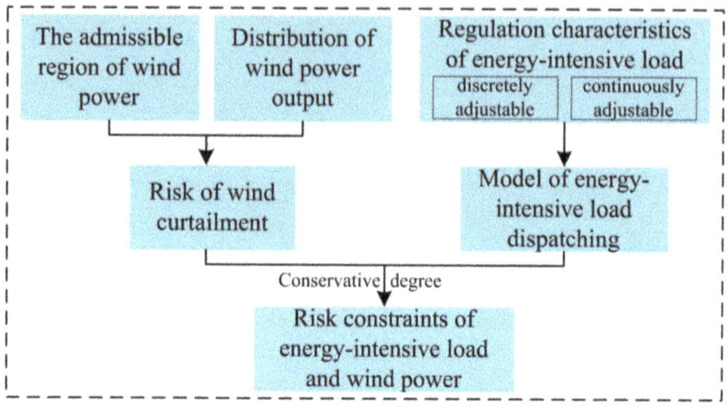

Figure 4. Schematic diagram of risk constraints of energy-intensive load and wind power.

3.1. Regulation Characteristics of Energy-Intensive Load

Energy-intensive load are divided into continuously adjustable load and discretely adjustable load. In this section, the regulation characteristics of two typical energy-intensive load of an electrolytic aluminum and titanium alloy are analyzed as examples [25].

(1) Electrolytic aluminum production load

Electrolytic aluminum production uses cryolite-alumina as raw materials, and direct current is applied to electrolysis in its molten salt until, finally, aluminum is obtained. Under normal circumstances, the load of electrolytic aluminum is stable, and adjustment within a certain range only affects the output and does not affect product quality and equipment safety. However, due to the limited impact tolerance of electrolytic aluminum equipment, stable production is required for a period of time after one adjustment, and frequent adjustments are not allowed. The schematic diagram of electrolytic aluminum load adjustment is shown in Figure 5.

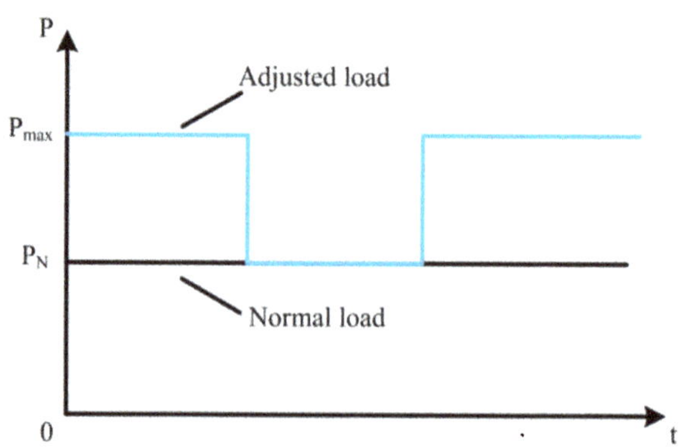

Figure 5. Schematic diagram of electrolytic aluminum load regulation characteristics.

(2) Titanium alloy production load

Titanium alloy production uses alloy oxide charge as raw material to reduce to titanium alloy at high temperature. Titanium alloys generally adopt uninterrupted production methods, and their production load fluctuates slightly, basically stable, with continuous adjustment capabilities, and flexible adjustments, which are not affected by stable production time. The schematic diagram of titanium alloy load adjustment is shown in Figure 6.

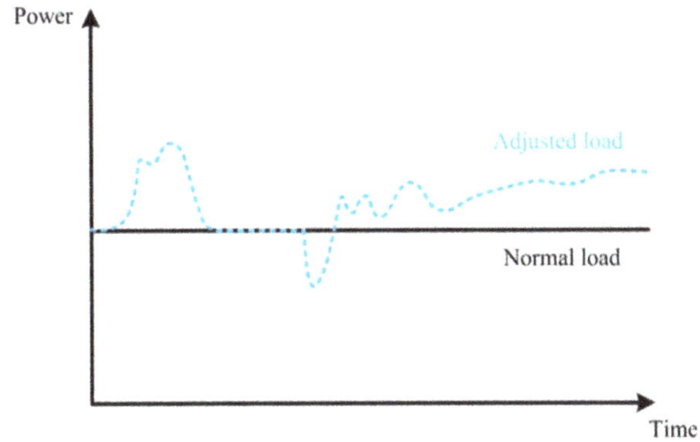

Figure 6. Schematic diagram of titanium alloy load regulation characteristics.

Summarizing the above load regulation characteristics, energy-intensive load can be divided into continuously adjustable loads and discretely adjustable loads. Various types of energy-intensive load regulation characteristics are shown in Table 1.

Table 1. Various types of energy-intensive load regulation characteristics.

Load Type	Typical Load	Power Stability Duration/h	Continuous Regulation
discretely adjustable load	Electrolytic aluminum	≥2	No
continuously adjustable load	Titanium alloy	0	Yes

3.2. *Model of Energy-Intensive Load Dispatching*

From the analysis in Section 3.1, it can be seen that the continuously adjustable load can be adjusted in real time according to the fluctuation of wind power, and the risk is relatively small. However, the discretely adjustable load cannot be adjusted continuously in a short period of time. After one adjustment, it needs to run stably for a period of time before the next adjustment can be carried out. The time period is longer. If the load regulation is large, the predicted output of wind power during this period is higher but the actual output is lower, which will cause the problem of a mismatch between the load increment and the power generation increment, resulting in a higher risk of load shedding. Conversely, if the load regulation amount is small, the flexibility of the energy-intensive load cannot be fully utilized, and a large wind curtailment may also occur. Therefore, the uncertainty of wind power and the regulation characteristics of the load should be fully considered when the discrete energy-intensive load participates in the wind power consumption.

In this paper, the discrete adjustable load is analyzed and studied [26]. Without losing generality, the mathematical model of a smelting furnace is used to represent the discrete adjustable load [27].

Other constraints of the electricity load and active power model of the smelting furnace are as follows:

$$\begin{cases} P_{j,t}^{EF} = P_j^{EF,\text{int}}(1 - x_{j,t}^{EF}) + P_j^{EF,on} x_{j,t}^{EF} + P_{j,t}^{EF,adj} \\ -P_j^{EF,d} x_{j,t}^{EF} \le P_{j,t}^{EF,adj} \le P_j^{EF,u} x_{j,t}^{EF} \\ -M(u_{j,t}^{EF} + 1 - x_{j,t}^{EF}) \le P_{j,t}^{EF,adj} - P_{j,t-1}^{EF,adj} \le M u_{j,t}^{EF} \end{cases} \tag{5}$$

Formula (5) is the active power constraints of the smelting furnace. Where $P_{j,t}^{EF}$, $P_{j,t}^{EF,adj}$, $x_{j,t}^{EF}$, $u_{j,t}^{EF}$ are the total active power, continuous regulation, state variable and start flag of smelting furnace j at time t, respectively; $P_j^{EF,\text{int}}$, $P_j^{EF,on}$, $P_j^{EF,d}$ and $P_j^{EF,u}$ are the oven power, normal production power, maximum down-regulated power and maximum up-regulated power of smelting furnace j, respectively.

$$\begin{cases} x_{j,t}^{EF} - x_{j,t-1}^{EF} \le u_{j,t}^{EF} \\ u_{j,t}^{EF} \le x_{j,t}^{EF} \\ u_{j,t}^{EF} \le 1 - x_{j,t-1}^{EF} \\ x_{j,t-1}^{EF} - x_{j,t}^{EF} \le 1 - x_{j,\tau}^{EF} \ \forall \tau \in \left[t+1, \min\left(t + T_j^{EF,on} - 1, T\right)\right] \\ u_{j,t}^{EF} \le 1 - x_{j,t+T_j^{EF,on}}^{EF} \\ \sum\limits_{\tau}^{\tau+T_j^{EF,\text{int}}-1} x_{j,\tau}^{EF} \ge 1 \ \forall \tau \in \left[1, T - T_j^{EF,\text{int}} + 1\right] \end{cases} \tag{6}$$

Formula (6) is the logical constraints of smelting furnace j, which are used to describe the discrete operating characteristics of smelting furnaces. Where $T_j^{EF,on}$ and $T_j^{EF,\text{int}}$ are the maximum smelting time and the maximum oven time of smelting furnace j, respectively.

3.3. Risk Constraints of Energy-Intensive Load

Energy-intensive load has a large load capacity. In order to make the energy be used efficiently, this paper introduces an energy-intensive load to participate in wind power consumption. When energy-intensive loads participate in wind power consumption, and considering that wind farms have obviously volatility, the risk constraint adjustment of energy-intensive load can modify the admissible range of wind power in Section 2.1 (so as to control the risk of wind curtailment).

When energy-intensive loads participate in wind power consumption, energy-intensive load enterprises can purchase electric energy from wind farms at a relatively low price. If the output of wind power is lower than expected after load adjustment, the interests of load enterprises may be harmed, thus dampening the enthusiasm of energy-intensive load to participate in wind power consumption. Therefore, in a dispatch cycle, the wind farm's curtailment expectations and load increase should meet certain risk constraints to ensure the abundance of wind power. This paper defines the conservative degree of load participating in coordinated dispatch as: the expected wind power curtailment before the load participates in the regulation can meet the minimum proportion of the load's increased power consumption after the load participates in the coordination. The concept of conservativeness can form the risk constraints when energy-intensive loads participate in the consumption of wind power.

$$\sum_{i=1}^{W} \sum_{t=1}^{T} E_{i,t}^{u}(w_{i,t}^{u}) \ge \beta^{\text{adj}} \sum_{j=1}^{E} \sum_{t=1}^{T} \max(0, P_{j,t}^{EF\prime} - P_{j,t}^{EF}) \tag{7}$$

where β^{adj} represents the degree of conservation; $P_{j,t}^{EF\prime}$ and $P_{j,t}^{EF}$ represent the electricity consumption plan before the adjustment of the energy-intensive load j at time t and the electricity consumption plan after the adjustment, respectively; W represents the number of wind farms; and E represents the number of energy-intensive load. This formula shows that the total wind curtailment expectation of the wind farm before the energy-intensive load participates in the mediation is greater than β^{adj} times the energy-intensive load adjustment.

Since the purpose of energy-intensive load is to consume wind power, after energy-intensive load participates in wind power consumption, the change of electric energy caused by the adjustment of the upper boundary of wind power curtailment shall be greater than or equal to β^{adj} times of the energy-extensive adjustment. This process meets the following requirements:

$$\begin{cases} w_{i,t}^{u,add} \geq w_{i,t}^{p} \\ \sum_{t=1}^{T} \sum_{i=1}^{W} (w_{i,t}^{u,add} - w_{i,t}^{u}) \geq \beta^{\text{adj}} \sum_{j=1}^{E} \sum_{t=1}^{T} \max(0, P_{j,t}^{EF\prime} - P_{j,t}^{EF}) \end{cases} \tag{8}$$

where $w_{i,t}^{u,add}$ represents the adjusted upper boundary of ARWP. It can be seen from the above calculation formula that when the upper boundary of wind power admissible interval is adjusted and changed, the risk of wind curtailment of the wind farm will be reduced, and the risk constraint of wind curtailment of the wind farm is further realized.

4. Bi-Level Optimization Model Considering Congested Wind Power Consumption

Based on the above analysis of the wind power uncertainty and energy-intensive load dispatching model, the upper model aims at the lowest investment cost of the energy storage system, and establishes an energy storage capacity optimization configuration model on the basis of ensuring the system power balance. The lower model aims at the maximum wind power consumption and the lowest operation cost of the system. Combined with the capacity configuration's results of the energy storage system obtained from the upper optimization model, a coordinated dispatching model of energy-intensive load and energy storage system is constructed. The bi-level optimization model considering congested wind power consumption is shown in Figure 7.

(1) Upper optimization model. According to the energy-intensive load data, the energy-intensive load model is self-dispatch with the goal of minimizing power consumption, and the initial energy-intensive load's electricity plan is obtained. At the same time, according to the wind power prediction data, system load prediction data, combined with the initial energy-intensive load's electricity plan, the configuration of energy storage capacity is optimized to minimize the investment cost of energy storage system.

(2) Lower optimization model. The uncertainty of wind power output follows the normal distribution, combined with the initial electricity plan of energy-intensive load, the upper boundary of ARWP before coordination, the expectation of wind curtailment before coordination and the increment of energy-intensive load are obtained by the risk constraint of energy-intensive load. Bring the above results into the coordinated dispatching model, aiming at the maximum wind power consumption and the lowest comprehensive operating cost of the system, using NSGA-II to solve the problem, and finally the pareto solution set is obtained.

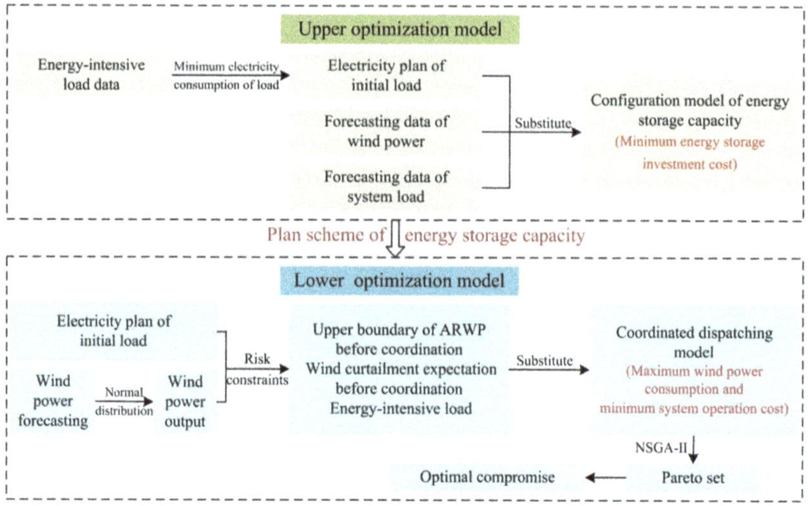

Figure 7. Flowchart of bi-level optimization model considering congested wind power consumption.

4.1. Model of Energy Storage Capacity Configuration

In order to improve the level of wind power consumption, this paper establishes the model by means of effective cooperation between the energy-intensive load and energy storage system. Wind farms are equipped with energy storage systems [28], relying on the peak-load shifting of the energy storage systems to improve system flexibility and reduce wind curtailment rate.

4.1.1. Objective Function

Configuring energy storage capacity with the goal of minimizing energy storage system investment, operation and maintenance costs, the expression is shown in (9):

$$\begin{cases} \min C = \frac{1}{365}[aC_{inv} + C_{on}] \\ a = \frac{\tau(1+\tau)^{\gamma}}{(1+\tau)^{\gamma}-1} \\ C_{inv} = (k_S P_b + k_E E_b) \\ C_{on} = k_S k_M P_b \end{cases} \tag{9}$$

where a is the equal-year system coefficient; τ is the annual interest rate; γ is the service life of the energy storage system; C_{inv} and C_{on} are the investment and construction cost and operation and maintenance cost of the energy storage system respectively; k_S and k_E are the unit power cost and unit capacity cost of the energy storage system respectively; k_M is the operation and maintenance cost rate of the energy storage system; and P_b and E_b are the investment power and investment capacity of the energy storage system, respectively.

4.1.2. Constraints

$$\begin{cases} E_b^{\min} \le E_b \le E_b^{\max} \\ P_b^{\min} \le P_b \le P_b^{\max} \end{cases} \tag{10}$$

where $E_b^{\max}$ and $E_b^{\min}$ are the upper and lower limits of the investment capacity of the energy storage system; $P_b^{\max}$ and $P_b^{\min}$ are the upper and lower limits of the investment power of the energy storage system, respectively.

4.2. Coordinated Dispatching Model of Energy-Intensive Load and Energy Storage System

4.2.1. Objective Function

The coordinated operation of energy-intensive load and energy storage system can enable the system to consume more wind power within the existing regulation capacity. However, using this method will increase the operation cost of the system. Therefore, how to maximize wind power consumption with the lowest operating cost is the key to cooperative operation. In this paper, a multi-objective optimization model is established with the goal of maximizing wind power consumption and minimizing system operating cost, and the expression is as follows:

$$\min F = \sum_{t=1}^{T} (C_t^{Con} + C_t^B + C_t^G + C_t^W) \tag{11}$$

where C_t^{Con}, C_t^B, C_t^G and C_t^W are the operating cost of conventional units, the charge and discharge management cost of energy storage system, the dispatching cost of energy-intensive load, and the penalty cost of curtailment wind, respectively. The specific calculation formula for each cost is as follows:

(3) The operating cost of conventional units C_t^{Con}

$$C_t^{Con} = \sum_{k=1}^{N} (a_k p_{k,t}^2 + b_k p_{k,t} + c_k d_{k,t} + d_{on,k,t} C_{u,k,t}) \tag{12}$$

where N is the number of thermal power units; a_k, b_k and c_k are the cost coefficient of thermal power units; $p_{k,t}$ is the output of the k-th thermal power unit at time t; $C_{u,k,t}$ is the start-up cost of the thermal power unit; $d_{k,t}$ is a 0–1 variable, which is used to indicate the current on/off state of the unit; $d_{on,k,t}$ is a 0–1 variable, which is used to indicate the starting state.

(4) The charge and discharge management cost of energy storage system C_t^B

$$C_t^B = \lambda_{\mathrm{b,dis}} \mathrm{P}_{\mathrm{dis,t}} + \lambda_{\mathrm{b,ch}} \mathrm{P}_{ch,t} \tag{13}$$

where $\lambda_{b,dis}$ is the discharging cost coefficient of the energy storage system; $\lambda_{b,ch}$ is the charging cost coefficient of the energy storage system; $P_{dis,t}$ is the discharge power of the energy storage system at time t; $\mathrm{P}_{ch,t}$ is the charging power of the energy storage system at time t.

(5) The dispatching cost of energy-intensive load C_t^G

$$C_t^G = (\beta_{MI} + \beta_{Wr}) E_{i,t}^u(w_{i,t}^u) + \pi N_f + \varepsilon(E_{i,t}^u(w_{i,t}^u)) c_{Lr} \tag{14}$$

where N_f is the number of power changes of energy-intensive load; π is the corresponding equipment loss cost in case of single power change; β_{MI} is the raw material cost coefficient per unit energy consumption; β_{Wr} is the equipment loss cost coefficient of unit regulated power; $E_{i,t}^u(w_{i,t}^{u,add})$ is the expected curtailment of wind during the dispatching period; c_{Lr} is the increased labor cost of participating in the consumption of congested wind power during the control period; $\varepsilon(E_{i,t}^u(w_{i,t}^u))$ is shown in Formula (15).

$$\varepsilon(E_{i,t}^u(w_{i,t}^u)) = \begin{cases} 0 & E_{i,t}^u(w_{i,t}^u) = 0 \\ 1 & E_{i,t}^u(w_{i,t}^u) > 0 \end{cases} \tag{15}$$

(6) The penalty cost of curtailment wind C_t^W

$$C_t^G = (\beta_{MI} + \beta_{Wr}) E_{i,t}^u(w_{i,t}^u) + \pi N_f + \varepsilon(E_{i,t}^u(w_{i,t}^u)) c_{Lr} \tag{16}$$

4.2.2. Constraints

Constraints include conventional unit constraints, system power balance constraints, energy storage system charging and discharging constraints, energy-intensive load constraints, etc.

(1) Conventional unit constraints

$$p_{\min,k} \leq p_{k,t} \leq p_{\max,k} \tag{17}$$

$$-p_{dn,k} \leq p_{k,t} - p_{k,t-1} \leq p_{up,k} \tag{18}$$

$$d_{on,k,t} \geq d_{k,t} - d_{k-1,t-1} \tag{19}$$

$$\begin{cases} (d_{k,t-1} - d_{k,t})(S_{on,k,t} - S_{on,\min,k}) \geq 0 \\ (d_{k,t} - d_{k,t-1})(S_{off,k,t} - S_{off,\min,k}) \geq 0 \end{cases} \tag{20}$$

$$\begin{cases} S_{on,k,t} = S_{on,k,t-1}d_{k,t} + d_{k,t} \\ S_{off,k,t} = S_{off,k,t-1}(1 - d_{k,t}) + (1 - d_{k,t}) \end{cases} \tag{21}$$

Formula (17) is the output constraint of the conventional unit, $p_{\max,k}$ and $p_{\min,k}$ are the upper and lower limits of the output of the k-th conventional unit respectively; Formula (18) is the ramp rate constraint of the conventional unit, $p_{dn,k}$ and $P_{up,k}$ are the maximum descent rate and maximum ascent rate of the active power output of the k-th conventional unit, respectively; Formula (19) is the 0–1 constraint for unit startup; Formula (20) is the minimum start-stop time constraint of the k-th conventional unit, which $S_{on,k,t}$ is the continuous start-up time of the k-th conventional unit, $S_{on,\min,k}$ is the minimum startup time of the k-th conventional unit, $S_{off,k,t}$ is the continuous shutdown time of the k-th conventional unit, and $S_{off,\min,k}$ is the minimum shutdown time of the k-th conventional unit; Formula (21) is the constraint of the continuous operation time and continuous shutdown time of the unit.

(7) System power balance constraints

$$\sum_{i=1}^{W} w_{i,t}^{p} + \sum_{k=1}^{G} p_{k,t}^{g} + P_{b,d}(t) = P_{load}(t) + \sum_{j=1}^{E} (P_{j,t}^{EF} + \Delta P_{j,t}^{EF}) + P_{b,c}(t) \tag{22}$$

where $P_{b,d}(t)$ and $P_{b,c}(t)$ represent the discharge and charging power of the battery at time t, respectively; $P_{load}(t)$ represents the conventional load power at time t; $\Delta P_{j,t}^{EF}$ represents the active power of the energy-intensive load j at time t.

(8) Energy storage system charging and discharging constraints are as follows:

$$\begin{cases} 0 \leq p_{ch,t} \leq (1 - ESS_t)p_{ch,\max} \\ 0 \leq p_{dis,t} \leq ESS_t p_{dis,\max} \end{cases} \tag{23}$$

$$\begin{cases} SOC_{t-1} + \eta_{ESS,ch}p_{ch,t}/E_{ESS} + ESS_tD \leq SOC_t \leq SOC_{t-1} + \eta_{ESS,ch}p_{ch,t}/E_{ESS} + ESS_tD \\ SOC_{t-1} - p_{dis,t}/(\eta_{ESS,dis}E_{ESS}) - ESS_tD \leq SOC_t \leq SOC_{t-1} - p_{dis,t}/(\eta_{ESS,dis}E_{ESS}) - ESS_tD \end{cases} \tag{24}$$

$$0.1 \leq SOC_t \leq 0.9 \tag{25}$$

Formula (23) is the constraint equation for the charge and discharge power of the energy storage system, $p_{ch,t}$ and $p_{dis,t}$ are the charge and discharge power of the energy storage system; $p_{ch,\max}$ and $p_{dis,\max}$ are the upper limit of the charge and discharge power of the energy storage device; ESS_t is a 0–1 variable indicating the state of energy storage: when $ESS_t = 0$ is in the charging state, when $ESS_t = 1$ is in the discharging state; Formula (24) is the energy storage state of charge constraint, $\eta_{ESS,ch}$ and $\eta_{ESS,dis}$ represent the charging and discharging efficiency of the energy storage system; SOC_t is the energy storage state of charge; and E_{ESS} is the upper limit of the capacity of the energy storage device; Formula (25) is the range constraint of the state of charge of energy storage. The D appearing in the model is a sufficiently large parameter introduced.

(9) Constraints of energy-intensive load are shown in Formulas (5) and (6).

5. Case Analysis

5.1. Basic Data and Scene Settings

This paper takes a wind farm in Gansu as an example to simulate and verify the effectiveness of the proposed model. The system includes 3 wind farms with installed capacities of 300 MW, 500 MW, and 700 MW; the energy-intensive load consists of 12 smelting furnaces, with a single operating power of 17.5–21.5 MW, oven power of 10.5 MW, and longest oven time of 2 h; wind curtailment cost is 300 yuan/MW·h. The energy-intensive load, energy storage system, and related parameter information of conventional units are shown in Tables 2–4 and the system load curve and forecasting curve of wind farms are shown in Figure 8.

Table 2. Energy-intensive load parameters.

Parameter	β^{adj}	β_{MI}	β_{Wr}	c_{Lr}	N_f
Value	0.8	450 yuan/MW·h	120 yuan/MW·h	120 yuan/MW·h	300 yuan/time

Table 3. Energy-intensive load parameters.

Parameter	τ	γ	k_S	k_E	k_M	$\lambda_{b,dis}$	$\lambda_{b,ch}$
Value	0.02	20	100,000 yuan/MVA	150,000 yuan/MW·h	0.04	6 yuan/MW	4 yuan/MW

Table 4. Conventional unit parameters.

Parameter Unit	**1**	**2**	**3**
$P_{max,i}/MW$	200	300	380
$P_{min,i}/MW$	50	80	100
$P_{up,i}/MW$	130	140	150
$P_{dn,i}/MW$	130	140	150
a_i	0.0039	0.0030	0.0027
b_i	17.33	16.23	14.12
c_i	300	500	700
Minimum startup time/h	2	3	4
Minimum downtime/h	2	3	4
Start-up cost/yuan	1000	1250	1500

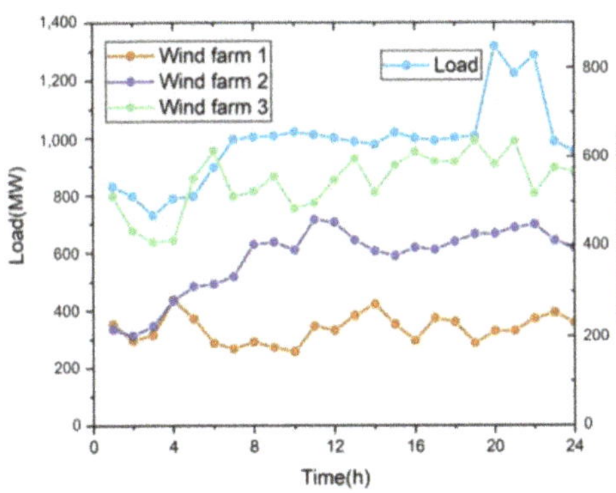

Figure 8. Curve of system load and forecasting power of wind farms.

In order to analyze the impact of wind power uncertainty and the addition of energy-intensive load and energy storage system on wind power consumption, this paper mainly considers the following four cases.

Case 1: Without considering the uncertainty of wind power, only energy-intensive load participates in regulation.
Case 2: Considering the uncertainty of wind power, only energy-intensive load participates in the regulation.
Case 3: Without considering the uncertainty of wind power, energy-intensive load and energy storage system work together.
Case 4: Considering the uncertainty of wind power, energy-intensive load and energy storage system work together.

5.2. Result Analysis

5.2.1. Analysis of Storage Capacity Configuration Results

For case 3 and 4, the energy storage capacity is configured with the lowest investment cost of the energy storage system as the goal, and the particle swarm algorithm is used to solve the problem [29]. The population size is 25, and the number of iterations is 50. The operation results are shown in Figure 9 and Table 5.

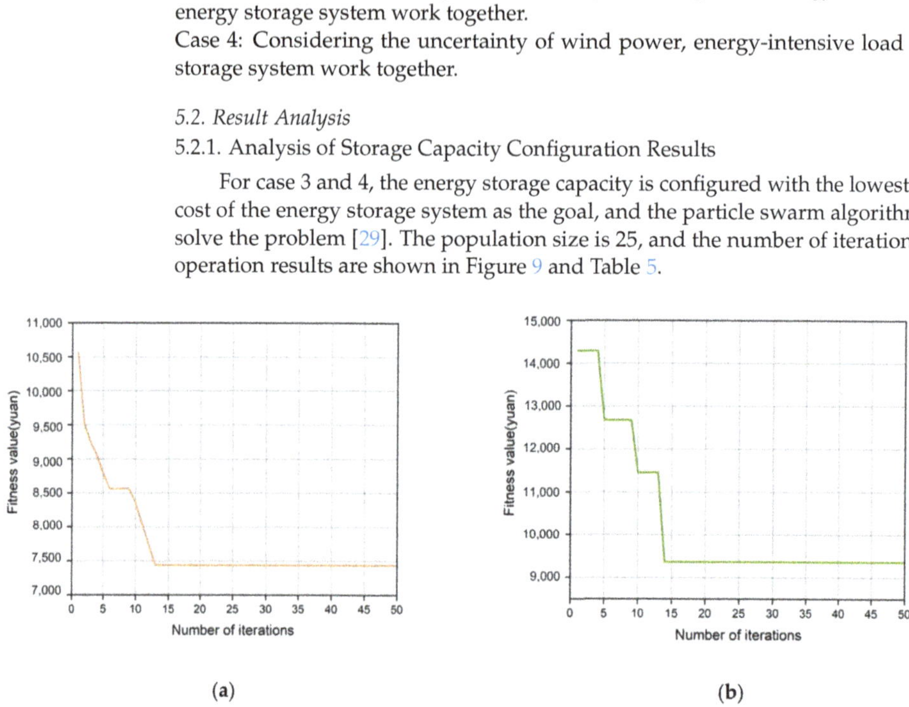

Figure 9. (**a**) Case 3 convergence result; (**b**) Case 4 convergence result.

Table 5. Energy storage system capacity configuration results.

Case	Configuration Parameter		Investment Cost/(Yuan·d^{-1})
	Energy Storage Capacity/MW·H	Energy Storage Power/MW	
3	166	118	7437
4	192	137	9366

It can be seen from Figure 9 that the operation results of case 3 and 4 converge in the 12th and 14th generations, respectively, and the best fitness values are 7437 and 9366, respectively. At the same time, the analysis in Table 5 shows that when case 4 considers the uncertainty of wind power, the energy storage capacity is increased compared to case 3 since, at this time, the risk constraints of wind power and energy-intensive load are considered, and the energy-intensive load regulation will be to reduce wind curtailment, and the capacity and power of energy storage will increase.

5.2.2. Analysis of Coordinated Dispatching Results

In this paper, the NSGA-II algorithm is used to solve the model established in the paper, and the distribution of the Pareto solution set in the four cases obtained in the objective function space is shown in Figure 10.

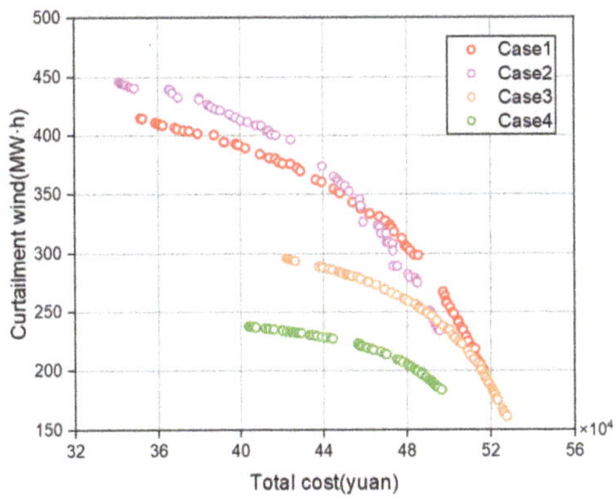

Figure 10. Pareto solution set distribution in different cases.

It can be seen from Figure 10 that the wind power curtailment volume and the operation cost of the system have previously shown an inverse proportional relationship. When the wind power curtailment volume decreases, the operation cost of the system will increase, which is not conducive to the economic indicators of the system. When the operation cost of the system decreases, the wind power curtailment volume will increase, which is not conducive to wind power consumption. Therefore, this paper selects the solution with the highest degree of satisfaction according to the multi-objective compromise strategy. Table 6 presents the two sets of solutions with the smallest wind power curtailment volume and the lowest system operating cost and the optimal compromise solution selected from the Pareto solution set.

Table 6. Comparison of Pareto optimal solutions in different cases.

	Goals	Minimum Expected Curtailment of Wind	Minimum Operation Cost	Optimal Compromise
Case1	Expected curtailment of wind/MW·h	203.5	418.4	325.8
	Operation cost/yuan	502,843	353,492	429,543
Case2	Expected curtailment of wind/MW·h	236.9	441.5	337.4
	Operation cost/yuan	499,201	349,363	405,563
Case3	Expected curtailment of wind/MW·h	162.5	284.7	223.9
	Operation cost/yuan	534,292	428,394	478,134
Case4	Expected curtailment of wind/MW·h	180.4	237.6	204.3
	Operation cost/yuan	509,278	409,021	459,272

In addition, Table 7 shows the overall system operating costs, energy-intensive load costs, energy storage costs, expected curtailment of wind, and energy-intensive load increments in the four cases.

Table 7. Comparison of results in different cases.

Case	1	2	3	4
System operating cost/yuan	429,543	405,563	478,134	459,272
energy-intensive load cost/yuan	281,219	255,923	253,842	246,438
Energy storage cost/yuan	0	0	128,165	122,695
Conventional unit cost/yuan	50,584	48,420	28,957	28,849
Expected curtailment of wind/MW·h	325.8	337.4	223.9	204.3
Load increment/MW·h	489.61	445.23	441.58	428.59

As can be seen from Table 7, the system operation cost in case 2 is reduced by 23,980 yuan compared with case 1. This is due to the fact that the introduction of risk constraints restricts the regulation of energy-intensive load and reduces the load increment, and the output of conventional units will also be reduced, so the system operation cost will be reduced. However, due to the impact of risk constraints, the expected curtailment of wind in case 2 has increased by 11.6 MW compared with case 1. Compared with case 1, the system operation cost of case 3 increased by 48,591 yuan. This is due to the fact that the energy storage system is introduced to participate in wind power consumption, and the energy storage cost is high, so the system operation cost increases. However, the energy storage system is adjusted flexibly and rapidly, the expectation of wind curtailment is significantly reduced, which is 31.27% lower than that in case 1, and the level of wind power consumption is significantly improved. The system operation cost of case 4 is slightly lower than that of case 3. This is due to the fact that the uncertainty of wind power has been taken into account, the risk of the load side has been further avoided, and the increment of energy-intensive load has been reduced. Meanwhile, the increase of energy storage capacity is conducive to the consumption of more wind power. It can be seen that the expected curtailment of wind is reduced by 19.4 MW·h compared with case 3. It can be seen from the comparison of different cases that through the effective cooperation between energy-intensive load and energy storage system, the expected curtailment of wind is significantly reduced and the consumption level of congested wind power is effectively improved. And through the risk constraints of energy-intensive load, enterprises can adjust the load in a targeted manner, which can effectively avoid the risk of mismatch between the adjustment increment of energy-intensive load and the wind power output, so as to greatly reduce the overall operating cost of the system. Moreover, with the reduction of the conventional units output, the startup and shutdown times of units are also relatively reduced, which increases the stability of unit operation. The specific operation conditions under different cases are analyzed below.

(1) Operation result analysis of case 1

In case 1, the uncertainty of wind power is not considered, and the wind power is consumed by adjusting energy-intensive load. Figure 11 shows the wind power curtailment expectation curve, load power plan curve, upper and lower boundaries of ARWP before and after energy-intensive load participates in the regulation, and system dispatching curve.

As can be seen from Table 7 and Figure 11, since the uncertainty of wind power output is not considered in case 1, in order to consume more wind power, energy-intensive load enterprises will increase load regulation as much as possible. It also increases the operating cost of energy-intensive load while reducing wind curtailment. Since the discretely adjustable energy-intensive load cannot be adjusted continuously in a short time, when the predicted output of wind power is higher but the actual output is lower, it will cause the problem of a mismatch between the load increment and the power generation increment, resulting in a higher risk of load shedding. In order to meet the constraints of system power balance and to try and avoid the risk of load shedding due to the uncertainty of wind power, conventional units will increase output, so the overall operating cost of the system will increase.

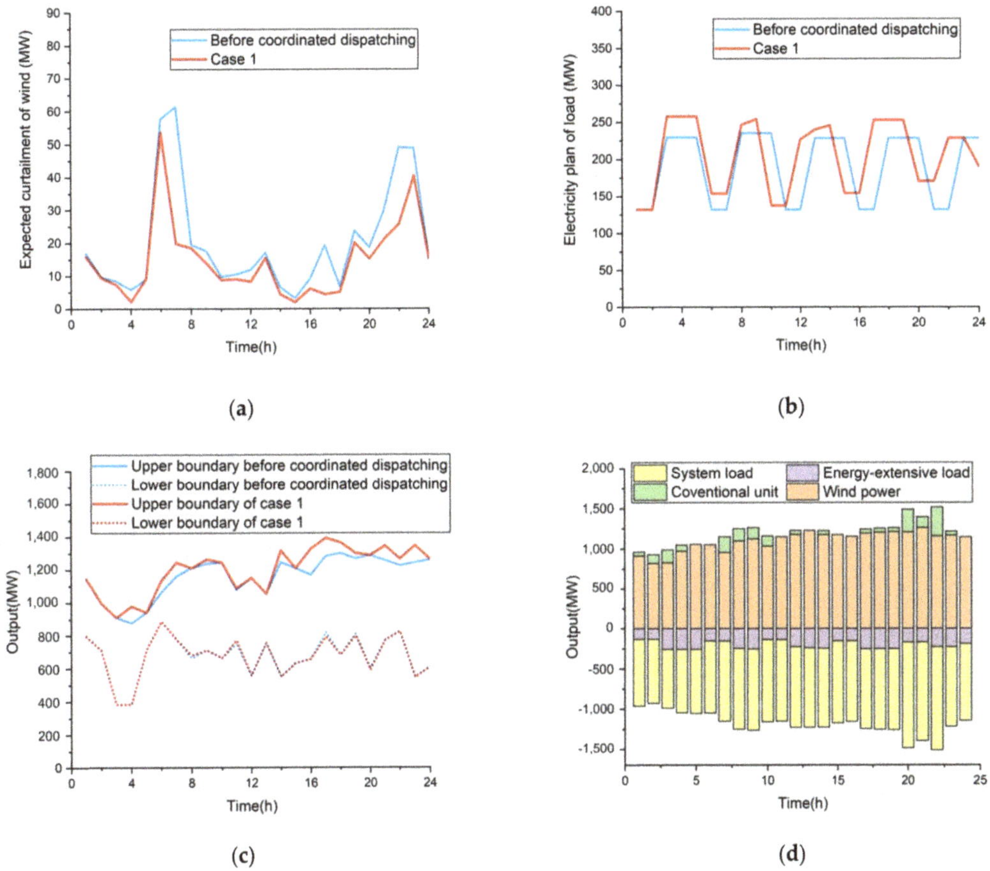

Figure 11. (**a**) Curves of expected curtailment of wind; (**b**) Curves of electricity plan of load; (**c**) ARWP boundary of power grid; (**d**) System coordination dispatching diagram.

(10) Operation result analysis of case 2

In case 2, the uncertainty of wind power is considered, and the wind power is consumed by adjusting the high load energy load. Figure 12 shows the wind power curtailment expectation curve, load power plan curve, upper and lower boundaries of ARWP before and after energy-intensive load participates in the regulation, and system dispatching curve.

In case 2, the risk constraints of wind power and energy-intensive load are considered. The increment of energy-intensive load is reduced by 44.38 MW·h compared with case 1, and the operation cost of corresponding energy-intensive load is reduced by 25,296 yuan. At the same time, the cost of conventional units is reduced by 2164 yuan, so the comprehensive operation cost of the system is slightly lower than that in case 1. It can be seen that due to the influence of risk constraints, the wind curtailment expectation of case 2 has increased by 11.6 MW·h compared with case 1. In addition, after the energy-intensive load participates in coordinated dispatch, the total ARWP upper boundary of the grid has increased by 499.29 MW·h the lower boundary of ARWP has increased by 9.96 MW·h, and the total amount of APWP has increased by 489.33 MW·h, which greatly improves the acceptance capacity of the power grid for wind power and effectively promotes the consumption of wind power.

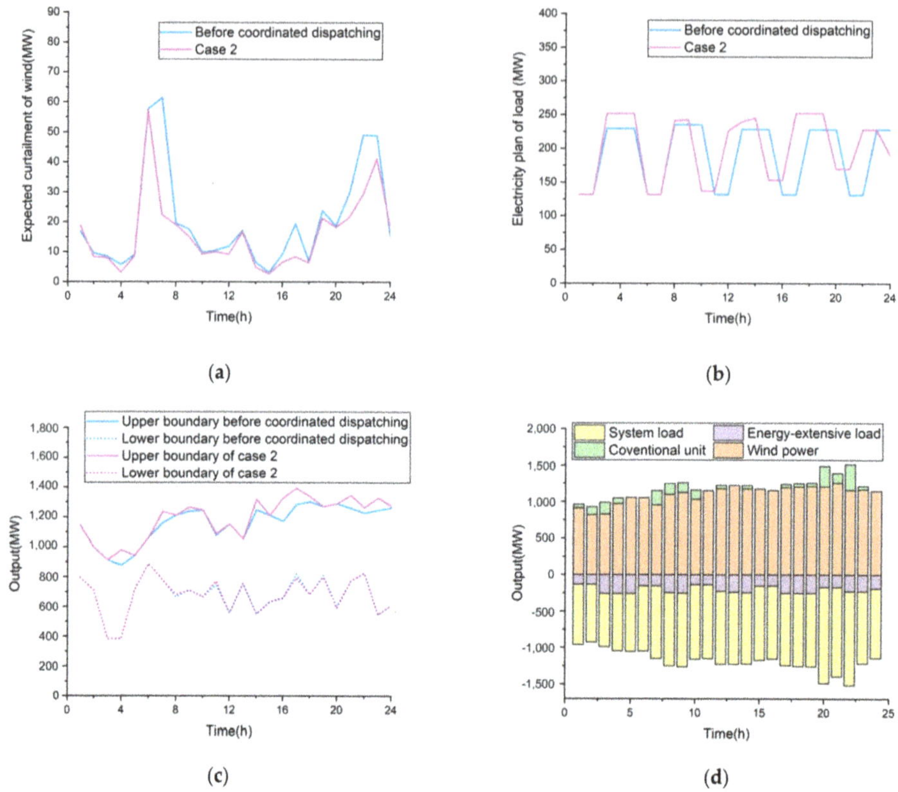

Figure 12. (**a**) Curves of expected curtailment of wind; (**b**) Curves of electricity plan of load; (**c**) ARWP boundary of power grid; (**d**) System coordination dispatching diagram.

It can be seen from Figure 12 that due to the increased risk constraints of wind power and energy-intensive load, compared with the original plan for energy-intensive load, the time period during which the adjusted high-energy load power increases generally corresponds to the time when wind power is relatively curtailed. This indicates that in the coordinated dispatching process of energy-intensive load and wind power, the addition of risk constraint makes the energy-intensive load tend to adjust electricity consumption in the period of more wind curtailment so as to avoid the economic risks brought by wind power shortage to the load side.

In addition, according to Table 7, the increment of energy-intensive load before and after coordination is not equal to the expected reduction of wind abandoning, nor is it equal to the increase of the upper and lower boundary width of the grid ARWP. The reason is that the three are not the same. The load increment is $E^u_{i,t}(w^u_{i,t})$, the expected reduction of wind curtailment is $E^u_{i,t}(w^{u,add}_{i,t}) - E^u_{i,t}(w^u_{i,t})$, and the upper boundary increase of ARWP is $w^{u,add}_{i,t} - w^u_{i,t}$.

(11) Operation result analysis of case 3

In case 3, the uncertainty of wind power is not considered, and wind power is consumed through the joint adjustment of energy-intensive load and energy storage system. Figure 13 show the wind power curtailment expectation curve, load power plan curve, upper and lower boundaries of ARWP before and after energy-intensive load participates in the regulation, charging and discharging conditions and energy storage system SOC change, and system dispatching curve.

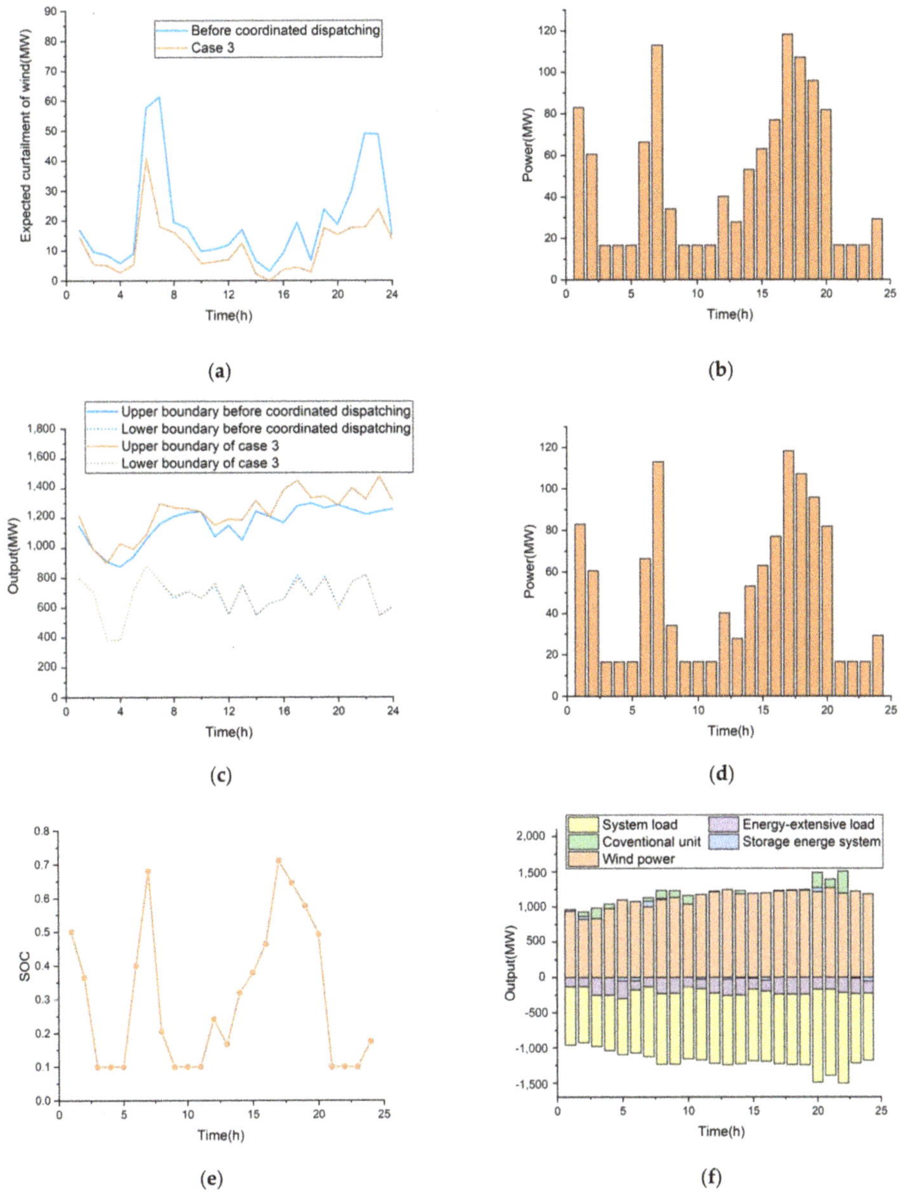

Figure 13. (**a**) Curves of expected curtailment of wind; (**b**) Curves of electricity plan of load; (**c**) ARWP boundary of power grid; (**d**) Charging and discharging of energy storage system; (**e**) Energy storage system SOC; (**f**) System coordination dispatching diagram.

After adding the energy storage system in case 3, it can be seen that the output of conventional units has been significantly decreased, and the start and stop of some conventional units have been reduced, saving the cost of conventional units. At the same time, when the system load is low, the excess electric energy can be stored in the energy storage system. When the load is high, the discharge of the energy storage system can make up for the insufficient wind power output. This shows that the energy storage system can assist the operation of the power system and optimit.

(12) Operation result analysis of case 4

In case 4, the uncertainty of wind power is considered, and wind power is consumed through the joint adjustment of energy-intensive load and energy storage system. Figure 14 shows the wind power curtailment expectation curve, load power plan curve, upper and lower boundaries of ARWP before and after energy-intensive load participates in the regulation, energy storage system SOC change and charging and discharging conditions, and system dispatching curve.

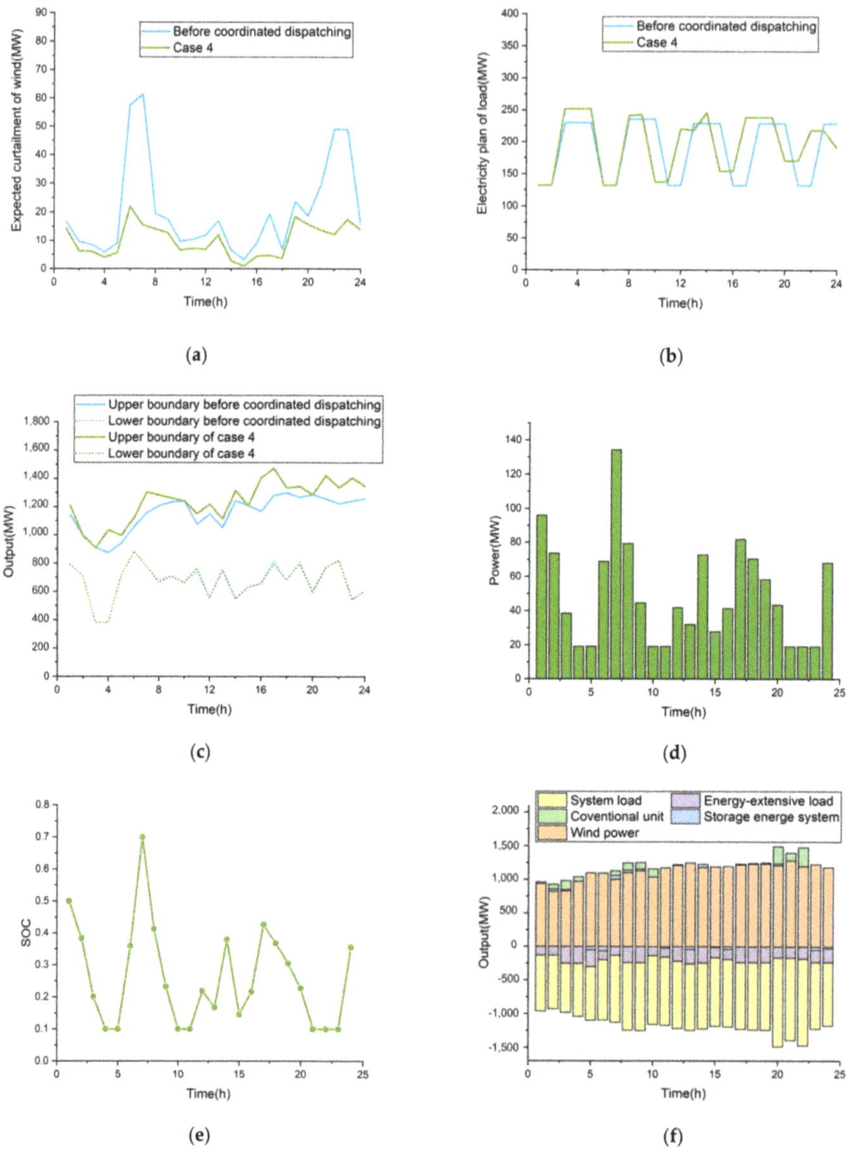

Figure 14. (**a**) Curves of expected curtailment of wind; (**b**) Curves of electricity plan of load; (**c**) ARWP boundary of power grid; (**d**) Charging and discharging of energy storage system; (**e**) Energy storage system SOC; (**f**) System coordination dispatching diagram.

It can be seen from Table 7 that when considering the uncertainty of wind power, the increase in energy-intensive load is reduced by 12.99 MW·h compared with case 3, and the capacity configuration of the energy storage system is increased by 26 MW. The energy storage system can convert part of the wind power waste into chemical energy and store it in the energy storage system when wind power is generated. Therefore, the conventional unit output in case 4 is the smallest among the four cases. In the comparison of different cases, case 4 is the optimal operation scenario, with the lowest expectation of wind curtailment and the most significant effect of wind power consumption.

Through the analysis of different cases, it can be seen that energy-intensive load and the energy storage system can effectively reduce the wind power curtailment volume, decrease the total system operation cost, and reduce the output fluctuation of conventional units while increasing the operation stability of the power system.

5.2.3. Influence of Conservative Degree Change of Energy-Intensive Load on Consuming Results

In this paper, the concept of conservatism is introduced to restrict the adjustment of energy-intensive load power under wind power uncertainty. The selection of conservatism parameters will affect the results of wind power consumption. Table 8 and Figure 15 show the results of energy-intensive load and energy storage system participating in wind power consumption in each case under different conservative parameters.

Table 8. Energy storage system capacity configuration results.

Case	Conservation Level	System Operating Costs/Yuan	Expected Curtailment of Wind/MW·h
1	1.3	401,638	355.68
	0.9	425,890	340.95
	0.5	439,105	316.03
	0.1	458,264	317.08
2	1.3	386,633	370.56
	0.9	401,640	351.63
	0.5	419,271	328.19
	0.1	437,434	327.21
3	1.3	462,743	262.29
	0.9	472,974	232.62
	0.5	489,671	218.01
	0.1	501,224	209.93
4	1.3	442,729	243.19
	0.9	451,964	215.05
	0.5	470,691	190.12
	0.1	489,830	185.44

From the analysis of Table 8 and Figure 15, it can be seen that when the wind curtailment expectation is certain, the lower the conservative level of the load, the greater the load increment. When the conservative degree is greater than or equal to 1, the load increment is less than or equal to the expected wind curtailment before coordination; when the conservative degree is less than 1, it is understood that the load side is willing to take a certain risk during the coordination process, so the load increment is greater than the expected wind curtailment before coordination. This shows that the introduction of risk constraints can help the load side choose the amount of risk it can bear according to its own characteristics (such that the dispatching method of power grid can meet the needs of load enterprises with different operation tendencies).

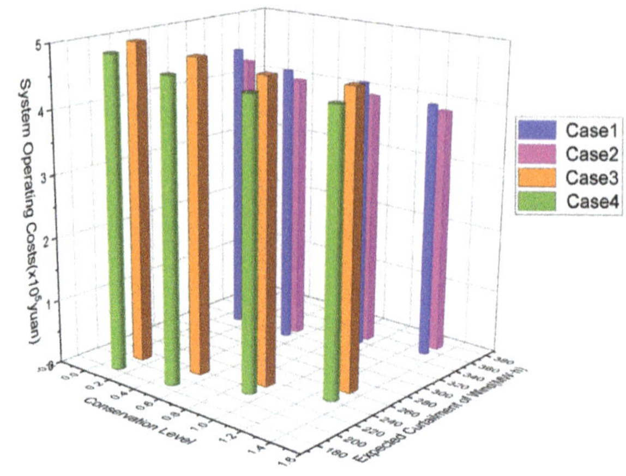

Figure 15. Results of congested wind power consumption under different conservative degrees.

6. Conclusions

In order to deal with the mismatch between the electricity plan of the load, the output caused by the uncertainty of wind power, and the fact that the discretely adjustable energy-intensive load cannot be continuously adjusted in a short time in the process of consuming congested wind power, a bi-level optimization model research for an energy-intensive load and energy storage system (considering congested wind power consumption) is proposed, and the effectiveness of this model is verified by a practical example. The specific results of his research are as follows:

(1) On the basis of analysis of a wind power uncertainty and energy-intensive load dispatching model, this paper establishes the risk constraints of energy-intensive load, which not only fully excavates the regulation potential of energy-intensive load but also solves the problem of mismatch between the regulation increment of energy-intensive load and wind power output.

(2) In order to maximize the consumption of congested wind power, a bi-level optimization model considering congested wind power consumption is established in this paper. The optimal configuration model of the energy storage capacity is established with the goal of minimizing the investment cost of the energy storage system, and the coordinated dispatching model of the energy-intensive load and energy storage system is established with the goal of maximizing the consumption of wind power and minimizing the comprehensive operation cost of the system (so as to achieve the purpose of maximizing the consumption of congested wind power by using an energy-intensive load and energy storage system).

(3) According to the example results, the minimum wind curtailment expectation of the proposed method is 204.3 MW·h, which is 65% lower than that before the coordinated dispatch, and the total admissible range of wind power increased by 572.5 MW·h, which effectively improve the ability of the power grid to integrate wind power and increased the overall level of consumption of congested wind power.

(4) This paper establishes a model of the conservative degree of energy-intensive load participating in the consumption of wind power such that the load side can adjust the load in a targeted manner to avoid the risk of excessive wind power uncertainty. It also analyzes the impact of different conservative values on the consumption of wind power. The results show that under the conditions of a certain wind curtailment expectation, the lower the conservative level of the load, the greater the load increment. energy-intensive load enterprises can choose the risks they can bear according to their own characteristics to ensure corporate benefits.

Author Contributions: Conceptualization, S.Z. and K.Z.; methodology, K.Z.; software, S.Z.; validation, S.Z., K.Z. and G.Z.; formal analysis, S.Z.; investigation, C.F. and W.B.; resources, G.Z.; data curation, J.W.; writing—original draft preparation, S.Z. and K.Z.; writing—review and editing, S.Z., K.Z. and G.Z.; visualization, T.X. and J.W.; supervision, T.X.; project administration, G.Z.; funding acquisition, G.Z. All authors have read and agreed to the published version of the manuscript.

Funding: This research was funded by Key Research and Development Plan of Shaanxi Province (2018-ZDCXL-GY-10-04), Natural Science Basic Research Program of Shaanxi (Program No.2019JLZ-15).

Institutional Review Board Statement: Not applicable.

Informed Consent Statement: Not applicable.

Data Availability Statement: Not applicable.

Conflicts of Interest: The authors declare no conflict of interest.

Abbreviations

ARWP	Admissible region of wind power
W	the number of wind farms
E	the number of energy-intensive load
N	the number of thermal power units
$w^{p}_{i,t}$	planned output
$\Delta\tilde{w}_{i,t}$	wind power output fluctuation
$w_{i,t}$	actual output
$w^{u}_{i,t}$	the upper boundary before coordinated dispatching
$w^{l}_{i,t}$	the lower boundary before coordinated dispatching
$w^{f}_{i,t}$	the predicted output of wind farm
σ_i	the initial standard deviation of wind farm i load forecasting
$\Delta\sigma_i$	the standard deviation increment of wind farm i load forecasting process with time scale
ρ^{u}	the penalty for wind curtailment
$w^{\max}_{i,t}$	the upper limit of output of wind farm i at time t
$P^{EF}_{j,t}$	the total active power
$P^{EF,adj}_{j,t}$	continuous regulation
$x^{EF}_{j,t}$	state variable
$u^{EF}_{j,t}$	start flag of smelting furnace
$P^{EF,\mathrm{int}}_{j}$	the oven power
$P^{EF,on}_{j}$	normal production power
$P^{EF,d}_{j}$	maximum down-regulated power of smelting furnace
$P^{EF,u}_{j}$	maximum up-regulated power of smelting furnace
$T^{EF,on}_{j}$	the maximum smelting time of smelting furnace
$T^{EF,\mathrm{int}}_{j}$	the maximum oven time of smelting furnace
β^{adj}	the degree of conservation
$P^{EF\prime}_{j,t}$	the electricity consumption plan before the adjustment
$P^{EF}_{j,t}$	the electricity consumption plan after the adjustment
$w^{u,add}_{i,t}$	the adjusted upper boundary of ARWP
a	the equal-year system coefficient
τ	the annual interest rate
γ	the service life of the energy storage system
C_{inv}	the investment and construction cost of the energy storage system
C_{inv}	the operation and maintenance cost of the energy storage system
k_S	the unit power cost of the energy storage system

k_E	the unit capacity cost of the energy storage system
k_M	the operation and maintenance cost rate of the energy storage system
P_b	the investment power of the energy storage system
E_b	the investment capacity of the energy storage system
$E_b^{\max}$	the upper limits of the investment capacity of the energy storage system
$E_b^{\min}$	the lower limits of the investment capacity of the energy storage system
$P_b^{\max}$	the upper limits of the investment power of the energy storage system
$P_b^{\min}$	the lower limits of the investment power of the energy storage system
C_t^{Con}	the operating cost of conventional units
C_t^B	the charge and discharge management cost of energy storage system
C_t^G	the dispatching cost of energy-intensive load
C_t^W	the penalty cost of curtailment wind
a_k	the cost coefficient of thermal power units
b_k	the cost coefficient of thermal power units
c_k	the cost coefficient of thermal power units
$p_{k,t}$	the output of thermal power unit
$C_{u,k,t}$	the start-up cost of the thermal power unit
$d_{k,t}$	0–1 variable
$d_{on,k,t}$	0–1 variable
$\lambda_{\mathrm{b,dis}}$	the discharging cost coefficient of the energy storage system
$\lambda_{\mathrm{b,ch}}$	the charging cost coefficient of the energy storage system
$\mathrm{P}_{\mathrm{dis,t}}$	the discharge power of the energy storage system
$\mathrm{P}_{ch,t}$	the charging power of the energy storage system
π	the corresponding equipment loss cost
β_{MI}	the raw material cost coefficient per unit energy consumption
β_{Wr}	the equipment loss cost coefficient of unit regulated power
c_{Lr}	the increased labor cost of participating in the consumption of congested wind power during the control period
$p_{\max,k}$	the upper limits of the output of the conventional unit
$p_{\min,k}$	the lower limits of the output of the conventional unit
$p_{dn,k}$	the maximum descent rate of active power output of the conventional unit
$p_{up,k}$	the maximum ascent rate of active power output of the conventional unit
$S_{on,k,t}$	the continuous start-up time of the conventional unit
$S_{on,\min,k}$	the minimum start-up time of the conventional unit
$S_{off,k,t}$	the continuous shutdown time of the conventional unit
$S_{off,\min,k}$	the minimum shutdown time of the conventional unit
$P_{b,d}(t)$	the discharge power of the battery
$P_{b,c}(t)$	the charging power of the battery
$P_{load}(t)$	the conventional load power
$p_{ch,\max}$	the upper limit of the charge power of the energy storage device
$p_{dis,\max}$	the upper limit of the discharge power of the energy storage device
$\eta_{ESS,ch}$	the charging efficiency of the energy storage system
$\eta_{ESS,dis}$	the discharging efficiency of the energy storage system
SOC_t	the state of charge

References

1. Shahriari, M.; Cervone, G.; Clemente-Harding, L.; Delle Monache, L. Using the analog ensemble method as a proxy measurement for wind power predictability. *Renew. Energy* **2020**, *146*, 789–801. [CrossRef]
2. Spokesperson of the National Energy Administration. Transcript of the online press conference of the National Energy Administration in the first quarter of 2021. *China Electr. Power* **2021**, *02*, 26–29.
3. Simla, T.; Stanek, W. Reducing the impact of wind farms on the electric power system by the use of energy storage. *Renew. Energy* **2020**, *145*, 772–782. [CrossRef]
4. Madaeni, S.H.; Sioshansi, R. Using Demand Response to Improve the Emission Benefits of Wind. *IEEE Trans. Power Syst.* **2013**, *28*, 1385–1394. [CrossRef]
5. Lamsal, D.; Sreeram, V.; Mishra, Y.; Kumar, D. Output power smoothing control approaches for wind and photovoltaic generation systems: A review. *Renew. Sustain. Energy Rev.* **2019**, *113*, 1–22. [CrossRef]

6. Tang, R.M. Promote the integration of power source network and load storage and multi-energy complementation—Policy interpretation of the "Guiding Opinions of the National Development and Reform Commission and the National Energy Administration on Promoting the Development of Power Source, Network, Load and Storage Integration and Multi-energy Complementation". *China Econ. Trade Her.* **2021**, *8*, 10–11.
7. Drovtar, I.; Uuemaa, P.; Rosin, A.; Kilter, J.; Valtin, J. Using demand side management in energy-intensive industries for providing balancing power—The Estonian case study. In Proceedings of the Power & Energy Society General Meeting, Vancouver, BC, Canada, 21–25 July 2013.
8. Paulus, M.; Borggrefe, F. The potential of demand-side management in energy-intensive industries for electricity markets in Germany. *Appl. Energy* **2011**, *88*, 432–441. [CrossRef]
9. Zhang, N.; Hu, Z.G.; Shen, B.; He, G.; Zheng, Y.N. An integrated source-grid-load planning model at the macro level: Case study for China's power sector. *Energy* **2017**, *126*, 231–246. [CrossRef]
10. Jonghe, C.D. Optimal Generation Mix with Short-Term Demand Response and Wind Penetration. *IEEE Trans. Power Syst.* **2012**, *27*, 830–839. [CrossRef]
11. Kiran, B.; Kumari, M.S. Demand response and pumped hydro storage scheduling for balancing wind power uncertainties: A probabilistic unit commitment approach. *Int. J. Electr. Power Energy Syst.* **2016**, *81*, 114–122. [CrossRef]
12. Daneshvar, M.; Mohammadi-Ivatloo, B.; Zare, K. Two-stage optimal robust scheduling of hybrid energy system considering the demand response programs. *J. Clean. Prod.* **2020**, *248*, 119267.1–119267.3. [CrossRef]
13. Yang, H.; Yu, Q.; Liu, J.P.; Jia, Y.W.; Yang, G.Y.; Ackom, E.; Dong, Z.Y. Optimal Wind-Solar Capacity Allocation with Coordination of Dynamic Regulation of Hydropower and Energy Intensive Controllable Load. *IEEE Access* **2020**, *8*, 110129–110139. [CrossRef]
14. Koohi-Fayegh, S.; Rosen, M.A. A review of energy storage types, applications and recent developments. *J. Energy Storage* **2020**, *27*, 101047. [CrossRef]
15. Li, X.; Cao, X.; Li, C.; Yang, B.; Cong, M.; Chen, D.W. A Coordinated Peak Shaving Strategy Using Neural Network for Discretely Adjustable Energy-Intensive Load and Battery Energy Storage. *IEEE Access* **2019**, *8*, 5331–5338. [CrossRef]
16. Khosravi, M.; Afsharnia, S.; Farhangi, S. Optimal sizing and technology selection of hybrid energy storage system with novel dispatching power for wind power integration. *Int. J. Electr. Power Energy Syst.* **2021**, *127*, 106660. [CrossRef]
17. Sulaeman, S.; Tian, Y.; Benidris, M.; Mitra, J. Quantification of Storage Necessary to Firm Up Wind Generation. *IEEE Trans. Ind. Appl.* **2017**, *53*, 3228–3236. [CrossRef]
18. Khalid, M.; Almuhaini, M.; Aguilera, R.P.; Savkin, A.V. Method for planning a wind–solar–battery hybrid power plant with optimal generation-demand matching. *IET Renew. Power Gener.* **2018**, *12*, 1800–1806. [CrossRef]
19. Dicorato, M.; Forte, G.; Pisani, M.; Trovato, M. Planning and Operating Combined Wind-Storage System in Electricity Market. *IEEE Trans. Sustain. Energy* **2012**, *3*, 209–217. [CrossRef]
20. He, H.; Peng, F.; Gao, Z.; Liu, X.; HU, S.; Zhou, W.; Sun, H. A Multi-Objective Risk Scheduling Model of an Electrical Power System-Containing Wind Power Station with Wind and Energy Storage Integration. *Energies* **2019**, *12*, 2153. [CrossRef]
21. Sun, R.F.; Zhang, T.; Liang, J. Evaluation and application of wind power integration capacity in power grid. *Autom. Electr. Power Syst.* **2011**, *35*, 70–76.
22. Li, P.; Yu, D.; Yang, M.; Wang, J. Flexible look-ahead dispatch realized by robust optimization considering cvar of wind power. *IEEE Trans. Power Syst.* **2018**, *33*, 5330–5340. [CrossRef]
23. Dvorkin, Y.; Lubin, M.; Backhaus, S.; Chertkov, M. Uncertainty Sets for Wind Power Generation. *IEEE Trans. Power Syst.* **2016**, *31*, 3326–3327. [CrossRef]
24. Ding, Y.; Shao, C.; Yan, J.; Song, Y.; Zhang, C.; Guo, C. Economical flexibility options for integrating fluctuating wind energy in power systems: The case of China. *Appl. Energy* **2018**, *228*, 426–436. [CrossRef]
25. Ge, S.; Yu, K.; Chen, X.Y.; Liao, Y.C.; Huang, X.S.; Zhao, J. Research on power loss reduction method based on continuous regulating features of energy-intensive industrial loads. In Proceedings of the 2016 IEEE International Conference on Power System Technology, Wollongong, NSW, Australia, 28 September–1 October 2016.
26. Ramlal, C.J.; Singh, A.; Rocke, S. Repetitive learning frequency control for energy intensive corporate microgrids subject to Cyclic Batch Loads. In Proceedings of the 2020 IEEE PES Innovative Smart Grid Technologies Europe (ISGT-Europe)—IEEE, Hague, The Netherlands, 26–28 October 2020.
27. Tovarovskii, I.G.; Merkulov, A.E. Features of Temperature and Concentration Fields During Pig and Cast Iron Smelting in a Blast Furnace Workspace. *Metallurgist* **2016**, *60*, 589–593. [CrossRef]
28. Psarros, G.N.; Dratsas, P.A.; Papathanassiou, S.A. A comparison between central- and self-dispatch storage management principles in island systems. *Appl. Energy* **2021**, *298*, 117181. [CrossRef]
29. Mahapatra, S.; Badi, M.; Raj, S. Implementation of PSO, it's variants and Hybrid GWO-PSO for improving Reactive Power Planning. In Proceedings of the 2019 Global Conference for Advancement in Technology (GCAT), Bangalore, India, 18–20 October 2019; pp. 1–6.

energies

Article

Heat Transfer Enhancement by Perforated and Louvred Fin Heat Exchangers

Miftah Altwieb [1], Rakesh Mishra [2], Aliyu M. Aliyu [2,*] and Krzysztof J. Kubiak [3]

1 Department of Mechanical and Industrial Engineering, University of Gharyan, Gharyan 010101, Libya; miftah.altwieb@hud.ac.uk
2 School of Computing and Engineering, University of Huddersfield, Huddersfield HD1 3DH, UK; r.mishra@hud.ac.uk
3 School of Mechanical Engineering, University of Leeds, Leeds LS2 9JT, UK; k.kubiak@leeds.ac.uk
* Correspondence: a.m.aliyu@hud.ac.uk

Abstract: Multi-tube multi-fin heat exchangers are extensively used in various industries. In the current work, detailed experimental investigations were carried out to establish the flow/heat transfer characteristics in three distinct heat exchanger geometries. A novel perforated plain fin design was developed, and its performance was evaluated against standard plain and louvred fins designs. Experimental setups were designed, and the tests were carefully carried out which enabled quantification of the heat transfer and pressure drop characteristics. In the experiments the average velocity of air was varied in the range of 0.7 m/s to 4 m/s corresponding to Reynolds numbers of 600 to 2650. The water side flow rates in the tubes were kept at 0.12, 0.18, 0.24, 0.3, and 0.36 m^3/h corresponding to Reynolds numbers between 6000 and 30,000. It was found that the louvred fins produced the highest heat transfer rate due to the availability of higher surface area, but it also produced the highest pressure drops. Conversely, while the new perforated design produced a slightly higher pressure drop than the plain fin design, it gave a higher value of heat transfer rate than the plain fin especially at the lower liquid flow rates. Specifically, the louvred fin gave consistently high pressure drops, up to 3 to 4 times more than the plain and perforated models at 4 m/s air flow, however, the heat transfer enhancement was only about 11% and 13% over the perforated and plain fin models, respectively. The mean heat transfer rate and pressure drops were used to calculate the Colburn and Fanning friction factors. Two novel semiempirical relationships were derived for the heat exchanger's Fanning and Colburn factors as functions of the non-dimensional fin surface area and the Reynolds number. It was demonstrated that the Colburn and Fanning factors were predicted by the new correlations to within ±15% of the experiments.

Keywords: heat exchanger; heat transfer; louvred fins; heat transfer effectiveness; Fanning friction factor; Colburn factor

Citation: Altwieb, M.; Mishra, R.; Aliyu, A.M.; Kubiak, K.J. Heat Transfer Enhancement by Perforated and Louvred Fin Heat Exchangers. *Energies* **2022**, *15*, 400. https://doi.org/10.3390/en15020400

Academic Editors: Alon Kuperman and Alessandro Lampasi

Received: 7 December 2021
Accepted: 4 January 2022
Published: 6 January 2022

1. Introduction

Heat exchanging devices are used to transfer thermal energy between two or more mediums, which could be fluid–fluid or fluid–gas systems. The heat transfer process is carefully considered in the design of heat exchangers, which may involve various modes of heat transfer. Heat exchangers are used widely in a wide range of industries where there may be a need for controlled heating or cooling of flow streams, controlled evaporation, or controlled condensation, such as ventilation and air conditioning systems (HVAC), power generation industries, process industries, and manufacturing plants [1,2].

There are specific guidelines and procedures for designing and predicting performance of the heat exchangers. Knowledge and adherence to these during a design process are of great importance for maintaining proper and efficient operation. The performance of heat exchangers depends on geometric, flow, and fluid variables. Thus, appropriate selection of these variables is very important for the optimum performance of the heat exchanger unit for a given duty. The heat exchanger geometry (flow paths and fins geometry) is often optimised to provide best heat transfer performance for minimum operational and capital costs.

Heat exchanger analysis and design processes have seen considerable improvement over the years because of extensive research in this area and currently significant focus is directed towards optimising such systems. The main aim of these investigations is directed towards improving the heat transfer rate and minimise pumping costs as well as reduce costs associated with size and weight of the heat exchanger. In general, optimisation approaches can be classified under either active or passive categories. In the first category, an external force is used to drive heat transfer performance. Conversely, inserts and other additional geometrical protrusions are used to modify the flow in the second category. In practice however, a combination of the two may be used to optimise the performance of a heat exchanger [3,4].

Wilson [5] developed an experimental technique to measure the effectiveness of various heat transfer processes. The process performance was quantified through the calculated convection coefficients. The overall thermal resistance was divided into three major categories: two convections (internal and external) and tube wall. The method has been extensively used and even adapted for use in modified systems i.e., for helical tubes and for pipe annuli. It assumes that the outside coefficient and the fouling resistance are constant and that the coefficients C_A, nA, and mA of the correlation devised are known:

$$Nu_A = C_A Re_A^{nA} Pr_A^{mA} \tag{1}$$

Wilson method was modified by different investigators such as Sieder-Tate [6], Colburn [7] and Dittus-Boelter [8]. The nature of parametric intercedence relating the Nusselt, Reynolds and Prandtl numbers in Equation (1) was the subject of most modifications.

Wang et al. [9] experimentally studied 15 plate, fin and tube heat exchangers having a 9.52 mm tube diameter with different geometries. The evaluated effects of parameters corresponding to fins (thickness, spacing) and tube (number of rows) on the typical flow and heat transfer characteristics. They found that the fin thickness and spacing have limited effect on the flow and heat transfer characteristics. Wang et al. [9] also found that the number of tube rows has a negligible influence on a friction factor behaviour.

Abu Madi et al. [10] assessed the performance behaviour of finned plate and tube heat exchangers. They correlated geometry of flat and corrugated fins with Colburn and friction factors. They found that the fin type had an effect on heat transfer and friction factor. However, the number of tube rows was found to be of much less significance. Furthermore, they found that the effect of the number of tube rows was influenced by the fin and tube geometries and the Reynolds number. It has been established by Webb et al. [11] and Wang et al. [12], that the most effective methods of enhancing the heat transfer performance is to extend the fin surface. Additionally, the plain fin is the most widely used due to its ease of manufacture, simplicity of assembly and has low pressure drop characteristics.

Wang et al. [13] analysed experimentally compact slit fins exchangers with plain and louvred fins. Similar to previous studies, it was seen that the frictional performance had been affected only minimally by the number of rows, whereas louvred fins had been found to increase the heat transfer. Fernández-Seara et al. [14] designed an experimental setup to measure the heat transfer coefficients in the processes of vapour generation and its condensation in heat exchanger tubes. They extended the use of underlying method to a number of convection heat transfer cases which they noted would be useful to design engineers handling thermal problems.

An experimental study was carried out by Wang et al. [15] and in this study the airside performance characteristics of plain, semi-dimpled vortex generator (VG) and louvred fin-and-tube heat exchangers were comparatively evaluated. They investigated the effect of the number of tube rows and the effect of different fins on the heat transfer coefficient. Their results showed that number of tubes in a row had a negligible effect on the heat transfer coefficients for the louvred and semi-dimpled VG fin geometry. Moreover, the heat transfer coefficients for the louvred fin geometry were found to be higher (in the range of 2–15%) than in the case of the semi-dimpled VG geometry. It is however noted that these findings are valid for heat exchangers with number of tubes rows of between 2 and 4.

Liu et al. [16] conducted CFD investigations on the effect of perforation size, fin spacing, and number of perforations, on the Colburn factor corresponding to the air side. The thermal characteristics of finned-tube heat exchangers were also studied. The thermal performance of perforated heat exchanger was compared with that of the plain fin heat exchanger, and they found that the Colburn factor (air side) increased by more than 3 and 8% for constant fin spacing, respectively when the Reynolds number (air side) increases from 750 to 2350. Conversely, it was found that the Colburn factor (air side) was lower for plain fin heat exchanger in comparison to perforated fins heat exchanger.

Kalantari et al. [17] carried out a parametric study to cover a wide range of design configurations, geometrical and operating parameters. They investigated Reynolds numbers of up to 12,000 and found that longer fins, fin pitch and smaller tube diameter result in higher heat transfer coefficients. A correlation for the conjugate heat transfer coefficient was developed that applies to gas–liquid finned tube heat exchangers. In the correlation, the Nusselt number was expressed in terms of the Prandtl number and non-dimensional geometrical parameters.

Altwieb et al. [18] assessed the thermal performance of a multi-tube heat exchanger with plain fins and having different geometrical modification using three-dimensional CFD simulations. Three enhancements were analysed. These include fin spacing and longitudinal as well as transverse pitch. This was done to determine their influence on the Colburn and Fanning factors. Validation experiments were carried out and compared with the CFD; and were in turn utilised to calculate the Fanning and Colburn factors; and the local fin efficiency for each of the geometrical modifications. Two empirical correlations were developed for the Fanning and Colburn friction factors and the authors demonstrated predictions within 10% the experimental data. Similarly, Altwieb & Mishra [19] reported an experimental and numerical study on the thermal response of multi tube and fin heat exchanger with plain, louvred and semi-dimple vortex generator. The heat transfer and pressure drop characteristics were extensively investigated in this work. Two new design equations were developed for the heat transfer rate and the pressure drop behaviour.

The scope of the work summarised above is quite limited since most investigations are focussed only on the simple arrangement of perforations in the plain fins and only limited information is available on complex perforations in plain fins. Also limited information is available on louvered fins. Perforations are used to provide passive heat transfer enhancements in the heat exchanger. The effect of fin perforation on local and global performance indicators is a key and ongoing area of research that requires deeper understanding. Furthermore, the majority of design equations proposed have very limited range and do not include varied geometric parameters such as fin pitch, spacing and the presence of perforations.

The aim of this paper is to experimentally investigate the steady state heat transfer and thermal performance of a wide range of fin configurations (plain, louvred and perforated fin) heat exchanger. Using the experimental data, new mechanistic prediction models for the Colburn factor (j) and the Fanning friction factor (f) as a function of Reynolds number and heat exchanger geometry were developed and their prediction error margins analysed. It is envisaged that these equations will contribute to improved design and operation of such heat exchanger configurations.

2. Materials and Methods

An experimental setup was designed and developed to study the thermal behaviour of a multi-tube multi-fin heat exchanger. Details of the setup, equipment, instrumentation and uncertainties are given in the following sections.

2.1. Experimental Rig

The experimental setup is composed of the following parts: a 5-litre water tank (wrapped with a reflector foil to minimise heat loss), a heater, circulation pump integrated with a 0.9 kW heater unit, flow meter (Flowmax 44i with measuring range of 0.3–21 L/min or 0.018–1.26 m^3/h), the heat exchanger testing unit, pressure transducers (IMP, with 0–10 V analogue output signal and range 0–4 bars), T-type thermocouples (with accuracy of ±0.15 °C), thermocouples data logger, RTD sensors (PT100, with range of −75–250 °C) and data acquisition system. Figure 1a–f includes various details including a schematic representation of the experimental setup and photos of the various test rig components. These have been use d for numerous previous experimental campaigns [20–23] with a high degree of consistency and reliability.

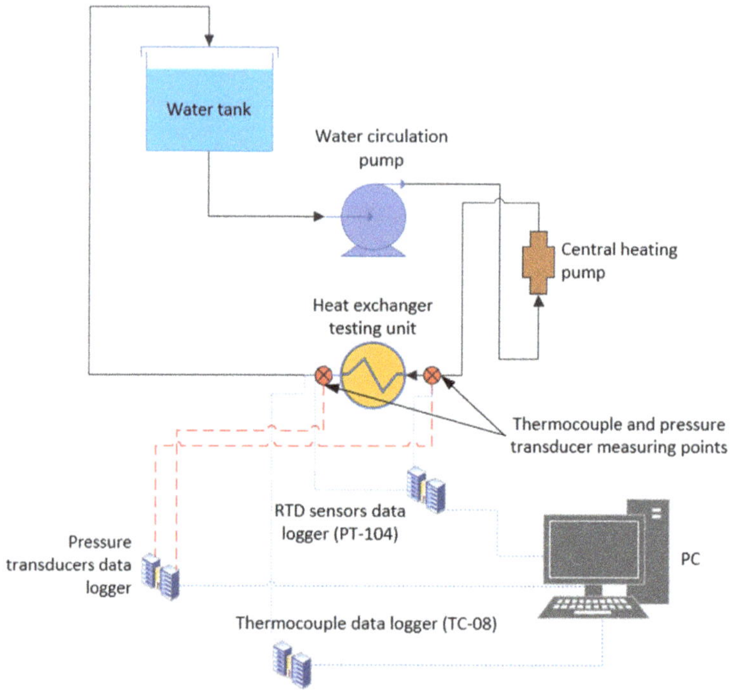

(a)

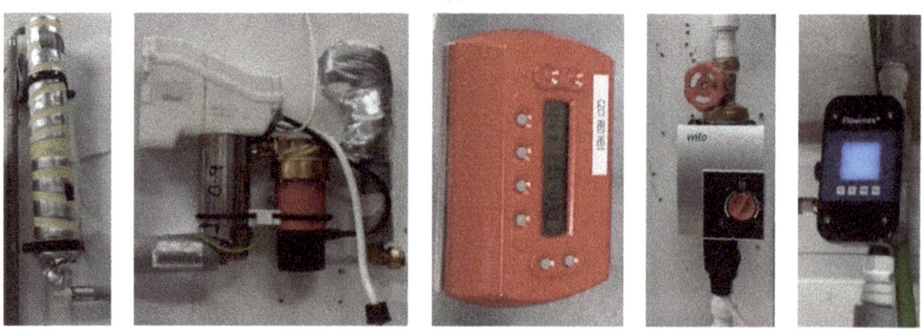

(b) (c) (d) (e) (f)

Figure 1. Schematic of the experimental setup: (**a**) overall schematic of a heater flow loop, (**b**) insulated 5 litre water tank, (**c**) water heater, (**d**) heater controller unit, (**e**) water pump, (**f**) water flow meter.

2.1.1. Heat Exchanger Testing Unit

Figure 2a illustrates a schematic diagram of the heat exchanger testing unit. The testing unit was made from a 2-mm thick galvanised steel sheet. The dimensions of the unit include a length of 0.650 m; and corresponding width of 0.165 m and a height of 0.175 m. Airflow is supplied to the testing unit using a single-sided centrifugal fan which has an incorporated electric motor. The fan has a power rating of 119 W and is able to deliver a maximum of 610 m^3/h airflow rate. The speed of the fan's electric motor was controlled using a potentiometer.

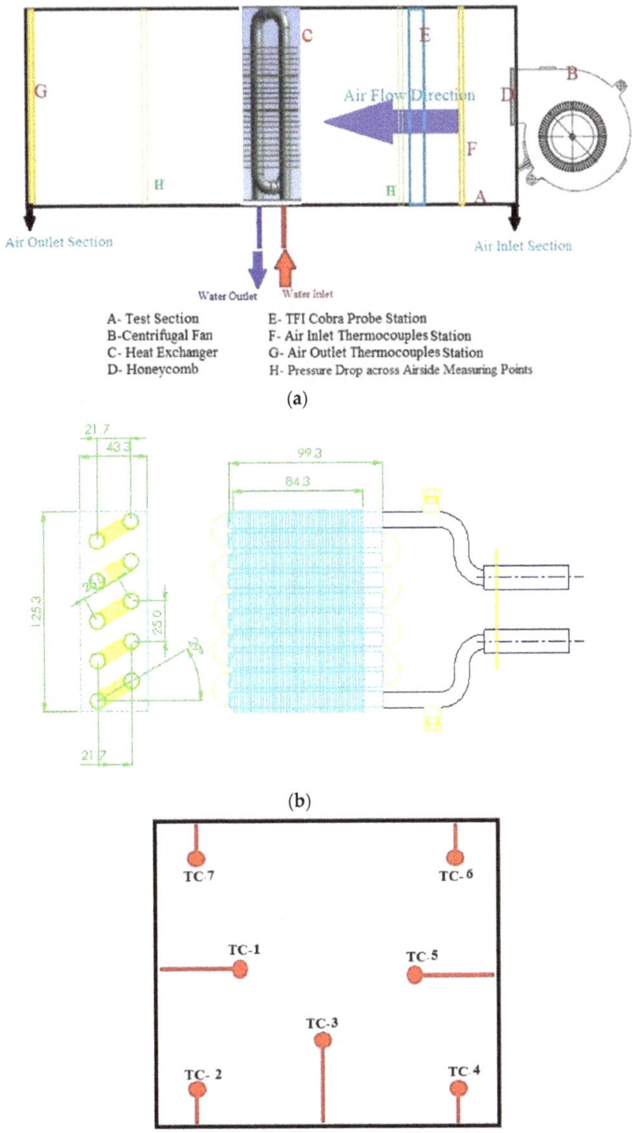

Figure 2. (**a**) Schematic representation of the heat exchanger testing unit (**b**) dimensions of the plain fin heat exchanger (**c**) Arrangement of thermocouples at the specific measurement location.

The incoming flow was conditioned by using a honeycomb structure and various velocity components associated with incoming flow were measure by TFI cobra probe [24] at the entry to the test section. A typical cobra probe has four pressure sensing ports, and these pressure values are used to find the three flow velocity components. At the inlet the flow velocity was measured at 25 points to obtain the average velocity using the ASHRAE standard 41.2 [15,25].

The upstream and downstream air temperatures within the test section were measured at two specific locations. At each of the specific locations seven T-type thermocouples were used. The thermocouples have exposed welded copper/constantan tips to minimise it thermal inertia [15]. There are benefits of using multiple thermocouples at each location. Due to this the accuracy is improved since more samples are available for averaging and automatic averaging can be carried out simultaneously. The arrangement of these thermocouples at each specific location is shown in Figure 2c. During the experimentation, the thermocouples repeatability was ensured by taking multiple measurements and individual thermocouples were calibrated using a laboratory grade thermometer. All temperatures measured were recorded using a data acquisition system [Pico Technology (PicoTech)]. The data from the thermocouples were logged were then averaged.

The inlet and outlet water temperatures in the tubes were measured by PicoTech temperature probes (RTD-PT100). The accuracy of these probes is ±0.03 °C during the testing and the repeatability in the measurements was ensured by having multiple measurements. Also, the probes were calibrated using a standard thermometer. The water flow rate was measured by using the Flowmax 44i water flowmeter which is an ultrasonic-based volumetric flow meter with a ±2% maximum error of measurement and its repeatability is within ±0.5%.

The heat exchanger's airside pressure drop was measured using a DPM TT550 micromanometer. It has the ability of measuring the static pressure within the range: 0.4–5000 Pa. The heat exchanger's water side pressure drop was measured using two pressure transducers. They were respectively placed at the water inlet and outlet tubes and were in turn connected to a PC via a USB-1616HS Series Data Acquisition interface. The voltage readings were then recorded. Using the calibration equations these voltage values were converted to corresponding pressure values.

2.1.2. Fin Geometries

Three main fin geometries were used to carry out this study. They are:

a. Plain fin
b. Perforated plain fin
c. Louvred fin

The plain fin heat exchanger used in the present investigation is a multi-tube multi-fin type. There are two tube rows provided, and each tube has a diameter of 9.52 mm. Each of the rows included five 0.26-mm thick copper tubes and the overall length of each tube is 0.130 m. The bend of each tube has a diameter of 16 mm. The heat exchanger has 21 staggered 0.12 mm aluminium plain fins which have a width of 43.3 mm and a height 125.3 mm. Along the heat exchanger, the fins are placed at a distance of 4.23 mm from each other. The dimensions of the heat exchangers are shown in Figure 2b.

The novel heat exchanger model (perforated plain fins) was manufactured by arranging twelve equal diameter holes (3 mm) in each plain fin material. This is shown in Figure 3a,b with the inclined distribution of the 3-mm perforated holes. Finally, the louvred fins heat exchanger used has the same dimensions as the plain and perforated plain fin models. It is shown in Figure 3c and is identical to the louvred fins used in Wang et al. [15].

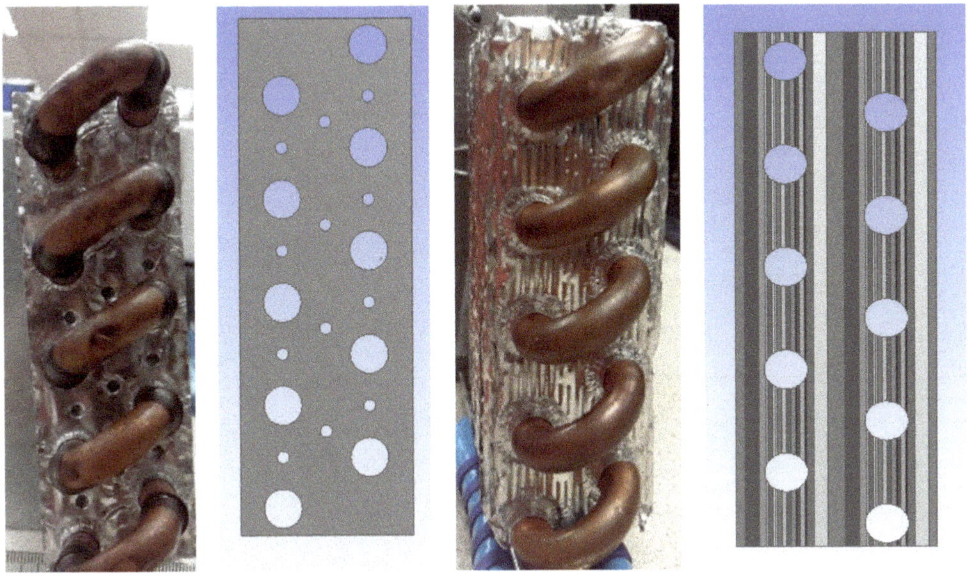

(a) (b) (c) (d)

Figure 3. (**a**) Picture of perforated plain fin type heat exchanger, (**b**) perforated holes' arrangement, (**c**) louvred fins heat exchanger, (**d**) louvred fin shape.

2.1.3. Experimental Procedure

Steady-state tests provide thermal performances of the heat exchanger that are time independent. The test process involved drawing an airflow within the heat exchanger on the fin-side and allowing hot water to circulate through the tubes within the heat exchanger. Both the average air velocity (from 0.7 m/s to 4 m/s) and water flow rates (0.12 to 0.36 m^3/h) were varied. The above indicates that the flow within the tubes is fully turbulent. The test matrix for the experiments carried out in this study are presented in Table 1. It shows that a total of 25 tests were carried out for up to 0.36 m^3/h water flow rate with 0.7–4 m/s air velocities for each water flow rate.

Table 1. Test matrix for the comparative steady-state tests conducted.

Test ID	Water Side		Air Side	
	Water Flow Rate (m^3/h)	**Water Inlet Temperature (°C)**	**Air Velocity (m/s)**	**Air Inlet Temperature (°C)**
Test case 1.1			0.705	
Test case 1.2			1.546	
Test case 1.3	0.12 ± 0.0018	60 ± 1	2.183	24 ± 1
Test case 1.4			3.177	
Test case 1.5			3.991	
Test case 2.1			0.705	
Test case 2.2			1.546	
Test case 2.3	0.18 ± 0.0018	60 ± 1	2.183	24 ± 1
Test case 2.4			3.177	
Test case 2.5			3.991	

Table 1. *Cont.*

Test ID	Water Side		Air Side	
	Water Flow Rate (m^3/h)	Water Inlet Temperature (°C)	Air Velocity (m/s)	Air Inlet Temperature (°C)
Test case 3.1			0.705	
Test case 3.2			1.546	
Test case 3.3	0.24 ± 0.0018	60 ± 1	2.183	24 ± 1
Test case 3.4			3.177	
Test case 3.5			3.991	
Test case 4.1			0.705	
Test case 4.2			1.546	
Test case 4.3	0.3 ± 0.0018	60 ± 1	2.183	24 ± 1
Test case 4.4			3.177	
Test case 4.5			3.991	
Test case 5.1			0.705	
Test case 5.2			1.546	
Test case 5.3	0.36 ± 0.0018	60 ± 1	2.183	24 ± 1
Test case 5.4			3.177	
Test case 5.5			3.991	

Each test point was run for about 2100 s (35 min) for the system to first attain steady conditions and overcome the initial high thermal inertia (as shown in Figure 4) before any readings were taken. The tests were further repeated twice to obtain representative measurements. An acceptable level of repeatability was obtained as measurements showed less than ±3% of deviation between each test condition.

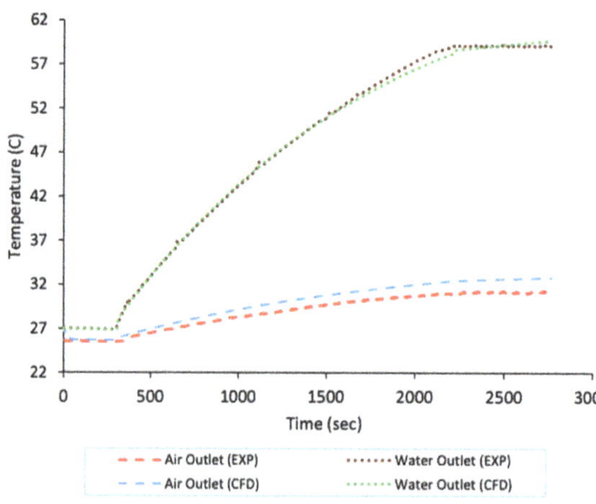

Figure 4. Starting up test showing experimental and transient CFD temperatures for air and water outlets.

2.2. Data Analysis

To compute heat transfer rates, first the temperatures of the flowing streams (air and hot water) were measured at both the entry and exit locations. Also, the respective pressure drops were also measured. The heat transfer rates then can be computed using the following equation both for water and air, respectively.

$$\dot{Q}_{hot=m_w}Cp_w(T_{wi} - T_{wo}) \quad (2)$$

$$\dot{Q}_{cold=m_a}Cp_a(T_{ao} - T_{ai}) \quad (3)$$

where the subscripts a and w indicate air and water; i and o indicate inlet and outlet, respectively. The average value of the heat transfer rate ($\dot{Q}_{avg}$) can be computed as follows:

$$\dot{Q}_{avg} = \frac{\dot{Q}_{hot} + \dot{Q}_{cold}}{2} \tag{4}$$

Furthermore, in order to carry out an assessment of heat transfer and pressure drop characteristics, the Colburn factor (j) and Fanning friction factor (f) were calculated and used for this purpose. The f-factor symbolises the pressure drop characteristics while the j-factor symbolises the heat transfer process and the j/f ratio is termed as the efficiency index. The Colburn factor j and the friction factor f are respectively calculated using:

$$j = St\mathrm{Pr}^{2/3} \tag{5}$$

$$f = \frac{A_c}{A_o}\frac{\rho_m}{\rho_1}\left[\frac{2\rho_1 \Delta P}{G_c^2} - \left(K_c + 1 - \sigma^2\right) - 2\left(\frac{\rho_1}{\rho_2} - 1\right) + \left(1 - \sigma^2 - K_e\right)\frac{\rho_1}{\rho_2}\right] \tag{6}$$

where A_c is the minimum free flow area of the air side; A_o is the total surface area of the air side; the variables K_c and K_e are the entrance and exit pressure loss coefficients. Equation (6) was developed by Kays and London [21] using the data from Figures 14–26 in McQuiston et al. [2]. Additionally, the Stanton and the Prandtl numbers used to define the Colburn j-factor in Equation (5) are, respectively, given as:

$$St = \frac{h_o}{\rho_a V_{a(max)} c_{pa}} \tag{7}$$

$$Pr = \frac{\mu c_{pa}}{\lambda} \tag{8}$$

where h_o is the heat transfer coefficient calculated using the total surface area of the air side; ρ_a is the density of air, $V_{a(max)}$ is the maximum air velocity; c_{pa} is the specific heat capacity of air; μ is the air dynamic viscosity and λ is its thermal conductivity. Since the f- and j-factors are most commonly used by researchers to assess the performance of heat exchanger fin strips, they will be used here for assessing the performance of the three geometries used in this study.

3. Results

In this section, the trend of the measured pressure drops, and heat transfer rates are studied and discussed in detail. These values were used to calculate the j- and f-factors, and the efficiency index for characterising the performance of the three fin and tube heat exchanger models.

3.1. Performance Comparison of Fin Geometry

The plots in Figure 5 show the trend of the mean heat transfer rate ($\dot{Q}_{avg}$) variation with respect to the average air velocity corresponding to the three heat exchangers (i.e., for perforated plain fins, ordinary plain fins and louvred fins) over a range of water flow rates namely 0.12, 0.18, 0.24, 0.3 and 0.36 m^3/h. The error bars represent the combined uncertainty of the thermocouples and deviation between repeated measurements. The uncertainties were determined to be ±5%.

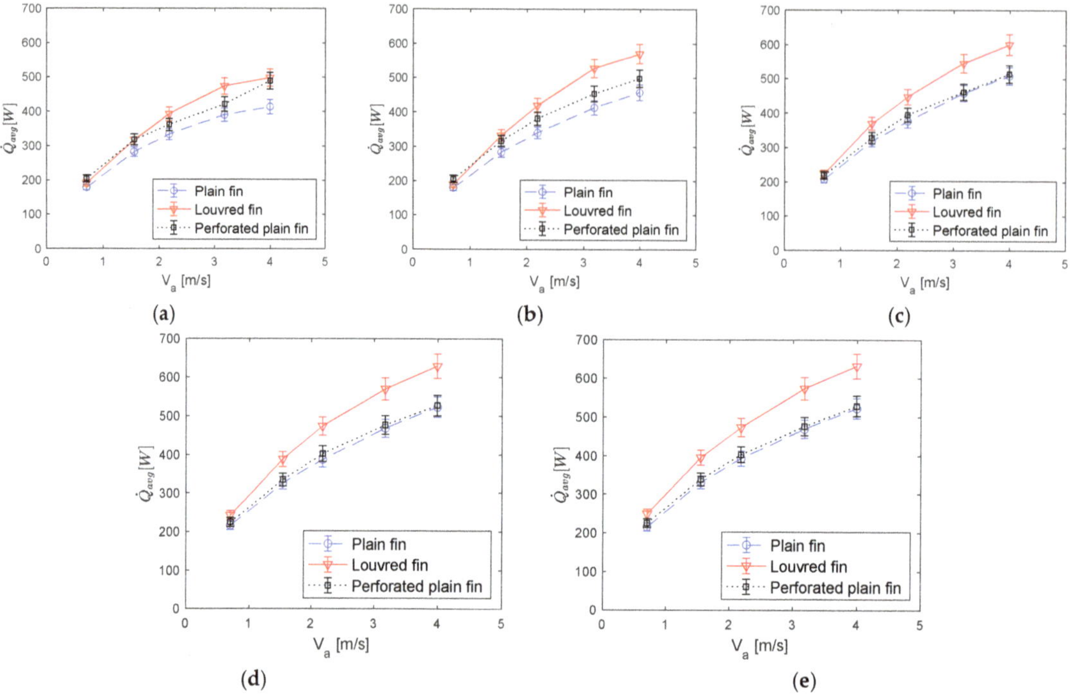

Figure 5. Variation of ($\dot{Q}_{avg}$ [W]) against V_a for various heat exchangers; (**a**) for water flow rate 0.12 m^3/h, (**b**) for water flow rate 0.18 m^3/h, (**c**) for water flow rate 0.24 m^3/h, (**d**) for water flow rate 0.3 m^3/h and (**e**) for water flow rate 0.36 m^3/h.

Figure 5a shows the variation of average heat transfer rate against the average air velocity for water flow rate of 0.12 m^3/h for different heat exchanger geometries. It can be seen from the figure that the louvred fins heat exchanger exhibited a higher mean heat transfer rate when compared with the perforated plain, and plain fin heat exchangers. For all cases, the average heat transfer rate increases as the water flow rate increases, first at a faster rate at lower air velocities and the rate of increase starts to decrease, more noticeably beyond 3 m/s. Figure 5b shows a similar trend but the rate of heat transfer increase beyond 3 m/s is now higher than the previous case and this is maintained for water flow rates of 0.24, 0.3 and 0.36 m^3/h as shown in Figure 5c–e, respectively. The louvred fin geometry produced correspondingly higher increase in the average heat transfer rates especially at 0.3 and 0.36 m^3/h water flow rates than at the lower water flow rates.

Figure 6a shows the variation of pressure gradient on air side with average air velocity at a given water flow rate. It can be seen that for louvered fins heat exchanger pressured drop increases almost linearly with air velocity. The louvred fin geometry exhibits much higher pressure drops than the other geometries. It is seen from the figure that the louvered fin produces up to 380% more pressure drop than the plain fin geometry at V_a = 4 m/s while the perforated plain fin geometry produced a maximum of 65% difference when compared with the plain fin geometry. The large pressure drop in the louvred fin heat exchanger is due to the uneven surface of the louvred fins which result in higher pressure losses. As expected, the pressure drop behaviour remains constant with changing water flow rates giving only marginal differences (which are within experimental uncertainties) as seen in Figure 6b–e.

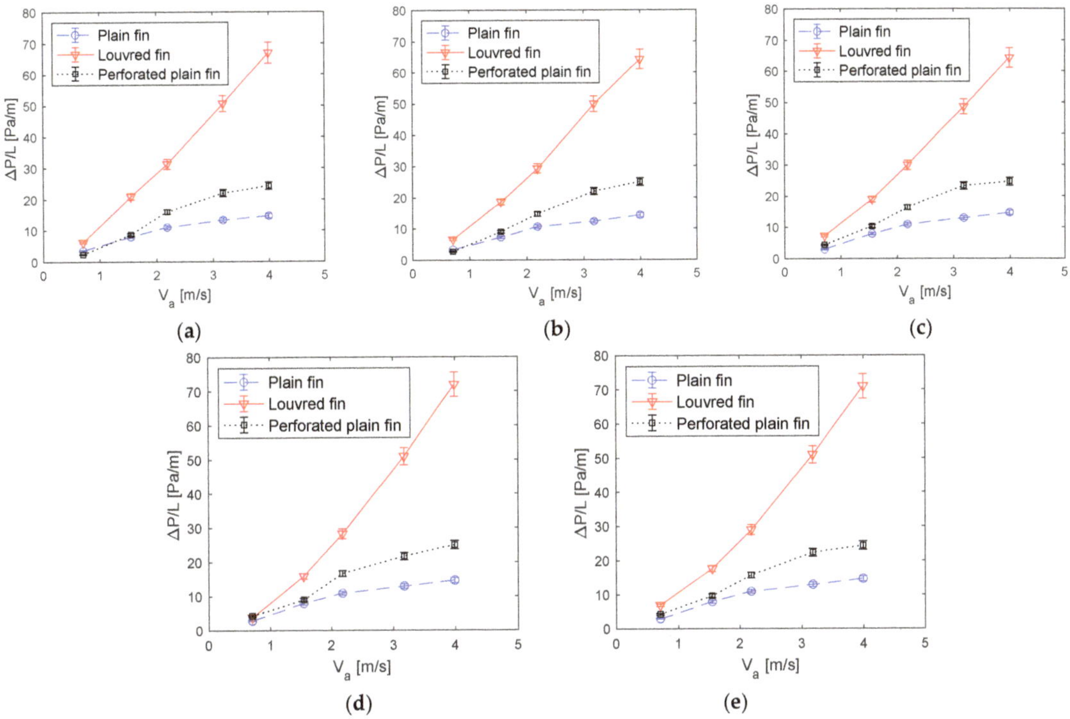

Figure 6. Variation of ($\Delta P/L$ [Pa/m]) against V_a for various heat exchanger; (**a**) at water flow rate of 0.12 m^3/h, (**b**) at water flow rate of 0.18 m^3/h, (**c**) at water flow rate of 0.24 m^3/h, (**d**) at water flow rate of 0.3 m^3/h and (**e**) at water flow rate of 0.36 m^3/h.

Looking at the heat transfer characteristics and corresponding pressure drops associated with different geometries (Figures 6 and 7), it can be seen that quantitatively, at the lowest water flow rates of 0.12 m^3/h, an increase in heat transfer rate was noticed for the louvred fins as compared to plain fins (about 13% lower) and plain perforated fins (2.7% lower). These values change to about 15% for plan fins and 6% for the perforated plain fin model at a water flow rate of 0.18 m^3/h. At flow rate of 0.24 m^3/h again the corresponding values were 13% and 11%. However, such significant increase of the heat transfer was made at the expense of a significant air pressure drop (Figure 6). It is assumed that the perforated fins increased the flow vorticity. At 0.3 m^3/h almost similar levels of change in heat transfer rate was noticed. In general, there is an increasing trend in heat transfer as the water flow rate increases for all types of heat exchangers. This is evidenced by an increase in the rate of heat transfer from about 500 to 600 W for louvered fins heat exchanger, 400 W to 523 W for plain fin heat exchanger, and 488 to 528 W for perforated plain fin heat exchanger at V_a = 4 m/s. Nevertheless, the perforated plain fin model appears to be least affected by an increase in water flow rate. Its heat transfer rate has only marginally increased at higher water flow rates. This may be because the perforations have reduced the heat transfer area below a certain threshold where the fluid flow rates can have a significant effect. As a result, in certain situations this aspect should be kept in view whilst designing a perforated plain heat exchanger.

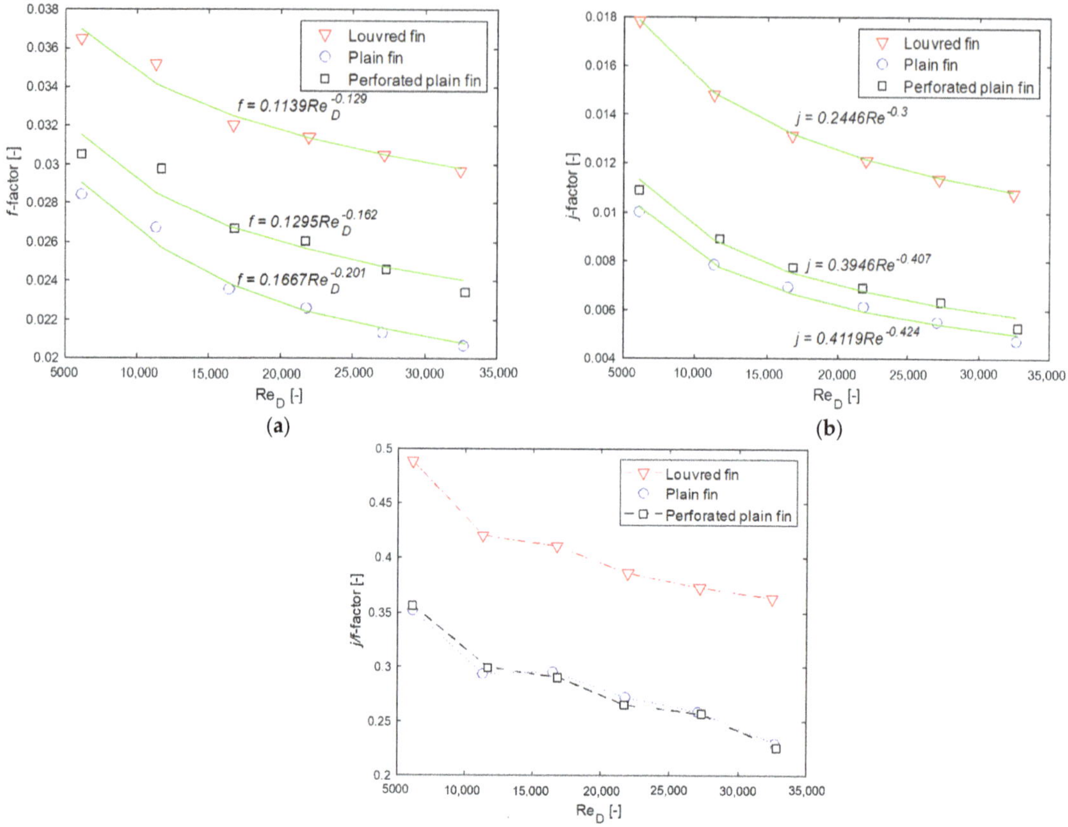

Figure 7. Variations of (**a**) friction f-factor (**b**) Colburn j-factor and (**c**) efficiency index j/f for different fin arrangements as a function of Reynolds number.

Figure 7a shows the variation of friction factor (f) for the heat exchangers as a function of Reynolds number. It shows that the friction factor decreases with increasing Reynolds number and is consistent with previous observations including those in the Moody chart for pipe flows. For each of heat exchanger, there is a steep decreasing slope in the curve between $Re = 11{,}000$ and 17,000 before slope of the curve becomes flatter. As is seen in the plots, power law curves were fitted to the three sets of data (louvred, plain and perforated fins). This was done to develop a prediction model for f as a function of Re. It can be seen that all the heat exchangers show slightly different slopes (power of Reynolds number are −0.129, −0.162 and −0.201). For the j-factor the corresponding values of indices are −0.300, −0.407 and −0.424 for the louvred, plain and perforated plain fin geometries, respectively.

Figure 7b shows the variation of the Colburn j-factor of the three heat exchangers as a function of Reynolds number. It shows that the Colburn j-factor decreases with increasing Reynolds number indicating a higher heat transfer rate at lower Reynolds numbers. The louvred fin heat exchanger gave the highest j values within the range 0.011–0.02 compared to 0.005–0.011 and 0.004–0.010 for the perforated and plain fin heat exchangers respectively. The fitted power law curves to the three sets of data give prediction models for j as a function of Re. Similar to the f-factor, for the j-factor, there is a slight decrease in slope from plain to perforated fin to louvred with the coefficients and indices of the equations being 0.2446, 0.3946, 0.4119 and −0.3, −0.407 and −0.424, respectively. Figure 7c depicts efficiency index (j/f) variations corresponding to three different heat exchangers as a function of Reynolds number with an inverse relationship existing between the efficiency index and *Re*. The plain and perforated models exhibited similar behaviour in j/f magnitudes of 0.022–0.036 across the entire *Re* range studied. However, the louvred fin gave much larger values of the efficiency index of between 0.37 and 0.49. While the louvred fin exhibits a far more superior efficiency than the other two geometries, the plain and perforated plain fin models gave near identical behaviour throughout the experimental range of *Re* investigated–with both geometries exhibiting a similar efficiency index across the experimental range. The nature of the curves in Figure 7a–c clearly establish that the friction, Colburn factors and the efficiency index asymptotically decrease with the Reynolds number for all the three different heat exchanger models. It can be seen that the Colburn and friction factor values are significantly higher for the louvred fins heat exchanger as compared to the plain and perforated plain fin geometries. This observation can be explained based on the higher surface area available for the louvred fin heat exchanger as compared to the plain and perforated plain fin heat exchangers. Due to this the heat transfer coefficient for louvered heat exchanger is higher which in turn leads to high Colburn j-factor values.

Based on the heat transfer measurements it can be shown that there is an improvement in the average heat transfer rate $(\dot{Q}_{avg})$ of nearly 10% and 20% for the perforated plain fin and louvred fin heat exchangers, respectively, when compared to the plain fin geometry. However, this improvement was accompanied by large increases in the corresponding pressure drop on the airside respectively as earlier highlighted. The data collected during the current experiments and after processing were used for the development of the two new design correlations for predicting the Fanning f- and Colburn j-factors as functions of the flow velocity and area available. To represent these in non-dimensional form, Reynolds number and an area ratio term (the ratio of the heat exchanger's fin surface area divided by the total surface area) have been used in the correlations.

3.2. Development of New Empirical Relations for Fanning *f* and Colburn *j*-Factor

For the complex geometries used in the present investigation, the experimental data obtained in the present investigation were used for developing a set of novel design equations for the prediction of Fanning *f*- and the Colburn *j*-factors. As previously stated, the *f*- and *j*-factors quantify the pressure drop and heat transfer characteristics of the heat exchanger units. Therefore, it is imperative to develop correlations that relate them with the flow, fluid, and geometrical parameters. The correlation was developed using multivariate regression analysis using the curve fitting tools in Microsoft Excel's Solver® which are based on the least squares' method. The dimensionless parameters used to develop the predictive correlation are, as mentioned earlier, the Reynolds number, Re_D and the ratio between total fin surface area to the total heat transfer surface area $\left(A_f/A_t\right)$ of the heat exchanger which is a design parameter used by Palmer et al. [26] that can be used for estimating heat transfer areas needed for a given amount of heat to be transmitted. Other authors have used similar dimensionless groupings to correlate the heat transfer properties of fin and tube heat exchangers [13,17,18]. The newly derived equations are as follows:

$$j = 10^{4.595}\left(\frac{A_f}{A_t}\right)^{29.918} Re_D{}^{-0.374} \tag{9}$$

$$f = 10^{1.203}\left(\frac{A_f}{A_t}\right)^{12.811} Re_D{}^{-0.139} \tag{10}$$

where Re_D is the Reynolds number calculated using the hydraulic diameter of the heat exchanger face's cross-section; D_c is the outside diameter of the fin's collar (m): A_f is the total surface area of the fins (m^2); A_t is the heat exchanger's overall heat transfer surface area (m^2). The Equations (9) and (10) show that the Colburn and Fanning factors are inversely proportional to the Reynolds numbers which are consistent with experimental observations (in Figure 7a,b). Additionally, the relatively large indices (29.218 and 12.811) for the $\left(A_f/A_t\right)$ parameter reflects the small magnitude of the $\left(A_f/A_t\right)$ ratio. It is advised that the equations are valid within the Reynolds number range: $6 \times 10^3 \leq Re_D \leq 30 \times 10^3$ and for the heating cycle in forced convection heat transfer.

In order to estimate the accuracy of predictions from the developed equations Figure 8a,b have been prepared to depict the relationship between the experimentally evaluated values and the predicted values of Colburn factor (j) and Fanning friction factor (f), respectively. From the Figure 7a,b, it can be concluded that there is a good match between the experimental and the computed values and almost 100% of the data lie within ±15% error band. Furthermore, the correlation coefficients which compare the goodness of fit between the experimentally evaluated and predicted data for Equations (9) and (10) are 0.853 and 0.811, respectively. As such, the newly developed correlations can be used with confidence.

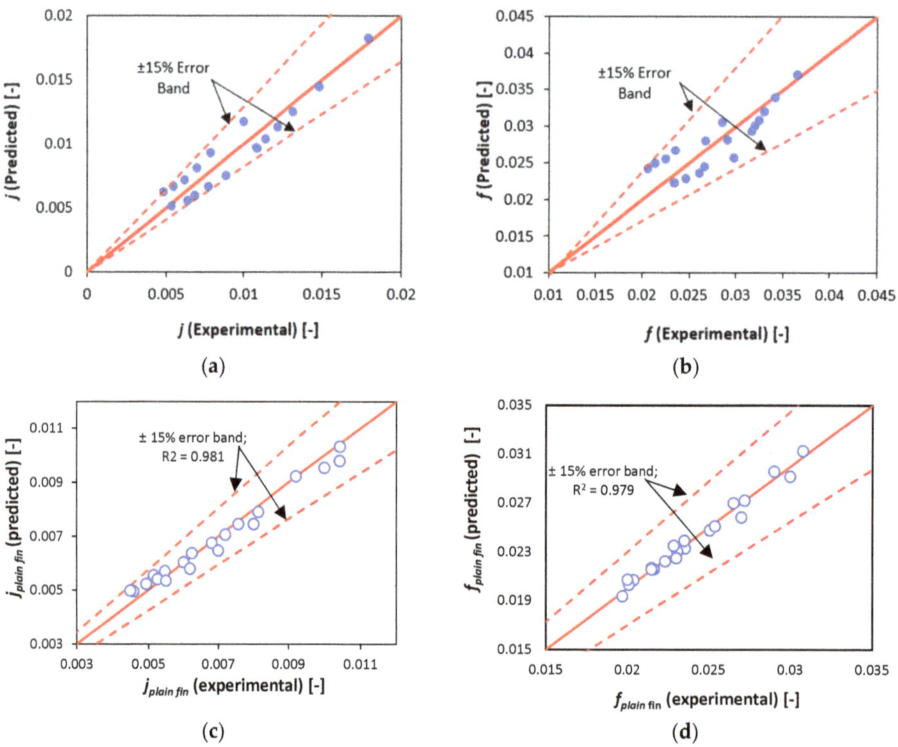

Figure 8. Comparisons of predicted (by Equations (9) and (10)) and experimental values of (**a**) Colburn j-factor and (**b**) Fanning f-factor (**c**) plain fin Colburn factor (**d**) plain fin Fanning factor.

In order to account for geometric variations of the fins (their spacing F_p, longitudinal pitch L_p and transverse pitch T_p), validated CFD data (previously reported in ref. [18]) was combined with the experimental data obtained in this study to generate new empirical correlations for j and f for the plain fin model using nonlinear least squares regression. The longitudinal pitches considered were 20, 22 and 24 mm; transverse pitches were 23.5, 25 and 26.5 mm; while the fin spacings investigated were 3.7, 4.2 and 4.7 mm at air Reynolds numbers of 6000 to 32,000. There were a combined total of 21 points (15 CFD and 6 experimental). Figure 7c,d show that the correlations have very high goodness of fit R^2 values of 0.931 and 0.979 for the j and f factors, respectively, and the predictions of the new correlations are well within ±15% of the underlying CFD and experimental data. The two new correlations are given below as follows:

$$j = 0.173\ Re_D^{-0.388}\left(\frac{F_p}{D_c}\right)^{-0.199}\left(\frac{L_p}{F_w}\right)^{-0.297}\left(\frac{T_p}{F_H}\right)^{-0.089} \quad (11)$$

$$f = 0.084\ Re_D^{-0.213}\left(\frac{F_p}{D_c}\right)^{-0.334}\left(\frac{L_p}{F_w}\right)^{-0.151}\left(\frac{T_p}{F_H}\right)^{-0.262} \quad (12)$$

To summarise, it can be concluded that the developed equations are very much capable of predicting the Fanning *f*- and Colburn *j*-factors of these heat exchangers having the stated fin geometries with sufficient accuracy. Consequently, the equations can be used during the design and evaluation of existing multi-tube multi-fin heat exchanger with plain, perforated or louvred fins.

4. Conclusions

This study has presented novel geometric configurations for multi-tube multi-fin heat exchanger. The configurations were designed in order to conduct a robust experimental investigation with three heat exchanger geometries namely plain, perforated plain and louvred fin heat exchangers. Some important observations were made during the experiments and analysis of the pressure drop and heat transfer data. It was found that for all inlet air and water flow rates and hence velocities, the louvred fins produced the highest heat transfer rate. This was attributed to increased surface area available for heat transfer. Conversely, it also produced the highest pressure losses when compared to the other two designs. Also, while the new perforated design produced a slightly higher pressure drop than the plain fin design, due to the vortices generated by the perforations, an enhancement in its heat transfer characteristics was observed when comparing with the plain and louvred fin models. This enhancement is relatively high at a small water flow rate. The experimental results were subsequently used to generate a set of novel empirical equations for design optimisation which can be used to predict the heat transfer and pressure drop characteristics of the heat exchangers represented by the Colburn and Fanning factors. The empirical equations were developed as functions of the heat exchangers' geometrical parameters, and we have shown that the performance of the equations are well within acceptable ±15% error margins in relation to the experimental data.

Author Contributions: M.A.: Experimentation, Data curation, formal analysis, Investigation, methodology, validation, visualization, writing—original draft; R.M.: Conceptualisation, supervision, writing—review and editing; A.M.A.: Writing—review & editing, formal analysis; K.J.K.: supervision, formal analysis, writing—review and editing. All authors have read and agreed to the published version of the manuscript.

Funding: This research did not receive any external funding.

Data Availability Statement: Data is available upon reasonable request.

Conflicts of Interest: The authors declare no conflict of interest.

References

1. Shah, R.K.; Sekulic, D.P. *Fundamentals of Heat Exchanger Design*; John Wiley & Sons: Hoboken, NJ, USA, 2003.
2. Mcquiston, F.C.; Parker, J.D.; Spilter, J.K. *Heating, Ventilating, and Air Conditioning: Analysis and Design*; Wiley: Hoboken, NJ, USA, 2004.
3. Naphon, P.; Wongwises, S. A review of flow and heat transfer characteristics in curved tubes. *Renew. Sustain. Energy Rev.* **2006**, *10*, 463–490. [CrossRef]
4. Thulukkanam, K. *Heat Exchanger Design Handbook*, 2nd ed.; Taylor & Francis: Abingdon, UK, 2013.
5. Wilson, E.E. A basis for rational design of heat transfer apparatus. *J. Am. Soc. Mech. Eng.* **1915**, *37*, 546–551.
6. Sieder, E.N.; Corporation, G.E.T.A.T.E.F.W.; York, N. *Heat Transfer and Pressure Drop of Liquids in Tubes O Oil A, Heating Oil B, Heating Oil A, Cooling Oil C, Cooling*; American Chemical Society: New York, NY, USA, 2021.
7. Colburn, A.P. A method of correlating forced convection heat-transfer data and a comparison with fluid friction. *Int. J. Heat Mass Transf.* **1964**, *7*, 1359–1384. [CrossRef]
8. Dittus, F.W.; Boelter, L.M.K. Heat transfer in automobile radiators of the tubular type. *Int. Commun. Heat Mass Transf.* **1985**, *12*, 3–22. [CrossRef]

9. Wang, C.-C.; Chang, Y.-J.; Hsieh, Y.-C.; Lin, Y.-T. Sensible heat and friction characteristics of plate fin-and-tube heat exchangers having plane fins. *Int. J. Refrig.* **1996**, *19*, 223–230. [CrossRef]
10. Abu Madi, M.; Johns, R.; Heikal, M. Performance characteristics correlation for round tube and plate finned heat exchangers. *Int. J. Refrig.* **1998**, *21*, 507–517. [CrossRef]
11. Webb, R.; Kim, N.-H. Advances in Air-Cooled Heat Exchanger Technology. *J. Enhanc. Heat Transf.* **2007**, *14*, 1–26. [CrossRef]
12. Wang, C.-C. A survey of recent patents of fin-and-tube heat exchangers from 2001 to 2009. *Int. J. Air-Cond. Refrig.* **2010**, *18*, 1–13. [CrossRef]
13. Wang, C.-C.; Lee, W.-S.; Sheu, W.-J. A comparative study of compact enhanced fin-and-tube heat exchangers. *Int. J. Heat Mass Transf.* **2001**, *44*, 3565–3573. [CrossRef]
14. Fernández-Seara, J.; Uhía, F.J.; Sieres, J.; Campo, A. A general review of the Wilson plot method and its modifications to determine convection coefficients in heat exchange devices. *Appl. Therm. Eng.* **2007**, *27*, 2745–2757. [CrossRef]
15. Wang, C.-C.; Chen, K.-Y.; Liaw, J.-S.; Tseng, C.-Y. An experimental study of the air-side performance of fin-and-tube heat exchangers having plain, louver, and semi-dimple vortex generator configuration. *Int. J. Heat Mass Transf.* **2015**, *80*, 281–287. [CrossRef]
16. Liu, X.; Yu, J.; Yan, G. A numerical study on the air-side heat transfer of perforated finned-tube heat exchangers with large fin pitches. *Int. J. Heat Mass Transf.* **2016**, *100*, 199–207. [CrossRef]
17. Kalantari, H.; Ghoreishi-Madiseh, S.A.; Kurnia, J.C.; Sasmito, A.P. An analytical correlation for conjugate heat transfer in fin and tube heat exchangers. *Int. J. Therm. Sci.* **2021**, *164*, 106915. [CrossRef]
18. Altwieb, M.; Kubiak, K.J.; Aliyu, A.M.; Mishra, R. A new three-dimensional CFD model for efficiency optimisation of fluid-to-air multi-fin heat exchanger. *Therm. Sci. Eng. Prog.* **2020**, *19*, 100658. [CrossRef]
19. Altwieb, M.O.; Mishra, R. Experimental and Numerical Investigations on the Response of a Multi Tubes and Fins Heat Exchanger under Steady State Operating Conditions. In Proceedings of the 6th International and 43rd National Conference on Fluid Mechanics and Fluid Power, Allahabad, India, 15–17 December 2016; pp. 1–3.
20. Singh, D.; Aliyu, A.; Charlton, M.; Mishra, R.; Asim, T.; Oliveira, A. Local multiphase flow characteristics of a severe-service control valve. *J. Pet. Sci. Eng.* **2020**, *195*, 107557. [CrossRef]
21. Singh, D.; Charlton, M.; Asim, T.; Mishra, R.; Townsend, A.; Blunt, L. Quantification of additive manufacturing induced variations in the global and local performance characteristics of a complex multi-stage control valve trim. *J. Pet. Sci. Eng.* **2020**, *190*, 107053. [CrossRef]
22. Asim, T.; Mishra, R.; Oliveira, A.; Charlton, M. Effects of the geometrical features of flow paths on the flow capacity of a control valve trim. *J. Pet. Sci. Eng.* **2019**, *172*, 124–138. [CrossRef]
23. Asim, T.; Charlton, M.; Mishra, R. CFD based investigations for the design of severe service control valves used in energy systems *Energy Convers. Manag.* **2017**, *153*, 288–303. [CrossRef]
24. TFI. Cobra Probe. 2021. Available online: https://www.turbulentflow.com.au/Products/CobraProbe/CobraProbe.php (accessed on 6 December 2021).
25. ASHRAE. *HVAC Design Manual for Hospitals and Clinics*; ASHRAE: Atlanta, GA, USA, 2013.
26. Palmer, E.; Mishra, R.; Fieldhouse, J. An optimization study of a multiple-row pin-vented brake disc to promote brake cooling using computational fluid dynamics. *Proc. Inst. Mech. Eng. Part D J. Automob. Eng.* **2009**, *223*, 865–875. [CrossRef]

energies

MDPI

Article

Analysis of Stand-Alone Photovoltaic—Marine Current Hybrid System and the Influence on Daily and Seasonal Energy Storage

Jorge Olmedo-González [1,*], Guadalupe Ramos-Sánchez [2], Erika Paola Garduño-Ruiz [3] and Rosa de Guadalupe González-Huerta [1,*]

1 Electrochemistry Laboratory, Instituto Politécnico Nacional-ESIQIE, UPALM, Av. Instituto Politécnico Nacional S/N, Mexico City 07738, Mexico
2 Chemistry Department, Universidad Autónoma Metropolitana-Iztapalapa, Av. San Rafael Atlixco 186, Mexico City 09340, Mexico; gramossa@conacyt.mx
3 Instituto de Ingeniería, Universidad Nacional Autónoma de México, Ciudad Universitaria, Building 17, Circuito Exterior S/N, Coyoacán, Mexico City 04510, Mexico; paola.quimar@gmail.com
* Correspondence: jorgeolmedog@outlook.com (J.O.-G.); rgonzalezh@ipn.mx (R.d.G.G.-H.)

Abstract: Stand-alone systems in remote regions require the utilization of renewable resources; however, their natural intermittence requires the implementation of energy-storage systems that allow a continuous power supply. More than one renewable source is usually available at the same site. Thus, the choice of a hybrid system seems viable. It is relevant to study hybrid systems as they could reduce energy storage; however, sizing the hybrid system might have several implications, not only for the available daily energy, but also for the required daily energy storage and surplus seasonal energy. In this work, we present a case study of a stand-alone, conventional household powered by photovoltaic and marine-current-energy systems in Cozumel, Mexico. The analysis of different hybridization degrees serves as a guidance tool to decide whether hybrid systems are required for a specific situation; in contrast to previous approaches, where ideal consumption and generation profiles have been utilized, yearlong profiles were utilized here. The renewable potential data were obtained on site at an hourly resolution; requirements such as size of and cycles in the daily and seasonal energy storage were analyzed according to the degree of participation or hybridization of the proposed renewable systems through an algorithm that evaluates power generation and daily consumption throughout the year. A further analysis indicated that marine-current-energy implementation reduces the size of the daily energy-storage system by 79% in comparison to the use of only a photovoltaic system due to the similarity between the energy-demand profile and the marine-current-energy production profile. The results indicate that a greater participation of marine currents can help decrease daily storage while increasing seasonal storage by 16% compared to using only solar energy. On the other hand, hybridization enabled a reduction in the number of daily charge and discharge cycles at 0.2 hybridization degrees. It also allowed us to reduce the seasonal energy storage by 38% at 0.6 hybridization degrees with respect to only using energy from marine currents. Afterwards, energy-storage technologies were evaluated using the TOPSIS Multi-Criteria Decision Analysis to validate the best-suited technology for the energy-storage system. The evaluation considered the characteristics of the technology and the periods of energy storage. In this work, hybrid storage systems were mandatory since, for daily storage, lithium-ion batteries are better suited, while for seasonal storage, hydrogen-producing systems are more suitable to manage the amount of energy and the storage duration due to the high seasonal renewable-energy variations.

Keywords: renewable-energy hybrid system; marine-current system; solar PV system; energy-storage technology; multi-criteria decision analysis (MCDA); battery–hydrogen energy-storage system

Citation: Olmedo-González, J.; Ramos-Sánchez, G.; Garduño-Ruiz, E.P.; González-Huerta, R.d.G. Analysis of Stand-Alone Photovoltaic—Marine Current Hybrid System and the Influence on Daily and Seasonal Energy Storage. *Energies* **2022**, *15*, 468. https://doi.org/10.3390/en15020468

Academic Editors: Alon Kuperman, Alessandro Lampasi and Surender Reddy Salkuti

Received: 23 October 2021
Accepted: 2 January 2022
Published: 10 January 2022

1. Introduction

Sun, wind, and marine energies are defined as Variable Renewable Energies (VREs) due to their inherent intermittence, irregularity, and dispatchability, although VREs might have several small- to large-scale applications in stand-alone power systems where energy storage is essential. Marine Renewable Energy (MRE) is broadly available in different regions. Estimations indicate that ocean energy could contribute from 500 to 1000 MW of the installed capacity by 2030 [1,2]. Ocean currents or marine currents can produce energy from tidal movements and/or ocean circulation due to thermal and salinity gradients [3]. In Mexico, this type of energy is attractive due to its natural occurrence in certain regions such as the Gulf of California and the Cozumel Current in the Yucatan Peninsula where the potential for producing energy is as high as 100 W/m^2 [4]. However, even with the technology to harvest ocean energy, significant challenges for renewable energy include bringing the energy into the coast, storing it, and using it in a cost-effective manner.

Stand-alone renewable energy systems are off-grid systems that are able to provide electricity for regions lacking power grids in specific remote applications. Energy supply is usually provided by VREs where the energy-storage systems (ESSs) play a very important role in balancing and controlling the generation and consumption of electricity in deferred periods of time [5]. Nowadays, the cost of energy-storage systems is high; therefore, the selection of an appropriate ESS requires studies with multiple approaches. Determining the correct one among the many options, such as Pumped Hydro Storage (PHS), Compressed Air Energy Storage (CAES), Pb batteries, li-ion batteries, flow batteries, flywheels, supercapacitors, and hydrogen, should be carried out whilst taking into account criteria such as, storage capacity, response time, lifetime, cycle life, efficiency, cost, power, energy density, and power rating [6,7]. These criteria are also important as ESSs may play an important role in the environmental impact of the system [8]. The large variety of options and complex characteristics make it difficult to choose a specific ESS to take full advantage of their properties [9]. Therefore, the first step to deciding the most appropriate ESS technology is to consider its benefits, advantages, disadvantages, and maturity [10]. Moreover, stand-alone systems require a higher degree of energy availability to provide autonomy and comfort related to the daily electricity consumption. Hybrid renewable energy systems could be an effective way to integrate different renewable energy sources, for instance, sun and marine energy sources or any other viable combination. Hybrid systems can provide major electricity-generation availability, improving the reliability of the energy supply and the overall efficiency due to a lower dependency on ESSs [11]. Likewise, they could allow energy-storage-installed capacity to be reduced while having a profound effect on the durability of the ESS since the cycle life is affected by the daily availability of the renewable source [12]. The study of the potential of renewable energies is the first stage of the development of a hybrid system where the analyses in different periods of time, from daily to seasonal, can provide substantial information on the performance and energy storage needs in these periods [13].

Cozumel, Mexico, has been considered as a potential zone for stand-alone systems for households due to its ecotourism attractions. Although many studies of ESS implementation in the main VRE systems (solar and wind energy) can be found elsewhere, only few studies for MRE exist, especially in potential regions of Mexico or other regions with an abundance MRE resources; Figure 1 shows MRE projects where ESSs are considered.

The relevance of incorporating an ESS relies on its inherent environmental and economic benefits, especially in stand-alone systems where the surplus energy can be used for other purposes such as mobility. Stand-alone systems and microgrids are quite promising in Mexico due to approximately two million people having no access to electricity, especially in isolated regions and coastal zones [14]. As MRE and photovoltaic energy are broadly available in coastal touristic zones, this work could serve as a guide for the implementation of hybrid systems where the choice of primary source might have further implications. The different hybridization degrees not only imply continuous household energy supply but also the great benefits of seasonal energy storage, thus providing further economic advantages.

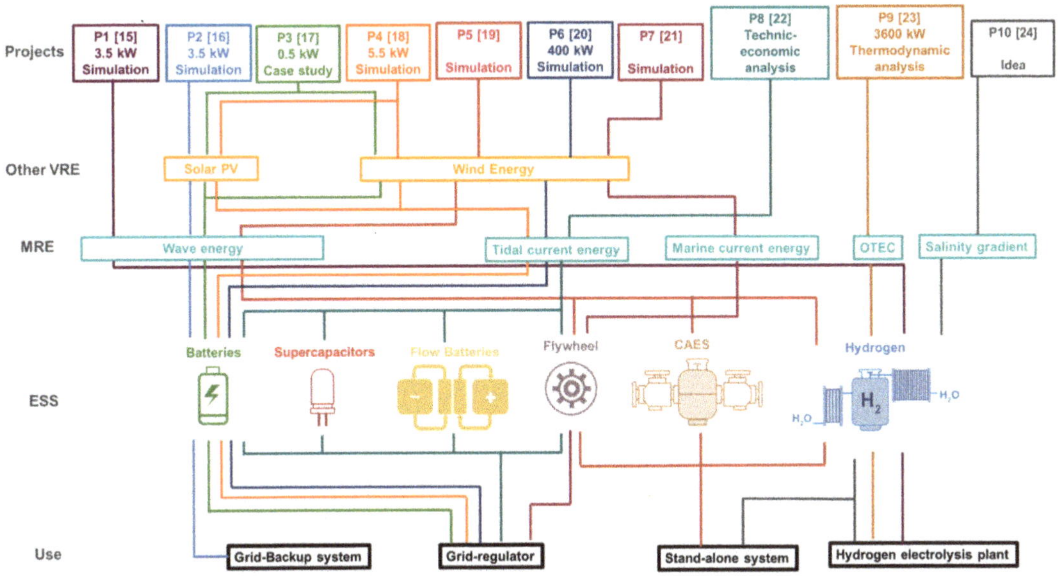

Figure 1. Different MRE projects with ESSs and their use [15–24].

2. Methods

This paper proposes a case study to evaluate the hybridization degree (HD) of a Solar–Marine Hybrid Energy System (SMHES) in the Cozumel region. The method comprises the following activities: (1) determination of daily and monthly energy storage assessment of the SMHES in the region; (2) comparison of home-load profiles to determine the average energy consumption; (3) identification of the appropriate ESS using Multi-Criteria Decision Analysis (MCDA) with a combination of AHP-TOPSIS methods.

2.1. Solar PV and Marine-Current-Energy Potential in Cozumel

Figures 2 and 3 show the potential profile to size the solar photovoltaic system (PVS) and the marine-current system (MCS), respectively. For this study, 1 h data resolution was considered for both VREs; solar data were obtained from the Photovoltaic Geographical Information System (PVGIS), and the data analysis was based on Yunez-Cano et al. 2016 [25]. A higher daily potential of solar energy can be obtained using solar-tracking systems as power and efficiency can increase mainly in large-scale solar energy applications [26]; in this work, as the information on this is insufficient, their effect is not included. Marine-current-energy data were provided by the CEMIE-Océano project Mexico [27]; for both systems, data from 2014 were considered. The Köppen–Geiger climate classification of the Cozumel is Aw, which presents a tropical savanna climate characterized by an extensive dry season that is more marked and prolonged than in the monsoon climate and which contrasts with a rainy season with intense rainfall [28,29]. The seasonality analysis considered the spring months (March–May), summer months (June–August), fall months (September–November), and winter months (December–February) [30].

Figures 2a and 3a show the daily standard behavior of solar energy and marine-current energy, where the variations in solar irradiance and marine-current speed are evident at different times (daily variability). Solar potential was obtained via the integration of irradiance data (W/m^2) during the day to obtain the irradiation potential (kWh/m^2) (Figure 2b). The marine-current seasonal potential is defined as the average of daily measurements of marine-current speed (m/s) for each month (Figure 3b).

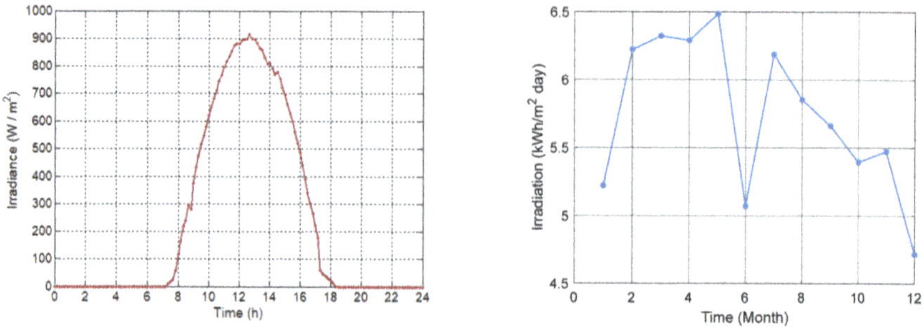

(a) (b)

Figure 2. Standard behavior of irradiance and irradiation in Cozumel: (**a**) daily solar-irradiance profile (W/m^2); (**b**) daily irradiation average per month (kWh/m^2).

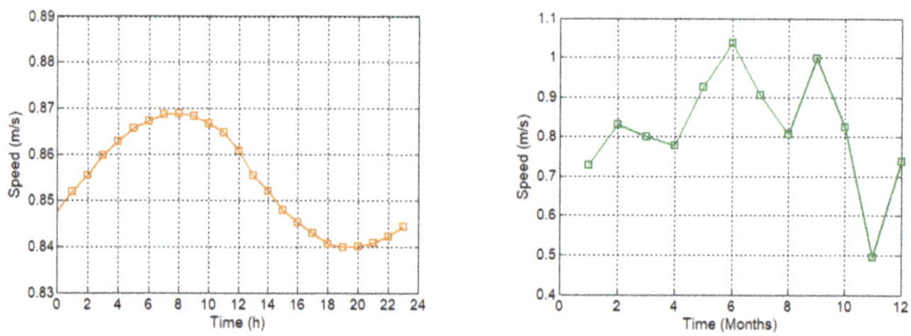

(a) (b)

Figure 3. Marine-current-speed profile in Cozumel: (**a**) daily standard marine-current-speed profile; (**b**) daily current speed average per month.

2.2. Home-Load Profile

The home-load profile with 1 h resolution represents an average energy consumption (E_C) of 7.5 kWh/day for a home on the Mexican coast [31,32]. Energy consumption was obtained from the statistical yearbook Quintana Roo state in Mexico in 2017, developed by the National Institute of Statistics and Geography [33]; Figure 4 shows the daily standard-load profile.

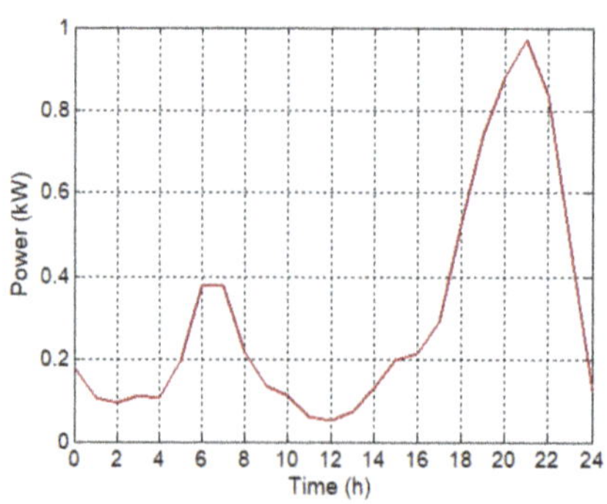

Figure 4. Daily standard-load profile of coastal house in Cozumel.

2.3. Sizing and Design of the Renewable Energy Hybrid System

SMHES sizing is based on the analysis of the profiles of solar irradiance (Figure 5) and marine-current speed (Figure 6). The sizing of both renewable energies was carried out based on the assumption that the energy supplies are provided individually and independently of each other. Figures 5a and 6a show the daily solar irradiance and marine-current-speed variations in a one-year timelapse. Figures 5b and 6b show the minima, maxima, and year averages to identify the maximum variations during the year; this shows the occurrence of days with higher or lower power generation than that estimated for sizing. The SMHES was dimensioned with respect to the seasonal potential minima [34]. In the case of the PVS, it was dimensioned with the minimum daily irradiation in December (4.72 kWh/day) (Figure 5c), while the SCM dimensioning was carried out with the minimum speeds in November (0.65 m/s) (Figure 6c).

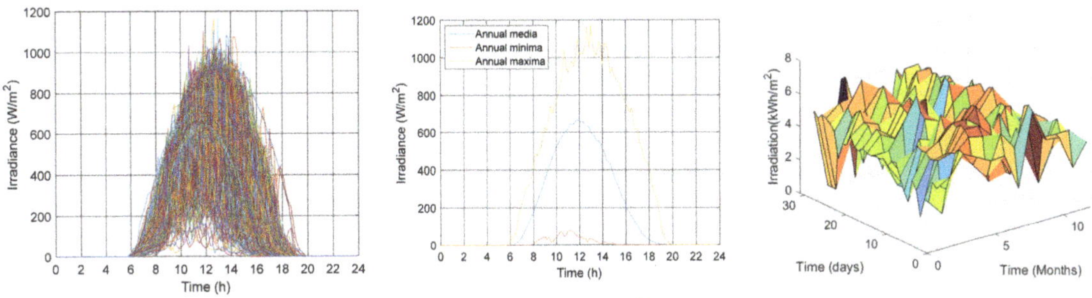

(a) (b) (c)

Figure 5. Solar potential in Cozumel: (**a**) solar-irradiance daily variations; (**b**) maximum, minimum, and average solar potential; (**c**) daily solar irradiation in different months of the year.

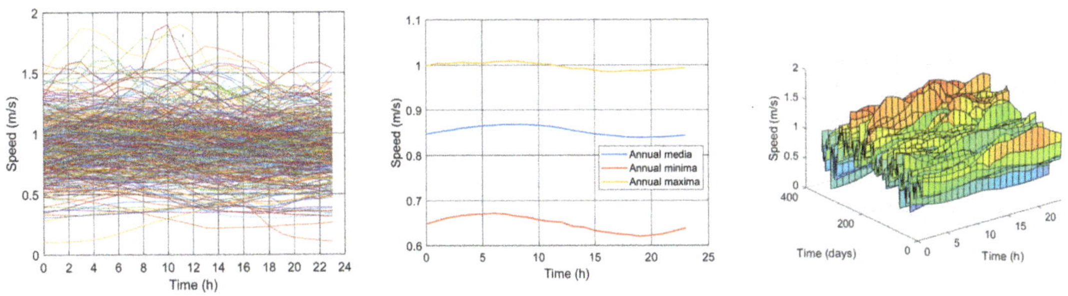

(a) (b) (c)

Figure 6. Marine-current potential in Cozumel; (**a**) marine-current-speed daily variations; (**b**) maximum, minimum, and average marine-current potential; (**c**) daily speed variation of the different days of the year.

2.3.1. PV Solar System Sizing

The sizing of the PVS was carried out according to annual irradiation minimums (Equation (1)) [34,35].

$$P_{PVS} = \frac{E_{PVOut} G_{CEM}}{G_{dm} \eta} \quad (1)$$

where P_{PVS} is the minimum power-installed capacity (kW) of the photovoltaic system, which was obtained directly through energy consumption per day (E_{PVOut}, kWh/day); G_{CEM} is the solar standard-test irradiance (1 kW/m^2); G_{dm} is the solar irradiation, which in this case is proposed as the media in the month of lower irradiation, December (4.72 kWh/m^2 day); and η is the overall efficiency of the auxiliary equipment (DC–DC controller regulator). Note that solar panel efficiency is not considered in Equation (1); the installed capacity is evaluated through commercial parameters where 300 W is proposed for solar panels (tested at 1 kW/m^2).

2.3.2. Marine-Current-Energy System Sizing

The marine-current system (MCS) is mainly composed of the marine-current turbine (MCT) and the permanent magnet synchronous generator (PMSG). The MCT instantaneous power P_{MCT} (kW) can be calculated using Equation (2) [36].

$$P_{MCT} = \frac{1}{2}\rho C_p A\, V^3 \tag{2}$$

where ρ is the seawater density, V is the current speed in m/s, and A is the cross-area section of the turbine. C_p is the power coefficient (dimensionless), which is a function of the tip speed ratio and pitch angle; for the typical marine-current turbine (MCT), C_p values are considered in the 0.35–0.5 range. A turbine with a diameter of 2 m is proposed, based on that reported by Shirasawa et al., 2016 [37], which allows the turbine to operate at the current speeds of Cozumel (~0.7 m/s).

To obtain the daily energy production, the integration of the power-generation profile was carried out. The daily marine-current energy (E_{MCS}) supplied by the MCS is given by Equation (3), where t is the time and η is the overall efficiency of the equipment (PMSG and AC–DC controller regulator).

$$E_{MCS} = \left(\int_0^t P_{MCT}\, dt\right)\eta \tag{3}$$

2.4. Solar PV–Marine Current and Energy-Storage System Hybridization

The SMHES hybridization analysis consisted of evaluating the energy supply with different PVS and MCS degrees of utilization for the 365 days of the year. A 0 HD considers that energy is provided only from the PVS and 1 HD when the energy is provided completely from the MCS. Therefore, hybridization supposes a decrease in the installed capacity of the PVS or MCS. In this study, the hybridization analysis was carried out using an algorithm developed in Matlab® software (Figure 7). The analysis procedure consisted of four steps; first, the power of each system was integrated as a function of time intervals of 1 h in a whole year; second, the HD was proposed from zero to one; third, the energy generated and consumed was obtained; and fourth, the energy storage as well as the number of charge and discharge cycles of the storage system were calculated considering the points of intersection between the generation and consumption profiles. In Section 3.1 the results are analyzed.

Daily electricity-power supply (E_{SMHES}) was determined by the HD, which defines the PVS (E_{Hpvs}) and MCS (E_{HMCS}) installed capacity (Equations (4)–(6)).

$$E_{HPVS} = E_{PVS}\,(1 - HD) \tag{4}$$

$$E_{HMCS} = E_{MCS}\, HD \tag{5}$$

$$E_{SMHES} = E_{Hpvs} + E_{HMCS} \tag{6}$$

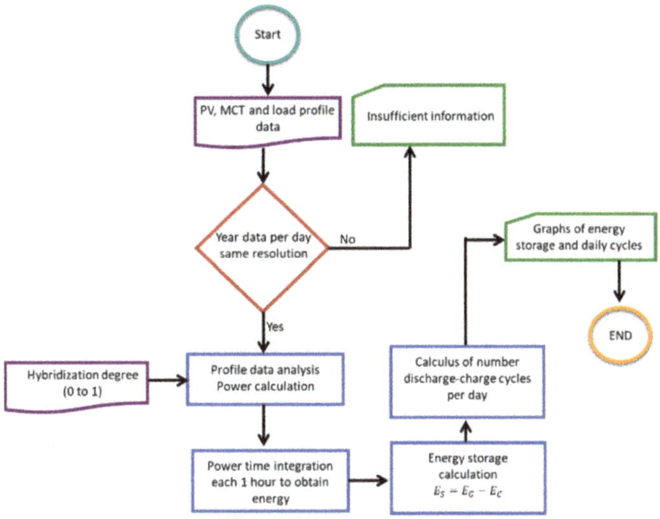

Figure 7. Energy-balance analysis algorithm to evaluate the daily and seasonal energy-storage system.

2.4.1. Hybrid Energy Storage

A hybrid battery–hydrogen storage system was proposed, and the sizing was carried out based on hybridization analysis, while its relevance was validated through the TOPSIS method. Figure 8 shows the SMHES diagram. The energy balance from Equation (7) determined the daily energy surplus–deficit that was proposed to be covered by the hydrogen energy-storage system (HESS). For hourly fluctuations, a battery energy-storage system (BESS) was proposed, where time fluctuations determined the energy balance; thus, a dynamic model was required. In this analysis, energy balance was carried out at different time intervals (Δt) with 1 h resolution (Equation (8)). The variation between the SMHES generation and the energy consumption (E_C) determines the moments in which surplus energy is available to store and when energy deficits are present, and thus, the stored energy must be used.

$$E_{HESS} = (E_{SMHES} - E_C)\, \eta_{HESS} \tag{7}$$

$$E_{BESS} = (E_{SMHES} - E_C)\, \eta_{BESS} \tag{8}$$

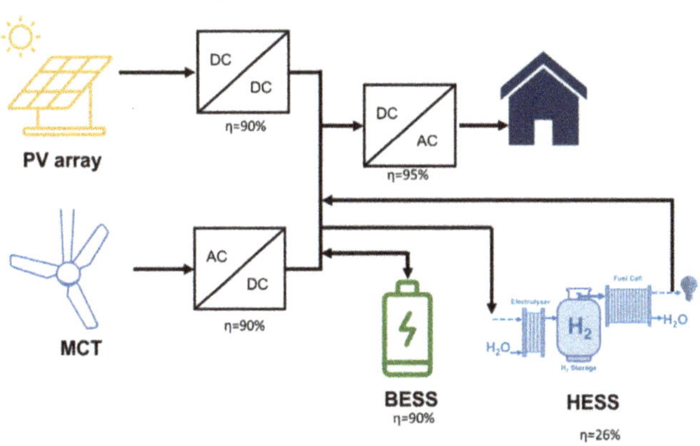

Figure 8. Solar–MarineCurrent Energy Hybrid System diagram.

2.4.2. Energy-Storage Selection

Due to the variety of ESS options to evaluate the relevance of the hybrid battery–hydrogen storage system, a MCDA analysis was proposed. In this work, ESSs were evaluated based on three objectives: (a) the Regulation Energy-Storage System (RESS), (b) the Post-Consumption Energy-Storage System (PCESS) for daily, monthly, or seasonal periods; and (c) the Regulation & Post-Consumption Energy-Storage System (RPCESS) for uses where the ESS is considered for the regulation and control of energy supply autonomy on certain days, e.g., batteries [9,38,39]. The proposed MCDA is a combination of the Analytic Hierarchy Process (AHP) and the Technique for Order Performance by Similarity to Ideal Solution (TOPSIS); this method provides solutions to problems involving conflicting and multiple objectives [10]. TOPSIS was developed by Hwang and Yoon (1981) [40] and is based on the concept that the best alternative should have the shortest geometric distance from the ideal positive solution but the largest geometric distance from the ideal negative solution [41–43].

The study evaluated eight ESSs with four classifications, (1) mechanical, (2) electrical, (3) electrochemical, and (4) chemical, and ten criteria, lifetime (C1), cycle life (C2), energy efficiency (C3), power rating (C4), response time (C5), storage duration (C6), power density (C7), energy density (C8), installed system cost (C9), and maturity (C10). The TOPSIS method was conducted according to Garduño-Ruiz et al. (2021) [44] and consisted of feeding a decision matrix with a set of alternatives and criteria (Figure 9) and then assigning levels of importance or weightings to each criterion through the AHP technique using the method of Saaty (1990) [45]. To determine the best alternative, the TOPSIS tool was used by means of a Python algorithm.

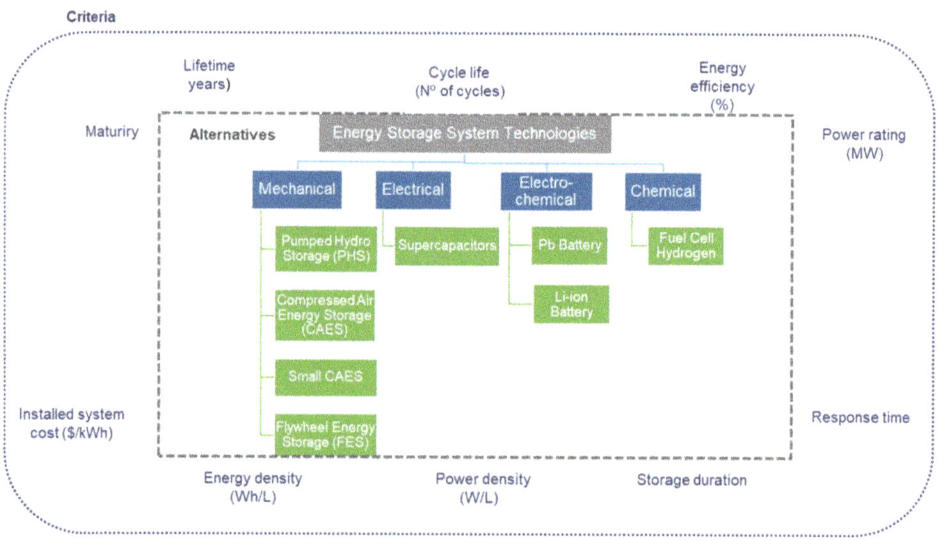

Figure 9. Classification of energy-storage technologies and alternatives for decision matrix.

3. Results

3.1. Renewable-Energy Hybrid System

The SMHES sizing was carried out for 0 HD (only PVS) and 1 HD (only MCS). Table 1 shows the parameters for the PVS and MCS.

Table 1. SMHES parameters in case study.

PVS		MCS	
Parameter	**Value**	**Parameter**	**Value**
P_{PVS}	2.1 kW	P_{MCS}	0.458 kW
G_{dm}	4.72 kWh/day	V_{min}	0.65 m/s
Panel array	7 modules (300 W)	MCT	3 turbines (153 W)
Panel area	1.95 m^2/module	Rotor area	3.14 m^2/turbine
		Temperature	25 °C
		Seawater density	1030 kg/m^3

The generation profile was analyzed with respect to the home power-demand profile, which allows the hours of the day, when it is possible to cover the energy demand, the range where there is an energy surplus, or where it is necessary to use energy storage to cover the demand to be determined. Figure 10a shows a consumption–generation profile in a representative day in the year (day 355, December) when each renewable system produces 7.5 kWh/day in the seasonal minimums. Under this scenario, the amount of energy extracted from the MCS is higher than that from the PVS due to a higher daily availability; therefore, the MCS requires lower installed capacity, 0.458 kW, while solar requires an installation of 2.1 kW. It is worth mentioning that although the MCS generation profile shows fewer hourly variations, there are fluctuations that are less visible on the scale compared to the PVS generation profile. Figure 10b shows the total generation. The system considers only output power. From this perspective, the intersection of the generation curve with the demand curve allows us to visualize the points where the hybrid system is not able to cover the demand.

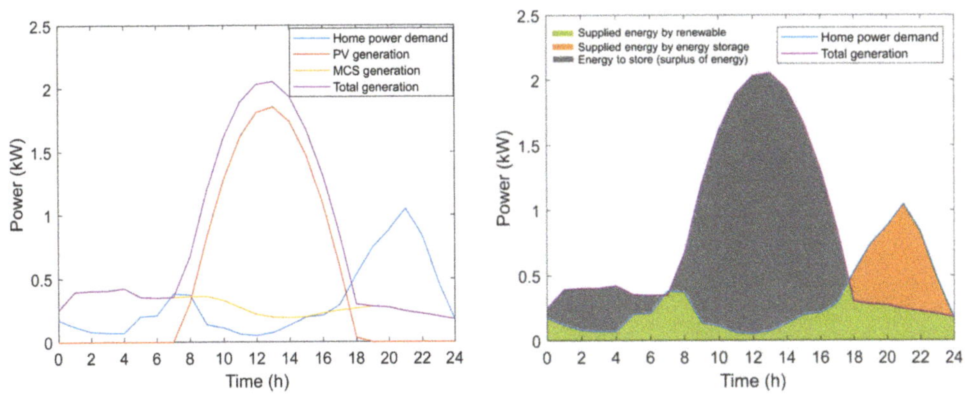

(a) (b)

Figure 10. (**a**) Generation–consumption profile of SMHES supplied by an installed capacity PVS of 2.1 kW or an MCS of 0.458 kW; (**b**) total generation–consumption profile of SMHES indicating the energy supplied by renewable resources, energy supplied by energy storage, and surplus of energy.

3.2. Hybridization Analysis Results

Applying the algorithm developed in the Matlab® program, the curves of the hybridization process were obtained at different HDs for the 365 days of the year; daily variations during the year are shown in Appendix A in Figure A1 and includes daily variations in energy consumption and supply in the 12 months. Figure 11 shows a representative day (355 day) of generation–consumption profiles with different HDs for comparison. For each case, an energy-balance analysis was carried out for different HDs. Table 2 shows the power (kW) and energy (kWh/day) generation of the SMHES, RPCESS, percentage reduction in the RPCESS for different HDs, and the average charge–discharge cycles.

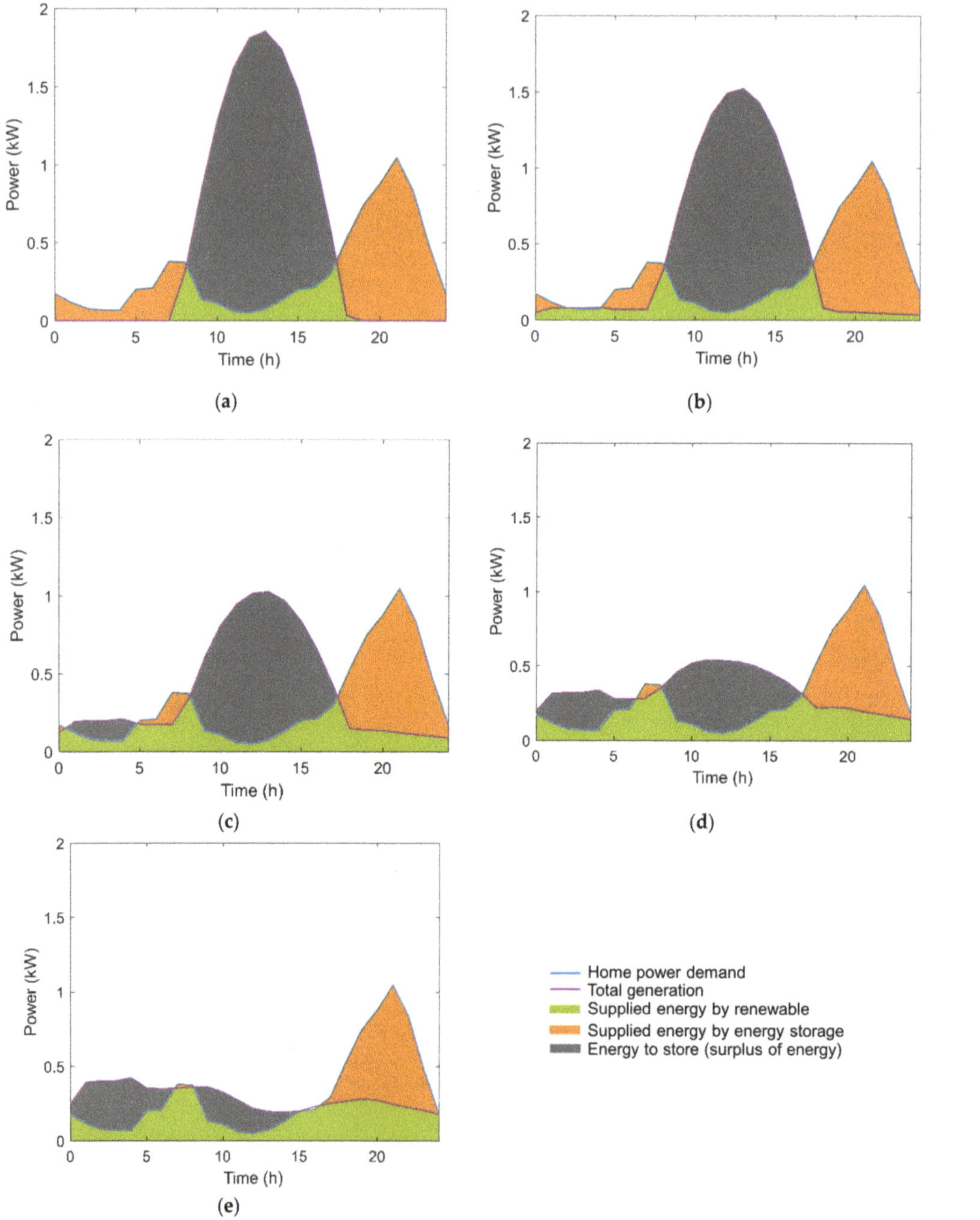

Figure 11. Generation–consumption profiles with different hybridization degrees: (**a**) 0 HD; (**b**) 0.2 HD; (**c**) 0.5 HD; (**d**) 0.8 HD; (**e**) 1 HD.

Table 2. Solar–Marine-Current Hybrid System (average results).

Case	HD	PV System		MC System		RPESS (kWh/day)	% Reduction RPESS **	Average Cycles/Day (Charge–Discharge)
		Power-Installed Capacity (kW)	Energy (kWh/day)	Power-Installed Capacity (kW)	Energy (kWh/day)			
1	0	2.1	11.55	0	0	5.39	-	1
2	0.2	1.68	9.24	0.09	5.65	3.17	59%	3.5
3	0.5	1.05	5.77	0.23	12.4	1.75	55%	3
4	0.8	0.42	2.31	0.36	22.6	1.07	61%	1.5
5	1	0	0	0.46	28.26	0.85	79%	1
6 *	-	2.1	11.55	0.46	28.26	-	-	1

* In Case 6, PVS and MCS provide the maximum power, double the energy that the system requires, which is included for comparison purposes. ** % Reduction in RPESS with respect to Case 1, where there is the largest energy-storage system.

The graphs in Figure 11 show the total generation of the PVS and the MCS for different HDs. The amount of energy that can be covered by the SMHES can be observed by comparing the area under the curve of the consumption profile of the house to the total generation curve. Similarly, the variation in the maximum power with respect to the degree of hybridization can be observed.

The hybridization mainly affects the total amount of energy stored per day. In Case 1 (HD 0), where the energy is totally supplied by the PVS, the energy required to be stored is 5.39 kWh/day, while for Case 5 (HD 1), where the energy is totally supplied by the MCS, the energy required to be stored is lower than in Case 1 at only 0.85 kWh/day. The number of cycles that the RPESS has to perform to achieve this goal is one cycle for Case 1 and three cycles at 0.5 HD. It is worth mentioning that these cycles do not imply full charge or discharge. The RPESS can have important implications in the cycle life as the number of cycles increase, with triple the charge–discharge cycles than with 0.5 HD. However, the capacity of the RPESS can be reduced by 55% for HD = 0.5 and 79% for HD = 1. This capacity reduction is due to the similarity of the demand and MCS profiles during the daytime to the PVS.

3.3. Evaluation System Hybridization in the Year

The annual analysis allowed us to determine the total number of required charge and discharge cycles according to the HD (Figure 12). The analysis showed that 0 HD requires more daily energy storage, 5.5 kWh/day, than 1 HD requiring 0.85 KWH/day, while the number of cycles per year is higher at 0.2 HD that the other HD.

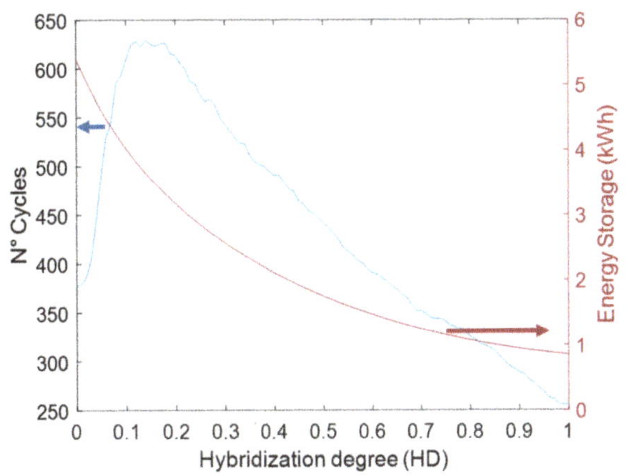

Figure 12. Number of cycles in a year in the different cases of hybridization and the comparison with daily energy storage.

According to the analysis in Section 3.1, the daily consumption (7.5 kWh/day) can be easily covered by the production of the PVS and MCS at different HDs. The estimated amount of energy lost with the full participation of the PVS (HD 0) is 1477 kWh/year (Figure 13a), while for a system with the full participation of the MCS (HD 1), it is 7576 kWh/year (Figure 13b). This surplus energy can be stored for days with a higher consumption than the estimated value (base load) and for days where the amount of energy produced by the SMCHS cannot cover the energy demand.

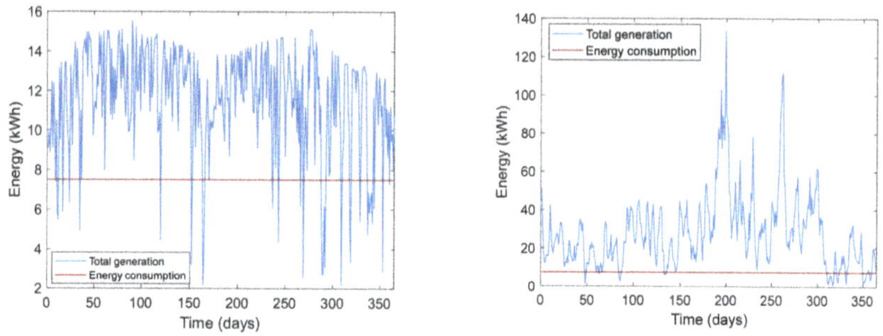

(a) (b)

Figure 13. PVS and MCS daily generation: (**a**) 0 HD and (**b**) 1 HD, with respect to the minimum seasonal generation corresponding to the base load (7.5 kWh/day).

These results show that a greater participation of the PVS has advantages in terms of lower seasonal storage (lower seasonal losses). However, seasonal storage also depends on the HD (Figure 14). In the case of 0.2 HD, the energy surplus is 2697 kWh/year; for 0.5 HD, it is 4527 kW/year; and for 0.8 HD, it is 6296 kWh/year.

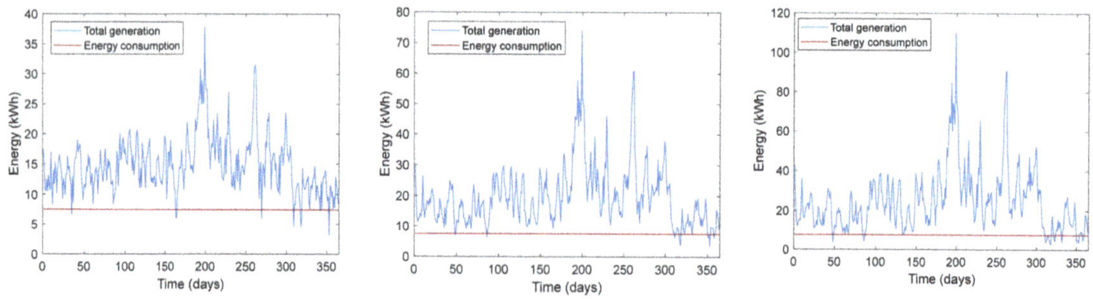

(a) (b) (c)

Figure 14. PVS and MCS daily generation: (**a**) 0.2 HD, (**b**) 0.5 HD, and (**c**) 0.8 HD, with respect to the minimum seasonal generation corresponding to the base load (7.5 kWh/day).

It was necessary to evaluate the seasonal variability of the system since it would help to complement the results obtained from the daily analysis. From this analysis, the percentage of days of the year in which the demand can be totally covered was found to be 91% for 0 HD and 92% for 1 HD, while the highest availability was found for 0.3 HD at 98.5% (Figure 15).

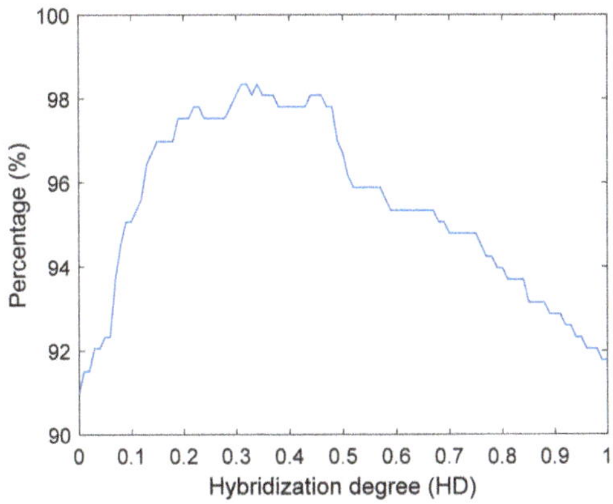

Figure 15. Minimum generation availability at different HDs.

To cover the deficit of energy, different alternatives can be used, from oversizing the SMCHS to the use of an auxiliary internal combustion system, increasing the PCEES capacity or using a second ESS; for this purpose, a PCESS was used. Figure 15 shows that despite having an adequate coverage of days, most of the ESSs have significant surpluses of energy that are not used, as shown in Figures 13 and 14. Figure 16 shows the amount of seasonal minimum storage required for days, which is unable to cover the base demand. The results indicate that at 0 HD, it is 5.4 kWh/day, while at 1 HD, it is 6.4 kWh/day. The analysis showed that the HD with the lowest seasonal storage for these days is 0.6 HD with 3.86 kWh/day.

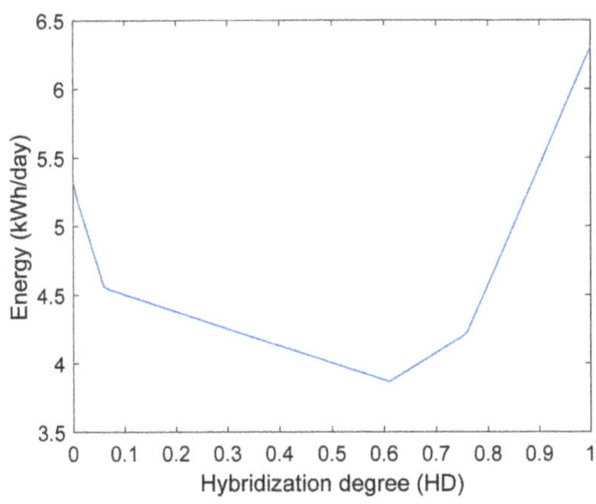

Figure 16. Seasonal energy storage at different HDs; additional energy storage per day with less generation than base load (7.5kWh/day).

3.4. Energy-Storage Selection

Renewable-energy hybridization analysis determines the feasibility of the degree of participation of the PVS and MCS based on the daily number of cycles of the storage system. The determination of the best RPCESS and PCESS options for a specific application is an important task, which requires the analysis of several factors (described in Table 3). Many aspects should be considered for the evaluation of a storage system; the most critical factors in this study were power rating (C4), response time (C5), and storage duration (C6), as the objective was to validate the technology.

Table 3. Decision matrix evaluating the different alternatives and criteria.

	Lifetime Years (C1)	Cycle Life, No. of Cycles (C2)	Energy Efficiency % (C3)	Power Rating MW (C4)	Response Time (C5)	Storage Duration (C6)	Power Density W/L (C7)	Energy Density Wh/L (C8)	Installed System Cost USD/kWh (C9)	Maturity (C10)
PHS (A1)	50	20,000	80	3000	minutes	hours–months	1.5	2	100	Mature
CAES (A2)	30	10,000	65	1000	minutes	hours–months	2	6	50	Commercialized
Small CAES (A3)	23	1200	80	10	seconds	hours–months	2	6	200	Early commer-cialized
Pb Battery (A4)	15	1500	85	40	milliseconds	minutes–days	200	100	300	Mature
Li-ion Battery (A5)	15	3000	90	100	milliseconds	minutes–days	3000	400	1000	Commercialized
Flywheel (A6)	20	20,000	85	20	seconds	seconds–minutes	1500	80	3000	Early commer-cialized
Supercapacitors (A7)	30	50,000	90	0.3	milliseconds	seconds–hours	100,000	30	1750	Deve-loping
Fuel cell Hydrogen (A8)	20	10,000	26	100	seconds	hours–months	500	600	1000	Develop-ing

According to the methodology of Saaty, the weighting carried out is considered reasonable, given that the consistency ratio (CR) was less than 0.1 (Table 4). On the other hand, the evaluations carried out with the TOPSIS method found that the electrochemical classification of EESs was better than the mechanical classification since the criterion power rating was minimized to find an EES with a low power rating. The best options were as follows: (1) li-ion batteries and (2) Pb batteries for the RPCESS (Table 5); (1) hydrogen and (2) small CAES for the PCEES; (1) supercapacitors and (2) li-ion batteries for the REES based on the best response time.

Table 4. Weighting of the Multi-Criteria Decision Analysis.

Criteria	Type (Min/Max)	Weighting REES	Weighting PCEES	Weighting RPCESS
Lifetime (years)	max	0.07	0.07	0.04
Cycle life	max	0.08	0.08	0.07
Energy efficiency (%)	max	0.10	0.09	0.09
Power rating (MW)	min	0.13	0.13	0.12
Response time	max	0.36	0.18	0.21
Storage duration	max	0.17	0.35	0.29
Power density (W/L)	max	0.03	0.03	0.03
Energy density (Wh/L)	max	0.03	0.03	0.05
Installed system cost (USD/kWh)	min	0.01	0.01	0.03
Maturity	max	0.02	0.02	0.06
		nmax = 11.41, CI = 0.15, CR = 0.09	nmax = 11.41, CI = 0.15, CR= 0.09	nmax = 10.91, CI = 0.10, CR= 0.06

Note: nmax is an eigenvalue of the decision matrix; CI is the consistency index; and CR is the consistency ratio.

Table 5. Ranking of EES using the TOPSIS method.

ESS	Hierarchy RESS	Hierarchy PCESS	Hierarchy RPCESS
PHS (A1)	8	8	8
CAES (A2)	7	6	6
Small CAES (A3)	5	2	5
Pb Battery (A4)	3	5	2
Li-ion Battery (A5)	2	4	1
Flywheel (A6)	6	7	7
Supercapacitors (A7)	1	3	3
Hydrogen (A8)	4	1	4

The obtained results consider the fluctuations between generation and demand in time intervals from hours to months (seasonal). The importance of daily storage was found to be able to cover the energy-demand profile, and variations in seasonal energy expedients were shown. Often, in stand-alone systems, the ESS is oversized in order to cover days with a higher consumption or lack of renewable generation. However, this might compromise the cycle life due to the number of daily cycles that the ESS performs. Therefore, a reserve (backup) storage system that exploits seasonal surpluses may be a technically viable alternative. The storage viability for daily and seasonal storage was evaluated through a multi-criteria method.

For energy storage for the RPCESS (daily fluctuations), it was found that the most important criteria are response time followed by the number of cycles and efficiency. In the case of the PCESS (seasonal fluctuations), the most important criteria are the duration of storage, followed by the response time; as it is used for low-power systems, a low-power range is better. The results of the Multi-Criteria Decision Analysis (TOPSIS) for the selection of the best ESS are shown in Table 5, which are listed from one to eight at the feasibility level. The RESS, PCESS, and RPCESS are compared, and a hierarchy is shown: (1) is the best option, and (8) is the worst storage system option. For the REES, supercapacitors were found to be adequate for periods of time shorter than days, while for longer periods, PCESS hydrogen storage is adequate, and RPCEES lithium-ion batteries are the best alternative. This information allows us to validate the importance of developing lithium-ion batteries and hydrogen energy-storage systems.

4. Conclusions

Sun and marine currents are broadly available renewable resources on coastal zones, and at first sight, choosing one or the other seems unimportant; however, the generation profiles in this work indicate that MCSs have lower daily variability and, consequently, have lower energy-storage requirements. For hybrid photovoltaics–MCSs, this study indicates that for a greater participation of the MCS, the amount of required daily energy storage is lower (79%). On the other hand, the hybridization degree affects the number of daily charge–discharge cycles that the system can stand; for 0 HD in comparison to 1 HD, the cycles increase up to 46%, while for 0.2 HD, the number of cycles increases to 156%. Although this work does not include the characteristics of the ESS, depending on the type of battery or storage system, the number of cycles and the power can influence the durability of the system. However, for energy surpluses, the SMCHS with a higher MCS share can have five times higher losses in comparison to hybrid systems with higher solar shares; these losses can be avoided by using a lower dimensioning factor, although the number of days that cannot meet the demand might increase. The analysis indicates that a HD between 0.3 and 0.5 has a greater effect on the increase in days of minimum generation availability, while seasonal storage at 1 HD increases by 16% compared to using only solar energy (0 HD). Finally, the TOPSIS method for the selection of the best ESS demonstrates the relevance of electrochemical storage (batteries) in stand-alone systems, where the response time between minutes and days is more relevant than for faster technologies; on the other hand, the surplus energy generated due to seasonal variations is ideal for chemical storage in the form of hydrogen. Further studies aiming to improve the limitations of this work, such as analyses of industrial and building consumption profiles, the inclusion of the characteristics and limitations of ESSs, and economic analysis, will be published elsewhere.

Author Contributions: Conceptualization, J.O.-G., R.d.G.G.-H. and G.R.-S.; methodology, J.O.-G., R.d.G.G.-H. and E.P.G.-R.; software, J.O.-G. and E.P.G.-R.; validation, R.d.G.G.-H. and G.R.-S.; formal analysis, J.O.-G., R.d.G.G.-H. and G.R.-S.; investigation, J.O.-G.; data curation, J.O.-G. and R.d.G.G.-H.; writing—original draft preparation, J.O.-G., E.P.G.-R., R.d.G.G.-H. and G.R.-S.; writing—review and editing, J.O.-G., R.d.G.G.-H. and G.R.-S.; visualization, G.R.-S.; supervision, R.d.G.G.-H. and G.R.-S.; project administration, R.d.G.G.-H. and G.R.-S.; funding acquisition, R.d.G.G.-H. All authors have read and agreed to the published version of the manuscript.

Funding: This research was developed under the framework of CEMIE-Océano (Mexican Centre for Innovation in Ocean Energy). Project FSE-2014-06-249795 financed by CONACYT-SENER-Sustentabilidad Energética.

Data Availability Statement: The solar irradiance database used in this paper was from database PVGIS-NSRDB provided by Solar radiation tool from Photovoltaic Geographical Information System, European Commission, (https://re.jrc.ec.europa.eu/pvg_tools/es/#MR, 10 March 2021).

Acknowledgments: The authors would like to thank the Centro Mexicano de Innovación en Energía del Océano (CEMIE-Océano), the Integration to the electricity grid and energy management resources line (I-LT1), the Currents and tidal energy line for the marine current database. IPN multidisciplinary project SIP-2024 (2019-2021) and Ciencia Basica Project A1-S-15770. Jorge Olmedo-Gonzalez acknowledges to CONACYT for the scholarship.

Conflicts of Interest: The authors declare no conflict of interest.

Nomenclature

A	Cross-area section of the turbine	m^2
C_p	Power coefficient	Dimensionless
E_C	Energy consumption	kWh
E_{HMCS}	Electricity-power supply from MCT	kWh
E_{HPVS}	Electricity-power supply from PVS	kWh
E_{MCS}	Daily marine-current energy	kWh
E_{PVOut}	Energy consumption per day	kWh/day
E_{SMHES}	Daily electricity-power supply	kWh
G_{CEM}	Solar standard-test irradiance	1 kW/m^2
G_{dm}	Solar irradiation	kWh/m^2 day
P_{MCT}	MCT instantaneous power	kW
P_{PVS}	Minimum power-installed capacity	kW
t	Time	h days/months/years
V	Current speed	m/s
η	Overall efficiency of auxiliary equipment	Dimensionless
η_{BESS}	BESS efficiency	Dimensionless
η_{HESS}	HESS efficiency	Dimensionless
ρ	Seawater density	kg/L

Abbreviations

AHP	Analytic Hierarchy Process
BESS	Battery Energy-Storage System
CAES	Compressed-Air Energy Storage
ESS	Energy-Storage Systems
HD	Hybridization degree
HESS	Hydrogen Energy-Storage System
MCDA	Multi-Criteria Decision Analysis
MCS	Marine-current system
MCT	Marine-current turbine
MRE	Marine Renewable Energy
PCESS	Post-Consumption Energy-Storage System
PHS	Pumped Hydro Storage
PMSG	Permanent magnet synchronous generator
PVGIS	Photovoltaic Geographical Information System
PVS	Photovoltaic System
RESS	Regulation Energy-Storage System
RPCESS	Regulation & Post-Consumption Energy-Storage System
SMHES	Solar–Marine Hybrid Energy System
VRE	Variable Renewable Energies

Appendix A

The appendix includes daily variations in the energy consumption and supply in the twelve months.

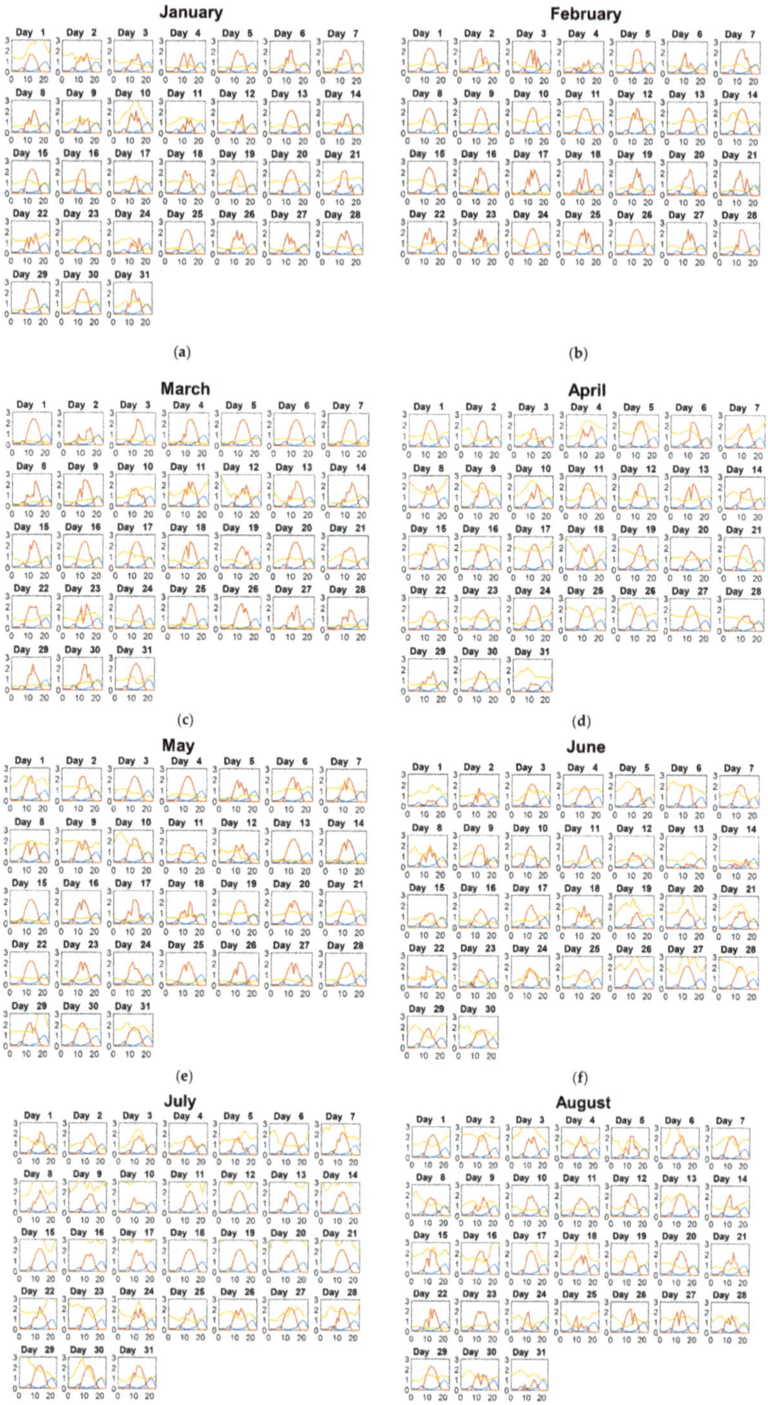

Figure A1. *Cont.*

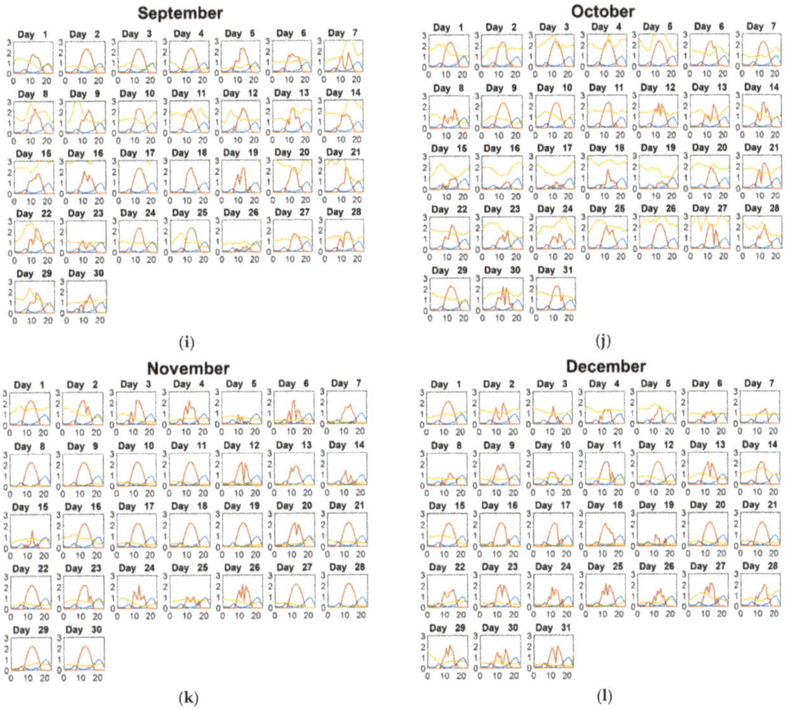

Figure A1. Variations in the SMHES for the different months of the year. The variations of solar power generation (orange curve) and marine current power generation (yellow curve) are shown for each day of the year and compared with the daily demand profile (blue curve), (**a**) January; (**b**) February HD; (**c**) March; (**d**) April; (**e**) May; (**f**) June; (**g**) July; (**h**) August; (**i**) September; (**j**) October; (**k**) November; (**l**) December.

References

1. Uihlein, A.; Magagna, D. Wave and tidal current energy—A review of the current state of research beyond technology. *Renew. Sustain. Energy Rev.* **2016**, *58*, 1070–1081. [CrossRef]
2. IRENA. *Renewable Energy Prospects: Mexico, REmap 2030 Analysis*; IRENA: Abu Dhabi, Saudi Arabia, 2015; Available online: http://www.irena.org/remap (accessed on 24 October 2021).
3. Rourke, F.O.; Boyle, F.; Reynolds, A. Marine current energy devices: Current status and possible future applications in Ireland. *Renew. Sustain. Energy Rev.* **2010**, *14*, 1026–1036. [CrossRef]
4. Hernández-Fontes, J.V.; Felix, A.; Mendoza, E.; Cueto, Y.R.; Silva, R. On the marine energy resources of Mexico. *J. Mar. Sci. Eng.* **2019**, *7*, 191. [CrossRef]
5. Cebulla, F.; Naegler, T.; Pohl, M. Electrical energy storage in highly renewable European energy systems: Capacity requirements, spatial distribution, and storage dispatch. *J. Energy Storage* **2017**, *14*, 211–223. [CrossRef]
6. Guney, M.S.; Tepe, Y. Classification and assessment of energy storage systems. *Renew. Sustain. Energy Rev.* **2017**, *75*, 1187–1197. [CrossRef]
7. Luo, X.; Wang, J.; Dooner, M.; Clarke, J. Overview of current development in electrical energy storage technologies and the application potential in power system operation. *Appl. Energy* **2015**, *137*, 511–536. [CrossRef]
8. Sinclair, K.; Copping, A.E.; May, R.; Bennet, F.; Warnas, M.; Perron, M.; Elmqvist, Å.; DeGeorge, E. Resolving environmental effects of wind energy. *WIREs Energy Environ.* **2018**, *7*, e291. [CrossRef]
9. Chen, H.; Cong, T.N.; Yang, W.; Tan, C.; Li, Y.; Ding, Y. Progress in electrical energy storage system: A critical review. *Prog. Nat. Sci.* **2009**, *19*, 291–312. [CrossRef]
10. Kim, J.; Suharto, Y.; Daim, T.U. Evaluation of Electrical Energy Storage (EES) technologies for renewable energy: A case from the US Pacific Northwest. *J. Energy Storage* **2017**, *11*, 25–54. [CrossRef]
11. Bajpai, P.; Dash, V. Hybrid renewable energy systems for power generation in stand-alone applications: A review. *Renew. Sustain. Energy Rev.* **2012**, *16*, 2926–2939. [CrossRef]

12. Barelli, L.; Bidini, G.; Bonucci, F.; Castellini, L.; Fratini, A.; Gallorini, F.; Zuccari, A. Flywheel hybridization to improve battery life in energy storage systems coupled to RES plants. *Energy* **2019**, *173*, 937–950. [CrossRef]
13. Shoaib, M.; Siddiqui, I.; Rehman, S.; Khan, S.; Alhems, L.M. Assessment of wind energy potential using wind energy conversion system. *J. Clean. Prod.* **2019**, *216*, 346–360. [CrossRef]
14. SENER Datos Abiertos de México-Regiones sin Electricidad. Available online: https://datos.gob.mx/busca/dataset/regiones-sin-electricidad (accessed on 22 June 2020).
15. Sánchez-Dirzo, R.; González-Huerta, R.G.; Mendoza, E.; Silva, R.; Sandoval Pineda, J.M. From wave to jet and from jet to hydrogen: A promising hybrid system. *Int. J. Hydrogen Energy* **2014**, *39*, 16628–16636. [CrossRef]
16. Ahmad, S.; Uddin, M.J.; Nisu, I.H.; Ahsan, M.M.; Rahman, I.; Samrat, N.H. Modeling of grid connected battery storage wave energy and PV hybrid renewable power generation. In Proceedings of the 2017 International Conference on Electrical, Computer and Communication Engineering (ECCE), Cox's Bazar, Bangladesh, 16–18 February 2017; pp. 375–380.
17. Wang, L.; Lee, D.J.; Lee, W.J.; Chen, Z. Analysis of a novel autonomous marine hybrid power generation/energy storage system with a high-voltage direct current link. *J. Power Sources* **2008**, *185*, 1284–1292. [CrossRef]
18. Murray, D.B.; Gallagher, P.; Duffy, B.; McCormack, V. Energy storage solutions for offshore wave and tidal energy prototypes. In Proceedings of the 2017 Twelfth International Conference on Ecological Vehicles and Renewable Energies (EVER), Monte Carlo, Monaco, 11–13 April 2017; pp. 1–8.
19. Zhou, Z.; Benbouzid, M.; Charpentier, J.F.; Scuiller, F.; Tang, T. Energy storage technologies for smoothing power fluctuations in marine current turbines. *IEEE Int. Symp. Ind. Electron.* **2012**, 1425–1430. [CrossRef]
20. Askarzadeh, A. Electrical power generation by an optimised autonomous PV/wind/tidal/battery system. *IET Renew. Power Gener.* **2017**, *11*, 152–164. [CrossRef]
21. Shirai, Y.; Minamoto, S.; Yonemura, K.; Rahman, M.L. Output power control of hybrid off-shore-wind and tidal turbine generation system with battery storage system. In Proceedings of the 2016 19th International Conference on Electrical Machines and Systems (ICEMS), Chiba, Japan, 13–16 November 2016; pp. 1–6.
22. Wang, L.; Yu, J.-Y.; Chen, Y.-T. Dynamic stability improvement of an integrated offshore wind and marine-current farm using a flywheel energy-storage system. *IET Renew. Power Gener.* **2011**, *5*, 387. [CrossRef]
23. Banerjee, S.; Musa, M.N.; Jaafar, A.B. Economic assessment and prospect of hydrogen generated by OTEC as future fuel. *Int. J. Hydrogen Energy* **2017**, *42*, 26–37. [CrossRef]
24. Tufa, R.A.; Pawlowski, S.; Veerman, J.; Bouzek, K.; Fontananova, E.; di Profio, G.; Velizarov, S.; Goulão Crespo, J.; Nijmeijer, K.; Curcio, E. Progress and prospects in reverse electrodialysis for salinity gradient energy conversion and storage. *Appl. Energy* **2018**, *225*, 290–331. [CrossRef]
25. Yunez-Cano, A.; González-Huerta, R.d.G.; Tufiño-Velázquez, M.; Barbosa, R.; Escobar, B. Solar-hydrogen hybrid system integrated to a sustainable house in Mexico. *Int. J. Hydrogen Energy* **2016**, *41*, 19539–19545. [CrossRef]
26. Hafez, A.Z.; Yousef, A.M.; Harag, N.M. Solar tracking systems: Technologies and trackers drive types—A review. *Renew. Sustain. Energy Rev.* **2018**, *91*, 754–782. [CrossRef]
27. Alcérreca-Huerta, J.C.; Encarnacion, J.I.; Ordoñez-Sánchez, S.; Callejas-Jiménez, M.; Barroso, G.G.D.; Allmark, M.; Mariño-Tapia, I.; Casarín, R.S.; O'Doherty, T.; Johnstone, C.; et al. Energy yield assessment from ocean currents in the insular shelf of Cozumel Island. *J. Mar. Sci. Eng.* **2019**, *7*, 147. [CrossRef]
28. Mazzeo, D.; Matera, N.; De Luca, P.; Baglivo, C.; Maria Congedo, P.; Oliveti, G. Worldwide geographical mapping and optimization of stand-alone and grid-connected hybrid renewable system techno-economic performance across Köppen-Geiger climates. *Appl. Energy* **2020**, *276*, 115507. [CrossRef]
29. Cui, D.; Liang, S.; Wang, D. Observed and projected changes in global climate zones based on Köppen climate classification. *WIREs Clim. Chang.* **2021**, *12*, e701. [CrossRef]
30. Felix, A.; Mendoza, E.; Chávez, V.; Silva, R.; Rivillas-Ospina, G. Wave and wind energy potential including extreme events: A case study of Mexico. *J. Coast. Res.* **2018**, *85*, 1336–1340. [CrossRef]
31. Cruz González, G.; Alejandro, M.; Saldívar, D. El consumo de energía eléctrica en los hogares de México por nivel de ingresos, 2012. *Tiempo Económico* **2015**, *X*, 43–56.
32. Taufiqul, M.T.; Oo, A.M.; Shawkat Ali, A.B.M. Estimation of Energy Storage and Its Feasibility Analysis. In *Energy Storage-Technologies and Applications*; Zobaa, A.F., Ed.; IntechOpen: Rijeka, Croatia, 2013; pp. 41–78.
33. INEGI. *Anuario Estadístico y Geográfico de Quintana Roo 2017*; Instituto Nacional de Estadística y Geografía México: Aguascalientes, Mexico, 2017; Available online: https://www.inegi.org.mx/contenido/productos/prod_serv/contenidos/espanol/bvinegi/productos/nueva_estruc/anuarios_2017/702825095130.pdf (accessed on 30 October 2021).
34. Fernández Salgado, J.M. *Compendio de Energía Solar: Fotovoltaica, Térmica y Termoeléctrica*; Mundi-Prensa AMV Ediciones: Madrid, España, 2010.
35. Khatib, T.; Ibrahim, I.A.; Mohamed, A. A review on sizing methodologies of photovoltaic array and storage battery in a standalone photovoltaic system. *Energy Convers. Manag.* **2016**, *120*, 430–448. [CrossRef]
36. Zhou, Z.; Benbouzid, M.; Frédéric Charpentier, J.; Scuiller, F.; Tang, T. A review of energy storage technologies for marine current energy systems. *Renew. Sustain. Energy Rev.* **2013**, *18*, 390–400. [CrossRef]
37. Shirasawa, K.; Tokunaga, K.; Iwashita, H.; Shintake, T. Experimental verification of a floating ocean-current turbine with a single rotor for use in Kuroshio currents. *Renew. Energy* **2016**, *91*, 189–195. [CrossRef]

38. Maghami, M.R.; Hassani, R.; Gomes, C.; Hizam, H.; Othman, M.L.; Behmanesh, M. Hybrid energy management with respect to a hydrogen energy system and demand response. *Int. J. Hydrogen Energy* **2019**, *45*, 1499–1509. [CrossRef]
39. Ibrahim, H.; Belmokhtar, K.; Ghandour, M. Investigation of usage of compressed air energy storage for power generation system improving—Application in a microgrid integrating wind energy. *Energy Procedia* **2015**, *73*, 305–316. [CrossRef]
40. Hwang, C.-L.; Yoon, K. *Multiple Attribute Decision Making: Methods and Applications, a State-of-the-Art Survey*, 1st ed.; Springer: Berlin, Heidelberg, 1981; ISBN 9783540105589.
41. Kacprzak, D. An extended TOPSIS method based on ordered fuzzy numbers for group decision making. *Artif. Intell. Rev.* **2020**, *53*, 2099–2129. [CrossRef]
42. Ma, X.; Li, N.; Tao, X.; Xu, H.; Peng, F.; Che, Y.; Guo, S. The optimal selection of electrochemical energy storage using Bayesian BWM and TOPSIS method. In Proceedings of the 2019 6th International Conference on Information Science and Control Engineering (ICISCE), Shanghai, China, 20–22 December 2019; pp. 610–614. [CrossRef]
43. Krohling, R.A.; Pacheco, A.G.C. A-TOPSIS—An approach based on TOPSIS for ranking evolutionary algorithms. *Procedia Comput. Sci.* **2015**, *55*, 308–317. [CrossRef]
44. Garduño-Ruiz, E.P.; Silva, R.; Rodríguez-Cueto, Y.; García-Huante, A.; Olmedo-González, J.; Martínez, M.L.; Wojtarowski, A.; Martell-Dubois, R.; Cerdeira-Estrada, S. Criteria for optimal site selection for ocean thermal energy conversion (Otec) plants in Mexico. *Energies* **2021**, *14*, 2121. [CrossRef]
45. Saaty, T.L. How to make a decision: The analytic hierarchy process. *Eur. J. Oper. Res.* **1990**, *48*, 9–26. [CrossRef]

energies

MDPI

Review

Review of Carnot Battery Technology Commercial Development

Vaclav Novotny [1,2,*], Vit Basta [1,2], Petr Smola [1] and Jan Spale [1,2]

1 Faculty of Mechanical Engineering, Czech Technical University in Prague, Technicka 4, 16607 Prague 6, Czech Republic; Vit.Basta@fs.cvut.cz (V.B.); Petr.Smola@fs.cvut.cz (P.S.); Jan.Spale@cvut.cz (J.S.)
2 University Centre for Energy Efficient Buildings, Czech Technical University in Prague, Trinecka 1024, 27343 Bustehrad, Czech Republic
* Correspondence: Vaclav.Novotny@cvut.cz

Abstract: Carnot batteries are a quickly developing group of technologies for medium and long duration electricity storage. It covers a large range of concepts which share processes of a conversion of power to heat, thermal energy storage (i.e., storing thermal exergy) and in times of need conversion of the heat back to (electric) power. Even though these systems were already proposed in the 19th century, it is only in the recent years that this field experiences a rapid development, which is associated mostly with the increasing penetration of intermittent cheap renewables in power grids and the requirement of electricity storage in unprecedented capacities. Compared to the more established storage options, such as pumped hydro and electrochemical batteries, the efficiency is generally much lower, but the low cost of thermal energy storage in large scale and long lifespans comparable with thermal power plants make this technology especially feasible for storing surpluses of cheap renewable electricity over typically dozens of hours and up to days. Within the increasingly extensive scientific research of the Carnot Battery technologies, commercial development plays the major role in technology implementation. This review addresses the gap between academia and industry in the mapping of the technologies under commercial development and puts them in the perspective of related scientific works. Technologies ranging from kW to hundreds of MW scale are at various levels of development. Some are still in the stage of concepts, whilst others are in the experimental and pilot operations, up to a few commercial installations. As a comprehensive technology review, this paper addresses the needs of both academics and industry practitioners.

Keywords: medium duration energy storage; long duration energy storage; Carnot battery; pumped thermal energy storage; power to heat to power; electricity storage; thermal energy storage; ETES; PTES; accumulation

Citation: Novotny, V.; Basta, V.; Smola, P.; Spale, J. Review of Carnot Battery Technology Commercial Development. *Energies* **2022**, *15*, 647. https://doi.org/10.3390/en15020647

Academic Editors: Alon Kuperman and Alessandro Lampasi

Received: 7 December 2021
Accepted: 11 January 2022
Published: 17 January 2022

1. Introduction

The share of renewable generation in electricity production is ever increasing with the feasibility of a 100% renewable supply supported by multiple studies [1–3]. The intermittent nature of these sources puts increasing requirements on electricity storage and system flexibility. Lithium batteries are a well-established technology within this field, provide high efficiency (95%, though in real operation, auxiliaries and performance decay by wearing and ageing can notably decrease this value [4,5]) and relatively low cost per unit power (€/kW). For grid scale medium and long duration applications, however, they are economically well fitted to no more than several hours of capacity due to the high cost per unit capacity (€/kWh). Investigation of a hypothetical 100% of renewable scenario for the UK has found, that, apart from the required installation of certain over-generation, it is the medium duration energy storage in the range of multiple hours to days, through which the majority of the stored electricity needs to flow [6,7]. The lifetime of electrochemical batteries, typically below 10 years, furthermore stresses the need to search for other solutions [8].

Currently, pumped hydro energy storage (PHES) largely dominates the installed storage capacity in comparison to other solutions, as we can see in Figure 1. Even though a rapid growth is experienced in electrochemical batteries and is also expected across other technologies [9], the logarithmic scale provides an idea of the overall storage power demand when the total installed capacity will need to increase several fold to accommodate the renewable production variation. The values are compiled from the database [10] with further added Carnot battery projects known to the authors, with respect to the year 2020. It might not be exhaustive, especially in case of the electro-chemical batteries, though it provides a good overview of the situation and the scale of storage for future energy systems. One can also observe that some systems classified as Carnot batteries are already operational, though only in the megawatt scale.

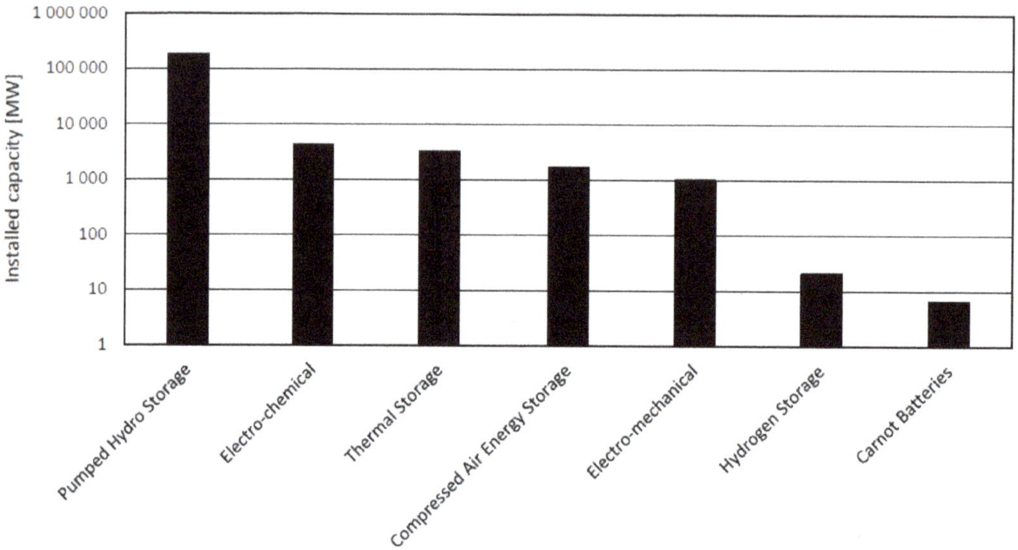

Figure 1. Overview of global installed grid scale electricity storage systems power rating in 2020.

The PHES is also the most commonly employed large scale storage (>100 MW) for medium to long durations. The PHES has major advantages such as high roundtrip efficiency, fast response time, long duration of operation and low self-discharging effect. However, the PHES suffers from the requirements of a suitable geographical location, impact on the environment and low energy density [11,12]. Compressed air energy storage (CAES) technology utilizes mostly underground caverns for storing large volumes of compressed air. Together with PHES it is therefore dependent on geographically suitable locations [13]. Nowadays, only two commercial large scale CEAS facilities are operating, Huntorf in Germany and McIntosh in Alabama, USA, having installed a capacity of 290 MW and 110 MW and reaching an efficiency of 42% and 54% respectively [14,15]. Both commercial CAES plants come under the first stage of development, but advanced CAES systems have been developed such as adiabatic-CAES (A-CAES), advanced adiabatic-CAES (AA-CAES) and isothermal-CAES (I-CAES), which aim for higher efficiency by being more sophisticated and complex [16]. AA-CAES has been technologically experimentally proven on the MW scale [17] and current development and construction plans are given for up to 2.3 GW and 28 GWh in the coming years [17,18]. Flow electrochemical batteries aim to eliminate some drawbacks of classical batteries, especially in capacity scaling, while retaining their advantages. Owing to notably lower efficiency, low energy density, degradation, still requiring toxic and scarce materials and technical flaws and shortcomings, there is a lot of research needed for actual widespread application [19]. Gravity storage systems are also either limited by geographical location or capacity. An increasing number of systems are progressing from conceptual to pilot and commercial stages [20,21]. Conversion to hydrogen and other synthetic fuels remains costly with very low efficiency and suitable for rather very long duration to seasonal applications [22].

1.1. Carnot Battery Principles

Carnot batteries (CB) comprise a set of multiple technologies which have a common underlying principle of converting the electricity to thermal exergy, storing it in thermal energy storage (TES) systems, and in a time of need converting the heat back to electricity. Based on this principle, alternative terms are also used as power to heat to power (P2H2P) or electric thermal (or electro-thermal) energy (electricity) storage (ETES). An excellent review work [23] provides a general overview of CB principles and therefore the reader is referred to this work for details. Prospects of PTES system are then provided in [24]. Here the general aspects will be therefore summarized rather briefly. A general principle of the CB is illustrated in Figure 2.

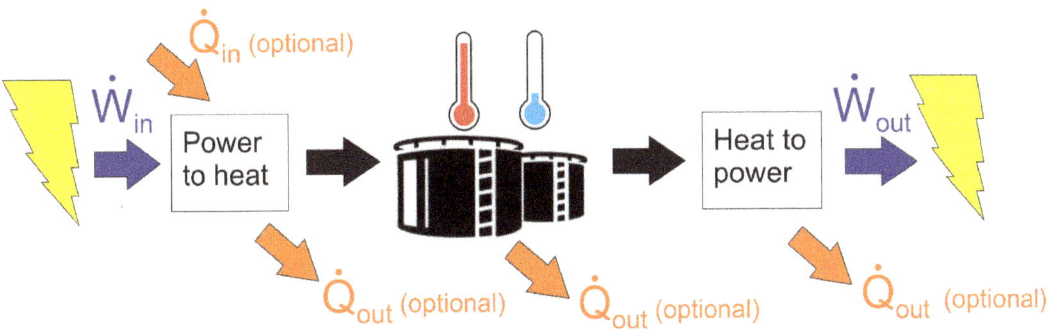

Figure 2. General principle of Carnot battery systems.

Carnot Batteries use surplus electricity as an input of a power to heat (P2H) system to create a temperature gradient (thermal exergy). It can have a form of hot and cold storage systems, or just one of those (hot or cold) with the temperature gradient defined against the environment. During the discharging process, the thermal exergy is converted back to work (electricity) by heat to power (H2P) system, in principle a heat engine. Various concepts of CB can be illustrated regarding the P2H and H2P conversion processes and thermal integration of the heat source in Figure 3.

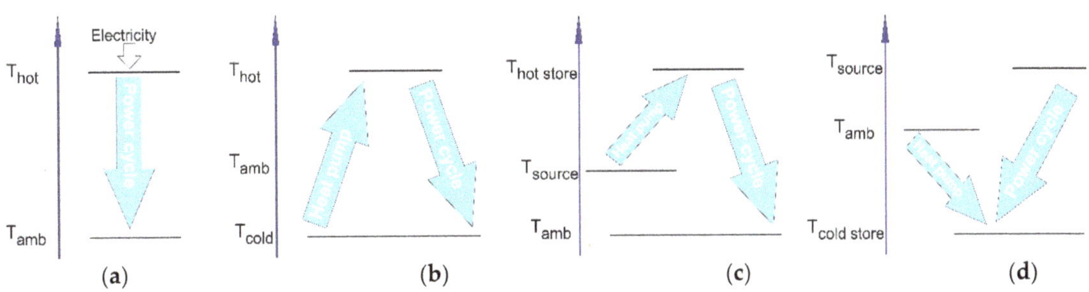

Figure 3. CB concepts regarding thermal integration of heat sources and conversion systems. (**a**) direct heat to power conversion, (**b**) reversible thermodynamic cycle, (**c**) with heat source integration and hot storage or (**d**) cold storage.

The simplest concept is the direct conversion of electricity to heat, which is then stored before its conversion back to power during discharging, typically by a power cycle. Other systems can be considered so called compressed heat energy storage (CHEST) or pumped thermal energy storage (PTES) as they utilize the thermodynamic cycle (in principle a heat pump) for the P2H conversion. The first PTES in Figure 3b works mostly between two distinct temperature levels arbitrarily chosen and (theoretically) independent of the environment, having separate hot and cold storage systems. A specific aspect of CB is the possibility of thermal integration, both on the side of energy input as well as output, providing many possibilities for sector coupling. Regarding the heat input, the heat source can be either upgraded to a higher temperature, which is then used as a heat input of the power cycle or the charging system can prepare cold, which is stored and subsequently used as a heat sink of the power cycle during discharging, increasing the overall temperature gradient of the heat source [25]. A specific case can be defined when the T_{source} is identical to the environment. Regarding hot storage, such systems are not really considered due to their low roundtrip efficiency. Regarding cold storage, it could be considered a highly simplified representation of liquid air energy storage, when the air is liquefied and stored at cryogenic temperatures. All real thermodynamic conversions depart from the ideal ones (e.g., minimum temperature differences in heat exchangers, efficiency of compressors and expanders, pressure drop in components). As a result, a portion of the heat needs to be rejected into the environment due to the irreversibilities [26]. The first concept (Figure 3a) minimizes the losses by converting and storing the heat at highest possible temperature, maximizing the power cycle efficiency. The second concept (Figure 3b) then optimizes the charging and discharging cycles to minimize the irreversibilities. In the heat source integrated concepts, efficiency is also a function of the temperature lift of the heat pump. With a very low lift, the roundtrip efficiency (defined below) can theoretically reach values above unity, in case of a zero lift (and work) of the heat pump, even going towards infinity.

As Dumont et al. [23] or Steinmann et al. [27] mention, there are many possible technological variations of CB. Charging can be realized besides direct conversion (joule heating) by any thermodynamic heat pump cycle. Discharging offers a similar range of options with heat engines to those of Brayton, Rankine or Stirling cycle, their combinations and also direct conversion as thermoelectric, thermi-ionic and thermophotovoltaic systems.

The first examples of these technologies can be traced to 1924, when Fritz Marguerre patented his own solution of thermal energy storage [28] or even to 1833 to work of Erricsson [23], but it has not been until the recent decade, when high volatility of electricity production and its mismatch with demand, it attracted wider interest in this technology. It typically provides relatively low efficiency in the range of 30% to 70% but also low cost for medium and long duration electricity storage. The widely increasing interest in CB was also a reason for establishing an IEA Task 36 on Carnot Batteries [29] with an aim of providing and unifying clear definitions, key performance indicators and classification of CB. Results of a short bibliographic study from Google Scholar using the main keywords of CB technologies shown in Figure 4 also confirms the trend of an increased interest in these technologies. At the same time, it shows that the unifying notation of these technologies as a Carnot battery is not yet in mainstream use.

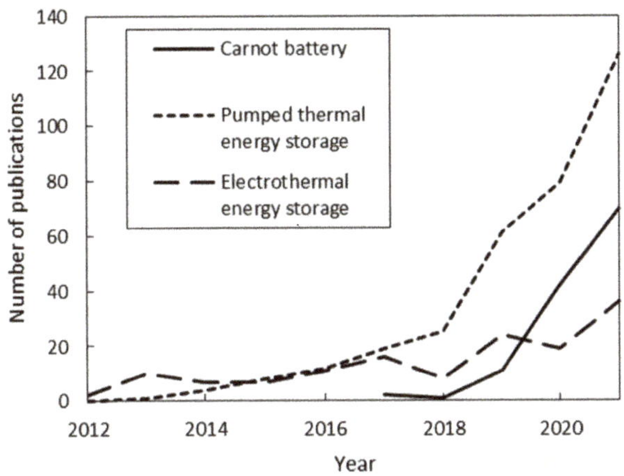

Figure 4. Bibliographic study on a yearly number of scientific works with main CB keywords.

CB as any electricity storage system is specified by its roundtrip efficiency (RTE), which is defined as a ratio between electricity produced during discharging and electricity consumed during charging, see Equation (1).

$$\eta_{RTE} = \frac{electricity\ discharged}{electricity\ charged} = \frac{W_{e,discharge}}{W_{e,charge}}. \tag{1}$$

Alternatively, useful energy efficiency (also referred to as total efficiency) can be defined by Equation (2) as total useful energy output in the case of sector coupling, especially also providing heat and/or eventually cold. This heat can be used from the CB system in the charging phase, separately drawn from storage, but the highest efficiency is obtained if it is a by-product or rejected heat (possibly also its part) from the discharging phase.

$$\eta_{UE} = \frac{W_{e,discharge} + Q_{useful}}{W_{e,charge}}. \tag{2}$$

Note that no heat input is considered even though in some CB concepts it is present. It can be argued that other formulations of efficiency may be also used. From thermodynamic standpoint, exergy efficiency explains the loss of potential and quality of all inputs in the best manner. This work however aims rather at an overview of CB technologies with respect to their prospective application and progress of scientific findings towards much needed commercialization. The heat input is moreover mostly considered as a low or zero cost input, typically in the case of waste heat with no alternative utilization.

1.2. Purpose of the Review

Throughout the increasingly extensive scientific research of the Carnot Battery technologies, commercial development is present as well. Examples of the main technologies were provided in the CB review [23] along with nine identified prototypes built between 2011 and 2020. Only some companies are publishing their findings and information about the CB systems directly in scientific publications as in [30,31]. A comprehensive summary of the CB technologies under commercial development is however missing. The present work addresses this gap, while the commercially developed technologies and presented parameters are also put into context with the approach and results in scientific publications. This provides both qualitative and quantitative views of this field of technology as a certain approach shared by multiple companies can be identified in some cases, while a certain approach is not, on the other hand, explored in the scientific literature. As a comprehensive technology review, it addresses the needs of both academics and industry practitioners, while it might in some cases bridge between these two, especially when pointing out technologies that are insufficiently addressed in scientific research or, vice versa, highly academically studied concepts with very limited industrial development.

1.3. Classification of CB Technologies

As was shown above, many classifications of CB concepts, principles and used technologies are available. The method of charging, discharging, thermal energy storage technology or the conversion system used can be named as examples. In this manuscript, classification by the discharging system is used, specifically as the Rankine cycle systems, Brayton cycle systems and other & hybrid systems. This is in accordance with the categories in the IEA Task 36 on Carnot Batteries [29]. Within each of these groups, further distinction is made between the systems with direct P2H conversion and with the PTES systems using a heat pump principle for charging.

Additionally, there are TES systems under commercial development, which are considered specifically for the CB application, mostly for direct P2H conversion and standard power cycle (Rankine or Brayton) technology. These specific technologies are therefore included as a separate section of this paper.

Within the list of the experimental systems, there are several levels reported. The first one is a proof of concept, largely scaled down system for demonstrating technical feasibility, often with poor roundtrip efficiency or other parameters. A demonstrator is then typically at a larger scale and better parameters but works at the manufacturer site with limited benefit of the operation or operated just for purposes of technical tests. A pilot is already installed on site and providing its designed services to its customer, though installation is largely or fully subsidized. Note, however, that this difference does not need to always be very clear in real systems. The last level is finally a fully commercial system.

2. Carnot Battery Development by Technologies

2.1. Rankine Cycle Systems

Rankine cycle systems can be further divided into steam Rankine cycles, which are predetermined to work efficiently at higher heat source temperatures and high power outputs typically in the order of dozens to hundreds MW and organic Rankine cycles (ORC), which are the domain of smaller power outputs and lower heat source temperatures up to several MW and possibly down to the kW scale. CO_2 cycles, typically with transcritical operation, make up a specific category. Considering the underlying principle, even systems, where the Rankine cycle thermodynamic changes are separated to different time periods, are included here. A typical example is storing heat in a liquefied gas or vapor (latent heat), while the gas itself is a working fluid in the thermodynamic conversion.

Table 1 comprehensively summarizes the CB systems utilizing the Rankine cycle for its discharge phase in the scope of commercial development (or having some commercially oriented aspects). They are sorted according to the power to heat conversion method to the resistance heated systems and fully reverse systems; then by working fluids from water to organic fluid, air and CO_2 and finally to systems with a Rankine cycle heat pump as a full cycle and the systems with part of the heat pump cycle with liquefaction and storage of the working fluid. Note that eight out of 13 reported systems are conducting experimental, demonstration or even pilot operations of their systems.

Table 1. List of commercial development projects in CB using Rankine cycle discharging.

Company, System	Charging Method	TES	Discharging Method	Power Output	Storage Capacity/ Duration	Roundtrip Efficiency	State	Ref.
Siemens Gamesa, ETES	Resistance heaters to air	Volcanic rock bed~600 °C	steam Rankine cycle	units to 100s MW	24 h	25% to >40%	Demo	[32–34]
RWE, Store2Power	Resistance heaters	molten salt	steam Rankine cycle	100s MW	hours	~40%	n.a.	[35]
E2S Power	Resistance heaters	Graphite-aluminium alloy at 700 °C	steam Rankine cycle	1–100s MW	hours	25–40%	lab proof of concept	[36]
Spilling	Steam compression & liquefaction	Saturated water (steam accumulators)	Steam expander (steam engine)	up to MW	hours	n.a.	n.a.	[37,38]
GE, AMSESS	CO_2 Brayton + el. heating	Molten salt, water tank	Steam cycle with extraction	20–100 MW	8 h	42–62%	Concept	[39]
Consortium CHESTER	Heat pump (organic fluid)	PCM and water	ORC	MW scale (8 kW exper.)	hours to days	n.a.	lab proof of concept	[40]
Climeon	Heat pump (organic fluid)	water (e.g., district heating system)	ORC	80 kW to MW	hours	25–60%	concept with existing ORC	[41]
TC Mach	Heat pump (organic fluid)	Stone dust TES	ORC	kW	hours	n.a.	construction of proof of concept	[42]
Future Bay	Heat pump (organic fluid)	water (hot) and PCM (cold)	ORC	10s kW	hours	n.a.	Demo	[43]
Highview	Air liquifaction	Liquid air + other TES	Vaporization, expansion turbine	50–350 MW	about 6	60–70%	Pilot, full scale construction	[44–46]
MAN/ABB, ETES	CO_2 heat pump	120 °C water + cold (ice) storage	CO_2 Rankine cycle	several MWe	~5 h	~45%	lab demo	[31,47–49]
Echogen, ETES	CO_2 heat pump, fluidized bed heat exchange	Sand (hot) and ice (cold)	CO_2 Rankine cycle	25 MW	250 MWh	~60%	design	[50–54]
Energydome, CO_2 battery	CO_2 compression & liquefaction	Liquid CO_2 + other TES	Vaporization, expansion turbine	10–80 MW modules	20–200 MWh	77%	Pilot construction	[55,56]

2.1.1. Electrically Heated RC Systems

Joule heating makes most technical sense in combination with a steam Rankine cycle. No other working fluid is considered in these directly heated systems. The reason, except for being the industrial standard, can be found in its higher efficiency compared to ORC systems and the fact that CO_2 cycles at high temperatures have seen only very few demonstrators built so far (e.g., CSP application [57] or Allam cycle for gas combustion [58]). Technically, the direct heating with steam cycle can be considered as the simplest technology. Furthermore, all electrically heated systems under development include the possibility of converting existing coal fired power plants to storage systems, utilizing the existing infrastructure and saving the typically costliest components of the CB, which is the power cycle unit. The system of this type is referred to as a straight forward one in [59] and the possible scale in the case of a fossil fired plant's partial refurbishment for increased flexibility or complete refurbishment can be found in [23].

Regarding the power output, electrically heated systems are well scalable, with the limitations provided by steam cycle systems. The smallest systems in the MW scale can be applied for example in CHP plants; as solely power to power CB, Siemens Gamesa with an output of 1.2 MW_e serves only as a demo. The largest systems can theoretically consist of blocks of several hundreds of MW each. Electrically heated storage is limited by relatively low roundtrip efficiency, which is determined mainly by the H2P conversion technology (power cycle efficiency) since the electric heater is employed for charging. Furthermore, the other losses occur during operation, namely heat loss into the environment and the pressure drops.

The first small scale Gamesa ETES test rig was built in Bergedorf in 2014 with 5 MWh storage capacity and a 750 kW gas burner. Its purpose was the testing of various storage concepts, materials and setups. This testing unit has run for over 2500 h [32]. Owing to the constructed full system demonstrator, the ETES system of Siemens Gamesa commissioned in 2019 in Hamburg is perhaps the best known; first and so far, the biggest constructed system of this kind. The system uses a horizontal flow packed bed of volcanic rock and air as a heat transfer fluid. With a maximum storage temperature of up to 750 °C the system has 130 MWh_{th} storage capacity. Volumetric storage capacity reaches approximately 0.2 MWh_{th}/m^3 (depending on charge-discharge temperature spread) and its great advantage is in high technology readiness level (TRL) due to the use of commercially available components, the resistive heaters from process industry and heat recovery steam generators (HRSG) utilized in combined cycle power plants. The storage concrete construction vessel is filled with about 1000 tons of crushed volcanic rocks and thermally insulated by aerated concrete and rock wool insulation [32]. The system was designed for 24 h charge and 24 h discharge operation. The rather low roundtrip efficiency (around 25%) is due to the steam cycle efficiency (small scale, low pressure) and auxiliary loads, especially air fans.

Thermal losses can be minimized by optimizing the amount of thermal insulation according to the economics. Pressure drops can be mitigated by packed bed construction and operation (air flow velocity, particle shape and diameter and vessel length to diameter ratio). Nominal storage capacity can further decrease by de-stratification effects due to heat transfer in a packed bed in a charged state [60] (reported for horizontal flow stores). Hence the storage unit must be oversized, which leads to an increase of thermal and pressure losses. Unlike in coal plants, parasitic loads related to its handling or boiler efficiency are however excluded here. The efficiency is expected to significantly increase in larger scale installations to over 42% [33,34,60]. Currently, Gamesa is ready for building the first series of commercial pilots in a range of 10–100 MW power; 100–2000 MWh storage capacity and 300–720 °C steam temperature [32].

The ETES system offers various applications, such as integration into existing power plants or combined heat and power plants, rebuilding conventional fossil fuel power plants into storage units or electrification of process heating. Therefore, Siemens Gamesa has been developing three market types of storage application, for the conversion of fossil power plants, adding storage to existing thermal systems (industrial plants for process heat electrification and heat recovery) and a whole stand-alone system supplying process steam, electricity or district heating [60].

Looking at the chosen horizontal configuration, much smaller scale investigations of rock bed TES found very strong effects of buoyancy, having a detrimental effect on temperature stratification and the utilization of the thermal capacity of the entire volume, further confirming the de-stratification effects. Either the TES capacity is then limited or additional horizontal air tight layers need to be added to the storage tank to maintain air flow uniformly through the entire volume [61].

Similar is the intended application scale of RWE's development, which is, with its project StoreToPower, focusing on an alternative technology of molten salt, adopted primarily from concentrated solar power plants (CSP) and, secondarily, also on solid materials with air as a heat transfer fluid [35]. However, no further public information provides any insight on the ongoing status. Focus on molten salt appears to be a logical choice as in the CSP plants it is a well proven commercial technology [62,63]. With the molten salts, special attention needs to also be paid to corrosion, especially at temperatures above 500 °C [64,65], which are typically found in coal fired power plants.

A P2H2P system with direct electrical heating primarily suggested for refurbishment of coal fired plants is offered by E2S Power [36]. The TES system is using an alloy (miscibility gap alloy, MGA) composed of graphite and aluminum and developed by a partner MGA Thermal. The MGAs are commonly proposed as a novel type of TES material with some unique properties such as macroscopically solid materials safely embedding PCM, providing combination of sensible and latent heat and with high thermal conductivity [66,67].

Blocks of the MGA are electrically heated during charging, half up to 700 °C while the second half only to temperature required by the steam turbine. The colder blocks then, during discharging and steam generation, serve for self-regulation of the steam outlet temperature. A laboratory system has been built demonstrating the TES with charging and steam generation, while plans exist for a 50 MWh_{th} pilot at a coal power plant in Montenegro.

2.1.2. Reversible RC Systems

Reversible systems use the principle of a vapor compression heat pump in the charging phase and standard RC during discharging. Various configurations were proposed theoretically for steam systems with a suggestion to further improve the roundtrip efficiency by integration of heat sources to the heat pump input [68]. Together with the additional possibility of various working fluids (steam, organic fluids, CO_2), direct or indirect use of the cycle working fluid for storage and choice between latent or sensible TES, there is a wide range of available technologies.

Starting from the most typical RC using water as a working fluid, the company Spilling, manufacturer of steam turbines, compressors and engines, entered into the development of CB. The system consists of a reversible steam engine/compressor and two steam accumulators, one at low pressure and other at high pressure [37,38]. A major advantage of such system is a combination of existing technology of steam engine from the company portfolio, while steam accumulators are a well-established technology over hundred years old. The accumulators however require relatively large, pressurized tanks for the saturated water. This makes the technology of steam accumulators suitable rather for industries and small power plants; one of the largest applications is a 20 MW_{th} (50 min at 5.5 MW_e) application for a CSP plant. The addition of PCM to the tanks is proposed to improve the thermal capacity at the same volume. A drop in the pressure during the discharging of the tanks is another disadvantage [69,70]. For the CB applications, a similar size can be expected as that for similar systems for small steam parabolic trough flexible plants as 2 MW/6–24 MWh units [71]. The pressure drop loss might be partly compensated for by sliding pressure in both charging and discharging process. Investigation of parameters of this CB concept are not, to the authors' knowledge, supported by any scientific literature.

GE has in the past proposed a system using a recuperated CO_2 Brayton cycle operating as a heat pump for charging with a steam Rankine cycle discharging as an advanced molten salt electrical storage system (AMSESS). High temperature heat is stored as a molten salt while low temperature heat input is provided from a water tank and the water is thus cooled down. An additional electric heater can be used to increase the salt temperature. During discharging, heat is transferred to the steam cycle, where most of the rejected heat goes into the environment, while a smaller portion such as a turbine bleed heats up the water tank. A design with off-the-shelf components has been performed for 20–100 MW_e power output over 8 h period with RTE ranging based on size and presence of the heater between 42% and 62%. An interesting note is that the space requirements are about one third of the same capacity in containerized lithium batteries [39].

Another set of technologies builds on standard heat pumps and subcritical organic Rankine cycles. These systems are specifically considered as well fitted to waste heat sources, or other low temperature heat sources, such as solar collectors or geothermal heat. The availability of the heat source then decreases temperature lift of the heat pump (or provides a heat source in less considered concept, where the heat sink is below ambient by cold energy stored and prepared in time of excess electricity by refrigeration cycle). Roundtrip efficiency can, as a result of heat input, theoretically exceed 100% [72]. Regarding the size and cost, storage tanks need to be carefully considered as in the case of low temperature glide; even a small power output requires large tanks, which might be costly for long duration storage. The power output can be expected from dozens kW up to no more than several MW as is the size of large CHP or geothermal ORC systems [73]. This concept gained much research attention, where theoretical investigations are performed in many institutes and several experimental systems are in experimental operation or under construction [74–76], all in kW scale.

Looking at the commercial development, the list is, however, rather short. Within the CHESTER research project [40], universities join research institutes and companies to focus on business models and market opportunities for such application, meanwhile one industry within the project, for example, focuses on the development of an isobaric expansion device considered for one of the concepts [77]. The main CB concept utilizes both sensible and latent heat storage to maximize efficiency and match temperature profiles during charging and discharging, while providing thermal integration to district heating. A kW scale experimental system is under development. Similarly, ORC unit manufacturer Climeon offers its units of output around 100 kW_e for CB applications with storage also proposed as district heating infrastructure including large seasonal storage tanks and up to MW_e size total output. Additionally, greenhouses and geothermal systems are suggested as application cases for CB. Except for the ORC conversion unit, the system is however in a conceptual state [41]. Another example is a heat pump manufacturer, TC Mach, developing together with a university a high temperature heat pump, TES. made of compacted stone dust and an ORC, where the constructed proof of concept unit is going to be sub-kW scale [42].

The only larger scale ORC CB system in state of a pilot application is developed by Futurebay [43]. This system is considered in several applications, with or without thermal integration of waste heat source. Without any external heat source, a heat pump is used to charge both hot and cold storage during a period of excess electricity and these stores are then utilized separately. Heat is stored in hot water tank while PCM storage (probably ice based) is used for cold storage. Both hot and cold can be used directly or the ORC can run between present temperature gradients—hot store and ambient (eventually space heating temperature) or waste heat to cold store. The system is considered primarily for integration into thermal systems with both cooling and heating requirements. A containerized demonstrator delivering 50 kW_e/200 kWh_e and cooling capacity 2 MWh_{th} has been built and operated; scalability is however suggested up to a 12 MW_e/72 MWh_e grid scale system.

Liquid air energy storage (LAES) is in principle also a Rankine cycle as it utilizes the phase change of the working fluid for storage and especially in the discharging phase, a typical cycle's heat input/output takes place. The working fluid is at the same time also a storage media in an open cycle. The air is compressed and liquefied during the charging by a relatively standard industrial technology. Special attention is paid to the conservation of thermal energy during the liquefaction process, both cold and hot, as it is essential for the high roundtrip efficiency of the system. Thermal energy is in this concept stored therefore threefold—in the liquefied air and near-ambient pressure as latent heat, thermal energy recovered after compressors' outlets and cold energy from the gas cooling before expansion and from recycled streams.

The advantage of LAES lies in relatively small storage volumes (in the order of 700 times smaller than those required for CAES) due to the higher energy density in liquid air. The LAES system also relies on commonly used components in industry (compressor, liquefier, turbine, etc.). Today's studies show that it is possible to achieve roundtrip efficiencies around 70%, with a specific investment cost of 1270–2090 €/kW [78]. To date, several studies have been published on various LAES configurations. In [79] they proposed integration with a conventional combined cycle power plant; in [80] they studied the LAES system integrated with a nuclear power plant, and many publications paid attention to the recovery of waste heat from LAES using ORC, which increased RTE by up to 12% [81–83].

The only commercial development is carried out by the British company Highview power. In 2011 their first pilot plant was launched (350 kW/2.5 MWh), which was tested on a biomass plant site and is now located at the University of Birmingham. The achieved roundtrip efficiency was only 8%, therefore the second Pilsworth Grid Scale pilot power plant (5 MW/15 MWh) was built in 2018. Highview power is now developing a 50 MW commercial plant in Carrington Village with a storage capacity of 250 MWh in cooperation with MAN Energy Solutions. The commissioning is expected in 2022 [44–46]. Another project is planned with the expected start of construction in 2023 in Chile with a power rating of 50 MW and a capacity of 500 MWh [84]. In the future, Highview power plans to offer LAES systems in a relatively large range of outputs from 20 MW/80 MWh to more than 200 MW/1.2 GWh [85].

Rather favorite among CB appears to be the utilization of CO_2 cycles, which otherwise struggle to find their place in other energy systems. Development of the ETES system of MAN and ABB aims at a very peculiar and rather low temperature system with a reversible transcritical CO_2 Rankine cycle. The system has the highest storage temperature only around 120–150 °C utilizing pressurized water and cold storage, and using ice as PCM. The hot storage system is divided into four tanks at different temperatures into which the heat is transferred via three separate heat exchangers. It is to balance the optimal mass flow rate and heat exchange temperature profiles due to the change of the supercritical CO_2 heat capacity, while the temperature differences along the whole length of the heat exchangers are designed in order of several Kelvins. Only one hot water tank needs to be pressurized, while the pressure is still moderate. During charging, part of the energy in high pressure CO_2 is recovered by a hydraulic turbine while rest below the saturation line is flashed to exclude the issues of two-phase expander. The resulting system is then a result of many techno-economical optimizations under constraints of isentropic efficiency of key turbomachinery components (the company's state-of-the-art) with the roundtrip efficiency reaching values of 38%–50%. The target size is an 8.5 MW_e system with 8 h charging. Development of this system is well documented in [31,47–49]. Currently, an MW scale laboratory demonstrator has been developed to test the system and equipment. The commercialization strategy aims at the possibility to use also standalone system of only heat pump or heat pump with storage, able to provide 5–50 MW_{th} of heat (3–30 MW_{th} cooling) with 2–15 MW_e power input. As such, a 50 MW_{th} seawater heat pump for district heating using this technology is to be built in Denmark by 2023 [86].

The MAN–ABB consortium is not the only one representative of CO_2 cycle utilization, even though they are clearly closest to application. Echogen [50], known mainly for its waste heat recovery 8 MW unit with CO_2 as a working fluid is also working on a CB system for more than 4 h and 10 MW_e size. Available information discloses a concept utilizing the CO_2 reversible recuperated cycle, low temperature storage also being ice/water storage (as brine for −2 °C to −10 °C) and high temperature storage being at 300–350 °C in the form of sand in silos or alternatively concrete blocks. The temperature has been selected to be within the limits of standard construction materials [51–53]. A proof-of-concept in a 100 kW_{th} scale heat pump and sand storage and heat transfer system has been developed while a 25 MW_e 8 h prototype system is in a design phase [54].

The CO_2 has been also proposed for a system using a similar principle to the LAES by the company Energy Dome [55]. The gaseous CO_2 is stored at ambient pressure in a large container (dome). During charging, it is compressed, liquefied and stored in tanks, while the heat (from intercooling and condensation) is separately stored in a TES. The discharging process then reverses the flow, high pressure CO_2 is evaporated by the stored thermal energy and expanded in the turbines back to the low pressure gas store to produce the electricity. The sequence of charging and discharging can be considered as thermodynamic changes in an open Rankine cycle. To maintain constant conditions at the compressor inlet, a flexible membrane is employed within the dome. TES is separated into five sections at different temperatures in configuration with a multiple section compressor and turbine with intercooling and reheating. Main advantages of this concept are mentioned as high energy density at moderate pressures. The calculated net roundtrip efficiency can reach about 77% [56]. Note that similar systems were proposed before, but with liquid CO_2 storage at a low pressure and with a maximum roundtrip efficiency of 57% [87]. The system has recently progressed from a concept with basic sizing and costing to the construction of a 2.5 MW_e/4 MWh pilot system scheduled to be finished in 2022 and in the same year there is planned the start of a full-scale 20 MW_e/100 MWh system construction [88].

2.2. Brayton Cycle Systems

Brayton cycles are favored for applications with sensible heat TES materials as during the isobaric heat addition and rejection, it is possible to obtain a well matching temperature profile, minimizing exergetic losses. Other advantage can be the maturity of the gas turbine industry, from which many systems are derived. The summary of the commercial development in Table 2 shows that the extent of the projects is smaller than in the case of the Rankine cycle systems (opposite trend to the publications numbers in academic research). It can be argued that one reason is in the requirement of very high efficiency of compressors and expanders to which the systems are highly sensitive, while such components have a limited industrial supply.

Table 2. List of commercial development projects in CB using Brayton cycle discharging.

Company, System	Charging Method	TES	Discharging Method	Power Output	Storage Capacity/ Duration	Roundtrip Efficiency	State	Ref.
247Solar, Heat2Power Turbine	Electric resistance heaters	Silica sand	Gas turbine (Brayton cycle)	200 kWe–100s MW	6–20 h	30%	Concept, design	[89,90]
1414Degrees, TESS	Electric resistance heaters	Silicon based alloy, melting temperature 1414 °C	Gas turbine (also steam turbine, Stirling engine, direct heat)	10 MW-GW	n.a.	n.a.	Demos (done) Planning grid scale pilot	[91]
Peregrine Turbine Technologies	Electric resistance heaters	Graphite-aluminium alloy (MGA), 800 °C	CO_2 Brayton cycle	1 MW	8 MWh	45%	CO_2 turbine/ compressor tests	[92]
Isoentropic	Heat pump (Brayton cycle -reciprocating devices)	Crushed rock packed bed	Brayton cycle (reciprocating devices)	2 MW (exper. 150 kW)	16 MWh	72%	Demo (bankruptcy)	[93]
Malta, Pumped Heat Energy Storage	Heat pump (reverse Brayton cycle)	Molten salt + hydrocarbon antifreeze	Recuperated Brayton cycle	10–100 MW	80 MWh–1 GWh	n.a.	Concept, design	[26]
Stiesdal, GridScale	Heat pump (reverse Brayton cycle)	Crushed basalt rock packed bed	Brayton cycle	2 MW–1 GW	100,000 MWh	35–60%	Pilot construction	[94]
Enolcon, OPTES	Heat pump (reverse Brayton cycle), N_2 or Ar	Silica sand packed bed (silica sand, iron based sand, basalt)	Brayton cycle, N_2 (or Ar)	~8 MW	~80 MWh	58–66%	Concept, design (Pilot designing/ constructing)	[30]
WindTP	Heat pump (reverse Brayton cycle)	Gravel bed, indirect heat transfer	Brayton cycle	3–20 MW	up to 100 h	up to 85%	component demo	[95]

2.2.1. Electrically Heated BC Systems

In order for the electrically heated Brayton cycle to be efficient, very high storage temperatures need to be reached, namely exceeding 1000 °C. However, the need of reaching such a high temperature for increase in roundtrip efficiency also leads to material constraints. Considering a parallel to the scientifically reported development in Brayton based PTES, interestingly, a very small number of publications is focusing on this concept.

A U.S. company, 247 Solar, provides a concept of an electrically heated CB, which combines subsystems named the 247Solar Heat2Power Turbine and the 247Solar Thermal Storage System. These subsystems are proven in CSP applications (Brayton based solar towers), where there is a specific solar irradiation receiver and vertically orientated thermal storage, both using air as a heat transfer fluid operating under conditions of near ambient pressure and high temperatures (970 °C) [90].

During charging, the electric heater (resistance coils) is employed for heating the air flow by using a blower, both supplied by surplus electricity, and afterwards the air passes the heat to the sand thermal storage. The discharging process employs a classic open loop Brayton cycle, meaning that the inlet ambient air is compressed, then heat is supplied in two heat exchangers (low and high temperature) and next the air expands in the turbine. To increase the overall system dispatchability, a burner for various types of fuel (including hydrogen) can be added. The presented roundtrip efficiency of this system amounts to 30%, nevertheless the combined heat and power production efficiency can reach more than 90%. 247 Solar considers standard configuration with nominal 200 kW_e power output and thermal storage capacity 1.8 MWh that corresponds to 8–10 h duration. This project is evolving in collaboration with Capstone Green Energy, the provider of customized microgrid solutions and on-site energy technology systems [96].

1414 Degrees Limited is an Australian company, which develops Thermal Energy Storage Systems (TESS). The novelty of this system lies in using molten silicon PCM with melting temperature 1414 °C, which enables reaching high energy densities and potentially high efficiencies of conversion back to electricity and usable heat [89].

For the charging of TES, an electric resistive heater is utilized. Another option for heating the molten silicone storage is combusting gas, for instance, methane generated from wastewater plants. The advantage of using silicon is in its high latent heat capacity in comparison to other PCMs. The silicon's sensible heat capacity below the melting point is up to 300 kWh/t; nevertheless, the latent heat supplied during the solid to liquid phase change at the constant temperature of 1414 °C is about 500 kWh/t. The operation of TES at such a temperature brings many material challenges. However, several well-established industrial processes are treating material with very high temperatures, for example the cement industry temperature level exceeds 1400 °C. Therefore, 1414 Degrees implements materials and technology knowledge from these proven technologies. The discharging process can be achieved by several conventional heat engines, where the most effective plants achieve over 60% thermal efficiency. Hence, 1414 Degrees considers four options of discharging. In terms of combined power and heat production for reaching the maximum overall roundtrip efficiency, the gas turbine, steam turbine or a Stirling engine can be employed; lastly, a steam generator or heat exchanger can provide steam or hot clean air for industrial processes [89].

The TESS is designed as a modular system; when connected in series it increases the power output and in parallel configuration the storage capacity rises. 1414 Degrees plans four main commercial applications; one oriented on bulk medium-long term energy storage, one providing medium scale energy storage for industries and residential developments requiring power and heat. The last one is developed to accommodate combustion waste gas (e.g., sewage treatment) and store the thermal energy for later recovery (for electricity and heat). As such, it is similar to the industrial application but with modified charging. Lastly, the TESS industrial system producing steam is best fitted for industries with high thermal rather than electricity demand, especially for replacing natural gas with renewable sources. Using a steam turbine for electricity generation is only secondary [91], but possible for example using extraction turbines providing also heat at lower parameters.

1414 Degrees built up the first demonstration plant in 2016 and is still operating it in its R&D facility in Lonsdale, South Australia. The demo comprises of 200 kg silicon PCM storage charged by 50 kW heater and was using 43 kW Stirling engine for discharging, in 2018 replaced by a gas turbine. In 2018, a 10 MWh_{th} GAS-TESS module was also put in operation at the Glenelg Wastewater Treatment Plant. The next step of 1414 Degrees is to acquire SolarReserve Australia II, which owns the Aurora solar project [97]. The current plans of the Aurora project, led by the 1414 Degrees company, are to build up a 70MW PV farm and 150 MW CSP plant, coupled with several thousand MWh_{th} storage capacity [98].

The use of a supercritical CO_2 in a BC has been proposed by Peregrine Turbine Technologies as a CB along several other use cases [92]. The system uses the MGA from MGA Thermal as already considered in Rankine cycle systems from E2S, here with temperatures of 800 °C. The power cycle uses a partially recuperated configuration with compression from right above the CO_2 critical point [99] and the whole system is designed as a 1 MW_e

modular system with a storage capacity of either 8 MWh_e or modular with 5 MWh_e increments. The unit has an expected CAPEX of 280 $/kWh and a lifetime of 20 years. The company has tested their compact turbine with a CO_2 compressor and plans full scale pilot construction for 2022 [92,100].

Contrary to these commercially developed systems, in the scientific literature the only application of the electric heaters with respect to the Brayton cycle has been considered in the form of additional upgrade and control of the gas temperature at the compressor outlet in works such as [13]. The purpose of adding an electric heater is the increase of the maximum storage temperature of the hot reservoir that increases the efficiency of the reconversion of heat into work during discharging and also significantly increases the energy storage density. Nevertheless, the implementation of an electric heater during charging leads to a reduction of the COP of the Brayton cycle heat pump, which may cause a decrease in roundtrip efficiency. As such, it is later shown to be proposed in the conceptual phase of one commercial reversible system development.

2.2.2. Reversible BC Systems

The development of a company, Isentropic Ltd., focused on a Brayton cycle using a reciprocating reversible piston machine capable of working as both compressor and expander together with a fine gravel storage in hot pressurized tank and a cold tank at ambient pressure. The company itself, established in 2012, was placed in administration in 2016, before finishing its "grid scale" 150 kW experimental demonstration system. It serves as an example of the consequences when the market is not yet ready for the developed systems. Many lessons and information can be learned, as much of the technological development has been scientifically reported. First, three small prototypes and the conceptualization of a commercial size unit were reported in [93]. The development, originally conducted with Newcastle University, was consequently picked up by Durham University, where the 150 kW demonstrator was built and experimentally tested [101].

The piston compressor/expander component utilizes a uniquely developed sliding valve approach, which is designed to minimize the pressure losses, reduce dead space, provide a high quality seal without excessive precision requirements and a very fast valve actuation speed [102], which under electrical control allows for a rapid change from charging to discharging regimes in less than one second. The storage is composed of a stack of grates with magnetite sand (~1 mm) with actively controlled flow only through the selected ones and bypassing others to limit pressure drop. The designed working fluid, argon, has been for the tests substituted by nitrogen due to leaks and the system was operated at reduced pressure, speed and power. The 150 kW_e designed system was operated at about 10 kW_e charging and discharging power, still reaching roundtrip thermodynamic efficiency (drive excluding mechanical and electrical losses) around 77%, while the prospect of likely final system performance with argon at full speed has been determined from the data as 73% roundtrip efficiency at about 125 kW_e charging and 100 kW_e discharging power [101]. The originally intended scaled up 2 MW_e/16 MWh_e commercial (modular) unit was analysed to have a LCOS of 9 to 11 €ct/kWh including given purchase electricity price and 2 to 5 €ct/kWh excluding them, while CAPEX were 350–797 €/kW_e of power and 13–21 €/kWh_e stored [103].

The concept developed by the company Malta follows an approach detailed by Laughlin [26]. The described concept is based on a closed Brayton cycle, which is transporting heat from a cryogenic fluid storage to a molten salt reservoir, when it operates in the charging regime. In discharging mode, the heat flow is reversed and runs the gas turbine, so the electricity is supplied back to the grid, when demanded. The Laughlin's storage system consists of a compressor, an expander, two molten salt hot storage tanks and two hexane coolant cold storage tanks, two heat exchangers for heat transfer between the heat transfer and working fluid (HTF) and a recuperator.

This system was developed in the early stages by Professor Laughlin, the company Brayton energy and Nick Cizek, who designed the baseline of closed cycle gas turbine with

Ar (or He) as a working fluid and later sold the rights to Google's (Alphabet) moon-shot factory X in 2016 [104]. After two years of incubation in X, when the potential of the grid scale storage technology cleared up, Malta became an independent company [105].

Nowadays, the novel Malta system, named Pumped Heat Energy Storage, is based on a closed loop, recuperated Brayton cycle. The purpose of recuperation is to get better minimum and maximum storage temperature spread (based on industry experience) and an increase of efficiency. The maximum temperature 570 °C is limited by material creep and corrosion, while the minimum temperature of −70 °C is set by material embrittlement and air (working fluid) drying requirements [106]. Malta's final goal is a full scale 10–100 MW_e facility in 2024, enabling the low-cost storage (unit CAPEX less than $100/kWh). Malta is currently developing a kW-scale demonstration facility, which should help with understanding the limitations of operational modes (start-up and shut down), especially the validation of transient analysis. This demo is characterized by lower costs and risks, due to its small scale. Contrarily, the small scale of turbomachinery and temperature limits of storage media causes significant roundtrip efficiency disruption, so the prediction of RTE is 10% [107]. Malta has also signed a term sheet with Canadian utility NB Power about establishing an Energy Storage Benefits Agreement to advance the first long-duration energy storage facility in the New Brunswick province. The planned 1000 MWh facility would help to achieve New Brunswick's emission reductions, improve grid stabilization, increase the integration of renewables, and bring new jobs into the region. However, the project of the storage facility is still in the early stages, the target year for the start of operation is 2024 [108].

GridScale is the official name of the concept developed by Danish company Stiesdal. Stiesdal's goal is to develop a storage facility for cost-effective electric storage, which is scalable and based on industrialized components. The application field of GridScale is a medium duration storage, which more specifically balances day-to-day production of photovoltaics (12–18 h) and smooths the wind farms production on several days (3–7) periods. GridScale is a reversible Brayton cycle Carnot battery, which uses low cost crushed rock packed beds as thermal storage and offers relatively high roundtrip efficiency without location limitations [94].

In the charging regime one pair of compressor-expander is employed for pumping the thermal energy from cold reservoir into the hot storage, due to the use of surplus electricity from the grid. The maximum temperature of hot reservoir is 600 °C and the minimum temperature of cold tank is designed at −30 °C. The COP of Stiesdal's configuration is determined approximately on 2.5 (depending on the temperature levels). During discharging, the system is operating as a regular gas turbine (Brayton cycle) with another similar compressor-turbine system for charging. The predicted efficiency of the gas turbine during discharging is 20%–25%, also depending on the temperature ranges. The range of GridScale's roundtrip efficiency, derived from COP and Brayton cycle efficiency, is 55%–60%. The useful energy efficiency could be increased to roughly 90% by using the rejected heat from the cycle during charging for district heating [94].

The project GridScale has been supported by grants and subsides and is developed in a partnership with Andel, Aarhus University, Technical University of Denmark, Welcon, BWSC, Energi Danmark and Energy Cluster Denmark [109]. The numerical models of GridScale have been validated under various conditions since 2019 at the test facility with a 1:10 scale steel tank. This demo facility reached similar heat transfer from heat transfer fluid (air) to storage media (basalt crushed rock) as theoretically modelled, and achieved even lower pressure drops than expected [109]. In terms of the serial manufacturing of storage units, Stiesdal can utilize extensive experience from welding the tubular steel towers of wind turbines. The company Stiesdal is preparing a demonstrational project, which will be placed in Rødby on the Danish Island of Lolland and installed next to Rødby's district heat and power plant. This demo plant is planned to operate in commercial regime with a 4 MW_e/2 MW_e charge/discharge power and 10 MWh_e storage capacity [94].

Enolcon GmbH in [30] presented their own Brayton based Carnot battery called OPTES Battery. The baseline design of OPTES consists of common closed Brayton cycle components: expander; compressor and two sand packed bed storage units and two heat exchangers. Alternatively, the bottoming ORC cycle using otherwise rejected heat after expansion in discharging regime can be added, in order to increase roundtrip efficiency. In the case of sufficiently cheap electricity input, an electric heater is a feasible solution for "boosting" temperature before inlet to the hot tank during charging. That enables a high level of temperature regulation, nevertheless it also causes a drop of roundtrip efficiency. The silica sand packed bed is developed by Enolcon's partner STORASOL GmbH and has been demonstrated together with an ORC unit, named the ORCTES-plant, at the University of Bayreuth in Germany, where it has been in operation since 2015 [110]. The parameters of this demonstrator are 1.5 MWh_{th} storage capacity, 1.8 MW_{th} of charging power and 600 °C operation temperature of the storage [111].

Currently, Enolcon is developing a proof of concept of the OPTES Battery power plant, which they plan to launch in 2023. Their CB should be able to reach approx. 40%–45% roundtrip efficiency even with off-the-shelf equipment. But the goal is to develop companders (devices combining turbocompressor and turboexpander) with nitrogen as a working fluid in a cooperation with Atlas Copco, which sustains higher pressures and temperatures above 715 °C and therefore reach higher efficiency. Hence, the CB with advanced turbomachinery achieves roundtrip efficiency roughly around 63%–67%. However, it is necessary to mention that the considered isentropic efficiencies in the described model are relatively high, specifically 93% for the expander and 91% for the compressor in a "realistic case" [30].

The capital expenditures (CAPEX) of OPTES Battery with a storage capacity of 80 MWh_e and electrical storage power of 7.6 MW_e were specified for 16–20 million EUR, meaning that the specific installation costs of electricity storage of such a system are estimated at 200–250 €/kWh_e. In comparison, the specific costs of electricity storage of current commercial large-scale Li-on battery storage in Germany are according to [30] approximately 450 EUR/kWh_e. Specifically, the Li-on battery project in south-eastern Brandenburg with 53 MWh_e storage capacity and CAPEX of 25 million EUR, which was under construction in 2020, is mentioned. The determination of CAPEX is coming from the market prices based on Enolcon's knowledge in project management, engineering and design of large technical facilities (process systems, power plans, etc.) [30].

Lastly a CB system designed specifically for integration into the wind turbine is a system designed by WindTP [95] in cooperation with the University of Nottingham and the University of Leeds [112]. The system is thus primarily charged not by electrical power but instead by the mechanical power of the wind turbine shaft. The work can be transferred directly to generator via subsequent expander as aerodynamic coupling. When energy storage is required, shaft drives a low speed multiple stage high efficiency piston compressor to compress the working gas from 2 to 50 MPa and around 600 °C. A separate optional compressor supplied with electricity from the grid may be included as well. After heat transfer to the hot storage, expansion takes place and cold energy is stored as well. Rock bed is assumed as both hot and cold TES. Research has shown a feasibility of this solution for wind turbines when the stored electricity accounts for at least 25% of the produced power. A specific multistage slow speed piston compressor concept is proposed for high efficiency [113], which is currently in experimental development [7] and can be utilized not just for wind direct driven CB but also CAES systems [114].

2.3. *Other Cycles & Hybrid Systems*

Other and hybrid systems classification can be also divided into the electrically heated ones and ones using reversible cycles. Their commercial development is comprehensively summarized in Table 3. One can notice a presence of proofs of concept and pilot experimental systems and even already started commercial applications, surely a positive indication towards CB applications. A specific aspect of this group of technologies is that the heat to power conversion is performed not only by thermodynamic cycles, but also by systems of direct conversion.

Table 3. List of commercial development projects in CB in the other and hybrid discharging cycle category.

Company, System	Charging Method	TES	Discharging Method	Power Output	Storage Capacity/ Duration	Roundtrip Efficiency	State	Ref.
Azelio	Electric resistance heaters	Aluminium based PCM (600 °C)	Stirling engine	13 kW	13 h	~30%	Multiple pilots, production line, commercial	[115]
CCT Energy Storage	Electric resistance heaters	Silicon based PCM (1400 °C)	Stirling engine	5–100 kW	up to 1.2 MWh per module	n.a.	Pilot	[116,117]
TEXEL Energy Storage	Electric resistance heaters	Metal hydrides (MH)/metal carbonates	Stirling engine	30 kW	15–720 kWh	40%	commercial installation effort	[118]
Kraftlagenn München	Electric resistance heaters/ waste heat	Ceramic system (1000 °C)	Stirling engine and ORC	60 kW (demo)	1.4 MWh_{th} (demo)	n.a.	demo	[119]
NREL ENDURING LDES (GE, PEI, Allied)	Electric resistance heaters	Fluidized packed bed with solid materials (1100 °C)	Combined Brayton and Rankine cycle	50–400 MW	10–100 h	50–55%	Components prototypes, demo preparation	[120]
Pintail Power, Liquid Salt Combined Cycle	Electric resistance heaters	molten salt	Combustion combined cycle integration	from dozens MW	24 h	82–96% (+ fuel)	Patented concepts	[121]
Pintail Power, Liquid Air Combined Cycle	Compressor for air liquefaction	liquid air	Combustion combined cycle integration	from dozens MW	24 h	118% (+ fuel)	Patented concepts	[122]
Antora Energy	Electric resistance heaters	Graphite blocks (1500 °C)	Thermophotovoltaic cell	0.1–1 MW	10 MWh_e	40%	Proof of concept	[123]
NaCompEx	Compresed heat followed by desorption	NaOH-H_2O solutions via concentration difference	Expansion followed by absorption	10–100 MW	60 kWh/m^3 storage	80%	Concept, design	[124,125]

2.3.1. Electrically Heated Other & Hybrid Systems

As with electrically heated RC or BC systems, electric resistance heaters are used in these systems for charging. The discharge can be realized either by thermodynamic cycles, for example by a Stirling engine or a combined Brayton with Rankine cycle, or by technology based on the direct conversion of thermal radiation to power (thermos-photovoltaic cell, multi-junction photovoltaic cells or hybrid thermionic photovoltaic cells). Thermal energy in these systems is stored at temperature levels from 600 °C up to even 2400 °C. All the systems can be therefore classified as high temperature ones.

Systems using a Stirling motor as the discharge method can be an interesting alternative to electrically heated RC and BC systems specifically for low power output in range of several to dozens kW and providing combined heat and power output. The development is probably furthest at the company Azelio, which has, for example, pilot installations in Morocco and Abu Dhabi and is already delivering first commercial units [126]. The system is using a sodium circuit as a heat transfer fluid to deliver the heat from recycled aluminum-silicon alloy PCM to the Stirling engine. The unit provides 13 kW_e and heat at temperatures 55 to 65 °C over 13 h of storage capacity. Systems up to 100 MW are suggested to be comprised of multiple modules [115]. Development of this system is well documented from previous application of this Stirling engine technology and TES for CSP systems [127,128], while the Stirling engine development can be traced even to submarine systems [129]. Based on these sources, the roundtrip efficiency of this CB system can be estimated at around 30%–40%. In CHP application, the useful energy efficiency is around 90%.

A similar system has also been suggested for balancing small scale renewables in [130], where the considered engine has efficiency of only 21% at 2.5 kW_e output without thermal output (cold cylinder at 30 °C). Thermal oil, molten salt and sand were explored as a storage material and economics was evaluated under a selected operation regime. The development of similar CSP applications using Stirling engine systems, often including various TES types, can be found in a number of other scientific publications [131,132].

This can be seen as the reason the Stirling engine also became the subject of quite a number of other commercial CB developments. Among them is the CCT Energy Storage, which unveiled its pilot project in March 2019 and was on track to install its first commercial facility [133]. The system uses phase changes of silicon to store thermal energy (PCM), which is then used to generate electricity using a Stirling engine. The installation can scale from 5 kW_e applications to a theoretically unlimited size and the estimated cost is only 60%–80% of the cost of equivalent storage with lithium-ion batteries [117].

Stirling engine can be coupled also to suitably sized thermochemical (sorption) TES and work as a CB as shows the company TEXEL Energy Storage. Its system is based on the V4-Stirling converter, originally developed by Ford Motors. The Stirling converter produces approximately 40% of electricity and 50% of thermal energy, which adds up to 90% of total energy efficiency [118]. The second basic element of the TEXEL system is the thermochemical thermal energy storage. The company uses technology based on metal hydrides. The advantage of metal hydrides is the high energy density and the ability to work with temperatures above 600 °C. The TES consists of two parts, a hot part containing high temperature metal hydride (HTMH) and a cold part containing low temperature metal hydride (LTMH). When charging the TES system, hydrogen desorbs from HTMH and the desorbed hydrogen is then stored in LTMH. During discharge, hydrogen is absorbed back on the HTMH side, releasing heat to drive the Stirling converter [134]. For example, according to [135], it is possible to use a material pair of CaH_2 and TiFe with the possibility of storage in the temperature range of 800 to 1000 °C and with a TES price of \$ 15/kWh_{th}. Useful metal hydrides can furthermore be, for example, $NaMgH_3$ or TiH_2. TEXEL also started the development of TES technology based on metal carbonates MCO_3 where M is the metal cation (Mg, Ca, Sr, Ba, or Pb). This system works similarly to the metal hydride technology but uses CO_2 as its working gas instead of H_2 [118,136].

Another proposal of Stirling engine is a project of Kraftanlagen München in cooperation with Solar—Institut Jülich, Germany. It is a combined heat and power project with the useful energy efficiency of up to 80% (roundtrip efficiency unspecified). Electricity production is to be ensured through a combination of Stirling engine and ORC. The multi-functionality of the device should consist in the choice of heat source-either electric heaters or waste heat. The system is using a ceramic TES system with intended more versatile application [119,137]. The TES system is also marketed and described separately among the TES systems for CB in following section.

Only one scientific work focuses on an investigation of CB systems involving Stirling engine on a system level [130]. Separate description of major components as TES or conversion systems is not uncommon, though originally for different applications. This shows also the dynamics of the CB development, which can effectively repurpose technologies from different applications, typically, but not only, CSP.

A technology using electrical heating with TES temperatures around 1200°C, which uses a combined cycle of gas turbine and steam cycle for discharging is investigated in an NREL research project, ENDURING, with the strong involvement of GE and Babcock and Wilcox on a power cycle and the component manufacturing side and Allied Mineral Products, Inc. focusing on TES. The TES consists of a silica sand, which requires specific particulate handling system and heat exchangers based on fluidized bed principle. Techno-economic analyses are performed for a 50–400 MW_e and 10–100 h of duration with the LCOS estimate below 0.05 \$/kWh. With the combined cycle and high temperatures, the roundtrip efficiency has set a baseline at 50% with a technological prospect of around 55% gross efficiency. Hardware-wise, laboratory prototypes of the components are being developed and plans aim towards MW scale pilot demonstration [120,138]. The concept of storage in hot sand is also based on previous projects aimed at the CSP application [139].

A specific approach with the integration of the CB principle into a combustion gas turbines combined cycle has been taken by Pintail power. Its first system uses electrical heating of a molten salt, which is then, during discharging, used for the evaporation of the water in the bottoming steam cycle [121]. This approach can substantially increase the steam cycle power output and provide a good temperature match between the flue gas and water in the heat recovery steam generator. The second system is a hybrid LAES system, where the air is liquefied during charging. During discharging the cold fluid is then used first as a heat sink of a bottoming ORC, then to pre-cool the combustion air before it is finally heated by the flue gas and expanded in an air turbine [122].

The last group of technologies in this section are technologies that do not use a thermodynamic cycle for discharge, but various types of direct heat to power conversion-thermophotovoltaic cell, specifically modified multi-junction photovoltaics or hybrid thermionic photovoltaic converter. These systems work with very high temperatures, typically from 1500 to 2400 °C, placing new technological demands on the systems, which developers have to manage. Antora has developed a thermophotovoltaic cell working with solid carbon heated to temperatures just above 1000 °C with 40% efficiency. While the cells are separately marketed products, a 100 kWh_e prototype of CB system has been built. The company aims towards efficiency over 50% and manufacturing CB as 10 MWh_e blocks at 0.1–1 MW_e and at the cost of less than 10 $/$kWh_e$ as a containerized solution [123,140].

Antora is not alone in development of these high temperature direct conversion technologies. A European project Amadeus focuses on the development of an extremely compact energy storage devices based on the use of a silicon and boron alloy as a high temperature PCM. The combination of a thermionic and photovoltaic cell as a new hybrid thermionic photovoltaic converter with an efficiency of approximately 40% is considered for discharging. The system should also have a high variable power in the range of kW to MW [141–143]. Similar is the "Sun in a box" project at MIT [144]. The system works with molten silicon, which cycles between 1900 °C (discharged tank) and 2400 °C (charged tank). When discharging the system, specialized solar cells, known as multi-junction photovoltaic cells with a conversion efficiency of 29%, are used to generate electricity [145,146]. It is also considered that the thermal energy storage technology developed in these projects could be used in concentrated solar power plants (CSP plants) [141].

2.3.2. Reversible Cycle and Other Hybrid Systems

A reversible thermodynamic cycle for charging can be achieved theoretically with a Stirling engine operating as a heat pump. This system has however not been found in any commercial or scientific research. A thermochemical conversion called the Lamm–Honigmann process, originally invented in the 19th century, which uses physical or chemical reactions with water for heat storage, is another CB concept. The system was developed with several working fluids though the most notable was water with caustic soda (NaOH) used to drive a "fireless" locomotive [147].

A German company NaCompEx has been developing a 10–100 MW_e storage system with a reported storage capacity of 60 kWh_e/m3 and the roundtrip efficiency of 80% [125,148]. When charging the storage, desorption of water from lean solution vapor is realized using heat from a condenser while the high-pressure vapor is provided by compressors. Condensed liquid water and concentrated solution are then stored separately. When discharging, the rich solution is absorbing the water vapor at low pressure, providing high temperature heat and turning high pressure water into a vapor. This vapor then produces work in an expander when flowing from high to low pressure.

No information on experimental work has been however reported and recently the company announced work on a hydrogen storage system instead [124]. The Lamm–Honigmann process is, however, a subject of a number of scientific research such as investigating efficiency limits [147] or conducting of a lab scale experiments with a system providing isentropic power potential of several hundred watts and expander power in the range of a dozen watts [149].

2.4. TES Systems for CB

Lastly, as many TES systems are developed separately but their application is specifically suggested for CB, they are summarized in this chapter. These are mostly high temperature technologies, where the heat might be utilized directly in the industry, but also drive a steam turbine, eventually in some high temperature TES systems also in gas turbines. As it can be seen from the summary in Table 4, various concepts and principles are employed. The review includes only entire TES systems (technical solutions) which comprise auxiliary systems for heat transfer and connection to heat demand and source; considering also separate development of suitable materials would be a topic for a separate extensive review work.

Table 4. Overview of the TES systems for Carnot battery applications.

Company, System	TES Type (Temperature Limit)	Thermal Capacity	State	Ref.
EnergyNest	High temperature concrete modules (up to 380 °C)	Scalable to hundreds of MWh	Pilot plant, preparing commercial project	[150]
Storworks power, BolderBloc	High temperature concrete modules (up to 600 °C)	from 30 MWh	200 kWh demo, construction of 10 MWh	[151]
Kraftblock	High temperature concrete granules (up to 1300 °C)	Scalable	Pilot industrial applications	[152]
Magaldi Green Energy	Silica sand fluidized bed (up to 1000 °C)	5 to 100 MWh	full size demonstrator	[153]
Dürr & Kraftanlagen	Ceramic blocks with gas heat exchange	MWh to GWh	Demonstrator as CB	[154]
Carboclean	Ceramic blocks with direct heating (above 1000 °C)	up to 1 GWh	Laboratory proof of concept	[155]
Joule Hive (Electrified Thermal Solutions)	Ceramic firebricks (up to 1700 °C)	n.a.	n.a.	[156]
BrenmillerEnergy	Rock bed	6–750 MWh	Pilot applications	[157]
Lumenion	Steel rods (up to 650 °C)	up to 500 MWh	Pilot applications	[158]
Kyoto, Heatcube	Molten salt (525 °C produced steam)	4–100 MWh	Pilot applications, commercial order	[159]
Hyme	Molten hydroxide up to 700 °C	Scalable	Patents, laboratory experiments	[160]
SaltX	Thermochemical	from 10 MWh	Pilot applications	[161]

Starting from the systems using sensible heat storage, a specifically developed high temperature energy storage concrete named Heatcrete from EnergyNest [162] works on the principle of having HTF in pipes embedded into the cylindrical high temperature concrete blocks and together coupled as multiple blocks creating a thermal battery module. Suggested applications are primarily industrial heat, CSP, but CB with steam turbine is also suggested. Such is the situation for most of the TES systems listed here. The modules have undergone various pilot testing [163]. The Thermal Battery pilot has been tested at Masdar Institute Solar Platform (MISP) in Abu Dhabi, UAE with 2 × 500 kWh_{th} thermal capacity at temperatures up to 380 °C over a period of more than 20 months. The measured demonstrator behaved as was predicted from numerical simulations and multi-cycle operation have proved the integrity and operational feasibility of concrete TES. The measurement of HTF inlet and outlet TESS temperature over 279 charge and discharge cycles (6000 h) show a stable and repetitive performance, which demonstrates that the concrete storage medium stays stable with no sign of degradation. The cylindrical storage element was cut into smaller sections and, following inspection, revealed no degradation, for instance spalling or cracking. Moreover, on the samples no separation between steel pipes and concrete storage material has been revealed [150].

A similar system was developed by Storworks, also consisting of steel pipes within concrete modules, which are factory built and can be assembled on site. The material utilizing waste fly ash is however claimed to withstand higher temperatures with operation up to 600 °C. A 10 MWh pilot system is being built on a working power plant site [151].

Specifically developed thermal storage high temperature concrete can also take the form of granules, where heat transfer with gaseous fluid is secured by fluidized bed technology, as developed by Kraftblock [152,164]. The material is at least 85% recycled and should withstand up to 1300 °C. The technology is currently tested in several pilot industrial applications with containerized units of 4 to 60 MWh_{th}.

The use of silica sand has also been developed separately as a TES system utilizing air fluidized bed heat transfer by Magaldi Green Energy [153,165]. Maximal temperatures of the system can reach up to 1000 °C and the design can be performed as 5 to 100 MWh thermal capacity modules, which can then be connected both in parallel and series. One significant advantage of this system is the experience from the pilot application of a very similar TES system applied to a concentrated solar mirror field.

Ceramic materials are another explored option. Structured ceramic blocks using air or other gases for charging and discharging were developed by Dürr together with Kraftanlagen München [154,166]. The technology has been adopted from more complex high temperature applications, such as regenerators in regenerative thermal oxidation processes. These ceramic blocks can withstand up to 1000 °C and 350 bar in dynamic thermal cycling. Experience from long term cycling is available as it was tested at Jülich solar plant [167,168]. The suggested sizing of CB using this technology is up to hundreds MW and MWh to GWh capacity.

Alternatively, direct insertion of heating element into the ceramic blocks was proposed as a high temperature TES system by a consortium led by Carboclean with EEW and University of Darmstadt. Storage material is expected to also withstand temperatures above 1000 °C with the storage capacity systems expected up to 1 GWh_{th} and volumetric capacity up to 1 MWh/m^3. A laboratory system was developed to test this technology with plans for integration primarily into waste incineration combined heat and power plants [155,169,170].

Ceramic material in the form of a firebrick has been additionally suggested and marketed by Electrified Thermal Solutions under trademark Joule Hive as a high temperature heat storage material up to 1700 °C [156]. A specifically developed electrically conductive ceramics has the advantage of generating the heat from electricity directly in the entire volume of the material. The temperatures are also well suited to gas turbine applications. The outlet air temperature in the proposed configuration, however, varies during discharging and its control is suggested by mixing with a cold air [171,172].

Use of natural materials of rock beds is also being developed as a standalone TES system, specifically by BrenmillerEnergy [157]. Maximum declared storage temperatures reach 750 °C and in the case of steam production, its temperature then reaches 500 °C. The TES system combines the storage material with charging and discharging heat exchanger (possibly usable also as steam generator) as integrated compact blocks that are then connected for larger capacity and power as containerized solution, which can be further combined. Alternatively, for electrical charging, heating elements are also directly incorporated. Sizing of the units is distinguished between industrial and utility scale. Pilot applications include projects with charge by waste heat or flue gas and smoothing profile or shifting the energy for high thermal demand in industrial or heating systems and one solar thermal project for up to 24 h a day electricity production.

Last representative of sensible heat TES system here is Lumenion. In this system, steel rods are used together with inert (nitrogen) gas circulation as heat transfer fluid, heated to temperatures up to 650 °C. A complete plug and play system TES is developed with pilot applications including 2.4 MWh system for district heating in Berlin operated by Vattenfal (charge 720 kW, discharge 100 kW) as a follow-up of a smaller 450 kWh system. Intended application scale is 0.2 to 20 MW systems [158] A cogeneration regime is proposed with electric efficiency around 25% and thermal 70%. Storage installation costs is estimated at €25/kWh and plans exist for up to 500 MWh_{th} systems [173].

Molten salt is also considered for standalone TES units. Norwegian company Kyoto has developed a modular system named Heatcube fitting separate salt tanks into a 20 feet containers, providing 4–100 MWh systems with a discharge rate of up to 25 MW and around 90% energy efficiency. Charging can take place from electricity or directly from a heat source. A primary application is for industrial heat but production of up to 525 °C steam offers the possibility of CHP steam plant application with electric roundtrip efficiency predicted in the range 15%–25%. Two pilot projects below 1 MWh were completed in 2021 while commercial orders for 10–20 MWh units are present [159].

Molten hydroxides were proposed for the storage of heat at up to 700°C for periods of up to 14 days by company Hyme [160]. It is a spinoff of a small modular reactor company Seaborg considering the hydroxides as a heat transfer and storage fluid. Major know-how of this solution is in a specific corrosion control technology. Compared to molten salt the hydroxides are expected to be substantially cheaper [174].

Chemical high temperature reactions are representative of another two technologies. Reversible reaction of CaO with water to form Ca $(OH)_2$ is used by SaltX, where the granules of salt use nano-size layer coating to secure robust behavior in cycling and prevent agglomeration. Charging is performed by heat at a temperature of 550°C and higher while discharging process releases heat up to 450 °C. A pilot system of 0.5 MW/10 MWh is operated as a pilot in Berlin for steam district heating, again by Vattenfall. It is a follow-up on a small 20 kW experimental system. Further pilots are planned with power to steam or heat to heat applications. The general technology concept is, however, planned to be changed from the heated screw reactor type to fluidized bed reactors [161]. Note that this process has been proposed for energy storage a long time ago [175] and much of the research is still ongoing, focusing for example on kinetics and stability [176], which the SaltX appears to have solved. The process is also similar to the more scientifically explored Calcium looping, which is also usable for high temperature TES application [177].

Most of the TES systems can also be utilized in the utility scale to improve the flexibility of the thermal fossil fired power plants. The various levels and complexity of TES integration, as well as technologies of storage, were proposed. Often the impact of storage integration on flexibility can, however, be considered as too low in the hundred MW scale plants. The maximum load variation is below 5% in [70], for around 13% in [178] and with more extensive TES integration separately for high temperature reheated steam and lower temperature feed water regeneration up to 50% in [179]. Only in the smaller plants, such as industrial CHP, can the examples of larger load variation up to 100% be found as in the case of thermochemical storage in [180]. These works can serve as a basis for the analysis of the current thermal plants' conversions to full CB as such works are absent in the available literature. A large part of the original analyses of increasing plant flexibility, such as TES integration, can remain, though the target operation regime needs to be more dynamic with frequent shutdowns, as found, on the other hand, rather commonly in concentrated solar plants. The major difference would also be in focusing on longer duration storage and on the method of charging, utilizing a surplus of renewable electricity instead of flue gases and steam.

3. Discussion and Conclusions

In the increasing need of medium and long duration energy storage, Carnot batteries (CB) offer a potentially cost-effective solution with systems ranging from large grid scale applications down to even dozens of kW. Therefore, the concept has attracted not only academic, but already also considerable industrial, research and development. Among many concepts and systems, this work provides an extensive overview of commercially developed technologies, their classification by the level of maturity and experimental development. These technologies are then put into the perspective of the scientific research, providing either background for the technologies or pointing at research opportunities. The general composition of references with actual dates points out how recent and rapidly developing the topic of CB is.

There is a total 30 CB systems reported, the technologies and states of which are described in detail. They cover a large range of concepts including both direct power to heat (P2H) Joule heating conversion and heat pump based conversion, thermodynamic cycles, such as the Rankine cycle (with steam, organic or CO_2 fluid), the Brayton cycle, as well as their combination, Stirling cycle or direct heat to power (H2P) conversion with systems such as thermophotovoltaics. The range of the power output and storage capacity (respectively storage duration in hours) summarized from all collected systems is presented in Figure 5. When exact data were not available, an engineering estimate was made. It shows that the CB covers the range from kW to GW systems. The smallest ones are based on ORC and Stirling engines while the largest are generally based on current thermal power plants' technologies. The storage duration confirms the application range between about 4 and 24 h, the time range of so-called medium duration energy storage systems. With the higher maturity of these systems and the increased storage duration required, it might be feasible to add further storage capacity at a relatively low cost.

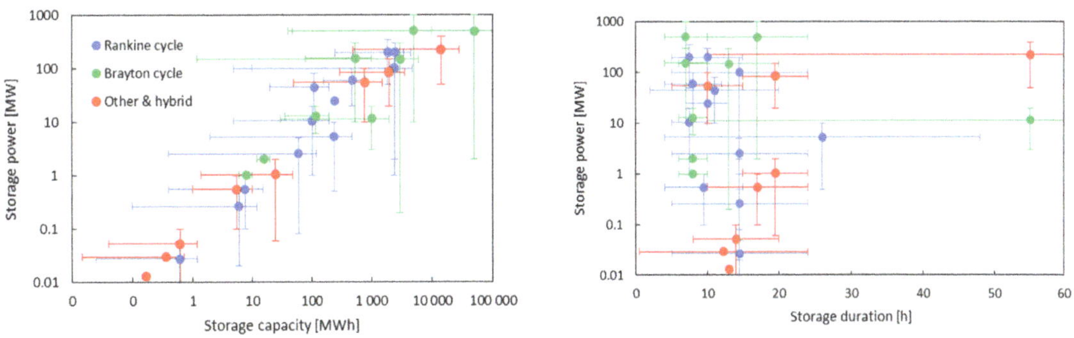

(a) (b)

Figure 5. Storage power output and capacity (**a**) and discharge duration (**b**) for the commercially developed CB systems.

Out of them, the largest CB commissioned so far is a liquid air energy storage (LAES) pilot 5 MW system of Highview, where the company is furthermore constructing a full-scale unit with 50 MW and 5 h of operation. The first-of-a-kind challenge is being overcome as contracts are prepared for other LAES units and thus the technology is on a clear path for commercial applications. The second largest is Siemens Gamesa ETES, using direct power to heat conversion, the storage of heat in a rock packed bed and using a steam cycle for power to heat conversion with a 1.2 MW_e and 29 MWh_e scale pilot system. The only existing commercial installations are, however, much smaller-scale units of Azelio with a 13 kW_e output provided by the Stirling engine, while charging is also performed by direct conversion and thermal storage comprised of an aluminum alloy phase change material. Many other systems using direct electricity to heat conversion are then at least in a state of experimental demonstration.

Systems using the heat pump principle (mostly vapor compression or Brayton cycle) for charging are much more favored in the scientific literature due to their potential of higher efficiency. Except for the Highview LAES system, the commercial development is significantly less advanced with many more systems staying at the conceptual state and yet unrealized design. This might however change in the near future as Stiesdal is planning a construction of an MW scale pilot for 2022 with a reversible Brayton and rock bed storage system. A peculiar approach to commercialization is explored by MAN, where the transcritical Rankine CO_2 heat pump used for charging can be operated separately and as such is planned to be built at the 50 MW_{th} scale.

For the reversible Brayton cycle CB, there is however also an example of the bankruptcy of the company Isentropic, stressing the need for real CB systems to have a low unit cost with a simple and robust design and the fact that their feasibility is expected to improve with future, more renewable-based grids. The collected data can be displayed from the point of round trip efficiency, which is shown in Figure 6. A wide spread of the values between about 25% and 80% can be seen, with an exception in the fuel co-fired systems reaching numerically values even over 100%. High roundtrip efficiency is often reported in conceptual systems, which is similar to theoretical results with rather optimistic estimates of system assumptions. Therefore, caution must be taken, especially for the systems existing only in the conceptual or early design states, without the support of experimental data. Even though demonstration and pilot plants exist, the values are still mostly projected for full scales and thus are usually yet to be verified.

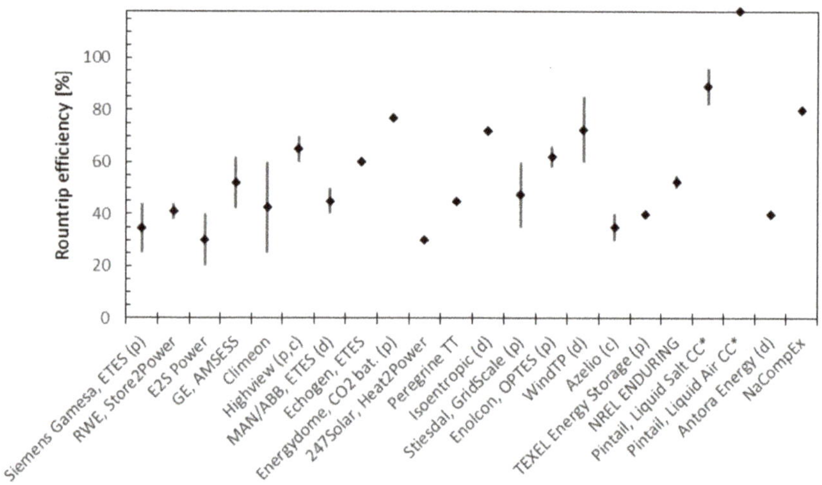

Figure 6. Overview of round trip efficiency in the commercially CB systems (mostly declared as experimental values are limited). In notation (d) stands for demo, (p) pilot, (c) commercial units (built or under construction), * for systems with additional fuel firing.

Lastly, a brief timeline of the CB commercial development considering systems from laboratory demonstrators and proof of concept systems to commercial installations and large scale pilots (often operating in commercial regime) is presented in Figure 7. It shows how the development is gaining momentum, mostly in recent years, with the first major grid scale pilot coming online in 2019, full commercialization of 13 kW modular unit in 2021 and multiple pilot and commercial systems scheduled for commissioning in 2022 and 2023. Once the experience from these systems is obtained and renewable installations with energy prices' volatility further increase, we may witness even more rapid development.

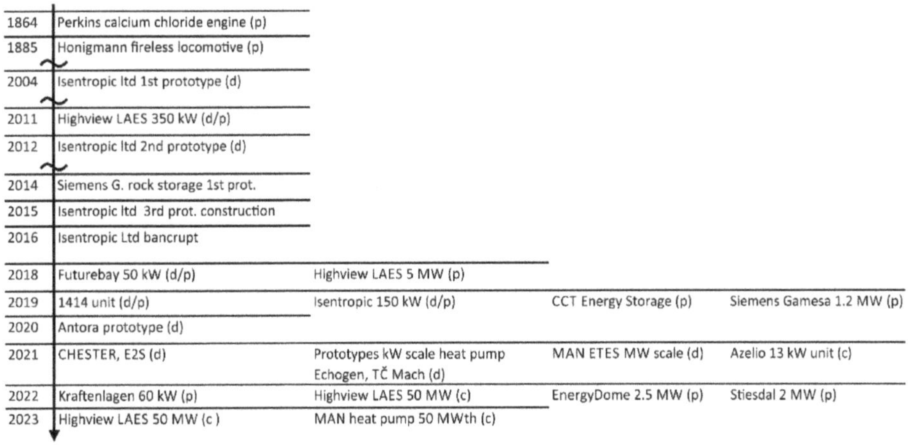

Figure 7. Timeline of experimental commercial CB development. In notation (d) stands for demo or laboratory prototype, (p) pilot, (c) commercial units.

Furthermore, there is a considerable development of thermal energy storage systems where the CB application is specifically suggested as one of the use cases. As the market appears to not yet be fully ready for CB applications due to the still lower added value compared to the production processes, industrial and process heat are the primary target customers. As the situation might change with the larger adoption of intermittent renewables, CB-ready TES systems are listed here as well. They often include electric charging and would require adding a relatively standard, typically steam, heat to power units.

Finally, one interesting trend in CB systems with high temperature storage is a re-adoption of technologies previously considered for concentrated solar power plants and industrial process heat. This confirms that CB is not that much of a novel technology, but is rather a novel use case and a combination of existing technologies. As such, perhaps the simplicity causes the electrically heated systems to be mostly absent from the scientific literature although they have the prospect of a low cost and a simple/robust system. Techno-economic studies evaluating actual costs and LCOE might especially be a suitable area for future studies.

Author Contributions: Conceptualisation, V.N.; methodology, V.N.; investigation, V.N., V.B., P.S.; resources, V.N., V.B., P.S., J.S.; data curation, V.N., V.B., P.S.; writing—original draft preparation, V.N. V.N., V.B., P.S.; writing—review and editing, V.N., J.S.; visualisation, V.N.; supervision, V.N., J.S.; project administration, V.N., J.S.; funding acquisition, V.N., J.S. All authors have read and agreed to the published version of the manuscript.

Funding: This research was funded by the Grant Agency of the Czech Technical University in Prague, grant No. SGS21/111/OHK2/2T/12 and by the Technology Agency of the Czech Republic, grant number TJ04000326.

Data Availability Statement: All data presented in this study are contained within the article or its references.

Conflicts of Interest: The authors declare no conflict of interest. The funders had no role in the design of the study; in the collection, analyses, or interpretation of data; in the writing of the manuscript, or in the decision to publish the results.

Nomenclature

Symbols	
$\dot{Q}$	heat flow [kW]
$\dot{W}$	work power [kW]
T	thermodynamic temperature [K]
Q	heat [kJ]
W	work [kJ]
η	efficiency [-]
Subscripts	
amb	*ambient*
in	*Inlet*
out	*outlet*
UE	*useful energy*
Abbreviations	
AA-CAES	advanced adiabatic compressed air energy storage
A-CAES	adiabatic compressed air energy storage
BC	Brayton cycle
CAES	compressed air energy storage
CAPEX	capital expenditures
CB	Carnot battery/batteries
COP	coefficient of performance
CSP	concentrated solar power plants
ETES	electric thermal energy storage/electro—thermal energy storage
H2P	heat to power
HRSG	heat recovery steam generators
HTF	heat transfer fluid
HTMH	high temperature metal hydride
CHEST	compressed heat energy storage
CHP	combined heat and power production
I-CAES	isothermal compressed air energy storage
IEA	international energy agency
LAES	liquid air energy storage
LCOE	levelized cost of energy
LCOS	levelized cost of storage
LTMH	low temperature metal hydride
MGA	miscibility gap alloy
MISP	Masdar Institute Solar Platform
OPTES	optimized pumped thermal energy storage
ORC	organic Rankine cycles
P2H	power to heat
P2H2P	power to heat to power
PCM	phase change material
PHES	pumped hydro energy storage
PHS	pumped hydro storage
PTES	pumped thermal energy storage
PV	photovoltaic
R&D	research and development
RC	Rankine cycle
RTE	roundtrip efficiency
TES	thermal energy storage
TESS	thermal energy storage system
TRL	technology readiness level

References

1. Hansen, K.; Mathiesen, B.V.; Skov, I.R. Full energy system transition towards 100% renewable energy in Germany in 2050. *Renew. Sustain. Energy Rev.* **2019**, *102*, 1–13. [CrossRef]
2. Brown, T.W.; Bischof-Niemz, T.; Blok, K.; Breyer, C.; Lund, H.; Mathiesen, B.V. Response to "Burden of proof: A comprehensive review of the feasibility of 100% renewable-electricity systems". *Renew. Sustain. Energy Rev.* **2018**, *92*, 834–847. [CrossRef]
3. Aghahosseini, A.; Bogdanov, D.; Breyer, C. Towards sustainable development in the MENA region: Analysing the feasibility of a 100% renewable electricity system in 2030. *Energy Strat. Rev.* **2020**, *28*, 100466. [CrossRef]
4. Redondo-Iglesias, E.; Venet, P.; Pelissier, S. Efficiency Degradation Model of Lithium-Ion Batteries for Electric Vehicles. *IEEE Trans. Ind. Appl.* **2019**, *55*, 1932–1940. [CrossRef]
5. Schimpe, M.; Naumann, M.; Truong, N.; Hesse, H.C.; Santhanagopalan, S.; Saxon, A.; Jossen, A. Energy efficiency evaluation of a stationary lithium-ion battery container storage system via electro-thermal modeling and detailed component analysis. *Appl. Energy* **2018**, *210*, 211–229. [CrossRef]
6. Cárdenas, B.; Swinfen-Styles, L.; Rouse, J.; Hoskin, A.; Xu, W.; Garvey, S. Energy storage capacity vs. renewable penetration: A study for the UK. *Renew. Energy* **2021**, *171*, 849–867. [CrossRef]
7. Garvey, S.D. Medium-Duration Energy Storage Explained. Mediu. Durat. Energy Storage Net Zero UK, Nottingham: Supergen Energy Storage. 2021. Available online: https://www.era.ac.uk/Medium-Duration-Energy-Storage (accessed on 10 November 2021).
8. Smith, K.; Saxon, A.; Keyser, M.; Lundstrom, B.; Cao, Z.; Roc, A. Life prediction model for grid-connected Li-ion battery energy storage system. In Proceedings of the 2017 American Control Conference (ACC), Seattle, WA, USA, 24–26 May 2017. [CrossRef]
9. Moore, J.; Henbest, S. New Energy Outlook 2020. Available online: https://assets.bbhub.io/professional/sites/24/928908_NEO2020-Executive-Summary.pdf (accessed on 22 November 2021).
10. Gyuk, I. DOE Global Energy Storage Database. Sandia Natl Lab 2021. Available online: https://sandia.gov/ess-ssl/gesdb/public/index.html (accessed on 22 November 2021).
11. Benato, A.; Stoppato, A. Pumped Thermal Electricity Storage: A technology overview. *Therm. Sci. Eng. Prog.* **2018**, *6*, 301–315. [CrossRef]
12. Rehman, S.; Al-Hadhrami, L.M.; Alam, M. Pumped hydro energy storage system: A technological review. *Renew. Sustain. Energy Rev.* **2015**, *44*, 586–598. [CrossRef]
13. Benato, A. Performance and cost evaluation of an innovative Pumped Thermal Electricity Storage power system. *Energy* **2017**, *138*, 419–436. [CrossRef]
14. Aghahosseini, A.; Breyer, C. Assessment of geological resource potential for compressed air energy storage in global electricity supply. *Energy Convers. Manag.* **2018**, *169*, 161–173. [CrossRef]
15. Sterner, M.; Stadler, I. *Handbook of Energy Storage*; Springer: Berlin/Heidelberg, Germany, 2019.
16. Budt, M.; Wolf, D.; Span, R.; Yan, J. A review on compressed air energy storage: Basic principles, past milestones and recent developments. *Appl. Energy* **2016**, *170*, 250–268. [CrossRef]
17. Projects—Hydrostor 2021. Available online: https://www.hydrostor.ca/projects/ (accessed on 15 October 2021).
18. King, M.; Jain, A.; Bhakar, R.; Mathur, J.; Wang, J. Overview of current compressed air energy storage projects and analysis of the potential underground storage capacity in India and the UK. *Renew. Sustain. Energy Rev.* **2021**, *139*, 110705. [CrossRef]
19. Lourenssen, K.; Williams, J.; Ahmadpour, F.; Clemmer, R.; Tasnim, S. Vanadium redox flow batteries: A comprehensive review. *J. Energy Storage* **2019**, *25*, 100844. [CrossRef]
20. Botha, C.; Kamper, M. Capability study of dry gravity energy storage. *J. Energy Storage* **2019**, *23*, 159–174. [CrossRef]
21. Fyke, A. The Fall and Rise of Gravity Storage Technologies. *Joule* **2019**, *3*, 625–630. [CrossRef]
22. Abdon, A.; Zhang, X.; Parra, D.; Patel, M.K.; Bauer, C.; Worlitschek, J. Techno-economic and environmental assessment of stationary electricity storage technologies for different time scales. *Energy* **2017**, *139*, 1173–1187. [CrossRef]
23. Dumont, O.; Frate, G.F.; Pillai, A.; Lecompte, S.; De Paepe, M.; Lemort, V. Carnot battery technology: A state-of-the-art review. *J. Energy Storage* **2020**, *32*, 101756. [CrossRef]
24. Frate, G.F.; Ferrari, L.; Desideri, U. Energy storage for grid-scale applications: Technology review and economic feasibility analysis. *Renew. Energy* **2021**, *163*, 1754–1772. [CrossRef]
25. Dumont, O.; Lemort, V. Mapping of performance of pumped thermal energy storage (Carnot battery) using waste heat recovery. *Energy* **2020**, *211*, 118963. [CrossRef]
26. Laughlin, R.B. Pumped thermal grid storage with heat exchange. *J. Renew. Sustain. Energy* **2017**, *9*, 044103. [CrossRef]
27. Steinmann, W.-D. Thermo-mechanical concepts for bulk energy storage. *Renew. Sustain. Energy Rev.* **2017**, *75*, 205–219. [CrossRef]
28. Marguerre, F. Thermodynamic Energy Storage. U.S. Patent US2065974A, 29 December 1936.
29. iea Energy Storage Task 36—Carnot Batteries 2021. Available online: https://www.eces-a36.org/ (accessed on 25 October 2021).
30. Schneider, G.; Maier, H.; Häcker, J.; Siegele, S. Electricity Storage with a Solid Bed High Temperature Thermal Energy Storage System (HTTES)—A Methodical Approach to Improve the Pumped Thermal Grid Storage Concept. In *Atlantis Highlights in Engineering*; Atlantis Press: Paris, France, 2021; Volume 6, pp. 26–33.
31. Jacquemoud, E. Electro-Thermal Energy Storage System (ETES) Based on CO_2 Cycles. *Ref. Modul. Earth Syst. Environ. Sci.* **2021**. [CrossRef]

32. ETES-Electric Thermal Energy Storage-Technology and Commercial Proposition. Siemens Gamesa Renew Energy GmbH Co KG n.d. Available online: https://www.siemensgamesa.com/en-int/-/media/siemensgamesa/downloads/en/products-and-services/hybrid-power-and-storage/etes/siemens-gamesa-etes-general-introduction-3d.pdf (accessed on 11 October 2021).
33. Barmeier, T. Electric Thermal Energy Storage (ETES) Transition of Energy Supply. 26. Wind. Warn., Warnemünde, Germany. 2017. Available online: https://windenergietage.de/wp-content/uploads/sites/2/2017/11/26WT0811_F11_1120_Dr_Barmeier.pdf (accessed on 22 November 2021).
34. Electric Thermal Energy Storage (ETES) ETES:SWITCH 2021. Available online: https://www.siemensgamesa.com/en-int/products-and-services/hybrid-and-storage/thermal-energy-storage-with-etes-switch (accessed on 4 November 2021).
35. Witold, A. *StoreToPower Pilot Plant for the Development of a Heat Storage Power Plant. 2nd Int. Work. Carnot Batter*; DLR: Stuttgart, Germany, 2020.
36. E2S Power | 2020. Available online: http://e2s-power.com/ (accessed on 4 November 2021).
37. Schäfer, T. Thermo-MechanischerSpeicher (CARNOT-Batterie). Zittau, Germany. 2020. Available online: https://www.cleantech-ost.de/fileadmin/user_upload/chefredakteur1/AK_Energiespeicher/IV_IPM_Thermomechanischer_Speicher_Schaefer.pdf (accessed on 15 November 2021).
38. Spilling Technologies: About Spilling 2021. Available online: https://www.spilling.de/about-spilling/about-spilling.html (accessed on 21 October 2021).
39. Aga, V.; Conte, E.; Carroni, R.; Burcker, B.; Ramond, M. Supercritical CO_2-Based Heat Pump Cycle for Electrical Energy Storage for Utility Scale Dispatchable Renewable Energy Power Plants. In Proceedings of the 5th International Symposium—Supercritical CO2 Power Cycles, San Antonio, TX, USA, 28–31 March 2016.
40. Chester Project | Compressed Heat Energy Storage for Energy from Renewable Sources n.d. Available online: https://www.chester-project.eu/ (accessed on 21 October 2021).
41. Karthäuser, J. *The Low Tempeature (80–120 °C) Carnot Battery and its Potential for the Integrated Energy System. 2nd Int. Work. Carnot Batter*; DLR: Stuttgart, Germany, 2020.
42. Novotny, V.; Mach, S.; Rathan, M.; Basta, V.; Spale, J. Modelling and experimental development of a waste heat upgrade integrated ORC Carnot battery with stone dust thermal storage. In Proceedings of the 6th International Seminar on ORC Power Systems, Munich, Germany, 11–13 October 2021.
43. FutureBay—Energy Storage, Hvac 2021. Available online: https://futurebay.uk.com/ (accessed on 22 October 2021).
44. Plants | Highview Power 2021. Available online: https://highviewpower.com/plants/ (accessed on 2 October 2021).
45. Morgan, R.; Nelmes, S.; Gibson, E.; Brett, G. Liquid air energy storage—Analysis and first results from a pilot scale demonstration plant. *Appl. Energy* **2015**, *137*, 845–853. [CrossRef]
46. Ross, K. Highview Hires MAN Energy for UK Cryogenic Energy Storage Project. *Power Eng. Int.* 2021. Available online: https://www.powerengineeringint.com/smart-grid-td/energy-storage/highview-hires-man-energy-for-uk-cryogenic-energy-storage-project/ (accessed on 24 November 2021).
47. Sanz Garcia, L.; Jacquemoud, E.; Jenny, P. Thermo-Economic Heat Exchanger Optimization for Electro-Thermal Energy Storage Based on Transcritical CO_2 CYCLES. In Proceedings of the 3rd European supercritical CO_2 Conference, Paris, France, 19–20 September 2019; p. 2019-sCO2.eu-151. [CrossRef]
48. Morandin, M.; Mercangöz, M.; Hemrle, J.; Maréchal, F.; Favrat, D. Thermoeconomic design optimization of a thermo-electric energy storage system based on transcritical CO2 cycles. *Energy* **2013**, *58*, 571–587. [CrossRef]
49. Mercangöz, M.; Hemrle, J.; Kaufmann, L.; Z'Graggen, A.; Ohler, C. Electrothermal energy storage with transcritical $_{CO2}$ cycles. *Energy* **2012**, *45*, 407–415. [CrossRef]
50. ETES System Overview | Echogen Power Systems. 2021. Available online: https://www.echogen.com/energy-storage/etes-system-overview (accessed on 3 November 2021).
51. Echogen Energy Storage. Supercritical CO_2-Based Long-Duration Electrical Energy Storage Technical Overview. 2019. Available online: https://www.echogen.com/_CE/pagecontent/Documents/Echogen-TechnicalOverview7.18.19-3.pdf (accessed on 15 November 2021).
52. Echogen Storage Systems | Distributed Systems | F6S Profile. 2022. Available online: https://www.f6s.com/echogenstoragesystems (accessed on 14 January 2022).
53. Held, T. Systems and Methods for Generating Electricity Via a Pumped Thermal Energy Storage System. U.S. Patent US10883388B2, 5 January 2021.
54. Held, T. Low-Cost, Long-Duration Electrical Energy Storage Using a CO_2-Based Electro Thermal Energy Storage (ETES) System. In *Arpa-E DAYS Annu. Meet.*; 2021. Available online: https://arpa-e.energy.gov/sites/default/files/2021-03/03Day1-TimHeld_Echogen.pdf (accessed on 5 October 2021).
55. Energy Dome | Groundbreaking Long Duration Energy Storage 2021. Available online: https://energydome.com/ (accessed on 2 November 2021).
56. Astolfi, M.; Rizzi, D.; Macchi, E.; Spadacini, C. A Novel Energy Storage System Based on Carbon Dioxide Unique Thermodynamic Properties. *Proc. ASME Turbo Expo.* **2021**, *4*. [CrossRef]
57. Moore, J.; Cich, S.; Day-Towler, M.; Mortzheim, J. Development and Testing of a 10 MWe Supercritical CO_2 Turbine in a 1 MWe Flow Loop. *Proc. ASME Turbo Expo.* **2020**, *11*. [CrossRef]

58. Allam, R.; Martin, S.; Forrest, B.; Fetvedt, J.; Lu, X.; Freed, D.; Brown, G.W.; Sasaki, T.; Itoh, M.; Manning, J. Demonstration of the Allam Cycle: An Update on the Development Status of a High Efficiency Supercritical Carbon Dioxide Power Process Employing Full Carbon Capture. *Energy Procedia* **2017**, *114*, 5948–5966. [CrossRef]
59. Steinmann, W.-D.; Jockenhöfer, H.; Bauer, D. Thermodynamic Analysis of High-Temperature Carnot Battery Concepts. *Energy Technol.* **2020**, *8*, 1900895. [CrossRef]
60. von der Heyde, M.; Schmitz, G. Electric Thermal Energy Storage Based on Packed Bed. *Ref. Modul. Earth Syst. Environ. Sci.* **2021**. [CrossRef]
61. Soprani, S.; Marongiu, F.; Christensen, L.; Alm, O.; Petersen, K.D.; Ulrich, T.; Engelbrecht, K. Design and testing of a horizontal rock bed for high temperature thermal energy storage. *Appl. Energy* **2019**, *251*, 113345. [CrossRef]
62. Prieto, C.; Osuna, R.; Fernandez, A.I.; Cabeza, L.F. Thermal storage in a MW scale. Molten salt solar thermal pilot facility: Plant description and commissioning experiences. *Renew. Energy* **2016**, *99*, 852–866. [CrossRef]
63. Relloso, S.; Gutiérrez, Y. SENER molten salt tower technology. Ouarzazate NOOR III case. *AIP Conf. Proc.* **2017**, *1850*, 030041. [CrossRef]
64. Bell, S.; Steinberg, T.; Will, G. Corrosion mechanisms in molten salt thermal energy storage for concentrating solar power. *Renew. Sustain. Energy Rev.* **2019**, *114*, 109328. [CrossRef]
65. Kruizenga, A.M.; Gill, D.D.; LaFord, M.E. Corrosion of High Temperature Alloys in Solar Salt at 400, 500, and 680 °C. 2013. Available online: https://www.osti.gov/biblio/1104752/ (accessed on 5 October 2021).
66. Copus, M.; Fraser, B.; Reece, R.; Hands, S.; Cuskelly, D.; Sugo, H.; Reed, S.; Bradley, J.; Post, A.; Kisi, E. On-sun testing of Miscibility Gap Alloy thermal storage. *Sol. Energy* **2019**, *177*, 657–664. [CrossRef]
67. Reed, S.; Sugo, H.; Kisi, E.; Richardson, P. Extended thermal cycling of miscibility gap alloy high temperature thermal storage materials. *Sol. Energy* **2019**, *185*, 333–340. [CrossRef]
68. Steinmann, W. The CHEST (Compressed Heat Energy STorage) concept for facility scale thermo mechanical energy storage. *Energy* **2014**, *69*, 543–552. [CrossRef]
69. Steinmann, W.-D.; Eck, M. Buffer storage for direct steam generation. *Sol. Energy* **2006**, *80*, 1277–1282. [CrossRef]
70. Krüger, M.; Muslubas, S.; Loeper, T.; Klasing, F.; Knödler, P.; Mielke, C. Potentials of Thermal Energy Storage Integrated into Steam Power Plants. *Energies* **2020**, *13*, 2226. [CrossRef]
71. Terrajoule | 24/7 Clean Energy 2021. Available online: http://www.terrajoulecorp.com/ (accessed on 21 October 2021).
72. Frate, G.F.; Ferrari, L.; Desideri, U. Rankine Carnot Batteries with the Integration of Thermal Energy Sources: A Review. *Energies* **2020**, *13*, 4766. [CrossRef]
73. Macchi, E.; Astolfi, M. *Organic Rankine Cycle (ORC) Power Systems: Technologies and Applications*; Woodhead Publishing: Sawston, UK, 2016.
74. Dumont, O.; Charalampidis, A.; Lemort, V.; Karellas, S. Experimental Investigation of a Thermally Integrated Carnot Battery Using a Reversible Heat Pump/Organic Rankine Cycle. In Proceedings of the 18th International Refrigeration and Air Conditioning Conference at Purdue, West Lafayette, IN, USA, 24–28 May 2021.
75. Trebilcock, F.; Ramirez, M.; Pascual, C.; Weller, T.; Lecompte, S.; Hassan, A.H. Development of a compressed heat energy storage system prototype. *IIR Rank Conf.* **2020**, 400–409. [CrossRef]
76. Steger, D.; Karl, J.; Schlücker, E. Launch and first experimental results of a reversible heat pump-orc pilot plant as carnot battery. In Proceedings of the 6th International Seminar on ORC Power Systems, Munich, Germany, 11–13 October 2021.
77. Glushenkov, M.; Kronberg, A.; Knoke, T.; Kenig, E.Y. Isobaric Expansion Engines: New Opportunities in Energy Conversion for Heat Engines, Pumps and Compressors. *Energies* **2018**, *11*, 154. [CrossRef]
78. O'Callaghan, O.; Donnellan, P. Liquid air energy storage systems: A review. *Renew. Sustain. Energy Rev.* **2021**, *146*, 111113. [CrossRef]
79. Chino, K.; Araki, H. Evaluation of energy storage method using liquid air. *Heat Transfer-Asian Res.* **2000**, *29*, 347–357. [CrossRef]
80. Li, Y.; Cao, H.; Wang, S.; Jin, Y.; Li, D.; Wang, X.; Ding, Y. Load shifting of nuclear power plants using cryogenic energy storage technology. *Appl. Energy* **2014**, *113*, 1710–1716. [CrossRef]
81. Mikolajczak, A.; Wolowicz, M.; Kurkus-Gruszecka, M.; Badyda, K.; Krawczyk, P. Improving the efficiency of Liquid Air Energy Storage by organic rankine cycle module application. In Proceedings of the 2018 International Interdisciplinary PhD Workshop (IIPhDW), Świnoujście, Poland, 9–12 May 2018. [CrossRef]
82. Peng, X.; She, X.; Li, Y.; Ding, Y. Thermodynamic analysis of Liquid Air Energy Storage integrated with a serial system of Organic Rankine and Absorption Refrigeration Cycles driven by compression heat. *Energy Procedia* **2017**, *142*, 3440–3446. [CrossRef]
83. She, X.; Peng, X.; Zhang, T.; Ding, Y. Configuration optimization of stand-alone Liquid Air Energy Storage for efficiency improvement. *IOP Conf. Ser. Mater. Sci. Eng.* **2019**, *502*, 012015. [CrossRef]
84. Highview Enlasa Developing 50MW/500MWh Liquid Air Energy Storage Facility in the Atacama Region of Chile | Highview Power 2021. Available online: https://highviewpower.com/news_announcement/highview-enlasa-developing-50mw-500mwh-liquid-air-energy-storage-facility-in-the-atacama-region-of-chile/ (accessed on 20 October 2021).
85. Highview Power to Develop Multiple Cryogenic Energy Storage Facilities in the UK and to Build Europe's Largest Storage System | Highview Power 2019. Available online: https://highviewpower.com/news_announcement/highview-power-to-develop-multiple-cryogenic-energy-storage-facilities-in-the-uk-and-to-build-europes-largest-storage-system/ (accessed on 20 September 2021).

86. Jacquemoud, E. CO_2 *Basierte Industrielle Heatpump und Carnot—Batterie. Carnot-Batterien Thermische Stromspeicher für die Energiewende*; Energie Campus Nürnberg: Nurnberg, Germany, 2021. Available online: https://www.encn.de/fileadmin/user_upload/Flyer_Carnot-Batterie-Workshop_final_V2.pdf (accessed on 25 October 2021).
87. Wang, M.; Zhao, P.; Yang, Y.; Dai, Y. Performance analysis of energy storage system based on liquid carbon dioxide with different configurations. *Energy* **2015**, *93*, 1931–1942. [CrossRef]
88. Colthorpe, A. 'CO2 Battery' Startup Energy Dome Gets US$11m Financing, First Agreement to Deploy 100MWh Project-Energy Storage News. *Energy Storage News*. 2021. Available online: https://www.energy-storage.news/co2-battery-startup-energy-dome-gets-us11m-financing-first-agreement-to-deploy-100mwh-project/ (accessed on 7 December 2021).
89. Parham, J.; Vrettos, P.; Levinson, N. Commercialisation of ultra-high temperature energy storage applications: The 1414 Degrees approach. *Ultra-High Temp. Therm. Energy Storage Transf. Convers.* **2021**, *2021*, 331–346. [CrossRef]
90. Anderson, B.N. Modular Solar Systems for 24/7 Scalable, Flexible, Affordable Electricity. In Proceedings of the ASME Power Conference, Charlotte, NC, USA, 26–30 June 2017; Volume 2, p. V002T09A002.
91. Moriarty, K. Prospectus 1414 DEGREES 2018. Available online: http://1414degrees.com.au/wp-content/uploads/2018/12/1414Degrees-Limited-Replacement-Prospectus-14D.pdf (accessed on 7 November 2021).
92. Technology—Peregrine Turbine Technologies 2021. Available online: https://www.peregrineturbine.com/technology/ (accessed on 6 December 2021).
93. Howes, J. Concept and Development of a Pumped Heat Electricity Storage Device. *Proc. IEEE* **2012**, *100*, 493–503. [CrossRef]
94. The GridScale Technology Explained | Stiesdal n.d. Available online: https://www.stiesdal.com/storage/the-gridscale-technology-explained/ (accessed on 31 October 2021).
95. WindTP—Wind Driven Thermal Pumping 2021. Available online: https://www.wind-tp.com/ (accessed on 6 December 2021).
96. Heatstore. 247Solar 2020. Available online: https://www.247solar.com/247Solar/heatstore.html (accessed on 1 November 2021).
97. Maisch, M. 1414 Degrees to Power Abandoned Aurora CSP Site with PV and Thermal Storage—Pv Magazine Australia. Pv Mag 2019. Available online: https://www.pv-magazine-australia.com/2019/11/29/1414-degrees-to-power-abandoned-aurora-csp-site-with-pv-and-thermal-storage/ (accessed on 9 November 2021).
98. Aurora Energy Project. 1414 DEGREES 2021. Available online: https://1414degrees.com.au/auroraenergyproject/ (accessed on 9 November 2021).
99. Stapp, D.S. A Novel SCO2 Primary Cycle for Air-Combustible Fuels. 4th Int. Symp. CO2 Power Cycles, Pittsburgh, Pennsylvania: 2014. Available online: http://sco2symposium.com/papers2014/systemConcepts/72-Stapp.pdf (accessed on 9 November 2021).
100. Dewis, D. PTT's Thermal Energy Storage (TES) Application Utilizes the Power Block Comprising Proprietary Turbo-Machinery and PCHE Technology. 2021. Available online: https://www.peregrineturbine.com/wp-content/uploads/2021/10/Peregrine-white-paper-EPRI.pdf (accessed on 9 November 2021).
101. Smallbone, A. Experimental Results and Modelling of a Grid-scale Pumped Heat Energy Storage Demonstrator. *Int. Work. Carnot Batter.* **2020**.
102. Ameen, M.T.; Smallbone, A.; Roskilly, A.P.; Carpenter, E. The development of a screen valve for reciprocating heat pump/engine applications. *J. Renew. Sustain. Energy* **2020**, *12*, 054101. [CrossRef]
103. Smallbone, A.; Jülch, V.; Wardle, R.; Roskilly, A.P. Levelised Cost of Storage for Pumped Heat Energy Storage in comparison with other energy storage technologies. *Energy Convers. Manag.* **2017**, *152*, 221–228. [CrossRef]
104. Laughlin Battery—Google Recuperator. Brayt Energy 2021. Available online: https://www.braytonenergy.net/our-projects/malta-program-google-recuperator/ (accessed on 31 July 2020).
105. Our Story—From Theory to a Successful Proof of Concept. Malta Inc. 2021. Available online: https://www.maltainc.com/our-story/ (accessed on 30 October 2021).
106. Bollinger, B.R. Malta Pumped Heat Electricity Storage (PHES) for Coal Exit and Energy Transition from Fossil to Renewable Energies. 2nd Int. Work. Carnot Batter. 2020. Available online: https://iwcb2020.welcome-manager.de/archiv/web/userfiles/iwcb2020/Downloads/Presentations/IWCB20_VI_3_MaltaPHESCoal_BOLLINGER.pdf (accessed on 17 October 2020).
107. Smith, N. Small-Scale PHES Demonstration. Arpa-E DAYS Annu. Meet. 2021. Available online: https://arpa-e.energy.gov/sites/default/files/2021-03/04Day1-NatalieSmith_SwRI.pdf (accessed on 30 October 2021).
108. Fredericton, N.B.; Cambridge, M.A. Malta and NB Power Announce Collaboration Towards Development of Province's LDES. Malta Inc. 2021. Available online: https://www.maltainc.com/maltanbpowerannouncement (accessed on 30 October 2021).
109. Andel and Stiesdal Join Forces on Large-Scale Energy Storage. Andel 2021. Available online: https://andel.dk/en/news/andel-and-stiesdal-join-forces-on-large-scale-energy-storage/# (accessed on 31 October 2021).
110. König-Haagen, A.; Höhlein, S.; Brüggemann, D. Detailed exergetic analysis of a packed bed thermal energy storage unit in combination with an Organic Rankine Cycle. *Appl. Therm. Eng.* **2020**, *165*, 114583. [CrossRef]
111. Schlipf, D.; Faust, E.; Schneider, G.; Maier, H. First operational results of a high temperature energy storage with packed bed and integration potential in CSP plants. *AIP Conf. Proc.* **2017**, *1850*, 080024. [CrossRef]
112. Garvey, S.D.; Pimm, A.; Buck, A.J.; Woolhead, S.; Liew, K.W.; Kantharaj, B.; Garvey, E.J.; Brewster, B.D. Analysis of a Wind Turbine Power Transmission System with Intrinsic Energy Storage Capability. *Wind. Eng.* **2015**, *39*, 149–173. [CrossRef]
113. Garvey, S.D. Multi-Stage High-Efficiency Adiabatic Compressor/Expander Uses Covered Liquid Pistons. U.S. Patent US20160356268A1, 8 December 2016. [CrossRef]

114. Swinfen-Styles, L.; Garvey, S.D.; Giddings, D. Combining wind-driven air compression with underwater compressed air energy storage. In Proceedings of the 2019 Offshore Energy and Storage Summit (OSES), Brest, France, 13–12 July 2019. [CrossRef]
115. Azelio—Clean Power When & Where You Need It 2021. Available online: https://www.azelio.com/ (accessed on 15 November 2021).
116. CCT Energy | Thermal Energy Storage Specialists, n.d. Available online: https://www.cctenergystorage.com/ (accessed on 18 October 2021).
117. Blain, L. 'World's First Working Thermal Battery' Promises Cheap, Eco-Friendly, Grid-Scalable Energy Storage. New Atlas 2019. Available online: https://newatlas.com/cct-silicon-energy-battery-thermal-energy-storage/59098/ (accessed on 30 September 2021).
118. TECHNOLOGY—TEXEL. Texel Energy Storage n.d. Available online: https://www.texeles.com/technology/ (accessed on 7 October 2021).
119. Herrmann, U. Projects Energy Storage: FH Aachen. Solar-Institut Jülich n.d. Available online: https://www.fh-aachen.de/en/research/solar-institute-juelich/focus-areas/projects-energy-storage (accessed on 5 September 2021).
120. Ma, Z.; Hofer, D.; Tallman, J.; Zhang, R.; Lambert, M.M. Economic Long-Duration Electricity Storage Using Low-Cost Thermal Energy Storage and a High-Efficiency Power Cycle (ENDURING). Arpa-E DAYS Annu. Meet, Advanced Research Projects Agency—Energy. 2021. Available online: https://arpa-e.energy.gov/sites/default/files/2021-03/07%20Day1-Zhiwen%20Ma_NREL.pdf (accessed on 4 October 2021).
121. Liquid Salt Combined Cycle. Pint Power 2019. Available online: https://www.pintailpower.com/technology/liquid-salt-combined-cycle/ (accessed on 7 December 2021).
122. Liquid Air Combined Cycle. Pint Power 2019. Available online: https://www.pintailpower.com/technology/liquid-air-combined-cycle/ (accessed on 7 December 2021).
123. Briggs, J. Antora Energy: Solid State Thermal Battery. Arpa-E DAYS Annu. Meet., Advanced Research Projects Agency—Energy. 2021. Available online: https://arpa-e.energy.gov/sites/default/files/2021-03/09Day1-JustinBriggs_AntoraEnergy-reducedfilesize.pdf (accessed on 4 October 2021).
124. Nacompex. 2021. Available online: https://feredox.de/ (accessed on 11 October 2021).
125. Nacompex Thermal Heating Power Plant 2019. Available online: http://nacompex.com/ (accessed on 30 May 2020).
126. Colthorpe, A. Long-Duration Thermal Energy Storage Startup Azelio Wins First Commercial Order—Energy Storage News 2020. Available online: https://www.energy-storage.news/long-duration-thermal-energy-storage-startup-azelio-wins-first-commercial-order/ (accessed on 4 October 2021).
127. Nilsson, M.; Abou-Taouk, A.; Sandberg, H.; Lindh, J. A Stirling engine for thermal energy storage. *AIP Conf. Proc.* **2019**, *2126*, 140005. [CrossRef]
128. Lindquist, T.; Karlsson, J.; Wallmander, J.; Guedez, R.; Hedlund, M.-L.; Jamot, J.; Gloss, D.; Lindh, J.; Hertin, A.; Nilsson, M.; et al. A novel modular and dispatchable CSP Stirling system: Design, validation, and demonstration plans. *AIP Conf. Proc.* **2019**, *2126*, 06005. [CrossRef]
129. Nilsson, H. Submarine Power Systems Using the V4-275R Stirling Engine. *Proc. Inst. Mech. Eng. Part A Power Process. Eng.* **1988**, *202*, 257–267. [CrossRef]
130. Tetteh, S.; Yazdani, M.R.; Santasalo-Aarnio, A. Cost-effective Electro-Thermal Energy Storage to balance small scale renewable energy systems. *J. Energy Storage* **2021**, *41*, 102829. [CrossRef]
131. Acharya, S.; Bhattacharjee, S. Stirling engine based solar-thermal power plant with a thermo-chemical storage system. *Energy Convers. Manag.* **2014**, *86*, 901–915. [CrossRef]
132. Rea, J.E.; Oshman, C.; Singh, A.; Alleman, J.; Buchholz, G.; Parilla, P.A.; Adamczyk, J.M.; Fujishin, H.-N.; Ortiz, B.R.; Braden, T.; et al. Prototype latent heat storage system with aluminum-silicon as a phase change material and a Stirling engine for electricity generation. *Energy Convers. Manag.* **2019**, *199*, 111992. [CrossRef]
133. Thermal Battery Producer Heats up Storage Market n.d. Available online: https://theleadsouthaustralia.com.au/industries/mining-resources/thermal-battery-producer-heats-up-storage-market/ (accessed on 18 October 2021).
134. Griffond, A.C.M.; Sofianos, M.V.; Sheppard, D.A.; Humphries, T.D.; Sargent, A.L.; Dornheim, M.; Aguey-Zinsou, K.-F.; Buckley, C.E. High-temperature thermochemical energy storage using metal hydrides: Destabilisation of calcium hydride with silicon. *J. Alloys Compd.* **2021**, *858*, 158229. [CrossRef]
135. Corgnale, C.; Hardy, B.; Motyka, T.; Zidan, R.; Teprovich, J.; Peters, B. Screening analysis of metal hydride based thermal energy storage systems for concentrating solar power plants. *Renew. Sustain. Energy Rev.* **2014**, *38*, 821–833. [CrossRef]
136. Miller, M.A.; Petrasch, J.; Randhir, K.; Rahmatian, N.; Klausner, J. Chemical energy storage. *Therm. Mech. Hybrid Chem. Energy Storage Syst.* **2021**, 249–292. [CrossRef]
137. IEA ES TCP Newsletter September 2021. Available online: https://iea-es.org/news/es-tcp-newsletter-september-2021/ (accessed on 5 October 2021).
138. Ma, Z.; Wang, X.; Davenport, P.; Gifford, J.; Martinek, J. Economic Analysis of a Novel Thermal Energy Storage System Using Solid Particles for Grid Electricity Storage: Preprint. 15th Int. Conf. Energy Sustain., Virtual, Online: The American Society of Mechanical Engineers. 2021. Available online: https://www.nrel.gov/docs/fy21osti/79014.pdf (accessed on 6 October 2021).

139. Ma, Z.; Mehos, M.; Glatzmaier, G.; Sakadjian, B.B. Development of a Concentrating Solar Power System Using Fluidized-bed Technology for Thermal Energy Conversion and Solid Particles for Thermal Energy Storage. *Energy Procedia* **2015**, *69*, 1349–1359. [CrossRef]
140. Patel, P. White-Hot Blocks as Renewable Energy Storage?—IEEE Spectrum 2021. Available online: https://spectrum.ieee.org/could-storing-electricity-in-whitehot-blocks-give-supercheap-renewables-storage (accessed on 19 October 2021).
141. Datas, A.; Zeneli, M.; Cañizo, C.; Malgarinos, I.; Nikolopoulos, A.; Nikolopoulos, N.; Karellas, S.; Marti, A. Molten silicon storage of concentrated solar power with integrated thermophotovoltaic energy conversion. *AIP Conf. Proc.* **2018**, *2033*, 090005. [CrossRef]
142. Ramos, A.; Datas, A.; Cañizo, C.D.; Martí, A. Molten Silicon at the hearth of a novel energy storage system. *Silicon Chem. Sol. Ind. XI Svolvaer Norwa* **2018**. [CrossRef]
143. Datas, A.; Vaillon, R. Thermionic-enhanced near-field thermophotovoltaics. *Nano Energy* **2019**, *61*, 10–17. [CrossRef]
144. Bullis, K. Molten Salts Might Provide Half-Price Grid Energy Storage—MIT Technology Review. MIT Technol Rev 2014. Available online: https://www.technologyreview.com/s/525121/molten-salts-might-provide-half-price-grid-energy-storage/ (accessed on 26 June 2018).
145. Amy, C.; Seyf, H.R.; Steiner, M.A.; Friedman, D.J.; Henry, A. Thermal energy grid storage using multi-junction photovoltaics. *Energy Environ. Sci.* **2019**, *12*, 334–343. [CrossRef]
146. Schulte, K.L.; France, R.M.; Friedman, D.J.; LaPotin, A.D.; Henry, A.; Steiner, M.A. Inverted metamorphic AlGaInAs/GaInAs tandem thermophotovoltaic cell designed for thermal energy grid storage application. *J. Appl. Phys.* **2020**, *128*, 143103. [CrossRef]
147. Thiele, E.; Jahnke, A.; Ziegler, F. Efficiency of the Lamm–Honigmann thermochemical energy storage. *Therm. Sci. Eng. Prog.* **2020**, *19*, 100606. [CrossRef]
148. EW GmbH—Technology, n.d. Available online: http://www.o-ew.de/technologie/ (accessed on 11 October 2021).
149. Jahnke, A. Untersuchung des Honigmann-Prozesses zur Thermochemischen Energiespeicherung [Investigation of the Honigmann Process for Thermochemical Energy Storage]. Ph.D. Thesis, Technische Universität Berlin, Berlin, Germany, 2019. [CrossRef]
150. Hoivik, N.; Greiner, C.; Barragan, J.; Iniesta, A.C.; Skeie, G.; Bergan, P.; Blanco-Rodriguez, P.; Calvet, N. Long-term performance results of concrete-based modular thermal energy storage system. *J. Energy Storage* **2019**, 24. [CrossRef]
151. Storworks Power—Our Technology 2020. Available online: https://storworks.com/our-technology/ (accessed on 7 December 2021).
152. Electrical Power | Kraftblock n.d. Available online: https://kraftblock.com/en/applications/electrical-power.html (accessed on 10 November 2021).
153. Green Energy Solutions | Magaldi Green Energy. Magaldi Green Energy n.d. Available online: https://www.magaldigreenenergy.com/en/technology (accessed on 10 November 2021).
154. Thermal Processes for Air Pollution Control—Dürr n.d. Available online: https://www.durr.com/en/products/environmental-technology/exhaust-gas-and-air-pollution-control/oxi-x-thermal-processes (accessed on 10 November 2021).
155. Philipp, R. Carbonclean and EEW Strengthen Collaboration in Carnot Batteries: Developing Thermal Waste Utilisation Plants into Flexible Storage Power Plants—EEW Energy from Waste 2021. Available online: https://www.eew-energyfromwaste.com/en/news/media/detail/news/carbonclean-und-eew-vertiefen-zusammenarbeit-bei-carnot-batterie-ausbau-energetischer-abfallverwertungsanlagen-zu-flexiblen-speicherkraftwerken/ (accessed on 10 November 2021).
156. Electrified Thermal Solutions | Joule Hive 2021. Available online: https://www.electrifiedthermal.com/ (accessed on 5 November 2021).
157. Brenmiller Energy | Carbon-Free Thermal Energy Storage n.d. Available online: https://bren-energy.com/ (accessed on 10 November 2021).
158. LUMENION Builds Storage for an Efficient CO_2-Free Energy Supply 2021. Available online: https://lumenion.com/?lang=en (accessed on 5 November 2021).
159. The Norwegian Thermal Battery Company. Kyoto Gr AS 2021. Available online: https://www.kyoto.group/ (accessed on 24 November 2021).
160. Hyme. Energy Storage in Molten Hydroxides 2021. Available online: https://www.hyme.energy/ (accessed on 4 January 2022).
161. SaltX Technology—Energy Storage with Nano Coated Salt. SaltX Technol Hold 2021. Available online: https://saltxtechnology.com/ (accessed on 5 November 2021).
162. Thermal Battery Technology | Modular, Scalable, Recycable. EnergyNest 2018. Available online: https://energy-nest.com/technology/ (accessed on 20 September 2021).
163. Hoivik, N.; Greiner, C.; Tirado, E.B.; Barragan, J.; Skeie, G.; Calvet, N.; Bergan, P.; Blanco, P. Demonstration of EnergyNest thermal energy storage (TES) technology. *AIP Conf. Proc.* **2017**, *1850*, 080011. [CrossRef]
164. Jacqui, G. Innovative Eenergy Storage. Compass Mag—Renew Energy Innov 2021. Available online: https://compassmag.3ds.com/innovative-energy-storage/?fbclid=IwAR1_GNU5PDIPFoF0rYmT7Sj9MtTKaPX0EjcvsKXxCXJyx92ZLbquF-dAwSI (accessed on 10 November 2021).
165. Acri, R.; Bassetti, F.; Falvo, M.C.; Magaldi, L.; Manganelli, M.; Romagnoli, L.; Santi, F.; Scafuri, A. A new thermal energy storage technology for power system services. In Proceedings of the 2021 IEEE International Conference on Environment and Electrical Engineering and 2021 IEEE Industrial and Commercial Power Systems Europe (EEEIC/I&CPS Europe), Bari, Italy, 7–10 September 2021. [CrossRef]

166. Green Heat Module—Kraftanlagen 2021. Available online: https://www.kraftanlagen.com/en/solutions/energy/green-heat-module/ (accessed on 10 November 2021).
167. Baumann, T.; Göhring, F.; Stadler, H.; Doerbeck, T. Jülich solar power tower—System behavior during downtime. *AIP Conf. Proc.* **2017**, *1850*, 030004. [CrossRef]
168. Kronhardt, V.; Alexopoulos, S.; Reißel, M.; Sattler, J.; Hoffschmidt, B.; Hänel, M.; Doerbeck, T. High-temperature Thermal Storage System for Solar Tower Power Plants with Open-volumetric Air Receiver Simulation and Energy Balancing of a Discretized Model. *Energy Procedia* **2014**, *49*, 870–877. [CrossRef]
169. Präsentation Carnot Batterien für Kraftwerke. Carbonclean 2019. Available online: https://www.innovationsforum-energiewende.de/fileadmin/user_upload/2019-01-28-Firmenpraesentation-carbonclean-Speicherkraftwerk.pdf (accessed on 10 November 2021).
170. Preiss, S. EEW cooperates with Carbon Clean on electricity storage | EUWID New Energy News 2021. Available online: https://www.euwid-energie.de/eew-kooperiert-mit-carbon-clean-bei-stromspeicherung/ (accessed on 10 November 2021).
171. Forsberg, C.W.; Stack, D.C.; Curtis, D.; Haratyk, G.; Sepulveda, N.A. Converting excess low-price electricity into high-temperature stored heat for industry and high-value electricity production. *Electr. J.* **2017**, *30*, 42–52. [CrossRef]
172. Stack, D.C.; Curtis, D.; Forsberg, C. Performance of firebrick resistance-heated energy storage for industrial heat applications and round-trip electricity storage. *Appl. Energy* **2019**, *242*, 782–796. [CrossRef]
173. Ondrey, G. Storing Excess Electricity as High-Temperature Heat. *Chem. Eng.* **2021**, *1*. Available online: https://www.chemengonline.com/storing-excess-electricity-high-temperature-heat/#disqus_thread (accessed on 5 November 2021).
174. Billing. Mimi. Salt Breakthrough Could Halve the Cost of Storing Solar Energy. Sifted 2021. Available online: https://sifted.eu/articles/salt-energy-storage-seaborg-hyme/ (accessed on 4 January 2022).
175. Kanzawa, A.; Arai, Y. Thermal energy storage by the chemical reaction augmentation of heat transfer and thermal decomposition in the CaOCa$(OH)_2$ powder. *Sol. Energy* **1981**, *27*, 289–294. [CrossRef]
176. Dai, L.; Long, X.F.; Lou, B.; Wu, J. Thermal cycling stability of thermochemical energy storage system $Ca(OH)_2/CaO$. *Appl. Therm. Eng.* **2018**, *133*, 261–268. [CrossRef]
177. Alovisio, A.; Chacartegui, R.; Ortiz, C.; Valverde, J.M.; Verda, V. Optimizing the CSP-Calcium Looping integration for Thermochemical Energy Storage. *Energy Convers. Manag.* **2017**, *136*, 85–98. [CrossRef]
178. Li, D.; Wang, J. Study of supercritical power plant integration with high temperature thermal energy storage for flexible operation. *J. Energy Storage* **2018**, *20*, 140–152. [CrossRef]
179. Wang, Y.; Lou, S.; Wu, Y.; Wang, S. Flexible Operation of Retrofitted Coal-Fired Power Plants to Reduce Wind Curtailment Considering Thermal Energy Storage. *IEEE Trans Power Syst.* **2020**, *35*, 1178–1187. [CrossRef]
180. Angerer, M.; Djukow, M.; Riedl, K.; Gleis, S.; Spliethoff, H. Simulation of Cogeneration-Combined Cycle Plant Flexibilization by Thermochemical Energy Storage. *J. Energy Resour. Technol. Trans. ASME* **2018**, *140*. [CrossRef]

energies

MDPI

Article

Grid and PV Fed Uninterruptible Induction Motor Drive Implementation and Measurements †

Ruben Rafael Boros * and István Bodnár

Institute of Physics and Electrical Engineering, University of Miskolc, 3515 Miskolc, Hungary; vegybod@uni-miskolc.hu

* Correspondence: elkruben@uni-miskolc.hu

† This paper is an extended version of our paper published in 2021 IEEE International Carpathian Control Conference, Ostrava, Czech Republic, 31 May–1 June 2021.

Abstract: Motors powered directly from solar panels are becoming more and more popular in pump applications. However, solar panels can be the source of operational issues due to varying irradiance, ambient temperature, weather. This paper shows how it is worth expanding a solar induction motor drive to provide an uninterrupted flow of electricity to the motor. In addition, the main components of the uninterruptible induction motor drive are presented, including the LLC (inductor-inductor-capacitor) converter, the three-phase inverter, and the three-phase rectifier. LLC converters that can increase the voltage from 25–40 V to 330 V cannot be bought directly from manufacturers. Therefore, a custom LLC converter was made for the research. It was necessary to build a custom converter to avoid the use of solar panel strings. This way, solar panels connected in parallel can be used. A low-voltage (25–40 V) supply was implemented from the solar side, while the induction motor requires 230 V AC three-phase voltage in delta connection. For this reason, a voltage boost is required from the low voltage side. The grid feeds the universal DC link through the three-phase rectifier. This allows the motor to consume varying amounts of electricity from the grid or the solar panel. The study also presents in detail the LLC converter that performs the voltage boost. Measuring the entire motor drive, switching transients and efficiencies can be observed at different input voltages for different supplies as well as loads.

Keywords: solar panel; induction motor drive; LLC converter; uninterruptible

Citation: Boros, R.R.; Bodnár, I. Grid and PV Fed Uninterruptible Induction Motor Drive Implementation and Measurements. *Energies* **2022**, *15*, 708. https://doi.org/10.3390/en15030708

Academic Editors: Alon Kuperman and Alessandro Lampasi

Received: 24 December 2021
Accepted: 17 January 2022
Published: 19 January 2022

1. Introduction

Nowadays, the world's demand for electricity is constantly increasing, while renewable energy sources are also becoming more and more widespread [1]. Solar panels allow efficient electricity generation over a relatively small area. Typically, the direct current and direct voltage produced by a solar panel are connected to a solar inverter that feeds electricity back into the grid or supplies electricity to the household. In some cases, solar panels are also used in island mode with batteries, in which case loads are fed directly from the solar panel and optionally from the battery through power electronic converter. Dynamic loads, such as electric motors, can cause control problems in electronic converters when powered solely by solar panels. Therefore, it is generally not advisable to power electric motor drives directly from solar panels, although it is worth to consider using them in lower power applications, such as in geothermal heat pumps, water pumps, heating circulating pumps, where three-phase or one-phase induction motors are usually used [2].

In Europe, dwarf three-phase induction motors (<1 kW) are in most cases designed for 230/400 V (delta/wye connection). The operating voltage of the solar panel is a few tens of volts, depending on the type. If voltage boost LLC converters are used, there is no need to form a solar panel string where several panels are connected in series to increase the voltage. Certain problems can occur when using solar panel strings if one of the solar panels is shaded and/or damaged. Shading a solar cell alone reduces the power of the entire string to a fraction when connected in series. Shading a single solar cell result in a much smaller power reduction when they are connected in parallel, than shading a cell of multiple cells connected in series. Due to this phenomenon, microinverters are used to feed back to the grid without needing to form a solar panel string. It is worth to mention that it is possible to form strings for low power motors (a few hundred watts), but not as cost-effective, since the unit cost of lower power solar panels are higher. Thus, it is worth to use higher power panels connected in parallel. An LLC converter is also required for solar panels connected in series and in parallel as well.

The voltage of a solar panel depends on many factors, such as the load, the solar irradiance, the temperature etc. On top of that, induction motors are AC loads; thus, an inverter is required to drive the motor from DC voltage and make it possible to adjust the speed of the motor. The electronic converter circuit must include a boost converter and an inverter. The boost converter generates stable high voltage from the fluctuating low-voltage of the solar panel, and then the inverter converts it to a three-phase alternating voltage for the motor with variable frequency. Depending on the size of the induction motors, the starting inrush current can be several times higher than the full load current, which means that the operating voltage of the solar panel changes abruptly. It is necessary to use a boost converter that can perform the formerly mentioned task with high efficiency. Using resonant LLC switching power supplies with good efficiency is sensible.

Efficiency is a crucial aspect of solar systems. Modern electronic converters can convert electricity with an acceptable efficiency; however, the efficiency of solar panels currently available on the market is not higher than approximately 20% [3]. Their efficiency, however, is increasing steadily as they are constantly being developed. The efficiency of dwarf induction motors is around 60–75%. Overall, it is worth installing solar fed motor drives in pump applications.

An induction motor drive powered directly from a solar panel cannot be considered reliable, since there is no constant sunlight in any particular area, and the solar irradiance itself changes dynamically. Operational safety can be greatly increased with batteries integrated into the system. However, even for low-power motors, several batteries are required for the pump to run overnight. It is reasonable to combine batteries with the use of electricity from the power grid, if available. This also increases operational safety since in the event of insufficient irradiance, the motor can just draw power from the grid. In addition, in the event of a grid failure, the motor can run on solar and/or battery power, so this system provides a versatile uninterrupted power supply to the motor [4]. Installing batteries in the system incurs an additional cost, as an MPPT controller is now required within the system. The battery system should be designed in a way that the solar panels charge the battery through the MPPT controller, and the LLC converter is powered from the battery.

Voltage boost LLC converters from 25–40 V to 330 V are not widespread, so it was necessary to build a custom converter for this research. Another advantage of a custom converter is that it does not require the use of solar panel strings. In this way, solar panels can be connected in parallel, if necessary. Moreover, this paper examines a solar and grid powered induction motor drive (Figure 1) in terms of electrical and mechanical efficiency and versatility. In theory, the following system can also be used for dwarf motors and high-performance motors. However, the size of the LLC converter increases greatly as its power increases, including the elements of the LLC tank. For this reason, it is worth using up to a few 10 kW.

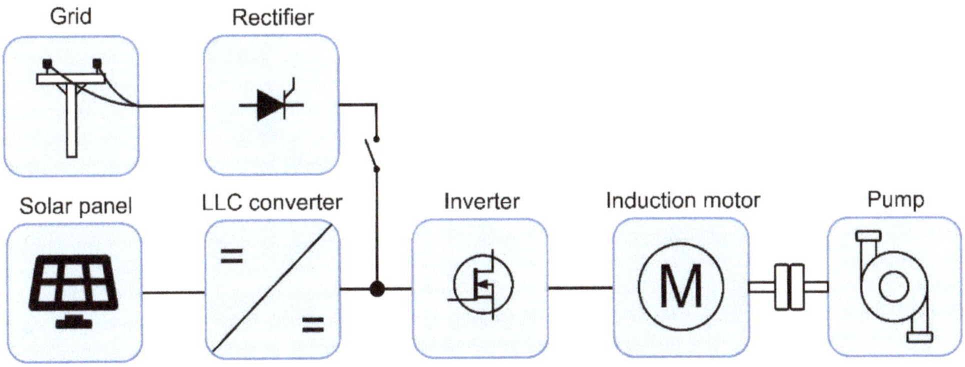

Figure 1. Grid and PV fed uninterruptible induction motor drive.

The design of the LLC converter and the uninterruptible induction motor drive is presented in Section 2. The definition of the converters parameters is presented in detail. Section 3. details the hardware and software implementation of the LLC converter with PCB layout. The experimental setup used for the measurements is described in Section 4. Measurements are also presented where the LLC converter is powered by solar panel, from the power grid and from battery, as well as power supply. NI 9215 data logger with 0.2% accuracy was used with LabView software to measure the efficiencies. The torque and speed of the motor were measured by a SE2662-3S instrument and SE2662-1R magnetic powder brake from Lucas-Nülle. The efficiency of the motor was calculated from the ratio of the mechanical output power to the electrical power input. Mechanical performance was derived from the product of torque and angular velocity measured by the Lucas-Nülle device.

2. Designs

2.1. LLC Converter in General

The LLC converter is a resonant switching power supply that is widespread due to its high efficiency and relatively small size. LLC converters include a high-frequency transformer and external inductors, depending on the type. The LLC converters can also be used as a boost and buck converter as well. By increasing the switching frequency, the size of the transformer can be reduced, although the semiconductors operate with a higher switching heat loss, which results in higher cooling demand, a larger heat sink [5].

With resonant converters, the switching loss can be greatly reduced. In case of resonant converters, the semiconductors are switched when the voltage drop across the switching element is zero (ZVS: Zero Voltage Switching) or the current flow is zero (ZCS: Zero Current Switching). These switching modes result in soft switching on the semiconductor, so the dynamic power loss on the semiconductor is greatly reduced [5–8]. The LLC converters used operate in ZVS mode. This mode can be achieved in only inductive region. ZCS switching mode should be avoided. In this case, the current leads the voltage, so the current on the MOSFET flows in the reverse direction before the MOSFET turns off. After switching off, a large current would flow through the body diode. Hard commutation would then occur when a MOSFET in the bridge was turned on. This would result in a large recovery loss, noise and high current spike, which can result in device failure [7].

It is advisable to use the LLC full-bridge converter shown in Figure 2 for the task. A major advantage of the LLC converter is that in the case of a wide range of varying loads, the switching frequency only needs to be adjusted at a narrow interval, so it can be well controlled at a very low loads, or even without a load. It also has the additional advantage of low EMI noise due to soft turn off [5–8]. However, it is not always advisable to use resonant converters. It is worth using in medium power applications (few 10 kW), but above that the elements of the LLC tank would be very large and heavy.

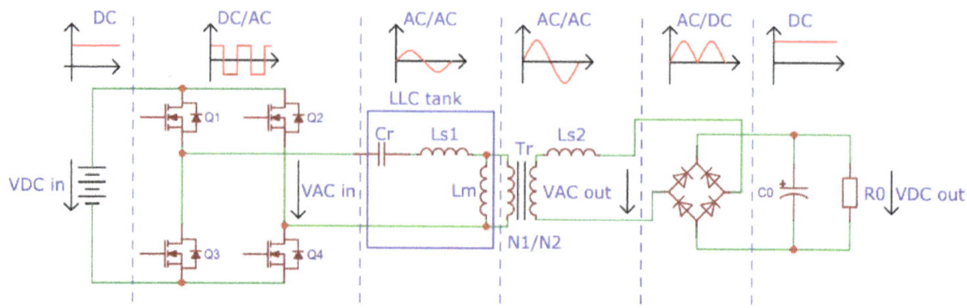

Figure 2. LLC full-bridge converter.

A constant direct current (DC) is connected to the input of the LLC converter, which is applied to a single-phase inverter (DC/AC). The inverter consists of four (Q1, Q2, Q3, Q4) MOSFETs that produce alternating voltage with a 50% duty cycle. At the output of the inverter is the LLC tank, which converts the square wave voltage into sinusoidal voltage (AC/AC). The LLC tank circuit and the transformer increase or decrease together the voltage to the desired value (AC/AC). The AC voltage is applied to a full-bridge rectifier that provides DC voltage at its output (AC/DC) with a high ripple. Capacitor C_0 smooths the ripple of the DC voltage. The load (R_0) connected to the constant DC voltage [5–8].

Two magnetic components are required for the resonant circuit: the inductances L_r and L_m. L_r denotes the series resonant inductance, L_m the magnetizing inductance, which is shunt-like. C_r is the series resonant capacitance. It is also necessary to install a transformer for high gain. With the help of an integrated transformer, the inductances L_r and L_m can be implemented [5–8].

Consider Figure 3, which substitutes Figure 2 for the actual parameters of the transformer, i.e., the leakage primary and secondary and magnetizing inductances. It further simplifies Figure 2. At the bottom of the figure, in the simplified network, the transformer can already be considered ideal. The operation of LLC converters is also greatly influenced by leakage inductances (L_{s1}, L_{s2}). Although measuring leakage inductances is very difficult in practice, this is no longer necessary after simplification of the network. The parameter L_p can be determined by measuring the inductance on the primary winding of the transformer while the secondary side is open. The parameter L_r is similarly measured for a short-circuited secondary coil [6,7].

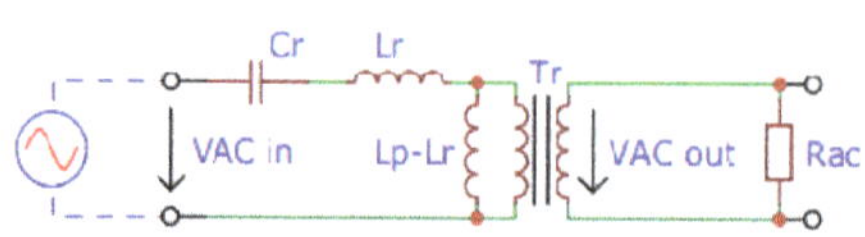

Figure 3. Simplified LLC full-bridge converter.

The appearance of the parameter L_r (resonant inductance) therefore stems from the simplification that can be determined according to Equation (1) [5–8].

$$L_r = L_{s1} + L_m \times \left(n^2 \cdot L_{s2}\right) = L_{s1} + \frac{L_m \cdot L_{s1}}{L_m + L_{s1}} \tag{1}$$

The parameter L_p (primary inductance) is also created, which combines the primary leakage inductance and the magnetizing inductance (Equation (2)) [5–8].

$$L_p = L_{s1} + L_m \tag{2}$$

The magnetizing inductance can be given according to Equation (3), only two values of the result of the measurement on the transformer are required [5–8].

$$L_m = L_p - L_r \tag{3}$$

An important static parameter in the design of LLC converters is the m parameter, which can be changed to optimize the converter (Equation (4)). Reducing this parameter entails a reduction in the efficiency of the converter, a higher gain at the output, a narrower frequency range for control. Increasing the parameter m results in higher efficiency of the converter by a smaller magnetizing circulating current [5–8].

$$m = \frac{L_p}{L_r} = \frac{L_r + L_m}{L_r} \tag{4}$$

Figure 3 also shows a R_{AC} resistor corresponding to the load tested from the AC side (Equation (5)). Here, R_0 stands for DC load and A_V for virtual gain [5–8].

$$R_{AC} = \frac{8}{\pi^2}\cdot\frac{R_0}{{A_v}^2} = \frac{8\cdot n^2}{\pi^2}\cdot\frac{{V_{DC\_OUT}}^2}{P_0} \tag{5}$$

There are two resonant frequencies, these are derived from the Thomson formula (Equations (6) and (7)). The first resonant frequency is determined by the values of L_r, C_r and the second by the values of L_p and C_r [5–8].

$$f_r = \frac{1}{2\pi\sqrt{L_r\cdot C_r}}. \tag{6}$$

$$f_p = \frac{1}{2\pi\sqrt{L_p\cdot C_r}} \tag{7}$$

The quality factor of the resonant circuit is described by (Equation (8)) [5–8].

$$Q = \frac{\sqrt{\frac{L_r}{C_r}}}{R_{AC}} \tag{8}$$

Equation (9) gives the transfer function of the simplified network shown in Figure 3 [5–8].

$$G(Q,\ m,\ F_n) = \frac{F_n^2(m-1)}{\sqrt{\left(m\cdot F_n^2 - 1\right)^2 + F_n^2\cdot\left(F_n^2 - 1\right)^2\cdot(m-1)\cdot Q^2}} \tag{9}$$

The coefficient F_n represents the normalized frequency, which is the quotient of the frequency of the voltage supplying the resonant circuit and the resonant frequency (Equation (10)) [5–8].

$$F_n = \frac{f_s}{f_r} \tag{10}$$

The graphical representation of the transfer function is shown in Figure 4, where the static parameter m = 8 as an example. At different loads, we get various curves of the magnitude of the gain as a function of the normalized frequency.

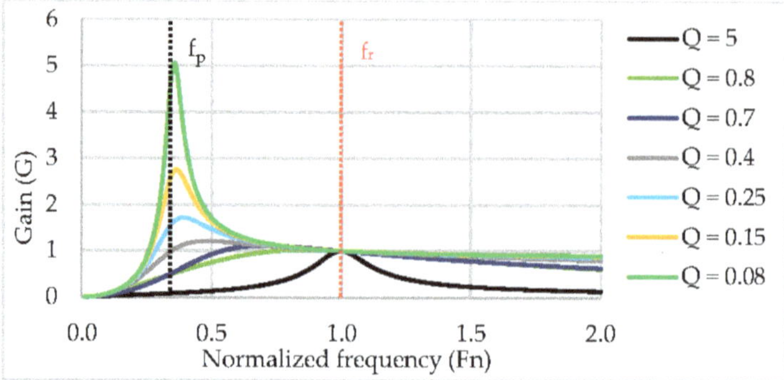

Figure 4. Gain as a function of normalized frequency under different loads.

As the load increases (R_{AC} decreases), Q also increases. As the load increases the maximum of the curve (gain) decreases. The gain of the resonant circuit fed at the resonant frequency will be G = 1 regardless of the load. However, as the load decreases, the maximum of the curves are converging to the primary resonant frequency (f_P). The gain G = 1 at the resonant frequency is only true if the secondary leakage inductance of the transformer is not considered, otherwise the gain changes slightly. The resonant frequency displays the virtual gain, denoted by A_V (Equation (11)) [5–8].

$$A_V = \sqrt{\frac{L_P}{L_P - L_r}} = \sqrt{\frac{m}{m-1}} \tag{11}$$

2.2. Uninterruptible Induction Motor Variable-Frequency Drive

Figure 5 shows the schematic diagram of the uninterruptible induction motor variable-frequency drive. The half-controlled three-phase rectifier is located at the top of the figure. It is advisable to use a three-phase rectifier due to the smoother DC voltage. An alternative solution is the single-phase full-bridge rectifier, which generates approx. 320 V DC at its output. The DC link can be considered universal, as different rectifiers, DC/DC converters and several inverters can be connected to it [9,10].

If the SW switch is closed, the DC link's ground is not isolated anymore from the grid, but it is not a problem because the LLC converter's transformer isolates the grid's ground from the solar panel and the battery, and the grid can supply the DC link through the rectifier. From the DC voltage, the three-phase inverter generates variable-frequency voltage for the motor, thus controlling the speed of the motor. The DC link is also fed by the LLC converter which is located under the three-phase rectifier of the figure. With this circuit, three operating states can be achieved:

1. The LLC converter is turned off, only the rectifier is feeding the motor;
2. The rectifier is turned off, only the LLC converter is feeding the motor;
3. Both the rectifier and the LLC converter are simultaneously feeding the motor at a certain rate. The ratio depends on the magnitude of the voltage generated by the converters.

Since the motor is a dynamic load, it is necessary to monitor the voltage of the DC link which the single-phase inverter controls in the LLC converter [11,12]. Due to the load being dynamic, the quality factor and the gain also change fast. This is monitored by the resonant circuit control circuit through an isolation amplifier. The controller controls the frequency so that the output is set to 330 V. Meanwhile, the operating voltage of the solar panel also changes dynamically due to the changing irradiance as well as the load impedance [13].

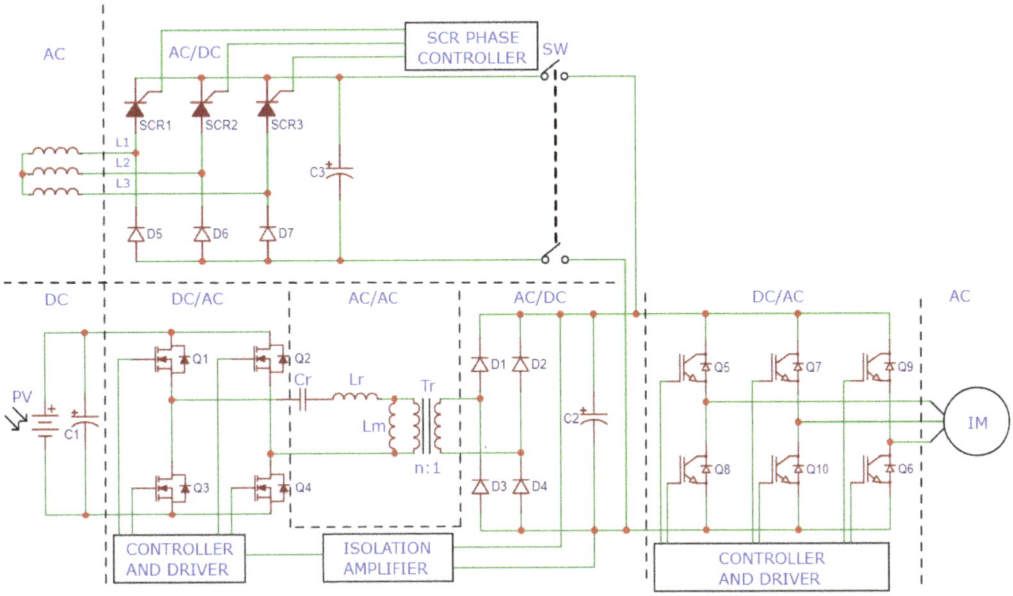

Figure 5. Schematic diagram of the uninterruptible induction motor variable-frequency drive.

2.3. LLC Converter Design for Induction Motor Variable-Frequency Drive

The first step in designing the LLC converter is to specify the input parameters. The amount of power required at the output must be declared. The aim of the study is to power dwarf three-phase induction motors of max. 400 W electrical power (Equation (12)).

$$P_0 = 400\ \text{W} \tag{12}$$

The output voltage in the DC link must be 330 V to produce an effective voltage of 230 V at the output of the inverter for the motor, including losses. The motor must then be connected in a delta connection. The magnitude of the current flowing on the DC side and the magnitude of the DC load can be calculated (Equations (13) and (14)).

$$I_{DC\_OUT} = \frac{P_0}{V_{DC\_OUT}} = \frac{400\ \text{W}}{330\ \text{V}} = 1.2121\ \text{A} \tag{13}$$

$$R_0 = \frac{V_{DC\_OUT}}{I_{DC\_OUT}} = \frac{330\ \text{V}}{1.2121\ \text{A}} = 272.26\ \Omega \tag{14}$$

The maximum and minimum of the input voltage are both required input parameters. The circuit is powered by a solar panel with a maximum operating (nominal) voltage of around 29 V and a maximum power of 500 W at 1000 W/m^2 irradiance. The lower limit was set at 25 V and the upper limit at 40 V. Based on these, the transformer ratio can be determined (Equation (15)). It requires the value of virtual gain already mentioned in Equation (11) of the study. The result is given by Equation (16). The m parameter was chosen to be 8 in this research for good efficiency. V_F is the voltage drop of a diode in the rectifier bridge circuit (approx. 3 V).

$$n = \frac{V_{DC\_IN\_MAX}}{V_{DC\_OUT} + 2 \cdot V_F} \cdot A_{min} = \frac{40\ \text{V}}{330\ \text{V} + 2 \cdot 1.5\ \text{V}} \cdot 1.0691 = 0.128414 \tag{15}$$

$$A_{min} = A_V = \sqrt{\frac{m}{m-1}} = \sqrt{\frac{8}{8-1}} = 1.0691 \tag{16}$$

The maximum gain must also be determined when the input voltage is the lowest (25 V in this case). Equation (17) can be used for this.

$$A_{max} = A_{min} \cdot \frac{V_{DC\_IN\_MAX}}{V_{DC\_IN\_MIN}} = 1.0691 \cdot \frac{40\ V}{25\ V} = 1.71 \quad (17)$$

The quality factor (Q) is determined in the following way to reach the value of the maximum gain. It is necessary to substitute the initial values into the transfer function, plot the obtained function and then find the maximum of this (executed in Excel). Until then, the value of Q is iterated, even if the maximum of the function exceeds the value of A_{max}. However, it is important to mention that at the switching frequency used at the maximum of the function, the semiconductor switching elements do not operate in ZVS mode, so the efficiency is greatly reduced. In this case, it is worth increasing the value of A_{max} by 10%. Through the iterations, a value of Q = 0.232 was obtained, thus, covering the 10% oversizing. With a quality factor Q = 0.232 and a static parameter m = 8, the transfer function shown in Figure 6 is obtained. The figure also shows the minimum and maximum gain required (without oversizing). The maximum of the function is displayed at $F_n = 0.39$ (39 kHz), but at maximum load—due to the oversizing—47.6 kHz is required.

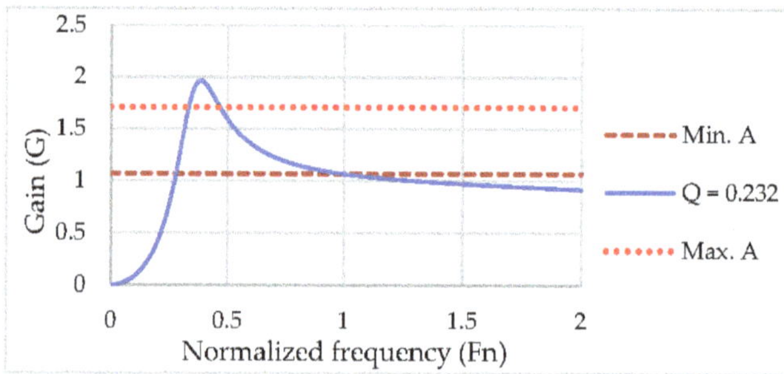

Figure 6. The transfer function with the designed values at maximum load.

The value of the resonant frequency (f_r) must also be specified. This value is limited by the properties of the semiconductors and the capabilities of the control circuit. A frequency of up to 100 kHz is still acceptable, with SiC MOSFETs, suitable drives and a high-speed microcontroller, henceforth f_n = 100 kHz. Thus, the maximum gain is at 47.6 kHz. Equations (18)–(20) are required to determine the capacitive and magnetic parameters, as well as the value of the load from the AC side (Equation (22)). The magnetizing inductance is given by Equation (21).

$$C_r = \frac{1}{2 \cdot \pi \cdot Q \cdot f_r \cdot R_{AC}} = \frac{1}{2 \cdot \pi \cdot 0.232 \cdot 10^5\ Hz \cdot 3.1842\ \Omega} = 2154\ nF \quad (18)$$

$$L_r = \frac{1}{(2 \cdot \pi \cdot f_r)^2 \cdot C_r} = \frac{1}{(2 \cdot \pi \cdot 10^5\ Hz)^2 \cdot 2154 \cdot 10^{-9}\ F} = 1.1756\ \mu H \quad (19)$$

$$L_p = m \cdot L_r = 8 \cdot 1.1756\ \mu H = 9.405\ \mu H \quad (20)$$

$$L_m = L_p - L_r = 9.405\ \mu H - 1.1756\ \mu H = 8.2294\ \mu H \quad (21)$$

$$R_{AC} = \frac{8 \cdot n^2}{\pi^2} \cdot \frac{V_{DC\_OUT}{}^2}{P_0} = \frac{8 \cdot 0.1284^2}{\pi^2} \cdot \frac{330^2\ V}{400\ W} = 3.184\ \Omega \quad (22)$$

3. LLC Converter Implementation

The implemented LLC converter is shown in Figure 7. The single-phase inverter MOSFETs (1.), power supply (2.), microcontroller (3.), transformer with external inductors

(4.), full-bridge rectifier (5.), LLC tank capacitors (6.), voltage measurement circuits (7.), MOSFET's gate driver with fault detection (8.) can be found on the printed circuit board.

Figure 7. The implemented LLC converter.

3.1. Single-Phase Inverter

The single-phase inverter contains four NTHL060N090SC1 SiC MOSFETs with a static drain-source on-resistance of 60 mΩ (if V_{GS} = 15 V). SiC MOSFETs are assembled on a heat sink. The relatively large heat sink is required because a high current (approx. 20 A at 25 V) flows at the input of the converter at full load, resulting in a high static power loss on the MOSFETs. The MOSFETs are controlled by an intelligent optodriver circuits (ACPL-352J) and isolated DC/DC converters (MGJ2D051505SC). This drive method also allows stable control above the resonant frequency (100 kHz). This method is necessary at the energy-free turn on.

3.2. LLC Tank and Transformer

The LLC tank circuit contains 27 pieces 82 nF in parallel connection, 1000 V foil capacitor, 6 pieces external inductor in parallel connection (PCV-0-472-10L, 4.7 μH) and the transformer as well. A total of 27 capacitors connected in parallel are needed to be able to operate at high current, as one capacitor can conduct max. 1.1 A_{RMS}. The transformer must be sized for the lowest operating frequency (47.6 kHz), as this will be the maximum flux in the iron core. The type of iron core is EE65/65/27A ferrite. The transformer has n = 0.129 turns ratio. The primer winding has N_1 = 4; the secondary winding has N_2 = 31 number of turns. The iron core of the transformer consists of two pieces. The inductance L_p of the transformer can be adjusted with the air gap; however, the inductance L_r changes slightly. The inductance L_r depends to a large extent on the stray inductance of the transformer. The stray inductance must be made very large (compared with a normal transformer) in this case. The primary and secondary coils should be placed as far apart as possible. However, if this is not achieved, the stray inductance can be increased by inserting external coils. Therefore, 6 coils connected in parallel were installed. The air gap was set while measuring L_p inductance, so a very accurate result can be achieved. The plastic spacer and the iron core were assembled with ferrite glue. The following results were achieved: L_p = 9.4 μH, L_r = 0.39 μH. The resulting inductance of the six coils is 0.783 μH. This gives a total inductance of L_r = 1.173 μH. These values deviate very slightly from the design values, but it does not affect the operation.

3.2.1. Output Rectifier

At the transformer output, a four-piece SiC diode (STPSC15H12) creates a full-bridge rectifier. One diode has approx. V_F = 1.5 V voltage drop. A buffer capacitor (330 µF) smooths the voltage at the rectifier output. A heatsink is not mounted on the diodes due to the output current being low (max. 1.22 A) and the diode recovery time is negligible.

3.2.2. Output Voltage Measurement Circuit and Microcontroller

The output voltage is monitored by the microcontroller using analog input. With ACPL-C87A optically isolated amplifier the voltage can measure safely. The amplifier circuit requires low-pass filter and isolated DC/DC converter. At the output of the isolated amplifier, a differential operational amplifier conditions the signal to the analogue input of the microcontroller. The type of microcontroller is STM32F103C8T6. Using the microcontroller timers, the control signals of the single-phase inverter can be easily generated. The timer frequency is set by a discrete PID controller. The setpoint of the PID controller is 330 V, for which the output is controlled by the microcontroller. The frequency is set between 47.6 kHz and 100 kHz depending on the magnitude of the load and the input voltage. The PID controller is set to turn off the single-phase inverter when the voltage reaches 390 V. This is required when the load is slight and the input voltage is nearly 40 V. As soon as the output drops below 340 V, the inverter switches on again. Filter capacitors are electrolytic capacitors with a high leakage current, so the output cannot be considered unloaded. Thus, the voltage oscillates between 390 and 340 V with a large time constant when no load connected. Figure 8 shows the flowchart of the implemented microcontroller's algorithm.

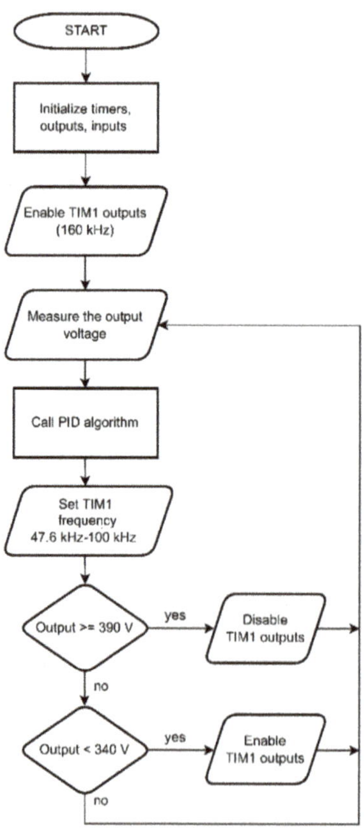

Figure 8. Flowchart of the implemented algorithm.

4. Uninterruptible Induction Motor Drive Implementation

The uninterruptible induction motor drive and the experimental setup are shown in Figure 9. Several voltage sources were connected to perform different measurements. The figure shows the measurement powered by the battery and the grid. The motor has a mechanical power of 370 W, which can be loaded dynamically with any torque with a magnetic brake pad. The brake pad controller instrument indicates the actual speed and torque. Furthermore, the instrument has analogue output, thus, with four channel datalogger the actual mechanical and electrical power can measure at same time. From these, the efficiency can be determined.

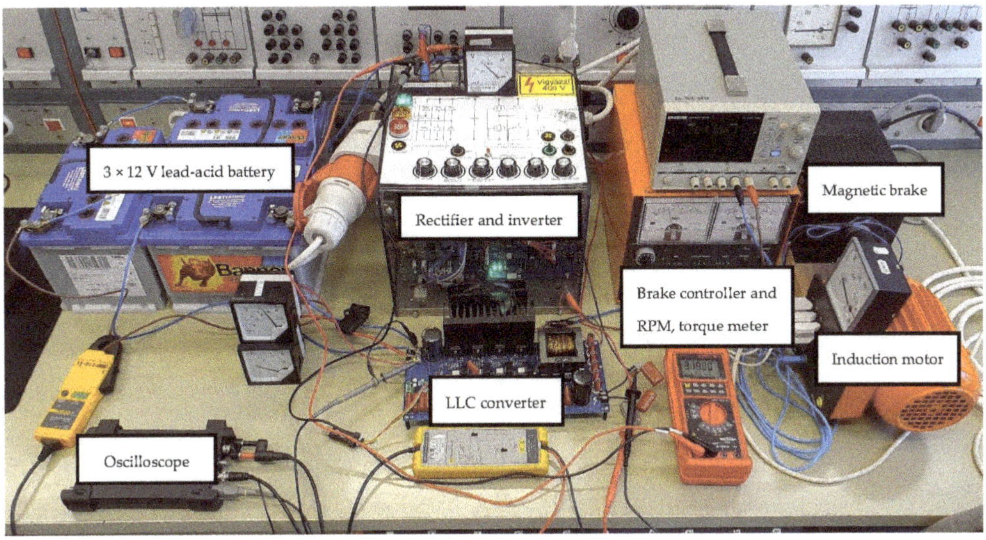

Figure 9. The implemented uninterruptible induction motor drive with mechanical load.

The three-phase rectifier is half-controlled converter. It contains three thyristors (SCR) and three diodes. This rectifier is designed for previous research (max. output power is 10.56 kW), but it can also be inserted into the low power uninterruptible induction motor drive. The firing circuit is made up of pulse transformers. Thyristors are turned on by a microcontroller-timer circuit as a function of the desired firing angle. The firing angle directly controls the average value of the DC voltage. By changing the firing angle, the induction motor draws different proportions of electrical power from the grid and the LLC converter.

The three-phase inverter has already been built for previous research purposes. The inverter is designed to be universal. It contains 1200 V, 100 A IGBT modules with the corresponding driver circuit. The output of the LLC converter can be connected to the inverter's DC link circuit. Potentiometers can be used to adjust the modulation index, which controls the motor voltage, the switching frequency, the first-order frequency, as well as ramp speed.

5. Measurements

In this section, the measurements of implemented circuits are described. The measurements confirm the correctness of the calculated values, examine the transients in time; thus, the stability of the PID control circuit can be examined. In addition, the efficiency of the LLC converter and the entire motor drive are determined.

5.1. LLC Converter Energy-Free Turn on Transient

Turning on the energy-free LLC tank and output filter capacitor results in a large inrush current, as can be seen in Figure 10. At the moment of switching on, the microcontroller switches on the inverter with 160 kHz and then controls it at a frequency of 100 kHz for a short time, this ensures the lowest possible inrush current. During this time, the PID controller algorithm also starts and takes over the control. At this measurement, the LLC converter was fed by a 3 × 12 V lead-acid battery.

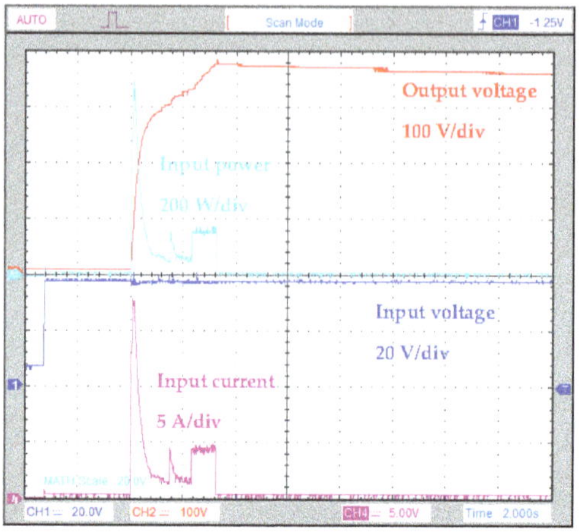

Figure 10. Energy-free turn on transient when no load is applied at the output.

5.2. LLC Converter Normal Operation

The voltage and the current at the output of the inverter of the LLC converter are shown in Figure 11. In the figure on the left, there is a light load on the output, the inverter is controlled above the resonant frequency (>100 kHz). In the figure on the right, the inverter is controlled at a frequency close to the maximum gain when heavy load is applied at the output. The high oscillation in the voltage is due to the parasitic oscillation of the SiC MOSFETs. Oscillation occurs because the surge voltage resonates with the MOSFET's drain-source parasitic capacitor (C_{DS}), with stray inductance (L_S) of printed circuit board wires and with the input filter capacitor. This oscillation is reduced by a snubber capacitor connected close to the MOSFETs to the input DC link. At this measurement, the LLC converter was fed by a 3 × 12 V lead-acid battery.

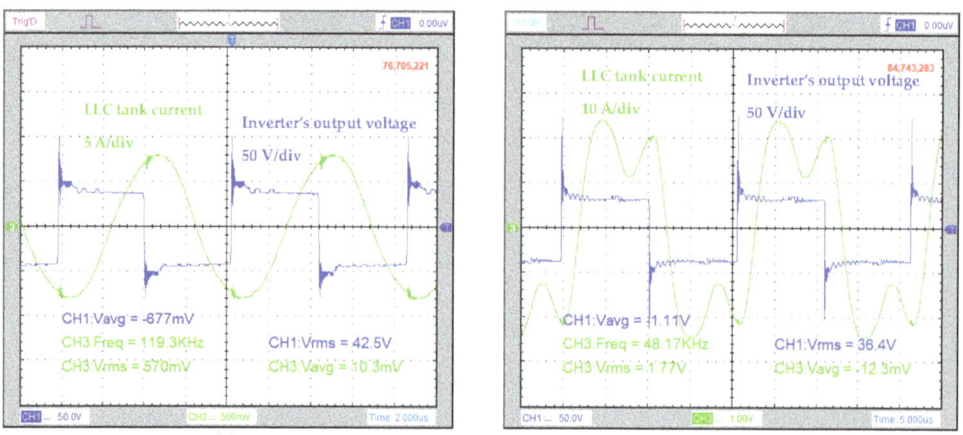

(**a**) (**b**)

Figure 11. LLC converter at normal operation: (**a**) at resonant frequency with light load when f = 119.3 kHz; (**b**) at heavy load when f = 48.17 kHz.

Figure 12 shows the output voltage at constant load. The voltage is kept stable by the PID controller. At this measurement, the LLC converter was fed by a 2 × 12 V lead-acid battery (23.9 V).

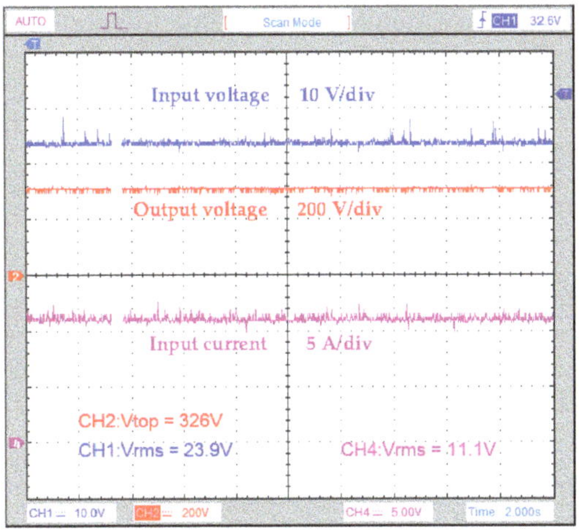

Figure 12. Input voltage, current and the output voltage at constant mechanical load.

5.3. LLC Converter Fed by Solar Panel

An LLC converter powered directly with a solar panel is not preferred. On one hand, as the irradiance decreases, the maximum power that can be extracted from the solar panel also decreases. As the maximum power decreases, the voltage of the solar panel decreases. If the load is still constant, more current flows from the solar panel to the converter. This event is still well handled by the PID controller (Figure 13a). Sudden high-power consumption can cause the solar panel to reach its maximum current. In this case, the voltage of the solar panel is greatly reduced (below 25 V). This is handled by the PID controller by setting the frequency for maximum gain, however, the output voltage does not reach the setpoint. Once the overload has ceased, the output voltage stabilizes rapidly (Figure 13b). It is also possible to observe that the input current oscillates slightly under constant load. This is because the PID controller never controls the inverter with one frequency value, as the controllers also have oscillations. The oscillation amplitude depends on the value of the maximum power that can be extracted from the solar panel, so it depends on the irradiance. The oscillation of the PID controller changes the load on the solar panel, so the input voltage also changes.

5.4. LLC Converter Fed by Batteries and Solar Panel

The measurement of the battery-powered LLC converter is shown in Figure 14. In the figure on the left, the motor torque changes only slightly, while in the figure on the right, it changes suddenly. In the case of measurement, Figure 14a, the output voltage varied to a lesser extent than in the case of solar power supply, because at high input current, the battery terminal voltage decreases only slightly. This allows the PID controller to hold the output voltage more stably. Furthermore, in the event of an overload, the voltage will not drop due to battery power as in the case of solar power, as shown in Figure 14b. The output voltage decreases greatly under dynamic load because the output filter capacitor is slightly discharged. When the output voltage exceeds 390 V, the converter is off. If the induction motor is started without a soft start, the output voltage can be greatly reduced. In this case, the PID controller needs more time to charge the output capacitor to the setpoint. Once the converter is turned on, it responds more quickly to dynamic loads. A decrease in output voltage will have a negligible effect on motor operation. In this case, the motor voltage also decreases, with the result that the flux and, thus, the torque also decrease. A large drop in output voltage can be avoided if the motor is started softly and no mechanical load occurs instantaneously as shown in Figure 14a.

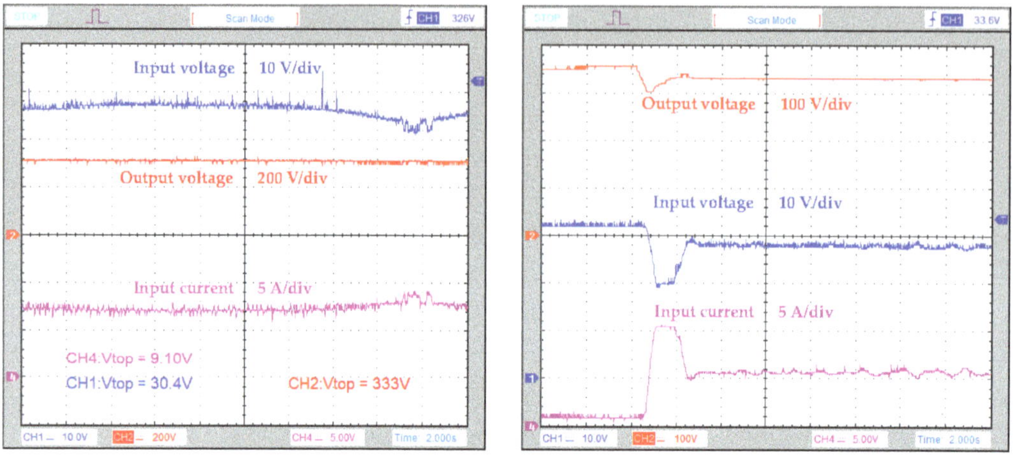

(a) (b)

Figure 13. LLC converter fed by solar panel: (**a**) irradiance slightly decreasing; (**b**) solar panel overloaded operating condition.

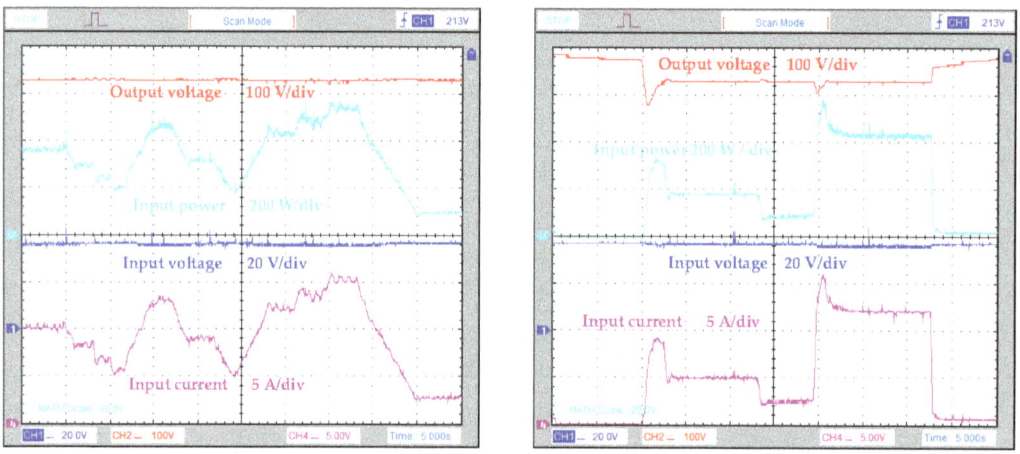

(a) (b)

Figure 14. LLC converter fed by batteries: (**a**) the torque changes slowly; (**b**) the torque changes suddenly.

5.5. Uninterruptible Induction Motor Drive

The measurement of the uninterruptible induction motor drive are shown in Figure 15. During the measurement, the motor is loaded with constant torque at a constant speed. Figure 15a shows the three operating states:

1. The LLC converter is turned off, only the rectifier is feeding the motor;
2. The rectifier is turned off, only the LLC converter is feeding the motor;
3. Both the rectifier and the LLC converter are simultaneously feeding the motor at a certain rate.

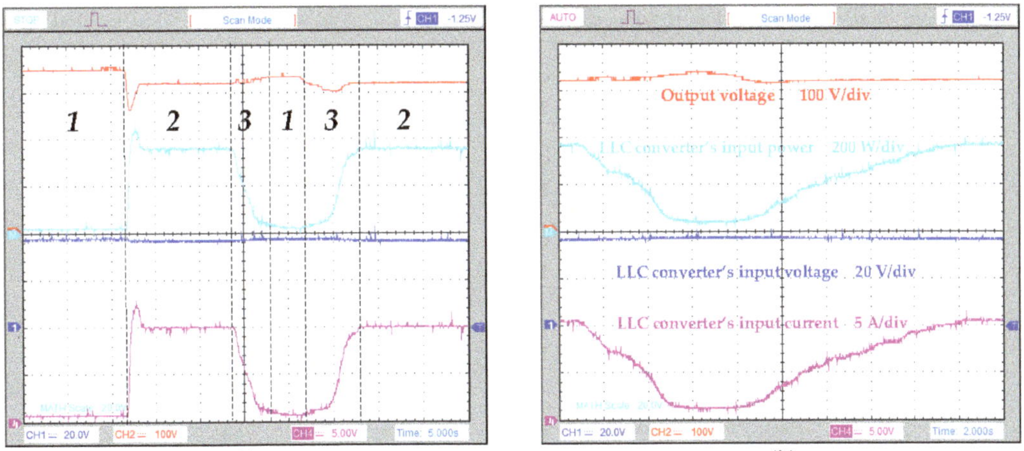

(a) (b)

Figure 15. Uninterruptible induction motor drive measurement: (**a**) three operating states; (**b**) the rectifier's firing angle changes slowly.

At the beginning of the measurement, the rectifier supplies a DC voltage of 350 V to the DC link. The LLC converter was then set to a 320 V setpoint, so its output is turned off by the control circuit, so electricity only comes from the grid to the three-phase inverter. The three-phase inverter monitors the voltage of the DC link, so it is not a problem that the DC voltage rises above 330 V. The inverter uses pulse width modulation to reduce the motor voltage so that the iron core will not be saturated. Operating state 1 can only be achieved if the rectifier produces a higher voltage than the LLC converter. In contrast, diodes and thyristors do not open.

The measurement examines the sudden change from the first operating state to the second operating state. The rectifier will then turn off completely, simulating a grid power outage. The LLC converter immediately turns on and charges the output capacitor and maintains the voltage at 320 V.

The rectifier then turns on and the firing angle decreases slowly, causing the output voltage to increase. The rectifier and the LLC converter feed the three-phase inverter in different proportions. This ratio can be adjusted with the firing angle. If the voltage rises above 320 V, the LLC converter continuously increases the frequency of the LLC tank, thus decreasing the amount of voltage generated by the LLC converter. Figure 15b shows this process back and forth over a longer period (operating stages: 2 → 3 → 1 → 3 → 2).

5.6. *Efficiency Measurements*

The left side of Figure 16a shows the efficiency measurement of the standalone LLC converter. The figure on the right Figure 16b shows the efficiency between the input of the LLC converter and the mechanical power measured on the motor shaft. During measurement Figure 16a, the LLC converter was operated at approximately 505 W output power, when the input power is 580 W (at V_{in} = 40 V). The converter can be overloaded due to the large heat sink. There is a difference between the efficiencies of a voltage boost LLC converter and a voltage drop LLC converter with the same power. A higher efficiency can be achieved with a voltage reducing LLC converter [14–16]. This is because their input voltage is higher, but the input current is lower. At lower currents, the power loss on the MOSFETs is significantly lower. The efficiency of the presented voltage boost LLC converter is lower due to the relatively high current at the input and the high channel resistance of the MOSFETs. These result in a large static power loss.

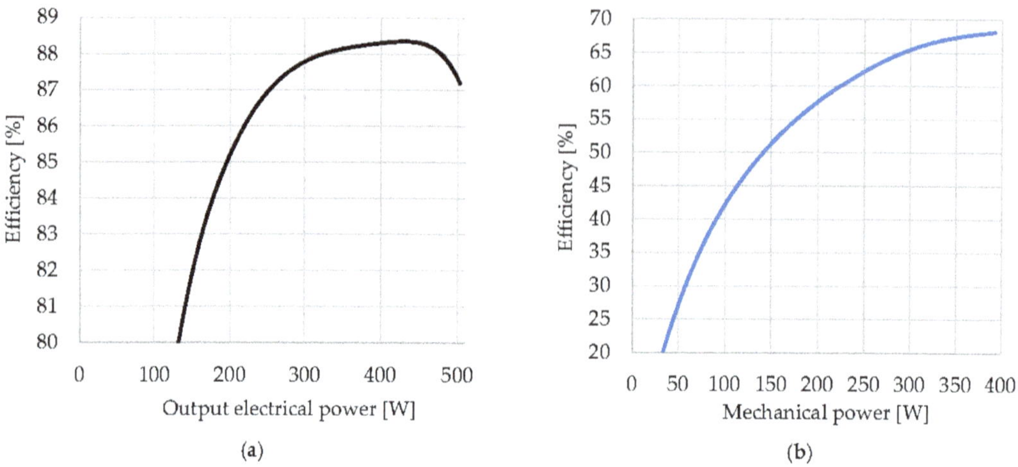

Figure 16. Efficiency: (**a**) between LLC converter input power and output power; (**b**) between LLC converter input power and motor's mechanical power.

The efficiency of the 370 W induction motor used during the measurement was approx. 80.4% at the maximum load. As the load torque decreases, the efficiency also decreases. In the following, it can be seen in Figure 17 how the efficiency varies between the mechanical power of the motor and the input of the LLC converter at different input voltages. During the measurement, the rated torque was not applied to the motor, only 1.15 Nm. The speed varied from 0 to 2000 RPM, which was set by the variable frequency drive. As the input voltage decreases, the efficiency also decreases, as a higher input current is required to achieve the same mechanical power. Due to the higher current consumption, the static power loss increases, resulting in reduced efficiency.

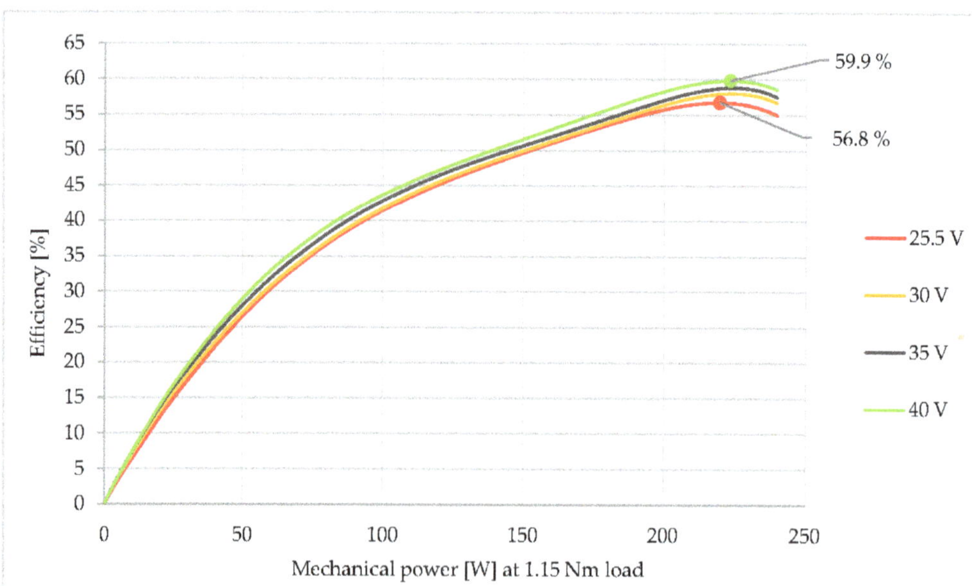

Figure 17. The total efficiency for different input voltages when the torque constant 1.15 Nm (the RPM changes 0–2000).

6. Discussion and Conclusions

In summary, the uninterruptible induction motor drive is working properly and stable. The measurements were made under different conditions, as the LLC converter was also powered by a battery, solar panel and power supply. However, this does not affect the result of the demonstration that the uninterruptible induction motor drive can be used reliably at low power. During the measurements, no condition occurred when the LLC converter was powered by a solar panel and a battery at the same time. This is because an electronic converter that charges the battery with the DC voltage generated by the solar panel has not yet been completed. Research is currently underway to create an MPPT (Maximum Power Point Tracking) control circuit that can handle dynamic loads well. By implementing the MPPT circuit, it is possible to power the LLC converter from a solar panel and a battery at the same time. In terms of measurements, it seems that it is not worthwhile to power the LLC converter from a solar panel alone, as achieving continuous rated power is not guaranteed. The solar panel would be able to supply the battery and the LLC converter simultaneously if an MPPT control circuit was also installed. In the event of a sudden power consumption, the batteries will withstand this without any problems, so that the voltage of the solar panel will not drop greatly. Conventional MPPT controllers do not fit well with this system, as sudden power consumption can occur in this case, this control algorithm cannot handle it properly, so research is focused on developing a new type of MPPT controller in the future. In the previous research, where the LLC converter was designed in theory and the parameters were checked by simulation, an efficiency of 90.6% was obtained. The implemented converter achieved an efficiency of 88.3%. In the simulation, the capacitors, inductors, the transformer was ideal, so in reality it is therefore less efficient. Furthermore, the power supply of the LLC converter electronics is also covered from the input and this was not set in the simulation. The driver circuit of the power MOSFETs consume the most power from the auxiliary power supply.

The efficiency of the entire drive could otherwise be improved if the LLC converter was supplied from a higher input voltage. However, an additional cost would occur as more batteries and solar panels would be required. The efficiency would not increase significantly though, as the lowest efficiency in the system is the motor. Another possible way to increase the efficiency is replacing the voltage-hertz control algorithm in the three-phase inverter with Fuzzy-logic Field-Oriented Control [17,18].

Author Contributions: Conceptualization, R.R.B. and I.B.; software, R.R.B.; validation R.R.B. and I.B.; formal analysis, R.R.B.; investigation, R.R.B.; resources, R.R.B. and I.B.; data curation, R.R.B.; writing—original draft preparation, R.R.B.; writing—review and editing, R.R.B.; visualization, R.R.B.; supervision, I.B.; project administration, I.B. All authors have read and agreed to the published version of the manuscript.

Funding: This research received no external funding.

Institutional Review Board Statement: Not applicable.

Informed Consent Statement: Not applicable.

Data Availability Statement: Not applicable.

Conflicts of Interest: The authors declare no conflict of interest.

References

1. Hasanuzzaman, M.; Zubir, U.S.; Ilham, N.I.; Seng Che, H. Global electricity demand, generation, grid system, and renewable energy polices: A review. *WIREs Energ. Environ.* **2016**, *6*, 18. [CrossRef]
2. Frivaldsky, M.; Kascak, S.; Morgos, J.; Prazenica, M. From Non-Modular to Modular Concept of Bidirectional Buck/Boost Converter for Microgrid Applications. *Energies* **2020**, *13*, 3287. [CrossRef]
3. Derbeli, M.; Napole, C.; Barambones, O.; Sanchez, J.; Calvo, I.; Fernández-Bustamante, P. Maximum Power Point Tracking Techniques for Photovoltaic Panel: A Review and Experimental Applications. *Energies* **2021**, *14*, 7806. [CrossRef]
4. Adamczyk, M.; Orlowska-Kowalska, T. Virtual Current Sensor in the Fault-Tolerant Field-Oriented Control Structure of an Induction Motor Drive. *Sensors* **2019**, *19*, 4979. [CrossRef] [PubMed]

5. Rafael, R.B.; Istvan, B. LLC Resonant Converter Design and Simulation for PV Motor Drives. In Proceedings of the 2021 22nd International Carpathian Control Conference (ICCC), Ostrava, Czech Republic, 31 May–1 June 2021; pp. 1–5.
6. Escudero, M.; Kutschak, M.-A.; Pulsinelli, F.; Rodriguez, N.; Morales, D.P. On the Practical Evaluation of the Switching Loss in the Secondary Side Rectifiers of LLC Converters. *Energies* **2021**, *14*, 5915. [CrossRef]
7. Sam, A.-R. Resonant LLC Converter: Operation and Design. *Infineon* **2012**, *1*, 19.
8. Hang-Seok, C. *Half-Bridge LLC Resonant Converter Design Using FSFR-Series Fairchild Power Switch (FPS^{TM})*; Fairchild: Sunnyvale, CA, USA, 2007; p. 20.
9. Shukla, S.; Singh, B. Single Stage PV-Grid Interactive Induction Motor Drive with Improved Flux Estimation Technique for Water Pumping with Reduced Sensors. *IEEE Trans. Power Electron.* **2020**, *35*, 12988–12999. [CrossRef]
10. Shukla, S.; Singh, B. Grid-PV Array Fed Induction Motor Drive for Water Pumping. In Proceedings of the 2018 IEEE International Conference on Power Electronics, Drives and Energy Systems (PEDES), Madras, India, 18–21 December 2018; p. 6.
11. Abdelwanis, M.I.; Rashad, E.M.; Taha, I.B.M.; Selim, F.F. Implementation and Control of Six-Phase Induction Motor Driven by a Three-Phase Supply. *Energies* **2021**, *14*, 7798. [CrossRef]
12. Wu, Z.; Gao, C.; Tang, T. An Optimal Train Speed Profile Planning Method for Induction Motor Traction System. *Energies* **2021**, *14*, 5153. [CrossRef]
13. Cortés, B.; Tapia, R.; Flores, J.J. System-Independent Irradiance Sensorless ANN-Based MPPT for Photovoltaic Systems in Electric Vehicles. *Energies* **2021**, *14*, 4820. [CrossRef]
14. Yang, C.; Liang, T.; Chen, K.; Li, J.; Lee, J. Loss analysis of half-bridge LLC resonant converter. In Proceedings of the 1st International Future Energy Electronics Conference (IFEEC), Tainan City, Taiwan, 2–6 November 2013; p. 6.
15. Wang, Q.; Zhang, X.; Burgos, R.; Boroyevich, D.; White, A.; Kheraluwala, M. Design considerations for a high efficiency 3 kW LLC resonant DC/DC transformer. In Proceedings of the IEEE Energy Conversion Congress and Exposition (ECCE), Montreal, QC, Canada, 20–24 September 2015; pp. 5454–5461.
16. Schmidt, O.R.; Myhre, E. 380Vdc/48Vdc/ 3kw DC/DC converter with 98.2% efficiency. In Proceedings of the IEEE International Telecommunications Energy Conference (INTELEC), Osaka, Japan, 18–22 October 2015; pp. 1–6.
17. Eltamaly, A.M.; Alolah, A.I.; Badr, B.M. Fuzzy controller for three phases induction motor drives. In Proceedings of the International Conference on Autonomous and Intelligent Systems, Povoa de Varzim, Portugal, 21–23 June 2010; p. 6.
18. Eltamaly, A.M.; Alolah, A.I.; Hamouda, R.; Abdulghany, M.Y. A novel digital implementation of AC voltage controller for speed control of induction motor. *Int. J. Power Energy Convers.* **2010**, *2*, 78–94. [CrossRef]

processes

Article

Planning Method and Principles of the Cloud Energy Storage Applied in the Power Grid Based on Charging and Discharging Load Model for Distributed Energy Storage Devices

Junfang Li [1,*], Yue Xing [2] and Donghui Zhang [2]

[1] Department of System Planning and Research, Grid Planning & Research Center, Guangdong Power Grid Corporation, CSG, Guangzhou 510080, China
[2] Department of System Analysis and Hybrid Simulation, Sichuan Energy Internet Research Institute, Tsinghua University, Chengdu 610042, China; xy_sophie@163.com (Y.X.); zhangdonghui@tsinghua-eiri.org (D.Z.)
* Correspondence: jfliitwo@163.com; Tel.: +86-138-2220-8406

Abstract: The cloud energy storage system (CES) is a shared distributed energy storage resource. The random disordered charging and discharging of large-scale distributed energy storage equipment has a great impact on the power grid. This paper solves two problems. On one hand, to present detailed plans for designing an orderly controlled CES system in a realistic power system. On the other hand, Monte Carlo simulation (MCS) is used for analyzing the load curves of five types of distributed energy storage systems to manage and operate the CES system. A method of its planning and the principles of CES for applied in a power grid, are presented by analyzing the impact based on five load curves including the electric vehicle (EV), the ice storage system, the demand response, the heat storage system, and the decentralized electrochemical energy storage system. The MCS simulates the random charging and discharging of the system over a five-year planned scaling of distributed energy storage from 2021 through 2025. The influence of distributed energy storage systems on power grid capacity, load characteristics, and safety margins is researched to summarize the applicable fields of CES in supporting large power grids. Finally, important conclusions are summarized and other research possibilities in this field are presented. This paper represents a significant reference for planners.

Keywords: cloud energy storage system; demand response; electric vehicles; ice storage system; load modelling; Monte Carlo simulation; power system planning; shared distributed energy storage system

Citation: Li, J.; Xing, Y.; Zhang, D. Planning Method and Principles of the Cloud Energy Storage Applied in the Power Grid Based on Charging and Discharging Load Model for Distributed Energy Storage Devices. *Processes* **2022**, *10*, 194. https://doi.org/10.3390/pr10020194

Academic Editors: Alon Kuperman and Alessandro Lampasi

Received: 15 December 2021
Accepted: 31 December 2021
Published: 20 January 2022

Publisher's Note: MDPI stays neutral with regard to jurisdictional claims in published maps and institutional affiliations.

1. Introduction

1.1. Motivation

In the future, power systems will operate with a high proportion of renewable energy, which needs more flexible operational resources to compensate for power imbalances that are currently scarce. Significant intermittence in renewable resources for power systems causes great changes in real-time price, and so requires energy storage to balance power. Distributed, shared energy storage technology has high application value and will be an important and widely used resource in future power systems. At present, large-scale, pumped-storage power stations are the main energy storage resource in power systems, with costs lower than those of battery energy storage systems. High cost causes a scarcity applied battery energy storage technology in power grids. To reduce the cost of energy storage services, cloud energy storage (CES) technology, presented in [1,2], is one strategy for centralizing all distributed energy storage devices from consumers into a cloud service center, as virtual energy storage capacity, instead of real devices.

Depending on the interactive value of the consumer and power grid, a CES can be an opportunity to promote a shared business model, setting the precedent for consumers as energy resource suppliers. Such a shared business model can provide users more opportunity to be involved in the power market and to make energy storage more flexible. The advantages thereof are as follows: (1) having no limits to capacity from time, location, or demand; (2) having lower costs than traditional energy storage technology, only paying for leasing; (3) large-scale, flexible, shared, and distributed energy storage devices from load-side consumers and concentrated energy storage equipment from professional providers, working together in the CES environment, to huge social benefit and improved spare energy usage; (4) the consumer who has distributed energy storage participating in their business can reduce their total electricity expenses paid to the grid.

The major consumers are residents and small businesses, who are sensitive to electricity prices. Additionally, some substations within a power grid need distributed energy storage devices for emergency service. At present, to multiple energy systems (MES) a CES can be a flexible resource for end-user demand responsiveness.

Technical support for such a system includes power-load forecasting, planning optimization, communication technology, data and economic analysis, and so on. Buyers send information, such as their needed storage capacities and payments, to sellers, transferred based on the power market, communications companies, and banks. Buyers who need energy storage capacity can purchase the right of use for a given period. After obtaining the right to its use, these consumers can charge and discharge the cloud battery, according to their demand circumstances. Consumers can buy virtual energy storage capacity in a cloud network instead of building physical energy storage to reduce cost.

A CES is used as a distribution network, wherein users benefit from each other. When its battery needs charging, CES suppliers pay for the power. When its batteries discharge, CES suppliers collect fees from the consumers and the distributed power grid. Cloud energy storage suppliers need to make optimization decisions, considering cost and profit under the constraints of consumers' demand for charging and discharging the cloud battery. Then, energy storage device suppliers control real-world equipment through the building, operating, and maintenance to provide good service.

In summary, the research goal of this paper is as follows.

1. The modelling of distributed energy storage equipment, such as electric vehicles (EV), ice storage systems, and distributed energy storage, based on demand response and electrochemical energy storage systems in different application scenarios, such as on the power-supply side, the distribution-network side, and the user side.
2. The predicting of future charging and discharging behaviors of distributed energy storage users based on the charging-and-discharging load model established in this paper, considering the key factors of temporal and spatial distributional characteristics of charging and discharging loads.
3. Research on the impact of distributed energy storage on power grid capacity, load characteristics, and safety margins in order to summarize the fields involved in the application of a CES supporting a large power grid.

1.2. Paper Innovation

This paper is to solve two problems: (1) how to plan and design an orderly, controlled CES system in a realistic power system; (2) how to manage and operate the CES system. To solve the first problem, this paper presents a detailed planning drawing. To solve the second problem, this paper analyzes the load curve of five types of distributed energy storage system using Monte Carlo simulation (MCS). A planning method and principles of the CES applied in the power grid are presented. In this paper, the main innovations are as follows.

(1) The traditional research on summarizing the main factors and probabilistic models of load distribution of different types of distributed energy storage in one paper are seldom. We establish the timely and spatial distribution of charge and discharge load model considering multi-variable series of factors such as electricity price and user

demand, after summarizing the impact factors from the aspects of energy storage type, battery capacity, state of charge (SOC), charging mode and user behavior.

(2) The traditional research on the impact of distributed cloud energy storage on grid seldom focuses on comprehensive analysis considering all common seen energy storage device kinds. We model the electric vehicle (EV), ice storage system, distributed energy storage based on demand response and electrochemical energy storage system in different application scenarios.

(3) We simulated the application scenarios of five kinds of distributed energy storage system through MCS, combined with the preliminary design scheme of a provincial five-year plan, analyzed the load curve of peak load day, and presented some suggestions. This paper designs an integrated energy sharing management platform and market trading mode for the next five years, in order to manage the integrated energy charging orderly and disorderly.

1.3. Structure of the Article

The structure of this article is as follows: Section 2 summarizes our literature review. Section 3 presents the planning method and establishes the charging and discharging load model for distributed energy storage behaviors. Section 4 analyzes the impact of the orderly or disorderly charging and discharging of different energy storage behaviors on power grid capacity, load characteristics, and safety margin, in order to summarize the application fields of CES in supporting large power grids. Section 5 concludes the paper.

2. Literature Review

Over the last ten years, multi-type energy storage technologies for micro grids [3,4] have been the focus of a large number of studies. With the development of large-scale renewable energy resources that easily integrate distributed energy resources into power systems with high power quality and operational control, the building of new transfer lines needs significant investment to meet end-user demand response; to this end, installing suitable scales of energy storage devices is being adopted as a better planning solution. In such a solution, the operation department adopts an active control strategy, such as energy storage systems, to compensate the renewable energy sources' volatility and smooth peak–valley load differences. For producing maximum comprehensive economic benefit, virtual power plants (VPP) [5,6] aggregate scheduled and non-scheduled units, including renewable and non-renewable energy resources, storage devices, and flexible loads, to operate as a single entity participating in the power market. Much literature has considered frameworks for and the modelling of components and operational systems [7]. To reduce the imbalance between the power generation and consumption due to intermittently generating units, scientifically adopting a bidding strategy for a VPP to elicit maximum benefit is a decidedly primary issue.

Electric vehicles (EV) represent mobile charging loads, as either plug-in electric vehicles or plug-in hybrid electric vehicles. Due to EVs' ability to stabilize the grid and provide significant battery storage capacity without upfront capital cost to grid infrastructure, research on the integration of vehicle-to-grid (V2G) energy storage units and cooperation with intermittent wind/PV in a VPP is a future social focus [8,9], but it's presently unrealistic. However, research on vehicle-to-grid aggregators for frequency regulation [10] and operational modeling of electric vehicle charging stations [11] provides novel ideas about distributed energy storage consumption.

The cloud energy storage (CES) systems presented in [1,2] in 2017 centralize all distributed energy storage devices from consumers into the cloud service center as a virtual energy storage capacity, belonging to the energy storage units of such VPP. The framework of the CES system for power grids is presented in [12] and built by the consumer, the CES operator, the energy storage supplier, and the distribution grid. With the information and cash flow calculated through distributed computing, the CES operator, as a centralized agency, is an intermediary service provider in contact with the consumers, the storage supplier, and the grid. For the centralized system, considering the hierarchical architecture of the power grid, cloud–edge intelligence [13–15] for wide-area load frequency control, substation simulation and protection control, and load modeling and management, is suitable for application in CES systems. In recent years, distributed storage coordination control strategies [16] and application–research scenarios on large-scale battery energy storage systems [17] are two common fields of energy storage system research.

As for renewable energy resources and EVs, the power grid cannot afford a large volume of electric vehicles (EVs) charging at peak load time because they greatly influence the load curve [18]. To deal with the peak load regulation and demand response for a VPP supporting EVs, three solutions are: optimization dispatch [19,20], control strategies [21], and trade mechanisms [22]. Distributed computing is the main solution for solving the distributed EVs integrated into the power market, e.g., multi-agent intelligent computing [23] and distributed online algorithm [24]. Blockchain represents realistic market and dispatch mechanisms for EVs [25–29] by its technical characteristics e.g., transparent, untampered with, privacy protection, and smart contract, allowing the tracing of EV charging and the extension of its leasing activity.

In recent three years, blockchain technology has led to a complete set of distributed energy trading and supply systems by connecting energy producers and consumers directly, greatly reducing the transactional cost of electricity and improving transactional efficiency. This enables power producers, transmission grid operators, distribution grid operators, and retail energy service providers to trade at different levels, simplifying the complex multilevel structure of current power systems. Blockchain technology can also deploy its tamper-proof characteristics in identifying the certification of carbon power and renewable energy power, directly recording renewable power, and providing the convenience of credible transactions of renewable energy. Building a power-trading platform with blockchain technology will be a technological upgrade to the current power-trading market.

Blockchain, a distributed database technology, has changed some major application fields of the internet of things (IoT) network over the last five years [30,31]. Introduced by Satoshi Nakamoto's Bitcoin in 2008 [32] as a peer-to-peer (P2P) system for distributed computing and decentralized data sharing, blockchain is composed of a distributed series of blocks, linked together by their hash values, that has the characteristics of decentralization, time traceability, autonomy, openness, and tamper-proof information by using time stamps, asymmetric cryptography, distributed consensus, and flexible programming technique. The application domains of blockchain technologies in IoT, e.g., the internet of vehicles, the internet of energy, the internet of cloud, fog computing, etc., are surveyed in [33]. Blockchain techniques, applied in the Chinese energy internet, is surveyed in [34,35]. The architecture and functionality of the blockchain groups in the intelligent distributed electrical energy systems presented in [36,37] is a realizable model for future implementation. At present, realistic engineering samples are so few that most research focuses on the blockchain's framework design concerning power system planning [37–41] and market transaction infrastructure [42–48].

3. Planning Method and Modelling

This section presents the architecture design for a CES and then builds a load model of the charging and discharging of distributed energy storage.

3.1. Planning Method

3.1.1. Architecture Design for CES

Figure 1 shows the architecture design for the CES market trade for local and cross-region power grid. Figure 2 is the unified model architecture for trade participators, a flowchart, and trade codes. The business participators are as follows:

(1) Consumers: The consumers include the users with wind or PV resources, small commercial users (e.g., load or electric vehicles), distributed generation (e.g., wind farms or PV stations), and the power plant. Distributed generation represents a certain scale

of "wind farm /PV station and battery storage (BS)". Small commercial users include the common industrial/commercial loads or batteries. Fossil-fuel power plants output power to charge energy storage devices through the distribution grid. They send fixed addresses—an ID number—and administrator information as block data.

(2) CES operators: Information concerning user demand is sent, through the CES operator, to the power market. Energy flows between consumers and the distribution grid. The CES operator is a software platform based on dispatching infrastructure and information technology, not an organization, optimizing dispatch and connected to the grid to send dispatched demand to the power market and receiving basic information about the available storage devices from the power market. In managing discharging and charging, energy flow is the main work of the CES system, that is, to decide how much and when to transfer energy from consumers to the distribution grid under the constraint of balancing supply and demand.

(3) Distributed energy storage suppliers: Distributed energy storage deploys bi-directional power conversion systems (PCS) and energy management systems (EMS) to different places for flexible scheduling. Its address and company information is fixed in block data. Its operation rules are as follows: (1) when the power plants or the distributed generation companies generate more energy, the energy storage devices store it for use during peak load demand; (2) due to the high cost of these devices, the price of providing energy storage service is higher than paid to the consumer for charging their own batteries; (3) the maximum energy storage capacity needs to be more than the maximum energy output of the distributed generation in a day, deducting the daily power energy supply in order to store excess power; (4) the minimum energy storage capacity needs to be more than the demand for energy during an emergency in the nearby area.

(4) Distribution grids: Distribution grids are the medium of power, energy, information, and cash flow. The power market trade center and dispatch center are two main operation organizations, including four functional systems, e.g., consumer monitoring centers, forecasting systems, contract management, and transaction settlement systems.

Local consumption of renewable energy is better than long-distance transmission to the power grid. The cost of centralized labor management of operation and maintenance is so high that the smart distributed management system needs user-end intelligence.

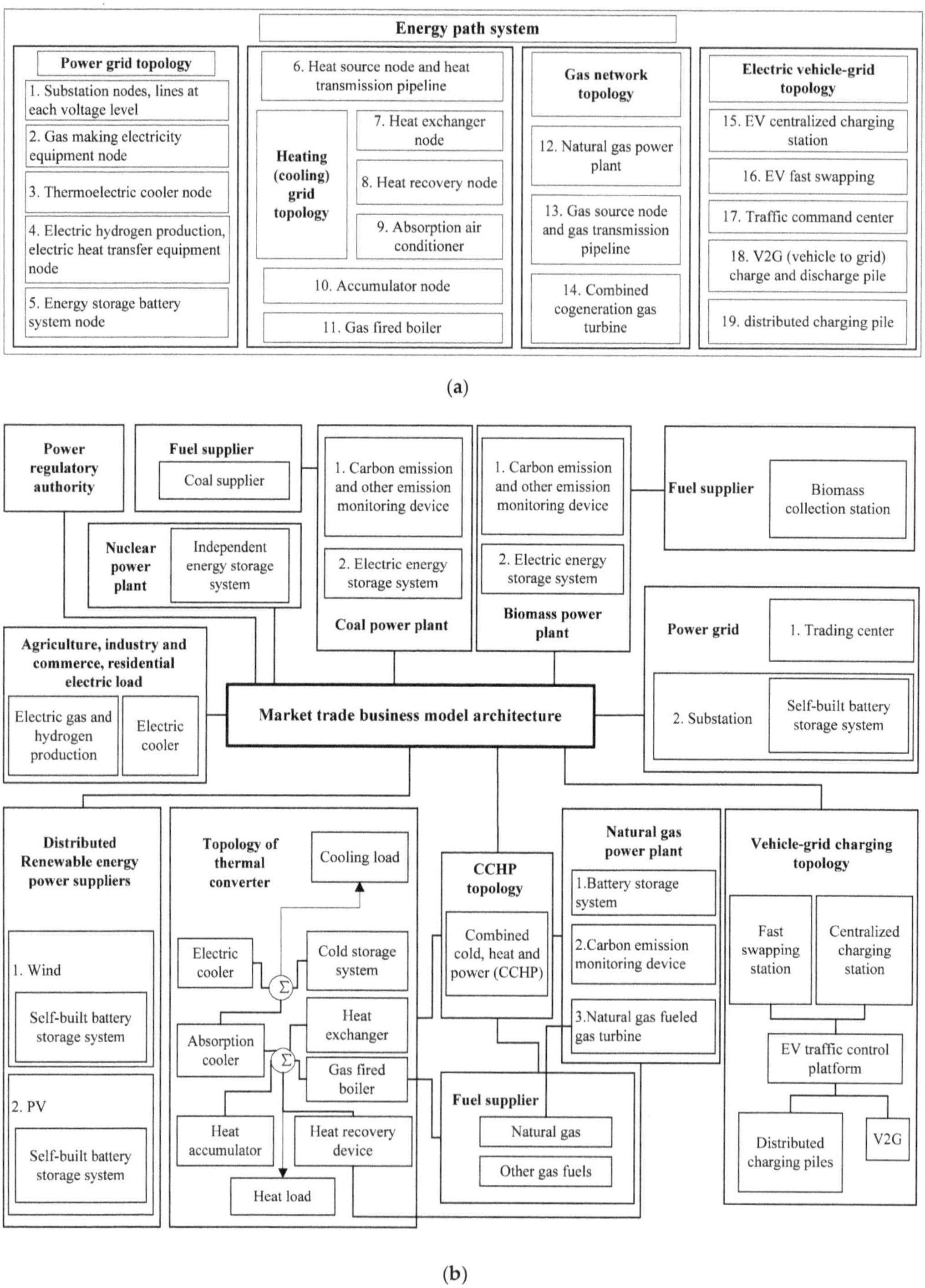

Figure 1. The architecture of a market trade business: (**a**) energy path system; (**b**) market trade business model architecture.

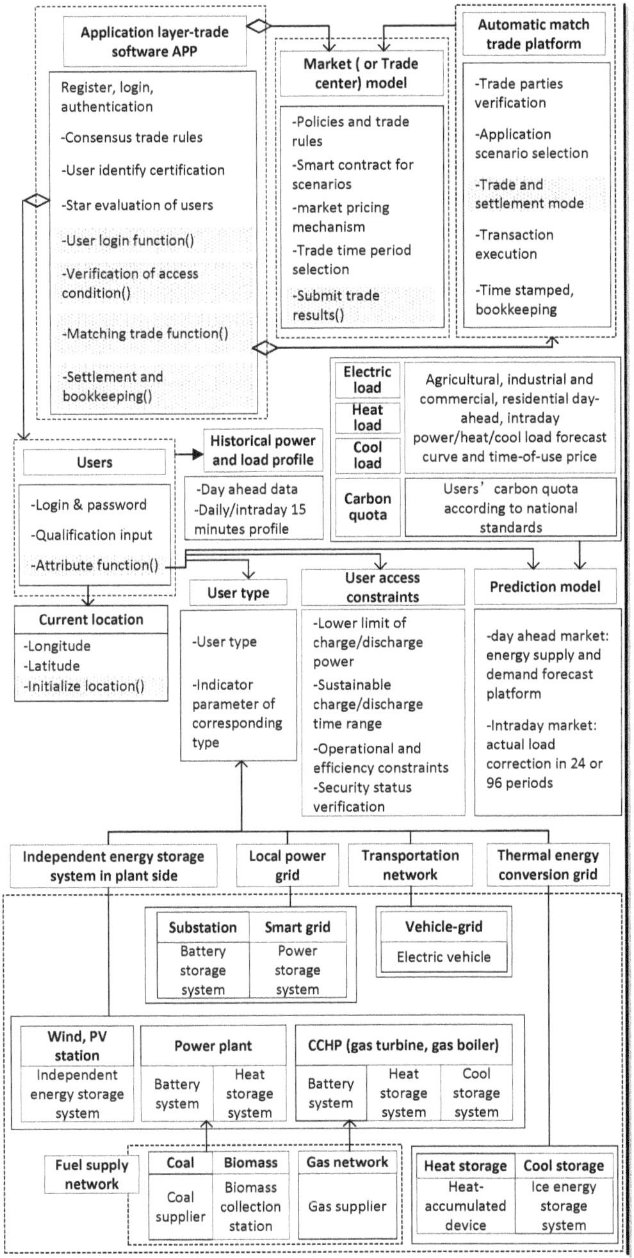

Trade code	Scene	Trade mount and price	Trade party A	Trade party B	Trade time	Trade place
	x x	x x x	x x x	x x x	x x x	x x x

Trade code		Content	Code
Energy storage application scenarios		1. Wind/PV generation leases independent energy storage in proportion for priority to stabilizing output	X
		2. Peak load shifting by energy storage in user side	X
		3. Peak load shifting by energy storage in power plant side	X
		4. Peak load shifting by independent energy storage	X
		5. Valley load filling by electric vehicle-grid	X
		6. Valley load filling by electric thermal-storage, electric cool-storage device	X
Trade mount and price		1. Power storage capacity, price per capacity satisfying charging electricity load	X
		2. Trade mount, electricity price satisfying discharging electricity load	X
		3. Heat storage capacity, price per capacity satisfying electric heat-transfer load	X
		4. Cool storage capacity, price per capacity satisfying electricity-driven refrigeration load	X
Users providing charging and discharging services by self-built energy storage		1. Self-built or independent electric energy storage devices lessor at the power plant side	X
		2. Electric energy storage system in industrial and commercial user side	X
		3. Electric vehicle	X
		4. Lessor of heat storage devices	X
		5. Lessor of cool storage devices	X
Regulator		1. Electricity market access certification authority	X
		2. Power regulation agency	X
Power supplier		1. Power plant	X
		2. Power grid, smart grid	X
Load user		Electricity/heat/cooling load, residual carbon quota	X
Time period		1. Day ahead (24 hours)	X
		2. Intraday per 1 hour/ per 15 minutes transaction micro-regulation	X
Location		1. Address of lessor as starting point	X
		2. Address of lessee as destination	X

Figure 2. Unified model architecture for trade participators, flowchart and trade code.

3.1.2. Planning of the CES System in a Power Grid

To build the CES system in a realistic power system, this paper presents the planning ideas as follows:

The cloud energy storage operator only needs to build a software platform that has a basic distribution grid with energy storage units. During the planning process, the most important concern is the power/energy flow, information flow, and cost flow. Due to the separation of current dispatching and transaction platforms, the cost flow can be in the market transaction platform. The realistic terminal devices that use these three kinds of flows are the consumers' electrical appliances and all kinds of energy storage device, which are connected by the distribution grid. There are two kinds of energy storage devices, namely the fixed energy storage stations and distributed mobile energy storage devices. The fixed energy storage stations, because of their large capacities and high cost, need to be built by the power grid or by some significant entrepreneurial investment. Mobile batteries feature location flexibility, without a demand to be located, as they can be anywhere. Thus, no realistic building construction cost in the CES system that only the mobile battery vehicles and the charge/discharge interface stations need to be invested instead.

The CES operator is responsible for the following tasks: building an information cloud detailing all service providers and energy storage suppliers, receiving consumers' charge/discharge capacities and energy demands, receiving electric vehicles' charging demands and locations and navigating them to the nearest charging stations, optimizing dispatch suggestions, load shifting from the peak to the valley, smoothing output power, tracking planned output power, ancillary services, primary frequency control in wind/PV and storage systems, meter monitoring, solving the problem of abandoning wind and PV power in order to build a friendly power supply, and improving system operational flexibility. These tasks can be set as functional modules in the desktop platform to be chosen by consumers.

The aim is to operate economically, reliably, with high quality, and efficiently. Costs needs to consider initial investment in the platform and the operation of charging stations based on users' reported installation of system capacity and energy demands. Then the cost is converted into the service price (currency/kWh). The platform needs to manage payment of all functional modules.

3.1.3. Trade Method

We present trade adopting a bidirectional auction mechanism, as follows. The platform sets the access conditions of energy storage power stations and matches the trade between the supply and demand sides of incorporated energy storage devices. The application scenarios are described next, with each scenario representing the purchase and sale of rental services.

(1) Wind/PV generation is given proportional priority in the rental energy storage services. Wind/PV generation leases shared energy storage facilities and electric vehicle charging to store extra electric output. The rental capacity is regarded as similar to its virtual self-built energy storage capacity, at a proportion of no less than 10% of the total capacity, for which the continuous charging time is no less than 2 h.

(2) In peak-load shifting, as performed by energy storage devices on the user side, the user side builds electric energy storage devices for self-use and participation in the peak-load shift market trade. During peak-load periods, the user side needs the electric energy storage devices and V2G electric vehicles to discharge for load demand. The charged electricity is traded by the previous valley-load or current market electricity price. When discharging, energy storage devices and V2G can sell electricity, similarly to distributed power generation, to nearby users. The discharge price is settled according to the independent energy storage price stipulated by the territorial government.

(3) In peak-load shifting by energy storage devices on the power-plant side, during valley-load time periods, electric storage devices built/owned by the power plant are leased to other power plants preparing for peak-load shifting by valley-load charging by signing transaction contracts, or as an independent energy storage entity participating in the peak-load shifting market to improve the frequency regulation performance of other power units. The amount of discharged electricity is equal to the amount of electricity generated at the power plant, and settled according to national standards of price.

(4) In peak-load shifting by independent electric energy storage, smaller independent electric energy storage suppliers participate in the peak-load shifting transactions of energy storage capacity as the main body of the market, and submit information to nearby local power dispatch departments. The charging and discharging state of energy storage devices are uniformly regulated by the local power dispatch department. The charging electricity is settled by peak and valley electricity price, or purchased at an appointed valley price. When discharging, it is similar to distributed power generation, selling electricity to users in the near area, and the discharge price is settled according to the independent energy storage price stipulated by national standards.
(5) Electric vehicles fill grid valleys. In night valley-load periods, electric vehicle charging is required to fill the valley. The vehicle network system regulates and displays the valley price information and signals the surplus power of the geographically nearest charging station or pile on the platform, so as to guide the electric vehicle to the nearest station or pile for charging. The charging amount complies with the charging price of electric vehicle stipulated by the government. For electric vehicles with special V2G functions, the independent energy storage price settlement stipulated by the national standards is used when discharging.
(6) Valley-load electric heat storage, electric cold storage and thermal cold storage. The valley-load electric heat storage, electric cold storage and a heat-and-cold storage structured heat (cold) network topology is composed of heat-source nodes, heat-recovery device nodes, heat-exchange device nodes, electric heating equipment nodes, absorption refrigerator nodes, electric refrigerator nodes, micro gas turbine nodes, gas boiler node and heat (cold) transmission pipelines. Electric heat storage, electric refrigeration, and other heat storage (cooling) should be used when the electricity price is high during the day, so as to reduce the electricity cost of direct heating and cooling during the day.

Concerning carbon emission rights, the remaining carbon quota is measured according to standards of national carbon emission quotas and participates in carbon quota trading according to the requirements of the carbon-trading market. The smaller independent energy storage devices, who set quotes based on self-operating profit, can offer quotes within the scope of a national floating standard.

We present the trading capacity matching method of day-ahead and intraday markets, in which appointments are made for each scenario in the day-ahead market after uploading the supply and demand capacities and quoted energy storage suppliers, including electricity/hot/cold energy storage suppliers, and two or three energy storage demands at the same time. Micro-adjusting the actual supply and demand capacity occurs in 1-h and 15-min increments each day. If the capacity on the supply side is less than the actual capacity on the current demand side, when the supply side cannot provide new capacity, the system will continue to provide at least two options in the near area for the buyer to choose until all transactions on the demand side are completed; if the reserved capacity on the supply side is greater than the actual capacity on the current demand side, only the actual used capacity will be measured. In order to avoid the demand side increasing the forecast demand in the day-ahead market, the range of acceptable error rate and penalty are formulated in the contract. The error rate = (predicted value − actual value)/actual value × 100%, and the range of acceptable error rate is set to ±10%. For the period exceeding the error rate, a penalty of 5‰ should be paid in 24hours of the day.

3.2. Load Model of Charging and Discharging Distributed Energy Storage

3.2.1. Electric Vehicle (EV)

There have been 2,356,657 EVs recorded in the national regulatory platform between 2017 and 2019, of which blade electric vehicles (BEV) account for 84.6%, plug-in hybrid electric vehicles (PHEV) account for 15.3%, and fuel cell electric vehicles (FCEV) account for 0.01% [49]. The disorderly charging behavior of EVs will create harmonics, aggravating the peak–valley load difference and increasing grid loss. The factors affecting the temporal and spatial distribution characteristics of EVs' charging and discharging are as follows: different EVs types, fast- or normal-charging mode of charging facilities, EVs users' behavior, number of different types of EVs, environment temperature, and time-of-use price. The EV charging power is assumed to be uniformly distributed U(a, b). The initial state of charge (SOC) is assumed to be normal distribution N(μ, σ). The detail modelling process is as follows.

(1) Respectively obtain the charging time probability of buses, private cars, taxis, and official vehicles, charging power, charging time, and discharge probability and time within the preset time period. Obtain a daily 15-min cumulative charging load and discharge power of all types of EVs.

According to the national standard of China GB/T 20,234, shown in Table 1, fast-charging mode adopts DC, with a maximum voltage of 750/1000 V and the maximum currents of 80/125/200/250 A, i.e., the charging powers from 60 kW through 250 kW. The normal charging mode adopts AC, with the maximum voltage 440 V and the maximum current 16/32/63 A, i.e., the charging powers from 7.04 kW through 27.72 kW.

Table 1. Different EV charging modes according to national standard of China GB/T 20,234.

Charging Mode	AC/DC	Maximum Voltage (V)	Maximum Current (A)	Maximum Power (kW)
L1 (slow charging)	AC	250	10/16/32	2.5/4/8
L2 (normal charging)	AC	440	16/32/63	7.04/14.08/27.72
L3 (fast charging)	DC	750/1000	80/125/200/250	60/93.75/150/187.5 or 80/125/200/250

Taking the taxi operation in Beijing, China, as examples, large-class taxi drivers shift every 24 h and small-class taxi drivers shift every 12 h. The ratio of large class to small class is about 4:1. The large class taxi drivers have a long rest and can choose normal charging, satisfying U(14.08, 27.72) (referring to Table 1) at public charging piles in residential or commercial areas. The small-class taxi drivers have a short rest time, and generally charge in fast mode, satisfying U(80, 250) (referring to Table 1). They charge twice a day, respectively, 2:00–5:00 am (the 8th through 20th time period) and 12:00–14:00 pm (the 48th through 56th time period) for large class taxi, and 2:00–4:00 am and 12:00–14:00 pm for small-class taxi. The battery capacity is assumed to be uniformly distributed under U(80, 100). The SOC is assumed to be normally distributed under N(0.5, 0.1). The charging time length, T (number of time periods), is calculated as follows, where C_{EV} is the battery capacity (kW·per time period), η is the charging efficiency 0.8–0.9, and P_{charge} is the charging power (kW).

$$T = (1 - \mathrm{SOC}) \cdot \frac{C_{\mathrm{EV}}}{\eta \cdot P_{\mathrm{charge}}} \tag{1}$$

(2) According to the probability distribution simulating the type of EV and its charging behavior, randomly sample the charge power, battery capacity, SOC, charge starting time of each EV. Then calculate charging time duration and charging end time. Finally, calculate cumulative charge load curve for 96 time periods.

The battery capacity of private cars follows a uniform distribution, U(20, 30), as does slow-charging power, U(4, 6). Considering that the expectation of the probability distribution at the beginning of charging of private cars is 17.6 h, the variance is 3.4. The expectation of the initial state of charge is 0.6, while the variance is 0.2. The charging efficiency is 0.9. The mean value of the probability distribution of mileage is 3.2, and the variance is 0.88. Figure 3a depicts the load curves of private cars of different sizes during disorderly charging as the number of vehicles grows from 200,000 in 2020 to 1,000,000 in 2025.

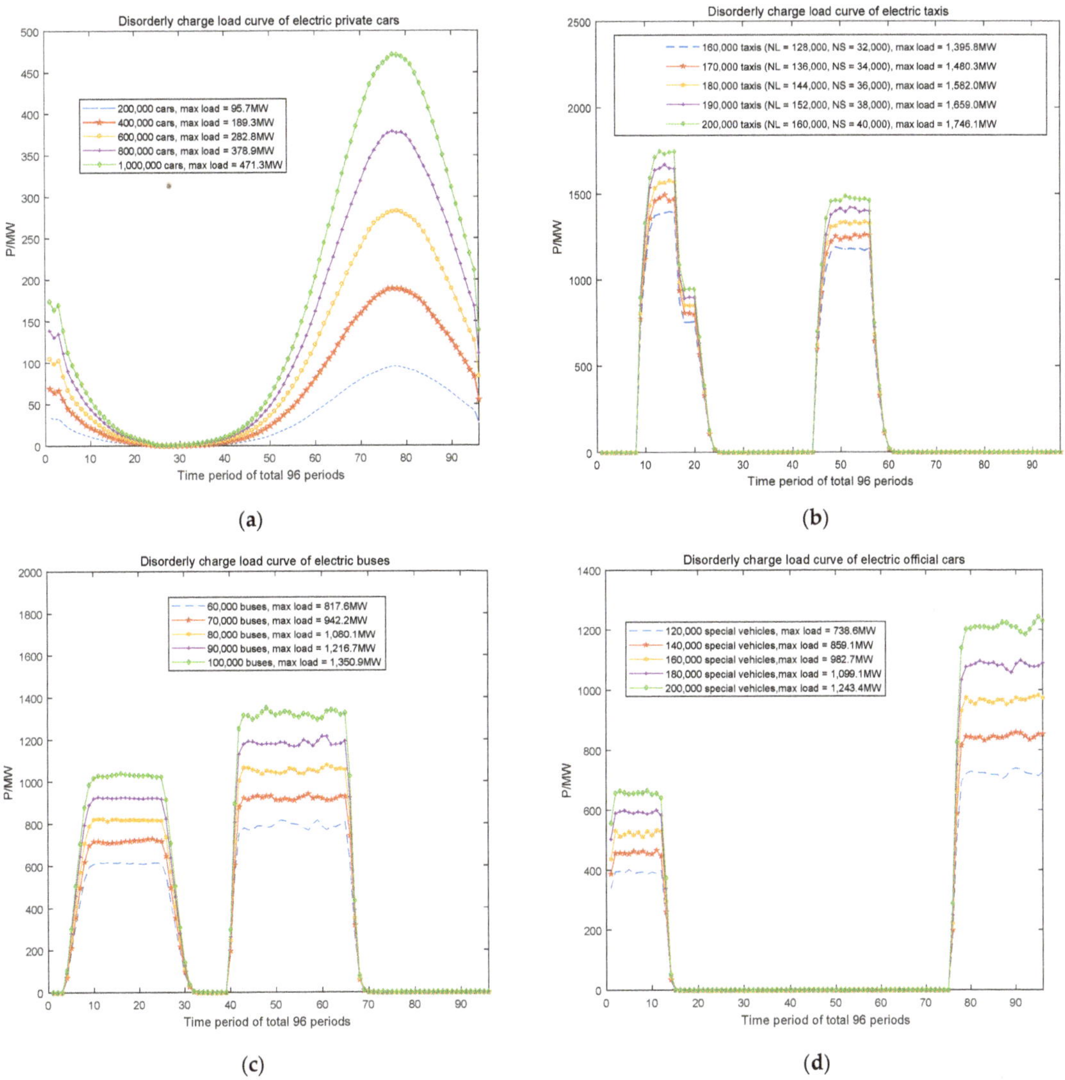

Figure 3. The disorderly charging load curve of EVs: (**a**) private cars; (**b**) taxis; (**c**) buses; (**d**) official and special vehicles.

Similarly, the disorderly charging load curves of electric taxis, buses, and special vehicles are respectively shown in Figure 3b–d.

The total charge load curve of the four types of EVs shown in Figure 4 includes the cars in Figure 3a, taxis in Figure 3b, buses in Figure 3c, and official and special vehicles in Figure 3d. The maximum charging power, 3425 MW, of EVs in 2025 in Figure 4 accounts for no more than 2.8% of current provincial peak load, about 123,810 MW in 2020.

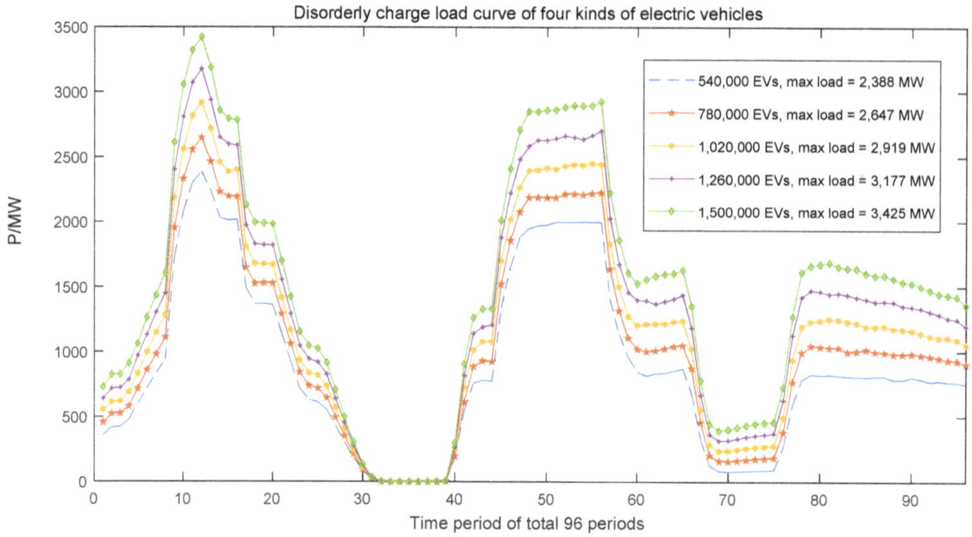

Figure 4. The total charging load curve of private cars, taxis, buses, and official and special vehicles.

3.2.2. Ice Storage System

Ice storage systems make use of the low electricity price at night to make ice and store it in an ice storage device. During the peak period of power consumption in the daytime, the ice melting releases its cooling capacity, which reduces the system cooling pressure during the peak load period and reduces the system's operational cost. The main equipment of an ice storage system includes a refrigerator and an ice storage tank. Its working mode can be divided into separate cooling by the electric refrigerator, separate cooling by the ice storage tank, simultaneous cooling by the electric refrigerator and ice storage tank, and combined cooling by the electric refrigerator and ice storage tank. The electric load and cooling power curves of an ice storage system are shown in Figure 5.

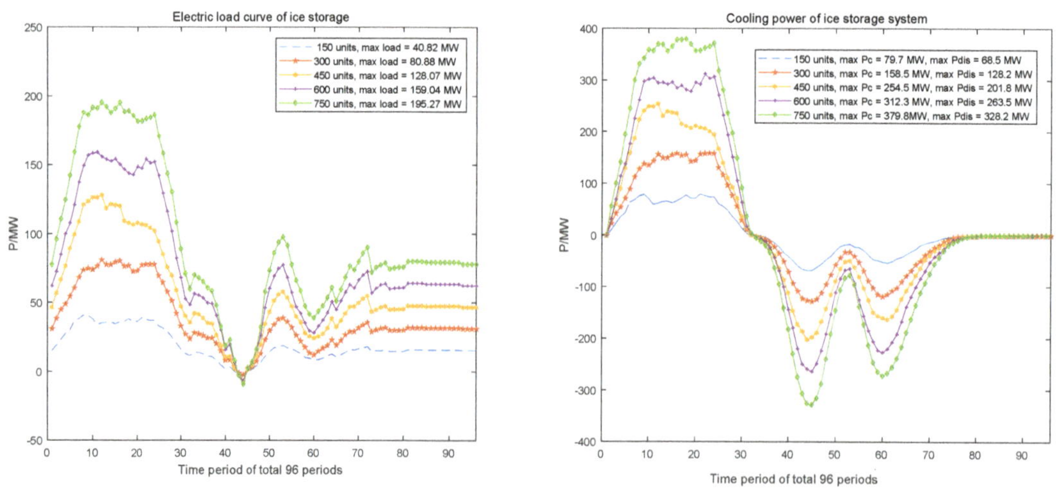

(**a**) (**b**)

Figure 5. An ice storage system: (**a**) electric load curve of ice storage system; (**b**) cooling power of an ice storage system.

Ice storage units can effectively use surplus power for refrigeration at night, transferring part of the power load at the peak of the day to the low-cost power period at night. The refrigeration host does not turn on, or turns on less, during peak times to reduce power load and improve the load composition of the power grid, which is conducive to the stable operation of the power grid.

3.2.3. Considering Demand Response

According to the power consumption characteristics of users and the operation time and energy consumption demand of different electrical equipment, power loads can be divided into uncontrollable loads, transferable loads, and interruptible loads.

Uncontrollable loads have no energy storage characteristics; the power consumption time is relatively fixed or may use power at any time according to the users' wishes, such as lighting load, or load from TVs, computers, etc. Their operation time or power fluctuation range are small and basically without ability to transfer load. Transferable loads owe to equipment with or without energy storage characteristics, but whose power consumption periods are flexible. Washing machines and timed rice cookers do not have energy storage, but they use electricity flexibly, and, as their total power consumption is certain, they have load transfer capacity. Interruptible loads, obtain from equipment such as air conditioners and water heaters, have flexible power consumption characteristics and short-term power-off operation without great impact on users, but they may deviate from the original power consumption habits, such as by causing room and water temperature changes. For more on the operational model of integrated controllable load, we refer the reader to [50]. The response curve of a controllable load is shown in Figure 6, below.

According to the Chinese national standard of residential electricity, each family uses maximum 6 kW as the basic design capacity for ordinary residential living in a 61–100 m^2 house. The power of the washing machines is generally 0.7–1 kW. The power of the timed rice cooker for a family having six persons is generally 0.8–1 kW. The power of an air conditioner with a heating function is generally 1.1–1.5 kW and there are typically 2–3 air conditioners for a large family. The power of the water heater is generally 1.2–2 kW. All household appliances always don't use electricity at the same time, to avoid tripping circuit breakers. Thus, for a family, the transferable load, including a washing machine or a timed rice cooker, is about 0.7–1 kW. The interruptible load including an air conditioner or a water heater is about 1.1–1.5 kW.

Supposing there are 93 million people in a province, and every six persons form a family, we can deduce that there are about 15.5 million households in a province. Assuming the maximum load transfer of 20,000 households is 9 MW during the 45th–48th period, 14 MW during the 66th–68th period, and 8.4 MW during the 79th–84th period, we deduce that, ideally, there are about 6975 MW, 10,850 MW, and 6510 MW load transfers at three peak load periods, respectively. For residential interruptible loads, from 10:00 to 12:00 am, shutting down two air conditioners, or only one from 15:00–19:00 pm can reduce peaking load. In Figure 6c,d, if the maximum interrupted load of 20,000 households is 22 MW in the 47th–48th period and 7.5 MW in the 74th–76th period, we deduce that, ideally, there are about 17,050 MW and 5813 MW for a province. Thus, there are about maximum 24,025 MW over 11:30–12:00 am, 10,850 MW over 16:00–16:30 pm, 5813 MW over 18:30–19:00 pm, and 6510 MW over 19:45–21:00 pm in terms of demand response. During the maximum demand response load, the transferable and interrupted load accounts for 29% and 71%, respectively.

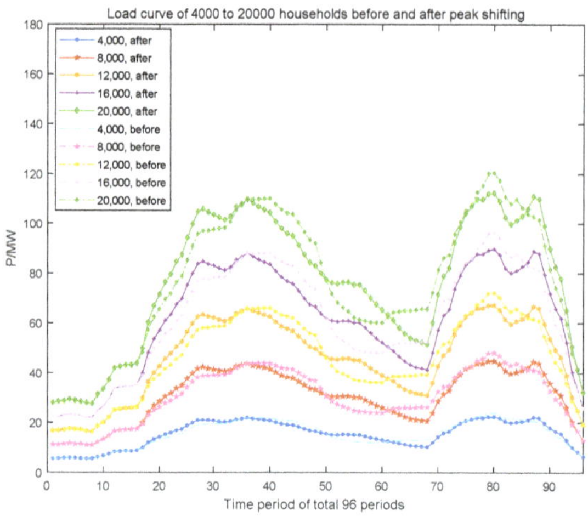

(**a**)

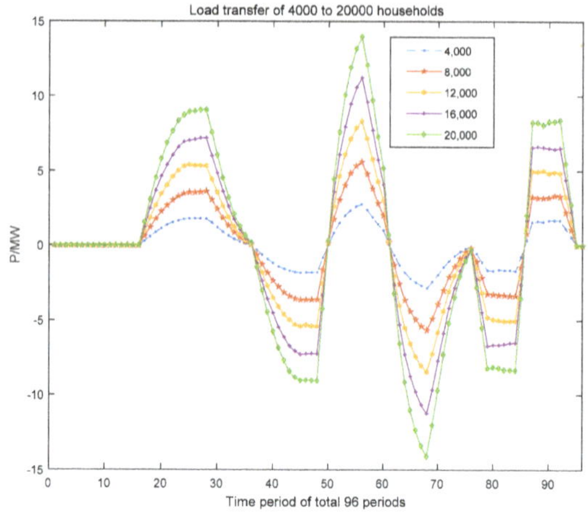

(**b**)

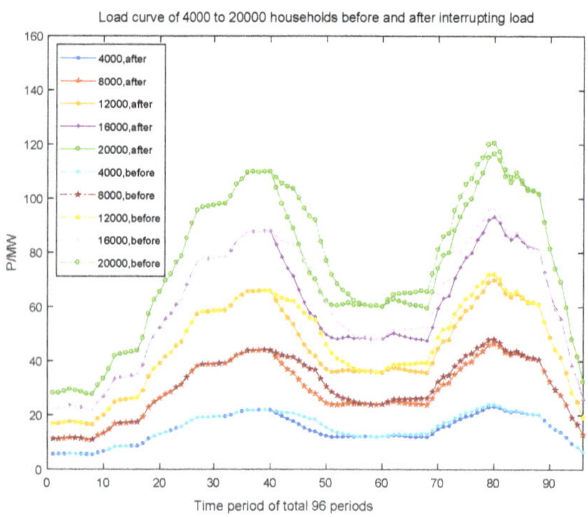

(**c**)

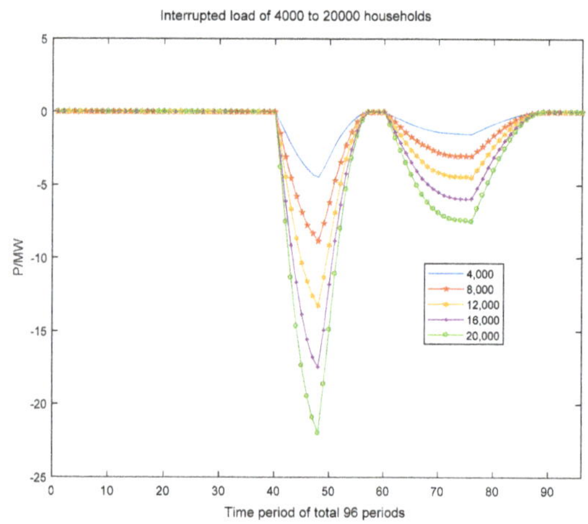

(**d**)

Figure 6. The response curve of 4000 to 20,000 household loads. (**a**) Transferable load curve before and after peak shifting and valley filling; (**b**) load transfer curve; (**c**) interrupted load curve before and after interrupting load; (**d**) interrupted load curve.

3.2.4. Heat Storage System

Heat storage devices can adjust the heating load through the process of repeatedly cycling heat storage and heat release.

The general energy storage system adopts the two charging and discharging cycles of "valley charging and peak discharging" and "flat charging and peak discharging", every day, to reduce the times of charging and discharging to ensure the service life of the energy storage device. Three different types of heat storage devices are selected for the project, and the heat storage power and capacity meet the normal distribution. The expected values are configured as 1 MW/4 MWh, 2 MW/8 MWh, and 4 MW/12 MWh accounting for 1/3. At the same time, the heat storage exothermic conversion rate of the heat storage device is 0.92, and its heat storage/exothermic efficiency is 0.90. The initial heat storage state and the initial time of heat storage/exothermic efficiency follow different uniform distributions. By changing the total number of heat storage devices to 100, 200, 300, 400, or 500, the disordered charge–discharge curves of heat storage devices in different scales can be obtained from Figure 7a–d, respectively.

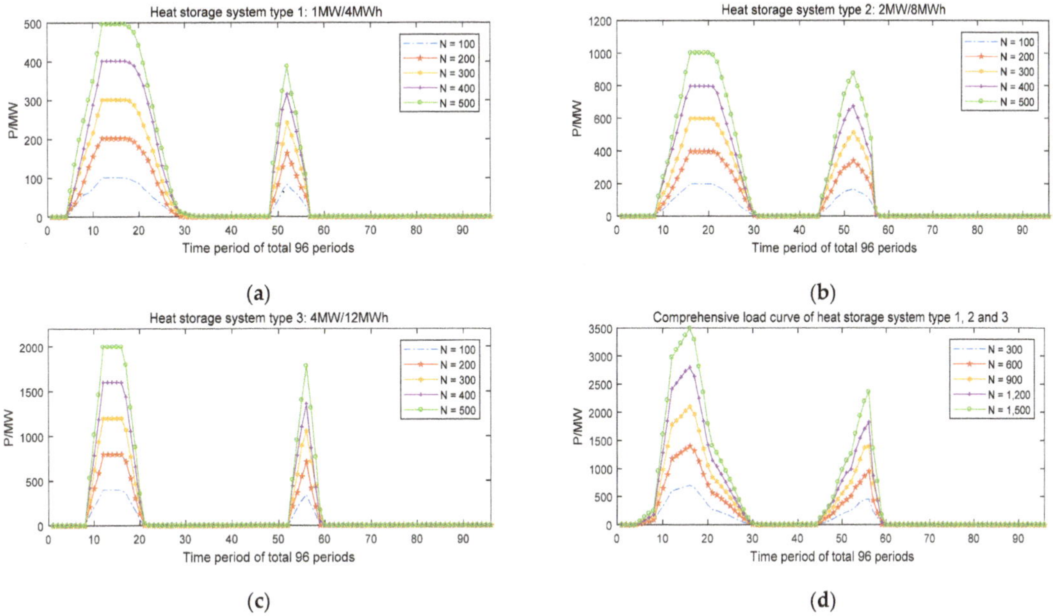

Figure 7. Electrical load curve of a heat storage system: (**a**) curve of a type 1 (1 MW/4 MWh) heat storage device; (**b**) curve of a type 2 (2 MW/8 MWh) heat storage device; (**c**) curve of a type 3 (4 MW/12 MWh) heat storage device; (**d**) comprehensive heat storage load curve of a heat storage device.

3.2.5. Decentralized Electrochemical Energy Storage

Under the time-of-use price proposed by the power grid company, industrial users adopt electrochemical energy storage devices because of their advantages of high energy density, fast response, and low maintenance cost [51]. The user-side energy storage uses the peak valley price difference to obtain income, namely, charging in the low electricity price period and discharging in the peak period to profit between peak and valley electricity prices. The total installed capacity of electrochemical energy storage in a province from 2021 to 2025 is about 480 MW, 1680 MW, 3840 MW, 6000 MW, and 8400 MW, respectively by year, and a decentralized system of 1 MW/2 MWh accounts for 40%, of 2 MW/5 MWh account for 40%, and of 6 MW/36 MWh accounts for 20%. In order to make use of the price differences between peaks and valleys, under two-charge and two-discharge modes, the optimal strategy is to charge the energy storage in the valley period of 0:00–8:00 am every morning and in the normal period of 12:00–14:00 pm, and discharge with a total duration of 4h during the peak period of 10:00–12:00 am and 14:00–16:00 pm every day. The 35% of full charging and discharging load curves of 200, 700, 1600, 2500, and 3500 devices are shown in Figure 8a,b, respectively.

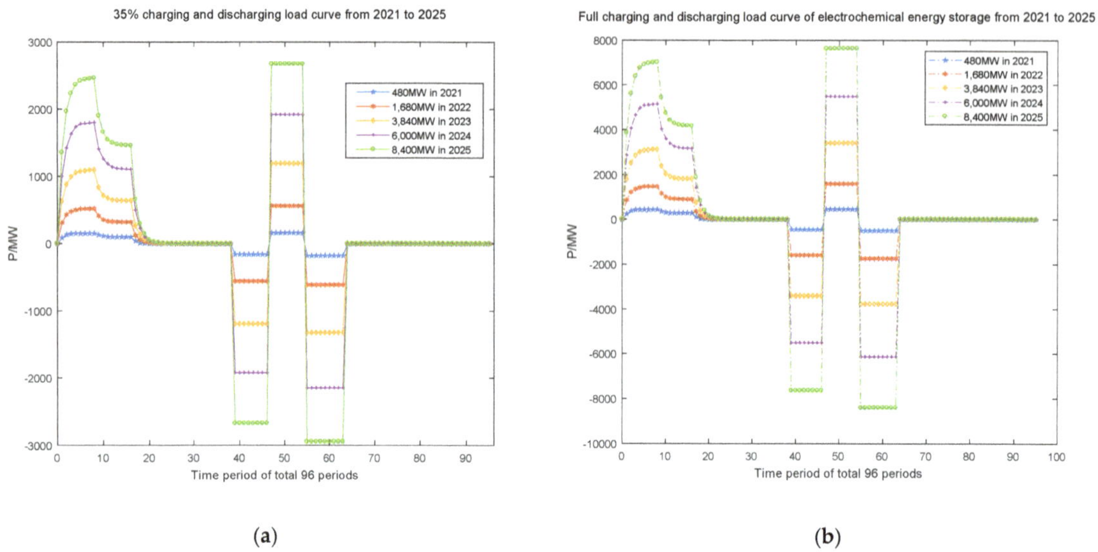

Figure 8. The charging and discharging load curves of the planned capacity of electrochemically stored energy from 2021 to 2025: (**a**) 35% charging and discharging; (**b**) full charging and discharging.

In a sample case, the maximum discharging load is 8243 MW and the maximum charging load is 7462 MW in full charging and discharging mode, resulting in reduced battery life. In a realistic case, to protect battery life, a 30–40% charging and discharging mode is adopted. In the sampled 35% case, the maximum discharging load is 2942 MW and the maximum charging load is about 2671 MW.

3.2.6. Distributed Storage Aggregation Provider (DSAP)

A DSAP is an independent organization that integrates all kinds of distributed energy storage resources and provides them to market buyers. As an intermediate organization in the distributed energy storage and power grid for providing power resources, it extracts, evaluates, and integrates distributed energy storage resources through professional means, integrates decentralized energy storage into system-schedulable resources, and participates in the operation of the power system.

The grid regulation takes reducing the variance of load curve and reducing the peak load as its charging and discharging objectives when dispatching and managing distributed energy storage. In order to reduce the negative impact of distributed energy storage on the power grid, combined with the time of use price mechanism, the goals of reducing the peak-to-valley difference of loads and reducing peak load can be set. A day is divided into 24 time periods-hours-and the optimization variable is the charge and discharge power of energy storage equipment in each time period.

In Figure 9a, EVs are the main charging load for the basic load curve in the period of 2:00–6:00 a.m. and 12:00–14:00 p.m., accounting for 65% and 52%, respectively, leading to an increased maximum load of 5167–5260 MW at 2:15–3:00 a.m. and of 5526–5670 MW, without a heat storage system, at 12:00–13:45 p.m., as shown in Figure 9b, corresponding to the electricity prices of 0.3052 yuan/kWh and 0.6104 yuan/kWh in Table 2, respectively. Until 2025, considering the heat storage system in summer, an extreme and unrealistic case for evaluating extreme maximum loads, the total increased maximum load of five energy storage systems between 6226–8209 MW at 2:15–3:00 a.m. and 6012–7829 MW at 12:00–13:45 p.m. accounts for 5–7% of Guangdong's unified regulation load of 123,810 MW in 2020.

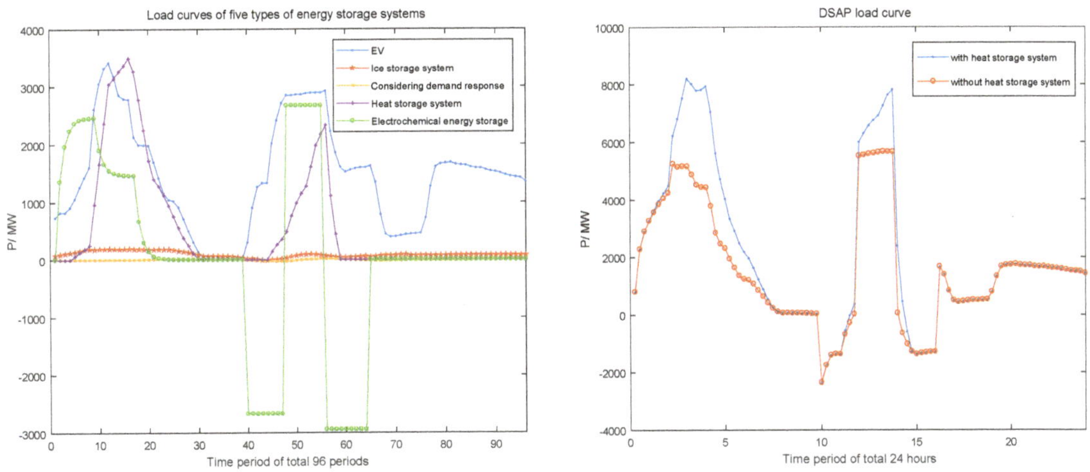

(a) (b)

Figure 9. The load curve of distribution energy storage systems in 2025: (**a**) charging electric load curves of five types of energy storage systems; (**b**) DSAP total charging power load curve including EVs and ice storage, considering demand response and a decentralized shared system on a day using heat storage systems (in winter) and without the use of heat storage systems (in summer).

Table 2. The electricity price of large industrial power in Guangzhou, China, in its distribution grid.

Peak and Valley Time Period	Classification of Load Voltage	Price (RMB yuan/kWh)
flat load (8:00–10:00, 12:00–14:00, 19:00–24:00)	1–10 kV	0.6104
	20 kV	0.6072
	35–110 kV	0.5854
peak load (10:00–12:00, 14:00–19:00)	1–10 kV	1.0072
	20 kV	1.0019
	35–110 kV	0.9659
valley load (0:00–8:00)	1–10 kV	0.3052
	20 kV	0.3036
	35–110 kV	0.2927

4. Impact on Power Grid Capacity, Load Characteristics, and Safety Margins

4.1. Distributed Energy Storage System on Load Side

4.1.1. Impact on Peak Shifting, Energy Efficiency, and Economic Benefit for Consumers

The construction of power grid needs to meet the demand of peak load, resulting in low asset utilization and operational efficiency. Especially in summer, more than 40% of the grid load is air-conditioning load, while its annual power consumption is less than 10%. In Figure 9b, without considering heat storage systems until 2025, if the basic load is 123,810 MW in 2020, according to energy storage planning, using distributed energy storage for peak shifting of 1359–2353 MW at 10:00–12:00 a.m. and of 1024–1338 MW at 14:00–16:00 p.m., valley filling of 2314–5260 MW at 1:00–5:00 a.m. and a flat period fulfilling 5526–5670 MW at 12:00–14:00 p.m. can transfer no more than 1–2% of peak load, which increases to 4–5% load in valley and flat periods.

If the power consumption on a peak-load day is 2,000,000 MWh, the total charging and discharging power consumption of four energy storage systems without heat storage systems, depicted in Figure 9b, are 43,224 MWh and 4691 MWh, accounting for 2.2% and 0.23% of the daily consumption, respectively. The net increased daily power consumption is 38,533 MWh, accounting for 2% of the daily consumption. The net increased daily power consumption includes rigid power demand of EVs and energy loss of electrochemical battery system, may cause reduced energy efficiency.

According to rough estimation, the total power fee in the peak load day is 20.73 million yuan at the charging period, and a maximum 4.30 million yuan at the discharging period, as decided by the proportions at the user side, the power grid, and the supplier side. If the electrochemical battery capacity of the user side accounts for 7% of all distributed electrochemical energy storage in 2025, the total saving cost for users on a peak-load day is about 0.3 million yuan.

4.1.2. Impact on Safety Margin

The disorderly charging behavior of large volumes of EVs increases in risk until 2025, although the maximum charging power, 3425 MW, of the EVs simulated in Figure 4 accounts for no more than 2.8% of current provincial peak load, and 52–65% of energy storage load. For five types of energy storage, although the whole energy storage system flattens the single peak at 10:00–12:00 a.m. and 14:00–16:00 p.m., it forms multiple sub-peaks at 1:00–5:00 a.m. and 12:00–14:00 p.m. Although the risk level, at peak periods, of exceeding the limit decreases, the number of multiple sub-peak risks increases.

The orderly charging behavior of electrochemical energy storage invested in by the power grid, in case of emergency, can avoid overloading some equipment and of having low voltage of some nodes. In this case, the energy storage equipment put into operation at such critical moments can play the role of emergency support.

4.1.3. Impact on Planning and Construction Considering EVs

The power demand of EV batteries, accounting for 65% of charging power in the valley period and influenced by their work mode, affects the distribution grid planning, substation capacity, and equipment selection. The site selection of a substation is selected in combination with the distribution of EV charging load. The fixed capacity of the substation refers to the determination of the main transformer capacity of the substation, and the appropriate transformer capacity load ratio shall be considered. The simultaneous rate of conventional load peaks and EV charging load peak have an important impact on the determination of transformer capacity load ratio. With the growth of EVs, the uncertainty of charging load leads to the maximum load prediction deviation, requiring great changes of total substation capacity and layout.

If a large number of EVs are charged at widely distributed charging piles, the voltage waveform of a 380 V public bus will be seriously distorted because the piles are distributed in a 400 V low-voltage distribution system. The short-term fast charging of charging equipment may cause too fast a load change, produce impulse voltage, and endanger the safety of the power grid. In order to ensure power quality, on the one hand, corresponding active-filter and reactive power-compensation devices can be equipped. On the other hand, when planning to build a large-capacity charging station, it may not share the same section of bus with loads sensitive to power quality. The uncertain characteristics of EV charging time add uncertain factors to the power flow calculation of distribution system. In order to meet safe transmission requirements under various operation modes, a certain margin can be reserved for the line capacity. The weak links can be strengthened to ensure that the line current does not exceed the limit. When considering the impact of EV development on distribution network planning, the analysis of planning operational cost is more important. It is necessary to optimize the economic model in the planning process, so that the actual planning scheme can take into account adaptability and economy.

4.2. Distributed Energy Storage System on Power Resource Side

4.2.1. Impact on Stabilizing Output Power of Renewable Energy

In order to reduce the impact of clean energy output fluctuation on the power grid, installing appropriate distributed energy storage can increase the controllability of clean energy output power and stabilize power fluctuation to ensure the safe, stable, and economic operation of the clean energy grid with high permeability.

Based on power grids' operational requirements, the objectives of stabilizing the output fluctuations of energy storage devices fall into three categories. The first category has the lowest requirements. When the power system's operational conditions and regulation capacity are certain, it is expected that the grid connection of clean energy will not affect dynamic stability or cause frequency regulation pressure, that is, the output can meet the fluctuation constraints and frequency regulation requirement. The second type has an additional and more stringent requirement that the power generated by clean energy be able to track the power generation plan in real time. The third category has the highest requirements; the expected output must not only quickly respond to changes in power grid frequency and have the abilities of system frequency modulation and peak shaving, but also have a certain schedulability and the ability to coordinate large-scale clean energy generation.

4.2.2. Improving Clean Energy Consumption

We next consider the impact on the clean energy consumption of the power grid under various application modes, such as the joint configuration of a distributed energy storage system and flexible interconnection devices, and on participation in the demand response.

(1) The joint configuration of distributed energy storage system and flexible interconnection devices. Compared with separate construction, the integrated construction mode of soft open point (SOP) and energy storage realizes the effective reuse of two groups of high-capacity power electronic converters in SOP, improves the utilization rate of SOP equipment, and greatly reduces the system construction investment and operation costs. From the perspective of operation, the addition of energy storage elements improves the operational inertia of SOP devices and enhances the ability of SOP devices to deal with transient disturbances in maintaining system energy balance for adapting to more complex operational scenarios and control requirements. From the perspective of the distribution grid, SOPs containing energy storage will have the energy transfer capability in both spatial and temporal dimensions, which can not only realize the real-time adjustment of transmission power between different feeders or stations, but also realize the functions of stabilizing fluctuation, peak shifting, and valley filling within a given time period, further strengthening the dispatching control capability of SOPs and improving the level of intermittent energy consumption. They will play an important role in improving power supply quality and optimizing the operation level of the distribution network. The dual role of SOPs with energy storage means they face higher technical requirements. In the planning stage, the investment cost and operation cost of SOP including energy storage will be closely related to the capacity and power of energy storage elements and the operation life under different charge and discharge strategies, which need to be fully considered; in terms of operation control, the charging and discharging of energy storage elements need to be completed by the cooperation of two groups of converters in the SOPs, which greatly increases the complexity of the control strategy. In addition, when the DC-side voltage level of SOPs is high, the energy storage components may need to be connected through DC chopper boosting, and the coordination among multiple power electronic converters will also become one of the key issues.

(2) Distributed power generation combined with a distributed energy storage system participating in demand response. Photovoltaic output is concentrated in the daytime, the fluctuation amplitude and frequency of which are significantly greater. Wind power output fluctuates all day, but the overall output at night is large, having a

negative effect in valley periods. Wind and solar grid connections reduce peak load, greatly reduce the overall time series curves, and increases the capacity margin. The net load curve has multiple valleys, and the net load in valley periods can even be negative. In order to avoid the phenomenon of abandoning wind and PV, the main network and other distributed energy sources need to have downward regulation capacity and reserve capacity. In some periods, the wind power output and load trends are inconsistent, resulting in large fluctuation times and amplitude of the net load curve, which requires the system to have a flexible climbing ability.

In order to improve the consumption level of clean energy and avoid the phenomenon of abandoning wind and PV when the net load is negative, distributed energy storage can be charged during valley periods and discharged during peak load periods. Through the time sequence transfer of the net load curve, the energy utilization rate of uncontrollable distributed generation can be increased, to avoid the phenomena of abandoning wind and PV power and high-price power purchasing under the low acceptance capacity of the system.

5. Conclusions

There is a developing trend of establishing a hierarchical and partitioned-energy internet sharing operation platform to gather wind power, photovoltaic, energy storage, flexible load, heat and cold energy systems, and energy suppliers to realize consumer energy services. The energy interconnection and sharing platform of Dongguan's local dispatching was established in 2017 and put into operation in 2019. Based on cloud–edge computing technology, it has realized the pioneering construction of distributed cloud energy storage with access to a local power grid management platform. Although the current market policy has not been liberalized, the platform performs only power monitoring and operation and the maintenance of power load equipment and does not have the functions of heat energy collection, carbon emission monitoring, or market transaction. However, with the progress of technology and the attention of the national industry, it will further realize such a platform with ideal functionality.

Building a local dispatching platform of the provincial demand side's response platform has great significance for consolidating new power system infrastructure for carbon peaking and carbon neutralization and for improving the efficiency of comprehensive energy management. Our suggestions are as follows:

(1) Though energy storage has many advantages, there are some focuses: single-peak load shifting transferring to multiple sub-peak loads; lower energy efficiency, as electricity increase in valley periods are much greater than the reductions of peak loads; the high cost of electrochemical batteries; EV load demand accounting for more than 60% of increased energy storage load and needing a regulated work mode; unnecessary heat load and replaceable and interruptible load implemented by setting enterprise rules.
(2) The grid company should propose standards of energy storage configuration, improve the features of the proportion of storage capacity, such as location and operational and maintenance measures, and configure energy storage access to the sharing platform.
(3) Electrochemical energy storage has the fastest response and highest cost. At present, in order to protect battery service life, realistic operation requires shallow charging and discharging, of about 30–40% in one charging. It is not necessary to plan a too-large overall capacity scale on the power-supply side, power-grid side, or user side in the next five years; rather, we recommend such installed capacity only when other measures cannot solve overload problems.
(4) Control the power demand of users. It is necessary to make mandatory management rules for energy conservation and emission reduction, such as controllable air conditioning and heating load.

Author Contributions: Conceptualization, J.L. and Y.X.; methodology, J.L., Y.X. and D.Z.; software, Y.X., D.Z. and J.L.; validation, D.Z., Y.X. and J.L.; formal analysis, Y.X., J.L. and D.Z.; investigation, J.L. and D.Z.; resources, Y.X.; data curation, Y.X. and J.L.; writing—original draft preparation, J.L. and Y.X.; writing—review and editing, J.L.; supervision, J.L. and D.Z.; project administration, J.L. and Y.X.; funding acquisition, J.L. All authors have read and agreed to the published version of the manuscript.

Funding: This research was funded by China Southern Power Grid Company Limited Science and Technology Project Foundation (030000KK52190010 (GDKJXM20198116)).

Institutional Review Board Statement: Not applicable.

Informed Consent Statement: Not applicable.

Acknowledgments: Thanks to the researchers in Sichuan Energy Internet Research Institute, Tsinghua University, who helped provide the source code and some analysis contents. We would like to express our gratitude to them for their help and guidance.

Conflicts of Interest: The authors declare no conflict of interest.

References

1. Liu, J.; Zhang, N.; Kang, C.; Kirschenb, D.; Xia, Q. Cloud energy storage for residential and small commercial consumers: A business case study. *Appl. Energy* **2017**, *188*, 226–236. [CrossRef]
2. Liu, J.; Zhang, N.; Kang, C. Research framework and basic models for cloud energy storage in power system. *Proc. CESS* **2017**, *37*, 3361–3371.
3. Zhou, L.; Huang, Y.; Guo, K.; Feng, Y. A survey of energy storage technology for micro grid. *Power Syst. Prot. Control* **2011**, *39*, 147–152.
4. Chen, W.; Shi, J.; Ren, L.; Tang, Y.; Shi, Y. Composite usage of multi-type energy storage technologies in microgrid. *Autom. Electr. Power Syst.* **2010**, *34*, 112–115.
5. Awerbuch, S.; Preston, A. *The Virtual Utility: Accounting, Technology & Competitive Aspects of the Emerging Industry*; Springer: New York, NY, USA, 2012.
6. Mashhour, E.; Moghaddas-Tafreshi, S.M. Bidding strategy of virtual power plant for participating in energy and spinning reserve markets—Part I: Problem formulation. *IEEE Trans. Power Syst.* **2011**, *26*, 949–956. [CrossRef]
7. Ghavidel, S.; Li, L.; Aghaei, J.; Yu, T.; Zhu, J. A Review on the Virtual Power Plant: Components and Operation Systems. In Proceedings of the 2016 IEEE International Conference on Power System Technology (POWERCON), Wollongong, Australia, 28 September–1 October 2016; pp. 1–6.
8. Kempton, W.; Tomić, J. Vehicle-to-grid power implementation: From stabilizing the grid to supporting large-scale renewable energy. *J. Power Sources* **2005**, *144*, 280–294. [CrossRef]
9. Tuttle, D.P.; Baldick, R. The evolution of plug-in electric vehicle-grid interactions. *IEEE Trans. Smart Grid* **2012**, *3*, 500–505. [CrossRef]
10. Han, S.; Sezaki, K. Development of an optimal vehicle-to-grid aggregator for frequency regulation. *IEEE Trans. Smart Grid* **2010**, *1*, 65–72.
11. Tan, X.Q.; Yang, S.C.; Fang, Y.P.; Xue, D. Discussion on operation model to the electric vehicle charging station. *AMR* **2014**, *875–877*, 1827–1830. [CrossRef]
12. Li, J.; Zhou, S.; Lu, X. Planning method and principles of the cloud energy storage applied in the power grid. In Proceedings of the 2020 IEEE Sustainable Power and Energy Conference (ISPEC), Chengdu, China, 23–25 November 2020; pp. 2040–2046.
13. Mu, L.; Cui, L.; An, N. Research and Practice of Cloud Computing Center for Power System. *Power Syst. Technol.* **2011**, *35*, 171–175.
14. Rao, W.; Ding, J.; Lu, Q. Cloud Computing Platform Construction for Smart Grid. *East China Electr. Power* **2011**, *39*, 1493–1496.
15. Bai, Y.; Huang, Y.; Chen, S.; Zhang, J.; Li, B.; Wang, F.-Y. Cloud-edge Intelligence: Status Quo and Future Prospective of Edge Computing Approaches and Applications in Power System Operation and Control. *Acta Autom. Sin.* **2020**, *46*, 397–410.
16. Jin, W. Research on Distributed Storage Convergence and Coordination Control Strategy. Master's Thesis, China Electric Power Research Institute, Beijing, China, June 2018.
17. Guo, B.; Niu, M.; Lai, X.; Chen, L. Application research on large-scale battery energy storage system under Global Energy Interconnection framework. *Glob. Energy Interconnect.* **2018**, *1*, 79–86.
18. Jin, R.; Zhang, X.; Wang, Z.; Sun, W.; Yang, X.; Shi, Z. Blockchain-Enabled Charging Right Trading Among EV Charging Stations. *Energies* **2019**, *12*, 3922. [CrossRef]
19. Yao, W.; Zhao, J.; Wen, F.; Xue, Y.; Xin, J. A charging and discharging dispatching strategy for electric vehicles based on bi-level optimization. *Autom. Electr. Power Syst.* **2012**, *36*, 30–37.
20. Hu, W.; He, L.; Chen, J. A bi-layer optimization based schedule considering large-scale electric vehicles. *Power Syst. Prot. Control* **2016**, *44*, 22–28.

21. Zhang, Q.D.; Huang, X.L.; Chen, Z.; Chen, L.X.; Xu, Y.P. Research on control strategy for the uniform charging of electric vehicle battery swapping station. *Trans. China Electrotech. Soc.* **2015**, *30*, 447–453.
22. Liu, W.; Wen, F.; Ma, L.; Xue, S. Demand-side Transactive Energy Mechanism Considering Electric Vehicles and Controlable Loads. *Electr. Power Constr.* **2019**, *40*, 24–30.
23. Vasirani, M.; Kota, R.; Cavalcante, R.L.G.; Ossowski, S. An Agent-Based Approach to Virtual Power Plants of Wind Power Generators and Electric Vehicles. *IEEE Trans. Smart Grid* **2013**, *4*, 1314–1322. [CrossRef]
24. Fan, S.; Liu, J.; Wu, Q.; Cui, M.; Zhou, H.; He, G. Optimal coordination of virtual power plant with photovoltaics and electric vehicles: A temporally coupled distributed online algorithm. *Appl. Energy* **2020**, *277*, 115583. [CrossRef]
25. Wang, H.; Chen, S.; Yan, Z.; Ping, J. Blockchain-enabled charging right trading among EV charging stations: Mechanism, model, and method. *Proc. CSEE* **2020**, *40*, 425–435.
26. Jin, Z.; Wu, R.; Li, G.; Yue, S. Transaction model for electric vehicle charging based on consortium blockchain. *Power Syst. Technol.* **2019**, *43*, 4362–4369.
27. Zhang, F.; Li, G.; Wang, T. Electric vehicle charging chain based on blockchain technology. *Comput. Technol. Dev.* **2020**, *30*, 161–166.
28. Yang, Y.; Peng, D.; Wang, W. Block-chain based Energy Tracing Method for Electric Vehicles Charging. In Proceedings of the 2020 IEEE Sustainable Power and Energy Conference (iSPEC 2020), Chengdu, China, 23–25 November 2020; pp. 2622–2627.
29. Zhang, X.; Yang, Y.; Su, S. Study on Electric Vehicle Sharing and Leasing Business Model for Group Users based on Blockchain. In Proceedings of the 2020 IEEE Sustainable Power and Energy Conference (iSPEC 2020), Chengdu, China, 23–25 November 2020; pp. 2628–2633.
30. Tapscott, D.; Tapscott, A. *Blockchain Revolution*; Tap Publications Inc.: New York, NY, USA, 2016.
31. Di Silvestre, M.L.; Gallo, P.; Guerrero, J.M.; Musca, R.; Sanseverino, E.R.; Sciumè, G.; Vásquez, J.C.; Zizzo, G. Blockchain for power systems: Current trends and future applications. *Renew. Sustain. Energy Rev.* **2020**, *119*, 109585. [CrossRef]
32. Nakamoto, S. Bitcoin: A Peer-to-Peer Electronic Cash System. Consulted. 2009, pp. 1–9. Available online: https://www.bitcoin.org or http://www.spacepirates.com/bitcoin.pdf (accessed on 30 March 2009).
33. Ferrag, M.A.; Derdour, M.; Mukherjee, M.; Derhab, A.; Maglaras, L.; Janicke, H. Blockchain Technologies for the Internet of Things: Research Issues and Challenges. *IEEE Internet Things J.* **2019**, *6*, 2188–2204. [CrossRef]
34. Yang, D.; Zhao, X.; Xu, Z.; Li, Y.; Li, Q. Developing Status and Prospect Analysis of Blockchain in Energy Internet. *Proc. CSEE* **2017**, *37*, 3664–3671.
35. Wang, S.; Guo, C.; Feng, B.; Zhang, H.; Du, Z. Application of blockchain technology in power systems: Prospects and ideas. *Autom. Electr. Power Syst.* **2020**, *44*, 10–24.
36. Yuan, Y.; Wang, F. Blockchain: The state of the art and future trends. *Acta Autom. Sin.* **2016**, *42*, 481–494.
37. Zhang, J.; Gao, W.Z.; Zhang, Y.C.; Zheng, X.H.; Yang, L.Q.; Hao, J.; Dai, X.X. Blockchain based intelligent distributed electrical energy systems: Needs, concepts, approaches and vision. *Acta Autom. Sin.* **2017**, *43*, 1544–1554.
38. Zhang, N.; Wang, Y.; Kang, C.; Cheng, J.; He, D. Blockchain Technique in the Energy Internet: Preliminary Research Framework and Typical Applications. *Proc. CSEE* **2016**, *36*, 4011–4022.
39. Zeng, M.; Cheng, J.; Wang, Y.; Li, Y.; Yang, Y.; Dou, J. Primarily research for multi module cooperative autonomous mode of energy internet under blockchain. *Proc. CSEE* **2017**, *37*, 3672–3681.
40. Dong, Z.; Luo, F.; Liang, G. Blockchain: A secure, decentralized, trusted cyber infrastructure solution for future energy systems. *J. Mod. Power Syst. Clean Energy* **2018**, *6*, 1–10. [CrossRef]
41. Cui, J.; Wang, S.; Xin, Y. Research on technical framework of smart grid data management from consortium blockchain perspective. *Proc. CSEE* **2020**, *40*, 836–848.
42. Tai, X.; Sun, H.; Guo, Q. Electricity transactions and congestion management based on blockchain in energy internet. *Power Syst. Technol.* **2016**, *42*, 2630–3638.
43. Ouyang, X.; Zhu, X.; Ye, L.; Yao, J. Preliminary applications of blockchain technique in large consumers direct power trading. *Proc. CSEE* **2017**, *37*, 3737–3745.
44. Ma, T.; Peng, L.L.; Du, Y.; Gou, Q.F.; Wang, C.; Guo, X.F. Competition game model for local multi-microgrid market based on block chain technology and its solution algorithm. *Electr. Power Autom. Equip.* **2018**, *38*, 191–203.
45. Li, B.; Cao, W.; Zhang, J.; Chen, S.; Yang, B.; Sun, Y.; Qi, B. Transaction system and key technologies of multi-energy system based on heterogeneous blockchain. *Autom. Electr. Power Syst.* **2018**, *42*, 183–193. [CrossRef]
46. Wang, B.; Li, Y.; Zhao, S.; Chen, H.; Jin, Y.; Ding, Y. Key technologies on blockchain based distributed energy transaction. *Autom. Electr. Power Syst.* **2019**, *43*, 53–64.
47. Qi, B.; Xia, Y.; Li, B.; Li, D.; Zhang, Y.; Xi, P. Photovoltaic trading mechanism design based on blockchain-based incentive mechanism. *Autom. Electr. Power Syst.* **2019**, *43*, 132–139.
48. Wang, Y.; Li, J.; Hu, Y. Research on Integrated Energy Trading Mechanism Based on Blockchain Smart Contract Technology. In Proceedings of the 2020 IEEE Sustainable Power and Energy Conference (iSPEC 2020), Chengdu, China, 23–25 November 2020; pp. 2634–2639.
49. National Big Data Alliance of New Energy Vehicles (NDANEV). The National Regulatory Platform for New Energy Vehicles Has Accessed 2.357 Million New Energy Vehicles, with a Driving Mileage of 27.42 Billion Kilometers. 2019. Available online: http://www.china-nengyuan.com/news/141934.html (accessed on 9 July 2019).

50. Gao, H.; Wang, L.; Liu, J.; Wei, Z. Integrated day-ahead scheduling considering active management in future smart distribution system. *IEEE Trans. Power Syst.* **2018**, *33*, 6049–6061. [CrossRef]
51. Jiang, K.; Li, H.; Li, W.; Cheng, S. On several battery technologies for power grids. *Autom. Electr. Power Syst.* **2013**, *37*, 47–53.

energies

MDPI

Article

Thermal Performance Improvement for Different Strategies of Battery Thermal Management Systems Combined with Jute—A Comparison Study †

Rekabra Youssef [1,2,*], Md Sazzad Hosen [1], Jiacheng He [1], Mohammed AL-Saadi [1,2], Joeri Van Mierlo [1] and Maitane Berecibar [1]

1 Mobility, Logistics & Automotive Technology Research Group (MOBI), Electrical Engineering and Energy Technology (ETEC) Department, Vrije Universiteit Brussel, 1050 Brussels, Belgium; md.sazzad.hosen@vub.be (M.S.H.); jiacheng.he@vub.be (J.H.); Mohammed.al-saadi@vub.be (M.A.-S.); Joeri.Van.Mierlo@vub.be (J.V.M.); Maitane.Berecibar@vub.be (M.B.)

2 Flanders Make, 3001 Heverlee, Belgium

* Correspondence: Rekabra.Youssef@vub.be

† This article is a revised version of our paper published in 2021 IEEE the 9th International Conference on Smart Grid and Clean Energy Technologies (ICSGCE), Sarawak, Malaysia, 15–17 October 2021, pp. 68–71.

Abstract: Jute is a cheap, eco-friendly, widely available material well-known for its cooling properties. In electric vehicles (EVs), dissipating a huge amount of the heat generated from lithium-ion batteries with an efficient, light, and low-power consumption battery thermal management system (BTMS) is required. In our previous study, jute fibers were proposed and investigated as a novel medium to enhance the cooling efficiency of air-based battery thermal management systems. In this paper, as the first attempt, jute was integrated with a phase change material (PCM) passive cooling system, and the thermal performance of a 50 Ah prismatic battery was studied. Temperature evolution, uniformity, and cooling efficiency were investigated. A comparison between the thermal behavior of the air-based BTMS and PCM-assisted cooling system was performed. The results indicated that adding jute to the BTMS increased the cooling efficiency and especially decreased the temperature development. Furthermore, the temperature difference (ΔT) efficiency was enhanced by 60% when integrating jute with PCM, and temperature uniformity improved by 3% when integrating jute with air-based BTMS. This article compared the integration of jute with active cooling and passive cooling; thus, it shed light on the importance of jute as a novel, eco-friendly, lightweight, cheap, available, and nontoxic material added to two strategies of BTMS. The setup was physically made and experimentally studied for the purpose of BTMS optimization.

Keywords: jute; electric vehicles; thermal management; evaporative cooling; passive cooling; air cooling; phase change material (PCM); lithium-ion battery

Citation: Youssef, R.; Hosen, M.S.; He, J.; AL-Saadi, M.; Van Mierlo, J.; Berecibar, M. Thermal Performance Improvement for Different Strategies of Battery Thermal Management Systems Combined with Jute—A Comparison Study †. *Energies* **2022**, *15*, 873. https://doi.org/10.3390/en15030873

Academic Editors: Alon Kuperman and Alessandro Lampasi

Received: 17 November 2021
Accepted: 18 January 2022
Published: 25 January 2022

1. Introduction

The electric vehicle (EV) market has matured over the last few decades because of the future it promises of environmental and green transportation with zero emissions [1]. The lithium-ion battery (LIB) is the most preferred energy storage system used by the majority of EVs [2] because of their large energy, high cycle life, light weight, and low rate of self-discharge [3]. On the other hand, the performance characteristics of LIBs are strongly limited by their operating temperature [4,5]. High charge and discharge produce a huge amount of heat [6] and nonuniform surface temperature distribution [6,7]. Moreover, heat generation can diminish battery performance [8,9], leading to a quick thermal runway [10]. Therefore, solutions to this issue are demanded to exceed the last stage of large-energy LIB applications. Maintaining the operating temperature within the proper range requires an effective battery thermal management system (BTMS) [11]. As pointed out by many researchers [12], BTMSs can be classified into air cooling (active and passive cooling) [13], liquid cooling [14], phase change material (PCM) cooling [15], heat pipe cooling [16], and hybrid cooling [17], which is a combination of two or more of the mentioned cooling strategies.

A liquid cooling system is the most common cooling strategy used in electrified vehicles [18]. Because of high thermal conductivity and heat transfer efficiency [19], it can control battery temperature development and distribution [20]. However, liquid cooling systems have complexity in structure, water leakage risk, and additional weight and energy consumption [21].

The heat pipe has better thermal performance and higher thermal conductivity [22] compared to the other passive cooling strategies, but the contact area of the heat pipe system with the battery is small, and the system structure is volumetric [23,24].

Air cooling systems are considered one of the most traditional, common, and widely embraced cooling strategies in industrial applications [25]. This is due to their simplicity in structure, flexibility in maintenance and packaging, water leakage avoidance, and low weight and power consumption [26]. However, the main concerning points are low cooling efficiency and nonuniform temperature distribution [27]. Nevertheless, respectable research efforts have been employed to optimize the air cooling strategy [28]. Na et al. [29] proposed an optimization for an air cooling BTMS with a reverse airflow design. They compared the results with unidirectional airflow by computational fluid dynamics (CFD) and declared that the maximum average temperature differences were enhanced with a reverse airflow design. Chen et al. [30] improved the BTMS with U-type airflow and concluded that battery temperature difference and power consumption were minimized. Wang et al. [31] experimentally found that by reciprocating airflow, uneven temperature distribution caused by heat accumulation could be mitigated. Xie et al. [32] experimentally and numerically researched the effects of inlet and outlet angles and the channel width between battery cells on the heat dissipation rate. They revealed that airflow channels had a remarkable influence on the maximum temperature rise and the temperature difference. In terms of safety, alternative BTMSs have been studied and proposed, such as helium-based [33] and ammonia-based [34] battery thermal management.

A phase change material (PCM) is a substance that can absorb or release heat at a specific temperature value when it reaches the melting point and transforms from one physical phase to another [35,36]. The PCM absorbs the heat accumulated on the battery surface by conduction [37]. It has large latent heat and constitutes a low-cost thermal management strategy [38]. On the other hand, PCMs suffer from low thermal conductivity, and thermal management systems based on them are considered volumetric and massive. Therefore, many studies have been recorded to overcome these disadvantages [39]. Hussain et al. [40] used paraffin as a phase change material and infiltrated it into graphene-coated nickel foam. They claimed that the battery temperature rise was decreased by 17%. Li et al. [41] proposed an optimized design to reduce the mass and volume of PCMs. Huo et al. [42] employed a porous medium for PCM heat transfer improvement. Zhang et al. [43] suggested a combination of PCMs and designed a composite of olin/expanded graphite (EG)/paraffin. They found that the complex could control the battery temperature rise at 45 °C ambient conditions and a high discharge rate.

A respectable research effort has been performed to optimize battery thermal management either experimentally [42,43] or by thermal modeling [44,45].

All the mentioned studies were carried out targeting battery thermal management system optimization. Most of the optimization techniques did not take environmental aspects into consideration and even ignored them at some stages. Moreover, leakage, bulkiness, volumetric, uneven temperature distribution, and complexity in design are the most common issues between all existing battery thermal management systems. Therefore, in this paper, environmental and novel design optimization is proposed to enhance battery thermal management performance and design. Jute, as a cheap, eco-friendly, lightweight, available, and novel cooling medium, was integrated as a first attempt into a BTMS [46]. An active air cooling strategy was chosen as the common BTMS used in industrial applications. Then, jute was integrated into the system, and the results indicated an improvement in battery thermal performance [46].

In this study, a promising passive cooling strategy, a PCM with high latent heat, was chosen to combine with jute, since the PCM strategy is not enough to maintain the temperature within the preferable scope, especially for the demands of future EVs with fast charging. Therefore, merging further cooling mediums such as jute with PCM was hypothesized to contribute to enhancing thermal performance and lowering the temperature increase.

Therefore, the PCM cooling strategy was built with and without jute. Then, the thermal behavior of a 50 Ah prismatic battery cell was investigated. Maximum temperature, temperature differences, and distribution with cooling efficiency were analyzed. A comparison among different BTMSs (natural convection, active air cooling, and passive PCM cooling) integrated with jute was studied. The results indicated that integrating jute with BTMSs improved thermal performance with less weight, cost, and equipment, with respect of the environmental side, and kept the main purpose of EVs within the environmental aspect. The remainder of the paper is arranged as follows: Section 2 describes the experimental setup and the proposed design, Section 3 discusses and analyzes the thermal performance of integrating jute with PCM, and Section 4 presents the analysis for integrating jute with air-based BTMS. A comparison among BTMSs integrated with jute is represented in Section 5 and the experimental uncertainty analysis is given in Section 6. Finally, the conclusion and future work are drawn in Section 7.

2. Methodology and Development

A prismatic lithium-ion battery holding a large capacity of 50 Ah was used; its specifications are reported in Table 1.

Table 1. Battery properties declared by the manufacture.

Item	Parameter Specification
Nominal capacity	50 Ah
Nominal voltage	3.7 V
Battery weight	900 ± 25 g
Battery dimension	148 × 98 × 27 mm

With the aim of generating as much heat as possible, a high, constant current of 125 A (meaning a 2.5 °C discharge rate) was injected into the battery, first in natural convection without any cooling system. Then, jute fibers were prepared to be integrated into the following cooling strategies: active air cooling using fans and passive cooling with PCM assistance. Afterward, battery temperature increase, temperature distribution, and cooling efficiency were investigated for each of the cooling strategies integrated with jute. In the end, a comparison between the proposed BTMSs was made. Figure 1 gives an overview of the experiments carried out and result analysis.

2.1. Experimental Setup

In order to conduct trustworthy and accurate tests, an experimental test bench was built in the MOBI (mobility, logistics, and automotive technology research center) laboratory at Vrije Universitiet Brussel. The equipment and setup used in this study can be divided into general setup and specific setup.

The specific setup is illustrated and clarified separately in the following section for each cooling strategy. However, the general test bench and setup is shown in Figure 2. It consisted of a battery tester (PEC manufactured ACT0550 model) with an accuracy of ±0.005% for the voltage reading, used to cycle the battery; a CTS manufactured climate chamber, to control battery surface temperature; four thermocouples with accuracy of ±3%, used for temperature measurements; and a microcontroller device, used to record the thermocouples' temperature readings. Finally, all the mentioned equipment were linked to a computer to monitor the parameters (voltage, current, and temperature) and acquire the data.

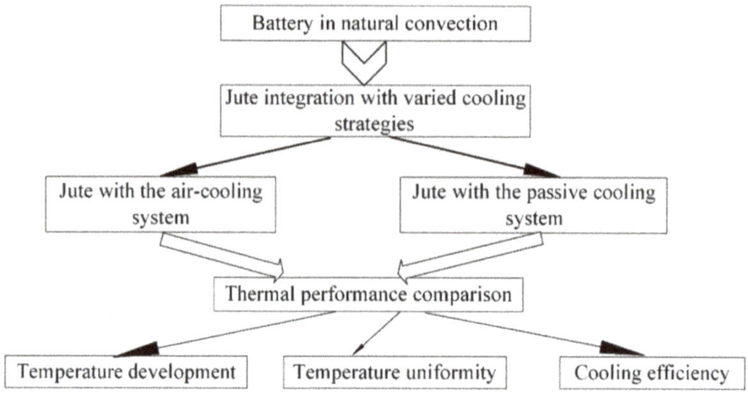

Figure 1. Design of the study and work flowchart.

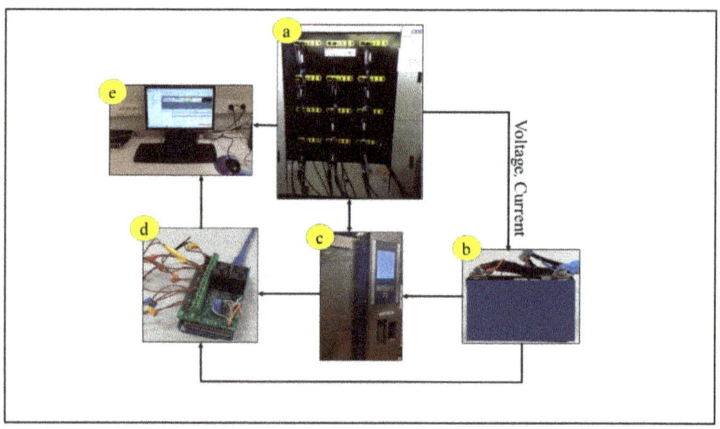

Figure 2. Schematic of the test bench: (**a**) battery tester; (**b**) 50 Ah battery; (**c**) climate chamber; (**d**) microcontroller; (**e**) computer.

2.2. *Description of the Proposed Designs*

Jute is considered a low-cost, natural fiber [47] and cellulosic material [48]. Because of its attractive cooling efficiency, light weight, low cost, and environmental benefits [49–51], it was chosen to insert into a battery thermal management system. Dedicated test setups were designed and constructed to evaluate the performance of jute fibers used in two strategies of BTMS, which are described in the following subsections.

2.2.1. Passive Cooling PCM-Assisted BTMS

An illustrative image of the proposed design as it was tested in this study is shown in Figure 3. Four thermocouples (glass-encapsulated sensors, standard type, B57560G, B57560G1) with accuracy of ±3% were used for temperature measurement. Three thermocouples were attached to the front side of the battery at the locations shown in Figure 3, and the fourth was used to record the ambient temperature. Then, the battery was surrounded with jute mesh and placed in a plexiglass case constructed to fit with the battery, jute, PCM, and thermocouples. A composite paraffin\graphite PCM was prepared, representing the passive cooling system. Its thermophysical properties are given in Table 2.

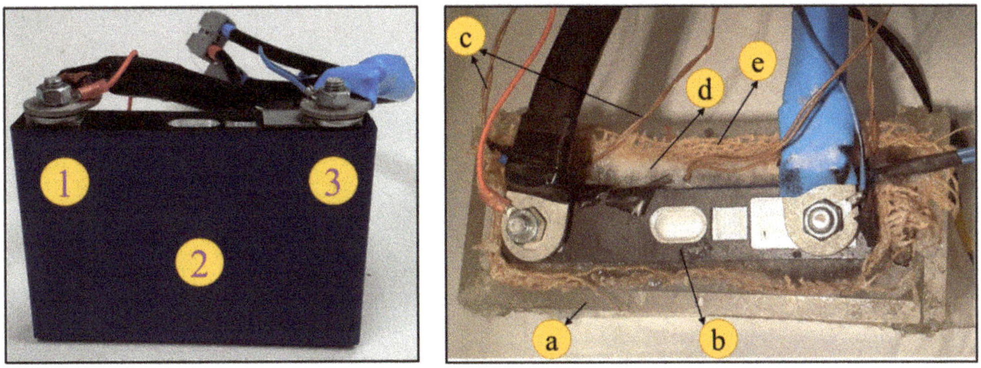

Figure 3. Thermocouple locations on the battery surface (**left**); PCM case with jute (**right**): (**a**) plexiglass case; (**b**) battery; (**c**) thermocouples; (**d**) PCM; (**e**) jute.

Table 2. Specifications and thermophysical properties for the PCM used.

Properties	Value
Thermal conductivity	2.32 W/m·K
Heat capacity	1430 J/kg·K
Density	1305 kg/m^3
Melting temperature	27–28 °C
Latent heat	150,000 J/kg

The PCM complex was poured into the plexiglass casing at an environmental temperature of 25 °C. The battery was discharged from a 100% state of charge (SoC) to a 0% state of charge, and the temperature was recorded at three cooling conditions. First, battery temperature measurements were taken inside the case without any PCM or jute. Then, PCM was poured into the case, and the measurement was recorded in the presence of the paraffin/graphite PCM. Finally, jute was wrapped around the battery periphery and PCM was poured into the plastic casing.

2.2.2. Active Air Cooling BTMS

An illustrative image of the test bench for air cooling with jute is shown in Figure 4. A casing was prototyped from plexiglass, and the battery was placed in cross-section inside the casing. Four fans were set on both sides of the case. Two fans on one side were used to blow the air with a speed of 7 m/s, and air circulation surrounded the battery before exiting from the suction fans with airflow of 2.5 m/s on the opposite side of the case.

A frame of jute fibers was prepared to fit with the case side. Furthermore, a filling water system was designed to feed the jute with water and drain the extra water from the cooling system.

For the purpose of validating the proposed air-based cooling design, a high discharge current (125 A) was applied to the battery with four conditions of airflow for the proposed design. The first condition featured two functional suction fans while the opposite side fans were in off mode and not operated. In the second condition, the air cooling case was operated with four fans; two fans were used as inlets, and the airflow traveled through the case to the other two suction fans (outlets). Then, the jute frame with the water filling system was fixed to the cooling case, replacing the inlet fans, in the third condition. In this condition, the jute was saturated with water and revealed to the exhaust fans, where the airflow evaporated the water from the jute's surface and cooled down the battery temperature. Therefore, the jute was wet at the beginning of the test and was then dehydrated. In the fourth condition, the jute frame was kept in its place, facing the suction fans, but it was refilled with water and maintained as wet during the entire test time.

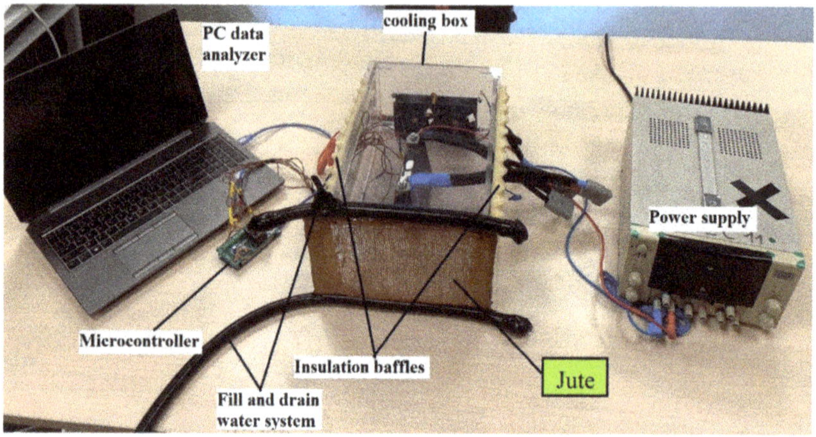

Figure 4. Proposed air cooling design with jute (Reprinted with permission from ref. [46]).

3. Thermal Performance of Jute Integration with PCM-Assisted Cooling System

The amount of thermal energy that could be stored in the jute fabric increased by applying PCMs on its textiles. This was due to the nature of PCMs and latent heat. When the battery was wrapped with jute fabric integrated with PCMs, the battery grew cooler when its temperature rose and hotter when the battery temperature decreased. This was because the heat from the battery was used to melt the PCMs in the jute fabric, essentially cooling the battery down. Furthermore, when the battery temperature decreased again, the energy stored as latent heat was released again, which in turn kept the battery warmer for a longer duration.

3.1. Temperature Development

Temperature development (evolution) and increase were calculated as the average surface temperature among the three thermocouples attached to the battery. Figure 5 compares the temperature increase in three tests, which were performed to represent the cases of no cooling (Box), cooling assisted with PCM, and cooling with jute merged into PCM. As expected, the use of PCM resulted a noteworthy reduction in the surface temperature. The maximum temperature for the battery surface reaches 46 °C in the no-cooling condition, but applying the PCM as a cooling medium decreased the maximum temperature at the end of discharge to 40 °C. Furthermore, an additional reduction in temperature occurred when jute was merged with PCM; a maximum temperature of 36 °C was obtained when the battery was fully discharged. This was because the thermal properties and cooling efficiency of jute were enhanced by merging it with the PCM [51–53], which led to better cooling performance.

3.2. Uniformity

Temperature uniformity can be defined as the temperature distribution throughout the cell surface. This distribution becomes more nonuniform at critical discharge current, which leads to heat accumulation and thermal runaway [52,53]. To measure the cooling uniformity for the proposed cooling strategy, the temperature nonuniformity (temperature distribution) over the surface of the battery was analyzed in this research. It was studied as the average of the differences between every two contiguous points of temperature obtained at the three thermal sensor locations. Figure 6 shows the temperature nonuniformity (temperature distribution) versus time for all three cases studied. The maximum nonuniformity of the temperature distribution on the battery surface was 2.7 °C, 2.4 °C, and 7.14 °C for the no-cooling (box), PCM cooling, and PCM+jute conditions, respectively. The pure PCM cooling case achieved the best temperature uniformity, whereas the PCM+jute response of heat dissipation was slower, and the temperature distribution was mostly nonuniform. The deficiency in thermal performance could have been due to the fiber cumulation and the absence of uniformity in the distribution the of jute [51–55].

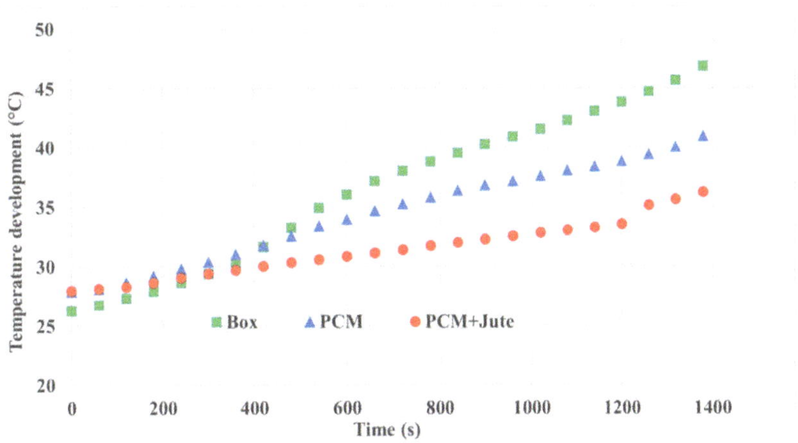

Figure 5. Temperature increase over time during fast discharge with a constant high current of 125 A.

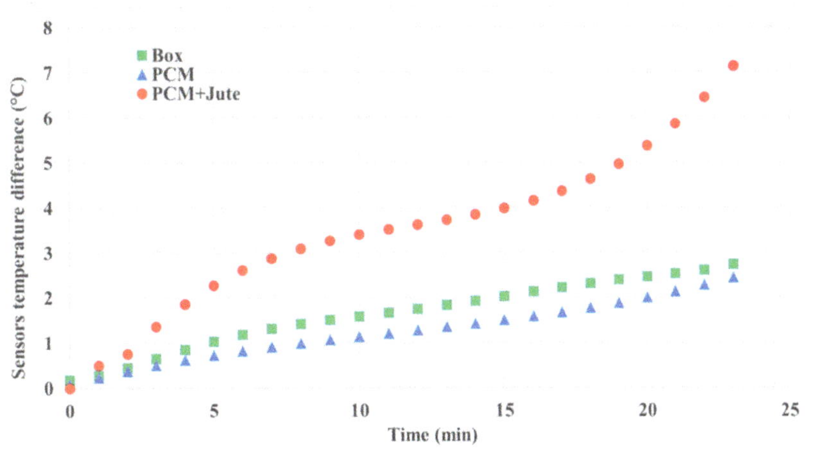

Figure 6. Temperature nonuniformity (temperature distribution) as a function of time for the passive cooling strategy.

3.3. Efficiency

Three parameters were considered to determine the cooling efficiency: maximum temperature (T_{max}) efficiency, temperature increase (ΔT) efficiency, and uniformity efficiency. The maximum temperature for every cooling scenario, $T_{max(k)}$, was compared with the maximum temperature that the battery reached without any cooling system, $T_{max(Box)}$ (Equation (1)). The same was done for battery uniformity efficiency (Equation (2)). Then, because the initial temperature for the three cases did not have an equal value, it was trustworthy to give a comparison of delta T for both cooling cases. Delta T (ΔT) represents the difference between the maximum temperature the battery reached and the initial temperature at every time interval. Then, ΔT efficiency was calculated the same way as T_{max} efficiency and uniformity efficiency (Equation (3)).

$$T_{max}\text{efficiency} = \frac{T_{max(Box)} - T_{max(k)}}{T_{max(Box)}} \times 100 \quad (1)$$

$$T_{max}\text{efficiency} = \frac{T_{max(Box)} - T_{max(k)}}{T_{max(Box)}} \times 100 \quad (2)$$

$$\Delta T \text{ efficiency} = \frac{\Delta T_{max(Box)} - \Delta T_{max(k)}}{\Delta T_{max(Box)}} \times 100 \quad (3)$$

The efficiency compression between the PCM cooling strategy and the PCM+jute cooling technique is clarified in Figure 7. Uniformity efficiency was not achieved with the PCM+jute technique. However, the maximum temperature efficiency was enhanced with the PCM+jute technique by 23%, which indicated better performance than the PCM cooling strategy (13% T_{max} efficiency). Moreover, a notable improvement in ΔT efficiency occurred when jute was embedded in the PCM. Increases in ΔT efficiency of 37% and 60% were obtained for the PCM and PCM+jute cooling strategies, respectively.

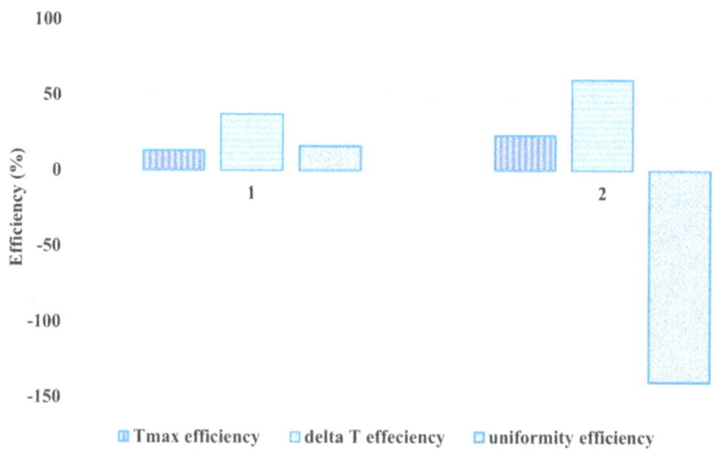

Figure 7. Efficiency comparison for cooling strategies: (1) PCM, (2) PCM+jute.

4. Thermal Performance of Jute Integration with Air Cooling Strategy

When moistened jute fabric is affected by a hot environment, the latent heat is absorbed by some of the water droplets, which then evaporate and change their phase from liquid to vapor [53]. Moreover, the remaining nonabsorbed water withdraws from the jute fabric to the outside air.

4.1. Temperature Development (Evolution)

Temperature development and increase were calculated as the average of the temperatures recorded at the three thermocouple locations. The average temperature trend is plotted in Figure 8, which displays the curves of average surface temperatures for the battery cell over time for the following cooling five cases: no-cooling (NC) condition; a forced air cooling strategy including two cases, first with two outflow fans and second with both inflow and inflow fans; then combining air cooling with jute for evaporative cooling; and the fourth cooling case evaporative cooling through entirely wet jute. The most aggressive temperature rise was recorded for the case of no cooling (NC), in which the temperature reached 48 °C when the battery fully discharged. Then, the temperature was diminished impressively by the case of using two exhaust fans, reaching approximately 40.5 °C. A significant reduction in temperature occurred in the evaporative cooling case, where jute was integrated into the forced cooling. With this integration, an interesting improvement in the cooling was obtained during the discharge process of about two degrees on average. This was due to the vacuum position of the jute, where it enhanced and cooled down the airflow temperature. Although jute helped with cooling, air humidity contributed to a further increase in temperature, and this was noticed with both cases of integrating jute (wet until getting dehydrated and wet throughout the discharge time). However, the final temperature reached with evaporative cooling was about 38.5 °C. Nonetheless, with the use of four fans, the highest temperature obtained was around 37.5 °C, which was the lowest maximum temperature that occurred among all the forced cooling cases. This discussion resulted in a novel outcome, that the same rate of cooling improvement occurred by using four fans or two fans. However, by utilizing jute, further reduction in weight, power consumption, and cost can be gained.

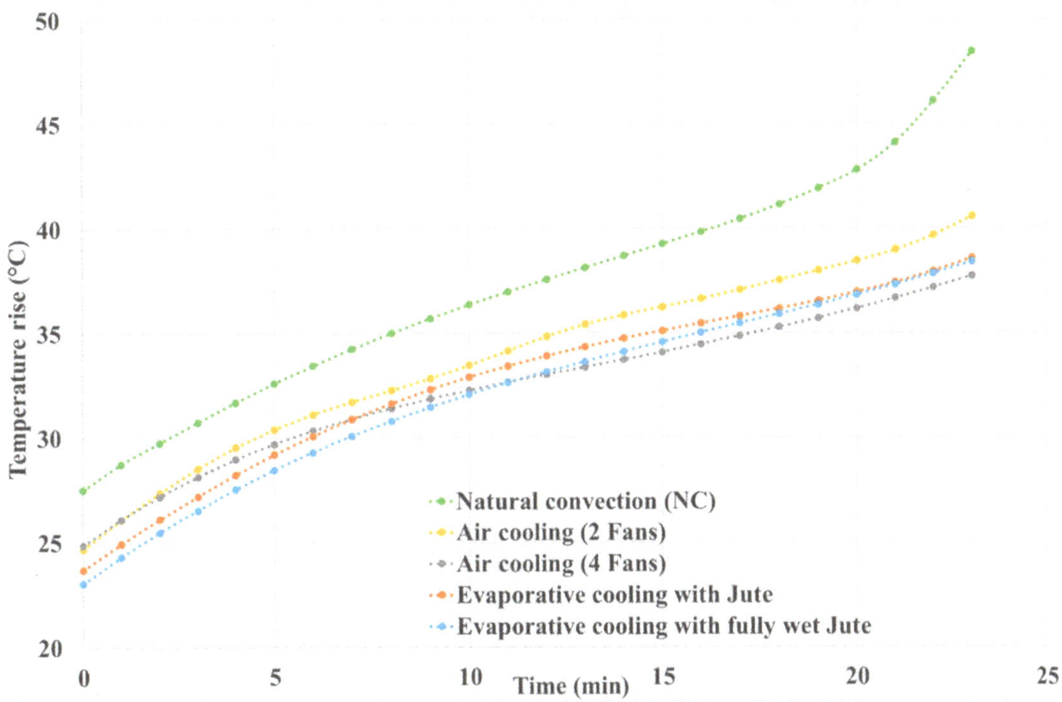

Figure 8. Battery surface temperature as a function of time for all the tests (Reprinted with permission from ref. [46]).

4.2. Uniformity

The temperature nonuniformity (distribution) over the battery surface was studied and calculated the same way as described in Section 3.2. Figure 9 shows the nonuniformity temperature versus time for all active cooling strategies. Each second, the measurement was recorded twice; therefore, the horizontal axis represents the time field and is the discharge time duplicated. It was found that except for evaporative cooling with fully wet jute, the temperature uniformity for all cooling strategies increased progressively to meet the peak when the battery was completely discharged. The maximum nonuniformity on the battery surface was 1.29 °C, 1.42 °C, 1.40 °C, and 1.39 °C for the natural air convection with no cooling (NC), air cooling with two fans, air cooling with four fans, and evaporative cooling with jute, respectively. However, in the fully wet jute cooling scenario, the uniformity characteristic grew at the early part of the battery discharge process, then arrived at a mostly stationary status and kept semi steady uniformity characteristics between 0.5 and 1 °C. Furthermore, close to the highest temperature, the temperature distribution for evaporative cooling with the fully wet jute strategy was the lowest among the active cooling strategies.

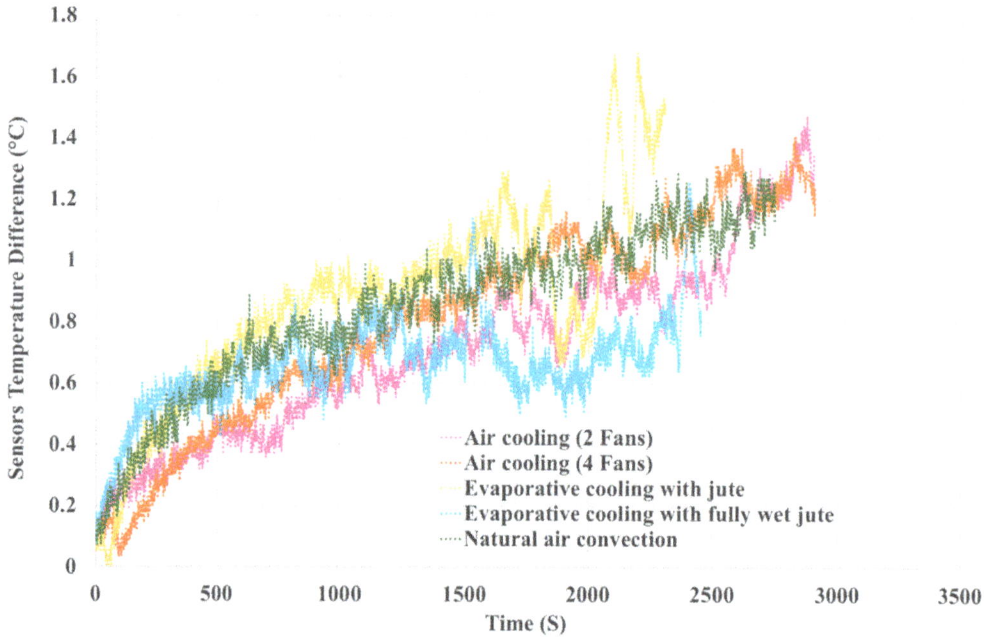

Figure 9. The nonuniformity temperature distribution versus time for different air-based cooling strategies (Reprinted with permission from ref. [46]).

4.3. *Efficiency*

All the active cooling strategies' efficiencies, in terms of maximum temperature (T_{max}) efficiency, temperature increase (ΔT) efficiency, and uniformity efficiency, are summarized in Figure 10. T_{max}, ΔT, and uniformity efficiency were calculated the same way as explained in Section 3.3. The T_{max} efficiency for all active cooling strategies fluctuated between 14% and 20%. The best cooling system in terms of T_{max} efficiency was the cooling scenario with four fans; in second place, the cooling scenario with jute achieved an efficiency of 18.7%. Furthermore, ΔT efficiency reached 44.45% with the four-fan cooling strategy and 33.06% with the wet jute strategy. Uniformity efficiency was accomplished exclusively in the wet jute strategy by 2.5%. Last but not the least, it is noteworthy that integrating wet jute with an active cooling strategy could contribute to improving the cooling efficiency in terms of T_{max}, ΔT, and uniformity simultaneously.

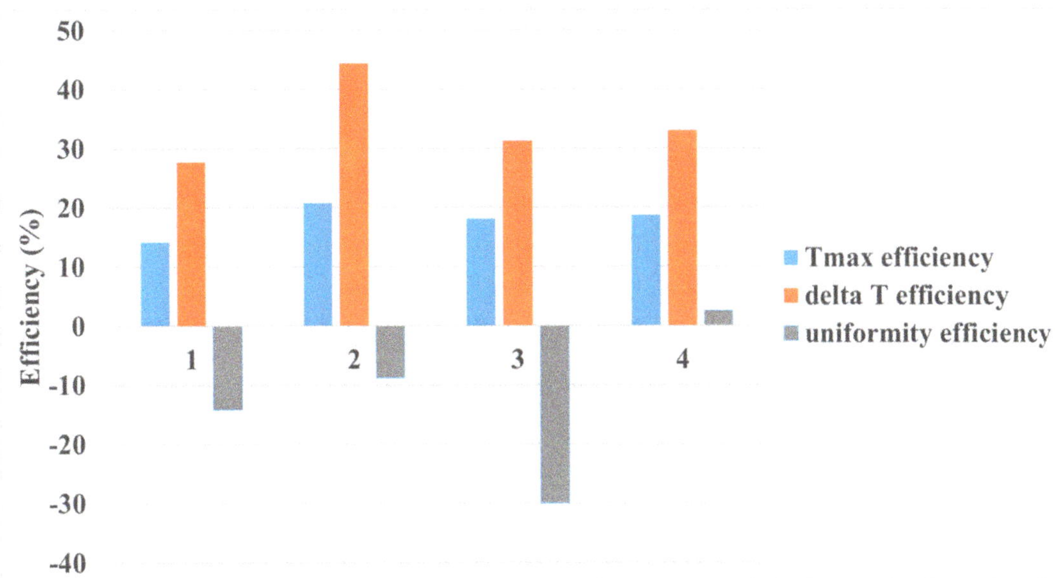

Figure 10. Efficiency comparison for cooling strategies: (1) air cooling (two suction fans); (2) air cooling (suction and blow fans); (3) evaporative cooling with jute; (4) evaporative cooling with fully wet jute (Reprinted with permission from ref. [46]).

5. Thermal Performance Comparison of Combining Jute into the (Active Cooling) Air-Based Cooling Strategy and (Passive Cooling) PCM-Assisted Cooling Strategy

5.1. Temperature Development

Figure 11 compares the changes in the average battery surface temperatures measured by thermocouples under different cooling conditions. The thermal schematic shows that the battery temperature reached the highest value, around 48 °C, in natural air convection with no cooling (NC). The temperature increment decreased by applying different cooling strategies: cooling with fans (active cooling), evaporative cooling (EC) with combined jute, and passive cooling assisted with PCM. Comparing active cooling and EC, EC with wet jute obtained a relatively close result to active cooling with four fans, but with less equipment, weight, and power consumption. Moreover, comparing EC with passive cooling, an additional reduction in temperature occurred with passive cooling integrated with jute. It can be concluded that jute had a positive impact on both environmental aspects and temperature enhancement for cooling strategy optimization.

5.2. Uniformity

According to Figure 12, the use of the PCM+jute technique decreased the temperature uniformity more than the rest of the cooling strategies. However, with the use of jute integrated into the active cooling with fans, the uniformity was enhanced, and an improvement in temperature distribution was achieved.

5.3. Efficiency

In Figure 13, all the performed characterizations are abbreviated to a single plot, which describes and compares the efficiency for every cooling scenario. Maximum temperature (T_{max}) and ΔT efficiency were generally achieved under most of the cooling strategies, but integrating jute with PCM led to the highest T_{max} and ΔT efficiency at 23% and 60%, respectively. Uniformity efficiency was not accomplished under most of the cooling strategies, excepting PCM and EC with wet jute. Therefore, it can be concluded that jute had a remarkable impact on the cooling efficiency in terms of T_{max}, ΔT, and uniformity and could be integrated into BTMS design optimization.

Time (min)	NC	4fans	2 fans	EC with Jute	EC with wet Jute	PCM	PCM+Jute
23	48.47	37.71	40.59	38.56	38.39	40.94	36.21
22	46.08	37.17	39.66	37.94	37.83	40.04	35.61
21	44.12	36.65	38.96	37.41	37.31	39.42	35.11
20	42.80	36.16	38.44	36.94	36.82	38.84	33.53
19	41.92	35.70	37.98	36.53	36.35	38.40	33.24
18	41.15	35.27	37.53	36.16	35.90	38.06	33.01
17	40.46	34.86	37.07	35.81	35.46	37.60	32.83
16	39.84	34.46	36.63	35.47	35.02	37.15	32.53
15	39.25	34.09	36.25	35.11	34.58	36.83	32.24
14	38.69	33.73	35.87	34.75	34.12	36.39	31.99
13	38.13	33.37	35.42	34.35	33.65	35.82	31.72
12	37.56	33.01	34.84	33.91	33.15	35.25	31.39
11	36.98	32.64	34.16	33.43	32.62	34.67	31.14
10	36.36	32.26	33.47	32.89	32.06	33.97	30.84
9	35.70	31.85	32.83	32.29	31.45	33.41	30.59
8	35.00	31.40	32.25	31.62	30.79	32.57	30.33
7	34.24	30.90	31.70	30.89	30.08	31.80	30.03
6	33.44	30.34	31.10	30.08	29.30	31.05	29.70
5	32.59	29.70	30.40	29.20	28.46	30.41	29.40
4	31.69	28.98	29.54	28.25	27.55	29.82	29.05
3	30.75	28.15	28.53	27.22	26.56	29.24	28.63
2	29.77	27.20	27.38	26.12	25.48	28.68	28.29
1	28.76	26.11	26.09	24.95	24.32	28.14	28.14
0	27.54	24.86	24.70	23.72	23.06	27.92	27.97

Color scale: 50, 45, 40, 35, 30, 25, 20

Figure 11. Comparison of the average surface temperature rise for all the discussed cooling strategies.

Time (min)	NC	4fans	2 fans	EC with Jute	EC with wet Jute	PCM	PCM+ Jute
23	1.27	1.31	1.48	1.58	1.12	2.44	7.14
22	1.18	1.31	1.39	1.21	0.98	2.28	6.44
21	1.15	1.28	1.25	1.12	0.93	2.13	5.85
20	1.12	1.26	1.22	1.09	0.85	2.00	5.36
19	1.08	1.22	1.17	1.07	0.80	1.88	4.96
18	1.08	1.21	1.08	1.07	0.79	1.77	4.63
17	1.04	1.15	0.97	1.04	0.78	1.67	4.36
16	0.97	1.10	0.95	0.98	0.77	1.58	4.15
15	0.97	1.10	0.95	0.98	0.77	1.50	3.98
14	0.93	1.09	0.94	0.94	0.77	1.42	3.85
13	0.88	1.09	0.93	0.93	0.77	1.35	3.73
12	0.84	1.06	0.91	0.93	0.77	1.28	3.62
11	0.82	1.05	0.91	0.93	0.75	1.21	3.52
10	0.82	1.04	0.89	0.89	0.71	1.14	3.40
9	0.81	1.04	0.88	0.89	0.71	1.07	3.26
8	0.75	1.00	0.88	0.87	0.70	0.99	3.09
7	0.75	0.98	0.88	0.84	0.64	0.91	2.87
6	0.73	0.98	0.86	0.84	0.64	0.82	2.60
5	0.70	0.98	0.80	0.84	0.64	0.73	2.27
4	0.55	0.96	0.79	0.83	0.62	0.62	1.86
3	0.50	0.96	0.77	0.78	0.60	0.51	1.36
2	0.42	0.89	0.77	0.76	0.59	0.38	0.76
1	0.18	0.87	0.73	0.74	0.58	0.24	0.54
0	0.11	0.80	0.71	0.69	0.57	0.08	0.42

Color scale: 8, 7, 6, 5, 4, 3, 2, 1, 0

Figure 12. Comparison of nonuniformity temperature distributions for all the discussed cooling strategies.

Figure 13. Efficiency comparison for cooling strategies: (1) PCM, (2) PCM+jute, (3) 2 fans, (4) 4 fans, (5) EC with jute, (6) EC with wet jute.

6. Experimental Uncertainty Analysis

Using the method presented by Moffat [56], the experiment error analysis was estimated. Assuming that the result, S, of the experiment is acquired from a set of measurements as follows:

$$S = S(x_1, x_2, x_3, \ldots, x_n)$$

The uncertainty is specified by the following equation:

$$\delta S = \left\{ \sum_{k=1}^{n} \left(\frac{\partial R}{\partial x_k} \delta x_k \right)^2 \right\}^{1/2} \tag{4}$$

where δS is the total uncertainty and δx_k is the uncertainty of every singular measurement. The uncertainties of experimental equipment were assumed as the absolute bias given by the aperture specification and described in Section 2.1. However, the uncertainty of the average temperature measurement was acquired based on the average temperature of the thermocouples attached to the cell. Thus, the total uncertainties were calculated following Equation (5) and estimated as 3.6%.

$$\frac{\Delta T_R}{\mathbf{T}_R} = \sqrt{\sum_{i=1}^{4} (\Delta T_{Ri}/T_{Ri})^2} \tag{5}$$

7. Conclusion and Future Work

This study attempted to investigate and analyze a novel and environmental optimization for LIB thermal management systems by integrating jute fiber with an active cooling strategy and a passive cooling strategy with PCM. It concluded with interesting results, which are summarized as follows:

1. Integrating jute fabrics with PCM (passive cooling strategy) achieved the desired effect on the battery thermal behavior and enhanced the cooling efficiency; temperature difference (ΔT) efficiency was enhanced by 60%. Furthermore, less PCM and

nonenvironmental cooling material was used. Therefore, system weight reduction and environmental material enhancement were achieved.

2. By adding jute to active cooling, high cooling efficiency was delivered with the wet jute in terms of maximal temperature, ΔT, and temperature uniformity; the latter improved by 3%. Merging jute into BTMS design would boost the opportunities for electric vehicles to work properly in hot environments.
3. Comparing the integration of jute with active cooling (air-based) and passive cooling (PCM), the results indicated that uniformity efficiency with the PCM+jute technique was mostly not achieved, while it was enhanced with the active cooling strategy. Therefore, integrating jute into an active BTMS has great potential to improve BTMS cooling efficiency in terms of T_{max}, ΔT, and uniformity together.
4. Last but certainly not least, jute, as a novel, eco-friendly, weightless, cheap, available, and nontoxic material, was added to two strategies of BTMS. The setup was physically made and experimentally studied for the purpose of BTMS design optimization. This work will be extended in the future to the modeling stage and simulation. Jute fabric thermal properties will be studied more, and further investigation with analysis will be added to optimize battery thermal management environmentally.

Author Contributions: Conceptualization, Methodology, Software, Validation, Formal analysis, Investigation, Resources, Data Curation, Writing—Original Draft, Visualization, and Writing—Review & Editing, R.Y.; Writing—Review & Editing, M.S.H.; Writing—Review & Editing, J.H.; Supervision & Project administration, M.A.-S.; Supervision & Project administration, J.V.M.; Supervision & Project administration, M.B. All authors have read and agreed to the published version of the manuscript.

Funding: European Union Horizon 2020 research and innovation program under Grant Agreement No. 824311, the study was mainly developed within the framework of the ACHILES project.

Institutional Review Board Statement: Not applicable.

Informed Consent Statement: Not applicable.

Data Availability Statement: Not applicable.

Acknowledgments: This study was mainly developed within the framework of the ACHILES project, which has received the funding from the European Union Horizon 2020 research and innovation program under Grant Agreement No. 824311. The authors also wish to acknowledge Flanders Make for support to the MOBI research group.

Conflicts of Interest: The authors declare no conflict of interest.

References

1. Muratori, M.; Alexander, M.; Arent, D.; Bazilian, M.; Cazzola, P.; Dede, E.M.; Farrell, J.; Gearhart, C.; Greene, D.; Jenn, A.; et al. The rise of electric vehicles—2020 status and future expectations. *Prog. Energy* **2021**, *3*, 022002. [CrossRef]
2. Sanguesa, J.; Torres-Sanz, V.; Garrido, P.; Martinez, F.; Marquez-Barja, J. A Review on Electric Vehicles: Technologies and Challenges. *Smart Cities* **2021**, *4*, 372–404. [CrossRef]
3. Yoshio, M.; Brodd, R.J.; Kozawa, A. *Lithium-Ion Batteries*; Springer: Berlin/Heidelberg, Germany, 2009.
4. Bandhauer, T.M.; Garimella, S.; Fuller, T.F. A Critical Review of Thermal Issues in Lithium-Ion Batteries. *J. Electrochem. Soc.* **2011**, *158*, R1. [CrossRef]
5. He, J.; Youssef, R.; Hosen, S.; Akbarzadeh, M.; Van Mierlo, J.; Berecibar, M. A novel methodology to determine the specific heat capacity of lithium-ion batteries. *J. Power Sources* **2021**, *520*, 230869. [CrossRef]
6. Youssef, R.; Hosen, S.; He, J.; Jaguemont, J.; Akbarzadeh, M.; De Sutter, L.; Van Mierlo, J.; Berecibar, M. Experimental and Numerical Study on the Thermal Behavior of a Large Lithium-Ion Prismatic Cell With Natural Air Convection. *IEEE Trans. Ind. Appl.* **2021**, *57*, 6475–6482. [CrossRef]
7. Lin, J.; Liu, X.; Li, S.; Zhang, C.; Yang, S. A review on recent progress, challenges and perspective of battery thermal management system. *Int. J. Heat Mass Transf.* **2021**, *167*, 120834. [CrossRef]
8. Ma, S.; Jiang, M.; Tao, P.; Song, C.; Wu, J.; Wang, J.; Deng, T.; Shang, W. Temperature effect and thermal impact in lithium-ion batteries: A review. *Prog. Nat. Sci.* **2018**, *28*, 653–666. [CrossRef]
9. Youssef, R.; He, J.; Akbarzadeh, M.; Jaguemont, J.; De Sutter, L.; Berecibar, M.; Van Mierlo, J. Investigation of Thermal Behavior of Large Lithium-Ion Prismatic Cell in Natural Air Convection. In Proceedings of the 2020 9th International Conference on Renewable Energy Research and Application (ICRERA), Glasgow, UK, 27–30 September 2020; pp. 43–47. [CrossRef]

10. Feng, X.; Ren, D.; He, X.; Ouyang, M. Mitigating Thermal Runaway of Lithium-Ion Batteries. *Joule* **2020**, *4*, 743–770. [CrossRef]
11. Kim, J.; Oh, J.; Lee, H. Review on battery thermal management system for electric vehicles. *Appl. Therm. Eng.* **2019**, *149*, 192–212. [CrossRef]
12. Jaguemont, J.; Van Mierlo, J. A comprehensive review of future thermal management systems for battery-electrified vehicles. *J. Energy Storage* **2020**, *31*, 101551. [CrossRef]
13. Zhang, X.; Li, Z.; Luo, L.; Fan, Y.; Du, Z. A review on thermal management of lithium-ion batteries for electric vehicles. *Energy* **2022**, *238*, 121652. [CrossRef]
14. Thakur, A.K.; Prabakaran, R.; Elkadeem, M.; Sharshir, S.W.; Arıcı, M.; Wang, C.; Zhao, W.; Hwang, J.-Y.; Saidur, R. A state of art review and future viewpoint on advance cooling techniques for Lithium–ion battery system of electric vehicles. *J. Energy Storage* **2020**, *32*, 101771. [CrossRef]
15. Liu, C.; Xu, D.; Weng, J.; Zhou, S.; Li, W.; Wan, Y.; Jiang, S.; Zhou, D.; Wang, J.; Huang, Q. Phase Change Materials Application in Battery Thermal Management System: A Review. *Materials* **2020**, *13*, 4622. [CrossRef] [PubMed]
16. Zhang, Z.; Wei, K. Experimental and numerical study of a passive thermal management system using flat heat pipes for lithium-ion batteries. *Appl. Therm. Eng.* **2020**, *166*. [CrossRef]
17. Kong, D.; Peng, R.; Ping, P.; Du, J.; Chen, G.; Wen, J. A novel battery thermal management system coupling with PCM and optimized controllable liquid cooling for different ambient temperatures. *Energy Convers. Manag.* **2020**, *204*, 112280. [CrossRef]
18. Liu, H.; Wei, Z.; He, W.; Zhao, J. Thermal issues about Li-ion batteries and recent progress in battery thermal management systems: A review. *Energy Convers. Manag.* **2017**, *150*, 304–330. [CrossRef]
19. Yang, W.; Zhou, F.; Zhou, H.; Wang, Q.; Kong, J. Thermal performance of cylindrical lithium-ion battery thermal management system integrated with mini-channel liquid cooling and air cooling. *Appl. Therm. Eng.* **2020**, *175*, 115331. [CrossRef]
20. Deng, Y.; Feng, C.; Jiaqiang, E.; Zhu, H.; Chen, J.; Wen, M.; Yin, H. Effects of different coolants and cooling strategies on the cooling performance of the power lithium ion battery system: A review. *Appl. Therm. Eng.* **2018**, *142*, 10–29. [CrossRef]
21. Xia, G.; Cao, L.; Bi, G. A review on battery thermal management in electric vehicle application. *J. Power Sources* **2017**, *367*, 90–105. [CrossRef]
22. Yang, X.; Yan, Y.Y.; Mullen, D. Recent developments of lightweight, high performance heat pipes. *Appl. Therm. Eng.* **2012**, *33*, 1–14. [CrossRef]
23. Zou, H.; Wang, W.; Zhang, G.; Qin, F.; Tian, C.; Yan, Y. Experimental investigation on an integrated thermal management system with heat pipe heat exchanger for electric vehicle. *Energy Convers. Manag.* **2016**, *118*, 88–95. [CrossRef]
24. Wang, Q.; Jiang, B.; Xue, Q.; Sun, H.; Li, B.; Zou, H.; Yan, Y. Experimental investigation on EV battery cooling and heating by heat pipes. *Appl. Therm. Eng.* **2015**, *88*, 54–60. [CrossRef]
25. Volkswagen E-Up! Air Cooled Battery, Featured in ViaVision Magazine | My Electric Car Forums n.d. Available online: https://www.myelectriccarforums.com/volkswagen-e-up-air-cooled-battery-featured-in-viavision-magazine/ (accessed on 17 June 2021).
26. Zhao, G.; Wang, X.; Negnevitsky, M.; Zhang, H. A review of air-cooling battery thermal management systems for electric and hybrid electric vehicles. *J. Power Sources* **2021**, *501*, 230001. [CrossRef]
27. Numerical Investigation of Thermal Behaviors in Lithium-Ion Battery Stack Discharge | Elsevier Enhanced Reader n.d. Available online: https://reader.elsevier.com/reader/sd/pii/S0306261914007053?token=28CCD1C09F370C552B30C563E03D4360789A389587DCD03455AF641EB4997591FB5E9493D6F3F3F6629A58C1933B719B&originRegion=eu-west-1&originCreation=20210623085040 (accessed on 23 June 2021).
28. Akinlabi, A.H.; Solyali, D. Configuration, design, and optimization of air-cooled battery thermal management system for electric vehicles: A review. *Renew. Sustain. Energy Rev.* **2020**, *125*, 109815. [CrossRef]
29. Na, X.; Kang, H.; Wang, T.; Wang, Y. Reverse layered air flow for Li-ion battery thermal management. *Appl. Therm. Eng.* **2018**, *143*, 257–262. [CrossRef]
30. Chen, K.; Song, M.; Wei, W.; Wang, S. Structure optimization of parallel air-cooled battery thermal management system with U-type flow for cooling efficiency improvement. *Energy* **2018**, *145*, 603–613. [CrossRef]
31. Wang, S.; Li, K.; Tian, Y.; Wang, J.; Wu, Y.; Ji, S. Improved thermal performance of a large laminated lithium-ion power battery by reciprocating air flow. *Appl. Therm. Eng.* **2019**, *152*, 445–454. [CrossRef]
32. Xie, J.; Ge, Z.; Zang, M.; Wang, S. Structural optimization of lithium-ion battery pack with forced air cooling system. *Appl. Therm. Eng.* **2017**, *126*, 583–593. [CrossRef]
33. Alipour, M.; Hassanpouryouzband, A.; Kizilel, R. Investigation of the Applicability of Helium-Based Cooling System for Li-Ion Batteries. *Electrochem* **2021**, *2*, 135–148. [CrossRef]
34. Al-Zareer, M.; Dincer, I.; Rosen, M.A. Electrochemical modeling and performance evaluation of a new ammonia-based battery thermal management system for electric and hybrid electric vehicles. *Electrochimica Acta* **2017**, *247*, 171–182. [CrossRef]
35. Siddique, A.R.M.; Mahmud, S.; Van Heyst, B. A comprehensive review on a passive (phase change materials) and an active (thermoelectric cooler) battery thermal management system and their limitations. *J. Power Sources* **2018**, *401*, 224–237. [CrossRef]
36. Souayfane, F.; Fardoun, F.; Biwole, P.-H. Phase change materials (PCM) for cooling applications in buildings: A review. *Energy Build.* **2016**, *129*, 396–431. [CrossRef]
37. Landini, S.; Leworthy, J.; O'donovan, T. A Review of Phase Change Materials for the Thermal Management and Isothermalisation of Lithium-Ion Cells. *J. Energy Storage* **2019**, *25*. [CrossRef]

38. Bose, P.; Amirtham, V.A. A review on thermal conductivity enhancement of paraffinwax as latent heat energy storage material. *Renew. Sustain. Energy Rev.* **2016**, *65*, 81–100. [CrossRef]
39. Verma, A.; Shashidhara, S.; Rakshit, D. A comparative study on battery thermal management using phase change material (PCM). *Therm. Sci. Eng. Prog.* **2019**, *11*, 74–83. [CrossRef]
40. Hussain, A.; Abidi, I.H.; Tso, C.Y.; Chan, K.; Luo, Z.; Chao, C.Y. Thermal management of lithium ion batteries using graphene coated nickel foam saturated with phase change materials. *Int. J. Therm. Sci.* **2018**, *124*, 23–35. [CrossRef]
41. Li, Y.; Du, Y.; Xu, T.; Wu, H.; Zhou, X.; Ling, Z.; Zhang, Z. Optimization of thermal management system for Li-ion batteries using phase change material. *Appl. Therm. Eng.* **2018**, *131*, 766–778. [CrossRef]
42. Lyu, P.; Huo, Y.; Qu, Z.; Rao, Z. Investigation on the thermal behavior of Ni-rich NMC lithium ion battery for energy storage. *Appl. Therm. Eng.* **2020**, *166*, 114749. [CrossRef]
43. Zhang, X.; Liu, C.; Rao, Z. Experimental investigation on thermal management performance of electric vehicle power battery using composite phase change material. *J. Clean. Prod.* **2018**, *201*, 916–924. [CrossRef]
44. Li, Y.; Wei, Z.; Xiong, B.; Vilathgamuwa, D.M. Adaptive Ensemble-Based Electrochemical-Thermal-Degradation State Estimation of Lithium-ion Batteries. *IEEE Trans. Ind. Electron.* **2021**, 1. [CrossRef]
45. Wu, J.; Wei, Z.; Liu, K.; Quan, Z.; Li, Y. Battery-Involved Energy Management for Hybrid Electric Bus Based on Expert-Assistance Deep Deterministic Policy Gradient Algorithm. *IEEE Trans. Veh. Technol.* **2020**, *69*, 12786–12796. [CrossRef]
46. Youssef, R.; Hosen, S.; He, J.; Jaguemont, J.; De Sutter, L.; Van Mierlo, J.; Berecibar, M. Effect analysis on performance enhancement of a novel and environmental evaporative cooling system for lithium-ion battery applications. *J. Energy Storage* **2021**, *37*, 102475. [CrossRef]
47. Shah, S.S.; Shaikh, M.N.; Khan, M.Y.; Alfasane, A.; Rahman, M.M.; Aziz, A. Present Status and Future Prospects of Jute in Nanotechnology: A Review. *Chem. Rec.* **2021**, *21*, 1631–1665. [CrossRef]
48. Chattopadhyay, D.; Patel, B.H.; Chattopadhyay, D.P.; Patel, B.H. Imparting water repellency to jute fabric by nano paraffin and nano copper colloid treatment Functional finishing of textiles with natural resources. View project Imparting water repellency to jute fabric by nano paraffin and nano copper colloid treatment. n.d. Available online: https://www.researchgate.net/publication/308334126 (accessed on 15 November 2021).
49. Jahan, A. The environmental and economic prospects of jute with a connection to social factors for achieving Sustainable Development. Independent Master thesis Advanced level, Uppsala University, Uppsala, Sweden, 2019.
50. El Nemr, A. From natural to synthetic fibers. In *Textiles: Types, Uses and Production Methods*; Nova Science Publishers, Inc.: Hauppauge, NY, USA, 2012.
51. Brend, D.E.C.; Eva, B.; Ignacio, M.; Pablo, D.; Jaime, G.; Valérie, D.E.V. Structural influence on the thermal behavior of jute fabrics treated with pcms. n.d. pp. 43–46. Available online: http://textile.webhost.uoradea.ro/Annals/Vol%20XIX-No%201-2018/Textile/Art.%20no.%20290-pag.%2043-46.pdf (accessed on 20 January 2022).
52. Al-Sulaiman, F. Evaluation of the performance of local fibers in evaporative cooling. *Energy Convers. Manag.* **2002**, *43*, 2267–2273. [CrossRef]
53. Feng, X.; Ouyang, M.; Liu, X.; Lu, L.; Xia, Y.; He, X. Thermal runaway mechanism of lithium ion battery for electric vehicles: A review. *Energy Storage Mater.* **2018**, *10*, 246–267. [CrossRef]
54. Babapoor, A.; Azizi, M.; Karimi, G. Thermal management of a Li-ion battery using carbon fiber-PCM composites. *Appl. Therm. Eng.* **2015**, *82*, 281–290. [CrossRef]
55. Yuen, M.; Chen, L. Heat-transfer measurements of evaporating liquid droplets. *Int. J. Heat Mass Transf.* **1978**, *21*, 537–542. [CrossRef]
56. Moffat, R.J. Describing the uncertainties in experimental results. *Exp. Therm. Fluid Sci.* **1988**, *1*, 3–17. [CrossRef]

energies

MDPI

Article

Polarization Voltage Characterization of Lithium-Ion Batteries Based on a Lumped Diffusion Model and Joint Parameter Estimation Algorithm

Bizhong Xia, Bo Ye * and Jianwen Cao

Division of Advanced Manufacturing, Shenzhen International Graduate School, Tsinghua University, Shenzhen 518055, China; xiabz@sz.tsinghua.edu.cn (B.X.); jw-cao18@mails.tsinghua.edu.cn (J.C.)
* Correspondence: yb18@mails.tsinghua.edu.cn; Tel.: +86-186-1385-9890

Abstract: Polarization is a universal phenomenon that occurs inside lithium-ion batteries especially during operation, and whether it can be accurately characterized affects the accuracy of the battery management system. Model-based approaches are commonly adopted in studies of the characterization of polarization. Towards the application of the battery management system, a lumped diffusion model with three parameters was adopted. In addition, a joint algorithm composed of the Particle Swarm Optimization algorithm and the Levenberg-Marquardt method is proposed to identify model parameters. Verification experiments showed that this proposed algorithm can significantly improve the accuracy of model output voltages compared to the Particle Swarm Optimization algorithm alone and the Levenberg-Marquardt method alone. Furthermore, to verify the real-time performance of the proposed method, a hardware implementation platform was built, and this system's performance was tested under actual operating conditions. Results show that the hardware platform is capable of realizing the basic function of quantitative polarization voltage characterization, and the updating frequency of relevant parameters can reach 1 Hz, showing good real-time performance.

Keywords: battery polarization; lumped diffusion model; parameter identification; particle swarm optimization; Levenberg-Marquardt method

Citation: Xia, B.; Ye, B.; Cao, J. Polarization Voltage Characterization of Lithium-Ion Batteries Based on a Lumped Diffusion Model and Joint Parameter Estimation Algorithm. *Energies* **2022**, *15*, 1150. https://doi.org/10.3390/en15031150

Academic Editor: Alon Kuperman

Received: 12 January 2022
Accepted: 31 January 2022
Published: 4 February 2022

1. Introduction

Due to the high operating voltage and high energy density of lithium-ion batteries, both grid and off-grid applications using lithium-ion batteries have gained a lot of attention [1], especially in the field of electric vehicles [2]. In addition to the advancement of battery manufacturing technology, a sufficiently accurate and agile battery management system (BMS) is important for the further application of lithium-ion batteries [3].

During battery charge/discharge cycles, the external characteristics of batteries behave as a time-varying nonlinear system due to the nonlinear relationship between electric potential versus State of Charge (*SOC*), as well as the voltage drop due to various polarization phenomena (e.g., ohmic polarization, activation polarization, concentration polarization) [4]. It is important to clarify that polarization is a universal phenomenon that occurs inside the lithium-ion battery during operation, with different types of polarization occurring at different locations inside the battery. Furthermore, the relative proportion of each type of polarization varies with the change of external excitation [5]. Polarization hinders lithium intercalation and deintercalation kinetics, leading to a decline in energy efficiency and performance of the battery [6]. Therefore, polarization is of concern during the entire phase from battery material development to the end-of-life phase of batteries. In addition, higher charge/discharge rates, extremely low ambient temperatures, and increased cycles, which are common scenarios encountered during the use of power batteries, can all lead to increased battery polarization [7].

Polarization is concerned in different stages of battery development and application.

(i.) In the cell design phase, improvements are made for the internal composition of the cell. Zheng et al. pointed out that polarization leads to uneven distribution of electrode active material, which causes performance decline for lithium-ion batteries [6]. The separator is a key component inside a lithium-ion battery cell. Feng et al. redeveloped the separator material to mitigate the polarization phenomenon during battery operation, which enhanced the power performance and cycling stability of the battery [8]. Kim et al. improved the anode materials for lithium-ion batteries to mitigate the polarization phenomenon of lithium-ion batteries during operation [9]. In [10], Shi et al. prepared an electrical conductive graphene nanosheet with hybrid lithium titanate nanoparticles dispersed on it as an anode, which shortened the ion transport path, greatly improving the ion and electron transport efficiency at the particle/electrolyte coupling interface, and remarkably reducing the charge transfer impedance, and improving the battery cycling performance under large rate conditions. The DRT (Distribution of Relaxation Times) method has been used to analyze the impedance spectrum of lithium-ion batteries in the frequency domain to precisely characterize various types of polarization, and used to develop new cathode materials [11]. Song et al. developed a ferroelectric polyvinylidene difluoride (PVDF) polymer as a binder material and demonstrated by the galvanostatic intermittent titration technique (GITT) measurement and in situ galvanostatic electrochemical impedance spectroscopy (GS-EIS) analysis that this new material can significantly reduce the lithium-ion diffusion impendence as a binder inside cells, compared to the paraelectric PVDF binder material, thereby improving the battery performance under large rate conditions. This provided new inspiration for the design of high-performance lithium-ion batteries [12].

(ii.) Polarization has been considered an influential factor in numerous studies related to the thermal management system of lithium-ion batteries. In [13], the respective proportion of ohmic internal resistance and polarization internal resistance under different discharge rate conditions were explored to summarize the contribution of polarization in the accumulation of internal battery temperature at the early stage of thermal runaway, which provides a theoretical basis for safer battery design. As discussed in [14,15], the heat of polarization is the main component of the internal heat production of the cell and consists of four parts: ohmic heat, polarization heat, reaction heat, and side-reaction heat. Taheri P, Mansouri A et al. developed a two-dimensional analytical model of the lithium-ion battery, and a concentration-independent polarization voltage was derived to explore the application for battery thermal management through the thermal coupling model [16]. Goonetilleke D. et al. found that increasing the ambient temperature increased the reaction rate inside the battery and reduced polarization inside battery cells [17]. In [18], the optimal alternating current (AC) frequency was determined based on the Thevenin-thermal coupling model in the frequency domain, and an internal heating strategy based on a constant polarization voltage at low ambient temperatures was developed, which achieved a good balance between cell aging and heating efficiency.

(iii.) State of Charge and State of Health (SOH) are two very important state parameters in BMS, and the influence of polarization on estimation accuracy cannot be ignored. Li et al. pointed out that polarization resistance contributes significantly to the battery capacity decay, so the accurate characterization of polarization is important for developing the estimation method of SOH [19]. Marino C. et al. quantified the electrode polarization resistance and established a functional relationship with the external current excitation at different cycle rates to estimate the aging state of the battery to determine battery failure [20]. SOH decays with increasing cycle number, and Xia and Chen et al. defined the concept of Degree of Polarization (DOP) to correct SOH estimation results to improve the accuracy of *SOC* estimation [21].

(iv.) In battery charging technology, polarization is also considered a controlled variable. Zhang et al. used the voltage drop from polarization as a controlled variable in the charging process to make a balance between charging time and temperature rise and combined it with a Genetic Algorithm (GA) to find the optimal charging current trajectory. Verification experiments of battery aging proved that this charging method has similar capacity retention to the 0.5C Constant current Constant voltage (CC-CV) charging method, but with improved charging efficiency [22,23].

The main characterization methods of battery polarization can be roughly divided into experiment-based methods and model-based methods. For model-based methods, the electrochemical model and equivalent circuit model are two of the most common models adopted.

The experiment-based method probes the polarization characteristics of a battery cell using invasive or noninvasive methods through physical and chemical probing techniques. Ex-situ X-ray diffraction technology is adopted to observe polarization behavior during the initial lithiation process in the oxide of WO_3 and derivatives, and polarization arises from conversion reaction alleviated using the WO_{3-x} oxide compared to the WO_3 oxide, due to higher electrical conductivity [9]. Xu et al. used synchrotron X-ray tomography analysis and microstructure-resolved computational modeling to analyze the morphological defects of electrodes from multiple spatial scales, combined with battery frequency domain analysis, to explore the correlation between polarization and morphological defects of electrode structures [24]. The GITT method uses the subtraction of the quasi-open-circuit voltage (QOCV) from the closed-circuit voltage (CCV) as a characterization of electrode polarization [9,12,25]. Noelle D.J. et al. imposed abuse conditions on a battery and direct current (DC) internal resistance analysis was used to quantitatively characterize the ohmic and polarization resistance during thermal runaway [13]. Furthermore, a destructive intrusion test was conducted to investigate the relationship between electrolyte concentration and the polarization internal resistance of the battery in Noelle's research.

An electrochemical model is a common tool in battery modelling and simulation [26]. Nyman A. et al. adopted the Pseudo-2D model to locate and quantify the various types of polarization occurring inside the cell, but in this study only the polarization occurrence at specific *SOC* levels (40% and 80%) was investigated [5]. Huang et al. developed a coupled electrochemical-thermal model based on a one-dimensional electrochemical model with COMSOL Multiphysics software to study the effect of discharge rate on heat production of the cell, including the heat of polarization [14]. Li et al. improved the traditional Single Particle Model (SPM) and identified the model parameters by excitation response analysis and conducted experiments to confirm the model's performance at large charge/discharge C-rate conditions (up to 4C discharge condition for $LiCoO_2$ batteries, and up to 5C discharge for $LiFePO_4$ batteries) [3]. Yan et al. developed a 3D model that preserves the effect of inhomogeneous geometrical characteristics on global polarization as well as the local polarization in the electrodes, which can provide more information on the polarization characteristics of the real cell compared to the above-mentioned Pseudo-2D model [27]. Qiu et al. investigated the effects of ambient temperature, charge/discharge rate, and the number of cycles on the polarization characteristics of batteries based on an electrochemical-thermal coupling model [7]. In [16], for planar electrodes of pouch-type lithium-ion batteries, an analytical model was established and the concentration-independent polarization expression was derived. In addition, the potential and current distribution of the electrodes during the constant-current discharge process were studied, but limited only to the constant-current discharge condition. However, for the Pseudo-2D model, SPM, and their derivatives, parameter identification of the models often requires customization of specific cycle data and cannot be based on real battery cycle data. As in [28], the parameters in improved SPM are identified by frequency response analysis. As well as in [29], seven model parameters are to be identified after rederivation of the original SPM, those seven parameters are divided into three groups, and different customized cycle conditions are designed to identify these three groups of model parameters. In addition, the control equations of this type of model are of very high order and require high calculation power, which is not suitable for real-time applications [30].

An equivalent circuit model which consists of passive components such as resistors, capacitors, inductors, and W-resistors is widely used in BMS [4,18,19,31], but fails to reveal the essential information since passive components do not directly correspond to the internal components or reaction processes inside real batteries [32]. Parameter identification methods for ECMs can be divided into two categories: the time-domain approach and the frequency-domain approach. For example, Li X. et al. used a second-order equivalent circuit model and correlated the small time-constant RC component and the activation polarization and the large time constant RC component with the concentration polarization [19]. The problem is that the two RC components couple with each other, and such a simple distinction does not strictly distinguish between these two types of polarization in a physical sense and on a time scale when using time-domain identification methods. As for the frequency domain approach, the identification of equivalent circuit model parameters using electrochemical impedance spectra is a common approach [11,33–35]. In addition, the DRT approach can distinguish more precisely voltage drops caused by each type of polarization under the frequency domain, and then assign a reasonable time constant to each polarization loss. Furthermore, Zhou et al. combined the DRT method and a physics-based impedance model to separate the solid-phase diffusive polarization voltage drop and the liquid-phase diffusive polarization voltage drop [36]. However, the frequency domain analysis method requires a large input excitation frequency span, and the current BMS on board is not able to meet the requirement. In addition, electrochemical impedance spectroscopy involves a specific *SOC* level [11], and continuous identification during battery cycling is not possible.

Characterizing polarization by changes in cell potential under battery operating conditions is a common method [17], in which the directly measurable cell terminal voltage is used as a measure of characterization accuracy. In this paper, we propose a quantitative battery polarization characterization tool based on a lumped diffusion model (LDM) [32,37] with a joint parameter identification algorithm consisting of the Particle swarm optimization (PSO) algorithm and Levenberg-Marquardt (L-M) method and demonstrate the effectiveness of the proposed method through accuracy verification experiments. Furthermore, a hardware platform was built to demonstrate that the proposed method is capable of quantitative real-time characterization of three types of polarization voltage drops, and has a good tracking performance for the terminal voltage. Compared with the equivalent circuit model, this model preserves the internal physicochemical processes of the battery. Compared with other electrochemical models, this model has fewer parameters to be identified and has good prospects for online applications. The remainder of this paper is organized as follows. First, LDM and the joint parameter identification algorithm are introduced, then the accuracy and effectiveness of the proposed method are verified based on two sets of real-world testing data. Finally, the implementation and results analysis of the hardware platform for online applications are introduced.

2. Battery Modeling and Joint Algorithm Scheme

2.1. Battery Model Description

A Pseudo-2D model was proposed by Doyle et al. in 1993 to describe the behavior of lithium-ion batteries based on porous electrode theory and concentrated solution theory [38]. The physicochemical equations that constitute the Pseudo-2D model are the electrochemical reaction process at the critical surface between the active particle surface and the electrolyte solution in both electrode regions, the solid-phase diffusion process, the liquid-phase diffusion process, the solid-phase ohmic resistance, and the liquid-phase ohmic resistance [39,40]. A single particle model was proposed by B. Haran in 1998, which was obtained by further simplifying the assumptions based on the Pseudo-2D model, neglecting the differences in the liquid-phase concentration distribution in the thickness dimension of the electrode sheet, so that one spherical particle can be used to represent the whole electrode [37,41].

The model adopted in this study is LDM, which is a further simplified version of the SPM and Pseudo-2D model, and no longer distinguishes the difference in the spatial distribution of the same type of polarization at different locations of the cell while retaining the control equations reflecting the real physicochemical processes inside the cell. The voltage drop under the battery operating condition is attributed to three types of polarization impedance: ohmic polarization impedance, activation polarization impedance, and concentration polarization impedance. The battery terminal voltage E_{cell} can be obtained by:

$$E_{cell} = E_{OCP}(SOC) + \eta_{ohm} + \eta_{act} + \eta_{con} \tag{1}$$

where $E_{OCP}(SOC)$ is the open-circuit voltage, which is a function of the average SOC of electrodes. η_{ohm}, η_{act}, and η_{con} are ohmic polarization overpotential, activation polarization overpotential, and concentration polarization overpotential, respectively. SOC at a certain moment $SOC(t)$ can be obtained by the ampere-hour integral method:

$$SOC(t) = SOC(t_0) + \frac{\int_{t_0}^{t} i(t)dt}{Q_n} \tag{2}$$

where Q_n is the battery capacity, and $i(t)$ is the instantaneous current. The time-discrete expression of the above equation takes the form:

$$SOC(k) = SOC(0) + \frac{\sum_{t=0}^{t=k} I(t) \cdot T_s}{Q_n} \tag{3}$$

where T_s is the sampling period, and $I(t)$ is the applied current at time point t. Ohmic polarization overpotential is defined as:

$$\eta_{ohm} = R_{ohm} \cdot I \tag{4}$$

where R_{ohm} is the ohmic resistance. Under the lithium deintercalation/intercalation kinetics assumption on the electrode particle surface in the Pseudo-2D model, the relationship between current density, lithium concentration, and intercalation overpotential is given by the Butler-Volmer formula [42] is expressed in the form:

$$\frac{I}{I_{1C}} = J_0\left(\exp\left(\frac{(1-\alpha)F}{RT}\eta_{act}\right) - \exp\left(\frac{-\alpha F}{RT}\eta_{act}\right)\right) \tag{5}$$

where J_0 is the dimensionless charge exchange current, used to describe the charge transfer reaction rate on the surface of both electrodes, I_{1C} is the value of the applied current taken at 1 C rate, which is related to the cell capacity, R is the molar gas constant, F is the Faraday constant, T is the reference temperature, and α is the charge transfer coefficient. In LDM, the difference in the spatial distribution of current density is neglected and the charge transfer coefficients of both electrodes take the value of 0.5. The overpotential of the two electrodes is considered as a whole, and a single equation expresses the reaction overpotential of the whole cell about the input current excitation:

$$\eta_{act} = \frac{2RT}{F}\operatorname{asinh}\left(\frac{I}{2J_0 I_{1C}}\right) \tag{6}$$

In this model, the electrode is idealized as a spherical particle, and the electrode local State of Charge $iSOC(X,t)$ varies with both time t and the dimensionless spatial variable X. The partial differential control equation is obtained by reformulation of Fick's law and solved using a spherically symmetric solution, expressed as:

$$\tau\frac{\partial iSOC}{\partial t} = -\frac{\partial}{\partial X}\left(-\frac{\partial}{\partial X}(iSOC)\right) \tag{7}$$

where τ is the diffusion time constant in s, and $X \in [0,1]$ denotes the dimensionless spatial variable in particle size scale. For this partial differential equation, the initial value condition is

$$iSOC(X,0) = SOC_0, \quad t = 0 \tag{8}$$

Both the left and right side boundary conditions are Neuman boundary conditions:

$$\frac{\partial}{\partial X}(iSOC(0,t)) = 0, \quad X = 0 \tag{9}$$

$$\frac{\partial}{\partial X}(iSOC(1,t)) = \frac{\tau I}{N_{shape} Q_n}, \quad X = 1 \tag{10}$$

where, for spherical particles, the dimension number N_{shape} takes the value of 3. When State of Charge of the electrode particle surface is defined as SOC_{surf}:

$$SOC_{surf}(t) = iSOC(1,t), \quad X = 1 \tag{11}$$

The electrode particle average State of Charge SOC_{ave} reflected the molarity of lithium ions inside the particle, which can be obtained by integrating the local State of Charge $iSOC$ over the particle volume with:

$$SOC_{ave}(t) = \frac{\int_0^1 iSOC(X,t) \cdot 4\pi X^2 dX}{\int_0^1 4\pi X^2 dX} \tag{12}$$

Therefore, the concentration polarization overpotential is expressed as:

$$\eta_{con} = E_{OCP}\left(SOC_{surf}\right) - E_{OCP}(SOC_{ave}) \tag{13}$$

After introducing the above equation, the battery terminal voltage E_{cell} can be reformulated as:

$$E_{cell} = E_{OCP}\left(SOC_{surf}\right) + \eta_{ohm} + \eta_{act} \tag{14}$$

2.2. *Joint Parameter Estimation Algorithm Design*

Three parameters can be identified in the above-mentioned LDM, namely, the ohmic resistance R_{ohm} associated with ohmic polarization, the dimensionless charge exchange current J_0 associated with activation polarization, and the diffusion time constant τ associated with concentration polarization. The objective of the parameter identification algorithm is to find the optimal solution of the state parameters by solving for the minimum of the objective error function so that the model output voltage is as close as possible to the real-world terminal voltage.

Many algorithms were adopted for the identification of model parameters, which in general can be divided into gradient-free methods (i.e., PSO algorithm) and gradient methods (i.e., L-M method). The fitting accuracy of the PSO algorithm is often inferior to that of the L-M method, while the initial value of the L-M method is crucial to determine will fall into a local optimum [43]. Therefore, in this case, we first used the PSO algorithm for the prediction of the model parameters and used the identification results as the initial values of the L-M method to establish a joint algorithm for the accurate estimation of the parameters in LDM.

The PSO algorithm was proposed by Eberhart and Kennedy in 1995 [44]. The population contains a certain number of particles, each of which represents a possible solution. The initial positions of the particles are generally determined randomly. The fitness function associated with those model parameters to be optimized represents the distance of each particle from the optimal solution. The particle velocity determines the direction and distance of each particle's motion during each iteration. Based on this set of rules, particles in the population search for the optimal solution in the solution space.

In the PSO algorithm, the fitness function established based on the above LDM is defined as:

$$f = \text{abs}(u_t(k) - \hat{u}_t(k)) \tag{15}$$

where u_t is the real-world terminal voltage and $\hat{u}_t$ is the model output voltage.

In the three-dimensional search space $S \subseteq R^3$, the population contains P particles $[1,2, \ldots ,p, \ldots ,P]$ with the maximum iteration number G. The position vector $x_p = [x_{p,\tau}, x_{p,invJ0}, x_{p,\eta_{IR,1C}}] \in S$, and the velocity vector is $v_p = [v_{p,\tau}, v_{p,invJ0}, v_{p,\eta_{IR,1C}}] \in S$ for the pth particle. Note that to avoid a divide-by-zero error in the calculations, the dimensionless exchange current density J_0 is used in the calculations using its inverse $invJ_0$ for the operation and the ohmic resistance R_{ohm} is replaced using the ratio of ohmic overpotential at 1 C rate to 1 C rate current $\frac{\eta_{IR,1C}}{I_{1C}}$. Define history optimum $pbest_p$ as $[p_{p,\tau}, p_{p,invJ0}, p_{p,\eta_{IR,1C}}] \in S$ for the pth particle, and the global optimum particle $gbest$ as $[g_\tau, g_{invJ0}, g_{\eta_{IR,1C}}] \in S$ for the whole population, then the velocity and position update rules for a certain particle are:

$$\begin{cases} v_p^{kg+1} = \omega * v_p^{kg} + c_1 rand_1\left(pbest_p - x_p^{kg}\right) + c_2 rand_2\left(gbest - x_p^{kg}\right) \\ x_p^{kg+1} = x_p^{kg} + v_p^{kg+1} \end{cases} \tag{16}$$

where ω is the inertia weight, which functions to scale the feasible domain. c_1 and c_2 are the local learning factor and the global learning factor, respectively. $rand_1$ and $rand_2$ are random numbers uniformly distributed in (0,1). kg is the current iteration number. Setting the velocity and position bounds for each particle in the population to ensure that the current velocity and position are restricted to the preset range after each iteration:

$$v_p^{kg+1} = \begin{cases} LB_v & v_p^{kg+1} < LB_v \\ UB_v & v_p^{kg+1} > UB_v \end{cases} \tag{17}$$

$$x_p^{kg+1} = \begin{cases} LB_x & x_p^{kg+1} < LB_x \\ UB_x & x_p^{kg+1} > UB_x \end{cases} \tag{18}$$

When facing constrained optimization problems with multi-peak distribution, the following two improvements are applied to this case to avoid the identification algorithm falling into local optimum, and to improve the global search capability.

(i) The decreasing time-varying inertia weight $\omega_{kg} \in [\omega_{min}, \omega_{max}]$ is introduced, and $\omega_{kg} v^{kg}$ represents the momentum of particle motion in the population. In this case, the inertia weight decreases uniformly as the number of iterations increases. The purpose of this is to make sure that the particles have good global search ability at the beginning to avoid falling into a local optimum, and at the end of the iteration to facilitate local search and accelerate convergence, which achieves a good balance between convergence efficiency and global searchability.

$$\omega_{kg} = \omega_{max} + kg * \frac{\omega_{max} - \omega_{min}}{G} \tag{19}$$

Furthermore, the local learning factor and global learning factor are set as a function of the time-varying inertia weights, defined as:

$$\begin{cases} c_{1,kg} = k_1 * \left(1 - \omega_{kg}\right) \\ c_{2,kg} = k_2 * \left(1 - \omega_{kg}\right) \end{cases} \tag{20}$$

where k_1 and k_2 are the learning factor gain, generally take $k_2 > k_1$. $c_{1,kg}$ and gradually increase with the increase of the number of iterations. The motion of particles in the early stage is less influenced by the history and other particles to enhance the global searchability, while in the later stage particles are increasingly influenced by the history and other particles to accelerate convergence. Then, update the velocity and position of the pth particle by the rules below:

$$\begin{cases} v_p^{kg+1} = \omega_{kg} * v_p^{kg} + c_{1,kg} rand_1 \left(pbest_p - x_p^{kg} \right) + c_{2,kg} rand_2 \left(gbest - x_p^{kg} \right) \\ x_p^{kg+1} = x_p^{kg} + v_p^{kg+1} \end{cases} \tag{21}$$

(ii) In each iteration, a certain probability of particle mutation occurrence is set, and the position information of the selected particles is reassigned to ensure that some of the particles in the population can jump out of the local optimum trap and continue to search for other possible global optimum solutions.

$$\begin{cases} x_p^{kg}(dim) = LB_x(dim) + (UB_x(dim) - LB_x(dim)) * r_1 \\ dim = ceil(3 * r_2) \end{cases} \tag{22}$$

where $x_p^{kg}(dim) \subseteq [x_{p,\tau}^{kg}, x_{p,invJ0}^{kg}, x_{p,R}^{kg}]$ represent the randomly selected dimension among those three dimensions in the position vector for the pth particle, and r_1 and r_2 are random numbers uniformly distributed on the interval (0,1), respectively. Among the population, update $pbest_p$ and $gbest$ after each iteration by the equations below, until the maximum iteration number is met and the value of $gbest$ will be the final solution.

$$\begin{cases} pbest_p = \min\left[f_p, pbest_p\right] \\ gbest = \min\left[f_p, gbest\right] \end{cases} \tag{23}$$

L-M method [45,46] is a classical numerical solution method for solving the minimum of nonlinear equations, which retains both the stability of the steepest descent method and the fast convergence property of the Gaussian Newton method. In this paper, we set the parameter vector $\theta = [\tau, invJ_0, \eta_{IR,1C}]$, the original data is M sets of battery cycling data (u_m, i_m), $m = 1, 2, \ldots, M$, and the output voltage of LDM is $\hat{u}(i_m, \theta)$. Then, the error function of the L-M method is expressed as:

$$E(\alpha) = \sum_{m=1}^{M} e^2 = \sum_{m=1}^{M} (u_m - \hat{u}(i_m, \theta))^2 \tag{24}$$

where e is the voltage error for one set of data. The optimal model parameter vector θ is obtained by iteratively solving for the minimum of the above error function, and the iterative expression of the L-M method is:

$$\theta_{k+1} = \theta_k + \left[J^T{}_k \cdot J_k + \mu L \right]^{-1} \cdot J_k \cdot e(k) \tag{25}$$

where μ is the damping factor, L is the identity matrix, and k is the current iteration number. J_k is the Jacobian matrix:

$$J_k = \begin{bmatrix} \frac{\partial e_1}{\partial \theta[1]} & \frac{\partial e_1}{\partial \theta[2]} & \frac{\partial e_1}{\partial \theta[3]} \\ \frac{\partial e_2}{\partial \theta[1]} & \frac{\partial e_2}{\partial \theta[2]} & \frac{\partial e_2}{\partial \theta[3]} \\ \vdots & \vdots & \vdots \\ \frac{\partial e_M}{\partial \theta[1]} & \frac{\partial e_M}{\partial \theta[2]} & \frac{\partial e_M}{\partial \theta[3]} \end{bmatrix} \tag{26}$$

The iteration termination conditions are:

$$(k > k_{max}) OR (e(k) < \varepsilon) \tag{27}$$

where k_{max} is the maximum iteration number, and ε is the tolerance. The procedure of the proposed joint algorithm is summarized in Figure 1.

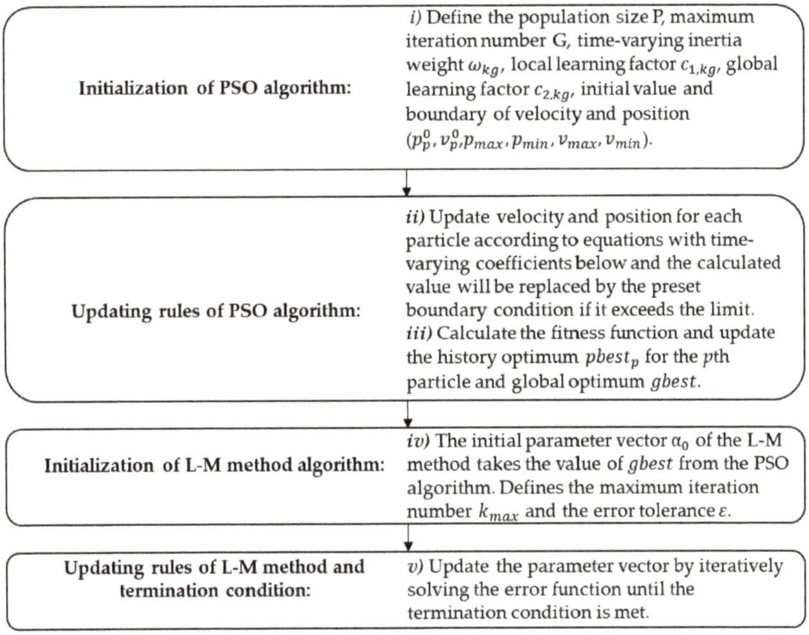

Figure 1. Major steps of the joint parameter identification algorithm.

3. Experiment Verification

3.1. Introduction of the Experiment Bench

A Samsung INR18650-33 G battery (Cell Business Division, Samsung SDI Co., Ltd., Yongin, Korea) with nominal capacity 2700 mAh (0.2C, 2.50 V discharge), nominal voltage 3.6 V, charging end voltage 4.1 V, and discharge cut-off voltage 2.5 V, was adopted as the sample battery. An Arbin BT-5HC (Arbin Instruments, LLC, College Station, TX, USA) with voltage range 0–5 V DC, and maximum current ±30 A was adopted for calibration, driving schedule simulation, and temperature monitoring. A Sanwood SC-80-CC-2 (Sanwood, Dongguan, China) thermal cabinet provided a controlled temperature and humidity environment during experiments. Arbin Mits Pro Software (v7 PV.202103) and MATLAB R2019b (MathWorks, Natick, MA, USA) were adopted for driving schedule file editing and application, data logging, model establishing, and data processing [47,48]. The configuration of the offline test bench is shown in Figure 2.

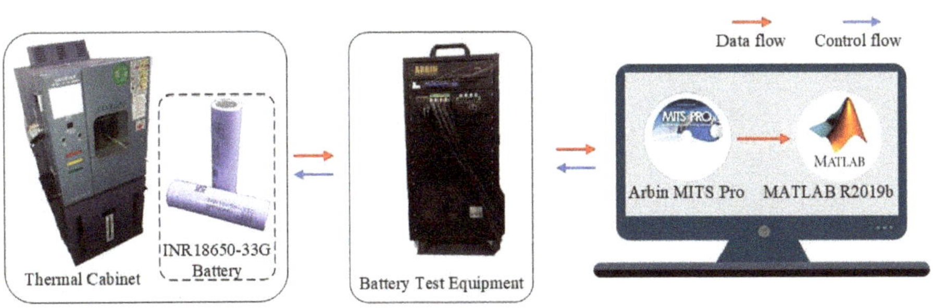

Figure 2. Outline of the test bench: offline calibration and schedule condition application.

In addition, to test the online application performance of the proposed method, a hardware platform for real-time quantitative characterization on polarization voltage of lithium-ion batteries was built, as shown in Figure 3. A Chroma DC electronic load

63206E and programmable DC power supply 62050H (Chroma Electronics (Shenzhen) Co., Ltd., Shenzhen, China) were adopted for schedule condition application. A battery fixture (homemade) was used to hold battery cells and connect the circuit. Batteries were connected in series in this case. A data acquisition board (homemade) was used to acquire current and voltage signals, where the sampling frequency was 1 Hz. MATLAB R2019b (MathWorks, Natick, MA, USA) was adopted for driving schedule file editing and application, data communication, logging, and data processing.

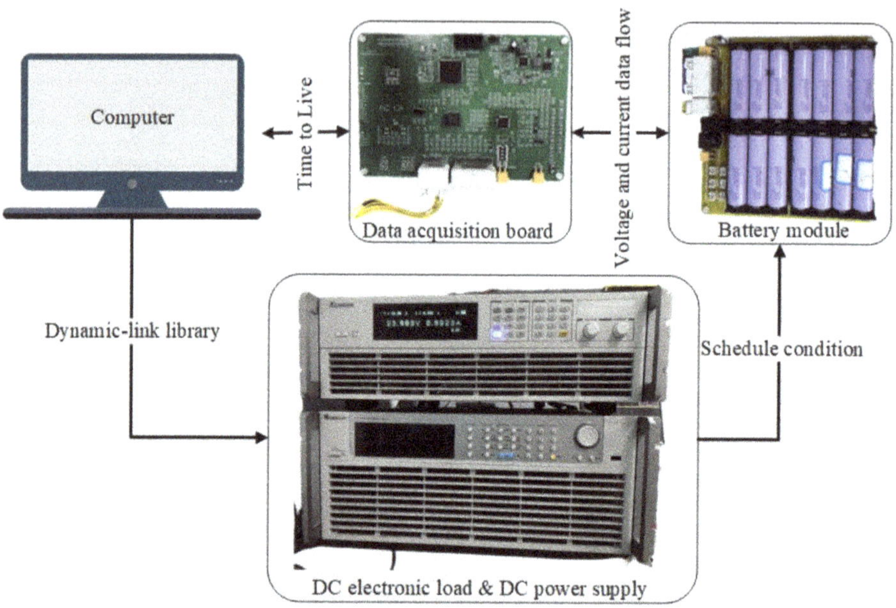

Figure 3. Outline of the hardware platform: online characterization of polarization voltage.

3.2. Experiment Configuration

The experimental flow was divided into two main parts: the offline validation of the proposed method and the online hardware implementation of the polarization characterization, as shown in Figure 4.

In preliminary work, the battery was calibrated for relevant parameters, including the actual capacity of the battery, *SOC-OCV* curve, and offline identified model parameters. The temperature dependence was not considered, and all experiments were conducted in the thermal cabinet at 25 °C and temperature fluctuations on the cell surface were monitored using thermocouples.

Battery capacity was obtained using standard capacity testing methods at 25 °C. The battery was first fully charged using the Constant current Constant voltage (CC-CV) method and rest for 2 h, then discharged to the lower cut-off voltage at 0.2 C constant rate (1 C is 2.7 A in this case). The above steps were repeated three times and the average of three discharge capacities was used as the exact value of the actual battery capacity. The actual capacity of the cell in this case was obtained by the above method is 2.5907 Ah.

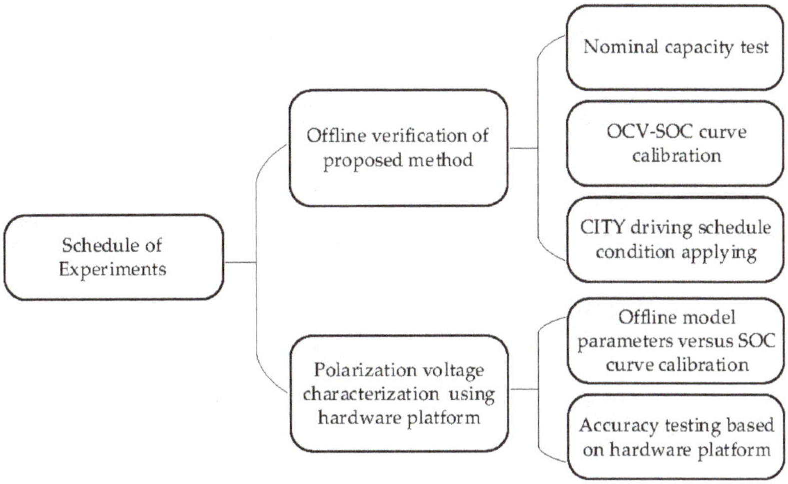

Figure 4. Schedule of experiments.

The *SOC-OCV* curve was obtained by a series of discharge pulses with different spacing to calibrate a series of points and then fit. The specific steps were: (1) use the CC-CV method to fully charge the battery, and rest for 2 h, record the end voltage as the open-circuit voltage (*OCV*) of the battery at 100% *SOC*; (2) discharge the battery to 98% *SOC* using a constant current rate of 0.2 C and rest for two hours, and record the end voltage as the open-circuit voltage at *SOC* level of 98%; (3) repeat step (2) and measure the open-circuit voltage of *SOC* at levels 95%, 90%, 80%, 70%, 60%, 50%, 40%, 30%, 20%, 10%, 8%, 5%, 3%, 1% and 0%. The *SOC-OCV* points obtained at different *SOC* levels were recorded and the relationship between *OCV* and *SOC* was described using a sixth-order polynomial. And the recorded result is shown in Table 1 and the recorded data points and fitted curve are shown in Figure 5.

$$OCV = \sum_{n=0}^{6} a_n SOC^n$$

Table 1. Polynomial coefficients of *SOC-OCV* curve.

Coefficients	a_1	a_2	a_3	a_4	a_5	a_6	a_7
Values	−5.0010	20.8142	−34.1273	27.8136	−11.4670	2.8176	3.2400

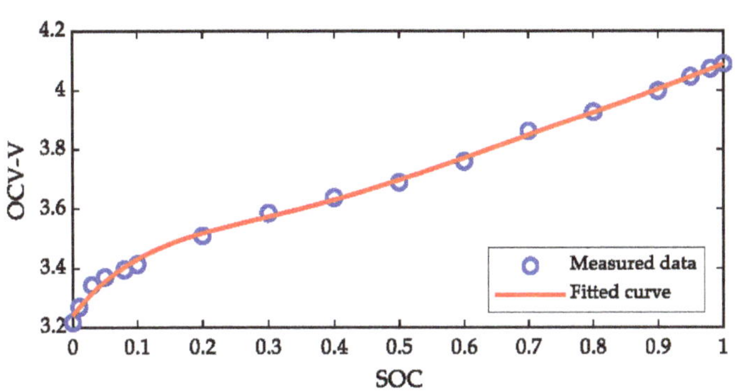

Figure 5. The measured data points and fitted curve of *SOC* versus *OCV*.

3.3. Offline Verification of Proposed Method

A CITY driving cycle was applied to the battery at an ambient temperature of 25 °C to obtain realistic battery cycle data to verify that the model could accurately describe the battery behavior or not. Part of the battery cycle data was intercepted as input data for the improved PSO algorithm, L-M method, and the joint algorithm, respectively. In this case, $SOC_0 = 0.69692$ was used as the starting point, and 1408 subsequent data points (i.e., one CITY cycle) were intercepted. The current curve is shown in Figure 6 as the input data for the parameter identification algorithm. The test data were fed into those three algorithms separately, and the identification results of the three model parameters were obtained, as shown in Table 2.

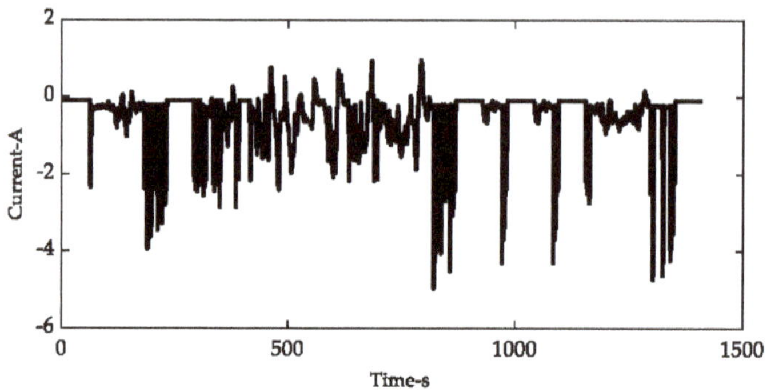

Figure 6. Current profile of the CITY operating condition test.

Table 2. Model parameter identification results of three types of algorithm.

Algorithm	τ	$invJ_0$ (Dimensionless)	$\eta_{IR,1C}$/mV
PSO Algorithm	17,447	2.6007	49.279
L-M Method	10,163	1.0765	69.642
Joint Estimation Algorithm	10,034	1.1412	69.62

To verify the superiority of the parameter identification algorithm proposed in this paper, the model parameters identified by PSO algorithm alone, L-M method alone, and the joint algorithm were substituted into the model, and the model output voltage was compared with the sampling voltage data, as shown in Figure 7. The mean error (ME) and root mean square error (RMSE) were used to describe the deviation between the model output voltage and the sampling voltage. The results are listed in Table 3. The joint algorithm significantly improved the fit accuracy of the model to the sampling voltage, in terms of voltage ME or voltage RMSE, compared to the PSO algorithm or L-M method alone for the intercepted cycle data. Specifically, the joint algorithm reduced voltage ME by 86.3% compared to the PSO algorithm and 83.2% compared to the L-M method, and reduced voltage RMSE by 77.1% compared to the PSO algorithm and 72.3% compared to the L-M method.

To further verify the effectiveness of the proposed method, the second set of test data ($SOC_0{}' = 0.621165046$, data length 1408) was fed into the model which adopted model parameters obtained from the first set of test data, and the error of the model output voltage from the sampling voltage was compared, as shown in Figure 8. The voltage RMSE was 0.0095320142 V and the voltage ME was 0.0082487339 V. Based on the above experiments, it can be concluded that: (1) the LDM describes the nonlinear characteristics of the battery under high dynamic driving cycles, and (2) the same model parameters are used in test data from different but adjacent *SOC* stages, which can still describe, relatively well, the terminal voltage characteristics of the battery, indicating that the proposed method reflects the real physicochemical processes inside the battery to a certain extent.

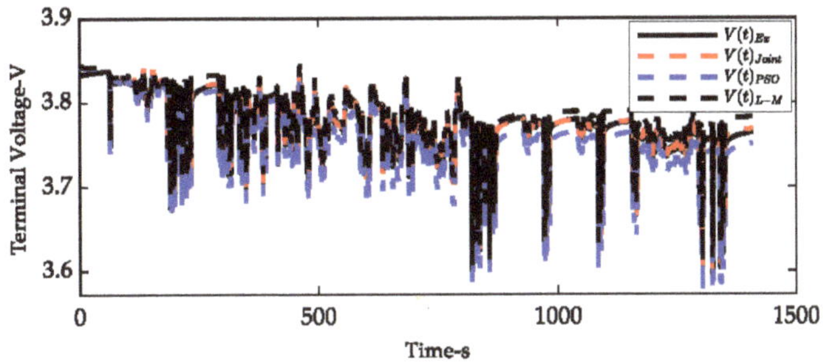

(a)

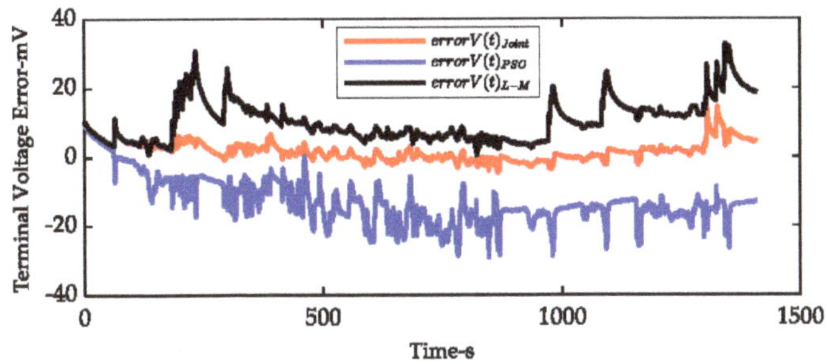

(b)

Figure 7. Comparison of LDM output voltage curve based on three types of parameter identification algorithm during the CITY test at 25 °C ambient temperature. (**a**) Voltage curve; (**b**)voltage error curve.

Table 3. Voltage error statistics based on model parameters from three algorithms.

Algorithm	PSO Algorithm	L-M Method	Joint Estimation Algorithm
Mean Voltage Error/V	0.0128780611	0.0104785487	0.0017621340
Voltage RMSE/V	0.0145556022	0.0120200028	0.0033290570

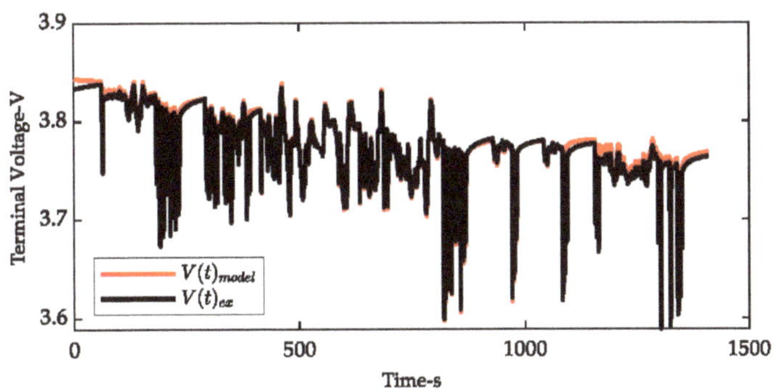

Figure 8. Model output voltage error curve based on the second group of testing data.

After solving the partial differential equation in LDM, the distribution of the local *SOC* inside the electrode particle can be obtained, as shown in Figure 9. At the particle size dimension taken as X = 1, the distribution of *SOC* on the electrode particle surface with time is obtained, as shown in Figure 10. Based on LDM, the variation curves of activation polarization voltage, ohmic polarization voltage, and concentration polarization voltage with time can be obtained, respectively, as shown in Figure 11. A conclusion can be drawn that the activation polarization and ohmic polarization respond quickly to the change of input current excitation; compared with the other two, and the concentration polarization responds more slowly to the change of input current excitation. When a non-zero current was applied to the cell system, a gradient in the concentration of the active material in the solid and liquid phases was gradually formed, and the voltage drop from concentration polarization gradually increased, while the time constant of this process was much larger than that of the ohmic and activation polarization. This conclusion is consistent with that obtained in [5] using the Pseudo-2D model under EUCAR driving conditions. The superposition of these three types of polarization phenomena is reflected in the output terminal voltage of LDM, which determines whether the proposed method can accurately describe the cell behavior. The results of the terminal-voltage accuracy comparison above justify the proposed method.

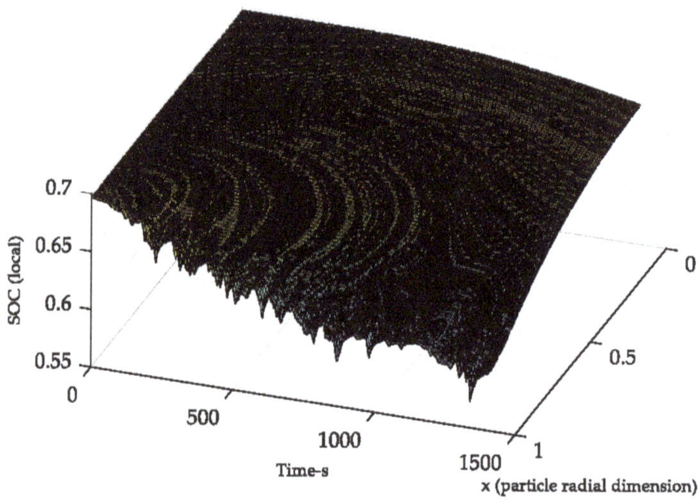

Figure 9. Temporal and spatial distribution of *iSOC* inside the electrode particle during the CITY test at 25 °C ambient temperature.

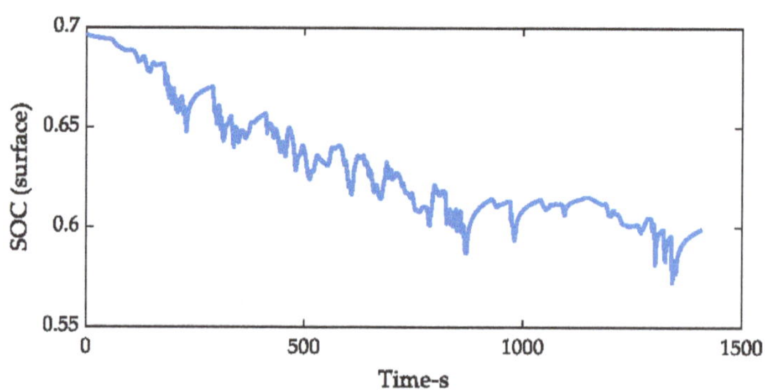

Figure 10. Variation curve of *SOC* on the surface of electrode particles with time.

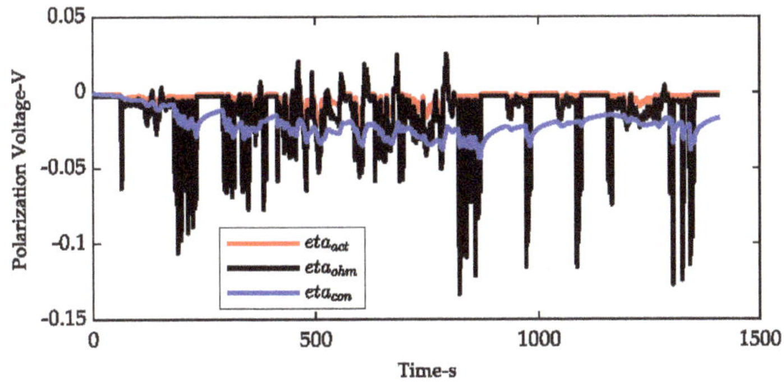

Figure 11. Quantitative characterization curves of three types of polarization voltage based on LDM.

3.4. Online Polarization Voltage Characterization Using a Hardware Platform

To realize the online quantitative characterization of the polarization voltage drop based on LDM, the model parameters at different *SOC* levels need to be identified offline in advance. Discharge pulses were applied to the battery at 25 °C at different *SOC* levels and rest for 2 h after each discharge pulse, and current versus voltage data were recorded throughout. The *SOC* levels were selected as 98%, 95%, 90%, 80%, 70%, 60%, 50%, 40%, 30%, 20%, 10%, 8%, 5%, 3%, 1% and 0%. A portion of the data before and after each discharge pulse, containing the zero-state response and zero-input response phases, was intercepted as input data for the parameter identification algorithm. A 9th order polynomial was used to fit the parameter points. The fitting curves for three model parameters are shown in Figures 12–14.

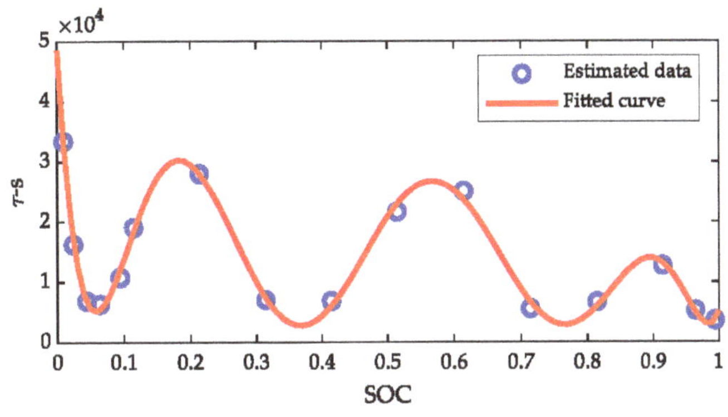

Figure 12. Variation curve of diffusion time constant τ with *SOC*.

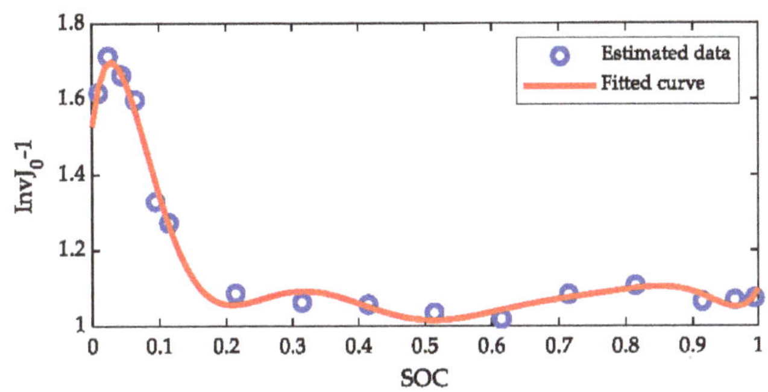

Figure 13. Variation curve of dimensionless charge exchange current $invJ_0$ with *SOC*.

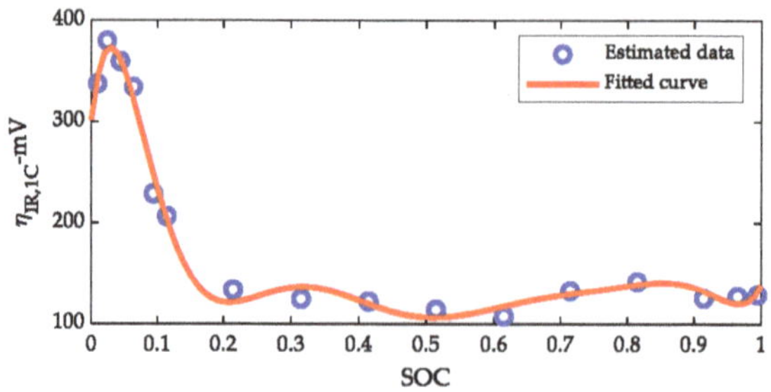

Figure 14. Variation curve of ohmic overpotential at 1 C rate $\eta_{IR,1C}$/mV with *SOC*.

At an ambient temperature of 25 °C, the New European Driving Cycle (NEDC) data were used as the test data for real-time quantitative characterization of polarization voltage based on LDM. The data were fed into the characterization platform when *SOC* = 0.8468 and stopped when *SOC* = 0.7006. Input current update frequency and terminal voltage acquisition frequency were 1 Hz. The values of the model parameters at specific *SOC* level were obtained by interpolation of the previously calibrated curves. The current vs. terminal voltage curves (Figure 15), *SOC* curves (Figure 16), and polarization voltages (Figure 17) were plotted dynamically during cycling. Based on the hardware platform used, real-time characterization of polarization voltage drops at a frequency of 1 Hz could be achieved using the proposed method (actual calculation time consumption for each time step is less than 500 ms). The model output voltage maintained good tracking performance by comparing with the battery terminal voltage data obtained from the acquisition board. However, it was observed that the voltage tracking error increased when the current increased. The possible sources of error are (1) LDM does not include the electrolyte concentration polarization, (2) errors from the identification algorithm or the curve fitting, which are expected to be further improved. It can be seen that the voltage drop from all three types of polarization was positively correlated with the current applied to the cell, which is consistent with the findings of previous studies [14,17,49]. In summary, the proposed method achieves the function of quantitative characterization of polarization voltage, and the algorithm computation efficiency can meet a good real-time performance.

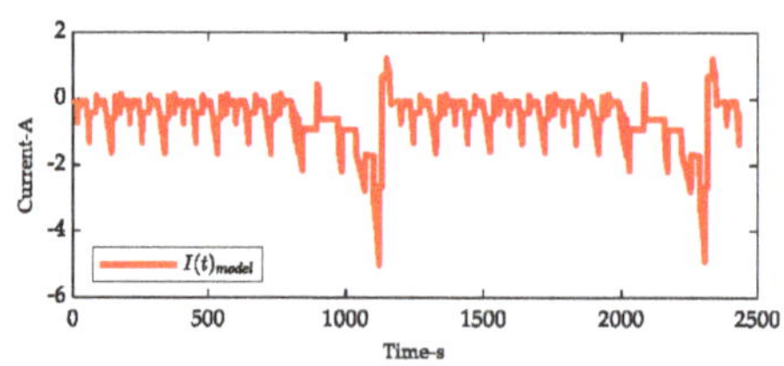

(a)

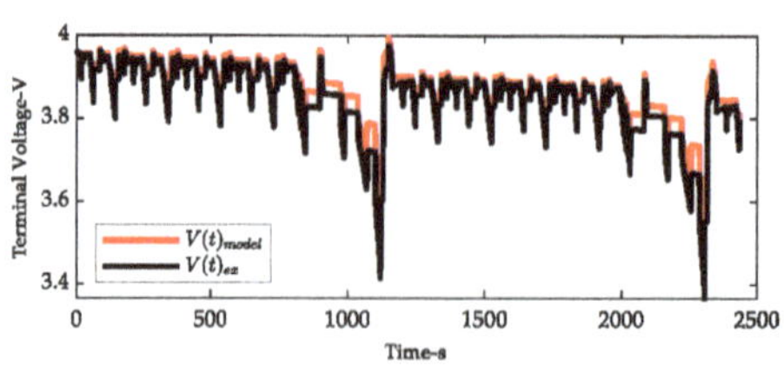

(b)

Figure 15. Accuracy test of hardware implementation platform under NEDC schedule at 25 °C ambient temperature. (**a**) NEDC current profile. (**b**) Comparison of model output voltage and actual voltage.

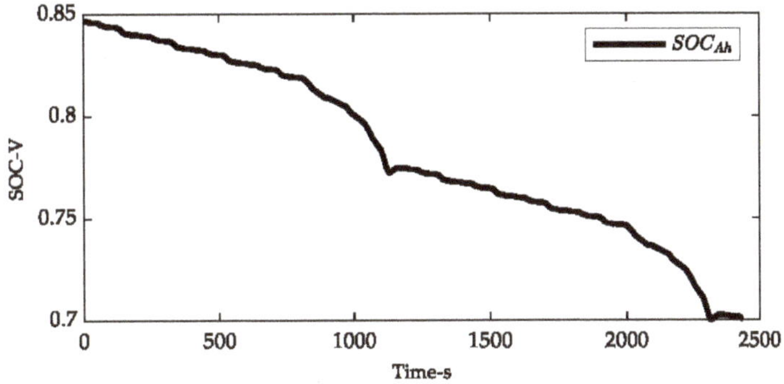

Figure 16. SOC curve based on ampere-hour integral method under the NEDC condition.

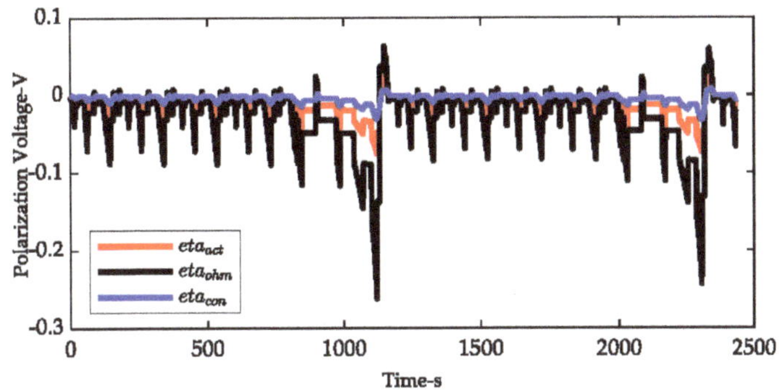

Figure 17. Curves of three kinds of polarization voltage under the NEDC condition.

4. Conclusions

Based on LDM, this study characterizes all three types of polarization voltage of Lithium ion batteries under operating conditions. Three model parameters were used: dimensionless charge exchange current J_0, ohmic resistance R_{ohm}, and diffusion time constant τ to characterize activation polarization, ohmic polarization, and concentration polarization, respectively. A joint algorithm consisting of the PSO algorithm and the L-M method was used to identify the model parameters. The deviation of the model output terminal voltage from the actual terminal voltage was used as the accuracy criterion, and the proposed algorithm was compared with the PSO algorithm alone and the L-M method alone. In terms of the intercepted battery testing data, RMSE as the criterion, the voltage error of the joint algorithm was reduced 77.1% compared to the PSO algorithm only, and 72.3% compared to the L-M method only. To further test the effectiveness of the model and the identification algorithm, the identified model parameters were substituted into the second battery test data. It can be concluded that (1) the proposed scheme describes the nonlinear characteristics of the battery cell under the excitation of high dynamic driving conditions, (2) the model reflects the real physical-chemical processes inside the battery to a certain extent. To test the real-time performance of the proposed method, a hardware implementation platform for the real-time quantitative characterization of the polarization voltage of lithium-ion batteries was built, and the model parameters were calibrated and fitted using an offline method. The hardware platform was capable of realizing the basic function of quantitative polarization voltage characterization, and the update frequency of relevant parameters was 1 Hz, with good real-time performance. It has the potential for further development for BMS applications.

Author Contributions: B.X. and B.Y. proposed the novel method of polarization voltage characterization for lithium-ion batteries; B.Y. designed the experiment scheme and conducted experiments. B.Y. and J.C. designed and implemented the hardware platform; B.Y. wrote the paper. All authors have read and agreed to the published version of the manuscript.

Funding: This research was funded by National Natural Science Foundation of China (Grant No. 51877120).

Conflicts of Interest: The authors declare no conflict of interest.

Nomenclature

Acronyms	
AC	Alternating current
BMS	Battery management system
CC-CV	Constant current Constant voltage
CCV	Closed-circuit voltage
DC	Direct current
DOP	Degree of Polarization
DRT	Distribution of Relaxation Times
GA	Genetic Algorithm
GITT	Galvanostatic intermittent titration technique
GS-EIS	Galvanostatic electrochemical impedance spectroscopy
LDM	Lumped diffusion model
L-M method	Levenberg-Marquardt method
Pseudo-2D	Pseudo 2 dimensional
PSO	Particle swarm optimization
PVDF	Ferroelectric polyvinylidene difluoride
QOCV	Quasi-open-circuit voltage
SOC	State of Charge
SOH	State of Health
SPM	Single particle model
Symbols	
a_n	Coefficients of polynomial describing SOC-OCV curve
c_1	Local learning factor
c_2	Global learning factor
dim	Randomly picked dimension among three dimensions
E_{OCP}	Open-circuit voltage
E_{cell}	Battery terminal voltage
e	Voltage error
F	Faraday constant
f	Fitness function of PSO algorithm
G	Maximum iteration number
$gbest$	$= [g_\tau, g_{invJ0}, g_{\eta_{IR,1C}}] \in S$, global optimum particle
I_{1C}	Applied current taken at 1C rate
I	Applied current under discrete time domain
i	Instantaneous current
$iSOC$	Local SOC in electrode particle
J_0	Dimensionless charge exchange current
J_k	Jacobian matrix
k_1, k_2	Learning factor gain
k_{max}	Maximum iteration number
k	Current iteration number
LB_v	Lower velocity bound
LB_x	Lower position bound
L	Identity matrix
N_{shape}	Dimension number of the particle
$pbest_p$	$= \left[p_{p,\tau}, p_{p,invJ0}, p_{p,\eta_{IR,1C}}\right] \in S$, history optimum for the pth particle

P	Population size
Q_n	Battery capacity
R	Molar gas constant
R_{ohm}	Ohmic resistance
$rand_1$, $rand_2$, r_1, r_2	Random numbers uniformly distributed in (0,1)
SOC_{ave}	Average SOC of electrode particle
SOC_{surf}	Surface SOC of electrode particle
SOC_0	SOC at initial time point
S	Search space
T	Reference temperature
t_0	Initial time
t	Current time
UB_v	Upper velocity bound
UB_x	Upper position bound
$\hat{u}$	Model output voltage
u	Real world terminal voltage
v_p	$= \left[v_{p,\tau}, v_{p,invJ0}, v_{p,\eta_{IR,1C}}\right] \in S$, velocity vector for the pth particle
x_p	$= \left[x_{p,\tau}, x_{p,invJ0}, x_{p,\eta_{IR,1C}}\right] \in S$, position vector for the pth particle
X	Dimensionless space variable in particle size scale
Greek symbols	
α	Charge transfer coefficient
ε	Tolerance
η_{act}	Activation polarization overpotential
η_{con}	Concentration polarization overpotential
η_{ohm}	Ohmic polarization overpotential
θ	Parameter vector in the L-M method
τ	Diffusion time constant
ω_{max}	Maximum inertia weight
ω_{min}	Minimum inertia weight
ω	Inertia weight
Subscripts/Superscripts	
$+$	Positive electrode
$-$	Negative electrode
p	The pth particle
kg	Iteration number

References

1. Bizeray, A.M.; Kim, J.-H.; Duncan, S.R.; Howey, D.A. Identifiability and Parameter Estimation of the Single Particle Lithium-Ion Battery Model. *IEEE Trans. Control Syst. Technol.* **2019**, *27*, 1862–1877. [CrossRef]
2. Xiaosong, H.; Li, S.; Peng, H. A Comparative Study of Equivalent Circuit Models for Li-Ion Batteries. *J. Power Sources* **2012**, *198*, 359–367.
3. Junfu, L.; Wang, D.; Pecht, M. An Electrochemical Model for High C-Rate Conditions in Lithium-Ion Batteries. *J. Power Sources* **2019**, *436*, 226885.
4. Shunli, W.; Stroe, D.; Fernandez, C.; Yu, C.; Zou, C.; Li, X. A Novel Energy Management Strategy for the Ternary Lithium Batteries Based on the Dynamic Equivalent Circuit Modeling and Differential Kalman Filtering under Time-Varying Conditions. *J. Power Sources* **2020**, *450*, 227652.
5. Andreas, N.; Zavalis, T.G.; Elger, R.; Behm, M.; Lindbergh, G. Analysis of the Polarization in a Li-Ion Battery Cell by Numerical Simulations. *J. Electrochem. Soc.* **2010**, *157*, 11.
6. Jiaxin, Z.; Lu, J.; Amine, K.; Pan, F. Depolarization Effect to Enhance the Performance of Lithium Ions Batteries. *Nano Energy* **2017**, *33*, 497–507.
7. Changsheng, Q.; He, G.; Shi, W.; Zou, M.; Liu, C. The Polarization Characteristics of Lithium-Ion Batteries under Cyclic Charge and Discharge. *J. Solid State Electrochem.* **2019**, *23*, 1887–1902.
8. Guanhua, F.; Li, Z.; Mi, L.; Zheng, J.; Feng, X.; Chen, W. Polypropylene/Hydrophobic-Silica-Aerogel-Composite Separator Induced Enhanced Safety and Low Polarization for Lithium-Ion Batteries. *J. Power Sources* **2018**, *376*, 177–183.
9. Hyun-seung, K.; Kim, J.; Jang, J.; Kim, N.; Ryu, J.H.; Yoon, S.; Oh, S.M. A Comparative Study of Polarization During the Initial Lithiation Step in Tungsten-Oxide Negative Electrodes for Lithium-Ion Batteries. *Solid State Ion.* **2017**, *311*, 1–5.

10. Ying, S.; Wen, L.; Li, F.; Cheng, H. Nanosized Li4ti5o12/Graphene Hybrid Materials with Low Polarization for high rate lithium ion Batteries. *J. Power Sources* **2011**, *196*, 8610–8617.
11. Balasundaram, M.; Ramar, V.; Yap, C.; Balaya, P. Investigation of Physico-Chemical Processes in Lithium-Ion Batteries by Deconvolution of Electrochemical Impedance Spectra. *J. Power Sources* **2017**, *361*, 300–309.
12. Woo-Jin, S.; Joo, S.H.; Kim, D.H.; Hwang, C.; Jung, G.Y.; Bae, S.; Son, Y.; Cho, J.; Song, H.; Kwak, S.K.; et al. Significance of Ferroelectric Polarization in Poly (Vinylidene Difluoride) Binder for High-Rate Li-Ion Diffusion. *Nano Energy* **2017**, *32*, 255–262.
13. J, N.D.; Wang, M.; Le, A.V.; Shi, Y.; Qiao, Y. Internal Resistance and Polarization Dynamics of Lithium-Ion Batteries Upon Internal Shorting. *Appl. Energy* **2018**, *212*, 796–808.
14. Yingxia, H.; Lai, H. Effects of Discharge Rate on Electrochemical and Thermal Characteristics of Lifepo4/Graphite Battery. *Appl. Therm. Eng.* **2019**, *157*, 113744.
15. Noboru, S. Thermal Behavior Analysis of Lithium-Ion Batteries for Electric and Hybrid Vehicles. *J. Power Sources* **2001**, *99*, 70–77.
16. Peyman, T.; Mansouri, A.; Yazdanpour, M.; Bahrami, M. Theoretical Analysis of Potential and Current Distributions in Planar Electrodes of Lithium-Ion Batteries. *Electrochim. Acta* **2014**, *133*, 197–208.
17. Damian, G.; Faulkner, T.; Peterson, V.K.; Sharma, N. Structural Evidence for Mg-Doped Lifepo4 Electrode Polarisation in Commercial Li-Ion Batteries. *J. Power Sources* **2018**, *394*, 1–8.
18. Haijun, R.; Jiang, J.; Sun, B.; Zhang, W.; Gao, W.; Wang, l.; Ma, Z. A Rapid Low-Temperature Internal Heating Strategy with Optimal Frequency Based on Constant Polarization Voltage for Lithium-Ion Batteries. *Appl. Energy* **2016**, *177*, 771–782.
19. Xiaokang, L.; Wang, Q.; Yang, Y.; Kang, J. Correlation between Capacity Loss and Measurable Parameters of Lithium-Ion Batteries. *Int. J. Electr. Power Energy Syst.* **2019**, *110*, 819–826.
20. Cyril, M.; Fullenwarth, J.; Monconduit, L.; Lestriez, B. Diagnostic of the Failure Mechanism in Nisb2 Electrode for Li Battery through Analysis of Its Polarization on Galvanostatic Cycling. *Electrochim. Acta* **2012**, *78*, 177–182.
21. Bizhong, X.; Chen, G.; Zhou, J.; Yang, Y.; Huang, R.; Wang, W.; Lai, Y.; Wang, M.; Wang, H. Online Parameter Identification and Joint Estimation of the State of Charge and the State of Health of Lithium-Ion Batteries Considering the Degree of Polarization. *Energies* **2019**, *12*, 2939.
22. Caiping, Z.; Jiang, J.; Gao, Y.; Zhang, W.; Liu, Q.; Hu, X. Polarization Based Charging Time and Temperature Rise Optimization for Lithium-Ion Batteries. *Energy Procedia* **2016**, *88*, 675–681.
23. Zhang, C.; Jiang, J.; Gao, Y.; Zhang, W.; Liu, Q.; Hu, X. Charging Optimization in Lithium-Ion Batteries Based on Temperature Rise and Charge Time. *Appl. Energy* **2017**, *194*, 569–577. [CrossRef]
24. Rong, X.; Yang, Y.; Yin, F.; Liu, P.; Cloetens, P.; Liu, Y.; Lin, F.; Zhao, K. Heterogeneous Damage in Li-Ion Batteries: Experimental Analysis and Theoretical Modeling. *J. Mech. Phys. Solids* **2019**, *129*, 160–183.
25. Hosang, P.; Yoon, T.; Kim, Y.; Ryu, J.H.; Oh, S.M. Li2nio2 as a Sacrificing Positive Additive for Lithium-Ion Batteries. *Electrochim. Acta* **2013**, *108*, 591–595.
26. Moya, A.A.; Castilla, J.; Horno, J. Ionic Transport in Electrochemical Cells Including Electrical Double-Layer Effects. A Network Thermodynamics Approach. *J. Phys. Chem.* **1995**, *99*, 1292–1298. [CrossRef]
27. Bo, Y.; Lim, C.; Song, Z.; Zhu, L. Analysis of Polarization in Realistic Li Ion Battery Electrode Microstructure Using Numerical Simulation. *Electrochim. Acta* **2015**, *185*, 125–141.
28. Junfu, L.; Wang, L.; Lyu, C.; Liu, E.; Xing, Y.; Pecht, M. A Parameter Estimation Method for a Simplified Electrochemical Model for Li-Ion Batteries. *Electrochim. Acta* **2018**, *275*, 50–58.
29. Emil, N.; Torregrossa, D.; Cherkaoui, R.; Paolone, M. Parameter Identification of a Lithium-Ion Cell Single-Particle Model through Non-Invasive Testing. *J. Energy Storage* **2017**, *12*, 138–148.
30. Thanh-Son, D.; Vyasarayani, C.P.; McPhee, J. Simplification and Order Reduction of Lithium-Ion Battery Model Based on Porous-Electrode Theory. *J. Power Sources* **2012**, *198*, 329–337.
31. Xiaowei, Z.; Cai, Y.; Yang, L.; Deng, Z.; Qiang, J. State of Charge Estimation Based on a New Dual-Polarization-Resistance Model for Electric Vehicles. *Energy* **2017**, *135*, 40–52.
32. Henrik, E.; Fridholm, B.; Lindbergh, G. Comparison of Lumped Diffusion Models for Voltage Prediction of a Lithium-Ion Battery Cell During Dynamic Loads. *J. Power Sources* **2018**, *402*, 296–300.
33. Andre, D.; Meiler, M.; Steiner, K.; Wimmer, C.; Soczka-Guth, T.; Sauer, D.U. Characterization of High-Power Lithium-Ion Batteries by Electrochemical Impedance Spectroscopy. I. Experimental Investigation. *J. Power Sources* **2011**, *196*, 5334–5341. [CrossRef]
34. Zhang, S.; Xu, K.; Jow, T. Electrochemical Impedance Study on the Low Temperature of Li-Ion Batteries. *Electrochim. Acta* **2004**, *49*, 1057–1061. [CrossRef]
35. Moya, A.A. Identification of Characteristic Time Constants in the Initial Dynamic Response of Electric Double Layer Capacitors from High-Frequency Electrochemical Impedance. *J. Power Sources* **2018**, *397*, 124–133. [CrossRef]
36. Xing, Z.; Huang, J.; Pan, Z.; Ouyang, M. Impedance Characterization of Lithium-Ion Batteries Aging under High-Temperature Cycling: Importance of Electrolyte-Phase Diffusion. *J. Power Sources* **2019**, *426*, 216–222.
37. Santhanagopalan, S.; Guo, Q.; Ramadass, P.; White, R.E. White. Review of Models for Predicting the Cycling Performance of Lithium Ion Batteries. *J. Power Sources* **2006**, *156*, 620–628. [CrossRef]
38. Doyle, M.; Fuller, T.F.; Newman, J. Modeling of Galvanostatic Charge and Discharge of the Lithium/Polymer/Insertion Cell. *J. Electrochem. Soc.* **1993**, *140*, 1526–1533. [CrossRef]

39. Srinivasan, V.; Wang, C.Y. Analysis of Electrochemical and Thermal Behavior of Li-Ion Cells. *J. Electrochem. Soc.* **2003**, *150*, A98. [CrossRef]
40. Botte, G.G.; Subramanian, V.R.; White, R.E. White. Mathematical Modeling of Secondary Lithium Batteries. *Electrochim. Acta* **2000**, *45*, 2595–2609. [CrossRef]
41. Haran, B.S.; Popov, B.N.; White, R.E. White. Determination of the Hydrogen Diffusion Coefficient in Metal Hydrides by Impedance Spectroscopy. *J. Power Sources* **1998**, *75*, 56–63. [CrossRef]
42. Diwakar, V.D. Towards Efficient Models for Lithium Ion Batteries. Ph.D. Thesis, Tennessee Technological University, Ann Arbor, MI, USA, 2009.
43. Shen, W.-J.; Li, H.-X. Parameter Identification for the Electrochemical Model of Li-Ion Battery. In Proceedings of the 2016 International Conference on System Science and Engineering (ICSSE), Puli, Taiwan, 7–9 July 2016.
44. Kennedy, J.; Eberhart, R. Particle Swarm Optimization. In Proceedings of the ICNN'95—International Conference on Neural Networks, Perth, WA, Australia, 27 November–1 December 1995.
45. Levenberg, K. A Method for the Solution of Certain Non-Linear Problems in Least Squares. *Q. Appl. Math.* **1944**, *2*, 164–168. [CrossRef]
46. Marquardt, D.W. An Algorithm for Least-Squares Estimation of Nonlinear Parameters. *J. Soc. Ind. Appl. Math.* **1963**, *11*, 431–441. [CrossRef]
47. Lyshevski, S.E. *Engineering and Scientific Computations Using Matlab*; John Wiley & Sons: Hoboken, NJ, USA, 2005.
48. Bizhong, X.; Huang, R.; Lao, Z.; Zhang, R.; Lai, Y.; Zheng, W.; Wang, H.; Wang, W.; Wang, M. Online Parameter Identification of Lithium-Ion Batteries Using a Novel Multiple Forgetting Factor Recursive Least Square Algorithm. *Energies* **2018**, *11*, 11.
49. Myounggu, P.; Zhang, X.; Chung, M.; Less, G.B.; Sastry, A.M. A Review of Conduction Phenomena in Li-Ion Batteries. *J. Power Sources* **2010**, *195*, 7904–7929.

energies

MDPI

Article

Stress Coupling Analysis and Failure Damage Evaluation of Wind Turbine Blades during Strong Winds

Kangqi Tian [1], Li Song [1,2,3,*], Yongyan Chen [1,2,3], Xiaofeng Jiao [4], Rui Feng [5] and Rui Tian [1,2,3]

1 College of Energy and Power Engineering, Inner Mongolia University of Technology, Hohhot 010051, China; 20181800177@imut.edu.cn (K.T.); yongyanchen@aliyun.com (Y.C.); tianr@imut.edu.cn (R.T.)
2 Key Laboratory of Wind Energy and Solar Energy Technology, Ministry of Education, Hohhot 010051, China
3 Key Laboratory of Renewable Energy in Inner Mongolia, Hohhot 010051, China
4 Inner Mongolia Power Science Research Institute, Hohhot 010051, China; jxf_0329@sina.com
5 Guoshui Group Huade Wind Power Co., Ltd., Ulanqab 012000, China; feng_rui@ctg.com.cn
* Correspondence: songli@imut.edu.cn; Tel.: +86-139-4812-0724

Abstract: Blades in strong wind conditions are prone to various failures and damage that is due to the action of random variable amplitude loads. In this study, we analyze the failure of 1.5 MW horizontal axis wind turbine blades. The computational fluid dynamics unsteady calculation method is used to simulate the aerodynamic load distribution on the blade. Fluid–structure coupling methods are applied to calculate the blade stress. The results show that the equivalent stress of the blade is the largest when the azimuth angle is 30°, and the maximum equivalent stress is 20.60 MPa. There are obvious stress peaks in six sections, such as r/R = 0.10 (the span length of blade/the full length of the blade = 0.10). The frequency of damage that is caused by the stress in each area of the blade is determined based on the blade damage. The frequency of gel coat cracking in the blade tips and leaves is 77.78% and 22.22%, respectively, and the frequency of crack occurrence is 87.75%, 10.20% and 2.05%, respectively. By combining the stress concentration area and the damage results, the cause of blade damage is determined, which can replace the traditional inspection methods and improve the inspection efficiency.

Keywords: wind turbine blade; fluid–solid coupling; azimuth; stress concentration; failure location; failure damage evaluation

Citation: Tian, K.; Song, L.; Chen, Y.; Jiao, X.; Feng, R.; Tian, R. Stress Coupling Analysis and Failure Damage Evaluation of Wind Turbine Blades during Strong Winds. *Energies* **2022**, *15*, 1339. https://doi.org/10.3390/en15041339

Academic Editor: Andrés Elías Feijóo Lorenzo

Received: 7 January 2022
Accepted: 1 February 2022
Published: 12 February 2022

1. Introduction

During the operation of a wind turbine, the blades are subjected to alternating effects of aerodynamic, centrifugal, and gravity loads. Wind turbines that have been operating for nearly 20 years are generally at the end of their service; the blades are prone to various failures and damage that is due to the action of random variable amplitude loads. Especially in strong wind conditions, a variety of damage can occur in the blades, such as in the leading edge of the blades. This includes paint peeling, multi-directional cracks on the trailing edge, and blade spanwise damage. Wind turbine blades are more prone to damage under strong wind conditions. Therefore, studying the force acting on the wind turbine blades during strong winds has important guiding significance for the analysis of failure locations and failure modes during the entire life cycle of the turbine.

Scholars worldwide have explored the load characteristics, stress, and displacement of the wind turbine blades under strong winds. Santo et al. [1] used the transient fluid–structure coupling method to consider the influence of strong winds with wind speeds that are greater than 18 m/s on the dynamic load characteristics of the wind turbine. Zhu and his colleagues [2] analyzed the stress and deformation characteristics of the 1.5 MW horizontal axis wind turbine blades under the ultimate load. In addition, they analyzed the vibration shape of the blades to design the structure of large wind turbine blades in order to provide a certain reference value. Scholars such as Fernandez [3] introduced an automated program to calculate the overall and local stress on wind turbine blades under different wind conditions. The accurate estimation of wind load depends on the deformation of the blades achieving good consistency. A stress–strain test system was proposed by Jiang et al. [4] to test the strain and dynamic response of the wind turbine blades during operation; this system provided real-time load monitoring and a means to diagnose damages in wind turbines. Kim et al. [5] proposed a real-time shape prediction technology that is based on a multi-point strain measurement. Under high wind speeds, the deformation of the blades was captured by a stereo pattern recognition system that consisted of eight cameras. At the same time, a finite element model of the blade was established to describe the relationship between the displacement and the stress of the blade. Through comparison, it was determined that the simulation results are in good agreement with the experimental results. Zhang et al. [6] calculated the equivalent stress and displacement response of the blade under strong winds based on a numerical simulation method that relies on fluid–structure coupling for large offshore wind turbine blades. Cheng et al. [7] used co-simulation technology to monitor the multi-parameters of the coupling structure of large wind turbine blades; they analyzed the response under different working conditions. The fluid–solid coupling method was adopted to analyze the action of the wind shear and blades. Zhang et al. [8] used finite element analysis software to calculate the blade displacement and stress distribution under strong winds. The results show that the amplitude of the entire blade along the span shows a nonlinear growth trend. When the tip amplitude is the largest, the maximum stress appears in the middle of the blade. A three-dimensional parametric finite element model was proposed by Zhou et al. [9] to explore the influence of the web on the blade; they studied the stress distribution under different web offsets. By performing orthogonal experiments, the most important factors that affect the stress distribution were determined; the results have a certain guiding significance for the design and manufacturing of blades. Choudhury et al. [10] explored the influence of aeroelasticity on the output power, and a method to reduce the stress coupling was discovered. Dimitrov et al. [11] explored the fatigue characteristics of the wind turbine blades under strong winds. The results show that the reduction of the wind shear index is beneficial to reduce the fatigue damage equivalent load of the wind turbine blades. Mesfin et al. [12] took NREL(National Renewable Energy Laboratory) 5 MW blades as their research object; they designed the ply and explored the effects of the methods on the results of the blade stress and displacement based on the calculation methods of applying one-way fluid–solid coupling and fluid–solid coupling. The study noted that under the action of strong winds, the fluid–solid coupling can solve the stress distribution of the blade more accurately. Zhu et al. [13] used Workbench to calculate the fluid–structure interaction of the blade under rated and extreme wind loads; they analyzed the force and strength characteristics of the blade under strong winds. Bae et al. [14] established a complete fluid-structure coupling model of a 2 MW horizontal axis wind turbine blade and calculated the structural strength and load characteristics of the blade under various loads. Ullah et al. [15] generated a three-dimensional (3D) surface model of the wind turbine blades and analyzed the structural response of composite blades under extreme loads by using the software program ANSYS (American ANSYS, Pittsburgh, Pennsylvania). By optimizing the blade structure, the overall power characteristics of the wind turbine were improved.

Shen et al. [16] studied the cyclical unsteady characteristics of the wind shear and determined that fatigue loads can cause local damage to the wind turbine blades and ultimately lead to failure. Fu et al. [17] numerically calculated the dynamic response of the wind turbine that is based on the actual wind load; thus, they could easily obtain the reliability of the structure under random loads. Guo et al. [18] established a 5 MW large-scale wind turbine finite element model to analyze the displacement response and vibration frequency of the wind turbine blades under random wind loads. Wang et al. [19] systematically analyzed the aerodynamic load and dynamic response of large-scale wind turbines at different stages of a typhoon. In addition, they obtained the vibration characteristics of the wind turbines at different stages of the typhoon. Boujleben et al. [20] constructed a 3D model of the flexible NREL 5 MW wind turbine blades and performed highly iterative calculations through fluid–structure coupling. From this, they accurately calculated the aerodynamic load distribution on the blades and analyzed the aerodynamic load of the wind turbine blades. The fatigued state and stress distribution law of the model was used to verify the validity of the calculation model. According to the magnitude of the wind load, Mathijs and his team [21] applied equal tension to 43 m long wind turbine blades to conduct bending experiments; they compared the stress and displacement characteristics of the blades. Relying only on numerical simulation to analyze the dynamic response of large wind turbine blades cannot meet the actual needs of the project. It is necessary to supplement the effectiveness of the actual application in the project with a more practical experimental research method.

Herein, in order to explore the stress distribution law of the wind turbine blades under strong wind loads and the damage of the blades after applying a force, this study focuses on 1.5 MW horizontal axis wind turbine blades. This investigation applies the fluid–solid coupling calculation method and it analyzes the wind turbine under strong wind conditions. The equipment stress of the blade is analyzed, the stress distribution law during the operation of the blade is explored and the failure area of the wind turbine blade is explored. By combining the results of the numerical simulation with the observational experiments, this study defines the typical failure areas. According to the frequency of the damage in the various parts of the wind turbine blade, the damage rules of the blade are explored and the failure mode of each area of the blade is finally determined.

2. Materials and Methods

2.1. Theories

2.1.1. Fluid–Solid Coupling Control Equation

Wind turbine blades have a large windward area and the wind turbine flow field environment is very complex. Therefore, the fluid–solid coupling method is adopted and the weak coupling method and sequential solution method are used for the fluid–solid coupling. Fluid–solid coupling follows the most basic principles of conservation. Thus, the fluid and solid stress τ, displacement d, heat flow q, temperature T, and other variables should be equal or conserved [22].

$$\tau_f n_f = \tau_s n_s \tag{1}$$

$$d_f = d_s \tag{2}$$

$$q_f = q_s \tag{3}$$

$$T_f = T_s \tag{4}$$

2.1.2. Theoretical Basis of the Turbulence Model

The shear stress transport model is referred to as the SST k-ω model, and they are used to represent the wind flowing through the blades. The governing equation is as follows [23]:

$$\frac{\partial}{\partial t}(\rho k) + \frac{\partial}{\partial x_i}(\rho k u_t) = \frac{\partial}{\partial x_i}(\Gamma_k \frac{\partial_k}{\partial x_j}) + G_k - Y_k \tag{5}$$

$$\frac{\partial}{\partial t}(\rho \omega) + \frac{\partial}{\partial x_i}(\rho \omega u_t) = \frac{\partial}{\partial x_j}(\Gamma_k \frac{\partial_\omega}{\partial x_j}) + G_\omega - Y_\omega + D_\omega \tag{6}$$

where G_ω is the turbulent kinetic energy generation term that is caused by the velocity gradient, Γ_k and Γ_ω are the convection terms of and ω, k is turbulent kinetic energy, ω is specific dissipation rate, and Y_k and Y_ω are the effective diffusion terms of k and ω that are caused by the turbulence.

2.1.3. Sliding Mesh Theory

The basic principle of the sliding mesh is to divide the geometric model mesh into several regions; the meshes on both sides of the interface slide against each other [24]. The number of grid nodes on both sides of the interfaces should not be too different, and the fluxes on both sides should be equal, as shown in Figure 1.

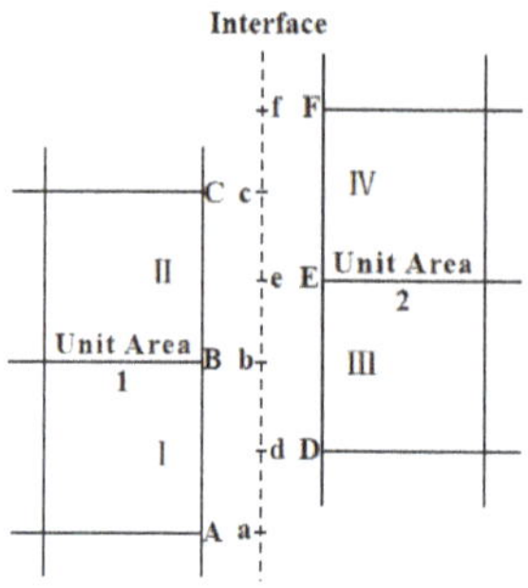

Figure 1. Schematic diagram of the sliding grid principle [24].

For the generalized scalar Φ in any control body V with a moving boundary, the integral conservation equation [24] is written as follows:

$$\frac{d}{dt}\int_V \Phi dV + \int_{\partial V} \Phi(v - v_g)dA = \int_{\partial V} \Gamma \bigtriangledown \Phi dA + \int_V SdV \tag{7}$$

where ∂V is the boundary of the control body V, and V_g is the mesh velocity vector.

2.2. Physical Model and Numerical Settings

2.2.1. Physical Model

Our research object is the 1.5 MW horizontal axis wind turbine and data about the 1.5 MW wind turbine comes from Guoshui Group Huade Wind Power Co., Ltd. The main parameters are shown in Table 1.

Table 1. Parameters of the 1.5 MW wind turbine.

Name	Specifications	Name	Specifications
Airfoil	Provided by manufacturer	Rated power P/MW	1.5
Number of blades /N	3	Rated speed n/rpm	19.8
Hub center height H/m	65	Rated wind speed v/m·s^{-1}	12
Hub diameter d/m	2	Rated tip speed ratio λ	8.5
Wind wheel diameter D/m	77	Cut-in wind speed v/m·s^{-1}	3
Wind wheel quality m/t	30.1	Cut-out wind speed v/m·s^{-1}	25

Herein, considering the structure of the 1.5 MW wind turbine blades and using the ACP (ANSYS Composite PrepPost) module (American ANSYS, Pittsburgh, Pennsylvania), the 1.5 MW wind turbine blade layering design is divided into five areas, as shown in Figure 2 made by ANSYS. Based on the layering method offered by reference [25] of the wind turbine blade, we adjusted the layering parameters appropriately. The blade is subjected to shear, bending moment, and torque. The force characteristic is that it decreases from the root to the tip of the blade. Therefore, in terms of the number of material layers, the thickness of the blade root material should be greater than that of the middle blade. Moreover, it is larger than that of the blade tip; thus, the layering shows a descending trend along the blade length. The blade layer material is considered to be composite material glass fiber reinforced plastic (FRP) with density ρ = 2100 kg/m^3. The other material parameters are presented in Table 2.

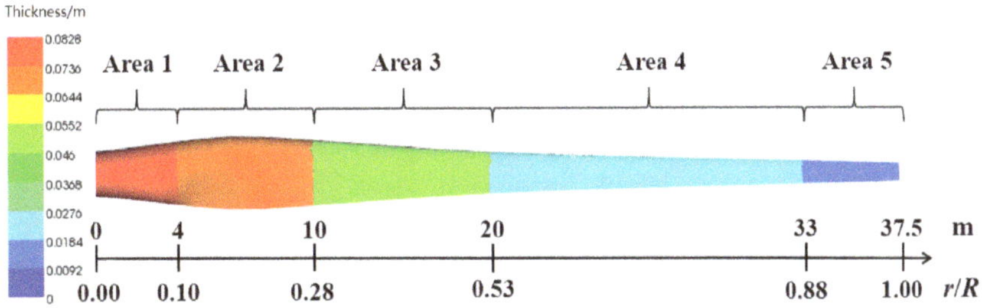

Figure 2. Layering design of the 1.5 MW wind turbine blade used in this study.

Table 2. Material properties.

Material	Elastic Modulus/MPa			Poisson's Ratio			Shear Modulus/MPa		
	E11	E22	E33	v12	v23	v13	G12	G23	G13
FRP	39,000	8600	8600	0.28	0.47	0.28	3800	2930	3800

As shown in Figure 3, a suitable cuboid calculation domain is established according to the model size, which is 150 m high, 150 m wide, and 300 m long. A physical model that had a 1:1 scale with the actual wind wheel was constructed. The hub center is 65 m from the inlet and 235 m from the outlet. It is also 65 m from the ground; the center of the rotation domain coincides with the center of the wind wheel, and the domain diameter is 82 m. The computational domain uses unstructured tetrahedral grids, and through grid independence verification, the total number of divided grids was determined to be 15.8 million. The non-steady-state calculation method of the sliding grid was used to simulate the aerodynamic performance of the wind turbines. This method solves the problem of the grid mismatch on both sides of the interface between the flow field and the rotation field [24].

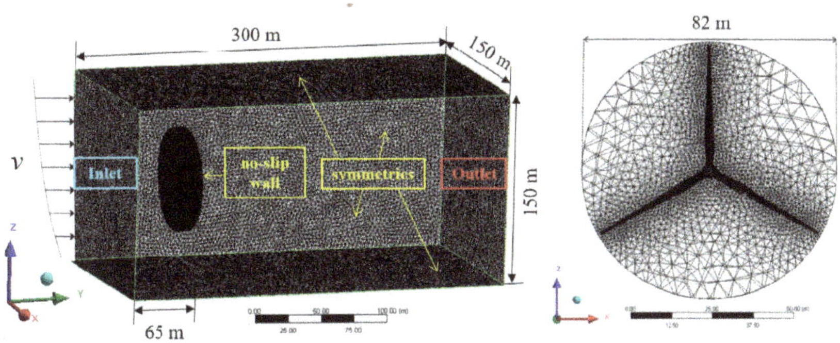

Figure 3. Computational domain mesh comprising unstructured tetrahedral grids to construct the physical model of the wind wheel.

The geometric center of the wind wheel along the center of gravity of the rotation axis is defined as the coordinate origin O point, the direction of the incoming flow parallel to the ground is the positive direction of the Y-axis, and the Z-axis is vertically upward. The model outlet is set as a pressure outlet, the domain wall is set as a symmetrical boundary, and the wind wheel is set as a non-slip wall. In this study, the SST k-ω turbulence model is selected as it takes into account the influence of the turbulent shear stress; it can better simulate the gas flow and blade pressure distribution [23].

As for the structural model of the wind turbine, due to the irregular aerodynamic shape of the wind turbine blade, based on the Transient Structural module of ANSYS Workbench 19.2, the unstructured tetrahedral grid is used to divide the solid model, and the grid is properly encrypted to ensure the accuracy of subsequent fluid-structure coupling data transmission. The structural model and meshing of the blade are referred to [25]. Here, the wind turbine startup wind speed of 3.6 m/s, shutdown wind speed of 21.1 m/s and strong wind speed of 19 m/s are selected to verify the grid independence of the CFD model. The grid independence verification about the CFD model is as follows in Table 3.

Table 3. Grid independence verification about CFD model.

Wind Speed	Grid Numbers/Ten Thousand							
3.6 m/s	340	430	520	620	700	850	900	1070
Maximum pressure/Pa	164	165	167	169	172	175	175	175
19 m/s	340	430	520	620	700	850	900	1070
Maximum pressure/Pa	1573	1587	1604	1624	1646	1674	1674	1674
21.1 m/s	340	430	520	620	700	850	900	1070
Maximum pressure/Pa	1618	1643	1668	1694	1720	1743	1743	1743

According to the calculation, it was found that when the grid number exceeds 8.5 million, the maximum pressure of the blade under three working conditions no longer changes, and the value is highly consistent with the results calculated by Wang, Quant, and Kolios [22], so the grid independence verification is accurate and effective, and the total number of grids divided is 8.5 million.

Unstructured tetrahedral meshes were adopted in the solid structure model. The grid independence verification of the structural solid model is shown in Table 4. After the analysis, the grid number of the blade model is suitable at 1.09 million.

Table 4. Grid independence verification about CFD model.

Wind Speed	Grid Numbers/Ten Thousand							
19 m/s	75	81	90	98	105	109	109	109
Maximum Equivalent stress /Mpa	19.2	19.5	19.7	20.3	20.6	20.6	20.6	20.6
Maximum displacement/m	0.56	0.58	0.59	0.61	0.63	0.63	0.63	0.63

2.2.2. Boundary Conditions

The wind speed distribution of the Huade Wind Farm of the Inner Mongolia Guoshui Group was considered for this study. We used the exponential wind shear function to describe the distribution. This study integrated calculations that are based on the measured data of the 65 m high wind measurement tower in the wind field; the incoming wind speed can be determined by applying the following equation:

$$v = v_{ref}\left(\frac{Z}{Z_{ref}}\right)^{\alpha} \tag{8}$$

where v is the wind speed at height z, m/s; v_{ref} is the known wind speed at Z_{ref}, m/s; Z_{ref} is the height at the center of the hub, m; α is the wind shear coefficient, which is fitted according to the measured data and is 0.2.

Because the influence of strong wind load on the structural strength of wind turbine blades can not be ignored and combined with the local strong wind conditions of the wind farm, it is necessary to analyze the unsteady stress characteristics of wind turbine blades under strong wind conditions. According to the literature, it is necessary to define strong winds. According to the reference [26], winds with instantaneous wind speeds of 17 m/s or more (or visually estimated winds of eight or more) are strong winds. Therefore, v_{ref} is 19 m/s, Z_{ref} is 65 m, and the blade pitch angle is 28.04°. From this, the UDF (User Defined Function) of the shear wind at the entrance of the calculation domain is compiled to simulate the shear and flow of the wind turbine blade during operation. By combining the observations in Figure 4 and Equation 8, the maximum and minimum heights during the blade tip rotation are 103.5 m and 26.5 m, respectively. This corresponds to wind speeds of 20.84 m/s and 15.88 m/s. Based on the meteorological parameters of the wind field in December 2018, the air density is 1.147 kg/m^3, the ambient temperature is 256.45 K, the atmospheric pressure is 86.4 KPa, and the turbulence is 0.07.

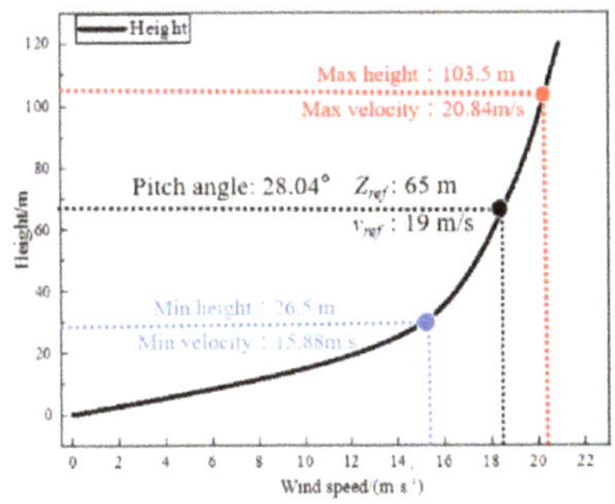

Figure 4. Inflow wind speed profile.

This investigation selected the Fluent double precision and SIMPLE algorithm for performing the calculations. The residual value is 10^{-4}, and the difference method is a second-order upwind. The calculation selects a strong wind with a wind condition of 19 m/s. When the wind speed is greater than the rated wind speed and less than the maximum wind speed, the wind wheel rotates at the rated speed; then, the set speed is 19.8 rpm, and it takes 3.029 s for each rotation of the wind wheel. The calculation is performed every time the blade rotates 30°, and the corresponding transient calculation time step is 0.252 s. The wind wheel is set to rotate for 20 cycles. When the torque coefficient is stable, the 16th cycle aerodynamic load is introduced into the transient structure field, and the fluid–solid coupling method is used to solve the blade surface stress distribution. The consistency of the coupling solution is ensured by keeping the step length and total time of the transient structural field consistent with the calculation time of the fluent flow field.

3. Results and Discussion

3.1. Analysis of the Blade Equivalent Stress under Different Azimuth Angles

The vertical upwards of the blade tip is defined as an azimuth angle of 0°, and the clockwise direction is considered to be the positive direction of the rotation. The rotation period is equally divided into 12 parts, where each part represents 30°. After the calculation is completed, the equivalent stress cloud diagram and the maximum equivalent stress periodic change are extracted.

It can be observed from Figure 5 that under the same rotation period, the maximum equivalent stresses on the windward and leeward sides at an azimuth angle of 30° are 20.60 MPa and 8.09 MPa, respectively. The maximum equivalent stresses on the windward and leeward sides have an azimuth angle of 240° and they are 19.20 MPa and 7.23 MPa. The equivalent stress value of the first half cycle is generally greater than the second half cycle. Because of the shear flow, the stress increases with an increase in height; gravity does positive work in the first half cycle and negative work in the second half cycle. The equivalent stress on the windward surface is concentrated near the root and the middle of the blade; this concentration is caused by the force characteristics of the windward surface and the structure of the blade. The equivalent stress on the leeward surface appears in the transition zone at each layer section, which is due to the stress concentration and layering that is caused by the segmented position.

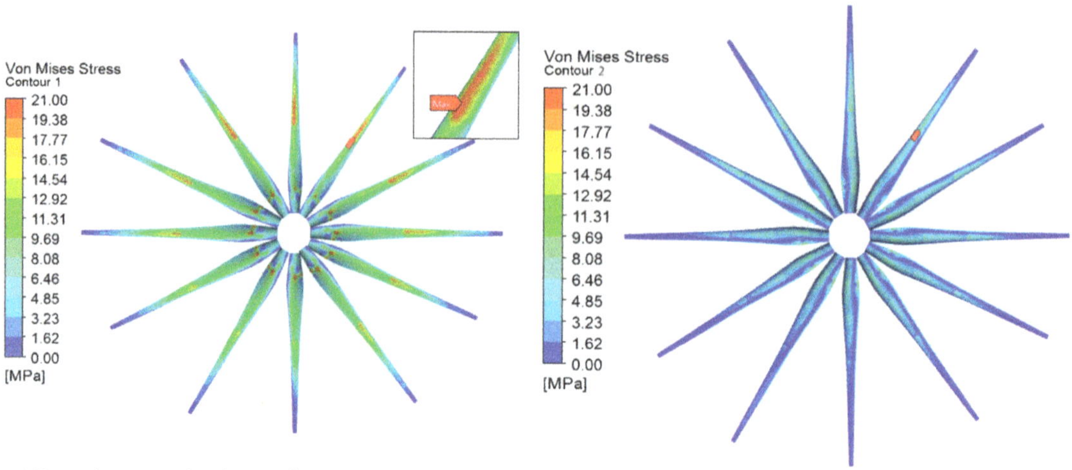

(**a**) Equivalent stress distribution cloud diagram on the windward surface.

(**b**) Equivalent stress distribution cloud diagram on the windward surface.

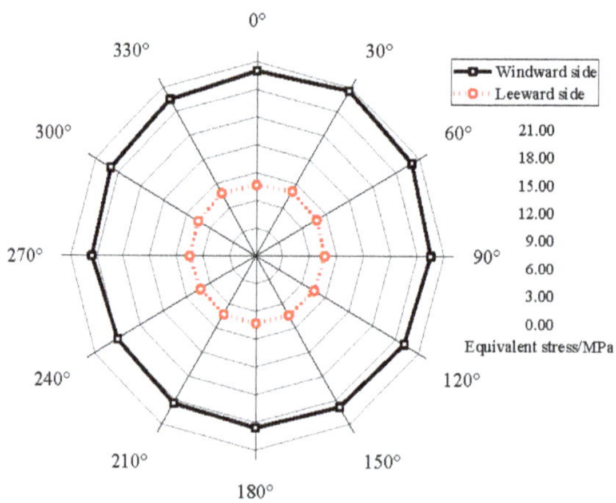

(**c**) Maximum equivalent stress change of a single blade during the rotation period.

Figure 5. Equivalent stress distribution during one cycle.

This study set the wind speed as v, the centrifugal force load as q, gravity as G, the spanwise component of gravity as G_z, and gravity along the blade rotation direction as G_x. The blade load distribution is shown in Figure 6. The wind speed v and G_x at a certain point *ri* (r/R = 0.5) of the monitoring blade are plotted in Figure 7, and the two change in the form of sine and cosine. When the blade rotates to the azimuth angle θ, the wind speed slowly decreases from 0° to 45°. G_x increases rapidly from 0° to 45° and reaches the maximum at 45°. The blade receives the largest force at 30°, which is reflected in the maximum equivalent stress of the blade at an azimuth angle of 30°, which is the largest.

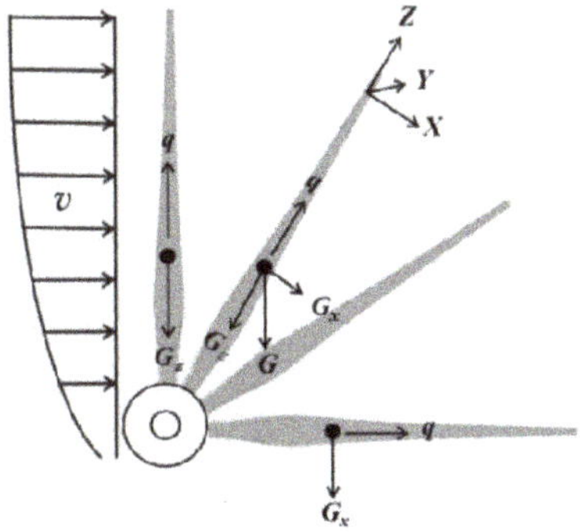

Figure 6. Spanwise load distribution.

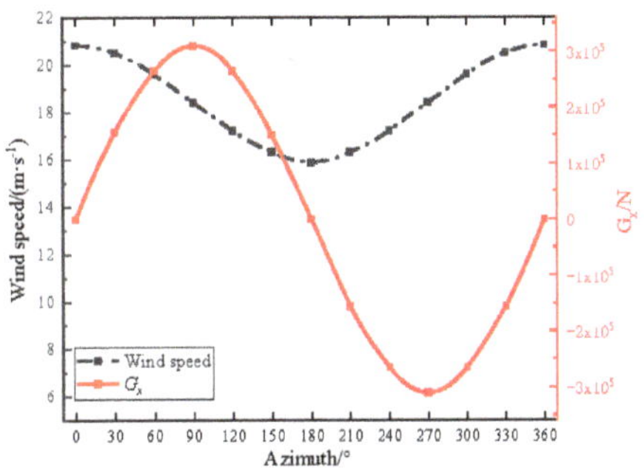

Figure 7. Distribution of the gravity and wind speed at r_i.

3.2. Load Analysis of a Typical Blade Failure Area

The blade is susceptible to fatigue due to large aerodynamic loads on the windward side of the leading edge and weak tail edge. Therefore, it is important to explore the distribution of equivalent stress on the leading edge and the trailing edge of the blade. Under the condition of a 19 m/s wind speed, this study first carried out a spanwise equivalent stress analysis on the leading edge and trailing edge of the blade under an azimuth angle of 30° to find the typical stress concentration section.

It can be observed from Figure 8 that the equivalent stress at the trailing edge line of the blade is generally greater than the leading edge line. The span length of the blade/the full length of the blade is set to r/R. The equivalent stress at the trailing edge of the blade is generally higher than that at the leading edge, and it includes an r/R = 0.10 section, an r/R = 0.28 section, an r/R = 0.53 section, and r/R = 0.88 at the blade layer section. The abrupt peak value of the equivalent stress is due to the sharp change in the number of layers at the blade layer section, and the poor structural stability of the section, which makes the position more prone to stress concentration and damage that is attributed to sudden changes in the stress. At the same time, there are also stress mutation peaks in the r/R = 0.16 section and r/R = 60 section, which is due to the special aerodynamic shape of the blade. This leads to the distribution of the blade spanwise stress.

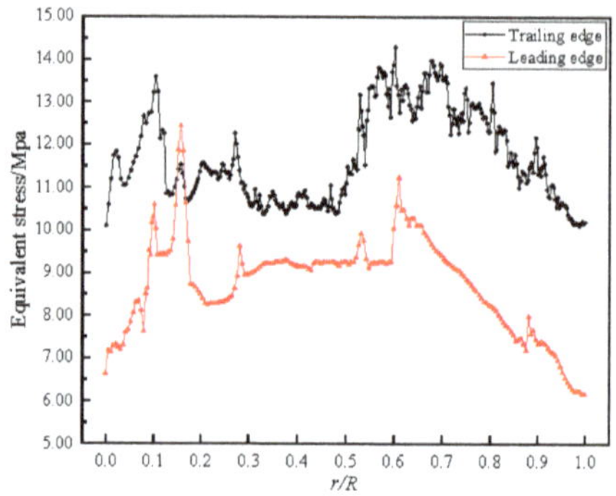

Figure 8. Equivalent stress distribution when it is spanwise at a 30° azimuth angle.

There are obvious equivalent stress peaks in the aforementioned six sections. In order to further explore the chordal distribution of the equivalent stress, the chordal equivalent stress analysis under one rotation cycle is performed on the six sections, as shown in Figure 9.

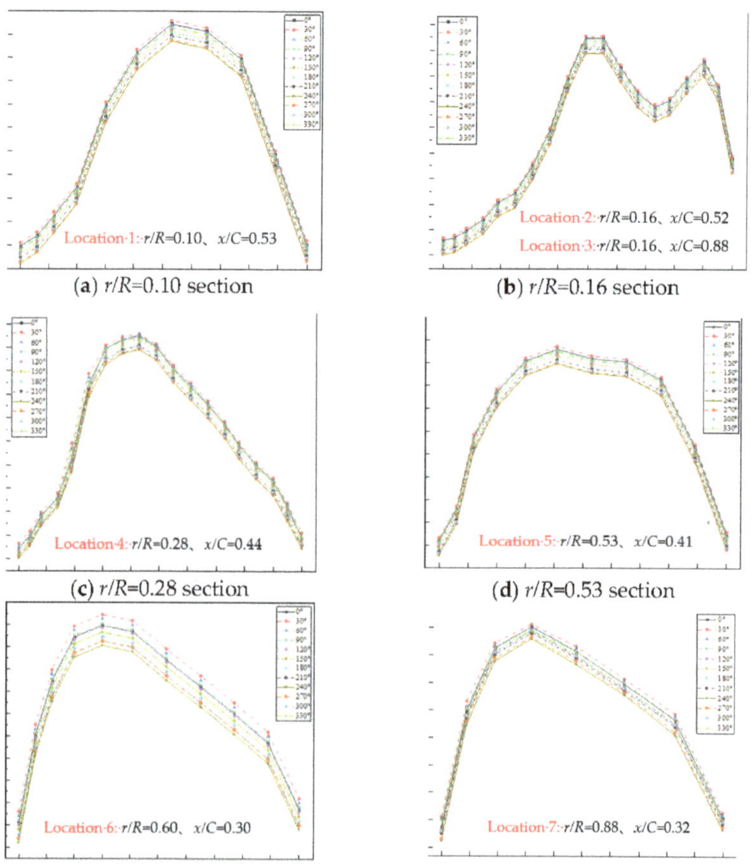

(**a**) r/R=0.10 section (**b**) r/R=0.16 section

(**c**) r/R=0.28 section (**d**) r/R=0.53 section

(**e**) r/R=0.60 section (**f**) r/R=0.88 section

Figure 9. Stress of a typical failure section at different azimuth angles.

It can be observed from Figure 9 that each section has the maximum equivalent stress at an azimuth angle of 30°, and the changing trend of the equivalent stress along the chord direction of each section first increases and then decreases. The r/R = 0.10, 0.28, 0.53, and 0.88 sections are located in the pavement section, the number of plies has a sharp transition, and the superimposition of the multiple loads results in stress concentration peak in four sections. The equivalent stress at the section position is too concentrated, and the curve has a sudden peak. In particular, the r/R = 0.16 section is located in the transition area that is between the blade root circle and the blade root airfoil. The relatively large transition of the relative thickness results in large equivalent stress in the middle of the airfoil chord and near the trailing edge; the curve has two obvious peaks. The equivalent stress value of the r/R = 0.60 section is larger than the other five sections. The aerodynamic shape and structural characteristics of the blade at this position indicate that the section bears a large load and is prone to failure. According to the graph, six typical failure locations can be found in Table 5.

Table 5. Failure location and stress.

Location	Failure Location	Max Equivalent Stress/MPa	Azimuth Angle of the Maximum Equivalent Stress
Location 1	r/R = 0.10 x/C = 0.53	10.60	Around 30°
Location 2	r/R = 0.16 x/C = 0.52	17.10	Around 30°
Location 3	r/R = 0.16 x/C = 0.88	15.80	Around 30°
Location 4	r/R = 0.28 x/C = 0.44	9.36	Around 30°
Location 5	r/R = 0.53 x/C = 0.41	9.74	Around 30°
Location 6	r/R = 0.60 x/C = 0.30	20.60	Around 30°
Location 7	r/R = 0.88 x/C = 0.32	5.28	Around 30°

From Figure 10 and Table 5, it can be observed that the load when the blade rotates to different positions is quite different. The maximum value of the equivalent stress at each failure location is around a 30° azimuth angle, and the maximum equivalent stress value at Location 6 is 20.60 MPa. Compared with the incoming positive wind, under the action of a strong wind, when the blade runs from 0–60° and 300–360°, not only is the equivalent stress value large but the influence of the shear stress and normal stress cannot be ignored. Considering this, we explored the typical shear stress and normal stress distribution of the section.

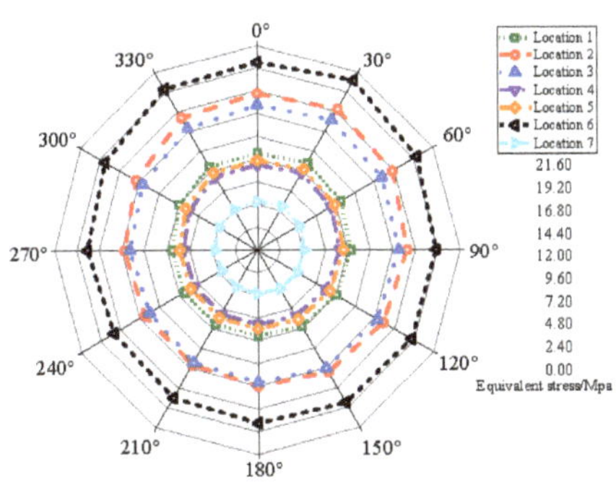

Figure 10. Stress distribution at a typical position.

By analyzing the relationship between the positive and shear stress in Figure 11, it was determined that during the transition from the root to the tip, the normal stress shows a gradual decrease, whereas the change in the shear stress is relatively stable. The normal stress at $r/R = 0.16$ section near the blade root circle and the transition area of the airfoil is the maximum at 4.85 MPa. The normal stress at $r/R = 0.88$ is rapidly reduced to 0.35 MPa, and the maximum shear stress is 2.60 MPa, which is 7.5 times the maximum normal stress of the section. Based on the aforementioned analysis, it can be observed that excessive shear stress is an important factor that leads to the failure of blade roots and blade tips.

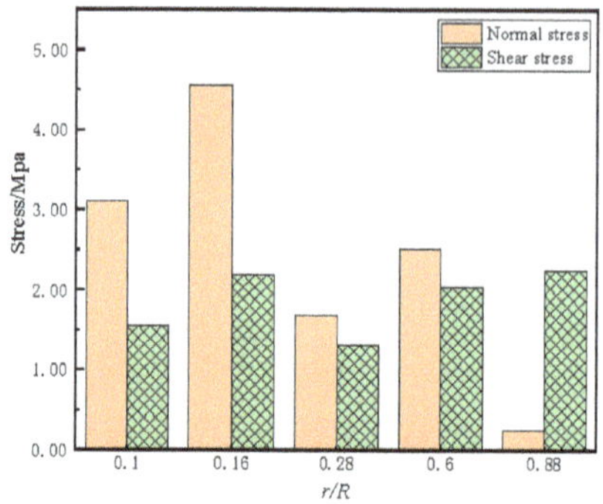

Figure 11. Stress under a 30° azimuth angle.

The composite material that is used for the blade is glass fiber reinforced plastic, and the strength of the blade ensures that the equivalent stress of the blade under a load cannot exceed the material damage limit [21].

$$\sigma_{\max} \leq [\sigma] = \frac{\sigma_s}{\gamma} \tag{9}$$

When the safety factor γ is 3, $\sigma_{\max}$ is the maximum stress of the blade; (σ) is the allowable stress of the blade, in which the value is 73.33 MPa; and σ_s is the yield stress of the blade, in which the value is 220 MPa.

According to Equation (9), the strength of the wind turbine blade is checked based on the maximum stress criterion. The results of the evaluation are shown in Table 6.

Table 6. Strength check of the blade.

Location	$\frac{\sigma_{\max}}{(\sigma)}$
Location 1	14.45%
Location 2	23.32%
Location 3	21.55%
Location 4	12.76%
Location 5	13.28%
Location 6	28.09%
Location 7	7.20%

According to the calculated data in Table 6, it can be determined that under strong wind conditions where the speed is 19 m/s, the maximum blade stress value is 28.09% of the allowable stress. It can be observed that the wind turbine blades that are used in this study are theoretically safe and reliable.

3.3. Research and Investigations of Failure and Damage of the Typical Area Outside the Blade

There are so many techniques for wind turbine damage detection, such as based on strain measurement, vibration, thermography, etc. However, Strain and vibration measurements rely on sticking strain gauges to the blades, and the operation is difficult and dangerous for operators. Although thermal imaging technology has the advantage of not contacting wind turbine blades, it is vulnerable to environmental interference [27–29]. So the method of UAV(Unmanned aerial vehicle) damage detection is more safe and efficient.

When the wind turbine is in a shutdown state, an unmanned aerial vehicle is used to inspect the blades of the wind turbine in service, and the typical external problems of the blades are highlighted through image processing technology. The experimental wind farm is the Huade wind farm that belongs to the Guoshui Group. The farm is located in Ulanqab City in the Inner Mongolia Autonomous Region. All wind turbine models consisted of WGTS1500A, and there were a total of 132 wind turbines. The total capacity of the fourth phase is 200 MW. The scale of each phase is 33 × 1.5 MW. The terrain of the site consists of low mountain hills, gentle slope hills, and there are a few buildings and trees at the site. By taking the first phase as an example, the layout of the wind turbines in the wind farm is shown in Figure 12.

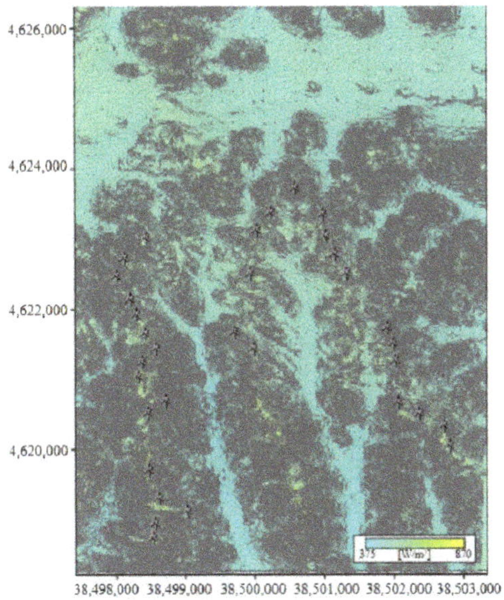

Figure 12. Schematic diagram of the layout of the wind turbine units.

As shown in Figure 12, it is significant that most wind turbines are distributed in places with a high wind power density. Before the experiment, the wind turbines that were to be tested were shut down, and the 132 wind turbines that consist of the WTGS1500(Wind turbine generator system1500) model and belong to the wind farm were inspected by drones. When the wind turbine is in a shutdown state, an unmanned aerial vehicle is used to inspect the blades of the wind turbine in service, and the typical external problems of the blades are highlighted through image processing technology. Finally, a series of blade damage pictures were collected. By combining the results of the blade stress concentration area that were obtained by a numerical simulation analysis and comparing the damage results that were observed on-site, we can comprehensively determine the cause of the blade damage, and provide guidance for the maintenance of the blade.

As shown in Figure 13, according to the distribution of the failure locations, this investigation divides the wind turbine blade into three regions along the span direction, namely the root region r/R = 0.00–0.16, the middle region r/R = 0.28–0.60, and the tip region r/R = 0.88–1.00. By taking a wind farm in western Inner Mongolia as a research sample, this study used an unmanned aerial vehicle that was equipped with a 60-megapixel lens.

The blades of 132 wind turbines were inspected sequentially, and the common failure and damage forms in each area of the wind turbine blades at strong winds were summarized.

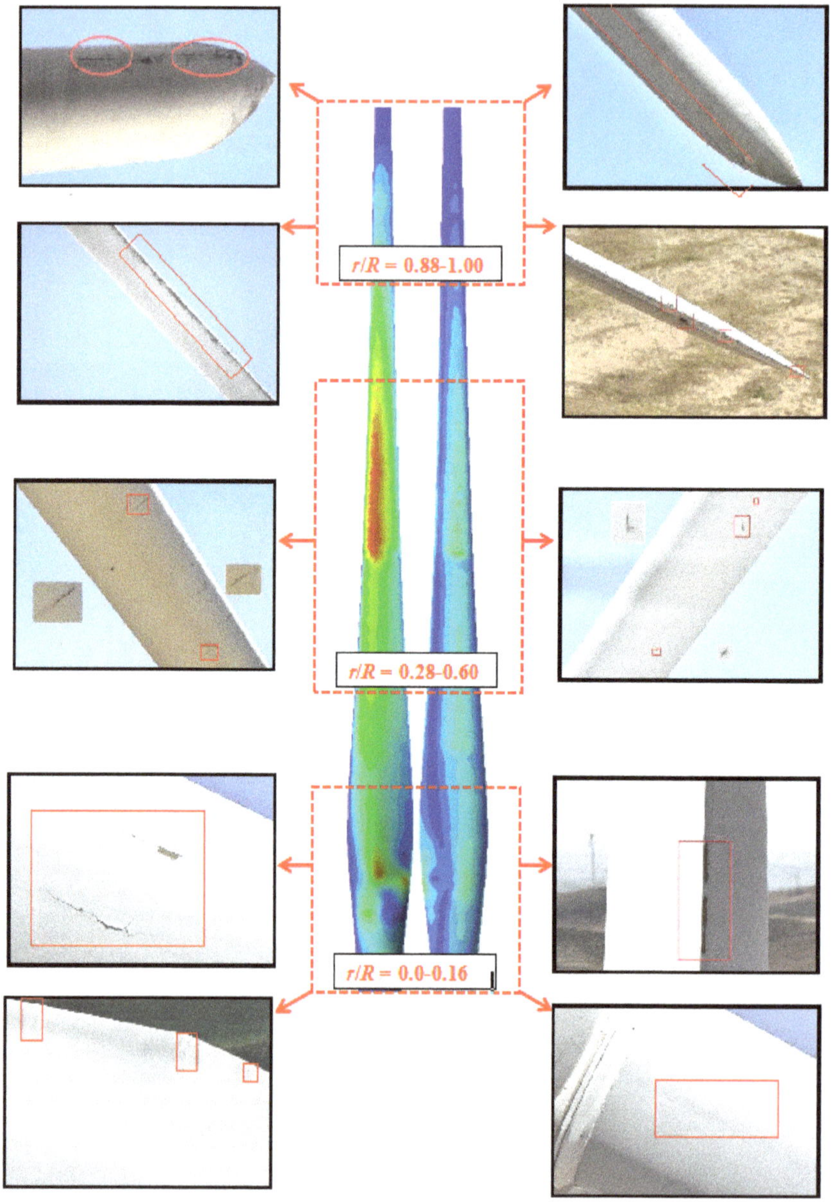

Figure 13. Common damage and failure of the blades.

After inspecting a total of 396 blades from the 132 wind turbines in this wind farm, the failure damage was assessed. Figure 13 is the cloud diagram of the equivalent stress distribution of the wind turbine blade at 30° azimuth calculated by the fluid-structure coupling numerical simulation, which is compared and correlated with the actual damaged area of the blade. The types of failure damage that appear in order from light to heavy are: paint peeling, oil pollution, scratching injuries, bulges, cracks, cracks in the gel coat, and lightning damage. The afore-mentioned damages are the most common and frequent failure damage modes for this type of 1.5 MW wind turbine blade in this wind farm.

The types of blade damage and the common types of wind turbine blade damage are summarized. In this batch of wind turbine blade inspections, a total of 8917 images were taken, which include 2180 blade damage images and 6737 images without blade damage. According to the calculations, damaged images account for 24.45% of the total number of images (most of them were paint peeling and oil stains, etc.), which contain 1988 images, that account for 22.29% of the total number of images; the rest are scratches, cracks, gel coat cracking, lightning strikes, and bulges, which consist of 192 images. Among them, the damage that is caused by stress includes cracks and gel coat cracking in about 58 images, which accounts for 0.65% of the total number of images.

Based on the stress distribution of the wind turbine blades in the previous subsection, after identifying and classifying the captured damage images, the damage of the blade that is caused by stress includes cracks, gel coat cracks; the various types of damage that were counted. Figure 14 shows the gel coat cracking phenomenon that is caused by stress; Figure 15 displays the crack image of the blade that is caused by stress.

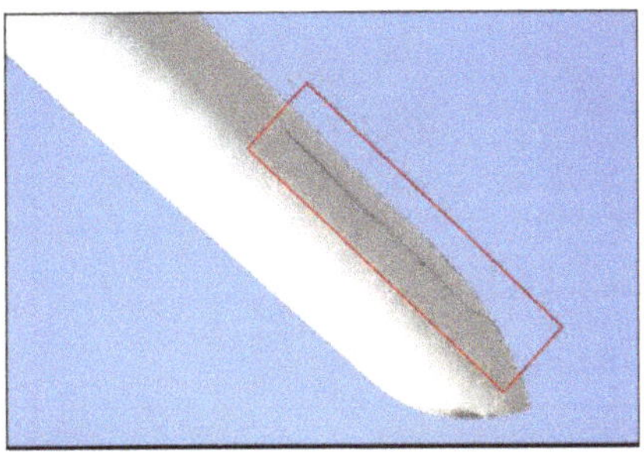

(**a**) Gel coat cracking in the blade tip area

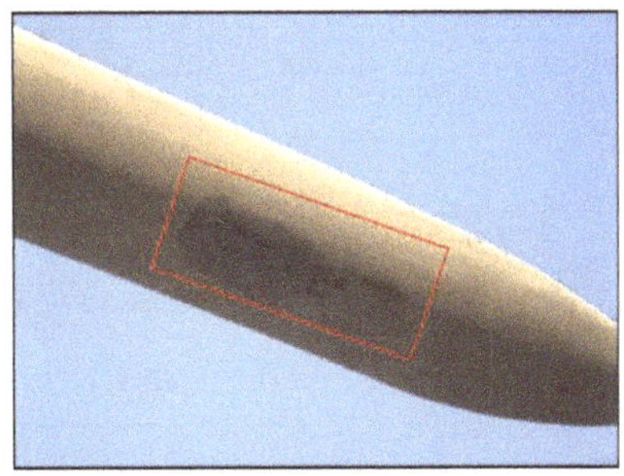

(**b**) Gel coat cracking in the blade tip area

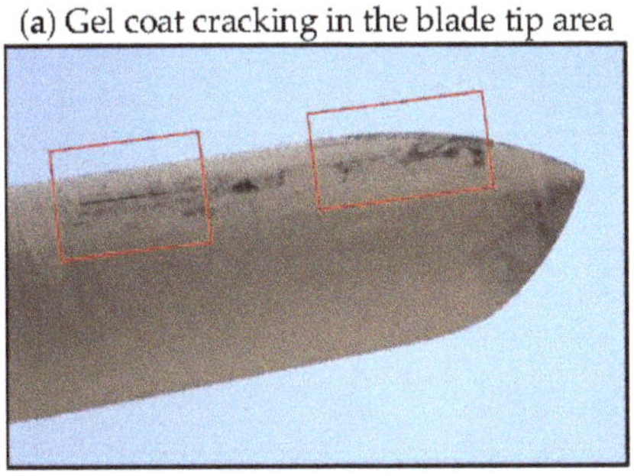

(**c**) Gel coat cracking in the blade tip area

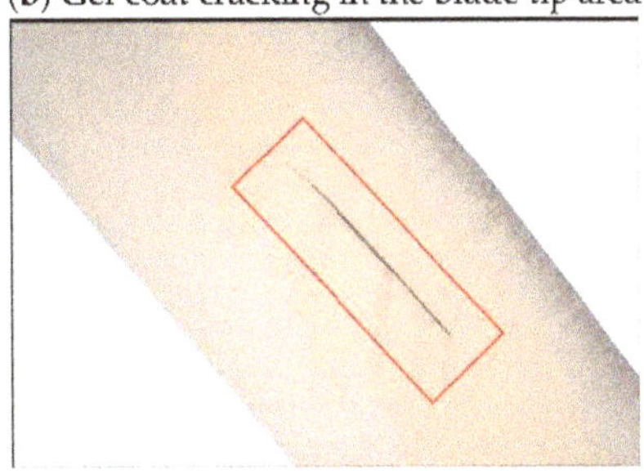

(**d**) Gel coat cracking in the blade middle area

Figure 14. Cracking of the gel coat caused by stress.

Among them, the frequency of the various types of damage = the number of various types of damage/the total number of damage. The frequency for a certain type of damage in each area = the number of certain types of damage in each area/the total number of damage for a certain type. The failure conditions are shown in Tables 7 and 8.

It can be observed from Table 7 that by screening the damage images of the wind turbine blade failure that is due to stress, the most frequent failure type are cracks. The number of images is 49, which accounts for 84.48% of the damage that is caused by stress. The gel coat crack is another form of failure caused by stress. The number of images is nine, which accounts for 15.52% of the damage that is caused by stress.

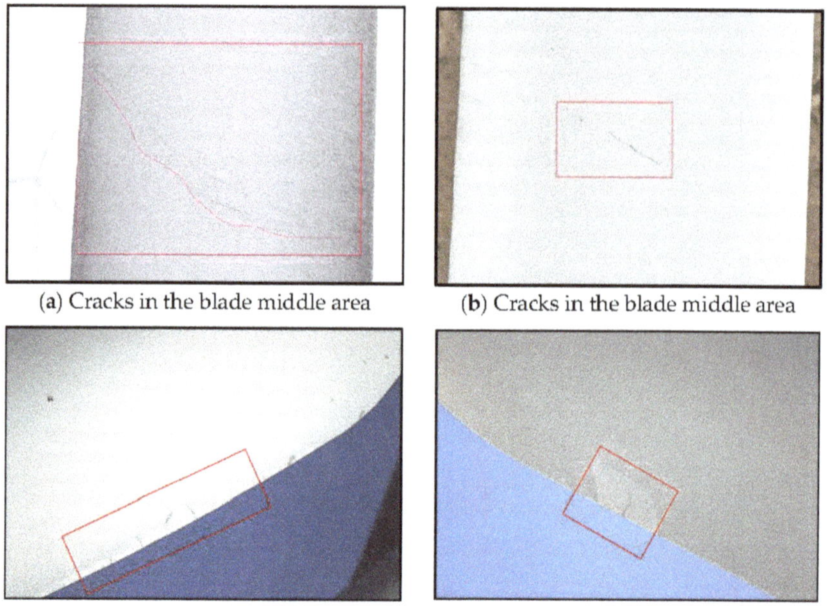

(**a**) Cracks in the blade middle area (**b**) Cracks in the blade middle area

(**c**) Cracks in the blade root area (**d**) Cracks in the blade root area

Figure 15. Cracks caused by stress.

Table 7. Failures of the blade caused by stress.

Damage Types Caused by Stress	Gel Coat Cracking	Cracks
Damage numbers	58	
Damage type numbers	9	49
Frequency	15.52%	84.48%

Table 8. Frequency of most common types of failurs of blades.

Damage Types	Gel Coat Cracking		Cracks	
Typical area	Number	Frequency	Number	Frequency
Blade root $r/R = 0.00$–0.16	0	0.00%	43	87.75%
Blade middle $r/R = 0.28$–0.60	2	22.22%	5	10.20%
Blade tip $r/R = 0.28$–0.60	7	77.78%	1	2.05%

According to the classification of the typical areas, we can summarize the frequency of the main failure damage that is caused by the stress in each area, and this is summarized in Table 8. There were nine cases of gel coat cracking; seven times, it occurred along the tip of the blade, and the frequency was 77.78%. It occurred two times along the middle of the blade with a frequency of 22.22%; however, no gel coat cracking occurred at the root of the blade. A total of 49 cracks appeared in the root of the blade, and the number and frequency were 43 times and 87.75%, respectively. There were three cracks in the blade, and the frequency of occurrence was 10.20%. The number of cracks appeared less at the tip of the blade, in which the frequency was once, and the probability was 2.05%. It can be observed that the main failure mode of the blade tip area is gel coat cracking, the root area of the blade is the main failure mode, and gel coat cracking and cracking in the blade area occur, but the frequency is low.

According to the image statistics, when considering all the failure types that are caused by stress, the frequency of cracks in the root area (r/R = 0.00–0.16) is as high as 87.75%. However, the cracks in this area are relatively slight and they mostly exist in the surface layer of the trailing edge of the blade root area. These cracks are insufficient to bring destructive hazards to the wind turbine blade. Regular repairs for this crack type can maintain the normal operation of the wind turbine, the maintenance is convenient, and the cost is low. However, the frequency of cracks in the mid-blade area (r/R = 0.28–0.60) also reached 10.20%. According to the results of the field survey, once a chordal crack in the mid-blade area appears, it may be fatal to the blade and the whole machine. Because of the damage, the maintenance is inconvenient and high-cost.

This investigation compared the results of the blade stress concentration area that was obtained by the numerical simulation analysis with the blade damaged area that was obtained from the field observation. It was determined that the actual damage position of the blade is mostly near the stress concentration area that is calculated by the numerical simulation. In addition, the damage failure mode can confirm the blade force characteristics. The results can guide drone inspections.

4. Conclusions

This paper discusses the fluid–structure coupling calculation method in order to explore the stress distribution of a 1.5 MW wind turbine during strong winds. By combining the numerical simulation results and the UAV observation experiments, we can comprehensively determine the failure area and failure mode of the blade and draw the following conclusions:

1. The maximum stress of the blade near a 30° azimuth angle is 20.6 MPa. This is because the resultant force on the blade is the largest when the azimuth angle is 30°. By having an azimuth angle of 30° there are obvious equivalent stress peaks in the six sections of the blade, and the most vulnerable position is r/R = 0.60 and x/C = 0.30. The maximum stress value of the blade under strong wind conditions is 28.09% of the allowable stress. Theoretically, the blade structure is safe.
2. When considering the UAV's inspection of 132 wind turbines in the wind farm, 2145 failure images of the blades were collected. The damaged images account for 24.45% of the total number of images (i.e., paint peeling and oil pollution are the most frequent failure types at the end of the wind farm's service, which appear 1988 times and account for 22.29% of the total number of image samples). The rest include scratches, cracks, bulges, gel coat cracks and so on, which resulted in a total of 192 photos.
3. Cracks appeared 49 times in total, with more occurrences at the root of the blade, 43 times with a frequency of 87.75%, respectively. There were three occurrences along the middle of the blade, with a frequency of 10.20%; and there were fewer occurrences at the tip of the blade, which occurred only once and the probability was 2.05%. Therefore, the main failure mode of the blade tip is gel coat cracking, the root of the blade is mainly cracked, and gel coat cracking and cracking in the blade occur, but the frequency is low.
4. It was determined that the actual damage position of the blade is mostly near the stress concentration area that is calculated by the numerical simulation. The damage failure mode can confirm the force characteristics of the blade. These results can provide guidance for drone inspections.

Author Contributions: K.T. contributed to the conception of the study and wrote the manuscript. L.S. reviewed and edited the manuscript. Y.C. was responsible for preparing the methodology. X.J. was responsible for the technical support. R.F. provided the experimental data support. R.T. helped perform the analysis with constructive discussions. All authors have read and agreed to the published version of the manuscript.

Funding: This research was funded by the Inner Mongolia Science and Technology Project Plan (2019), which was funded by the Department of Science and Technology in the Inner Mongolia Autonomous Region. (No grant number).

Data Availability Statement: Not applicable.

Acknowledgments: The wind speed model and wind field environmental data that were selected for the numerical simulation in this article were provided by the Inner Mongolia Electric Power Research Institute Branch of the Inner Mongolia Electric Power (Group) Co., Ltd. and the Guoshui Group Huade Wind Power Co., Ltd. We would like to express our sincere thanks to them.

Conflicts of Interest: The authors declare no conflict of interest.

Abbreviations

ACP	ANSYS Composite PrepPost
α	the wind shear coefficient
E	Elastic Modulus
FRP	Fiber-reinforced plastic
G	Shear modulus
G	Gravity
x	Chord length along chordwise
C	Chord length
r	Blade length along spanwise
R	Blade length
UDF	User-Defined Function
v	Poisson's ratio
v	Wind speed
G_z	Spanwise component of gravity
G_x	Gravity along the blade rotation direction
v_{ref}	Known wind speed at Z_{ref}
Z_{ref}	Height at the center of the hub
q	Centrifugal force load
θ	Azimuth angle

References

1. Santo, G.; Peeters, M.; Van Paepegem, W.; Degroote, J. Dynamic load and stress analysis of a large horizontal axis wind turbine using full scale fluid-structure interaction simulation. *Renew. Energy* **2019**, *140*, 212–226. [CrossRef]
2. Zhu, J.; Cai, X.; Pan, P.; Gu, R.R. Static and Dynamic Characteristics Study of Wind Turbine Blade. *Adv. Mater. Res.* **2012**, *433-440*, 438–443. [CrossRef]
3. Fernandez, G.; Usabiaga, H.; Vandepitte, D. An efficient procedure for the calculation of the stress distribution in a wind turbine blade under aerodynamic loads. *J. Wind Eng. Ind. Aerodyn.* **2018**, *172*, 42–54. [CrossRef]
4. Jiang, X.; Gao, Z.; Wang, J.; Bai, Y.; Yan, M.; Wang, X. Experiment on correlation of wind turbine strain and tower vibration. *J. Drain. Irrig. Mach. Eng.* **2017**, *35*, 685.
5. Kim, H.-I.; Han, J.-H.; Bang, H.-J. Real-time deformed shape estimation of a wind turbine blade using distributed fiber Bragg grating sensors. *Wind Energy* **2014**, *17*, 1455–1467. [CrossRef]
6. Zhang, J.; Guo, L.; Wu, H.; Zhou, A.; Hu, D.; Ren, J. The influence of wind shear on vibration of geometrically nonlinear wind turbine blade under fluid–structure interaction. *Ocean Eng.* **2014**, *84*, 14–19. [CrossRef]
7. Cheng, Y.; Xue, Z.; Jiang, T.; Wang, W.; Wang, Y. Numerical simulation on dynamic response of flexible multi-body tower blade coupling in large wind turbine. *Energy* **2018**, *152*, 601–612. [CrossRef]
8. Zhang, J.P.; Li, D.L.; Liu, Y.; Wu, H.L.; Ren, J.X.; Pan, W.G. Dynamic Response Analysis of Large Wind Turbine Blade Based on Davenport Wind Speed Model. *Adv. Mater. Res.* **2011**, *347-353*, 2330–2336. [CrossRef]
9. Zhou, B.; Wang, X.; Zheng, C.; Cao, J.; Zou, P. Finite Element Analysis for the Web Offset of Wind Turbine Blade. *IOP Conf. Ser. Earth Environ. Sci.* **2017**, *63*, 12011. [CrossRef]
10. Choudhury, S.; Sharma, T.; Shukla, K. Effect of orthotropy ratio of the shear web on the aero-elasticity and torque generation of a hybrid wind turbine blade. *Renew. Energy* **2017**, *113*, 1378–1387. [CrossRef]
11. Dimitrov, N.; Natarajan, A.; Kelly, M. Model of wind shear conditional on turbulence and its impact on wind turbine loads. *Wind Energy* **2014**, *18*, 1917–1931. [CrossRef]
12. Ageze, M.B.; Hu, Y.; Wu, H. Comparative Study on Uni- and Bi-Directional Fluid Structure Coupling of Wind Turbine Blades. *Energies* **2017**, *10*, 1499. [CrossRef]

13. Zhu, R.-S.; Zhao, H.-L.; Peng, J.-Y.; Li, J.-P.; Wang, S.-Q.; Zhao, H. Numerical Investigation on Fluid-structure Coupling of 3 MW Wind Turbine Blades. *Int. J. Green Energy* **2015**, *13*, 241–247. [CrossRef]
14. Bae, S.-Y.; Kim, Y.-H. Structural design and analysis of large wind turbine blade. *Mod. Phys. Lett. B* **2019**, *33*, 1940032. [CrossRef]
15. Ullah, H.; Ullah, B.; Riaz, M.; Iqbal, M.; Badshah, A. Structural Analysis of a Large Composite Wind Turbine Blade under Extreme Loading. In *2018 International Conference on Power Generation Systems and Renewable Energy Technologies, Islamabad, Pakistan, 10–12 September 2018*; IEEE: Piscataway, NJ, USA, 2018.
16. Shen, X.; Zhu, X.; Du, Z. Wind turbine aerodynamics and loads control in wind shear flow. *Energy* **2011**, *36*, 1424–1434. [CrossRef]
17. Fu, B.; Zhao, J.; Li, B.; Yao, J.; Teifouet, A.R.M.; Sun, L.; Wang, Z. Fatigue reliability analysis of wind turbine tower under random wind load. *Struct. Saf.* **2020**, *87*, 101982. [CrossRef]
18. Guo, S.; Li, Y.; Chen, W. Analysis on dynamic interaction between flexible bodies of large-sized wind turbine and its response to random wind loads. *Renew. Energy* **2020**, *163*, 123–137. [CrossRef]
19. Wang, H.; Ke, S.T.; Wang, T.G.; Zhu, S.Y. Typhoon-induced vibration response and the working mechanism of large wind turbine considering multi-stage effects. *Renew. Energy* **2020**, *153*, 740–758. [CrossRef]
20. Boujleben, A.; Ibrahimbegovic, A.; Lefrançois, E. An efficient computational model for fluid-structure interaction in application to large overall motion of wind turbine with flexible blades. *Appl. Math. Model.* **2020**, *77*, 392–407. [CrossRef]
21. Peeters, M.; Santo, G.; Degroote, J.; Van Paepegem, W. Comparison of Shell and Solid Finite Element Models for the Static Certification Tests of a 43 m Wind Turbine Blade. *Energies* **2018**, *11*, 1346. [CrossRef]
22. Wang, L.; Quant, R.; Kolios, A. Fluid structure interaction modelling of horizontal-axis wind turbine blades based on CFD and FEA. *J. Wind Eng. Ind. Aerodyn.* **2016**, *158*, 11–25. [CrossRef]
23. Tran, T.; Kim, D.; Song, J. Computational Fluid Dynamic Analysis of a Floating Offshore Wind Turbine Experiencing Platform Pitching Motion. *Energies* **2014**, *7*, 5011–5026. [CrossRef]
24. Steijl, R.; Barakos, G. Sliding mesh algorithm for CFD analysis of helicopter rotor-fuselage aerodynamics. *Int. J. Numer. Methods Fluids* **2008**, *58*, 527–549. [CrossRef]
25. Tian, K.; Song, L.; Jiao, X.; Feng, R.; Chen, Y.; Tian, R. Aeroelastic stability analysis and failure damage evaluation of wind turbine blades under variable conditions. *Energy Sources Part A Recover. Util. Environ. Eff.* **2021**, 1–18. [CrossRef]
26. Song, L.; Li, Q.; Chen, W.; Qin, P.; Huang, H.H.; Yuncheng, H. Wind characteristics of a strong typhoon in marine surface boundary layer. *Wind Struct.* **2012**, *15*, 1–15. [CrossRef]
27. Mao, Y.; Wang, S.; Yu, D.; Zhao, J. Automatic image detection of multi-type surface defects on wind turbine blades based on cascade deep learning network. *Intell. Data Anal.* **2021**, *25*, 463–482. [CrossRef]
28. Xu, D.; Wen, C.; Liu, J. Wind turbine blade surface inspection based on deep learning and UAV-taken images. *J. Renew. Sustain. Energy* **2019**, *11*, 053305. [CrossRef]
29. Toft, H.S.; Branner, K.; Berring, P.; Sorensen, J.D. Defect distribution and reliability assessment of wind turbine blades. *Eng. Struct.* **2011**, *33*, 171–180. [CrossRef]

energies

MDPI

Article

Modeling and Design of High-Power Enhanced Leakage-Inductance-Integrated Medium-Frequency Transformers for DAB Converters

Xinsheng Zhang, Fei Xiao, Ruitian Wang *, Wei Kang and Beichao Yang

National Key Laboratory of Science and Technology on Vessel Integrated Power System, Naval University of Engineering, Wuhan 430033, China; marvinzxs@163.com (X.Z.); xfeyninger@gmail.com (F.X.); kangxiaodouXJTU@163.com (W.K.); yangbeichao199405@163.com (B.Y.)
* Correspondence: wangrt4321@163.com

Abstract: For dual active bridge (DAB) converters, integrating the phase-shifting inductance (PSI) in the medium-frequency transformer (MFT) is an effective way to improve the overall power density. Different from the existing leakage-inductance-integrated (LII) structure, a concentric-winding (CW) enhanced leakage-inductance-integrated (ELII) structure, which includes an additional core, is proposed in this paper. In order to explain the operating mode of CW ELII MFT, a magnetic circuit model is established, and the analysis is carried out under the typical DAB excitation. The total leakage inductance of CW ELII MFT is divided into the winding leakage inductance and the additional leakage inductance for calculation. The integrated structure makes the heat dissipation of the MFT challenging. Therefore, the air–water combined cooling method is adopted in the design. A thermal resistance model is built for the winding air channel under forced convection. On this basis, MFT designs with different integration structures for different leakage inductance requirements are compared. Finally, a 200 kW/4 kHz/200 μH MFT prototype was designed and manufactured, which achieved the power density of 5.16 kW/dm^3 and the efficiency of 99.30%. The prototype was tested in a DAB converter, which is a module of a 2 MW modular multilevel converter-bidirectional DC–DC converter (MMC-BDC).

Keywords: medium-frequency transformer; inductor integration; DAB converter; optimal design

Citation: Zhang, X.; Xiao, F.; Wang, R.; Kang, W.; Yang, B. Modeling and Design of High-Power Enhanced Leakage-Inductance-Integrated Medium-Frequency Transformers for DAB Converters. *Energies* **2022**, *15*, 1361. https://doi.org/10.3390/en15041361

Academic Editors: Alon Kuperman and Alessandro Lampasi

Received: 27 January 2022
Accepted: 10 February 2022
Published: 14 February 2022

1. Introduction

In recent years, benefiting from advances in switching devices, magnetic materials, and control methods, power conversion systems based on MFTs have gradually increased. With the increase of operating frequency, the volume and weight of the transformer can be reduced, thus improving the power density of the converter. Therefore, the MFT has broad application prospects in many fields—such as flexible DC transmission, renewable energy grid connection, and electrified transportation [1–3]. A typical application occasion is vessel integrated power system (IPS) [4], and a modular multi-level converter-bidirectional DC/DC converter (MMC-BDC) for IPS is shown in Figure 1 [5]. The converter adopts multiple DAB converters to realize DC/DC energy conversion and bidirectional flow, which can serve DC loads or energy storage elements.

In the topology of DAB, MFT provides voltage matching and galvanic isolation. The phase-shifting inductance (PSI) is determined by the power capacity of the DAB converter and provides the instantaneous energy storage in the conversion process [6]. In the reported high-power DAB prototypes, an auxiliary inductor and an MFT is usually used. The volume and weight of the auxiliary inductor may even be close to the MFT [3]. In [7], the integrated/non-integrated MFT designs are compared. Although the integrated design has obvious advantages in power density, the non-integrated design is finally chosen for the sake of heat dissipation and installation flexibility. In [8–10], the design method for leakage-inductance-integrated (LII) MFT has been investigated. However, in order to achieve a larger leakage inductance, the LII MFT requires more winding turns or a greater primary-secondary distance, and the unconstrained leakage flux will cause additional losses in tape wound cores. Therefore, it is hard for the LII MFT to meet the requirements of the high inductance and the high power.

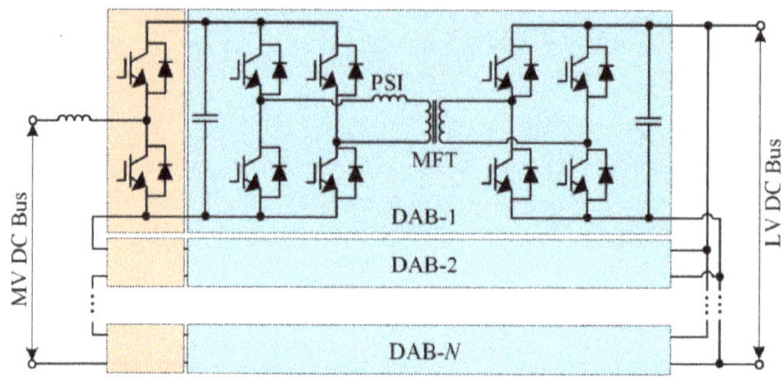

Figure 1. The modular multilevel converter–bidirectional DC–DC converter.

Thermal performance is one of the decisive factors for the success of MFT design. Natural air cooling is the most common heat dissipation method for MFTs. In [11] and [12], the thermal network modeling method for natural air-cooled MFT is comprehensively described. In order to increase power density, water cooling is adopted in [13] and [14], and forced air cooling is adopted in [3,15]. In [3], the forced convection formula for circular tubes is adopted for thermal modeling without considering the structural characteristics of the winding air channel.

Many achievements have emerged in the optimal design of MFT in recent years, which further pushes the MFT to large-scale industrial applications. In [16,17], the brute force grid search method is adopted to optimize the design of a 100 kW/10 kHz shell-type MFT. The sensitivity analysis of design parameters is carried out based on the large-scaled calculation data. In [18], the researchers attempt to establish a unified mathematical model for transformer/inductor optimal design based on existing classical parameter models and carried out the optimal design with the goal of minimum core and minimum loss. A 200 kVA/10 kHz shell-type MFT is designed based on the genetic algorithm in [10], and the optimal design of the heat sink is embedded in the design process. The above work is carried out for typical core-type or shell-type MFT. The design of inductor-integrated MFT is more challenging due to the parameter coupling and the integration structure.

Compared with the previous work, the main contributions of the work in this paper include the following: (1) Different from the conventional leakage-inductance-integrated (LII) MFT, a concentric-winding enhanced leakage-inductance-integrated (CW ELII) MFT, which includes an additional core, is proposed in this paper. Moreover, a developed example of 200 kW, 4 kHz, 200 μH MFT with CW ELII structure is presented specifically. (2) The air channel in the forced air-cooled winding is simplified to the parallel-plate structure, and the thermal resistance model is established based on classic models. (3) For different leakage inductance requirements, the power density of the CW LII and CW ELII solutions are quantitatively compared.

This paper is organized as follows. Section 2 discusses different integration structures and focuses on the magnetic circuit analysis of the CW ELII MFT in DAB. Section 3 introduces the calculation method of the loss and the leakage inductance. Section 4 presents the thermal modeling method of air–water combined cooled MFT. In Section 5, the design method and quantitative comparison of the CW LII and CW ELII MFT are presented. The developed 200 kW, 4 kHz, 200 μH MFT prototype, and experimental results are also presented in Section 5. Finally, Section 6 lays out the conclusions and future work possibilities.

2. Integration Structures and Operating Mode

2.1. Integration Structures

The inductor-integrated MFT can be divided into three categories: detached-inductor integrated (DII), leakage-inductance-integrated (LII), and enhanced leakage-inductance-integrated (ELII) MFT. The characteristic of DII MFT is that it has independent inductor windings. The inductor core and the transformer core are integrated or just packaged together. The DII MFT will not be focused on in this paper.

LII MFT features no additional winding or core. The target leakage inductance is achieved only through the design of the winding structure. Figure 2a, b show typical separate-winding (SW) and concentric-winding (CW) structures. It is usually easier to achieve a greater leakage inductance by the SW structure [3]. However, since the magnetic motive force (MMF) on the core limb of the SW MFT is unbalanced, the leakage flux is widely distributed in the external space, which may lead to eddy losses in the adjacent metal structure. Therefore, the SW MFT should be used in compact high-power converters with caution. No matter the SW or the CW structure, part of the leakage flux will pass through the core tape vertically and cause apparent eddy current loss of the wound core [19].

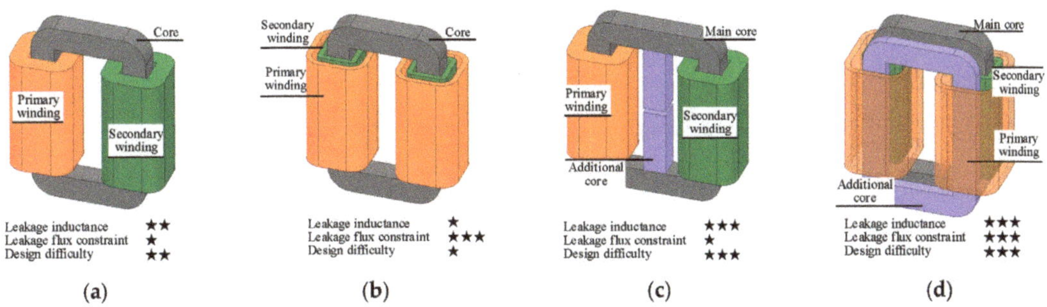

Figure 2. Comparison of different integration structures. (**a**) SW LII, (**b**) CW LII, (**c**) SW ELII, (**d**) CW ELII.

The characteristic of ELII MFT is that the additional core is used instead of the air in the leakage flux path of the LII MFT. Due to the higher equivalent permeability of the additional core, it is possible to achieve a higher power density. Figure 2c shows the SW ELII structure, which will cause the external leakage flux like the SW LII structure. The flux from the main core to the additional core limb will pass through the core tape vertically, which will cause local additional core loss [19]. Figure 2d shows the CW ELII structure, in which the additional independent core restricts the flow path of the leakage flux. Therefore, the CW ELII structure may be more suitable for the requirements of the high power and high inductance.

2.2. CW ELII MFT Operating Mode in DAB

The circuit topology of DAB is shown in Figure 3. The MFT is excited by H-bridges on both sides simultaneously, and the operating condition is determined by the duty cycle and the phase-shifting duty of the two bridges. In order to analyze the operating mode of the CW ELII MFT under typical DAB excitation, the magnetic circuit model is established. As shown in Figure 4, the voltage excitations u_p and u_s at the primary and the secondary sides work as flux sources Φ_p and Φ_s in the same direction in the magnetic circuit.

$$\begin{cases} N_p\Phi_p(t) = \int u_p(t)dt \\ N_s\Phi_s(t) = \int u_s(t)dt \end{cases} \tag{1}$$

where N_p and N_s are the number of turns of the primary and secondary windings.

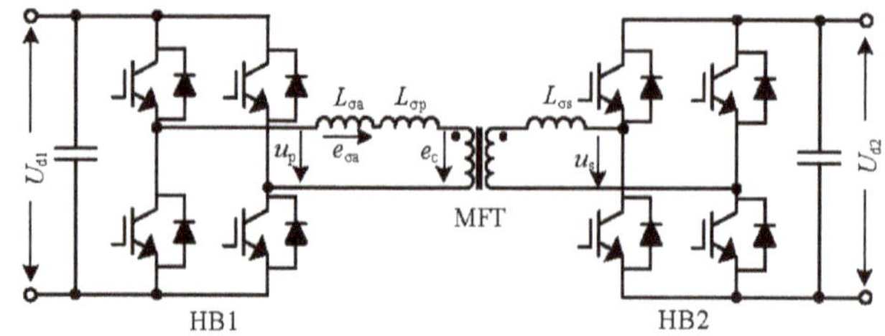

Figure 3. DAB circuit topology.

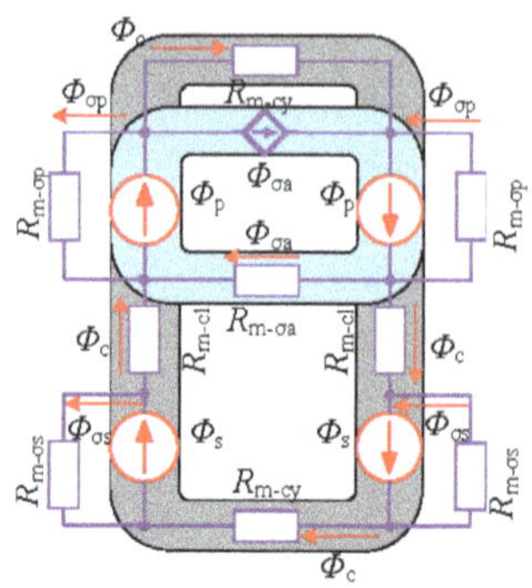

Figure 4. Magnetic circuit model of CW ELII MFT.

In Figure 4, $R_{\text{m-cl}}$ and $R_{\text{m-cy}}$ are respectively the reluctance of the core limb and core yoke; $R_{\text{m-}\sigma\text{p}}$ and $R_{\text{m-}\sigma\text{s}}$ are respectively the air leakage reluctance of the primary winding and the secondary winding; $R_{\text{m-}\sigma\text{a}}$ is the reluctance of the additional core. The magnetic circuit equation is given as

$$\begin{cases} \Phi_{\sigma a} R_{m-\sigma a} - 2\Phi_{\sigma p} R_{m-\sigma p} + \Phi_c R_{m-cy} = 0 \\ \Phi_{\sigma a} R_{m-\sigma a} - 2\Phi_c R_{m-cl} + 2\Phi_{\sigma s} R_{m-\sigma s} - \Phi_c R_{m-cy} = 0 \\ \Phi_{\sigma a} + \Phi_{\sigma p} + \Phi_c = \Phi_p \\ \Phi_{\sigma s} + \Phi_c = \Phi_s \end{cases} \tag{2}$$

According to the reluctances, the additional leakage inductance $L_{\sigma a}$, primary leakage inductance $L_{\sigma p}$, and secondary leakage inductance $L'_{\sigma s}$ (in primary) can be defined as

$$\begin{cases} L_{\sigma a} = N_p^2 / R_{m-\sigma a} \\ L_{\sigma p} = N_p^2 / 2R_{m-\sigma p} \\ L'_{\sigma s} = N_p^2 / 2R_{m-\sigma s} \end{cases} \tag{3}$$

Since the main core is without air gaps, the reluctances $R_{\text{m-cl}}$ and $R_{\text{m-cy}}$ are much smaller than those of the air or the additional core and can be assumed to be zero. Thus, according to (1) and (2), the magnetic fluxes can be obtained.

$$\begin{cases} \Phi_c = \frac{L'_{\sigma s}\Phi_p + (L_{\sigma p} + L_{\sigma a})\Phi_s}{L'_{\sigma s} + L_{\sigma p} + L_{\sigma a}} \\ \Phi_{\sigma a} = \frac{L_{\sigma a}(\Phi_p - \Phi_s)}{L'_{\sigma s} + L_{\sigma p} + L_{\sigma a}} \\ \Phi_{\sigma p} = \frac{L_{\sigma p}(\Phi_p - \Phi_s)}{L'_{\sigma s} + L_{\sigma p} + L_{\sigma a}} \\ \Phi_{\sigma s} = \frac{-L'_{\sigma s}(\Phi_p - \Phi_s)}{L'_{\sigma s} + L_{\sigma p} + L_{\sigma a}} \end{cases} \tag{4}$$

Take typical DAB excitation as an example: the duty of two H-bridges $D_h = 0.5$; phase-shifting duty $0 < D_s < 0.5$. Then the voltage excitations of MFT, u_p and u_s, are given as

$$\begin{aligned} u_p(t) &= \begin{cases} U_{d1}, [0, 0.5T] \\ -U_{d1}, [0.5T, T] \end{cases} \\ u_s(t) &= \frac{U_{d2}}{U_{d1}} u_p(t - D_s T) \end{aligned} \tag{5}$$

where U_{d1} is the DC bus voltage of the primary side, and U_{d2} is the DC bus voltage of the secondary side.

The induced electromotive force $e_c(t)$ of the main core and $e_{\sigma a}(t)$ of the additional core can be calculated according to Equations (1), (4), and (5).

$$-e_c(t) = \begin{cases} \frac{L'_{\sigma s} - L_{\sigma p} - L_{\sigma a}}{L'_{\sigma s} + L_{\sigma p} + L_{\sigma a}} U_{d1}, [0, D_s T] \\ U_{d1}, [D_s T, 0.5T] \\ \frac{-L'_{\sigma s} + L_{\sigma p} + L_{\sigma a}}{L'_{\sigma s} + L_{\sigma p} + L_{\sigma a}} U_{d1}, [0.5T, (0.5 + D_s)T] \\ -U_{d1}, [(0.5 + D_s)T, T] \end{cases} \tag{6}$$

$$-e_{\sigma a}(t) = \begin{cases} \frac{2L_{\sigma a}}{L'_{\sigma s} + L_{\sigma p} + L_{\sigma a}} U_{d1}, [0, D_s T] \\ 0, [D_s T, 0.5T] \\ -\frac{2L_{\sigma a}}{L'_{\sigma s} + L_{\sigma p} + L_{\sigma a}} U_{d1}, [0.5T, (0.5 + D_s)T] \\ 0, [(0.5 + D_s)T, T] \end{cases} \tag{7}$$

For example, the turns ratio of the MFT satisfies $N_p:N_s = U_{d1}:U_{d2}$, and the leakage inductances satisfies $L_{\sigma a} + L_{\sigma p} + L'_{\sigma s} = 5L_{\sigma p} = 5L'_{\sigma s}$. Then the voltage and flux waveforms with the phase-shifting duty of 0 and 0.25 are shown in Figure 5.

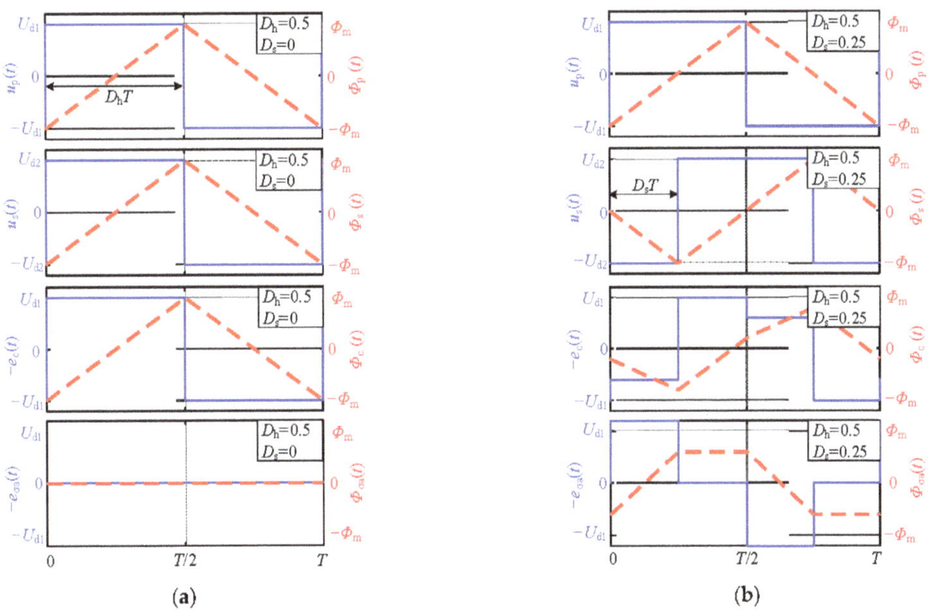

Figure 5. Voltage, induced electromotive force, and flux waveforms of CW ELII MFT under typical DAB excitation. (**a**) $D_h = 0.5$, $D_s = 0$; (**b**) $D_h = 0.5$, $D_s = 0.25$.

3. Loss and Leakage Inductance Calculation

Loss and inductance are important electrical parameters of MFT. In view of the structure particularity of the CW ELII MFT, this section discusses the calculation methods of the loss and the leakage inductance based on the magnetic circuit analysis above.

3.1. Core Loss of CW ELII MFT

The core loss of CW ELII MFT contains the main core loss and the additional core loss. Core loss under non-sinusoidal excitation can be calculated by the improved generalized Steinmetz equation (iGSE) [20], in which the peak value and the differential of the flux density are needed. Based on the magnetic circuit analysis and the assumption that $L_{\sigma p} = L'_{\sigma s}$, the peak value and the differential of the flux densities can be derived from (6) and (7) as

$$B_{\mathrm{m}} = \frac{U_{\mathrm{d1}}}{2N_{\mathrm{p}} f A_{\mathrm{cv}}}\left[\frac{1}{2} - \frac{L_{\sigma\mathrm{w}}}{L_{\mathrm{int}}}D_{\mathrm{s}}\right] \tag{8}$$

$$B_{\mathrm{ma}} = \frac{U_{\mathrm{d1}}}{N_{\mathrm{p}} f A_{\mathrm{acv}}}\frac{L_{\sigma\mathrm{a}}}{L_{\mathrm{int}}}D_{\mathrm{s}} \tag{9}$$

$$\left|\frac{\mathrm{d}B_{\mathrm{m}}}{\mathrm{d}t}\right| = \begin{cases} \frac{L_{\sigma\mathrm{a}}}{L_{\mathrm{int}}}\frac{U_{\mathrm{d1}}}{N_{\mathrm{p}}A_{\mathrm{cv}}}, [0, D_{\mathrm{s}}T] \\ \frac{U_{\mathrm{d1}}}{N_{\mathrm{p}}A_{\mathrm{cv}}}, [D_{\mathrm{s}}T, 0.5T] \\ \frac{L_{\sigma\mathrm{a}}}{L_{\mathrm{int}}}\frac{U_{\mathrm{d1}}}{N_{\mathrm{p}}A_{\mathrm{cv}}}, [0.5T, (0.5+D_{\mathrm{s}})T] \\ \frac{U_{\mathrm{d1}}}{N_{\mathrm{p}}A_{\mathrm{cv}}}, [(0.5+D_{\mathrm{s}})T, T] \end{cases} \tag{10}$$

$$\left|\frac{\mathrm{d}B_{\mathrm{ma}}}{\mathrm{d}t}\right| = \begin{cases} \frac{2L_{\sigma\mathrm{a}}}{L_{\mathrm{int}}}\frac{U_{\mathrm{d1}}}{N_{\mathrm{p}}A_{\mathrm{acv}}}, [0, D_{\mathrm{s}}T] \\ 0, [D_{\mathrm{s}}T, 0.5T] \\ \frac{2L_{\sigma\mathrm{a}}}{L_{\mathrm{int}}}\frac{U_{\mathrm{d1}}}{N_{\mathrm{p}}A_{\mathrm{acv}}}, [0.5T, (0.5+D_{\mathrm{s}})T] \\ 0, [(0.5+D_{\mathrm{s}})T, T] \end{cases} \tag{11}$$

Based on the iGSE model [20], the engineering calculation model for the core losses under DAB excitation (D_{h} =0.5) is given as

$$P_{\mathrm{c}} = M_{\mathrm{c}}k_{\mathrm{i}}f^{\alpha-\beta}\left(\frac{U_{\mathrm{d1}}}{N_{\mathrm{p}}A_{\mathrm{cv}}}\right)^{\beta}\left(\frac{1}{2} - \frac{L_{\sigma\mathrm{w}}D_{\mathrm{s}}}{L_{\mathrm{int}}}\right)^{\beta-\alpha}\left[1 + 2D_{\mathrm{s}}\left(\frac{L_{\sigma\mathrm{a}}^{\alpha}}{L_{\mathrm{int}}^{\alpha}} - 1\right)\right] \tag{12}$$

$$P_{\mathrm{ca}} = M_{\mathrm{ac}}k_{\mathrm{i}}2^{\beta+1}f^{\alpha-\beta}\left(\frac{U_{\mathrm{d1}}}{N_{\mathrm{p}}A_{\mathrm{acv}}}\frac{L_{\sigma\mathrm{a}}}{L_{\mathrm{int}}}\right)^{\beta}D_{\mathrm{s}}^{1-\alpha+\beta} \tag{13}$$

$$k_{\mathrm{i}} = \frac{k}{(2\pi)^{\alpha-1}\int_0^{2\pi}|\cos\theta|^{\alpha}2^{\beta-\alpha}d\theta} \tag{14}$$

where P_{c} is the loss of the main core; P_{ca} is the loss of the additional core; M_{c} is the effective weight of the main core; M_{ac} is the effective weight of the additional core; f is frequency, k_{i} is the iGSE coefficient. k, α, and β are Steinmetz coefficients, which are extracted from sinusoidal measurement data of a gapped nanocrystalline core sample, which is shown in Figure 6.

The experiment is carried out with the core sample, and the calculated value and the measured value are compared in Figure 6, verifying the effectiveness of the model.

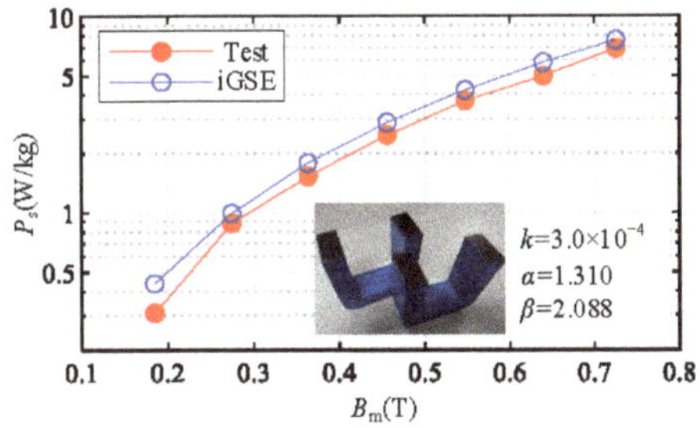

Figure 6. Comparison of experimental and calculated losses of the main core (D_h = 0.5, D_s = 0).

3.2. Winding Loss

In order to calculate the winding losses under non-sinusoidal excitation, the current is usually decomposed by Fourier transform, and the winding loss is given as

$$P_{Cu} = \sum_{i=1}^{\infty} I_{RMS-i}^2 R_{AC-i} = \sum_{i=1}^{\infty} I_{RMS-i}^2 F_{R-i} R_{DC} \tag{15}$$

where $I_{RMS\text{-}i}$ is the i_{th} harmonic components of the load current; the ac resistance coefficient F_R can be obtained by the Dowell model [21] as

$$F_R = \Delta\left[\frac{\sin h2\Delta + \sin 2\Delta}{\cos h2\Delta - \cos 2\Delta} + \frac{2(m^2-1)}{3}\frac{\sin h\Delta - \sin \Delta}{\cos h\Delta + \cos \Delta}\right] \tag{16}$$

$$\Delta = d_c\sqrt{\pi f \sigma \mu_0 \frac{nh_c}{h_{wd}}} \tag{17}$$

where, m is the number of winding layers, d_c is the thickness of conductor, h_c is the height of single conductor, Δ is the ratio of conductor thickness to penetrated depth; σ is conductivity, f is frequency, μ_0 is the vacuum permeability, n is the number of conductors per layer, and h_{wd} is the window height.

3.3. Leakage Inductance of CW ELII MFT

Figure 7 shows the simplified magnetic field distribution of CW ELII MFT. The integrated leakage inductance of CW ELII MFT consists of additional leakage inductance $L_{\sigma a}$, primary leakage inductance $L_{\sigma p}$, and secondary leakage inductance $L_{\sigma s}$. Since the leakage inductance of the primary winding and secondary winding are difficult to be separated physically, the two are usually combined as the winding leakage inductance $L_{\sigma w}$, so the total leakage inductance of the transformer is given as

$$L_\sigma = L_{\sigma a} + L_{\sigma p} + L'_{\sigma s} = L_{\sigma a} + L_{\sigma w} \tag{18}$$

The Rogowski model and Dowell model are classical calculation methods for transformer leakage inductance. The Rogowski model is a 2D static model in which the internal characteristics of the winding (for example, conductor thickness and inter-turn distance) and the effect of eddy current in the conductor are ignored. The Dowell model is a 1D dynamic model, and describes the influence of the eddy current on the leakage inductance based on the assumption of parallel flux and uniform arrangement of conductors [21]. The Dowell model is suitable for cases with eddy current effects but has an obvious error in cases with low utilization of the core window height [22]. Researchers introduced Rogowski coefficient into the Dowell model, making up for the deficiency of Dowell model [22,23]. The hybrid model is given as [22]

$$L_{\sigma w-H} = \frac{N^2\mu_0 l_w}{h_{eq}}\left[\frac{d_{cs}m_s}{3}F_{sw} + \frac{d_{cp}m_p}{3}F_{pw} + d_d + \frac{d_{is}(m_s-1)(2m_s-1)}{6m_s} + \frac{d_{ip}(m_p-1)(2m_p-1)}{6m_p}\right] \tag{19}$$

$$F_w = \frac{\left[(4m^2-1)\frac{\sinh 2\Delta - \sin 2\Delta}{\cosh 2\Delta - \cos 2\Delta} - 2(m^2-1)\frac{\sinh\Delta - \sin\Delta}{\cosh\Delta - \cos\Delta}\right]}{2m^2\Delta} \tag{20}$$

$$\Delta = d_c\sqrt{\mu_0\pi f\sigma\frac{nh_c}{h_{wd}}} \tag{21}$$

$$h_{eq} = \frac{h_w}{K_R} \tag{22}$$

$$K_R = 1 - \frac{1-e^{-\pi h_w/d_w}}{\pi h_w/d_w} \tag{23}$$

where N is the turns of the winding; l_w is the average length of turns; m_p and m_s are respectively the numbers of primary and secondary winding layers; Δ is the ratio of conductor thickness to skin depth; σ is conductivity; f is frequency; μ_0 is the vacuum permeability; n is the number of conductors per layer; K_R is Rogowski coefficient; and h_{eq} is the corrected magnetic circuit length. As shown in Figure 8, d_d is the main insulation distance; d_{wp}, d_{ws}, and d_w, are respectively the width of the primary, secondary, and the entire winding; h_w is the height of winding; h_{wd} is the window height; d_{ip} and d_{is} are inter-layer distances; d_{cp} and d_{cs} are the widths of single conductors; h_{cp} or h_{cs} are the heights of conductors.

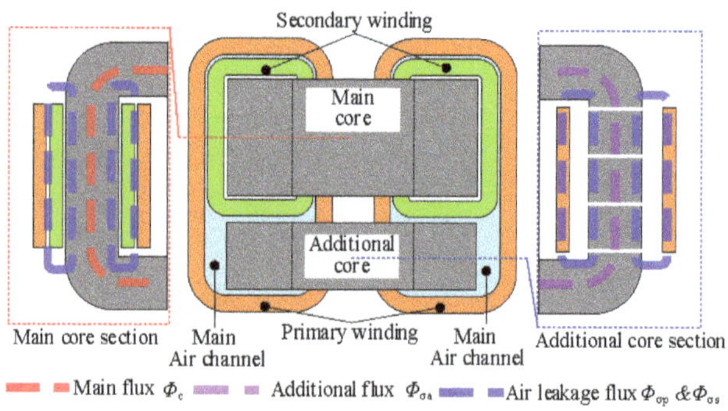

Figure 7. Simplified magnetic field distribution of CW ELII MFT.

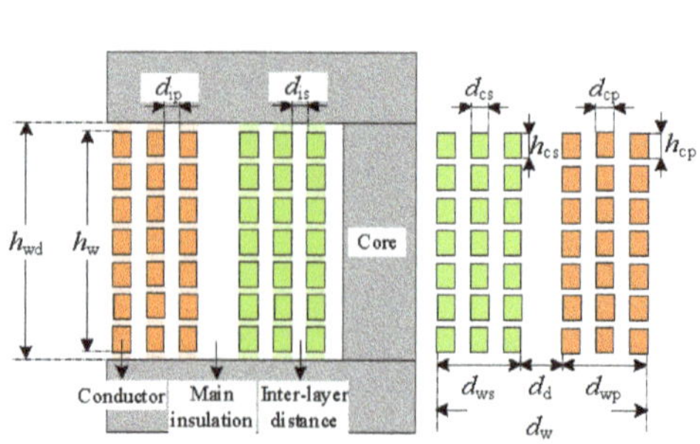

Figure 8. Winding parameters.

According to the assumption of the Dowell model, the main insulation layer has the same width d_d in the perimeter direction; conductor layers are arranged uniformly with the same inter-layer distance, i.e., d_{is} and d_{ip}. As shown in Figure 7, the width of the main insulation in the CW ELII MFT is not equal in the circumference direction. In addition, when there is an internal air channel in the winding, the inter-layer distance will not be equal. Therefore, the original model needs to be improved to enable it to be used in the CW ELII MFT.

Assuming that the cross-sectional area of the main insulation layer is S_d, the inductance of the main insulation layer can be obtained based on the magnetic field energy

$$L_d = \frac{N^2 S_d \mu_0}{h_{eq}} \tag{24}$$

Assuming that conductor layers are arranged non-uniformly with the inter-layer distances $d_{i\text{-}1}, d_{i\text{-}2}, \ldots, d_{i\text{-}q}, \ldots, d_{i\text{-}(m-1)}$, then the inter-layer inductance is given as

$$L_i = \frac{\mu_0 N^2 l_w}{h_{eq} m_p^2} \sum_{q=1}^{q=m-1} q^2 d_{i-q} \tag{25}$$

Replace the static terms in (19) by (24) and (25), and the mean length of turns of the primary and secondary windings, l_{wp} and l_{ws}, are used to calculate the leakage inductance. The improved hybrid model is given as

$$L_{\sigma w-HI} = \frac{N^2 \mu_0}{h_{eq}} \left[\frac{l_{wp} d_{cp} m_p}{3} F_{wp} + \frac{l_{ws} d_{cs} m_s}{3} F_{ws} + S_d + \frac{l_{wp}}{m_p^2} \sum_{q=1}^{q=m_p-1} q^2 d_{ip-q} + \frac{l_{ws}}{m_s^2} \sum_{r=1}^{r=m_s-1} r^2 d_{is-r} \right] \tag{26}$$

where, $d_{ip\text{-}q}$ and $d_{is\text{-}r}$ refer to the width of the q_{th} and r_{th} inter-layer distance of the primary and secondary winding from the outside to the inside.

A winding model with a non-uniform arrangement of conductors is shown in Figure 9. The model parameters are: 18 turns of the primary winding, 11 turns of the secondary, the width of the main insulating layer is 8 mm, the thickness of the copper foil is 1 mm, the height is 220 mm, the height of the window is 245 mm. An 8 mm-wide air channel is located between the q layer and the q + 1 layer in the primary, and the other turns of the insulating layer have a width of 0.2 mm.

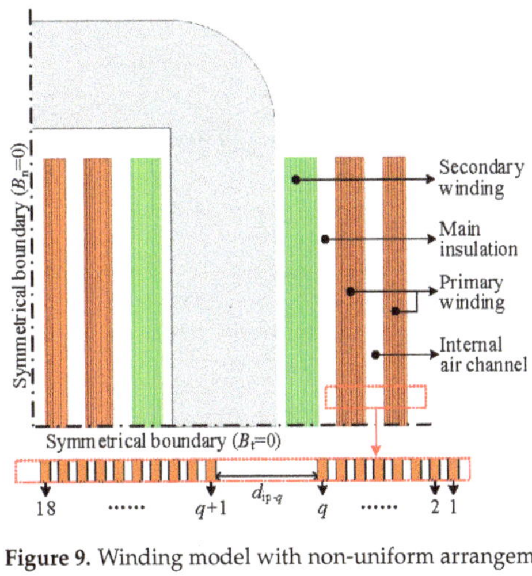

Figure 9. Winding model with non-uniform arrangement of conductor layers.

As shown in Figure 10, different models and FEM are used to calculate the leakage inductance of the example winding with different air channel locations. In the original and hybrid models, the average inter-layer width is used for the calculation. Both the FEM results and the calculation results of the proposed model increase as the position of the channel gradually approaches the main insulating layer, while the original model and the hybrid model cannot reflect the influence of the air channel on the leakage inductance. Based on the FEM results, the average error of the proposed model is 1.9%, and the maximum error is 2.4%, both smaller than the other two models; when q = 10–12, the errors of the three models are relatively close, all less than 5%.

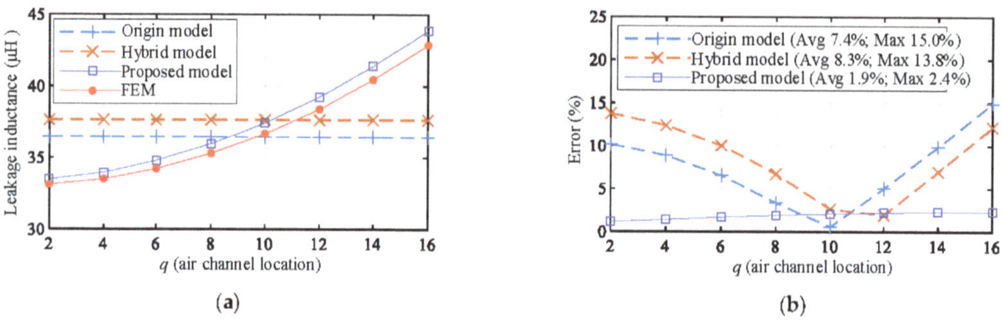

Figure 10. Comparison of FEM and analytical calculation results. (**a**) Comparison of results, (**b**) Comparison of errors.

The additional leakage inductance is composed of multiple air-gap inductances. In [24], an empirical formula is provided for calculating the inductance of the lumped air-gap inductor. In [25], the 2D expressions of air-gap reluctance are derived based on the Schwarz–Christoffel transformation.

A simplified air gap is shown in Figure 11, and its reluctance is composed of the internal reluctance R_{in} and the fringing reluctance R_{fr} in parallel. The expressions are given as [26]

$$R_{\text{in}} = \frac{d_g}{\mu_0 A_{\text{ac}}} \quad (27)$$

$$R_{\text{fr}} = \frac{\pi}{\mu_0 C \ln\left(\frac{2h+d_g}{d_g}\right)} \quad (28)$$

$$R_{g-i} = R_{\text{fr}} || R_{\text{in}} \quad (29)$$

where d_i is the height of the air gap, A_{ac} is the cross-sectional area of the core, μ_0 is air permeability, and C is the perimeter of the cross-section.

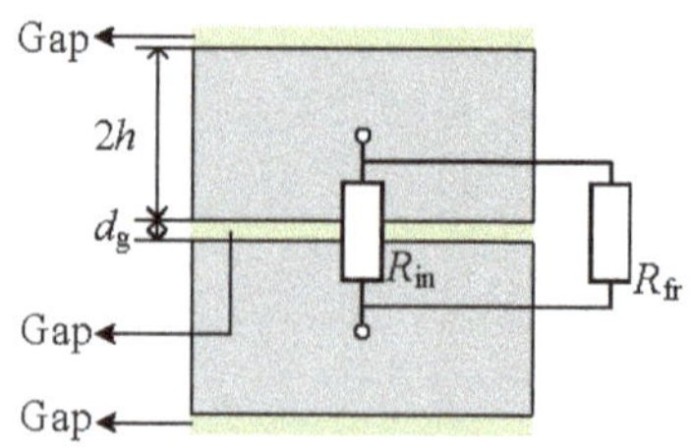

Figure 11. Simplified model of the air gap.

Thus, the additional leakage inductance is given as

$$L_{\sigma a} = \frac{N^2}{\sum R_{g-i}} \quad (30)$$

where N is the number of turns of the winding.

The FEM simulation is carried out to verify the calculation method for the additional leakage inductance, and the model and parameters are shown in Figure 12a. As shown in Figure 12b, as the air gap height increases, the inductance value is greatly reduced. Therefore, in the MFT design process, the height and number of the air gap are used to adjust the inductance value.

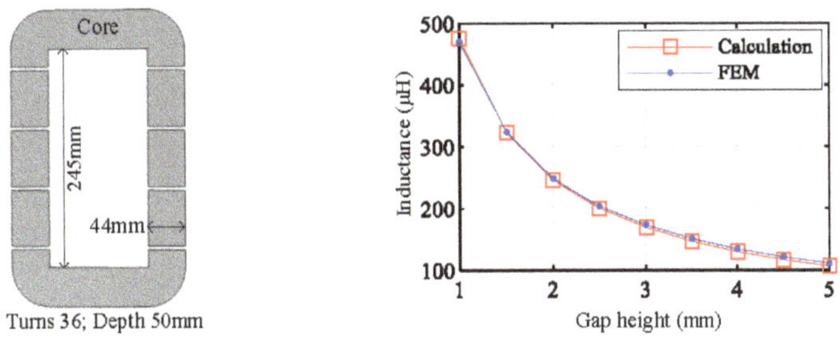

(a) (b)

Figure 12. Calculation and simulation of the additional leakage inductance. (**a**) FEM model (**b**) comparison of the calculated and simulated values.

4. Thermal Modeling and Analysis

For the design of the air–water combined cooling MFT, a thermal model is necessary. The steady-state thermal model mainly includes the heat source, thermal resistance, and network topology. The heat source, namely the core and winding losses, has been discussed above. Therefore, this section mainly discusses the modeling of thermal resistance and network topology.

4.1. Convective Thermal Resistance of the Air Channel

The air channel is an important heat dissipation structure of transformer windings. The air channel can be simplified as a parallel-plate channel for thermal modeling. As shown in Figure 13, the length of the air channel is L, the cross-section is $W \times S$, and $W >> S$. The hydraulic diameter is given as [27]

$$d_{\rm h} = \frac{4A}{P} = \frac{4WS}{2(W+S)} \approx 2S \tag{31}$$

where A is the cross-section area and P is the inner perimeter.

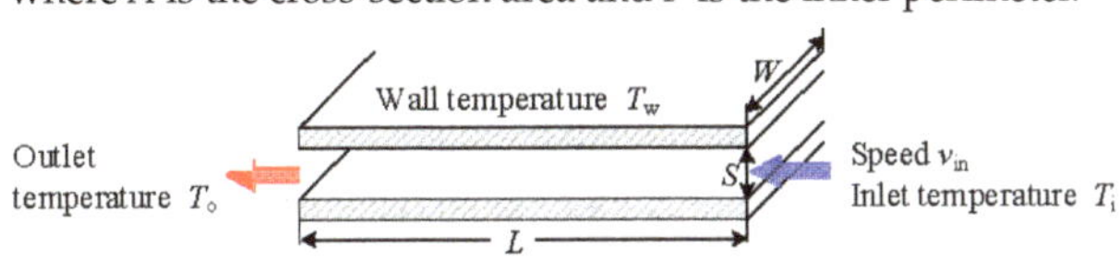

Figure 13. Parallel-plate air channel.

Since the convection model of the parallel-plate channel in literature is not complete, a combination of the laminar-flow model [28] for the parallel-plate channel and a turbulent-flow model [28] for the non-circular tube is adopted in this paper. At the same time, linear interpolation is used in the transition period between laminar flow and turbulent flow. The Nusselt number calculation formula of the parallel-plate air channel is given as

$$N_{\rm u} = \begin{cases} \left[N_1^3 + N_2^3 + N_3^3\right]^{1/3}, \ (R_{\rm e} \le 2300) \\ \dfrac{\left(\frac{f_{\rm r}}{8}\right) R_{\rm e} P_{\rm r} \left[1 + \left(\frac{d_{\rm h}}{L}\right)^{2/3}\right] \left(\frac{T_{\rm f}}{T_{\rm w}}\right)^{0.45}}{1 + 12.7\sqrt{\frac{f_{\rm r}}{8}}\left(P_{\rm r}^{2/3} - 1\right)}, \ (R_{\rm e} \ge 10^4) \\ (1-r)N_{\rm u2300} + rN_{\rm u10000}, \ (2300 < R_{\rm e} < 10^4) \end{cases} \tag{32}$$

$$R_e = \frac{\rho v_{in} d_h}{\eta} \tag{33}$$

$$N_1 = 7.541 \tag{34}$$

$$N_2 = 1.841(R_e P_r d_h/L)^{1/3} \tag{35}$$

$$N_3 = \left(\frac{2}{1+22P_r}\right)^{1/6} (R_e P_r d_h/L)^{1/2} \tag{36}$$

$$f_r = (1.8\log_{10} R_e - 1.5)^{-2} \tag{37}$$

$$P_r = \frac{c_p \eta}{\lambda_f} \tag{38}$$

$$T_f = \frac{T_i + T_o}{2} \tag{39}$$

$$r = \frac{R_e - 2300}{10^4 - 2300} \tag{40}$$

where R_e is the Reynolds number; N_1, N_2, and N_3 are the laminar flow model coefficients; P_r is the Prandtl number; T_f is the average fluid temperature; T_i is the inlet temperature, which is equal to the ambient environment temperature T_{am}; T_o is the outlet temperature; T_w is the wall temperature; f_r is the friction coefficient; r is the interpolation coefficient in the transition period; ρ is the fluid density; η is dynamic viscosity; c_p is fluid specific heat capacity; and λ_f is fluid thermal conductivity.

After the Nusselt number is obtained, the surface heat transfer coefficient is given as

$$h = \frac{\lambda_f N_u}{d_h} \tag{41}$$

The FEM simulation is carried out to verify the accuracy of the above calculation methods. Figure 14 shows the temperature distribution inside the air channel under different wind velocities. The simulated and calculated results are compared in Figure 15. As shown in Figure 15a, the average heat transfer coefficient increases with the higher wind velocity. The reason is that the higher wind velocity reduces the thickness of the boundary layer. As shown in Figure 15b, the shorter channel can achieve higher average heat transfer coefficients. The reason is that the thickness of the boundary layer increases with the channel length. The error between the calculated and simulated result is less than 10%.

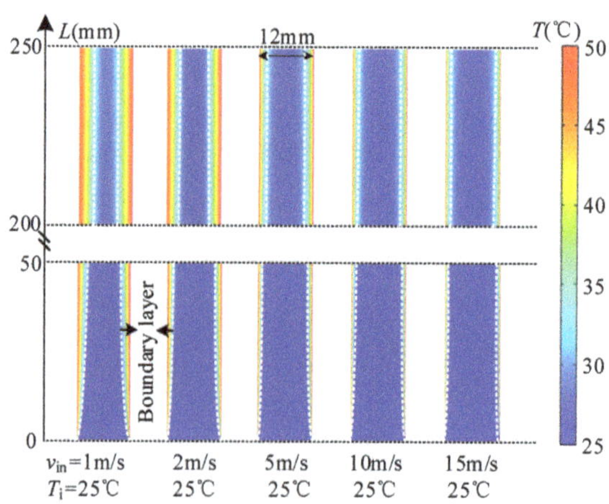

Figure 14. Temperature distribution of the air channel under different wind velocities.

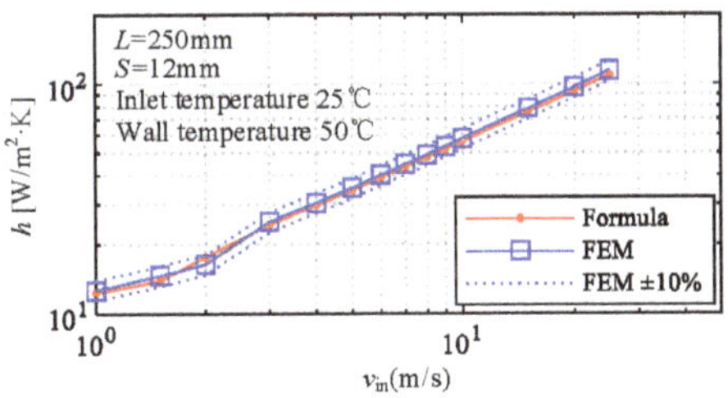

(a)

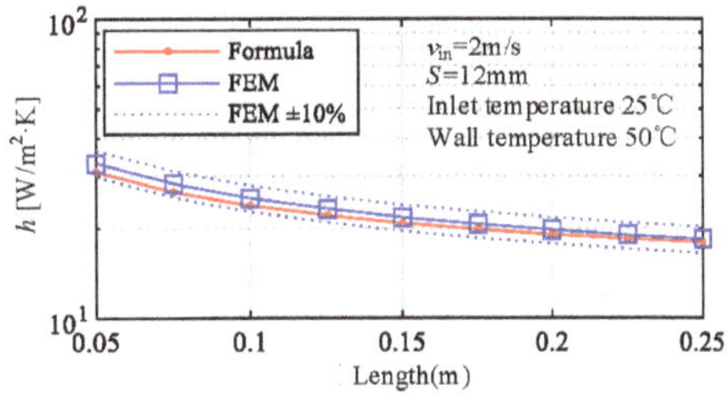

(b)

Figure 15. Comparison of the simulated and calculated heat transfer coefficients. (**a**) Different wind velocities; (**b**) different channel lengths.

After obtaining the heat transfer coefficient, the convective thermal resistance can be obtained according to the Newton cooling formula of the internal convection.

$$Q = hA\left(T_\mathrm{w} - \frac{T_\mathrm{i} + T_\mathrm{o}}{2}\right) = \frac{(T_\mathrm{w} - T_\mathrm{f})}{R'_\mathrm{inconv}} \tag{42}$$

In (42), the temperature difference is between the surface temperature T_w and the fluid temperature T_f. As the fluid temperature T_f is related to the internal convection process, the thermal resistance R'_inconv defined by (42) cannot be directly connected to the external environment nodes in the thermal network. Therefore, transform (42) as

$$Q = \frac{hA\left(T_\mathrm{w} - \frac{T_\mathrm{i}+T_\mathrm{o}}{2}\right)}{(T_\mathrm{w} - T_\mathrm{i})}(T_\mathrm{w} - T_\mathrm{i}) = \frac{(T_\mathrm{w} - T_\mathrm{i})}{R_\mathrm{inconv}} \tag{43}$$

The thermal resistance R_inconv defined by (43) can be directly connected to the external environment nodes for the temperature difference is between surface temperature T_w and inlet temperature T_i. The proposed thermal resistance definition for the internal convection is more convenient for thermal network modeling.

4.2. Conduction and Radiation Thermal Resistance

Conduction: for a uniform cuboid with length l, thermal conductivity λ, and cross-sectional area A, the thermal resistance is given as [27]

$$R_\mathrm{cond} = \frac{l}{\lambda A} \tag{44}$$

Radiation: for a surface with the area A_sur, the surface temperature T_sur, and the ambient temperature T_am, the thermal radiation resistance is given as [27]

$$R_\mathrm{rad} = \frac{T_\mathrm{sur} - T_\mathrm{am}}{\varepsilon_\mathrm{i}\sigma A_\mathrm{sur}(T_\mathrm{sur}^4 - T_\mathrm{am}^4)} \tag{45}$$

where σ is Boltzmann constant and ε_i is emissivity factor.

4.3. Thermal Network Topology

The structure of the air–water cooled CW ELII MFT is shown in Figure 16. As shown in Figure 16b, to reduce the eddy current loss caused by the fringing flux, the tooth-slot structure is adopted in water-cooling plates adjacent to the additional core.

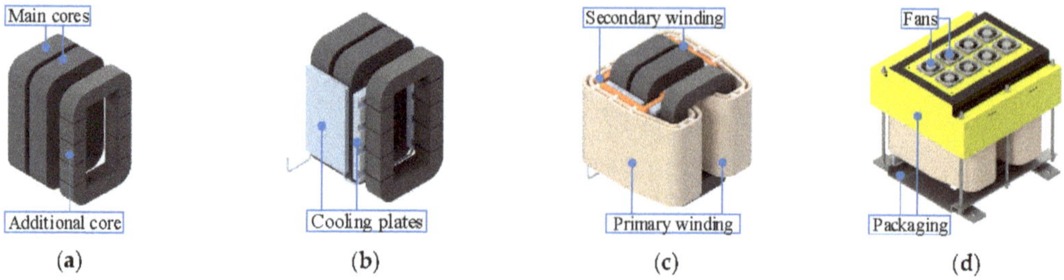

Figure 16. The structure of CW ELII MFT. (**a**) Cores; (**b**) water-cooling system; (**c**) windings; (**d**) packaging and fans.

As shown in Figure 17, the thermal network model has ten nodes. N_1~N_3 denote the upper yoke, the limb, and the lower yoke of the main core, respectively. N_4 represents the secondary winding. N_5 represents the primary winding. N_6~N_8 represent the upper yoke, the limb, and the lower yoke of the additional core. N_9 represents the external air. N_{10} represents the water-cooling plate. The additional nodes n_1~n_8 represent the convection and radiation surfaces of nodes N_1~N_8 and are used to provide the wall temperature for convection and radiant thermal resistance calculations. Q_1–Q_8 represent the losses of each node, respectively; T_{am} and T_{co} represent the temperature of the environment and the water-cooling plate. In the above model, the internal forced convection process of the water-cooling plate is ignored. The implication of $R_{xy\text{-Cond/Conv/Rad}}$ is conduction/convection/radiation thermal resistance from node x to node y. In a steady-state, the governing equation is given as [12]

$$0 = \boldsymbol{AT} + \boldsymbol{BU} \tag{46}$$

where $\boldsymbol{A}$ describes the thermal resistance network of internal nodes; $\boldsymbol{B}$ describes the thermal resistance network between the internal nodes and the external nodes (air and water-cooling plates); $\boldsymbol{T}$ is the node temperature vector; $\boldsymbol{U}$ is the external excitation vector.

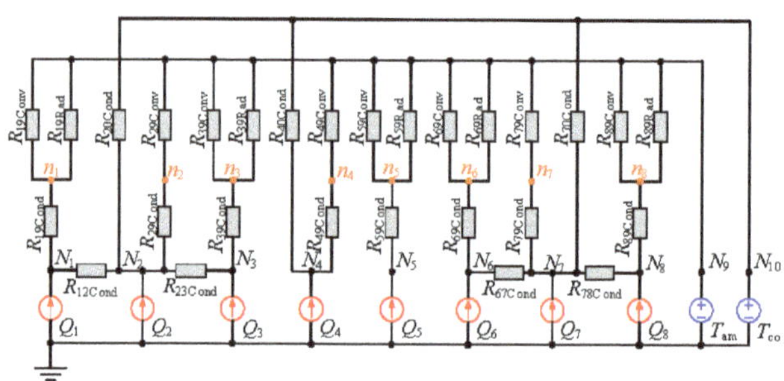

Figure 17. Thermal network model of the CW ELII MFT.

5. Optimal Design, Analysis, and Example

The MFT design is a non-linear, non-convex, and non-continuous problem. The optimization design and the comparative evaluation of the inductor-integrated MFT is presented in this section.

5.1. Design Method

The critical factors of optimal design include: design inputs, constraints, and optimization objectives.

5.1.1. Design Inputs

As shown in Figure 18, there are many variables in the optimal design of MFT, and it is unrealistic to optimize all variables. Therefore, seven free variables are selected in this paper, including the flux density of main core B_{m}, the flux density of additional core B_{ma}, turns of the primary winding N_{p}, the height of foils h_{c}, the thickness of primary foils d_{cp}, the thickness of secondary foils d_{cs}, and the width of core tape D_{cs}. The free variables constitute the input vector of the optimization design as

$$x = [B_{\mathrm{m}}, B_{\mathrm{ma}}, N_{\mathrm{p}}, h_{\mathrm{c}}, d_{\mathrm{cp}}, d_{\mathrm{cs}}, D_{\mathrm{cs}}] \tag{47}$$

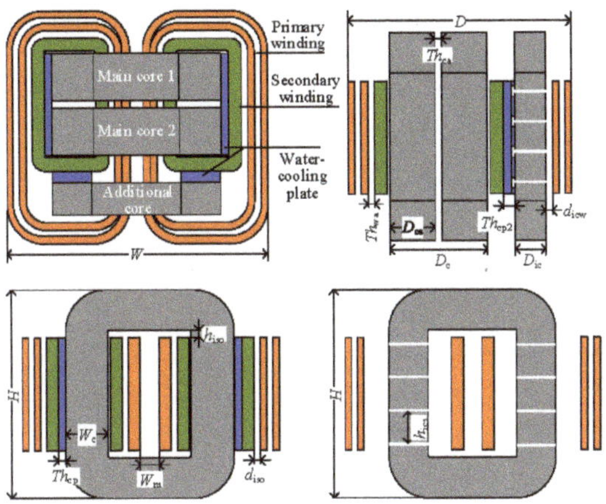

Figure 18. Sketch of the CW ELII MFT structure.

In addition, inputs design includes many fixed parameters about operating conditions and material properties, such as the DC voltage U_{dc1} and U_{dc2}, the turns ratio of the transformer k_{PS}, frequency f, and electrical and thermal properties of materials.

5.1.2. Constraints

The design constraints include temperature rise ΔT_{max}, leakage inductance error e_{Lmax}, and the spatial size H_{max}, W_{max}, D_{max}.

$$s.t.\begin{cases} \Delta T < \Delta T_{\mathrm{max}} \\ H < H_{\mathrm{max}} \\ W < W_{\mathrm{max}} \\ D < D_{\mathrm{max}} \\ \left|\frac{L_{\sigma}-L_{\sigma t}}{L_{\sigma t}}\right| < e_{L\mathrm{max}} \end{cases} \tag{48}$$

where ΔT is the temperature increase; H, W, and D are spatial sizes; and $L_{\sigma\mathrm{t}}$ is the target leakage inductance.

5.1.3. Objectives

In general, efficiency and power density are the main parameters for evaluating the design results.

$$\begin{aligned} f_1(x) &= \max \eta(x) = \max\left(\frac{P_{\mathrm{out}}}{P_{\mathrm{out}}+P_{\mathrm{Fe}}+P_{\mathrm{Cu}}}\right) \\ f_2(x) &= \max P_d(x) = max\left(\frac{P_{\mathrm{out}}}{V}\right) \end{aligned} \tag{49}$$

where η is efficiency, P_{out} is the output power, P_{Fe} is the core loss, P_{Cu} is the winding loss, V is the volume of MFT (ignoring structural components and terminals).

The design procedures of inductor-integrated MFT are shown in Figure 19. The brute force grid search method, which is extremely robust, is adopted on the premise that the calculation time is acceptable. According to the free parameters of the design input, a design space is constructed, which contains all possible designs. Designs in the space are calculated, and the results are saved. According to the constraints, the qualified designs are obtained. The Pareto front is solved according to the optimization objectives, and the designer can determine the final design according to the front.

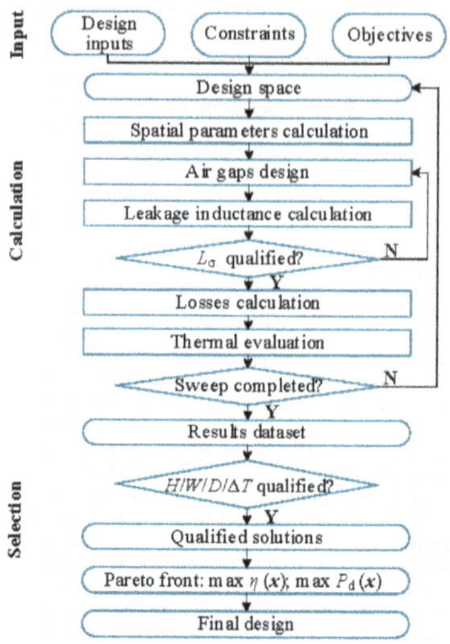

Figure 19. Design procedures of the CW ELII MFT.

5.2. Design Results and Comparative Analysis

The 4 kHz, 200 kW, 200 μH inductor-integrated MFT for a DAB converter is taken as an example of optimal design, and the main parameters and constraints are shown in Table 1.

Table 1. Parameters and constraints.

Parameters	Description	Value
P	rated power	200 kW
U_{d1}/U_{d2}	MV/LV DC voltage	1.6 kV/1 kV
k_{PS}	turns ratio	1.6
D_h	H-bridge duty	0.5
D_s	phase-shifting duty	0.0875
f	fundamental frequency	4 kHz
$L_{\sigma t}$	target leakage inductance	200 μH
e_{Lmax}	max leakage error	5%
$H_{max} \times W_{max} \times D_{max}$	space limit	0.4 × 0.4 × 0.4 m

The optimal designs of the CW LII MFT and the CW ELII MFT are carried out, respectively. The CW LII MFT is also combined air–water cooling, the difference is that the CW LII structure has no additional core, and the target leakage inductance is achieved by adjusting the width of the main insulation layer. The SW LII structure and the SW ELII structure are not discussed in this paper due to the inevitable external leakage flux.

Figure 20 shows the comparison of qualified designs of the two integration structures. The designs with the maximum power density in the CW LII and the CW ELII qualified solutions are shown in Table 2, and only the transformer body and internal cooling plates are considered in the power density specification. The CW ELII design has a lower core loss, but the eddy current of the water-cooling plate makes the efficiency of the CW ELII solution 0.09% lower than that of the CW LII solution. The volume power density of the CW LII solution is 2.24 kW/dm^3, and that of the CW ELII solution is 5.16 kW/dm^3, which is about 130% higher. Meanwhile, the weight power density of the CW ELII solution is only 9.9% higher than that of the CW LII solution, indicating that the main advantage of the CW ELII structure is the more compact structure. Considering the core material and wire material, the unit cost of the CW ELII design is 7.7% lower than that of the CW LII design.

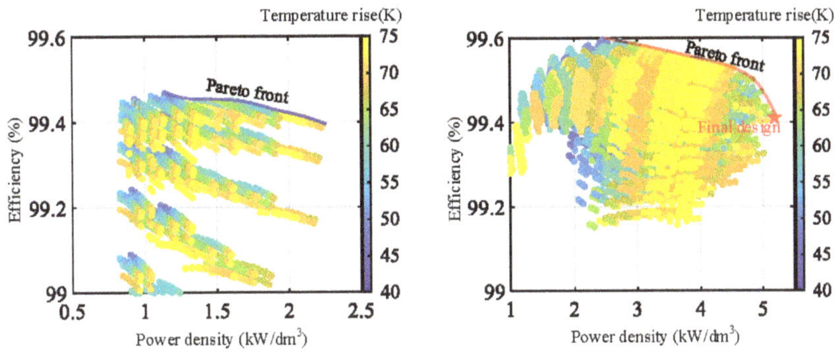

(a) (b)

Figure 20. Comparison of CW LII and CW ELII qualified designs. (**a**) CW LII qualified designs; (**b**) CW ELII qualified designs.

Table 2. Comparison of CW LII and CW ELII solutions.

Parameters		CW LII	CW ELII
Power (kW)		200	200
Leakage inductance (μH)		200	200
Winding	Turns	32/20	36/22
	Foil size (mm)	0.4 × 200	0.4 × 200
	Air channel width (mm)	86.3	12
	Winding loss (W)	913.7	916.8
Main core	Flux density (T)	0.65	0.65
	No-load core loss (W)	305.4	200.9
Additional core	Flux density (T)	/	0.4
	Load core loss (W)	/	56.8
	Cooling plate loss (W)	/	244
Temperature rise (K)		66	64.1
Efficiency (%)		99.39	99.30
Unit cost (RMB/kW)		78	72
Power density (kW/kg)		2.63	2.88
Power density (kW/dm^3)		2.24	5.16

In order to make a more comprehensive comparison of the two structures, the optimal designs for different leakage inductance values are carried out. It should be noted that other design inputs remain unchanged among the designs. The comparison of the Pareto fronts is shown in Figure 21. As shown in Figure 21a, when the leakage inductance is 10% p.u. (40 μH), the Pareto fronts of the CW LII solutions and the CW ELII solutions are very close. However, with the increase of leakage inductance, the power density of CW LII decreases continuously, while the CW ELII has not changed significantly.

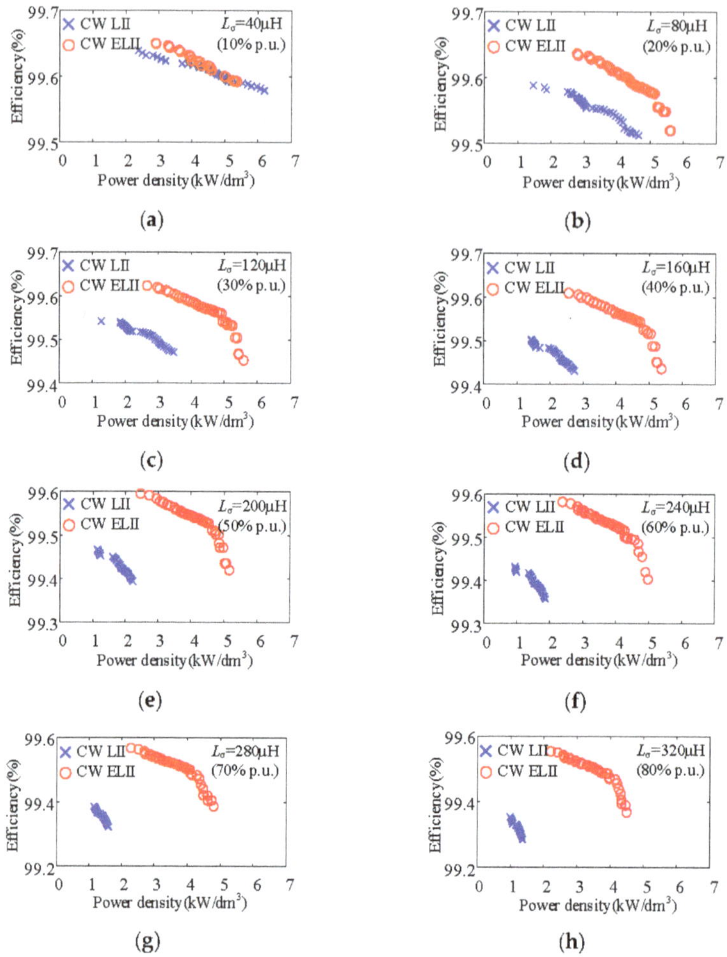

Figure 21. Pareto fronts of CW LII and CW ELII solutions with different leakage inductances. (**a–h**) leakage inductance: 10% p.u. to 80% p.u. (40 μH to 320 μH).

Based on Figure 21, the CW LII and CW ELII solutions with the highest power density in each group are selected for comparison, as shown in Figure 22. As shown in Figure 22a, the power density of the CW LII solution decreases with larger leakage inductance, and the maximum power density is 6.2 kW/dm^3 at 10% p.u. (40 μH). The CW ELII solution achieves the maximum power density of 5.8 kW/dm^3 at 20% p.u. (80 μH). When the leakage inductance per unit value exceeds 20% (p.u.), the CW ELII solution can achieve a higher power density than the CW LII solution. It becomes more evident as the leakage inductance increases. As shown in Figure 22b, as the leakage inductance increases, the efficiency of the two solutions is gradually reduced. When the leakage inductance per unit value is greater than 40% (p.u.), the CW ELII solution has advantages.

While keeping other parameters unchanged, the rated current in the design input is changed to analyze the influence of the power on the design result. In all groups, the leakage inductance in per-unit is 50%. Figure 23 shows the comparison of the Pareto fronts of CW ELII MFT designs with different power requirements. As the power increases, it is easier to achieve higher power densities. A possible reason is that the insulation distance in the MFT is a fixed parameter under the same insulation requirement. A possible reason is that the insulation distance in the MFT is a fixed parameter under the same insulation requirement.

With the increase of rated current (80 kW to 280 kW), the proportion of windings in the core window increases and the proportion of insulating space decreases, which makes the MFT power density increase. When the rated current is large enough (280 kW and 320 kW), the proportion of insulating space is small, and this effect is no longer obvious.

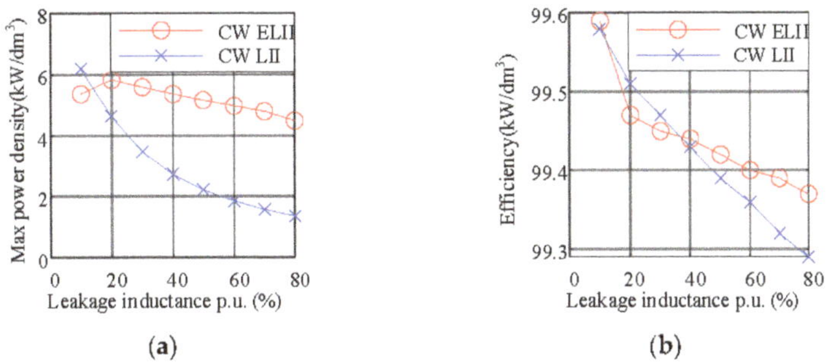

Figure 22. Comparison of solutions for different inductance requirements. (**a**) Max power density; (**b**) efficiencies.

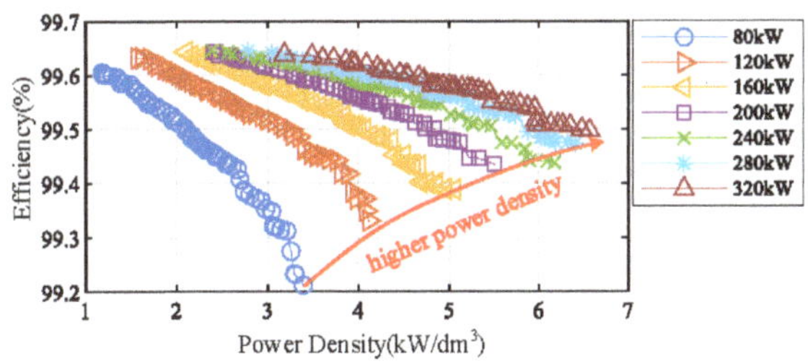

Figure 23. Comparison of Pareto fronts of CW ELII MFT designs under different power requirements (different current amplitudes; same leakage inductance in per unit, 50%).

5.3. Prototype and Experimental Verification

According to the final design of the CW ELII MFT, a prototype was manufactured and tested in a 200 kW DAB converter, which is a module of a 2 MW MMC-BDC. The MFT prototype and the experimental platform are shown in Figure 24.

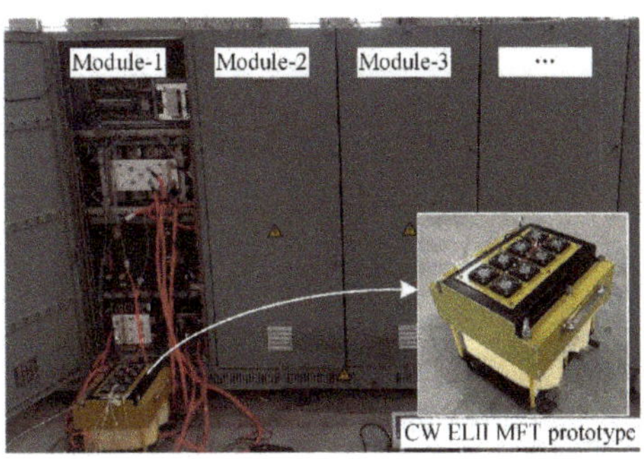

Figure 24. Experimental platform and the MFT prototype.

Figure 25 shows the waveforms of the no-load voltage and current under LV side excitation. The voltage was measured by Tek P5210A voltage probe (bandwidth 50 MHz), and the current was measured by Tek TCP303 + TCPA300 current probe (bandwidth 15 MHz). The no-load loss can be calculated by the no-load voltage and current.

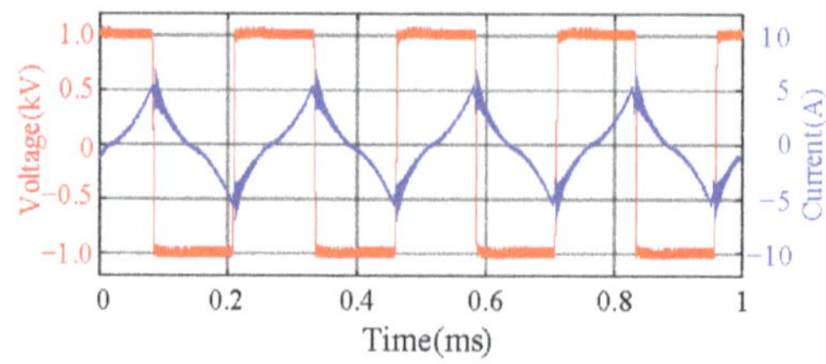

Figure 25. Voltage and current waveforms of the no-load condition.

Figure 26 shows the voltage and current waveforms under the load condition with an output power of 200 kW, in which the DC voltages is 1.6 kV/1 kV and the phase-shifting duty is 0.0875. The temperature rise experiment was carried out under this condition, where the inlet water temperature was 35.5 °C, the ambient temperature was 39 °C, and the average inlet wind speed was 1.0 m/s (measured by KIMO VT110). Figure 27 shows the MFT surface temperature distribution in thermal steady state. Contact sensors were placed on the core surface to monitor the core temperature. At the same time, to avoid the influence of internal wind on winding temperature measurement, the resistance method was used to measure the average winding temperature.

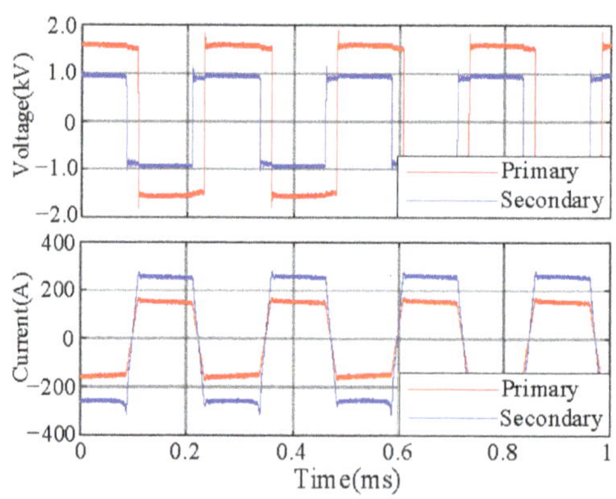

Figure 26. Voltage and current waveforms of the load condition.

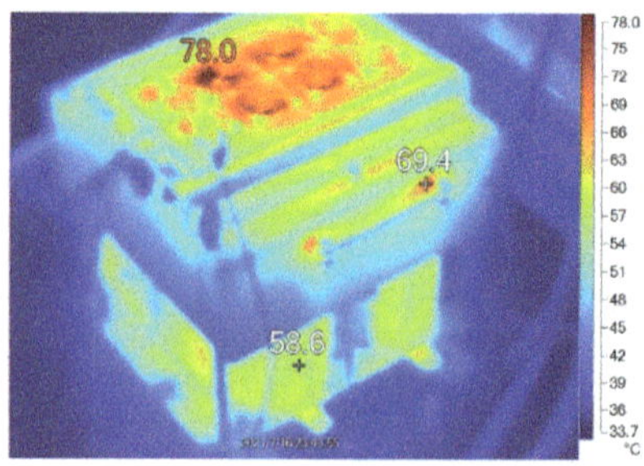

Figure 27. Temperature distribution measured by the infrared thermal camera.

The experimental results are summarized as in Table 3. The experimental results of the inductance, no-load loss, and temperature increase are basically consistent with the calculated values, verifying the validity of the parameter models and the optimization design method.

Table 3. Experimental results.

Parameters		Calculated	Measured	Error
Leakage inductance@4 kHz (μH)		200	209	4.5%
No-load loss (W)		200.9	208.1	3.5%
Temperature (°C)	Primary winding	93.7	92	1.8%
	Secondary winding	103.1	108.5	5.0%
	Main core	81.3	78.2	4.0%
	Additional core	95.9	96.8	0.9%

6. Conclusions

The design and analysis of the inductor-integrated MFT is a challenging multi-physical problem. Facing the application requirements of the DAB converter, the research on the inductor-integrated MFT is carried out. The main conclusions are drawn as follows:

(1) By comparing different integration structures, the CW ELII structure is adopted. The operating mode of CW ELII MFT under typical DAB excitation is analyzed based on the magnetic circuit model, then obtaining the magnetic flux expressions at different phase-shifting duties. Aiming at the characteristics of the irregular cross-section of the main insulation layer and the non-uniform arrangement of conductor layers, the leakage inductance model is improved. The results show that the maximum error of the proposed model is 2.4%, and that of the classical model is 15.0% for the cases with the non-uniform arrangement of conductor layers.

(2) The thermal network model of air–water combined cooled CW ELII MFT is established. Based on the combination of classical models, the thermal resistance model for the winding air channel under forced convection is proposed, and the relative error with FEM results is less than 10%. The analysis shows that a higher average heat transfer coefficient can be obtained using a higher wind speed or a shorter channel.

(3) The 200 kW, 4 kHz, 200 μH MFT for DAB converter is chosen as an example, and the optimal design is carried out using two structures of CW LII and CW ELII. The volume power density of the CW ELII solution is about 130% higher than that of the CW LII solution with the similar efficiency and cost. The weight power density is 9.9% higher. A CW ELII MFT prototype was manufactured with the power density of 5.16 kW/dm^3 and the efficiency of 99.30%. The prototype was tested in a 2 MW DC MMC-BDC prototype verifying the electrical and thermal performance.

(4) According to a more extensive comparison and analysis, the power density of the CW LII solution decreases with the increase of the leakage inductance. In contrast, the CW ELII solution achieves a max power density of 5.8 kW/dm^3 when leakage inductance is 20% (p.u.). The CW ELII solution can achieve a higher power density when the per unit value of leakage inductance exceeds 20% (p.u.). Both structures achieve the highest efficiency when the leakage inductance is the smallest. The CW ELII solution has an efficiency advantage when the leakage inductance exceeds 40% (p.u.).

There are still some limitations in this work, which is also the direction of the future research. In this work, the operating frequency of the research object is relatively low. With the popularization of SiC devices, the operating frequency of MFT is also gradually increasing. Therefore, it is necessary to further develop research of the inductor-integrated MFT with higher operating frequency, focusing on the high-frequency winding loss and the fringing flux loss of the additional core. In this work, only the optimization design of the MFT body is carried out; however, parameters of the converter will significantly affect the MFT design. Therefore, it is necessary to further carry out the system-level optimal design of the converter. The modeling and design method of the CW ELII MFT presented in this paper lays a foundation for further research.

Author Contributions: Conceptualization, X.Z., R.W. and F.X.; Methodology, X.Z. and R.W.; Software, X.Z.; Validation, X.Z.; Writing—original draft preparation, X.Z., W.K. and B.Y. All authors have read and agreed to the published version of the manuscript.

Funding: This research was funded by the National Natural Science Foundation of China, grant no. 51907199.

Conflicts of Interest: The authors declare no conflict of interest. The funders had no role in the design of the study; in the collection, analyses, or interpretation of data; in the writing of the manuscript, or in the decision to publish the results.

References

1. Dujic, D.; Zhao, C.; Mester, A.; Steinke, J.K.; Weiss, M.; Lewdeni-Schmid, S.; Chaudhuri, T.; Stefanutti, P. Power electronic traction transformer-low voltage prototype. *IEEE Trans. Power Electron.* **2013**, *28*, 5522–5534. [CrossRef]
2. Zhang, L.; Zhang, D.; Shui, H.; Yuan, Y.; Pei, Q.; Zhu, J. Optimization design of medium frequency transformer for the offshore dc grid based on multi-objective genetic algorithm. *IET Power Electron.* **2017**, *10*, 2157–2162. [CrossRef]
3. Guo, Z.; Yu, R.; Xu, W.; Feng, X.; Huang, A. Design and optimization of a 200-kw medium-frequency transformer for medium-voltage SiC PV inverters. *IEEE Trans. Power Electron.* **2021**, *36*, 10548–10560. [CrossRef]
4. Chen, P.; Xiao, F.; Liu, J.; Zhu, Z.; Ren, Q. Unbalanced Operation Principle and Fast Balancing Charging Strategy of a Cascaded Modular Multilevel Converter–Bidirectional DC–DC Converter in the Shipboard Applications. *IEEE Trans. Transp. Electrif.* **2020**, *6*, 1265–1278. [CrossRef]
5. Liu, J.; Zhu, Z.; Xiao, F.; Chen, P.; Ren, Q. A modular three-port DC-DC converter for vessel integrated power system. *Trans. China Electrotech.* **2020**, *35*, 4085–4096. [CrossRef]
6. Zhao, B.; Song, Q.; Liu, W.; Sun, Y. Overview of dual-active-bridge isolated bidirectional DC-DC converter of high-frequency-link power-conversion system. *IEEE Trans. Power Electron.* **2014**, *29*, 4091–4106. [CrossRef]
7. Yao, P.; Jiang, X.; Xue, P.; Li, S.; Lu, S.; Wang, F. Design optimization of medium frequency transformer for DAB converters with dc bias capacity. *IEEE J. Emerg. Sel. Top. Power Electron.* **2020**, *9*, 5043–5054. [CrossRef]
8. Garcia-Bediaga, A.; Villar, I.; Rujas, A.; Mir, L.; Rufer, A. Multiobjective optimization of medium-frequency transformers for isolated soft-switching converters using a genetic algorithm. *IEEE Trans. Power Electron.* **2017**, *32*, 2995–3006. [CrossRef]
9. Chen, B. Design optimisation of an inductor-integrated MF transformer for a high-power isolated dual-active-bridge DC–DC converter. *IET Power Electron.* **2019**, *12*, 2912–2922. [CrossRef]
10. Chen, B.; Liang, X.; Wan, N. Design methodology for inductor-integrated litz-wired high-power medium-frequency transformer with the nanocrystalline core material for isolated DC-link stage of solid-state transformer. *IEEE Trans. Power Electron.* **2020**, *35*, 11557–11573. [CrossRef]
11. Peng, S. Optimal Design of Highly Efficient, Compact and Silent Medium Frequency Transformers for Future Solid-State Transformers. Ph.D. Thesis, ETH Zurich, Zurich, Switzerland, 2017.
12. Villar, I. Multiphysical Characterization of Medium-Frequency Power Electronic Transformers. Ph.D. Thesis, EPFL, Lausanne, Switzerland, 2010.
13. Leibl, M.; Ortiz, G.; Kolar, J. Design and experimental analysis of a medium frequency transformer for solid-state transformer applications. *IEEE J. Emerg. Sel. Top. Power Electron.* **2017**, *5*, 110–123. [CrossRef]
14. Zhang, X.; Wang, R.; Xiao, F.; Fan, X.; Yang, G.; Yang, B. Design of highly compact short-time duty power inductors for supercapacitor charging applications. *IET Power Electron.* **2020**, *13*, 4377–4385. [CrossRef]
15. Kauder, T.; Belgrand, T.; Lemaître, R.; Thul, A.; Hameyer, K. Medium-frequency power transformer using GOES for a three-phase dual active bridge. *J. Magn. Magn. Mater.* **2020**, *504*, 166672. [CrossRef]
16. Marko, M.; Dujic, D. 100 kW, 10 kHz medium-frequency transformer design optimization and experimental verification. *IEEE Trans. Power Electron.* **2019**, *34*, 1696–1708. [CrossRef]
17. Marko, M.; Drazen, D. Sensitivity Analysis of Medium-Frequency Transformer Designs for Solid-State Transformers. *IEEE Trans. Power Electron.* **2019**, *34*, 8356–8367. [CrossRef]
18. Yu, X.; Su, J.; Lai, J.; Guo, S. Analytical optimization of nonsaturated thermally limited high-frequency transformer/inductor design considering discreteness of design variables. *IEEE Trans. Power Electron.* **2019**, *35*, 6231–6250. [CrossRef]
19. Cougo, B.; Kolar, J.W. Integration of leakage inductance in tape wound core transformers for dual active bridge converters. In Proceedings of the 7th International Conference on Integrated Power Electronics Systems (CIPS), Nuremberg, Germany, 6–8 March 2012.
20. Venkatachalam, K.; Sullivan, C.; Abdallah, T.; Tacca, H. Accurate prediction of ferrite core loss with nonsinusoidal waveforms using only Steinmetz parameters. In Proceedings of the IEEE Workshop on Computers in Power Electronics, Mayaguez, PR, USA, 3–4 June 2002.
21. Dowell, P. Effects of eddy currents in transformer windings. *Proc. Inst. Elect. Eng.* **1966**, *113*, 1387–1394. [CrossRef]
22. Mogorovic, M.; Dujic, D. Medium frequency transformer leakage inductance modeling and experimental verification. In Proceedings of the Energy Conversion Congress and Exposition (ECCE), Cincinnati, OH, USA, 1–5 October 2017.
23. Lv, F.; Guo, Y.; Li, P. Calculation and correction method for leakage inductance of high-power medium-frequency transformer. *High Volt. Eng.* **2016**, *42*, 1702–1707. [CrossRef]
24. Mclyman, C.W.T. *Transformer and Inductor Design Handbook*; China Elect. Power Press: Beijing, China, 2008.

25. Balakrishnan, A.; Joines, W.; Wilson, T. Air-gap reluctance and inductance calculations for magnetic circuits using a Schwarz-Christoffel transformation. *IEEE Trans. Power Electron.* **1997**, *12*, 654–663. [CrossRef]
26. Zhang, X.; Xiao, F.; Wang, R.; Fan, X.; Wang, H. Improved Calculation method for inductance value of the air-gap inductor. In Proceedings of the 1st China International Youth Conference on Electrical Engineering (CIYCEE), Wuhan, China, 1–4 November 2020.
27. Incropera, F.; DeWitt, D.; Bergman, T.; Lavine, A. *Fundamentals of Heat and Mass Transfer*; Chemical Industry Press: Beijing, China, 2007.
28. VDI. *VDI Heat Atlas*; Springer: Berlin/Heidelberg, Germany, 2010.

energies

MDPI

Article

State of Charge Dual Estimation of a Li-ion Battery Based on Variable Forgetting Factor Recursive Least Square and Multi-Innovation Unscented Kalman Filter Algorithm

Hongyuan Yuan [1,2,*], Youjun Han [1,3], Yu Zhou [1], Zongke Chen [4], Juan Du [1] and Hailong Pei [1,*]

[1] School of Automation Science and Engineering, South China University of Technology, Guangzhou 510640, China; hanyoujun@gziit.ac.cn (Y.H.); zhouyu2@gedi.com.cn (Y.Z.); dujuan@scut.edu.cn (J.D.)
[2] China Energy Construction Group Guangdong Electric Power Design and Research Institute Co., Ltd., Guangzhou 510700, China
[3] Guangzhou Institute of Industrial Technology, Guangzhou 511458, China
[4] School of Electric Power Engineering, South China University of Technology, Guangzhou 510640, China; tsungkechen@163.com
* Correspondence: hongyuan_yuan@163.com (H.Y.); auhlpei@scut.edu.cn (H.P.)

Abstract: Battery management is the key technical link for electric vehicles. A good battery management system can realize the balanced charge and discharge of batteries, reducing the capacity degradation and the loss of health caused by battery overcharge and discharge, which all depend on the real-time and accurate estimation of the battery's state of charge (SOC). However, the battery's SOC has highly complex nonlinear time-varying characteristics related to the complex chemical and physical state and dynamic environmental conditions, which are difficult to measure directly, and this has become a difficulty in design and research. According to the characteristics of ternary lithium-ion batteries of electric vehicles, a battery SOC dual estimation algorithm based on the Variable Forgetting Factor Recursive Least Square (VFFRLS) and Multi-Innovation Unscented Kalman Filter (MIUKF) is proposed in this paper. The VFFRLS algorithm is used to estimate battery model parameters, and the MIUKF algorithm is used to estimate the battery's SOC in real time. The two algorithms are coupled to update battery model parameters and estimate the SOC. The experiment results show that the algorithm has high accuracy and stability.

Keywords: equivalent circuit model; multi-innovation; Unscented Kalman Filter; variable forgetting factor recursive least square; SOC online estimation; battery management system

Citation: Yuan, H.; Han, Y.; Zhou, Y.; Chen, Z.; Du, J.; Pei, H. State of Charge Dual Estimation of a Li-ion Battery Based on Variable Forgetting Factor Recursive Least Square and Multi-Innovation Unscented Kalman Filter Algorithm. *Energies* **2022**, *12*, 1529. https://doi.org/10.3390/en15041529

Academic Editor: Carlos Miguel Costa

Received: 4 January 2022
Accepted: 17 February 2022
Published: 18 February 2022

Publisher's Note: MDPI stays neutral with regard to jurisdictional claims in published maps and institutional affiliations.

1. Introduction

With the reduction in fossil energy reserves and the intensification of environmental pollution, in order to achieve the goal of "carbon neutralization" and optimize the industrial structure and energy structure, countries all over the world have increased their investment in the new energy industry. Among these new technologies, research on electric vehicles is one of the main directions of study. In many aspects of electric vehicles' battery technology, battery management technology is extremely important. A good battery management system can prevent the battery from overcharging and discharging and realize balanced management. Accurate battery SOC estimation is the basis of battery management system design, but a battery's SOC comprises highly complex nonlinear time-varying characteristics, which are difficult to measure directly. Therefore, it has become the focus of design and research.

At present, SOC estimation methods mainly include the ampere hour integration method, the open circuit voltage method, and machine learning-based algorithms such as the neural network algorithm [1], Kalman filter (KF) and its extension based on parameter estimation model and system identification.

The estimation accuracy of the ampere hour integration method mainly depends on the initial SOC value and sensor error accumulation—the initial SOC value is difficult to determine, and the errors will accumulate when the online estimation time is long, resulting in low accuracy. The open circuit voltage method uses the functional mapping relationship between the battery open circuit voltage (OCV) and the SOC to estimate the SOC. The error of this method is large when it is used for real-time estimation, because the determined functional relationship requires the battery to be at rest for a long time. The algorithm based on machine learning needs a large number of labeled sample data. Benchmark SOC, which is necessary for use as labeling data, is difficult to obtain through online metrics [2,3], while the cost of obtaining a large number of labeled data in the experimental environment is high, meaning that the research and applications are limited at present. KF uses the recursive method of "prediction-actual measurement-correction", which is more suitable for the dynamic system state estimation such as battery SOC estimation. By establishing the battery models and estimating the parameters, the prior state probability determined by the system model parameters is updated according to the measured a posteriori state probability. Therefore, it has been widely used and has become the mainstream research direction in the field of SOC estimation in recent years.

The KF algorithm assumes that the system is a linear system, but the SOC estimation of the battery is the state estimation of a typical nonlinear time-varying system, so the KF algorithm will introduce linear errors. To solve this problem, many improved methods have been proposed, mainly extending and expanding within the framework of the KF algorithm. Among them, the extended Kalman filter (EKF) and Unscented Kalman Filter (UKF) are widely used. The EKF linearizes the nonlinear system through Taylor series expansion. After ignoring the second-order and higher-order terms, it is transformed into the KF algorithm. The EKF algorithm involves the heavy calculation workload of the Jacobian matrix and ignores the high-order terms. It is suitable for systems with weak nonlinearity because it is easy to cause filter divergence to systems with strong nonlinear systems in addition to introducing linear errors [4–6]. For a strong nonlinear system, based on the idea that the probability density function distribution of the approximate nonlinear function is easier to obtain than the approximate nonlinear function, the UKF algorithm is proposed. This algorithm analyzes the probability density of the functional relationship in the battery model and obtains the set of sampling points around the estimated value through unscented transformation. The statistical characteristics of the variables to be estimated are approximated by a series of sampling points, which avoids the complex operation of Jacobian matrix and considers the influence of higher-order terms on the matrix [7–9]. Related research shows that the accuracy of the UKF reaches the second-order Taylor series at least and is higher than the approximate linearization of EKF.

At the same time, the KF algorithm framework assumes that the system is a Markov process, and the process has no memory—that is, the state at the current time is only affected by the state at the previous time—and has nothing to do with the state before the previous time—that is, the system is a complete information system at any time—but the battery's nonlinear time-varying characteristics, dynamic environments and working conditions determine that the SOC estimation cannot strictly meet this assumption. In order to further improve the performance of the estimation algorithm and break through the boundary assumptions of the KF framework, the multi-innovation (MI) identification theory is proposed. When the algorithm updates the parameters, the errors of multiple historical moments are introduced, and the performance is further improved [10].

The KF algorithm framework and its extension are based on the battery model to estimate the a priori probability of SOC. The a priori probability determined by the system model parameters is then updated by the measured a posteriori probability. The variation of the battery model's parameters is one of the main characteristics of the nonlinear time-varying characteristics of battery SOC estimation. Therefore, the accurate and real-time identification of model parameters is the key to the success of the algorithm. Most of the early research used the offline parameter identification method. Offline parameter identification refers to identifying the parameters of the battery model in advance in the laboratory. During the operation of the battery, the corresponding model parameters are transferred for SOC estimation according to the dynamic working state or working environment [11,12]. Offline parameter identification struggles to cover all of the dynamic characteristics in different environments and working conditions, such as the changes in temperature, the health condition of the battery and other factors. In order to solve this problem, an online parameter identification method is proposed. This method identifies the parameters of the battery model in real time and synchronously with the SOC online estimation through the online parameter identification algorithm according to the characteristic parameters such as current, voltage and temperature collected by the sensor during the working process of the battery system. Online parameter identification needs to adapt to the changing environment and operating state and has high requirements for real-time performance [13,14]. Generally, the amount of data used is smaller than offline parameter identification, which may result in lower accuracy. However, due to its ability to better adapt to the characteristics of a nonlinear time-varying battery, it has become the mainstream research direction in recent years. Among various online identification algorithms for battery model parameters, recursive least square (RLS) and its extended algorithm have become the mainstream online identification algorithms due to their advantages of small storage space, a small amount of computation and being suitable for real-time control [15]. There is an important forgetting factor parameter in the RLS algorithm, which is used to determine the influence weight of the previous time on the identification result of the current time. The selection of this parameter has a great impact on the convergence speed and accuracy of the battery's parameters. In the early stage, the trial-and-error method was mainly used to obtain a fixed value. Later, the Variable Forgetting Factor RLS (VFFRLS) algorithm was proposed, which associated the forgetting factor with the algorithm estimation error in the current time window and realized the design of the dynamic forgetting factor [16].

At present, extensive and in-depth research on the SOC estimation algorithm and the battery model parameter identification algorithm has been conducted based on KF framework and its extension, but most of the research is focused on one aspect, and there is less research on dual or joint estimation algorithms integrating SOC estimation and parameter identification. Recently, a multi-scale EKF joint estimation SOC algorithm [17] and a UKF and VFFRLS joint estimation algorithm [18] were proposed, but there is no dual or joint estimation method that further integrates MI, UKF and VFFRLS. This paper attempts to integrate advantageous algorithms of MI, UKF and VFFRLS and creatively realize the dual estimation experiment of MIUKF + VFFRLS. The experiment results show that the algorithm has obvious advantages in accuracy and stability compared with offline parameter + EKF, offline parameter + UKF and offline parameter + MIUKF. Compared with UKF + VFFRLS, it has advantages in convergence speed, accuracy and stability. The overall performance of the fused algorithm is outstanding.

2. Battery Model Establishment and Parameter Identification

2.1. Battery Model Establishment

There are two kinds of battery models: the electrochemical model and equivalent circuit model. The electrochemical model abstracts the complex physical and chemical reaction process inside the battery to describe the dynamic characteristics of the battery. Due to the fundamental process experiment, the accuracy of the model is high, but the model is complex, and the amount of calculation is large, which is difficult to meet the requirements of real-time tasks. The equivalent circuit model uses circuit elements such as resistance, capacitance and power supply to form a circuit network to describe the dynamic characteristics of the battery. It can clearly reflect the electrical characteristics of the battery, and the model is relatively simple, meaning that it is suitable and widely used for tasks requiring real-time calculation such as SOC.

A common equivalent circuit model is the n*RC circuit model. The dynamic characteristics of the battery are simulated by N groups of circuits connected in parallel and then connected in series. It can be divided into zero-order, first-order, second-order and multi-order models according to the number of resistance and capacitor groups. The zero-order equivalent circuit network model is also called the Rint model, which is only composed of a voltage source and a resistor in series. The Rint model can only represent the static process of the battery and cannot describe the dynamic characteristics of the battery. The first-order equivalent circuit network model is also called the Thevenin model [19]. This model consists of a group of capacitors and resistors in a parallel circuit, a voltage source and a resistor which are connected in series. The resistance in series simulates ohmic internal resistance, and the combination of capacitors and resistors in parallel simulates polarization internal resistance. The second-order equivalent circuit network, also known as the DP model [20], is composed of two sets of capacitor resistance parallel circuits, one voltage source and one resistance which are connected in series. The series resistance simulates ohmic internal resistance. The two sets of capacitor resistance parallel circuits are used to describe the short-term electrochemical polarization effect and long-term concentration polarization effect of the battery, respectively, which can describe the dynamic characteristics more accurately.

With the increase in model order, the accuracy of the model will increase, but the complexity of the model will also increase. Considering the balance of accuracy and calculation, the DP model is adopted in this paper, and its circuit structure is shown in Figure 1.

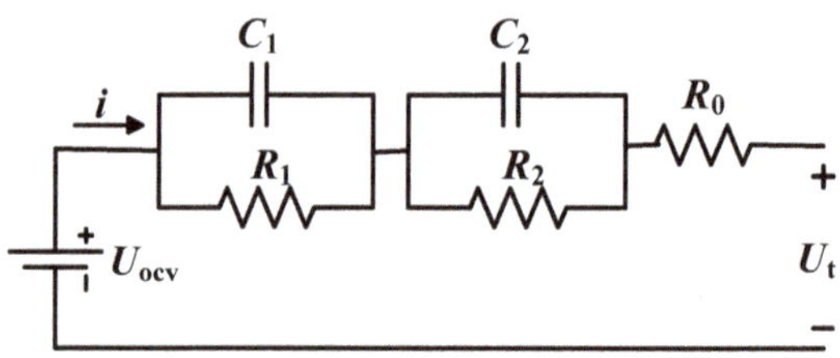

Figure 1. DP equivalent circuit model.

It can be seen from the figure that the DP model is composed of a voltage source, ohmic internal resistance and RC networks. U_{ocv} is the battery open circuit voltage. R_0 is the battery ohmic internal resistance. U_t is the battery terminal voltage. The parallel network constructed by R_1 and C_1 is used to reflect the gradual change in the battery terminal voltage, and its time constant is relatively large, which is used to describe the long-term concentration polarization effect. The parallel network constructed by R_2 and C_2 is used to reflect the sudden change in the battery terminal voltage, and its time constant is small, which is used to describe the short-term electrochemical polarization effect.

According to Kirchhoff's law, the output voltage relation of the DP model is as follows:

$$\begin{cases} U_t = U_{ocv} - IR_0 - U_1 - U_2 \\ \dot{U}_1 = -\frac{1}{C_1R_1}U_1 + \frac{1}{C_1}I \\ \dot{U}_2 = -\frac{1}{C_2R_2}U_2 + \frac{1}{C_2}I \end{cases} \tag{1}$$

where $\dot{U}_1$ and $\dot{U}_2$ are the derivatives of U_1 and U_2 with respect to time, respectively.

2.2. Open Circuit Voltage Parameter Identification of Battery Model

The open circuit voltage U_{ocv} is an ideal power supply in the battery model. It is not affected by the resistance capacitance parameters of the battery. It can be measured by the terminal voltage when the battery discharge current is close to zero, and the internal chemical reaction is stable because when the chemical reaction is stable, the dynamic characteristics of the reaction are close to zero. According to Formula (1), U_{ocv} is equal to U_t.

There is a relatively stable functional relationship between the open circuit voltage U_{ocv} and SOC. The functional relationship between SOC and U_{ocv} can be obtained by function fitting (generally polynomial fitting) of SOC and U_{ocv} measured at different SOC value points after sufficiently long relaxation.

This functional relationship can be used in battery resistance capacitance parameter identification and SOC estimation.

2.3. Resistance Capacitance Parameter Identification of Battery Mode

Offline parameter identification and online parameter identification can be adopted in the resistance capacitance parameter identification of the battery model. The principles are to use the resistance capacitance parameters of the battery to fit the terminal voltage curve measured in the dynamic charge and discharge process and to optimize the error between the fitting curve and the measured curve through the adjustment of the resistance capacitance parameters. Offline parameter identification is limited by the experimental conditions and cannot cover all situations such as different temperatures, different charge and discharge currents, etc. A large amount of data can be collected for fitting because the real-time performance is not required in the experimental environment, which generally brings high accuracy under the corresponding experimental conditions but may bring sharply dropping accuracy under different conditions. On the contrary, online parameter identification can be dynamically identified according to the real-time operating conditions, but the real-time requirements limit the data amount for fitting, which generally causes lower accuracy, meaning that proper algorithm design is critical. The online parameter identification method is generally modified and innovated based on the RLS method.

2.3.1. Recursive Least Squares Parameter Identification

The RLS algorithm is developed from the least square (LS) algorithm, and the basic principle is as follows [15].

For discrete linear systems

$$y_{k+1} = \theta_k \Phi^{\theta}_{k+1} + e^{\theta}_{k+1} \tag{2}$$

where y_{k+1} is the output vector of the system, and θ_k is the model parameter vector to be identified; Φ^{θ}_{k+1} is the input data matrix of the system, and e^{θ}_{k+1} is the error vector.

The optimal value estimated by minimizing the sum of squares of e^{θ}_{k} errors is the LS algorithm. The resource consumption of the algorithm will continue to rise until the resources are exhausted if all data sequences are used for the estimation of the length of Φ^{θ}_{k+1}. RLS is proposed to solve this problem, in which the identification results of the current time and the system input of the next time are used to recursively obtain the system parameter value of the next time. The specific methods are as follows:

$$\begin{cases} K^{\theta}_{k+1} = \frac{P^{\theta}_{k}\Phi^{\theta}_{k+1}}{\lambda + (\Phi^{\theta}_{k+1})^T P^{\theta}_{k}\Phi^{\theta}_{k+1}} \\ \theta_{k+1} = \theta_k + K^{\theta}_{k+1} e^{\theta}_{k+1} \\ P^{\theta}_{k+1} = \lambda^{-1}P^{\theta}_{k} - \lambda^{-1}K^{\theta}_{k+1}(\Phi^{\theta}_{k+1})^T P^{\theta}_{k} \end{cases} \tag{3}$$

where K is the algorithm gain, and P is the error covariance matrix of the identification parameters. The λ is the forgetting factor, which represents the forgetting degree to the previous identification result and determines the confidence comparison between the old and new sampling data. Its value range is between 0 and 1. When the forgetting factor is 1, it means that the algorithm has no forgetting function, and all data points are used in parameter identification; then, the RLS algorithm degenerates to the LS algorithm. When the forgetting factor is 0, it means that the algorithm will forget all the previous identification results and only use the data of the current time for parameter identification. The selection of forgetting factor has a great impact on the accuracy of parameter identification results.

In the RLS method, the parameter needs to be preset as a fixed value, which has a poor effect in the battery parameter identification with complex working conditions.

2.3.2. Parameter Identification of Variable Forgetting Factor Recursive Least Square Method

The VFFRLS method is an improved algorithm based on the recursive least square method (RLS) used to find the optimal value of the forgetting factor adaptively according to the estimation error in the process of parameter identification [16].

The calculation formula of the variable forgetting factor is as follows.

$$\begin{cases} e_{k+1} = y_{k+1} - \hat{y}_{k+1} \\ L(k+1) = -\rho \frac{\sum_{i=k-S+2}^{k+1} e_i e_i^T}{S} \\ \lambda = \lambda_{\min} + (\lambda_{\max} - \lambda_{\min}) 2^{L(k+1)} \end{cases} \tag{4}$$

where e_{k+1} is the estimation error at $k + 1$ time, λ is the forgetting factor, and $\lambda_{\min}$ and $\lambda_{\max}$ are the minimum and maximum forgetting factors, respectively. The larger the λ is, the smaller the influence of the system fluctuation on the estimation accuracy of the algorithm; the smaller the λ is, the stronger the tracking ability and convergence ability of the algorithm. ρ is the sensitivity factor, and S is the window size. In this paper, we set the value of $\lambda_{\min}$ as 0.9, the value of $\lambda_{\max}$ as 0.995 [18], the value of ρ as 200, and the value of S as 22.

3. Battery State of Charge Estimation based on Multi-Innovation and Kalman Filter Framework Algorithm

3.1. Definition and Characteristics of a Battery's State of Charge

SOC refers to the ratio of the remaining charge margin in the battery to the rated charge capacity of the battery. The calculation formula is as follows.

$$SOC = SOC_0 - \frac{1}{Q_N} \int_{\tau=t_0}^{\tau=t} \mu I d\tau \tag{5}$$

where SOC_0 is the initial SOC value, Q_N is the maximum discharge capacity, μ represents coulomb efficiency and generally is set to 1, and I is the load current, and its discharge is specified as positive.

SOC can only be estimated through indirect methods with, e.g., voltage and current as inputs, which is related to resistance capacitance parameters and working conditions, and so the estimation is related to the ohmic internal resistance, polarization internal resistance, the temperature and health state of the battery, etc. The system has highly complex nonlinear characteristics, and so the estimation error is large if only using the measurement method. If the data of the two dimensions—system state estimation and measurement results—can be used at the same time, the state of the system model over time can be used as an a priori estimation value, the measured value is updated as an a posteriori estimate, and the algorithm should have a very efficient performance improvement in theory. A Bayesian filter has such characteristics. A Bayesian filter can be derived from the Bayesian full probability formula without any omission of the nonlinear characteristics of the system, but it is difficult to obtain the prior probability distribution of the system in practical use, and the calculation is also too complex. In order to facilitate calculation, in practical application, it is generally assumed that the probability distribution is normally distributed, and the system is a complete information system—that is, the state of the next time is only related to the current time. In this case, the algorithm is simplified to the KF algorithm.

KF and its extended algorithm adopt the recursive filtering algorithm, which has the characteristics of simple calculation. At the same time, the algorithm uses the data of two relatively independent dimensions—state equation and measurement equation—for verification and updating, which can effectively improve the accuracy of the algorithm. Therefore, it is widely used in SOC estimation.

3.2. Kalman Filter

The basic formula of KF is as follows:

$$x_{k+1} = A_{k+1}x_k + B_{k+1}u_{k+1} + \omega_{k+1} \tag{6}$$

$$y_{k+1} = H_{k+1}x_{k+1} + v_{k+1} \tag{7}$$

where Equation (6) is called the state equation, and Equation (7) is called the measurement equation. x_k is the state vector at time k, x_{k+1} is the state vector at time $k + 1$, y_{k+1} is the observation vector at time $k + 1$, u_{k+1} is the system input vector at time $k + 1$, A is the state transition parameter matrix, B is the input control parameter matrix, H is the observation parameter matrix, $\omega_{k+1} \sim N(0, Q_{k+1})$ is the system noise vector, and $v_{k+1} \sim N(0, R_{k+1})$ is the measurement noise vector. ω_{k+1} and v_{k+1} are independent of each other.

The recursive process of the algorithm is as follows.

(1) Predict the system state at the next time.

$$\hat{x}_{k+1}^{(-)} = A_{k+1}\hat{x}_k^{(+)} + B_{k+1}u_{k+1} \tag{8}$$

(2) Predict the system covariance at the next time.

$$P_{k+1}^{(-)} = A_{k+1}P_k^{(+)}A_{k+1}{}^T + Q_{k+1} \tag{9}$$

(3) Calculate the Kalman gain of the measurement update.

$$K_{k+1} = \frac{P_{k+1}^{(-)}H_{k+1}{}^T}{H_{k+1}P_{k+1}^{(-)}H_{k+1}{}^T + R_{k+1}} \tag{10}$$

(4) Update the system status by the measured values.

$$\hat{x}_{k+1}^{(+)} = \hat{x}_{k+1}^{(-)} + K_{k+1}(y_{k+1} - H_{k+1}\hat{x}_{k+1}^{(-)}) \tag{11}$$

(5) Update system covariance by the measured values.

$$P_{k+1}^{(+)} = (I - K_{k+1}H_{k+1})P_{k+1}^{(+)} \tag{12}$$

where $\hat{x}_{k+1}^{(-)}$ represents an a priori estimate of x at time k + 1, $\hat{x}_{k+1}^{(+)}$ represents an a posteriori estimate of x at time k + 1, $\hat{x}_k^{(+)}$ represents an a posteriori estimate of x at time k, $P_{k+1}^{(-)}$ represents an a priori estimate of P at time k + 1, $P_{k+1}^{(+)}$ represents an a posteriori estimate of P at time k + 1, and $P_k^{(+)}$ represents an a posteriori estimate of P at time k. K_{k+1} represents the Kalman gain at time k + 1.

KF requires that the state equation and measurement equation of the system are linear, which is difficult to apply to nonlinear tasks such as the battery state of charge. Instead, the EKF and UKF algorithms are generally used in such tasks.

3.3. Extended Kalman Filter Algorithm

The EKF algorithm is transformed into the KF algorithm by expanding the Taylor series of state and measurement functions and ignoring the second-order and higher-order terms. The basic formula is as follows:

$$x_{k+1} = f(x_k, u_k, k, \omega_k) \tag{13}$$

$$y_{k+1} = h(x_{k+1}, k) + v_{k+1} \tag{14}$$

where u_k is the system input vector at time k, $\omega_k \sim N(0, Q_k)$ is the system noise vector, and $v_k \sim N(0, R_k)$ is the measurement noise vector. ω_k and v_k are independent of each other.

The recursive process of the algorithm is as follows.

(1) Predict the system state at the next time.

$$\hat{x}_{k+1}^{(-)} = f(\hat{x}_k, u_k, k) \tag{15}$$

(2) Predict the system covariance at the next time.

Let

$$F_k = \left.\frac{\partial f}{\partial x}\right|_{x_k = \hat{x}_k^{(+)}} \tag{16}$$

Get

$$P_{k+1}^{(-)} = F_k P_k^{(+)} {F_k}^T + Q_k \tag{17}$$

(3) Calculate the Kalman gain of the measurement update.

Let

$$H_{k+1} = \left.\frac{\partial h}{\partial x}\right|_{x_{k+1} = \hat{x}_{k+1}^{(-)}} \tag{18}$$

Get

$$K_{k+1} = \frac{P_{k+1}^{(-)} {H_{k+1}}^T}{H_{k+1} P_{k+1}^{(-)} {H_{k+1}}^T + R_{k+1}} \tag{19}$$

(4) Update the system status by the measured values.

$$\hat{x}_{k+1}^{(+)} = \hat{x}_{k+1}^{(-)} + K_{k+1}\left\{y_{k+1} - h\left[\hat{x}_{k+1}^{(-)}, k+1\right]\right\} \tag{20}$$

(5) Update system covariance by the measured values.

$$P_{k+1}^{(+)} = P_{k+1}^{(-)} - K_{k+1}\left[H_{k+1} P_{k+1}^{(-)} H_{k+1}^T + R_{k+1}\right] \tag{21}$$

where $\hat{x}_{k+1}^{(-)}$ represents an a priori estimate, and $\hat{x}_{k+1}^{(+)}$ represents an a posteriori estimate of x at time $k + 1$; $\hat{x}_k^{(+)}$ represents an a posteriori estimate of x at time k; $P_{k+1}^{(-)}$ represents an a priori estimate of P at time $k + 1$; $P_{k+1}^{(+)}$ represents an a posteriori estimate of P at time $k + 1$, and $P_k^{(+)}$ represents an a posteriori estimate of P at time k. K_{k+1} represents the Kalman gain at time $k + 1$.

The EKF algorithm omits the influence of second-order and higher-order terms and is only applicable to the case of weak nonlinearity. For the case of heavy nonlinearity, the error is large, and more complex nonlinear filtering algorithms are often required, such as the UKF.

3.4. Unscented Kalman Filter Algorithm

The UKF takes the KF as the basic framework—the basic formula is consistent with that of the EKF, and unscented transformation is applied to realize nonlinear application scenarios. The recursive algorithm flow is as follows.

(1) Let the a posteriori state estimation and covariance at time k be $x_k^{(+)}$ and $P_k^{(+)}$, respectively.

(2) Calculate sampling points.

$$\begin{cases} \chi_k^0 = x_k^{(+)} \\ \chi_k^i = x_k^{(+)} + \sqrt{(L+\eta)P_{xx}}, i = 1, 2 \ldots L \\ \chi_k^i = x_k^{(+)} - \sqrt{(L+\eta)P_{xx}}, i = L+1, L+2, \ldots 2L \end{cases} \tag{22}$$

where L is the length of the state vector, and the weight value is calculated as follows:

$$\begin{cases} \eta = \alpha^2(L + k_i) - L \\ W_m^0 = \frac{\eta}{L+\eta}, W_m^i = \frac{1}{2(L+\eta)}, i = 1, 2 \ldots 2L \\ W_c^0 = \frac{\eta}{L+\eta} + 1 - \alpha^2 + \beta, W_c^i = \frac{1}{2(L+\eta)}, i = 1, 2 \ldots 2L \end{cases} \tag{23}$$

where subscript c represents the weight of covariance; subscript m represents the weight of mean square deviation; η represents the scaling ratio, and α represents the distribution state of sampling points—when α is large, it indicates a greater weight of sigma points at the average value. β is a weight, which is used to combine the dynamic differences of higher-order terms in the equation. In this paper, we set $L = 3$, $\alpha = 0.01$, $k_i = 0$, $\beta = 2$.

(1) Update the a priori state value $x_{k+1}^{(-)}$ and a priori variance value $P_{k+1}^{(-)}$.

$$\chi_{k+1}^i = f(\chi_k^i, u_k, k) \tag{24}$$

$$x_{k+1}^{(-)} = \sum_{i=0}^{2L} W_m^i \chi_{k+1}^i \tag{25}$$

$$P_{k+1}^{(-)} = \sum_{i=0}^{2L} (W_c^i(\chi_k^i - x_{k+1}^{(-)})(\chi_k^i - x_{k+1}^{(-)})^T) + Q_{k+1} \tag{26}$$

where Q_k is the system noise covariance matrix; in this paper, we set $Q_k = 0.00000001 * \begin{bmatrix} 1 & 0 & 0 \\ 0 & 1 & 0 \\ 0 & 0 & 1 \end{bmatrix}$.

(2) Calculate the observation estimation value $\hat{y}_{k+1}$, observation variance prediction value P_{k+1}^{yy} and estimated covariance difference P_{k+1}^{xy}.

$$y_{k+1}^i = h(\chi_{k+1}^i, k+1) \tag{27}$$

$$\hat{y}_{k+1} = \sum_{i=0}^{2L} W_m^i y_{k+1}^i \tag{28}$$

$$P_{k+1}^{yy} = \sum_{i=0}^{2L} (W_c^i(y_{k+1}^i - \hat{y}_{k+1})(y_{k+1}^i - \hat{y}_{k+1})^T + R_{k+1} \tag{29}$$

$$P_{k+1}^{xy} = \sum_{i=0}^{2L} W_c^i(\chi_{k+1}^i - \hat{x}_{k+1}^{(-)})(y_{k+1}^i - \hat{y}_{k+1})^T \tag{30}$$

where R_k is the measurement noise covariance matrix; in this paper, we set $R_k = 1$.

(3) Update the a posteriori state value $x_{k+1}^{(+)}$ and a posteriori state error covariance $P_k^{(+)}$ using the measured value y_{k+1}.

$$K_{k+1} = \frac{P_{k+1}^{xy}}{P_{k+1}^{yy}} \tag{31}$$

$$x_{k+1}^{(+)} = x_{k+1}^{(-)} + K_{k+1}(y_{k+1} - \hat{y}_{k+1}) \tag{32}$$

$$P_{k+1}^{(+)} = P_{k+1}^{(-)} - K_{k+1} P_{k+1}^{yy} K_{k+1} \tag{33}$$

where K_{k+1} represents Kalman gain at time $k + 1$.

Compared with the EKF algorithm, the UKF can adapt to systems with stronger nonlinearity and can achieve third-order approximation accuracy in the case of Gaussian distribution and second-order approximation accuracy in the case of non-Gaussian distribution; the UKF does not need to calculate the Jacobian matrix, but the number of sampling points is $2n + 1$, and the overall amount of calculation is larger than the EKF algorithm.

3.5. Application of Multi-Innovation in Kalman Filter Framework Algorithm

In the traditional KF framework algorithm, only the error of the current time is used to update the state of the next time. The model is simple and easy to calculate, but it also brings problems. For highly complex nonlinear time-varying battery operating conditions, the state of the next time is likely to be related not only to the current time but also to several times before the current time, resulting in a decline in accuracy. In order to solve the problem and further improve the estimation progress, the multi-innovation identification theory is introduced into the measurement equation. The calculation formula of multi-innovation identification is as follows.

Expand a single innovation e_k into an innovation matrix $e_{p,k}$.

$$e_{p,k} = \begin{bmatrix} e_k \\ e_{k-1} \\ e_{k-2} \\ \vdots \\ e_{k-p+1} \end{bmatrix} \tag{34}$$

At the same time, the gain k_k is extended to the gain matrix $k_{p,k}$

$$k_{p,k} = \left[k_k, k_{k-1}, \cdots, k_{k-\mathrm{p}+1}\right] \tag{35}$$

Therefore, the status measurement update needs to be modified as follows:

$$y_k = \hat{y}_k + \left[\mathrm{k}_k, k_{k-1}, \cdots k_{k-\mathrm{p}+1}\right] e_{p,k} \tag{36}$$

Namely,

$$y_k = \hat{y}_k + \sum_{i=0}^{p} \gamma_{\mathrm{i}} k_{i,k} e_{k-i} \tag{37}$$

where

$$\begin{cases} \gamma_1 = 1 \\ \gamma_2 = \gamma_3 = \cdots \gamma_p = \frac{a}{M-1}, 0 \leq a \leq 1, M \geq 2 \end{cases} \tag{38}$$

where M is the innovation length and a is the adjustable coefficient. This paper takes M as 22and takes a as 0.5. A detailed discussion about the selection of the two parameters will be conducted in Section 5.3.1.

4. Dual Estimation of VFFRLS Battery Model Parameters and MIUKF Battery SOC

4.1. Union of Basic Equations

It can be obtained from Equations (2) and (14) that

$$\theta_k \Phi_{k+1}^{\theta} + e_{p,k+1} = h(x_{k+1}^{(-)}, k+1) + v_{k+1} \tag{39}$$

Then, it can be obtained from Equation (13) that

$$\theta_k \Phi_{k+1}^{\theta} + e_{p,k+1} = h(f(x_k^{(+)}, u_k, \omega_k, k)) + v_{k+1} \tag{40}$$

Finding partial derivatives on both sides of the equation,

$$\Phi_{k+1}^{\theta} + \frac{\partial}{\partial \theta_k} e_{p,k+1} = \frac{\partial}{\partial \theta_k}\left(h(f(x_k^{(+)}, u_k, \omega_k, k)\right) + \frac{\partial}{\partial \theta_k} v_{k+1} \tag{41}$$

where $e_{p,k}$,ω_{k-1}, and v_k are defined as being independent of θ_k. Equation (42) can be simplified to

$$\Phi_{k+1}^{\theta} = \frac{\partial}{\partial \theta_k}\left(h(f(x_k^{(+)}, u_k, k)\right) \tag{42}$$

Thus, the basic equations of VFFRLS battery model parameter estimation and MIUKF SOC estimation are combined.

4.2. Setting of Estimation Period

Compared with the time-varying characteristics of SOC, the time-varying characteristics of battery parameters are relatively flat, and the time interval for updating the estimation can be relatively long. In the dual estimation, the SOC estimation is carried out for each sampling period while the estimation of battery parameters adopts multiple sampling intervals as T_θ, which is taken as 60 in this paper.

4.3. Battery Parameters Transmission

Every time the battery parameters are updated, the VFFRLS part of the algorithm passes the current battery parameters to the SOC estimation and starts the timer. Before reaching the sampling interval set by the timer, the battery parameters remain unchanged—that is, if the update time parameter is θ_k, the subsequent parameters θ_{k+1}, $\theta_{k+2} \ldots \theta_{k+T_\theta-1}$ are equal to θ_k.

4.4. Forgetting factor Transmission

The timer starts after the battery parameter is updated. When the sampling interval set by the timer is reached, the battery parameter will be updated again. At this time, the SOC estimation part of the algorithm transmits the error in the previous M sampling cycle time window to the VFFRLS part of the algorithm to calculate the forgetting factor λ, which can then be substituted into the parameter in the equation.

4.5. Algorithm Flow

The algorithm flow is shown in Figures 2–4.

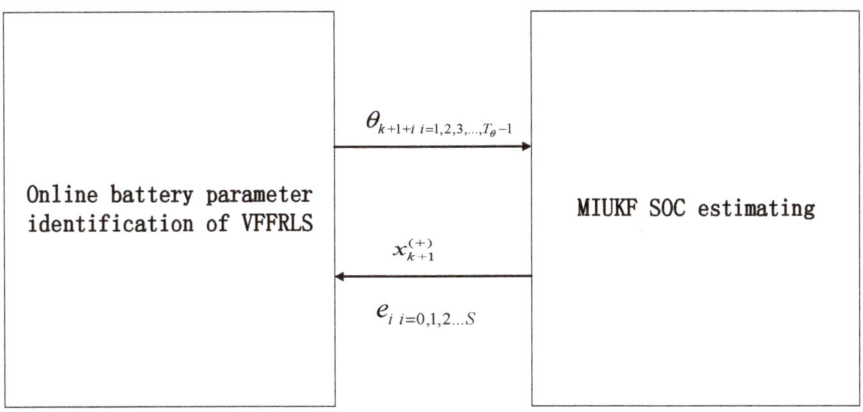

Figure 2. Overall block diagram of algorithm.

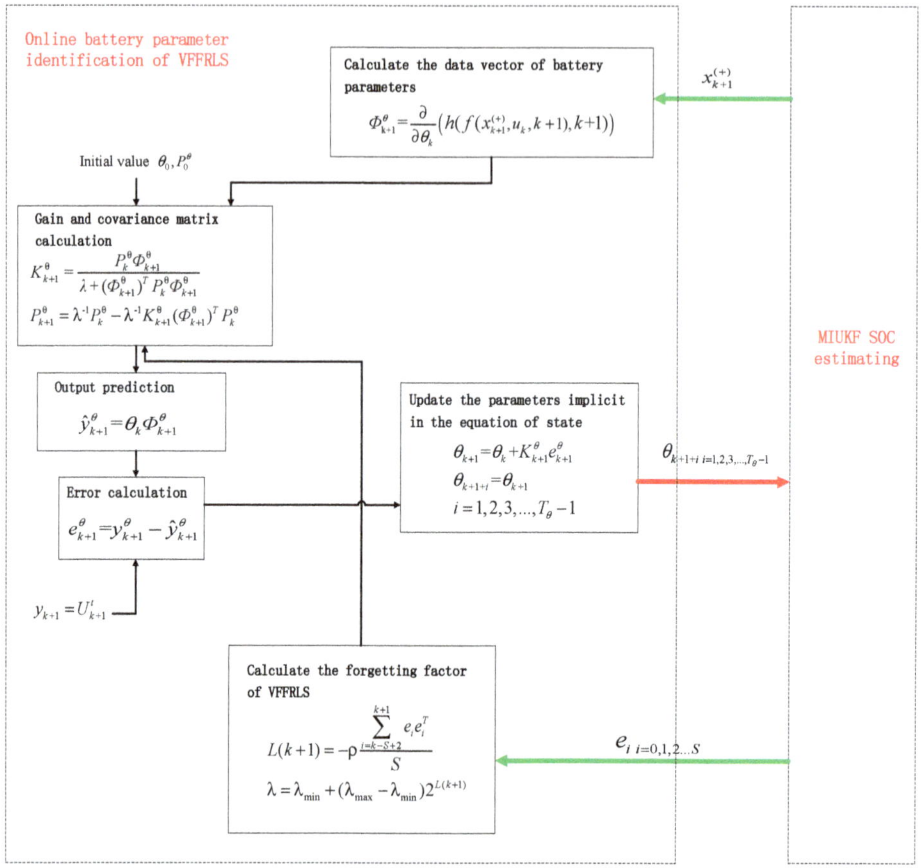

Figure 3. The algorithm flow of VFFRLS part.

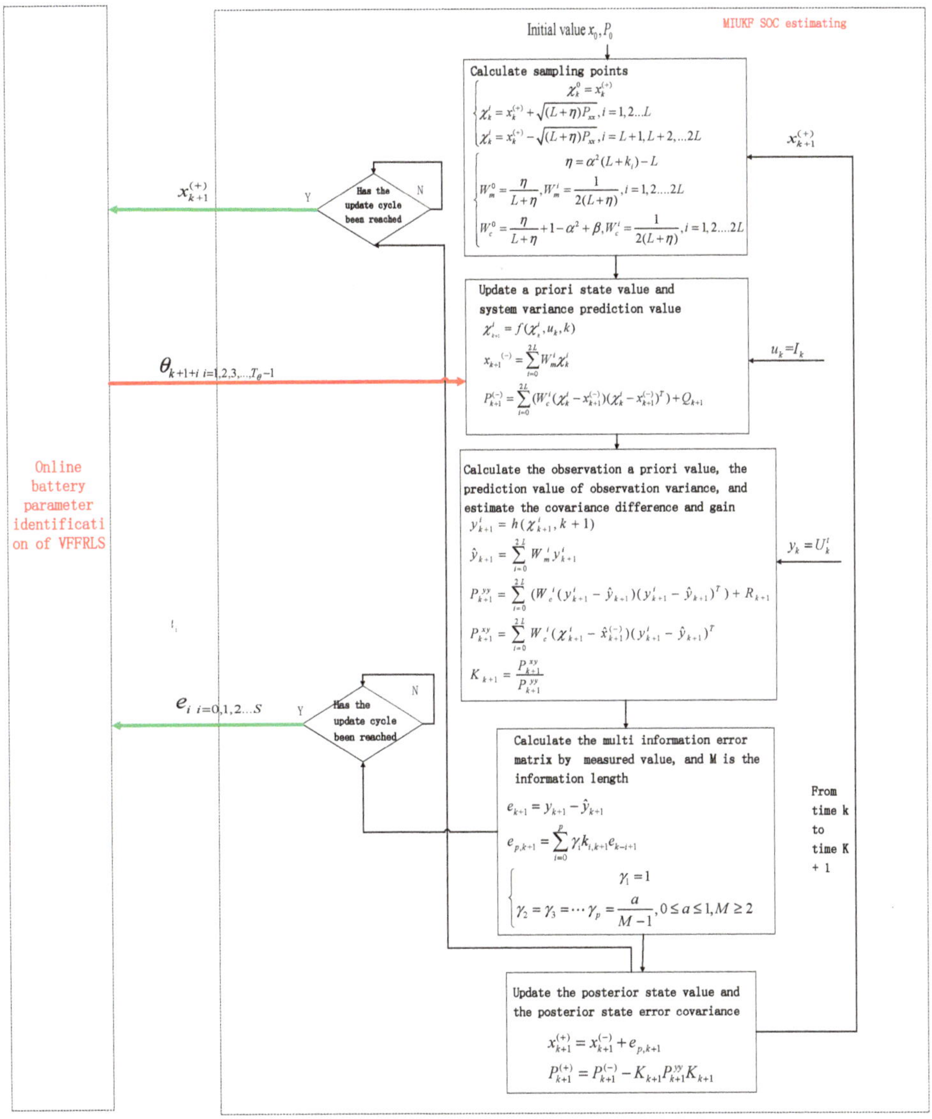

Figure 4. The algorithm flow of MIUKF part.

5. Analysis of Experiment Results

5.1. Data Sources and Tools

The battery used in this paper is a 32 Ah/3.7 V square ternary material lithium power battery produced by Ningde Times. The test platform consists of a battery cell, a power battery test system and a high- and low-temperature damp heat alternating test chamber. In this paper, the effectiveness of the algorithm is verified by using the Urban Dynamometer Driving Schedule (UDDS) [21] cycle test data. The test conditions are set to 25 °C. After the battery is fully charged and left to stand for half an hour, it is discharged for 13 UDDS cycle cycles, the SOC is reduced from 100% to 1.2%, the voltage is reduced from 4.18 V to 3.21 V, the experimental data acquisition interval is set as 1 s, and a total of 20,000 data samples are collected. The algorithm is simulated by MATLAB R2019b based on the collected data.

5.2. Model Substitution

Let $\left[U_1^k, U_2^k, SOC^k\right]^T$ be the basic form of the state vector x_k and the observation vector y_k at time k, and the battery parameter matrix $\theta_k = [R_0^k, R_1^k, C_1^k, R_2^k, C_2^k]^T$ is embedded into the state equation as an intermediate variable through the functional relationship with U_1^k, U_2^k, SOC^k and I_k. The initial setting value of $x_k = \left[U_1^k, U_2^k, SOC^k\right]^T$ is $[0, 0, 0.8]^T$, and the initial value of θ_k is obtained from offline identified parameters, that is $[0.002, 0.0012, 7.23e+04, 0.0011, 4.49e+04]^T$, while the initial variance of θ_k is set as $P_0 = [0.0001, 0.0001, 10000, 0.0001, 10000]^T$.

The measured data are a matrix composed of vectors $\left[k, I^k, U_t^k, Soc_r^k\right]^T$. I_k is substituted into the measurement equation as the input vector at time k. U_t^k is substituted into the measurement equation through the functional relationship with U_1^k, U_2^k and U_{ocv}^k, and the functional relationship between U_{ocv}^k and Soc^k. The data provided by the original factory for fitting the functional relationship between SOC and OCV are shown in Table 1.

Table 1. Relationship data between SOC and OCV.

Uocv	3.423	3.521	3.596	3.644	3.696	3.784	3.88	3.948	4.02	4.075	4.181
SOC	0	0.1	0.2	0.3	0.4	0.5	0.6	0.7	0.8	0.9	1

Under the experimental condition, one discharge cycle is tested, so there are no long-term cumulative error problems. Then, Soc_r^k, which is obtained from the ampere hour integral method, has high accuracy and can be used as the benchmark for the comparison of the algorithm results.

The data are then substituted into the Kalman filter framework algorithm. The essence of the error is the difference between the a priori state terminal voltage determined by the battery resistance capacitance parameter and the measured terminal voltage, which is characterized by the difference between the a priori state value and the measured value. The task of this paper is to update the resistance capacitance parameters and Kalman filter parameters through the iterative process of the VFFRLS + MIUKF algorithm to minimize the terminal voltage error.

5.3. Experiment Result

5.3.1. Experiment Results of Different Parameters of MI

According to Formula (38), the parameters M and a are the key parameters affecting the multi-innovation model, and their value and influence need to be analyzed. In the following, nine parameters are selected for an interval 20 of $M \in$ [2162], and eleven parameters are selected for an interval 0.1 of $a \in [0,1]$ for combined analysis. When $a = 0$, the multi-innovation model does not work. At this time, MIUKF + VFFRLS degenerates into the UKF + VFFRLS algorithm. Therefore, the comparison of the two algorithms can be transformed into the comparison of algorithms when a is non-zero and a is zero.

The nine values of M are classified, and different values of a are taken in each classification, substituted into the model, and the final SOC error is calculated to draw a three-dimensional diagram (in order to ensure the image display effect, the data at 41 to 20,000 time points are intercepted in t dimension). The results are shown in Figure 5.

At the same time, the average and standard deviation of the absolute value of SOC error under different parameter values are calculated. Among them, the average value of the absolute value of SOC error under different parameter values is shown in Table 2.

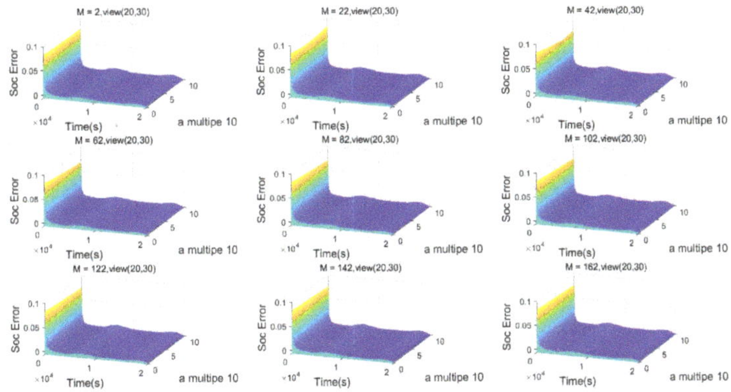

Figure 5. Variation of SOC error with time under different value combinations of M and a.

Table 2. Average value of the absolute value of SOC error under different parameter values.

	$M = 2$	$M = 22$	$M = 42$	$M = 62$	$M = 82$	$M = 102$	$M = 122$	$M = 142$	$M = 162$
$a = 0$	0.247%	0.247%	0.247%	0.247%	0.247%	0.247%	0.247%	0.247%	0.247%
$a = 0.1$	0.230%	0.229%	0.228%	0.227%	0.226%	0.224%	0.223%	0.221%	0.220%
$a = 0.2$	0.224%	0.223%	0.222%	0.220%	0.218%	0.216%	0.214%	0.212%	0.210%
$a = 0.3$	0.223%	0.221%	0.220%	0.218%	0.216%	0.214%	0.211%	0.209%	0.207%
$a = 0.4$	0.223%	0.222%	0.221%	0.219%	0.217%	0.215%	0.213%	0.211%	0.209%
$a = 0.5$	0.226%	0.225%	0.224%	0.223%	0.221%	0.219%	0.217%	0.215%	0.213%
$a = 0.6$	0.231%	0.230%	0.229%	0.227%	0.225%	0.223%	0.221%	0.220%	0.218%
$a = 0.7$	0.235%	0.234%	0.234%	0.232%	0.230%	0.228%	0.226%	0.225%	0.223%
$a = 0.8$	0.240%	0.239%	0.239%	0.237%	0.235%	0.233%	0.231%	0.230%	0.229%
$a = 0.9$	0.245%	0.244%	0.244%	0.242%	0.240%	0.238%	0.236%	0.235%	0.234%
$a = 1$	0.250%	0.249%	0.248%	0.248%	0.246%	0.243%	0.242%	0.240%	0.240%

The standard deviation of absolute value of SOC error under different parameter values is shown in Table 3.

Table 3. Standard deviation of the absolute value of SOC error under different parameter.

	$M = 2$	$M = 22$	$M = 42$	$M = 62$	$M = 82$	$M = 102$	$M = 122$	$M = 142$	$M = 162$
$a = 0$	0.735%	0.735%	0.735%	0.735%	0.735%	0.735%	0.735%	0.735%	0.735%
$a = 0.1$	0.716%	0.716%	0.715%	0.714%	0.712%	0.711%	0.710%	0.710%	0.710%
$a = 0.2$	0.700%	0.700%	0.699%	0.696%	0.693%	0.691%	0.690%	0.690%	0.690%
$a = 0.3$	0.686%	0.687%	0.685%	0.682%	0.678%	0.675%	0.674%	0.674%	0.674%
$a = 0.4$	0.674%	0.676%	0.675%	0.671%	0.665%	0.662%	0.660%	0.660%	0.661%
$a = 0.5$	0.664%	0.667%	0.666%	0.662%	0.655%	0.651%	0.649%	0.650%	0.651%
$a = 0.6$	0.655%	0.659%	0.659%	0.655%	0.647%	0.642%	0.640%	0.641%	0.642%
$a = 0.7$	0.648%	0.653%	0.653%	0.649%	0.641%	0.634%	0.632%	0.633%	0.635%
$a = 0.8$	0.641%	0.647%	0.649%	0.645%	0.635%	0.628%	0.625%	0.626%	0.629%
$a = 0.9$	0.636%	0.643%	0.645%	0.642%	0.631%	0.623%	0.620%	0.621%	0.624%
$a = 1$	0.631%	0.639%	0.642%	0.640%	0.629%	0.619%	0.615%	0.616%	0.619%

It can be seen from Tables 2 and 3 that MIUKF + VFFRLS has advantages over the UKF + VFFRLS algorithm in a wide range of parameters. Nevertheless, it is still necessary to consider setting reasonable parameters to make the algorithm reach the state-of-the-art.

It also can be seen from Figure 5, Tables 2 and 3 that, as the value of a increases, the standard deviation of the absolute value of SOC error decreases, but the average value of the absolute value of SOC error fluctuates from large to small and then to large. The fluctuation from large to small in the first part is characterized by the curve becoming more smooth in the figure, while the fluctuation from small to large in the second partis characterized by the curve becoming less smooth in the figure. Comparing the three-dimensional diagrams with different pitch angles when $M = 162$, as shown in Figure 6, it can be seen that the curve is less smooth when $a = 1$ than when $a = 0$, and there is obvious jitter at the time point close to convergence.

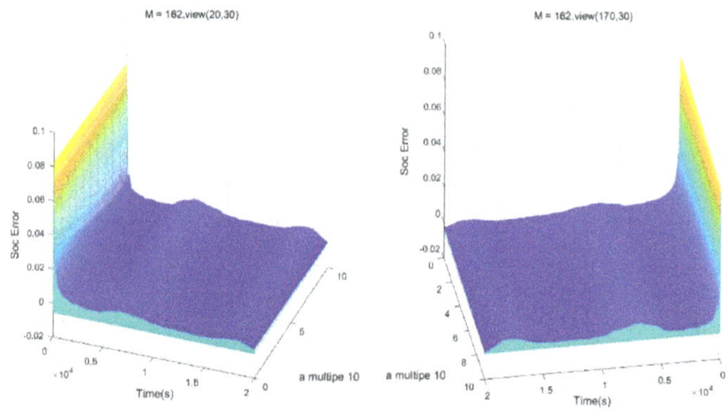

Figure 6. Three-dimensional SOC error diagrams with different pitch angles when $M = 162$.

The selection of MI model parameters is needed to improve the accuracy and stability of the algorithm as much as possible in order to reduce the mean and standard deviation of the estimated absolute error value. At the same time, it also needs to consider reducing the consumption of algorithm resources in order to reduce the value of M, which represents the time sliding window length. Considering the data comprehensively, $a = 0.5$ and $M = 22$ are selected in this paper.

5.3.2. Experiment Results of VFFRLS + MIUKF

(1) Current condition

The dynamic characteristics of the UUDS cycle are very strong, and the current changes almost every second. In the experimental process, the discharge is set at a positive number, the maximum discharge current is 62 A, and the maximum charging current is 41 A. The curve of the current over time is shown in Figure 7.

(2) Comparison between measured and estimated terminal voltage

The estimated terminal voltage of battery parameters is close to the measured terminal voltage, and the gap at the end of battery discharge is slightly enlarged, which is also consistent with the intuitive feeling that the internal polarization reaction of the battery tends to be intense at the end of battery discharge, resulting in the enhancement of nonlinear characteristics. The maximum error of terminal voltage is 20.3%, the minimum value is −7.87%, and the average value of the absolute error is 0.57%. The time-varying terminal voltage curve of the measured value and estimated value and the time-varying curve of the terminal voltage error are shown in Figures 8 and 9, respectively.

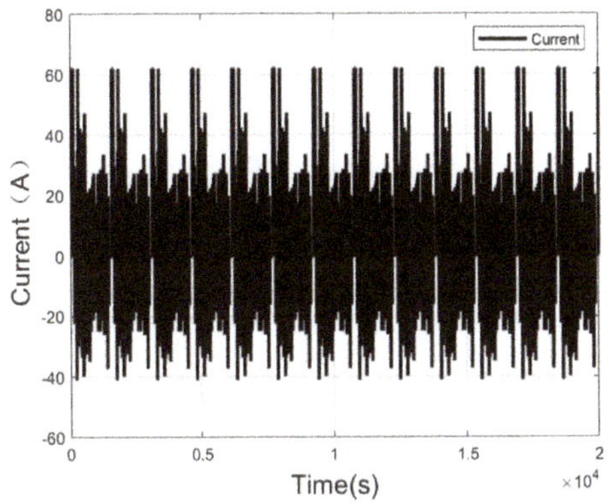

Figure 7. Discharge current over time.

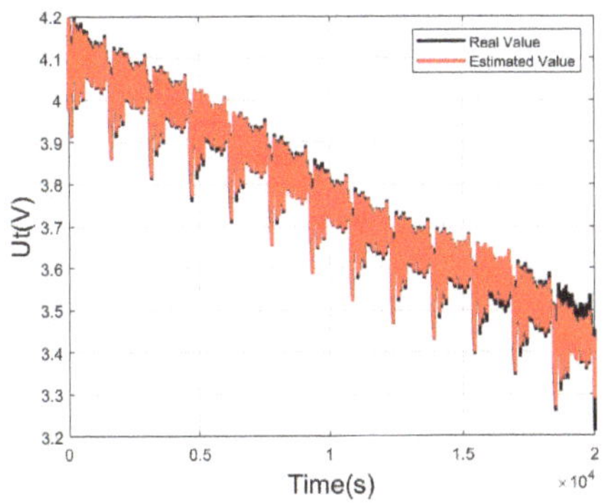

Figure 8. Curve of measured and estimated terminal voltage values over time.

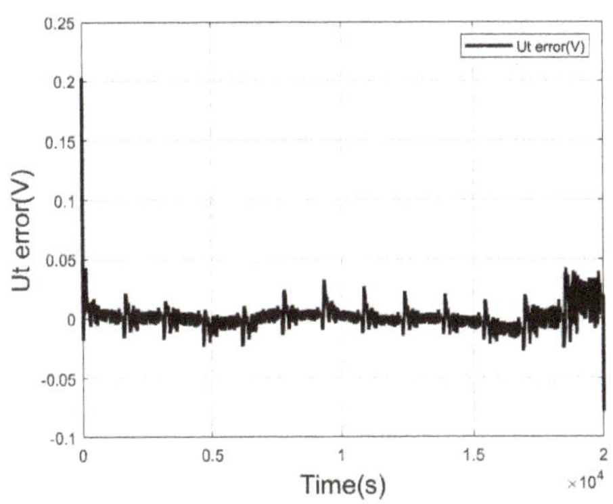

Figure 9. Curve of terminal voltage error over time.

(3) Identification results of battery resistance and capacitance parameters

The parameters related to the resistance and capacitance of the battery identified by the algorithm are shown in the Figure 10.

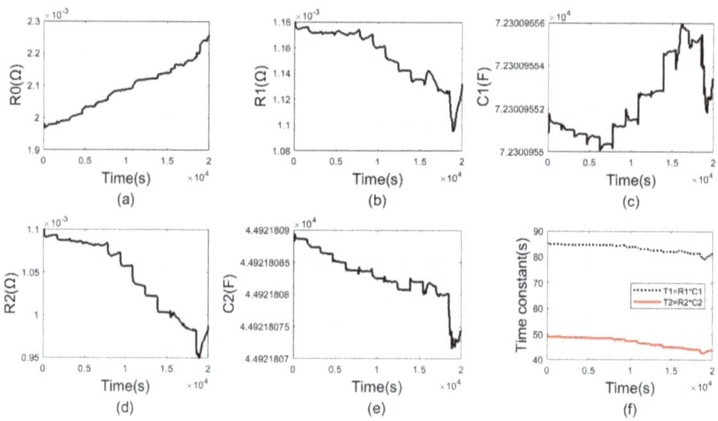

Figure 10. Identification results of battery resistance and capacitance parameters.: (**a**) R0 result, (**b**) R1 result, (**c**) C1 result, (**d**) R2 result, (**e**) C2 result, (**f**) T1 and T2 results.

R0 represents ohmic internal resistance, and its size depends on the activation degree of the electrode and active material which decreases with the decrease in SOC. Therefore, generally speaking, R0 shows a gradual increase trend with the passage of discharge time—that is, it gradually increases with the decrease in SOC. The identified R0 parameters conform to this physical characteristic.

R1 and C1 are represented as slow reaction polarization phenomena in the model, and R2 and C2 are represented as fast reaction polarization phenomena in the model. Their respective products are called time constants, which should conform to R1 * C1 > R2 * C2, and the identified parameters also conform to this physical feature.

It is worth noting that, similarly to R0, R1 and R2 generally increase gradually with the decrease in SOC. The parameters identified in this paper are inconsistent with this. Considering that this goal is not present in the model, and the inconsistency here does not cause significant deviation from the preset objectives of the model, the identification results are still successful.

(4) SOC estimate vs. baseline

The maximum value of SOC error is 19%, the minimum value is −0.44%, and the average value of absolute value is 0.225%. The SOC estimation value of the algorithm can quickly adjust the gap with the reference value. After 114 sampling cycles—that is, 114 s—the error of the algorithm decreases from 19% to less than 5%, and after 416 s, it decreases to less than 1%, after which it remains below 1%.

The time-varying curves of SOC estimated value and reference value and the time-varying curves of error are shown in Figures 11 and 12, respectively.

5.3.3. Comparison of Algorithm Experiment Results

Based on the same data, the online parameter estimation algorithm VFFRLS + UKF and the offline parameter estimation algorithm MIUKF, the UKF and EKF, are used for the experiment. The comparison between the online and offline algorithms is based on similar cost baselines, as it is difficult to compare them in other aspects, so the same offline resistance and capacitance parameters are used as initial value for online parameters, and the same system and measurement noise covariance matrix are used as well. The comparison results show that the online parameter estimation has obvious advantages in accuracy and stability—the average value of the absolute value of the error is small, and the error curve is stable and close to zero over time. Compared with the VFFRLS + UKF algorithm of the estimation of the same online parameter, VFFRLS + MIUKF is superior in convergence speed, accuracy and stability, which shows that the error converges to zero faster, and the average absolute value of the error and the standard deviation of the error are small.

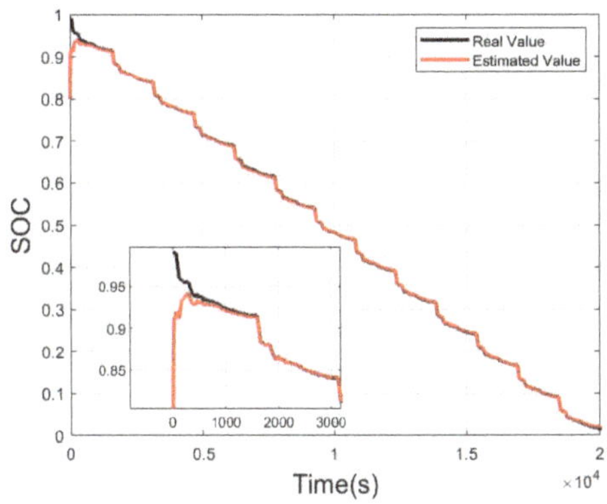

Figure 11. Curve of SOC estimated value and reference value over time.

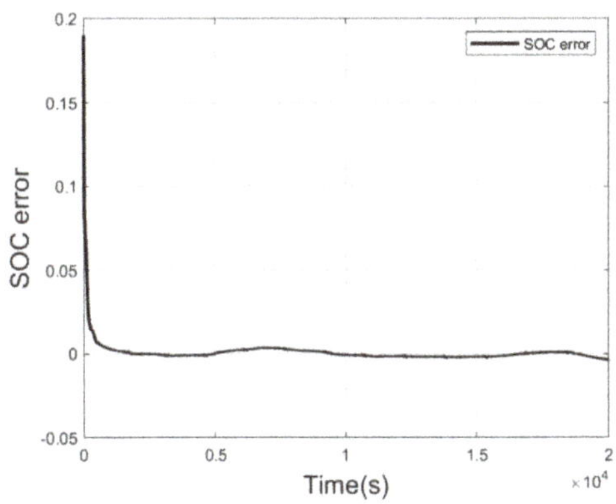

Figure 12. Curve of SOC estimation error over time.

The index statistics related to the stability, accuracy and convergence speed of the SOC prediction error results of each algorithm are shown in Table 4.

Table 4. The index statistics related to the stability, accuracy and convergence speed of the SOC prediction error results of each algorithm.

Algorithm	Maximum Error Value	Minimum Error Value	Average Absolute Values of Error	Standard Deviation of Error Value	The First Time When the Absolute Value of the Error Starts to Be Less than the Average Value	The Second Time When the Absolute Value of the Error Begins to Be Greater than the Average Value
VFFRLS + MIUKF	19.00%	−0.44%	0.23%	0.67%	1138 s	6012 s
VFFRLS + UKF	19.00%	−0.34%	0.25%	0.74%	1835 s	6014 s
MIUKF	19.00%	−1.14%	0.78%	1.22%	408 s	3331 s
UKF	19.00%	−1.92%	1.02%	0.88%	25 s	3310 s
EKF	19.00%	−2.65%	1.20%	0.49%	7 s	158 s

The curve of the SOC predicted value and reference value of each algorithm over time is shown in Figure 13.

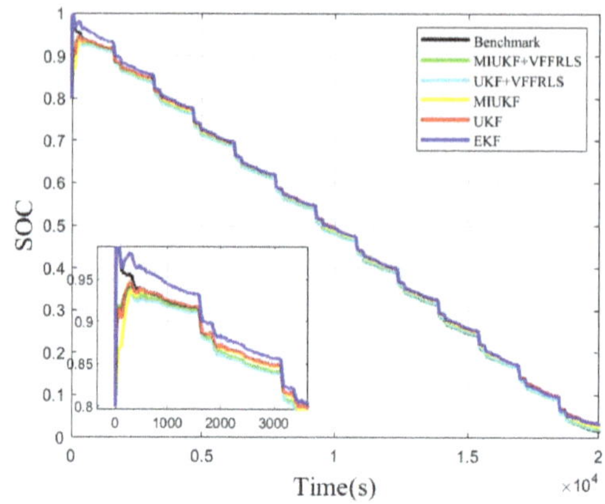

Figure 13. The curve of the SOC predicted value and reference value of each algorithm over time.

The time-varying curve of the SOC prediction error of each algorithm is shown in Figure 14.

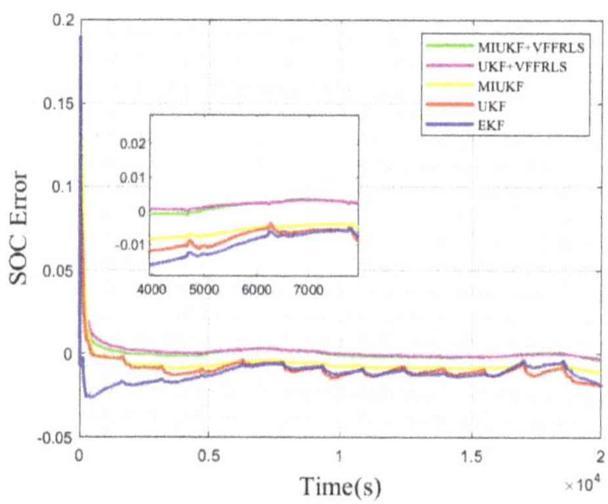

Figure 14. The time-varying curve of SOC prediction error of each algorithm.

6. Conclusions

Accurate and real-time SOC estimation is the basis and key to realizing balanced battery management, which can reduce battery internal resistance loss and the possibility of battery overcharge and discharge. Due to the complex internal chemical and physical reactions and dynamic environmental conditions, the SOC of a battery has obvious nonlinear and time-varying characteristics, which has always been the focus of and main difficulty in battery management system research.

In this paper, a joint SOC estimation algorithm based on online parameter identification and a second-order RC equivalent circuit model is proposed, which innovatively realizes the dual estimation of MIUKF + VFFRLS. The experimental results based on UDDS test data show that the algorithm has obvious advantages in stability and accuracy compared with offline parameter + EKF, offline parameter + UKF and offline parameter +MIUKF; compared with UKF + VFFRLS, it has advantages in convergence speed, accuracy and stability. The overall performance of the fused algorithm is outstanding.

Through the above research work, the SOC estimation accuracy can be effectively improved, the battery consistency management ability can be improved, and the theoretical value and practical value can be reflected, but there are still limitations and deficiencies.

The tuning of the KF is critical to the SOC estimation results, and optimization methods can be further discussed. The accuracy of RLS algorithm is very sensitive to measurement noise, and the associated noise-compensation methods can be further studied. The model-based SOC estimation also depends on accurate estimation of the battery capacity, and data-based capacity estimation can be further studied.

The applicability of the algorithm is also related to the efficiency of the algorithm. The calculation time of the algorithm is not compared in this experiment because the calculation time is strongly related to factors such as the type of program language and the method of coding. The battery type used in the algorithm is a ternary lithium battery with a relatively strong linear relationship between the U_{ocv} and the SOC curve. The duration of experimental data is short. The factors of battery capacity attenuation and temperature change are not considered.

The follow-up research can work in the following directions: making noise-compensated methods research [22,23], optimizing the tuning of the KF [24], considering the factors of battery capacity attenuation [25] and temperature change, designing an analogous algorithm efficiency model to compare the calculation time of different types of SOC estimation algorithms, performing experiments to collect data for a longer time or to seek a larger public dataset, further verifying the effectiveness of the algorithm based on larger datasets with different batteries, further studying the effectiveness of intelligent algorithms such as neural networks and the algorithm proposed by this paper in large data sets and exploring more efficient SOC estimation methods suitable for more scenarios.

Author Contributions: Conceptualization, H.Y. and H.P.; methodology, H.Y. and Y.H.; software, H.Y.; validation, H.Y., Y.H. and Y.Z.; formal analysis, J.D.; investigation, H.Y. and Y.H.; resources, Y.Z.; data curation, Z.C.; writing—original draft preparation, H.Y.; writing—review and editing, Y.H.; visualization, H.Y.; supervision, H.P.; project administration, H.P.; funding acquisition, J.D. All authors have read and agreed to the published version of the manuscript. H.Y. and Y.H. contributed equally to this work.

Funding: This research was funded by the project 2021 Natural Science Foundation of Guangdong Province, Grant No. is 2021A1515011851.

Conflicts of Interest: The authors declare no conflict of interest.

References

1. Rivera-Barrera, J.P.; Muñoz-Galeano, N.; Sarmiento-Maldonado, H.O. SoC estimation for lithium-ion batteries: Review and future challenges. *Electronics* **2017**, *6*, 102. [CrossRef]
2. Hong, J.; Wang, Z.; Chen, W.; Wang, L.-Y.; Qu, C. Online joint-prediction of multi-forward-step battery SOC using LSTM neural networks and multiple linear regression for real-world electric vehicles. *J. Energy Storage* **2020**, *30*, 101459. [CrossRef]
3. Zhao, F.; Li, Y.; Wang, X.; Bai, L.; Liu, T. Lithium-ion batteries State of Charge prediction of electric vehicles using RNNs-CNNs neural networks. *IEEE Access* **2020**, *8*, 98168–98180. [CrossRef]
4. Liu, X.; Li, K.; Wu, J.; He, Y.; Liu, X. An extended Kalman filter based data-driven method for State of Charge estimation of Li-ion batteries. *J. Energy Storage* **2021**, *40*, 102655. [CrossRef]
5. Al-Gabalawy, M.; Hosny, N.S.; Dawson, J.A.; Omar, A.I. State of Charge estimation of a Li-ion battery based on extended Kalman filtering and sensor bias. *Int. J. Energy Res.* **2021**, *45*, 6708–6726. [CrossRef]
6. Zhao, Y.; Xu, J.; Wang, X.; Mei, X. The Adaptive Fading Extended Kalman Filter SOC Estimation Method for Lithium-ion Batteries. *Energy Procedia* **2018**, *145*, 357–362. [CrossRef]
7. Wang, W.; Wang, X.; Xiang, C.; Wei, C.; Zhao, Y. Unscented kalman filter-based battery SOC estimation and peak power prediction method for power distribution of hybrid electric vehicles. *IEEE Access* **2018**, *6*, 35957–35965. [CrossRef]
8. Xing, J.; Wu, P. State of Charge estimation of lithium-ion battery based on improved adaptive unscented kalman filter. *Sustainability* **2021**, *13*, 5046. [CrossRef]
9. Peng, X.; Li, Y.; Yang, W.; Garg, A. Real-Time State of Charge Estimation of the Extended Kalman Filter and Unscented Kalman Filter Algorithms Under Different Working Conditions. *J. Electrochem. Energy Convers. Storage* **2021**, *18*, 041007. [CrossRef]
10. Li, W.; Yang, Y.; Wang, D.; Yin, S. The multi-innovation extended Kalman filter algorithm for battery SOC estimation. *Ionics* **2020**, *26*, 6145–6156. [CrossRef]

11. Zahid, T.; Li, W. A Comparative Study Based on the Least Square Parameter Identification Method for State of Charge Estimation of a LiFePO4 Battery Pack Using Three Model-Based Algorithms for Electric Vehicles. *Energies* **2016**, *9*, 720. [CrossRef]
12. Lin, C.; Zhang, X.H. Modeling and Simulation Research on Lithium-Ion Battery in Electric Vehicles Based on Genetic Algorithm. *Appl. Mech. Mater.* **2014**, *494–495*, 246–249. [CrossRef]
13. He, H.; Zhang, X.; Xiong, R.; Xu, Y.; Guo, H. Online model-based estimation of state-of-charge and open-circuit voltage of lithium-ion batteries in electric vehicles. *Energy* **2012**, *39*, 310–318. [CrossRef]
14. He, H.; Xiong, R.; Guo, H. Online estimation of model parameters and state-of-charge of $LiFePO_4$ batteries in electric vehicles. *Appl. Energy* **2012**, *89*, 413–420. [CrossRef]
15. Tong, S.; Klein, M.P.; Park, J.W. On-line optimization of battery open circuit voltage for improved state-of-charge and state-of-health estimation. *J. Power Sources* **2015**, *293*, 416–428. [CrossRef]
16. Ouyang, T.; Xu, P.; Chen, J.; Lu, J.; Chen, N. Improved parameters identification and State of Charge estimation for lithium-ion battery with real-time optimal forgetting factor. *Electrochim. Acta* **2020**, *353*, 136576. [CrossRef]
17. Xiong, R.; Sun, F.; Chen, Z.; He, H. A data-driven multi-scale extended Kalman filtering based parameter and state estimation approach of lithium-ion polymer battery in electric vehicles. *Appl. Energy* **2014**, *113*, 463–476. [CrossRef]
18. Lao, Z.; Xia, B.; Wang, W.; Sun, W.; Lai, Y.; Wang, M. A novel method for lithium-ion battery online parameter identification based on variable forgetting factor recursive least squares. *Energies* **2018**, *11*, 1358. [CrossRef]
19. Waag, W.; Fleischer, C.; Sauer, D.U. On-line estimation of lithium-ion battery impedance parameters using a novel varied-parameters approach. *J. Power Sources* **2013**, *237*, 260–269. [CrossRef]
20. He, H.; Xiong, R.; Fan, J. Evaluation of Lithium-Ion Battery Equivalent Circuit Models for State of Charge Estimation by an Experimental Approach. *Energies* **2011**, *4*, 582–598. [CrossRef]
21. US EPA. Urban Dynamometer Driving Schedule, Emission Standards Reference Guide [EB/OL]. USA. 2017. Available online: https://www.epa.gov/vehicle-and-fuel-emissions-testing/dynamometer-drive-schedules (accessed on 25 November 2021).
22. Wei, Z.; Zou, C.; Leng, F.; Soong, B.H.; Tseng, K.-J. Online model identification and state-of-charge estimate for lithium-ion battery with a recursive total least squares-based observer. *IEEE Trans. Ind. Electron.* **2018**, *65*, 1336–1346. [CrossRef]
23. Wei, Z.; Zhao, J.; Xiong, R.; Dong, G.; Pou, J.; Tseng, K.J. Online estimation of power capacity with noise effect attenuation for lithium-ion battery. *IEEE Trans. Ind. Electron.* **2019**, *66*, 5724–5735. [CrossRef]
24. He, J.; Wei, Z.; Bian, X.; Yan, F. State-of-Health Estimation of Lithium-Ion Batteries Using Incremental Capacity Analysis Based on Voltage–Capacity Model. *IEEE Trans. Transp. Electrif.* **2020**, *6*, 417–426. [CrossRef]
25. Bian, X.; Wei, Z.; He, J.; Yan, F.; Liu, L. A Two-Step Parameter Optimization Method for Low-Order Model-Based State-of-Charge Estimation. *IEEE Trans. Transp. Electrif.* **2021**, *7*, 399–409. [CrossRef]

energies

MDPI

Article

A Robust Algorithm for State-of-Charge Estimation under Model Uncertainty and Voltage Sensor Bias

Yang Guo and Ziguang Lu *

School of Electrical Engineering, Guangxi University, Nanning 530004, China; zzguoyang@163.com
* Correspondence: luzg@gxu.edu.cn

Abstract: Accurate estimation of the state of charge (SOC) of zinc–nickel single-flow batteries (ZNBs) is an important problem in battery management systems (BMSs). A nonideal electromagnetic environment will usually cause the measured signal to contain nonnegligible noise and bias. At the same time, due to the influence of battery ageing, environmental temperature changes, and a complex reaction mechanism, it is difficult to establish a very accurate system model that can be applied to various complex working conditions. The unscented Kalman filter (UKF) is a widely used SOC estimation algorithm, but the UKF will reduce the estimation accuracy and divergence under the influence of inaccurate model and sensor errors. To improve the performance of the UKF, a robust desensitized unscented Kalman filter (RDUKF) is proposed to realize an accurate SOC estimation of batteries in the context of different disturbances. Then, the proposed method is applied to cases of error interference, such as Gaussian noise, voltage sensor drift, an unknown initial state, and inaccurate model parameters. The simulation and experimental results show that compared with the standard UKF algorithm, the proposed estimation algorithm can effectively suppress the influence of various errors and disturbances and achieve higher accuracy and robustness.

Keywords: real-time estimation; robust desensitized unscented Kalman filter; state of charge; zinc–nickel single-flow batteries

Citation: Guo, Y.; Lu, Z. A Robust Algorithm for State-of-Charge Estimation under Model Uncertainty and Voltage Sensor Bias. *Energies* **2022**, *15*, 1537. https://doi.org/10.3390/en15041537

Academic Editors: Alon Kuperman and Alessandro Lampasi

Received: 29 November 2021
Accepted: 16 February 2022
Published: 19 February 2022

Publisher's Note: MDPI stays neutral with regard to jurisdictional claims in published maps and institutional affiliations.

1. Introduction

Renewable energy has been considered an alternative to coal-fired energy because its use can result in reduced pollution and greenhouse gas emissions. However, the fluctuation caused by the wide application of intermittent renewable energy, such as wind energy and solar energy, will increase the burden of the power grid. Integrating rechargeable batteries into the power grid as an energy storage system can effectively improve power quality and system reliability [1]. Redox flow batteries (RFBs) have become one of the most promising grid-connected energy storage technologies on the market because of their long life cycle, good scalability, high efficiency, and low spatial requirements [2,3]. The main difference between a typical RFB and other types of batteries is that the active materials of its positive and negative electrodes are stored in an external electrolyte storage tank in the form of electrolytes and circulated into the battery stack chamber by two independent auxiliary pumps. The RFB separates the two electrolytes through the ion exchange membrane to prevent the mixing of redox substances and to maintain the electrical neutrality of the energy storage system [4,5]. Since the concept of RFBs was put forward, different kinds of RFBs have rapidly been developed, but the high cost of ion exchange membranes and the interference of electrolyte cross pollution are still the main bottlenecks hindering the development of RFBs. A novel RFB structure was proposed in [6]. This battery dissolves zincate in high-concentration potassium hydroxide as the electrolyte. As only one electrolyte is used, there is no need for an expensive ion exchange membrane in the system design, and cross pollution of the electrolyte is avoided. At the same time, this RFB has a high open-loop voltage and energy density, is nontoxic, has a simple structure, and is low-cost [7–9]. This makes its commercialization easier.

Zinc–nickel single-flow batteries (ZNBs) have been developed rapidly since they were first proposed. However, there are still some important problems to be solved in their practical engineering. State of charge (SOC), an important state monitored in battery management systems (BMSs), is very important to consider in attempting to avoid overcharging and/or overdischarging. At present, SOC estimation has three main research directions: the Coulomb counting method, the machine learning method, and the state observer method [10]. The Coulomb counting method is an open-loop method that is simple to implement and has a low calculation cost. However, its accuracy is greatly affected by the sensor accuracy and initial errors; thus, it is necessary to periodically use the charge–discharge cut-off voltage or open-circuit voltage to correct the estimated value [11–13]. Machine learning methods do not need to understand the internal electrochemical or external circuit characteristics of a battery. They estimate the SOC by fitting the nonlinear relationship between relevant factors such as the current, voltage, temperature, and SOC. The support vector machine (SVM) [14], neural network (NN) [15], fuzzy logic (FL) [16], and long short-term memory (LSTM) network [17] have been implemented for SOC estimation. If the appropriate training dataset is given, the machine learning method can estimate the SOC of a battery more accurately. When the training data and test data correspond to different working conditions, the robustness of the state estimation algorithm will worsen, and machine learning will require a longer training time and higher computing cost, which restricts the wide application of machine learning in actual SOC estimation. The observer method usually assumes that the battery is a dynamic system. The battery model is established to describe the change in battery state, and various filter algorithms are introduced to estimate the state variables. Among the different types of observers, the Kalman filter is the most widely used. The extended Kalman filter (EKF) was first used to solve the problem of battery state estimation. However, as the battery state space model is usually a nonlinear model and the EKF uses first-order Taylor series expansion to linearize a nonlinear model, there are large errors in estimating battery SOC using the EKF; in contrast, using the unscented Kalman filter (UKF) to estimate battery SOC results in higher robustness and accuracy [18,19].

The Kalman filter algorithm is based on a basic premise: that the established model is accurate enough to accurately track the change in battery terminal voltage under dynamic current conditions. This assumption is usually difficult to fully satisfy because the actual parameters of the model will change in accordance with the change in ambient temperature and battery life, and it is very difficult to establish a model considering all influencing factors. In [20], the error sources of SOC estimation are systematically analyzed from the perspectives of the measurement signal, model error, and state estimation algorithm. In [21], the model-based SOC estimation algorithm was shown to be more affected by voltage measurement errors and model errors than by current measurement errors and initial errors. To improve the accuracy of SOC estimation under different error sources, many different improved algorithms have been proposed.

In [22], the bias term of the model error was taken as a new state vector, which was jointly estimated with the SOC through an estimation algorithm to solve the problems of open-circuit voltage drift and voltage sensor drift [23]; a proportional integral (PI) state observer and a current integrator were used to estimate the SOC of the battery; and a drift current corrector was used to suppress the influence of the drift current. For the situation in which the battery model parameters are uncertain due to battery ageing and there is instrument measurement noise, [24] proposed a smooth variable structure filter (SVSF) to estimate the SOC and health state of lithium batteries. Compared with the EKF, the SVSF can produce more accurate SOC estimation results. In [25], the extended Kalman smoothing variable structure filter was proposed, which is a new algorithm combining the EKF and SVSF technologies. The experimental results showed that it has strong robustness with respect to inaccurate models and can improve the accuracy of SOC estimation. In [26], an observer with multiple feedback gains was designed to estimate the SOC. In the case of inaccurate modeling and current sensor drift, the SOC of the battery can be estimated robustly, and the amount of calculation is low. In [27], an algorithm combining the strong tracking UKF and adaptive UKF was proposed, and the uncertainty of the model was identified through statistical information, which can effectively suppress the state estimation error caused by sensor drift. In [28], the observability of the second-order RC nonlinear model was analyzed, and it was proposed that accurate voltage measurement is more important for SOC estimation than current measurement accuracy, which is consistent with the conclusion of [21]. At the same time, Zhao et al. [28] proposed using the extended model to address the reduction in the accuracy of SOC estimation when errors occur in sensors and pointed out that when errors occur in voltage and current sensors, the method used may not accurately estimate the SOC and bias at the same time. In [29,30], the state estimation problem was studied when noise and bias occur in voltage sensors or current sensors. The DEKF was used to estimate the battery SOC, and the results showed that the proposed estimation method has good convergence and robustness. The two-layer state estimator proposed in [31] estimated the SOC of a battery, limiting the state estimation error to $\pm 4\%$ when the model is inaccurate and current deviation exists.

In summary, although voltage offset and model error are the main factors affecting the reduction in SOC estimation accuracy, few studies have considered the SOC estimation problem when voltage offset or model error exists. Moreover, the BMS is affected by external electromagnetic interference in harsh operating environments. The signals received from voltage and current sensors are usually polluted by random noise, which reduces the SOC estimation accuracy. To solve the above problems, a robust UKF algorithm is proposed to improve the robustness of the SOC estimation of ZNBs under Gaussian noise, inaccurate modeling, or voltage sensor drift. The proposed method has the following advantages: (1) it can suppress the influence of Gaussian measurement noise in a sensor, (2) it has good robustness to model parameter uncertainty, and (3) it improves the accuracy of battery SOC estimation in the presence of voltage sensor drift.

The structure of the paper is as follows: In Section 2, the equivalent circuit model of a ZNB is introduced. In Section 3, the application of the proposed robust UKF algorithm in battery SOC estimation is introduced. In Section 4, the estimation results of a robust UKF in the presence of model parameter disturbance or voltage bias are verified and compared with those of a UKF. Section 5 offers the main conclusions of the paper.

2. Equivalent Circuit Model

The SOC of ZNBs can be defined as the ratio of residual capacity to maximum capacity, which can be expressed by (1). SOC_t and SOC_0 are the initial SOC at time t and at the beginning of estimation, respectively, Q_n is the maximum capacity of the battery, and $i(\tau)$ is the load current. It is assumed that the current is positive when the battery is discharged.

$$SOC_t = SOC_0 - \frac{1}{Q_n}\int_0^t i(\tau)d\tau \tag{1}$$

At present, the commonly used battery models mainly include neural network models, electrochemical models, and equivalent circuit models (ECMs). Among the electrochemical models, the one-dimensional (1D) model [32] and pseudo-two-dimensional (P2D) model [33,34] are widely used. As the electrochemical model is the first-principles model and has clear physical significance, it can not only describe the relationship between current and voltage at the macroscale but also analyze the concentration, current and potential distribution of reaction products in the electrode and electrolyte at the microscale. However, the complex coupled nonlinear partial differential equations and the model parameters that are difficult to measure directly cause the electrochemical model to have high requirements regarding the amount of calculation. The neural network model is the opposite of the electrochemical model. This model does not consider the mechanism of a battery and uses only the data-driven method to approximate the dynamic characteristics of the battery. Compared with other models, it requires more data to train the network model. ECM uses circuit elements such as resistance, capacitance, and voltage source to build a circuit network to simulate the dynamic response of voltage under different working conditions. The commonly used equivalent circuit models include the Thevenin model and second-order RC equivalent circuit model, which adds a group of RC networks. Therefore, the polarization and diffusion characteristics inside the battery can be better simulated.

A typical second-order-equivalent circuit model is shown in Figure 1. R_1 and R_2 represent the polarization resistance, and C_1 and C_2 represent the polarization capacitance. The equivalent circuit model uses parallel RC branches to characterize the electrochemical polarization and concentration polarization of ZNBs during charge and discharge. V_{OCV} represents the open-circuit voltage of the battery, which is the difference in the equilibrium electrode potential between the positive and negative electrodes when the battery is in a reversible equilibrium state during the open-circuit period. R_0 is the ohmic resistance, which is used to characterize the energy loss caused by the resistance between the battery electrode and electrolyte. The terminal voltage and current of the battery are expressed as V_t and I, respectively, and the current during discharge is defined as a positive number. According to Kirchhoff's law and (1), the discrete state space equation of a dynamic battery can be easily deduced, as shown in (2) and (3). It is worth noting that for a ZNB, because the open-circuit voltage, polarization capacitance, and polarization resistance are functions related to the SOC, the state equation and output equation of this state space equation are nonlinear equations.

$$\begin{bmatrix} SOC(k) \\ U_1(k) \\ U_2(k) \end{bmatrix} = \begin{bmatrix} 1 & 0 & 0 \\ 0 & \exp(-\Delta t/R_1/C_1) & 0 \\ 0 & 0 & \exp(-\Delta t/R_2/C_2) \end{bmatrix} \begin{bmatrix} SOC(k-1) \\ U_1(k-1) \\ U_2(k-1) \end{bmatrix} + \begin{bmatrix} -\eta \frac{\Delta t}{Q_n} \\ [1-\exp(-\Delta t/\tau_1)]R_1 \\ [1-\exp(-\Delta t/\tau_2)]R_2 \end{bmatrix} I(k-1) \tag{2}$$

$$V_t(k) = V_{OCV}(SOC(k)) - U_1(k) - U_2(k) - R_0 I(k) \tag{3}$$

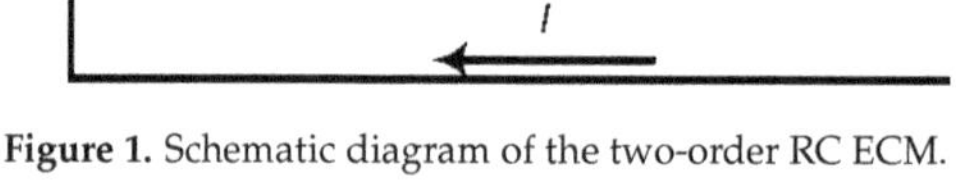

Figure 1. Schematic diagram of the two-order RC ECM.

To simplify the model, the parameters are assumed to be related only to the SOC, but in fact, the model parameters are also affected by the ambient temperature, current, and battery ageing state. Ignoring these factors in modeling reduces the accuracy of the model in a complex working state and affects the accuracy of the algorithm when used to estimate the SOC.

3. Robust Unscented Kalman Filter

For nonlinear systems such as ZNBs, the corresponding discrete state space representation is as follows:

$$x_k = f(x_{k-1}, u_{k-1}) + w_{k-1} \tag{4}$$

$$y_k = h(x_k, u_k) + v_k \tag{5}$$

where $x_k = [SOC\ U_1\ U_2]^T$ is the state at time k, and u_k and y_k are the input current and output voltage at time k, respectively. $f(x_{k-1}, u_{k-1})$ is the state transfer function, $h(x_k, u_k)$ is the measurement function, and w_{k-1} and v_k are the process noise and the measurement noise, which are usually assumed to be zero-mean white noise with covariance matrices Q and R, respectively.

Unlike the EKF, which uses only Taylor series expansion to approximate the nonlinear state space model [35], the UKF uses a traceless transformation (UT) to determine the sampling points near the estimation points and uses these sample points to approximately represent the random distribution of the nonlinear function [36]; thus, the UKF algorithm does not need to calculate the Jacobian matrix of the system model (compared with the EKF). At the same time, there is no need for random sampling (compared with the particle filter (PF) algorithm); therefore, the UKF can be regarded as a compromise between the EKF and PF. The SOC estimation process of a ZNB based on the UKF can be summarized as follows:

(1) Initialize the value of the state variable and the covariance:

$$\begin{cases} \bar{x}_0 = E(x_0) \\ P_0 = E[(x_0 - \bar{x}_0)(x_0 - \bar{x}_0)^T] \end{cases} \tag{6}$$

where $E(\cdot)$ is the expected mean value.

(2) Generate $2n + 1$ sigma vectors and corresponding weights at the $k - 1$ step, where n is the dimension of the state variable:

$$\begin{cases} \hat{\chi}_{0,k-1} = \hat{x}_{k-1} \\ \hat{\chi}_{i,k-1} = \hat{x}_{k-1} + \left(\sqrt{(n+\lambda)P_{k-1}}\right)_i, i = 1, 2, \ldots, n \\ \hat{\chi}_{i,k-1} = \hat{x}_{k-1} - \left(\sqrt{(n+\lambda)P_{k-1}}\right)_i, i = n+1, \ldots, 2n \end{cases} \tag{7}$$

$$\begin{cases} \omega_0^m = \frac{\lambda}{(n+\lambda)} \\ \omega_0^c = \frac{\lambda}{(n+\lambda)} + (1 + \beta - \alpha^2) \\ \omega_j^m = \omega_j^c = \frac{1}{2(n+\lambda)}, j = 1, 2, \ldots, 2n \end{cases} \tag{8}$$

Here, λ is a scaling parameter, which can be expressed as $\lambda = \alpha^2(n+h) - n$, and α determines the spread of the sigma points around $\hat{x}_{k-1}$ and is usually set to a small positive value. ω^m and ω^c are covariance weights, and β is a parameter that is used to incorporate prior knowledge of the distribution of the state variable. For Gaussian distributions, one can assume $\beta = 2$.

(3) State prediction

Update the sample point and calculate the propagated mean:

$$\overline{\chi}_{j,k} = f\left(\hat{\chi}_{j,k-1}, u_{k-1}\right), j = 0, 1, \ldots, 2n \tag{9}$$

$$\bar{x}_k = \sum_{j=0}^{2n} \omega_j^{(m)} \overline{\chi}_{j,k} \tag{10}$$

Update the priori error covariance of state:

$$\overline{P}_k = \sum_{j=0}^{2n} \omega_j^{(c)} [\overline{\chi}_{j,k}^x - \bar{x}_k][\overline{\chi}_{j,k}^x - \bar{x}_k]^T + Q_k \tag{11}$$

where Q_k is the covariance of system process noise.

(4) Measurement update

Predict the measurement from the propagated sigma points and estimate the mean of the predicted measurement.

$$\mathcal{Y}_{j,k} = h\left(\overline{\mathcal{X}}_{j,k}, u_k\right), j = 0, \ldots, 2L \tag{12}$$

$$\overline{y}_k = \sum_{j=0}^{2n} \omega_j^{(m)} \mathcal{Y}_{j,k} \tag{13}$$

Calculate the covariance of the measured variables and associated covariance matrices, where R_k is the covariance of measurement noise.

$$\overline{P}_{yy,k} = \sum_{j=0}^{2n} \omega_j^{(c)} \left[\mathcal{Y}_{j,k} - \overline{y}_k\right] \left[\mathcal{Y}_{j,k} - \overline{y}_k\right]^T + R_k \tag{14}$$

$$\overline{P}_{xy,k} = \sum_{j=0}^{2n} \omega_j^{(c)} \left[\overline{\mathcal{X}}_{j,k} - \overline{x}_k\right] \left[\mathcal{Y}_{j,k} - \overline{y}_k\right]^T \tag{15}$$

Calculate the posterior estimate and the covariance:

$$\hat{x}_k = \overline{x}_k + K_k(y_k - \overline{y}_k) \tag{16}$$

$$\hat{P}_k = \overline{P}_k - \overline{P}_{xy,k} K_k^T - K_k \overline{P}_{xy,k}^T + K_k \overline{P}_{yy,k} K_k^T \tag{17}$$

where K_k is the Kalman gain, which can be expressed as $K_k = \underline{P}_{xy,k} \underline{P}_{yy,k}^{-1}$.

The standard UKF has been successfully applied to battery SOC estimation [37,38]. It is easy to prove that the UKF is an unbiased state estimator when all system model parameters and noise statistics are accurately known. However, when the parameters of the system model are uncertain, the theoretical behavior and actual behavior of the filter will be inconsistent. Compared with the actual state value, the state estimation value will have a large deviation and even a serious divergence problem. Therefore, the RDUKF is then applied to the SOC estimation of ZNBs [39] to obtain a more robust SOC state estimation scheme.

Before using the RDUKF to estimate the battery SOC, the state equation and observation equation need to be changed to the following form:

$$x_k = f(x_{k-1}, u_{k-1}, \alpha) + w_{k-1} \tag{18}$$

$$y_k = h(x_k, u_k, \alpha) + v_k \tag{19}$$

where α is the uncertainty parameter in the established model.

The influence of model disturbance parameters on the accuracy of state estimation is measured by the sensitivity matrix of the state estimation error relative to the disturbance parameters. Therefore, (20) is used to evaluate the influence of uncertain parameters on a priori estimation and a posteriori estimation in the UKF algorithm, and α_i is the I component of uncertain parameters.

$$\overline{\sigma}_{i_k} = \frac{\partial \overline{x}_k}{\partial \alpha_i}, \hat{\sigma}_{i_k} = \frac{\partial \hat{x}_k}{\partial \alpha_i} \tag{20}$$

The sensitivity matrix can be generated by calculating the partial derivatives of the parameters in (10) and (16), where l is the number of uncertain parameters.

$$\overline{\sigma}_{i_k} = \sum_{j=0}^{2L} \omega_j^{(m)} \frac{\partial \overline{\mathcal{X}}_{j,k}}{\partial \alpha_i}, i = 1, \ldots, \ell \tag{21}$$

$$\sigma i_k = \overline{\sigma}_{i_k} - K_k \frac{\partial \overline{y}_k}{\partial \alpha_i}, i = 1, \ldots, \ell \tag{22}$$

Corresponding to the UKF algorithm, the calculation of the sensitivity of the RDUKF can be divided into four steps:

(1) Calculate the sensitivities of the sigma points for each uncertain parameter.

$$\begin{cases} \frac{\partial \hat{\chi}_{0,k-1}}{\partial \alpha_i} = \hat{\sigma}^T_{i_{k-1}} \\ \frac{\partial \hat{\chi}_{j,k-1}}{\partial \alpha_i} = \hat{\sigma}^T_{i_{k-1}} + \sqrt{(n+\lambda)} \left(\frac{\partial \sqrt{\hat{P}_{k-1}}}{\partial \alpha_i} \right)_j, j = 1,2,\ldots,n \\ \frac{\partial \hat{\chi}_{j,k-1}}{\partial \alpha_i} = \hat{\sigma}^T_{i_{k-1}} + \sqrt{(n+\lambda)} \left(\frac{\partial \sqrt{\hat{P}_{k-1}}}{\partial \alpha_i} \right)_j, j = 1,2,\ldots,n \end{cases} \tag{23}$$

(2) State prediction

Calculate the sensitivities of the sigma points with the state equation and evaluate the sensitivity of the a priori state estimation and a priori covariance matrix.

$$\frac{\partial \overline{\chi}_{j,k}}{\partial \alpha_i} = \frac{\partial f\left(\hat{\chi}_{j,k-1}, u_{k-1}, \alpha\right)}{\partial \alpha_i} + \frac{\partial f\left(\hat{\chi}_{j,k-1}, u_{k-1}, \alpha\right)}{\partial \hat{\chi}_{j,k-1}} \frac{\partial \hat{\chi}_{j,k-1}}{\partial \alpha_i}, \ j = 0,\ldots,2L \tag{24}$$

$$\overline{\sigma}_{i_k} = \frac{\partial \overline{x}_k}{\partial \alpha_i} = \sum_{j=0}^{2L} \omega_j^{(m)} \frac{\partial \overline{\chi}_{j,k}}{\partial \alpha_i} \tag{25}$$

$$\frac{\partial \overline{P}_k}{\partial \alpha_i} = \sum_{j=0}^{2L} \omega_j^{(c)} \left\{ \left[\frac{\partial \overline{\chi}_{j,k}}{\partial \alpha_i} - \overline{\sigma}_{i_k} \right] \left[\overline{\chi}_{j,k} - \overline{x}_k \right]^T + \left[\overline{\chi}_{j,k} - \overline{x}_k \right] \left[\frac{\partial \overline{\chi}_{j,k}}{\partial \alpha_i} - \overline{\sigma}_{i_k} \right]^T \right\}, i = 1,\ldots,\ell \tag{26}$$

(3) Measurement update

Calculate the sensitivity of the sigma points with the observation equation and the sensitivity of the predicted measurement.

$$\frac{\partial \mathcal{Y}_{j,k}}{\partial \alpha_i} = \frac{\partial h\left(\overline{\chi}_{j,k}, u_k, \alpha\right)}{\partial \alpha_i} + \frac{\partial h\left(\overline{\chi}_{j,k}, u_k, \alpha,\right)}{\partial \overline{\chi}_{j,k}} \frac{\partial \overline{\chi}_{j,k}}{\partial \alpha_i}, j = 0,\ldots,2L \tag{27}$$

$$\frac{\partial \overline{y}_k}{\partial \alpha_i} = \sum_{j=0}^{2L} \omega_j^{(m)} \frac{\partial \mathcal{Y}_{j,k}}{\partial \alpha_i} \tag{28}$$

Calculate the sensitivity of the innovation covariance matrix and the cross-covariance matrix.

$$\frac{\partial \overline{P}_{yy,k}}{\partial \alpha_i} = \sum_{j=0}^{2L} \omega_j^{(c)} \left\{ \left[\frac{\partial \mathcal{Y}_{j,k}}{\partial \alpha_i} - \frac{\partial \overline{y}_k}{\partial \alpha_i} \right] \left[\mathcal{Y}_{j,k} - \overline{y}_k \right]^T + \left[\mathcal{Y}_{j,k} - \overline{y}_k \right] \left[\frac{\partial \mathcal{Y}_{j,k}}{\partial \alpha_i} - \frac{\partial \overline{y}_k}{\partial \alpha_i} \right]^T \right\}, = 1,\ldots,\ell \tag{29}$$

$$\frac{\partial \overline{P}_{xy,k}}{\partial \alpha_i} = \sum_{j=0}^{2L} \omega_j^{(c)} \left\{ \left[\frac{\partial \overline{\chi}_{j,k}}{\partial \alpha_i} - \overline{\sigma}_{i_k} \right] \left[\mathcal{Y}_{j,k} - \overline{y}_k \right]^T + \left[\overline{\chi}_{j,k} - \overline{x}_k \right] \left[\frac{\partial \mathcal{Y}_{j,k}}{\partial \alpha_i} - \frac{\partial \overline{y}_k}{\partial \alpha_i} \right]^T \right\}, i = 1,\ldots,\ell \tag{30}$$

Calculate the Kalman gain and sensitivity of posterior covariance of the posterior state estimation.

$$K_k \grave{P}_{yy,k} + \sum_{i=1}^{\ell} W_{i_k} K_k \gamma_{i_k} \gamma_{i_k}^T = \grave{P}_{xy,k} + \sum_{i=1}^{\ell} W_{i_k} \overline{\sigma}_{i_k} \gamma_{i_k}^T \tag{31}$$

Here, $\gamma_{i_k} = \partial \overline{y}_k / \partial \alpha_i$:

$$\frac{\partial \hat{P}_k}{\partial \alpha_i} = \frac{\partial \overline{P}_k}{\partial \alpha_i} - \frac{\partial \overline{P}_{xy,k}}{\partial \alpha_i} K_k^T - K_k \frac{\partial \overline{P}_{xy,k}^T}{\partial \alpha_i} + K_k \frac{\partial \overline{P}_{yy,k}}{\partial \alpha_i} K_k^T, i = 1,\ldots,\ell \tag{32}$$

$$\hat{\sigma}_{i_k} = \overline{\sigma}_{i_k} - K_k \frac{\partial \overline{y}_k}{\partial \alpha_i}, i = 1,\ldots,\ell \tag{33}$$

Similar to the standard UKF, the trace minimizing the error covariance is used to define the cost function, and the RDUKF uses the trace of the sensitivity matrix product of all disturbance parameters to measure the impact of disturbance parameters on the accuracy of state estimation. Therefore, a new cost function (34) is constructed. The Kalman gain can be set as $\partial J_k/\partial K_k = 0$ with (17) and (33) and is obtained by algebraically solving linear equations.

$$J_k = \min_{K_k} \mathrm{tr}\left(\hat{P}_k\right) + \sum_{i=1}^{\ell} \hat{\sigma}_{i_k}^T W_{i_k} \hat{\sigma}_{i_k} \tag{34}$$

The sensitivity of the root square matrix $\sqrt{\hat{P}_k}$ needs to be calculated when the sensitivities of the sigma points are calculated. $\partial\hat{P}_k/\partial\alpha_i$ can be obtained by taking the partial derivative of $\hat{P}_k = \sqrt{\hat{P}_k}\sqrt{\hat{P}_k}$.

$$\frac{\partial \hat{P}_k}{\partial \alpha_i} = \frac{\partial\sqrt{\hat{P}_k}}{\partial \alpha_i}\sqrt{\hat{P}_k} + \sqrt{\hat{P}_k}\frac{\partial\sqrt{\hat{P}_k}}{\partial \alpha_i} \tag{35}$$

Thus, $\partial\sqrt{\hat{P}_k}/\partial\alpha_i$ can be calculated using (36):

$$\mathrm{vec}\left(\frac{\partial\sqrt{\hat{P}_k}}{\partial \alpha_i}\right) = \left(\sqrt{\hat{P}_k}^T \otimes I + I \otimes \sqrt{\hat{P}_k}\right)^{-1} \times \mathrm{vec}\left(\frac{\partial \hat{P}_k}{\partial \alpha_i}\right), i = 1, \ldots, l \tag{36}$$

where vec $(\cdot)$ and $\otimes$ are the column straightening operator and Kronecker product operator, respectively, and I is the identity matrix. The RDUKF approach is summarized in Table 1.

Table 1. Summary of the RDUKF approach for SOC estimation.

	Initialization
	Initial state $\underline{x}_0$; covariance matrix P_0; sensitivity parameters $\hat{\sigma}_0$; and $\partial P_0/\partial\alpha_i, i = 1, \ldots, l$
	For $k = 1, 2, \ldots$ Time update
1.	Generate sigma points via (7) and their sensitivities via (23).
2.	Propagate the sigma points and their sensitivities via (9) and (24).
3.	Compute the mean and covariance of the predicted state via (10) and (11).
4.	Compute the sensitivities of the prior estimate using (26) and the prior sensitivity using (25).
5.	Measurement update
6.	Propagate the sigma points of the measurement and their sensitivities via (12) and (27).
7.	Calculate the predicted measurements and their sensitivities using (13) and (28).
8.	Estimate the covariance of the measurement and the sensitivities using (14) and (29).
9.	Estimate the cross-covariance and the sensitivities using (15) and (30).
10.	Compute the Kalman gain via (31).
11.	Update the posterior mean of the state and the sensitivities by (16) and (33).
12.	Calculate the posterior covariance matrix and the sensitivities using (17) and (32). $k = k + 1$.

4. Experimental Results and Discussion

In this section, the experimental verification method is used to further evaluate the robustness of the proposed state estimation algorithm when the model parameters are inaccurate and there is voltage sensor bias.

In this study, a prototype of ZNBs was fabricated and tested. The nominal capacity was 56 Ah, and the charging cut-off voltage and discharging cut-off voltage were 2.1 V and 1.2 V, respectively. The test platform was a BTS-3000 programmable battery test system produced by Neware (Shenzhen, China). All experimental data (including current, voltage, and charge–discharge capacity) were collected at a sampling frequency of 1 Hz. As the experiment was completed in a laboratory environment with precise and accurate equipment, noise sequences with a random normal distribution and zero mean whose variance equals 0.36 A^2 and 0.004 V^2 were added to the collected voltage and current data to simulate the influence of a harsh electromagnetic environment. The voltage and current with noise under the dynamic stress test (DST) cycle are shown in Figure 2.

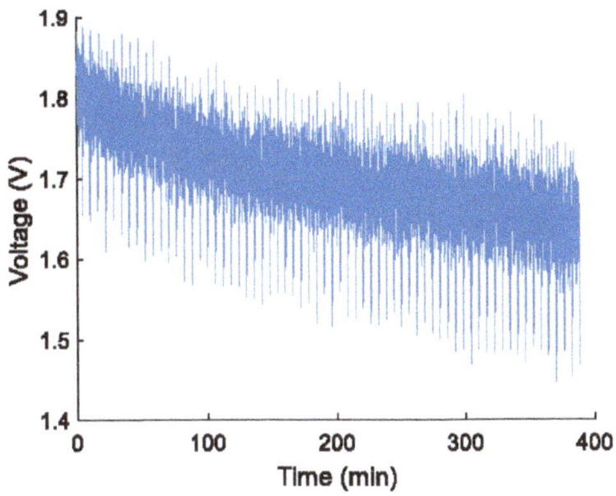

(a)

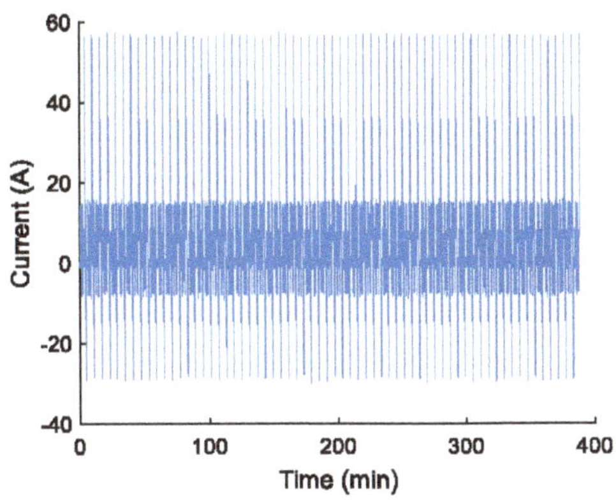

(b)

Figure 2. (**a**) Voltage with noise; (**b**) current with noise.

As the laboratory current sensor has high accuracy, after the initial SOC was determined by the open-circuit voltage, the SOC calculated by using the Coulomb count was taken as the accurate value to verify the state estimation performance of the different algorithms. The *MAE* and *RMSE* were used to evaluate the error of deviation between the estimated value and true value and are defined by (37) and (38), where *K* is the data length.

$$MAE = \frac{1}{K}\sum_{k=1}^{K}\left|\hat{x}_k - x_{ture,k}\right| \tag{37}$$

$$RMSE = \sqrt{\frac{1}{K}\sum_{k=1}^{K}(\hat{x}_k - x_{ture,k})^2} \tag{38}$$

In practical engineering applications, the BMS may not be able to obtain an accurate initial SOC. To verify the sensitivity and accuracy of different algorithms relative to the initial error, we set the initial SOC to 50% and the actual value to 90%. The estimation results and errors are shown in Figure 3. Both the UKF and RDUKF can still converge to the real SOC without knowing the real initial SOC. The RMSE and MAE obtained from the experiment are shown in Table 2. The RDUKF can more accurately track the real SOC changes.

Voltage drift may occur when the sensor is affected by drastic changes in ambient temperature or electromagnetic interference. As the Kalman filter algorithm relies on the feedback of the measured data to correct the state estimation value, the sensor drift will reduce the estimation accuracy. To evaluate the influence of voltage sensor drift on the UKF and RDUKF algorithms, in addition to setting the initial SOC error to 40%, the maximum drift voltage was set to $\pm$10 mV. The SOC estimation results and errors of the two methods are shown in Figure 4. To clarify the differences between the different algorithms, the figures display only the SOC changes that occur after 100 s. Due to the influence of sensor drift, the SOC value estimated by the UKF algorithm gradually deviates from the real value with the increase in the number of measured values. Although the estimation error of the RDUKF algorithm is larger than that without sensor drift, it still converges to a fixed value. The MAE and RMSE values obtained by the two methods under different drift voltages are shown in Figure 5. With increasing drift voltage, the MAE and RMSE of the two algorithms tend to increase. Under the same bias voltage, except when the voltage drift is +2 mV, the MAE and RMSE values of the method based on the RDUKF are smaller; thus, this method is more robust.

It is difficult to establish a battery model that covers the entire life cycle. Affected by ageing and ambient temperature, the established model has difficulty tracking the real voltage change, and the mismatched model usually causes divergence of the Kalman filter. To test the accuracy of the UKF and RDUKF algorithms under incorrect model parameters, it is also necessary to set the initial SOC to an incorrect value, i.e., a $\pm$ 50% deviation is added to the real model parameters. Compared with R_2, C_2, and C_1, when R_1 and R_0 change, the terminal voltage of the model undergoes greater changes under the same working conditions. Therefore, only the two models in which only R_1 is changed and only R_0 is changed are considered. It is worth noting that these two assumptions are used only to analyze the robustness of the state estimation algorithm and may not reflect practical applications. Figure 6 shows the SOC estimation results and errors of the two methods, when R_1 has a deviation of $\pm$50%. Similarly, only the change in SOC after 100 s is shown. Compared with the RDUKF, the UKF algorithm exhibits obvious divergence. The MAE and RMSE values obtained by the two methods under different model errors are shown in Figure 7. The RDUKF algorithm is obviously more robust.

Table 2. Estimated error under the DST.

	MAE	RMSE
RDUKF	0.0009	0.0029
UKF	0.0044	0.0056

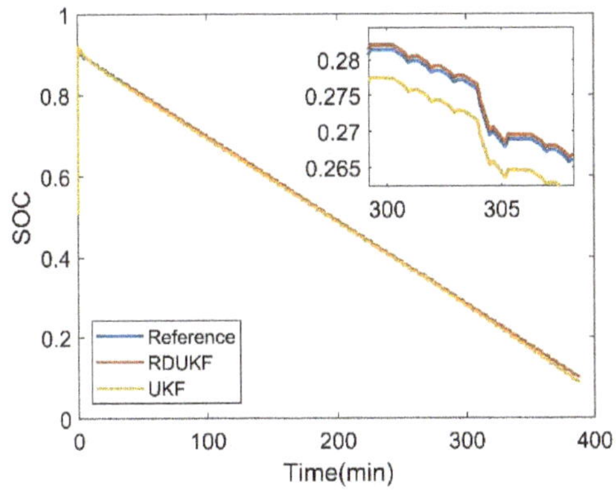

(a)

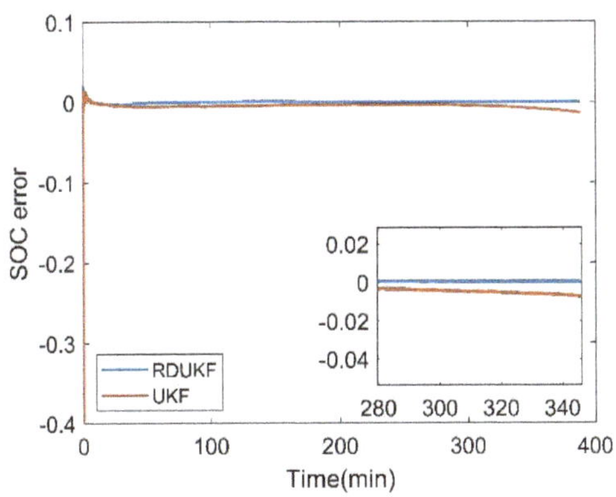

(b)

Figure 3. (**a**) SOC of the estimation results; (**b**) SOC errors of the estimation results.

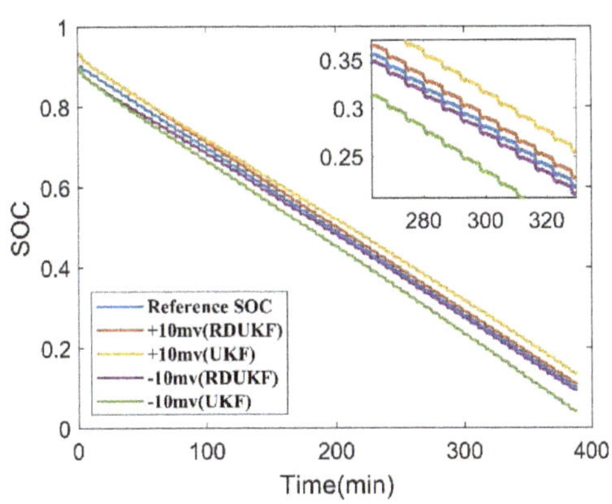

(a)

Figure 4. *Cont.*

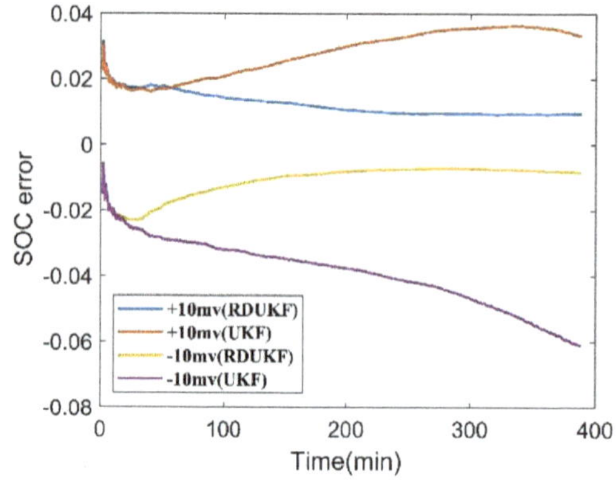

(b)

Figure 4. (**a**) SOC estimation results under different drift voltages; (**b**) SOC errors of the estimation results under different drift voltages.

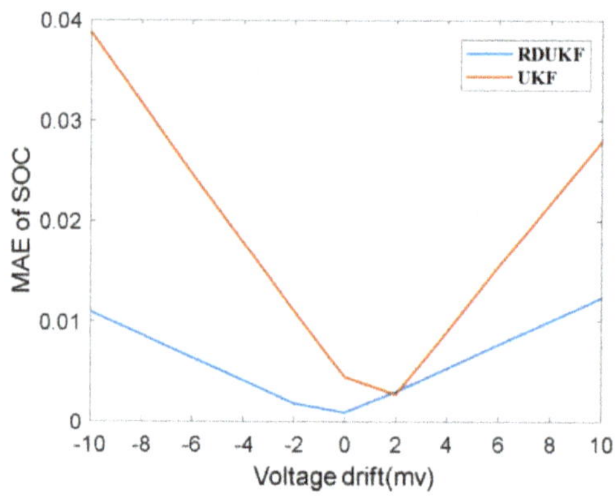

(**a**)

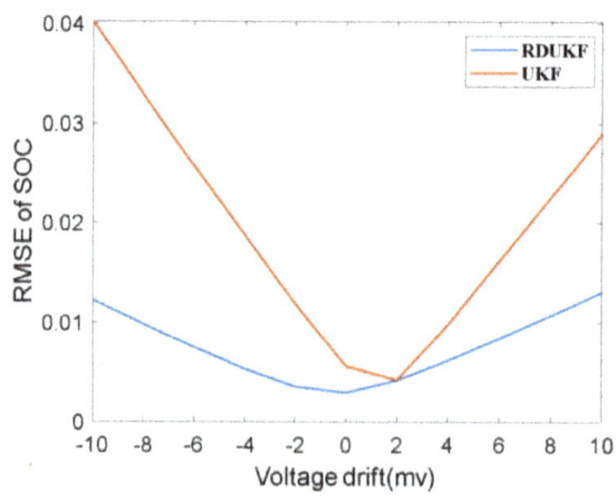

(**b**)

Figure 5. (**a**) MAE under different drift voltages; (**b**) RMSE under different drift voltages.

Figure 8 shows the SOC estimation results and errors of the two methods, when R_0 undergoes a deviation of ±50% after 100 s. The UKF algorithm also diverges. The MAE and RMSE values obtained by the two methods under different model errors are shown in Figure 9. The RDUKF algorithm has a higher estimation accuracy, but under the same deviation ratio, it can be seen that the RDUKF is affected to a greater extent by the R_0 model parameters.

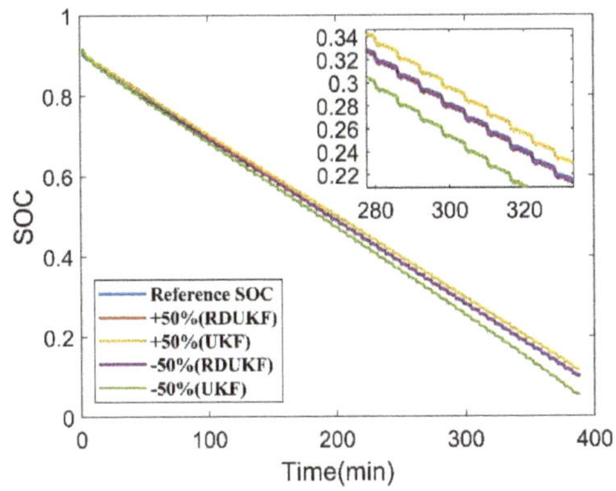

(a)

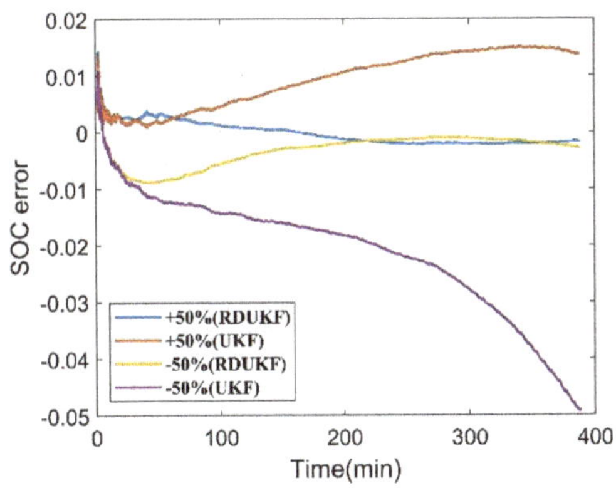

(b)

Figure 6. (**a**) SOC estimation results according to deviations in R_1; (**b**) SOC error according to deviations in R_1.

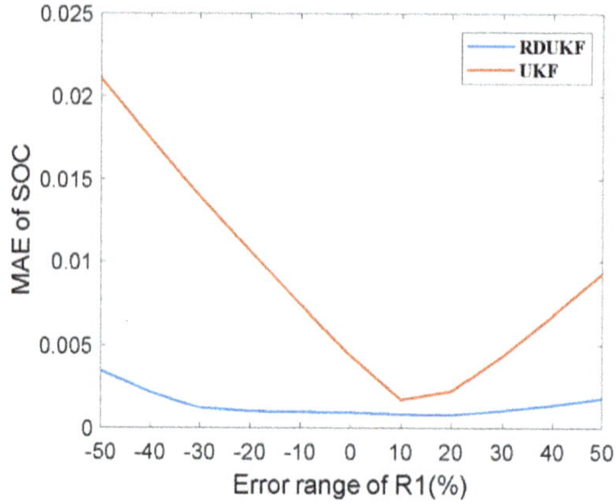

(a)

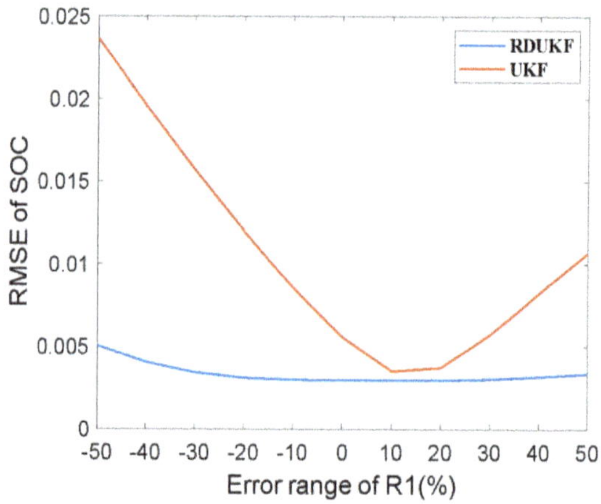

(b)

Figure 7. (**a**) RMSE according to deviations in R_1; (**b**) MAE according to deviations in R_1.

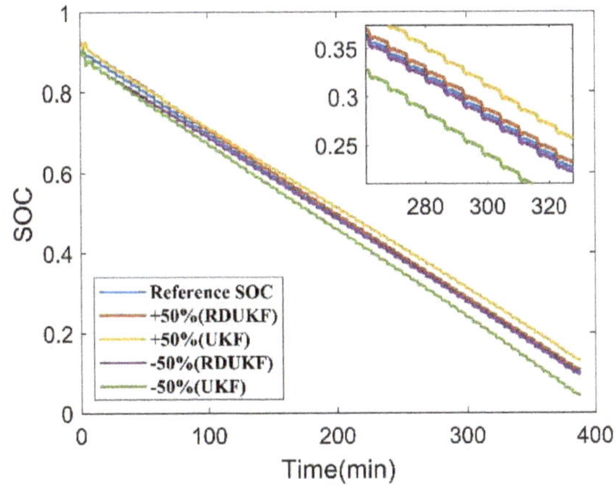

(a)

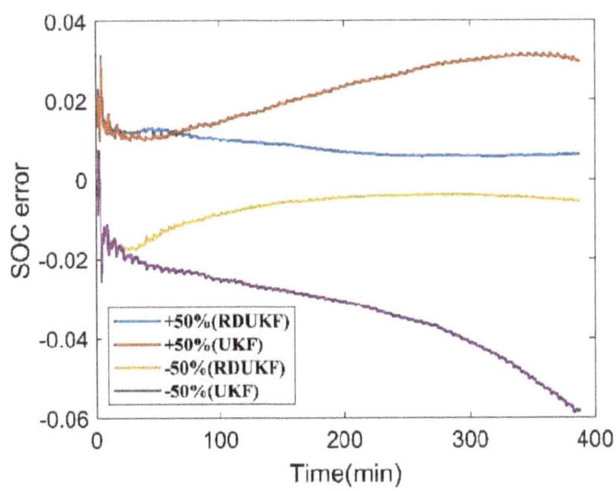

(b)

Figure 8. (**a**) SOC estimation results according to deviations in R_0; (**b**) SOC error according to deviations in R_0.

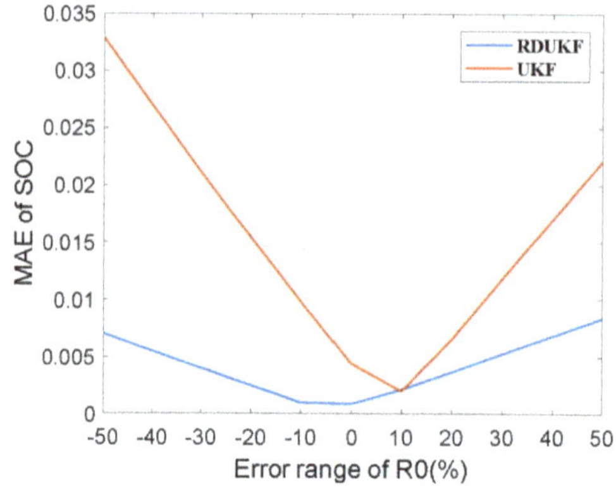

(a)

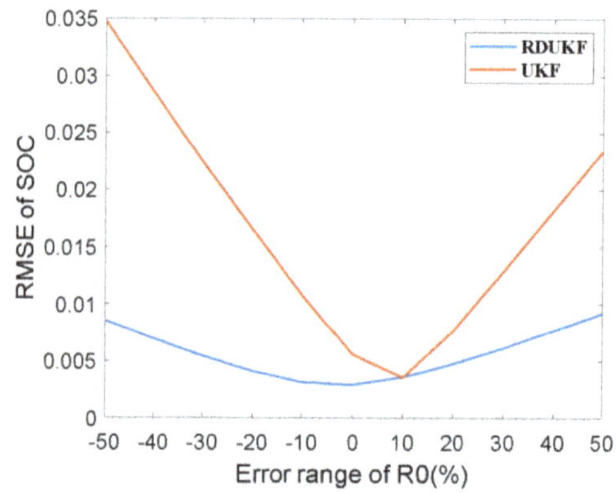

(b)

Figure 9. (**a**) MAE according to deviations in R_0; (**b**) RMSE according to deviations in R_0.

5. Conclusions

As a commonly used SOC estimation algorithm, the UKF has high accuracy in an ideal working environment, but over a complete life cycle, it is difficult to fully grasp the impacts of battery ambient temperature, ageing degree, and working conditions on the battery model and sensor drift, which makes it difficult for the UKF to accurately estimate the battery SOC. To improve the estimation accuracy of the SOC in complex environments, this paper discusses the application of the RDUKF to battery state estimation. By comparing the accuracy of the UKF and RDUKF state estimation under different assumptions, it is verified that the proposed method has high robustness in the presence of model uncertainty and sensor voltage drift. For changing parameters or $\pm$10% voltage deviation, the estimation accuracy can be maintained in the error range of $\pm$1.5%. Although the RDUKF has good robustness, it still requires accurate information on the statistical characteristics of system noise. Therefore, the future research goal is to study an estimation algorithm that can accurately estimate the SOC when both a priori noise is statistically unknown and model uncertainty exists.

Author Contributions: Conceptualization, Y.G.; methodology, Y.G.; software, Y.G.; validation, Y.G.; formal analysis, Y.G.; investigation, Y.G.; resources, Y.G.; data curation, Y.G.; writing—original draft preparation, Z.L. and Y.G.; writing—review and editing, Z.L. and Y.G.; visualization, Z.L.; supervision, Z.L.; project administration, Z.L. All authors have read and agreed to the published version of the manuscript.

Funding: This work was supported by the National Natural Science Foundation of China (51767005).

Institutional Review Board Statement: Not applicable.

Informed Consent Statement: Not applicable.

Data Availability Statement: Not applicable.

Conflicts of Interest: The authors declare no conflict of interest.

References

1. Luo, X.; Wang, J.; Dooner, M.; Clarke, J. Overview of current development in electrical energy storage technologies and the application potential in power system operation. *Appl. Energy* **2015**, *137*, 511–536. [CrossRef]
2. Leung, P.; Li, X.; de León, C.P.; Berlouis, L.; Low, C.T.J.; Walsh, F.C. Progress in redox flow batteries, remaining challenges and their applications in energy storage. *RSC Adv.* **2012**, *2*, 10125–10156. [CrossRef]
3. Weber, A.Z.; Mench, M.M.; Meyers, J.P.; Ross, P.N.; Gostick, J.T.; Liu, Q. Redox flow batteries: A review. *J. Appl. Electrochem.* **2011**, *41*, 1137–1164. [CrossRef]
4. Trovo, A.; Di Noto, V.; Epoupa Mengou, J.; Gamabaro, C.; Guarnieri, M. Fast response of kW-class vanadium redox flow batteries. *IEEE Trans. Sustain. Energy* **2021**, *12*, 2413–2422. [CrossRef]
5. Sankaralingam, R.K.; Seshadri, S.; Sunarso, J.; Bhatt, A.I.; Kapoor, A. Overview of the factors affecting the performance of vanadium redox flow batteries. *J. Energy Storage* **2021**, *41*, 102857. [CrossRef]
6. Cheng, J.; Zhang, L.; Yang, Y.S.; Wen, Y.H.; Cao, G.P.; Wang, X.D. Preliminary study of single flow zinc–nickel battery. *Electrochem. Commun.* **2007**, *9*, 2639–2642. [CrossRef]
7. Li, S.; Li, K.; Xiao, E.; Zhang, J.; Zheng, M. Real-time peak power prediction for zinc nickel single flow batteries. *J. Power Sources* **2020**, *448*, 227346. [CrossRef]
8. Ito, Y.; Nyce, M.; Plivelich, R.; Klein, M.; Steingart, D.; Banerjee, S. Zinc morphology in zinc–nickel flow assisted batteries and impact on performance. *J. Power Sources* **2011**, *196*, 2340–2345. [CrossRef]
9. Lai, Q.; Zhang, H.; Li, X.; Zhang, L.; Cheng, Y. A novel single flow zinc–bromine battery with improved energy density. *J. Power Sources* **2013**, *235*, 1–4. [CrossRef]
10. Yang, B.; Wang, J.; Cao, P.; Zhu, T.; Shu, H.; Chen, J.; Zhang, J.; Zhu, J. Classification, summarization and perspectives on state-of-charge estimation of lithium-ion batteries used in electric vehicles: A critical comprehensive survey. *J. Energy Storage* **2021**, *39*, 102572. [CrossRef]
11. Aylor, J.H.; Thieme, A.; Johnso, B.W. A battery state-of-charge indicator for electric wheelchairs. *IEEE Trans. Ind. Electron.* **1992**, *39*, 398–409. [CrossRef]
12. Gismero, A.; Schaltz, E.; Stroe, D.-I. Recursive state of charge and state of health estimation method for lithium-ion batteries based on coulomb counting and open circuit voltage. *Energies* **2020**, *13*, 1811. [CrossRef]
13. Zhang, H.; Zhao, L.; Chen, Y. A lossy counting-based state of charge estimation method and its application to electric vehicles. *Energies* **2015**, *8*, 13811–13828. [CrossRef]
14. Li, J.; Ye, M.; Meng, W.; Xu, X.; Jiao, S. A novel state of charge approach of lithium ion battery using least squares support vector machine. *IEEE Access* **2020**, *8*, 195398–195410. [CrossRef]
15. Hannan, M.A.; Lipu, M.S.H.; Hussain, A.; Saad, M.H.; Ayob, A. Neural network approach for estimating state of charge of lithium-ion battery using backtracking search algorithm. *IEEE Access* **2018**, *6*, 10069–10079. [CrossRef]
16. Awadallah, M.A.; Venkatesh, B. Accuracy improvement of SOC estimation in lithium-ion batteries. *J. Energy Storage* **2016**, *6*, 95–104. [CrossRef]
17. Bian, C.; He, H.; Yang, S.; Huang, T. State-of-charge sequence estimation of lithium-ion battery based on bidirectional long short-term memory encoder-decoder architecture. *J. Power Sources* **2020**, *449*, 227558. [CrossRef]
18. Lin, X.; Tang, Y.; Ren, J.; Wei, Y. State of charge estimation with the adaptive unscented Kalman filter based on an accurate equivalent circuit model. *J. Energy Storage* **2021**, *41*, 102840. [CrossRef]
19. Yang, F.; Xing, Y.; Wang, D.; Tsui, K.-L. A comparative study of three model-based algorithms for estimating state-of-charge of lithium-ion batteries under a new combined dynamic loading profile. *Appl. Energy* **2016**, *164*, 387–399. [CrossRef]
20. Zheng, Y.; Ouyang, M.; Han, X.; Lu, L.; Li, J. Investigating the error sources of the online state of charge estimation methods for lithium-ion batteries in electric vehicles. *J. Power Sources* **2018**, *377*, 161–188. [CrossRef]
21. Shen, P.; Ouyang, M.; Han, X.; Feng, X.; Lu, L.; Li, J. Error analysis of the model-based state-of-charge observer for lithium-ion batteries. *IEEE Trans. Veh. Technol.* **2018**, *67*, 8055–8064. [CrossRef]

22. Wang, W.; Mu, J. State of charge estimation for lithium-ion battery in electric vehicle based on kalman filter considering model error. *IEEE Access* **2019**, *7*, 29223–29235. [CrossRef]
23. Tang, X.; Wang, Y.; Chen, Z. A method for state-of-charge estimation of $LiFePO_4$ batteries based on a dual-circuit state observer. *J. Power Sources* **2015**, *296*, 23–29. [CrossRef]
24. Afshari, H.H.; Attari, M.; Ahmed, R.; Delbari, A.; Habibi, S.; Shoa, T. Reliable state of charge and state of health estimation using the smooth variable structure filter. *Control Eng. Pract.* **2018**, *77*, 1–14. [CrossRef]
25. Wadi, A.; Abdel-Hafez, M.; Hussein, A.; Alkhawaja, F. Alleviating dynamic model uncertainty effects for improved battery SOC estimation of EVs in highly dynamic environments. *IEEE Trans. Veh. Technol.* **2021**, *70*, 6554–6566. [CrossRef]
26. Tang, X.; Liu, B.; Lv, Z.; Gao, F. Observer based battery SOC estimation: Using multi-gain-switching approach. *Appl. Energy* **2017**, *204*, 1275–1283. [CrossRef]
27. Li, Y.; Wang, C.; Gong, J. A combination Kalman filter approach for State of Charge estimation of lithium-ion battery considering model uncertainty. *Energy* **2016**, *109*, 933–946. [CrossRef]
28. Zhao, S.; Duncan, S.R.; Howey, D.A. Observability analysis and state estimation of lithium-ion batteries in the presence of sensor biases. *IEEE Trans. Control Syst. Technol.* **2017**, *25*, 326–333. [CrossRef]
29. Al-Gabalawy, M.; Hosny, N.S.; Dawson, J.A.; Omar, A.I. State of charge estimation of a Li-ion battery based on extended Kalman filtering and sensor bias. *Int. J. Energy Res.* **2021**, *45*, 6708–6726. [CrossRef]
30. Ren, L.; Zhu, G.; Wang, J.V.; Luo, B.; Kang, J. Comparison of robustness of different state of charge estimation algorithms. *J. Power Sources* **2020**, *478*, 228767. [CrossRef]
31. He, J.; Feng, D.; Hu, C.; Wei, Z.; Yan, F. Two-layer online state-of-charge estimation of lithium-ion battery with current sensor bias correction. *Int. J. Energy Res.* **2019**, *43*, 3837–3852. [CrossRef]
32. Rahimian, S.K.; Rayman, S.; White, R.E. State of charge and loss of active material estimation of a lithium ion cell under low earth orbit condition using kalman filtering approaches. *J. Electrochem. Soc.* **2012**, *159*, A860–A872. [CrossRef]
33. Han, X.; Ouyang, M.; Lu, L.; Li, J. Simplification of physics-based electrochemical model for lithium ion battery on electric vehicle. Part II: Pseudo-two-dimensional model simplification and state of charge estimation. *J. Power Sources* **2015**, *278*, 814–825. [CrossRef]
34. Jokar, A.; Rajabloo, B.; Désilets, M.; Lacroix, M. Review of simplified Pseudo-two-Dimensional models of lithium-ion batteries. *J. Power Sources* **2016**, *327*, 44–55. [CrossRef]
35. Julier, S.J.; Uhlmann, J.K. New extension of the Kalman filter to nonlinear systems. In *Signal Processing, Sensor Fusion, and Target Recognition VI*; SPIE: Orlando, FL, USA, 1997; Volume 3068, pp. 182–193.
36. Tian, Y.; Xia, B.; Sun, W.; Xu, Z.; Zheng, W. A modified model based state of charge estimation of power lithium-ion batteries using unscented Kalman filter. *J. Power Sources* **2014**, *270*, 619–626. [CrossRef]
37. Peng, S.; Zhu, X.; Xing, Y.; Shi, H.; Cai, X.; Pecht, M. An adaptive state of charge estimation approach for lithium-ion series-connected battery system. *J. Power Sources* **2018**, *392*, 48–59. [CrossRef]
38. Zhu, R.; Duan, B.; Zhang, J.; Zhang, Q.; Zhang, C. Co-estimation of model parameters and state-of-charge for lithium-ion batteries with recursive restricted total least squares and unscented Kalman filter. *Appl. Energy* **2020**, *277*, 115494. [CrossRef]
39. Shen, H.; Karlgaard, C.D. Sensitivity reduction of unscented Kalman filter about parameter uncertainties. *IET Radar Sonar Navig.* **2015**, *9*, 374–383. [CrossRef]

energies MDPI

Article

Analysis of a Wind-Driven Air Compression System Utilising Underwater Compressed Air Energy Storage

Lawrie Swinfen-Styles *, Seamus D. Garvey, Donald Giddings, Bruno Cárdenas and James P. Rouse

Department of Mechanical, Materials and Manufacturing Engineering, University of Nottingham, Nottingham NG7 2RD, UK; seamus.garvey@nottingham.ac.uk (S.D.G.); donald.giddings@nottingham.ac.uk (D.G.); bruno.cardenas@nottingham.ac.uk (B.C.); james.rouse@nottingham.ac.uk (J.P.R.)

* Correspondence: lawrie.swinfen-styles@nottingham.ac.uk

Abstract: The increasing push for renewable penetration into electricity grids will inevitably lead to an increased requirement for grid-scale energy storage at multiple time scales. It will, necessarily, lead to a higher proportion of the total energy consumed having been passed through storage. Offshore wind is a key technology for renewable penetration, and the co-location of energy storage with this wind power provides significant benefits. A novel generation-integrated energy storage system is described here in the form of a wind-driven air compressor feeding underwater compressed air energy storage. A direct drive compressor would require very high intake swept volumes. To overcome this difficulty, some prior compression is introduced. This paper discusses the constituent technologies for this concept, as well as the various configurations that it might take and the logic behind operating it. Special consideration has been given to the differences resulting from utilising a near-isothermal wind-driven compressor versus a near-adiabatic one. Multiple iterations of the system have been simulated. This has been done using a price-matching algorithm to optimise the system operation and using volumetric air flow rates to calculate exergy flow. Simulated operation has been performed for a year of real wind and electricity price data. This work has been performed in order to clarify the relationships between several key parameters in the system, including pressure and work ratios, volumetric flowrates, storage costs and profit rates. An additional objective of this paper was to determine whether the system has the potential for economic viability in some future energy grid, especially when compared with alternative wind and energy storage solutions. The results of the simulation indicated that, with proper sizing, the system might perform competitively with these alternatives. Maximum one-year return on investment values of 9.8% for the isothermal case and 13% for the adiabatic case were found. These maxima were reached with ~15–20 h of output storage. In all cases, it was found that maximising the power of the wind-driven compressor compared with the initial compressor was favourable.

Keywords: generation integrated energy storage; wind-integrated energy storage; compressed air energy storage; underwater compressed air energy storage; wind-driven air compression; intake swept volumes; capture value; alternative wind technology; wind system simulation; optimal operation of energy stores

Citation: Swinfen-Styles, L.; Garvey, S.D.; Giddings, D.; Cárdenas, B.; Rouse, J.P. Analysis of a Wind-Driven Air Compression System Utilising Underwater Compressed Air Energy Storage. *Energies* **2022**, *15*, 2142. https://doi.org/10.3390/en15062142

Academic Editors: Alon Kuperman and Alessandro Lampasi

Received: 12 January 2022
Accepted: 7 March 2022
Published: 15 March 2022

1. Introduction

Large-scale energy storage will likely play a significant role in any future energy system that wishes to implement high levels of renewable penetration [1]. This is due to its ability to provide much-needed flexibility in grids whose generation portfolio will become increasingly inflexible as a result of the natural variability of renewable sources. This widely accepted notion has led to a surge in the number and scope of energy storage technologies, both proposed and implemented.

Most of the work done in the energy storage sector has focused on standalone storage systems, which absorb electricity and convert to and from some other storable form of exergy before re-supplying (a proportion of) the electricity at a later time. Standalone storage can be developed and implemented entirely separately from the generation, enabling traditional renewable energy technologies to be used for generation. There are also non-grid applications for energy storage, such as the use of Li-ion batteries in electric vehicles, that are, by definition, standalone.

In contrast to standalone storage, generation-integrated energy storage (GIES) technologies store energy before it is converted to electricity [2]. Figure 1 contrasts the energy conversions seen in GIES systems with those in standalone storage.

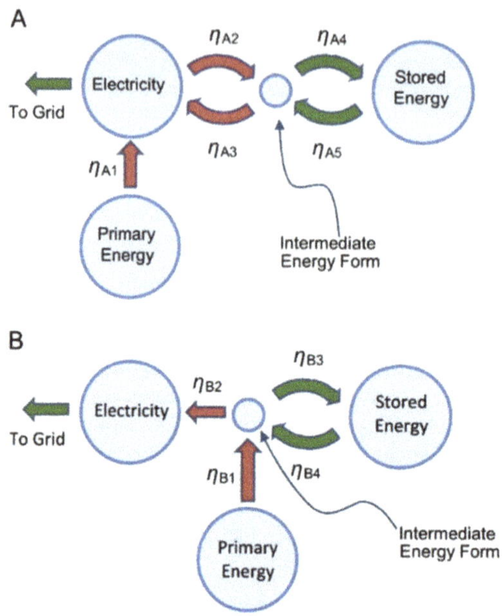

Figure 1. Comparison of energy conversions between (**A**) standalone and (**B**) generation-integrated energy storage configurations. From [2].

GIES systems minimise the conversions between the primary energy collected and the consumed output electricity for any energy that passes through storage. A familiar example of a GIES system is natural hydro, where the energy from flowing water is converted into a storable form (gravitational potential via damming) before it is ever passed through turbines to generate electricity. Although pure transmission efficiencies (that is, the output exergy divided by the input exergy for exergy that does not pass through storage) may be slightly lower than in conventional generation systems, GIES systems have the potential to achieve significantly higher efficiencies for energy that has passed through storage. Such GIES systems may therefore achieve higher overall throughput efficiencies than systems with independent generation and storage, if a sufficient proportion of the total output energy is required to pass through storage, as represented in Figure 2 (from [2]). GIES systems therefore have the potential to be relatively cheap forms of energy storage. Additional examples of well-established GIES technologies include many concentrated solar plants [3,4]. GIES systems have also been proposed for wind [5–8] and nuclear power [9,10].

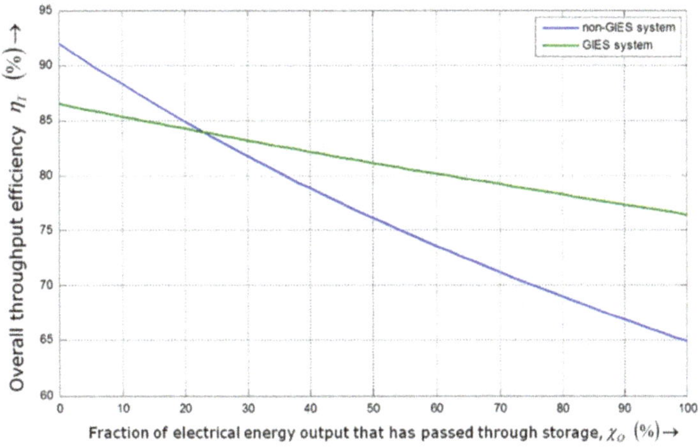

Figure 2. Overall throughput efficiency against fraction of output energy that passed through storage for standalone and GIES stores. From [2].

An offshore wind-driven GIES system that exercises multiple stages of air compression, with integrated compressed air energy storage, is proposed in [11]. This paper clarifies and expands upon that idea. In Section 2, the technologies that constitute the system are explained and examples of other systems utilising them are given. In Section 3, the system from [11] is described, with particular note to possible alternative configurations. Section 4 describes simulations of the system that were performed in order to understand the relationships between parameters such as storage size and pressure ratio. The economic benefits of such a system are then discussed, along with the barriers impeding it.

2. Constituent Technologies

2.1. Isothermal and Adiabatic Compressed Air Energy Storage

Compressed air energy storage (CAES) is a set of thermomechanical energy storage technologies. In CAES, work is done on ambient air in order to compress it to a higher pressure, at which point it is stored (at near-ambient temperature) for some period of time before being expanded back to ambient conditions to generate electricity. Two grid-scale CAES plants in Huntorf, Germany, and Alabama, USA, are diabatic (D-CAES), meaning that the heat from compression is discarded and the air is stored cool, before being reheated prior to expansion. In these cases, the pre-heating of air is done through the burning of natural gas. For several reasons, not least of which is the desire not to use fossil fuels in energy storage, modern CAES proposals tend towards Isothermal (I-CAES) or Adiabatic (A-CAES) configurations, which do not consume any combustible fuel, although CAES systems that use some carbon-neutral fuel (such as green hydrogen) have also been posited [12–14].

In I-CAES systems, air is compressed and expanded near-isothermally with the use of effective heat transfer. In a successful implementation, almost no exergy is lost, because any heat taken from the air is close to ambient temperature and therefore has negligible exergetic content. Assuming ideal gas behaviours, the work done on a volume of gas compressed isothermally is given by:

$$W_{ISO} = p_1 V_1 \ln\left(\frac{p_2}{p_1}\right) =: p_1 V_1 \ln(r), \quad (1)$$

where p is the gas pressure, V is the gas volume and subscripts 1 and 2 denote inlet and outlet, respectively.

Conversely, A-CAES systems purposely minimise the heat transfer between the air and its surroundings. In A-CAES, the heat is stripped from the air after compression and is then stored in thermal energy stores. For adiabatic compression, the work done on a gas is:

$$W_{ADIA} = p_1 V_1 \frac{\left(r^{((\gamma-1)/\gamma)} - 1\right)}{((\gamma-1)/\gamma)} =: p_1 V_1 \frac{(r^{\chi} - 1)}{\chi}, \tag{2}$$

where γ = 1.4 for diatomic gases. As before, r represents the pressure ratio p_2/p_1.

2.2. Wind-Driven Energy Storage

Integrating energy storage into wind power generation has been proposed in various formats. The simplest form of this is the co-location of a conventional energy store with wind turbines, used to smooth the intermittency of renewable generation [15–17]. If sufficiently inexpensive, this has some economic advantages for the operator, as electricity can be supplied to the grid during periods of higher value. There is also a benefit to the grid by reducing transmission losses. However, such systems cannot be described as GIES. Wind farms providing energy in the form of electricity to some co-located energy store could be described as "wind-powered".

By contrast, "wind-driven" systems are defined here as those that initially convert the energy provided by the wind into a form of energy other than electricity. The majority of wind-driven GIES systems in the literature involve some form of gas compression, although the use of induction or other forms of direct heating have also been suggested [18].

Of the systems utilising gas compression, D-CAES [5], adiabatic compression [6] and offshore I-CAES [19] configurations have been proposed. In all cases, the gearbox of a conventional geared wind turbine—a source of considerable capital and maintenance cost—is removed, and the generator is relocated close to the energy stores at ground- or sea-level [20–22]. This leaves the nacelle of the wind turbine relatively free for compression machinery. In [19], a liquid piston compressor is used, with seawater performing the duty of the liquid and also acting as an excellent heat sink. Liquid piston technology is also used in [6], but is designed to perform near-adiabatically, with a radial piston configuration employing multiple stages to minimise the temperature gradients and therefore heat transfer.

2.3. Direct Drive Air Compression and the Problem of Intake Swept Volumes

The wind-driven CAES systems mentioned are direct drive, meaning that the compressor runs synchronous with the wind turbine itself. For most modern turbines, the tip-speed ratio (TSR) relates the tangential speed of the blade tips (u_{TIP}) to the oncoming wind speed (u):

$$TSR = \frac{u_{TIP}}{u} = \frac{\omega D}{2u}, \tag{3}$$

where ω is the rotational speed and D is the diameter of the turbine's swept area. The rotational frequency of the rotor, f, is:

$$f = \frac{\omega}{2\pi} = \frac{uTSR}{\pi D}. \tag{4}$$

The parameters ω and f have rated values for any wind speed at or above the turbine's rated wind speed u_R (and below its maximum wind speed). In this paper, the MHI Vestas V164-8.0 MW turbine will be used as the reference turbine. Data for this turbine is summarised in Table 1.

Table 1. Data for reference turbine used in this paper. From [23].

MHI Vestas V164-8.0 MW Turbine			
Name of Parameter	**Parameter**	**Value**	**Units**
Swept Diameter	D	164	m
Tip-speed	-	104	m/s
Rated windspeed	u_R	13	m/s
Cut-in speed	-	4	m/s
Cut-out speed	-	25	m/s
Tip-speed ratio	TSR	8	-
Rotational speed	ω	1.27	rad/s
Rotational frequency	f	0.20	rotations/s

For wind-driven compression, it is useful to find the intake swept volume of air, V_{swept}, that must be absorbed by the compressor in each rotation. This can be done by rearranging Equations (1) and (2), using the fact that the work done per rotation is equal to the rated power of the turbine $P_{turbine}$ divided by its rotational frequency. For isothermal compression, this gives:

$$V_{swept,ISO} = \frac{P_{turbine}}{f p_1 \ln(r)}, \tag{5}$$

and for adiabatic compression:

$$V_{swept,ADIA} = \frac{P_{turbine} \chi}{f p_1 (r^{\chi} - 1)}. \tag{6}$$

For a set upper pressure $p_2 = p_1 r$, a minimum swept volume is found in an isothermal compressor when $r = e \approx 2.7182$ and in an adiabatic compressor when $r = 1.4^{1/\chi} \approx 3.2467$. For pressure ratios above these optima, swept volumes increase much more slowly than for pressure ratios below these values (with V_{swept} tending to infinity as $r \to 1$).

For illustration, if air were to be compressed by the reference turbine from ambient pressure to 74 bar, the upper pressure of the D-CAES plant at Huntorf, an adiabatic compressor would need to intake 47.2 m^3 of air per revolution and an isothermal compressor would need to intake 92.9 m^3. It should be noted that the swept volume is relatively insensitive to the upper pressure. It has also been assumed here that the adiabatic compression would take place over a single stage (with regard to heat transfer). A multi-stage compressor might be used in a scenario where the upper temperature would be too high for conventional materials.

To estimate the acceptable swept volume for a compressor housed in a wind turbine, consider that the volume of both the gearbox and generator is freed up in a wind-driven system.

Assuming the use of a radial piston compressor design similar to that described in [6], the machine can be split into two distinct parts: here defined as the "displacer" and the "converter". The displacer houses a camshaft and radial pistons and the converter contains the compression cylinders. It is therefore reasonable to make the approximation that the displacer replaces the gearbox and the converter replaces the generator in the turbine nacelle. The nacelle dimensions of the reference turbine are 24 × 12 × 7.5 m and the generator takes up approximately 1/8th of the total volume, or ~270 m^3. A prototype compressor utilising the technology described in [6], in development at the University of Nottingham, has a ratio between the inlet swept volume and overall converter volume of roughly 0.004. Scaling up to the reference nacelle size, this relates to an intake swept volume of ~1.1 m^3. Accounting for slightly larger allowable nacelles, as well as more optimised compressor geometries and nacelle layouts, it is likely that intake swept volumes of ~5–15 m^3 might be achievable in the reference turbine. Even at the upper end of this estimate, a turbine compressing to 74 bar would have a power of 2.54 MW (adiabatic) or 1.29 MW (isothermal)—only a fraction of the 8 MW reference turbine power.

The problem of large intake swept volumes is therefore significant for all direct drive systems. The I-CAES system in [19] solves this problem by moving the liquid piston to sea-level and instead utilising a variable displacement hydraulic pump in the nacelle. This combines well with the use of seawater as the liquid piston (thereby not requiring seawater to be pumped up to the nacelle). By contrast, Ref. [6] solves the problem of intake swept volumes by increasing the inlet air pressure p_1. Rather than a CAES system, Ref. [6] is a pumped thermal energy storage system, wherein all exergy stored is in the form of high-grade heat (or coolth) that is stripped from the compressed gas [24]. This involves a closed gas loop, meaning that the inlet and outlet pressures for the compressor can be freely set.

2.4. Underwater Compressed Air Energy Storage

Underground caverns are often considered the default method of air storage when discussing CAES plants. The CAES plants in Huntorf and Alabama both use this method of storage, as do many other CAES proposals.

Such storage is generally isochoric (constant volume), meaning that emptying a proportion of the air in the store reduces the pressure of the remaining air. Expanders used in an isochoric CAES system must therefore run over a series of pressures, limiting their efficiency, as they cannot be optimised for a single pressure. As there will be a pressure below which an expander cannot run at an acceptable efficiency, this also significantly limits the proportion of the store that can be flexed. Cavern stability also limits operational pressure ranges, as a certain volume of "cushion gas" is required to ensure that the caverns do not collapse.

In contrast, isobaric (constant pressure) storage allows for the flexing of the entire storage volume, at a single pressure for which the expander can be optimised. Isobaric storage can be achieved in combined hydro-CAES plants where a water reservoir is used to keep constant pressure [25]. However, most isobaric storage designs revolve around the store being placed deep underwater, in a lake or ocean. This is known as underwater compressed air energy storage (UWCAES). In UWCAES, each 10 metres of water depth provides approximately 1 bar of additional pressure on the store. By filling the store with air of equivalent pressure to the hydrostatic pressure supplied by the water, the store itself does not need to provide any confining force. A UWCAES system utilising flexible canvas bags (energy bags) is described in [26,27] and expects storage capacity costs in the region of 25 \$/kWh (dependent on the storage depth). A representation of this design can be seen in Figure 3.

The components comprising an energy bag system are anchoring ballast (AB), tension cables (TC) and the canvas surface area itself (SU). The role of the tension cables is to counteract the buoyant force created by the air volume. The cables are therefore the sole structural element of the energy bag. It is explained in [26] that the cost of the anchoring ballast is proportional to the buoyancy force created by the energy bag volume and is therefore proportional to the cube of the characteristic dimension (the diameter, D) of the energy bag:

$$C_{AB} \propto D^3. \tag{7}$$

The tension cable cost is proportional to the product of its length L and tension force F_T, which is also proportional to the buoyancy force:

$$C_{TC} \propto LF_T \propto D^4. \tag{8}$$

The cost of the canvas is clearly proportional D^2 and so the total energy bag cost can be written as:

$$C_{BAG} = C_{AB} + C_{TC} + C_{SU} = aD^4 + bD^2 + cD^3, \tag{9}$$

where a, b and c are constants. By noting that the energy stored in the bag, E, is equal to another constant, k, multiplied by the energy bag volume,

$$\frac{C_{BAG}}{E} = \frac{aD^4 + bD^2 + cD^3}{kD^3}. \tag{10}$$

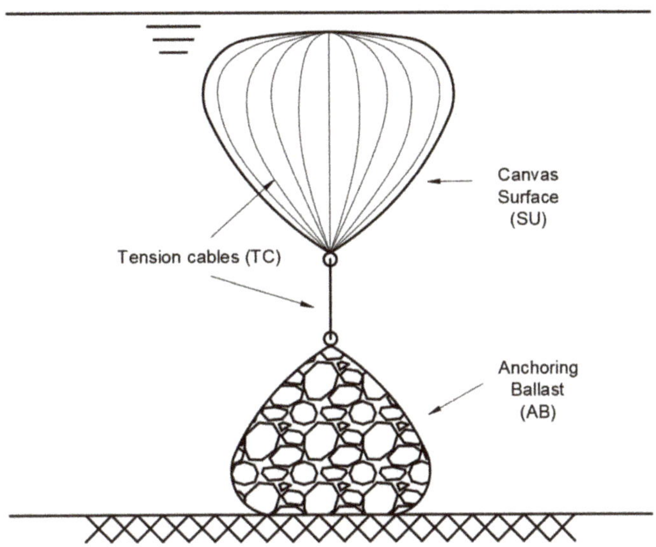

Figure 3. Energy bag underwater compressed air energy storage (UWCAES) system. Shown without pipework for simplicity.

Differentiating this to find the optimum and substituting into Equation (9) then shows that the optimum diameter is achieved when C_{SU} is equal to C_{TC}. Multiple energy bags of this optimal diameter would therefore be better value for money than a single energy bag of equivalent volume. The use of multiple smaller bags also has the benefit of reducing the proportion of storage that is out of service if damage to a single bag occurs. Assuming that any storage system employs multiple optimally sized energy bags (with the volumetric capacity of V_{opt}), the total cost of the system is the number of energy bags used multiplied by the cost of a single bag. Therefore, for a system with a volumetric capacity of nV_{opt}, where n is an integer, the total cost of the storage system is directly proportional to its volume.

The technologies discussed in this section have a clear synergy. The use of UWCAES specifically for the balancing of wind energy has been investigated previously [28,29]. However, iterations of this concept assume a conventional wind turbine with standalone UWCAES. Thus far, no literature has discussed the combination of wind-driven air compression with UWCAES. The possibility of utilising UWCAES in a GIES-type system is unexplored. The concept of wind-integrated energy storage itself (beyond the co-location of standalone energy storage) is underexplored.

Furthermore, no wind-driven air compression system attempts to solve the problem of large swept volumes with a prior stage of electrically driven air compression. It may initially seem counterintuitive to add electrically driven air compression to a wind-driven system. However, there is significant advantage to be had by removing the wind turbine gearbox and allowing direct drive air compression. There are also potential grid-balancing advantages to such a system, discussed later in this paper.

The system described in the remainder of this paper is not intended as a catch-all solution to the problem of inflexible wind generation. Indeed, it is a complimentary technology that might be used in tandem with a grid utilising any subset of systems previously mentioned, conventional wind turbines, and large-scale standalone storage (for example, in the form of CAES or green hydrogen).

3. System Overview

3.1. Description

The system described in [11] is named *Wi-DACIS*, standing for wind-driven air compression and isobaric storage. It utilises the technologies previously mentioned in Section 2. A wind-driven, direct drive air compressor replaces the gearbox system in a conventional geared offshore wind turbine. The high-pressure air exhausted from the compressor can either be immediately expanded to generate electricity or passed into an energy bag air storage system. This provides flexibility to the otherwise intermittent renewable energy supply from the turbine. Because primary energy from the wind is used directly to compress air before it is ever converted into electricity, this constitutes a GIES system.

Being direct drive, the problem of intake swept volume must be considered. In a similar manner to the pumped heat system described in [6], the solution is to supply air to the primary compressor at a pressure higher than ambient. However, instead of the closed loop utilised by [6], *Wi-DACIS* employs a prior stage of compression, which absorbs air from the environment, compressing it from ambient conditions to a suitable pressure for the wind-driven compressor.

In [11], the *Wi-DACIS* system was originally described with two pressures of air store: a high-pressure store fed by the wind-driven compressor, and an additional medium-pressure store fed by the first stage compressor. While the authors believe that there could be significant merit to the additional flexibility provided by the medium-pressure store, the full cost/benefit analysis of such a store is beyond the scope of this paper and will be addressed in future work.

The classification of *Wi-DACIS* as a purely GIES system is dependent on the form that the first stage of compression takes. If it is driven by electricity from the grid, as is assumed for the analysis in Section 4, then it acts as a hybridisation of standalone and GIES systems, as some fraction of the energy in storage has been absorbed from the grid. The ratio of standalone to GIES utility is then dependent on the relative powers of the two compressor stages.

However, if the first stage of compression is run by some non-grid method (i.e., one in which electricity from the grid is not required, such as solar or wave power), then *Wi-DACIS* performs as a wholly GIES system. The possibility of using a form of trompe compressor as a method of offshore first-stage compression is also discussed in [11], using the water head created by wave heights to push volumes of water to a depth, pressurising the air entrained within that water.

Figure 4 shows the *Wi-DACIS* system in its hybrid GIES-standalone configuration, utilising a single high-pressure air store.

Important components and parameters for the *Wi-DACIS* system are:

- COMG—the grid-connected first stage of compression.
- COMW—the main stage, wind-driven compressor, which can be either isothermal or adiabatic (discussed in Section 3.2).
- HPEB—the high-pressure energy bag system (shown as a single bag for simplicity).
- EXP—the expander/generator set.
- HXU/TES—the heat-exchanger unit and thermal energy stores, relevant only if the main stage compressor is adiabatic.
- r_1 and r_2—the pressure ratios of COMG and COMW. The pressure of the high-pressure store is $r_1 r_2$.
- P_G and P_W—the powers of COMG and COMW. P_{EXP} is the expander power.
- V_1, V_2, V_3—the volumetric flowrate into COMG, out of COMG/into COMW and out of COMW/into EXP. p_1, p_2, p_3 are the respective pressures.

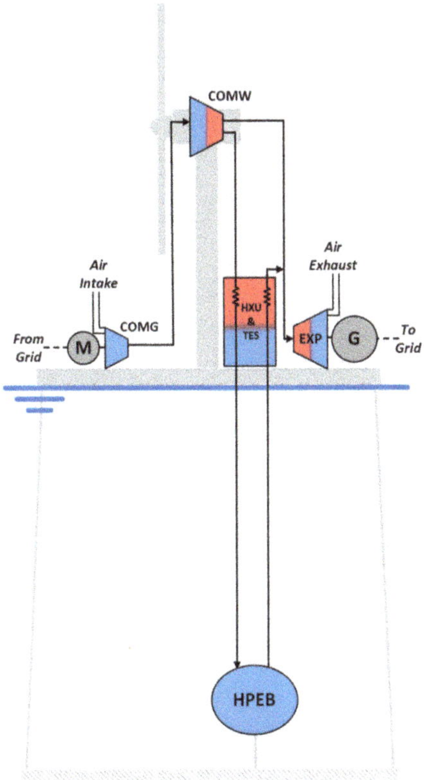

Figure 4. The *Wi-DACIS* system in an adiabatic main stage configuration. In an isothermal configuration, HXU and TES are not present.

3.2. Possible Configurations

An important consideration in a *Wi-DACIS* system is whether the main stage compressor is adiabatic or isothermal. This decision dictates several system parameters, including the relative powers (or relative pressure ratios, depending on which values are set) of the two compressors. Because the product of p and V remains constant throughout isothermal compression, if the main stage compressor is isothermal, the relative powers of the two stages can be described as:

$$\frac{P_G}{P_W} = \frac{\ln(r_1)}{\ln(r_2)}. \tag{11}$$

By comparison, if the main stage compressor is adiabatic, the relative powers can be described as:

$$\frac{P_G}{P_W} = \frac{\chi \ln(r_1)}{(r_2)^{\chi} - 1}. \tag{12}$$

Figure 5 shows that, for a set upper pressure $r_1 r_2$, an adiabatic main stage will result in a smaller grid-powered compressor. Given that the *Wi-DACIS* system intends the main-stage compressor to have a reasonably high power compared to the first-stage compressor (certainly $P_G/P_W < 1$), this limits the acceptable relative pressure ratios. In the example given in Figure 5, where $r_1 r_2$ = 200, r_2 would therefore need to be greater than 9 for an adiabatic main stage or greater than 14 for an isothermal one.

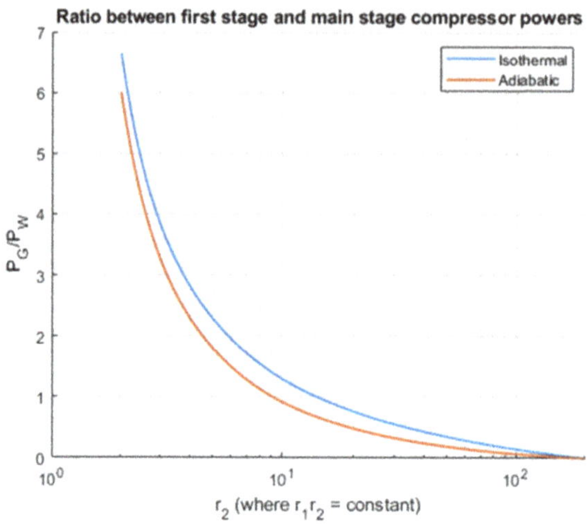

Figure 5. Power ratios for an isothermal and an adiabatic main stage of compression, where the upper pressure, $r_1 r_2$, is set.

The choice between isothermal and adiabatic compression also affects the required inlet swept volume, as is described in Section 2.3. For any given intake pressure and pressure ratio, an adiabatic compressor will require a lower intake swept volume than an isothermal one. This is due to the additional work done in adiabatic compression to heat the gas. As previously mentioned, the upper temperature is a limiting factor in the adiabatic case, but multi-stage compression is possible to keep this temperature within reasonable limits (by taking heat from the air at several points during the compression process).

With both of these effects taken into consideration, an adiabatic main stage appears to be the clear choice. However, this is also dependent on the relative costs of the compressors. Regardless of whether the main stage compressor is isothermal or adiabatic, heat exchange of some kind will be required, either in the form of inter-stage cooling, or to strip the heat from the air for storage. Assuming, then, that the isothermal and adiabatic compressors and heat exchangers are of similar costs, an important measure of the viability of adiabatic main stage compression is whether the thermal energy storage (TES) required by an adiabatic compressor is cost-competitive with the high-pressure energy bag system. Given that the costs calculations for the energy bags in Section 2.4 only include material costs, the same will be applied to the TES.

4. System Analysis

4.1. Simulation Data and Assumptions

Simulations of the *Wi-DACIS* system have been run for both the isothermal and adiabatic main-stage compressor configurations. These will provide insight into the relationship between various *Wi-DACIS* parameters, as well as providing a reasonable minimum size for the energy store and some sense of the profitability of such a system.

Wind speed data for 2015 were obtained from the UK Met Office Integrated Data Archive System (MIDAS) [30]. This took the form of sea-level wind speed data, which were converted to wind speed at turbine-height with:

$$u(z) = \frac{u_*}{\kappa}\left[\ln\left(\frac{z}{z_0}\right)\right], \tag{13}$$

where $u(z)$ is the wind speed at altitude z, u_* is the friction velocity, κ is the Von Kármán constant (~0.4) and z_0 is the roughness length (~0.0002 m for open sea) [31,32].

Hourly data from 12 locations were used (Figure 6). These sites were chosen due to the availability of sufficient wind speed data over the given period. Several of these sites would not necessarily be suitable for a *Wi-DACIS* system, due to their location on the continental shelf implying relatively shallow water. However, the data at these sites are sufficiently representative of generic offshore wind data to be suitable in the analysis. Figure 6 includes the approximate outline of the continental shelf, showing the closeness of the sites to relatively deep water. Figure 7 shows the worldwide water depth map. It is clear that a depth of 2000 m is closely accessible for much of the world's coastline.

Figure 6. Map of the 12 wind data point locations. Letters used to differentiate between sites. Adapted from [33] (Map Data: ©2022 Google, GeoBasis-DE/BKG ©2009).

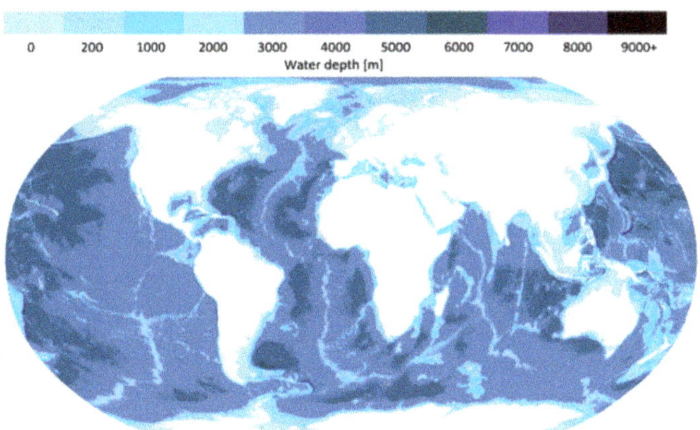

Figure 7. Map of worldwide water depth. Adapted from [34].

The wind speed data were converted to energy generation data using the real power curve of the reference turbine, shown in Figure 8 [23]. The energy generation data could then be used in the simulation to calculate the energy flow in the system via air volumetric flow rates.

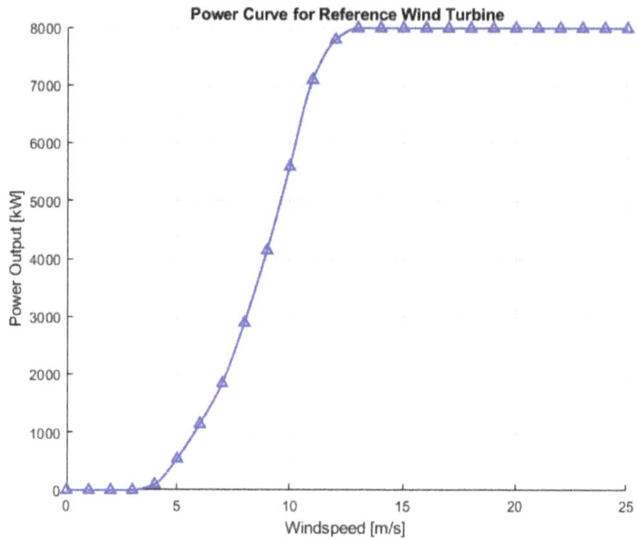

Figure 8. Power curve for reference wind turbine. Adapted from [23].

UK day-ahead electricity price data over the same 2015 time period were taken from [35]. It is assumed that the system operates with a merchant contract, using the hourly day-ahead electricity prices as a basis for decision-making. This is in contrast to many UK wind farms, which operate on the Contracts for Difference (CfD) scheme, wherein a single price is agreed upon per unit of electricity supplied to the grid, regardless of instantaneous electricity value. The revenue of a conventional wind turbine on the CfD scheme will also be considered for the sake of comparison.

The simulation assumes that the *Wi-DACIS* wind-driven compressor is capable of performing at part-load to the same extent as the conventional reference turbine. Optimal part-load performance would be made possible with a compressor design similar to the one described in [6], which allows individual compression cylinders to be deactivated via automated active valve control. It is expected that these operations could be performed at much shorter timescales than is required for the turbine blades to adjust pitch, meaning that overall part-load performance should not be affected.

The grid-powered isothermal compressor is assumed to be of sufficient power to provide the necessary volumetric flowrate to the main stage compressor at its nameplate capacity. It is also assumed to have no variation in efficiency at partial load.

In order to gain some measure of the cost of each iteration, unit costs were applied to each element of the system. Reasonable values for a, b and c from Equation (9) were found that satisfied the condition of a single optimum at $D \approx 20$ m and gave a cost of roughly 50 £/m^3, indicated by [26]. The total energy bag storage was calculated as multiples of these optimal bags, with an additional smaller bag (if necessary) at a less optimal cost. It is assumed that each energy bag can store a parcel of air for as long as is required between the charging and discharging cycles. This is a reasonable assumption given the store sizes considered (<200 h).

4.2. Methodology

The initial inputs for the simulation can be split into system parameters and unit costs. The system parameters tested in the simulation are presented in Table 2.

Table 2. System parameters for *Wi-DACIS* simulations.

Name of Parameter	Parameter	Value(s)	Units
Main compressor swept volume	V_{swept}	~5–15	m^3/rev
Maximum pressure	p_2	200	bar
High-pressure store size	-	1–200	hours

For each combination of store capacity and swept volume, the system powers, inlet and outlet volumetric flowrates, pressure ratios and energy bag sizes were calculated, both for the isothermal and adiabatic configurations. The initial state of charge (SoC) assumed for the store was 50% of the maximum charge. Operation over the year time period was then simulated for each system and at each site.

The simulation was performed assuming perfect knowledge of a year of day-ahead electricity prices and wind availability. A modified version of the price-matching algorithm found in [36] was used to find the optimal store and compressor power usage at each hour time interval.

The income (revenue) was calculated as a summation of the hour-wise multiplication of instantaneous electricity price and instantaneous expander power. Similarly, the expenditure was calculated as a summation of the hour-wise multiplication of instantaneous electricity price and the instantaneous power of the first-stage, grid-powered compressor.

For the configuration where the wind-driven stage is adiabatic, the required capacity of the thermal energy store was calculated using the work done by the adiabatic compressor from Equation (2), and the temperature was found by:

$$T_{store} = T_{ref} r_2^{X}, \tag{14}$$

where T_{ref} is the ambient temperature of the air. In a packed gravel bed TES, using a thermal medium (quartzite) with a specific heat capacity of 850 J/kg·K, the required mass was found:

$$m_{quartzite} = \frac{W_{Heat}}{850 \cdot \left(T_{store} - T_{ref}\right)}. \tag{15}$$

The thermal medium was given a material cost of 0.1 GBP/kg [37]. A mass overrating factor of 2 was then applied.

An estimate of the steel required for the thermal store was obtained using a random sphere packing density ~0.64, quartzite and steel densities of 2700 kg/m^3 and 8000 kg/m^3, respectively, and assuming a TES store with a height to diameter aspect ratio of 2 and 20 mm thick walls (considered to be very conservative):

$$V_{store} = \frac{m_{quartz}}{0.64 \times 2700}, \tag{16}$$

$$m_{steel} \approx 0.646 \left(m_{quartz}\right)^{2/3}. \tag{17}$$

The cost of worked steel for wind turbines is in the order of 2 GBP/kg, based on assertions made by Henrik Stiesdal on several occasions, including at Floating Offshore Wind UK (30 October 2018) and the EWEA Floating Wind Power Debate (18 November 2015). Steel was therefore costed at 2 GBP/kg. The capital costs per unit power for the compressors and expanders were chosen partly from values in [2,13,38–40] and partly on the idea that a conventional wind turbine costs ~1 GBP/W and replacing the expensive gearbox element with an expensive compressor element will not change that value too significantly. These costs are summarised in Table 3. It should be noted that the cost (GBP/kW) for compressor and expander machinery is dependent on several factors, including its rated power—larger machines generally cost less per unit power [39–41]. In order to take advantage of the scalability in the *Wi-DACIS* system, it is likely that several turbines would be fed by a single grid-powered compressor and would feed into a single expander. Although a single turbine has been simulated, this scalability has been accounted for in the capital costs chosen.

Table 3. Per-unit power capital costs for the compressor/expander machinery. Based on values in [2,13,38–40].

Machine	Cost Range (Nominal Value) (GBP/kW)
COMG	200–400 (300)
COMW	800–1200 (1000)
EXP	300–700 (500)

To reasonably account for losses in the system, the efficiencies shown in Table 4 were applied in the simulation. As with cost, the efficiency of machinery can be affected by its rated power [42,43]. The large scale of the machines used in *Wi-DACIS* means that the upper range of efficiencies is certainly accessible.

Table 4. Efficiencies of the system machinery used. Based on values in [44–48].

Machine	Efficiency Range, η (Nominal Value)
COMG	0.8–1.0 (0.9)
COMW	1 *
EXP	0.8–1.0 (0.9)
HPEB	0.98

* The efficiency value of the main stage compressor is given as 1 because it is a relative efficiency, which will be used to compare with alternative wind + storage systems in Section 4.5. The result of applying the COMG efficiency is that the pure transmission efficiency of the Wi-DACIS system will be 90% of the transmission efficiency of a conventional turbine.

The range of cost and efficiency values shown were used in data validation testing, whereas the nominal values were used in the main simulation.

The efficiency losses in the grid-powered compressor and the expander were accounted for by increasing the expenditure and decreasing the income by a factor of η, while keeping the actual exergy absorbed/sold the same.

Efficiency losses in the energy bags are expected to be extremely small. In this paper, a representative value of 0.98 was chosen. This was applied by increasing the cost of the stores by this factor.

The outputs of the simulations were: instantaneous compressor and expander powers over time, state of charge for the store over time, capital costs, revenue and expenditure (the difference of which is the profit), a one-year return on investment and potentially limiting parameters such as TES temperature and ratio of P_G/P_W.

4.3. Results and Discussion

Each of the 12 sites used in the simulation exhibited the same behaviours, although the exact values were slightly different due to local wind data. The averaged results of the 12 sites are used in the following figures. Maximum and minimum values given in the text cover all 12 sites.

The primary effect of increasing the inlet swept volume is that the main stage compressor, COMW, becomes a larger proportion of the total system power; alternatively, COMG becomes a smaller proportion. This is caused by the increased ratio r_2/r_1, seen in Figure 9.

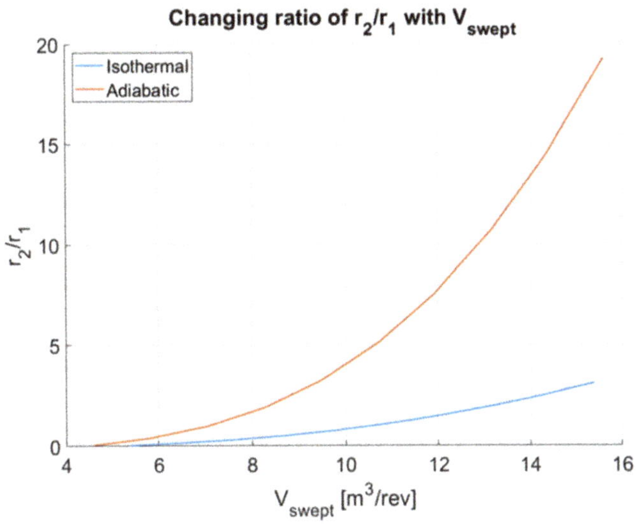

Figure 9. The relationship between the relative power of the main-stage and first-stage compressors and the intake swept volume, in the form of pressure ratio r_2/r_1.

Increasing the proportional power of COMW in this way serves to significantly improve the yearly profit (income minus expenditure) of the system, in all iterations. This can be seen in Figure 10, which highlights the improvement of the system with increased V_{swept} (and reduced proportional power of COMG).

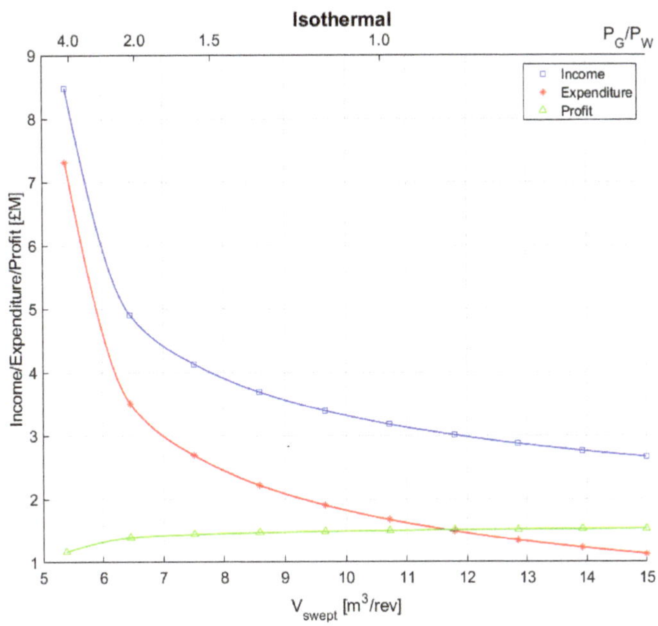

Figure 10. *Cont.*

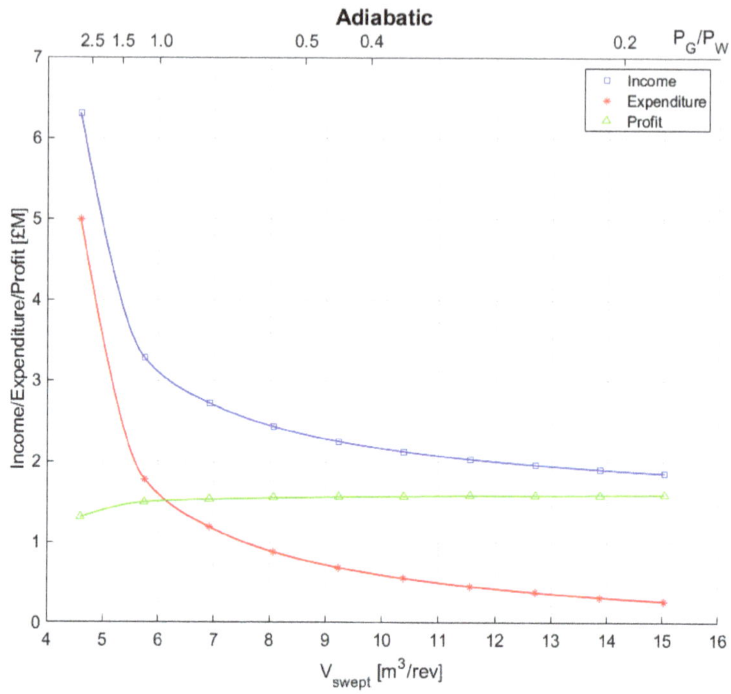

Figure 10. Income, expenditure and profit for representative isothermal (**above**) and adiabatic (**below**) systems with 200 h of storage. Also shown on the upper x-axes is the proportional power of COMG compared with COMW. Decreasing this ratio improves financial performance.

While the income of the system is reduced by increasing V_{swept}—a function of the total system power reducing as the proportional power of COMW, which is set at 8 MW, increases—the expenditure of the system decreases more quickly, resulting in greater profit. The expenditure is further reduced in the adiabatic case compared with the isothermal case, due to the further increase in the ratio between COMW and COMG powers, a result of the work of Equations (1) and (2).

Due to the effect the swept volume has on the system powers and pressure (and therefore costs), it is useful to show a breakdown of the cost elements with changing V_{swept} for both the isothermal and adiabatic cases. Figure 11 shows a breakdown of the cost elements for a system with 200 h of storage. Figure 12 normalises these costs by the power of the system at each value of V_{swept}.

For the adiabatic case, the reduction in the cost of the thermal energy store with increasing V_{swept} is due to the increasing pressure ratio of COMW. This results in a higher outlet air temperature; the thermal mass is therefore used more effectively. Likewise, the reduction in the cost of the air store is due to a larger proportion of the exergy being stored as heat, as V_{swept} increases.

Given that system capital costs are reduced and profits increased with increased V_{swept}, it is clear that the one-year return on investment (ROI) is significantly affected by increasing the inlet swept volume of the system. This can be seen in Figure 13, using a system with 200 h of storage for demonstration.

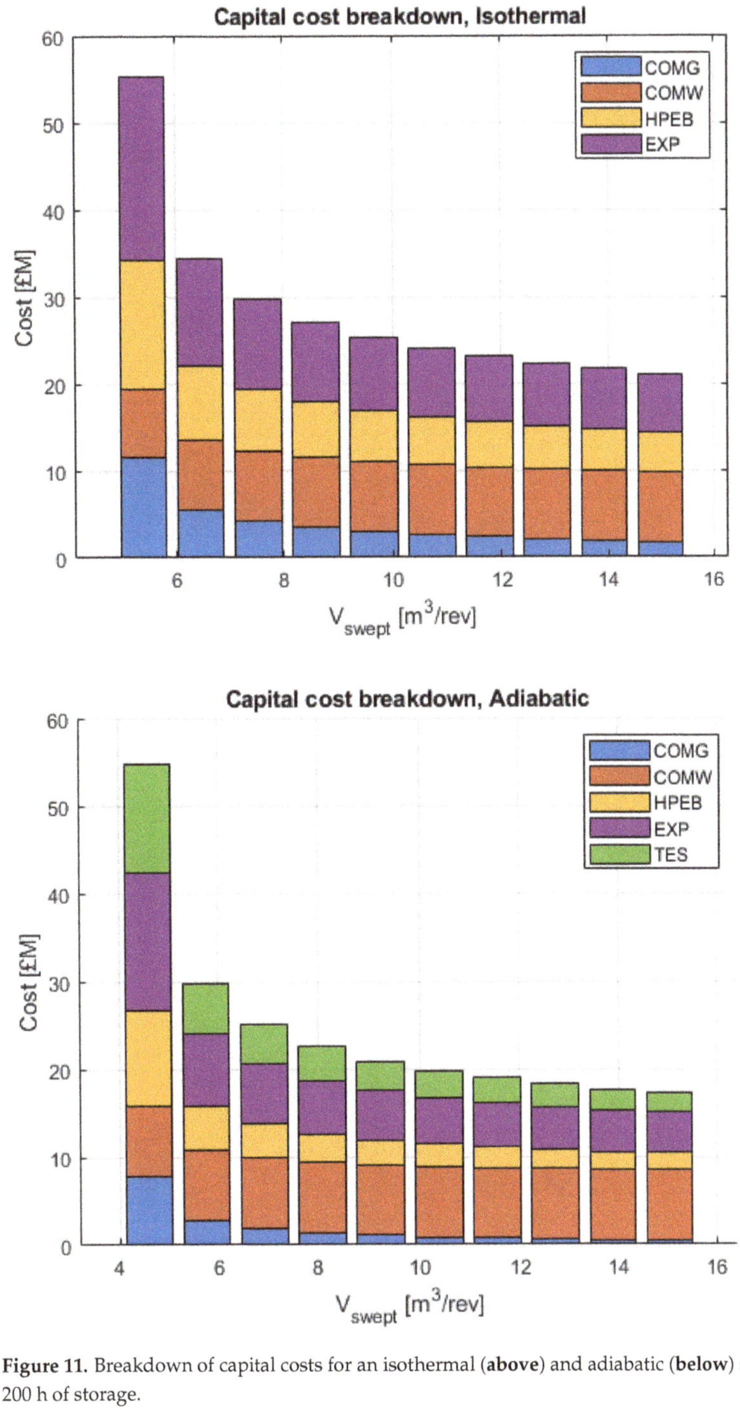

Figure 11. Breakdown of capital costs for an isothermal (**above**) and adiabatic (**below**) system with 200 h of storage.

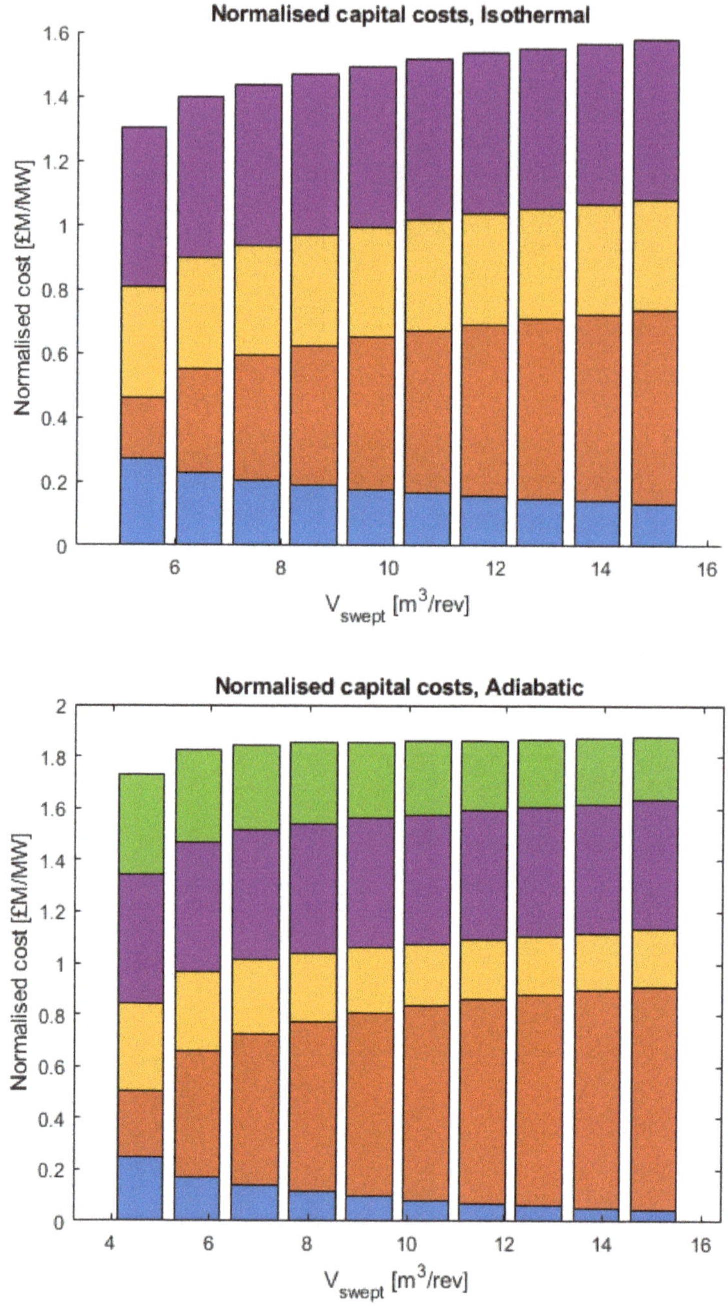

Figure 12. Breakdown of capital costs shown in Figure 10, normalised by the total discharge power of the system. Isothermal (**above**) and adiabatic (**below**) configurations shown.

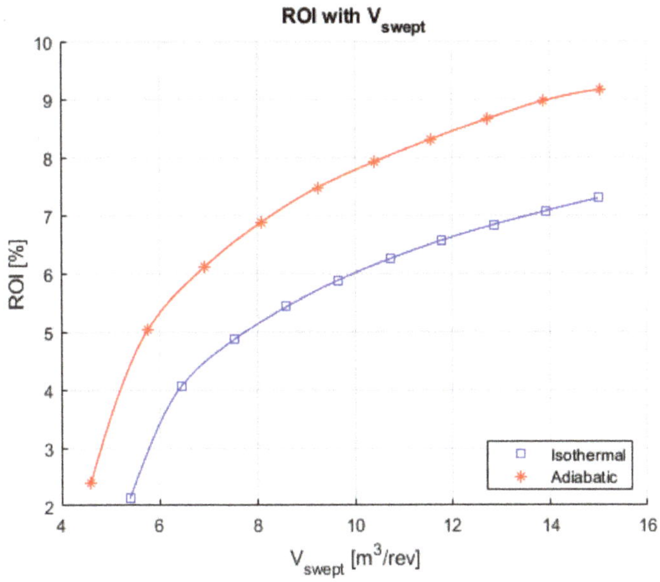

Figure 13. The relationship between one-year return on investment and inlet swept volume for a representative system with 200 h of storage.

It is therefore desirable for the profitability of the *Wi-DACIS* system to maximise V_{swept} to its upper practical limit.

It is important to determine whether there is an optimal energy storage capacity for the system. Figure 14 shows the ROI of a system with a constant swept volume and changing store size, from 1 to 200 h.

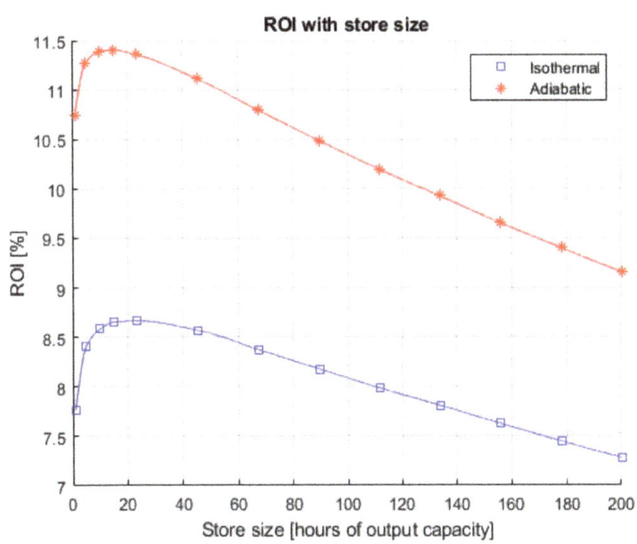

Figure 14. The relationship between one-year return on investment and storage capacity for a system with the maximum allowable inlet swept volume.

In fact, a maximum ROI for the system is reached in all cases. At the highest value of V_{swept} (shown in Figure 14), this occurs at ~20 h for the isothermal case and ~15 h for the adiabatic case. Before this point, there is very little value provided by the stores, despite the increased cost of the system compared with a conventional turbine (through the addition of the first-stage compressor and the expander). After this maximum, the increased value offered by the store does not make up for the climbing costs of such a system at present. This is summarised in Figures 15 and 16.

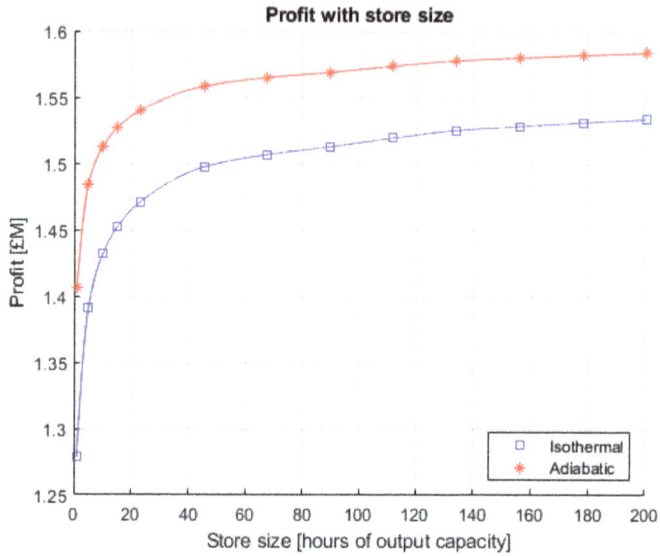

Figure 15. The relationship between profit and storage capacity for a representative system with the maximum allowable inlet swept volume.

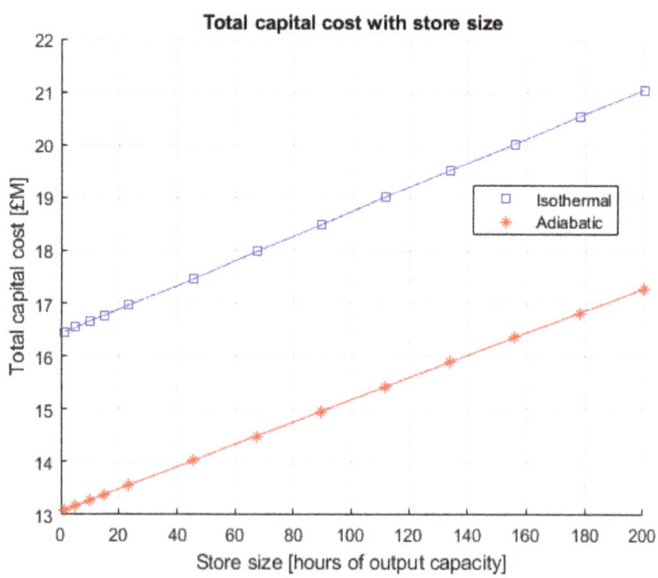

Figure 16. The relationship between total system cost and storage capacity for a representative system with the maximum allowable inlet swept volume.

For the isothermal case, the maximum ROI at each site ranged from 6.4% to 9.8%. For the adiabatic case, the maximum ROI ranged from 8.1% to 13.0%. At all sites, this maximum occurred at the highest value of swept volume, with the optimal store size varying only slightly.

The lowest ROI values (between 0.8% and 1% for the isothermal case and between 1.8% and 2.4% for the adiabatic case) are found when the swept volume is at its lowest and the storage capacity is close to zero, due to the high system costs and minimal benefit provided by the store.

In the adiabatic case, at higher values of swept volume, the air reaches temperatures nearing the limits of sensible heat storage capabilities (~900 K). This could be accounted for with multi-stage adiabatic compression utilising inter-stage cooling.

Using the method and values described in this paper, it seems likely that an adiabatic wind-driven compressor is significantly financially favourable over an isothermal one for the *Wi-DACIS* system. It is advantageous to design the system such that the wind-driven compressor performs the majority of the total work, achieved by designing for maximum swept volume and adiabatic operation. The existence of an optimal store size in each case is also of interest, as it creates a clear system design goal for store size—something that is not seen to be the case for standalone storage in Section 4.5.

4.4. Validation

In addition to the real wind and electricity price data used in this paper, and the nominal costs and efficiencies based on prior literature, data validation was performed in the form of a Monte Carlo simulation. The ranges of cost and efficiency shown in Tables 3 and 4 were used to create triangular probability distributions for each parameter. For each iteration of the *Wi-DACIS* system (characterised by the values of V_{swept} and store size), random weighted samplings of these distributions were created, and the cost, income, expenditure, profit and ROI values of that iteration were adjusted accordingly. With 10^5 runs in each iteration, a normal distribution was then fit to the result, and the coefficient of variation (that is, the standard deviation normalised by the mean and presented as a percentage) of the ROI was found, as shown in Figure 17.

The coefficient of variation is considerably higher at low values of V_{swept}. At its highest, it reaches 141% in the isothermal case and 51% in the adiabatic case. This is due to the low mean ROI values and higher proportional power of COMG at the minimum swept volume. For some values of cost and efficiency, the ROI at these low V_{swept} iterations can be negative, a result of the losses in the system outweighing the benefits of absorbing and storing the electricity. Clearly a system with such parameters would not be economically viable.

However, at middling and higher values of V_{swept}, the coefficient of variation quickly becomes tolerable, plateauing at around 10–15% for the isothermal case and 7–10% for the adiabatic case. As with the effects of V_{swept} on the variability, the isothermal case is more variable due to the larger proportional power of COMG, resulting in lower profit margins and more chance for an expensive compressor to impact the ROI of the system.

Given that the iterations that have been shown to be more favourable are those with higher values of V_{swept}, the variability calculated due to possible variations in the input parameters is considered to be acceptable.

4.5. Comparison with Other Systems

Comparing the ROI and capital costs of the *Wi-DACIS* system with alternative wind–storage options is useful in determining whether the system described here has a potential place in the energy grid.

A conventional 8 MW wind turbine (capital cost ~£8 M) on the CfD scheme with a strike price of 40 GBP/MWh has a maximum ROI of 22% using the same wind and price data. Approximately the same maximum ROI was found for a free-market wind turbine over the same time period. The difference in the ROI between these turbines and the

Wi-DACIS system supports the generally held belief that it is much more expensive to store a unit of energy than it is to generate it.

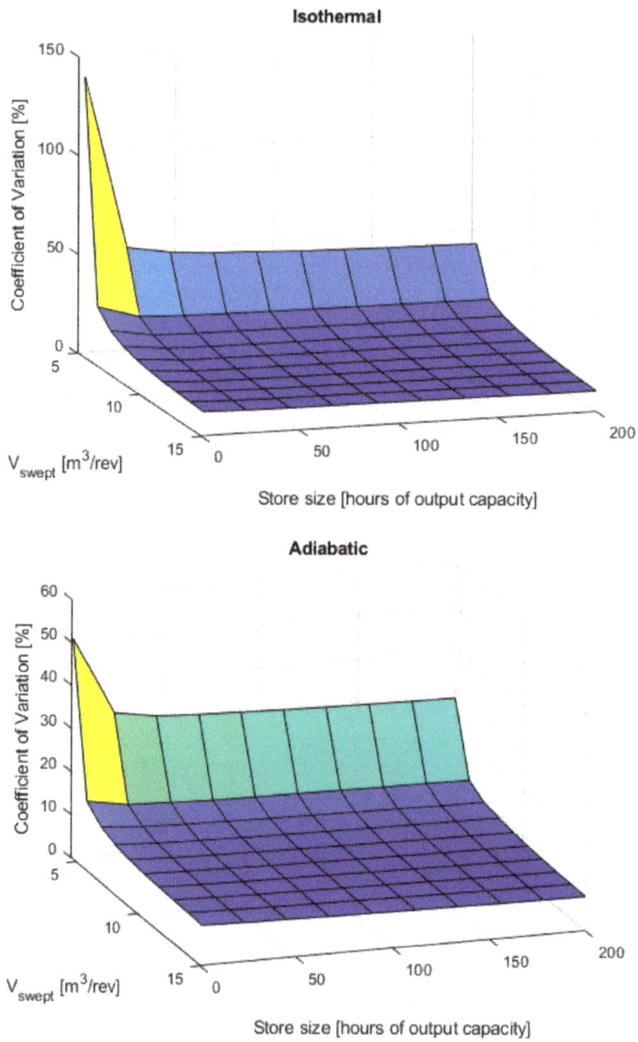

Figure 17. Results of data validation for isothermal (**above**) and adiabatic (**below**) systems, presented as the coefficient of variation of the ROI for each iteration.

However, given the necessity for large-scale storage in future energy grids, regardless of its relative profitability, it is useful to compare like-for-like. Therefore, two other wind–storage systems have been simulated, using the same initial logic: a wind turbine with co-located Li-ion battery storage, and a wind turbine connected to both a compressed air energy store (CAES) and the grid.

The efficiencies and costs assumed for these systems are summarised in Table 5.

Table 5. Efficiencies and costs of the other wind–storage systems. Costs estimated from [49,50].

Name of Parameter	Li-Ion Store	CAES Store	Units
Efficiency	0.9	0.65	-
Cost (capacity)	150	20	£/kWh
Scale (capacity)	8–1600	8–1600	MWh
Cost (power)	1500	300	£/kW

The efficiencies were applied in a similar approach as the *Wi-DACIS* simulation. During costing, the store was considered larger by a factor $\sqrt{\eta}$ (to apply the same efficiency loss to the beginning and the end of the process), and during revenue calculations, income was reduced by the same factor.

The wind turbine was considered to be acting on the free market in both cases, and the same algorithm (based on [36]) was used to decide whether to sell or store energy, where possible. The power costs provided in Table 5 are the total costs for charging and discharging power. As with the *Wi-DACIS* system, the storage capacity was calculated in hours, from 1 to 200. Figure 18 shows the representative ROI curves for the two systems.

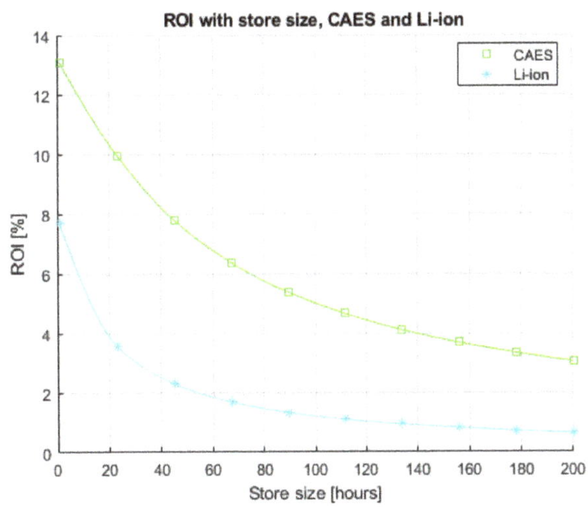

Figure 18. The relationship between one-year return on investment and storage capacity for the Li-ion and CAES simulations.

For a set storage capacity, the Li-ion system achieves higher revenue in all cases (due, obviously, to the only simulated difference in the income of the two systems being their efficiencies). However, the CAES system achieves significantly higher ROI values. The maximum ROI values for CAES were between 9.7% and 14.3% and were between 4.9% and 8.4% for Li-ion. These all occur during the lowest possible store size. Comparing Figures 14 and 18, it appears as though the *Wi-DACIS* system is extremely competitive with alternative storage systems, especially at higher storage capacities. Section 4.6 discusses an interesting quirk of *Wi-DACIS* that may give it further viability, from a whole grid perspective.

4.6. Capture Value

The capture value (CV) of a wind turbine is a measure of how effectively the turbine can achieve the average electricity value over a given time period. Wind power generally has a negative correlation with electricity price—during times of high wind, the wind fleet of a grid is generating more electricity, increasing supply, and thereby reducing the electricity price. Capture value is calculated as:

$$CV = \frac{\sum_{i=1}^{n} output_i \times price_i}{\overline{price} \times \sum_{i=1}^{n} output_i}, \tag{18}$$

where *output* and *price* are the wind electricity output and electricity price at time *i*, and $\overline{price}$ is the average electricity price over the time period [51].

The average capture values are expected to decrease in the future, due to significantly increased wind generation capacity. This is a concern for wind farm operators, due to the uncertainty it creates [51,52]. Less certainty in wind farm output may result in reduced investment in wind farms in the future.

Any system that stores exergy prior to sending electricity to the grid has the potential to lessen the negative impact of this capture value, both for itself (by only selling during times of high electricity price) and for the whole grid (by reducing the wind supply during times of high wind). *Wi-DACIS* is an example of such a system, but so is any co-located wind–storage system. What sets *Wi-DACIS* apart is that it also draws electricity from the grid when the wind turbine is working, via its grid-powered compressor. Because the instantaneous power of this compressor is directly proportional to the instantaneous wind resource, *Wi-DACIS* would serve to draw electricity mainly at times of high wind. When this coincides with times of low electricity value (which will occur increasingly more frequently), *Wi-DACIS* will store this energy. Through this mechanism, *Wi-DACIS* further lessens the capture value problem. This improved capture value could reduce uncertainty in future wind farm investments. However, any system intending to make a significant impact on the capture value of a whole grid would need to account for a significant proportion of the wind generation capacity of that grid.

5. Summary

The wind-driven GIES system described in this paper has the potential to offer significant generation flexibility to wind power generation. Further costing work must be done before the economic case is fully made. By including some electrically driven compression prior to supplying air to the wind-driven compressor, a solution is found to the problem that large intake swept volumes would otherwise be needed. This allows direct drive machinery to replace the expensive gearbox system in a conventional turbine.

The use of underwater compressed air energy storage in the form of flexible energy bags has natural synergy with floating offshore wind turbines, which may be situated over extremely deep water. It has been shown in a previous paper that these energy bags have an optimum diameter, which was suggested to be approximately 20 m. Any energy store utilising these bags should therefore employ multiple optimally sized bags.

For multiple configurations of the system, simulations were run for both an adiabatic and an isothermal wind-driven compressor. For a set wind turbine and upper pressure, as the allowable volumetric swept intake of the main stage compressor increases:

- The pressure ratio of the grid-powered compressor reduces and the pressure ratio of the wind-driven compressor increases;
- The power ratio between the wind-driven and grid-powered compressors increases; and
- The one-year return on investment of the system increases.

It is therefore desirable to accommodate the maximum possible swept volume.

It was also found that the adiabatic main stage compressor achieved higher ROI values than the isothermal configuration. This is partly due to the relatively cheap thermal energy storage used in the adiabatic case. The adiabatic case also results in a more favourable ratio of wind-driven to grid-powered compressor powers, resulting in a configuration closer to a pure generation-integrated energy storage system.

There is an optimum value of store size for ROI with the *Wi-DACIS* system, which is generally between 10 and 20 h of rated output. With proper design and sizing, ROI values in the region of 10–15% are certainly possible. A Monte Carlo simulation validated the assumed costs of the system by showing low variability (in the order of 10%) at higher values of V_{swept}, where these ROI values were found.

With the wind data used in this paper, a conventional wind turbine achieves higher ROI values than the *Wi-DACIS* system. However, under the same assumptions, the other wind–storage systems simulated also achieved lower ROI values than the conventional turbine, and the optimum values for these systems were found at the minimum storage capacity, which most closely resembles a standalone wind turbine. The *Wi-DACIS* system performed as well as or better than the alternative storage solutions, especially at higher storage capacities.

The addition of a medium-pressure store may serve to add flexibility to the system, reducing the expenditure by allowing *Wi-DACIS* to draw energy from the grid at times of lower electricity value. This additional store should be analysed in future work.

Further analysis should also be performed on the relative costs of the system elements, as these will dictate proper sizing. The effect that the power ratio between the two compressors has on the overall efficiency of the system is of interest, as an iteration that is closer to standalone storage than GIES may result in a less efficient system. The benefit of overrating of the expander, allowing for quicker discharge of the high-pressure store, should also be investigated.

In this paper, the efficiency of the wind-driven compressor was given in relative terms, with the assumption being that it would be as efficient as a traditional wind turbine transmission system. This should be verified as the development of the adiabatic compressor in [6] continues.

With a sufficient grid presence, *Wi-DACIS* may also act as a capture value balancer, due to its dual mechanisms of storing energy during times of low electricity price (which will increasingly frequently occur during times of high wind) and drawing electricity from the grid during times of high wind availability, thereby increasing demand. With a significant market share of the wind generation capacity of a grid, *Wi-DACIS* could greatly reduce the uncertainty surrounding capture value in wind farms. Further work should be done to determine the value of the *Wi-DACIS* and similar wind-driven systems in an expected future energy network, as the negative correlation between electricity price and wind availability becomes more pronounced in the coming years. This negative correlation may heavily impact conventional wind farms that lack co-located energy storage, at which point systems such as *Wi-DACIS* may have an extremely promising economic case.

Author Contributions: Conceptualization, L.S.-S., S.D.G. and D.G.; methodology, L.S.-S. and S.D.G.; software, L.S.-S. and B.C.; validation, L.S.-S. and B.C.; formal analysis, L.S.-S.; investigation, L.S.-S.; writing—original draft preparation, L.S.-S.; writing—review and editing, L.S.-S., S.D.G., D.G. and J.P.R.; visualization, L.S.-S.; supervision, S.D.G., D.G. and J.P.R.; project administration, S.D.G.; funding acquisition, S.D.G. All authors have read and agreed to the published version of the manuscript.

Funding: The authors would like to thank the United Kingdom's Engineering and Physical Sciences Research Council (EPSRC) for funding this work through the following research grants: 'Multi-scale Analysis for Facilities for Energy Storage' (Manifest: EP/N032888/1) and 'Generation Integrated Energy Storage - A Paradigm Shift' (EP/P023320/1).

Data Availability Statement: Restrictions apply to the availability of wind speed data. Data was obtained from NCAS British Atmospheric Data Centre and the post-processed data are available from the authors with permission of NCAS British Atmospheric Data Centre. Publicly available day-ahead electricity price data were analysed in this study. This data can be found here: https://transparency.entsoe.eu/dashboard/show. The outcome data presented in this study are available upon reasonable request from the corresponding author.

Conflicts of Interest: The authors declare no conflict of interest.

Nomenclature

C	cost [GBP]
D	diameter [m]
E	energy [J]
f	frequency [1/s]
F	force [N]
L	length [m]
m	mass [kg]
p	pressure [bar]
P	power [W]
r	ratio of pressures
T	temperature [K]
u	(wind) speed [m/s]
V	volume [m^3]
W	work [J]
z	altitude [m]
γ	ratio of specific heats
η	efficiency
κ	Von Kármán constant
ω	rotational speed [rad/s]

Subscripts and Superscripts

0	roughness length
1	inlet stage 1 (pressure, volume)
2	outlet stage 1/inlet stage 2 (pressure, volume)
3	outlet stage 2 (pressure, volume)
ADIA	adiabatic
G	grid (pressure ratio)
Heat	heat (energy)
ISO	isothermal
opt	optimal (volume)
quartz	quartz (mass)
ref	reference (temperature)
steel	steel (mass)
store	thermal store (temperature, volume)
swept	swept (volume) per rev
T	tension (force)
TIP	wind turbine blade tip (speed)
turbine	turbine (power)
W	wind (pressure ratio)
*	friction velocity

Abbreviations

AB	anchoring ballast
A-CAES	adiabatic compressed air energy storage
CAES	compressed air energy storage
CfD	contracts for difference
COMG	compressor (grid-driven)
COMW	compressor (wind-driven)
CV	capture value
D-CAES	diabatic compressed air energy storage
EXP	expander
GIES	generation-integrated energy storage
HPEB	high-pressure energy bag
HXU	heat exchanger unit
I-CAES	isothermal compressed air energy storage

MIDAS	Met Office Integrated Data Archive System
ROI	return on investment
SoC	state of charge
SU	canvas surface
TC	tension cables
TES	thermal energy storage
TSR	tip-speed ratio
UWCAES	underwater compressed air energy storage
Wi-DACIS	wind-driven air compression and isobaric storage

References

1. Cárdenas, B.; Swinfen-Styles, L.; Rouse, J.; Garvey, S.D. Short-, Medium-, and Long-Duration Energy Storage in a 100% Renewable Electricity Grid: A UK Case Study. *Energies* **2021**, *14*, 8524. [CrossRef]
2. Garvey, S.D.; Eames, P.C.; Wang, J.H.; Pimm, A.J.; Waterson, M.; MacKay, R.S.; Giulietti, M.; Flatley, L.C.; Thomson, M.; Barton, J.; et al. On Generation-Integrated Energy Storage. *Energy Policy* **2015**, *86*, 544–551. [CrossRef]
3. Dunn, R. A global review of concentrated solar power storage. In Proceedings of the Solar2010, the 48th AuSES Annual Conference, 1–3 December 2010; p. 10.
4. Bergan, P.G.; Greiner, C.J. A New Type of Large Scale Thermal Energy Storage. *Energy Proc.* **2014**, *58*, 152–159. [CrossRef]
5. Ingersoll, E.; Marcus, D.R. Wind Turbine System. U.S. Patent Application 11/804,704, 28 February 2008.
6. Garvey, S.D.; Pimm, A.J.; Buck, J.A.; Woolhead, S.; Liew, K.W.; Kantharaj, B.; Garvey, J.E.; Brewster, B.D. Analysis of a Wind Turbine Power Transmission System with Intrinsic Energy Storage Capability. *Wind. Eng.* **2015**, *39*, 149–173. [CrossRef]
7. Salter, S.H.; Rea, M. Hydraulics for Wind. In Proceedings of the European Wind Energy Conference, Hamburg, Germany, 22–26 October 198; pp. 534–541.
8. Lee, J.L. On-Demand Generation of Electricity from Stored Wind Energy. U.S. Patent 8,664,793, 4 March 2014.
9. Ren, Z.; Wang, H.; Chen, G.; Li, X.; Esfarjani, K. Metallic Composite Phase-Change Materials and Methods of Using the Same. U.S. Patent US2014109895A1, 24 April 2014.
10. Denholm, P.; King, J.C.; Kutcher, C.F.; Wilson, P.P.H. Decarbonizing the Electric Sector: Combining Renewable and Nuclear Energy Using Thermal Storage. *Energy Policy* **2012**, *44*, 301–311. [CrossRef]
11. Swinfen-Styles, L.; Garvey, S.D.; Giddings, D. Combining Wind-Driven Air Compression with Underwater Compressed Air Energy Storage. In Proceedings of the Offshore Energy and Storage Summit (OSES), Brest, France, 10–12 July 2019. [CrossRef]
12. Wang, J.; Lu, K.; Ma, L.; Wang, J.; Dooner, M.; Miao, S.; Li, J.; Wang, D. Overview of Compressed Air Energy Storage and Technology Development. *Energies* **2017**, *10*, 991. [CrossRef]
13. Rogers, A.; Henderson, A.; Wang, X.; Negnevitsky, M. Compressed Air Energy Storage: Thermodynamic and Economic Review. In Proceedings of the IEEE PES General Meeting | Conference & Exposition, National Harbor, MD, USA, 27–31 July 2014. [CrossRef]
14. Safaei, H.; Aziz, M.J. Thermodynamic Analysis of Three Compressed Air Energy Storage Systems: Conventional, Adiabatic, and Hydrogen-Fueled. *Energies* **2017**, *10*, 1020. [CrossRef]
15. Dufo-López, R.; Bernal-Agustín, J.L.; Domínguez-Navarro, J.A. Generation Management Using Batteries in Wind Farms: Economical and Technical Analysis for Spain. *Energy Policy* **2009**, *37*, 126–139. [CrossRef]
16. Cavallo, A. Controllable and Affordable Utility-Scale Electricity from Intermittent Wind Resources and Compressed Air Energy Storage (CAES). *Energy* **2007**, *32*, 120–127. [CrossRef]
17. Bennett, J.A.; Simpson, J.G.; Qin, C.; Fittro, R.; Koenig, G.M.; Clarens, A.F.; Loth, E. Techno-Economic Analysis of Offshore Isothermal Compressed Air Energy Storage in Saline Aquifers Co-Located with Wind Power. *Appl. Energy* **2021**, *303*, 117587. [CrossRef]
18. Matsuo, T.; Okazaki, T. A Basic Theory of Induction Heating for a Wind-Powered Thermal Energy System. *IEEE Trans. Magn.* **2017**, *53*, 1–5. [CrossRef]
19. Li, P.Y.; Loth, E.; Simon, T.W.; Van de Ven, J.D.; Crane, S.E. Compressed air energy storage for offshore wind turbines. In Proceedings of the International Fluid Power Exhibition (IFPE), Las Vegas, NV, USA, 23–25 March 2011.
20. Ribrant, J.; Bertling, L. Survey of Failures in Wind Power Systems with Focus on Swedish Wind Power Plants during 1997–2005. In Proceedings of the IEEE Power Engineering Society General Meeting, Tampa, FL, USA, 24–28 June 2007. [CrossRef]
21. Ragheb, A.; Ragheb, M. Wind Turbine Gearbox Technologies. In Proceedings of the 1st International Nuclear & Renewable Energy Conference (INREC), Amman, Jordan, 21–24 March 2010. [CrossRef]
22. Daneshi-Far, Z.; Capolino, G.A.; Henao, H. Review of Failures and Condition Monitoring in Wind Turbine Generators. In Proceedings of the XIX International Conference on Electrical Machines—ICEM, Rome, Italy, 6–8 September 2010. [CrossRef]
23. Bauer, L.; Matysik, S. Wind-Turbine-Models. Available online: https://en.wind-turbine-models.com/turbines/1419-mhi-vestas-offshore-v164-8.0-mw (accessed on 10 February 2022).
24. Smallbone, A.; Jülch, V.; Wardle, R.; Roskilly, A.P. Levelised Cost of Storage for Pumped Heat Energy Storage in Comparison with Other Energy Storage Technologies. *Energy Convers. Manag.* **2017**, *152*, 221–228. [CrossRef]
25. Kim, Y.M.; Shin, D.G.; Favrat, D. Operating Characteristics of Constant-Pressure Compressed Air Energy Storage (CAES) System Combined with Pumped Hydro Storage Based on Energy and Exergy Analysis. *Energy* **2011**, *36*, 6220–6233. [CrossRef]

26. Pimm, A.J.; Garvey, S.D.; Drew, R.J. Shape and Cost Analysis of Pressurized Fabric Structures for Subsea Compressed Air Energy Storage. *Proc. Inst. Mech. Eng. Part C J. Mech. Eng. Sci.* **2011**, *225*, 1027–1043. [CrossRef]
27. Pimm, A.J.; Garvey, S.D.; de Jong, M. Design and Testing of Energy Bags for Underwater Compressed Air Energy Storage. *Energy* **2014**, *66*, 496–508. [CrossRef]
28. Moradi, J.; Shahinzadeh, H.; Khandan, A.; Moazzami, M. A Profitability Investigation into the Collaborative Operation of Wind and Underwater Compressed Air Energy Storage Units in the Spot Market. *Energy* **2017**, *141*, 1779–1794. [CrossRef]
29. Astolfi, M.; Guandalini, G.; Belloli, M.; Hirn, A.; Silva, P.; Campanari, S. Preliminary Design and Performance Assessment of an Underwater CAES System (UW-CAES) for Wind Power Balancing. In Proceedings of the Volume 3: Coal, Biomass, Hydrogen, and Alternative Fuels; Cycle Innovations; Electric Power; Industrial and Cogeneration, Organic Rankine Cycle Power Systems, Phoenix, AZ, USA, 17–21 June 2019. [CrossRef]
30. Met Office (2012): Met Office Integrated Data Archive System (MIDAS) Land and Marine Surface Stations Data (1853-Current; 2015 Used); NCAS British Atmospheric Data Centre. Available online: https://catalogue.ceda.ac.uk/uuid/dbd451271eb04662beade68da43546e1 (accessed on 20 March 2018).
31. Holmes, J.D.; Paton, C.; Kerwin, R. *Wind Loading of Structures*; CRC Press: Boca Raton, FL, USA, 2007. [CrossRef]
32. Wieringa, J. Updating the Davenport Roughness Classification. *J. Wind. Eng. Ind. Aerodyn.* **1992**, *41*, 357–368. [CrossRef]
33. Google. Google Maps. Available online: https://www.google.com/maps (accessed on 20 March 2018).
34. Databayou. World Ocean Depth Map. 2021. Available online: https://databayou.com/ocean/depth.html (accessed on 25 November 2021).
35. ENTSO-E Transparency Platform. UK Day-Ahead Electricity Prices. Online Dataset. Available online: https://transparency.entsoe.eu/dashboard/show (accessed on 20 June 2020).
36. Connolly, D.; Lund, H.; Finn, P.; Mathiesen, B.V.; Leahy, M. Practical Operation Strategies for Pumped Hydroelectric Energy Storage (PHES) Utilising Electricity Price Arbitrage. *Energy Policy* **2011**, *39*, 4189–4196. [CrossRef]
37. Pelay, U.; Luo, L.; Fan, Y.; Stitou, D.; Rood, M. Thermal Energy Storage Systems for Concentrated Solar Power Plants. *Renew. Sustain. Energy Rev.* **2017**, *79*, 82–100. [CrossRef]
38. Safaei, H.; Keith, D.W.; Hugo, R.J. Compressed Air Energy Storage (CAES) with Compressors Distributed at Heat Loads to Enable Waste Heat Utilization. *Appl. Energy* **2013**, *103*, 165–179. [CrossRef]
39. Arsalis, A. Thermoeconomic Modeling and Parametric Study of Hybrid SOFC–Gas Turbine–Steam Turbine Power Plants Ranging from 1.5 to 10MWe. *J. Power Sources* **2008**, *181*, 313–326. [CrossRef]
40. Akrami, E.; Chitsaz, A.; Nami, H.; Mahmoudi, S.M.S. Energetic and Exergoeconomic Assessment of a Multi-Generation Energy System Based on Indirect Use of Geothermal Energy. *Energy* **2017**, *124*, 625–639. [CrossRef]
41. Luyben, W.L. Capital Cost of Compressors for Conceptual Design. *Chem. Eng. Processing Process Intensif.* **2018**, *126*, 206–209. [CrossRef]
42. Mashimo, T.; Ariga, I.; Sakai, T.; Watanabe, I. On the Performance Prediction of a Centrifugal Compressor Scaled Up. In Proceedings of the Volume 1: Turbomachinery, London, UK, 18–22 April 1982. [CrossRef]
43. Chen, H. Size and Reynolds Number Effects on Compressor Performance and Scaling. In Proceedings of the Volume 2E: Turbomachinery, Virtual, Online, 21–25 September 2020. [CrossRef]
44. Ebadollahi, M.; Rostamzadeh, H.; Ghaebi, H.; Amidpour, M. Exergoeconomic Analysis and Optimization of Innovative Cascade Bi-Evaporator Electricity/Cooling Cycles with Two Adjustable Cooling Temperatures. *Appl. Therm. Eng.* **2019**, *152*, 890–906. [CrossRef]
45. Aali, A.; Pourmahmoud, N.; Zare, V. Exergoeconomic Analysis and Multi-Objective Optimization of a Novel Combined Flash-Binary Cycle for Sabalan Geothermal Power Plant in Iran. *Energy Convers. Manag.* **2017**, *143*, 377–390. [CrossRef]
46. Weiner, D.; Shnaid, I. Second-Law Analysis of a Compressed Air Energy Storage (CAES) System. In Proceedings of the Volume 4: Heat Transfer; Electric Power; Industrial and Cogeneration, Cologne, Germany, 1–4 June 1992. [CrossRef]
47. Courtois, N.; Najafiyazdi, M.; Lotfalian, R.; Boudreault, R.; Picard, M. Analytical Expression for the Evaluation of Multi-Stage Adiabatic-Compressed Air Energy Storage (A-CAES) Systems Cycle Efficiency. *Appl. Energy* **2021**, *288*, 116592. [CrossRef]
48. Houssainy, S.; Janbozorgi, M.; Kavehpour, P. Thermodynamic Performance and Cost Optimization of a Novel Hybrid Thermal-Compressed Air Energy Storage System Design. *J. Energy Storage* **2018**, *18*, 206–217. [CrossRef]
49. Steward, D.; Saur, G.; Penev, M.; Ramsden, T. *Lifecycle Cost Analysis of Hydrogen Versus Other Technologies for Electrical Energy Storage*; Technical Report; NREL: Golden, CO, USA, 2009.
50. Argyrou, M.C.; Christodoulides, P.; Kalogirou, S.A. Energy Storage for Electricity Generation and Related Processes: Technologies Appraisal and Grid Scale Applications. *Renew. Sustain. Energy Rev.* **2018**, *94*, 804–821. [CrossRef]
51. Jansen, M.; Staffell, I.; Kitzing, L.; Quoilin, S.; Wiggelinkhuizen, E.; Bulder, B.; Riepin, I.; Müsgens, F. Offshore Wind Competitiveness in Mature Markets without Subsidy. *Nat. Energy* **2020**, *5*, 614–622. [CrossRef]
52. Impacts of Inter-Annual Wind and Solar Variations on the European Power System. *Joule* **2018**, *2*, 2076–2090. [CrossRef]

energies

Article

Solidified-Air Energy Storage: Conceptualization and Thermodynamic Analysis

Sandro Hiller [1], Christian Hartmann [1], Babette Hebenstreit [1,2] and Stefan Arzbacher [1,*]

[1] Research Centre Energy, Vorarlberg University of Applied Sciences, 6850 Dornbirn, Austria; sandro.hiller@fhv.at (S.H.); chartmann@gmx.ch (C.H.); babette.hebenstreit@fhv.at (B.H.)
[2] Division of Energy Science, Department of Engineering Sciences and Mathematics, Luleå University of Technology, 971 87 Luleå, Sweden
* Correspondence: stefan.arzbacher@fhv.at

Abstract: Grid-scale electrical energy storage (EES) is a key component in cost-effective transition scenarios to renewable energy sources. The requirement of scalability favors EES approaches such as pumped-storage hydroelectricity (PSH) or compressed-air energy storage (CAES), which utilize the cheap and abundant storage materials water and air, respectively. To overcome the site restriction and low volumetric energy densities attributed to PSH and CAES, liquid-air energy storage (LAES) has been devised; however, it suffers from a rather small round-trip efficiency (RTE) and challenging storage conditions. Aiming to overcome these drawbacks, a novel system for EES is developed using solidified air (i.e., clathrate hydrate of air) as the storable phase of air. A reference plant for solidified-air energy storage (SAES) is conceptualized and modeled thermodynamically using the software CoolProp for water and air as well as empirical data and first-order approximations for the solidified air (SA). The reference plant exhibits a RTE of 52% and a volumetric storage density of 47 kWh per m^3 of SA. While this energy density relates to only one half of that in LAES plants, the modeled RTE of SAES is comparable already. Since improved thermal management and the use of thermodynamic promoters can further increase the RTEs in SAES, the technical potential of SAES is in place already. Yet, for a successful implementation of the concept—in addition to economic aspects—questions regarding the stability of SA must be first clarified and challenges related to the processing of SA resolved.

Keywords: electrical energy storage; thermal energy storage; air energy storage; air clathrate hydrate

Citation: Hiller, S.; Hartmann, C.; Hebenstreit, B.; Arzbacher, S. Solidified-Air Energy Storage: Conceptualization and Thermodynamic Analysis. *Energies* **2022**, *15*, 2159. https://doi.org/10.3390/en15062159

Academic Editors: Alon Kuperman and Alessandro Lampasi

Received: 22 February 2022
Accepted: 14 March 2022
Published: 16 March 2022

1. Introduction

With the increasing expansion of renewable energy sources for power generation, the integration of their volatile production capacity into the electrical grid infrastructure is becoming an ever greater challenge. To address this challenge, a variety of flexibility measures and electrical energy storage (EES) have been discussed in the literature. See, e.g., the review of Zehrran et al. [1], the discussions about storage demands by Sterner and Stadler [2], as well as recent simulation results from Child et al. [3] and Schill et al. [4]. In general, for low-to-medium (up to 70%) penetration of renewable energy into the electricity supply mix, flexibility measures such as demand-side management and flexible power plants appear most cost effective. However, in a 100% renewable energy system, EES with storage periods of some hours up to a day play a major role.

In this respect, a plethora of solutions for EES have been developed, deployed, or are at the demonstration stage (see the reviews by Chen et al. [5], Gür [6], and Hameer and van Niekerk [7]). All of them aim to meet one or more of the following requirements: high (volumetric) storage density, low storage losses, fast charge and discharge rates, cheap to build, cheap to operate, and cheap to scale. While there are a few solutions that can meet several of these requirements, none of them perform well in all of these aspects. The requirement of scalability is particularly hard to meet, as it demands the utilization of materials which are both cheap and sufficiently abundant on Earth. Next to an increasing number of materials which are discussed as solutions, particularly for thermo- and electrochemical energy storage systems (see the comprehensive review by Gür [6]), these two criteria are also prominently met by (1) water and (2) air.

(1) Water is predominantly used as an easy-to-transport carrier of potential energy in pumped-storage hydroelectricity (PSH) plants which exhibit superior cost-effectiveness and the highest round-trip efficiency (RTE), between roughly 70% and 80% [8]. As the technology for PSH is now over a hundred years old, mature, and well-tested, PSH provides the largest capacity for grid-scale energy storage to date [8]. However, the expansion of PSH capacity is becoming ever more challenging as the number of sites which are technically and economically feasible for PSH is reducing, as Rehman et al. [8], Kucukali et al. [9], and Lu et al. [10] report independently.

(2) Air, on the other hand, is a ubiquitous gas which serves as the storage medium for internal energy in compressed air energy storage (CAES) systems (see the reviews by Budt et al. [11] and Chen et al. [12]). Excess electricity is used to compress the air and raise its internal energy which can be converted back to electricity by expanding the air in a gas turbine. Due to the low volumetric energy density of CAES (0.5 kWh to 25 kWh per m^3 of compressed air storage [11]), storage is done most cost-effectively underground in large salt caverns. This implies that here, too, suitable geological structures are required for large scale EES, hence limiting the deployment of the technology to a small number of sites, such as the Huntdorf and the McIntosh plant, both described in detail by Budt et al. [11] and Chen et al. [12].

One way to get around the geographic dependence while still benefiting from the advantages of the ubiquity of air is liquid-air energy storage (LAES), a cryogenic energy storage (CES) technology, which has recently been implemented for the first time in a pilot-scale demonstration plant [13]. The idea is to simply liquefy the air and store it in the liquid instead of the compressed state. The 600 volumes of air at standard temperature and pressure (STP) which are contained within one volume of liquid can eventually be used to drive a gas turbine when the liquid is evaporated using low temperature heat. As the liquid air can be readily stored in well-insulated cryogenic containers which minimize the boil-off losses, LAES is free from geological constraints. Moreover, almost all processes and components in an LAES plant are well-known from cryotechnology (e.g., the liquefaction of air), making the design and operation of the plant calculable and low-risk for the operators. Yet, the liquefaction of air is a heat-intensive process which results in only mediocre RTEs of less than about 50%, even when the heat is stored and recycled. This follows from independent thermodynamic analyses of idealized and simplified plants [13–16]. In practice, though, RTEs can also be significantly lower. For instance, Ameel et al. [14] and Borri et al. [17] report RTEs of only 20% and 10%, respectively. This is in part because of the extremely low temperature of the liquid air (<77 K) which makes the storage and recovery of cold energy prone to large losses, particularly for extended periods of storage (e.g., many hours to several days).

To overcome the drawbacks of site restrictions and low round-trip efficiencies attributed to CAES and LAES, respectively, this work introduces a novel concept for the large-scale storage of electrical energy using solidified air as the storage medium. Solidified air, i.e., the clathrate hydrate of air, or air hydrate for short (see Koh et al. [18] for a general overview on clathrate hydrates and Miller [19], Pauer et al. [20], and Shoji et al. [21] for details on naturally occurring air hydrate), is an inclusion compound of water and air in which around 170 volumes of air at STP can be stored in one volume of the solid. This makes up a molecular storage tank for compressed air with an achievable air-based volumetric storage density of about one-quarter that of LAES systems. While the utilization of hydrogen-bonded water cavities as the high-pressure containment for air allows for cheap and scalable storage, the disadvantage of a slightly lower storage density could be set off by milder storage conditions. Together this could yield a CES system which alleviates the heat storage-loss problem of LAES but retains and extends its advantages.

Although clathrate hydrates of natural gas, also termed solidified natural gas (SNG) by Linga and coworkers [22–24], are widely discussed as alternative solutions for the storage of chemical energy in the form of natural gas, the usage of air hydrates or solidified air (here used in analogy to SNG) in a CES system has not been considered so far. Based on equilibrium thermodynamics and a well justified assumption, the utilization of solidified air (SA) for EES seems feasible on first thought. Yet, due to the complexity of the phase diagram and a few unknowns with respect to the kinetics and thermodynamics of SA, careful evaluation of the new approach to CES is needed.

Therefore, in this paper, the concept for a SAES system is laid out for the first time, examined, and compared with the state of the art of LAES using a thermodynamic analysis. While the main goal of our analysis is to clarify whether the concept of SAES is technically feasible at all, it is additionally used to provide an initial recommendation as to whether the concept is worth developing further and—if so—where future research must be directed at.

2. Conception

The principle idea behind SAES is to use excess electricity, water, and air, to form SA and store it in heaps or skips on or in ground, respectively. In periods of high electricity demand, the SA is dissociated to recover the water and air, which, similar to LAES, is used to drive a turbine and generate electricity.

The term SA refers to air hydrate [19], an inclusion compound in which the molecules of the air (predominantly N_2 and O_2) occupy cavities formed by a hydrogen-bonded network of water molecules (see Figure 1a for an illustration). The composition of SA is nonstoiciometric, i.e., not all cavities of the structure need to be occupied. However, the maximum occupancy is 5.75 mol of water per mole of air. That relates to approximately 170 volumes of air at STP per volume of hydrate, which is about one-quarter the air storage capacity of liquefied air (600 vol/vol), but relates to approximately 17 MPa, a pressure much larger than that applied in CAES (typically <10 MPa [11,12]).

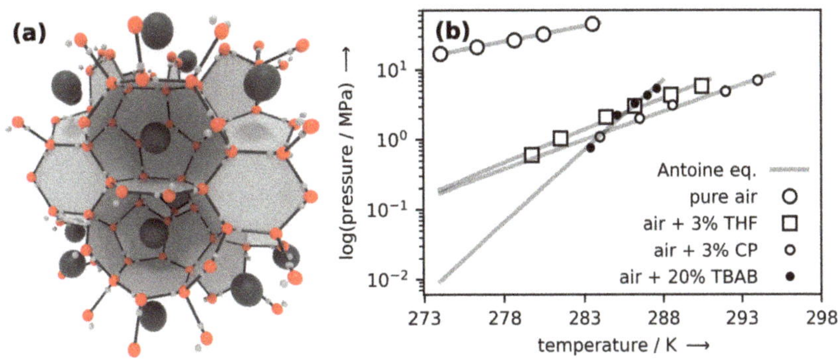

Figure 1. (**a**) Artistic illustration of the molecular structure of SA, i.e., air hydrate. The large grey balls denote air molecules (N_2 and O_2), the red balls denote the oxygen atoms of water molecules which form a hydrogen-bonded network of cavities around the air molecules. (**b**) Hydrate–liquid–vapor equilibrium data for pure air hydrate (data from Mohammadi et al. [25]) and for hydrates of air with the growth-promoting substances Tetrahydrofuran (THF), Cyclopentane (CP), Tetrabutylammonium bromide (TBAB) (data from Yang et al. [26]). The solid grey lines fit the data and denote the phase boundary above which the hydrate is thermodynamically stable (also see Lipenkov and Istomin [27] and Mohammadi and Richon [28] for a discussion of the phase diagram).

Due to its water-rich composition, many of the SA's properties, such as its appearance or density (approx. 1 g/cm^3 [29]), resemble those of ice. Likewise, the heat of their dissociation is comparable to the heat of fusion of ice (334 kJ/kg [30]). While due to their large heat of fusion, both ice and gas hydrates can be utilized for thermal energy storage (see the reviews by Saito [31] for ice and Wang et al. [32] for gas hydrate thermal energy storage), gas hydrates can additionally be used to store electrical energy if their ability to act as molecular storage vessel for gases is exploited. SA can be formed by bringing together liquid water and air at conditions above the hydrate–liquid–vapor equilibrium curve (the solid gray lines in Figure 1b), which means at moderately low temperatures but high pressures. It can be dissociated by exposing it to conditions below that curve, that is, by superheating or underpressurizing with respect to stability conditions. The equilibrium curve and its variation with temperature is very important in the context of this paper, because, by varying the temperature, the SA formation and dissociation pressure can be modified drastically. For instance, one could form SA at low temperature/pressure and dissociate it at higher temperature/pressure to minimize the work of compression while maximizing that of turbine expansion. Furthermore, as shown in Figure 1b, the addition of small amounts of stability-promoting substances such as THF, CP, and TBAB, can have a dramatic effect on the equilibrium curve and thus on both formation and dissociation conditions [26]. Although promoters and their effects must be addressed because of their potential to improve the RTE of a future SAES plant, for ease of understanding, stability-promoting substances are left out in our analysis of a very first reference case.

A key aspect of our concept for SAES is the storage of the synthesized SA at the most cost-efficient conditions, i.e., preferably at ambient conditions. Yet, as can be seen from the phase diagram in Figure 1b, SA is far from thermodynamic stability at ambient conditions. Therefore, storage in the stable form is unsuitable for our purposes. However, a kinetic anomaly termed "self-preservation" (SP) [33], which results in very low rates of dissociation at conditions far outside the stability region, is likely to allow for a medium-term (days to weeks) storage of SA at ambient pressure and temperatures just below 273 K with minute amounts of "boil-off" losses. While SP has been reported numerous times for hydrates of CH_4 [34,35] and CO_2 [36,37], little is known for hydrates of air, probably because it seemed not relevant enough to this date. However, several reports [38–40] of SP for O_2 and N_2—the major constituents of air—render the occurrence of SP in SA plausible. In case SP in SA is similarly effective as in hydrates of CH_4, SA could be stored at subzero temperatures and ambient pressure for months with losses of only a few percent [41]. Additionally, due to the strongly endothermic transition, every dissociation of SA cools down the environment. Therefore, when the stored SA is thermally well insulated, it does not have to be actively cooled.

For the rest of this work we build on the assumption of pronounced SP with the full knowledge that its refutation will render the whole concept purely theoretical. Even then, though, this work can still be regarded a long-missing potential application needed to stimulate research on SP in hydrates of air.

3. Materials and Methods

In this section, a theoretical process for SAES is outlined before a thermodynamic model for the steady-state operation of a reference plant is developed. Besides, performance parameters are defined to allow for comparisons with competing EES technologies.

3.1. Plant Description

The SAES plant is conceptualized in the process scheme in Figure 2.

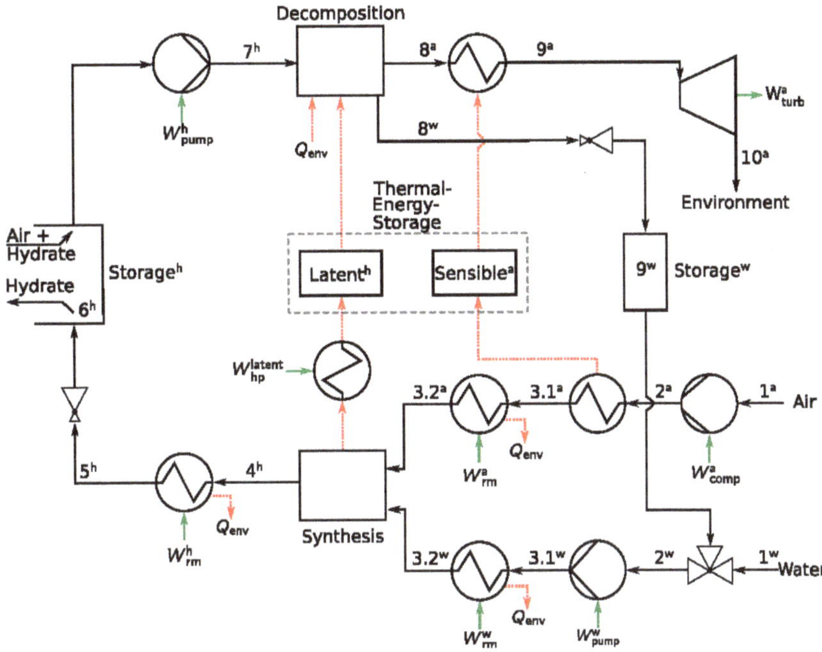

Figure 2. Simplified process scheme of a reference plant for SAES. Solid black lines denote pathways for media transport. Dashed red lines are used to indicate heat transport. Short green arrows are used to label energy flows. In the state labeling, consecutive numbers are used according to a "energy charging → energy storage → energy discharging" order. Superscripts "a", "w", and "h" denote air, water and hydrate (i.e., SA), respectively.

Similar to CAES and LAES, electrical energy is used to compress air (1^a to 2^a). However, instead of storing the pressurized air directly, like in CAES, or liquefying the air, as in LAES, SA is formed from the pressurized air and water (3 to 4). Before entering the reactor, air and water are brought to formation conditions (1 to 3.2). After its formation in the reactor, the SA is extracted and pelletized (as described by, e.g., Rehder et al. [42]) for ease of transport. Note that this step is neither shown in the process scheme nor taken into account in the calculation. Other options for SA processing, such as in a slurry or via "dry water" [43], complicate the process scheme and are therefore not considered for now. The pelletized SA is cooled below 0 °C and the pressure is released (4^h to 6^h). At this state, i.e., at ambient pressure and moderately below 0 °C, the SA reaches SP conditions and can be stored in the SA storage. During discharge, the process is reversed. To this end, the SA is first pressurized by a high-density solids pump, e.g., a single or two-cylinder piston pump (6 to 7). The dissociation of the SA to air and water takes place in the dissociation reactor (7 to 8). Eventually, the air is heated (8^a to 9^a) and expanded in a turbine (9^a to 10^a) to recover electricity.

Both processes, the compression/expansion of air, as well as the formation/dissociation of SA, are heat intensive. Therefore, three thermal energy storage (TES) units are integrated: a sensible heat storage, which is charged by the hot air after the compressor (2^a to 3.1^a) and discharged by the cool air before the turbine (8^a to 9^a); a latent heat storage, which is charged during the exothermic SA formation (3 to 4) and discharged during the endothermic SA dissociation (7 to 8); and a water tank, to reuse the cold water exiting the dissociation reactor (8^w) in the next cycle. Water losses are compensated by mixing with water at ambient conditions (1^w to 2^w). Where no heat integration is possible, refrigeration machines and heat pumps are used. To account for minimum temperature difference constraints as well as heat losses in the storage units, two refrigeration machines are integrated for air and water, to ensure the attainability of the formation conditions (3.1 to 3.2). Another refrigeration machine is used to cool the SA after formation (4 to 5) to the storage temperature. A heat pump, integrated between the formation reactor and the latent heat storage unit, is operated only in case the minimum temperature difference constraint cannot be met for charging the latent heat storage unit. During discharge, no further refrigeration machines or heat pumps are used. When the heat stored in the latent heat storage unit is not sufficient to completely dissociate the SA, low-temperature ambient heat can be used in addition. The sensible heat storage unit enables the reuse of the heat of compression and determines the inlet temperature at the turbine.

In this very first analysis of SAES, one-stage compression and expansion are assumed for the air. This leads to unrealistically high temperatures and compression rates. It is believed, though, that this approach still allows for valuable insights into the SAES concept. Clearly, upon further developing the idea, multistep compression and expansion have to be considered.

3.2. Plant Modeling

A few simplifications had to be made in the thermodynamic modeling of the SA due to lack of better data. While there are thermodynamic models and simulations available for the prediction of the hydrate–liquid–vapor curve (see the compilation by Khan et al. [44]), accurate predictions of enthalpy changes due to combinations of both phase transition and change of state are still missing. Therefore, the specific enthalpy change Δh_{tr} of phase transition (formation and dissociation) is modeled using the available data of the hydrate–liquid–vapor equilibrium (cf. Figure 1) and the Clapeyron equation

$$\Delta h_{\mathrm{tr}} = T\Delta v(\mathrm{d}p/\mathrm{d}T). \tag{1}$$

The exact procedure is described in great detail by Anderson [45,46]. The change of enthalpy caused by a change of state without phase transition can be described by

$$\mathrm{d}h = \left(\frac{\partial h}{\partial T}\right)_p \mathrm{d}T + \left(\frac{\partial h}{\partial p}\right)_T \mathrm{d}p = c_p\,\mathrm{d}T + \left[v - T\left(\frac{\partial v}{\partial T}\right)_p\right]\mathrm{d}p. \tag{2}$$

Both the variation of c_p [29] as well as the thermal expansivity [47] are small over the range of temperatures considered. Further, SA is nearly incompressible [48]. Consequently, Equation (2) simplifies to

$$\mathrm{d}h \approx c_p\,\mathrm{d}T + v\,\mathrm{d}p, \tag{3}$$

where both $c_p = 2.1$ kJ/kg/K [29] and $v = 1.07$ cm^3/g [49] are assumed to be constant for all conditions considered.

The modeling of air and water does not require such simplifications and is done using the Python wrapper to the software CoolProp [50], an open-source database for fluid and humid air properties. The discrepancy between the simplifications necessary in the modeling of SA and the accurate modeling of water and air naturally results in uncertainties with respect to both the thermodynamic states and the energy balances. In view of other unknowns of much larger magnitude (e.g., the degree of SP, losses due to technical challenges in the handling of SA, etc.) these can be neglected, though.

The modeling of the plant components is done using enthalpy balances according to the parameters listed in Table 1.

Except for the storage units, heat losses are generally neglected. Compressors, pumps, and turbines are described via the isentropic efficiencies η_{comp}, η_{pump}, and η_{turb}, respectively. For ease of comparison, the isentropic efficiencies are chosen in agreement with those used for the modeling of an LAES plant [13]. The expansion valve is assumed to be isenthalpic. The refrigeration machines and heat pumps are calculated with fixed coefficients of performance β_{rm} and β_{hp}, respectively. Heat exchangers are modeled isobarically with fixed minimum temperature differences $\Delta T_{\mathrm{sensible}}$ and $\Delta T_{\mathrm{latent}}$ for the sensible and the latent heat transfer, respectively. The heat losses of the sensible and the latent heat storage are modeled using the storage efficiencies η_{sensible} and η_{latent}. h_{latent} and c_{sensible} denote the specific phase change enthalpy for the latent heat storage and the specific heat capacity for the sensible heat storage, respectively. The storage efficiencies and volumetric capacities are loosely calibrated to gravel beds for the sensible heat storage and to paraffin for the latent heat storage (see ref. [2] for typical values). In general, a pronounced degree of SP for the storage of SA is assumed, but some "boil-off" losses due to an intake of heat cannot be avoided. These losses are modeled by the hydrate storage efficiency η_{hs}. The hydration number efficiency η_{hn} allows deviations from the stoichiometric hydration number to be accounted for and reduces the air storage capacity of the SA. The void fraction η_{vs} is used to model the fraction of air-filled void space in the SA pellet with respect to its volume.

Table 1. Plant parameters of reference case.

Parameter	Value	Parameter	Value	Parameter	Value
T_{env}	298 K	T_{form}	278 K	T_{diss}	274 K
T_{storage}	273 K	$\Delta T_{\mathrm{sensible}}$	20 K	$\Delta T_{\mathrm{latent}}$	2 K
p_{env}	0.1 MPa	p_{form}	25.3 MPa	p_{diss}	16.6 MPa
p_{storage}	0.1 MPa				
η_{comp}	0.85	η_{pump}	0.90	η_{turb}	0.85
η_{hs}	0.975	η_{hn}	1.0	η_{vs}	0.2
η_{sensible}	0.80	η_{latent}	0.95	$\beta_{\mathrm{rm}}^{\mathrm{a}}$	7
$\beta_{\mathrm{rm}}^{\mathrm{w}}$	6	$\beta_{\mathrm{rm}}^{\mathrm{h}}$	6	$\beta_{\mathrm{hp}}^{\mathrm{latent}}$	8
h_{latent}	154 MJ m^{-3}	c_{sensible}	1.5 MJ m^{-3} K^{-1}		

In the simulations of SAES plant operation, all parameters listed in Table 1 can be selected within physically acceptable ranges, while all other states are derived from the enthalpy balances as described above. Typically, efficiencies are fixed before the start of the simulation and the input parameters environment temperature T_{env}, formation temperature T_{form}, dissociation temperature T_{diss}, and SA storage temperature T_{storage} are varied to find optimal conditions of operation.

3.3. Key Indicators

One of the most important indicators of an energy-storage system is the RTE. For the SAES system, the RTE is calculated by

$$\eta_{\mathrm{rt}} = \frac{W_{\mathrm{out}}}{W_{\mathrm{in}}} = \frac{W_{\mathrm{turb}}^{\mathrm{a}} - W_{\mathrm{pump}}^{\mathrm{h}}}{W_{\mathrm{comp}}^{\mathrm{a}} + W_{\mathrm{pump}}^{\mathrm{w}} + W_{\mathrm{rm}}^{\mathrm{a}} + W_{\mathrm{rm}}^{\mathrm{w}} + W_{\mathrm{rm}}^{\mathrm{h}} + W_{\mathrm{hp}}^{\mathrm{latent}}}. \tag{4}$$

This relates the net electricity W_{out}, obtained upon discharging, to the electricity that had to be used for charging.

Another important parameter for characterizing stationary EES systems is the volumetric energy density which expresses the amount of energy stored per unit volume. Considering only the air stored in the SA, the volumetric (solidified) air-based energy density is given by

$$w_{\mathrm{v}}^{\mathrm{a}} = \frac{W_{\mathrm{out}}}{V_{\mathrm{storage}^{\mathrm{h}}}}. \tag{5}$$

To account for the size of the TES system, which is a crucial part of any energy-storage system in which a gas is compressed and expanded, the total energy density

$$w_{\mathrm{v}}^{\mathrm{tot}} = \frac{W_{\mathrm{out}}}{V_{\mathrm{storage}^{\mathrm{h}}} + V_{\mathrm{sensible}^{\mathrm{a}}} + V_{\mathrm{latent}^{\mathrm{h}}} + V_{\mathrm{storage}^{\mathrm{w}}}} \tag{6}$$

is defined in addition.

4. Results and Discussion

This section provides a thermodynamic analysis of the SAES reference plant, followed by the determination of optimal operating conditions and a sensitivity analysis.

4.1. Thermodynamic Analysis

The SAES plant outlined above is modeled using the reference parameters provided in Table 1. As the following analysis will show, these parameters optimize the RTE of the reference plant. An overview of the corresponding states of air, water, and SA is provided in Table 2 for all components of the plant. Key indicators are summarized and compared to a LAES plant in Table 3.

While the RTEs are fairly similar, there is a significant difference in the energy densities. That is mainly because of the more than three-fold density of air in the liquid state compared to SA. The total energy density of LAES including TES has not yet been reported, though, a similar reduction with respect to the air-based value seems reasonable. The big difference between w_v^a and w_v^{tot} already indicates the importance of TES in the SAES plant. This is highlighted even more by the following figures: To generate a net electricity W_{out} of 1.000 kWh, roughly 1.754 kWh and 2.205 kWh of heat have to be stored in the sensible and latent heat storage unit, respectively. For comparison, the compressor (W_{comp}^a) and turbine work (W_{turb}^a) are 1.729 kWh and 1.107 kWh , respectively.

Table 2. States for the reference case.

State	1	2	3.1	3.2	4	5	6	7	8	9	10
T^a/K	298	1474	314	278	-	-	-	-	274	1243	468
p^a/MPa	0.1	25.3	25.3	25.3	-	-	-	-	16.6	16.6	0.1
T^w/K	298	278	278	278	-	-	-	-	274	278	278
p^w/MPa	0.1	0.1	25.3	25.3	-	-	-	-	16.6	0.1	0.1
T^h/K	-	-	-	-	278	261	273	273	-	-	-
p^h/MPa	-	-	-	-	25.3	25.3	0.1	16.6	-	-	-

Table 3. Key indicators for the reference case of SAES and for LAES.

Indicator	SAES	LAES [15]
Round trip efficiency η_{rt}	52 %	54 %
Energy density w_m^h	0.047 kWh kg^{-1}	0.12 kWh kg^{-1}
Air-based energy density w_v^a	47 kWh m^{-3}	104 kWh m^{-3}
Total energy density w_v^{tot}	19 kWh m^{-3}	-

To better illustrate these relations, Figure 3 shows a detailed analysis of the works during charging and discharging, again normalized to a net work output of 1.000 kWh.

While W_{comp}^a and W_{turb}^a dominate the picture, the works necessary for the compression of water (W_{pump}^w) and SA (W_{pump}^h) stand out as well. As together they sum up to about 25% of the net work output, potential improvements of the RTE are possible by either storing at high pressures instead of the two expansions in the system or by using a water turbine in the expansion process. The arrow in the storage stage termed "different losses" is used to close the enthalpy balance. It combines all heat and storage losses, but also all other losses associated with the various processes executed during charging.

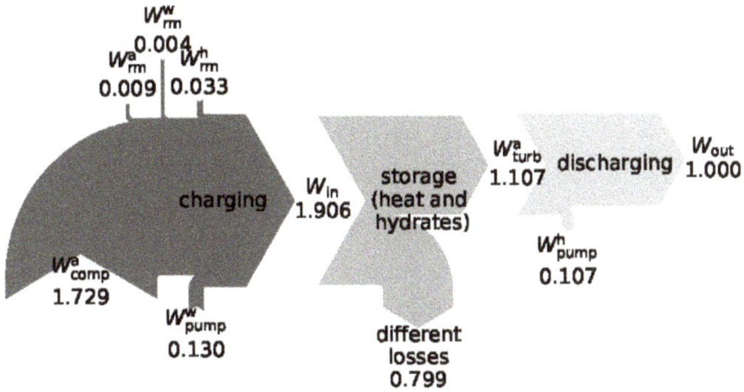

Figure 3. Sankey diagram to illustrate the relation between works (normalized to W_{out}) occurring in the SAES reference case.

4.2. Optimum Operating Conditions

A further analysis of the SAES plant aims at exploring the RTE of the plant while varying the main operational parameters. Figure 4a displays the RTE at different formation and dissociation temperatures.

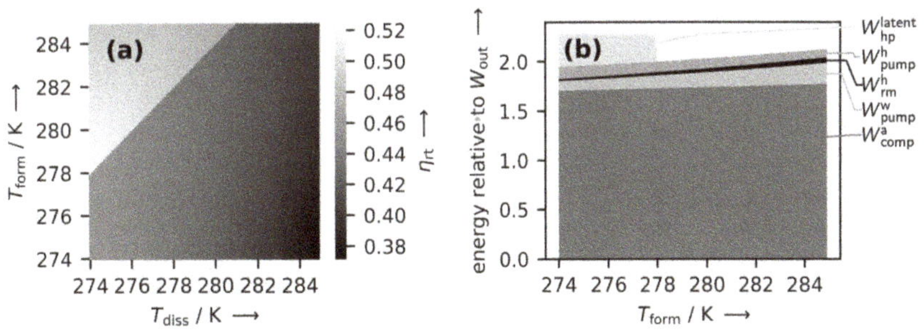

Figure 4. (**a**) Heat map of the RTE as a function of the formation and dissociation temperature. Bright spots indicate a high efficiency, dark spots a low efficiency. (**b**) Sum of process works, normalized to a net work output of 1.000 kWh, as a function of formation temperature. The dissociation temperature is kept constant at 274 K. Note that W^a_{rm} and W^w_{rm} are negligible on the scale of the graph and omitted for better readability.

The highest RTE is attained at the lowest possible dissociation temperature of 274 K and a formation temperature of 278 K, i.e., conditions also used in the reference case. This indicates that lower formation/dissociation temperatures and thus lower pressures are, in general, favorable. The sudden increase in RTE along a diagonal step can be explained with the help of Figure 4b. Therein, all energies to be expended are displayed as a function of formation temperature while the dissociation temperature is kept constant at 274 K. This can be interpreted as a line plot from the leftmost lower corner in Figure 4a to the top. Following this line, the RTE jumps abruptly as soon as a formation temperature of 278 K is reached. This is because at 278 K the latent heat pump (W^{latent}_{hp}) is no longer needed to ensure heat flow from the formation to the dissociation reactor. Consequently, the expended energy drops and the RTE increases. According to this observation, the dissociation pressure is best kept below the formation pressure to obtain high RTE. This is in contrast to the naive idea to use higher pressures at the turbine inlet than at the compressor outlet to both decrease compressor work and increase turbine work for maximal efficiency, an idea that the authors of this paper also initially fell for. According to the phase diagram in Figure 1, this would imply a higher dissociation than formation temperature. When a latent heat pump is to be avoided for enhanced efficiency, this is only possible with an external heat supply for the dissociation, an option not further followed in this paper.

4.3. Sensitivity Analysis

In addition to the effect of different temperatures, the sensitivity of the RTE to heat storage and other parameters is investigated in Figure 5a,b, respectively.

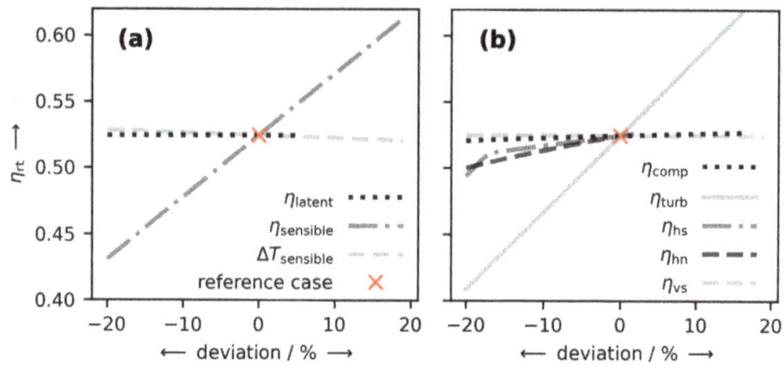

Figure 5. Sensitivity of the RTE to (**a**) heat storage and (**b**) other parameters with respect to a deviation from the reference case. The reference case (red cross) corresponds to the parameters of Table 1.

The parameters are varied by reducing/increasing them by fractions in the range of −20 % to 20 % with respect to the reference case values of Table 1 (indicated by a red cross). Note that to avoid unphysical values, some parameters do not extend over the full *x*-axis range. For example, all efficiency parameter lines were plotted only up to an efficiency of 100 %. From Figure 5a it becomes apparent that an efficient sensible heat storage system is crucial for the attainment of high RTE. As this holds in general for any storage system which includes the compression and expansion of gases, this observation is to be expected. Conversely, as can be seen from the influence of $\Delta T_{sensible}$, the heat transfer in the sensible storage unit is of only minor importance. Similarly, η_{latent}, the efficiency of latent heat storage, exhibits hardly any effect on the RTE. That is because in the simulations, ambient heat (at $T_{env} \gg T_{diss}$) is used to provide the missing heat of dissociation.

Regarding the performance parameters in Figure 5b, special attention has to be paid to the insentropic turbine efficiency η_{turb}. Its reduction impairs the RTE significantly. Not as dramatic as the effect of η_{turb}, but still important for an efficient plant operation, are the parameters η_{hn} and η_{hs}. These imply the importance of the degree of SP as well as of the air capacity of the SA structure. Remarkably, the slope of η_{hs} changes distinctly at around −17 %. Naturally, a decrease in η_{hs} results in a decrease in RTE, because the mass flow through the turbine decreases. At the same time, though, the sensible heat storage can heat the air to a higher temperature resulting in a higher W^{a}_{turb}. Yet, beyond a certain point (at around −17 % in Figure 5b), the temperature of the air cannot be increased any further (even though the required amount of heat would be available) because at this point, the minimum temperature difference $\Delta T_{sensible}$ becomes the limiting factor. As the effect of η_{vs} is negligible, the additional work because of the void space in and between the SA pellets is of no particular interest. The influence of η_{comp}, though minor, is more interesting because its variation affects the RTE in two opposite directions. On the one hand, with a poorer isentropic efficiency, more work is needed for compression. On the other hand, a poorer efficiency also leads to more heat being released. This additional heat can be used to heat the air to higher temperatures before expansion, which in turn results in a larger turbine work. Ultimately, this can almost compensate the additional work needed for compression. Yet, from a technical point of view, it is to be doubted if this additional heat can be fully utilized. Hence, in a more realistic scenario, the influence of η_{comp} on the RTE will most likely be larger than here.

5. Conclusions

Aiming for the development of a novel CES system which overcomes the drawbacks of low RTE and extremely low temperatures in LAES, the most popular CES system to date, a concept for EES in SA, i.e., the clathrate hydrate of air, has been developed and thermodynamically analyzed. Here, the SA acts as a water-based molecular storage vessel for compressed air, which, when released upon dissociation of the SA, drives a gas turbine.

Similar to LAES plants and due to the large heat of SA formation/dissociation and the heat generated during compression, the storage and recovery of heat is crucial for an efficient operation of an SAES plant. Assuming high rates of heat recovery but no utilization of waste heat from external processes, an RTE of 52% is calculated. While this reference case value is already comparable to the maximum RTE envisaged for mature large-scale LAES plants [13], the achievable energy density (47 kWh per m^3 of solidified air) is only half of that of the LAES plants. This is mainly due to a more than three-fold density of air in the liquid, when compared to that in SA.

Nevertheless, as the only components of SA are water and air, the concept of SAES is very friendly to the environment. Moreover, as practically all water can be recycled, water usage is negligible. Additionally, SAES offers some advantages which have the potential to become decisive in a comparison between SAES and LAES. Due to the possible reuse of the water, the addition of thermodynamic promoters is harmless but allows for the careful design of operating conditions. This means that conditions can be adapted to the environmental conditions at the site of the plant or to the conditions needed for the integration of SAES in a larger process. Both options are likely to enable the achievement of significantly higher RTE. Moreover, under the assumption of a pronounced degree of SP, the SA can be readily stored in huge heaps or skips. Therefore, storage tanks are unnecessary and capital expenditures can be reduced.

Though the concept for SAES is well substantiated, for now, the whole concept is purely theoretical. Before proofing the concept in a pilot plant, several questions have to be answered. With respect to the fundamentals, the phenomenon of SP in air hydrates has to be investigated and thoroughly ascertained. The practical goal is to find a route of SA processing which results in the most effective inhibition of SA dissociation at the conditions of storage. Similarly, fast routes for SA formation and dissociation have to be established for typical conditions in an SAES plant. With respect to process engineering, methods for the efficient transport, compression, and expansion of SA have to be studied. These analyses must include the options of processing in a slurry, in pellets, or in dry water hydrate. To better represent a real SAES plant, future models will also need to include multistage compression and expansion of the air, as well as more detailed heat transfer analyses.

In summary, SAES has the potential to compete with its closest competitor LAES, at least in terms of RTEs. This follows from our thermodynamic analysis and potentially milder storage conditions in the case of SAES, of course, given a pronounced degree of SP can be established for hydrates of air. The option to optimize RTEs by adjusting process parameters to the conditions at the plant's site further adds to the appeal of the concept. One particularly interesting option is the idea of combining subsea SAES with off-shore wind power generation. There, water and air are abundant. Additionally, hydrostatic pressure aids in compression and allows for storage at thermodynamically stable conditions. Moreover, SA can be readily stored at the seafloor because the buoyancy of SA in water is close to zero.

However, since at the moment the concept is purely theoretical, technical difficulties associated with the processing and storage of SA are hard to foresee. Even if the assumption of a pronounced degree of SP holds, it will be technically very challenging to implement SAES on a large scale. In terms of economics, it will be similarly challenging to even compete with LAES, which has yet to prove itself. Nevertheless, as engineering developments favoring LAES will likely also favor SAES, it seems to early to ultimately decide on the future of SAES. To enable the decision-making on solid grounds, at least the most pressing open question, namely, that regarding the existence of a pronounced degree of SP in air hydrate, must be clarified first.

Author Contributions: Conceptualization, C.H., B.H. and S.A.; methodology, C.H., B.H., S.H. and S.A.; software, S.H. and S.A.; validation, S.H., B.H. and S.A.; formal analysis, S.H. and S.A.; investigation, S.A.; resources, S.A.; data curation, S.H.; writing—original draft preparation, S.H. and S.A.; writing—review and editing, S.H., B.H. and S.A.; visualization, S.H. and S.A.; supervision, S.A.; project administration, S.A. All authors have read and agreed to the published version of the manuscript.

Funding: The funding from the illwerke vkw Endowed Professorship for Energy Efficiency as well as the Open Access Funding by the Vorarlberg University of Applied Sciences are greatly acknowledged.

Institutional Review Board Statement: Not applicable.

Informed Consent Statement: Not applicable.

Conflicts of Interest: The authors declare no conflict of interest.

Abbreviations

The following abbreviations are used in this manuscript:

CAES	Compressed-air energy storage
CES	Cryogenic energy storage
CP	Cyclopentane
EES	Electrical energy storage
LAES	Liquid-air energy storage
PSH	Pumped-storage hydroelectricity
RTE	Round-trip efficiency
SA	Solidified air
SAES	Solidified-air energy storage
SNG	Solidified natural gas
SP	Self-preservation
STP	Standard temperature and pressure
TBAB	Tetrabutylammonium bromide
THF	Tetrahydrofuran

References

1. Zerrahn, A.; Schill, W.P. Long-run power storage requirements for high shares of renewables: Review and a new model. *Renew. Sustain. Energy Rev.* **2017**, *79*, 1518–1534. [CrossRef]
2. Sterner, M.; Stadler, I. (Eds.) *Handbook of Energy Storage: Demand, Technologies, Integration*; Springer: Berlin/Heidelberg, Germany, 2019. [CrossRef]
3. Child, M.; Kemfert, C.; Bogdanov, D.; Breyer, C. Flexible electricity generation, grid exchange and storage for the transition to a 100% renewable energy system in Europe. *Renew. Energy* **2019**, *139*, 80–101. [CrossRef]
4. Schill, W.P.; Zerrahn, A. Long-run power storage requirements for high shares of renewables: Results and sensitivities. *Renew. Sustain. Energy Rev.* **2018**, *83*, 156–171. [CrossRef]
5. Chen, H.; Cong, T.N.; Yang, W.; Tan, C.; Li, Y.; Ding, Y. Progress in Electrical Energy Storage System: A Critical Review. *Prog. Nat. Sci.* **2009**, *19*, 291–312. [CrossRef]
6. Gür, T.M. Review of Electrical Energy Storage Technologies, Materials and Systems: Challenges and Prospects for Large-Scale Grid Storage. *Energy Environ. Sci.* **2018**, *11*, 2696–2767. [CrossRef]
7. Hameer, S.; van Niekerk, J.L. A Review of Large-Scale Electrical Energy Storage. *Int. J. Energy Res.* **2015**, *39*, 1179–1195. [CrossRef]
8. Rehman, S.; Al-Hadhrami, L.M.; Alam, M.M. Pumped Hydro Energy Storage System: A Technological Review. *Renew. Sustain. Energy Rev.* **2015**, *44*, 586–598. [CrossRef]
9. Kucukali, S. Finding the Most Suitable Existing Hydropower Reservoirs for the Development of Pumped-Storage Schemes: An Integrated Approach. *Renew. Sustain. Energy Rev.* **2014**, *37*, 502–508. [CrossRef]
10. Lu, B.; Stocks, M.; Blakers, A.; Anderson, K. Geographic Information System Algorithms to Locate Prospective Sites for Pumped Hydro Energy Storage. *Appl. Energy* **2018**, *222*, 300–312. [CrossRef]
11. Budt, M.; Wolf, D.; Span, R.; Yan, J. A Review on Compressed Air Energy Storage: Basic Principles, Past Milestones and Recent Developments. *Appl. Energy* **2016**, *170*, 250–268. [CrossRef]

12. Chen, L.; Zheng, T.; Mei, S.; Xue, X.; Liu, B.; Lu, Q. Review and Prospect of Compressed Air Energy Storage System. *J. Mod. Power Syst. Clean Energy* **2016**, *4*, 529–541. [CrossRef]
13. Morgan, R.; Nelmes, S.; Gibson, E.; Brett, G. Liquid Air Energy Storage–Analysis and First Results from a Pilot Scale Demonstration Plant. *Appl. Energy* **2015**, *137*, 845–853. [CrossRef]
14. Ameel, B.; T'Joen, C.; De Kerpel, K.; De Jaeger, P.; Huisseune, H.; Van Belleghem, M.; De Paepe, M. Thermodynamic Analysis of Energy Storage with a Liquid Air Rankine Cycle. *Appl. Therm. Eng.* **2013**, *52*, 130–140. [CrossRef]
15. Guizzi, G.L.; Manno, M.; Tolomei, L.I.; Vitali, R.I. Thermodynamic Analysis of a Liquid Air Energy Storage System. *Energy* **2015**, *93*, 1639–1647. [CrossRef]
16. Sciacovelli, A.; Vecchi, A.; Ding, Y. Liquid Air Energy Storage (LAES) with Packed Bed Cold Thermal Storage – From Component to System Level Performance through Dynamic Modelling. *Appl. Energy* **2017**, *190*, 84–98. [CrossRef]
17. Borri, E.; Tafone, A.; Romagnoli, A.; Comodi, G. A Review on Liquid Air Energy Storage: History, State of the Art and Recent Developments. *Renew. Sustain. Energy Rev.* **2021**, *137*, 110572. [CrossRef]
18. Koh, C.A.; Sloan, E.D.; Sum, A.K.; Wu, D.T. Fundamentals and Applications of Gas Hydrates. *Annu. Rev. Chem. Biomol. Eng.* **2011**, *2*, 237–257. [CrossRef] [PubMed]
19. Miller, S.L. Clathrate Hydrates of Air in Antarctic Ice. *Science* **1969**, *165*, 489–490. [CrossRef]
20. Pauer, F.; Kipfstuhl, S.; Kuhs, W.F.; Shoji, H. Air Clathrate Crystals from the GRIP Deep Ice Core, Greenland: A Number-, Size- and Shape-Distribution Study. *J. Glaciol.* **1999**, *45*, 22–30. [CrossRef]
21. Shoji, H.; Langway, C.C. Air Hydrate Inclusions in Fresh Ice Core. *Nature* **1982**, *298*, 548–550. [CrossRef]
22. Veluswamy, H.P.; Kumar, S.; Kumar, R.; Rangsunvigit, P.; Linga, P. Enhanced Clathrate Hydrate Formation Kinetics at near Ambient Temperatures and Moderate Pressures: Application to Natural Gas Storage. *Fuel* **2016**, *182*, 907–919. [CrossRef]
23. Veluswamy, H.P.; Kumar, A.; Kumar, R.; Linga, P. An Innovative Approach to Enhance Methane Hydrate Formation Kinetics with Leucine for Energy Storage Application. *Appl. Energy* **2017**, *188*, 190–199. [CrossRef]
24. Veluswamy, H.P.; Kumar, A.; Seo, Y.; Lee, J.D.; Linga, P. A Review of Solidified Natural Gas (SNG) Technology for Gas Storage via Clathrate Hydrates. *Appl. Energy* **2018**, *216*, 262–285. [CrossRef]
25. Mohammadi, A.H.; Tohidi, B.; Burgass, R.W. Equilibrium Data and Thermodynamic Modeling of Nitrogen, Oxygen, and Air Clathrate Hydrates. *J. Chem. Eng. Data* **2003**, *48*, 612–616. [CrossRef]
26. Yang, H.; Fan, S.; Lang, X.; Wang, Y.; Sun, X. Hydrate Dissociation Conditions for Mixtures of Air + Tetrahydrofuran, Air + Cyclopentane, and Air + Tetra- *N* -Butyl Ammonium Bromide. *J. Chem. Eng. Data* **2012**, *57*, 1226–1230. [CrossRef]
27. Lipenkov, V.Y.; Istomin, V. On the stability of air clathrate-hydrate crystals in subglacial Lake Vostok, Antarctica. *Mater. Glyatsiol. Issled* **2001**, *91*, 2001.
28. Mohammadi, A.H.; Richon, D. Ice-Clathrate Hydrate-Gas Phase Equilibria for Air, Oxygen, Nitrogen, Carbon Monoxide, Methane, or Ethane + Water System. *Ind. Eng. Chem. Res.* **2010**, *49*, 3976–3979. [CrossRef]
29. Sloan, E.D.; Koh, C.A. *Clathrate Hydrates of Natural Gases*, 3rd ed.; CRC Press: Boca Raton, FL, USA, 2007.
30. Hobbs, P.V. *Ice Physics*; Oxford Classic Texts in the Physical Sciences; Oxford University Press: New York, NY, USA, 2010.
31. Saito, A. Recent Advances in Research on Cold Thermal Energy Storage. *Int. J. Refrig.* **2002**, *25*, 177–189. [CrossRef]
32. Wang, X.; Zhang, F.; Lipiński, W. Carbon Dioxide Hydrates for Cold Thermal Energy Storage: A Review. *Sol. Energy* **2020**, *211*, 11–30. [CrossRef]
33. Yakushev, V.S.; Istomin, V. Gas Hydrates Self-Preservation Effect. In *Physics and Chemistry of Ice*; Maeno, N., Hondoh, T., Eds.; Hokkaido Univ. Press.: Sapporo, Japan, 1992; pp. 136–140.
34. Stern, L.A.; Circone, S.; Kirby, S.H.; Durham, W.B. Anomalous Preservation of Pure Methane Hydrate at 1 Atm. *J. Phys. Chem. B* **2001**, *105*, 1756–1762. [CrossRef]
35. Falenty, A.; Kuhs, W.F.; Glockzin, M.; Rehder, G. 'Self-Preservation' of CH_4 Hydrates for Gas Transport Technology: Pressure–Temperature Dependence and Ice Microstructures. *Energy Fuels* **2014**, *28*, 6275–6283. [CrossRef]
36. Falenty, A.; Kuhs, W.F. 'Self-Preservation' of CO_2 Gas Hydrates—Surface Microstructure and Ice Perfection. *J. Phys. Chem. B* **2009**, *113*, 15975–15988. [CrossRef] [PubMed]
37. Circone, S.; Stern, L.A.; Kirby, S.H.; Durham, W.B.; Chakoumakos, B.C.; Rawn, C.J.; Rondinone, A.J.; Ishii, Y. CO_2 Hydrate: Synthesis, Composition, Structure, Dissociation Behavior, and a Comparison to Structure I CH_4 Hydrate. *J. Phys. Chem. B* **2003**, *107*, 5529–5539. [CrossRef]
38. Takeya, S.; Ripmeester, J.A. Dissociation Behavior of Clathrate Hydrates to Ice and Dependence on Guest Molecules. *Angew. Chem. Int. Ed.* **2008**, *120*, 1296–1299. [CrossRef]
39. Hallbrucker, A.; Mayer, E. Unexpectedly Stable Nitrogen, Oxygen, Carbon Monoxide and Argon Clathrate Hydrates from Vapour-Deposited Amorphous Solid Water: An X-Ray and Two-Step Differential Scanning Calorimetry Study. *J. Chem. Soc. Faraday Trans.* **1990**, *86*, 3785–3792. [CrossRef]
40. Mayer, E.; Hallbrucker, A. Unexpectedly Stable Nitrogen and Oxygen Clathrate Hydrates from Vapour Deposited Amorphous Solid Water. *J. Chem. Soc. Chem. Commun.* **1989**, *12*, 749–751. [CrossRef]
41. Zhang, G.; Rogers, R.E. Ultra-Stability of Gas Hydrates at 1atm and 268.2K. *Chem. Eng. Sci.* **2008**, *63*, 2066–2074. [CrossRef]
42. Rehder, G.; Eckl, R.; Elfgen, M.; Falenty, A.; Hamann, R.; Kähler, N.; Kuhs, W.F.; Osterkamp, H.; Windmeier, C. Methane Hydrate Pellet Transport Using the Self-Preservation Effect: A Techno-Economic Analysis. *Energies* **2012**, *5*, 2499–2523. [CrossRef]

43. Melnikov, V.P.; Podenko, L.S.; Nesterov, A.N.; Drachuk, A.O.; Molokitina, N.S.; Reshetnikov, A.M. Self-Preservation of Methane Hydrates Produced in 'dry Water'. *Dokl. Chem.* **2016**, *466*, 53–56. [CrossRef]
44. Khan, M.N.; Warrier, P.; Peters, C.J.; Koh, C.A. Advancements in Hydrate Phase Equilibria and Modeling of Gas Hydrates Systems. *Fluid Phase Equilibria* **2018**, *463*, 48–61. [CrossRef]
45. Anderson, G.K. Enthalpy of Dissociation and Hydration Number of Carbon Dioxide Hydrate from the Clapeyron Equation. *J. Chem. Thermodyn.* **2003**, *35*, 1171–1183. [CrossRef]
46. Anderson, G.K. Enthalpy of Dissociation and Hydration Number of Methane Hydrate from the Clapeyron Equation. *J. Chem. Thermodyn.* **2004**, *36*, 1119–1127. [CrossRef]
47. Hester, K.C.; Huo, Z.; Ballard, A.L.; Koh, C.A.; Miller, K.T.; Sloan, E.D. Thermal Expansivity for sI and sII Clathrate Hydrates. *J. Phys. Chem. B* **2007**, *111*, 8830–8835. [CrossRef] [PubMed]
48. Kuhs, W.; Chazallon, B.; Radaelli, P.; Pauer, F. Cage Occupancy and Compressibility of Deuterated N_2-Clathrate Hydrate by Neutron Diffraction. *J. Incl. Phenom. Mol. Recognit. Chem.* **1997**, *29*, 65–77. [CrossRef]
49. Takeya, S.; Honda, K.; Yoneyama, A.; Hirai, Y.; Okuyama, i.; Hondoh, T.; Hyodo, K.; Takeda, T. Observation of Low-Temperature Object by Phase-Contrast x-Ray Imaging: Nondestructive Imaging of Air Clathrate Hydrates at 233 K. *Rev. Sci. Instrum.* **2006**, *77*, 053705. [CrossRef]
50. Bell, I.H.; Wronski, J.; Quoilin, S.; Lemort, V. Pure and Pseudo-pure Fluid Thermophysical Property Evaluation and the Open-Source Thermophysical Property Library CoolProp. *Ind. Eng. Chem. Res.* **2014**, *53*, 2498–2508. [CrossRef] [PubMed]

processes MDPI

Article

A Techno-Economic Assessment of Fischer–Tropsch Fuels Based on Syngas from Co-Electrolysis

Ralf Peters [1,*], Nils Wegener [1,2], Remzi Can Samsun [1], Felix Schorn [1,3], Julia Riese [2], Marcus Grünewald [2] and Detlef Stolten [3,4]

1 Institute of Energy and Climate Research-Electrochemical Process Engineering (IEK-14), Forschungszentrum Jülich GmbH, Wilhelm-Johnen-Str., 52428 Jülich, Germany; n.wegener@fz-juelich.de (N.W.); r.c.samsun@fz-juelich.de (R.C.S.); f.schorn@fz-juelich.de (F.S.)
2 Laboratory for Fluid Separations, Faculty of Mechanical Engineering, Ruhr University Bochum, 44721 Bochum, Germany; julia.riese@fluidvt.ruhr-uni-bochum.de (J.R.); marcus.gruenewald@fluidvt.ruhr-uni-bochum.de (M.G.)
3 Institute of Energy and Climate Research-Techno-Economic System Analysis (IEK-3), Forschungszentrum Jülich GmbH, Wilhelm-Johnen-Str., 52428 Jülich, Germany; d.stolten@fz-juelich.de
4 Chair for Fuel Cells, RWTH Aachen University, 52072 Aachen, Germany
* Correspondence: ra.peters@fz-juelich.de; Tel.: +49-2461-61-4260

Abstract: As a part of the worldwide efforts to substantially reduce CO_2 emissions, power-to-fuel technologies offer a promising path to make the transport sector CO_2-free, complementing the electrification of vehicles. This study focused on the coupling of Fischer–Tropsch synthesis for the production of synthetic diesel and kerosene with a high-temperature electrolysis unit. For this purpose, a process model was set up consisting of several modules including a high-temperature co-electrolyzer and a steam electrolyzer, both of which were based on solid oxide electrolysis cell technology, Fischer–Tropsch synthesis, a hydrocracker, and a carrier steam distillation. The integration of the fuel synthesis reduced the electrical energy demand of the co-electrolysis process by more than 20%. The results from the process simulations indicated a power-to-fuel efficiency that varied between 46% and 67%, with a decisive share of the energy consumption of the co-electrolysis process within the energy balance. Moreover, the utilization of excess heat can substantially to completely cover the energy demand for CO_2 separation. The economic analysis suggests production costs of 1.85 €/l_{DE} for the base case and the potential to cut the costs to 0.94 €/l_{DE} in the best case scenario. These results underline the huge potential of the developed power-to-fuel technology.

Keywords: CO_2 electrolysis; co-electrolysis; electrofuels; power-to-fuel; power-to-liquid; solid oxide electrolysis; synthetic diesel; synthetic kerosene; water electrolysis

Citation: Peters, R.; Wegener, N.; Samsun, R.C.; Schorn, F.; Riese, J.; Grünewald, M.; Stolten, D. A Techno-Economic Assessment of Fischer–Tropsch Fuels Based on Syngas from Co-Electrolysis. *Processes* **2022**, *10*, 699. https://doi.org/10.3390/pr10040699

Academic Editors: Alon Kuperman and Alessandro Lampasi

Received: 23 February 2022
Accepted: 30 March 2022
Published: 4 April 2022

1. Introduction

In order to mitigate anthropogenic climate change, a considerable reduction in the emission of climate-damaging emissions is necessary. For this purpose, it is essential to either electrify sectors or convert them to the use of alternative fuels in order to minimize dependence on fossil fuels. As a result of the introduction of various sustainable technologies, the energy and household sectors, for example, have seen the first reductions in greenhouse gas emissions [1]. The reduction of emissions in the transport sector, however, presents a bigger challenge. Heavy haulage, ship, and air traffic, in particular, can only be converted to electrified drivetrains to a limited extent. As a result, it can be expected that the demand for liquid fuels with high energy densities will remain high in the future [2]. One way of achieving CO_2-neutrality for the transport sector involves the power-to-liquid concept. 'Power-to-liquid' is a collective term for various technologies employed in the production of liquid energy carriers through the use of renewable electrical energy with the addition of carbon dioxide (CO_2) [3]. The energy carriers produced in this way can be converted back into electricity and thus function as electricity storage systems that can be used for other applications. In the event that the energy sources are used as fuels in the transport sector, this is referred to as 'power-to-fuel'. If both the required electrical energy and CO_2 are obtained from renewable sources, fuels can be produced in a CO_2-neutral manner using power-to-fuel processes. Thus, these represent a possibility for effectively defossilizing the transport sector [4]. Various power-to-fuel concepts already exist. On one hand, the production of alternative fuels such as methanol or dimethyl ether (DME) is a focus of research [5]. However, on the other hand, researchers are also exploring the synthesis of traditional fuels such as gasoline, kerosene, and diesel, as the existing infrastructure and greater applicability of these represent an advantage over alternatives [6]. For the structured processing of these tasks, this paper was divided into the following sections:

- Section 2 provides some insights into the motivation to apply power-to-fuel processes. An overview of already implemented and future planned power-to-liquid or power-to-fuel projects is given.
- Section 3 explains the basics of the individual components used in the developed power-to-fuel process. At the end, the degree of technological maturity of the individual components of the developed fuel synthesis is examined.
- The topic of Section 4 is the development and design of the selected power-to-fuel process. For this purpose, the more precise framework conditions and resulting structure of the fuel synthesis are presented first. Then, the modeling and simulation of the process in Aspen Plus is explored.
- Section 5 presents and discusses the results of the process analysis simulations. First, it is determined whether the fuels produced meet all of the requirements and then the material and energy balance of the process is examined. Finally, based on the efficiency, an energetic comparison between the developed fuel synthesis and alternative power-to-liquid processes is carried out.
- Section 6 analyzes the economic aspects of the developed power-to-fuel process. First, the manufacturing costs of the fuels produced are determined. Then, the influence of various factors on the production costs is examined through a sensitivity analysis. Finally, the production costs of the developed fuel synthesis route are compared with those of alternative power-to-liquid processes.
- In Section 7, the results of the work are summarized and an outlook on the main research areas are given.

2. Background

In 2012, the German transport sector consumed 2772 PJ ($\approx$770 TWh) of energy. Around 26.9% of this energy requirement was accounted for by gasoline, 51% by diesel, and 15.7% by aviation fuels [7].

On one hand, as vehicles with alternative drivetrains such as battery- or fuel cell-based ones are increasingly being used, it is expected that gasoline will lose its importance in the long term. On the other hand, due to the lack of alternatives in freight and air traffic, it can be assumed that the fuels diesel and kerosene will also be of great importance over the longer term [2].

Accordingly, the production of renewable diesel and kerosene is of both academic and industrial interest. A major advantage of synthetically-produced diesel and kerosene is that they are compatible with existing infrastructures and can be used in current vehicles [4]. The prerequisite for use is the fulfillment of the fuel specifications, which are set out in the relevant standards. Synthetically-produced diesel must comply with EN 15940 in Europe. ASTM 7566 applies to Jet A-1 kerosene used in civil aviation. This allows conventionally-produced Jet A-1 to be mixed with up to 50% synthetically-produced kerosene, depending on the synthesis route [8]. This synthetic kerosene is called synthesized paraffinic kerosene (SPK). An extract of the most important parameters for diesel (class A) according to EN 15940 and for SPK produced via a Fischer–Tropsch synthesis according to ASTM 7566 is discussed in more detail in Section 5.1. Class A describes diesel with an increased cetane number, which is a characteristic value of the ignitability of diesel fuels. The higher the cetane number, the more readily ignitable the diesel fuel [9].

One possibility of producing renewable, synthetic fuels is via the power-to-fuel concept. Schemme et al. [10] discussed the power-to-fuel concept as a solution to the present challenges of the transport sector in terms of the energy transition by coupling the energy and transport sectors. According to this concept, renewable electricity is used to produce hydrogen via water electrolysis, offering a storage possibility for volatile renewable energy sources. In the following synthesis step, renewable fuels are synthesized in a reaction or a series of reactions and further treatment steps combining the produced hydrogen with carbon dioxide from various possible sources. Different electrolysis technologies can be utilized for the generation of hydrogen.

In 2014, Sunfire GmbH commissioned the "Fuel 1" demonstration plant in Dresden, Germany. At this facility, high-temperature water electrolysis (SOEC) is used to provide hydrogen. The hydrogen is then mixed with carbon monoxide, which is generated in a reverse water–gas shift reactor and converted into so-called blue crude by means of Fischer–Tropsch synthesis. Blue crude is a renewable crude oil that can be further processed in a conventional refinery into synthetic gasoline, kerosene, or diesel, for example [11]. The plant was run for 1500 h [12] and produced one barrel (159 L) of blue crude per day [13].

In 2017, the VTT Technical Research Center of Finland and the Lappeenranta University of Technology operated a demonstration plant in Lappeenranta (Finland) for around 300 operating hours as part of the "SOLETAIR" project [14]. Fischer–Tropsch synthesis was also used in this system. CO_2 was obtained through direct air capture and converted into carbon monoxide in a reverse water–gas shift reactor. The hydrogen was provided via PEM electrolysis [15].

As part of the Kopernikus project "Power-to-X", funded by the Federal Ministry of Education and Research, the so-called SUNFIRE-SYNLINK was put into operation in Karlsruhe (Germany) in 2019 [16]. A co-electrolysis system from Sunfire GmbH was used to produce synthesis gas. This was combined with a Fischer–Tropsch reactor from INERATEC and a hydrocracker unit from the Karlsruhe Institute of Technology to produce synthetic fuels. The CO_2 required was obtained using Climeworks' direct air capture (DAC) technology. The co-electrolysis currently in use has an output of 10 kW, but Sunfire plans to upscale the process to an industrial scale [17].

An industrial-scale plant is being planned by the Norwegian company, Nordic Electrofuel (formerly Nordic Blue Crude). The plant is to be built and commissioned in Herøya Industripark (Norway) by 2022. The use of high-pressure alkaline electrolysis is planned and the plant is set to achieve an initial production capacity of 10 million liters per year. The required CO_2 is to be supplied from both industrial sources and via DAC technology. Originally, the use of a co-electrolysis unit from Sunfire GmbH was planned for the plant planned by Nordic Electrofuel [17]. However, the business partners separated in 2020 and Sunfire GmbH founded the industrial consortium, Norsk e-Fuel [18,19], together with Climeworks, Paul Wurth (SMS Group), and Valinor. Norsk e-Fuel also plans to build a plant capable of producing 10 million liters of synthetic kerosene per year at Herøya Industrial Park (Norway). This facility is scheduled to be commissioned in 2023 and expanded to a production capacity of 100 million liters of renewable fuels by 2026 [20]. Amongst other things, the co-electrolysis technology from Sunfire GmbH and the DAC technology from Climeworks are to be used in it [18,20].

In Rotterdam, the Hague Airport announced a study in 2019 in which, in collaboration with several European partners, a demonstration plant for the production of aviation fuel was to be developed. The plant is expected to produce around 1000 L of renewable fuel, but there has not yet been a specific date for its commissioning [21]. Based on this study, the two startups Synkero and Zenid were presented on 8 February 2021 [22,23]. Both of these plan to build a plant for the production of synthetic kerosene, but each is pursuing different concepts for providing the required CO_2. Although Zenid's goal is a plant that obtains the CO_2 exclusively by means of DAC technology [24], Synkero also considers other CO_2 sources such as industrial exhaust gases or biogenic sources [25].

Figure 1 illustrates a special power-to-fuel concept for the production of synthetic fuels, which is to be examined in more detail within the scope of this work. Initially, synthesis gas consisting of hydrogen (H_2) and carbon monoxide (CO) is produced from water and CO_2 using renewable electrical energy. The resulting synthesis gas is then converted into liquid fuels.

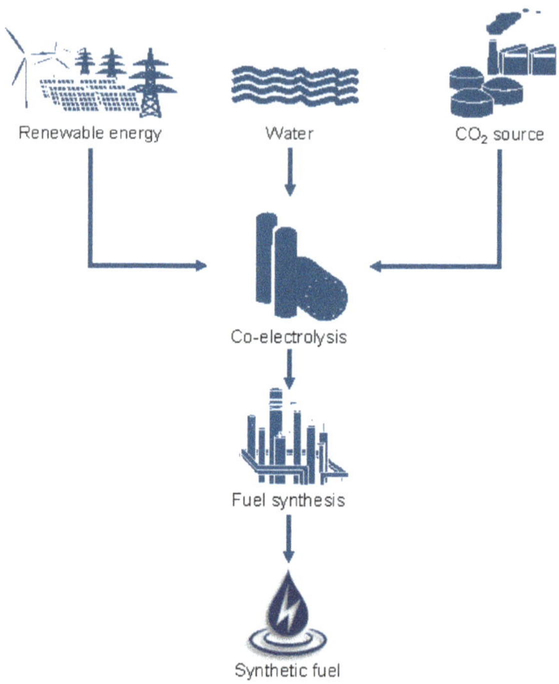

Figure 1. Schematic of a power-to-fuel process with co-electrolysis.

What is special about the power-to-fuel concept shown in Figure 1 is that the synthesis feed gas is produced using what is known as co-electrolysis. This allows for the production of synthesis gas in a single step, and so there is no need to produce hydrogen and carbon monoxide separately. In addition, co-electrolysis offers some energetic advantages over other synthesis gas production routes. For example, it is possible to substitute some of the electrical energy required for electrolysis with excess thermal energy from fuel synthesis. Therefore, the combination of co-electrolysis with Fischer–Tropsch synthesis represents an interesting basis for power-to-fuel processes. In this route, synthesis gas is converted into hydrocarbons by means of strongly exothermic reactions, which are then processed into fuels such as gasoline, diesel, or kerosene. The heat of reactions that occur during Fischer–Tropsch synthesis can be used to operate co-electrolysis, which opens up the possibility of improving the overall efficiency of power-to-fuel processes.

The aim of this work was to develop a power-to-fuel process based on co-electrolysis, in combination with a Fischer–Tropsch synthesis, and to model and simulate the process in the Aspen Plus program. The entire process chain, starting with CO_2 and water through to fuel, should be considered. Then, the developed power-to-fuel process should be techno-economically analyzed and compared to alternative power-to-fuel processes.

3. Basic Process Units of PtF System Design

The following section presents the technical basics of the most important components of the modeled power-to-fuel process. The power-to-fuel system developed within the scope of this work consists of a water electrolysis, co-electrolysis, Fischer–Tropsch synthesis, hydrocracker, reformer, and carrier steam distillation. This section deals with the basics of these system components.

3.1. Electrolysis and Co-Electrolysis

Depending on the electrolyte or ionic charge carrier used, a distinction is made between three different electrolysis methods. Alkaline electrolysis and PEM electrolysis are already available on the megawatt scale, and hydrogen thus produced can achieve high purities of over 99% [6,26]. In addition, alkaline and PEM electrolysis can be operated under pressures of 60 to 80 bar, thus reducing the need for compressor power for downstream processes [6].

In contrast, the solid oxide electrolyzer cell (SOEC) is not yet commercially available in the megawatt range. Operation under pressure is also still in the development phase. However, due to their high operating temperature, SOECs offer thermodynamic advantages over other electrolysis types. Therefore, it is theoretically possible to achieve electrical efficiencies of over 100% based on the calorific value of the products [6]. Typical operating temperatures of a SOEC are 800–1000 °C [27] or 700–1000 °C [28], depending on the literature source. In addition, due to its high operating temperature, it has improved kinetics [29] and can be operated as a so-called co-electrolysis unit [30]. In co-electrolysis, not only water, but also CO_2 is broken down. This electrolysis process is especially interesting for power-to-liquid (PtL) and power-to-fuel (PtF) processes as it makes it possible to produce synthesis gas consisting of hydrogen and carbon monoxide in a single process step [6,30].

The electrochemical reactions that take place at the cathode and anode are as follows. Cathode:

$$H_2O + 2e^- \rightarrow H_2 + O^{2-} \quad (1)$$

$$CO_2 + 2e^- \rightarrow CO + O^{2-} \quad (2)$$

Anode:

$$2O^{2-} \rightarrow O_2 + 4e^- \quad (3)$$

The minimum energy expenditure for the decomposition of water or CO_2 corresponds to the enthalpy of reaction ΔH_R, which in the case of the usual operating conditions of co-electrolysis (860 °C, 1 bar) is 249 kJ/mol for water and 283 kJ/mol for CO_2 [30]. According to the second law of thermodynamics, the reaction enthalpy is composed of the free Gibbs energy ΔG_R and the reaction entropy ΔS_R, multiplied by the temperature T as follows [31]:

$$\Delta H_R = \Delta G_R + T \cdot \Delta S_R \quad (4)$$

Here, ΔG_R is the part of the reaction enthalpy that must be provided in the form of electricity during electrolysis, whereas $T \cdot \Delta S_R$ can be supplied to the reaction in the form of heat [32]. An advantage of high-temperature electrolysis compared to other electrolysis types is that if the water evaporates outside the electrolysis cell, this energy must no longer be introduced into the cell in the form of electricity [29]. After evaporation with increasing temperature, the total energy requirement of the reaction remains almost constant, but the amount of electrical energy that is absolutely necessary significantly decreases. On the basis of these two facts, valuable electrical energy can be saved with high-temperature electrolysis in comparison to, for example, PEM electrolysis. Considering the thermal energy that is exchanged, the internal thermal losses are of particular importance in electrolysis, as they can be used to provide the heat of the reaction. The case in which the thermal losses correspond precisely to the heat of the reaction is referred to as the thermoneutral operating point, and the electrical voltage applied to the SOEC at this operating point is correspondingly referred to as the thermoneutral voltage [3]. At the thermoneutral operating point, the entire electrical energy E_{el} supplied to the electrolysis cell is converted into chemical potential energy and the following applies to cell efficiency at thermoneutral point $\eta_{cell,TN}$ [31]:

$$\eta_{cell,TN} = \frac{\Delta H_R}{E_{el}} = 1 \tag{5}$$

Due to the simultaneous presence of H_2O, CO_2, H_2, and CO at the cathode of the SOEC and the high operating temperatures, in addition to the reactions listed in Equations (1)–(3), the so-called water–gas shift reaction (WGS) or reverse water–gas shift reaction (RWGS) must also be considered in the co-electrolysis [33]. The reverse water–gas shift reaction is favored at the high operating temperatures of a SOEC [34].

$$CO + H_2O \leftrightarrow CO_2 + H_2 \tag{6}$$

In addition, methanation reactions (Equations (7) and (8)) can occur at the cathode of a SOEC [33]:

$$CO + 3H_2 \leftrightarrow CH_4 + H_2O \tag{7}$$

$$CO_2 + 4H_2 \leftrightarrow CH_4 + 2H_2O \tag{8}$$

Finally, the Boudouard reaction (Equation (9)) must be taken into account, and can lead to the precipitation of solid carbon under certain operating conditions of a SOEC [33]:

$$2CO \leftrightarrow CO_2 + C_{(s)} \tag{9}$$

According to Equation (5) the efficiency of co-electrolysis at the thermo-neutral operating point corresponds to 100%. According to Peters et al. [22], however, SOECs are not usually operated with an exactly thermoneutral voltage, so the heat must be either added or removed. In addition, the efficiency of the entire system is influenced by other factors such as the power required for the compression and storage of the products or the losses of the voltage converter to rectify the alternating current. Therefore, in order to assess the efficiency of the overall system (η_{SOEC}), the degree of efficiency is usually defined through the calorific value of the product (*Mass flow rate of product* $m_{product}$ *multiplied by its lower heating value* H_u^0) in relation to the electrical ($E_{el,\text{total}}$) and thermal energy ($E_{th,\text{total}}$) used [22]:

$$\eta_{SOEC} = \frac{m_{product} \cdot H_u^0}{E_{el,\text{total}} + E_{th,\text{total}}} \tag{10}$$

3.2. Fischer–Tropsch Synthesis

Fischer–Tropsch synthesis is a process in which a synthesis gas consisting of hydrogen and carbon monoxide is converted into liquid hydrocarbons [9]. This results in a wide range of hydrocarbon chains of different lengths with chain length n, according to the following equation [33]:

$$(2n+1) \cdot H_2 + n \cdot CO \rightarrow C_nH_{2n+2} + n \cdot H_2O \tag{11}$$

Another reaction that occurs in the Fischer–Tropsch synthesis process is the water–gas shift reaction already described in Equation (6). The methanation reactions (Equations (7) and (8)) and the Boudouard reaction described in Equation (9) can also occur [35]. Which reactions take place to which extent during the Fischer–Tropsch synthesis and also the product distribution of it are determined by the process parameters [35].

A distinction is made between high-temperature Fischer–Tropsch synthesis (HTFT) and low-temperature Fischer–Tropsch synthesis (LTFT) [36]. High temperatures (300–350 °C), with iron as a catalyst, favor short chain lengths and therefore shift the product distribution in the direction of (liquid) gases (n = 1–4) and synthetic gasoline (n = 5–12) [6,35]. Lower temperatures (200–240 °C) in combination with iron or cobalt catalysts, in contrast, favor the formation of longer hydrocarbon chains (i.e., middle distillates such as kerosene (n = 8–16) [37] and diesel (n = 10–23) as well as long-chain ones (n > 22)) [6,35]. In addition, cobalt catalysts suppress the water–gas shift reaction, and low temperatures reduce the formation of methane and solid carbon. The operating pressure also has a direct influence on the product distribution [35]. Typical operating pressures in Fischer–Tropsch synthesis are between 1 and 40 bar, with higher pressures resulting in longer average hydrocarbon chain lengths [35]. Another influencing factor on Fischer–Tropsch synthesis is the ratio of H_2 to CO. Typically, H_2/CO ratios of around two are used, with the average chain length of the product decreasing with higher ratios and increasing with lower ones [33].

The product distribution can be approximately determined using the Anderson–Schulz–Flory distribution [36]. With this, both the mass fractions w_n (Equation (12)) and molar fractions x_n (Equation (13)) of the respective hydrocarbon chains with chain length n can be determined [35]:

$$w_n = \alpha^{n-1} \cdot (1-\alpha)^2 \cdot n \tag{12}$$

$$x_n = \alpha^{n-1} \cdot (1-\alpha) \tag{13}$$

Here, α stands for the chain growth probability, which is determined by reactor design and operating conditions. The chain growth probability can be either determined empirically or taken from the literature. The influence of chain growth probability on the product distribution is discussed further in Appendix A. In Vervloet et al. [38], the approach in Equation (14) was given for α for the low-temperature Fischer–Tropsch synthesis assumed in this work and the use of a cobalt catalyst. In this model, the chain growth probability is determined through the ratio between the chain growth rate and chain growth termination rate:

$$\alpha = \frac{1}{1 + k_\alpha \left(\frac{c_{H2}}{c_{CO}}\right)^{\beta} \exp\left(\frac{\Delta E_\alpha}{R}\left(\frac{1}{493.15} - \frac{1}{T}\right)\right)} \tag{14}$$

where

k_α is the ratio of the speeds of the chain growth rate and chain growth termination rate ($k_\alpha = 0.0567$);
c_{H2} is the hydrogen concentration in mol/m^3;
c_{CO} is the carbon monoxide concentration in mol/m^3;
β is the exponential parameter for selectivity ($\beta = 1.76$);
ΔE_α is the difference of activation energies for chain growth and chain growth termination ($\Delta E_\alpha = 120.4 \frac{kJ}{mol}$);
R is the ideal gas constant ($R = 8.314 \frac{J}{mol}$); and
T is the reactor temperature in K.

The influence of the chain growth probability on the product distribution is illustrated in (Figure A1 in Appendix A, which shows the product distribution of a Fischer–Tropsch synthesis for α = 0.88 and α = 0.92 for C_1 to C_{60}.

3.3. Hydrocrackers

Hydrocracking refers to a chemical process in which long-chain, higher-molecular hydrocarbon chains are split into shorter ones through the addition of hydrogen. The distribution of the chain lengths of the products is strongly influenced by the catalyst used and the selected process conditions. Therefore, these must always be adapted to the respective application [39,40]. For power-to-fuel processes, chain lengths in the range from n = 5 to n = 20 are of great importance, as these hydrocarbon chains are required for the production of synthetic gasoline, kerosene, and diesel [36]. One possibility for maximizing these fractions in the product of the hydrocracker is the use of so-called "ideal hydrocracking" [41,42]. As a power-to-fuel process is to be modeled and simulated within the scope of this work, the focus in the following was on ideal hydrocracking.

The most important properties of ideal hydrocracking are defined as follows, drawing on Bouchy et al. [41]. If C_n-molecules are cracked, the selectivity to all C_4- to C_{n-4} hydrocarbons is identical, the selectivity to C_3 and C_{n-3} is half of that, and C_1, C_2, C_{n-1}

and C_{n-2} cannot be formed. Furthermore, only primary cracking occurs. Exclusively primary hydrocracking means that the shorter hydrocarbons that are created after a longer hydrocarbon chain has been cracked cannot be cracked any further [39]. In the case of non-ideal hydrocracking, the proportion of middle distillates is significantly lower than in ideal hydrocracking. In addition, a large peak of the C_3 to C_5 hydrocarbons was identified by Wegener [43]. This course is due to the occurrence of secondary cracking, which cracks the middle distillates, with the proportion of short-chain hydrocarbons increasing.

According to Bouchy et al., ideal bifunctional catalysts with a hydrogenation/dehydrogenation function and an acid function can be used, whereby it must be ensured that the reaction taking place at the acid function is the limiting one [41]. In addition, it is important to ensure that the pore structure of the catalyst is correct, so that no undesired increased cracking occurs at the ends of the hydrocarbon chains, which can lead to the stronger formation of short-chain gases [41,44].

As already described, the operating conditions also have a major influence on the product distribution of a hydrocracker. Conventional hydrocracking, depending on the literature source, is carried out at temperatures ranging from 350–430 °C and pressures of 100–200 bar [41], or in the upper range of 290–445 °C and 10–200 bar [40]. Ideal hydrocracking takes place under significantly milder process conditions, with temperatures in the range of 324-372 °C and pressures of 35–70 bar [41].

In addition, in order to suppress soot formation and catalyst deactivation, it must be ensured that the proportion of H_2 in the feed stream is high enough [39]. In the literature, values of 6–15% by weight are recommended [45]. With hydrocracking, conversions of up to 99% can be theoretically achieved [39]. However, according to Bouchy et al. [41], conversions that are too high for ideal hydrocracking become problematic, as secondary cracking inevitably occurs, even with ideal hydrocracking at very high conversions.

3.4. Reformers–Steam Reforming and Partial Oxidation

Reformers make it possible to convert hydrocarbons into synthesis gas. Various reactions and side reactions take place simultaneously in a reformer, with steam reforming and partial oxidation playing the greatest role [4].

In steam reforming, hydrocarbons are converted into carbon monoxide and hydrogen with the addition of steam. The general reaction equation for steam reforming is as follows:

$$C_nH_{2n+2} + n{\cdot}H_2O \rightarrow n{\cdot}CO + (2n+1){\cdot}H_2 \tag{15}$$

This is a strongly endothermic reaction which, with the exception of methane (n = 1), can be regarded as irreversible at the normal operating temperatures of over 500 °C for reformers [46]. The partial oxidation of hydrocarbons is an exothermic reaction with the general reaction equation:

$$C_nH_{2n+2} + \frac{n}{2}{\cdot}O_2 \rightarrow n{\cdot}CO + (n+1){\cdot}H_2 \tag{16}$$

If the supply of oxygen is regulated, the degree of reaction of the partial oxidation can also be adjusted. Accordingly, if both reactions are carried out at the same time, the required heat of reaction for the steam reforming can be provided via partial oxidation. A reformer can be operated endothermically, exothermically, or autothermically through the oxygen supply in the overall balance [46]. As already noted, numerous side reactions occur in a reformer. Due to the high operating temperatures and the simultaneous presence of water, hydrogen, carbon monoxide, and CO_2, the water–gas shift reaction (Equation (6)) takes place [46]. In addition, soot can form due to various reaction mechanisms. These reactions are, in particular, the Boudouard reaction (Equation (9)), methane splitting (Equation (17)), and CO or CO_2 hydrogenation (Equations (18) and (19)) [46]. These soot formation reactions are undesirable during operation of the reformer, and can be suppressed by means of a suitable starting material composition. According to Rostrup-Nielsen [47], H_2O/C rates of 0.6 are suitable for this:

$$CH_4 \leftrightarrow H_2 + C_{(s)} \tag{17}$$

$$CO + H_2 \leftrightarrow H_2O + C_{(s)} \tag{18}$$

$$CO_2 + 2H_2 \leftrightarrow 2H_2O + C_{(s)} \tag{19}$$

3.5. Carrier Steam Distillation

In the petrochemical industry, what is known as carrier steam distillation is usually used to separate hydrocarbon mixtures. Carrier steam distillation constitutes a special case of distillation that enables mixtures to be gently separated. It uses the addition of the vapor pressures of immiscible liquids. The mixture to be separated is evaporated with a low-boiling entrainer (often water) so that the boiling temperature is reduced. The desired fractions can then be drawn off from the carrier steam distillation column via side draws, and the entrainer can then be separated off again [48].

At this point, it should be noted that the petrochemical products are rarely specific chemicals, but are usually mixtures of different components with different properties. To characterize these mixtures and design separation processes, therefore, boiling point ranges or specific temperatures along these boiling curves are generally used. The temperature at which the mixture begins to evaporate is referred to as the initial boiling point (IBP), and the temperature at which the mixture has completely evaporated is called the final boiling point (FBP). The temperature at which a certain volume, for example, 10% of the liquid has evaporated is called the 10% point or T10. More information on the characteristic points is presented by Wegener [43] concerning the boiling curve of jet A-1 aircraft fuel [48].

3.6. Technology Readiness Level

In this section, the well-known technology readiness level (TRL) method is used to evaluate the power-to-fuel process developed in this paper. The TRL method indicates the maturity of a technology on a scale from 1 to 9. The method was originally developed by NASA [49] and is now used with adapted definitions in various areas [50]. In this work, the definitions established by the European Commission for the renewable energy sector are employed [51].

The TRLs of CO_2 capture technologies range from medium to very high. According to Schmidt et al. [8], for example, CO_2 separation from industrial waste gases by means of amine scrubbing (MEA) is already in use and has a TRL of 9. The separation of CO_2 from the air, however, is still at an earlier stage of development and assigned a TRL of 6 [8]. As described in Section 3.1, high-temperature electrolysis was used as part of "Fuel 1" by Sunfire GmbH in a larger demonstration plant and therefore assigned a TRL of 5 [13]. High-temperature co-electrolysis has only just started its test phase using the SUNFIRE-SYNLINK technology in 2019, and accordingly has not yet reached the same level of maturity as high-temperature water electrolysis. The remaining technologies used as part of the developed power-to-fuel process are highly developed technologies that are already in use on an industrial scale. The TRLs of these are correspondingly high. There are no specific statements regarding the values for the TRL of carrier steam distillations and reformers, but Luyben [48] describes the industrial use of carrier steam distillations and, it is well known that reformers are used in large-scale processes.

The TRLs of the individual components of the power-to-fuel process considered in this work are, with the exception of high-temperature electrolysis and CO_2 separation from the ambient air, very high. However, according to Schmidt et al. [8] and Marchese et al. [33], the TRL of a power-to-fuel process automatically falls to the lowest TRL in the process chain. According to this, the TRL of the developed power-to-fuel process in this work was assessed as 3 due to the low TRL level of the co-electrolysis step.

4. Modeling and Simulation in ASPEN PLUS

This section is dedicated to the modeling of the power-to-fuel process in Aspen Plus. For this purpose, the material data and property data models used are first presented. Then, the procedural design of the individual process components as well as the respective selected operating conditions are explained in more detail. For the sake of clarity, the relevant sections of the process flow diagram created in Aspen Plus are shown in Sections 4.2–4.6. When modeling the process, care is taken to ensure that the simulation is suitable for any mass flows.

4.1. Material Property Data and Material Property Data Models

As part of the power-to-fuel process developed, a low-temperature Fischer–Tropsch synthesis was used. Hence, with respect to de Klerk [36,52] and Dry [53], only straight-chain, unbranched alkanes with the empirical formula C_nH_{2n+2} were considered. The material data of the hydrocarbons C_1 to C_{29} were obtained from the database integrated in Aspen Plus. Based on Schemme [54], the C_{30+}-hydrocarbons were viewed as three groups of pseudo-components, with the C_{30-35}-, the C_{36-47}-, and C_{48+}-hydrocarbons grouped together. A representative molecular structure was selected for each of the three pseudo-components. The American Petroleum Institute (API) method, with the data given in Table 1, was used to calculate the thermodynamic properties of the pseudo-components. The Aspen Plus database was used for the material data of components H_2, H_2O, CO, and CO_2. The material data models used were selected on the basis of the general recommendations of Carlson [55] and other application-specific literature. A total of four different material data models were used to simulate the process.

Table 1. Properties of the pseudo-components [56].

	Pseudo-Components		
	C_{30-35}	C_{36-47}	C_{48+}
Representative molecular structure	$C_{32}H_{66}$	$C_{41}H_{84}$	$C_{61}H_{124}$
Molar mass in g/mol	454.9	572.2	861.7
Relative density at 60 °F (≈15.6 °C)	0.818	0.827	0.839
Boiling point at 1 atm in °C	469.3	528.1	624.0

The equation of state of Soave–Redlich–Kwong is widely used in the field of gas processing. In the context of this work, the equation of state with reference to Marchese et al. [33] was used as a material data model for the modeled electrolysis types. In order to be able to more precisely calculate the phase equilibrium between gas and liquid in the presence of hydrocarbons and light gases such as CO_2 and H_2 in the supercritical range, the Soave–Redlich–Kwong equation of state can be extended to the RKS–BM material data model with the Boston–Mathias alpha function. Drawing on Schemme [54], this model was used to calculate the reformer and the parts of the product separation with very low proportions of pseudo-components. The material model Braun K-10 was used to calculate the material flows with higher proportions of pseudo-components. This material data model was especially developed for calculating hydrocarbon mixtures with both real and pseudo-components, and was used in simulations for the Fischer–Tropsch reactor and hydrocracker. The NRTL model enables the description of gas–liquid equilibria as well as liquid–liquid equilibria of strongly non-ideal mixtures. The activity coefficients of the liquid phases are calculated on the basis of experimentally-determined binary interaction parameters. The calculation of CO, CO_2, and H_2 also takes Henry's law into account. In the simulation, the NRTL–RK material data model was used to calculate the carrier steam distillation. In general, the calculation of the gas phase is by default carried out using ideal gas law. In this work, the Redlich–Kwong equation of state was used instead to describe the gas phase. The NRTL–RK material data model was also used for heat exchangers in which a large proportion of water is in liquid form.

4.2. Co-Electrolysis and Water Electrolysis

As Aspen Plus does not have a stored model for the simulation of high-temperature electrolysis, a combination of different Aspen Plus blocks, also known as "Unit Operations", must be used to calculate co- and water electrolysis. In addition, so-called design specs are employed to establish the required process conditions. Similar configurations for simulating high-temperature electrolysis have already been used by Cinti et al. and Marchese et al. [31,33]. The simulation flow diagram for co-electrolysis is shown in Figure 2. It should be noted that not all heat flows of the process can be seen directly in the flow diagram. Various operating resources were used in the simulation to provide the required heating or cooling capacity. For instance, W-5 is an air cooler and W-6 a water cooler, each of which uses the corresponding operating media "air" and "cooling water". The balancing of the resources used was carried out through the "utilities" function integrated in Aspen Plus and is discussed in greater detail in Section 5 The same applies to all of the following flow diagrams.

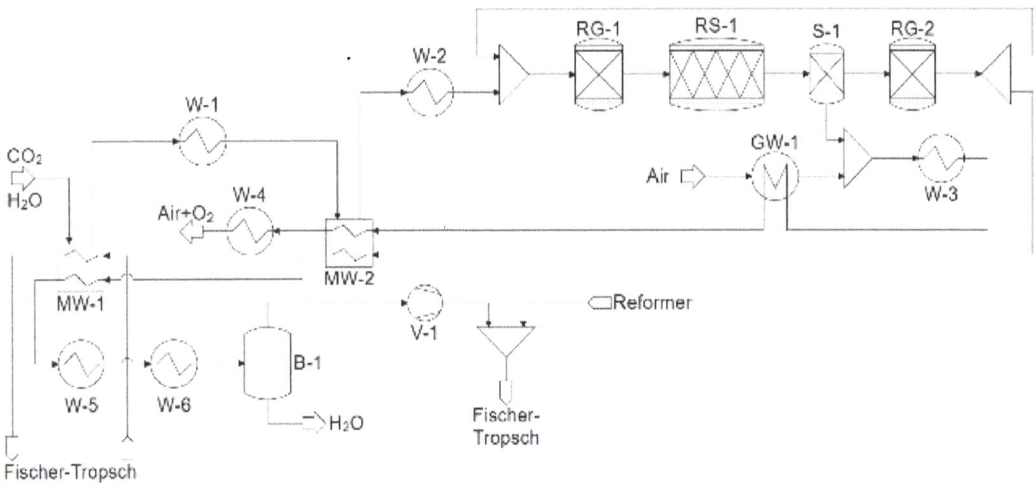

Figure 2. Excerpt from the process flow diagram: co-electrolysis.

First, the mixture of CO_2 and water, which is present at 1 bar, is warmed up over several stages in a heat exchanger and the water is evaporated. For this purpose, both the waste heat from co-electrolysis and the product flow of the Fischer–Tropsch synthesis as well as part of the waste heat from the Fischer–Tropsch reactor itself, was used. In addition, an electric heater was used with W-2, which ensured that the feed stream reached the electrolysis inlet temperature of 780 °C. The hot gas stream consisting of water vapor and CO_2 then enters the electrolysis cell. As already described, the electrolysis cell consisted of several unit operations with which the cathode (RG-1, RS-1, RG-2), the electrolyte (S-1), and anode (GW-1, W-3) were modeled. The feed stream was combined with a recycling stream and passed into a first equilibrium reactor (RG-1). Taking into account the equilibrium reactions that occur (WGS or RWGS: Equation (6), methanation: Equations (8) and (9)), this determines the composition of the gas flow by minimizing the Gibbs energy. Due to the high temperatures, it was assumed that the reaction equilibrium would be reached quickly (see Sun, et al. [57]). The Boudouard reaction (Equation (9)) was not taken into account in the simulation because, according to Sun, Chen, Jensen, Ebbesen, Graves and Mogensen [57], there is no deposition of solid carbon under the selected operating conditions of 800 °C and 1 bar. In the next step, the gas flow passes into the stoichiometric reactor RS-1, in which the electrolysis reactions (Equations (1)–(3)) take place. The conversion of RS-1 was automatically set with a design spec so that the total conversion of the electrolysis cell reactant utilization (RU) corresponded to the specified RU (see Equation (20) [33]). The total turnover was set at 70% with reference to Sun, Chen, Jensen, Ebbesen, Graves and Mogensen [57] and Marchese, Giglio, Santarelli and Lanzini [33]:

$$RU = \frac{\dot{n}_{react,\ in} - \dot{n}_{react,\ out}}{\dot{n}_{react,\ in}} \quad (20)$$

The electrolyte was modeled as a simple separator block (S-1), which separates the oxygen produced by the electrolysis reactions. The composition of the synthesis gas was then adjusted in parallel to RG-1 in a further equilibrium reactor (RG-2). While most of the synthesis gas then leaves the electrolysis segment, a portion of the stream is fed back. The size of the returned portion was set using a design spec so that the H_2 concentration in the feed of the electrolysis was at least 10 mol% in order to avoid oxidation of the nickel-based cathode [27,58]. Synthesis gas leaving the electrolysis cell is gradually cooled down for improved energetic utilization, and the unconverted water is condensed out. The synthesis gas is then compressed to 30 bar by a multi-stage compressor and merged with the reformed synthesis gas (see Section 4.6). Thereby, the H_2/CO ratio required for the Fischer–Tropsch synthesis was available after mixing of the synthesis gas streams and a further design spec was used that adjusts the ratio of water to CO_2 in the feed stream of the co-electrolysis system accordingly.

MW-1 and MW-2 represent multi-component counter-flow heat exchangers, which were used to recover the process heat for educt conditioning. W-1 is a heater used for further educt conditioning. V-1 represents a compressor. The separator block B-1 is used to separate the vapor phase and the liquid phase at equilibrium.

The oxygen stream separated by the separator block functioning as an electrolyte was mixed with an air stream on the anode side of the electrolytic cell. According to Cinti et al. [31], it is common practice to carry the oxygen through a stream of air in order to prevent the cell performance from being negatively influenced by a high concentration of the oxygen. The amount of air flow was set by a design spec so that the partial pressure of the oxygen after mixing with the air was 0.5 bar [31]. The air flow was warmed up as much as possible using a counter flow heat exchanger (GW-1) before it entered the electrolysis cell. As the cell temperature of 800 °C was not reached as a result, a heat exchanger block (W-3) was used to take into account the additional heating output that must be provided by the electrolysis cell. The air flow was then cooled in two steps and left the system air-enriched with oxygen.

A simplified flow chart of the water electrolysis for hydrogen provision for the hydrocracker is displayed in Figure 3. Its structure corresponded to that of co-electrolysis, but with the simplification that no equilibrium reactors were used. These were not required, as there was no CO_2 or CO present in the water electrolysis segment. Another difference is that the hydrogen was compressed to the operating pressure of the hydrocracker of 50 bar. Using the same logic as in the previous figure, W-7 and W-8 were heaters for steam generation, assisting the multi-component counter-flow heat exchanger MW-3. RS-3 was the stoichiometric reactor representing the water electrolysis. The separator block S-2 divided the products to anode and cathode sections. W-9 and W-10 were used to cool the product mixtures of the water electrolysis, GW-2 was a counter-flow heat exchanger used for pre-heating air feed, and V-2 was used to compress the produced hydrogen.

For the techno-economic analysis of the power-to-fuel process, it is necessary to determine the performance of the co- and water-electrolysis processes. For this purpose, the calorific value of the product flows could be output via Aspen Plus and the electrolysis output could then be calculated using a specified efficiency (see Equation (10)).

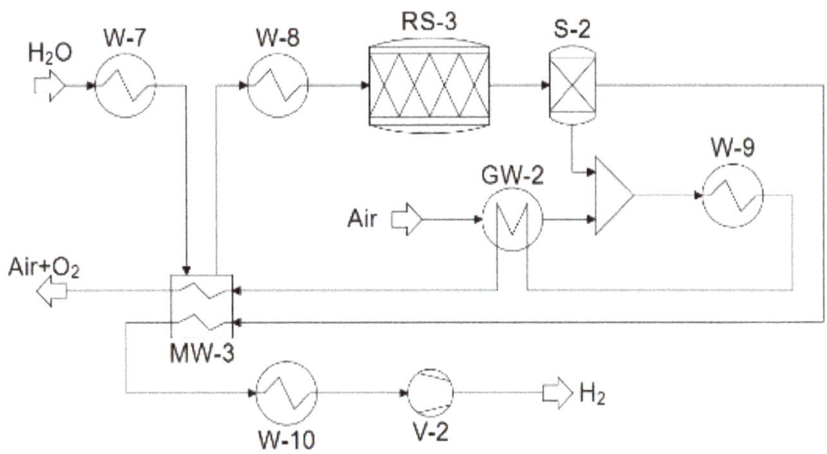

Figure 3. Excerpt from the process flow diagram: water electrolysis.

4.3. *Fischer–Tropsch Synthesis and Product Separation*

The flow chart of the Fischer–Tropsch synthesis and the subsequent product separation is shown in Figure 4. The synthesis gas from the co-electrolysis was mixed with the synthesis gas from the reformer and fed into the Fischer–Tropsch reactor (RS-2). This was modeled as an isothermal bubble column reactor with a pressure of 30 bar and a temperature of 210 °C. A stoichiometric reactor was used for modeling, in which the reaction equations for 32 parallel reactions according to Equation (11) were stored (for $n = 1$ to $n = 29$ and the pseudo components corresponded to $n = 32$, $n = 41$ and $n = 61$). The conversions of the various reactions were determined using a calculator block with an integrated Excel file, based on the ASF distribution (Equation (12)), with the chain growth probability α calculated using Equation (14). The modeling was therefore suitable for calculations with variable H_2/CO ratios; however, a ratio of 1.8 was chosen in the context of this work in order to maximize the proportion of middle distillates in the product of the Fischer–Tropsch synthesis. Under the selected operating conditions and the set H_2/CO ratio, there was a chain growth probability of approximately 0.92. According to de Klerk [36], this value is in the range of typical industrial low-temperature Fischer–Tropsch syntheses. The total conversion of carbon monoxide was set at 80%, reflecting the work of Becker, et al. [59], Trippe [35], and Schemme [54].

As the stoichiometric reactor used only had one product stream, in a first step (B-2) the product from the Fischer–Tropsch reactor was divided isothermally and isobarically into gas and liquid phases. The liquid phase, which consisted mainly of C_{20+}-hydrocarbons, was fed to the hydrocracker, while the gas phase was cooled isobarically in several steps to 40 °C, with part of the waste heat being used in the co-electrolysis process. After cooling, the light gases were separated from the middle distillates in the next container (B-3). Almost all of the C_{8+} hydrocarbons that could be used for kerosene and diesel were sent to the carrier steam distillation, and the short-chain hydrocarbons were sent to the reformer. In addition, the differences in density and polarity of the water and hydrocarbons in both tanks were used to separate the water produced by the Fischer–Tropsch synthesis. In order to avoid soot formation in the reformer, some of the separated water was fed into the reformer. The proportion fed into the reformer was determined with a design spec corresponding to the selected H_2O/C ratio (see Section 4.6).

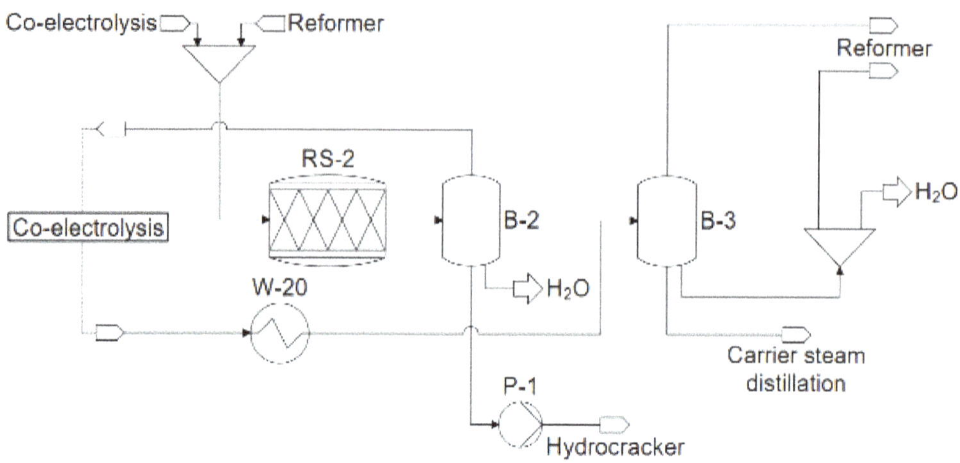

Figure 4. Excerpt from the process flow diagram: Fischer–Tropsch synthesis and product separation.

4.4. Hydrocracker

Figure 5 displays the modeled hydrocracker. The long-chain hydrocarbons were brought together with several re-circulations in which the required hydrogen was also added. The hydrogen was produced using water electrolysis and the total amount adjusted so that the mass fraction of hydrogen after mixing was around 8% (mass). This value was chosen as a compromise between the suppression of both soot formation and catalyst deactivation (see Section 3.3) and minimization of the required additional electrolysis performance.

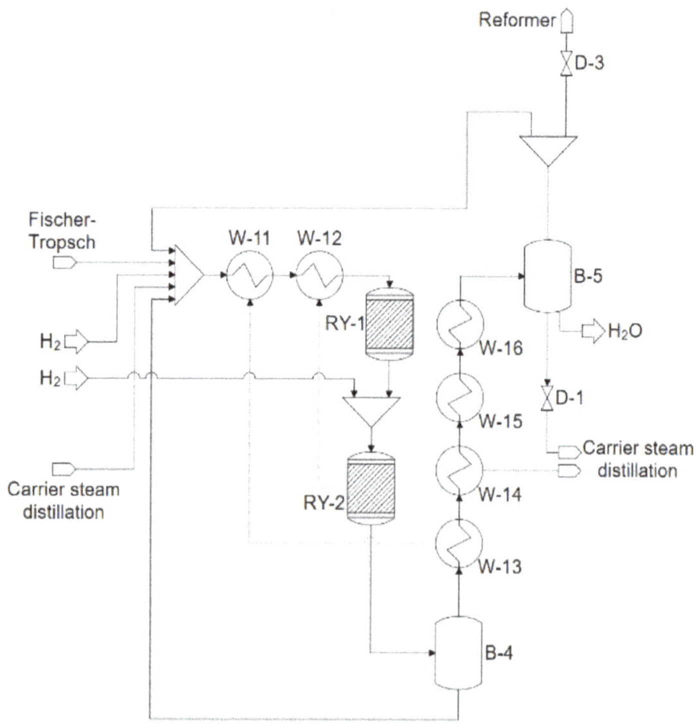

Figure 5. Excerpt from the process flow diagram: hydrocracker.

The mixed material flow was preheated and fed into the hydrocracker. The hydrocracker was modeled as a two-stage isothermal fixed bed reactor, whereby the two fixed beds were each modeled by a RYield block (i.e., a yield reactor). To ensure ideal hydrocracking, an operating pressure of 50 bar and a temperature of 350 °C were selected and a conversion per bed of 60% assumed. The yields of the respective cracking products were calculated in calculator blocks connected to Excel files, based on the uniform distribution of the cracking products of the ideal hydrocracking, which is described in more detail in Section 3.3 The heat required for the cracking process was provided by the exothermic reformer (see Section 4.6). The product stream was then separated into gas and liquid phases, and the long-chain hydrocarbons that had not been cracked were returned to increase the conversion of the hydrocracker. The gas stream was cooled in several stages for better energy utilization and the middle distillates were separated off, expanded to 30 bar, and passed to the carrier steam distillation. The gas stream separated in B-5 consisted mainly of hydrogen and C_{7-}-hydrocarbons, H_2O and CO_2, and was partly returned to the hydrocracker in order to reduce the need for fresh hydrogen and thus the required performance of the water electrolysis system. A purge flow of 10% was expanded to 30 bar and fed to the reformer to prevent CO_2 and short-chain hydrocarbons from accumulating in the hydrocracker.

4.5. Carrier Steam Distillation

The feed streams of the carrier steam distillation, which can be seen in Figure 6 were the product stream of the hydrocracker unit and the separated middle distillates from the product separation following the Fischer–Tropsch synthesis (see Figure 4). As the NRTL–RK material data model was to be used in the distillation column, which is not suitable for calculating mixtures with pseudo components, the remaining pseudo components (approximately 0.6% by weight) were assigned to the real component $C_{29}H_{60}$. A RYield block (RY-3) was used for this. In the next step, the feed was directed into a so-called pre-flash drum (B-6), in which part of the material flow was evaporated by releasing it to 1 bar. The use of a pre-flash drum is recommended in the literature [48] for the separation of some of the gases and thus simplify product separation in the column. In the simulation, almost all of the remaining CO_2, hydrogen, and methane as well as most of the C_2- to C_5-hydrocarbons were separated by the flash evaporation. The liquid phase was preheated by heat integration and fed to the vaporizer (furnace) of the column, which was modeled with a Petrofrac block (K-1). The energy required to evaporate the feed was provided by the exothermic reformer (see Section 4.6). Steam was fed to the column via the sump, which reduced the partial pressure of the hydrocarbons and thus lowered the boiling temperatures. The column had two strippers with four stages and their own steam supplies, each for kerosene and diesel. There was no need for a side exhaust for gasoline, which is common in industrial applications. The number of stages in the column as well as the position of the feed and the side draws were determined using a sensitivity analysis in such a way that as much kerosene and diesel as possible was produced, but the requirements of the standards (see Table 2 were still met. The requirements of the standards were taken into account in two design specs through which the specified T50 temperature of the respective product flow was set by automatically varying the amount of kerosene or diesel withdrawn. At this point, it should also be noted that the use of pump-arounds was dispensed with. Pump-arounds, through the recirculation of material flows within the column, represent a possibility for recovering thermal energy and reducing the column's energy requirement [48]. In the simulation, however, the use of pump-arounds led to considerable convergence problems. The bottom product, consisting of long-chain hydrocarbons, was returned to the hydrocracker, and the top product, together with the gas flow from the pre-flash drum, was compressed to 30 bar and sent to the reformer. A detailed analysis of the fuels produced is presented in Section 5.1.

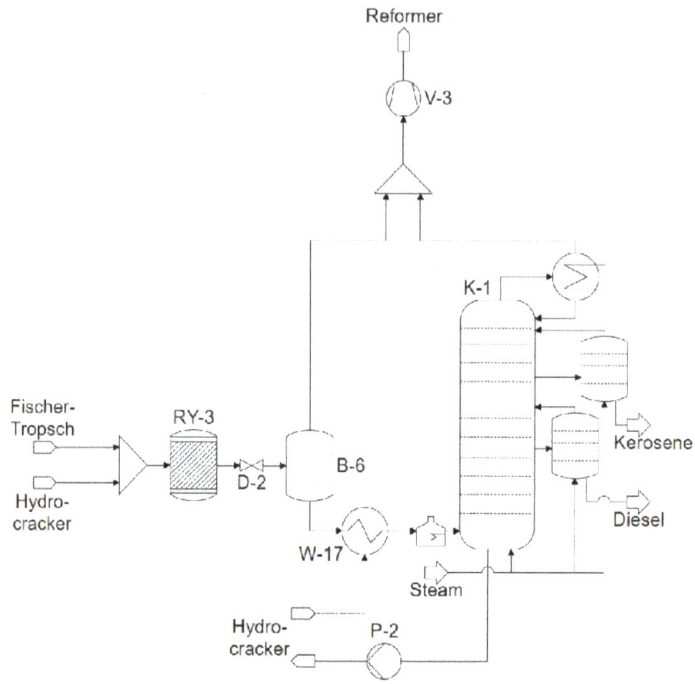

Figure 6. Excerpt from the process flow diagram: carrier steam distillation.

Table 2. Characteristic values of the synthetic kerosene and diesel produced.

Characteristic Value	Unit	Kerosene FT-SPK (ASTM 7566)	Syn. Kerosene (Simulation)	Diesel Class A (EN 15940)	Syn. Diesel (Simulation)
T10	°C	≤205	157	-	-
T95	°C	-	-	≤360	356
T90–T10	K	≥22	70	-	-
FBP	°C	≤300	274	-	-
Density @ 15 °C	kg/m^3	730–770	738	765–800	779
Cetane number	-	-	-	≥70	120 [1]
Freezing point	°C	≤−40	n/a	-	-
Heating value (LHV)	MJ/kg	-	44.17	-	43.85

[1] Probably overrated; for an explanation see text.

4.6. Reformer

The reformer was used to convert the unusable gases into synthesis gas and thus to recycle them. The corresponding section of the simulation flow diagram is shown in Figure 7. The gas streams available at 30 bar from the distillation column, the hydrocracker and the product separation were combined with the addition of water. The amount of water added was automatically set using a design spec so that there was an H_2O/C ratio of 0.6 after mixing. The feed stream was then heated to 900 °C with the waste heat from the product stream and passed into the reformer. The reformer was modeled with an RGibbs block (i.e., an equilibrium reactor). Due to the high operating temperature, the short residence time and gas phase reaction, the kinetics can be neglected [54] and the product composition can be determined by minimizing the Gibbs energy. The addition of oxygen was automatically controlled by a design spec so that a product temperature of 950 °C results. The reformer was exothermically-operated in order to provide the high-temperature heat required for the hydrocracker and the distillation column through the waste heat. The additional heat to be produced was taken into account by a cooling capacity imposed on the reformer from the outside. The reformed synthesis gas was cooled and then mixed with the synthesis gas from the co-electrolysis unit and fed into the Fischer–Tropsch reactor.

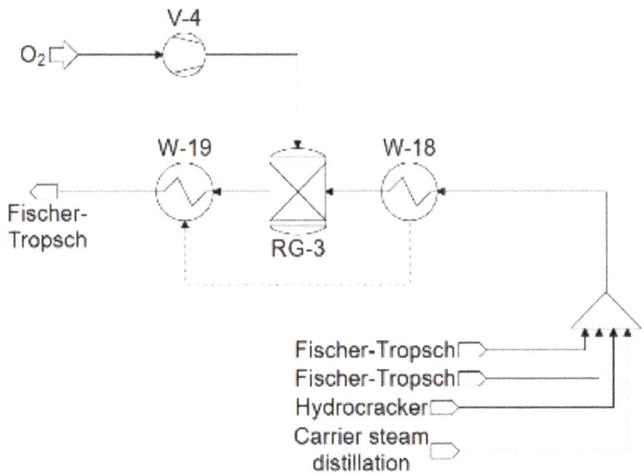

Figure 7. Excerpt from the process flow diagram: reformer.

5. Results from Process Analysis

In this section, the simulation results of the power-to-fuel process presented in Sections 3 and 4 are presented. First, it was determined whether the diesel and kerosene fuels produced met the requirements of the relevant standard. The composition of the fuels was considered, the boiling curves were calculated and compared with the real boiling curves of kerosene and diesel, and finally the fuel property data were checked. Then, the process balance was considered and the PtL efficiency of the process established. The next step was to examine the extent to which heat extraction for the provision of thermal energy for CO_2 separation is possible and useful. Finally, the developed process was compared with other power-to-fuel or PtL processes.

5.1. Fuel Property Analysis

The mass fractions of the various carbon chains in the kerosene and diesel produced as well as the mass-related average molecules of the respective fuels are shown in Figure 8. The kerosene withdrawn from the carrier steam distillation column contained almost no C_1- to C_7-hydrocarbons and approximately 8.3% by weight of C_8-hydrocarbons. The mass fraction of C_9 to C_{11} is between 18 and 19% by weight and the fraction of C_{12} is around 16.5% by weight. The proportion of longer hydrocarbons steadily decreased, with C_{13} being contained in kerosene at approximately 11.6% by weight, C_{14} at approximately 5.4% by weight, and C_{15} at approximately 1.8% by weight. Carbons with a chain length of 16 and longer are only contained in kerosene to a very small extent, with a total of around 0.7% by weight. The mass average chain length of the hydrocarbons in kerosene is 10.9. The diesel fuel produced consisted of approximately 0.7% by weight of carbons with a chain length of eleven or less. The mass fraction of longer hydrocarbon chains increased steadily from C_{12} at about 2% by weight, to C_{15} at 11.5% by weight. The curve of the mass fractions then flattened out and the C_{16}- to C_{20}-hydrocarbons were each contained in the diesel fuel at approximately 12.2% by weight. The mass fraction of longer hydrocarbons initially dropped sharply at 2.1% by weight for C_{21}, and then dropped uniformly with increasing chain lengths to approximately 0.3% by weight of C_{28} hydrocarbons. The proportion of C_{29}-hydrocarbons was around 1.2% by weight. The mass-related average chain length of diesel fuel was 17.5.

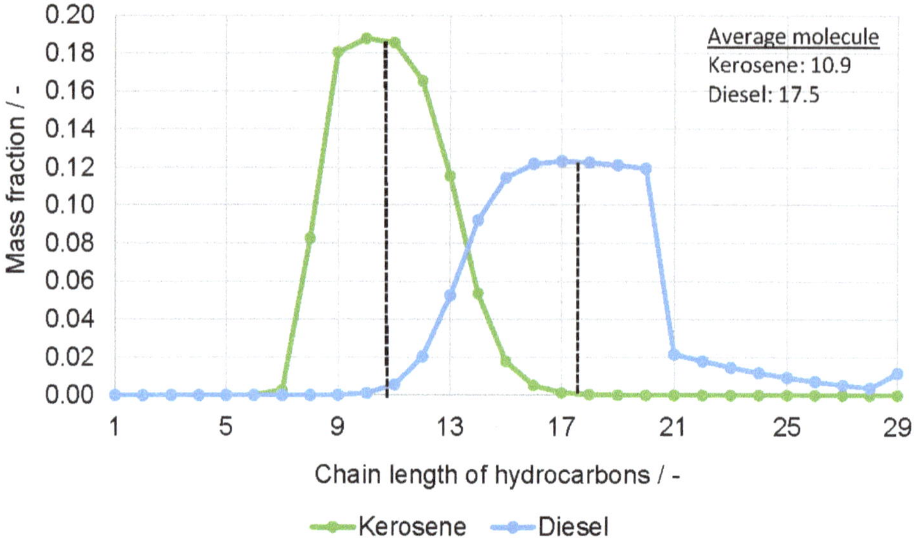

Figure 8. Product distribution of the produced kerosene and diesel fraction.

In the literature, chain lengths of between 8 and 16 carbon atoms were assigned to the kerosene fraction [37]. Carbons with these chain lengths made up more than 99.5% by weight of the kerosene produced in the simulation. The product distribution of synthetic kerosene was therefore in the range of typical aircraft fuels. The sharp drop in the mass fractions from C_{20} to C_{21} in the diesel produced can be explained by the fact that the proportion of C_{21+}-hydrocarbons in the feed of the distillation column was also low. This was, in turn, due to the majority of the C_{21+} hydrocarbons produced in the Fischer–Tropsch reactor being fed directly to the hydrocracker and cracked there. The comparatively high proportion of C_{29} can be attributed to the fact that, as described in Section 4.4, the pseudo-components were combined with the C_{29} hydrocarbons. Therefore, the 1.2 wt.% not only consisted of C_{29} hydrocarbons, but also contained all of the hydrocarbons with a chain length of 30 or more. Which hydrocarbon chains are assigned to diesel varies with the literature source. For example, Trippe [35] only counted the C_{13} to C_{20} hydrocarbons in the diesel fraction, whereas Bacha, et al. [60] and Dieterich et al. [6] assigned hydrocarbons with chain lengths of 10 to 22 or 23 to the diesel fraction. The synthetic diesel fuel produced in the simulated power-to-fuel process consisted of approximately 86.8% by weight of C_{13} to C_{20} and approximately 95% by weight of C_{13} to C_{23}. Accordingly, the product composition of the diesel produced fell within the range of typical diesel fuels. Both the product composition of the synthetic kerosene and that of the synthetic diesel were therefore within the acceptable range.

In the next step, the boiling curves of the fuels produced in the simulation were calculated and compared with the real boiling curves of kerosene Jet A-1 and diesel (see Figures 9 and 10). The boiling curves were calculated using the D86CRV method integrated in Aspen Plus, which calculates the boiling curve of a mixture of substances at atmospheric pressure.

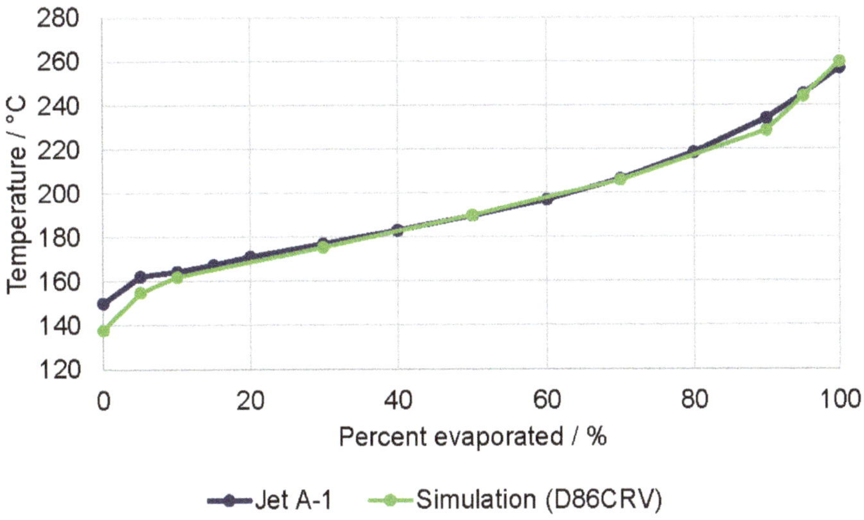

Figure 9. Boiling curves of jet fuel in comparison to Jet A-1 (boiling curve according to Edwards [61]).

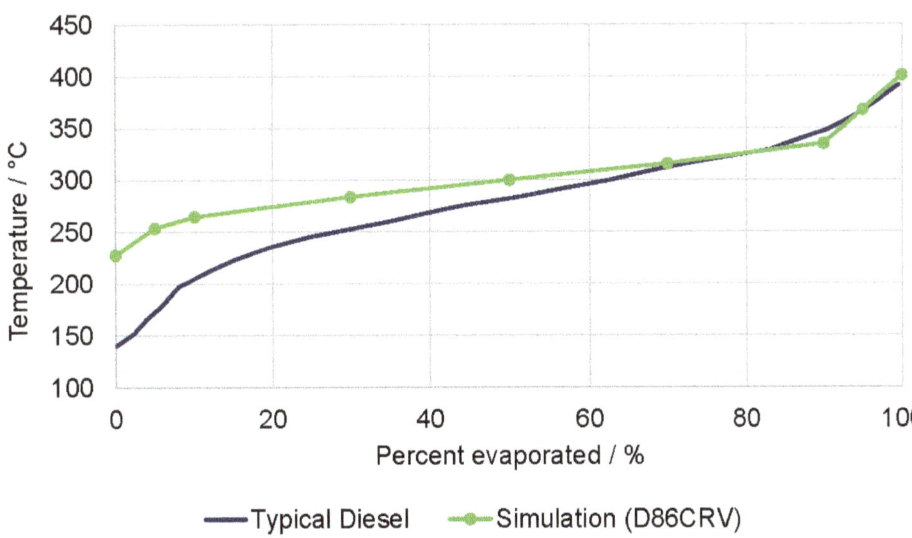

Figure 10. Boiling curves of diesel in comparison (boiling curve of "typical diesel fuel" according to Bacha et al. [4]).

The calculated boiling curve of synthetic kerosene and that of Jet A-1, according to Edwards [61], are shown in Figure 9. Both curves were almost identical, with the most noticeable difference being that the IBP of the synthetic kerosene of the simulation (137 °C) was below that of the reference kerosene of type Jet A-1 of approximately 150 °C. The T5 points of the boiling curves deviated by approximately 7 K at 155 °C for the calculated value and approximately 162 °C for the reference. With the exception of the T90 point, the boiling curves deviated from each other by less than 3 K in the area from T10 to the FBP, the deviation at the T90 point being approximately 5 K. Overall, it can be concluded that the boiling behavior of the kerosene produced in the simulated power-to-fuel process corresponded well to that of jet A-1 kerosene.

Figure 10 shows the calculated boiling curve of the synthetic diesel fuel produced and the boiling curve of a "typical" diesel fuel according to Bacha et al. [60]. A comparatively large difference could be seen between the IBP of the two boiling curves. The IBP of the reference diesel was around 140 °C, whereas that of the diesel fuel produced was around 228 °C; about 88 K higher. The boiling curves converged as the proportion of evaporated volume increased. The T10 points were about 55 K apart at about 210 °C for the reference diesel and 265 °C for the calculated boiling curve, whereas the T50 points differed by about 16 K at approximately 283 °C and 299 °C.

As the proportion of evaporated fuel increased, the boiling curves further converged. The FBP of the reference diesel was around 390 °C and that of the calculated boiling curve was around 400 °C. Overall, the deviations between the boiling curves of the reference and synthetic diesels were greater than those between the boiling curves of kerosene. Larger deviations could be found, especially in the area spanning the IBP to the T50 point. When comparing the compositions of the two diesel fuels, however, it was noticeable that the reference diesel according to Bacha et al. [60] had a comparatively higher proportion of short-chain light hydrocarbons. The average chain length of the hydrocarbons of the reference diesel was approximately 16, and accordingly, below the average chain length of the diesel fuel produced of 17.9. The deviations in the boiling curve can be explained by the correspondingly different compositions of diesel fuels. In order to determine whether the kerosene and diesel fraction produced is suitable for use as fuels, the requirements of the respective standards were checked in the next step. Table 2 compares the requirements of the standards and corresponding characteristic values of the fuels calculated in the simulation. In addition, the calorific values of synthetic kerosene and diesel are given, which were calculated using the Aspen Plus component properties (property sets) QVALNET.

The T10, T95, and T90 temperatures as well as the FBP of the synthetic kerosene were read from the boiling curves and were in the required range for both the synthetic kerosene and synthetic diesel. The densities of the two fuels were determined by the Aspen Plus component property RHOLSTD. The density of the FT SPK (738 kg/m^3) was at the lower end of the permitted range, but still met the requirements of ASTM 7566. The density of the synthetic diesel was also within the required range at 779 kg/m^3. The Aspen Plus component property CETANENO was used to determine the cetane number of synthetic diesel. This yielded a cetane number of 109.2 for pure n-hexadecane ($C_{16}H_{34}$). According to Dry [53], however, the actual cetane number of n-hexadecane is 100. Accordingly, it can be assumed that the actual cetane number of synthetic diesel is below the calculated value of 120. The freezing point of kerosene cannot be calculated using Aspen Plus. Due to the very high proportion of n-alkanes in the product of the Fischer–Tropsch synthesis [9], and therefore a very high proportion of n-alkanes in kerosene, there is the possibility that an after-treatment of the kerosene is necessary in order to improve the low-temperature properties. One possibility would be isomerization, as isoalkanes have a significantly lower freezing point [62]. On the basis of this sub-section, it should be noted that both of the synthetic fuels produced, namely kerosene and diesel, met the requirements of the respective standards. However, if the synthetic kerosene is to be used as a mixture component for Jet A-1, an after-treatment may be necessary to improve the low-temperature properties.

The following section deals with the balance of the process and presents the amount of energy and material required to produce the synthetic fuels.

5.2. Balancing the Power-to-Fuel Process

The simulation of the PtF process created in Aspen Plus is essentially suitable for calculating any mass flows. The process balance based on the production output is discussed in the following section. For this purpose, the unit liter diesel equivalent l_{DE} was defined as 35.9 MJ. First, the materials required for the production of synthetic fuels are discussed. Then, the process balance is considered from an energetic point of view, whereby the balance of the equipment used is also presented. The material balance is shown in Figure 11 as an overview. It should be noted that not all process-internal heat flows are shown in the figure. The heat flows shown in Figure 11 are those that have a major influence on the overall process, and are discussed in more detail below. If 1 l_{DE} is produced in the simulation, this liter consists of 38.9% synthetic kerosene and 61.1% synthetic diesel. A total of 3.99 kg of water per liter of diesel equivalent produced is required as a feed for the water and co-electrolysis as a material for the production of the fuels, of which approximately 0.18 kg of water is used for pure water electrolysis. In addition, CO_2 is required for the co-electrolysis process. The required amount is approximately 2.54 kg CO_2/l_{DE}. In addition to water and CO_2, the power-to-fuel process requires oxygen to operate the reformer. For each liter of diesel equivalent produced, around 0.34 kg of oxygen is required. Due to internal returns and recycling streams in the process, the only waste streams that arise are oxygen-enriched air streams from electrolysis and separated water. An overview of the material balance of the process is presented in Table A1 (Appendix B). In the developed power-to-fuel process, there are various heat sinks and heat sources. In order to minimize the energy requirement, an energy integration was carried out. Heat sinks and sources were partially coupled to each other via direct heat exchange between material flows, and the required heating or cooling capacity was partially provided by operating equipment.

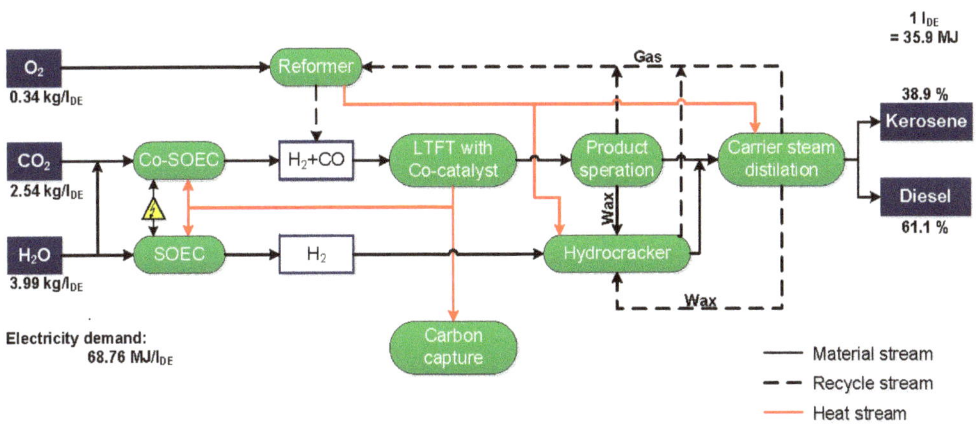

Figure 11. Energy-specific material balance of the developed fuel synthesis.

The low- and medium-pressure saturated steam were used as both a heating medium and coolant by generating the corresponding steam. For example, the Fischer–Tropsch reactor was cooled by generating medium-pressure saturated steam with the waste heat from the reaction. In addition, it was assumed that low-pressure saturated steam was used as an entrainer for the operation of the carrier steam distillation column, and this consumption was correspondingly taken into account in the balance. Electricity is required to operate the compressors and pumps as well as the two electrical heaters W-2 (see Figure 2) and W-8 (see Figure 3). An isentropic efficiency of 76% was assumed for the compressors and of 60% for the pumps. The electricity demand for water and co-electrolysis was considered separately.

Table A3 in Appendix B shows that the demand for both low-pressure and medium-pressure steam can not only be covered within the process, but there is even usable thermal energy left over. As displayed in Figure 11, this thermal energy can be discharged from the process and used, for instance, for carbon capture. This heat coupling is examined in greater detail in Section 6.3. The table does not include the heating power required to operate the carrier steam distillation column and the hydrocracker of 0.39 MJ/l_{DE} and 0.41 MJ/l_{DE}, respectively. As both the distillation column and hydrocracker are operated at over 300 °C, it is not possible to provide this heat demand using low- or medium-pressure steam. In order that no valuable electrical energy must be used to provide heating power, the reformer was operated exothermically instead of autothermically, as is common in many industrial applications [46]. The heat made available in this way can, as Figure 11 indicates, be used to operate the hydrocracker and distillation column. Thermal oils can be used for heat transfer as they enable heat transfer at temperatures of up to 400 °C [63]. As a result, with the exception of the high-temperature heat required to operate the water and co-electrolysis, all of the required process heat can be provided via internal heat integration. As described in Section 3.1, the electrical power required for the electrolysis can be calculated using the calorific value of the electrolysis products and determining the efficiency. The calorific value and output of the products can, in parallel to the calorific value of synthetic fuels, be determined using the Aspen Plus component property, QVALNET. For the electrical efficiency of high-temperature electrolysis, values from 60% to over 100% can be found in the literature, depending on the mode of operation of the SOEC and how the balance space of the electrolysis system is selected [64]. If, for example, the required heat is not included in the calculation, if this is potentially available free of charge as waste heat from another process, the efficiency of the electrolysis system improves accordingly. In the developed power-to-fuel process, the low-temperature heat for the evaporation of the water is provided through heat integration and therefore does not need to be taken into account when determining the electrolysis efficiency. However, the required high-temperature heat cannot be provided through heat integration but must be supplied to the system in the form of electrical energy. Therefore, it must be taken into account in the efficiency calculation. In this case, Peters et al. [64] specified an electrolysis efficiency η_{SOEC} of approximately 80% for approximately thermo-neutral operation. At this degree of efficiency, however, compression work for storing the electrolysis products is also included. Although this type of compression was not carried out in the developed power-to-fuel process, an electrolysis efficiency of 80% was selected to be on the safe side, in order not to underestimate the required electrical power of the electrolyzers. The amount of synthesis gas or hydrogen calculated in Aspen Plus and the specified efficiency result in the electrical energy required to operate the electrolyzers of approximately 3.09 MJ/l_{DE} for water electrolysis and 58.78 MJ/l_{DE} for co-electrolysis. One of the greatest advantages of energy integration, which is made possible by a combination of high-temperature electrolysis and Fischer–Tropsch synthesis, can also be made clear on the basis of these values. As shown in simplified form in Figure 12, the feed stream of the co-electrolysis is evaporated and preheated in a heat exchanger section (see Figure 2). About 16.4 MJ/l_{DE} of thermal energy was added to the feed stream. For this purpose, the waste heat from the Fischer–Tropsch synthesis was used, amongst other things, via the energy integration. Without energy integration, this amount of energy would need to be added to the high-temperature electrolysis. Conversely, this means that the energy requirement for co-electrolysis is reduced by more than 20% through the energy integration, and that excess thermal energy is also available and can be used for CO_2 separation (see Section 5.3).

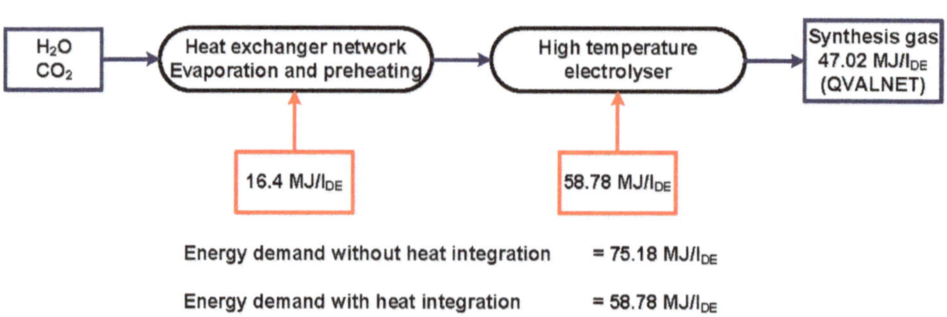

Figure 12. Simplified balance for co-electrolysis.

Once the power-to-fuel process has been fully balanced, the overall efficiency of the process, referred to as the PtF or PtL efficiency, can be determined. As the fuel synthesis does not need to be supplied with external thermal energy, the efficiency is calculated on the basis of the production output and electrical energy requirement. The electrical energy requirement was made up of the energy requirement of 58.78 MJ/l_{DE} for co-electrolysis, the energy requirement of 3.09 MJ/l_{DE} for water electrolysis, and the energy requirement for operating all other system components of 6.89 MJ/l_{DE} (see Table A3). It should be noted that CO_2 separation is not yet taken into account at this point. The following section deals with the influence of CO_2 separation on the PtL efficiency. This results in a power-to-fuel efficiency for fuel synthesis of:

$$\eta_{PTL} = \frac{35.9 \frac{MJ}{l_{DE}}}{6.89 \frac{MJ}{l_{DE}} + 3.09 \frac{MJ}{l_{DE}} + 58.78 \frac{MJ}{l_{DE}}} = 52.21\% \quad (21)$$

It becomes clear that the electrical power required for high-temperature electrolysis and so the electrolysis efficiency exerts the greatest influence on the power-to-fuel efficiency. To take a closer look at this effect, Figure 13 displays the overall efficiency of the process versus that of high-temperature electrolysis. The overall efficiency of the process is linearly dependent on the electrolysis efficiency. For an η_{SOEC} of 60%, the η_{PTL} drops to 40%. However, according to Peters et al. [64], electrolysis efficiencies of less than 70% only occur when the low-temperature heat for evaporation of the water must be provided by electrical energy. As this heat is available in the developed power-to-fuel process via the energy integration, an electrolysis efficiency of 70% was assumed for the worst case. Accordingly, the lowest possible power-to-fuel efficiency was around 46.26%. In the event that high-temperature heat is available and correspondingly does not need to be provided by electrical energy, electrolysis efficiencies of 100% and higher are possible. If the developed power-to-fuel process is to be built, for instance, in a network location where such high-temperature heat is available, a power-to-fuel efficiency of 63.67% is theoretically possible with an electrolysis efficiency of 100%. At this point, it should be noted that the assumed lower limit of the efficiency of 70% and also the 80% chosen for the base case represent conservative assumptions. Due to the energy integration carried out and the fact that in the developed fuel synthesis no compression of the electrolysis products for storage was carried out, the electrolysis efficiency tended to be in the range above 80%, rather than in that from 70% to 80% [64].

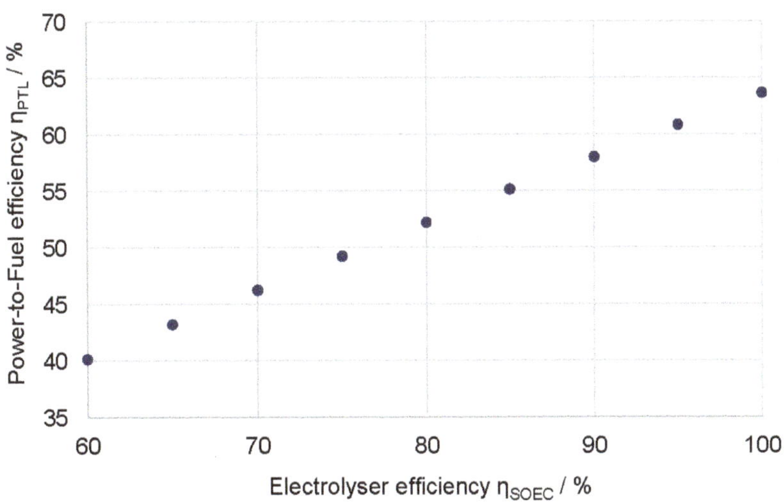

Figure 13. Power-to-fuel efficiency as a function of electrolysis efficiency.

5.3. Heat Recuperation and CO_2 Separation

The possibility of energy integration is often mentioned in the literature as a great advantage of PtL processes based on Fischer–Tropsch synthesis. Energy integration has already been carried out for the developed process in order to provide the required process heat. However, the values listed in Table A3 in Appendix B demonstrate that there are still large amounts of excess heat available that can be removed from the process in the form of low- and medium-pressure steam. In the following, it should be determined whether this heat can be used for CO_2 separation in order to provide the CO_2 required for co-electrolysis. In addition, the influence of such a process coupling on the efficiency of the entire process chain from CO_2 separation to synthetic fuel was examined. For the separation of CO_2 from biogases and from industrial exhaust gases, heat is required at a temperature level [65,66] that exceeds that of the excess low-pressure steam. Therefore, only the excess medium-pressure steam is considered in the following. For every liter of diesel-equivalent produced, the power-to-fuel process produces 9.191 MJ of medium-pressure steam. This energy can then be converted into the amount of energy released per unit of CO_2 consumed.

$$\frac{9.191\frac{MJ}{l_{DE}}}{2.54\frac{kg_{CO2}}{l_{DE}}} = 3.619\frac{MJ}{kg_{CO2}} = 1.005\frac{MWh}{t_{CO2}} \tag{22}$$

Accordingly, in the developed power-to-fuel process, 1.005 MWh of thermal energy was generated per ton of CO_2 consumed, which can in turn be used for CO_2 separation. The resulting thermal coverage for the separation of CO_2 from the CO_2 sources of biogas, industrial waste gas (cement works), and ambient air are given in Table 3. For the sake of completeness, the thermal and electrical energy requirements are also listed.

Table 3. Thermal coverage for CO_2 separation from various CO_2 sources.

CO_2 Source	Electricity Demand [MWh/t_{CO2}]	Heat Demand [MWh/t_{CO2}]	Thermal Coveragee
Biogas (amine washing) [66]	0.011	0.631	159%
Ambient air (direct air capture) [65]	0.5	1.5	67%
Cement production (amine washing) [66]	0.2	1.03	97.6%

The heat required to separate a ton of CO_2 from biogases is comparatively low, at 0.631 MWh, and so the entire thermal energy requirement can be covered by the excess medium-pressure steam and even remaining, unused medium-pressure steam. The separation of CO_2 from the ambient air requires the largest amount of thermal energy with 1.5 MWh per ton of CO_2. About 67% of this can be provided through the excess medium-pressure steam. As an example of the separation of CO_2 from industrial waste gases, separation from the waste gases of a cement plant by means of amine scrubbing was selected at this point. This technology requires 1.03 MWh of thermal energy to separate one ton of CO_2. Accordingly, 97.6% of this thermal energy can be provided through waste heat from the developed power-to-fuel process. The degrees of coverage listed in Table 3 make it clear that the developed PtF process can be coupled with CO_2 separation in order to provide the thermal energy required for CO_2 separation. In order to examine the influence of such a coupling in greater detail, Figure 14 shows the efficiencies of the entire process chain for the case of the coupling of CO_2 capture and the PtF process, and for the case that both processes are operated separately.

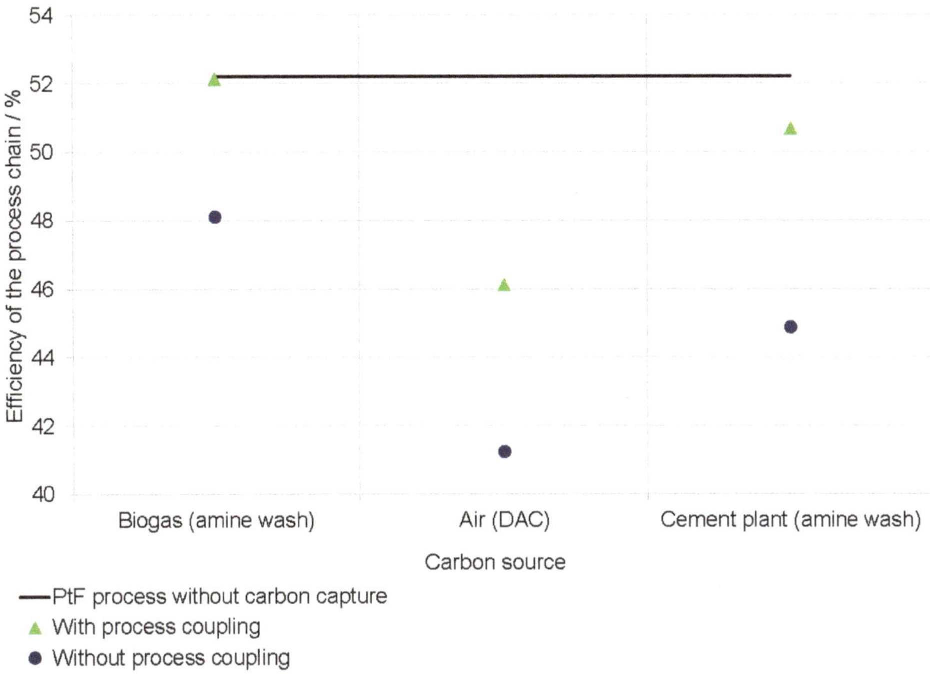

Figure 14. Coupling of the modeled synthesis with CO_2 capture technologies.

If the separation of CO_2 from biogas is conducted independently of the modeled PtF process, the efficiency of the process chain is approximately 48.1%. If both technologies are coupled, the entire thermal energy requirement can be provided by the fuel synthesis, so that the overall efficiency is 52.1% (i.e., almost the efficiency of the PtF process without taking CO_2 separation into account). The remaining difference in results from the electrical energy requirement for CO_2 separation are listed in Table 3. The efficiency of the entire process chain with CO_2 separation from ambient air was lower due to the high thermal and electrical expenditure of the technology (see Figure 14).

Nevertheless, by coupling the technologies, the efficiency can be increased from 41.3% to 46.1% (i.e., 4.8 percentage points). The greatest increase in efficiency can be seen in the coupling of the developed fuel synthesis with CO_2 separation from the exhaust gases of a cement works. By coupling the technologies, the efficiency can be increased by 5.8 percentage points, from 44.9% to 50.7%. Overall, it can be said that a coupling of the developed fuel synthesis with CO_2 separation technologies is both possible and advantageous. The excess heat incurred by the Fischer–Tropsch synthesis can be discharged from the process in the form of medium-pressure steam and used to operate various CO_2 separation technologies. This makes it possible to increase the efficiency of the entire process chain, from CO_2 to synthetic fuel, by up to around 4.87 percentage points, depending on the CO_2 source used.

5.4. Comparison to a Related Analysis for E-Fuels Based on Low-Temperature Electrolysis

In order to be able to better assess the calculated power-to-fuel efficiency of the developed fuel synthesis, it was compared with the efficiencies of alternative PtF or PtL processes. Three PtL processes developed and examined by Schemme [54] were considered. The comparison is possible, as Schemme [54] assumed the same boundary conditions and efficiencies for ancillary units as in this work. In two of the processes considered, alternative fuels were the target product, with methanol being produced in the first process and dimethyl ether (DME) in the second. In the third process considered, synthetic kerosene and synthetic diesel were also produced through a Fischer–Tropsch synthesis, with the synthesis gas for the Fischer–Tropsch reaction being generated via a reverse water–gas shift reactor. The hydrogen supply for all three processes was provided through PEM electrolysis, for which an electrolysis efficiency of 70% was assumed. The PtL efficiencies of the three processes are compared in Figure 15 with the PtL efficiencies of the fuel synthesis developed herein for different electrolysis efficiencies. It should be noted that efficiencies are only really comparable if all of the reference points for determining them are known [67].

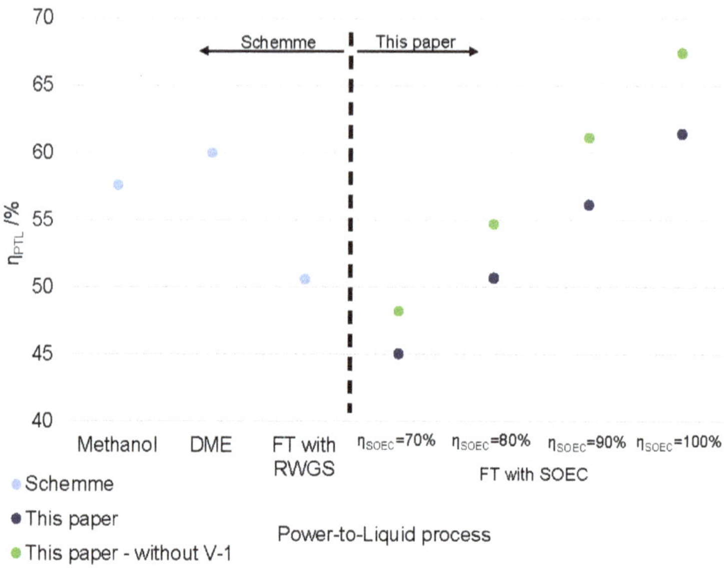

Figure 15. Comparison of the PtL efficiency of the developed fuel synthesis with those of alternative PtL processes (η_{PTl} of methanol, DME, and FT with the RWGS process according to Schemme [54]).

Therefore, the following comparison should not serve as a precise quantitative assessment of the processes, but represents a qualitative classification of the developed fuel synthesis. The efficiencies shown in Figure 15 are the PtL efficiencies for the case that the processes are coupled with CO_2 separation from industrial exhaust gases. With the three efficiencies according to Schemme [54], CO_2 separation from industrial exhaust gases was also assumed, but with the difference that only an electrical energy requirement of 0.333 MWh per ton of CO_2 was taken into account, and the thermal energy requirement was not. Accordingly, it can be assumed that the actual PtL efficiencies of the processes from Schemme [54], taking into account the thermal energy requirement, are below the specified values. The efficiencies of the methanol and DME processes in particular were probably overestimated, as significantly less excess heat is produced in these processes than in those that employ a Fischer–Tropsch synthesis (see Schemme [54]). The PtF efficiency of the developed fuel synthesis is shown in Figure 15 for two different cases. For the standard configuration shown in blue, the PtL efficiency of around 45% for an η_{SOEC} of 70% was below the efficiencies of all reference processes. With an electrolysis efficiency of 80%, the PtL efficiency of the developed fuel synthesis corresponded to that of the Fischer–Tropsch process with a RWGS reactor at around 51.1%. With increasing electrolysis efficiency, the PtL efficiency also further increased. With an η_{SOEC} of 100%, the PtL efficiency of the developed fuel synthesis was approximately 61.4% and above the power-to-liquid efficiencies of the methanol and DME syntheses of 57.6% and 60%, respectively. The conclusion for this is that co-electrolysis only achieves the same PtL efficiency with a 10 percentage point higher efficiency than the process configuration of PEM electrolysis, RWGS reactor, and Fischer–Tropsch synthesis, as assumed by Schemme [54]. However, the specified efficiencies for high-temperature electrolysis described by Peters et al. [64] partially take into account a compression of the electrolysis products to 70 bar for storage, which was not carried out in the developed fuel synthesis. Accordingly, it can be assumed that the required electrolysis power was overestimated and therefore the PtF efficiency was underestimated. To compensate for this effect, the PtF efficiency was recalculated, neglecting the compression of the synthesis gas to 30 bar by V-1 (cf. Figure 2). The resulting PtF efficiencies are highlighted in green in Figure 15. With an efficiency of the high-temperature electrolysis of 70%, the η_{PTL} was around 48.2% at a similar level to the Fischer–Tropsch process investigated by Schemme [54], with a PEM electrolysis efficiency of 70%. The PtF efficiency of the fuel synthesis developed in this work, neglecting V-1, further increased with η_{SOEC}, with a PtF efficiency of approximately 67.4% being calculated for an η_{SOEC} of 100%. Figure 15 thus shows the great potential of PtF processes based on co-electrolysis, and thus also the potential of the developed fuel synthesis. With an electrolysis efficiency of 70%, the high-temperature electrolysis-based process achieved similar efficiencies as the PEM electrolysis-based process. However, with high-temperature electrolysis, there is the possibility of substituting electrical energy with thermal energy and thus achieving very high electrolysis efficiencies and therefore also very high PtF efficiencies.

6. Techno-Economic Analysis

This section is dedicated to the analysis of the economic aspects of the developed PtF process. As the developed model is suitable for calculation with any feed stream, it is necessary to determine an order of magnitude for the process for investigation. According to a report by the Intergovernmental Panel on Climate Change [68], the cement industry emits around 932 megatons of CO_2 annually. For the number of cement works of 1175 given in the same source, assuming 8000 annual operating hours, this approximately corresponds to an average CO_2 production of 99.15 tons per hour and plant. Based on this value, a CO_2 feed stream of 100 tons per hour are defined for the techno-economic analysis in this work. This corresponds to a total production capacity of around 39,400 l_{DE} per hour of synthetic kerosene and diesel, with a total energy consumption of around 750 MW. In the following, the methodology described by Schemme et al. and Peters et al. [69,70] was used to calculate the costs of manufacturing, whereby the investment costs for the system as well as the material and personnel costs for operating it, were first determined. The cost of the product was then determined on the basis of the specific costs. A sensitivity analysis was carried out in order to examine the influence of various cost factors on the production costs in greater detail. Finally, the calculated production costs of the developed fuel synthesis were compared with alternative PtF or PtL processes. The cost of product to be expected for the reference year 2030 was also calculated as part of the economic analysis, with all costs being converted to 2019 equivalents. Accordingly, all cost information below relates to the year 2019, unless stated otherwise.

6.1. Investment Cost

The first step in calculating the cost of the product was to determine the system's investment costs. The calculation of the component costs, with the exception of the reactors and electrolysis, was conducted with the Turton method using the CAPCOST Excel tool; the methodology is explained in Schemme [54]. The investment costs for the reactors and electrolysis are calculated separately. Due to the large number of components required, the investment costs for each of the system components are not discussed individually below. A detailed breakdown of the investment costs is presented in Table 4. To calculate the component costs using the Excel tool CAPCOST, the material, operating pressure, and component-specific size parameters Z must be specified. In the cost accounting, it was assumed that all system components were made of stainless steel. The operating pressures were taken from the Aspen Plus simulation, whereby an additional safety factor of 1.5 was taken into account for all pressures. The size parameters Z were determined depending on the component group. For the pumps, compressors, and drives, the respective nominal capacities are required as size parameters for CAPCOST. These can be taken directly from Aspen Plus, taking the efficiency into account. For the cost calculation of the distillation column and the two strippers, the diameters and heights of the respective apparatuses are required. The Aspen Plus Tray sizing function was used to determine the diameter, assuming sieve trays with a distance of approximately 0.6 m (2 feet) from one another. The heights of the columns and strippers were calculated using the number of trays and the distance between them. An additional distance between the top floor and head or the lowest floor and the sump of 1.5 m was taken into account—a so-called disengagement space. The size parameter for calculating the heat exchanger corresponded to the heat transfer area of the respective heat exchanger.

Table 4. Investment costs for the modeled power-to-fuel process.

Component (-Group)	Investment Cost [Mil.-€]	Share [%]
Pumps	0.146	0.02%
Compressor	122.238	12.87%
Drives	1.680	1.12%
Columns & stripper	1.252	0.13%
Heat exchanger	135.524	14.27%
Vessels	3.593	0.38%
FT reactor	87.185	9.18%
Hydrocracker	70.186	7.38%
Reformer	1.862	0.20%
High-temperature water electrolyzer	25.809	2.72%
High-temperature co-electrolyzer	491.420	51.73%
Sum	949.896	100%

In the next step, the investment costs for the three reactors were determined using the corresponding equations published Baliban et al. [71]. The calculation method presented here requires the capacities of the Fischer–Tropsch reactor, the hydrocracker, and the reformer. These can be read from the Aspen Plus simulation and are listed in Table A4 in Appendix B. It should be noted that the capacity of the Fischer–Tropsch reactor was above the permitted s_{max} value and that two Fischer-Tropsch reactors were therefore operated in parallel. The determined investment costs for the reactors are shown in Table 4.

To calculate the investment costs for high-temperature electrolysis, the electrical performance of the water and co-electrolysis processes must be determined. For this purpose, as in Section 5.2, an electrolysis efficiency was calculated and the required electrolysis output determined through the calorific value of the electrolysis products. An electrolysis efficiency of 80% was assumed for the "base case." This resulted in an output of approximately 643.4 MW for the co-electrolysis and of about 33.8 MW for the water electrolysis. Table 4 gives the forecasts for the investment costs for high-temperature electrolysis for the year 2030. As part of this cost calculation, the data from Brynolf, et al. [72] was employed, as these also expressly apply to SOECs in co-electrolysis operation. The reference case was based on Brynolf et al. [72], given a mean value of €764 (≙700 € 2015) per kilowatt assumed for the investment costs for high-temperature electrolysis. The sum of the investment costs for the system components resulted in investment costs of around €949.9 mil. for the entire system. It can clearly be seen that the investment costs for the high-temperature electrolysis, totaling around €517 million and thus with a share of over 50%, made up by far the largest share of the total investment costs. Next up were the investment costs for the heat exchangers and compressors at around 14% and 13%, respectively. Costs for the Fischer–Tropsch reactor and hydrocracker made up a considerable part of the total investment costs at around 9% and 7%, respectively, whereas the costs for the remaining system components only played a subordinate role, at less than 2%. Due to the high share of investment costs for high-temperature electrolysis, the influence of the electrolysis efficiency and investment costs for the electrolysis against the cost of the product is discussed in greater depth in Section 6.4.3.

6.2. Material and Personnel Costs

The material costs were derived from the costs of the raw materials required for production and those for the required resources. The amounts of the respective materials can either be taken directly from the Aspen simulation or determined using the data shown in Figure 11. If the required quantities for raw materials and operating resources are known, the annual material costs are determined using a specific price for the respective material. An annual operating time of 8000 h was assumed for calculation of the costs. The calculated material costs for the raw materials are presented in Table 5. For the CO_2 price, the base case was assumed to be €70 per ton. This is the lowest price that can be expected in the short- to medium-term for CO_2 that is separated from the exhaust gases of a cement plant (see Schemme [54]). Overall, it can be seen that the costs for CO_2 make up the largest part of the raw material costs, at €56 million out of a total of around €64 million. The influence of the costs of CO_2 on the production costs was considered in the sensitivity analysis.

Table 5. Raw material costs for the PtF process.

Raw Material	Demand	Spec. Costs	Material Cost
Carbon dioxide	100.0 t/h	70.0 €/t [1]	56,000,000 €/a
Oxygen	13.4 t/h	70.0 €/t [2]	7,491,344 €/a
Process water	157.2 t/h	0.1 €/t [3]	125,753 €/a
C_R	-	-	63,617,097 €/a

[1] Brynolf et al. [72]. [2] Cheaper also in the literature: Fraunhofer-Institut [73]. 70 €/t was chosen to not underestimate the costs. [3] Cheaper also in the literature: INTRATEC [74]. 0.1 €/t was chosen to not underestimate the costs.

The annual material costs for the required equipment are presented in Table 6. For the base case of the cost calculation, an electricity price of €40 per MWh was assumed, which was the approximate average electricity exchange price in Germany for 2019 according to the 'energy charts' provided by the Fraunhofer Institute [75]. Table 6 shows that electricity costs made up the majority of operating costs. Therefore, Section 6.4.1 examines in greater detail how the price of electricity influences the cost of the product.

Table 6. Operating costs for the power-to-fuel process.

Raw Material	Consumption	Specific Costs	Material Costs
Cooling water	3324.6 t/h	0.1 €/t [1]	2,659,653 €/a
Cooling air	42,800.8 t/h	-	-
Electricity	752.65 MW	40 €/MWh [2]	240,848,000 €/a
C_B	-	-	243,507,653 €/a

[1] Cheaper also in the literature: INTRATEC [74]. 0.1 €/t was chosen not to underestimate costs. [2] 'Energy charts' [75].

To calculate the annual personnel costs C_p using Equation (23), the number of work steps with particulate solids P and the number of system components to be monitored or controlled N_{NP} must be determined. In the developed power-to-fuel process, no work steps were carried out with particulate solids, and P was accordingly zero. The number of system components to be monitored and the total N_{NP} were as follows: 65 compressors, one column, 181 heat exchangers, three reactors, and 14 electrolyzers.

The high number of compressors and heat exchangers compared to the modeling resulted from the fact that in the Excel tool CAPCOST, the highest possible compressor output of a single compressor was limited to 3000 kW, and the largest possible heat exchanger surface of a single heat exchanger to 1000 m^2. The required electrolysis power for the base case was determined to be approximately 643.4 MW for the co-electrolysis and 33.8 MW for the water electrolysis processes. According to Brynolf et al. [59], the maximum output by an SOEC to be expected by 2030 is 50 MW. Accordingly, a total of 14 SOEC units are required for the power-to-fuel process. This results in a value of 264 for N_{NP}. The average annual gross salary of a full-time employee in the chemical industry in Germany in 2019 was €58,896; see [76]. Taking into account the non-wage labor costs of approximately 23% (see [77]), the annual personnel costs can be calculated using Equation (23), whereby PAYROLL is determined as €76,488 according to [76,77].

$$\begin{aligned} C_P &= 5.38 \cdot \sqrt{6.29 + 31.7 \cdot \mathrm{P}^2 + 0.23 \cdot N_{NP}} \cdot \mathrm{PAYROLL} \\ &= 5.38 \cdot \sqrt{6.29 + 31.7 \cdot 0^2 + 0.23 \cdot 264} \cdot 76,488€ \\ &\approx 3,365,561€ \end{aligned} \tag{23}$$

6.3. Product Cost

Using Equation (24), annual production costs were calculated. However, a depreciation period t and an interest rate i must first be specified in order to determine the annuity. According to Brynolf et al. [59], lifetimes of between 10 and 20 years are to be expected for SOEC systems and maximum lifetimes of less than 90,000 operating hours for SOEC stacks. According to Schmidt et al. [78], for SOEC stacks, maximum operating times of over 100,000 h or, according to one of the experts questioned, of just 30,000 h, can be expected. A depreciation period of 12 years was assumed for the base case of the cost calculation, which corresponded to a total of approximately 96,000 operating hours with an annual operating time of 8000 h. Different values can be found for the interest rate in the literature for power-to-fuel processes such as an interest rate of 4% in Schmidt, et al. [79] and 8% in Buddenberg et al. [80]. An interest rate of 5% was selected for the base case of cost accounting. The influence of the selected depreciation period and selected interest rate was examined as part of the sensitivity analysis. If the depreciation period and interest rate are selected, the annual production costs can be calculated according to Equation (24):

$$\begin{aligned} COM &= 0.151 \cdot FCI + 2.284 \cdot C_P + 1.031 \cdot (C_R + C_B) \\ &+ FCI \cdot \left(\frac{i \cdot (1+i)^t}{(1+i)^t - 1} + i \cdot 0.15\right) \\ &= 582.1\ Mil.€ \end{aligned} \tag{24}$$

With:

Investment costs FCI = €949.9 million, see Table 4;
Personnel costs C_p = €3.37 million, see Equation (23);
Raw material costs C_R = €63.6 million, see Table 6;
Operating costs C_B = €243.5 million, see Table 6;
Depreciation period t = 12;
Interest rate i = 0.05.

For the specified feed stream of 100 tons of CO_2 per hour, the production output of the power-to-fuel process totaled around 39,400 L of diesel equivalent per hour. The specific production costs result from Equation (25):

$$com = \frac{582.1 \cdot 10^6 \frac{€}{a}}{39400 \frac{l_{DE}}{h} \cdot 8000 \frac{h}{a}} \approx 1.85 \frac{€}{l_{DE}} \tag{25}$$

Table 7 shows the cost allocation based on Otto [81]. As a result, the base case under consideration resulted in fuel production costs of around €1.85 per liter of diesel equivalent. Several important influencing factors on the cost of product could also be identified. At almost 42%, the operating resources made up the largest share of fuel production costs. It can be seen in Table 6 that the majority of operating costs were caused by electricity costs, and therefore the electricity price exerts a strong influence on the cost of the product.

Table 7. Distribution of the fuel production costs according to Otto [81].

Cost Component	Specific Production Cost [€/l_{DE}]	Share [%]
Raw materials	0.20	10.9
Resources	0.77	41.9
Overhead: transportation, storage, etc.	0.12	6.3
Manufacturing personnel	0.01	0.6
Surveillance and office staff	<0.01	0.1
Maintenance and repair work	0.19	9.8
Auxiliary materials	0.03	1.5
Laboratory costs	<0.01	0.1
Patent and license fees	0.04	2.4
Taxes and insurance	0.10	5.2
Administrative costs	0.03	1.6
Annuity	0.36	19.6
Total	1.85	100

The annuity accounted for the second largest share of fuel production costs, the amount of which depends on the selected interest rate and the selected depreciation period as well as on the investment costs FCI. In addition, the FCI have, according to the cost allocation by Otto [81], a direct impact on several other cost components such as maintenance and repair work. Over 50% of the total investment costs are made up by those for high-temperature electrolysis. Accordingly, it can be assumed that the investment costs for the electrolysis have a major influence on fuel production costs. The investment costs for high-temperature electrolysis depend, on one hand, on the specific investment costs per kilowatt of power and, on the other, on the efficiency of the high-temperature electrolysis. The last important cost factor can be identified as the CO_2 costs, as these account for almost 90% of the raw material costs (see Table 2) and so almost 10% of the total cost of the product. Overall, the electricity price, efficiency of high-temperature electrolysis, specific investment costs for high-temperature electrolysis, depreciation period, and interest rate as well as the price of CO_2 are identified as important cost items. The influence of these factors is examined in more detail in the following section.

6.4. Sensitivity Analysis

A sensitivity analysis is presented in this section to examine the influence of the cost factors identified in Sections 6.1–6.3. For this purpose, the assumptions for the electricity price, electrolysis efficiency, specific investment costs for the electrolysis, depreciation period, and interest rate as well as the CO_2 costs were varied and the fuel production costs calculated. The other cost factors were left at the values assumed for the base case (see Table 8). Finally, the influences of the respective cost factors were compared in the form of a so-called tornado diagram.

Table 8. Assumptions for the base case.

Cost Parameter	Assumption
Electricity price	40 €/MWh
η_{SOEC}	80%
Spec. investment electrolysis	764 €/kW
Depreciation period	12 years
Interest rate	5%
CO_2 cost	70 €/t

6.4.1. Influence of the Electricity Price

Figure 16 displays the production costs in relation to the assumed electricity price. For the sake of clarity, the cost components "production staff", "monitoring and office staff", "auxiliary materials", "laboratory costs", and "patent and license fees" were combined into "other production costs" in this and the following figures.

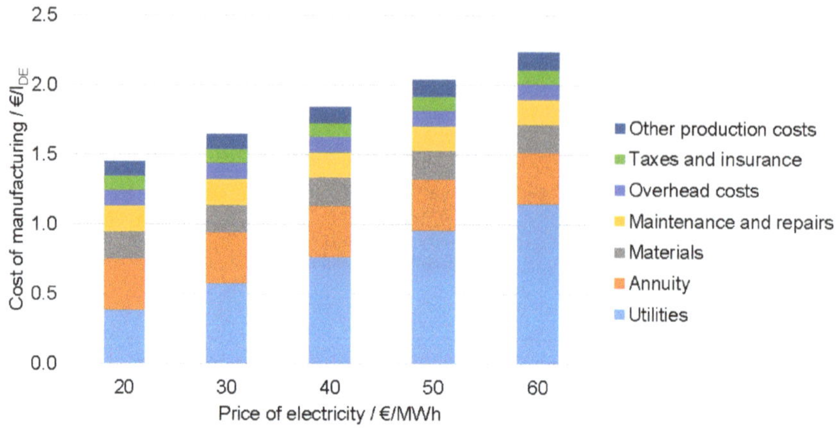

Figure 16. Production costs depending on the electricity price.

The electricity price was varied based on the €40 per megawatt hour assumed for the base case plus or minus 50%. For €20 per megawatt hour, the cost of the product was approximately €1.5/l_{DE}, and for €60 per megawatt hour, it was around €2.2/l_{DE}. The operating costs, which mainly consisted of electricity, made up around 25% at €20/MWh and over 50% of the cost of the product at €60/MWh. The share of electricity costs in the production costs and, accordingly, their influence on the production costs, was very high. Thus, the local electricity price should be considered as an important criterion when choosing locations.

6.4.2. Influence of Electrolysis Efficiency

Figure 17shows the fuel production costs as a function of the electrolysis efficiency. In Section 6.2, it was noted that electrolysis efficiencies of less than 70% are not to be expected for the developed fuel synthesis. Therefore, the efficiency of high-temperature electrolysis was varied from 70% to 100% for the sensitivity analysis. For an efficiency η_{SOEC} of 70%, the cost of product was around €2.0/l_{DE}. The levelized costs of the product decreased with increasing electrolysis efficiency, and for an efficiency η_{SOEC} of 100%, the levelized costs of the product were around €1.6/l_{DE}. It can be seen in Figure 17 that with increasing electrolysis efficiency, both the operating material costs and cost items that are dependent on the investment costs for the system fell. This was due to the fact that the required electrolysis power decreased with increasing electrolysis efficiency, which in turn reduced the investment costs required for the electrolysis and its power consumption. As stated above, the electrolysis efficiency is strongly influenced by the extent to which high-temperature heat is available for the operation of the water and co-electrolysis. Accordingly, when choosing a location for fuel synthesis, it should be determined whether such high-temperature heat is generated in an existing system, and if heat coupling is possible.

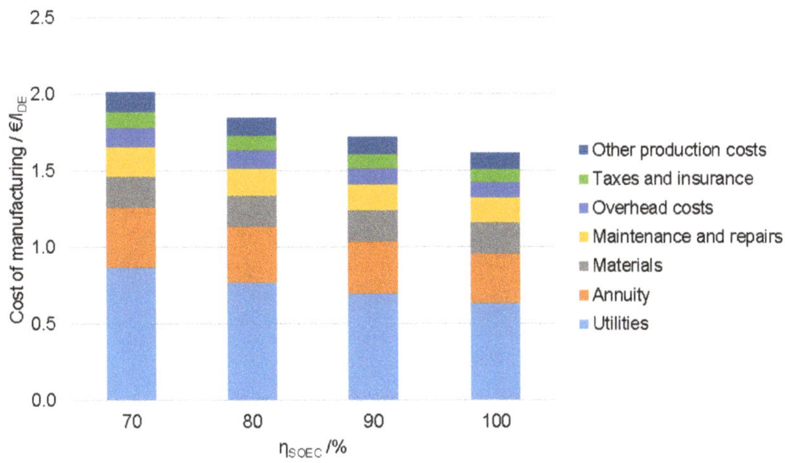

Figure 17. Product production costs depending on the electrolysis efficiency η_{SOEC}.

6.4.3. Influence of the Specific Investment Costs for Electrolysis

Figure 18 shows the cost of the product as a function of the specific investment costs for water and co-electrolysis. Brynolf et al. [72] provide a value of €436/kW (≙400 €/kW @ 2015) as the lower limit and a value of €1091/kW (≙1000 €/kW @ 2015) as the upper limit for the specific investment costs for high-temperature electrolysis expected by 2030 (see Table 4). For the sensitivity analysis, the specific investment costs were varied accordingly. The cost of product for €436/kW was around €1.7/l_{DE}, and for €1091/kW, there was a product cost of around €2.0/l_{DE}. This corresponded to a deviation from the fuel production costs for the base case of €764/kW of around ±10%. The influence of the specific investment costs was, accordingly, significantly less than that of the electricity price, for instance. However, it must be noted that the projection of Brynolf et al. [72] only applies in the event that major technical advances are made by 2030 and are therefore to be considered target values. It is important to invest in research and development so that the developed fuel synthesis or other power-to-fuel processes based on SOEC technology can be implemented in the future.

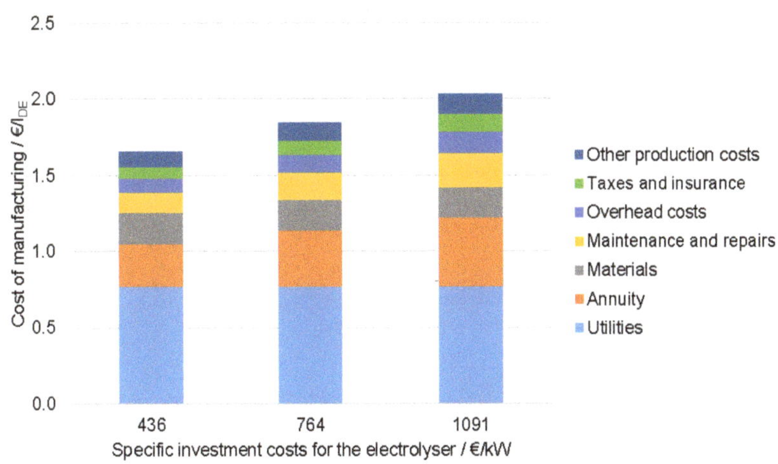

Figure 18. Product production costs as a function of electrolyzer investment costs.

6.4.4. Influence of the Depreciation Period and Interest Rate

The cost of product as a function of the selected depreciation period and selected interest rate are shown in Figure 19. The cost of product for a short depreciation period of nine years with a high interest rate of 7% and a long depreciation period of 15 years with a low interest rate of 3% were compared with the base case of 12 years and 5%, respectively. In the worst case, with a short depreciation period and high interest rate, the cost of product was €2.0/l_{DE}, and in the favorable case with a long depreciation period and low interest rate, the cost of product was €1.7/l_{DE}. This deviation from the base case resulted from the annuity, which was reduced by almost 45% from the unfavorable case to the favorable one. Accordingly, the location-dependent investment conditions should also be taken into account when choosing a location.

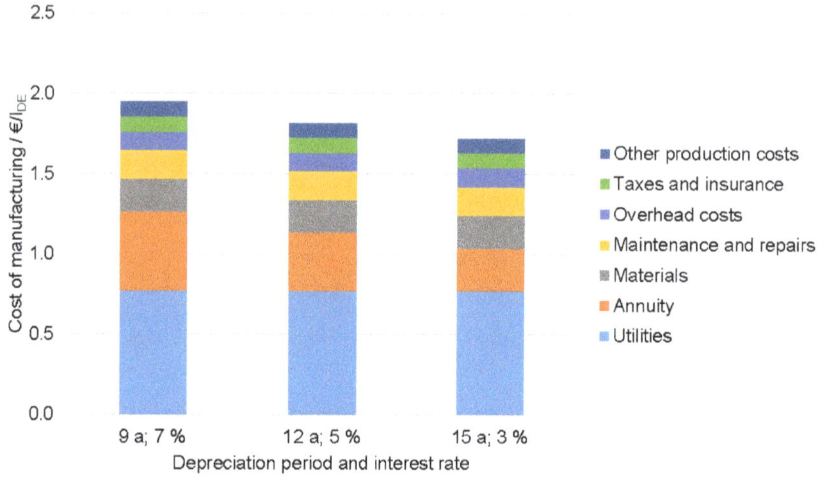

Figure 19. Product production costs depending on the depreciation period and interest rate.

6.4.5. Influence of CO_2 Costs

Figure 20 shows the cost of the product as a function of that of the required CO_2. According to Brynolf and Taljegard [37], the expected long-term costs for CO_2 that are separated from the exhaust gases of a cement plant are between €30 and €50 per ton. Accordingly, for the sensitivity analysis, production costs in this range were assumed. For the sake of completeness, the cost of the product was also calculated to be between €70 and €90 per ton of CO_2. The cost of the product was around €1.7/l_{DE} and €1.9/l_{DE}. It is clear that the CO_2 price has an influence on the production costs, but that it is significantly lower compared to the other cost factors examined. This means that a favorable CO_2 price in the range considered is advantageous for fuel synthesis, but not absolutely essential.

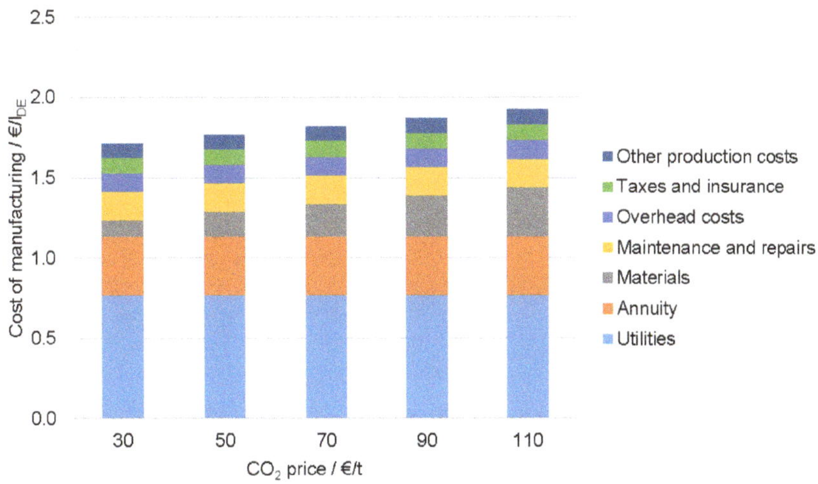

Figure 20. Product production costs as a function of the CO_2 price.

6.4.6. Influence of the Cost Factors in Comparison

A tornado diagram, which can be seen in Figure 21, was used to compare the influence of the respective cost factors on the levelized cost of electricity (LCOE). This diagram compares the areas in which the levelized product costs are dependent on the respective cost factors, based on the base case of €1.85/$l_{DE.}$ For this comparison, the best case and the worst case assumptions from the previous sections were used and highlighted together with the base case assumptions in Figure 21. In Figure 21, it is once again clear that the electricity price had the greatest influence on the fuel production costs and could influence fuel production costs up or down by up to 40 ct/l_{DE} for the price range examined. It is important to note that the tornado diagram only considered the cost parameters individually. In reality, it can be assumed that there will be a mixture of "best", "base", and "worst" cases presented. However, due to the sum of all possible savings when all of the best cases arrive, a lower limit for the fuel production costs was given at €0.94/l_{DE}. In the same manner, with the arrival of all worst cases with €2.95/l_{DE}, there was an upper limit for the cost of the product.

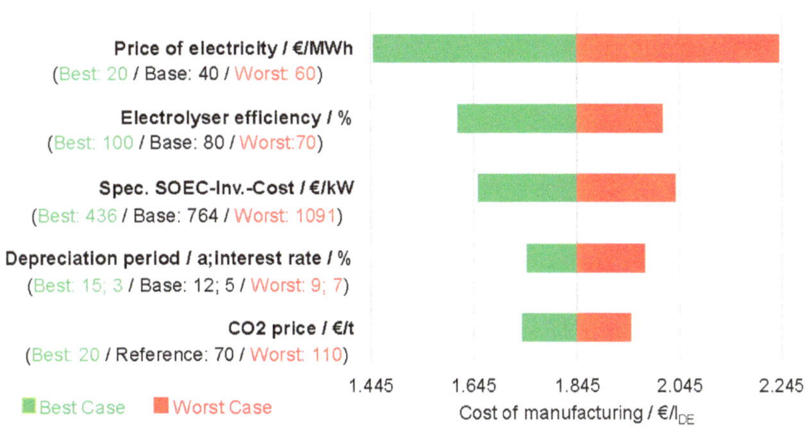

Figure 21. Tornado diagram of the sensitivity analysis of product generation costs.

6.5. Comparison with Alternative Power-to-Liquid Processes

In the following section, the calculated production costs are compared with those of alternative power-to-fuel processes. As noted in Section 6.4, the methanol and DME synthesis investigated by Schemme [54] and the Fischer–Tropsch synthesis in combination with a RWGS reactor were used for the comparison. The production costs of the three processes are compared in Figure 22 with the manufacturing costs for the worst, base, and best cases specified in Section 6.4.

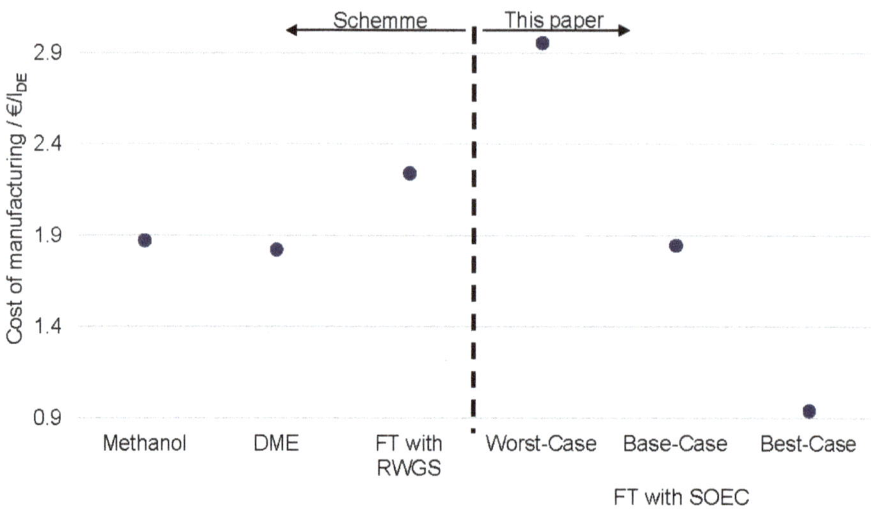

Figure 22. Comparison of the production costs of the developed fuel synthesis with those of alternative power-to-liquid processes (costs of methanol, DME, and FT with the RWGS process according to Schemme [54]).

It must be taken into account that with respect to Schemme [54], different framework conditions were selected in some cases than in the economic analysis carried out in this study. In the three comparison processes, for example, the required hydrogen was not produced in the process, but was obtained externally and taken into account in the cost calculation at a price of €4.6/kg of H_2. Therefore, the following comparison serves as a qualitative classification of the calculated manufacturing costs rather than an exact quantitative comparison of the production costs of the different power-to-liquid processes. In addition, the processes investigated by Schemme [54] were smaller than the fuel synthesis investigated herein.

For example, the Fischer–Tropsch reactor investigated by Schemme [54] was fed with a feed stream of approximately 245,000 Nm^3/h, whereas in the economic analysis carried out in this study, approximately 350,000 Nm^3/h flowed into the Fischer–Tropsch reactor. It is therefore possible that there is a potential for savings through scaling effects for the processes examined by Schemme [54]. However, Figure 22 clearly shows the great potential of the fuel synthesis that has been developed, also taking into account these potential savings. The production costs of €1.85/l_{DE} are already competitive for the base case with the fuel production costs of the methanol and DME synthesis of €1.87/l_{DE} and €1.82/l_{DE}, respectively. In the best case, the costs can even be undercut with €0.94/l_{DE}. An additional advantage of the developed fuel synthesis compared to methanol and DME synthesis is that the infrastructure for traditional fuels already exists and further costs can therefore be saved.

7. Conclusions

Power-to-fuel technology represents a promising possibility for making the transport sector CO_2-neutral in the future. An especially interesting power-to-fuel concept is the coupling of high-temperature co-electrolysis with Fischer–Tropsch synthesis, as this carries some thermodynamic advantages. The aim of this study was to develop such a power-to-fuel process, model the developed process in a process simulation program, and then carry out a techno-economic analysis of the overall process. In the developed fuel synthesis, the entire process chain was considered, starting with water and CO_2 and ending with the fuel according to specifications. First, water and CO_2 were converted into synthesis gas, consisting of hydrogen and carbon monoxide, by means of high-temperature co-electrolysis. In the next step, the synthesis gas was converted into hydrocarbons through a Fischer–Tropsch synthesis, and then processed into synthetic diesel according to EN 15940 and synthetic kerosene type FT-SPK according to ASTM 7566. The fuel preparation consisted of a hydrocracker, reformer, and carrier steam distillation. An additional high-temperature water electrolysis system was used to provide the hydrogen required for the hydrocracker. The process simulation was implemented in the simulation program Aspen Plus, whereby the model was designed for the calculation of any mass flows and so any system sizes. In addition, an energy integration analysis was conducted. The results of the process simulation provide information regarding the material and energetic balance of the process. In the developed fuel synthesis, 1 L of diesel equivalent (35.9 MJ) of synthetic fuels was produced, which was then broken down energetically into 38.9% kerosene and 61.1% diesel. An examination of the fuels produced indicated that both synthetic diesel and synthetic kerosene meet the requirements of the above standards. To produce one liter of diesel equivalent, 2.54 kg of CO_2, 3.99 kg of water, and 0.34 kg of oxygen are required. The energetic analysis of the process shows that the energy requirement of the high-temperature co-electrolysis was reduced by the energy integration from about 75 MJ/l_{DE} over 20% to about 59 MJ/l_{DE}. This makes it clear that the coupling of Fischer–Tropsch synthesis with high-temperature electrolysis represents an attractive power-to-fuel concept. In addition, it was found that the energy requirement of the process and so the power-to-liquid efficiency depends heavily on the efficiency of the electrolysis. The power-to-liquid efficiency for an electrolysis efficiency of 70% was approximately 46%, and with an electrolysis efficiency of 100%, the PtL efficiency was almost 67%. The assumed base case electrolysis efficiency of 80% resulted in a PtL efficiency of 52%, whereby the electrical energy for the co-electrolysis, with about 59 MJ/l_{DE}, made up more than 85% of the total energy requirement of about 69 MJ/l_{DE}. Accordingly, the co-electrolysis represents the critical element of the developed fuel synthesis and presents itself as a topic for further research in order to develop a better understanding of the technology as well as to identify possible energy-saving potentials. The energetic analysis also showed that the developed power-to-fuel process generated an excess heat of around 1.005 MWh per ton of CO_2 consumed. This heat can be used for CO_2 capture technologies. This study showed that the excess CO_2 could cover around 67% of the thermal energy required to separate a corresponding amount of CO_2 from the ambient air and around 97% of the thermal energy requirement for separating CO_2 from industrial waste gases (cement works). The thermal energy requirement of CO_2 separation from biogas can be fully covered. This option is very attractive since it offers a biogenic CO_2 source, resulting in a completely sustainable route. In the long-term, the further technical development of the direct separation from ambient air will surely enable a broad application possibility for the developed technology. The thermal coupling of the power-to-fuel process with CO_2 capture technologies therefore represents a good opportunity to improve the efficiency of the entire process chain, from CO_2 to synthetic fuel. If fuel synthesis is coupled with CO_2 separation from biogases, the overall efficiency can be increased from 48.1% to 52.1% (i.e., by four percentage points). In the case of CO_2 separation from the ambient air, the thermal coupling can increase the efficiency by 4.8 percentage points, from 41.3% to 46.1%. The largest increase in efficiency was found when the fuel synthesis was coupled with the separation of CO_2 from industrial exhaust gases. These values show the importance of the CO_2 capture technology and its relevance for the overall process. In order to maximize the efficiency of the future demonstration projects, it is therefore important to select the optimal site concerning the CO_2 potential. The process model developed in this study can be used to analyze the economic viability of the examined locations given that, as already described, it is suitable for calculating any system parameters. The further development of the outlined process model can also be the subject of future research. For example, the models developed for calculating the co-electrolysis or Fischer–Tropsch reactor could be enhanced by kinetic models.

In order to realize the potential of the developed fuel synthesis, two important considerations are necessary. On one hand, it is important to invest in the research and development of SOEC technology and to increase the low TRL of the SOEC and bring high-temperature electrolysis to a megawatt scale and market maturity. On the other hand, the choice of location plays a decisive role. In this way, the great potential of the developed fuel synthesis can, above all, be realized in locations where cheap electricity and high-temperature heat are available for the operation of the SOEC.

Author Contributions: Conceptualization: R.P., R.C.S. and N.W.; Methodology: R.P., F.S. and R.C.S.; Literature review and writing: N.W.; Process analysis ASPEN PLUS: N.W. and F.S.; Techno-economic analyses: F.S., N.W. and D.S.; Technical writing—original draft preparation: N.W.; Writing and visualization—review and editing: All authors; Supervision: R.P., J.R., M.G. and D.S. All authors have read and agreed to the published version of the manuscript.

Funding: This study received no external funding.

Acknowledgments: The authors would like to thank all members of the Institute for Electrochemical Process Engineering (IEK-14) for fruitful discussions on all of the technologies of the energy transition and their role in the future energy system. Special thanks to the research groups of Ro. Peters and Q. Fang on SOCs, systems engineering, and electrochemistry for valuable discussions related to technical data on SOECs. The exchange of information with D. Schäfer deserves special mention. Additionally, we thank C. Wood for the language editing.

Conflicts of Interest: The authors have no conflict of interest to disclose.

Abbreviations

SRK	Equation of State (EOS, cubic) Soave–Redlich–Kwong
RKS-BM	Soave–Redlich–Kwong EOS with Boston–Mathias Alpha-Function
BK10	Braun K-10
NRTL-RK	Non-Random-Two-Liquid, extended by EOS Redlich–Kwong for calculation of the gas phase

Appendix A. Effect of Chain Growth Probability on the Product Distribution

In order to illustrate the influence of the chain growth probability on the product distribution, Figure A1 compares the product distribution of a Fischer–Tropsch synthesis for $\alpha = 0.88$ and $\alpha = 0.92$ for C_1 to C_{60} according to the Anderson–Schulz–Flory distribution (Equation (12)). The course of the mass fractions of the various hydrocarbons was similar in both curves. First, the mass fractions increased with increasing chain length until a maximum was reached and the curve fell again. However, the curve for $\alpha = 0.88$ rose much more rapidly in the area of short hydrocarbon chains and already reached the maximum at a chain length of $n = 8$. In addition, it flattened out very steeply, which means that the proportion of long-chain hydrocarbons in the product was very low. In comparison, the curve for $\alpha = 0.92$ rose much more slowly and only reached the maximum at a chain length of $n = 12$, and then flattened out more slowly. Accordingly, the proportion of long-chain hydrocarbons in the product of the Fischer–Tropsch synthesis was significantly higher for $\alpha = 0.92$, or for high chain growth probabilities in general. It follows that to maximize kerosene and diesel production in a power-to-fuel process, the chain growth probability of the Fischer–Tropsch synthesis should also be maximized.

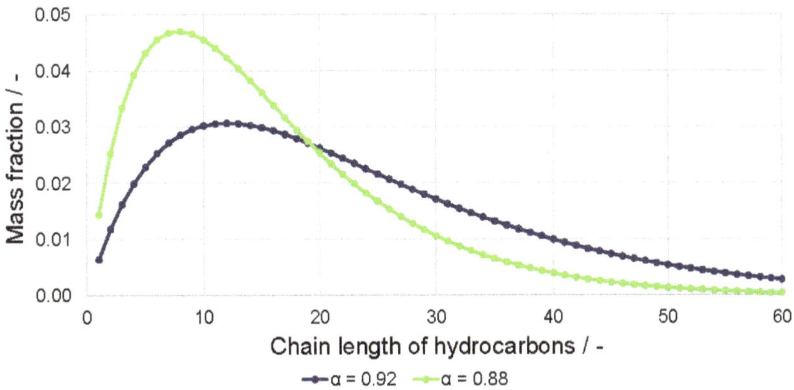

Figure A1. Product distribution of the Fischer–Tropsch synthesis for α = 0.88 and α = 0.92.

Appendix B. Key Data from Process Simulations

Table A1. Energy-specific material balance of the developed fuel synthesis.

Input		Output	
Carbon dioxide	2.54 kg	Synthetic fuel	1 l_{DE}
Water	3.99 kg	Share kerosene	38.9%
Oxygen	0.34 kg	Share diesel	61.9%
		Water	2.89 kg

Table A2. Resources used.

Resources	Temperature Range	Specific Heating-/ Cooling-Power
Low-pressure saturated steam	124–125 °C	2193 kJ/kg
Medium-pressure saturated steam	174–175 °C	2036 kJ/kg
Cooling water	20–25 °C	21 kJ/kg
Cooling air	30–35 °C	5 kJ/kg
Electricity	-	-

Table A3. Energy-specific resource balance of the developed fuel synthesis. Negative values in the sum mean that excess steam is available from the process. See Table A2 for details on the resources used.

Resources	Quantity [kg/l_{DE}]	Energy [MJ/l_{DE}]
	Low-pressure saturated steam	
Produced	−0.235	−0.515
Demand	0.196	0.429
Sum	−0.039	−0.086
	Medium-pressure saturated steam	
Produced	−6.059	−12.339
Demand	1.546	3.148
Sum	−4.513	−9.191
	Cooling water	
Demand	84.370	1.764
	Cooling air	
Demand	1086.192	5.431
	Electricity (w/o electrolysis)	
Demand	-	6.890

Table A4. Capacities of the Fischer–Tropsch reactor, the hydrocracker and the reformer.

	FT Reactor	Hydrocracker	Reformer
Educt-/product rate	350,322 Nm^3/h	1191 t/d	159,483 Nm^3/h
s_{max}**-value**	228,029 Nm^3/h	6256 t/d	9,438,667 Nm^3/h
Number of reactors	2	1	1

References

1. BMU. *Klimaschutz in Zahlen-Fakten, Trends und Impulse deutscher Klimapolitik*; Bundesministerium für Umwelt, Naturschutz und nuklear Sicherheit (BMU): Berlin, Germany, 2019.
2. Ausfelder, F.; Wagemann, K. Power-to-Fuels: E-Fuels as an Important Option for a Climate-Friendly Mobility of the Future. *Chem. Ing. Tech.* **2020**, *92*, 21–30. [CrossRef]
3. Sterner, M.; Stadler, I. *Energiespeicher-Bedarf, Technologien, Integration*, 2nd ed.; Springer: Berlin/Heidelberg, Germany, 2017.
4. Kramer, U. *Defossilierung des Transportsektors Optionen und Vorraussetzungen in Deutschland*; FVV Prime Movers. Technologies: Frankfurt am Main, Germany, 2018.
5. Marlin, D.S.; Sarron, E.; Sigurbjornsson, O. Process Advantages of Direct CO_2 to Methanol Synthesis. *Front. Chem.* **2018**, *6*, 446. [CrossRef] [PubMed]
6. Dieterich, V.; Buttler, A.; Hanel, A.; Spliethoff, H.; Fendt, S. Power-to-liquid via synthesis of methanol, DME or Fischer–Tropsch-fuels: A review. *Energy Environ. Sci.* **2020**, *13*, 3207–3252. [CrossRef]
7. Energiedaten: Gesamtausgabe. Available online: https://www.bmwi.de/Redaktion/DE/Artikel/Energie/energiedaten-gesamtausgabe.html (accessed on 31 October 2020).
8. Schmidt, P.; Weindorf, W.; Roth, A.; Batteiger, V.; Riegel, F. *Power-to-Liquids Potentials and Perspectives for the Future Supply of Renewable Aviation Fuel*; Umweltbundesamt, Ed.; German Environment Agency: Dessau-Roßlau, Germany, 2016.
9. De Klerk, A. *Fischer-Tropsch Refining*; Wiley-VCH Verlag & Co. KGaA: Weinheim, Germany, 2011.
10. Schemme, S.; Samsun, R.C.; Peters, R.; Stolten, D. Power-to-fuel as a key to sustainable transport systems–An analysis of diesel fuels produced from CO_2 and renewable electricity. *Fuel* **2017**, *205*, 198–221. [CrossRef]
11. Verdegaal, W.M.; Becker, S.; Olshausen, C.V. Power-to-Liquids: Synthetisches Rohöl aus CO2, Wasser und Sonne. *Chem. Ing. Tech.* **2015**, *87*, 340–346. [CrossRef]
12. GmbH, S. Blue Crude: Sunfire Produziert Nachhaltigen Erdölersatz. Available online: https://www.sunfire.de/de/unternehmen/news/detail/blue-crude-sunfire-produziert-nachhaltigen-erdoelersatz (accessed on 20 October 2020).
13. Bazzanella, A.; Krämer, D. (Eds.) *Technologies for Sustainability and Climate Protection-Chemical Processes and Use of CO2*; DECHEMA Gesellschaft für Chemische Technik und Biotechnologie e.V.: Frankfurt, Germany, 2017.
14. Vázquez, F.V.; Koponen, J.; Ruuskanen, V.; Bajamundi, C.; Kosonen, A.; Simell, P.; Ahola, J.; Frilund, C.; Elfving, J.; Reinikainen, M.; et al. Power-to-X technology using renewable electricity and carbon dioxide from ambient air: SOLETAIR proof-of-concept and improved process concept. *J. CO2 Util.* **2018**, *28*, 235–246. [CrossRef]
15. SOLETAIR-Technical Specifications. Available online: https://soletair.fi/technical-specifications/ (accessed on 20 October 2020).
16. Durchbruch Für Power-to-X. Available online: https://www.kopernikus-projekte.de/aktuelles?backRef=30&news=Durchbruch_fuer_Power_to_X (accessed on 20 October 2020).
17. Durchbruch Für Power-to-X: Sunfire Nimmt Erste Co-Elektrolyse in Betrieb und Startet die Skalierung. Available online: https://www.sunfire.de/de/unternehmen/news/detail/durchbruch-fuer-power-to-x-sunfire-nimmt-erste-co-elektrolyse-in-betrieb-und-startet-die-skalierung (accessed on 20 October 2020).
18. Norsk e-Fuel. Available online: https://www.norsk-e-fuel.com/en/ (accessed on 8 September 2020).
19. Translated: Two Companies Want to Make Synthetic Fuel on Herøya: Ex-Partner Becomes a Competitor and Settles on the Neighboring Plot. Available online: https://www.tu.no/artikler/to-bedrifter-vil-lage-syntetisk-drivstoff-pa-heroya-eks-partner-blir-konkurrent-og-slar-seg-ned-pa-nabotomta/493866?key=zbh08EHG (accessed on 8 September 2020).
20. Trotz Wasserstoffstrategie: Große Power-to-X-Anlage von Norsk e-Fuel Entsteht in Norwegen. Available online: https://www.cleanthinking.de/wasserstoffstrategie-powertox-norsk-e-fuel-norwegen/ (accessed on 8 September 2020).
21. The Hague Airport Initiates Study for the Production of Renewable Jet Fuel from Air. Available online: https://www.sunfire.de/de/unternehmen/news/detail/rotterdam-the-hague-airport-initiates-study-for-the-production-of-renewable-jet-fuel-from-air (accessed on 20 October 2020).
22. Schiphol and Sustainable Kerosene. Available online: https://news.schiphol.com/schiphol-and-sustainable-kerosene/ (accessed on 8 September 2020).
23. Zenid–Sustainable Aviation Fuel from Air. Available online: https://climeworks.com/news/christoph-gebald-co-ceo-and-co-founder-of-climeworks (accessed on 8 September 2020).
24. Zenid | Jet Fuel from Air. Available online: https://zenidfuel.com/ (accessed on 8 September 2020).
25. Homepage-Synkero. Available online: https://synkero.com/ (accessed on 8 September 2020).
26. Carmo, M.; Fritz, D.L.; Mergel, J.; Stolten, D. A comprehensive review on PEM water electrolysis. *Int. J. Hydrogen Energy* **2013**, *38*, 4901–4934. [CrossRef]

27. Andika, R.; Nandiyanto, A.B.D.; Putra, Z.A.; Bilad, M.R.; Kim, Y.; Yun, C.M.; Lee, M. Co-electrolysis for power-to-methanol applications. *Renew. Sustain. Energy Rev.* **2018**, *95*, 227–241. [CrossRef]
28. Zheng, Y.; Wang, J.; Yu, B.; Zhang, W.; Chen, J.; Qiao, J.; Zhang, J. A review of high temperature co-electrolysis of H2O and CO2 to produce sustainable fuels using solid oxide electrolysis cells (SOECs): Advanced materials and technology. *Chem. Soc. Rev.* **2017**, *46*, 1427–1463. [CrossRef]
29. Buttler, A.; Spliethoff, H. Current status of water electrolysis for energy storage, grid balancing and sector coupling via power-to-gas and power-to-liquids: A review. *Renew. Sustain. Energy Rev.* **2018**, *82*, 2440–2454. [CrossRef]
30. Dueñas, D.M.A.; Riedel, M.; Riegraf, M.; Costa, R.; Friedrich, K.A. High Temperature Co-electrolysis for Power-to-X. *Chem. Ing. Tech.* **2020**, *92*, 45–52. [CrossRef]
31. Cinti, G.; Baldinelli, A.; Di Michele, A.; Desideri, U. Integration of Solid Oxide Electrolyzer and Fischer-Tropsch: A sustainable pathway for synthetic fuel. *Appl. Energy* **2016**, *162*, 308–320. [CrossRef]
32. Frank, M.H. *Reversible Wasserstoffbetriebene Festoxidzellensysteme*; Forschungszentrum Jülich GmbH Zentralbibliothek: Jülich, Germany, 2019.
33. Marchese, M.; Giglio, E.; Santarelli, M.; Lanzini, A. Energy performance of Power-to-Liquid applications integrating biogas upgrading, reverse water gas shift, solid oxide electrolysis and Fischer-Tropsch technologies. *Energy Convers. Manag. X* **2020**, *6*, 100041. [CrossRef]
34. Peters, R.; Decker, M.; Eggemann, L.; Schemme, S.; Schorn, F.; Breuer, J.L.; Weiske, S.; Pasel, J.; Samsun, R.C.; Stolten, D. Thermodynamic and ecological preselection of synthetic fuel intermediates from biogas at farm sites. *Energy Sustain. Soc.* **2020**, *10*, 4. [CrossRef]
35. Trippe, F. *Techno-Ökonomische Bewertung Alternativer Verfahrenskonfigurationen Zur Herstellung Biomass-to-Liquid (BtL) Kraftstoffen und Chemikalien*; KIT: Karlsruhe, Germany, 2013.
36. de Klerk, A. Fischer–Tropsch fuels refinery design. *Energy Environ. Sci.* **2011**, *4*, 1177–1205. [CrossRef]
37. Holladay, J.; Albrecht, K.; Hallen, R. Renewable routes to jet fuel. *Curr. Opin. Biotechnol.* **2014**, *26*, 50–55.
38. Vervloet, D.; Kapteijn, F.; Nijenhuis, J.; van Ommen, J.R. Fischer–Tropsch reaction–diffusion in a cobalt catalyst particle: Aspects of activity and selectivity for a variable chain growth probability. *Catal. Sci. Technol.* **2012**, *2*, 1221–1233. [CrossRef]
39. Hsu, C.S.; Robinson, P.R. (Eds.) *Practical Advances in Petroleum Processing*; Springer: New York, NY, USA, 2006; Volume 1.
40. Speight, J.G. *Heavy and Extra-Heavy Oil Upgrading Technologies*; Gulf Professional Publishing: Houston, TX, USA; Elsevier Inc.: Amsterdam, The Netherlands, 2013. [CrossRef]
41. Bouchy, C.; Hastoy, G.; Guillon, E.; Martens, J.A. Fischer-Tropsch Waxes UpgradingviaHydrocracking and Selective Hydroisomerization. *Oil Gas Sci. Technol.* **2009**, *64*, 91–112. [CrossRef]
42. Calemma, V.; Gambaro, C.; Parker, W.O.; Carbone, R.; Giardino, R.; Scorletti, P. Middle distillates from hydrocracking of FT waxes: Composition, characteristics and emission properties. *Catal. Today* **2010**, *149*, 40–46. [CrossRef]
43. Wegener, N. Prozessmodellierung und Bewertung Synthetischer Kraftstoffe auf Basis von Wasserstoff und Kohlenstoffdioxid. Master's Thesis, University of Bochum, Bochum, Germany, 2020.
44. Lawson, K.H.; Jothimurugesan, K.; Espinoza, R.L. Catalyst for Hydroprocessing of Fischer-Tropsch Products. U.S. Patent 10/866,884, 15 December 2005.
45. Meyers, R.A. *Handbook of Petroleum Refining Processes*, 3rd ed.; McGraw-Hill: New York, NY, USA, 2004.
46. Pors, Z. *Eduktvorbereitung und Gemischbildung in Reaktionsapparaten zur Autothermen Reformierung von Dieselähnlichen Kraftstoffen*; Forschungszentrum Jülich GmbH Zentralbibliothek: Jülich, Germany, 2006.
47. Rostrup-Nielsen, J.R. Syngas in perspective. *Catal. Today* **2002**, *71*, 243–247. [CrossRef]
48. Luyben, W.L. *Distillation Design and Control Using Aspen(TM) Simulation*, 2nd ed.; John Wiley & Sons: Hoboken, NJ, USA, 2013.
49. Tzinis, I. Technology Readiness Level. Available online: https://www.nasa.gov/directorates/heo/scan/engineering/technology/technology_readiness_level (accessed on 22 February 2022).
50. Héder, M. From NASA to EU: The evolution of the TRL scale in Public Sector Innovation. *Innov. J.* **2017**, *22*, 1.
51. De Rose, A.; Buna, M.; Strazza, C.; Olivieri, N.; Stevens, T.; Peeters, L. *Technology Readiness Level: Guidance Principles for Renewable Energy Technologies*; European Comission: Brussels, Belgium, 2017.
52. de Klerk, A. Can Fischer−Tropsch Syncrude Be Refined to On-Specification Diesel Fuel? *Energy Fuels* **2009**, *23*, 4593–4604. [CrossRef]
53. Dry, M.E. High quality diesel via the Fischer-Tropsch process—A review. *J. Chem. Technol. Biotechnol.* **2002**, *77*, 43–50. [CrossRef]
54. Schemme, S. *Techno-Ökonomische Bewertung von Verfahren zur Herstellung von Kraftstoffen aus H2 und CO_2*; Forschungszentrum Jülich GmbH Zentralbibliothek: Jülich, Germany, 2020.
55. Carlson, E.C. Don't Gamble With Physical Properties For Simulations. *Chem. Eng. Prog.* **1996**, *92*, 35–46.
56. Auburn University. *Design of a Process for the Production if Fuels and/or Chemicals from Hydrocarbon Feestocks*; Department of Chemical Engineering, Auburn University: Auburn, AL, USA, 2013.
57. Sun, X.; Chen, M.; Jensen, S.H.; Ebbesen, S.D.; Graves, C.; Mogensen, M. Thermodynamic analysis of synthetic hydrocarbon fuel production in pressurized solid oxide electrolysis cells. *Int. J. Hydrogen Energy* **2012**, *37*, 17101–17110. [CrossRef]
58. Cinti, G.; Discepoli, G.; Bidini, G.; Lanzini, A.; Santarelli, M. Co-electrolysis of water and CO_2 in a solid oxide electrolyzer (SOE) stack. *Int. J. Energy Res.* **2016**, *40*, 207–215. [CrossRef]
59. Becker, W.L.; Braun, R.J.; Penev, M.; Melaina, M. Production of Fischer–Tropsch liquid fuels from high temperature solid oxide co-electrolysis units. *Energy* **2012**, *47*, 99–115. [CrossRef]

60. Bacha, J.; Freel, J.; Gibbs, A.; Gibbs, L.; Hemighaus, G.; Hoekman, K.; Horn, J.; Gibbs, A.; Ingham, M.; Jossens, L.; et al. *Diesel Fuels Technical Review*; Chevron: San Ramon, CA, USA, 2017.
61. Edwards, T. *Reference Jet Fuels for Combustion Testing*; Air Force Research Laboratory: Dayton, OH, USA, 2017.
62. Cookson, D.J.; Lloyd, C.P.; Smith, B.E. Investigation of the Chemical Basis of Kerosene (Jet Fuel) Specification Properties. *Energy Fuels* **1987**, *1*, 438–447. [CrossRef]
63. Martin, W.; Sandra, U.; Peter, M.; Friedrich, S.; Fabio, G. *Energietechnologien der Zukunft-Erzeugung, Speicherung, Effizienz und Netze*; Springer Vieweg: Wiesbaden, Germany, 2015. [CrossRef]
64. Peters, R.; Deja, R.; Blum, L.; Nguyen, V.N.; Fang, Q.; Stolten, D. Influence of operating parameters on overall system efficiencies using solid oxide electrolysis technology. *Int. J. Hydrogen Energy* **2015**, *40*, 7103–7113. [CrossRef]
65. Kuramochi, T.; Ramírez, A.; Turkenburg, W.; Faaij, A. Comparative assessment of CO2 capture technologies for carbon-intensive industrial processes. *Prog. Energy Combust. Sci.* **2012**, *38*, 87–112. [CrossRef]
66. Peter, V.; Juri, H.; Alexander, S.; Ole, Z. *Verfahren der CO_2-Abtrennung aus Faulgasen und Umgebungsluft innerhalb des Forschungsprojekts TF_Energiewende*; Wuppertal Institut für Klima, Umwelt, Energie gGmbH: Wuppertal, Germany, 2018.
67. Lettenmeier, P. *Wirkungsgrad-Elektrolyse-Whitepaper*; Siemens, Corporate Technology, Strategy & Business Development: Erlangen, Germany, 2019.
68. Metz, B.; Davidson, O.; de Coninck, H.; Loos, M.; Meyer, L. (Eds.) *Carbon Dioxide Capture and Storage*; Cambridge University Press: New York, NY, USA, 2005.
69. Peters, R.; Baltruweit, M.; Grube, T.; Samsun, R.C.; Stolten, D. A techno economic analysis of the power to gas route. *J. CO2 Util.* **2019**, *34*, 616–634. [CrossRef]
70. Schemme, S.; Breuer, J.L.; Köller, M.; Meschede, S.; Walman, F.; Samsun, R.C.; Peters, R.; Stolten, D. H2-based synthetic fuels: A techno-economic comparison of alcohol, ether and hydrocarbon production. *Int. J. Hydrogen Energy* **2020**, *45*, 5395–5414. [CrossRef]
71. Baliban, R.C.; Elia, J.A.; Floudas, C.A. Toward Novel Hybrid Biomass, Coal, and Natural Gas Processes for Satisfying Current Transportation Fuel Demands, 1: Process Alternatives, Gasification Modeling, Process Simulation, and Economic Analysis. *Ind. Eng. Chem. Res.* **2010**, *49*, 7343–7370. [CrossRef]
72. Brynolf, S.; Taljegard, M.; Grahn, M.; Hansson, J. Electrofuels for the transport sector: A review of production costs. *Renew. Sustain. Energy Rev.* **2018**, *81*, 1887–1905. [CrossRef]
73. Fraunhofer IKTS. *Dezentrale Sauerstoffproduktion*; Fraunhofer-Institut für Keramische Technologien und Systeme IKTS: Hermsdorf, Germany, 2020.
74. INTRATEC. Cooling Water Price. Available online: https://www.intratec.us/chemical-markets/cooling-water-price (accessed on 21 November 2020).
75. Energy-Charts. Available online: https://energy-charts.info/index.html?l=de&c=DE (accessed on 13 November 2020).
76. Bruttomonatsverdienste der Vollzeitbeschäftigten Arbeitnehmer in der Chemieindustrie in Deutschland in Den Jahren 2000 bis. Available online: https://de.statista.com/statistik/daten/studie/203407/umfrage/bruttomonatsverdienste-der-vollzeitbeschaeftigten-arbeitnehmer-in-der-chemischen-industrie-in-deutschland/ (accessed on 16 October 2020).
77. Bundesamt, S. Abeitskosten, Lohnnebenkosten-Detaillierte Zusammensetzung der Arbeitskosten im Produzierten Gewerbe und Dienstleistungsbereich. Available online: https://www.destatis.de/DE/Themen/Arbeit/Arbeitskosten-Lohnnebenkosten/Tabellen/struktur-kostenart.html;jsessionid=5CE90E6590FA9ECCACC6C3661523845C.internet8731 (accessed on 16 October 2020).
78. Schmidt, O.; Gambhir, A.; Staffell, I.; Hawkes, A.; Nelson, J.; Few, S. Future cost and performance of water electrolysis: An expert elicitation study. *Int. J. Hydrogen Energy* **2017**, *42*, 30470–30492. [CrossRef]
79. Schmidt, P.R.; Zittel, W.; Weindorf, W.; Raksha, T. *Renewables in Transport 2050-Empowering a Sustainable Mobility Future with Zero Emission Fuels from Renewables Electricity*; Fvv–Forschungsvereinigung Verbrennungskraftmaschinen E.V.: Frankfurt am Main, Germany, 2016.
80. Buddenberg, T.; Bergins, C.; Schmidt, S. Power to fuel as a sustainable business model for cross-sectorial energy storage in industry and power plants. In Proceedings of the 5th Conference of Carbon Dioxide as a Feedstock for Fuels, Chemistry and Polymers, Cologne, Germany, 6–7 December 2016.
81. Otto, A. *Chemische, Verfahrenstechnische und Ökonomische Bewertung von Kohlendioxid als Rohstoff in der Chemischen Industrie*; Forschungszentrum Jülich GmbH Zentralbibliothek: Jülich, Germany, 2015.

energies

MDPI

Article

Energy and Exergy Analysis of an Ammonia Fuel Cell Integrated System for Marine Vessels

Phan Anh Duong [1], Borim Ryu [1], Chongmin Kim [2], Jinuk Lee [1] and Hokeun Kang [3,*]

[1] Department of Marine System Engineering, Korea Maritime and Ocean University, 727 Taejong-ro, Yeongdo-gu, Busan 49112, Korea; anhdp@g.kmou.ac.kr (P.A.D.); ryuborim@g.kmou.ac.kr (B.R.); julee88@kmou.ac.kr (J.L.)

[2] System Safety Research Team, Korean Register, 36 Myeongji Ocean City 9-ro, Gangseo-gu, Busan 46762, Korea; ckim@krs.co.kr

[3] Division of Coast Guard Studies, Korea Maritime and Ocean University, 727 Taejong-ro, Yeongdo-gu, Busan 49112, Korea

* Correspondence: hkkang@kmou.ac.kr; Tel.: +82-51-410-4260; Fax: +82-51-404-3985

Abstract: In this paper, a new integrated system of solid oxide fuel cell (SOFC)–gas turbine (GT)–steam Rankine cycle (SRC)–exhaust gas boiler (EGB) is presented, in which ammonia is introduced as a promising fuel source to meet shipping decarbonization targets. For this purpose, an SOFC is presented as the main power-generation source for a specific marine propulsion plant; the GT and SRC provide auxiliary power for machinery and accommodation lighting, and steam from the waste heat boiler is used for heating seafarer accommodation. The combined system minimizes waste heat and converts it into useful work and power. Energy and exergy analyses are performed based on the first and second laws of thermodynamics. A parametric study of the effects of the variation in the SOFC current density, fuel utilization factor, superheat temperature, and SRC evaporation pressure is conducted to define the optimal operating parameters for the proposed system. In the present study, the energy and exergy efficiencies of the integrated system are 64.49% and 61.10%, respectively. These results serve as strong motivation for employing an EGB and SRC for waste heat recovery and increasing the overall energy-conversion efficiency of the system. The SRC energy and exergy efficiencies are 25.58% and 41.21%, respectively.

Keywords: SOFC; ammonia; integrated system; waste heat boiler; steam Rankine cycle

Citation: Duong, P.A.; Ryu, B.; Kim, C.; Lee, J.; Kang, H. Energy and Exergy Analysis of an Ammonia Fuel Cell Integrated System for Marine Vessels. *Energies* **2022**, *15*, 3331. https://doi.org/10.3390/en15093331

Academic Editors: Alon Kuperman and Alessandro Lampasi

Received: 15 April 2022
Accepted: 28 April 2022
Published: 3 May 2022

1. Introduction

Maritime transportation makes a significant contribution to global CO_2 emissions [1]. From 2012 to 2020, CO_2 emissions due to shipping activities increased from 962 million tons to 1056 million tons, which is equivalent to an increase of 2.76% to 2.89% [2]. To combat climate change and contribute to sustainable development, the International Maritime Organization (IMO) has adopted an initial strategy and related regulations to reduce CO_2 emissions per transportation work (carbon intensity) by 40% by 2030, and the total yearly GHG emissions by 50% by 2050 [3]. As a result, several technical and operational measures have been implemented to improve energy efficiency and reduce CO_2 emissions, including the use of alternative fuels, using renewable energy, increasing engine efficiency, implementing waste heat recovery systems, improving hull form, implementing speed reduction, and using alternative sea routes [4].

Hydrogen is a cost-effective and "clean and green" fuel option. However, its limitation is its low volumetric energy density (9.9 MJ/m^3 lower heating value) compared with that of fossil fuels; hence, larger vessels are required for storage and transportation [5]. Both physical and material-based technologies have been explored for the dense storage of hydrogen.

Ammonia, with the chemical formula NH_3, is a fuel with a hydrogen concentration of 17.6% by weight and has no carbon. It decomposes easily into a gas combination of 75% H_2 and 25% N_2, resulting in a high output of pure hydrogen with minimal carbon emissions [6]. Ammonia has emerged as an energy vector because of its ability to efficiently be stored and produced from renewable energy. It is a potential fuel for fuel cells because it is inexpensive, easy to store and carry, less combustible than other fuels, and relatively safe due to observable odor leaks [7]. With this important potential of ammonia as marine fuel, the IMO and international societies have adopted safety guidelines and operational regulations to minimize risks for ammonia utilization. In addition, fuel cells are considered possible alternatives to the aforementioned approaches, which have strong benefits over conventional combustion-based technologies [8]. Ammonia can be directly supplied and converted to electricity using technologies such as SOFCs [9] and internal combustion engines [10], or indirectly used through decomposition in polymer electrolyte membrane fuel cells (PEMFCs) [11]. In contrast to methanol, the ammonia fuel used in SOFCs undergoes thermal cracking at the anode, producing hydrogen and nitrogen; therefore, no steam is required, avoiding carbon formation. In addition, the technological infrastructure for ammonia use is well-established. As a result, there is growing interest in employing ammonia-fueled fuel cells, particularly SOFCs [6].

SOFCs are gaining more attention as a cutting-edge power-generation technology with high energy-conversion efficiency, low environmental impact, and flexibility of fuel supply [12]. Because of the high cost and poor volumetric energy density of hydrogen, portable SOFC applications seem unlikely, even though hydrogen has been widely used as a fuel for SOFCs. However, since SOFCs operate at a high temperature (approximately 600–1000 °C), their efficiency can be increased by combining them with thermodynamics cycles [13,14] to improve the energy-conversion efficiency.

Wojcik et al. [15] experimented with SOFCs using ammonia as the fuel and showed that ammonia can be directly used as fuel in a SOFC system. System performance is dependent on the current density and the catalysts used for the SOFC. The two recommended catalysts for ammonia SOFCs are nickel-and iron-based catalysts. Qianli et al. [16] tested ammonia with an SOFC cell, Ni-BCGO/BCGO/LSCO–BCGO, compared the performance results with hydrogen as the fuel, and proved the feasibility of using ammonia as a fuel for SOFCs. The decomposition of ammonia is efficient in the anode chamber, with a maximum open circuit voltage (OCV) and power density of 0.985 V and 355 $mWcm^{-2}$ at 700 °C, respectively. Baniasadi and Dincer [7] designed an SOFC system for vehicular applications employing ammonia as the fuel and showed that the energy and exergy efficiencies vary from 60% to 90% and 60% to 40%, respectively, depending on the current density of the SOFC. Ammonia can achieve 100% conversion to N_2 and H_2 at temperatures above 873 K. The largest exergy destruction rate was found in the SOFC stack, turbine, and heat exchanger. Al-Hamed and Dincer [17] proposed an integrated ammonia SOFC and hydrogen production system for locomotive applications and showed that hydrogen can be produced during the operation of a passenger locomotive. The energy and exergy efficiencies of the system are 61.2% and 66.3%, respectively. When more hydrogen is generated, the energy and exergy efficiencies are marginally improved by 0.28% and 0.24%, respectively. Ishak et al. [18] performed a thermodynamic analysis with the SOFC–ammonia system, and stated that 100% of ammonia conversion to N_2 and H_2 is achieved at a temperature of 700 K and atmospheric pressure. In a comparison between SOFC-H and SOFC-O, the peak power density of SOFC–H using ammonia is 20–30% higher than that of SOFC–O using ammonia. Ishak et al. [19] presented the energy and exergy analysis for a combined system of an SOFC-GT using ammonia as fuel and found that the highest energy and exergy efficiencies obtained were 81.1% and 74.3%, respectively. The configuration and layout of the system components, among other factors, impacted the total system performance. The highest exergy destruction rate was observed in the fuel cell and heat exchanger. Siddiqui et al. [20] designed and analyzed an integrated SOFC combined cycle system with ammonia as the fuel. The proposed system produces electricity, hot and cold water, and hydrogen for multiple purposes. The overall energy and exergy efficiencies of the system are 39.1% and 38.7%, respectively. Compared with a single-generation system, the multi-generation system provides a higher exergy efficiency of 18.7%. Thus, the developed multi-generation system is more efficient than a single system. The highest exergy destruction rate is found in the organic Rankine cycle (ORC) turbine.

In the present paper, a novel plant configuration is proposed to adapt to both the energy demands of the propulsion plant of the vessel and sustainable development in an environmentally friendly manner. The main objective is to use ammonia as a direct fuel for the SOFC integrated system for marine vessels. Additionally, the integration of an SRC and EGB is introduced to use the exhaust gas from the SOFC-GT subsystem. This combination has not been investigated previously. The novelty of this system is the inclusion of a steam boiler to provide steam for the heating of lubricating oil and seafarer accommodation in the system, and the enhancement of energy efficiency and waste heat recovery. In addition, the hot water generated during the cooling process of the SRC is used in seafarer accommodation. In summary:

- A novel integrated system with ammonia as the fuel in an SOFC for significantly improving the energy-conversion efficiency and green energy target for marine vessels is proposed. The SRC and EGB are integrated to serially recover the waste heat from the exhaust gas from the SOFC using an energy-cascade method.
- A thermodynamic model of the components, sub-system, and total system is built to estimate the working performance of the proposed system.
- A parametric study is carried out to investigate the impact of the main parameters on system performance.

The remainder of this paper is organized as follows. The novel integrated system and its operation are described in Section 2. The thermodynamic models developed to investigate the system energetically and exegetically are presented in Section 3. This is followed by a verification using the thermodynamic models in Section 4. The parametric study on the effect of key parameters on system performance is presented in Section 5. Lastly, the conclusions of the thermodynamic investigation are presented in Section 6.

2. Process Modelling of the Power Generation Plant for the Vessel

The proposed system is designed for an existing general cargo vessel (3000 DWT), which uses a type of electric propulsion unit (3800 kW) [21] powered by ammonia. SOFCs provide power for the main propulsion plant, while a GT and SRC supply power for the auxiliary machinery, lighting, and seafarer accommodation.

Figure 1 shows the flowsheet of the integrated SOFC-GT-SRC-EGB system for marine vessels using ammonia as fuel.

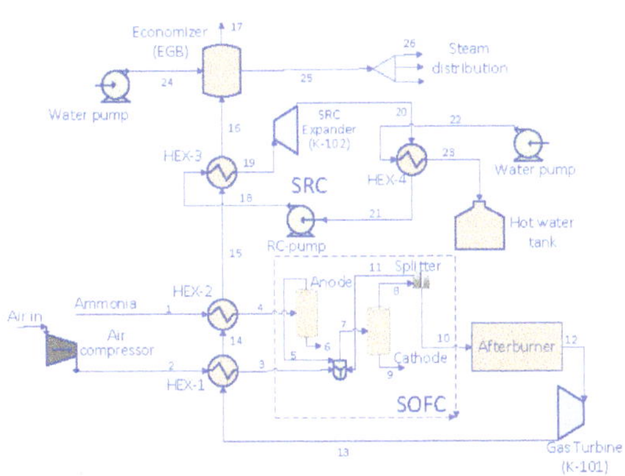

Figure 1. Schematic diagram of the SOFC-GT-SRC-EGB employing ammonia as fuel.

2.1. Design

The proposed multi-generation plant includes four main subsystems: (i) main power production based on SOFC technology, (ii) direct waste-heat utilization in a GT, (iii) energy harvested by the SRC, and (iv) steam generation by an EGB (economizer).

Ammonia is first preheated in the fuel regenerator (HEX-2) before being directly supplied to the SOFC anode and decomposed to H_2 and N_2 at high temperatures with the use of catalysts. Ni/YSZ is employed as the catalyst in this simulation study [9,13,22]. The decomposition of ammonia occurs by the dehydrogenation reactions [6], whereby (i) ammonia is absorbed by the catalyst's active area, (ii) N-H bond cleavage occurs to release hydrogen atoms, and (iii) N and H atoms are recombined to form nitrogen and hydrogen molecules. HEX-2 is needed for the fuel cell to attain its operational temperature. The temperature range for this SOFC is 700–900 °C.

After being compressed by an air compressor to attain the required pressure (stream 2), air is preheated by waste heat from the gas turbine in the heat exchanger (HEX-1). Then, the cathode side of the fuel cells is supplied with air at the required temperature and pressure, providing oxygen for the electrochemical reaction. After reacting in fuel cells, there are two exhaust streams generated. The main exhaust stream (stream 10) is supplied to the afterburner. Otherwise, the exhaust of a nitrogen-rich stream from the cathode (stream 9) and a mixed water and nitrogen stream (stream 6) from the anode, are discharged.

The afterburner burns the remaining (unused) ammonia fuel (stream 10) from the anode and cathode to produce a high-pressure, high-temperature exhaust gas. The exhaust gas (stream 12) then powers the turbine, which generates more electricity.

The integrated waste-heat recovery system includes a gas turbine (K-101), three recuperates (HEX-1, HEX-2, and HEX-3) that recover waste heat after the afterburner (stream 12) and expansion in the gas turbine (K-101) (stream 13). The recuperates pre-heat the feed air (via HEX-1), ammonia (via HEX-2), and the SRC working fluid (via HEX-3), after which heat is transferred to the water in the EGB.

In the SRC, the heat exchanger (HEX-3) is primarily responsible for recovering waste heat. The working fluid is water, which is initially pumped to a high pressure by a pump (RC-Pump) and then passed through a heat exchanger (HEX-3) to form superheated vapor. The high-pressure steam is then expanded and depressurized in the SRC expander (K-102) before being used to drive the reversible pump (RC-pump) and generate further electric power. The saturated vapor condensed in the heat exchanger (HEX-4) is transferred to cold water, releasing heat. The resulting hot water is supplied to seafarer accommodation.

Finally, the EGB uses waste heat, transferring it to fresh water and generating steam for multiple uses, such as the heating of lubricating oil and accommodation air conditioning.

The working performance of the integrated SOFC-GT-SRC-EGB hybrid system is strongly influenced by the SOFC efficiency, exhaust heat temperature, gas turbine cycle pressure, and GT pressure ratio. The SOFC power output/efficiency impacts the exergy of the exhaust gas, and the ORC turbine inlet pressure also affects the integrated system. For an increase in the power output of the SOFC, the mass flow of exhaust gas (stream 12) and heat provided to the gas turbine cycle increases, reducing the GT fuel requirement and improving the SOFC-GT efficiency.

2.2. Operating Data Performances and Assumptions

The proposed integrated SOFC with directly supplied ammonia is simulated with support from ASPEN-HYSYS V12.1 (AspenTech, Massachusetts, USA) [23–25]. Ammonia is defined as the conventional component of Aspen HYSYS; therefore, the Peng–Robinson (PR) equations of state are employed to estimate the thermodynamic properties and operating parameters of all components.

To simplify models and analysis, the below general assumptions are adopted:

- The supplied air is composed of 79% N_2 and 21% O_2 at 29.85 °C, 101.3 kPa;
- This simulation is accomplished in a steady state and thermodynamic equilibrium;
- The mass and pressure losses in all pipe components are negligible;

- The pressure losses on the tube side and shell side of the heat exchanger (HEX) are assumed to be 6.895 and 3.447 kPa, respectively;
- The flow temperatures of the air and fuel at the inlet and outlet of the SOFC are constant and equal to the working temperature of the SOFC.
- The boundary conditions of the simulation study are listed in Table 1 [17,26].

Table 1. Boundary conditions of the proposed system.

Component	Parameter	Value
SOFC	Operating pressure (bar)	4
	Operating temperature (°C)	814.1
	Number of single cells	14,202
	Fuel cell current density (A/m^2)	1455
	Active surface area (m^2)	0.22
	Anode thickness (cm)	0.002
	Cathode thickness (cm)	0.002
	Electrolyte thickness (cm)	0.004
	Stoichiometric rate of hydrogen	1.2
	Stoichiometric rate of oxygen	2
	Fuel utilization factor in SOFC (%)	85
Compressor	Isentropic efficiency (%)	82
Pumps	Isentropic efficiency (%)	82
Expanders	Isentropic efficiency (%)	80
Heat exchangers	Minimum temperature approach (°C)	7
Converter	DC-AC converter efficiency (%)	98

3. Material and Methodology

Ammonia is unstable at high temperatures and decomposes into H_2 and N_2 [19], as per Equation (1), at a temperature of 200 °C or greater [27]. At 425 °C, approximately 98–99% ammonia is decomposed, and if the temperature is above 600 °C (873 K), ammonia is entirely transformed [28]. However, the conversion rate is affected by the presence of catalysts.

$$2NH_3 \leftrightarrow N_2 + 3H_2 \tag{1}$$

$$2H_2 \leftrightarrow 4H^+ + 4e^- \tag{2}$$

The generated H_2 is used for the electrochemical generation of energy following the first cracking of NH_3 into N_2 and H_2 over the Ni/YSZ catalyst. Subsequently, oxide ions oxidize the hydrogen to H_2O and generate electricity [28].

$$O_2 + 4e^- + 4H^+ \rightarrow 2H_2O \tag{3}$$

$$O_2 + 2H_2 \rightarrow 2H_2O \tag{4}$$

Water (vapor state) and the remaining oxygen exit the fuel cell from the cathode side, while NH_3, N_2, and H_2 exit from the anode side. SOFCs rely primarily on an oxygen ion-conducting electrolyte to allow oxygen ions to move from the cathode to the anode, resulting in the production of nitrogen oxide (NO).

$$N_2 + 2O^{2-} \rightarrow 2NO + 4e^- \tag{5}$$

$$2NH_3 + 5O^{2-} \rightarrow 2NO + 3H_2O + 10e^- \tag{6}$$

By completely covering the YSZ electrolyte with a proton conductor and ammonia cracks, the production of NO may be avoided [15,16]. Ma et al. [29] showed that the partial pressure of NO increases with increasing of NH_3 conversion rate. However, the partial pressure of NO is only 10^{-12} atm when the ammonia conversion rate of ammonia is 0.99 and the working temperature of SOFC is 800 °C.

Reaction Equation (1) is endothermic and is influenced by pressure and temperature, in addition to other conditions. As the temperature rises, the reaction shifts to the right, producing more hydrogen, and as the pressure rises, it shifts to the left, producing less hydrogen. The overall Gibbs energy of the process can be decreased to attain thermodynamic equilibrium with ammonia [5,18,19]:

$$(\Delta G_{system})_{T,P} = 0 \tag{7}$$

The sum of the product of the number of moles of chemical species i multiplied by the appropriate standard Gibbs free energy at constant temperature and pressure is the Gibbs energy of the system:

$$G_{system} = \sum n_i \overline{g}_i \tag{8}$$

The Gibbs energy of one mole of the substance for a gas can be expressed in terms of this activity as per the following equation:

$$\overline{g}_i = \overline{g}^0_{fi} + RT\, ln \frac{f_i}{f_i^0} \tag{9}$$

The total Gibbs energy is equal to:

$$G_{system} = \left(\sum n_i [\overline{g}^0_{fi} + RT \ln(y_i P)]\right)_{gas} + \left(\sum n_i \overline{g}^0_{fi}\right)_{condensed} \tag{10}$$

3.1. Model of the SOFC

SOFCs [30] transform the chemical energy of the fuel into electrical power through electrochemical reactions at high temperatures (from 500 °C to 1100 °C). In the SOFC, ammonia participates in the reforming reaction according to Equation (1) under YSZ electrolytes [31].

3.1.1. Fuel and Oxidant Utilization

The utilization of fuel (ammonia) can be defined as follows [13,19,21,32,33]

$$U_f = \frac{(ammonia)_{react}}{(ammonia)_{in}} = \frac{(H_2)_{react}}{(H_2)_{in}} \tag{11}$$

where $(H_2)_{react}$ and $(H_2)_{in}$ denote the molar flow of the reacting hydrogen and the molar flow of hydrogen supplied to the fuel cells, respectively.

Similarly, the air utilization can be determined as the ratio of oxygen supply and usage because air is assumed to consist of a constant ratio of 21% oxygen and 79% nitrogen:

$$U_{air} = \frac{(Air)_{consumed}}{(Air)_{in}} = \frac{(O_2)_{consumed}}{(O_2)_{in}} \tag{12}$$

Air_{in} represents molar flow of inlet air.

With reference to Equation (4):

$$n_{O_2} = 0.5\, U_f n_{H_2} \tag{13}$$

n_{O_2} and n_{H_2} represent the molar flow rate of oxygen and hydrogen required for the fuel cells, respectively.

The net power output of an SOFC system can be determined using the component stack, as follows [34–36]:

$$W_{stack} = i{\cdot}a{\cdot}V_c \tag{14}$$

where i and V_c represent the current density (A/m^2) and the actual voltage of the stack (V), respectively, and a is the surface area (m^2) [26,30,35,37,38]

$$i = \frac{nFU_f q_{NH3}}{N_{cell} A_{cell}} \tag{15}$$

$$q_{NH3} = \frac{i\, N_{cell} A_{cell}}{U_f nF} \tag{16}$$

where n denotes the number of electrons transported by every oxygen atom of the electrochemical reaction (in this case, its value is 3); F denotes the Faraday constant, 96,485 C/mol; q_{NH_3} is the ammonia molar flow rate, mol/s; and N_{cell} and A_{cell} represent the number of cells and cell area (m^2), respectively.

$$V_c = V_R - V_{loss} \tag{17}$$

V_R is defined as the cell's ideal reversible voltage and V_{loss} is the voltage loss in the system. Based on the Nernst equation:

$$V_R = V^o + \frac{R_u T}{n_e F} \ln\left(\frac{P_{H_2} P_{O_2}^{\frac{1}{2}}}{P_{H_2O}}\right) \tag{18}$$

$$V_{loss} = V_{ohm} + V_{act} + V_{con} \tag{19}$$

where V_{ohm} denotes the ohmic losses (V), V_{act} denotes the activation losses (V), V_{con} denotes the concentration losses (V), and V^o denotes the electro-motoric force.

$$V_{ohm} = V_{ohm,a} + V_{ohm,c} + V_{ohm,e} + V_{ohm,int} \tag{20}$$

Furthermore, the I-V curve can be used to determine the actual voltage of the stack [17,29,32,39–41]

The generated power of the *SOFC*:

$$W_{SOFC} = W_{stack} N_{cell} \cdot \eta_{DA} \tag{21}$$

η_{DA} is the inverter efficiency.

Thermodynamically, the energy efficiency of the fuel cell [26,35,42–44]:

$$\eta_{en,SOFC} = \frac{\dot{W}_{SOFC}}{\dot{m}_{NH_3} LHV_{NH_3}} \tag{22}$$

where $\dot{m}_{NH_3}$ and LHV_{NH_3} represent the mass flow rate of NH_3 (kg/h) and the low heating value of NH_3 (KJ/kg), respectively.

3.1.2. Afterburner

The unreacted hydrogen and ammonia in the SOFC are combusted in the afterburner. A strong exothermic oxidation reaction occurs, increasing the temperature and pressure of the exhaust gas. The reactants are assumed to burn completely adiabatically [45].

3.2. *Model of the GT*

GT

The hot gaseous mixture expands and produces useful mechanical power as it exits the combustion chamber and enters the gas turbine. The outlet temperature of the gas turbine is

$$T_{out} = T_{in}\,(PR)^{\frac{(k-1)}{k}} \tag{23}$$

where $PR = \frac{P_{in}}{P_{out}}$ and $k = \frac{\sum_i y_i \bar{C}_{p,i}}{\sum_i y_i \bar{C}_{v,i}}$

The isentropic efficiency can be calculated as

$$\eta_{s,T} = \frac{\sum_i (\dot{n}_i \bar{h}_i)_{in} - \sum_i (\dot{n}_i \bar{h}_i)_{out}}{\sum_i (\dot{n}_i \bar{h}_i)_{in} - \sum_i (\dot{n}_i \bar{h}_i)_{s,\ out}} \tag{24}$$

For the *SOFC*-GT subsystem:

Energy efficiency can be determined as:

$$\eta_{en,SOFC,GT} = \frac{\dot{W}_{SOFC} + \dot{W}_{GT} - \dot{W}_{Air\ comp}}{\dot{m}_{NH3} LHV_{NH3}} \tag{25}$$

Exergy efficiency can be determined as:

$$\eta_{ex,SOFC,GT} = \frac{\dot{W}_{SOFC} + \dot{W}_{GT} - \dot{W}_{Air\ comp}}{\dot{m}_{NH3} ex_1} \tag{26}$$

Heat exchangers:

The following formulae are used to compute the hot (exhaust gas) and cold (fuel or air supply) streams through the heat exchanger:

$$\dot{Q} = \sum_i (\dot{n}_i \bar{c}_{p,i})_h \left(T_{h(c),in} - T_{h(c),out}\right) \tag{27}$$

3.3. Model of the SRC

The equation for the energy balance of the *SRC* turbines is (the subscript numbers in the equation represent the stream in the flow diagram):

$$\dot{m}_{wf,SRC} h_{20} = \dot{W}_{SRC,T} + \dot{m}_{wf,SRC} h_{19} \tag{28}$$

The waste heat absorbed by HEX-3 of the *SRC*:

$$\dot{Q}_{HEX-3} = \dot{m}_{wf,SRC}(h_{19} - h_{18}) = \dot{m}_{exhgas}(h_{15} - h_{16}) \tag{29}$$

The net power output of the *SRC*:

$$\dot{W}_{net,SRC} = \dot{W}_{SRC,Turbine} - \dot{W}_{P_{RC-Pump}} \tag{30}$$

The energy and exergy efficiency of the *SRC*:

$$\eta_{en,SRC} = \frac{\dot{W}_{net,SRC}}{\dot{m}_{15}\ (h_{15} - h_{16})} \tag{31}$$

$$\eta_{ex,SRC} = \frac{\dot{W}_{net,SRC}}{\dot{m}_{15}\ (ex_{15} - ex_{16})} \tag{32}$$

3.4. Model of the EGB

In this system, the EGB serves as a heat exchanger, and heat is transferred between the water and flue gas inside and outside the tube, resulting in the generation of steam in the boiler drum. The flue gas flow rate is 321 kgmole/h, at 359.4 °C. At the calculated steam production rate corresponding to the heat exchange facilitated in the boiler, steam is produced at 175 °C, 781.1 kPa, and 725 kg/h.

In this study, the EGB is divided into three working areas, as shown in Figure 2: a single-phase subcooled area, a two-phase evaporation area, and a single-phase superheated area. The heat balance equations between the hot and cold sides of each area are:

Superheated area:

$$\dot{m}_{16}(h_{16} - h_e) = \dot{m}_{24}(h_{25} - h_c) \tag{33}$$

Evaporation area:

$$\dot{m}_{16}(h_e - h_d) = \dot{m}_{24}(h_c - h_a) \tag{34}$$

Sub-cooled area:

$$\dot{m}_{16}(h_d - h_{17}) = \dot{m}_{24}(h_a - h_{24}) \tag{35}$$

The total heat transfer by the waste heat boiler [46] can be calculated as:

$$Q = U \cdot A \cdot LMTD \tag{36}$$

where $LMTD$ represents the log mean temperature difference, U is the heat exchange coefficient fixed at 30 $Wm^{-2}K^{-1}$, and A is the heat exchange area (m^2).

$$LMTD = \frac{\Delta T_{2,end} - \Delta T_{1,end}}{\ln\left(\frac{\Delta T_{2,end}}{\Delta T_{1,end}}\right)} \tag{37}$$

where $\Delta T_{2,end}$ and $\Delta T_{1,end}$ are the temperature differences between the hot and cold streams at the two ends of the heat exchanger, respectively.

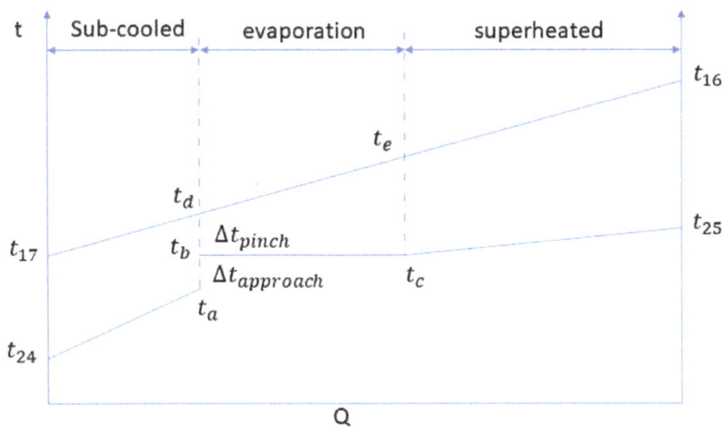

(a)

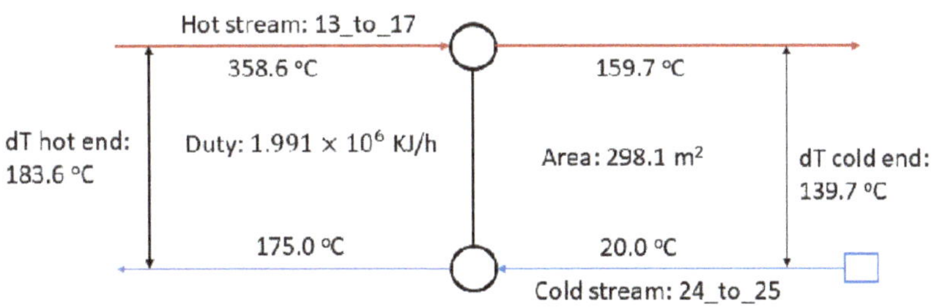

Figure 2. *Cont.*

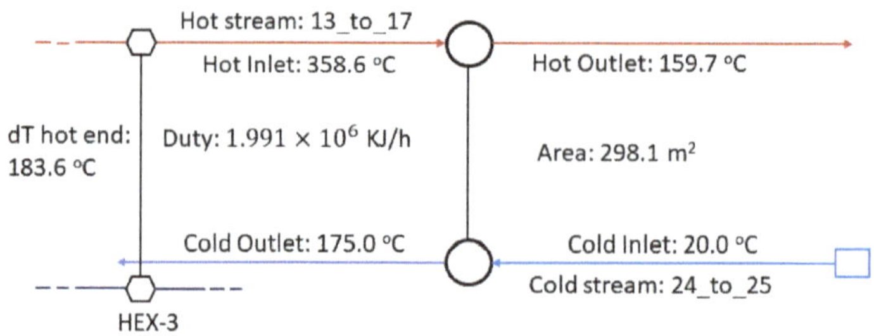

(b)

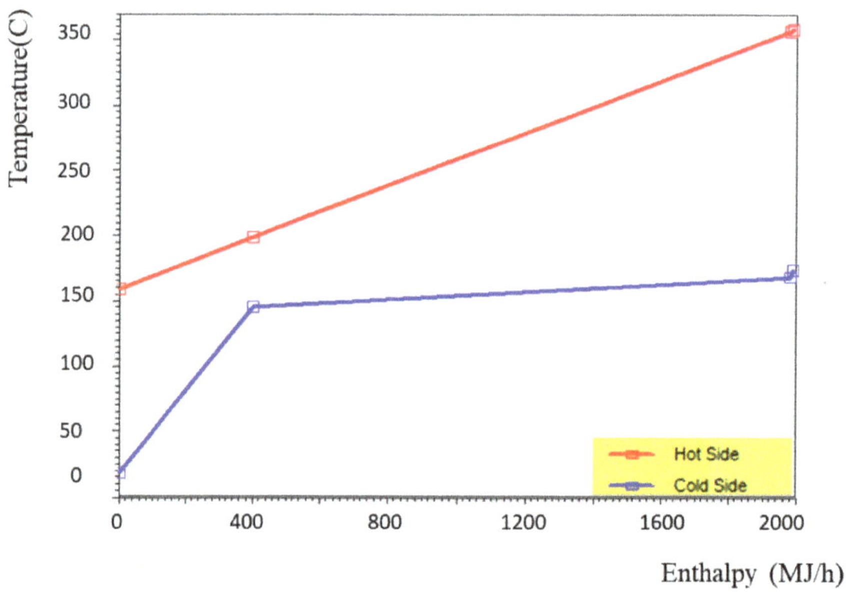

(c)

Figure 2. The temperature distribution of the EGB. (**a**) Temperature regions in the EGB. (**b**) Connectivity. (**c**) Temperature–enthalpy plot.

3.5. Exergy Efficiency of the Main Components

In the system, the exergy represents the maximum working capacity dissipated in the process. The exergy efficiencies of the main components of the system calculations are based on the second law of thermodynamics. Table 2 demonstrates the equilibrium relationship of the exergy destruction of the main components [38,47–50].

Table 2. Exergy destruction equations of the main components.

Components	Exergy Destruction Rate	
SOFC	$\dot{E}x_3 + \dot{E}x_4 + \dot{E}x_{11} - \dot{E}x_{10} + \dot{W}_s = \dot{E}x_{des}$	(38)
Afterburner	$\dot{E}x_{10} - \dot{E}x_{12} = \dot{E}x_{des}$	(39)
Gas Turbine	$\dot{E}x_{12} - \dot{E}x_{13} + \dot{W}_{Gas\ turbine} = \dot{E}x_{des}$	(40)
Air heat exchanger (HEX-1)	$\dot{E}x_2 + \dot{E}x_{13} - \dot{E}x_3 - \dot{E}x_{14} = \dot{E}x_{des}$	(41)

Table 2. *Cont.*

Components	Exergy Destruction Rate	
Fuel heat exchanger (HEX-2)	$\dot{E}x_1 + \dot{E}x_{14} - \dot{E}x_{15} - \dot{E}x_4 = \dot{E}x_{des}$	(42)
SRC heat exchanger (HEX-3)	$\dot{E}x_{15} + \dot{E}x_{18} - \dot{E}x_{19} - \dot{E}x_{16} = \dot{E}x_{des}$	(43)
Exhaust gas boiler	$\dot{E}x_{16} + \dot{E}x_{24} - \dot{E}x_{17} - \dot{E}_{x25} = \dot{E}x_{des}$	(44)

The overall energy and exergy efficiencies for the entire integrated system are written as [17,19,51]:

Energy balance equation at steady state:

$$\dot{Q} - \dot{W} + \sum_{in} \dot{m}_{in}\left(h_{in} + \frac{V_{in}^2}{2} + gZ_{in}\right) - \sum_{out} \dot{m}_{out}\left(h_{out} + \frac{V_{out}^2}{2} + gZ_{out}\right) = 0 \quad (45)$$

$$\eta_{en,overall} = \frac{\dot{W}_{elec,overall}}{\dot{m}_{NH3}\ LHV_{NH3}} \quad (46)$$

where $\dot{W}_{elec,overall}$ represents the net value of power production and consumption of the system:

$$\dot{W}_{elec,overall} = \dot{W}_{elec,SOFC} + \dot{W}_{GT} + \dot{W}_{SRC,\ turbine} - \dot{W}_{Air\ comp} - \dot{W}_{SRC,\ pump} \quad (47)$$

LHV_{NH3} is the lower heating value of ammonia (kJ/kg).

According to the second law of thermodynamics, the physical, chemical, potential, and kinetic exergy all contribute to the total exergy value. In this analysis, the potential and kinetic exergy are negligible.

$$ex_j = ex_j^{ph} + ex_j^{ch} = (h_j - h_0) - T_0(s_j - s_0) + \sum_k x_k\left(ex_j^{ch} - RT_0 x_k \ln(x_k)\right) \quad (48)$$

$$\eta_{ex,overall} = \frac{\dot{W}_{elec,overall}}{\dot{m}_{NH3}\ ex_{NH3}} \quad (49)$$

4. Model Verification

The values calculated using the integrated model employing ammonia as fuel and the corresponding values listed in the literature are shown in Table 3. The estimated values agree well with the data in the literature, and the discrepancy between the current data and the literature is within an acceptable range.

Table 3. Comparison of simulation results of the proposed integrated model with the corresponding values listed in the literature.

Parameter	Model Results	Reported [52]	Difference (%)
SOFC temperature (°C)	808.8	800	1
Cell voltage (V)	0.85	0.67	21
Fuel utilization factor (%)	85	80	5
Current Density (A/m^2)	1455	1450	0.3
SOFC efficiency	43.8	39	4.8

The proposed system ideally simultaneously supplies power for the propulsion plant and auxiliary equipment and hot water for seafarer accommodation.

Verification of the GT-SRC subsystem is necessary, as it is responsible for 26.87% of the integrated system's total power production. Chitgar et al. [36] used a multi-generation SOFC-GT system providing a total power generation of 4910.4 kW (1.8% difference), with energy and exergy efficiencies of 64.3% and 49.0%, respectively. The present model improves upon that energy efficiency by 0.19% and the exergy efficiency by 12.1%, demonstrating that the proposed model produces reasonable results.

5. Results and Discussion

Table 4 lists the thermodynamic parameters and state points of each node of the proposed integrated power system. The minimization of Equation (10) was performed using the optimization function in Aspen HYSYS V12.1.

Table 4. Simulation results of integrated system.

Node	Vapor Fraction	Temperature	Pressure	Molar Flow	Mass Enthalpy
Unit		C	kPa	kgmole/h	kJ/kg
Air in	1.00	29.85	101.30	187.62	4.63
1	1.00	29.85	400.00	80.10	−2687.22
2	1.00	203.36	400.00	187.62	183.31
3	1.00	492.60	396.55	187.62	495.48
4	1.00	492.60	396.55	80.10	−1518.26
5	1.00	152.64	396.55	105.22	−1518.21
6	0.00	152.64	396.55	0.00	−1518.21
7	1.00	383.66	396.55	309.58	90.18
8	1.00	814.11	396.55	334.66	90.18
9	0.00	814.11	396.55	0.00	90.18
10	1.00	814.11	396.55	317.93	90.18
11	1.00	814.11	396.55	16.73	90.18
12	1.00	1158.61	396.55	320.99	90.18
13	1.00	890.01	117.00	320.99	−358.61
14	1.00	736.20	110.11	320.99	−607.95
15	1.00	587.58	103.21	320.99	−843.23
16	1.00	359.39	96.32	320.99	−1192.75
17	1.00	159.71	89.42	320.99	−1486.54
18	0.00	72.12	19,000.00	56.62	−15,667.37
19	1.00	360.90	18,996.55	56.62	−13,345.07
20	0.73	74.69	38.00	56.62	−13,965.52
21	0.00	70.00	31.11	56.62	−15,693.37
22	0.00	20.00	100.00	444.07	−15,909.39
23	0.00	70.98	96.55	444.07	−15,689.09
24	0.00	20.00	784.53	40.24	−15,908.74
25	1.00	175.00	781.08	40.24	−13,162.57
26	1.00	814.10	396.55	16.73	90.21

The performance of the main components of the system are listed in Table 5. The results demonstrate an increase in the overall electrical efficiency of the proposed system to approximately 64.49% due to the recovery of the SOFC-GT exhaust gas via the SRC. Further improvements can be achieved via parameter optimization.

Table 5. Performance of integrated system.

Term	Value
SOFC power output (kW)	3800
Gas Turbine power (kW)	844.8
SRC Turbine power (kW)	175.8
SRC pump power (kW)	7.369
Air compressor power (kW)	268.7
SOFC electrical efficiency (%)	53.92
Electrical efficiency of entire system (%)	64.49
EGB mass flow rate (kg/h)	725
EGB superheated temperature (°C)	175

5.1. Energy and Exergy Efficiency of System

The target vessel requires 3800 kW of electric power for the main propulsion system, and additional power to run auxiliary machines, lighting, and seafarer accommodation requirements. The fuel cell should be operated at the best possible fuel utilization factor, while maintaining an acceptable reactant concentration and fuel cell efficiency. For a utilization factor of 0.85, the SOFC energy efficiency is 53.9%. With reference to the thermodynamic models built in Section 3, the total output power of the integrated system is 4820.6 kW, which is sufficient to drive the vessel and its auxiliary systems. The 1020.6 kW surplus over the power demand of the ship results almost exclusively from the waste heat recovery components, accounting for 21.17% of the total electrical output, which demonstrates that the subsystem was necessary for the running of auxiliary systems. The overall energy and exergy efficiencies of the proposed system are 64.49% and 61.10%, respectively. The energy and exergy efficiencies of the entire system and each subsystem are listed in Table 6.

Table 6. Energy and exergy efficiency of the proposed system.

Subsystem	Energy Efficiency	Exergy Efficiency
SOFC-GT	62.10	58.84
SRC	25.58	41.21
Total System	64.49	61.10

Figure 3 displays the exergy destruction of the key components of the proposed system as a result of the thermal processes within them. It is apparent that the system with the largest exergy destruction is the SOFC, with a value of 3136.25 kW, equivalent to 50.56% of the entire integrated system. The presence of electrochemical and chemical processes in this component are the primary sources of irreversibility.

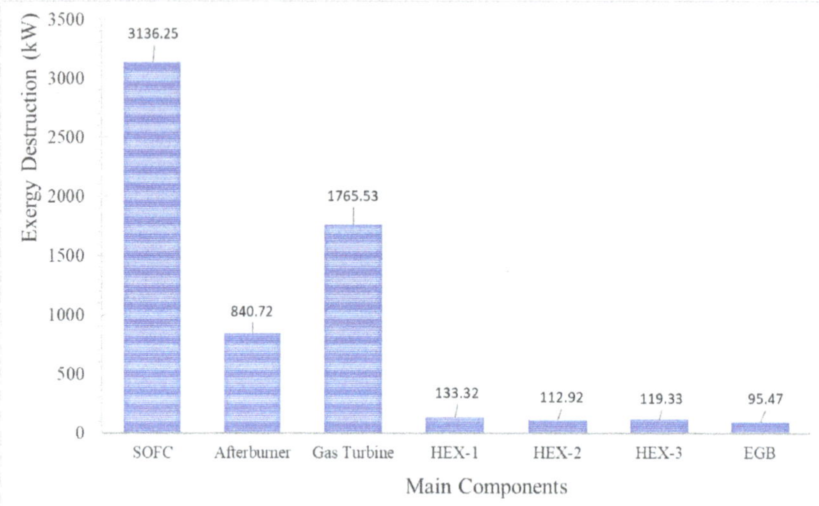

Figure 3. Exergy destruction of the main components of the proposed system.

The second largest sources of exergy destruction are the gas turbine and afterburner, at 28.46% and 13.55% of the whole system, respectively. The gas turbine has a comparatively high rate of exergy destruction, implying that there is a significant margin for development. The smallest exergies belong to the regenerators, HEX-1, HEX-2, and HEX-3, with values of 133.32 kW, 112.92 kW, and 119.33 kW, equivalent to 2.15%, 1.82%, 1.92% and 1.54% of the entire system, respectively. The phase transition process is the source of most of the thermal irreversibility, including internal and external irreversibility. The EGB had the lowest exergy destruction value of 95.47 kW due to the lower temperature of the flue gas provided to the boiler.

5.2. *Influence of the Key Parameters on the System's Performance*

A parametric investigation based on the first and second laws of thermodynamics was conducted to estimate the overall performance of the combined system.

Several variables affect the performance of the integrated SOFC-GT-SRC system. However, the main objective of the current study is to estimate the waste-heat recovery of the SOFC-GT system employing SRC and EGB to provide electric power to the propulsion plants of target vessels and other auxiliary electric equipment. Secondary system outputs include steam and hot water for machinery requirements and seafarer accommodation.

The parametric study examined how modifying the current density, fuel utilization factor, and turbine inlet temperature and pressure of the SRC affected the energy and exergy efficiency of the overall system.

5.2.1. Effect of Current Density

Current density is one of the key performance measures for the fuel cells and the entire system. Figure 4 demonstrates the effect of the current density on the energy efficiency of the system and other essential components. The current densities range from 955 A/m^2 to 2055 A/m^2. The cell operating pressure is 4 bar and the operating temperature is 800.8 °C. It can be seen that the efficiency of the SOFC-GT subsystem and the entire system decreases with increasing current density, whereas the energy efficiency of the SRC is independent of current density.

The highest net electrical cycle efficiency is 67.33% at 955 A/m^2, and the lowest is 60.24% at 2055 A/m^2. The energy efficiency of the SOFC-GT system decreases with an increase in current density. Increased current density necessitates an increase in air flow to provide more oxygen ions. Furthermore, decreased efficiency results in an increase in unconverted chemical energy, which is converted into heat; thus, heat dissipation at the air inlet must increase to maintain the operating temperature of the cells. Low-current-density operation results in greater efficiency. Additional capital (more cells) is required during low-current-density operation to generate the same amount of power as that generated during high-current-density operation.

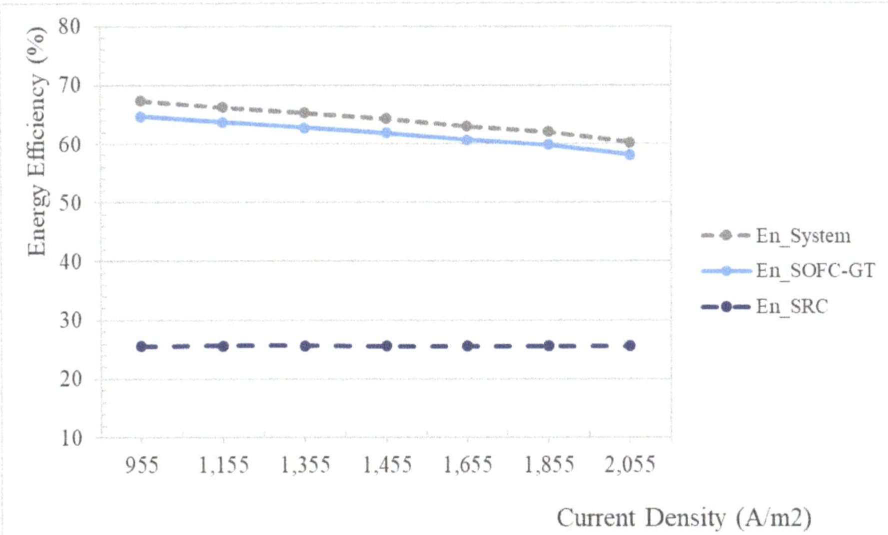

Figure 4. Effect of current density on the energy efficiency of system.

To maintain the power output of the system at predetermined levels, the mass flow rate was increased from 1247 kg/h to 1526 kg/h, and the current density was decreased from 2055 A/m^2 to 955 A/m^2. For a constant power scenario, it can be seen that lower current density requires more fuel for the SOFC. With reference to Equation (14), increasing current density also results in a decrease in cell voltage. As the current density increases, the Nernst potential and voltage losses decrease. Increases in the energy and exergy efficiency with a lowering in the current density were also reported in the literature [53] due to a large reduction in voltage losses. Based on these opposing requirements, an optimal value of current density was selected to maximize cell voltage. This maximum cell voltage was determined to be 0.85 V, corresponding to a current density of 1455 A/m^2.

The effect of the current density on the exergy efficiency is shown in Figure 5. With increasing current density, the net electrical exergy efficiency decreases, owing to a decrease in cell voltage, and as a result, a decrease in power production from the SOFC. In contrast, the exergy efficiency of the SRC tends to increase with increasing current density. This is because an increase in the flue gas temperature and mass exergy corresponds with an increase in the mass flow rate.

For the range of current densities investigated, the net electrical exergy efficiency of the entire system and SOFC-GT subsystem decreases with increasing current density from 63.79% to 57.8% and 61.32% to 55.05%, respectively.

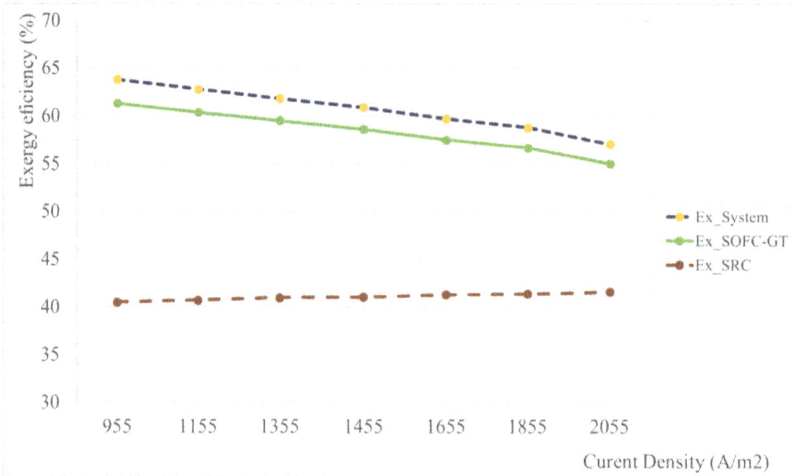

Figure 5. Effect of current density on the exergy efficiency of system.

5.2.2. Effect of the Turbine Inlet Pressure of the SRC

The parameters based on the first and second laws of thermodynamics are clear indicators of a system's performance, providing conclusive information about the current state of the system. Both exergy and thermal analyses are required to accurately represent the system's behavior as a function of the working pressure and temperature. In this section, the parametric optimization of the SRC is carried out in depth based on the turbine inlet pressure and temperature.

The turbine inlet pressure is a critical factor affecting the SRC performance. Figure 6 demonstrates the variation in the energy and exergy efficiency and power output of the SRC as a function of the turbine inlet pressure. The figure shows that the power output of the expander and the energy and exergy efficiency of the system fluctuate with the increasing evaporation pressure in the SRC in the range of 10,000 kPa to 30,000 kPa, whereas the power required form the SRC pump increases with an increase in evaporation pressure. In the observed range of evaporation pressures, the change in power output of SRC expander, energy, and exergy efficiency varies from 85.5 kW to 181.9 kW, 63.23% to 64.62%, and 59.91% to 61.23%, respectively.

It is noteworthy that the power output, energy efficiency, and exergy efficiency of the SRC decrease to a minimum at a turbine inlet pressure of 14,000 kPa; however, they reach a maximum at a turbine inlet pressure of 18,000 kPa. At a constant turbine inlet temperature, when the turbine inlet pressure is 14,000 kPa, liquid is formed at the inlet of the expander, influencing the results negatively. As the SRC uses two heat exchangers, it is important to consider the effects of turbine inlet pressure on the LMTD of the heat exchanger, as shown in Equations (36) and (37). Taking this into account, an evaporation pressure of 19,000 kPa was selected for this simulation.

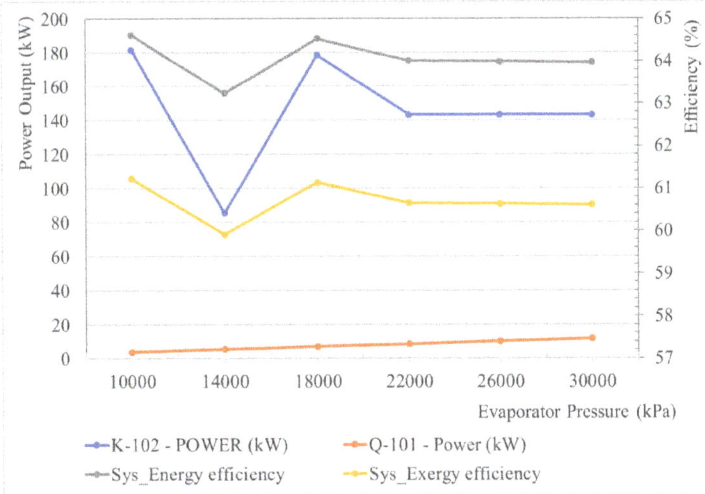

Figure 6. Effect of evaporator pressure on the SRC power output and efficiency.

5.2.3. Effect of the Evaporation Temperature of the SRC

The evaporation temperature of the SRC varies with the mass flow rate of the flue gas of the SOFC, the current density of the SOFC, and the compressor ratio of the exhaust gas turbine. Different superheat temperatures of the SRC working fluid were simulated, and the results are presented in Figure 7.

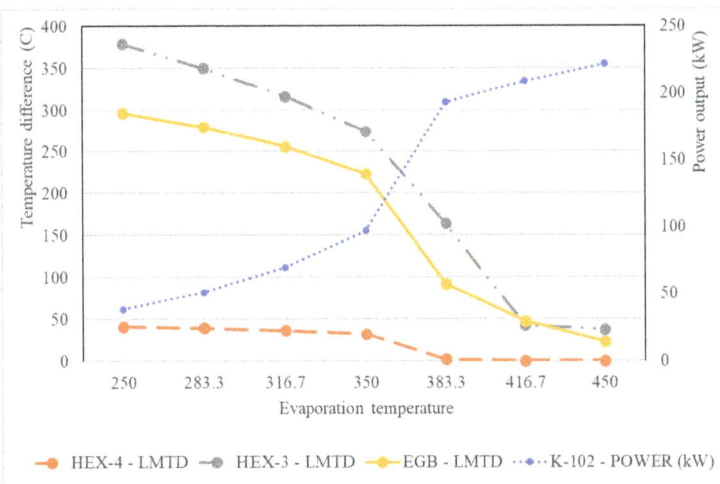

Figure 7. Effect of superheat temperature on SRC performance and heat exchanger efficiencies.

Figure 7 shows the effect of the superheat temperature, in the range 250 °C to 450 °C, on the SRC performance indicators and the temperature difference in the heat exchangers. With an increase in the superheat temperature, the power output of the SRC significantly increases, and the energy and exergy efficiency of the SRC are improved. The power output of the SRC changes from 38.46 kW to 221.9 kW with a superheat temperature increase from 250 °C to 450 °C. However, the temperature difference (LMTD) of the heat exchangers tends to decrease with an increase in the superheat temperature. As shown in Equations (36) and (37), with a decrease in LMTD, the temperature difference between the hot and cold sources of the heat exchangers decreases. This necessitates an increase in the heat-contacting area (A), which increases the cost of the design and operation. In contrast, a high value of LMTD necessitates a smaller heat exchange area with inherent manufacturing inefficiency. Therefore, the LMTD value selected must balance the heat exchange efficiency and cost of manufacturing.

The above analysis demonstrates the significant effect of the SRC superheat temperature on the SRC performance as well as the performance of heat exchangers (HEX-3, HEX-4, and EGB). Therefore, the selection of the stack cooling passage needs to be optimized to ensure that the SRC fluid can fully exchange heat and reach the desired superheat temperature.

5.2.4. Effect of the Fuel Utilization Factor

The performance of the SOFC system has been studied for fuel utilization factors (U_f) of 65%, 75%, and 85%, respectively. Figures 8 and 9 demonstrate the influence of U_f on the efficiency of the SOFC and the integrated system. It is observed that system efficiency increases with the fuel utilization ratio, if the concentration polarization is not significantly higher than that of other polarizations. When cells operate at a low current density, the influence of the fuel usage ratio becomes increasingly significant. When the SOFC-GT operates at lower current densities, the energy efficiency of the combined cycle improves. This is because the incoming fuel mass flow rate decreases faster than the output power. The calculated values for the energy efficiency of SOFC-GT are in agreement with the thermodynamic modeling results presented in References [7,54].

When the U_f value is 0.85, the energy efficiency of the system reaches a maximum value. This is because increasing the U_f further results in the consumption of more hydrogen in the SOFC stack, which simultaneously increases the current density and reduces the voltage due to internal irreversibility. In addition, the outlet temperature of the SOFC and afterburner is reduced, which further results in a decrease in the inlet temperature and power output of the GT.

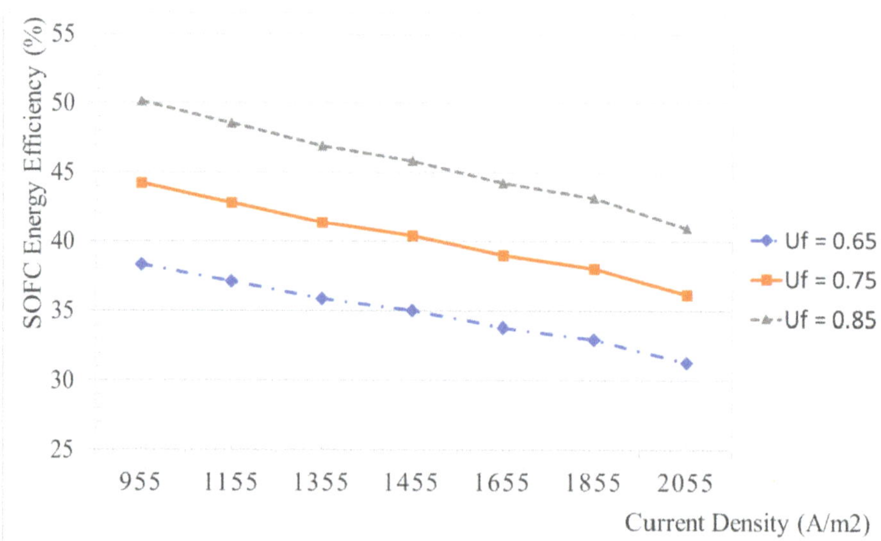

Figure 8. Effect of fuel utilization factor on the SOFC energy efficiency.

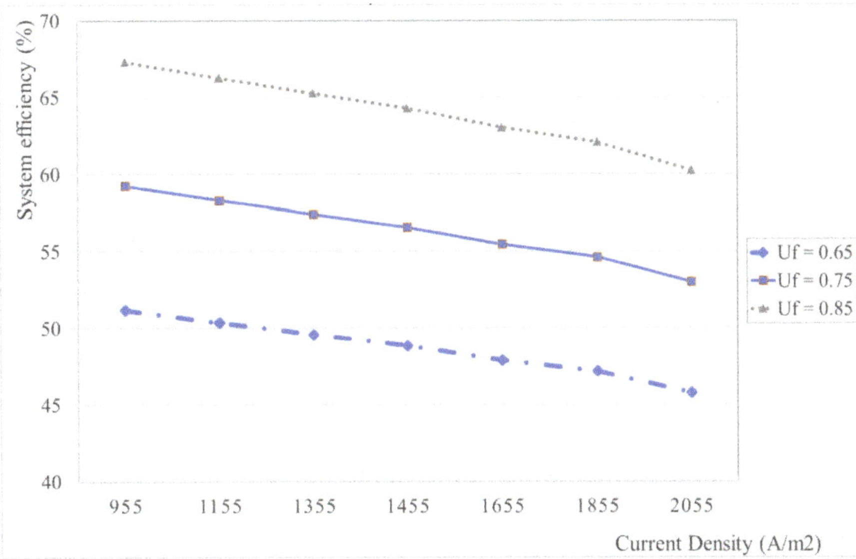

Figure 9. Effect of fuel utilization factor on the integrated system efficiency.

6. Conclusions

The energy and exergy performances of the system were assessed using the first and second laws of thermodynamics. The first law evaluates power, power density, current density, voltage, and electrical efficiency; the second law evaluates exergy efficiency and exergy losses.

In the present study, an SRC and EGB are employed to recover waste heat from an SOFC/GT system. A thermodynamic analysis predicted an increase of 175.8 kW of power due to the SRC, with energy and exergy efficiencies of 25.58% and 41.21%, respectively. Hot water and steam are generated for use in machinery and seafarer accommodation.

In addition, a parametric study showed the current density, fuel utilization factor, and turbine inlet pressures of the SRC to be the key variables affecting the system performance. Additional findings include:

As the current density increases, the exergy efficiency of the cycle decreases due to increased fuel consumption by the SOFC. As efficiency decreases, a greater amount of unconverted chemical energy is converted into heat, increasing the requirement for inlet air cooling to maintain the operating temperature of the cells.

For SRC, optimal turbine inlet pressures exist at which the net output power and exergy efficiency of the combined cycles can be maximized. The SRC is maximized for a power output of 178.5 kW, and the energy and exergy efficiencies of the entire system are 64.53%, and 61.14%, respectively. However, this results in a reduction in steam production and EGB efficiency due to the increase in the heat dissipation requirement in HEX-3.

The system efficiency increases with increasing fuel utilization factor. Within the testing range, at a U_f value of 0.85, the exergy efficiency of the hybrid system and combined cycles is maximized. On the other hand, the net output power of the cycles decreases as the fuel utilization factor increases.

Compared with the SOFC/GT system, combined cycles offer better exergy efficiency and provide an incentive to use the suggested combined cycles.

Utilizing the EGB, steam is produced at 175 °C, 781.1 kPa, and 725 kg/h. It is supplied to machinery and provides heating for seafarer accommodation.

Author Contributions: Formal analysis, P.A.D.; Investigation, B.R., C.K. and J.L.; Methodology, P.A.D.; Supervision, H.K.; Writing—original draft, P.A.D.; Writing—review and editing, H.K. All authors have read and agreed to the published version of the manuscript.

Funding: This work was supported by the project "Development of guidance for prevention of leaks and mitigation of consequences in hydrogen ships" (Grant No. 20200520), and the project "Test evaluation for LNG bunkering equipment and development of test technology (Grant No. 20180048)" funded by the Ministry of Oceans and Fisheries (Korea). This research was supported by BB21plus, funded by Busan Metropolitan City and Busan Institute for Talent and Lifelong Education (BIT).

Institutional Review Board Statement: Not applicable.

Informed Consent Statement: Not applicable.

Conflicts of Interest: The authors declare no conflict of interest.

References

1. Minutillo, M.; Perna, A.; Di Trolio, P.; Di Micco, S.; Jannelli, E. Techno-economics of novel refueling stations based on ammonia-to-hydrogen route and SOFC technology. *Int. J. Hydrogen Energy* **2021**, *46*, 10059–10071. [CrossRef]
2. IMO. MEPC Fourth IMO Greenhouse Gas Study. *Int. Marit. Organ.* **2021**, *11*, 951–952.
3. Hu, Q.; Zhou, W.; Diao, F. Interpretation of Initial IMO Strategy on Reduction of GHG Emissions from Ships. *Sh. Build. China* **2019**, *60*, 195–201.
4. Zhu, M.; Yuen, K.F.; Ge, J.W.; Li, K.X. Impact of maritime emissions trading system on fleet deployment and mitigation of CO2 emission. *Transp. Res. Part D Transp. Environ.* **2018**, *62*, 474–488. [CrossRef]
5. Rathore, S.S.; Biswas, S.; Fini, D.; Kulkarni, A.P.; Giddey, S. Direct ammonia solid-oxide fuel cells: A review of progress and prospects. *Int. J. Hydrogen Energy* **2021**, *46*, 35365–35384. [CrossRef]
6. Wan, Z.; Tao, Y.; Shao, J.; Zhang, Y.; You, H. Ammonia as an effective hydrogen carrier and a clean fuel for solid oxide fuel cells. *Energy Convers. Manag.* **2021**, *228*, 113729. [CrossRef]
7. Baniasadi, E.; Dincer, I. Energy and exergy analyses of a combined ammonia-fed solid oxide fuel cell system for vehicular applications. *Int. J. Hydrogen Energy* **2011**, *36*, 11128–11136. [CrossRef]
8. van Biert, L.; Godjevac, M.; Visser, K.; Aravind, P.V. A review of fuel cell systems for maritime applications. *J. Power Sources* **2016**, *327*, 345–364. [CrossRef]
9. Jeerh, G.; Zhang, M.; Tao, S. Recent progress in ammonia fuel cells and their potential applications. *J. Mater. Chem. A* **2020**, *9*, 727–752. [CrossRef]
10. Chiong, M.C.; Chong, C.T.; Ng, J.H.; Mashruk, S.; Chong, W.W.F.; Samiran, N.A.; Mong, G.R.; Valera-Medina, A. Advancements of combustion technologies in the ammonia-fuelled engines. *Energy Convers. Manag.* **2021**, *244*, 114460. [CrossRef]
11. Miyaoka, H.; Miyaoka, H.; Ichikawa, T.; Ichikawa, T.; Kojima, Y. Highly purified hydrogen production from ammonia for PEM fuel cell. *Int. J. Hydrogen Energy* **2018**, *43*, 14486–14492. [CrossRef]
12. Laval, A.; Haldor, T.; Siemens, G. Ammonfuel-an industrial view of ammonia as a marine fuel. *Hafnia* **2020**, *7*, 32–59.
13. Cinti, G.; Barelli, L.; Bidini, G. The use of ammonia as a fuel for transport: Integration with solid oxide fuel cells. *AIP Conf. Proc.* **2019**, *2191*, 020048. [CrossRef]
14. Parikhani, T.; Azariyan, H.; Behrad, R.; Ghaebi, H.; Jannatkhah, J. Thermodynamic and thermoeconomic analysis of a novel ammonia-water mixture combined cooling, heating, and power (CCHP) cycle. *Renew. Energy* **2020**, *145*, 1158–1175. [CrossRef]
15. Wojcik, A.; Middleton, H.; Damopoulos, I.; Van Herle, J. Ammonia as a fuel in solid oxide fuel cells. *J. Power Sources* **2003**, *118*, 342–348. [CrossRef]
16. Ma, Q.; Peng, R.; Lin, Y.; Gao, J.; Meng, G. A high-performance ammonia-fueled solid oxide fuel cell. *J. Power Sources* **2006**, *161*, 95–98. [CrossRef]
17. Al-Hamed, K.H.M.; Dincer, I. A novel ammonia solid oxide fuel cell-based powering system with on-board hydrogen production for clean locomotives. *Energy* **2021**, *220*, 119771. [CrossRef]
18. Ishak, F.; Dincer, I.; Zamfirescu, C. Thermodynamic analysis of ammonia-fed solid oxide fuel cells. *J. Power Sources* **2012**, *202*, 157–165. [CrossRef]
19. Ishak, F.; Dincer, I.; Zamfirescu, C. Energy and exergy analyses of direct ammonia solid oxide fuel cell integrated with gas turbine power cycle. *J. Power Sources* **2012**, *212*, 73–85. [CrossRef]
20. Siddiqui, O.; Dincer, I. Analysis and performance assessment of a new solar-based multigeneration system integrated with ammonia fuel cell and solid oxide fuel cell-gas turbine combined cycle. *J. Power Sources* **2017**, *370*, 138–154. [CrossRef]
21. Lee, H.; Jung, I.; Roh, G.; Na, Y.; Kang, H. Comparative analysis of on-board methane and methanol reforming systems combined with HT-PEM fuel cell and CO2 capture/liquefaction system for hydrogen fueled ship application. *Energies* **2020**, *13*, 224. [CrossRef]
22. Nafees, A.; Rasid, R.A. Study of natural gas powered solid oxide fuel cell simulation and modeling. *IOP Conf. Ser. Mater. Sci. Eng.* **2019**, *702*, 012017. [CrossRef]

23. Yoon-Ho, L. Thermo-economic analysis of a novel regasification system with liquefied-natural-gas cold-energy. *Int. J. Refrig.* **2019**, *101*, 218–229. [CrossRef]
24. Song, J.; Gu, C. wei Performance analysis of a dual-loop organic Rankine cycle (ORC) system with wet steam expansion for engine waste heat recovery. *Appl. Energy* **2015**, *156*, 280–289. [CrossRef]
25. Razmi, A.R.; Janbaz, M. Exergoeconomic assessment with reliability consideration of a green cogeneration system based on compressed air energy storage (CAES). *Energy Convers. Manag.* **2020**, *204*, 112320. [CrossRef]
26. Ezzat, M.F.; Dincer, I. Energy and exergy analyses of a novel ammonia combined power plant operating with gas turbine and solid oxide fuel cell systems. *Energy* **2020**, *194*, 116750. [CrossRef]
27. Lan, R.; Tao, S. Ammonia as a suitable fuel for fuel cells. *Front. Energy Res.* **2014**, *2*, 3–6. [CrossRef]
28. Larminie, J.; Dicks, A.; McDonald, M.S. *Fuel Cell Systems Explained*, 2nd ed.; John Wiley & Sons Ltd.: Hoboken, NJ, USA, 2003; ISBN 047084857X.
29. Ma, Q.; Peng, R.R.; Tian, L.; Meng, G. Direct utilization of ammonia in intermediate-temperature solid oxide fuel cells. *Electrochem. Commun.* **2006**, *8*, 1791–1795. [CrossRef]
30. Cui, Z.; Wang, J.; Lior, N. Thermodynamic Analysis of a Solid Oxide Fuel Cell Based Combined Cooling, Heating, and Power System Integrated with Biomass Gasification. *Entropy* **2021**, *23*, 1029. [CrossRef] [PubMed]
31. Siddiqui, O.; Dincer, I. A review and comparative assessment of direct ammonia fuel cells. *Therm. Sci. Eng. Prog.* **2018**, *5*, 568–578. [CrossRef]
32. Perna, A.; Minutillo, M.; Jannelli, E.; Cigolotti, V.; Nam, S.W.; Han, J. Design and performance assessment of a combined heat, hydrogen and power (CHHP) system based on ammonia-fueled SOFC. *Appl. Energy* **2018**, *231*, 1216–1229. [CrossRef]
33. Eveloy, V.; Karunkeyoon, W.; Rodgers, P.; Al Alili, A. Energy, exergy and economic analysis of an integrated solid oxide fuel cell—gas turbine—organic Rankine power generation system. *Int. J. Hydrogen Energy* **2016**, *41*, 13843–13858. [CrossRef]
34. Song, M.; Zhuang, Y.; Zhang, L.; Li, W.; Du, J.; Shen, S. Thermodynamic performance assessment of SOFC-RC-KC system for multiple waste heat recovery. *Energy Convers. Manag.* **2021**, *245*, 114579. [CrossRef]
35. Liu, Y.; Han, J.; You, H. Performance analysis of a CCHP system based on SOFC/GT/CO2 cycle and ORC with LNG cold energy utilization. *Int. J. Hydrogen Energy* **2019**, *44*, 29700–29710. [CrossRef]
36. Chitgar, N.; Moghimi, M. Design and evaluation of a novel multi-generation system based on SOFC-GT for electricity, fresh water and hydrogen production. *Energy* **2020**, *197*, 117162. [CrossRef]
37. Ghorbani, S.; Khoshgoftar-Manesh, M.H.; Nourpour, M.; Blanco-Marigorta, A.M. Exergoeconomic and exergoenvironmental analyses of an integrated SOFC-GT-ORC hybrid system. *Energy* **2020**, *206*, 118151. [CrossRef]
38. Zhao, H.; Hou, X.; Yang, Q. Thermodynamic Study and Exergetic Analysis of the Integrated SOFC-GT-Kalina Power Cycle. *Energy Power Eng.* **2018**, *10*, 43–64. [CrossRef]
39. Fuerte, A.; Valenzuela, R.X.; Escudero, M.J.; Daza, L. Ammonia as efficient fuel for SOFC. *J. Power Sources* **2009**, *192*, 170–174. [CrossRef]
40. Amiri, T.; Singh, K.; Sandhu, N.K.; Hanifi, A.R.; Etsell, T.H.; Luo, J.-L.; Thangadurai, V.; Sarkar, P. High Performance Tubular Solid Oxide Fuel Cell Based on Ba 0.5 Sr 0.5 Ce 0.6 Zr 0.2 Gd 0.1 Y 0.1 O 3-δ Proton Conducting Electrolyte. *J. Electrochem. Soc.* **2018**, *165*, F764–F769. [CrossRef]
41. Milewski, J.; Szczęśniak, A.; Szabłowski, Ł. A proton conducting solid oxide fuel cell—implementation of the reduced order model in available software and verification based on experimental data. *J. Power Sources* **2021**, *502*, 229948. [CrossRef]
42. Mehrpooya, M.; Dehghani, H.; Ali Moosavian, S.M. Optimal design of solid oxide fuel cell, ammonia-water single effect absorption cycle and Rankine steam cycle hybrid system. *J. Power Sources* **2016**, *306*, 107–123. [CrossRef]
43. Peters, R.; Deja, R.; Engelbracht, M.; Frank, M.; Nguyen, V.N.; Blum, L.; Stolten, D. Efficiency analysis of a hydrogen-fueled solid oxide fuel cell system with anode off-gas recirculation. *J. Power Sources* **2016**, *328*, 105–113. [CrossRef]
44. Ebrahimi, M.; Moradpoor, I. Combined solid oxide fuel cell, micro-gas turbine and organic Rankine cycle for power generation (SOFC-MGT-ORC). *Energy Convers. Manag.* **2016**, *116*, 120–133. [CrossRef]
45. Al-Hamed, K.H.M.; Dincer, I. A new direct ammonia solid oxide fuel cell and gas turbine based integrated system for electric rail transportation. *eTransportation* **2019**, *2*, 100027. [CrossRef]
46. Barelli, L.; Bidini, G.; Cinti, G. Operation of a solid oxide fuel cell based power system with ammonia as a fuel: Experimental test and system design. *Energies* **2020**, *13*, 6173. [CrossRef]
47. Aminyavari, M.; Mamaghani, A.H.; Shirazi, A.; Najafi, B.; Rinaldi, F. Exergetic, economic, and environmental evaluations and multi-objective optimization of an internal-reforming SOFC-gas turbine cycle coupled with a Rankine cycle. *Appl. Therm. Eng.* **2016**, *108*, 833–846. [CrossRef]
48. Chen, Y.; Wang, M.; Liso, V.; Samsatli, S.; Samsatli, N.J.; Jing, R.; Chen, J.; Li, N.; Zhao, Y. Parametric analysis and optimization for exergoeconomic performance of a combined system based on solid oxide fuel cell-gas turbine and supercritical carbon dioxide Brayton cycle. *Energy Convers. Manag.* **2019**, *186*, 66–81. [CrossRef]
49. Venkitesh, K.; Daniel, J.; Sreekanth, M. Thermodynamic Assessment of a Gas Turbine System Combined with A Solid Oxide Fuel Cell Powered by Synthetic Gas from Municipal Solid Waste. *IOP Conf. Ser. Mater. Sci. Eng.* **2021**, *1128*, 012036. [CrossRef]
50. Rangel-Hernández, V.H.; Niño-Avendaño, A.M.; Ramírez-Minguela, J.J.; Belman-Flores, J.M.; Elizalde-Blancas, F. Advanced Exergy Analysis of an Integrated SOFC-Adsorption Refrigeration Power System. *Appl. Energy* **2018**. [CrossRef]

51. Gholamian, E.; Zare, V. A comparative thermodynamic investigation with environmental analysis of SOFC waste heat to power conversion employing Kalina and Organic Rankine Cycles. *Energy Convers. Manag.* **2016**, *117*, 150–161. [CrossRef]
52. Saadabadi, S.A.; Patel, H.; Woudstra, T.; Aravind, P.V. Thermodynamic Analysis of Solid Oxide Fuel Cell Integrated System Fuelled by Ammonia from Struvite Precipitation Process. *Fuel Cells* **2020**, *20*, 143–157. [CrossRef]
53. Ranjbar, F.; Chitsaz, A.; Mahmoudi, S.M.S.; Khalilarya, S.; Rosen, M.A. Energy and exergy assessments of a novel trigeneration system based on a solid oxide fuel cell. *Energy Convers. Manag.* **2014**, *87*, 318–327. [CrossRef]
54. Ni, M.; Leung, D.Y.C.; Leung, M.K.H. Thermodynamic analysis of ammonia fed solid oxide fuel cells: Comparison between proton-conducting electrolyte and oxygen ion-conducting electrolyte. *J. Power Sources* **2008**, *183*, 682–686. [CrossRef]

energies

MDPI

Article

Valve Regulated Lead Acid Battery Evaluation under Peak Shaving and Frequency Regulation Duty Cycles

Nimat Shamim, Vilayanur V. Viswanathan *, Edwin C. Thomsen, Guosheng Li, David M. Reed and Vincent L. Sprenkle

Battery Materials & System Group, Pacific Northwest National Laboratory, Richland, WA 99352, USA; nimat.shamim@pnnl.gov (N.S.); edwin.thomsen@pnnl.gov (E.C.T.); guosheng.li@pnnl.gov (G.L.); david.reed@pnnl.gov (D.M.R.); vincent.sprenkle@pnnl.gov (V.L.S.)
* Correspondence: vilayanur.viswanathan@pnnl.gov; Tel.: +1-509-372-4745

Abstract: This work highlights the performance metrics and the fundamental degradation mechanisms of lead-acid battery technology and maps these mechanisms to generic duty cycles for peak shaving and frequency regulation grid services. Four valve regulated lead acid batteries have been tested for two peak shaving cycles at different discharge rates and two frequency regulation duty cycles at different SOC ranges. Reference performance and pulse resistance tests are done periodically to evaluate battery degradation over time. The results of the studies are expected to provide a valuable understanding of lead acid battery technology suitability for grid energy storage applications.

Keywords: valve regulated lead acid battery; grid energy storage; battery degradation; peak shaving; frequency regulation

Citation: Shamim, N.; Viswanathan, V.V.; Thomsen, E.C.; Li, G.; Reed, D.M.; Sprenkle, V.L. Valve Regulated Lead Acid Battery Evaluation under Peak Shaving and Frequency Regulation Duty Cycles. *Energies* **2022**, *15*, 3389. https://doi.org/10.3390/en15093389

Academic Editors: Alon Kuperman and Alessandro Lampasi

Received: 8 April 2022
Accepted: 3 May 2022
Published: 6 May 2022

1. Introduction

Enabling a more sustainable energy supply system requires the integration of renewable energy resources as well as energy storage systems (ESS) with the power grid. For a robust and resilient grid, ESS can provide various grid services such as load leveling, frequency regulation, energy management, backup power, voltage support, and grid stabilization [1,2]. To validate and identify the challenges related to the extensive use of ESS in the power grid, ESS is tested for reliability and safety as part of the "Grid Energy Storage Strategy" released by the U.S. Department of Energy (DOE) [3]. The lessons learned from these tests are expected to increase the deployment of ESS.

Many different energy storage technologies are being tested, evaluated, and deployed globally. Grid-scale lead-acid batteries were deployed as far back as the 1980s [4] and are a cost-effective and reliable option for different grid services [5–8]. Materials sustainability is one of the most important aspects of lead acid batteries since nearly a 100% recycling rate is achievable [2]. However, for the larger battery size required for grid scale battery energy storage systems (BESS), there are several challenges that need to be addressed.

The charge-discharge reactions for lead acid battery at the positive and negative electrodes are given in Equations (1) and (2) respectively, while Equation (3) shows the overall cell reaction [9,10].

$$PbSO_4(s) + 2H_2O \underset{\text{discharging}}{\overset{\text{charging}}{\rightleftarrows}} PbO_2(s) + H_2SO_4 + 2H^+ + 2e^- \tag{1}$$

$$PbSO_4(s) + 2H^+ + 2e^- \underset{\text{discharging}}{\overset{\text{charging}}{\rightleftarrows}} Pb(s) + H_2SO_4 \tag{2}$$

$$2PbSO_4(s) + 2H_2O \underset{\text{discharging}}{\overset{\text{charging}}{\rightleftarrows}} PbO_2(s) + Pb(s) + 2H_2SO_4 \tag{3}$$

The discharged species for both electrodes is lead sulfate ($PbSO_4$), while the charged species is lead dioxide (PbO_2) for the positive electrode and spongy lead (Pb) for the negative electrode. The formation of lead sulfate at both electrodes during discharge results in electrolyte dilution due to sulfuric acid consumption, with protons migrating from the negative to the positive electrode. Lead dioxide in the positive electrode is reduced while spongy lead in the negative electrode is oxidized to lead sulfate. The reverse process takes place during charge, with electrolyte specific gravity increasing due to sulfuric acid generation.

In a valve regulated lead acid battery (VRLA), the electrolyte is absorbed in a glass mat separator, also known as absorbed glass mat (AGM) separator, or absorbed into a gel, with no excess free-flowing electrolyte, unlike its flooded counterpart.

VRLA battery capacity is typically positive electrode limited, generating oxygen at the positive on overcharge (Equation (4)) before the negative electrode is fully charged. Gas channels are formed either as connected porosity in the AGM separator or as micro-cracks in the gel [11]. This creates a passage for oxygen gas to the negative, where it is electrochemically reduced in the presence of hydrogen ions to form water (Equation (5)). In addition to oxygen generation, there is an ongoing parasitic reaction related to positive grid corrosion during charge, which is accelerated during overcharge in the presence of oxygen. The depolarization of the negative electrode by the oxygen recombination reaction increases the positive electrode potential during constant voltage charge mode, further accelerating grid corrosion. This parasitic grid corrosion reaction has to be balanced at the negative electrode. Since the negative electrode is in excess, the balancing reaction is expected to be charging of the negative electrode active material from lead sulfate to spongy lead. If the oxygen generation rate exceeds the recombination rate at the negative, this excess oxygen generation also has to be balanced by charging of the negative electrode active material from lead sulfate to spongy lead. Ideally, there should be sufficient excess negative active material to account for positive grid corrosion and incomplete oxygen recombination, such that at the end of design life, hydrogen generation at the negative is avoided, thus preventing water loss. The reactions for oxygen generation at the positive during overcharge, followed by recombination with lead at the negative, and subsequent conversion to lead sulfate are given below in Equation (4) at the positive and Equation (5) at the negative, which is the reverse of Equation (4), thus keeping the positive and negative electrode state-of-charge (SOC) fixed [12]:

$$H_2O \rightarrow \frac{1}{2}O_2(g) + 2H^+ + 2e^- \tag{4}$$

$$\frac{1}{2}O_2(g) + 2H^+ + 2e^- \rightarrow H_2O \tag{5}$$

Common failure modes of VRLA batteries are discussed in Section 2, along with cell design to optimize performance and life. Degradation modes are proposed to explain the internal resistance increase and capacity degradation trends under peak shaving and frequency regulation duty cycles for VRLA battery modules in Section 4. Suggestions for post-test disassembly and analysis of individual components to validate the proposed failure mechanism for each duty cycle are provided in Section 4.4. Mapping has been done between the proposed degradation modes for each duty cycle with corresponding degradation modes for hybrid electric vehicles in Section 4.7.

2. Failure Modes for Lead Acid Batteries

The lead acid battery failure modes comprise the following, with electrolyte stratification and dry-out limited to VRLA [13–16]

- Positive grid corrosion
- Positive active material softening and shedding
- Electrolyte stratification (VRLA)
- Electrolyte dry-out (VRLA)

- Negative active material sulfation

2.1. Flooded Batteries

In flooded batteries positive grid corrosion is mitigated by using lead-antimony alloy grids [17,18], related to higher creep strength of this alloy [19]. Antimony also facilitates conversion of the highly resistive PbO corrosion product to the conductive PbO_2, which has a conductivity 17 orders of magnitude higher [20]. However, antimony dissolves in the electrolyte, migrates to the negative electrode, and promotes hydrogen evolution during standby, charge and overcharge. Hence for maintenance-free batteries, low antimony positive grid alloys are used, while for VRLA, antimony free alloys of Pb-Ca-Sn are used [19,21], with calcium providing mechanical strength and tin mitigating grid corrosion, while Pb-Ca alloy is used for the negative gird. Positive active material softening and shedding, and negative active material sulfation, common to flooded and VRLA batteries, are covered in detail in the next section. Electrolyte stratification is mitigated by overcharging to promote gas evolution with associated electrolyte mixing. Electrolyte dry-out is prevented by periodic addition of deionized water as part of standard maintenance. For flooded batteries, grid corrosion mitigation by use of lead-antimony alloy is expected to reduce positive active material shedding due to better grid-active material bond, while mitigation of electrolyte stratification and starting with a lower negative to positive active material ratio relative to VRLA are expected to reduce negative electrode irreversible sulfation.

2.2. VRLA Batteries

For VRLA batteries, there are areas that need further study. Oxygen recombination depolarizes the negative electrode, thus raising its potential. For batteries charged at a constant potential, this raises the positive electrode potential, leading to greater positive grid corrosion, with associated water loss [19,21] and additional oxygen evolution, which corrodes the grid further [12]. The higher sulfuric acid concentration due to water loss accelerates grid corrosion [19]. This leads to a synergistic effect of water loss on grid corrosion. Hence it is important to avoid water loss, not just to ensure the cell capacity is not electrolyte-limited, but to limit positive grid corrosion. Control of overcharge current such that it does not exceed the recombination rate ensures complete recombination of oxygen evolved from the positive electrode. For fresh VRLA batteries, there may be some excess electrolyte, which slows down oxygen transport to the negative, resulting in water loss till the oxygen recombination current can be supported. During this period, the oxygen generation current at the positive is supported by charging of the negative electrode active material. Cell design should take this into account and add sufficient excess negative capacity to ensure the negative electrode does not get fully charged during this period to avoid hydrogen evolution. The unintended effect of excess negative is that some uncharged lead sulfate crystals may eventually grow in size and become difficult to charge, increasing electronic resistance and preventing electronic and ionic access to parts of the electrode. Hence it is important to ensure the positive and negative electrode active mass and electrolyte content are optimized to avoid (1) hydrogen generation at the negative, (2) irreversible sulfation of negative electrode (3) electrolyte dry-out.

Upon discharge, there is a near doubling of volume in the positive electrode and about a 2.5× increase in the negative electrode volume [19] since the lead sulfate discharge product has a lower density. The negative spongy lead is compressible; hence this volume growth is partially compensated by a reduction in the volume of the remaining lead during discharge. For VRLA batteries using glass mat separators, while volume change in the in-plane direction due to positive grid corrosion has been mitigated by the use of antimony-free Pb-Ca-Sn alloy, the through plane increase in positive active material volume upon discharge is mitigated by applying a compressive force to the electrode assembly [19]. The glass mat is made of fibers with various diameters to optimize mechanical strength, pore size distribution, porosity, and wicking characteristics of the separator. Smaller fiber diameter lends itself to smaller pore size and greater wicking ability, while larger fiber diameter improves mechanical strength and resistance to compression [12,22]. With proper glass mat separator design, electrolyte stratification is avoided or mitigated by wicking, and positive active material growth is mitigated by compression, with the glass mat pressed against the positive active material at pressures up to 138 kPA [22]. In a gel VRLA battery, where sulfuric acid is mixed with fumed silica, forming an immobile electrolyte, stratification is less of an issue [22]. Note that for flooded batteries, as discussed earlier, electrolyte stratification is mitigated during overcharge, where the gases evolved stir the electrolyte.

VRLA batteries have a higher specific gravity at full charge compared to flooded batteries to compensate for the lower electrolyte content. The positive electrode has two oxide types in the fully charged state–α-PbO_2, which has lower activity, and β-PbO_2 which has higher activity. As long as the acid concentration is maintained in the 0.9–5 M (1.05–1.28 specific gravity) range, the active material is in the more active state, with the less active α-PbO_2 dominating outside this range [18]. Electrolyte stratification could result in greater α-PbO_2 formation at the bottom of the electrode, while water loss can result in α-PbO_2 dominating throughout the charged positive electrode. As discussed earlier, higher electrolyte specific gravity also accelerates positive grid corrosion, which manifests itself as a steady decrease in usable capacity at a fixed rate.

Electrolyte stratification also results in greater lead sulfate formation at the bottom. This uneven lead sulfate distribution across the electrode height results in lower electrode active mass utilization and associated capacity loss [12]. This is mitigated for tall cells and modules by placing the tall side horizontally [23].

Using the mitigation approaches described, the above failure modes have been reduced, resulting in the negative electrode failure being the main R&D topic for further work. The low surface area of the negative electrode and its low specific capacitance results in poor charge acceptance especially at high rates. The voltage range above which gassing occurs and below which charge is incomplete is quite narrow for VRLA batteries [23,24], which further contributes to poor negative electrode charge acceptance at high rates. The positive lead dioxide active material has an order of magnitude higher specific surface and three times higher specific capacitance relative to the negative electrode spongy lead [23,25]. This ensures uniform distribution of the charge transfer reaction across the bulk of the positive electrode, and formation of lead sulfate film at the negative electrode surface due to poor charge acceptance. This is especially the case for partial state of charge cycling of lead acid batteries [16,26], where the batteries are not fully charged at the end of each cycle (e.g., hybrid electric vehicle, frequency regulation). To overcome this, expanders are added to the negative electrode active mix during paste formulation. These consist of barium sulfate, which increases nucleation rate for the lead sulfate formation reaction [12,16,26], lignosulfonates with functionalities that promote the formation of lead sulfate uniformly across the electrode thickness by inhibiting lead sulfate crystal growth and formation of a passivating layer [12,24], and various forms of carbon such as graphite and carbon black that increase electronic conductivity, surface area and specific capacitance [24,27]. The higher electronic conductivity reduces isolation of negative active mass via electronic percolation, while the higher surface area allows uniform distribution of lead sulfate across the bulk of the electrode, with the higher capacitance improving charge acceptance at high rates [25]. The carbon is incorporated either in the paste mix or laminated onto the electrode to form a parallel lead-carbon hybrid, or simply replaces the negative spongy lead electrode or the grid [15,19]. For the batteries used in this work, high surface area carbon is introduced in the paste formulation step of the negative electrode [28]. Table 1 shows the summary of failure modes.

Table 1. Failure Modes of Absorbed Glass Mat Valve Regulated Lead Acid Batteries.

Failure Mode	Cause	Mitigation
Positive grid corrosion	Negative electrode depolarization during recombination for CV charge Water loss resulting in higher electrolyte concentration Electrolyte stratification, resulting in greater corrosion at electrode bottom	Mitigate water loss Reduce stratification by appropriate choice of separator fiber diameter Use of Pb-Ca-Sn alloy to increase creep strength and reduce grid corrosion
Water loss	Charge current exceeds recombination rate Charge voltage not temperature compensated Positive grid corrosion	Proper charge control Mitigate grid corrosion Adjust initial electrolyte amount, negative excess and optimize separator porosity to minimize hydrogen evolution through the design life
Positive active material shedding	Grid growth due to corrosion Active material expansion by 2× during charge	Mitigate grid corrosion Optimize separator fiber diameter distribution to balance mechanical strength (large diameter) versus wicking characteristics (small diameter)
Electrolyte stratification	Poor separator wicking	Optimize separator diameter distribution
α-PbO_2 formation	Electrolyte stratification Water loss	Mitigate stratification and water loss
Negative electrode sulfation	Low surface area leads to poor charge acceptance Partial state of charge cycling	Expanders that increase nucleation rate of lead sulfate formation and inhibit crystal growth High surface area carbon to increase surface area and electronic conductivity of partially charged electrode

3. Experimental Setup

3.1. Module Specification

A valve regulated lead acid battery (VRLA) module with six series-connected cells manufactured by C&D technology, Inc (Horsham, PA, USA), using absorbent glass mat (AGM) technology (model no. SHC 12-200FT), is used for testing. These batteries, also referred to as carbon lead batteries, use lead carbon composite in the active material paste of the negative electrode to reduce irreversible sulfation and improve charge acceptance of the negative plate. The battery has a capacity of 172 Ah when discharged at the C10 or 10-h (h) rate from a fully charged state to 1.8 V per cell at 25 °C, where C10 rate is the discharge current at which the battery requires 10 h to reach an end of discharge voltage of 1.8 V from a fully charged state at 25 °C. The battery specification sheet [28] provides battery weight (60 kg), dimensions, performance, and cycle life details.

3.2. Test Protocols

Four lead-acid battery modules are subjected to two peak shaving duty cycles and two frequency regulation duty cycles. The duty cycles are based on the "Protocol for Uniformly Measuring and Expressing the Performance of Energy Storage Systems" developed by energy storage industry stakeholders, led by Pacific Northwest National Laboratory and Sandia National Laboratory [29]. The tests are ongoing for over one year. The test data correspond to 28 months for one duty cycle and 18 months for three other duty cycles. Testing was performed using an N.H. Research Inc. (NHR, Irvine, CA, USA) battery pack test system (9200 series, model 4912), with a maximum DC voltage of 120 V and current of 200 amperes (A), with an 8 kW limit for charge and 12 kW limit for discharge [30]. Figure 1 shows the experimental setup for the BESS, where the lead acid battery module is the DC storage block, and the NHR battery tester performs the role of the grid and site controller. At the beginning of each test, the battery module is charged at a constant current (CC) of 17.5 A (10-h rate) until module voltage reaches 14.1 V (2.35 V/cell), followed by a constant voltage (CV) charge until current drops to 1 A. After charge, the battery is subjected to a rest period of three hours. Reference performance tests are done at the beginning of the test regime to establish a baseline and repeated periodically. These tests consist of a capacity test and a pulse resistance test to determine battery degradation during operation. The reference performance capacity test is conducted every month, corresponding to 58 peak shaving duty cycles or 30 frequency regulation duty cycles. A pulse resistance test is done every two months, after 116 peak shaving cycles and 60 frequency regulation cycles to measure internal resistance.

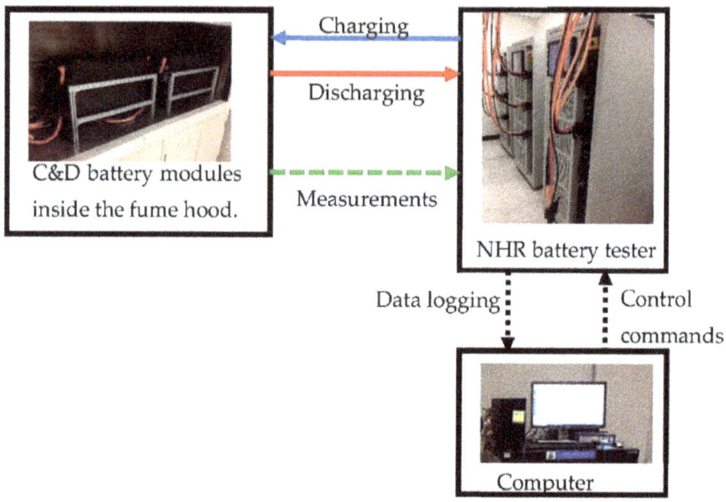

Figure 1. Experimental setup of the BESS, comprising battery module DC storage, and battery tester representing the grid and site controller.

3.2.1. Reference Performance Capacity Test

The battery is initially charged at a CC of 17.24 A (10-h rate) until the module voltage reaches 14.1 V, followed by a CV charge at 14.1 V until the charge current drops to 1 A. After the initial charge, the battery is discharged at the C5 rate of 31.34 A until the module voltage drops to 11.1 V (1.85 V/cell). Note that during this test, the end of discharge voltage is kept at 1.85 V/cell to avoid deep discharge-related degradation during the capacity test, since the main goal was to estimate duty cycle-related degradation. The battery is not subjected to rest after discharge to minimize duration at a low state of charge, where irreversible sulfation may occur. The CC/CV charge is repeated, with termination condition during CV charge corresponding to either the current decreasing to 1 A or charge capacity (Ah) reaching 103% of previous discharge Ah capacity. Note that the battery module is charged at a higher rate (5-h or C5 rate) for 1 h to minimize time spent at low SOC, followed by charge at the C10 rate. After charge, a 3 h rest period is incorporated. The discharge and charge steps are repeated for a total of two discharges, with the battery capacity calculated as the average of these two discharges. Capacity retention is calculated as the ratio of Ah capacity measured during reference performance test as the battery ages to the initial or baseline measured Ah capacity of the battery.

3.2.2. DC Current Pulse Test to Measure Internal Resistance

The lead acid battery modules' internal resistance can be measured using DC pulse current or electrical impedance spectroscopy (EIS). EIS employs multiple frequency sine waves to measure resistance over a wide range of frequencies [31]. However, EIS is not a viable option for a large battery module due to hardware limitations related to voltage and current. Per the battery cycler specifications [30], the current change time is <5 ms, while the internal resistance is reported every 10-millisecond (ms) time interval. However, it took

30–100 ms for the applied pulse current to stabilize, hence ohmic resistance is estimated at 100 ms, where there is a significant contribution expected from charge transfer.

The internal resistance of the battery module comprises ohmic resistance, charge transfer, and mass transport resistance. For AC impedance tests, ohmic resistance is estimated at 10–100 kHz [32], which corresponds to a duration of 10–100 μs, while charge transfer resistance is evaluated at 1–10 Hz (0.1–1 s), and mass transfer resistance or diffusion resistance to ion transport to and from the electrode pores is measured at 0.001–0.1 Hz (10–1000 s).

It has been shown that the pulse width at a fixed pulse current should be such that the ΔSOC is ≤0.1% [33]. This condition is met by using C1 or 1-h rate pulse of 115 A for 6 s, with a ΔSOC of 0.11%. The discharge and charge pulses are applied at every 10% SOC change from 100% SOC to 0% SOC, with only discharge pulse applied at 100% SOC and charge pulse at 0% SOC, with a 30-min rest period imposed after each pulse to allow the battery voltage to relax and the temperature to equilibrate. The battery's internal resistance is calculated from the voltage change during the pulse.

Total internal resistance, $R_{total} = \frac{\Delta V}{\Delta I}$, $\Delta V = (V_F - V_I)$ and $\Delta I = (I_P - I_I)$.

Where, V_I is the initial voltage before pulse is applied, V_F is the final voltage at the end of the pulse and I_I is the initial current, which is 0, and I_P is the pulse current. Figure 2 shows a discharge pulse current and voltage profile of the battery.

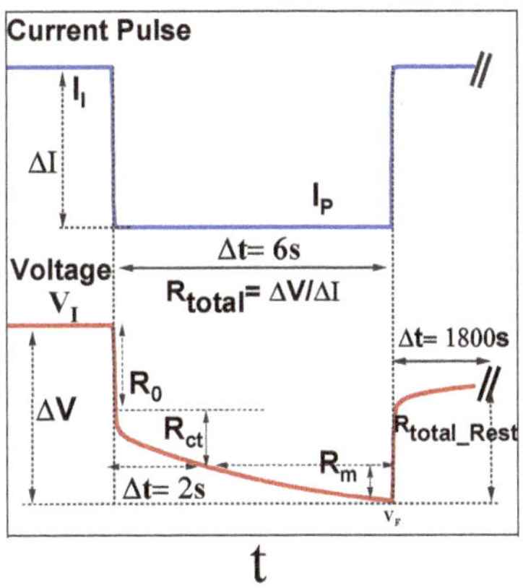

Figure 2. Measurement of internal resistance by DC pulse current.

In this work, the 100 ms data is used to calculate ohmic resistance, while the 2-s data is used to calculate the sum of ohmic (R_o) and charge transfer resistance (Rct), and the 6-s data is used to calculate total resistance (R_{total}) [34]. Mass transport resistance (R_m) is the difference in 6-s and 2-s data. Corresponding times for calculation of resistance from the voltage relaxation after charge or discharge are 0.1 s, 2 s, and 1800 s.

3.2.3. Peak Shaving Duty Cycle Test

Two battery modules are subjected to peak shaving duty cycles. A peak shaving (PS) duty cycle consists of a discharge at constant power for a duration ranging from 1–4 h during the daily on-peak period followed by a recharge during the off-peak period. Table 2 shows the peak shaving duty cycle operating parameters. The depth of discharge (DOD) of the battery module is 50% based on the initial Ah-capacity measurement. Discharge is done at two power levels (730 watts (W) and 246 W) to analyze the effect of power on performance and degradation. At the end of 58 peak shaving cycles, the battery is fully discharged, followed by a reference performance test.

For the 730 W, the discharge time is 1.2 h, with the highest current corresponding to the 2-h rate, and is referred to as PS 2-h. The corresponding numbers are 3.8 h and 8-h rate for discharge at 246 W, referred to as PS 8-h. The charge back power is kept the same for both modules at 265 W. The charge power is calculated by multiplying the voltage at the low end of SOC with the C10 charge current. The batteries are charged at a constant power of 265 W until the voltage is 14.1 V, followed by a CV charge until current decreases to 1 A. The total duration of both the peak shaving duty cycles is 12 to 14 h. The rest time for peak shaving is 2 h to accommodate close to 2 cycles per day.

Table 2. Peak Shaving Characteristics.

Parameter	Value	
Module	1	2
DOD	50%	50%
Initial SOC	100%	100%
Highest discharge current	2-h rate	8-h rate
Discharge power (W)	730	246
Discharge duration (h)	~1.2	~3.5
Rest after discharge (h)	1	1
Charge back power (W)	265	265
Charge time (h)	~8.5	~7.5
Rest after charge (h)	2	2
Total duration for each PS cycle (h)	~13	~14

3.2.4. Frequency Regulation Duty Cycle Test

Two modules have been subjected to the 24-h frequency regulation (FR) duty cycle. The 24-h duty cycle power is normalized with respect to the power corresponding to the C1 rate at 50% SOC of the battery module, estimated at 1090 W for both FR duty cycles. This corresponds to a discharge energy throughout of 3.9 kWh per cycle for this energy neutral signal. Note that positive power typically represents discharge from the battery module, and negative represents charge into the battery module from a grid perspective. In this work, since a battery cycler is used to charge and discharge the battery, charge is associated with a positive power signal, and discharge with a negative power signal. A reference to the raw data based on which the duty cycle is generated is given in Appendix B of [29]. Table 3 shows the operating parameters for the frequency regulation duty cycles, with the DOD for the 24-h duty cycle set at 20%.

Table 3. Frequency regulation characteristics.

Parameter	Value	
Module	1	2
DOD	20%	20%
Initial SOC	80%,	57%
Ending SOC	60%,	37%
Discharge duration (h)	24	24
Maximum power (W)	1090	1090
Charge back current	10-h rate	10-h rate
Total duration for each FR cycle (h)	~29	~29

Frequency regulation tests are done at two different initial SOC levels (80% and 57%) to determine the effect of different operating SOC ranges on performance and degradation. The charge acceptance of lead acid batteries is impacted by the preceding step. If the preceding step is a discharge step, the formation of small lead sulfate crystals offers sufficient surface area for the subsequent charge, whereas, if the preceding step is a charge step, the newly formed low surface area spongy lead reduces charge acceptance [35]. While optimization of expander properties is expected to mitigate this issue, for this work, starting an FR duty cycle after a preceding charge step to the initial SOC resulted in voltage excursions above the upper limit at the start of the duty cycle. Hence the batteries were brought to an SOC equal to target start SOC + 5% (62% for target SOC of 57% and 85% for target start SOC of 80%), and subsequently discharged to the target start SOC. This finding also indicates there is room for further improvement in expander research and development for duty cycles with volatile power signals.

Since the frequency regulation duty cycle is energy neutral, the total discharge Ah exceeds the total charge Ah for each cycle, since RTE < 1. That is, for the condition of charge energy equal to discharge energy, the average charge voltage is higher than average discharge voltage. Since energy in Wh is the product of Ah and battery voltage, the discharge Ah is greater than charge Ah for a 24-h FR duty cycle. At the end of each frequency regulation cycle, the battery modules are charged to bring them back to their starting SOC using a 10-h rate., followed by a constant voltage charge until the desired charge Ah is achieved. The desired charge Ah is calculated by measuring the difference between charge Ah and discharge Ah for the preceding frequency regulation cycle and applying a suitable overcharge to address gassing during charge. Note that no overcharge is applied for the FR duty cycle starting at 57% SOC, while a 0.75% overcharge is applied during charge back for the FR duty cycle starting at 80% SOC. The charge back step is followed by a rest period of 1 h before the next frequency regulation cycle. The total duration for frequency regulation, including charge back and rest, is about 29 h.

4. Results and Discussion

4.1. *Reference Performance Test Results*

Reference performance tests and pulse resistance tests are conducted at regular intervals to evaluate the performance stability of the batteries performing four different duty cycles. Performance stability assessment is done by monitoring changes in the battery capacity and internal resistance as a function of elapsed test time, number of cycles and cumulative energy throughput for each of the four duty cycles.

Figure 3a shows the voltage and current profile of one discharge and charge cycle for the reference performance capacity test. Figure 3b shows the current and temperature profiles for two charge and discharge cycles. Resistive and reversible heat contribute to heat generation during operation [36]. Lead acid batteries have a positive temperature coefficient for open circuit voltage of 0.2 mV/°C, which corresponds to endothermic contribution of reversible heat during discharge (and exothermic contribution of reversible heat during charge) [37,38]. The battery temperature rises during charge as expected, while during discharge, it rises during the first cycle and drops during the second cycle. At the start of the first discharge, the battery temperature was low at 22 °C, resulting in resistive heating that overwhelms the endothermic effect. In contrast, at the beginning of the second discharge, the temperature was higher at 24 °C, with associated lower resistive heating, such that the endothermic effect dominates. The exothermic behavior during charge has the potential for cell dry-out during constant voltage charging, hence it is prudent to set the CC current low to avoid significant temperature increase. None of the modules exhibited significant temperature increase during the course of testing, indicating that the charge profile is suitable, and the electrolyte dry-out is not a major failure mode.

Figure 4a shows battery capacity retention as a function of time elapsed since test start for different duty cycles. Capacity retention is calculated using the initial measured capacity as 100%. Over the same elapsed time, batteries subjected to FR duty cycles batteries degraded more than those subjected to PS duty cycles. The battery subjected to FR duty cycle operating in the 57–37% SOC range has the longest testing period and degraded the most. For PS duty cycle batteries, the battery capacity recovered after initial degradation. Performance enhancement can occur due to various factors such as improved electrode wetting during initial cycles, rearrangement of active material resulting in a greater active area, better electrolyte distribution among battery cell components [39], and reversal of irreversible sulfation. This recovery indicates that the initial capacity drop is not related to positive active mass shedding. Figure 4b shows capacity retention as a function of cycles, while Figure 4c shows capacity retention as a function of cumulative discharge energy. The sharp drop in capacity for FR (57–37)% in Figure 4b,c was observed after the battery was in an idle state for four months stored in a fully charged condition and is probably related to irreversible sulfation of the electrodes [40].

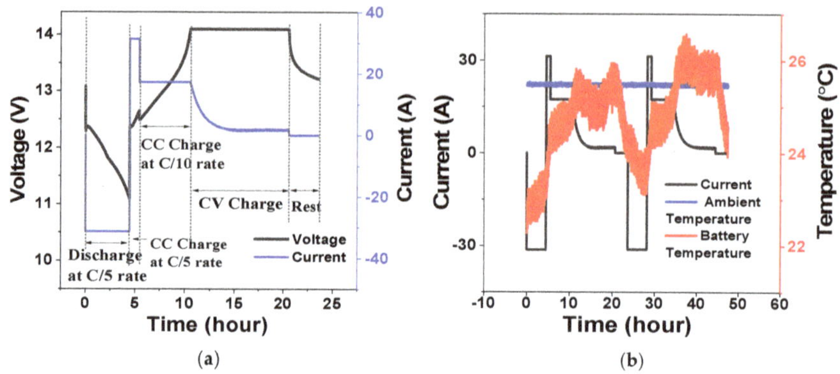

Figure 3. (**a**) Voltage and current profile of the capacity test; positive current is charge; (**b**) Temperature and current profile of the capacity test.

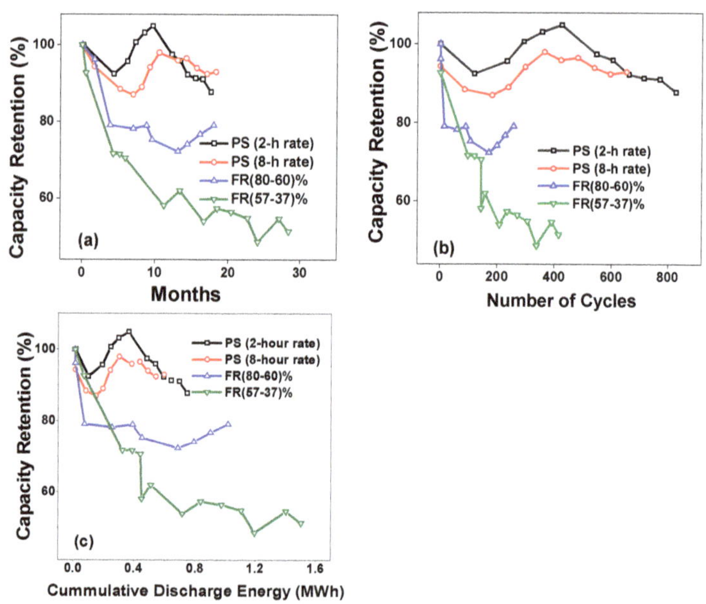

Figure 4. Capacity retention vs. (**a**) months (**b**) number of cycles and (**c**) cumulative discharge energy for four different duty cycles.

Typically, irreversible sulfation due to the growth of lead sulfate crystals occurs when batteries are stored for prolonged periods without being charged or subjected to low SOC during cycling [40,41]. The poor electronic conductivity of these lead sulfate crystals also isolate portions of the electrode, making them inaccessible, further reducing battery capacity. The battery subjected to FR (57–37)% duty cycle degraded more during the same period than FR (80–60)% duty cycle. The degradation may be related to irreversible sulfation of the battery electrodes caused by operation at lower SOC levels.

Since the frequency regulation duty cycle has higher discharge energy throughput per cycle than the peak shaving duty cycle, Figure 4b shows higher degradation per cycle for

FR, while the gap decreases when degradation is normalized per unit MWh discharged (Figure 4c), especially for the FR (80–60)% duty cycle. Note that the average SOC for this duty cycle is 70%, close to the average SOC of 75% for the PS duty cycles.

The greater rate of capacity loss for FR duty cycles, coupled with less of a capacity recovery appears to indicate that positive active material shedding, in addition to irreversible sulfation, may play a bigger role for the FR duty cycle, which involves multiple charge-discharges and the associated contractions and expansions of the active mass within a 24-h duty cycle. This needs to be taken into consideration while using VRLA batteries for frequency regulation. This failure mechanism appears to be absent for PS duty cycles. Note that converting cumulative energy throughput to number of 100% DOD equivalent cycles, FR (57–37)% has completed 728 100% DOD equivalent cycles, while the corresponding numbers for FR (80–60)%, PS (2-h) and PS (8-h) are 494, 362 and 289 respectively.

4.2. Pulse Resistance Test Results

Figure 5a shows internal resistance vs. SOC for discharge pulse durations from 100 (ms) to 6 s, while Figure 5b shows corresponding results for the charge pulse. The ohmic resistance, corresponding to the 100 ms measurement, includes contribution from charge transport resistance as well, and decreases as SOC increases. Ohmic resistance is determined by electronic conductivity of the current collector and the electrode active mass, along with electrolyte ionic conductivity and tortuosity of the separator and electrode pores. The discharge product at both electrodes, $PbSO_4$, has lower electronic conductivity than the corresponding charge states of PbO_2 at the positive and spongy Pb at the negative, while the electrolyte ionic conductivity increases with increasing state of charge [42]. While ohmic resistance should not depend on pulse direction (charge vs. discharge), for the 0.1-s discharge pulse, the ohmic resistance decreases more steeply with increasing state of charge compared to the charge pulse. The resistance at 10% SOC is nearly the same regardless of direction of the pulse, indicating the charge transfer resistance at this SOC is nearly the same for the charge and discharge pulse. As the SOC increases, the discharge pulse resistance decreases more than the charge pulse. This is expected, as the charge transfer resistance is expected to increase with state of charge for the charge pulse, negating the decrease in ohmic resistance at high SOC.

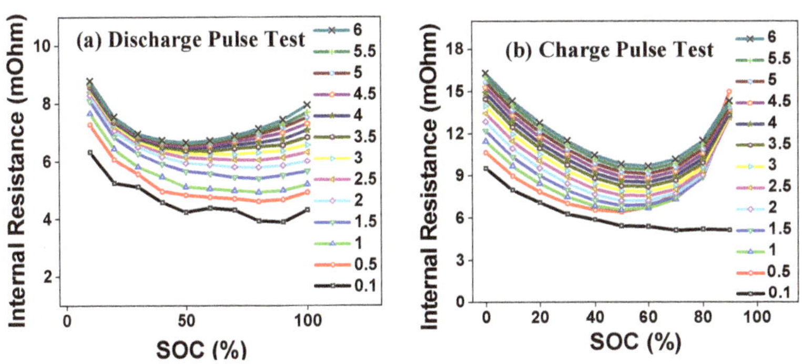

Figure 5. Internal resistance vs. SOC (%) profile (**a**) for discharge pulse (**b**) for charge pulse.

For the discharge pulse, at high SOC, the internal resistance increases with pulse width. While it is difficult to separate out charge transport and mass transport effects, assuming a 2-s cut-off for charge transport, it appears that mass transport effects can be estimated all the way up to 6-s pulse width. On the other hand, at low SOC, after 2–3 s, the resistance does not increase significantly. This indicates that due to low acid concentration at low SOC, mass transport effects are almost fully captured within 2–3 s. The data for the charge pulse is a mirror image. At low SOC, since there is plenty of water in the electrolyte, mass transport effects are observed for the entire 6-sec pulse duration. At high SOC, due to higher concentration of the sulfuric acid product, a majority of the mass transport related effects are captured within 2–3 s.

4.3. Degradation Analysis

Figure 6 shows all four batteries' capacity retention vs. total internal resistance. The total internal resistance is averaged from 80% SOC to 30% SOC for charge and discharge pulses. The capacity degrades linearly with an internal resistance increase. VRLA batteries are typically positive electrode limited. As the battery cycles, the positive electrode grids undergo corrosion, causing an in-plane mechanical stress on the grid, leading to a decrease in adhesion between the active material and the grid. The active material is separated from the grid because of the expansion and shrinkage of the active material during the discharge and charge cycles, reducing the capacity [43]. The capacity recovery for PS cycles in Figure 4c indicates that positive active material shedding is not the main reason for capacity loss. As stated earlier, use of appropriate alloys and compressive force appears to have mitigated this failure mode. This indicates that irreversible sulfation at the bottom of electrodes due to electrolyte stratification and at the low surface area negative during partial state of charge operation may be a reason for the linear decrease in capacity as resistance increases, while resistance increase may be associated with positive grid corrosion, water loss and irreversible sulfation. As irreversible sulfation at the negative increases, there is expected to be a transition from positive limit to negative limit, at which point gas generation increases, accompanied by further resistance increase due to water loss. The slight increase in capacity corresponding to an increase in resistance for FR duty cycles may be related to reversal of dense sulfation film (leading to increase in capacity), while grid corrosion and loss of water lead to an increase in resistance.

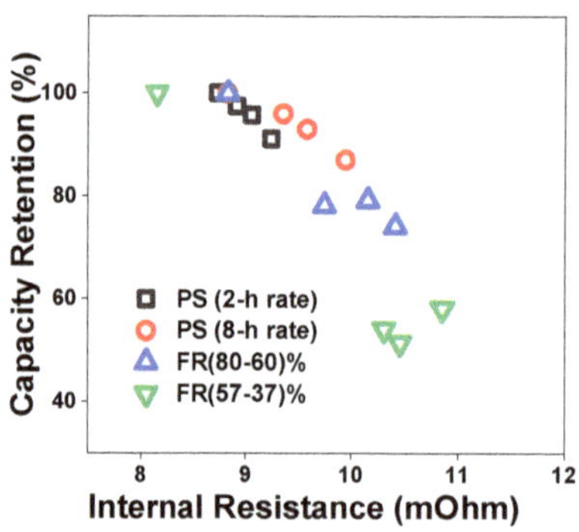

Figure 6. Capacity retention vs. internal resistance profile for four duty cycles.

Figure 7 shows internal resistance vs. elapsed test duration. Ohmic resistance increases with increasing duration. Some of the reasons could be (1) irreversible lead sulfate formation at the bottom of both electrodes related to electrolyte stratification and at the negative electrode due to partial state of charge cycling, (2) positive grid corrosion resulting in in-plane mechanical stress on grid and loss of contact with positive active material (3) Positive active material shedding due to active material volume change during cycling resulting in loss of electronic percolation in active mass (4) Electrolyte dry-out resulting in an increase in ionic resistance.

For all duty cycles, the ohmic resistance for charge pulse is greater than that for discharge pulse, which is due to contribution from charge transfer in the 100 ms pulse. Depending on the duty cycle, the reason for ohmic resistance increase may differ. For example, for peak shaving, since the charge rate is the same for both duty cycles, if electrolyte dry-out is the predominant mode, electrolyte loss is expected to be nearly the same for the 2-h and 8-h duty cycles. The slightly lower resistance for the 2-h discharge may be related to less time spent in the 75% to 50% SOC range, resulting in less irreversible sulfation at the negative electrode. The ohmic resistance for the FR (80–60)% run matches the results for the PS 8-h, whereas the corresponding values for the FR (57–37)% case is lower. Hence it appears that electrolyte dry-out could be the dominating reason for ohmic resistance increase for the PS duty cycles and the FR duty cycle operating in the 80–60% SOC range, with the PS duty cycles expected to have greater grid corrosion and water loss due to being subjected to a full charge after each cycle, while the FR (80–60)% duty cycle is expected to have more sulfation due to spending >3.5× duration at <70% SOC. Irreversible sulfation at the negative electrode could be the dominant reason for ohmic resistance increase for FR (57–37)%. Note that this battery was on open circuit for four months. During this time, some water loss is expected due to self-discharge, but the main degradation mechanism may be irreversible sulfate formation, which is corroborated by the steep increase in charge transfer resistance for the charge pulse at the 10-month mark (Figure 7b).

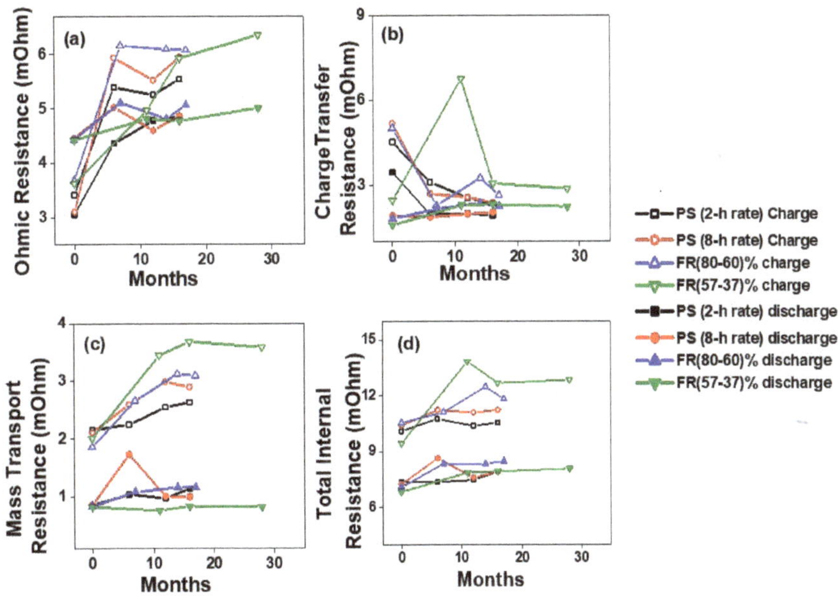

Figure 7. Resistance vs. elapsed test duration (**a**) ohmic, (**b**) charge transfer, (**c**) mass transport, (**d**) total.

The charge transfer resistance for the discharge pulse is relatively unchanged with a slight increase, while it decreases in the first few months for the charge pulse. This appears to indicate that the charge acceptance initially is not very high, probably related to sulfation formed over a long storage period of ~6 months prior to testing. The steep increase in charge transfer resistance for the charge pulse at the 10-month mark appears to be related to the 4-month rest on the open circuit, which leads to sulfation of the active material and the associated poor charge acceptance. As expected, the corresponding charge transfer resistance for the discharge pulse is not very high since the reacting species is the charged form of the active mass in each electrode.

Mass transport resistance for the charge pulse increases with time, while that for the discharge pulse increases to a much less degree (Figure 7c). During charge, sulfuric acid is produced in the pores, requiring transport of the acid away from the pores to allow reactant water access. It is difficult for bulk water to force its way into the pores while displacing the acid that is produced. During discharge, sulfuric acid is consumed at the electrode pores, which are now filled with water. It is easier for acid in bulk to enter the pores and displace the produced water. Irreversible sulfation results in the increase of active mass volume, with an associated decrease in porosity and pore size. Hence the pore size and porosity reduction associated with irreversible sulfate formation increase mass transport resistance, with more of an impact on the charge pulse.

This effect is maximum for FR (57–37)% duty cycle, where irreversible sulfation at the negative electrode is expected to be the largest, with corresponding greatest pore size and porosity reduction. The results are similar for FR (80–60)% duty cycle and PS 8-h, where both duty cycles spent significant time at <70% SOC and hence are expected to have a similar degree of irreversible sulfation, while the PS 2-h duty cycle spends less time at <70% SOC, and hence has less irreversible sulfation, accompanied by a lower increase in mass transport resistance.

It is our hypothesis that for the PS duty cycles and FR (80–60)% duty cycle, the main mode of degradation is water loss, with some irreversible sulfation at the bottom of both electrodes related to electrolyte stratification, with additional sulfation at the low surface area negative for FR (80–60%) and PS 8-h related to longer time spent at SOC < 70%, while for FR (57–37)% duty cycle, the main mode is irreversible sulfation at the negative electrode related to longer time spent at low SOC along with sulfation at the bottom of both electrodes related to electrolyte stratification, with water loss related effects expected to be lower. To validate these hypotheses, the aged battery modules, along with fresh modules to establish a baseline, need to be disassembled in the charged state and subjected to the recommended tests in the following section.

4.4. Recommended Tests to Validate the Proposed Failure Mechanism

(1) Determine mass of negative, positive and separator. The negative electrode needs to be transferred immediately in deionized water to avoid oxidation in air.
(2) Determine electrolyte content in positive, negative and separator by measuring wet and dry weight. The negative needs to be dried in a vacuum oven.
(3) Determine electrolyte specific gravity by titrating a known electrolyte volume obtained by squeezing electrolyte from separator.
(4) Determine electrolyte distribution at various heights of electrodes and separator (to estimate electrolyte stratification).
    a. Cut the electrode into three parts, top, middle, and bottom, measuring wet weight, followed by titration to determine acid content for each part, followed by measurement of dry weight.
(5) Determine acid content at the top, middle and bottom parts of the separator, followed by filtration and drying of the retained separator.
    a. Immerse each separator piece in deionized water after weighing and titrate vs. 1 M NaOH.
(6) Determine porosity and pore size distribution by BET and mercury porosimetry to correlate results with mass transport resistance.
(7) Determine elemental composition using energy dispersive X-ray analysis (EDX) for the top, middle and bottom part of positive and negative electrodes.

4.5. Internal Resistance vs. SOC Trends for the Duty Cycles

Figure 8 shows the internal resistance measured from discharge and charge pulse at different SOC levels for four different duty cycles. The internal resistance is averaged over time at the same SOC. From Figure 8a–d, it can be seen that the duty cycles affect the battery internal resistance similarly, with the shape of the resistance as f (SOC) similar for all duty cycles. For all four duty cycles, total internal resistance measured from charge pulse is higher than total internal resistance measured from discharge pulse. This is in line with our observation that charge transfer- and mass transport-related losses are higher for charge.

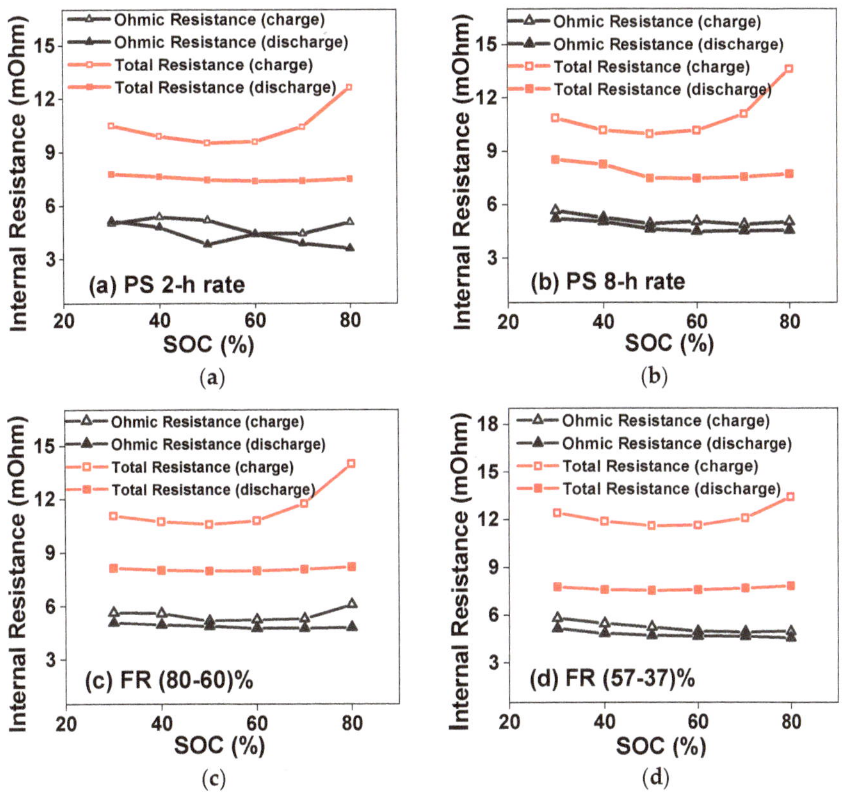

Figure 8. Internal resistance vs. SOC (%) for four duty cycles. (**a**) peak shaving 2-h rate (**b**) peak shaving 8-h rate (**c**) frequency regulation (80–60)% and (**d**) frequency regulation (57–37)%.

4.6. Round Trip Efficiency Results for the Duty Cycles

Figure 9 shows the round-trip efficiency (RTE) of (Figure 9a) peak shaving duty cycles and (Figure 9b) frequency regulation duty cycles. While the RTE is constant for 400 peak shaving duty cycles, it decreases subsequently, with significant fluctuations. This is due to a change in the charge procedure after 400 cycles. Initially, the charge back procedure consisted of a CC/CV charge with charge termination set at 3% overcharge or current decreasing to 0.8 A, whichever occurs first. Also, the constant current and constant voltage charge steps each had a 7-h time limit to reduce excessive gassing. After 400 cycles, the CC/CV charge is terminated when the current decreases to 0.8 A, with the 3% overcharge requirement removed. Note that in these subsequent cycles, the overcharge is more than 3% of the discharge capacity, thus reducing the RTE. Additionally, the charge capacity is different in each cycle as some of the charge back cycles reached the termination current of 0.8 A before the time limit, while for other cases, the time limit was reached before the current decreased to 0.8 A. Hence the decrease and fluctuation in RTE are not primarily due to battery degradation. The FR (80–60)% has slightly lower RTE than FR (57–37)%. For the energy neutral frequency regulation signal, as explained earlier, the total discharge Ah > total charge Ah for the FR duty cycle, since RTE < 1. The charge back Ah is determined by applying a 0.75% overcharge to the difference between total discharge Ah and total charge Ah for the FR (80–60)% duty cycle, while no overcharge was applied to the FR (57–37)% duty cycle since gassing is expected to be insignificant in this low SOC level. Hence FR (57–37)% has a slightly higher RTE.

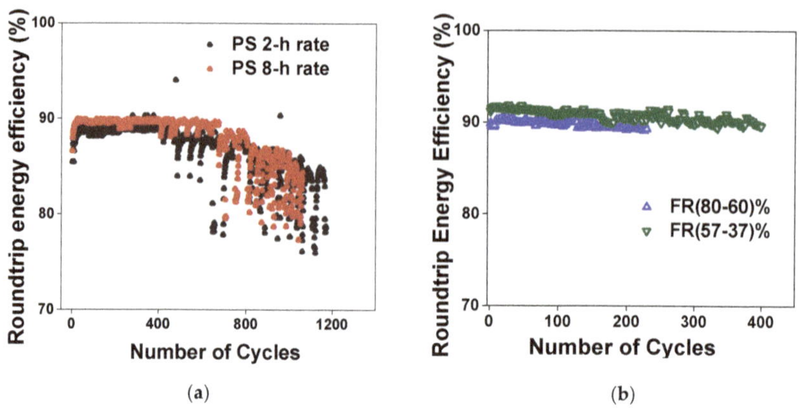

Figure 9. RTE (**a**) peak shaving duty cycle (**b**) frequency regulation duty cycle.

4.7. Mapping Hybrid Electric Vehicle (HEV) Degradation with Grid Services

It was determined that batteries subjected to a micro hybrid electric vehicle (HEV) drive cycle operating at 5% DOD in the 90–85% SOC range experienced no sulfation of the negative electrode, while batteries subjected to a mild HEV drive cycle operating in the 90–70% SOC range experienced mild sulfation, and batteries subjected to a full HEV drive cycle operating in the 80–30% SOC range experienced rapid sulfation [20]. The PS duty cycle appears to map with the mild HEV cycle, with a slightly lower average SOC, but getting fully charged after each cycle. The PS 8-h duty cycle spends more time at SOC < 75%, hence may have more sulfation at the negative. The FR (57–37)% duty cycle has a slightly lower average SOC than a full HEV drive cycle, is charged to only 57% SOC at the end of each cycle and is expected to show greater sulfation of the negatives. The FR (80–60)% duty cycle has an average SOC halfway between mild HEV and full HEV and is expected to experience sulfation greater than mild HEV and less than full HEV. The PS duty cycle and the FR (80–60)% duty cycle are expected to see greater water loss compared to the FR (57–37)% duty cycle. Table 4 below provides a mapping of grid services degradation with that experienced by various HEV cycles.

Table 4. Mapping of Grid Services Degradation with HEV Cycles.

Drive Cycle	**Micro HEV**	**Mild HEV**	**Full HEV**
SOC range	90–85%	90–70%	80–30%
Sulfation of negative	Negligible	Mild	Highest
Grid Corrosion	Medium	Lower	Lowest
Water loss	Medium	Lower	Lowest
Inactive α-PbO_2	Medium	Lower	Lowest
Grid services that map to drive cycles		PS 2-h lower sulfation, PS 8-h higher sulfation. Both expected to have similar grid corrosion, similar water loss. FR (80–60)% similar grid corrosion and water loss, greater sulfation	FR (80–60)% lower sulfation, FR (57–37)% higher sulfation, lower grid corrosion and water loss.

5. Conclusions

Leveraging upon work done on the degradation of lead acid batteries used for hybrid electric, degradation mechanisms of batteries used in micro, mild, and full HEV are mapped with the grid service duty cycles used in this work. Post-cycling tests have been proposed to validate the hypothesis for dominant failure modes for each duty cycle.

Assigning different durations within the charge or discharge current pulse, ohmic, charge transfer and mass transport resistances are estimated. Higher ohmic resistance for

PS 8-h relative to PS 2-h is explained by the longer duration spent at <70% SOC for the former, while similar ohmic resistance increase for the PS 8-h and FR (80–60)% is explained by higher water loss for the former balanced by the higher sulfation for the latter related to 3.5× longer time at <70% SOC for the FR (80–60)% grid service. The higher charge transfer resistance during the charge pulse at the start of testing may be related to extended storage prior to test, resulting in sulfation, which is subsequently reversed. This hypothesis is further validated by the spike in charge transfer resistance for the charge pulse for the FR (57–37)% duty cycle after a 4-month stand at open circuit, which is also reversed upon further testing. The higher mass transport resistance for the charge pulse also correlates with grid services with greater expected sulfation, with the value highest for FR (57–37)% and lowest for PS 2-h. Hence these measurements provide very useful insights into the degradation mechanisms for various grid services. The higher resistance during charge across all duty cycles places limitation on maximum power for energy neutral grid services such as frequency regulation.

Regardless of the duty cycle, capacity loss increased with an increase in internal resistance, with nearly the same slope for all duty cycles. For FR (57–37)%, while the capacity decreases as expected with increase in resistance, there is reversal in this trend with a slight increase in capacity as resistance increases, which appears to signal a switch from negative limit to positive limit related to reversal of the dense sulfation at the negative. Such signatures provide insights into battery management approaches for VRLA used in various grid services, taking into account the initial positive & negative electrode active material ratio and electrolyte content to maximize battery operating life.

While positive active material shedding does not appear to be the main failure mode for PS duty cycles. it may play a bigger role for the FR duty cycle, due to multiple contractions and expansions of the active mass within a 24-h duty cycle. This needs to be taken into consideration while using VRLA batteries for frequency regulation and other services such as renewable smoothing that are associated with volatile signals. The capacity loss per energy throughput is highest for grid services with the highest potential for sulfation—FR (57–37)% > FR (80–60)% > PS 8 h > PS 2 h—thus highlighting the need for the mitigation of sulfation to be a top priority. This is achieved by minimizing electrolyte stratification, which affects both electrodes, and by increasing negative electrode surface area to ensure uniform current distribution by addition of suitable expanders.

Author Contributions: Conceptualization, V.V.V., E.C.T., D.M.R. and V.L.S.; Formal analysis, N.S.; Funding acquisition, D.M.R. and V.L.S.; Methodology, N.S., V.V.V. and E.C.T.; Software, E.C.T.; Supervision, V.V.V., D.M.R. and V.L.S.; Writing—original draft, N.S. and V.V.V.; Writing—review & editing, N.S., V.V.V., E.C.T. and G.L. All authors have read and agreed to the published version of the manuscript.

Funding: This research was funded by the U.S. Department of Energy (DOE) Office of Electricity under contract number 57558. The APC was funded by the U.S. Department of Energy (DOE) Office of Electricity.

Institutional Review Board Statement: Not applicable.

Informed Consent Statement: Not applicable.

Data Availability Statement: Data is contained in the figures and tables of this article. Supporting data for the figures and tables are available from the corresponding author.

Acknowledgments: This work was funded by the U.S. Department of Energy (DOE) Office of Electricity under contract number 57558 through Pacific Northwest National Laboratory (Validated Safety & Reliability). Pacific Northwest National Laboratory is operated by Battelle for the DOE under contract number DE-AC05-76RL01830.

Conflicts of Interest: The authors declare that they have no known competing financial interest or personal relationship that could have appeared to influence the work reported in this paper.

References

1. Malhotra, A.; Battke, B.; Beuse, M.; Stephan, A.; Schmidt, T. Use cases for stationary battery technologies: A review of the literature and existing projects. *Renew. Sustain. Energy Rev.* **2016**, *56*, 705–721. [CrossRef]
2. May, G.J.; Davidson, A.; Monahov, B. Lead batteries for utility energy storage: A review. *J. Energy Storage* **2018**, *15*, 145–157. [CrossRef]
3. Choi, D.; Shamim, N.; Crawford, A.; Huang, Q.; Vartanian, C.K.; Viswanathan, V.V.; Paiss, M.D.; Alam, M.J.E.; Reed, D.M.; Sprenkle, V.L. Li-ion battery technology for grid application. *J. Power Sources* **2021**, *511*, 230419. [CrossRef]
4. Cook, G.; Spindler, W. Low-maintenance, valve-regulated, lead/acid batteries in utility applications. *J. Power Sources* **1991**, *33*, 145–161. [CrossRef]
5. Da Silva, V.C.; Gimenes, A.L.V.; Udaeta, M.E.M. Energy Storage and Multi-Source System for Reduction Energy Costs in the Consumer-Side. *J. Power Energy Eng.* **2021**, *9*, 80–105. [CrossRef]
6. Luo, X.; Wang, J.; Dooner, M.; Clarke, J. Overview of current development in electrical energy storage technologies and the application potential in power system operation. *Appl. Energy* **2015**, *137*, 511–536. [CrossRef]
7. Chen, H.; Cong, T.N.; Yang, W.; Tan, C.; Li, Y.; Ding, Y. Progress in electrical energy storage system: A critical review. *Prog. Nat. Sci.* **2009**, *19*, 291–312. [CrossRef]
8. Ibrahim, H.; Ilinca, A.; Perron, J. Energy storage systems—Characteristics and comparisons. *Renew. Sustain. Energy Rev.* **2008**, *12*, 1221–1250. [CrossRef]
9. Gandhi, K.S. Modeling of Sulfation in a Flooded Lead-Acid Battery and Prediction of its Cycle Life. *J. Electrochem. Soc.* **2020**, *167*, 13538. [CrossRef]
10. David, G.; Enos, S.N.L. *Lead-Acid Batteries and Advanced Lead-Carbon Batteries*; Sandia National Lab: Albuquerque, NM, USA, 2014.
11. Nelson, R. The basic chemistry of gas recombination in lead-acid batteries. *JOM* **2001**, *53*, 28–33. [CrossRef]
12. Culpin, B.; Rand, D. Failure modes of lead/acid batteries. *J. Power Sources* **1991**, *36*, 415–438. [CrossRef]
13. Moseley, P. Improving the valve-regulated lead–acid battery. *J. Power Sources* **2000**, *88*, 71–77. [CrossRef]
14. Cooper, A.; Moseley, P. Progress in overcoming the failure modes peculiar to VRLA batteries. *J. Power Sources* **2003**, *113*, 200–208. [CrossRef]
15. Moseley, P.T.; Rand, D.A.; Peters, K. Enhancing the performance of lead–acid batteries with carbon – In pursuit of an understanding. *J. Power Sources* **2015**, *295*, 268–274. [CrossRef]
16. Bullock, K.R. Carbon reactions and effects on valve-regulated lead-acid (VRLA) battery cycle life in high-rate, partial state-of-charge cycling. *J. Power Sources* **2010**, *195*, 4513–4519. [CrossRef]
17. Culpin, B. Separator design for valve-regulated lead/acid batteries. *J. Power Sources* **1995**, *53*, 127–135. [CrossRef]
18. Pavlov, D. Chapter 1—Invention and Development of the Lead–Acid Battery. In *Lead-Acid Batteries: Science and Technology*; Pavlov, D., Ed.; Elsevier: Amsterdam, The Netherlands, 2011; pp. 3–28.
19. Rand, D.A.J.; Moseley, P.T. 3—Lead–Acid Battery Fundamentals. In *Lead-Acid Batteries for Future Automobiles*; Garche, J., Karden, E., Moseley, P.T., Rand, D.A.J., Eds.; Elsevier: Amsterdam, The Netherlands, 2017; pp. 97–132.
20. Pavlov, D. Chapter 4—Lead Alloys and Grids. Grid Design Principles. In *Lead-Acid Batteries: Science and Technology*; Pavlov, D., Ed.; Elsevier: Amsterdam, The Netherlands, 2011; pp. 149–221.
21. Lam, L.; Ceylan, H.; Haigh, N.; Lwin, T.; Rand, D. Influence of residual elements in lead on oxygen- and hydrogen-gassing rates of lead-acid batteries. *J. Power Sources* **2010**, *195*, 4494–4512. [CrossRef]
22. Pavlov, D. Chapter 14—Valve-Regulated Lead–Acid (VRLA) Batteries. In *Lead-Acid Batteries: Science and Technology*, 2nd ed.; Pavlov, D., Ed.; Elsevier: Amsterdam, The Netherlands, 2017; pp. 593–620.
23. Wagner, R. 8—Positive Active-Materials for Lead–Acid Battery Plates. In *Lead-Acid Batteries for Future Automobiles*; Garche, J., Karden, E., Moseley, P.T., Rand, D.A.J., Eds.; Elsevier: Amsterdam, The Netherlands, 2017; pp. 235–267.
24. Pavlov, D. Chapter 7—Additives to the Pastes for Positive and Negative Battery Plates. In *Lead-Acid Batteries: Science and Technology*; Pavlov, D., Ed.; Elsevier: Amsterdam, The Netherlands, 2011; pp. 311–361.
25. Fernández, M.; Valenciano, J.; Trinidad, F.; Muñoz, N. The use of activated carbon and graphite for the development of lead-acid batteries for hybrid vehicle applications. *J. Power Sources* **2010**, *195*, 4458–4469. [CrossRef]
26. Pavlov, D.; Nikolov, P.; Rogachev, T. Influence of expander components on the processes at the negative plates of lead-acid cells on high-rate partial-state-of-charge cycling. Part I: Effect of lignosulfonates and $BaSO_4$ on the processes of charge and discharge of negative plates. *J. Power Sources* **2010**, *195*, 4435–4443. [CrossRef]
27. Boden, D.; Loosemore, D.; Spence, M.; Wojcinski, T. Optimization studies of carbon additives to negative active material for the purpose of extending the life of VRLA batteries in high-rate partial-state-of-charge operation. *J. Power Sources* **2010**, *195*, 4470–4493. [CrossRef]
28. C&D Technologies, Carbon Battery. Available online: http://www.energyinsight.co.za/wp-content/uploads/2015/06/Lead-Carbon-gen-brochure-en-1411-Final.pdf (accessed on 28 April 2022).
29. Conover, D.R.; Crawford, A.J.; Viswanathan, V.V.; Ferreira, S.; Schoenwald, D. *Protocol for Uniformly Measuring and Expressing the Performance of Energy Storage Systems*; PNNL-22010 Rev 2; Pacific Northwest National Lab: Richland, WA, USA, 2016.
30. NH Research Inc. *9200 Series Battery Module/Pack Test System*; NH Research Inc.: Irvine, CA, USA. Available online: https://nhresearch.com/power-electronics-test-systems-and-instruments/battery-module-pack-test-systems/high-power-battery-charge-discharge-test-systems-9200-series/ (accessed on 9 February 2022).

31. Research, P.A. Basics of Electrochemical Impedance Spectroscopy. In *Instruments, Complex Impedance in Corrosion;* In Application Note AC-1; Gamry Instruments Inc.: Warminster, PA, USA, 2006.
32. Alavi, S.M.M.; Birkl, C.R.; Howey, D.A. Time-domain fitting of battery electrochemical impedance models. *J. Power Sources* **2015**, *288*, 345–352. [CrossRef]
33. Schweiger, H.-G.; Obeidi, O.; Komesker, O.; Raschke, A.; Schiemann, M.; Zehner, C.; Gehnen, M.; Keller, M.; Birke, P. Comparison of Several Methods for Determining the Internal Resistance of Lithium Ion Cells. *Sensors* **2010**, *10*, 5604–5625. [CrossRef] [PubMed]
34. Barai, A.; Uddin, K.; Widanage, W.D.; McGordon, A.; Jennings, P. A study of the influence of measurement timescale on internal resistance characterisation methodologies for lithium-ion cells. *Sci. Rep.* **2018**, *8*, 21. [CrossRef] [PubMed]
35. Peters, K.; Rand, D.A.J.; Moseley, P.T. 7—Performance-Enhancing Materials for Lead–Acid Battery Negative Plates. In *Lead-Acid Batteries for Future Automobiles;* Garche, J., Karden, E., Moseley, P.T., Rand, D.A.J., Eds.; Elsevier: Amsterdam, The Netherlands, 2017; pp. 213–234.
36. Williford, R.E.; Viswanathan, V.V.; Zhang, J.-G. Effects of entropy changes in anodes and cathodes on the thermal behavior of lithium ion batteries. *J. Power Sources* **2009**, *189*, 101–107. [CrossRef]
37. Salkind, A.J.; Kelley, J.J.; Cannone, A.G. *Handbook of Batteries*, 2nd ed.; McGraw-Hill, Inc.: New York, NY, USA, 1995.
38. Electropaedia, Thermal Management. Available online: https://www.mpoweruk.com/thermal.htm#thermochemical (accessed on 1 February 2022).
39. Kohya, Y.; Takeda, T.; Takano, K.; Kohno, M.; Yotsumoto, K.; Ogata, T. A deterioration estimating system for 200-Ah sealed lead-acid batteries. In Proceedings of the Intelec 94, Vancouver, BC, Canada, 30 October–3 November 1994.
40. Hollenkamp, A.; Constanti, K.; Huey, A.; Koop, M.; Aputeanu, L. Premature capacity-loss mechanisms in lead/acid batteries. *J. Power Sources* **1992**, *40*, 125–136. [CrossRef]
41. Battery University. BU-804b: Sulfation and How to Prevent It. Last Updated 2 November 2021. Available online: https://batteryuniversity.com/article/bu-804b-sulfation-and-how-to-prevent-it (accessed on 9 February 2022).
42. Badeda, J.; Huck, M.; Sauer, D.U.; Kabzinski, J.; Wirth, J. 16—Basics of Lead–Acid Battery Modelling and Simulation. In *Lead-Acid Batteries for Future Automobiles*; Garche, J., Karden, E., Moseley, P.T., Rand, D.A.J., Eds.; Elsevier: Amsterdam, The Netherlands, 2017; pp. 463–507.
43. Kurisawa, I.; Iwata, M. Internal resistance and deterioration of VRLA battery-analysis of internal resistance obtained by direct current measurement and its application to VRLA battery monitoring technique. In Proceedings of the Power and Energy Systems in Converging Markets, Melbourne, Australia, 23–23 October 1997.

energies

Article

Preparation of a Honeycomb-like FeNi(OH/P) Nanosheet Array as a High-Performance Cathode for Hybrid Supercapacitors

Chenliang Li, Ruizhi Li * and Yingke Zhou *

The State Key Laboratory of Refractories and Metallurgy, Institute of Advanced Materials and Nanotechnology, College of Materials and Metallurgy, Wuhan University of Science and Technology, Wuhan 430081, China; lcl11905991@hotmail.com

* Correspondence: rzli@wust.edu.cn (R.L.); zhouyk@wust.edu.cn (Y.Z.)

Abstract: Polymetallic transition metal phosphides (TMPs) exhibit quasi-metallic properties and a high electrical conductivity, making them attractive for high-performance hybrid supercapacitors (HSCs). Herein, a nanohoneycomb (NHC)-like FeNi layered double hydroxide (LDH) array was grown in situ on 3D current collector nickel foam (NF), which is also the nickel source during the hydrothermal process. By adjusting the amount of NaH_2PO_2, an incomplete phosphated FeNi(OH/P) nanosheet array was obtained. The optimized FeNi(OH/P) nanosheet array exhibited a high capacity up to 3.6 C cm^{-2} (408.3 mAh g^{-1}) and an excellent long-term cycle performance (72.0% after 10,000 cycles), which was much better than FeNi LDH's precursor. In addition, the hybrid supercapacitor (HSC) assembled with FeNi(OH/P) (cathode) and polypyrrole (PPy/C, anode) achieved an ultra-high energy density of 45 W h kg^{-1} at a power density of 581 W kg^{-1} and an excellent cycle stability (118.5%, 2000 cycles), indicating its great potential as an HSC with a high electrochemical performance.

Keywords: polymetallic transition metal phosphides; hybrid supercapacitor; FeNi(OH/P) nanosheet array

Citation: Li, C.; Li, R.; Zhou, Y. Preparation of a Honeycomb-like FeNi(OH/P) Nanosheet Array as a High-Performance Cathode for Hybrid Supercapacitors. *Energies* **2022**, *15*, 3877. https://doi.org/10.3390/en15113877

Academic Editors: Alon Kuperman and Alessandro Lampasi

Received: 29 April 2022
Accepted: 22 May 2022
Published: 24 May 2022

1. Introduction

The energy crisis and global climate change have stimulated the search for sustainable energy. Numerous efforts have been made to develop energy storage devices that are efficient, economical, safe, and environmentally friendly in order to meet the future development of a low-carbon, sustainable economy. Batteries have a high energy density, but a low power density and poor cycling stability. Supercapacitors have a higher power density, faster charging time, and better long-term stability than batteries. However, supercapacitors have limited industrial applications because of their low energy density. By combining the advantages of batteries and supercapacitors, researchers proposed a hybrid supercapacitor (HSC) with a battery-type electrode and a capacitive electrode. Compared with carbon-based materials storing energy via a double electric layer, battery-type electrode materials can generate extreme energy density based on an active redox reaction [1,2]. Therefore, the performance of HSC is highly dependent on the efficiency and stability of the battery-type electrodes.

Recently, it has been demonstrated that transition-metal phosphides (TMPs) have garnered significantly more research because of their higher conductivity, electrochemical activity, and structural stability [3,4], all of which make them potentially useful in energy storage devices. TMP is a triangular prism made of metal bonds or covalent chemical bonds [5]. By increasing the number of metal atoms situated in the center of the prism's vertical plane, a nine-fold tetrakaidecahedron structure centered on phosphorus atoms is formed. The large channels and open framework of TMPs provide effective electron/ion transport. Because of the rapid reduction in ionic properties and the "P" dragging electron density in the sublattice presence of delocalization, TMPs have more free electrons, which is the fundamental reason for their higher conductivity. After the introduction of metals into the monometallic phosphide, the resulting polymetallic (or metal-rich) transition metal phosphide (e.g., MP or M_2P, where M is the transition metal) exhibits metallic properties (chemical stability), multiple redox centers, and the ability to carry more free electrons, resulting in a significantly increased electrical conductivity (compared with phosphorus-rich or monometallic phosphides) [6]. However, phosphorus-rich TMPs with abundant P−P bonds are unsuitable as energy storage materials because of the electron accumulation around the P atoms and decreased electrochemical performance [7]. Thus, understanding how to select appropriate metals and combine them to form multi-component metal phosphides with synergistic effects is critical for optimizing the electrochemical performance of electrochemical energy storage devices using TMPs as electrodes.

Recently, it was discovered that one of most effective tactics to improve the electrochemical performance is to combine Ni and Fe to form bimetallic phosphide, which improves the conductivity of Fe/Ni-based compounds [8,9]. The electrochemical properties of Ni and Fe are quite similar; the ionic radius of Fe^{3+}(0.65 Å) is close to that of Ni^{2+}(0.69 Å) [10,11]. As a result of the formation of hydroxyl oxides during charge storage, Ni and its compounds in various morphologies (e.g., nanoparticles, nanowires, and thin films) are typical supercapacitor electrode materials with an excellent redox behavior. Because of its high theoretical capacity (1951.2 C g^{-1}) and good electronic conductivity (1.2×10^{-3} S cm^{-1}) [12], Ni_2P is considered as a promising material for energy storage. However, because of structural degradation, the stability of FeNi-based materials is still unsatisfactory [9]. In recent years, many binder-free nanoarray electrodes with self-contained nano-units and porous structures have been investigated to minimize the volume expansion and improve the electrochemical performance. As a result of phosphorus's lower electronegativity and reduced bond ionicity, Fe/Ni bimetal phosphides give a series of various redox couples ($Fe^0/Fe^{2+}/Fe^{3+}$, $Ni^0/Ni^{1+}/Ni^{2+}$) with better metalloid properties [13]. Although there have been many reports about Ni/Fe bimetal phosphides as high-performance electrocatalysts for water splitting [14–16], their use as electrodes in SCs is unclear [17].

In this study, we designed, synthesized, and developed self-supported nanohoneycomb (NHC)-like FeNi(OH/P) electrodes using a simple two-step procedure, in which incomplete phosphated FeNi(OH/P) nanosheet arrays were oriented and anchored on 3D nickel foam without any bridging agent. The large contact areas and electron pathways of the unique interconnected NHC-like FeNi(OH/P) nanosheet array could provide high-efficiency electronic transmission for the charge storage. Furthermore, Ni/Fe bimetal phosphides have high density redox centers and a high conductivity. This study also describes a hybrid supercapacitor (HSC) with a cathode electrode made of bimetallic transition phosphides and an anode electrode made of polypyrrole/C (PPy/C). By optimizing gel electrolytes, the HSC exhibited excellent capacity retention. This work provides a new strategy for designing a cost-effective, reliable, and high-performance asymmetric supercapacitor for energy storage applications.

2. Materials and Methods

2.1. Materials

Nickel chloride hexahydrate ($NiCl_2{\cdot}6H_2O$), ferric chloride hexahydrate ($FeCl_3{\cdot}6H_2O$), sodium chloride (NaCl), and sodium hypophosphite ($NaH_2PO_2{\cdot}H_2O$) were purchased from Sinopharm Chemical Reagents Co., Ltd. All of the reagents were of analytical grade and were used without further purification.

2.2. Synthesis of Materials

2.2.1. Synthesis of FeNi LDH's Precursor

First, the nickel foam (NF, 5.2×3.7 cm^2) was cleaned with 2 M HCl, absolute ethanol, and deionized water by sonication for 15 min in sequence to remove the oxide layer on the surface. Then, 0.27 g $FeCl_3{\cdot}6H_2O$ [18,19] and 0.097 g NaCl were dissolved in 70 mL of deionized water by magnetic stirring for 30 min. The mixed solution and a piece of the pretreated NF were transferred to a 100 mL Teflon-lined stainless autoclave and kept at 120 °C for 10 h. The Ni foam coated with the FeNi LDH sample was washed alternately with ethanol and deionized water several times, and dried for 5 h at 60 °C. The mass density of the FeNi LDH nanosheet array (0.57 mg cm^{-2}) on the NF was determined by subtracting the weight of the NF (after acid etching) from the weight of the FeNi LDHs with the NF.

2.2.2. Synthesis of the FeNi(OH/P) Sheet Array Electrode

The NaH_2PO_2 and Ni foam-supported FeNi LDH were placed upstream and downstream of the tube furnace, respectively. The samples were heated to 300 °C at a heating rate of 2 °C min^{-1} for 2 h under an Ar flow. The FeNi(OH/P) was collected after the furnace was cooled to an ambient temperature. To explore the effect of the phosphorization degree, we considered NiFe(OH/P) with different amounts of NaH_2PO_2 (0 g, 0.6 g, 1.2 g, and 1.8 g, defined as FeNi(OH/P)-0, FeNi(OH/P)-0.6, FeNi(OH/P)-1.2, and FeNi(OH/P)-1.8, respectively). After phosphating, the mass increased by 1.87 mg cm^{-2} on average. The mass density of the FeNi(OH/P)-1.2 nanosheet array on the NF was 2.44 mg cm^{-2}.

2.2.3. Assembly of Hybrid Supercapacitor Device

All-solid-state hybrid supercapacitors (HSCs) were assembled by employing the as-fabricated FeNi(OH/P)-1.2 and PPy/C as the cathode and anode electrode, respectively, and PVA/KOH/PAAS as the electrolyte.

For the preparation of PPy/C, PPy aerogel was heated to 500 °C at a rate of 5 °C min^{-1} under an argon atmosphere. The preparation of PPy aerogel was mentioned in previous articles from our group [20]. PPy aerogel was synthesized through PPy hydrogels, which were obtained by oxidizing the pyrrole monomer with ammonium peroxysulfate (APS) in a Methyl orange (MO) solution. The mixture of PPy/C (80 wt%), acetylene black (10 wt%), and polyvinylidene fluoride (PVDF, 10 wt%) was coated onto an NF (1×1 cm^2) and then dried at 80 °C for 8 h to obtain the PPy/C electrodes. The mass loading of PPy/C powder attached to the NF was 8.4 mg cm^{-2}. The gel electrolyte was prepared as follows: 6 g PVA (1799 type) and 5 g KOH were dissolved in 60 mL deionized water at 90 °C, then 2 mL PAAS was added dropwise into the above solution with stirring and maintained at 65 °C for 15 min. The two electrodes were coated with a PVA/KOH gel electrolyte to obtain HSCs with a thickness of about 2.1 mm.

2.3. *Morphology and Structure Characterization*

The micromorphology and structure of the as-synthesized materials were analyzed using a scanning electron microscope (SEM, PHILIPS XL30 TMP) and transmission electron microscope (TEM, JEM-2000 UHR SETM/EDS). Energy dispersive X-ray spectroscopy (EDX), elemental mapping, and selected area electron diffraction (SAED) were performed on the same FE-TEM microscope. The phase characteristics of the as-synthesized materials were identified using X-ray diffraction (XRD) and X-ray photoelectron spectroscopy (XPS, AXIS SUPRA+). Mass weighting was recorded with a semi-micro balance (ESJ200-4B) with an accuracy of 0.01 mg.

2.4. *Electrochemical Measurements*

All of the electrochemical measurements (three electrodes/two electrodes) were performed using the CHI660D electrochemical workstation. The FeNi(OH/P) electrode was used directly as the working electrode, the Pt electrode (1×1 cm^2) was used as a counter electrode, the Ag/AgCl electrode was used as the reference electrode, and the electrolyte was 3 mol L^{-1} KOH. The scanning rates of the cyclic voltammetry (CV) were 1~100 mV s^{-1} within a potential window of 0~0.45 V, and the galvanostatic charge–discharge (GCD) test range was 0~0.4 V. Electrochemical impedance spectroscopy (EIS) was tested between 0.01 Hz and 100 kHz.

The areal capacity C_a (C cm^{-2}) was calculated from the CV curves using Equation (1):

$$C_a = \left(\int I(V)dV \right) / 2vA \tag{1}$$

The capacity C_s (C cm^{-2}) was calculated from the GCD curves using Equation (2):

$$C_s = \left(2I \int (V)dt \right) / \Delta V \tag{2}$$

where $I(V)$ refers to the current at V, dV is the differential potential, v is the scan rate, Δt is the discharge time, A is the electrode's surface area, m is the mass of active material, and ΔV is the potential window.

The HSC device was assembled using FeNi(OH/P)-1.2 as the cathode electrode and PPy/C as the anode electrode with a gel electrolyte. The charges stored between the cathode and anode electrodes were balanced (Q+ = Q−) based on the three-electrode system outcomes in order to achieve an excellent ASC device performance. According to Equation (3), the ratio of mass between FeNi(OH/P)-1.2 and PPy/C was 0.29.

$$m_+ / m_- = C_{a-} \times \Delta V_- / C_{a+} \times \Delta V_+ \tag{3}$$

where the subscripts "+" and "−" represent the cathode and anode electrodes, respectively. The energy densities (E) and the power density (P) were calculated using Equations (4) and (5):

$$E = \left(C \times \Delta V^2 \right) / 7.2 \tag{4}$$

$$P = (E \times 3600) / \Delta t \tag{5}$$

3. Results and Discussion

The synthesis process of the FeNi(OH/P)-1.2 nanosheet array on nickel foam is shown in Figure 1. First, as a 3D current collector, nickel foam (NF) with a 3D porous structure and high conductivity also provides the Ni source in the hydrothermal reaction. The honeycomb-like FeNi LDH nanosheet array was vertically grown on the 3D NF through a simple hydrothermal method (Figure 2a). Different from the traditional bimetallic co-precipitation method, the obtained FeNi LDHs grew robustly and did not fall off easily in the continuous testing work, which provided a good cycle performance. Additionally, NaCl acted as a structure directing agent, which contributed to the formation of the FeNi LDH nanosheet morphology. After heat treatment (without NaH_2PO_2), FeNi LDH was converted to FeNi(OH/P)-0 and the morphology remained largely unchanged (Figure S1a). After phosphating (1.2 g NaH_2PO_2), the color of the FeNi LDH precursor changed from orange to black, forming the FeNi(OH/P)-1.2 nanosheet array. The overall morphology of the FeNi(OH/P)-1.2 nanosheet array was maintained, which facilitated electron/ion transmission and promoted electrochemical energy storage. The surface of FeNi(OH/P)-1.8 (Figure S1c) was covered with a thick layer of phosphide and had a small number of nanotubes composed of P, and its specific surface area decreased, resulting in a reduction in the active reaction area [18].

When the phosphating reaction just occurred, P^{4-} reacted with metal ions on the surface of the material to produce a thin film of phosphide. Further reactions depended on the diffusion rate of the metal cations inside the material. Metal cation ions are relatively small and diffuse from the inside to the outside. P^{4-} anion ions are relatively large and diffuse from the outside to the inside. Therefore, when the reaction proceeded further, the outward diffusion of metal cations dominated, the phosphating reaction accumulated in the outer layer, and the metal phosphide shell thickened. Therefore, the morphology before and after phosphating was maintained. This was clearly evidenced in the TEM results. Figure 2d clearly shows the morphology of a single FeNi(OH/P)-1.2 nanosheet with a diameter of about 120–200 nm. The SAED (Figure 2e) presented typical diffraction rings, proving that the as-prepared FeNi(OH/P)-1.2 was polycrystalline.

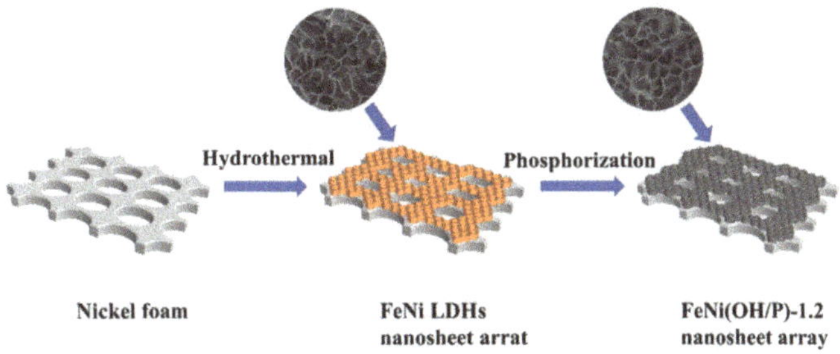

Figure 1. The synthesis procedure of the FeNi(OH/P)-1.2 nanosheet array on nickel foam. The inset shows the corresponding SEM images.

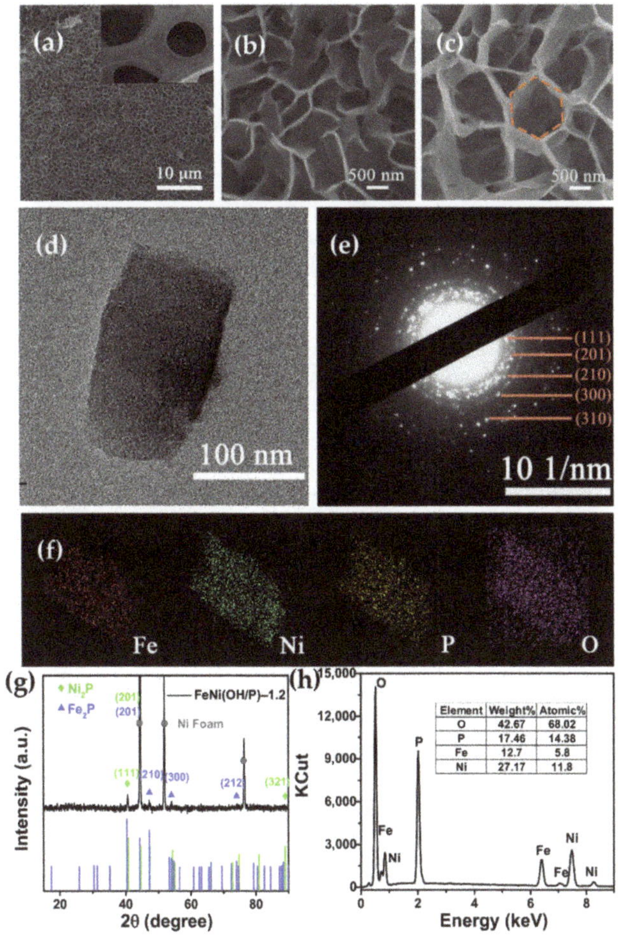

Figure 2. SEM images of (**a**,**b**) FeNi LDH nanosheet arrays and (**c**) a FeNi(OH/P)-1.2 nanosheet array. Inset (**a**) shows the SEM image of the FeNi LDH nanosheet array under low magnification. (**d**) TEM image; (**e**) SAED pattern; (**f**) the elements mapping of Fe, Ni, P, and O; (**g**) XRD pattern; and (**h**) EDX spectrum of the FeNi(OH/P)-1.2 nanosheet array. Inset of (**h**) shows the corresponding atomic/weight percentages of different elements: Fe, Ni, P, and O.

Figure 2g shows the XRD patterns of FeNi(OH/P)-1.2. Three broad peaks around 44.5°, 51.8°, and 76.4° could be clearly observed in the samples, which were assigned to the (111), (200), and (220) planes of the NF, respectively. The diffraction peaks of the sample at 2θ = 44.2°, 47.2°, 54°, and 73.9° corresponded to the (201), (210), (300), and (212) crystal planes of the hexagonal Fe_2P (JCPDS card no. 83-2337), respectively, while the peaks at 47.3°, 50.4°, and 88.9° were attributed to the (210), (300), and (321) planes of Ni_2P (JCPDS card no. 03–0953), respectively [21]. With the increase in the phosphorus source (the mass of NaH_2PO_2), the diffraction peaks were sharper and stronger, indicating that more FeNiP was successfully transformed (Figure S2). It can be seen from the XRD spectrum that the intensity of the diffraction peaks was weak because of the strong base Ni peak and incomplete phosphating of the FeNi(OH/P) material obtained. However, consistent with the XRD patterns, Fe, Ni, P, and O element peaks could also clearly be found from the element mapping (Figure 2f). The Fe, Ni, P, and O elements were uniformly distributed throughout the single nanosheet, confirming the formation of the FeNi(OH/P)-1.2 material. From the EDX spectrum (Figure 2g), the atomic ratios of Ni, Fe, P, and O were 0.12:0.06:0.14:0.68, which clearly indicate the existence of incomplete phosphating.

The elemental compositions and chemical valences of the samples were then revealed using XPS. Figure 3 shows the complete scan spectrum of FeNi(OH/P)-1.2. The appearance of Ni, Fe, P, and O peaks indicates the presence of these four elements. Figure 3b is the energy spectrum of Ni 2p. The two peaks at binding energies of 874.7 and 856.7 eV corresponded to Ni $2p_{1/2}$ and Ni $2p_{3/2}$, respectively. At the same time, the two satellite peaks at 880.9 and 861.9 eV confirmed the spin-orbit characteristics of Ni^{2+}, which is a typical feature of the combination of Ni^{2+} and oxygen. Note that the other peaks emerged at 852.0 eV ($2p_{3/2}$), demonstrating that Ni^{2+} ions were partially reduced to a metallic state, and the metallic characteristics typically offered fast electron transportation, thereby enhancing the electrochemical performance [14]. Figure 4c shows the energy spectrum of Fe 2p. The binding energies of 726.2 and 711.1 eV were accompanied by satellite peaks of 720.9 and 733.7 eV, which correspond to Fe $2p_{1/2}$ and Fe $2p_{3/2}$, respectively. The peak at 704.0 eV is related to Fe-P and indicates a positively charged Fe (δ+) [19]. Figure 4d shows the energy spectrum of the P 2p element, which can be fitted to two main peaks of 134.6 and 133.8 eV, corresponding to P $2p_{1/2}$ and P $2p_{3/2}$, respectively. The formation of the P−O bond is attributed to the oxidation on the surface of the metal phosphide [22]. It can be seen from Figure 4e that O 1s can be convoluted at 530.9 and 532 eV, which indicates the existence of metal hydroxides. Based on the XPS analysis, the percentages of Fe, Ni, P, and O in these samples were 5.8%, 11.6%, 13.8%, and 58.8%, respectively, which are consistent with the EDS results. In summary, the obtained FeNi(OH/P) material consisted of FeNi LDHs and FeNiP.

According to the above XPS results, the process of in situ growth of FeNi LDHs can be speculated. In the hydrothermal process, Fe^{3+} produced a redox attack on the NF and released Ni^{2+} on the surface of the NF, because $E^0_{Fe^{3+}/Fe^{2+}}$ (0.771 V) was much higher than $E^0_{Ni^{3+}/Ni^{2+}}$ (−0.257 V). In addition, as a result of the presence of oxygen, Fe^{2+} was oxidized to Fe^{3+}. Finally, Ni^{2+} and Fe^{3+} ions easily reacted with OH^- to form FeNi LDHs on the NF [23]. The equations used above are as follows:

$$2Fe^{3+} + Ni = 2Fe^{2+} + Ni^{2+} \tag{6}$$

$$4Fe^{2+} + O_2 + 2H_2O = 4Fe^{3+} + 4OH^- \tag{7}$$

$$Ni^{2+} + Fe^{3+} + OH^- = NiFe(OH)_x \tag{8}$$

NaCl as a structure directing agent did not participate in the formation of hydrothermal products. This was because the ionic radius of Na^+ (1.02 Å) [24] was significantly larger than those of Fe^{3+} (0.65 Å) and Ni^{2+} (0.69 Å), and LDHs usually consist of both divalent and trivalent cations [25]. Most notably, unlike the traditional bimetallic coprecipitation method, in situ grown interconnected FeNi nanosheet arrays are firmly attached to the NF, accompanied by ferric salts as catalysts, which gives the electrode materials excellent stability.

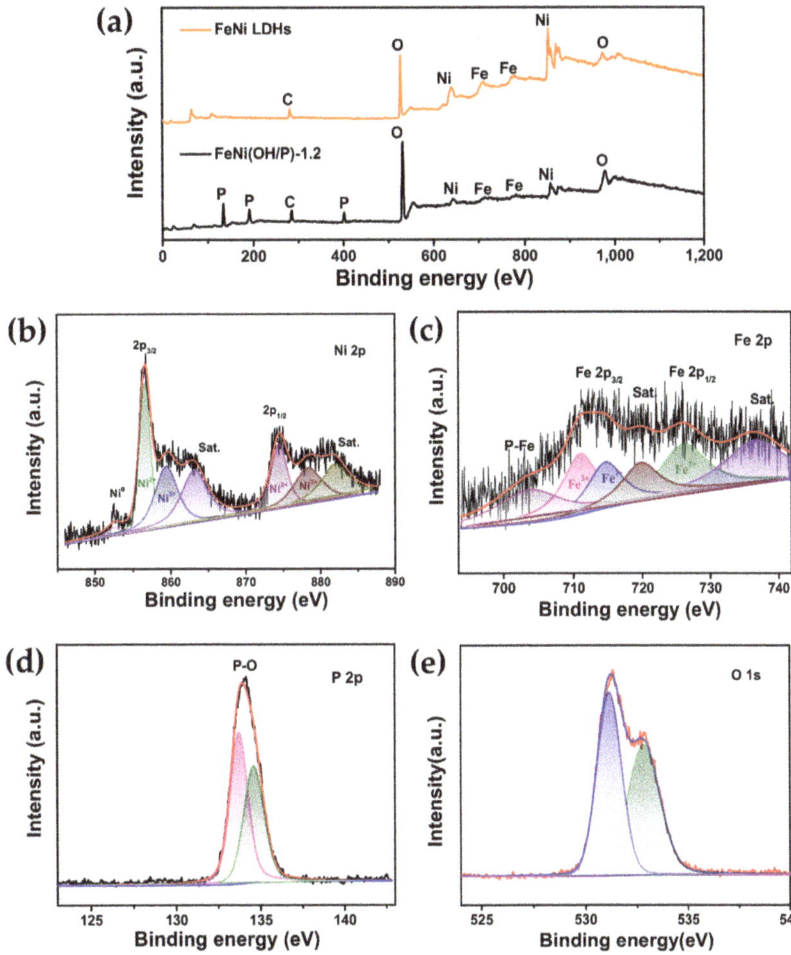

Figure 3. (**a**) Comparison of the XPS spectra for the FeNi LDH nanosheet array and FeNi(OH/P)-1.2 nanosheet array. (**b**) Ni 2p, (**c**) Fe 2p, (**d**) P 2p, and (**e**) O 1s XPS spectra of the FeNi(OH/P)-1.2 nanosheet array.

The electrochemical properties of the as-synthesized samples were first investigated in a three-electrode system using a 3.0 M KOH electrolyte. As shown in Figure 4a, CV tests of the FeNi(OH/P)-1.2 electrode between 0~0.45 V were obtained at 1~100 mV s^{-1}. As the scan rate increased, the closed area of the CV curve also increased, the response peak current also increased, and a faster redox reaction occurred. The peaks shifted towards a higher and lower potential, respectively. Two redox peaks were clearly observed on all CV curves, which demonstrated the characteristic battery-type behavior of the electrodes. This could be related to the highly reversible Faradaic reactions between Ni^{+}/Ni^{2+} and hydroxyl anions, as well as between Fe^{2+}/Fe^{3+} and hydroxyl anions. The possible redox reactions for the FeNi(OH/P) electrode involved in the KOH electrolyte can be summarized using Equations (9)–(11), as follows:

$$FeNiP + 2OH^{-} \leftrightarrow NiP_xOH + FeP_xOH + 2e^{-} \quad (9)$$

$$FeP_xOH + OH^- \leftrightarrow FeP_xO + H_2O + e^- \quad (10)$$

$$NiP_xOH + OH^- \leftrightarrow NiP_xO + H_2O + e^- \quad (11)$$

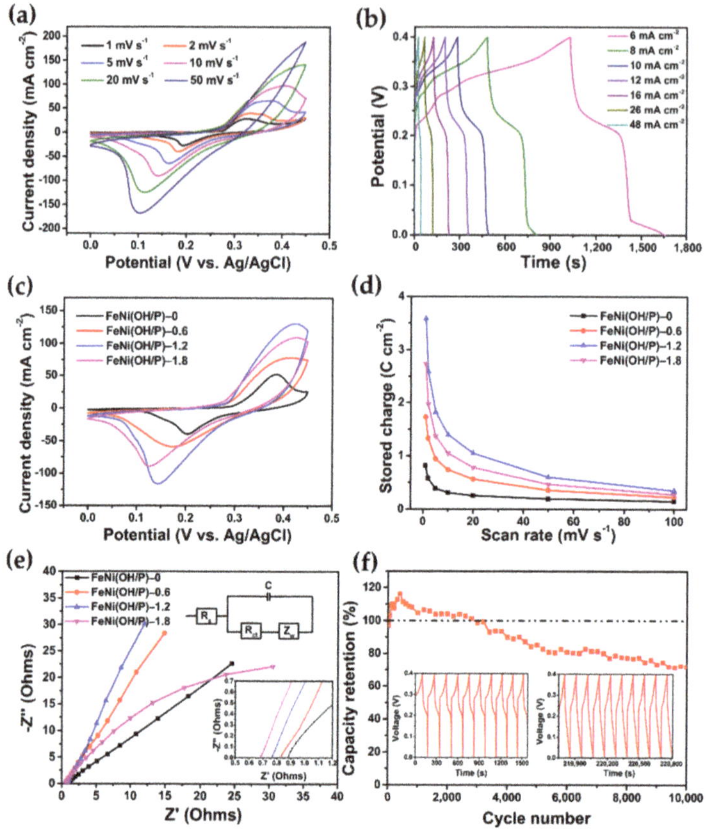

Figure 4. (**a**) CV curves at different scan rates and (**b**) GCD curves at different current densities for the FeNi(OH/P)-1.2 electrode. Comparison of (**c**) CV curves, (**d**) rate performance, and (**e**) EIS spectra for FeNi(OH/P) electrodes with different amount of NaH_2PO_2. The inset of (**e**) shows the equivalent circuit diagram. (**f**) Cycling performance of the FeNi(OH/P)-1.2 electrode; the inset of (**f**) shows the first 10 cycles and the last 10 cycles of the GCD curves of the FeNi(OH/P)-1.2. electrode at 24 mA cm^{-2}.

The chemical process of Equation (9) is related to the oxidation of TMP, and Equations (10) and (11) are related to the adsorption–desorption of OH^- ions. The adsorption capacity of the OH^- ions between the active material and the electrolyte affects the electrochemical performance of the electrode.

Figure 4b depicts the GCD curves of the FeNi(OH/P)-1.2 electrode at different current densities. There is an obvious platform in the discharge curve, indicating a more battery-type redox behavior, consistent with the CV curve results. The charging time was almost equal to the discharging time at all current densities, indicating that FeNi(OH/P)-1.2 has Coulombic efficiency and highly reversible redox reactions in GCD. The C_s of FeNi(OH/P)-1.2 was calculated to be 2740.9, 2393.0, 2196.6, 2045.0, 1649.9, and 1025.0 mC cm^{-2} at current densities of 6, 8, 10, 12, 26, and 48 mA cm^{-2}, respectively.

Figure S3a depicts the CV plots of the obtained FeNi LDHs and FeNi(OH/P)-1.2 electrode for 0~0.45 V. Notably, the FeNi(OH/P)-1.2 electrode demonstrates the larger CV area and peak current density. The capacities (Figure S3b) were then calculated from the CV curves. Compared with the FeNi LDH electrode, the FeNi(OH/P)-1.2 electrode had a better rate performance, which indicates significantly improved electrochemical activities and an enhanced capacity of phosphide samples. Figure 4c shows the CV curves of the obtained FeNi(OH/P)-0, FeNi(OH/P)-0.6, FeNi(OH/P)-1.2, and FeNi(OH/P)-1.8 electrodes at a scan rate of 20 mV s^{-1}. All of the curves exhibited similar pairs of redox peaks. To learn more about this, Figure 4d displays the rate performance for the FeNi(OH/P) electrode with different amounts of NaH_2PO_2. The result indicates that the FeNi(OH/P)-1.2 electrode demonstrated the highest specific capacity and best rate performance out of all of the as-prepared electrodes. Specifically, the FeNi(OH/P)-1.2 electrode possessed a specific capacity of up to 3.6 C cm^{-2} at 1 mV s^{-1}. This capacitive value was higher than that of the FeNi LDH electrode (1.0 C cm^{-2}), FeNi(OH/P)-0 electrode (0.82 C cm^{-2}), and FeNi(OH/P)-0.6 electrode (1.7 C cm^{-2}) at 1 mV s^{-1}. When the addition of NaH_2PO_2 was 1.8 g, the capacity of the FeNi(OH/P)-1.8 electrode decreased (2.7 C cm^{-2} at 1 mV s^{-1}) compared with FeNi(OH/P)-1.2, indicating that phosphorus-rich TMPs had a lower specific capacity as a result of electron accumulation around the P atoms. The highest capacity obtained of 3.6 C cm^{-2} ($\approx$1469.9 C g^{-1} and 408.3 mAh g^{-2}) at 1 mV s^{-1} was excellent compared with the reported metal phosphide electrodes.

To better explain the electrochemical properties, we need to clarify the electrochemical kinetics of the FeNi(OH/P) electrodes, which have two energy storage mechanisms: Faradaic and non-Faradaic reactions. The non-Faradaic reaction is the process of electrostatic adsorption in electric double-layer capacitors. In contrast, Faradaic reactions occur on both the surface (redox pseudocapacitance) and in the bulk of the material (typically as in conventional batteries; diffusion-controlled).

The capacitive effects can be characterized by analyzing the CV data at various sweep rates according to the following [26,27]:

$$i = av^b \tag{12}$$

where i is the current, ν is the scan rate, and a and b are constants. Here, b is determined from the slope of the plot of log i versus log v according to Equation (12). When b is close to 1, this indicates that it is predominantly capacitive. When b is equal to 0.5, this means that this is a diffusion-controlled process. The calculated b values for the FeNi(OH/P) electrodes are shown in Figure S4. It was found that the b value for all of the FeNi(OH/P) electrodes with different amounts of NaH_2PO_2 was always around 0.5, indicating that the current comes primarily from the diffusion-controlled process. This manifests the battery-type behavior of the FeNi(OH/P) electrode.

To further study the ion and electron transport kinetics of the FeNi(OH/P) electrodes, electrochemical impedance spectroscopy (EIS) measurements with equivalent circuits are shown in Figure 4e. The impedance spectra are composed of one semicircle and a linear segment in the high- and low-frequency region, respectively. The intercept at the real axis represents the equivalent series resistance (R_s), which is a combination of ionic resistance of the electrolytes, electronic resistance of the electrode materials, and interface resistance. The values of R_s for FeNi(OH/P)-0, FeNi(OH/P)-0.6, FeNi(OH/P)-1.2, and FeNi(OH/P)-1.8 are 0.88, 0.83, 0.76, and 0.68 Ω, respectively. As the content of NaH_2PO_2 increases, the R value decreases, demonstrating improved charge-transfer kinetics and rapid electron transport for redox reactions after phosphating. However, the FeNi(OH/P)-1.8 electrodes with the lowest slope in the low-frequency region demonstrate its higher diffusive resistance. Ion diffusion is affected because of the accumulation of electrons around the P atoms in the P-rich TMPs.

Figure 4f reveals the cycle performance of the FeNi(OH/P)-1.2 electrodes. After 10,000 cycles of GCD curves at 24 mA cm^{-2}, the capacitive retention of the FeNi(OH/P)-1.2 electrode was 72.0% and remained unchanged. The increase in capacity during the first 600 cycles could be attributed to the activation of FeNi(OH/P), which conformed with the characteristics of Ni(OH)2 [20] and NiP [18], indicating that they had standard metalloid characteristics. The diminishment of the capacity performance could be attributed to the following: (1) the formation of FeNi hydroxide from the surface of FeNi(OH/P) reacting with the electrolyte [28]. (2) In FeNi compounds, Ni was the active site for Faradaic reactions and it was easier to convert Ni^{2+} to Ni^{3+}, but not in the reverse process. After multiple charge/discharge processes, the irreversible reaction of Ni^{2+} increased, so the Faraday capacitance became lower and lower [9,29]. A detailed comparison between the electrochemical properties reported in the present study and in previously reported studies for composites in a three-electrode system is given in Table 1.

Table 1. Comparison of the electrochemical performances between the as-fabricated FeNi(OH/P) electrode and previously reported FeNi-based/TMP electrodes.

Electrode Materials	Current Collector	Electrolyte	Specific Capacitance	Capacitance Retention	Year Ref.
ZnNiFe(OH/P) nanosheet	Nickel foam	3 M KOH	1708 F g^{-1}/5.64 F cm^{-2} (1 A g^{-1})	83% (4000)	2021 [17]
FeCoP@NiCo nanosheet	Carbon foam	2 M KOH	795.5 C g^{-1} (1 A g^{-1})	89.7% (5000)	2021 [30]
FeNiP@CoNi LDHs nanosheet	Nickel foam	3 M KOH	2280 F g^{-1} (1 A g^{-1})	70% (5000)	2020 [31]
$NiFeP@NiCo_2S_4$ nanosheet	Carbon cloth	2 M KOH	874.4 C g^{-1} (1 A g^{-1})	85.6% (5000)	2020 [32]
NiCoP-MLG nanosheet (powder)	-	3 M KOH	1419.6 F g^{-1} (1 A g^{-1})	91% (3000)	2020 [33]
NiCoP nanoplate (powder)	-	2 M KOH	1306 F g^{-1} (1 A g^{-1})	-	2020 [34]
$Ni_2P@GO$ (powder)	-	2 M KOH	1526.66 F g^{-1} (1 A g^{-1})	86.3% (2500)	2020 [35]
$NiHPO_4$ nanowire	Nickel foam	1 M KOH	1472 F g^{-1} (1 A g^{-1})	90.3% (6000)	2019 [36]
NiCoP nanowire	Nickel foam	6 M KOH	19.9 F cm^{-2} (50 mA cm^{-2})	92% (2000)	2019 [37]
Ni_xP nanosphere	Nickel foam	6 M KOH	382.7 F g^{-1} (0.45 A g^{-1})	110.9% (4000)	2018 [11]
$NiCo_2O_4$-NiCoP nanosheet/nanowire	Nickel foam	3 M KOH	2288.8 F g^{-1} (1 A g^{-1})	71% (2000)	2018 [38]
NiCoP nanosheet	Nickel foam	6 M KOH	2143 F g^{-1} (1 A g^{-1})	73% (2000)	2018 [39]
NiP nanosheet	Nickel Sponge	6 M KOH	430.3 mAh g^{-1} (1 A g^{-1})	-	2016 [40]
Ni_2P nanosheet	Nickel foam	6 M KOH	2141 F g^{-1} (50 mV s^{-1})	-	2015 [21]
$CuFe_2O_4$ $NR@NiFe_2O_4$-NS nanosheet/nanowire	Stainless-steel mesh	6 M KOH	1366 C g^{-1} (1 A g^{-1})	94% (10,000)	2021 [41]
NiFe-LDH nanosheet	Nickel foam	1 M KOH	2708 F g^{-1} (5 A g^{-1})	42.6% (500)	2017 [9]
FeNi(OH/P)-1.2 nanosheet	Nickel foam	3 M KOH	3.6 C cm^{-2}/1469.9 C g^{-1}/ 408.3 mAh g^{-1} (1 mV s^{-1})	72.6% (10,000)	This work

It can be observed that the addition of NaH_2PO_2 in phosphating had a significant effect on the electrochemical performance of the as-prepared samples. This may be because of the following: (1) When the mass of NaH_2PO_2 was less than 1.2 g, there was no morphology change compared with the precursor, and the phosphating degree of the precursor was insufficient, resulting in insufficient active sites during the electrochemical reaction. (2) When the mass of NaH_2PO_2 was more than 1.2 g, the porosity decreased obviously owing to the aggregation of phosphide as a thick layer, resulting in a decrease in the specific surface area and ion/electron transport channels. Moreover, phosphorus-rich metal phosphides have a large number of P–P bonds, leading to free electron clusters and a reduced electrochemical performance. Therefore, when the addition of NaH_2PO_2 was 1.2 g, saturation phosphorization and nanostructure optimization of the as-prepared samples were achieved simultaneously. More importantly, the electrochemical performance was also optimized at the same time.

To further demonstrate the practical feasibility of the optimized FeNi(OH/P) electrode, an all-solid-state hybrid supercapacitor (HSC) was assembled with FeNi(OH/P)-1.2 as the cathode electrode and PPy/C as the anode electrode. Figure S5a displays the SEM image of the PPy/C anode, which is composed of a PPy/C nanorod. The XRD pattern is shown in Figure S3b. The diffraction peaks around 24° and 43° could be indexed to the (002) and (100) crystal planes of the carbon, respectively. The CV curves of the PPy/C electrode and FeNi(OH/P)-1.2 electrode at 20 mV s^{-1} are displayed in Figure S6a. The rectangular CV curve of the PPy/C anode reveals the electrical double-layer capacitance (EDLC) behavior, which is different from the electrochemical feature of the FeNi(OH/P) cathode. Figure 5a shows that the hybrid supercapacitor can work efficiently in a large operating voltage for 1.9 V. The GCD curves of the HSC at various current densities from 7 to 140 A cm^{-2} displayed an almost symmetrical triangular shape with a slight deviation, indicating its good electrochemical reversibility and ideal capacitance behavior (Figure 5b). Notably, the HSC reached 1.9 C cm^{-2}, 1.4 C cm^{-2}, 0.7 C cm^{-2}, and 0.3 C cm^{-2} at 7 mA cm^{-2}, 14 mA cm^{-2}, 70 mA cm^{-2}, and 140 mA cm^{-2}, respectively.

To study the influence of PAAS on electrolytes, we evaluated the electrochemical performance of the HSC with different gel electrolytes using the same electrodes. From the GCD curves (Figure 5a and Figure S6b), the HSC with PAAS had a longer discharge time. These results indicated the enhanced charge storage capacity of the FeNi(OH/P)//AC HSC after optimizing the electrolyte. As shown in Figure 5d, 2000 cycles were tested at 21 mA cm^{-2}. Impressively, after 2000 cycles, the capacity retention of the HSC with PAAS remained at 118.5%, which was better than the HSC without PAAS (51.3%). Sodium polyacrylate (PAAS) is a polyanionic electrolyte with good water solubility and a large number of hydrophilic groups [42]. Its aqueous solution has a good dissociation, wettability, and water retention. When added to the PVA/KOH electrolyte, it not only has good film-forming properties to encapsulate the two electrodes, but can also retain the water in the electrolyte well so as to reduce the water loss of the electrolyte and affect the electrochemical performance during electrochemical work. Moreover, PAAS is also a surfactant that can prevent the negative effect of metal ions. It can ameliorate the redox activity of battery-type electrodes and catalyze the reaction of the active material with OH^-. According to the above experimental results, PAAS, as a water-retaining agent/active agent in the electrolyte, can greatly improve the cycle performance of solid-state energy storage devices. The capacity retention of the HSC with PAAS is far superior to the reported supercapacitor; for instance, NiCoP@NF//AC (67.2%, 500 cycles) [31], $NiCo_2O_4$//$FeSe_2$ (90%, 1000 cycles) [43], CQDs/$NiCo_2O_4$//AC (101.9%, 5000 cycles) [44], $NiCo_2S_4$//C (76%, 5000 cycles) [45], and $NiCo_2O_4$/NF//$NiCo_2O_4$/NF (86%, 500 cycles) [46].

The kinetic factors for the enhanced electrochemical performance were analyzed using EIS spectra (Figure 5e). The electrolyte resistances (Rs) of the HSC with PAAS (Rs = 1.3 Ω) were smaller than those of the HSC without PAAS (Rs = 2.1 Ω), and the charge transfer resistance (Rct) of the HSC with PAAS (Rct = 0.7 Ω) was smaller than that of the HSC without PAAS (Rs = 1.06 Ω), indicating a fast electron transfer ability. Meanwhile, the slope of the HSC with PAAS in the low-frequency region was larger than that of the HSC without PAAS; thus, its ion diffusion kinetics were enhanced through the addition of PAAS in the electrolyte, which is consistent with the CV/GCD results. Furthermore, it can be seen that the Rs and Rct of the HSC with PAAS (Rs = 0.5 Ω, Rct = 0.61 Ω) were smaller after the cycles.

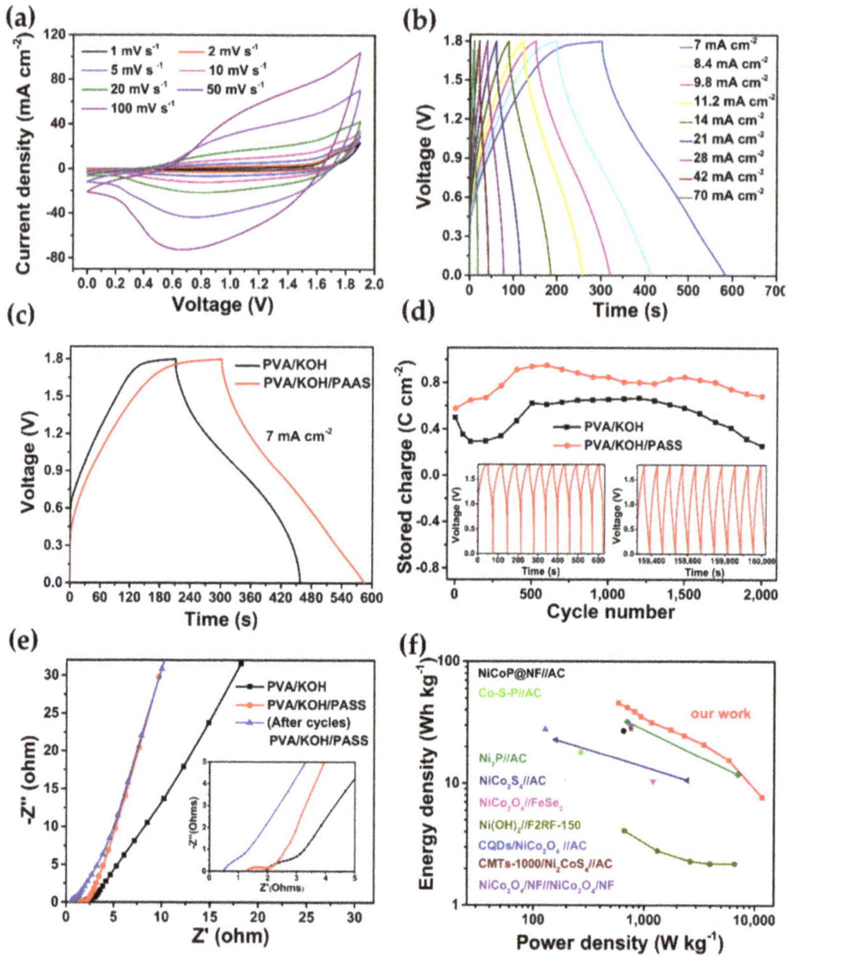

Figure 5. (**a**) CV curves and (**b**) GCD curves of the FeNi(OH/P)//PPy/C HSC with the PVA/KOH/PAAS gel electrolyte. Comparison of (**c**) GCD curves and (**d**) cycling performances of the FeNi(OH/P)//PPy/C HSC with different electrolytes. The inset of (**d**) shows the first 10 cycles and last 10 cycles of GCD curves of the FeNi(OH/P)//PPy/C HSC with PVA/KOH/PAAS gel electrolyte at 21 mA cm^{-2}. (**e**) EIS spectra for the FeNi(OH/P)//PPy HSC with PVA/KOH gel electrolyte, and the FeNi(OH/P)//PPy/C HSC with the PVA/KOH/PASS gel electrolyte before and after 2000 cycles. (**f**) Ragone plot.

The Ragone plots of the HSC in this research are compared with the reported literature in Figure 5f. The HSC exhibited a high energy density of 45 Wh kg^{-1} at 581 W kg^{-1}, which is superior to that reported in the literature, such as NiCoP@NF//AC (27 Wh kg^{-1} at 647 W kg^{-1}) [31], CQDs/$NiCo_2O_4$//AC (27.8 Wh kg^{-1} at 128 W kg^{-1}) [44], $NiCo_2O_4$//$FeSe_2$ (10.4 Wh kg^{-1} at 1200 W kg^{-1}) [43], NiCoP (30.2 Wh kg^{-1} at 891 W kg^{-1}) [27], $NiCo_2S_4$//C (22.8 Wh kg^{-1} at 160 W kg^{-1}) [45], $NiCo_2O_4$/NF//$NiCo_2O_4$/NF (30 Wh kg^{-1} at 750 W kg^{-1}) [46], Co-S-P//AC (17.9 Wh kg^{-1} at 750 W kg^{-1}) [47], CMTs-1000/Ni_2CoS_4//AC (28.1 Wh kg^{-1} at 753 W kg^{-1}) [48], and $Ni(OH)_2$//F2RF-150 (4.1 Wh kg^{-1} at 661.5 W kg^{-1}) [49].

4. Conclusions

In summary, a self-supported nanohoneycomb (NHC)-like FeNi(OH/P) nanosheet array was prepared through a simple two-step procedure. By adjusting the addition of

NaH_2PO_2, the FeNi(OH/P) electrode simultaneously achieved saturation phosphorization and nanostructure optimization. The as-fabricated FeNi(OH/P) electrode achieved a significantly enhanced specific capacity of 3.6 C cm^{-2} (≈408.3 mAh g^{-1}) at 1 mV s^{-1} with cyclic stability (72.0% retention after 10,000 cycles). Meanwhile, an all-solid-state hybrid supercapacitor device assembled by employing FeNi(OH/P) as the cathode electrode and PPy/C as the anode electrode exhibited a high capacity of 1.9 C cm^{-2} at 7 mA cm^{-2}, a maximum energy density of 45 Wh kg^{-1}, and excellent cyclic stability of 118.5% after 2000 cycles after optimizing the electrolyte. These excellent properties indicate that the FeNi(OH/P) nanosheet array has excellent prospects as an advanced electrode material for HSCs.

Supplementary Materials: The following supporting information can be downloaded at: https://www.mdpi.com/article/10.3390/en15113877/s1. Figure S1: SEM images of (a) FeNi(OH/P)-0, (b) FeNi(OH/P)-0.6, and (c) FeNi(OH/P)-1.8. Figure S2: Comparison of the XRD pattern for FeNi(OH/P) with different amount ofNaH_2PO_2. Figure S3: Comparison of (a) CV curves and (b) rate performance of FeNi LDHs and the FeNi(OH/P)-1.2 electrode. Figure S4: The value of b for FeNi(OH/P) electrodes with different amounts of NaH_2PO_2. Figure S5: (a) SEM images and (b) XRD pattern of the PPy/C. The inset of (a) shows SEM image of the PPy/C under high magnifications. Figure S6: (a) CV curves of the FeNi(OH/P)-1.2 cathode and PPy/C anode performed in 3 M KOH. (b) Comparison of the rate performance capacity for FeNi(OH/P)//PPy/C HSC with different electrolytes.

Author Contributions: Writing—original draft preparation, C.L.; formal analysis and data curation, C.L.; writing—review and editing, R.L.; supervision and funding acquisition, Y.Z. All authors have read and agreed to the published version of the manuscript.

Funding: This research was funded by the National Natural Science Foundation of China (no. 51802233) and the Natural Science Foundation of Hubei Province of China (no. 2018CFB154).

Institutional Review Board Statement: Not applicable.

Informed Consent Statement: Not applicable.

Data Availability Statement: Not applicable.

Conflicts of Interest: The authors declare no conflict of interest.

References

1. Gui, Q.; Ba, D.; Li, L.; Liu, W.; Li, Y.; Liu, J. Recent advances in materials and device technologies for aqueous hybrid supercapacitors. *Sci. China Mater.* **2022**, *65*, 10–31. [CrossRef]
2. Zuo, W.; Li, R.; Zhou, C.; Li, Y.; Xia, J.; Liu, J. Battery-Supercapacitor hybrid devices: Cecent progress and future prospects. *Adv. Sci.* **2017**, *7*, 1600539. [CrossRef] [PubMed]
3. Theerthagiri, J.; Murthy, A.P.; Lee, S.J.; Karuppasamy, K.; Arumugam, S.R.; Yu, Y.; Hanafiah, M.M.; Kim, H.; Mittal, V.; Choi, M.Y. Recent progress on synthetic strategies and applications of transition metal phosphides in energy storage and conversion. *Ceram. Int.* **2020**, *47*, 4404–4425. [CrossRef]
4. Zong, Q.; Liu, C.; Yang, H.; Zhang, Q.; Cao, G. Tailoring nanostructured transition metal phosphides for high-performance hybrid supercapacitors. *Nano Today* **2021**, *38*, 101201. [CrossRef]
5. Oyama, S.T.; Clark, P.; Wang, X.; Shido, T.; Iwasawa, Y.; Hayashi, S.; Ramallo-López, J.M.; Requejo, F.G. Structural characterization of tungsten phosphide (WP) hydrotreating catalysts by X-ray absorption spectroscopy and nuclear magnetic resonance spectroscopy. *J. Phys. Chem. B* **2002**, *106*, 1913–1920. [CrossRef]
6. Agarwal, A.; Sankapal, B.R. Metal phosphides: Topical advances in the design of supercapacitors. *J. Mater. Chem. A* **2021**, *9*, 20241–20276. [CrossRef]
7. Sun, M.; Liu, H.; Qu, J.; Li, J. Earth-Rich transition metal phosphide for energy conversion and storage. *Adv. Energy Mater.* **2016**, *6*, 1600087. [CrossRef]
8. Lu, Y.; Jiang, B.; Fang, L.; Ling, F.; Wu, F.; Hu, B.; Meng, F.; Niu, K.; Lin, F.; Zheng, H. An investigation of ultrathin nickel-iron layered double hydroxide nanosheets grown on nickel foam for high-performance supercapacitor electrodes. *J. Alloys Compd.* **2017**, *714*, 63–70. [CrossRef]
9. Li, M.; Jijie, R.; Barras, A.; Roussel, P.; Szunerits, S.; Boukherroub, R. NiFe layered double hydroxide electrodeposited on Ni foam coated with reduced graphene oxide for high-performance supercapacitors. *Electrochim. Acta* **2019**, *302*, 1–9. [CrossRef]
10. Chen, C.; Dong, C.; Wang, B.; Huang, J.; Wang, Y. Synthesis and room temperature ferromagnetism in Fe-doped $CuAlO_2$ semiconductor. *J. Wuhan Univ. Technol. Mater. Sci. Ed.* **2013**, *28*, 500–503. [CrossRef]

11. Varshney, D.; Kumar, A. Structural and optical properties of Ni substituted $CaCu_3Ti_{4-x}Ni_xO_{12}$. *Opt. Int. J. Light Electron Opt.* **2015**, *126*, 3437–3441. [CrossRef]
12. Verma, S.; Arya, S.; Gupta, V.; Mahajan, S.; Furukawa, H.; Khosla, A. Performance analysis, challenges and future perspectives of nickel based nanostructured electrodes for electrochemical supercapacitors. *J. Mater. Res. Technol.* **2021**, *11*, 564–599. [CrossRef]
13. Ma, L.; Sun, Y.; Guo, J. NiFeP nanocubes as advanced electrode material for hydrogen evolution and supercapacitor. *Colloid Interface Sci. Commun.* **2021**, *45*, 100520. [CrossRef]
14. Xiao, C.; Zhang, B.; Li, D. Partial-sacrificial-template Synthesis of Fe/Ni Phosphides on Ni Foam: A Strongly Stabilized and Efficient Catalyst for Electrochemical Water Splitting. *Electrochim. Acta* **2017**, *242*, 260–267. [CrossRef]
15. Liang, H.; Gandi, A.N.; Xia, C.; Hedhili, M.H.; Anjum, D.; Schwingenschlögl, U.N.; Alshareef, H. Amorphous NiFe-OH/NiFeP electrocatalyst fabricated at low temperature for water oxidation applications. *ACS Energy Lett.* **2017**, *2*, 1035–1042. [CrossRef]
16. Lv, Z.; Wang, K.; Si, Y.; Li, Z.; Jiang, L. High performance of multi-layered alternating Ni-Fe-P and Co-P films for hydrogen evolution. *Green Energy Environ.* **2020**, *7*, 75–85. [CrossRef]
17. Baasanjav, E.; Bandyopadhyay, P.; Saeed, G.; Lim, S.; Jeong, S.M. Dual-ligand modulation approach for improving supercapacitive performance of hierarchical zinc–nickel–iron phosphide nanosheet-based electrode. *J. Ind. Eng. Chem.* **2021**, *99*, 299–308. [CrossRef]
18. Zhang, S.; Yang, Y.; Tang, Y.; Chi, L.; Xu, B. Synthesis and electrochemical properties of high-activity Fe_2O_3@Ni composite electrode. *Inorg. Chem. Ind.* **2019**, *51*, 24–27.
19. Pan, C.; Liu, Z.; Huang, M. 2D iron-doped nickel MOF nanosheets grown on nickel foam for highly efficient oxygen evolution reaction. *Appl. Surf. Sci.* **2020**, *529*, 147201. [CrossRef]
20. Zhang, C.; Ma, B.; Zhou, Y. Three-dimensional polypyrrole derived N-doped carbon nanotube aerogel as a high performance metal free catalyst for oxygen reduction reaction. *ChemCatChem* **2019**, *11*, 5495–5504. [CrossRef]
21. Zhou, K.; Zhou, W.; Yang, L.; Lu, J.; Cheng, S.; Mai, W.; Tang, Z.; Li, L.; Chen, S. Ultrahigh-performance pseudocapacitor electrodes based on transition metal phosphide nanosheets array via phosphorization: A general and effective approach. *Adv. Funct. Mater.* **2015**, *25*, 7530–7538. [CrossRef]
22. Qu, M.; Jiang, Y.; Yang, M.; Liu, S.; Guo, Q.; Shen, W.; Li, M.; He, R. Regulating electron density of NiFe-P nanosheets electrocatalysts by a trifle of Ru for high-efficient overall water splitting. *Appl. Catal. B Environ.* **2020**, *263*, 118324. [CrossRef]
23. Li, J.; Lian, R.; Wang, J.; He, S.; Rui, Z. Oxygen vacancy defects modulated electrocatalytic activity of iron-nickel layered double hydroxide on Ni foam as highly active electrodes for oxygen evolution reaction. *Electrochim. Acta* **2019**, *331*, 135395. [CrossRef]
24. Luo, W.; Dahn, J.R. Preparation of $Co_{1-z}Al_z(OH)_2(NO_3)_z$ Layered Double Hydroxides and $Li(Co_{1-z}Al_z)O_2$. *Chem. Mater.* **2009**, *21*, 56–62. [CrossRef]
25. Wang, Q.; O'Hare, D. Recent advances in the synthesis and application of layered double hydroxide (LDH) nanosheets. *Chem. Rev.* **2012**, *112*, 4124–4155. [CrossRef]
26. Gogotsi, Y.; Penner, R.M. Energy Storage in Nanomaterials—Capacitive, Pseudocapacitive, or Battery-like? *ACS Nano* **2018**, *12*, 2081–2083. [CrossRef]
27. Okhay, O.; Tkach, A. Graphene/Reduced Graphene Oxide-Carbon Nanotubes Composite Electrodes: From Capacitive to Battery-Type Behaviour. *Nanomaterials* **2021**, *11*, 1240. [CrossRef]
28. Ji, J.; Zhang, L.L.; Ji, H.; Li, Y.; Zhao, X.; Bai, X.; Fan, X.; Zhang, F.; Ruoff, R.S. Nanoporous $Ni(OH)_2$ thin film on 3D ultrathin-graphite foam for asymmetric supercapacitor. *Acs Nano* **2013**, *7*, 6237–6243. [CrossRef]
29. Zhao, Y.; Zhao, M.; Ding, X.; Liu, Z.; Qiao, L. One-step colloid fabrication of nickel phosphides nanoplate/nickel foam hybrid electrode for high-performance asymmetric supercapacitors. *Chem. Eng. J.* **2019**, *373*, 1132–1143. [CrossRef]
30. Wan, L.; Wang, Y.; Zhang, Y.; Du, C.; Xie, M. FeCoP nanosheets@Ni-Co carbonate hydroxide nanoneedles as free-standing electrode material for hybrid supercapacitors. *Chem. Eng. J.* **2021**, *415*, 128995. [CrossRef]
31. Wan, L.; Chen, D.; Liu, J.; Zhang, Y.; Du, C. Construction of FeNiP@CoNi-layered double hydroxide hybrid nanosheets on carbon cloth for high energy asymmetric supercapacitors. *J. Power Sources* **2020**, *465*, 228293. [CrossRef]
32. Wan, L.; He, C.; Chen, D.; Liu, J.; Chen, J. In situ grown NiFeP@$NiCo_2S_4$ nanosheet arrays on carbon cloth for asymmetric supercapacitors. *Chem. Eng. J.* **2020**, *399*, 125778. [CrossRef]
33. Shuai, M.; Lin, J.; Wu, W.; Kuang, H.; Zhang, W.; Ling, Q.; Chen, H.; Komarneni, S. Metallic nickel–cobalt phosphide/multilayer graphene composite for high-performance supercapacitors. *New J. Chem.* **2020**, *44*, 8796–8804. [CrossRef]
34. He, S.; Li, Z.; Mi, H.; Ji, C.; Qiu, J. 3D nickel-cobalt phosphide heterostructure for high-performance solid-state hybrid supercapacitors. *J. Power Sources* **2020**, *467*, 228324. [CrossRef]
35. Parveen, N.; Hilal, M.; Han, J.I. Newly design porous/sponge red phosphorus@graphene and highly conductive Ni_2P electrode for asymmetric solid state supercapacitive device with excellent performance. *Nano Micro Lett.* **2020**, *12*, 1–16. [CrossRef] [PubMed]
36. Wang, Z.; Chen, F.; Kannan, P.; Ji, S.; Wang, H. Nickel phosphate nanowires directly grown on Ni foam as binder-free electrode for pseudocapacitors. *Mater. Lett.* **2019**, *257*, 126742. [CrossRef]
37. Zhou, C.; Gao, T.; Wang, Y.; Liu, Q.; Xiao, D. Through a hydrothermal phosphatization method synthesized NiCo and Fe-based electrodes for high-performance battery-supercapacitor hybrid device. *Appl. Surf. Sci.* **2019**, *475*, 729–739. [CrossRef]
38. Quan, Z.; Hui, Y.; Wang, Q.; Zhang, Q.; Shen, Q. $NiCo_2O_4$/NiCoP nanoflake-nanowire arrays: A homogeneous hetero-structure for high performance asymmetric hybrid supercapacitors. *Dalton Trans.* **2018**, *47*, 16320–16328.

39. Yong, Y.; Zhao, H.; Zong, Y.; Li, X.; Sun, Y.; Feng, J.; Wang, Y.; Zheng, X.; Du, Y. Phosphorization boosts the capacitance of mixed metal nanosheet arrays for high performance supercapacitor electrodes. *Nanoscale* **2018**, *10*, 11775–11781.
40. Shi, F.; Xie, D.; Zhong, Y.; Wang, D.H.; Xia, X.H.; Gu, C.D.; Wang, X.L.; Tu, J.P. Facile synthesis of self-supported Ni_2P nanosheet@Ni sponge composite for high-rate battery. *J. Power Sources* **2016**, *328*, 405–412. [CrossRef]
41. Bandgar, S.B.; Vadiyar, M.M.; Jambhale, C.L.; Ye, Z.; Kolekar, S.S. Construction of dual metal ferrite-based core-shell nanostructures as low-cost multimetal electrode for boosting energy density of flexible asymmetric supercapattery. *J. Energy Storage* **2021**, *36*, 102379. [CrossRef]
42. Liu, M.; Guo, T. Preparation and swelling properties of crosslinked sodium polyacrylate. *J. Appl. Polym. Sci.* **2010**, *82*, 1515–1520. [CrossRef]
43. Ji, C.; Liu, F.; Xu, L.; Yang, S. Urchin-like $NiCO_2O_4$ hollow microspheres and $FeSe_2$ micro-snowflakes for flexible solid-state asymmetric supercapacitors. *J. Mater. Chem. A Mater. Energy Sustain.* **2017**, *5*, 5568–5576. [CrossRef]
44. Zhu, Y.; Wu, Z.; Jing, M.; Hou, H.; Ji, X. Porous $NiCo_2O_4$ spheres tuned through carbon quantum dots utilised as advanced materials for an asymmetric supercapacitor. *J. Mater. Chem. A* **2014**, 3, 866–877. [CrossRef]
45. Kong, W.; Lu, C.; Zhang, W.; Pu, J.; Wang, Z. Homogeneous core-shell $NiCo_2S_4$ nanostructures supported on nickel foam for supercapacitors. *J. Mater. Chem. A* **2015**, *3*, 12452–12460. [CrossRef]
46. Cao, Z.; Liu, C.; Huang, Y.; Gao, Y.; Wang, Y.; Li, Z.; Yan, Y.; Zhang, M. Oxygen-vacancy-rich $NiCo_2O_4$ nanoneedles electrode with poor crystallinity for high energy density all-solid-state symmetric supercapacitors. *J. Power Sources* **2020**, *449*, 227571. [CrossRef]
47. Zhao, X.; Zhu, J.; Lu, Y.; Zhang, Z.; Hong, Y.; Wang, W.; Karimov, K.; Murtaza, I.; Wang, Q.; Dong, X. Three-dimensional Co–S–P nanoflowers as highly stable electrode materials for asymmetric supercapacitors. *ACS Sustain. Chem. Eng.* **2019**, *7*, 11448–11454. [CrossRef]
48. Wang, K.; Yan, R.; Tian, X.; Wang, Y.; Lei, S.; Li, X.; Yang, T.; Wang, X.; Song, Y.; Liu, Y.; et al. Multi-scale biomass-based carbon microtubes decorated with Ni-Co sulphides nanoparticles for supercapacitors with high rate performance. *Electrochim. Acta* **2019**, *302*, 78–91. [CrossRef]
49. Zhao, C.; Shao, X.; Zhang, Y.; Qian, X. Fe_2O_3/RGO/Fe_3O_4 composite in-situ grown on Fe foil for high performance supercapacitors. *ACS Appl. Mater. Interfaces* **2016**, *8*, 30133–30142. [CrossRef]

energies

MDPI

Article

Energy Consumption Estimation for Electric Buses Based on a Physical and Data-Driven Fusion Model

Xiaoyu Li [1,2], Tengyuan Wang [1], Jiaxu Li [1], Yong Tian [1] and Jindong Tian [1,3,*]

1 College of Physics and Optoelectronic Engineering, Shenzhen University, Shenzhen 518060, China; xiaoyu070220202@126.com (X.L.); 2070456016@email.szu.edu.cn (T.W.); lijiaxu188988@126.com (J.L.); ytian@szu.edu.cn (Y.T.)
2 National Engineering Laboratory for Electric Vehicles, School of Mechanical Engineering, Beijing Institute of Technology, Beijing 100081, China
3 Guangdong Laboratory of Artificial Intelligence and Digital Economy (SZ), Shenzhen University, Shenzhen 518060, China
* Correspondence: jindt@szu.edu.cn

Abstract: The energy consumption of electric vehicles is closely related to the problems of charging station planning and vehicle route optimization. However, due to various factors, such as vehicle performance, driving habits and environmental conditions, it is difficult to estimate vehicle energy consumption accurately. In this work, a physical and data-driven fusion model was designed for electric bus energy consumption estimation. The basic energy consumption of the electric bus was modeled by a simplified physical model. The effects of rolling drag, brake consumption and air-conditioning consumption are considered in the model. Taking into account the fluctuation in energy consumption caused by multiple factors, a CatBoost decision tree model was constructed. Finally, a fusion model was built. Based on the analysis of electric bus data on the big data platform, the performance of the energy consumption model was verified. The results show that the model has high accuracy with an average relative error of 6.1%. The fusion model provides a powerful tool for the optimization of the energy consumption of electric buses, vehicle scheduling and the rational layout of charging facilities.

Keywords: electric bus; energy consumption; physical model; CatBoost; fusion model

Citation: Li, X.; Wang, T.; Li, J.; Tian, Y.; Tian, J. Energy Consumption Estimation for Electric Buses Based on a Physical and Data-Driven Fusion Model. *Energies* **2022**, *15*, 4160. https://doi.org/10.3390/en15114160

Academic Editors: Alon Kuperman, Alessandro Lampasi and Michele Pastorelli

Received: 18 March 2022
Accepted: 2 June 2022
Published: 6 June 2022

1. Introduction

With energy shortages and environmental pollution problems becoming more pronounced, the global energy structure is gradually undergoing a transformation. Countries around the world are taking steps to achieve sustainable, green and efficient energy systems. New energy vehicles, especially electric vehicles, are gaining widespread attention due to their low pollution and high energy efficiency. By the end of 2021, the number of new energy vehicles worldwide had exceeded 10 million. In the Chinese urban bus system, many diesel buses have been replaced by energy-efficient and environmentally friendly electric buses [1]. Compared to traditional diesel buses, electric buses have more advantages in the public transport system. However, there are still some problems, such as difficulties in charging demand evaluation, vehicle route planning and battery energy storage system design. These issues are closely related to the range of electric buses and the energy consumption of vehicles under specific operating conditions [2]. Therefore, the study of an accurate vehicle energy consumption estimation model can solve the above problems, which is of great significance to the popularization of electric buses. The driving energy consumption of electric buses is affected by drivers' habits and working conditions. For electric buses on the same route, the difference in driving energy consumption under different working conditions can reach 40%. Therefore, information such as the temperature and departure time under current operating conditions need to be considered so as to accurately estimate the energy consumption of the vehicle.

The energy consumption of an electric vehicle is related to the vehicle powertrain dynamic performance, such as the windproof area of the vehicle, vehicle charging efficiency, etc. It is also influenced by the habits of drivers and environmental factors. As a result, it is difficult to achieve an accurate estimation of the energy consumption of a vehicle [3]. However, the national monitoring and management platform for new energy vehicles (NEVS) in China has enabled the aggregation of massive amounts of data on new energy vehicles. The platform can provide a large amount of vehicle driving data for the analysis and modeling of vehicle energy consumption.

Many related studies have been performed in this area. Yuan et al. [4] modeled vehicle powertrain dynamics by simulating driving data on a computer. An energy consumption model for electric vehicles was achieved. In the model, the energy of an electric vehicle is mainly consumed by rolling drag, air resistance and kinetic energy, and the error of the model is about 3%. However, the model only considers energy consumption under laboratory conditions and does not further consider possible congestion and air-conditioning factors in the actual driving process. Bracco et al. [5] used a simulation model to analyze the effects of different variables on energy consumption and the battery charging state. The results show that the number of passengers has the greatest impact on the energy consumption of electric vehicles. Qi et al. [6] used positive kinetic energy and negative kinetic energy to decompose the energy consumption under actual traffic congestion. Based on this decomposition, a data-driven model was established. The model is used to estimate the energy consumption of electric vehicles on the road, and the actual traffic conditions are taken into account. The model needs a lot of data input to obtain the accurate positive and negative kinetic energy of the vehicle. After simplifying the input to the average speed, the model error is 7%. Hao et al. [7] analyzed the energy consumption of electric buses, minibuses and taxis in Beijing. Through statistical analysis, vehicle energy consumption in different seasons and driving conditions was obtained. It was found that the energy consumption of electric vehicles per kilometer is lower than that at 5 °C. This shows that the energy transmission efficiency of the battery will change at different temperatures. Miraftabzadeh et al. [8] considered the driving route and weather conditions in the modeling process. Using a data-driven modeling method, the energy consumption prediction model of an electric taxi was established. In addition, by calculating the energy consumption of taxis on weekdays and weekends, the author compiled a taxi energy consumption table. The table shows that the month with the highest energy consumption of taxis in New York City is April, while July is the month with the lowest energy consumption. Al-Wreikat et al. [9] analyzed the effect of ambient temperature on the energy consumption of electric vehicles. They found that vehicles consumed 28% more energy at low temperatures of 0–15 degrees than at medium temperatures of 15–25 degrees. Björnsson et al. [10] designed a physical model of powertrain dynamics. The energy recovery performance in the braking process was analyzed. The research shows that under urban conditions, the energy regeneration potential per kilometer is higher under the condition of low average speed and multiple starts and stops. It lays a foundation for the study of the urban bus recovery coefficient. The energy consumption of electric vehicles has been modeled by many approaches.

According to the literature review, previous studies have considered physical modeling methods or artificial intelligence algorithms to obtain vehicle energy consumption models. However, vehicle energy consumption is the result of multiple factors. The energy consumption estimation results obtained by a single method or a single type of model are less reliable. Moreover, in practical applications, there is a lack of an energy consumption estimation model with limited input features for driving decision making. It is necessary to design a more reliable energy consumption estimation model with few input features. To address these issues, a physical and data-driven fusion model for vehicle energy consumption is proposed in this paper. Part of the energy consumed by the vehicle during driving can be expressed by physical formulas, such as the energy consumed by rolling resistance and air resistance [11]. The direct application of formula modeling will reduce the complexity of the model. Other driving factors also affect the driving energy consumption of vehicles, such as the vehicle departure time, the ambient temperature, etc. For these factors that cannot be expressed by the formula, the data-driven model is selected. Finally, the two models are fused to estimate vehicle energy consumption.

The content of each section is as follows: In Section 2, the statistical analysis of electric bus data is performed. The original data are preprocessed and reconstructed to obtain continuous data in the vehicle charging and driving cycle process. In Section 3, the energy consumption estimation model is designed. A physical vehicle energy consumption model is developed based on the powertrain dynamic performance of the electric bus. Model parameters are initially calibrated using the least-squares method. In addition, the factors affecting fluctuations in vehicle energy consumption, such as driving habits and environmental factors, are summarized and analyzed. The CatBoost decision tree model is used to characterize the effects. Finally, the two models are fused to obtain the final estimation result of vehicle energy consumption. In Section 4, the energy consumption estimation model is analyzed and validated. Some conclusions are presented in Section 5.

2. Data Statistics and Analysis

The electric bus data came from the National Monitoring and Management Platform for NEVS. The original data were collected by on-board terminals on electric buses and uploaded to the data platform. The dataset includes 38 items, such as the sampling time, battery management system (BMS) number, battery pack voltage, battery current, state of charge (SOC), minimum cell voltage, maximum cell voltage, minimum temperature, maximum temperature, etc. A detailed description of the items used in this paper is shown in Table 1. The data cover ten electric buses on the same bus route in one year. The travel distance is approximately 34 km. There are 24 stops along the route. After one round trip, the buses will be charged at the starting point of the bus stations. The purpose of this section is to process the original data and obtain the data on vehicle speed, data acquisition time, temperature, accelerator pedal value and deceleration pedal value. The data will be further processed for vehicle energy consumption modeling.

Table 1. A detailed description of the items used in this paper.

Item	Type	Explain
VIN	STRING	Vehicle identification number
State of charge	INT	Parking charging, driving charging, not charging, charging completed, abnormal, invalid
Speed	FLOAT	0~220 (km/h), abnormal
Cumulative travel	FLOAT	0~999,999.9 (km), abnormal
Total voltage	FLOAT	0~1000 (V), abnormal
Total current	FLOAT	−1000~1000 (A), abnormal
SOC	INT	0~100(%), abnormal
Temperature	INT	−40~210 (°C), abnormal
Accelerator pedal	INT	0~100, abnormal
Deceleration pedal	INT	0~100, abnormal
Data acquisition time	STRING	Time of each data acquisition

Data Processing and Analysis

The quality of the original electric bus data on the platform was not flawless. The acquired data often appeared to have outliers and missing values due to the influence of electromagnetic radiation and the unreliability of the circuit system. As a result, preprocessing of the original data was needed. During the process, data interpolation, outlier removal and data segmentation were performed.

The two main types of missing data are missing multiple rows and missing single features. For the first case, the data exhibit a discontinuity in specific intervals. Missing data were interpolated using the mean value [12]. For the latter case, the Lagrangian interpolation method was used to interpolate the data. Considering the outliers, the first quartile and the third quartile of the data were calculated by constructing a box plot. Values exceeding the upper and lower edges of the box plot were defined as outliers.

To facilitate the extraction of data features, the data were divided into short segments according to the state of charge in the dataset. The electric bus data transmission process and data preprocessing results are shown in Figure 1.

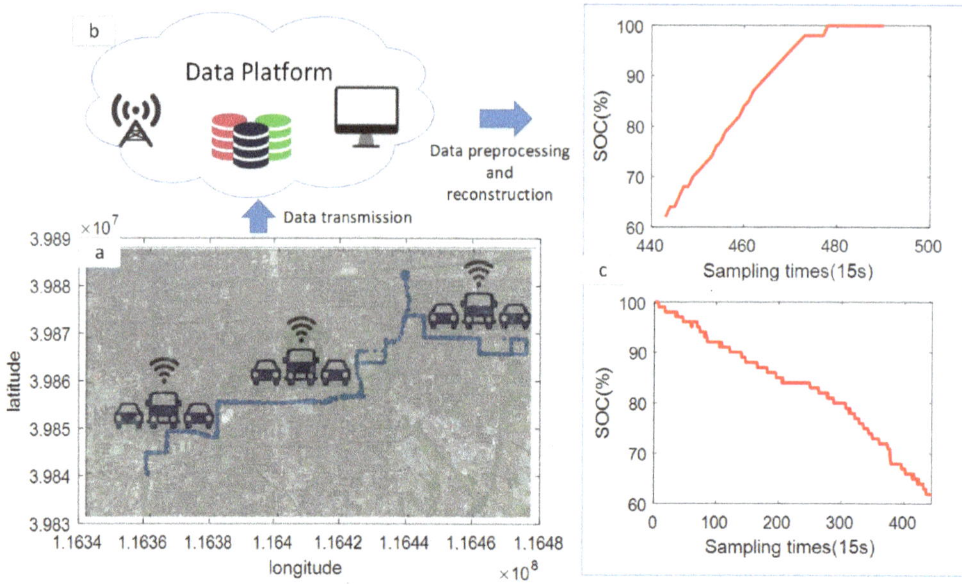

Figure 1. Electric bus operation data transmission and preprocessing results. (**a**) Driving route, (**b**) electric vehicle big data cloud platform and (**c**) preprocessing results of battery SOC.

After preprocessing the original data, the data were reconstructed according to the timestamp, vehicle velocity, etc. The vehicle driving season, departure time and velocity-related features were obtained. The statistics of the vehicle energy consumption under different operating conditions are shown in Figure 2. In the figure, the heading of the ordinate is the energy consumption. For brevity, it is abbreviated as EC. For electric vehicles, the operating temperature is related to the use of air conditioning, energy efficiency, etc. Figure 2a shows the relationship between energy consumption and operating temperature. It can be seen that the relationship can be approximated by a parabolic function. The minimum energy consumption occurs at a temperature of approximately 25 degrees. This phenomenon coincides with the fact that air conditioning and high energy transfer efficiency are rarely used [6]. Figure 2b shows the relationship between the variance of velocity and energy consumption. It can be seen that there is an approximately linear relationship between them. The higher the variance, the higher the energy consumption. As shown in Figure 2c, there is no obvious linear relationship between average speed and energy consumption, and this part of the analysis is described in detail later. Figure 2d shows the relationship between departure time and energy consumption. Departure times are related to road congestion and vehicle passenger weight. Specifically, vehicles consume more energy between 6 a.m. and 8 p.m. Vehicle energy consumption values differ by more than 20%.

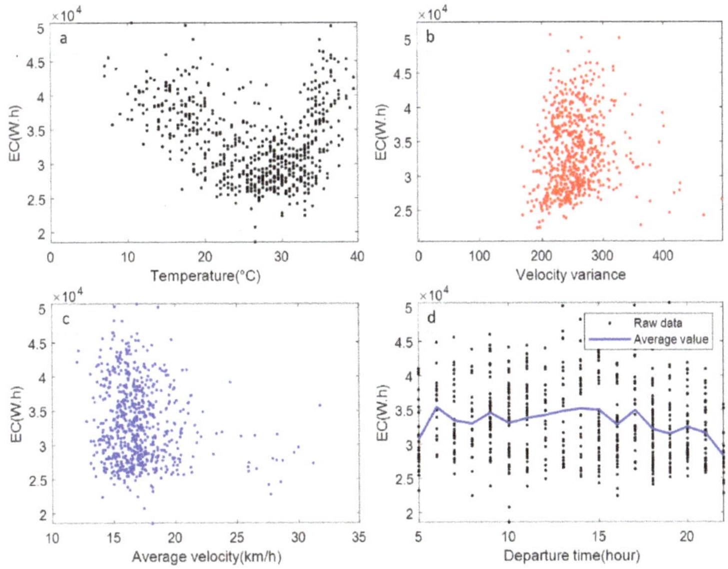

Figure 2. Statistical results of vehicle energy consumption after reconstruction. The relationships of (**a**) temperature, (**b**) vehicle velocity variance, (**c**) average velocity and (**d**) departure time with vehicle energy consumption.

3. Vehicle Energy Consumption Modeling

According to the powertrain dynamics of the vehicle, the energy consumption of the vehicle is mainly influenced by air resistance, rolling drag and kinetic energy changes during the driving process. Additionally, the energy consumption of the air-conditioning system in the electric bus should be taken into account [13]. The influence of these factors can be modeled with physics-based functions. However, the influences of driving habits and environmental factors are somewhat random and cannot be directly described by physical modeling. Therefore, the fluctuating energy consumption resulting from these factors is more suitable to be modeled by data-driven approaches. Data-driven modeling methods such as decision trees, support vector machines and neural networks are commonly used in many fields. The approaches have a good ability to solve complex, non-linear problems [14]. Based on the analysis, a fusion of the physical modeling and the data-driven modeling is proposed in this paper to achieve an accurate estimation of vehicle energy consumption.

3.1. The Physical Energy Consumption Model

As the energy source of the electric vehicle is attributed to charging stations, charging energy is regarded as the original energy of the vehicle in this paper. The charging efficiency of the battery pack is represented by η_{ch}. Due to the existence of the battery internal resistance, the value of the efficiency η_{ch} is less than 1. Additionally, the parameter fluctuates with the operating temperature. During the vehicle driving process, the discharging efficiency of the battery pack is also influenced by the battery internal resistance. Depending on the powertrain dynamics of the vehicle, the chemical energy of the battery pack is converted into electrical energy, which is further converted into mechanical energy to drive the vehicle. Meanwhile, vehicle energy is also consumed by the air-conditioning system. Therefore, vehicle energy consumption E can be expressed by:

$$E = \frac{1}{\eta_{ch}\eta_{mot}\eta_{bat}}(E_{roll} + E_{air} + (1-\eta_{re})E_{bra}) + \frac{(1-\eta_{re})}{\eta_{ch}\eta_{bat}}E_{ac} \tag{1}$$

As shown in Equation (1), the energy consumption consists of four main components: energy consumption from rolling drag E_{roll}, energy consumption from air resistance E_{air}, braking consumption E_{bra} and energy consumption from air conditioning E_{ac}. Energy transmission is also accompanied by motor efficiency η_{mot}. Energy recovery efficiency η_{re} shows that the change in kinetic energy during braking will reverse-charge the vehicle. Assuming that the energy recovery coefficient is η_{re}, the kinetic energy consumption lost during braking should be $(1-\eta_{re})E_{bra}$. Battery discharge efficiency is η_{bat}.

The energy consumption from rolling drag E_{roll} is influenced by the vehicle mass, velocity and other factors, and the equation is expressed as:

$$E_{roll}(i) = mgfv(i)t(i) \tag{2}$$

where m is the vehicle mass, g is the gravitational acceleration, f is the rolling drag coefficient, $v(i)$ is the speed at that time, and $t(i)$ is the sampling interval.

Considering the practicability of the energy consumption model, the number of model inputs should be as small as possible. Therefore, the parameters in Equation (1) need to be simplified and approximated. Herein, the velocity $v(i)$ is approximated as the average velocity, which is simplified as:

$$E_{roll} = mgf\overline{v}t_{total} \tag{3}$$

where t_{total} is the total time of travel, and $\overline{v}$ is the average velocity.

Energy consumption from air resistance is influenced by the vehicle velocity and windproof area of the vehicle. The energy consumption can be expressed by:

$$E_{air} = \rho CA\overline{v}^3t_{total} \tag{4}$$

where ρ is the air density, C is the air resistance coefficient, and A is the windproof area of the vehicle. However, according to the literature [8], there is a negative correlation between vehicle speed and energy consumption when the vehicle speed is lower than 45 km/h. With the increase in vehicle speed, the energy consumption should decrease slightly, which is inconsistent with Equation (4). As the average velocity of the electric bus used in this paper is very low, the energy consumed by air resistance was ignored in the energy consumption modeling process.

For the kinetic energy consumption of the vehicle, as the initial and end velocities of the vehicle in a driving cycle are both zero, it can be concluded that the deceleration kinetic energy and acceleration kinetic energy are roughly equal. As a result, an energy efficiency coefficient was added to the kinetic energy consumption to characterize the energy recovery performance during the driving cycle. Vehicle kinetic energy consumption can be expressed by Equation (5). In practical application scenarios, the velocity at each time point is unknown. For simplification, the variance of velocity is correlated with the change in kinetic energy, so the variance of velocity is used to replace the change in kinetic energy.

$$E_{bra} = \sum_{i=1}^{end} 0.5m\overline{v}(v(i+1) - v(i)) \approx 0.5m\mathrm{var}(v) \tag{5}$$

Due to the existence of a large passenger space, the energy consumption of air conditioning should be accounted for. According to the literature [15], the energy consumption of air conditioning during driving is directly proportional to the square of the temperature difference inside and outside the vehicle. Therefore, the energy consumption can be expressed as:

$$E_{ac} = c(T-25)^2 \tag{6}$$

where c is the air-conditioning coefficient, and T is the temperature.

Substituting Equations (3), (5) and (6) into Equation (1), the physical model can be obtained:

$$E = \beta_0 F_0 + \beta_1 F_1 + \beta_2 F_2 \\ \begin{cases} \beta_0 = \frac{mgft}{\eta_{ch}\eta_{mot}\eta_{bat}} \\ \beta_1 = \frac{0.5m(1-\eta_{re})}{\eta_{ch}\eta_{mot}\eta_{bat}} \\ \beta_2 = \frac{c}{\eta_{ch}\eta_{mot}\eta_{bat}} \end{cases} \quad (7)$$

where $F_0 = t_{total}\sum_{i=1}^{n}\overline{v}$, $F_1 = \mathrm{var}(v)$, and $F_2 = (T-25)^2$.

In the physical energy consumption model, the parameters related to the energy transmission efficiency are influenced by environmental factors. However, the model can be simplified by considering all energy efficiencies as fixed values. $\beta_0, \beta_1, \beta_2$ are constants and can be obtained using the least-squares fitting method based on statistical vehicle data [4].

3.2. The Data-Driven Energy Consumption Model

3.2.1. Analysis of Influencing Factors

The factors that cause energy consumption fluctuations can be summarized into three aspects, including driving habits, environmental factors and vehicle performance [16]. In terms of driving habits, vehicle velocity, acceleration and deceleration conditions can cause fluctuations in vehicle energy consumption. The conditions can be quantified accordingly as the average vehicle velocity, vehicle velocity variance and number of accelerator pedal presses. In terms of the environmental factors, temperature, which is related to the energy transfer efficiency and air-conditioning usage, is the main factor causing fluctuations in vehicle energy consumption. In addition, the road conditions can also cause fluctuations in energy consumption, such as whether the departure time is congested and whether the departure date is on the weekend [17]. In terms of vehicle performance, the energy efficiency of the battery storage system during the charging and discharging process can also cause energy consumption fluctuations. The energy efficiency is mainly affected by the internal resistance of the battery system and is directly related to the ambient temperature and battery aging. Based on the analysis, the statistical results of the impact of various energy consumption fluctuation factors on vehicle energy consumption are shown in Figure 3. In the figure, it should be noted that the fitting curves were obtained by fitting experimental data with polynomial functions.

It can be seen in Figure 3a,c,d that there are positive correlations between the velocity variance, acceleration pedal statistical parameter, deceleration pedal statistical parameter and vehicle energy consumption. In Figure 3b, there is no clear correlation between the average velocity and the change in vehicle energy consumption. According to a study in the literature [8], a negative correlation between vehicle speed and energy consumption is found when the vehicle speed is below 45 km/h. Due to the large number of stopping and idling situations during the driving of electric buses, when the speed is low, the energy consumption at low speed is greater than that at high speed. Figure 3e shows that the effect of temperature on vehicle energy consumption is relatively large, and the relationship can be approximated by a quadratic function. In Figure 3f,g, the difference in departure time and departure date affects vehicle energy consumption; however, at the same point in time, vehicle energy consumption fluctuates greatly. In Figure 3h, a positive correlation is shown between the internal resistance of the battery system and the energy consumption of the vehicle. Through further analysis of the internal resistance, it was found that there is a strong correlation between the internal resistance and the battery temperature. During the driving cycle, the battery internal resistance series data have great fluctuations. As a result, the internal resistance is not taken as an input feature.

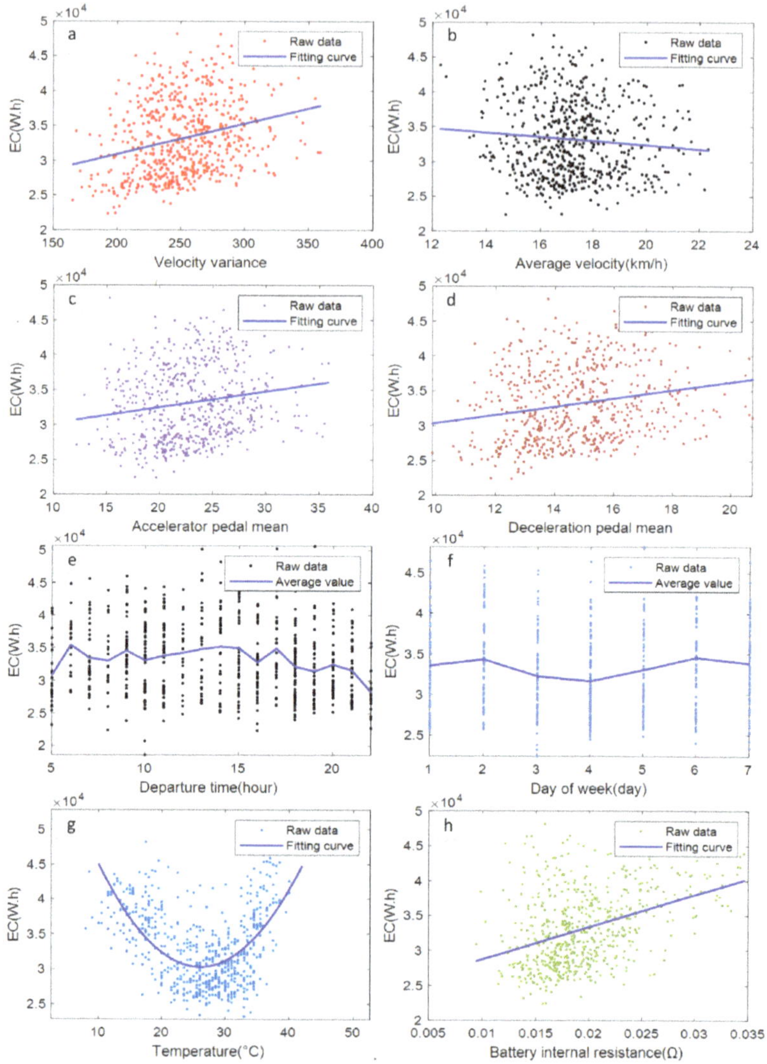

Figure 3. The statistical results of the impact of various energy consumption fluctuation factors on vehicle energy consumption. (**a–h**) The relationships between the velocity variance, average velocity, number of accelerator pedal presses, number of deceleration pedal presses, departure time, departure date, ambient temperature, internal resistance of the battery pack and energy consumption, respectively.

Based on the statistical analysis of the fluctuation factors of vehicle energy consumption, the main influencing features of vehicle energy consumption fluctuation are: velocity variance, average velocity, accelerator pedal parameter, deceleration pedal parameter, temperature and battery internal resistance. Considering that the internal resistance of the battery is mainly affected by the ambient temperature, the influencing features, except the internal resistance of the battery, are regarded as input features in the data-driven model.

3.2.2. Principle of CatBoost Modeling

In this paper, the CatBoost modeling approach is used to model fluctuations in vehicle energy consumption. CatBoost is an improvement of the gradient boosting decision tree

(GBDT) model [18]. The approach has the ability to improve the estimation accuracy with weak learners. Moreover, it has significant advantages in extracting important features and processing categorical features. In addition, the problem of poor model accuracy and overfitting can be avoided when the dataset is uneven. The main principle of this method is to construct many weak learners for training. The weights of the training samples are adjusted to focus on samples with large estimation errors and train the weak learners in turn. Finally, the weak learners are combined into a stronger learner model [19]. In the following content, the gradient boosting decision tree algorithm is introduced. Then, the optimization strategy of the CatBoost modeling approach is given. On the basis of the modeling approach, the vehicle's fluctuating energy consumption results can be obtained.

1. Gradient boosting decision tree

Gradient boosting decision tree is an iterative decision tree algorithm. The algorithm is composed of multiple decision trees, and the results of all trees are accumulated to obtain the final result [20]. Given a training dataset $D = \{(\mathbf{x}_i, y_i)\}_{i=1}^{n}$, $\mathbf{x}$ is the characteristic affecting energy consumption, and y is the predicted energy consumption of output. The goal of GBDT is to find a function $\hat{F}(\mathbf{x})$ that minimizes the given loss function $L(y, \hat{F}(\mathbf{x}))$. $\hat{F}(\mathbf{x})$ is accumulated by a series of decision trees $F(\mathbf{x})$. Each decision tree $F(\mathbf{x})$ is optimized as:

$$F_m(\mathbf{x}) = F_{m-1}(\mathbf{x}) + \rho_m h_m(\mathbf{x}) \tag{8}$$

where $h(\mathbf{x})$ is the decision tree function. ρ_m is the weight of the mth decision tree function $h_m(\mathbf{x})$. The initial value of $F(\mathbf{x})$ can be obtained by:

$$F_0(\mathbf{x}) = \underset{\alpha}{\operatorname{argmin}} \sum_{i=1}^{N} L(y_i, \alpha) \tag{9}$$

Subsequently, the optimization process of the model is achieved by minimizing the loss functions:

$$(\rho_m, h_m(\mathbf{x})) = \underset{\rho,h}{\operatorname{argmin}} \sum_{i=1}^{N} L(y_i, F_{m-1}(\mathbf{x}_i) + \rho h(\mathbf{x}_i)) \tag{10}$$

The gradient descent method is used to solve the above optimization problems. For each model $F(\mathbf{x})$, a new dataset $D = \{\mathbf{x}_i, r_{mi}\}_{i=1}^{N}$ is constructed and trained to obtain $h_m(\mathbf{x})$. r_{mi} can be obtained by:

$$r_{mi} = [\frac{\partial L(y_i, F(\mathbf{x}))}{\partial F(\mathbf{x})}]_{F(\mathbf{x})=F_{m-1}(\mathbf{x})} \tag{11}$$

The value of ρ_m is subsequently computed by solving a line search optimization problem. Its training process is shown in Figure 4.

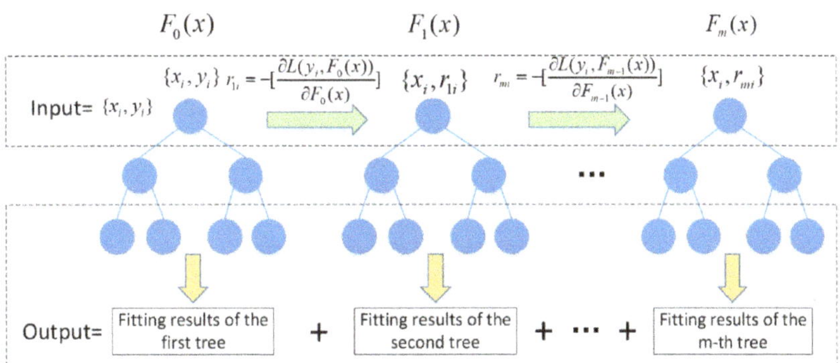

Figure 4. The training process of the gradient boosting decision tree modeling approach.

2. The CatBoost modeling approach

CatBoost is a kind of gradient-enhanced decision tree algorithm, which can handle category features well [21]. The variables extracted in this paper have certain category features. Therefore, CatBoost was selected for energy consumption modeling. This method differs from GBDT in the following ways [22]:

(1) CatBoost can process features during training [23]. First, the sample data are randomly sorted to generate multiple groups of random sequences. Then, for each random sequence, the average value of the same sample is calculated. When the sequence is $\Theta = [\sigma_1, \ldots, \sigma_n]_n^T$, it can be calculated by:

$$x_{\sigma_{p,k}} = \frac{\sum_{j=1}^{p-1} \left[x_{\sigma_{j,k}} = x_{\sigma_{p,k}}\right] \cdot y_{\sigma_j} + \beta \cdot P}{\sum_{j=1}^{p-1} \left[x_{\sigma_{j,k}} = x_{\sigma_{p,k}}\right] \cdot y_{\sigma_j} + \beta} \tag{12}$$

where P is an a priori value. For regression tasks, the prior value is the average value in the label. β is the weight of P.

(2) Feature combination. The numerical features calculated by Equation (12) may lose some information. Combining features can solve this problem and produce a more effective feature. CatBoost uses a greedy approach to consider feature combinations. The first segmentation does not consider the combination of category features, and the subsequent segmentation considers all feature combinations. CatBoost takes both groups of values after segmentation as category features to participate in the following combination.

In the previous sections, the fluctuation factors of vehicle energy consumption are analyzed. Seven features can be obtained that are related to vehicle energy consumption. Based on a full understanding of the factors and the CatBoost modeling approach, features such as average vehicle velocity, vehicle velocity variance, number of accelerator pedal presses, number of brake pedal presses, departure time, day of the week and temperature, mentioned in Section 3.2.1, are taken as the input features of the CatBoost decision tree model. The statistical range of input vehicle features is a round trip of the vehicle. After model parameter optimization, the data-driven model of energy consumption can be obtained.

3.3. *A Fusion of Physical and Data-Driven Models*

After physical and data-driven modeling of the basic energy consumption and fluctuating energy consumption of electric buses, the two parts needed to be fused to obtain a vehicle energy consumption model. In this study, the integrated learning approach in machine learning theory was used for model fusion. The reconstructed electric bus data were used to train the physical energy consumption model. The residual of the basic model was retrained as the training label of the data-driven model [24]. The flow chart of the energy consumption fusion modeling approach is shown in Figure 5.

The modeling approach can be divided into three steps:

(1) Data processing. In this process, the original data are interpolated. For different missing data types, the methods of average interpolation and Lagrange interpolation are adopted. For outliers in the data, the method of constructing quartile positions with a box plot is used to remove them. Then, the data are segmented according to the state of charge. After the specific driving segments are divided, data such as vehicle speed, data acquisition time, temperature, accelerator pedal value and deceleration pedal value can be obtained. The data are further processed for energy consumption modeling.

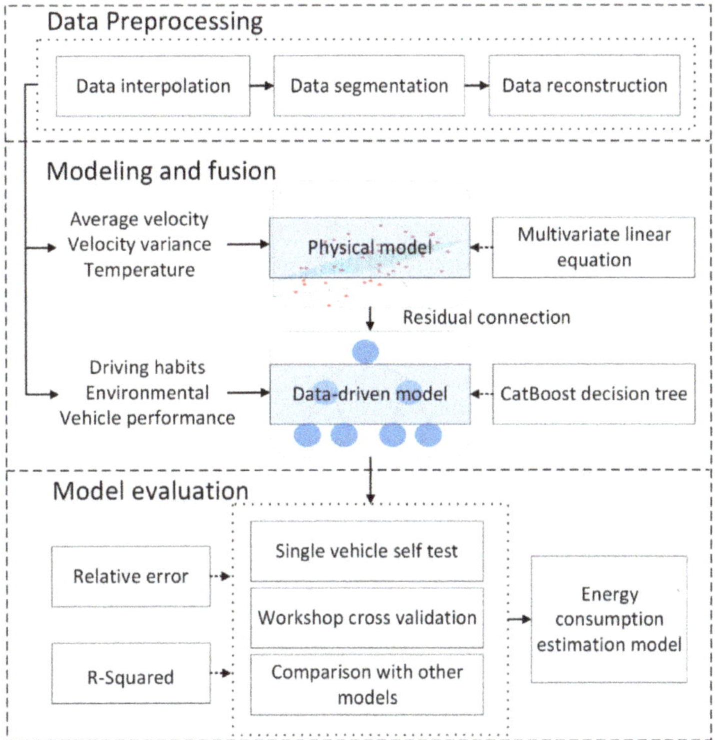

Figure 5. Flow chart of the energy consumption fusion modeling approach.

(2) Modeling and fusion. The original features obtained from the data processing step include vehicle speed, data acquisition time, temperature, accelerator pedal value and deceleration pedal value. These features need to be processed separately and input into the model. For the physical model, the vehicle driving distance, speed variance and the square of the difference between the temperature in the vehicle and the standard temperature are calculated as inputs. For the data-driven model, the departure time of the vehicle, whether it is a weekend, the temperature in the vehicle, the value of acceleration and deceleration pedal, the average speed and the speed variance are extracted and input into the CatBoost model. In engineering applications, many specific data in vehicle operation are unknown. Therefore, the input of the model needs to meet the following conditions: (1) The input parameters of the model can be obtained before the vehicle is driven. (2) The input parameters of the model need to include parameters that reflect the working condition information. This paper simplifies the input parameters according to this criterion and obtains the following input parameters: the mileage of the current route, the average speed, the speed variance, the temperature, the air-conditioning condition, the departure time, the departure day of the week and the average values of the accelerator pedal and the deceleration pedal. These parameters can be planned before driving. The aim of the fusion step of the model is to train the physical model to obtain the preliminary estimation results of energy consumption. Then, the residual of the physical model is retrained as the training label of the data-driven model to minimize the residual. The final energy consumption result is the sum of the results of the two models.

(3) Model evaluation. Two indicators are selected for the verification of model results, namely, the average relative error and the R-squared parameter. The verification is divided into a single vehicle division training set and a test set for verification. In order to test the robustness of the model, different vehicles are selected for verification.

4. Results and Discussions

4.1. Analysis of the Results of the Physical Model

The physical model to obtain the basic energy consumption was analyzed. In this work, data provided by the new energy vehicle big data platform were used for model training. The parameters of the physics-based basic energy consumption model of six electric buses were estimated. The results are shown in Table 2. Figure 6 shows the basic energy consumption estimation results of two electric buses (Bus 1 and Bus 2). As a large number of data points can cause the bar chart to be too small, only 20 points in Figure 6a,b were used to draw Figure 6c–f.

Table 2. Parameters of the physical model.

Vehicle Number	β_0	β_1	β_2
Bus 1	884.5	14.2	47.1
Bus 2	869.3	11.3	64.3
Bus 3	810.0	11.7	71.1
Bus 4	855.7	11.3	76.2
Bus 5	775.4	16.6	59.5
Bus 6	866.3	2.3	88.5
Average	843.6	11.2	67.7

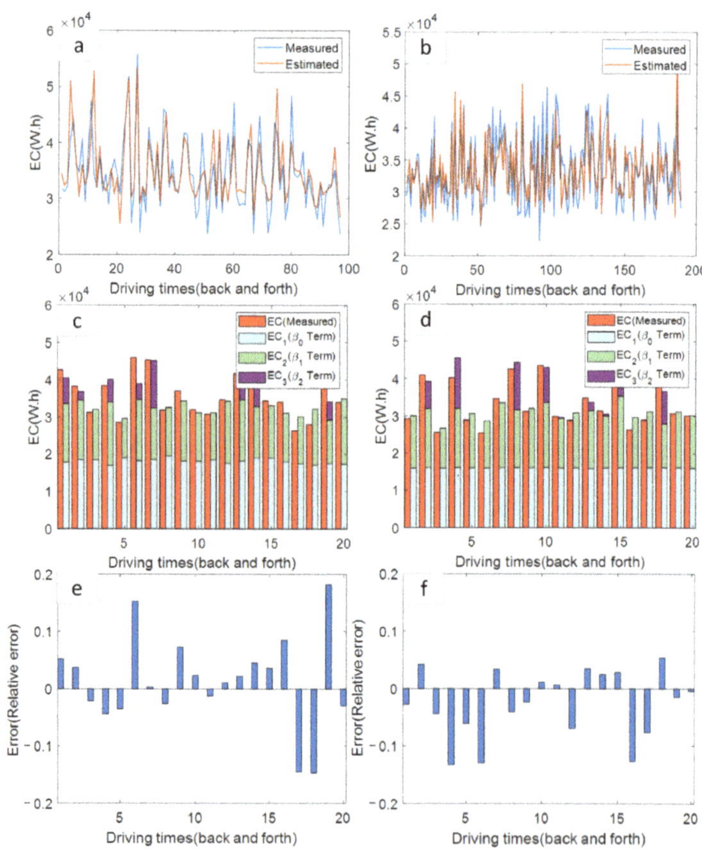

Figure 6. The basic energy consumption estimation results. (**a**,**b**) The fitting curves of two electric buses. (**c**,**d**) The energy consumption fitting results for each part. (**e**,**f**) The fitting errors.

It can be seen that the vehicle's rolling drag energy consumption coefficient is the largest. This shows that the main energy consumption of the vehicle during driving is consumed by the rolling drag. In the figure, Figure 6a,b compare the fitted values of the model with the real values of vehicle energy consumption. In the figure, the x-axis represents the number of vehicle round trips, and the y-axis represents the energy consumption. The model-fitting results are close to the values in terms of vehicle energy consumption data. Figure 6c,d show the proportion of the energy consumption of each component in the total energy consumption as a histogram. It is obvious that rolling drag and kinetic energy change dominate the energy consumption. Since the use of an air conditioner is closely related to the temperature difference between inside and outside the vehicle, the energy consumption data of the air conditioner fluctuate greatly. The errors of the basic energy consumption model are shown in Figure 6e,f. It can be seen that the average error of the estimation results is 7%. Since the model does not consider the influence of energy consumption fluctuations, there are large errors in the estimation of vehicle energy consumption.

4.2. Analysis of the Results of the Fusion Model

The energy consumption estimation results obtained from the physical energy consumption model only take into account energy consumption in ideal conditions. The model does not account for the influence of driving habits and environmental factors. In this context, the physical model and the data-driven model are fused to estimate vehicle energy consumption. The results of the vehicle energy consumption estimation of the fusion model are shown in Figure 7.

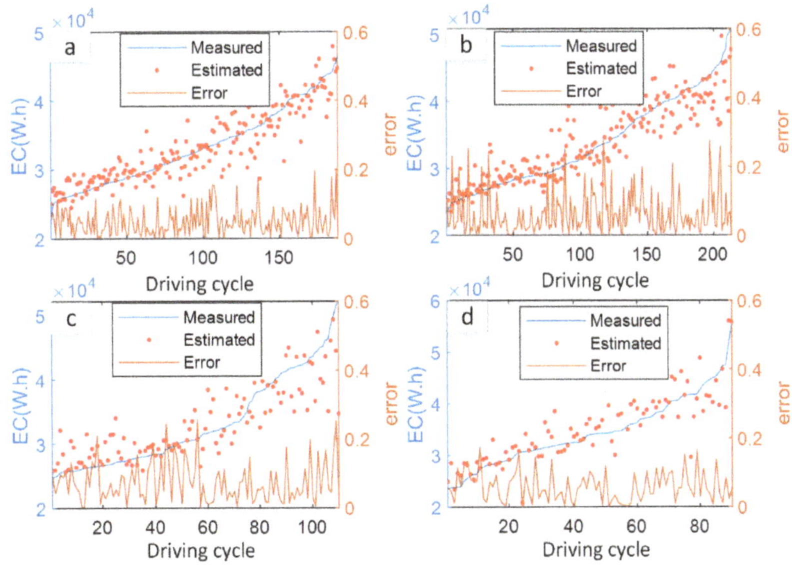

Figure 7. Energy consumption estimation results with the fusion model. (**a–d**) Bus 1 to Bus 4.

In Figure 7, the blue line represents the vehicle energy consumption obtained using the platform data, which can be considered a reference value for the vehicle energy consumption. The red points are the vehicle energy consumption obtained by the fusion model. It is clear that the estimation results of vehicle energy consumption are able to track changes in the real energy consumption of the vehicle with small estimation errors. The relative error of the fusion model on the Bus 1 dataset is 4.8%. However, not all results performed well. In some cases, the error reaches 15%. When the data were analyzed separately, it can be found that the larger errors occurred mainly during morning peaks and severe weather periods. Modeling these situations is complex and beyond the scope of this paper. To verify the generalizability capability of the model, two buses (Bus 1 and Bus 2) were selected as training samples, and other vehicle data were used as test data.

The energy consumption estimation results with multi-vehicle data and the fusion model are shown in Figure 8. The results show that the energy consumption estimation errors for multiple vehicles are within 8.1%. The statistics of the estimation results for the energy consumption of the fusion model are shown in Table 3. The average error of the vehicle energy consumption estimation results is 7.5%.

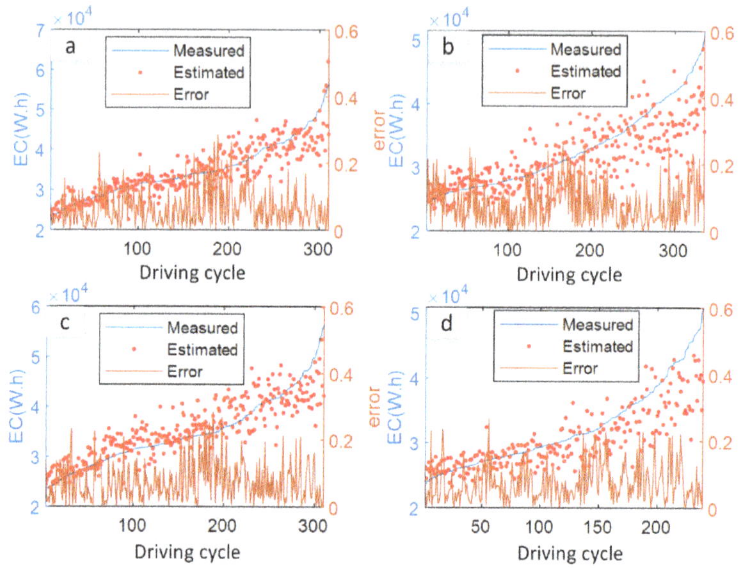

Figure 8. Energy consumption estimation results with multi-vehicle data and the fusion model. (**a**–**d**) Bus 1 to Bus 4.

Table 3. The statistics of the estimation results for energy consumption of the fusion model.

Vehicle Number	Relative Error of Vehicle Self-Test	R-Squared of Vehicle Self-Test	Relative Error of Other Vehicles	R-Squared of Other Vehicles
Bus 1	4.8%	0.83	Training vehicle	Training vehicle
Bus 2	5.5%	0.82	Training vehicle	Training vehicle
Bus 3	6.6%	0.75	7.5%	0.67
Bus 4	6.9%	0.76	7.7%	0.63
Bus 5	5.9%	0.77	7.0%	0.74
Bus 6	7.0%	0.78	8.1%	0.61
Average	6.1%	0.79	7.5%	0.66

4.3. Method Comparison and Verification

To further validate the effectiveness of the fusion model, several other vehicle energy consumption estimation models are introduced for comparison with the model proposed in this paper. The algorithms for comparison include the physical model, CatBoost decision tree model and fusion models with different approaches. In terms of validity assessment, the average relative error and the coefficient of determination are regarded as indicators for evaluation. The coefficient of determination is a correlation index that measures how well the data trend fits. Herein, the coefficient of determination is obtained using the R-squared method. The assessment indicators can be calculated by:

$$e_{Relative} = \frac{1}{n}(\sum_{i=1}^{n} \frac{|(y_i - \hat{y}_i)|}{y_i}) \tag{13}$$

$$R^2 = 1 - \frac{MSE(\hat{y}, y)}{\text{var}(y)} = 1 - \frac{\sum_i (\hat{y}_i - y_i)^2}{\sum_i (y_i - \overline{y}_i)^2} \tag{14}$$

where y_i and $\hat{y}_i$ represent the real value and the estimated value, respectively. n is the number of samples.

The results of the different energy consumption estimation models are shown in Table 4. These models were used to calculate the vehicle energy consumption in this paper. It can be seen that the physical model gives the worst energy consumption estimation results. In contrast, the CatBoost decision tree modeling approach has better estimation results. Ultimately, the physical-CatBoost decision tree model gives the best estimation results, with relative errors and coefficients of determination of 6.1% and 0.79, respectively.

Table 4. Vehicle energy consumption estimation results with different models.

Model	Average Relative Error of Vehicle	Average R-Squared of Vehicle
Physical model	8.4%	0.66
CatBoost model	7.0%	0.72
Physics-Neural network model	6.8%	0.72
Physics-Random forest model	6.8%	0.73
Physics-XGBoost model	6.9%	0.73
Physics-CatBoost fusion model	6.1%	0.79

The complexity of the Physics-CatBoost fusion model was tested. One million pieces of data were processed on a computer with Intel® core™ i5-10400 CPU @ 2.90 GHz running memory of 32 GB (Santa Clara, CA, USA). The data processing time was 6 s, and the model training time was only 0.9 s. It can be seen that the complexity of the model is low.

5. Conclusions

This research focused on the energy consumption estimation of electric buses based on a physical and data-driven fusion model. In terms of physical modeling, a basic energy consumption model was constructed. Rolling drag, kinetic energy consumption and air-conditioning factors were considered. In terms of data-driven modeling, the main factors affecting the fluctuation of vehicle energy consumption were studied. The input characteristics of the model were simplified so that the input of the model can be built before vehicle driving. A CatBoost decision tree modeling approach was employed to construct the model for estimating fluctuating energy consumption. In the model training process, the idea of integrated learning was utilized to optimize the model in a hierarchical iteration. The results show that the average relative error of the vehicle energy consumption estimation result is 6.1%. The coefficient of determination is 0.79. Compared with other energy consumption modeling methods, the fusion model performs best with the two indicators. The fusion model proposed in this paper has better accuracy and generalization ability than other models. It provides a reference basis for the optimization of the energy consumption of electric buses, vehicle scheduling and the rational layout of charging stations.

Based on the results, most of the points with large errors are concentrated in bad weather. In order to further improve the accuracy of the model, weather factors can be added to the model in the future. In addition, vehicle mass is regarded as a constant value in the driving process, which is also a reason for the model error. Therefore, the establishment of the dynamic estimation of vehicle mass in a follow-up work can improve the accuracy of the model.

Author Contributions: Conceptualization, X.L. and T.W.; methodology, X.L.; software, T.W.; validation, T.W. and J.L.; formal analysis, X.L. and T.W.; investigation, T.W. and J.L.; resources, Y.T. and J.T.; data curation, J.T.; writing—original draft preparation, X.L. and T.W.; writing—review and editing, X.L. and J.T.; visualization, T.W.; supervision, Y.T. and J.T.; project administration, J.T.; funding acquisition, X.L. All authors have read and agreed to the published version of the manuscript.

Funding: This work was supported by the National Natural Science Foundation of China (No. 52177219) and the Natural Science Foundation of Guangdong Province (2021A1515010525).

Data Availability Statement: The National Big Data Alliance of New Energy Vehicles (NDAEV) provided research data. The source of the data is the China National New Energy Vehicle Monitoring and Management Platform.

Conflicts of Interest: The authors declare no conflict of interest. The funders had no role in the design of the study; in the collection, analyses or interpretation of data; in the writing of the manuscript; or in the decision to publish the results.

References

1. Docherty, I.; Marsden, G.; Anable, J. The governance of smart mobility. *Transp. Res. Part A Policy Pract.* **2018**, *115*, 114–125. [CrossRef]
2. Babu, N. Review of the Estimation Methods of Energy Consumption for Battery Electric Buses. *Energies* **2021**, *14*, 7578. [CrossRef]
3. Pamuła, T.; Pamuła, W. Estimation of the energy consumption of battery electric buses for public transport networks using real-world data and deep learning. *Energies* **2020**, *13*, 2340. [CrossRef]
4. Yuan, X.; Zhang, C.; Hong, G.; Huang, X.; Li, L. Method for evaluating the real-world driving energy consumptions of electric vehicles. *Energy* **2017**, *141*, 1955–1968. [CrossRef]
5. Bracco, S.; Bianco, G.; Siri, S.; Barbagelata, C.; Casati, C.; Siri, E. Simulation Models for the Evaluation of Energy Consumptions of Electric Buses in Different Urban Traffic Scenarios. In Proceedings of the 2021 Sixteenth International Conference on Ecological Vehicles and Renewable Energies (EVER), Monte-Carlo, Monaco, 5–7 May 2021; pp. 1–6. [CrossRef]
6. Qi, X.; Wu, G.; Boriboonsomsin, K.; Barth, M.J. Data-driven decomposition analysis and estimation of link-level electric vehicle energy consumption under real-world traffic conditions. *Transp. Res. Part D Transp. Environ.* **2018**, *64*, 36–52. [CrossRef]
7. Hao, X.; Wang, H.; Lin, Z.; Ouyang, M. Seasonal effects on electric vehicle energy consumption and driving range: A case study on personal, taxi, and ridesharing vehicles. *J. Clean. Prod.* **2020**, *249*, 119403. [CrossRef]
8. Miraftabzadeh, S.M.; Longo, M.; Foiadelli, F. Estimation Model of Total Energy Consumptions of Electrical Vehicles under Different Driving Conditions. *Energies* **2021**, *14*, 854. [CrossRef]
9. Al-Wreikat, Y.; Serrano, C.; Sodré, J.R. Effects of ambient temperature and trip characteristics on the energy consumption of an electric vehicle. *Energy* **2022**, *238*, 122028. [CrossRef]
10. Björnsson, L.H.; Karlsson, S. The potential for brake energy regeneration under Swedish conditions. *Appl. Energy* **2016**, *168*, 75–84. [CrossRef]
11. Ritari, A.; Tammi, K.; Laitinen, H. Energy Consumption and Lifecycle Cost Analysis of Electric City Buses with Multispeed Gearboxes. *Energies* **2020**, *13*, 2117. [CrossRef]
12. Li, X.; Wang, T.; Wu, C.; Tian, J.; Tian, Y. Battery Pack State of Health Prediction Based on the Electric Vehicle Management Platform Data. *World Electr. Veh. J.* **2021**, *12*, 204. [CrossRef]
13. De Cauwer, C.; Van Mierlo, J.; Coosemans, T. Energy consumption prediction for electric vehicles based on real-world data. *Energies* **2015**, *8*, 8573–8593. [CrossRef]
14. Kanarachos, S.; Mathew, J.; Fitzpatrick, M.E. Instantaneous vehicle fuel consumption estimation using smartphones and recurrent neural networks. *Expert Syst. Appl.* **2019**, *120*, 436–447. [CrossRef]
15. Li, W. *Modeling and Energy Consumption Proportion Analysis of Pure Electric Vehicle Air Conditioning System*; Jilin University: Changchu, China, 2020. [CrossRef]
16. Vepsäläinen, J.; Otto, K.; Lajunen, A.; Tammi, K. Computationally efficient model for energy demand prediction of electric city bus in varying operating conditions. *Energy* **2019**, *169*, 433–443. [CrossRef]
17. Qin, W.; Wang, L.; Liu, Y.; Xu, C. Energy Consumption Estimation of the Electric Bus Based on Grey Wolf Optimization Algorithm and Support Vector Machine Regression. *Sustainability* **2021**, *13*, 4689. [CrossRef]
18. Truong, V.H.; Vu, Q.V.; Thai, H.T.; Ha, M.H. A robust method for safety evaluation of steel trusses using Gradient Tree Boosting algorithm. *Adv. Eng. Softw.* **2020**, *147*, 102825. [CrossRef]
19. Zhang, T.; He, W.; Zheng, H.; Cui, Y.; Song, H.; Fu, S. Satellite-based ground PM2.5 estimation using a gradient boosting decision tree. *Chemosphere* **2021**, *268*, 128801. [CrossRef] [PubMed]

20. Bentéjac, C.; Csörgő, A.; Martínez-Muñoz, G. A comparative analysis of gradient boosting algorithms. *Artif. Intell. Rev.* **2020**, *54*, 1937–1967. [CrossRef]
21. Jabeur, S.B.; Gharib, C.; Mefteh-Wali, S.; Arfi, W.B. CatBoost model and artificial intelligence techniques for corporate failure prediction. *Technol. Forecast. Soc. Chang.* **2021**, *166*, 120658. [CrossRef]
22. Huang, G.; Wu, L.; Ma, X.; Zhang, W.; Fan, J.; Yu, X.; Zhou, H. Evaluation of CatBoost method for prediction of reference evapotranspiration in humid regions. *J. Hydrol.* **2019**, *574*, 1029–1041. [CrossRef]
23. Prokhorenkova, L.; Gusev, G.; Vorobev, A.; Dorogush, A.V.; Gulin, A. CatBoost: Unbiased boosting with categorical features. *Adv. Neural Inf. Process. Syst.* **2018**, *31*, 1–11. Available online: https://proceedings.neurips.cc/paper/2018/file/14491b756b3a51daac41c24863285549-Paper.pdf (accessed on 1 December 2018).
24. Sivadas, N.; Okamoto, S.; Xu, X.; Fennie, C.J.; Xiao, D. Stacking-dependent magnetism in bilayer CrI3. *Nano Lett.* **2018**, *18*, 7658–7664. [CrossRef] [PubMed]

energies

Article

Optimization of Micro-CAES and TES Systems for Trigeneration

Paolo Maria Congedo [1], Cristina Baglivo [1,*], Simone Panico [1,2], Domenico Mazzeo [1,3] and Nicoletta Matera [1]

1 Department of Engineering for Innovation, University of Salento, 73100 Lecce, Italy
2 Institute for Renewable Energies, Eurac Research, Viale Druso 1, 39100 Bolzano, Italy
3 Department of Mechanical, Energy and Management Engineering (DIMEG), University of Calabria, 87036 Rende, Italy
* Correspondence: cristina.baglivo@unisalento.it

Abstract: Energy storage makes energy continuously available, programmable, and at power levels different from the original intensity. This study investigates the feasibility of compressed-air energy storage (CAES) systems on a small scale. In addition to the CAES systems, there are two TES (thermal energy storage) systems for the recovery of calories and frigories. The micro-CAES + TES system is designed for a single-family residential building equipped with a photovoltaic system with a nominal power of 3 kW. The system is optimized as a potential alternative to battery storage for a typical domestic photovoltaic system. The multi-objective optimization analysis is carried out with the modeFRONTIER software. Once the best configuration of the micro-CAES + TES system is identified, it is compared with electrochemical storage systems, considering costs, durability, and performance. The efficiency of CAES (8.4%) is almost one-tenth of the efficiency of the most efficient batteries on the market (70–90%). Its discharge times are also extremely short. It is shown that the advantages offered by the application of mechanical accumulation on a small scale are mainly related to the exploitation of the thermal waste of the process and the estimated useful life compared to the batteries currently on the market. The studied system proves to be non-competitive compared to batteries because of its minimal efficiency and high cost.

Keywords: CAES; TES; small-scale; battery storage; optimization

Citation: Congedo, P.M.; Baglivo, C.; Panico, S.; Mazzeo, D.; Matera, N. Optimization of Micro-CAES and TES Systems for Trigeneration. *Energies* **2022**, *15*, 6232. https://doi.org/10.3390/en15176232

Academic Editors: Alon Kuperman and Alessandro Lampasi

Received: 14 July 2022
Accepted: 24 August 2022
Published: 26 August 2022

1. Introduction

The gradual depletion of fossil fuels has driven research into renewable resources [1]. Many countries have made progress in promoting renewables within their energy mix, but obstacles remain, and further efforts are needed. Making a renewable system reliable over time requires provision from storage systems to avoid gaps in energy supply [2]. Energy storage is a key determinant in the energy transition process. Energy storage systems provide greater accessibility to renewable energy sources in the power grid, ensuring both energy savings and reduced impact on the environment [3,4], while reducing the gap between energy supply and demand [5]. The advantages offered by storage systems include increased stabilization of energy supply that can cope with high- and low-demand scenarios, relief of consumer bills by reducing and shifting peak electricity demand, and improved grid resilience [6].

The energy stored can be generated from a system (active storage) or can be stored passively in materials exploiting climatic fluctuations. In the latter case, some materials can store the solar energy directly in the building's walls using a sensible or latent process, i.e., with traditional or phase-change materials (PCMs) [7]. By the same principle, thermal solar energy can be stored in tanks integrated with PCMs to decrease the tank discharge time [8].

The classification of active storage systems can be made by considering the form of secondary energy in which the primary energy is stored.

In electrical energy storage systems (EESSs), electrical energy is converted into other types of potential energy, such as chemical, mechanical, elastic, and magnetic energy. Among them are the following:

- Electrochemical storage, which includes lead–acid and nickel–cadmium batteries;
- Mechanical energy storage, such as compressed-air energy storage (CAES);
- Electrical storage, such as supercapacitors or superconducting magnetic energy storage (SMES);
- Hydrogen storage coupled with fuel cells for its reconversion into electrical energy [9,10].

Although the importance of storage systems for energy efficiency is widely recognized, the range of energy storage techniques for microscale applications is very limited, and some mechanical and thermal energy storage systems include those that are applied in the high-power generation sector. In addition to being expensive [11], storage systems are often oversized [12], and their disposal and average lifespan are also challenging [13].

Electrochemical storage is currently the state of the art for small-scale energy storage. However, batteries are not yet considered to be a fully mature technology either technically or economically. The most promising technology—namely, the lithium-ion battery—depends on a strategic material that has limited uptake and is currently not recycled for economic reasons. However, in addition to cost reduction, technological improvements will also need to address increasing the specific energy and lifetime of storage batteries.

CAES technology enables the trigeneration of electrical, thermal, and cooling energy in the energy release process. Specifically, trigeneration systems simultaneously provide heat, power, and cooling using a single fuel source. Compared with conventional systems, cogeneration and trigeneration systems reduce fossil fuel demand and grid losses [14].

The trigeneration system of electrical, thermal, and refrigeration energy allows excess energy to be stored in the form of compressed air and thermal heat, and enables refrigeration energy to be produced through the direct expansion of compressed air. Modeling of compressed-air energy storage systems considering network-specific requirements has shown that optimal design leads to improved functionality and an overall reduction in system costs [15]. CAES systems provide several advantages over other storage systems, including high power and energy capacity, long service life, rapid response, and relatively low capital and maintenance costs [16].

CAES can be integrated with renewable energy systems, such as wind and solar power. This allows excess energy from renewable sources to be stored, effectively addressing the fluctuation of renewable sources (i.e., avoiding curtailment) [17]. Several articles in the literature have provided an overview of CAES in terms of scale, fuel utilization, and integration with other technologies (e.g., smart grid and energy internet), emphasizing its potential applications [18,19]. Wang et al. presented a discussion of the challenges and prospects of using CAES systems [20]. CAES is considered to be the most cost-effective technology, as well as being excellent for its scalability and ease of implementation when used on a utility scale [21].

Today, CAES is mainly used on a large scale (i.e., macro-CAES)—compressed air is stored, during the hours when the cost of energy is lowest, inside hermetically sealed underground cavities and at pressures generally around 70–100 bar; the same high-pressure gas is used in traditional turbo gas systems, or for pneumatic drives in production lines for a wide variety of needs and for automation in general. These systems have a good energy density—typically around 2–3 kWh/m^3 of storage, which is almost 10 times higher than the energy density of mechanical gravity storage used in hydroelectric power plants.

A potentially viable alternative to electrochemical systems for small-scale storage is micro-CAES. Among the main advantages of micro-CAES coupled with TES (thermal energy storage) is the possibility of recovering waste energy to make the micro electric generation system more competitive. The heat developed during the compression phase can be used for residential heating and/or domestic hot water production, while the expanding cold air can be used for space cooling.

The potential advantages of micro-CAES systems are as follows:

- Longer life than batteries, allowing a potentially unlimited number of charge and discharge cycles;
- High sustainability, as they do not require the use of toxic or rare chemicals;
- Flexibility in operation and installation, allowing micro-CAES to operate independently or in connection with the power grid, or even in combination with storage batteries.

Applications of small-scale CAES systems are not currently widespread. One reason that residential micro-trigeneration and trigeneration have received much less attention than large-scale systems is that small-scale applications provide a cooling load through a reverse Rankine cycle that requires high capital cost components, such as absorption chillers and boilers, to cool the load [22].

Cogeneration and trigeneration systems can operate more efficiently if electricity and heat production are decoupled using thermal energy storage, where unneeded heat is stored during the production period [23]. Achieving a highly feasible CAES system enables the design of a flexible energy system characterized by the optimal use of fluctuating renewable energy sources [24]. Different numerical and thermodynamic analyses have been conducted in the literature to highlight and evaluate the applicability of CAES systems [25,26]. Solutions to improve CAES systems' performance have been proposed as a result of optimization analyses [27]. Luo et al. [28] proposed a modeling and simulation tool for A-CAES (adiabatic compressed-air energy storage) system optimization to identify heat exchange and thermal storage units with suitable capacity and performance for air compression/expansion units, and then analyze system efficiency and identify potential improvement strategies.

Considering that the above, energy storage techniques for microscale applications are still very limited—especially micro-CAES applications. This study proposes an optimization of mechanical and thermal storage systems for small-scale trigeneration. The objectives are to identify solutions that are economical (e.g., components with affordable costs, technologies easily available on the market, and minimal maintenance) and adaptable to the spaces generally available in the residential area.

The present work is an extension of previous works [29,30]; the system is designed for a single-family residential building equipped with a photovoltaic system with a rated power of 3 kW.

First, this paper presents the case study to which the optimization for a mechanical and thermal storage system for small-scale trigeneration is applied. The purpose of the optimization was to improve the efficiency of the micro-CAES + TES system, along with attempting to simplify the structure as much as possible to make it economical and suitable for residential spaces. Finally, a comparison with the use of battery storage systems was conducted.

2. The Case Study: Micro-CAES for Trigeneration

The micro-CAES + TES system is designed for a single-family residential building equipped with a photovoltaic system with a nominal power of 3 kW. The average electricity demand of a family can be estimated at about 3000 kWh/year which, divided daily, becomes about 8.2 kWh/day. Assuming that at least 50% of household consumption takes place in the evening or at night when the photovoltaic system does not produce energy, about 4 kWh/day must be accumulated to have a good margin of autonomy from the grid. Therefore, on a typical day, the photovoltaic system can produce for immediate daytime consumption, or alternatively for night-time consumption by loading the mechanical storage system so that it can be discharged for the night.

The operation, as in any other storage technology, is divided into two phases deferred in time according to the time-shifting of the electricity demand of domestic users—a charging phase of the CAES's compressed air tank, with electricity absorption to the compressor, followed by a discharge phase of the same CAES tank with electricity generation to the expander. During the charging phase, the CAES system accumulates photovoltaic electrical energy in the form of energy elastic mechanics in the storage tank. At the same time, the HTTES (high-temperature thermal energy storage) system is also charged with the heat transfer fluid which, moving in a specially designed secondary circuit, recovers the thermal waste from the compression. During the discharge phase, the elastic energy is again converted into electrical energy using a turbine connected to the alternator. At the same time, the LTTES (low-temperature thermal energy storage) system is charged through the expanded cold air that cools the heat transfer fluid, which moves in its secondary circuit (distinct from that of the HTTES system). The heat developed during the compression phase can be used for residential heating and/or domestic hot water production, while the expanded cold air can be used for room cooling.

The problem is initially analyzed from a thermodynamic point of view. When filling the storage tank, the compressor sucks in atmospheric air and processes it to raise the storage pressure from the minimum operating value to the maximum. This process is non-stationary, and the type of compressor chosen must have suitable characteristics to keep the processed flow rate as constant as possible as the compression ratio varies. From this point of view, the best-performing compressors are the alternatives, which also offer a wide pressure range, long life, easy maintenance, and high reliability.

The choice of the type of expander, on the other hand, falls on the pneumatic reed motor as a technology widely used on the market and at affordable prices. It is chosen to power a motor with a nominal power of 5 kW with compressed air at 5 bar, obtaining a mechanical power of 3.85 kW on the shaft.

In this study, the aim was to optimize a micro-CAES system powered by a photovoltaic solar system with a nominal power of 3 kW for a domestic user, to which an electrical power of 3 kW was allocated through a pneumatic reed motor powered constantly at 5 bar. The CAES was combined with a high-temperature TES system for the recovery of thermal compression waste, which took place through a reciprocating volumetric compressor, and a low-temperature TES system for the recovery of the cooling capacity of the expanded air.

3. Optimization Setup

The goal of optimization is to maximize efficiency measures for mechanical and thermal storage. The charging and discharging processes of the micro-CAES were analyzed from a thermodynamic point of view to obtain the equations governing the problem in a previous study [29]. These equations were implemented in MATLAB and optimized with modeFRONTIER software.

Figure 1 shows the workflow of the optimization problem carried out in modeFRONTIER. The inputs, outputs, and constraints applied to the optimization model are described in detail in the following paragraphs (Sections 3.1 and 3.2).

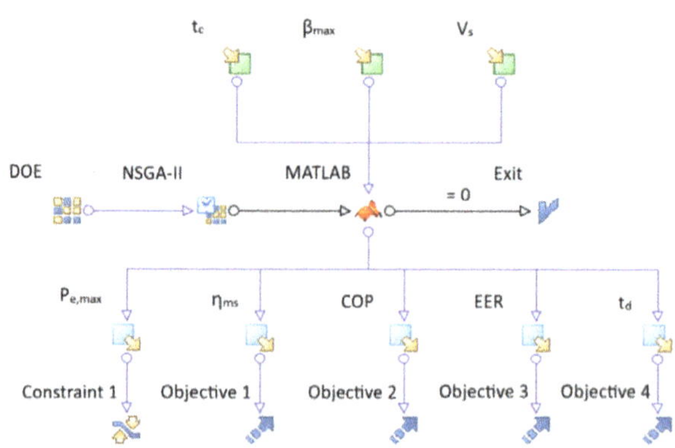

Figure 1. Workflow of the optimization problem on modeFRONTIER.

In the workflow, the first set of DOE (design of experiments) is a random sequence. The chosen NSGA-II (non-dominated sorting genetic algorithm II) optimization algorithm is an evolutionary algorithm suitable for solving optimization problems with multiple objectives [31,32].

3.1. Inputs

The thermodynamic equations were implemented in the MATLAB environment. Several operational coefficients and parameters were established in the preliminary analysis, as shown in Table 1.

Table 1. Thermodynamic coefficients and parameters that are not subject to optimization.

Parameters	Description	Values
c_p	Specific heat at constant dry air pressure	1005 J/kgK
k	Ratio of specific heats in dry air	1.4
n	Characteristic exponent of polytropic process in the compression cycle	1.25
p_0	Atmospheric pressure	1 bar
p_{min}	Minimum operating pressure	5 bar
P_E	Electric power generated by the alternator	3 kW
P_M	Power delivered by the pressure-fed reed motor	3.85 kW
R	Specific constant for dry air	287.05 J/kgK
T_0	Atmospheric temperature	20 °C
$T_{0'}$	Temperature of the air leaving the LTTES heat exchanger	10 °C
T_S	Temperature of the air stored in the CAES tank	20 °C
V_{exp}	Air consumption of the reed motor (FAD)	80 L/s
η_{al}	Minimum alternator efficiency	0.78
η_e	Efficiency of the electric motor of the compressor	0.8
η_g	Global compression efficiency	0.7
η_{is}	Isoentropic expansion efficiency	0.8

The optimization analysis was based on the variation, within a fixed range, of the parameters shown in Table 2. By simulating the variables in appropriate ranges, it is possible to obtain, among all possible combinations, those that optimize the predetermined objectives.

Table 2. Input variables.

Input Variable	Description	Lower/Upper Limit	Step	Number of Inputs Generated
t_c	Relative charging time	1/5 h	1	5
β_{max}	Maximum compression ratio	10/35	5	6
V_S	Volume of the CAES tank	1/10 m^3	1	10

For the maximum compression ratio (β_{max}) the lower limit of 10 was chosen, since for pressures below 10 bar the air compression has insignificant temperature variations, which would not justify the installation of the HTTES system. The upper limit, on the other hand, was taken to be 35, because the datasheets of manufacturers of reciprocating compressors show that with a maximum installed motor power of 3 kW (equal to the nominal power of the photovoltaic system) it is possible to compress air up to a maximum of 35 bar. The storage volume of the compressed air (V_S) had an upper limit of 10 m^3 for space and cost reasons. Finally, the duration of the charging phase of the CAES (t_c) must be consistent with the hours during which the solar photovoltaic system can operate during the day; for this reason, the upper limit of 5 h was set. The possible values assigned to each input variable result in up to 300 different experiments to be conducted.

3.2. Outputs

The storage efficiency of the system was optimized from both the mechanical and the thermal points of view. For the CAES system, the final purpose is electrical generation; the efficiency of mechanical energy storage (η_{ms}) is defined as the ratio between the electrical energy generated during the discharge phase ($L_{e,exp}$) and the electrical energy spent on the compressor during the charging phase ($L_{e,med}$).

$$\eta_{ms} = \frac{L_{e,exp}}{L_{e,med}} \quad (1)$$

For the HTTES system, the final purpose is the recovery of the thermal waste from the compressor, for which the coefficient of performance (COP) in calorific storage is defined as the ratio between the thermal energy stored in the system ($Q_{heating}$) and the net electrical energy spent for storage ($L_{e,med} - L_{e,exp}$).

$$\text{COP} = \frac{Q_{heating}}{L_{e,med} - L_{e,exp}} \quad (2)$$

For the LTTES system, the final purpose is the recovery of cooling capacity, so the energy efficiency ratio (EER) in refrigerated storage is defined as the ratio between the cooling capacity stored in the system ($Q_{cooling}$) and the net electrical energy spent to produce it ($L_{e,med} - L_{e,exp}$).

$$\text{EER} = \frac{Q_{cooling}}{L_{e,med} - L_{e,exp}} \quad (3)$$

For all three systems, the maximum electrical power consumption ($P_{e,med}$) was constrained, as shown in Table 3.

$$P_{e,med} = \frac{P_{max}}{\eta_g * \eta_e} \quad (4)$$

Table 3. Output variables and optimization choices in the workflow.

Output Variable	Objective Type
$P_{e,max}$	<3 kW
η_{ms}	Maximize
COP	Maximize
EER	Maximize
t_d	Maximize

For the CAES, the discharge time (t_d) was maximized.

$$t_d = \frac{M_{acc}}{G_{exp}} \quad (5)$$

where M_{acc} is the mass stored during the charging process and G_{exp} represents the mass flow rate of expanded air processed by the turbine [29].

The study of the optimization of the micro-CAES + TES system is not limited to simply maximizing the accumulation efficiency parameters; it is also necessary to combine other requirements that arise from the installation of such a system within a home unit, as follows:

- Reduce the space occupied by the facility, with particular reference to the CAES tank, which is the bulkiest element;
- Simplify the structure as much as possible to make it economical to purchase and maintain;
- Limit the maximum operating pressures to ensure greater safety;
- Power the households as long as possible so that the system is functional even in the evening hours when the PV system is not producing energy.

4. Results

The results show that of the 300 experiments generated, less than 34% meet the constraints imposed by the workflow. Unacceptable solutions were discarded from the design space table.

Figures 2 and 3 provide an estimation of the relationships between input and output factors. A positive value (red bar) indicates a direct relationship, while a negative value (blue bar) indicates an inverse relationship. In particular, Figure 2 represents the correlation between the maximum power absorbed by the compressor $P_{e,max}$ and the input variables (V_s, β_{max}, t_c), while Figure 3 shows the correlation between the CAES discharge time (t_d) and the input variables.

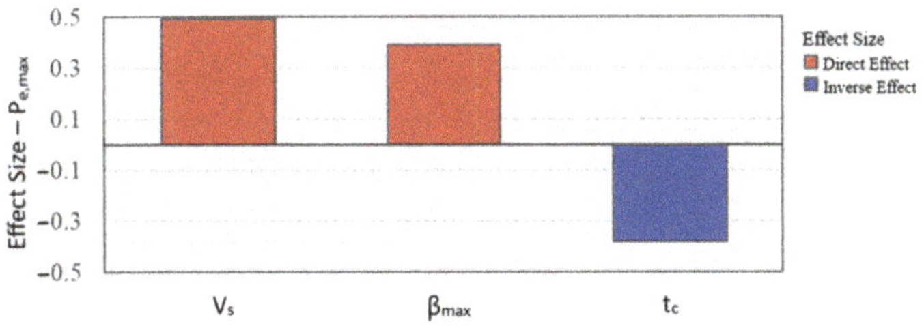

Figure 2. Relationship between the maximum power absorbed by the compressor $P_{e,max}$ and the input variables (V_s, β_{max}, t_c).

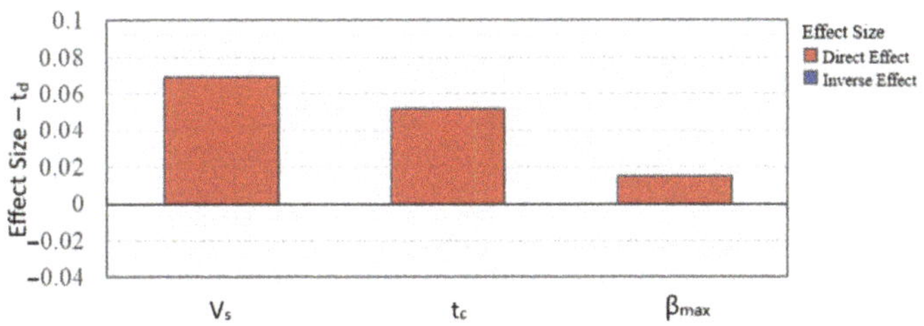

Figure 3. Relationship between the discharge time (t_d) and the input variables (V_s, β_{max}, t_c).

To present the relationship between the mechanical energy storage efficiency (η_{nm}) and the input variables, a 4D bubble chart (Figure 4) was used. This is particularly useful because it allows the values obtained from optimization for the four variables to be shown. In particular, the x and y axes represent β_{max} and η_{nm}, respectively, while the diameter indicates the t_c and the color indicates the V_s.

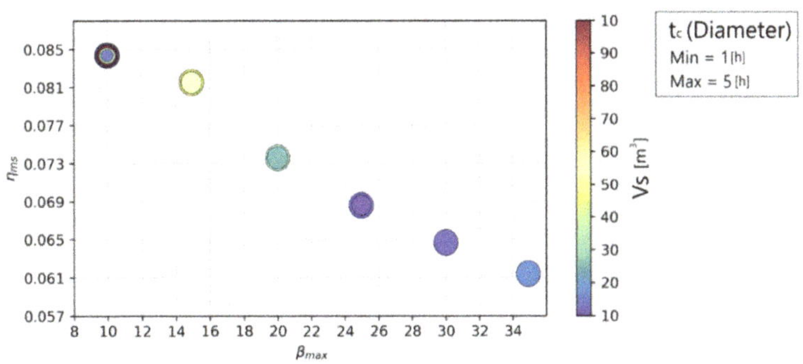

Figure 4. The x and y axes of the 4D bubble chart represent β_{max} and η_{nm}, respectively, while the diameter indicates the t_c and the color indicates the V_s.

The correlation matrix (Figure 5) summarizes all of the relationships concerning the mechanical accumulation system between the input factors (V_S, t_c, β_{max}) and the output variables $(P_{e,max}, t_d, \eta_{ms})$.

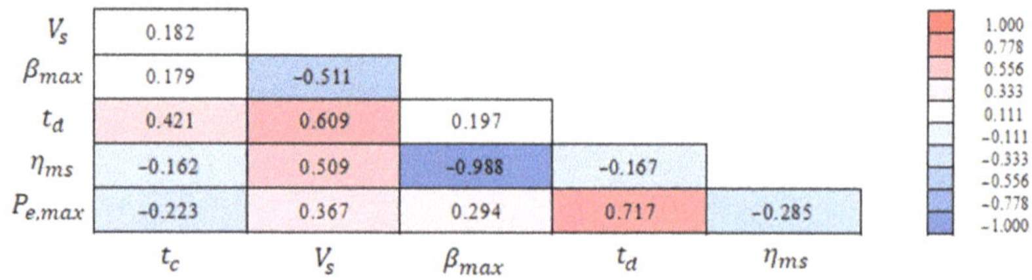

Figure 5. Correlation matrix to summarize the results referring to CAES. Values close to +1 indicate that the two variables are positively correlated, while values close to −1 indicate a negative correlation. If the value is close to 0, the variables are not correlated.

The results were also plotted with reference to TES—specifically, the 3D bubble chart in Figure 6 shows the relationships between COP, EER, and β_{max}.

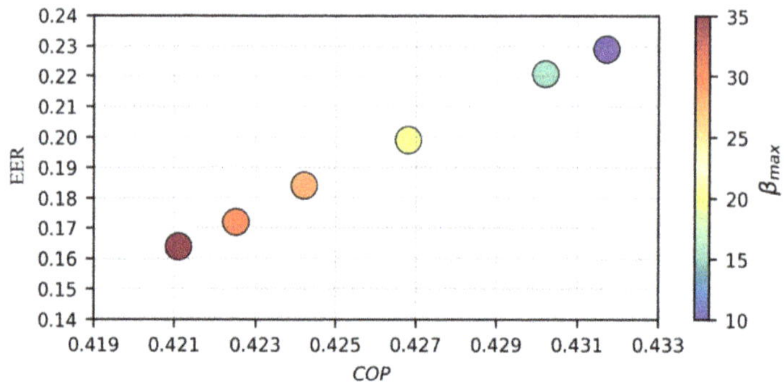

Figure 6. The x and y axes of the 3D bubble chart represent COP and EER, respectively, while color indicates the values of β_{max}.

Figures 7 and 8 show the parallel coordinate charts of the COP and EER output variables, respectively, in the optimized case. This is a particularly useful tool for plotting the input and output variables of each design simultaneously.

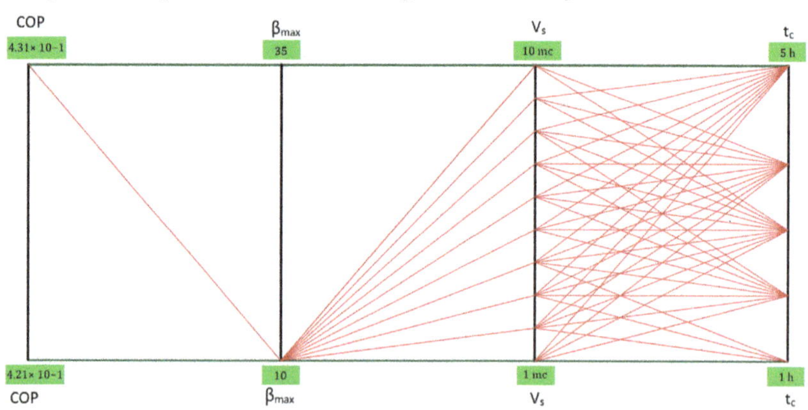

Figure 7. Parallel coordinate chart of the COP output variable in the optimized case.

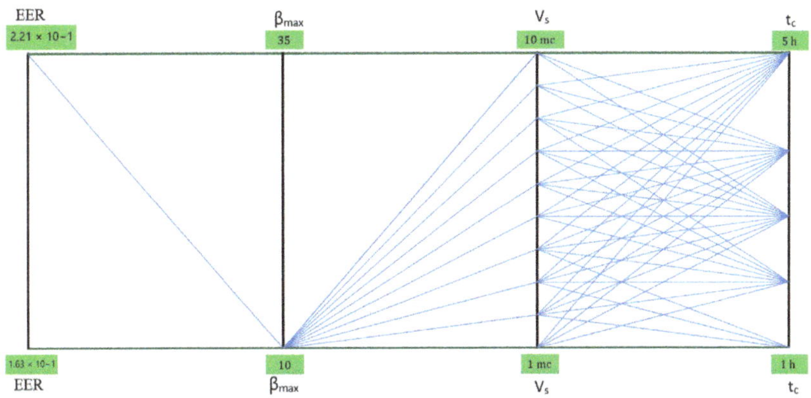

Figure 8. Parallel coordinate chart of the EER output variable in the optimized case.

5. Discussions

This section presents a discussion of the optimization process. As shown in Figure 2, the expanded charge times (t_c) reduce the power absorption to the compressor. Instead, there is a direct relationship between the power absorption to the compressor and the storage volumes (V_s) and compression ratios (β_{max}). Figure 3 indicates that the CAES discharge time (t_d) is directly related to all three input factors. The most influential factor is the volume of the CAES tank (V_s). It is beneficial to have the longest possible discharge times (t_d) in order to have a functional mechanical storage system for the power supply of the residential building.

Furthermore, increasing the storage volume (V_s) can be useful within certain limits, as it can be seen that as this factor increases, the power consumption to the compressor also increases significantly. Increasing the charging time (t_c) to the upper limit of 5 h, on the other hand, has a twofold positive effect: a reduction in power consumption and an increase in the discharge interval (the second most important factor after the storage volume).

It was assumed that the consumption of the building taken as a case study is concentrated in the evening hours, so during the day it accumulates electrical energy in elastic form for 5 h, and in the evening this energy is reconverted into electrical form to supply domestic users.

The maximum compression ratio (β_{max}) is almost irrelevant to the discharge time, but it is the second most important factor in terms of incidence with respect to the increase in compressor power. Therefore, within appropriate limits related to the space of the building and the purchase costs of the tanks, it is convenient to work with larger storage rather than higher storage pressures.

In Figure 4, it is possible to notice the inverse relation between η_{nm} and β_{max}. Therefore, in order to maximize the ratio between electrical energy produced during the discharge phase of the CAES system and electrical energy spent for the charge, it is necessary to reduce the compression ratio to a minimum. The reduction in the maximum pressure under which the CAES operates is a need in line with the considerations made previously for the other output variables. The efficiency of mechanical energy storage is substantially independent of the other input factors; it is not by chance that no trend emerges due to the diameter or color of the bubbles. With fixed β_{max}, there is the same efficiency for any combination of the factors V_s and t_c, provided that the relative experiments have not been discarded from the design table. For this reason, in the low-yield zone of the graph and, thus, at the high pressures of the CAES, only a few combinations of the two factors are possible (blue bubbles with large diameters, which indicate low storage volumes with extended charge times). The high-yield zone is the one to focus on for further analysis. This zone is also the one that offers the most choice in the configuration of the optimal setup for the system (bubbles of different colors with diameters of various sizes, meaning wider ranges of values in which to try to combine the input factors).

The resulting optimized CAES configuration, consistent with the interval thresholds set in the workflow of the problem, provides the following input factors for the solution:

- $V_S = 10\ m^3$;
- $t_c = 5$ h;
- $\beta_{max} = 10$.

Figure 6 highlights the fact that COP and EER are inversely proportional to the compression ratio. Therefore, for HTTES and LTTES systems, it is also convenient to work with $\beta_{max} = 10$.

To maximize COP and EER, a pair of V_S and t_c values is not specifically required, but it is essential that the compression ratio is set to a minimum. Therefore, the configuration adopted to optimize the CAES problem is also suitable for HTTES and LLTES systems. The fact that the trends of mechanical and thermal storage systems are not in contrast with one another is a very positive aspect. The input factor values of the experiment optimized for CAES can be confirmed for HTTES and LLTES systems as well.

The qualitative analysis of the trends allows us to define the optimized configuration for the micro-CAES + TES system under study.

In accordance with the equations reported in [29], starting from CAES, a single-stage reciprocating compressor (N = 1) is required to process an airflow rate of $G_{min} = 11.88$ kg/h $\approx$ 2.75 L/s (FAD) to a maximum pressure of 10 bar. The compressor requires a 1.5 kW electric motor. A comparison of the results with real models from manufacturers' catalogues was carried out, and the reliability of the study model was confirmed. The compressor takes 5 h to fill the CAES tank from the initial pressure of 5 bar to the final pressure of 10 bar, consuming about 6 kWh of electricity supplied by the solar photovoltaic system in the process. The compressed air storage tank has a volume of 10 m^3, equivalent to an overall length of 5.2 m by 1.65 m in diameter. During the loading phase, a total of almost 60 kg of air is stored to be used later for expansion.

Since the compressor is single-stage (N = 1), the HTTES system recovers the thermal waste from the charging phase exclusively through a heat exchanger located downstream of the compressor. The maximum temperature of the compressed air leaving the stage is estimated at about 190 °C, and the total heat recoverable in one charge is 8.6 MJ.

Moving on to the discharge phase of the CAES, this involves emptying the tank in order to supply the previously selected pneumatic reed valve motor, in order to deliver an electrical power of 3 kW to the user. The motor is supplied at a constant pressure of 5 bar, thanks to a pressure reducer installed downstream of the tank. Once this pressure threshold is reached in storage, the reducer interrupts the flow, and a mass of air equal to that discharged remains trapped in the CAES tank, i.e., about 60 kg. The discharge lasts a total of 0.17 h ($\approx$10 min). The minimum air temperature at the end of the expansion is just below −66 °C, and the cooling capacity that can be recovered through the heat exchanger of the LTTES system is about 4.6 MJ.

The mechanical and thermal storage values used for the optimized configuration return η_{ms} of 8.4%, COP of 0.43, and EER of 0.23.

Preliminary Analysis of Costs

Finally, a preliminary investigation of the system is presented, with the analysis of costs and relative comparison with the most common electrochemical storage systems on a small scale. Table 4 shows the purchase costs of the main components of the micro-CAES + TES system studied.

Table 4. The purchase costs of the main components of the micro-CAES + TES system.

Component	Cost
Tank (10 m^3–10 bar)	EUR 8000
Compressor (10 bar–1.5 kW)	EUR 500
Air heat exchangers	EUR 250 × 2 = EUR 500
Pneumatic reed motor	EUR 1800
Alternator (3 kW)	EUR 300
Total	EUR 11,100

To make the comparison with battery storage systems, it was considered that the average household consumption in the evening hours for a household amounts to approximately 4.5 kWh. Thus, for batteries, the cost analysis is summarized in Table 5.

Table 5. The purchase costs of storage batteries.

Battery	Cost
Lithium battery (5.7 kWh)	EUR 5000
Lead–acid battery (9 kWh)	EUR 2500
Lead-gel battery (9 kWh)	EUR 3000

For lead–acid and lead-gel batteries, it should be noted that about 50% of energy is retained in the charge–discharge cycle to avoid damaging them. For this, a 9 kWh battery is used. For lithium-ion technology, on the other hand, 80% of use is considered to be 5.7 kWh. In terms of purchase cost, the micro-CAES system is also cheaper than batteries.

Given that the CAES system under consideration can generate 0.51 kWh of electrical energy to the user at a constant power of 3 kW, it was determined what configuration the CAES system should have in order to generate a significant amount of energy, taking into account the purchase cost and available space.

To obtain 3 kWh from the discharge, it would be necessary for the discharge to last for 1 h. At this point, there are two options: increase the maximum compression pressure, which compromises the already low efficiency of mechanical energy storage; or increase the storage tank size.

In the first case, a compressor with a maximum air pressure of 35 bar and a 13 kW motor is required, which is incompatible with the nominal power of a typical domestic photovoltaic system. In addition, the mechanical energy storage efficiency would drop to 6.16%.

In the second case, a 60 m^3 tank would need to be filled to 10 bar in 5 h using a 9 kW compressor, and the efficiency (a function of β_{max}) would remain fixed at 8.4%.

The efficiency of CAES (8.4%) is almost one-tenth of the efficiency of the most efficient batteries present on the market (70–90%).

The only advantages that a micro-CAES + TES system such as the one studied can offer compared to state-of-the-art batteries lie in its longer service life and the possibility of recovering the thermal waste to be used for heating/cooling the building. The years of service life for the system studied, as shown in Table 6, can be estimated by taking into account the average life of the compressor, which is the component that is potentially the most prone to failures and which, moreover, requires more attention for maintenance.

Table 6. Comparison of estimated service life years for different storage systems.

Accumulation System	Estimated Service Life
Micro-CAES	20 years
Lithium-ion battery	10 years
Lead–acid battery	3 years
Lead-gel battery	5 years

6. Conclusions

This study investigated the feasibility of compressed-air energy storage (CAES) systems, with the objective of carrying out the optimization of mechanical and thermal storage systems for small-scale trigeneration, designed for a single-family residential building equipped with a photovoltaic system with a rated power of 3 kW.

When dealing with a multi-objective optimization problem, the optimal solution obtained is never unique. The results identify a set of potentially optimal solutions. Among these, it is up to the decision-maker to choose which objective to favor in the optimization strategy. In particular, for this case study, the best configuration for the CAES was obtained under the following conditions:

- Reduced compression ratios (β_{max}). This allows for maximizing the mechanical accumulation efficiency of the CAES and reducing the maximum power absorption to the compressor, which is a very important aspect in the case of storage systems powered by domestic photovoltaic systems, as in this case study. Finally, it is true that the discharge time t_d is reduced, but the effect is marginal when compared to V_S and t_c. This means that β_{max} can still be lowered, compensating for the effect on the discharge time by pushing slightly more on the other two input factors.
- High charging time (t_c). The power absorption is contained at compression, and the discharge time of the CAES is increased.
- High storage volumes (V_S). As this is the main factor influencing the increase in the discharge time of the CAES, even if the power consumption increases, the effect is still compensated by the choice to work with low compression ratios.

The results of the optimization showed some critical issues. A very small percentage of the electrical energy stored in elastic form is again convertible into electrical energy. The efficiency of CAES is nearly one-tenth of that of the most efficient batteries on the market, and the discharge times are also extremely short. At the same time, for TES systems, only 43% of the net electrical energy expended is recoverable as heat in the HTTES system, with the hot source (compressed air) at 190 °C; in the LTTES system, the recoverable cooling capacity reaches about 23% in relation to the net electrical energy expended, with the cold source (expanded air) at a temperature of −66 °C. A tank with a volume of 10 m^3 is not easy to install in a residential building.

The micro-CAES + TES system studied is inefficient, and is more expensive than batteries. The only advantages that micro-CAES + TES systems offer compared to batteries lie in their longer service life and in the possibility of recovering the thermal waste to be used for heating/cooling the building. The critical aspect of the system, optimized as much as possible in its configuration, can be attributed to the disproportion between the air consumption of the reed motor in the discharge phase and the flow rate that the compressor is able to process during the charging phase. The use of air motors, as widely accessible devices, has proven to be unsustainable for a micro-CAES system, even if optimized.

Therefore, in the future, this micro-CAES + TES system for small-scale trigeneration could be improved as a result of the optimization carried out in this work. If experimentation with new expanders is able to reduce the consumption of compressed air during discharge, the system could be configured as a viable alternative to batteries for the accumulation of photovoltaic energy. This is because the system offers indisputable advantages linked, as seen, to the exploitation of the thermal waste of the process to be used to heat/cool the building. Furthermore, the estimated useful life is double or even triple compared to some types of batteries currently on the market.

An interesting implementation of this study could involve the possibility of coupling the proposed system with an air–ground heat exchanger.

Author Contributions: Conceptualization, P.M.C., C.B., S.P. and D.M.; Data curation, P.M.C., C.B., S.P., D.M. and N.M.; Formal analysis, P.M.C., C.B., S.P., D.M. and N.M.; Investigation, P.M.C., C.B., S.P., D.M. and N.M.; Methodology, P.M.C., C.B., S.P. and D.M.; Resources, P.M.C., C.B., S.P. and D.M.; Software, P.M.C., C.B., S.P. and D.M.; Supervision, P.M.C., C.B., S.P. and D.M.; Validation, P.M.C., C.B., S.P. and D.M.; Visualization, P.M.C., C.B., S.P., D.M. and N.M.; Writing—original draft, P.M.C., C.B., S.P. and D.M.; Writing—review & editing, P.M.C., C.B., S.P. and D.M. All authors have read and agreed to the published version of the manuscript.

Funding: This research received no external funding.

Institutional Review Board Statement: Not applicable.

Informed Consent Statement: Not applicable.

Data Availability Statement: Not applicable.

Acknowledgments: We thank Francesco Tramonte of the Department of Engineering for Innovation at the University of Salento for support in the development of this study.

Conflicts of Interest: The authors declare no conflict of interest. The funders had no role in the design of the study; in the collection, analyses, or interpretation of data; in the writing of the manuscript; or in the decision to publish the results.

References

1. Mazzeo, D.; Matera, N.; De Luca, P.; Baglivo, C.; Congedo, P.M.; Oliveti, G. Worldwide geographical mapping and optimization of stand-alone and grid-connected hybrid renewable system techno-economic performance across Köppen-Geiger climates. *Appl. Energy* **2020**, *276*, 115507. [CrossRef]
2. Mazzeo, D.; Matera, N.; De Luca, P.; Baglivo, C.; Congedo, P.M.; Oliveti, G. A literature review and statistical analysis of photovoltaic-wind hybrid renewable system research by considering the most relevant 550 articles: An upgradable matrix literature database. *J. Clean. Prod.* **2021**, *295*, 126070. [CrossRef]
3. Beaudin, M.; Zareipour, H.; Schellenberglabe, A.; Rosehart, W. Energy storage for mitigating the variability of renewable electricity sources: An updated review. *Energy Sustain. Dev.* **2010**, *14*, 302–314. [CrossRef]
4. Gallo, A.B.; Simõões-Moreira, J.R.; Costa, H.K.M.; Santos, M.M.; Moutinho dos Santos, E.M. Energy storage in the energy transition context: A technology review. *Renew. Sustain. Energy Rev.* **2016**, *65*, 800–822. [CrossRef]
5. Rostamizadeh, M.; Khanlarkhani, M.; Sadrameli, S.M. Simulation of energy storage system with phase change material (PCM). *Energy Build.* **2012**, *49*, 419–422. [CrossRef]
6. Cheekatamarla, P.K.; Kassaee, S.; Abu-Heiba, A.; Momen, A.M. Near isothermal compressed air energy storage system in residential and commercial buildings: Techno-economic analysis. *Energy* **2022**, *251*, 123963. [CrossRef]
7. Mazzeo, D.; Oliveti, G. Thermal field and heat storage in a cyclic phase change process caused by several moving melting and solidification interfaces in the layer. *Int. J. Therm. Sci.* **2018**, *129*, 462–488. [CrossRef]
8. Nekoonam, S.; Ghasempour, R. Modeling and optimization of a thermal energy storage unit with cascaded PCM capsules in connection to a solar collector. *Sustain. Energy Technol. Assess.* **2022**, *52*, 102197. [CrossRef]
9. Pal, P.; Mukherjee, V. Off-grid solar photovoltaic/hydrogen fuel cell system for renewable energy generation: An investigation based on techno-economic feasibility assessment for the application of end-user load demand in North-East India. *Renew. Sustain. Energy Rev.* **2021**, *149*, 111421. [CrossRef]
10. Singh, A.; Baredar, P.; Gupta, B. Techno-economic feasibility analysis of hydrogen fuel cell and solar photovoltaic hybrid renewable energy system for academic research building. *Energy Convers. Manag.* **2017**, *145*, 398–414. [CrossRef]
11. Mazzeo, D.; Baglivo, C.; Matera, N.; Congedo, P.M.; Oliveti, G. A novel energy-economic-environmental multi-criteria decision-making in the optimization of a hybrid renewable system. *Sustain. Cities Soc.* **2020**, *52*, 101780. [CrossRef]
12. Baglivo, C.; Mazzeo, D.; Oliveti, G.; Congedo, P.M. Technical data of a grid-connected photovoltaic/wind hybrid system with and without storage battery for residential buildings located in a warm area. *Data Brief* **2018**, *20*, 587–590. [CrossRef] [PubMed]
13. Mazzeo, D.; Oliveti, G.; Baglivo, C.; Congedo, P.M. Energy reliability-constrained method for the multi-objective optimization of a photovoltaic-wind hybrid system with battery storage. *Energy* **2018**, *156*, 688–708. [CrossRef]
14. Angrisani, G.; Akisawa, A.; Marrasso, E.; Roselli, C.; Sasso, M. Performance assessment of cogeneration and trigeneration systems for small scale applications. *Energy Convers. Manag.* **2016**, *125*, 194–208. [CrossRef]
15. Rouindej, K.; Samadani, E.; Fraser, R.A. A comprehensive data-driven study of electrical power grid and its implications for the design, performance, and operational requirements of adiabatic compressed air energy storage systems. *Appl. Energy* **2020**, *257*, 113990. [CrossRef]
16. Mozayeni, H.; Wang, X.; Negnevitsky, M. Dynamic analysis of a low-temperature Adiabatic Compressed Air Energy Storage system. *J. Clean. Prod.* **2020**, *276*, 124323. [CrossRef]
17. Tong, Z.; Cheng, Z.; Tong, S. A review on the development of compressed air energy storage in China: Technical and economic challenges to commercialization. *Renew. Sustain. Energy Rev.* **2021**, *135*, 110178. [CrossRef]

18. Chen, L.; Zheng, T.; Mei, S.; Xue, X.; Liu, B.; Lu, Q. Review and prospect of compressed air energy storage system. *J. Mod. Power Syst. Clean Energy* **2016**, *4*, 529–541. [CrossRef]
19. Huang, B.; Qiu, X.; Wang, W.; Li, H.; Zhou, W. Overview of research situation and progress on compressed air energy storage technology. *IOP Conf. Ser. Earth Environ. Sci.* **2019**, *295*, 012020. [CrossRef]
20. Wang, J.; Ma, L.; Lu, K.; Miao, S.; Wang, D.; Wang, J. Current research and development trend of compressed air energy storage. *Syst. Sci. Control Eng.* **2017**, *5*, 434–448. [CrossRef]
21. Akinyele, D.O.; Rayudu, R.K. Review of energy storage technologies for sustainable power networks. *Sustain. Energy Technol. Assess.* **2014**, *8*, 74–91. [CrossRef]
22. Li, Y.; Wang, X.; Li, D.; Ding, Y. A trigeneration system based on compressed air and thermal energy storage. *Appl. Energy* **2012**, *99*, 316–323. [CrossRef]
23. Sala, J. Thermal energy storage (TES) systems for cogeneration and trigeneration systems. In *Advances in Thermal Energy Storage Systems*; Cabeza, L.F., Ed.; Woodhead Publishing Series in Energy; Woodhead Publishing: Sawston, UK, 2015; pp. 493–509. [CrossRef]
24. Kondoh, J.; Ishii, I.; Yamaguchi, H.; Murata, A.; Otani, K.; Sakuta, K.; Higuchi, N.; Sekine, S.; Kamimoto, M. Electrical energy storage systems for energy networks. *Energy Convers. Manag.* **2000**, *41*, 1863–1874. [CrossRef]
25. Manchester, S.; Swan, L. Compressed Air Storage and Wind Energy for Time-of-day Electricity Markets. *Procedia Comput. Sci.* **2013**, *19*, 720–727. [CrossRef]
26. Sciacovelli, A.; Li, Y.; Chen, H.; Wu, Y.; Wang, J.; Garvey, S.; Ding, Y. Dynamic simulation of Adiabatic Compressed Air Energy Storage (A-CAES) plant with integrated thermal storage—Link between components performance and plant performance. *Appl. Energy* **2017**, *185*, 16–28. [CrossRef]
27. Yu, H.; Engelkemier, S.; Gençer, E. Process improvements and multi-objective optimization of compressed air energy storage (CAES) system. *J. Clean. Prod.* **2022**, *335*, 130081. [CrossRef]
28. Luo, X.; Wang, J.; Krupke, C.; Wang, Y.; Sheng, Y.; Li, J.; Xu, Y.; Wang, D.; Miao, S.; Chen, H. Modelling study, efficiency analysis and optimisation of large-scale Adiabatic Compressed Air Energy Storage systems with low-temperature thermal storage. *Appl. Energy* **2016**, *162*, 589–600. [CrossRef]
29. Congedo, P.M.; Baglivo, C.; Carrieri, L. Hypothesis of thermal and mechanical energy storage with unconventional methods. *Energy Convers. Manag.* **2020**, *218*, 113014. [CrossRef]
30. Congedo, P.M.; Baglivo, C.; Carrieri, L. Application of an unconventional thermal and mechanical energy storage coupled with the air conditioning and domestic hot water systems of a residential building. *Energy Build.* **2020**, *224*, 110234. [CrossRef]
31. Sciacca, E. Algoritmi Evolutivi Multiobiettivo e Possibilistic Worst-Case Distance per l'ottimizzazione di circuiti elettronici, Università degli Studi di Catania. pp. 20–28. Available online: http://webusers.oact.inaf.it/esciacca/publications/Tesi_Sciacca.pdf (accessed on 10 May 2022).
32. Hanne, T. Global Multiobjective Optimization Using Evolutionary Algorithms. *J. Heuristics* **2000**, *6*, 347–360. [CrossRef]

energies

MDPI

Review

Solid–Gas Thermochemical Energy Storage Materials and Reactors for Low to High-Temperature Applications: A Concise Review

Anti Kur *, Jo Darkwa, John Calautit, Rabah Boukhanouf and Mark Worall

Buildings, Energy and Environment Research Group, Faculty of Engineering, University of Nottingham, Nottingham NG7 2RD, UK
* Correspondence: anti.kur@nottingham.ac.uk

Abstract: Thermochemical energy storage materials and reactors have been reviewed for a range of temperature applications. For low-temperature applications, magnesium chloride is found to be a suitable candidate at temperatures up to 100 °C, whereas calcium hydroxide is identified to be appropriate for medium-temperature storage applications, ranging from 400 °C up to 650 °C. For the high-temperature range (750–1050 °C), oxides of cobalt, manganese, and copper are found to have the redox behaviour required for thermochemical heat storage. However, some of these materials suffer from low thermal conductivities, agglomeration, and low cyclability and, therefore, require further improvements. The concept of enhancing thermal conductivities through additives such as nanomaterials has been encouraging. From an operational point of view, fluidized-bed reactors perform better than fixed- and moving-bed reactors due to better particle interactions. There is, however, a need for the reaction bed to be further developed toward achieving optimum heat and mass transfers. Agitated fluidized-bed reactors have shown encouraging results and are suggested for further exploration. A combination of appropriate computational tools can facilitate an in-depth understanding of bed dynamics.

Keywords: thermal energy storage; thermochemical energy storage; thermochemical reactors; solid–gas reactions; modelling; simulation

Citation: Kur, A.; Darkwa, J.; Calautit, J.; Boukhanouf, R.; Worall, M. Solid–Gas Thermochemical Energy Storage Materials and Reactors for Low to High-Temperature Applications: A Concise Review. *Energies* **2023**, *16*, 756. https://doi.org/10.3390/en16020756

Academic Editors: Alon Kuperman and Alessandro Lampasi

Received: 3 December 2022
Revised: 29 December 2022
Accepted: 4 January 2023
Published: 9 January 2023

1. Introduction

While energy supplies of various kinds and technologies underscore the economic development of societies, they also result in undesirable side effects [1]. It is common knowledge that the major environmental burden of global concern emanating from the energy sector is the greenhouse effect [2]. One of the prominent greenhouse gases (GHG) is carbon dioxide. It arises from the combustion of fossil fuels, natural gas, and coal in automobiles and power stations, as well as heating in buildings and industrial processes [2]. For instance, around 45–47% of the United Kingdom's (UK) total energy consumption is for heating purposes, and nearly 80% is from fossil fuel sources [3]. Moreover, domestic energy use alone has a share of more than a quarter of national GHG emissions, of which 75% is for space and water heating [4]. Hence, the decarbonization of heat is the major energy challenge that the world faces over the coming decades. One way to achieve 80% emissions reduction in 2050 is by decarbonizing industrial and domestic heat demand and improving resource efficiency [2]. Current choices are around district heating (DH) networks in combination with other technologies to electrify heat or 'green' the gas grid [5]. However, the variability of heat demand, with a predictable winter peak heat load, presents opportunities for thermal energy storage (TES) to manage supply requirements to meet specified demands. Presently, sensible heat storage, latent heat storage, and thermochemical heat storage are the three TES systems being explored by researchers.

Sensible heat storage (SHS) is the most mature and commercially used type of TES, available as tank thermal storage for hot water and electric storage heaters [3]. This technology is utilized for its material's convectional heat storage at very high temperature differences. Latent heat storage (LHS), on the other hand, is the heat absorbed or released by a substance during a change of phase. Typically, the heat is stored within a very narrow temperature range suitable only for applications requiring very small temperature differences [6]. The phase change materials (PCM) used for this purpose have the merit of achieving higher energy storage density, smaller volume requirements, and lower heat losses compared to sensible heat stores [6,7]. However, PCMs are unsuitable in comparison to SHS, which is more economical for applications requiring larger temperature differences. Thermochemical energy storage (TCES), which operates based on enthalpy change in reversible chemical reactions, is the most promising TES system [8] and has attracted greater interest in recent times.

TCES is recognized to have higher potential for energy stability and efficiency for reasons of high energy density (nearly 1000 kJ/L), smaller storage volume, minimal heat loss, long-term storage [9,10], high exergy efficiency [8], and lower charging temperature [10,11]. With respect to energy density, it is theoretically 5 to 10 times higher than LHS and SHS, respectively, when compared on the same scale basis [12,13]. As a result, the TCES system is more compact and could be effective where space constraints are significant. Additionally, TCES systems can be tuned to operate in a wide range of different temperatures and pressures [14], thus making them suitable for the storage of all grades of waste heat. This offers the possibility of being operated using various heat sources such as solar energy, aiming to operate a sustainable process. Moreover, the effective integration of TCES into district heating (DH) networks can lead to benefits such as increased energy efficiency and reliability, and reduction in energy use, costs, and GHG emissions. Thus, TCES is potentially useful in lowering fossil fuel consumption and related GHG emissions [15].

These attributes have attracted increasing interest in TCES research, albeit still at the experimental stage [3]. Many aspects of the technology are still unknown and yet to be discovered [14]. At present, TCES is hampered by a few problems, some of which include complexity in infrastructure [9] as well as low levels of thermal attainment in practical systems [9,16]. Therefore, a robust approach to operational control and understanding of the system must have a real-time model to predict its dynamics. For this reason, numerical or modelling studies are required for deeper theoretical insights and prediction of the system's behaviour. In addition, research still focuses on finding suitable materials with sufficient energy density, hydrothermal stability, and cyclability at conditions suitable for system operation [17]. Enhancement of materials' properties is crucial for integration into reactor systems. Additionally, it also requires suitable reactor optimization techniques and heat exchanger frameworks [14]. To date, experimental investigations of different reactor configurations have not overcome the dynamic limitations of heat and mass transfer [18]. In all reactors studied, nearly 70% are deployed for solid–gas reactions [19] because they show better values of chemical reaction efficiencies. However, the reaction bed must be further developed for optimum heat and mass transfers.

This paper aims to provide a concise review of the various TCES materials and reactors which could be used for low- to high-temperature applications. The review covers the fundamental concepts of TCES processes and materials, which are presented in Section 2. This includes a succinct discussion of the many experimental and theoretical efforts towards material property enhancement for application in different ranges of temperatures. The main types of solid–gas TCES reactors, together with their merits, challenges, and proposals for system improvement and optimization are discussed in Section 3. An overview of modelling and simulation of TCES systems is presented in Section 4, while the conclusion and recommendations are stated in Section 5.

We present a summary of the most recent reviews in which Desai et al. [20] reviewed sorption and chemical reactions for TCES. Experimental investigations and cyclic studies in low- and medium-temperature applications were discussed. Gbenou et al. [21] highlighted the many research issues hampering TCES implementation. They also discussed reactor prototypes, projects, and limitations as well as suggestions for better analysis of TCES problems. Efforts to link the practical and scientific aspects of TCES problems were the focus of Sadeghi [22]. In addition, findings from cutting-edge research and pertinent aspects of the TCES system consideration were reported. Marie et al. [23] presented advances in TCES with an overview of fluidized beds for low-temperature domestic applications, whilst Kant and Pitchumani [24] focused on open and closed reactors and prototypes for building applications. A perspective on the strengths and weaknesses of TCES materials and systems as well as a discussion on the evolution of TCES research has been provided in the work of Salgado-Pizarro et al. [25]. Gbenou et al. [26] reviewed TCES reactor prototypes and projects for low-temperature applications with analyses of the microscopic and macroscopic aspects of TCES systems.

However, the above reviews covered mainly low- to medium-temperature range applications, whereas, in this review, the coverage is from low- to high-temperature applications. Reviewing the entire spectrum will help to refocus on the broader issues of TCES. This might assist in honing the possibility of combining various techniques ideal for creating composites (hybrids) with novel properties. In addition, far-reaching proposals are made for the improvement of the thermophysical properties of TCES materials as well as the performance enhancement of TCES system integration. A succinct overview of the computational tools and resources available for the modelling and simulation of TCES systems provides motivation for researchers to gain valuable insights and understanding of these systems.

2. Thermochemical Energy Storage

Generally, thermochemical energy storage (TCES) uses a reversible system in which a source provides heat, for instance, to separate reactants (*AB*) into products (*A* and *B*). The products are stored separately at ambient temperatures, thereby eliminating the cost of insulation in storage containers. The separation route is endothermic and is the charging process. When heat release is required, the products (*A* and *B*) are recombined and pre-heated to an activation temperature to produce the reactant (*AB*) through a reversible exothermic route. The reversible exothermic path is the discharging process expressed as:

$$AB + \Delta H \leftrightharpoons A + B, \tag{1}$$

where ΔH is the molar heat (enthalpy) of the reaction. Essentially, a thermochemical energy storage cycle involves three main processes [27]: (i) charging, which is the endothermic reaction requiring a heat resource to dissociate the reactant AB; (ii) storing, where the products *A* and *B* are both stored separately; and (iii) discharging, which is the exothermic reaction where the products A and B are combined again to release heat energy.

The stored thermal energy (Q) depends on the molar reaction enthalpy (ΔH) and the number of moles (n) of one of the products [12], as in the equation:

$$Q = n\Delta H \tag{2}$$

Based on the storage mechanism, TCES processes are divided into sorption processes and reversible chemical reactions [9,13,28]. Figure 1 shows the classification chart for TCES systems. However, there appears to be a thin boundary for clear distinction between terminologies such as chemical storage, thermochemical storage, and sorption processes [27]. Often, a sorption process is considered a chemical reaction in which the chemical bonds are weak [29].

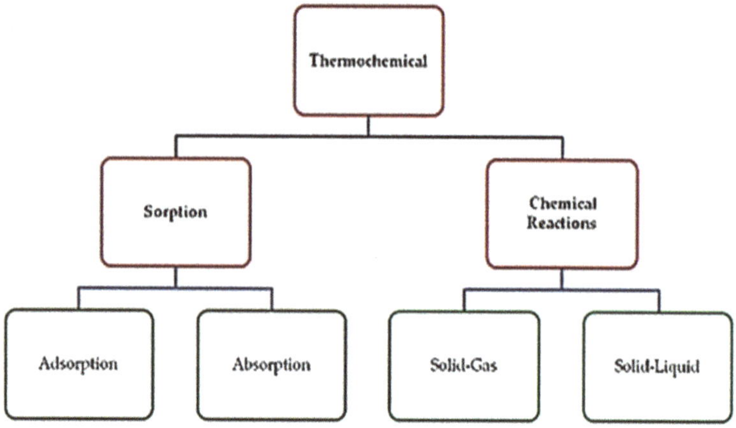

Figure 1. Classification of thermochemical energy storage systems [13].

2.1. Sorption Processes and Materials in Thermochemical Energy Storage

Sorption is described as a phenomenon by which a vapor or gas (sorbate) is captured by a denser substance (solid or liquid), called sorbent [27]. The reverse process, called desorption, requires heat to unbind the sorbate from the sorbent. Therefore, in a sorption process, heat storage is accomplished via a chemical potential when the force binding the sorbent and the sorbate is broken [30]. Essentially, absorption and adsorption are the two categories of sorption processes. Although dissimilar, they do, however, involve the physical transfer of a volume of mass or energy [27]. Absorption, in simple terms, is the process of one material (absorbent) retaining another material (absorbate). This takes place within the molecular enclave of the sorbent, resulting in alteration in its structure and morphology [17]. Examples of absorption materials for water include $MgSO_4$, LiCl, LiBr, $CaCl_2$, $MgCl_2$, KOH, and NaOH [29].

Unlike absorption, adsorption involves a very thin layer of atoms or molecules on the adsorbent surface without altering its structure [17]. Here, there is an accumulation of energy or matter (of adsorbate) onto a surface (of adsorbent) [27]. According to Srivastava and Eames [31], adsorption is a phenomenal occurrence at the periphery of two phases where both weak intermolecular and strong chemical bonds act between the molecules. Adsorption either proceeds as a physical process, physical adsorption (physisorption), or a chemical process, chemical adsorption (chemisorption), based on the type of bond between adsorbent and adsorbate. Generally, physisorption occurs whenever an adsorbate is brought into contact with the surface of the adsorbent and involves weak intermolecular forces (Van der Waals forces) [30]. Chemisorption, on the other hand, is due to strong chemical bonds (hydrogen bonds, charge-transfer interactions, covalent bonds) [32] in the same manner as in other chemical compounds. Both types of adsorptions involve the evolution of heat. The reason that chemical forces are stronger than physical forces is that the heat of chemisorption is larger than that of physisorption [33]. Though these are different processes, they often take place simultaneously at different sites or locations of the adsorbent [30,32]. Similarly, at the mesoscale, it is difficult to distinguish between absorption and adsorption [33] as both may occur simultaneously. In that case, the term sorption is generally used for both processes [31,34].

Materials for sorption processes are usually solids, liquids, and composite sorbents with solid/gas and liquid/gas systems as working pairs [13]. Moreover, reversible adsorption of vapours onto porous solid surfaces is a potential option for TCES, particularly for space heating applications [32]. According to Yu et al. [30], the most studied adsorbents are silica gels and zeolites using water as a working fluid. In particular, zeolite 13X is the commonest and most widely studied TCES material because of its hydrothermal and mechanical stability and corrosion behaviour [13], although aluminophosphates (AlPOs), silico-aluminophosphates (SAPOs), and metal-organic frameworks (MOFs) have recently been reviewed by Makhanya et al. [35] as promising materials for heat storage. Composite materials formed by a combination of salt hydrate and a porous additive with high thermal conductivity have also been studied [16]. Figure 2 shows a summary of typical sorption reactions and materials.

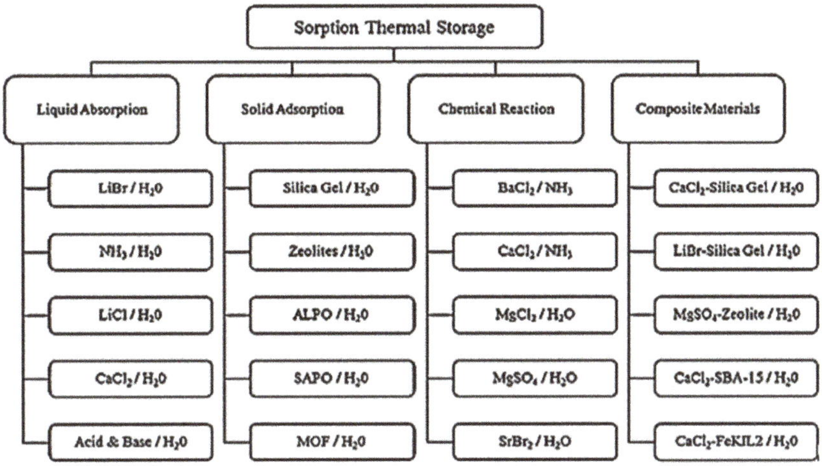

Figure 2. Sorption reactions and materials [10].

Sorption processes are essentially attractive for low-temperature applications due to their attribute of high kinetics at low temperatures [13]. Therefore, sorption reactions are typically not suitable for high-temperature applications [29].

2.2. *Chemical Reactions and Materials in Thermochemical Energy Storage*

TES based on chemical reactions is justifiably advantageous for seasonal storage [12]. These reaction systems store energy in the form of chemical potential, and the energy per mole required to break up chemical bonds is more than any other thermal storage system. These reactions are characterized by changes in the molecular composition of the reactants involved [13], and usually take place at temperatures above 400 °C [36]. High energy storage density and reversibility are key requirements for TCES materials, as it is challenging to find a suitable reversible reaction for a system. This is significant because the type of reaction has immense implications on the reactor design and system integration [37]. The difficult task for a reaction choice is the requirement for efficient heat and mass (HAM) transfer to and from the storage volume. This requirement, according to Aydin et al. [10] can be a limiting factor for the overall storage volume, unlike SHS and LHS, which allow higher volumes to be utilized. This volume limitation due to HAM transfer characteristics is the key area for current research in TCES systems.

In the literature [8,29,38], TCES reactions are classified into three categories, namely solid–gas, liquid–gas, and gas–gas reactions with regard to the nature of the reactants and products. However, for temperatures over 300 °C, only solid–gas and, in some cases, liquid–gas reactions remain practicable [29]. Furthermore, solid–gas reactions have been widely studied as a very promising heat storage method [39]. The interest in these reactions is due to their wide range of equilibrium temperatures and self-separation of reactants. Chemical reactions, including chemical sorption processes, premised on solid–gas systems are an encouraging method for the storage and conversion of heat energy for heating or cooling purposes [40]. While the sorption processes are used to store low (<100 °C) and medium (100–400 °C) grade heat with enthalpies in the range of 20–70 kJ/mol [40], chemical reactions are utilized for the storage of medium (100–400 °C) and high (>400 °C) grade heat and the enthalpies are in the range 80–180 kJ/mol [13,36].

Different kinds of solid–gas reactions are employed for TCES. These are categorized depending on the composition of the solid reactant as the most prominent [38]. The reactions include those based on hydrates, hydrides, hydroxides, carbonates, and oxides. It might be important that TCES materials be flanked by an appropriate reaction temperature and enthalpy for the application. For this reason, Bauer [29] characterized the solid–gas reactions according to reaction temperatures:

- Dehydration/hydration of metal salt hydrates (in the range of 40–260 °C);
- Dehydrogenation/hydrogenation of metal hydrides (in the range of 80–400 °C);
- Dehydration/hydration of metal hydroxides (in the range of 250–600 °C);
- Decarbonation/carbonation of metal carbonates (in the range of 100–950 °C);
- Deoxygenation/oxygenation of metal oxides (in the range of 600–1000 °C).

2.2.1. Dehydration/Hydration of Metal Salt Hydrates

TCES materials for low-temperature applications have attracted remarkable attention. Salt hydrates and composite sorbents based on salt hydrates belong to this category. They have become preferred materials for TCES in building applications [41] due to their high energy density and low turning temperatures [42]. The low turning temperatures are suitable for integration with sources such as solar energy or low-grade waste heat and make them fit for residential space heating applications [42]. Much literature is available on high-potential salts, and this includes chlorides—LiCl [43], $CaCl_2$ [44–46], and $MgCl_2$ [47]; bromides—$SrBr_2$ [48,49] and LiBr [43,50]; and sulphates—$MgSO_4$ [46,47,51], $Al_2(SO_4)_3$ [46,52], and $CuSO_4$ [53]. Furthermore, other promising hydrates such as Na_2S and K_2CO_3 were studied by de Jong et al. [54] and Gaeini et al. [55], respectively.

TCES materials must fulfil common conditions such as low cost, being non-poisonous, and non-corrosive, in addition to having sufficient energy density and suitable turning temperatures. These requirements are fulfilled by many salt hydrates [46]. However, numerous salts proposed for low-grade thermal energy storage have failed [42]. In a typical case, for instance, van Essen et al. [46] conducted a theoretical study of four salt hydrates, namely $MgSO_4.7H_2O$, $Al_2(SO_4)_3.18H_2O$, $CaCl_2.2H_2O$, and $MgCl_2.6H_2O$, using a thermogravimetry and differential scanning calorimetry (TG-DSC) apparatus. Based on the measured temperature lift under practical conditions, $MgCl_2$ was considered the most promising with a high theoretical energy density of 2.8 GJ/m^3. However, both hygroscopic chlorides under investigation tended to form a gel-like material (due to melting or formation of solution) during the hydration experiments, which prohibited further water uptake. Similarly, Donkers et al. [53] studied the cyclability of $CuCl_2$, $CuSO_4$, $MgCl_2$, and $MgSO_4$ in hydration/dehydration reactions. They observed the effect of fracturing to be greater in hydrates with larger volumetric changes. In conclusion, $CuCl_2$ was adjudged the most promising heat storage material.

A systematic evaluation of 125 salt hydrates was performed by N'Tsoukpoe et al. [52] using criteria such as safety, theoretical calculations, and thermogravimetry analysis (TGA). Out of 45 preselected salt hydrates, $SrBr_2.6H_2O$ and $LaCl_3.7H_2O$ appeared to be the most promising. However, the expected efficiency and net energy storage density (including water storage) remained low. Similarly, a review of 563 reactions was carried out [56] to evaluate the theoretical suitability of salt hydrates as seasonal heat storage materials. Up to 25 salt hydrates were identified. By considering cost, chemical stability, reaction kinetics, and safety, K_2CO_3 was determined to be the most promising candidate, but low energy density was noticed.

Table 1 gives the theoretical and experimental energy density, reaction temperature, and water vapour pressure of some salt hydrates (extracted from [57]). It is, however, noteworthy that in all these comparative investigations, the difference in behaviour is attributable to the intrinsic properties (crystal structure and thermodynamics) of the materials. Therefore, a general kinetic model of the sorption process in salt hydrates will require specific information on the material properties. Thus, besides the high potential shown by some salt hydrates, several associated issues are still obvious. These include poor hydrothermal stability, slow thermodynamics, high corrosivity, and toxicity [41]. Such attributes make it difficult for monomer salt hydrates to be used for TCES without modification of their properties. For this reason, researchers have experimented with composite materials. Fopah Lele et al. [48] evaluated four salt hydrates ($CaCl_2$, $MgCl_2$, $SrBr_2$, and $MgSO_4$) and host matrices (activated carbon, expanded natural graphite, and silica gel). The results on both systems for only salts gave thermal conductivity in the range of 0.3–1.3 W/mK with a measurement uncertainty of less than 14%. Zhao et al. [58] mixed $SrBr_2$ and expanded natural graphite treated with sulphuric acid. The composite with 10 wt% of $SrBr_2$ proved satisfactory with good mass transfer performance and no degradation in water uptake.

Table 1. Thermodynamic properties of some salt hydrates [57].

Reaction	Theoretical Energy Density (GJ/m^3)	Experimental Energy Density (GJ/m^3)	Temperature (Charging/Discharging) (°C)	Water Vapor Pressure (mbar)
$MgCl_2{\cdot}6H_2O \rightleftharpoons MgCl_2{\cdot}H_2O + 5H_2O$	2.5	0.71	150/30–50	13
$MgCl_2{\cdot}4H_2O \rightleftharpoons MgCl_2{\cdot}2H_2O + 2H_2O$	1.27	1.10	118/n.a.	13
$CaCl_2{\cdot}2H_2O \rightleftharpoons CaCl_2 + 2H_2O$	1.1	n.a.	95	n.a.
$Al_2(SO_4)_3{\cdot}6H_2O \rightleftharpoons Al_2(SO_4)_3 + 6H_2O$	1.9	n.a.	150	n.a.
$MgSO_4{\cdot}6H_2O \rightleftharpoons MgSO_4.H_2O + 5H_2O$	2.37	1.83	72/n.a.	13
$MgSO_4{\cdot}7H_2O \rightleftharpoons MgSO_4.H_2O + 6H_2O$	2.3	n.a.	150/105	n.a.
$CaSO_4{\cdot}2H_2O \rightleftharpoons CaSO_4 + 2H_2O$	1.4	n.a.	n.a./89	n.a.
$Na_2S{\cdot}5H_2O \rightleftharpoons Na_2S.1/2H_2O + 9/2H_2O$	2.7	n.a.	80/65	13
$SrBr_2{\cdot}6H_2O \rightleftharpoons SrBr_2.H_2O + 5H_2O$	2.3	2.08	n.a./23.5	20
$Li_2SO_4{\cdot}H_2O \rightleftharpoons LiSO_4 + H_2O$	0.92	0.80	103/n.a.	13
$CuSO_4{\cdot}5H_2O \rightleftharpoons CuSO_4.H_2O + 4H_2O$	2.07	1.85	92/n.a.	13

n.a.: not available.

Salt mixtures appear promising, but the general technical issue reported is mass transport within the matrix's pores due to deliquescence, overhydration (with possible leakage or pore blockage), and a low-temperature lift [41]. A proposal for pairing suitable salt hydrates according to different matrix materials, reactor analyses, and structural optimization methods for the enhancement of HAM transfer has recently been published [59]. Again, selecting a suitable binary salt mixture may increase the performance of each material and avoid its unique individual shortcomings. A double salt hydrate, $Na_2Zn(SO_4)_2{\cdot}4H_2O$, has been reported as having exhibited suitable stability at the first ten hydration/dehydration cycles, with an excellent energy storage density of 4.7 GJ/m^3 and theoretical efficiency up to 77.4% [60]. It might be necessary that the influence of material characterization and reaction parameters are considered to determine the optimum mixing pair and ratio, as well as optimize system controls under different operating conditions [41]. In spite of this, N'Tsoukpoe and Kuznik [34] assert that the performance achieved with salt-hydrate systems is not competitive and that the performance or advantages of the TCES materials have probably been overestimated.

2.2.2. Dehydrogenation/Hydrogenation of Metal Hydride

Metal hydrides (MHs) are compounds formed by the reversible reaction of hydrogen and metal or metal alloy, and this reversible absorption of hydrogen gas is exothermic [39]. The utilization of MHs for TCES is encouraged due to high energy efficiency, high volumetric energy density, and cost [61]. It also offers flexibility in its wide range of operating temperatures. On the other hand, one of the main disadvantages of MH systems is the need for hydrogen storage [37]. This means that the MH system can be a closed system with an intermediate hydrogen storage subsystem. It is suggested that by coupling a high temperature with a low-temperature metal hydride system, a self-regulating reversible metal hydride energy storage system can be established [62].

Lithium hydride (LiH), calcium hydride (CaH_2), and magnesium hydride (MgH_2) systems have been studied for their TCES potentials. However, more attention has been paid to MgH_2 [9]. It has a working temperature between 200–500 °C and decomposes into Mg metal, releasing hydrogen with a reaction enthalpy of 75 kJ/mol and a heat storage capacity of around 0.8 kWh/kg [39,63]. The hydrogen gas can be stored in a reservoir under the equilibrium pressure of MgH_2. For instance, the MgH_2/Mg equilibrium pressure of 10 bar at 350 °C and 20 bar at 400 °C is shown in Figure 3. According to Felderhoff et al. [39], if the pressure is lower than the equilibrium pressure at a given temperature (coloured area in Figure 3), MgH_2 decomposes until the pressure inside the system reaches the equilibrium pressure. At pressures higher than the equilibrium pressure, Mg metal can be hydrogenated (white area in Figure 3).

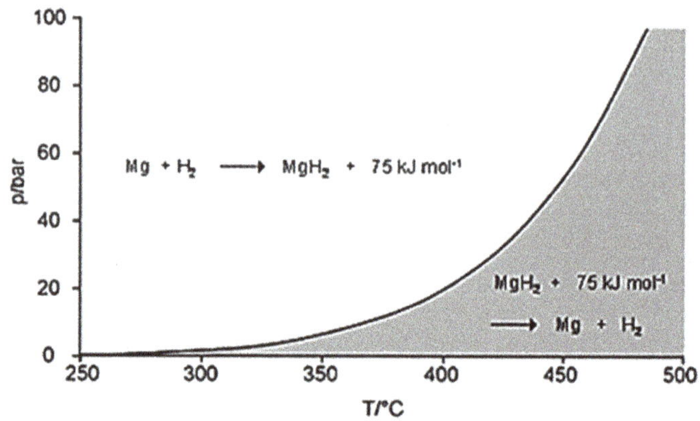

Figure 3. Equilibrium pressure of MgH_2 as a function of temperature [39].

Chen et al. [9] noted that the MgH_2/Mg pair suffers from poor reversibility. Its cyclic stability drops by 75% after 500 cycles [8] which is a limiting factor in large-scale application. Additionally, its high thermodynamic stability and sluggish sorption kinetics are the major obstacles to its extensive application [64]. Table 2 shows the thermodynamic properties of MgH_2 [65].

Table 2. Thermodynamic parameters and energy storage properties of MgH_2 [65].

Thermodynamic Parameters	Values
Formation enthalpy, kJ/(mol.H_2)	−74.5
Formation entropy, J/(mol.H_2.K)	−135
Hydrogen Storage Capacity (Theoretical)	
Gravimetric capacity, wt%	7.6
Volumetric capacity, g/(L.H_2)	110
Thermal Energy Storage Capacity (Theoretical)	
Gravimetric capacity, kJ/kg	2204
Volumetric capacity, kJ/dm^3	1763

Strategies employed to overcome these issues include the addition of nanostructures, alloying, and MgH_2-based composites. The catalytic addition of different transition metals or transition metal oxides can greatly accelerate the hydrogenation/dehydrogenation kinetics [39]. Khan et al. [66] investigated two nanostructured MgH_2 and cobalt (Co) powders. The hydrogen storage properties of the $2MgH_2$-Co powder and $2MgH_2$-Co compressed pellet were analysed. Fast hydrogenation was observed in the de-hydride $2MgH_2$-Co compressed pellet, with about 2.75 wt% absorbed in less than 1 min at 300 °C, and a maximum hydrogen storage capacity of 4.43 wt%. The hydrogen absorption activation energy of the $2MgH_2$-Co compressed pellet was also lower than in the $2MgH_2$-Co powder. Banrejee et al. [67] prepared nanocrystalline magnesium and compared it with micro-crystalline magnesium. The developed nanocrystalline Mg exhibited improved properties with a higher hydrogen storage capacity of 6.24 wt% at 300 °C. Prolonged ball milling led to faster hydrogenation kinetics (up to 90% of the saturation value in 15.5 min at 250 °C) and a substantial decrease in the activation barrier. Nanostructuring has also been studied [68,69] with remarkable improvements in Mg-based storage properties. However, nanostructuring could result in poor thermal conductivity [9]. Additionally, the drawback in powder materials is usually due to their tendency towards coarsening and sintering during dehydrogenation/hydrogenation cycles. This is susceptible in Mg, which has a relatively low melting temperature and, thus, displays significant atomic mobility at the cycle operating temperatures [66].

Alloying is an effective and easy-to-handle method of improving the sorption property of MgH_2/Mg [9]. Intermetallic compounds of transition metals are among the catalytic materials that can facilitate the thermal storage processes in MgH_2 [69]. Usually, intermetallic hydrides are composites of a hydride-forming element at high temperatures and a non-hydride-forming element, such as Mg_2NiH_4 and Mg_2FeH_6. Research efforts aimed at reducing the reaction temperatures of these composite hydrides have been achieved through the addition or substitution of existing elements [70]. There seems to be a consensus that increasing the number of 3D elements would improve the kinetics by decreasing the activation energy of hydrogen desorption [71]. For instance, the enthalpy change associated with the formation of Mg_2FeH_6 at 500 °C was measured to be 77.4 kJ/mol H_2, lower than the reported values of 98 kJ/mol H_2 [62]. Zhang et al. [64] also reported a decrease in the hydrogen desorption enthalpy and initial dehydrogenation temperature of MgH_2 through incorporation of either Ti or Ni. Sulaiman et al. [72] reported that a 5 wt% K_2NiF_6-doped MgH_2 sample started desorbing around 260°C, which was a reduction of about 95 °C and 157 °C compared with the as-milled and as-received MgH_2. Additionally, the de/absorption kinetics were also improved significantly compared to the un-doped MgH_2. In another approach, Majid et al. [73] selected $TiFe_{0.8}Mn_{0.2}$, graphite, and Fe as additives. Compared to pure milled MgH_2 powder, they found that the dehydrogenation peak temperatures were decreased by 90, 160, and 165 °C for Mg-$TiFe_{0.8}Mn_{0.2}$-graphite, Mg-Fe-graphite, and Mg-$TiFe_{0.8}Mn_{0.2}$-Fe-graphite composites, respectively. The co-addition of $TiFe_{0.8}Mn_{0.2}$, graphite, and Fe exhibited synergistic effects in improving the hydrogen desorption properties of MgH_2.

The roles of Ti-based catalysis and its consequent hydrogen storage effects on MgH_2 were reviewed by Zhou et al. [65]. They concluded that the doping technique via Ti-based catalysis is a viable approach to enhancing the reaction of Mg-based materials. A comprehensive compilation of Ti-based catalysis of MgH_2 systems, corresponding synthesis approaches, and kinetic behaviours is presented in their review. On the other hand, Kumar et al. [74] performed calculations based on the first principles to investigate the dehydrogenation kinetics, considering doping at various layers of MgH_2 (110) surface with Ca, Al, Ga, Sc, Ni, Ti, and V. Doping at the first and second layers of MgH_2 (110) had a significant role in lowering the H_2 desorption (from surface) barrier energy. The screening approach found Al and Sc to be the best possible dopants at lowering desorption temperature while preserving similar gravimetric density and bulk modulus to a pure MgH_2 system. By extending frontiers, Jain et al. [75] conducted an investigation on the role of alkaline metal fluoride (MgF_2) as a catalyst in the hydrogen-storage behaviour of MgH_2. For 5 mol% MgF_2 admixed into MgH_2 powder, hydrogenation measurements at 335 °C showed 92% of absorbed theoretical capacity in less than 20 min (compared to 70% by pure MgH_2). Sorption studies further point to the possibility of complete absorption at low temperatures down to 145 °C. Again, cyclic measurements made at 310 °C revealed an inconsequential loss in the total storage capacity. These results implied that the sensitivity of the material to atmospheric conditions is low, and it is easy to handle. Thus, it can be employed in applications where operation at relatively high temperatures is insignificant.

A variety of dopants for MgH_2 has been reported and the respective Mg-based hydride materials have been enhanced. Despite improvement in the material properties, thermodynamic tuning remains a major challenge [76]. Present approaches have been successful in addressing it, to some extent, but much is still desired for practical application.

2.2.3. Dehydration/Hydration of Metal Hydroxides

Thermochemical heat storage with metal hydroxides results from a reversible reaction of water (steam) and metal oxides at high temperatures (~500 °C) and near-atmospheric pressures [8]. The alkaline earth metal hydroxides such as $Mg(OH)_2$, $Ca(OH)_2$, $Sr(OH)_2$, or $Ba(OH)_2$ have been considered as storage materials [39]. The initial candidate hydroxide–oxide pairs are $Mg(OH)_2$/MgO, $Ca(OH)_2$/CaO, $Sr(OH)_2$/SrO, and $Ba(OH)_2$/BaO. The theoretical turning temperatures and thermodynamic data of these hydroxides are presented in Table 3. The reactions are in the range of 70–1005 °C, though most of the reactions are too low for high-temperature application [11], usually occurring at medium temperatures of 250 < T < 450 °C. The steam partial pressure and the temperature drive the hydration/dehydration reactions [12]. Figure 4 shows some of the couples which could be used for TCES application.

Table 3. Equilibrium temperature and heat storage capacity of metal hydroxides [11].

Material	Temperature (°C)	Reaction Enthalpy (kJ/mol)	Gravimetric Energy Density (kJ/kg)
$Ca(OH)_2$/CaO	515	100.177	1352
$Mg(OH)_2$/MgO	265	77.745	1333
$Be(OH)_2$/BeO	70	51.276	1191
$Mn(OH)_2$/MnO	190	67.072	754
$Sr(OH)_2$/SrO	755	88.581	728.3
$Ba(OH)_2$/BaO	1005	93.462	545.47
$Ni(OH)_2$/NiO	70	47.846	516
$Zn(OH)_2$/ZnO	55	49.609	498.96
$Cd(OH)_2$/CdO	125	59.952	409.4

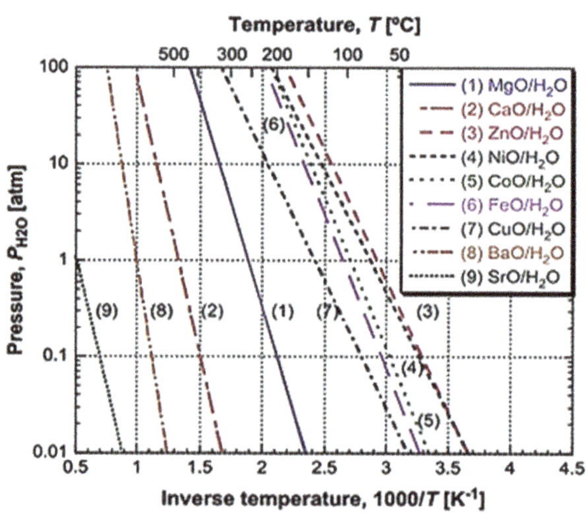

Figure 4. Reaction equilibrium lines for metal oxide/water reaction systems [8].

With reference to the high reaction enthalpies and energy storage densities (Table 3), mainly the hydroxides $Ca(OH)_2$ and $Mg(OH)_2$ have been extensively studied theoretically and experimentally [9,39]. However, the $Ca(OH)_2$/CaO system is more attractive [10]. It is the most-explored hydroxide system for thermochemical energy storage, prompting tests in both lab-scale reactors and TGA [14]. One reason is that the hydration of MgO is very slow in superheated steam and the rate of reaction drops with the rise in temperature [8,39]. Figure 5 shows the decomposition of the $Mg(OH)_2$ system at a relatively low temperature of around 330 °C [9]. The reaction enthalpy of $Mg(OH)_2$ degenerates with temperature up to 500 °C, unlike the enthalpy of $Ca(OH)_2$. In addition, CaO has a higher heat of adsorption over a short period [77] than MgO, as well as being much cheaper [8]. In view of these attractive attributes, a plethora of research efforts has been conducted for the potential application of $Ca(OH)_2$/CaO as a long-term thermal energy storage system.

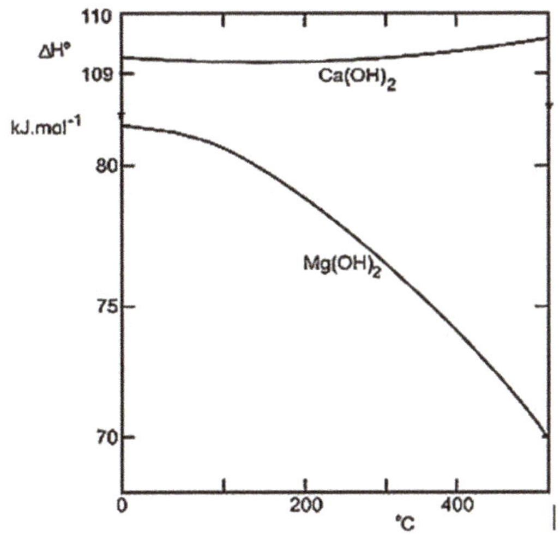

Figure 5. Enthalpy of dehydration reaction is dependent on temperature for $Ca(OH)_2$ and $Mg(OH)_2$ [39].

Schaube et al. [78] investigated a 10 mg sample in a $Ca(OH)_2$/CaO reversible system, and full conversion and cycling stability were reported over 100 cycles at a water partial pressure of 1 bar (even at 0.956 bar), with an equilibrium temperature of 505 °C and enthalpy of 104.4 kJ/mol. However, little success was achieved in the cycling stability of a 60 g sample of the reaction system as degradation was reported over 25 cycles. Agglomeration was also observed [79]. Additionally, Criado et al. [80] investigated $Ca(OH)_2$/CaO hydration/dehydration reaction and obtained higher rates than those reported in the literature at temperatures in the range of 400–560 °C and partial steam pressures between 0 and 100 kPa. However, particle attrition was observed for large particle sizes of the material. In another work, Dai et al. [81] investigated the cycling stability of the $Ca(OH)_2$/CaO system for 20 successive dehydration/hydration cycles. Existing problems relating to agglomeration, sintering, poor thermal conductivity, and irregularity in the rate of heat release were raised. There is a consensus that major problems encountered in $Ca(OH)_2$ reactors relate to particle agglomeration and sintering, poor heat transfer characteristics, and low permeability of the packed bed. Though the issue of permeability has been addressed in optimized reactors, other problems still exist [39]. As a result, much of the research has been around material enhancement through additives or composites, as well as reactor optimizations.

Criado et al. [82] synthesized composite materials using sodium silicate ($Na_2Si_3O_7$) to bind $Ca(OH)_2$/CaO particles for fluidized/fixed-bed application. The mechanical properties of the resulting CaO/Ca-silicate composites over hundreds of hydration/dehydration cycles were investigated. The results confirmed the primary role of $Ca(OH)_2$ anisotropic expansion as the main cause of the reduction in the crushing strength of the pellets. Another study by Funayama et al. [83] relates to a composite material using SiC/Si foam. The performance of the ~63 g of composite material with a packed-bed reactor was evaluated. The rate of heat output per volume under maximum hydration pressure of the composite was 1.3 kW/L-bed for the first 5 min, which is 1.4 times higher than previously reported for a bed of $Ca(OH)_2$ pellets. The composite material with pore size 400 μm of the $CaO/Ca(OH)_2$ samples maintained high reactivity and bulk volume during cycle reactions. This study was extended by using a ceramic honeycomb support composed of SiC/Si [84]. A volumetric energy density of 0.76 MJ/L-bed was obtained, and a heat output rate 1.8 times higher than the previously reported value for the pure $Ca(OH)_2$ pellet bed was achieved. In addition, the material also sustained high reactivity during the reaction cycles. Mixtures of expanded graphite (EG) with $Ca(OH)_2$ were also investigated by Kariya et al. [85] with the aim of enhancing the heat transfer and reactivity of the hydroxide. The results indicated that the maximum mean heat output of a sample mixture containing 11 wt% EG was twice as high as the heat of pure $Ca(OH)_2$. The decreasing effect of EG in the hydration reaction in the repetitive cycles was due to particle pulverization.

The doping of $Ca(OH)_2$ by hexagonal boron nitride (HBN) was approached by Huang et al. [86]. Analysis showed improvement in both thermal conductivity and dehydration enthalpy of the material. It also revealed a 15 wt% as optimal mass content of HBN-doped composite with improved activity after 10 dehydration/rehydration cycles. In addition, a 67% rehydration conversion and energy density greater than 1000 kJ/kg were achieved. Doping of $Ca(OH)_2$ with potassium nitrate, KNO_3, has also been reported by Shkatulov et al. [87]. With a 5 wt% KNO_3 addition to $Ca(OH)_2$, the dehydration temperature of the material was reduced and the reaction rates increased, but the material lost its dehydration heat by 7%. Wang et al. [88] obtained a similar result with a 10 wt% KNO_3 addition and the doped $Ca(OH)_2$ further showed good cycling stability in the nitrogen atmosphere, but failed in air. Gollsch et al. [89] modified $Ca(OH)_2$ powder with nanostructured flow agents to improve the powder's flowability. The additives consisted of nanostructured Si and/or Al_2O_3. Additives of weight fractions 6–12% improved the flowability of the powder. However, after cycling, the flowability of the mixtures decreased, while that of the pure powder increased. Analysis showed a correlation between growth in particle size and increased flowability. Additionally, the formation of phases in the additives led to a decrease in absolute heat release of up to 50%, although some of the side products seemingly added to the measured heat release by hydrating exothermally.

In tackling the problem of low conductivity and cohesive nature of powder bulk material, Mejia et al. [90] investigated ceramic-encapsulated CaO granules and $Ca(OH)_2$ granules coated with Al_2O_3 nanostructured particles. The results showed that both encapsulated materials did not change their shape after six-fold cycling. However, the Al_2O_3-coated sample exhibited volume expansion during hydration. There was a reduction in the reaction activity of the ceramic sample, whereas the performance of the Al_2O_3 sample was almost the same as the unmodified $Ca(OH)_2$ particles. Afflerbach et al. [91] investigated an encapsulated sample and good mechanical stability of the material was attained, followed by a considerably improved thermal conductivity. Again, over ten reaction cycles were attained in a lab-scale reactor. Additionally, issues concerning the poor flowability of the storage material and poor heat and mass transport with a strong agglomeration tendency could be overcome by persistent particle size stabilization. Of course, much is still required in terms of material enhancement, but research efforts so far show the $Ca(OH)_2/CaO$ system to have higher prospects for long-term TCES application.

2.2.4. Decarbonation/Carbonation of Metal Carbonates

Decarbonation/carbonation reactions of metal carbonates have also proven to be attractive high-temperature heat storage systems. In this case, heat is used to perform the endothermic breakdown of carbonate, and the products are CO_2 and metal oxide. The interest in carbonates is due to their relatively high operating temperatures (typically over 800 °C), high volumetric density, low operating pressure, non-toxicity, abundance, and cheapness [8,13]. The decomposition of $CaCO_3$, $SrCO_3$, $BaCO_3$, $MgCO_3$, and $PbCO_3$ has been studied [14]. Alas, the controversy about the high refractoriness of $MgCO_3/MgO$ and the toxicity of $PbCO_3/PbO$ [9] is a drawback for further research in these materials. Additionally, the carbonation reaction of BaO into $BaCO_3$ was hindered by the melting of the material during the decomposition step [92]. Therefore, among the carbonates, $CaCO_3$ is considered the most promising heat storage material [9,39] and the focus will be on the $CaCO_3/CaO$ system. It has been reported that after 40 high-temperature carbonation/decarbonation cycles with CaO, the carbonation (adsorptive) reaction significantly decreased because of the decrease in pore volume in the material [9]. This loss in porosity is caused by a decrease in the surface area of CaO due to the sintering of the particles [14,39], thereby inhibiting CO_2 access to the active sites within the material. Several techniques have been developed to minimize this loss in adsorption capacity. To increase the active surface area and stability of the pore structure, the use of additives, reduction in the particle size, and the synthesis of novel materials with the microporous structure were proposed [93].

To this end, Lu and Wu [94] doped nano CaO with Li_2SO_4 and showed that the Li_2SO_4-nano CaO adsorbent maintained a 51% conversion after 11 cycles, compared to pure nano CaO maintaining 27.3% under the same conditions. The superior performance of the Li_2SO_4-nano CaO adsorbent was attributed to pore enlargement and increase in macro-pore proportion through the Li_2SO_4 addition. Moreover, there were increased reaction rates and a lowering of the decomposition temperature by 15 °C in comparison with the pure material. In another work [95], different MgO concentrations were added to the CaO material. The additions with 5 and 10 wt% of MgO exhibited high CO_2 adsorption and retention capacity over multiple cycles. In particular, the CaO with 10 wt% MgO exhibited steady adsorption capacity over 30 cycles. Similarly, Wang [96] recently synthesized a porous MgO-stabilized nano CaO powder and realized highly effective long-term conversion because of its resistance to pore-plugging and sintering. Benitez-Guerrero [97] reported the synthesis of porous CaO/SiO_2 composites through a bio-template route using calcium nitrate, $Ca(NO_3)_2$, and rice husk as support. The morphology and composition of the biomorphic material improved the CaO multicycle activity, as it served to enhance CaO and inhibited pore-plugging effects. The influence of SiO_2 on $CaO/CaCO_3$ was also studied by Chen et al. [98]. The optimal 5 wt% SiO_2-doped $CaCO_3$ was demonstrated to enhance the reactivity and heat capacity, and led to a 28% enhancement of the reversibility, owing to the rise in grain boundary migration resistance. Table 4 shows the cycling stability achievements of some dopants, adapted from [98].

Table 4. Comparison of cycling stability of $CaO/CaCO_3$ after doping [98].

Doping Materials	Number of Cycles	Storage Conversion
SiO_2	20	62.09%
Li_2SO_4	11	51.0%
MgO	20	42.03%
$Ca_3Al_2O_6$	20	51.69%

Binary metallic elements and oxides have been experimented with recently. For instance, composites of CaO doped with Mn and Fe were reported to enhance the cycling stability of the TCES material [99]. A synergy between the small grain size and the reinforced skeletal structures prevented agglomeration of the composites, thereby enhancing their cycling stability. On the other hand, Sun et al. [100] reported that 5 wt% Al_2O_3 and 5 wt% CeO_2 co-doped on CaO showed the highest and most stable energy storage capacity under the carbonation pressure of 1.3 MPa during 30 cycles. In addition, the synthetic material possessed strong basicity and provided a large surface area and pore volume during the multicycle energy storage. Again, Raganati et al. [101] experimented with the application of an acoustic perturbation method that remarkably enhanced the carbonation performance of fine limestone particles. Indeed, it prevented agglomeration, which affects carbonation from both the gaseous (CO_2) and solid ($CaCO_3$) sides of the reaction, thus enhancing the fluidization quality, reactants' contact, and mass transfer coefficients. More information on in situ data of $CaCO_3$ doping samples, measurement parameters, and results can be accessed in the work of Moller et al. [102].

Calcium carbonate has the most economic advantage of being widespread, cheap, and having high gravimetric energy density (3029 kJ/kg). The high operating temperatures make this TCES system suitable for various applications such as integration with a solar furnace, and calcium-looping technology. However, this system is stable only up to 20 cycles without any degradation in the absorption capacity. The cycling stability and reversibility must, however, be improved up to 1000 cycles to make this system practical [8].

2.2.5. Deoxygenation/Oxygenation of Metal Oxides

Suitable transition metal oxides undergo a reduction reaction at high temperatures, through which thermal energy is absorbed. The reversible re-oxidation takes place below specific equilibrium temperatures and hence thermal energy will be delivered [103]. Thus, the reversible reduction/oxidation (redox) reactions of metal oxides show high potential as TCES materials. In comparison to the other TCES options, redox systems have the advantage of using air as both the heat transfer fluid (HTF) and the reactant. This eliminates the necessity for a different heat exchanger or gas storage needs. For this reason, TCES based on metal oxide redox reactions permits working with an open system [14]. In this case, it is important to investigate these systems in consistency with the control of oxygen partial pressure (pO_2) [14]. With lowering partial pressures of the reactive gas, the reduction temperature also decreases, as represented in the Van 't Hoff diagram in Figure 6.

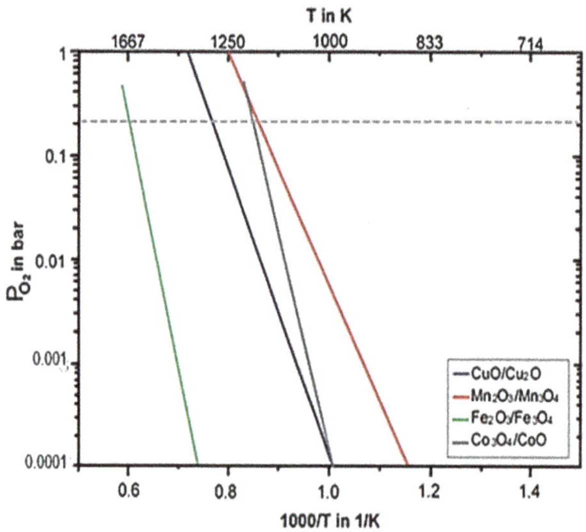

Figure 6. Variation of temperature with oxygen partial pressures for Cu, Mn, Fe, and Co oxides [103].

By comparing metal oxide systems, it was revealed that only cobalt oxide (Co_3O_4), iron oxide (Fe_2O_3), copper oxide (CuO), and manganese oxide (Mn_3O_4) showed befitting reaction temperatures, enthalpies, cycling stabilities, and material costs [103]. In another work, Silakhori et al. [104] assessed the redox reactions of CuO/Cu_2O, Co_3O_4/CoO, Mn_2O_3/Mn_3O_4, and Pb_3O_4/PbO using TGA. The results showed that CuO/Cu_2O and Co_3O_4/CoO were highly reversible under isothermal pressure-swing cycles, while Mn_2O_3/Mn_3O_4 exhibited slight signs of sintering, and Pb_3O_4 was unreactive up to 550 °C. The free Gibbs energy ($\Delta G°$) was determined for several oxides and PbO_2/PbO, PbO_2/Pb_3O_4, Pb_3O_4/PbO, CuO/Cu_2O, and Sb_2O_5/Sb_2O_3 were confirmed to show thermal storage attributes based on negative $\Delta G°$. Among these, CuO/Cu_2O displayed higher total enthalpy of 404.67 kJ/mol. However, the occurrence of phase transition was observed at temperatures near 1200 °C, and the molten state is prone to corrosiveness. Deutsch et al. [105] carried out kinetic investigations of the CuO/Cu_2O reaction cycle under isothermal and isokinetic conditions and used simultaneous thermal analysis (STA) and a lab-scale fixed-bed reactor. The outcome of the reaction resulted in substantial discrepancies between both analyses. In STA, outstanding stability of the reaction over 20 cycles was shown with some sintering occurring, whereas heavy sintering occurred in the reactor, which hampered the reaction as well as increased the reaction time three times higher than previously reported values in the literature. Alonso et al. [106] tested the suitability of CuO/Cu_2O in an argon atmosphere and the results indicated the reduction of CuO led to nearly 80% conversion. The reduction in air atmosphere was not favourable because of stronger coalescing particles that hindered the redox reactions. The synthesis of porous CuO-based granules with yttria-stabilized zirconia (YSZ) was also reported [107]. The synthesized granules exhibited high conversion over 100 consecutive cycles in air between 950 and 1050 °C. Stable cycling performances were also obtained in the reactor for 30 consecutive isobaric and isothermal operation modes.

Cobalt and manganese oxides have also been considered promising redox systems for TCES. The CoO/Co_3O_4 system has the potential to be the most suitable pure metal oxide system for TCES due to its fast reaction kinetics and complete reaction reversibility. However, cobalt oxide is also considered potentially toxic and would not be cost-effective for large-scale storage [108]. On the other hand, the Mn_2O_3/Mn_3O_4 redox couple is favoured in terms of minimal cost and toxicity in comparison to its alternatives and has been suggested as an appropriate material for TCES. However, several contentions have been singled out with respect to its capability of full energy storage sustainability with the required number of cycles necessary for this application [109]. In view of this, Bielsa et al. [109] studied several variables such as temperature and heating/cooling rates. A suitable choice of these variables was shown to enhance the heat storage capacity by 1.46 times in a 10-cycle test. The weight-change curves during the TGA are shown in Figure 7, although several levels of sintering were observed, proving the major drawback of this material.

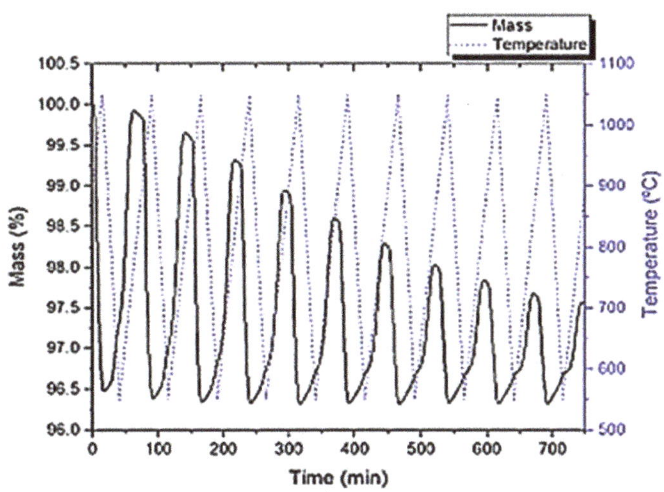

Figure 7. Weight-change curves during 10 Mn_2O_3/Mn_3O_4 TGA cycles between 500 and 1100 °C [109].

Andre et al. [110] studied the impact of Fe, addition which decreased the redox activity and energy storage capacity of Co_3O_4. However, the cycling stability of Mn_2O_3 was significantly improved with added Fe amounts above 20 mol% while the energy storage capacity was unchanged. Similarly, a mixed oxide of Co-Cu-O with low amounts

(<10 mol%) of Cu showed very good cycling stability and higher reaction enthalpy than the others (Mn-Cu-O and Co-Mn-O systems) [108]. Neises et al. [111] performed 30 cycles on a 5 wt% Al_2O_3-doped Co_3O_4 without any material degradation but yielding only a 50% conversion. This was attributed to the insufficient stirring and mixing of the metal oxide particle bed inside the reactor. Notwithstanding, about 400 kJ/Kg energy density was achieved per cycle. The graphical presentation of oxygen absorbed per mol of the doped Co_3O_4 during reduction is shown in Figure 8.

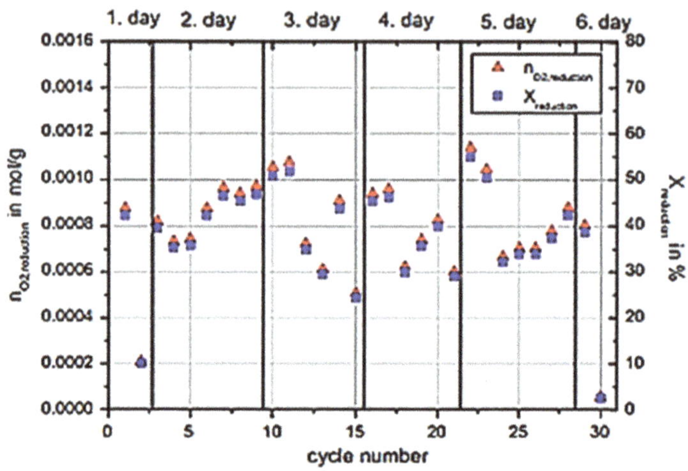

Figure 8. Amount of oxygen per mol of Co_3O_4 and conversion during reduction [111].

In contrast, Carrillo et al. [37] obtained results that indicated that operation with pure oxides (Mn_2O_3 and Co_3O_4) was more effective for TCES application compared to their mixed oxides. Although the values of heat released and absorbed for Mn_2O_3 were far from those obtained with Co_3O_4, its excellent cycling stability, low toxicity, and low cost make it an interesting candidate for heat storage applications. Relatively, the Cu_2O/CuO system has more prospects for TCES application than the CoO/Co_3O_4 system if the reaction is optimized in the reactor. In one experiment [112], isothermal runs at different oxygen partial pressures were carried out with TGA, and defined fractions of CuO samples were analysed. The results revealed that the oxygen partial pressure affects the kinetics, and the reparameterization of the pressure term influences the kinetic analysis of the oxidation reaction. It was concluded that the models described for various parameters provide indispensable prerequisites for the redox reactor designs [112].

So far, this section has presented focused discussions on the various aspects of TCES materials at low, medium, and high temperatures. It is, nevertheless, exigent to present a comparison of some techno-economic parameters of these materials, as summarized in Table 5. The materials in the table are representative candidates according to each temperature domain and could help in making preliminary considerations for suitable technology.

Table 5. Comparison of cost, efficiency, and charge/discharge cycles of TCES materials.

Material	Reaction Temperature (°C)	Energy Density (kJ/kg)	Price ($/kg)	Conversion Efficiency (%)	Number of Cycles	Reference
$MgCl_2.6H_2O$	130	940	0.18	87	25	[22,113]
MgH_2	300–480	2160	800–900	20	20	[8,12,22,113]
$Ca(OH)_2$	400–600	2000	0.1–0.18	49	100	[8,20,113]
$CaCO_3$	973–1273	3029	0.2–2.05	27.3	30	[8,9,12,22,113]
Co_3O_4	863–896	844	10–20	70	30	[8,104,113]

Ferchaud et al. [114] demonstrated that the heat stored (0.84 GJ/m^3) and released (0.71 GJ/m^3) in $MgCl_2.6H_2O$ was, respectively, 79% and 87% in the storage cycle. The small difference in the charging/discharging heat was attributed to the textural changes in the material. In another TGA/DSC test, the $MgCl_2.6H_2O$ material showed a 40% lower heat output after 25 cycles and a further 10% drop at the 28th cycle, compared to the first [115]. An open sorption system using $MgCl_2.6H_2O$ has also been tested, but the lower heat-recovery efficiency resulted in a power loss of nearly 70% [116]. It is common knowledge that the actual energy density obtained from experimental prototype tests differs significantly from the theoretical energy density [59].

For MgH_2, a multicycle system has been reported [117], where cycle 1 represented the activation phase at around 300 °C. The desorption kinetics showed a slight improvement from cycle 4, and stabilized in cycles 6–8 with 90% storage efficiency. However, performance degradation of the system was noticed after 10 cycles with a storage capacity reduction of 50% after 20 cycles. However, a solar-heated MgH_2 material at 420 °C showed the metal hydride to be thermally cycled more than 20 times with a minimal loss in hydrogen capacity [118]. A coupled TGA study of a modified MgH_2 powder also reported a cyclic conversion of 98.4% after 30 full cycles, with a calculated degradation rate of 0.00043 wt% per cycle [119].

In TGA experiments, a remarkable 100% conversion efficiency and cycling stability of $Ca(OH)_2$ has been proven over 100 cycles up to 95.6 kPa vapour pressure [78]. However, only 32 charge/discharge cycles could achieve 100% conversion efficiency at 100 kPa due to material structural failure [80]. Yan and Zhao [120] analysed the charging/discharging characteristics of $Ca(OH)_2$ and showed that heat storage efficiency increased with temperature (47% at 510 °C and 65% at 540 °C). However, higher heat-release efficiency could be achieved by reducing the temperature and increasing the vapour pressure. In particular, the conversion was 31.7%, 60.9%, and 72.8% under vapor pressures of 180 kPa, 240 kPa, and 320 kPa, respectively.

Generally, pure $CaCO_3$ does not show complete reversibility during decarbonation/carbonation cycles due to pore-plugging effects [9]. The best performance reported a 40-cycle high-temperature carbonation/decarbonation run. The initial cycle was always 100%, but the cyclic conversion progressively decreased due to the loss of pore volume [9]. A recent study on the multicyclic stability of different $CaCO_3$ minerals showed a limitation after 20 conversion cycles [97]. A comparative study of a sulphate-modified $CaCO_3$ found a 51% cyclic conversion over 11 decarbonation/carbonation cycles, which was higher than the 27.3% of pure $CaCO_3$ [121]. Thus, the addition of inert materials is seen as a viable option for improving the cycling stability of the $CaCO_3$. This is proven by CaO multicycle conversion data for different pretreated samples over 20 carbonation/decarbonation cycles carried out under calcium-looping-concentrated solar power (CaL−CSP) storage conditions.

For Co_3O_4, TGA results show rapid full thermal reduction and complete weight recovery for CoO oxidation to Co_3O_4 [11]. Hutchings et al. [122] reported 100 cycles of Co_3O_4 between 870 °C and 955 °C with 99% conversion efficiency and no evidence of degradation under these conditions. Again, Co_3O_4 powders exhibited long-term (30 cycles) performance with complete and reproducible cyclic redox performance within the temperature range of 800–1000 °C [123]. It has been observed that TGA results generally show higher oxidation fractions with lower cooling rates. Neises et al. [111] tested the solar-heated redox reaction of Co_3O_4 for 30 cycles in an integrated system and achieved a storage density of 400 kJ/kg per cycle. The system was only limited by insufficient mixing of the material.

3. Thermochemical Energy Storage Reactors for Solid–Gas Reactions

Reactors provide the platform for the operation of thermochemical storage systems. To guarantee efficiency in the charging and discharging process of the TCES material, an innovative reactor concept fashioned for particular operation and storage material is necessary [124]. That is, the nature of reactants or the type of reaction determines the type and design of the reactor and the system integration. Additionally, the art of designing efficacious reactors is as important as material enhancement itself. Different criteria have been used to classify reactors, such as operation mode, number of phases, reaction types, or a combination of these [125]. Based on the mode of operation, three major solid–gas reactors are implemented: fixed (packed)-bed, moving-bed, and fluidized-bed reactors [8], as shown in Figure 9.

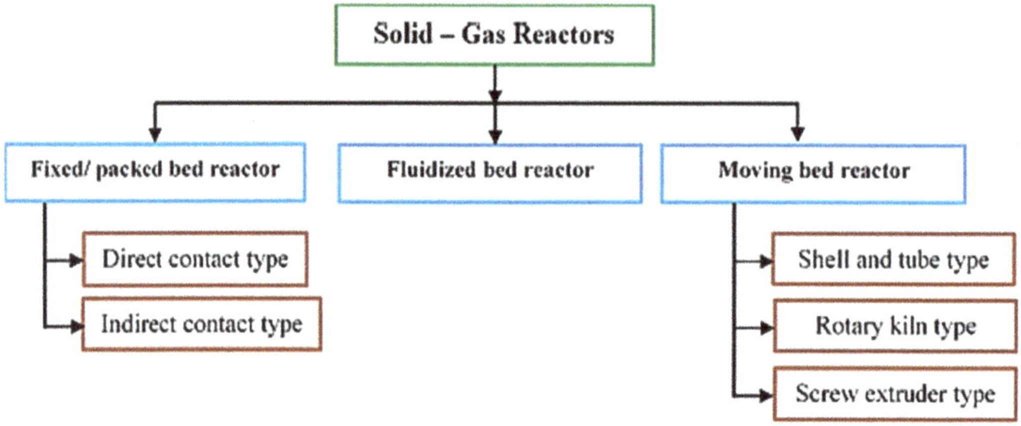

Figure 9. Classification of solid–gas reactors [8].

3.1. Fixed-Bed Reactors

Fixed or packed beds are hollow tubes, pipes, or other vessels filled with packing materials for chemical processing. The packing material may contain catalyst particles or adsorbents, and the purpose of such a bed is to improve contact between two phases in a chemical process. Thus, in packed-bed reactors, the solid reactants/products are arranged in the vessel's bed during heat storage and release, with the flux of reactants passing through the stationary bed. The HTF and reactants are not in contact with each other, resulting in heat transfer through the walls of heat exchangers.

Fixed-bed reactors are the most common lab-scale test rigs and have been investigated by Schmidt et al. [126]. In this work, several dehydration and hydration cycles were performed to study the charge and discharge characteristics of a 20 kg $Ca(OH)_2$ material. The dehydration was performed at 45 °C and the rehydration at about 55 °C. A conversion of 77% without degradation after 10 cycles was achieved. Investigation of different charging and discharging temperatures was then recommended to optimize the overall conversion of the material. Again, Ranjha et al. [127] modelled a two-dimensional rectangular bed filled with $CaO/Ca(OH)_2$ powder for different flow geometries and bed properties. They showed that increasing the porosity of the bed provided higher energy density but slowed the reaction, resulting in a lower average outlet temperature. A possible remedy for this deficiency would be to increase the bed dimensions, but this could result in a slower reaction owing to the poor thermal conductivity of the materials. This suggests that a systematic optimization process could be applied as a compromise between various parameters. However, this depends on the desired output temperature, energy requirements, and the rate of thermal energy storage and retrieval. Figure 10 shows the results of comparing varying bed porosities with total conversion time and the maximum outlet temperature in counter-current flow geometry.

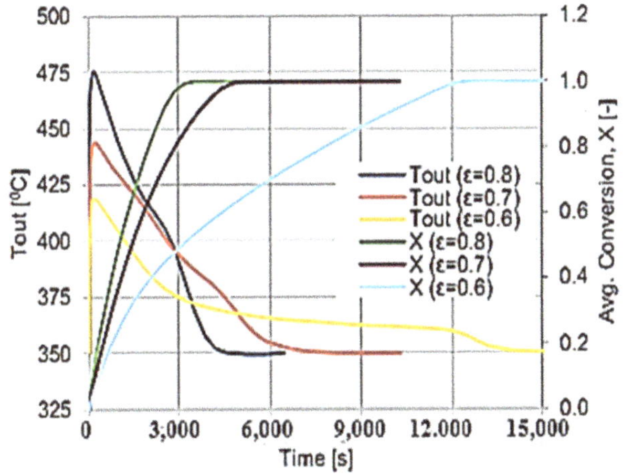

Figure 10. Outlet temperature of the HTF and the conversion as a function of time during hydration for bed porosity of 0.6, 0.7, and 0.8 in counter-current flow geometry [127].

On the other hand, Funayama et al. [128] evaluated a 60 g $Ca(OH)_2$ pellet in a packed-bed reactor. The heat storage density of the bed was 1.0 MJ/L-bed, and an average heat output rate of 0.71 kW/L-bed was observed for the first 10 min under a hydration pressure of 84.6 kPa. Although the bed showed a net expansion and formation of agglomerated lumps in the middle, the effects had a small influence on the reactivity. The stability of the reaction conversion of the bed was demonstrated during 17 cycles of experiments. Figure 11 shows the experimental setup for the test rig.

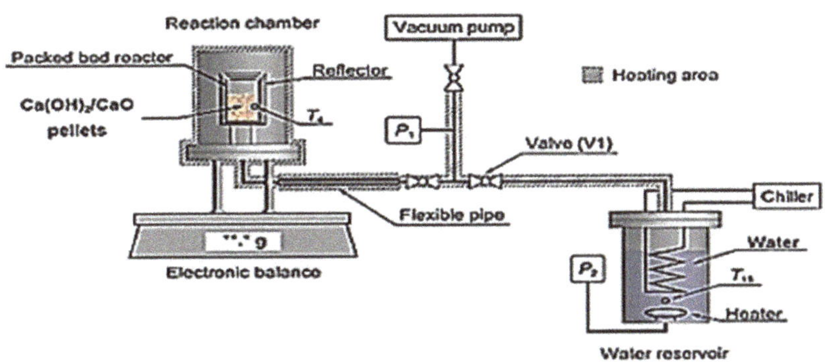

Figure 11. Schematic diagram of the experimental setup in Funayama et al. [128].

It is established that the heat transfer coefficients in indirectly heated fixed beds are generally limited by the low thermal conductivity of the reactants [129]. Therefore, a fixed-bed reactor in indirect operation at varying technical operating conditions was investigated by Schmidt et al. [130]. Thermal charging and discharging were experimentally demonstrated at vapour pressures between 1.4 kPa and 20 kPa. This indicated the possibility of operating the system at low vapour pressures, thus raising the total efficiency of the storage system. However, the range of operation of the $Ca(OH)_2$ system was constrained because of its efficient rate of reaction at low vapour pressure. Again, Peng et al. [131] simulated fixed-bed reactors under various operating conditions, and the impact of key process parameters was evaluated. An optimized model for the design was then used to compare the performance of three TCES reactors for Mn_2O_3, $Ca(OH)_2$, and $CaCO_3$. The results showed $CaCO_3$ and $Ca(OH)_2$ had more favourable conversion efficiencies.

Moreover, the HTF inlet and outlet temperature profiles also indicated the two aforementioned reactors rapidly reaching the endothermic reaction temperature. These results are shown in Figure 12.

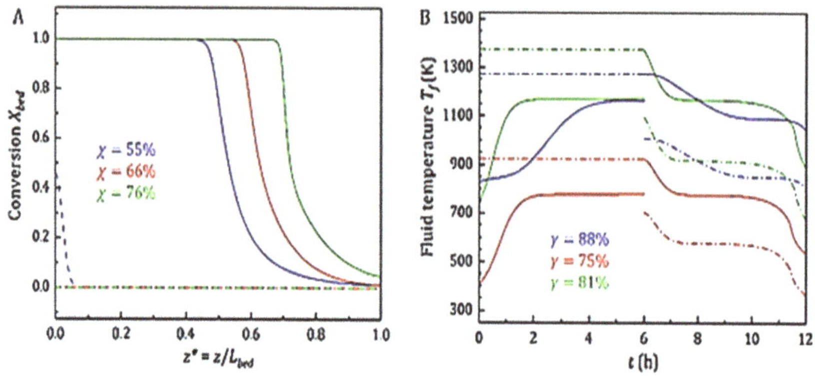

Figure 12. (**A**) Bed conversion efficiency and (**B**) fluid temperature of fixed-bed reactors. In (**A**), conversions at the start of charging (t = 0) are shown as dashed lines, and those at the end of charging (t = 6 h) as solid lines. In (**B**), fluid temperatures at the inlet are shown as dash-dotted lines, and those at the outlet as solid lines (blue: Mn_2O_3, red: $Ca(OH)_2$, green: $CaCO_3$) [131].

On the other hand, Schaube et al. [132] earlier investigated a reactor with direct heat transfer for the $CaO/Ca(OH)_2$ material. The simulated results of a 2D model developed showed good agreement with experiments. However, deviations were observed in temperature characteristics with an increasing flow rate of the HTF. This was due to the overestimation of the reaction rate. To forestall this in further research, a derived kinetic equation would be adapted to the actual conditions, for instance, particle size, to account for diffusion limitation. Then, the developed model could be used to predict the performance of a reactor with direct or indirect heat transfer.

Therefore, integrating these heat storage systems into industrial processes needs additional theoretical and experimental investigations [28]. Improving the low thermal conductivity, for instance, in an indirectly heated fixed bed requires a large heat exchanger, which adds cost. One way to surmount this challenge is to separate the heat exchanger (power) of the costly reactor from the storage material (capacity). A moving-bed concept in which the material moves through the reactor could be used to accomplish this [133].

3.2. Moving-Bed Reactors

In moving-bed reactors (MBRs), the bed can be shuffled in continuous or regular intervals in portions, and the flow of fluid is similar to what happens in a fixed bed. Moving-bed reactors were thought to have the advantage of improving thermal conductivity through enhanced solid–gas particle interaction. These solid–gas reactors can be categorized into two different regimes: (i) the axial-flow regime (concurrent and counter-current) and (ii) the crossflow pattern [134]. For instance, in the axial-flow solid–gas moving bed configuration, the advantages of the counter-current flow regime have been applied through the direct interaction of reactants. In this case, the solid flows downward while the gas moves upward as the chemical transformation occurs.

Notably, fine powders were observed to have very low flowability [135]. To improve this, a 1D simulation of a manganese-iron oxide (Mn-Fe oxide) TCES reactor was investigated [136]. An extension of the particle flow model was simulated for a counter-current MBR, and complete conversion was attainable only for low gas as well as low rates of solid flow. Thus, the oxidation kinetics of the redox transition was the limiting factor. Moreover, a contrasting trend was observed in which the energy density dropped, although the thermal power surged independently of the attained conversion efficiency in the given conditions of operation. (See Figure 13). Additionally, sensitivity analyses revealed the possibility of channels developing within the moving bulk material, which could slow down the heat transfer between solid and gas. Hence, a proposal for direct and indirect heat transfer coupling was recommended as a promising operational mode.

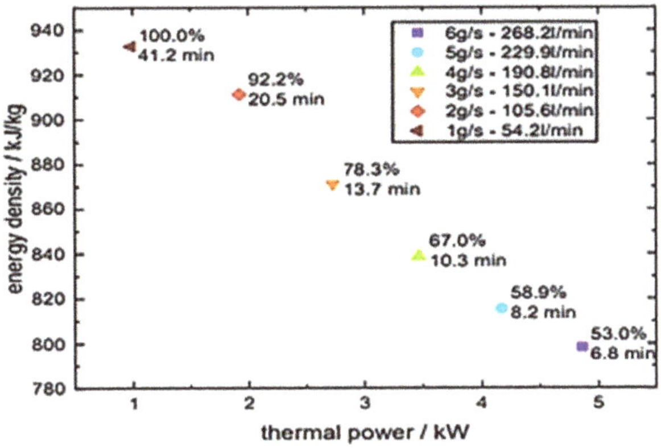

Figure 13. Thermal power of the MBR for the steady-state case and energy density of the material with different particle and gas flow rates [136].

Furthermore, Preisner and Linder [137] investigated the reaction behaviour of manganese-iron oxide (Mn-Fe oxide) at different temperatures and pressures. The TCES material was oxidized to an extent of 80.2% after a previous reduction of 77.1% at 20 kPa oxygen partial pressure. Moreover, a sufficiently stable oxidation/reduction reaction was established for two successive cycling tests (of 30 cycles each) in TGA. However, the flowability of the material particles was limited at high temperatures between 850 and 1050 °C due to mass loss, as shown in Figure 14.

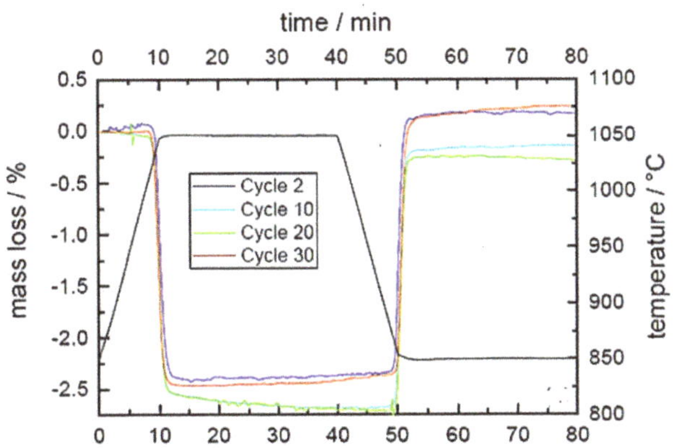

Figure 14. Relative mass loss of the Mn-Fe oxide during the first 30 redox cycles [137].

Huang et al. [138] investigated the complex physical and chemical transport phenomena in MBRs. The coupled heat/mass transfer and reactions were successfully modelled based on uniform flow and plug flow. Both models showed potency for simulating the transient flow processes. In particular, the plug flow model based on Ergun's equation successfully predicted the increased gas velocity inside the reaction zone. However, due to a lack of exact information on the heat flux input during the transient phase, the reaction zone temperature was overpredicted, and this led to overestimated oxygen concentration at the exit. In another concept, a rotary kiln was also tested for suitability in a thermochemical storage process [106]. Rotation movement helped to improve the reactiveness of the sample by mixing the particles. However, the reaction was handicapped because of strong coalescing particles. Additionally, the chemical conversion was significantly lower for oxidation than for reduction. For this reason, optimization of the operation parameters such as rotational speed, particle size, gas flow, and initial mass of reactant was recommended. This would increase the conversion of both reduction and oxidation and avoid the progressive loss of reactive fraction. So far, no successful operation of a moving-bed reactor over multiple cycles has been reported [135].

3.3. Fluidized-Bed Reactors

For fluidized-bed reactors, the fine solid particles are sustained in suspension by the fluid's upward flow [57]. Essentially, the base of a fluidized bed is similar to that of a fixed bed. With the increase in the fluid velocity, the solid particles are suspended at the minimum fluidization velocity. Better heat and mass transfer can be accomplished with fluidized-bed reactors (FBRs). FBRs require significantly lower gas velocities [135], therefore, they have been extensively suggested for more efficient TCES. However, getting the material fluidized, which in turn demands a huge increment in gas flow volume, has the consequence of decreasing the efficacy of the stowage mechanism. Notwithstanding, the advantages of fluidized beds over other reactors include better reaction efficiency due to enhanced fluid-particle interaction. Again, besides the limited pressure drop across the bed, fluidized beds efficient HAM transfer from the continuous particulate interactions and higher transfer coefficients [139]. In addition, the uniform temperature gradient ensures the elimination of dead zones and hot spots. A comparison of the three solid–gas reactors is summarized in Table 6.

Table 6. Comparison of fixed-, moving-, and fluidized-bed reactors [24].

Reactor	Advantages	Disadvantages
Fixed/Packed bed	• Easier modelling	• Low heat and mass transfer • High-pressure drops
Moving bed	• Direct solid–gas heat transfer	• Complex hydrodynamics
Fluidized bed	• Minimization of hotspots and thermal instability • High heat transfer coefficients	• Complex hydrodynamics and modelling • Internal component erosion

Generally, TCES technology is still at the infancy level, notwithstanding the increasing investment in research. So far, no known commercialization project has been launched, but a number of lab-scale projects and, in some cases, prototypes are reported [26]. We, again, present a focused outlook on the technology status of the potential TCES systems in the low to high-temperature range (Table 7).

Table 7. Examples of TCES system prototypes and projects.

Reference	Storage Material	Technology	Description
Zondag et al. [140]	$MgCl_2.6H_2O$	Packed bed	▪ Prototype for solar heat storage ▪ Produced heat for 40 h ▪ Power transfer efficiency of 33%
Delhomme et al. [141]	MgH_2	Tank rig	▪ Reactor coupled to an industrial heat source ▪ Degradation after 10 cycles ▪ Achieved 60% efficiency

Table 7. *Cont.*

Reference	Storage Material	Technology	Description
Schmidt et al. [126]	$Ca(OH)_2$	Fixed bed	▪ Reactor with indirect heat transfer ▪ Achieved 10 stable reversible cycles ▪ Attained 77% efficiency and 10 kW power
Meier et al. [142]	$CaCO_3$	Rotary kiln	▪ Solar-heated moving-bed reactor ▪ Achieved 90–98% conversion rate ▪ 10 kW power; thermal efficiency of 20%
Neises et al. [111]	Co_3O_4	Rotary kiln	▪ Solar-heated prototype reactor ▪ Achieved 30 stable cycles with 90% efficiency

It can be seen that current projects or prototypes involve mainly packed (or fixed) beds and moving beds (rotary kilns). Fixed-bed reactors can be operated continuously, are cost-effective, and offer high reaction conversion per unit mass of material. A rotary kiln, on the other hand, serves to raise materials to high temperatures. The rotation of the vessel ensures a continuous movement of materials between the entrance and exit of the kiln. In the literature, fixed-bed and moving-bed prototypes have a reactor efficiency of between 12 and 69%, whilst the whole (integrated) system efficiency is between 12 and 42% [21]. In terms of the overall cost of TCES technology, a probabilistic analysis is necessary to account for the variability and uncertainty of transient factors [143]. A method to estimate the capital costs and the profitability of TCES systems has been developed by Flegkas et al. [144].

Designing a gas–solid reactor is influenced by three factors: reaction system kinetics, particle size sequence, and particle flow model for solid and gas in the reactor [57]. This supposes a requirement for system modelling to deepen understanding and help in the simulation of processes. Moreover, the research effort is currently aiming to gain new insights into the reactants' basic molecular level, as well as scale up improvements and optimization of fluidized beds [139]. Admittedly, the modelling of these systems is complicated, and more so when storage materials are to be included. Many researchers have carried out work on the modelling of physical phenomena in TCES reactors. Thus far, numerical or experimental efforts have rarely investigated the physical and chemical processes at the particle scale [28]. Moreover, it is only recently that energy and exergy analyses on conceptual thermochemical heat storage systems are being implemented.

Flegkas et al. [145] proposed a method to model FBRs based on solid-state kinetics and fluidization hydrodynamics, using the $MgO/Mg(OH)_2$ reaction couple. The reaction enthalpy dropped significantly using water, hence the supply of steam constituted the main drawback. In another work, Angerer et al. [135] proposed a reactor concept featuring a bubbling fluidized bed with continuous, guided solid flow and immersed heat exchanger tubes. Fluidization of the $CaO/Ca(OH)_2$ powder proved herculean, but challenges were surmounted with the use of mild calcination settings and a peculiarly designed gas distributor plate. Analyses revealed that heat transfer between the reactor and the immersed heat exchangers had the largest influence on the system performance. The results of the first steady-state experiments in a new power plant implemented to investigate the idea under practical reactor conditions were reported by Rouge et al. [146]. The basic conceptual scheme of the integrated system is shown in Figure 15. The experimental results during dynamic and steady-state periods were fitted to a KL reactor bubbling-bed model. The reactor performance modes were sufficiently predicted by the model, as observed during the experiments under dynamic and steady-state conditions. This model would be a relevant tool for the future expansion of this energy storage technology.

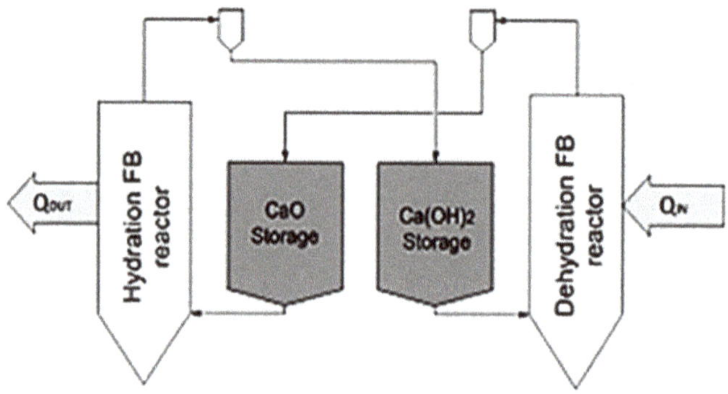

Figure 15. Basic conceptual scheme of the CaO/Ca$(OH)_2$ energy storage system [146].

In addition, Mu et al. [147] carried out a numerical simulation study of the thermal behaviour and hydrodynamics of FBR using the computational fluid dynamics-discrete element method (CFD-DEM). The effects of superficial gas velocity, bed height, and heat source distribution were analysed. The results showed that both the gas superficial velocity and the bed aspect ratio had a profound influence on fluidization behaviour and temperature distributions. Figure 16 shows the voidage as a function of the superficial velocity and aspect ratio. This work is similar to that of Hawwash et al. [18], in which the reactor design and area ratio were shown to impact the thermal performance and energy storage during the dehydration of a TCES material.

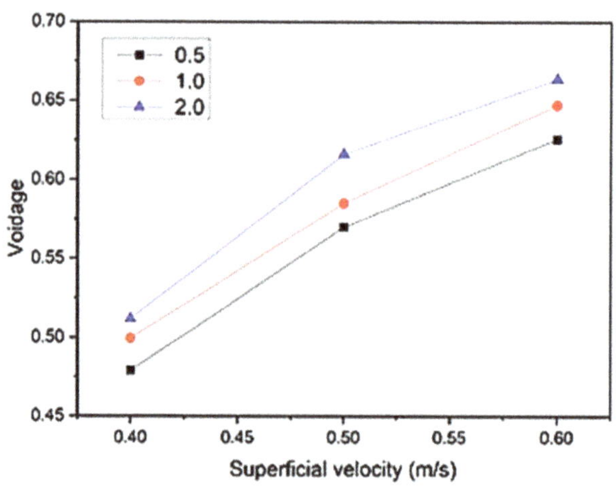

Figure 16. Temporally and spatially averaged bed voidage as a function of superficial velocity and aspect ratio [147].

Despite these milestones in understanding reactor dynamics, technical barriers still exist in conventional thermochemical energy storage systems. Efforts to enhance reactants with various additives or using different geometries of reaction beds have not abated the dynamic limitations of HAM transfers. For this reason, Darkwa et al. [148] proposed and investigated an agitated fluidized-bed reactor system. The model revealed substantial improvements in adsorption capacities and enhanced heat transfer rates. However, the thermophysical variables that influenced the minimal fluidization velocity in the adsorption column needed to be optimized. This would be necessary for efficient exothermic reaction and thermal exchange. The effect of agitation on the fluidization characteristics of fine materials was validated with different materials. Kim and Han [149] used fine particles

(22 μm) of material in a fluidized bed. The agitator was of the pitched-blade turbine type. The results showed that smoother fluidization was achieved with increasing agitation speed. Additionally, agglomeration and channelling were reduced by the mechanical effect of the agitation. Spectral analysis of the pressure drop fluctuation had the shape of a short-term correlation with different agitation speeds (See Figure 17). The void fraction also increased with the increasing speed of agitation at the constant fluidizing gas velocity, as shown in Figure 18.

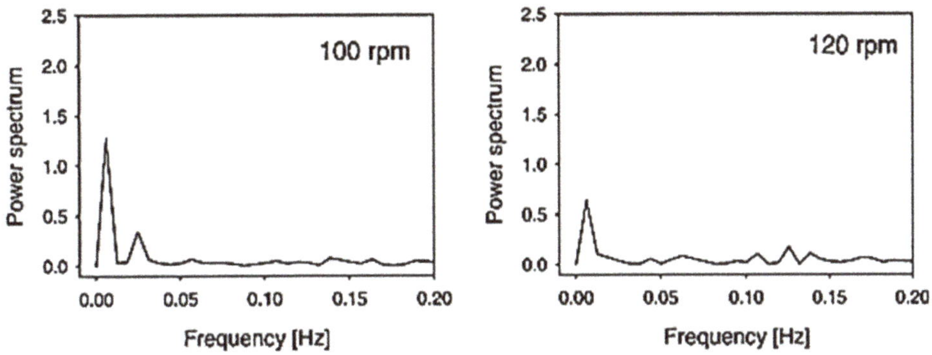

Figure 17. Power spectrum of the bottom zone of the bed for different agitation speeds [149].

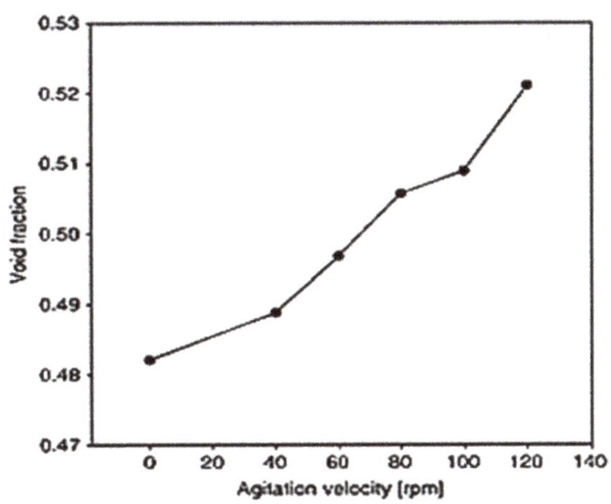

Figure 18. Graph showing an increase in the void fraction of powder material with agitation speed [149].

Again, the effect of agitation on fluidization in a solid–gas fluidized-bed reactor with a frame impeller was studied using a 3D unsteady CFD method [150]. The numerical method combined a two-fluid model and the kinetic theory of granular flow. The results showed that a substantially high agitation speed yielded higher performance in fluidization in addition to reduced bubble diameters and internal circulations of particles. Lv et al. [151] also reported the effectual performance of a fluidized bed under mechanical agitation, where the symmetrical motion pattern of the bed particles was identified. Similarly, the bubble diameter was also significantly reduced with increasing agitation speed, as presented in Figure 19. Furthermore, encouraging results showing the reduction in entrainment rate and channel flow and enhanced particulate interaction have been reported with an arch agitator [152]. Again, using a 3D CFD model, Shi et al. [153] established strong evidence of improved fluidization efficiency and reduction in the operation stability of a solid–gas fluidized bed.

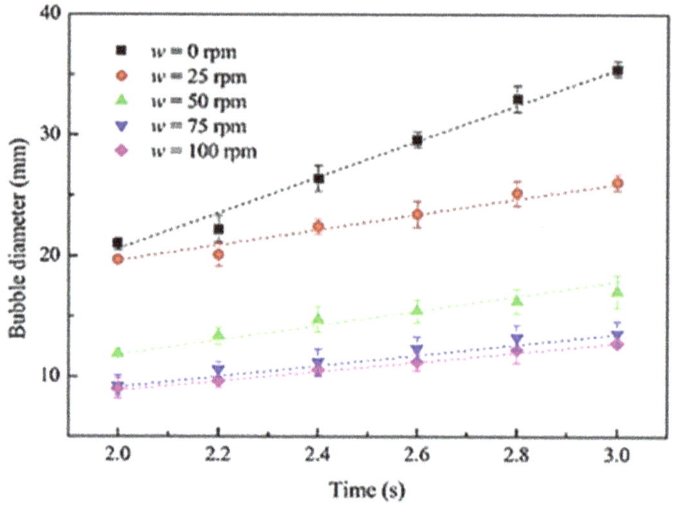

Figure 19. Graph of variation of bubble diameter with agitation speed [151].

A variety of reactor types have been developed, as found in the reviews of [19,28]. A concise summary of different reactor classifications, descriptions, and reactor and/or integrated system efficiencies is available in the review by Gbenou et al. [21]. Much research effort has been devoted to the development of thermochemical reactors, but few laboratory prototypes have been tested in large-scale pilot projects. In a recent review [26], the low technology-readiness level of TCES was attributed to the cost or performance of materials and reactor design. To improve on this rating, the reviewers canvassed a shift from microscopic views of numerical design processes to macroscopic comprehension of reactor design and analysis. Their paper, therefore, focused on low-temperature reactors and, again, reiterated the promise of FBRs. One significant highlight of the work was a proposal for computational resources (COMSOL Multiphysics and TRNSYS software) for the visualization of reactor phenomena for a better understanding of the transient processes.

4. Overview of Modelling and Simulation in TCES Systems

Modelling and simulation have the potential to improve understanding of the complex phenomena in TCES, either at the level of the material or in integrated systems. So far in this review, a good number of numerical studies have been cited. However, this section is intended to give an overview of the modelling and simulation approaches being explored by researchers for insights, analysis, or validation in thermal storage experiments. In the last couple of years, numerical modelling tools have been increasingly used, and the growing capability in computational power is increasing in reliability and accuracy, as well as lowering the time and cost invested in traditional experiments. Already, research efforts are currently aiming to gain new insights at the reactants' basic molecular level as well as scale up improvements and optimization of fluidized beds.

At the level of the material, for example, TCES can benefit from the achievements of molecular dynamics (MD) simulations in establishing good knowledge of thermophysical properties in materials such as phase change materials (PCMs) [154]. In particular, MD simulation with LAMMPS (Large-Scale Atomic/Molecular Massively Parallel Simulator) code has proven to be a capable tool for calculating the trajectories of particulate interactions in materials under some boundary conditions, and there is the possibility of obtaining desired thermodynamic information. For instance, to adequately decipher the metastability in a nitrate/MgO system regarding charging and discharging reactions, Shkatulov et al. [155] carried out MD simulation of $LiNO_3$-doped $Mg(OH)_2$ and pure $Mg(OH)_2$ systems to understand the difference in equilibrium temperatures for the dehydration process. The results showed striking proximity of the theoretical and experimental values. Again, agglomeration is one of the recurring challenges, especially in TCES with powder materials. In a quest for deeper insight into its occurrence, simulations of the agglomeration behaviour of the $CaO/Ca(OH)_2$ system were carried out [156]. The results revealed that the agglomeration rate was slower during discharge cycles of CaO in the presence of H_2O. In addition, the introduction of silica (SiO_2) particles into CaO was shown to effectively subdue agglomeration of the reactants. Thermophysical properties of CaO have also been investigated by Alvares et al. [157]. The obtained theoretical results provided accurate perspectives for calculating the thermodynamic properties in a temperature region difficult to access through experiments.

In complementarity to the atomistic scale simulation of TCES materials, suitable reactor concepts such as FBRs also require optimization of HAM flow, modelling, and simulation of the impact of transient effects. A large body of literature employs a variety of CFD-based models for the analysis and prediction of processes in fluidized-bed systems. The CFD-DEM model has been used to study the thermal behaviour and hydrodynamics in FBRs [147,158]. In one work, the fluidization velocity and bed aspect ratio impacted the temperature distribution [147], indicating that multiphase flow influences heat transfer in the FBR. In another, the particle size, density and bubble formation essentially affected the thermochemical properties of the fluidized material [158]. Shen et al. [159] carried out a feasibility study of a solar-integrated fluidized bed employing the CFD Eulerian model using ANSYS-FLUENT. Analysis of the results showed robust performance of the bed at high operating temperatures, but stabilization of the outlet temperature was required for optimal yield. Similarly, a 2D axisymmetric model of the Eulerian type was used to investigate the $MgO/Mg(OH)_2$ system [160]. The bed temperature was shown to be unaffected by the efficiency of the heat transfer, but the reactor throughput was limited by the kinetics of the hydration reaction. Essentially, the rate of gas flow and bed expansion ratio largely impacted the system heat stowage. The numerical results were validated experimentally by investigating the energy flow and dissipation optimization in the FBR. TRNSYS was also used by Li et al. [161] to test the $MgO/Mg(OH)_2$ system for solar thermal stowage. Analysis of the simulated charging/discharging phases of the TCES system indicated a contribution of 94.6% of the heating demand during the exothermic phase. Furthermore, AspenPlus was used to design, model, simulate, and optimize a combined charge/discharge process for a $CaO/Ca(OH)_2$ system coupled with concentrated solar power [162]. A sensitivity analysis of the results of the fluidized-bed-based process flowsheets was comparable to the reported models in the literature.

Although a plethora of literature exists on different models applicable to various fluidized bed systems, CFD provides a vast array of tools. In a comprehensive state-of-the-art review of CFD methods, Alobaid et al. [163] described the CFD-DEM model as becoming formidable if coupled with chemical reactions and HAM transfer, yet still requiring upgrading to be competitive. In particular, CFD-DEM simulations are reported to be computationally costly in tracking an avalanche of particles in solid–gas flows requiring the detection of particle–particle and particle–wall collisions. According to the reviewers, two-fluid models, for example, ANSYS-FLUENT, have the advantage of flexibility in the implementation of new algorithms and chemical or physical models, such as in the implementation of the so-called user-defined function (UDF). The major challenge with fluidized-bed modelling lies in the nonlinear and nonequilibrium behaviour of the solid–gas flows [163]. This implies a requirement for a combination of more than one model in the modelling and/or simulation processes. In this review, we emphasize the need to harmonize efforts from the viewpoint of material development processes to conceptual reactor design issues, by modelling and simulating the integrated systems. A specific potential material of interest could be enhanced according to its identified limitations and intended application. Molecular dynamics (MD) simulations, especially with LAMMPS code, can give useful numerical details of the kinetics and thermodynamics of the reactions. This molecular scale insight will be of enormous help in obtaining useful physical parameters for reaction conditions, design, and optimization of suitable reactor systems. MD simulations of this nature are, therefore, recommended for various TCES materials of interest in the desired temperature range of applications.

For the reaction bed, it is auspicious that HAM flow optimization, modelling and simulation of transient effects, and the charge/discharge reaction kinetics, among others, can be investigated in the FBR. We propose that modelling and simulating the FBR with an incorporated agitator mechanism will be a novel approach to enhancing HAM transfer coefficients via rigorous particulate interactions and shortened residency times. Together with a suitable heat exchanger framework, the overall efficiency of the system will be raised. The reactor geometry and the physical phenomena occurring within it could be visualized using capable computational tools such as AspenPlus software. There are good prospects for modelling such a transient state with ANSYS-FLUENT that can couple the multiphase flow and chemical reaction, on the premise of the Eulerian-Eulerian model, heat transfer, and kinetics equations. The influence of bed and fluid flow dynamics on thermal storage and release could then be established upon analysis of the reaction processes.

5. Conclusions

This review looked at research efforts in TCES from the viewpoint of energy storage materials to reactor systems. The work mainly focused on numerical and experimental studies carried out on promising storage materials from low to high temperatures. Additionally, the three conventional solid–gas reactor concepts, namely fixed-bed, moving-bed, and fluidized-bed reactors were reviewed. Metal hydroxides and metal carbonate systems, especially Ca-based, are currently under renewed investigation. Their attributes of high energy densities, relative cheapness, and non-toxicity have encouraged interest in developing compact and low-cost systems for industrial waste heat utilization.

Most promising materials investigated still suffer from agglomeration and low thermal conductivity, which has limited their performance in the reactor. Efforts have been made to improve the properties of the materials with various additives with some levels of success, but the problems of poor mass and heat transfer in different reactor configurations have remained. Coupled TGA-DSC procedures have been employed to characterize TCES materials, and these can be helpful for preliminary material selection. Nonetheless, for appropriate characterization of TCES materials for meaningful reactor design, the materials need to be characterized in an integrated lab-scale system. Generally, TCES materials are asserted to show better efficiencies in fluidized-bed reactors than fixed and moving-bed reactors. Therefore, experimental and numerical modelling of the system is an essential step in the right direction. Optimizations can be carried out to enhance HAM transfer in the fluidized-bed reactor, for instance, by increasing the contact between the solid and gas. We recommend:

1. Developing novel TCES materials with lower agglomeration tendency, improved thermal conductivity, cyclability, and storage temperatures. This can be achieved by exploring various composites and the addition of suitable nanomaterials;
2. Redesigning fluidized-bed reactors with novel heat exchangers with thermal properties for enabling efficient heat transport over the charge/discharge phases to the HTF;
3. Incorporating a suitable agitator mechanism in the fluidized-bed reactor to facilitate robust particulate interactions, for enhanced HAM transfer coefficients;
4. Finding appropriate simulation models to describe the kinetics of the TCES materials and coupled HAM transfer components of the fluidized-bed reactor. This could be one way to understand all aspects of the integrated thermochemical storage system, and hence maximize the storage potential of this technology.

Author Contributions: Conceptualization, A.K., J.D., J.C., R.B. and M.W.; writing—original draft preparation, A.K.; writing—review and editing, A.K., J.D., J.C., R.B. and M.W.; supervision, J.D.; project administration, J.D.; funding acquisition, J.D. All authors have read and agreed to the published version of the manuscript.

Funding: This research was supported by the Engineering and Physical Sciences Research Council (EPSRC), UK, [grant number EP/V041452/1].

Data Availability Statement: Not applicable.

Acknowledgments: The authors are grateful to the Tertiary Education Trust Fund (tetfund), Nigeria for Anti's studentship award.

Conflicts of Interest: The authors declare no conflict of interest.

References

1. Kumar, M. Social, Economic, and Environmental Impacts of Renewable Energy Resources. *Wind. Sol. Hybrid Renew. Energy Syst.* **2020**, *1*, 89494. [CrossRef]
2. Cooper, S.J.G.; Hammond, G.P. "Decarbonising" UK Industry: Towards a Cleaner Economy. *Proc. Inst. Civ. Eng. Energy* **2018**, *171*, 147–157. [CrossRef]
3. Eames, P.; Loveday, D.; Haines, V.; Romanos, P. The Future Role of Thermal Energy Storage in the UK Energy System. 2014. Available online: https://ukerc.ac.uk/publications/the-future-role-of-thermal-energy-storage-in-the-uk-energy-system/ (accessed on 4 January 2022).
4. Watson, S.D.; Lomas, K.J.; Buswell, R.A. Decarbonising Domestic Heating: What Is the Peak GB Demand? *Energy Policy* **2019**, *126*, 533–544. [CrossRef]
5. Whittle, B. Decarbonisation of Heat—A Crossroads—Energy Saving Trust. 2019. Available online: https://energysavingtrust.org.uk/decarbonisation-heat-crossroads/ (accessed on 4 January 2022).
6. Abedin, A.H.; Rosen, M.A. A Critical Review of Thermochemical Energy Storage Systems. *Open Renew. Energy J.* **2011**, *4*, 42–46. [CrossRef]
7. Shilei, L.; Guohui, F.; Neng, Z.; Li, D. Experimental Study and Evaluation of Latent Heat Storage in Phase Change Materials Wallboards. *Energy Build* **2007**, *39*, 1088–1091. [CrossRef]
8. Sunku Prasad, J.; Muthukumar, P.; Desai, F.; Basu, D.N.; Rahman, M.M. A Critical Review of High-Temperature Reversible Thermochemical Energy Storage Systems. *Appl. Energy* **2019**, *254*, 113733. [CrossRef]
9. Chen, X.; Zhang, Z.; Qi, C.; Ling, X.; Peng, H. State of the Art on the High-Temperature Thermochemical Energy Storage Systems. *Energy Convers Manag.* **2018**, *177*, 792–815. [CrossRef]
10. Aydin, D.; Casey, S.P.; Riffat, S. The Latest Advancements on Thermochemical Heat Storage Systems. *Renew. Sustain. Energy Rev.* **2015**, *41*, 356–367. [CrossRef]
11. André, L.; Abanades, S.; Flamant, G. Screening of Thermochemical Systems Based on Solid-Gas Reversible Reactions for High Temperature Solar Thermal Energy Storage. *Renew. Sustain. Energy Rev.* **2016**, *64*, 703–715. [CrossRef]
12. Pardo, P.; Deydier, A.; Anxionnaz-Minvielle, Z.; Rougé, S.; Cabassud, M.; Cognet, P. A Review on High Temperature Thermochemical Heat Energy Storage. *Renew. Sustain. Energy Rev.* **2014**, *32*, 591–610. [CrossRef]
13. Farulla, G.A.; Cellura, M.; Guarino, F.; Ferraro, M. A Review of Thermochemical Energy Storage Systems for Power Grid Support. *Appl. Sci.* **2020**, *10*, 3142. [CrossRef]
14. André, L.; Abanades, S. Recent Advances in Thermochemical Energy Storage via Solid–Gas Reversible Reactions at High Temperature. *Energies* **2020**, *13*, 5859. [CrossRef]
15. Yan, Y.; Wang, K.; Clough, P.T.; Anthony, E.J. Developments in Calcium/Chemical Looping and Metal Oxide Redox Cycles for High-Temperature Thermochemical Energy Storage: A Review. *Fuel Process. Technol.* **2020**, *199*, 106280. [CrossRef]
16. Xu, C.; Yu, Z.; Xie, Y.; Ren, Y.; Ye, F.; Ju, X. Study of the Hydration Behavior of Zeolite-$MgSO_4$ Composites for Long-Term Heat Storage. *Appl. Therm. Eng.* **2018**, *129*, 250–259. [CrossRef]
17. Scapino, L.; Zondag, H.A.; Van Bael, J.; Diriken, J.; Rindt, C.C.M. Sorption Heat Storage for Long-Term Low-Temperature Applications: A Review on the Advancements at Material and Prototype Scale. *Appl. Energy* **2017**, *190*, 920–948. [CrossRef]
18. Hawwash, A.A.; Hassan, H.; El Feky, K. Impact of Reactor Design on the Thermal Energy Storage of Thermochemical Materials. *Appl. Therm. Eng.* **2020**, *168*, 114776. [CrossRef]
19. Zsembinszki, G.; Sole, A.; Barreneche, C.; Prieto, C.; Fernández, A.I.; Cabeza, L.F. Review of Reactors with Potential Use in Thermochemical Energy Storage in Concentrated Solar Power Plants. *Energies* **2018**, *11*, 2358. [CrossRef]
20. Desai, F.; Sunku Prasad, J.; Muthukumar, P.; Rahman, M.M. Thermochemical Energy Storage System for Cooling and Process Heating Applications: A Review. *Energy Convers. Manag.* **2021**, *229*, 113617. [CrossRef]
21. Gbenou, T.R.S.; Fopah-Lele, A.; Wang, K. Recent Status and Prospects on Thermochemical Heat Storage Processes and Applications. *Entropy* **2021**, *23*, 953. [CrossRef]
22. Sadeghi, G. Energy Storage on Demand: Thermal Energy Storage Development, Materials, Design, and Integration Challenges. *Energy Storage Mater.* **2022**, *46*, 192–222. [CrossRef]

23. Marie, L.F.; Landini, S.; Bae, D.; Francia, V.; O'Donovan, T.S. Advances in Thermochemical Energy Storage and Fluidised Beds for Domestic Heat. *J. Energy Storage* **2022**, *53*, 105242. [CrossRef]
24. Kant, K.; Pitchumani, R. Advances and Opportunities in Thermochemical Heat Storage Systems for Buildings Applications. *Appl. Energy* **2022**, *321*, 119299. [CrossRef]
25. Salgado-Pizarro, R.; Calderón, A.; Svobodova-Sedlackova, A.; Fernández, A.I.; Barreneche, C. The Relevance of Thermochemical Energy Storage in the Last Two Decades: The Analysis of Research Evolution. *J. Energy Storage* **2022**, *51*, 104377. [CrossRef]
26. Gbenou, T.R.S.; Fopah-Lele, A.; Wang, K. Macroscopic and Microscopic Investigations of Low-Temperature Thermochemical Heat Storage Reactors: A Review. *Renew. Sustain. Energy Rev.* **2022**, *161*, 112152. [CrossRef]
27. Ding, Y.; Riffat, S.B. Thermochemical Energy Storage Technologies for Building Applications: A State-of-the-Art Review. *Int. J. Low-Carbon Technol.* **2013**, *8*, 106–116. [CrossRef]
28. Pan, Z.H.; Zhao, C.Y. Gas–Solid Thermochemical Heat Storage Reactors for High-Temperature Applications. *Energy* **2017**, *130*, 155–173. [CrossRef]
29. Bauer, T. Fundamentals of High-Temperature Thermal Energy Storage, Transfer, and Conversion. In *Ultra-High Temperature Thermal Energy Storage, Transfer and Conversion*; Elsevier: Amsterdam, The Netherlands, 2021; pp. 1–34.
30. Yu, N.; Wang, R.Z.; Wang, L.W. Sorption Thermal Storage for Solar Energy. *Prog. Energy Combust. Sci.* **2013**, *39*, 489–514. [CrossRef]
31. Srivastava, N.C.; Eames, I.W. A Review of Adsorbents and Adsorbates in Solid-Vapour Adsorption Heat Pump Systems. *Appl. Therm. Eng.* **1998**, *18*, 707–714. [CrossRef]
32. Wu, H.; Salles, F.; Zajac, J. A Critical Review of Solid Materials for Low-Temperature Thermochemical Storage of Solar Energy Based on Solid-Vapour Adsorption in View of Space Heating Uses. *Molecules* **2019**, *24*, 945. [CrossRef]
33. Kuznik, F.; Johannes, K. Thermodynamic Efficiency of Water Vapor/Solid Chemical Sorption Heat Storage for Buildings: Theoretical Limits and Integration Considerations. *Appl. Sci.* **2020**, *10*, 489. [CrossRef]
34. N'Tsoukpoe, K.E.; Kuznik, F. A Reality Check on Long-Term Thermochemical Heat Storage for Household Applications. *Renew. Sustain. Energy Rev.* **2021**, *139*, 110683. [CrossRef]
35. Makhanya, N.; Oboirien, B.; Ren, J.; Musyoka, N.; Sciacovelli, A. Recent Advances on Thermal Energy Storage Using Metal-Organic Frameworks (MOFs). *J. Energy Storage* **2021**, *34*, 102179. [CrossRef]
36. Cabeza, L.F.; Martorell, I.; Miró, L.; Fernández, A.I.; Barreneche, C. Introduction to Thermal Energy Storage (TES) Systems. In *Advances in Thermal Energy Storage Systems: Methods and Applications*; Woodhead Publishing: Sawston, UK, 2015.
37. Carrillo, A.J.; Moya, J.; Bayón, A.; Jana, P.; De La Peña O'Shea, V.A.; Romero, M.; Gonzalez-Aguilar, J.; Serrano, D.P.; Pizarro, P.; Coronado, J.M. Thermochemical Energy Storage at High Temperature via Redox Cycles of Mn and Co Oxides: Pure Oxides versus Mixed Ones. *Sol. Energy Mater. Sol. Cells* **2014**, *123*, 47–57. [CrossRef]
38. Carrillo, A.J.; González-Aguilar, J.; Romero, M.; Coronado, J.M. Solar Energy on Demand: A Review on High Temperature Thermochemical Heat Storage Systems and Materials. *Chem. Rev.* **2019**, *119*, 4777–4816. [CrossRef]
39. Felderhoff, M.; Urbanczyk, R.; Peil, S. Thermochemical Heat Storage for High Temperature Applications-A Review. *Green* **2013**, *3*, 113–123. [CrossRef]
40. Cot-Gores, J.; Castell, A.; Cabeza, L.F. Thermochemical Energy Storage and Conversion: A-State-of-the-Art Review of the Experimental Research under Practical Conditions. *Renew. Sustain. Energy Rev.* **2012**, *16*, 5207–5224. [CrossRef]
41. Lin, J.; Zhao, Q.; Huang, H.; Mao, H.; Liu, Y.; Xiao, Y. Applications of Low-Temperature Thermochemical Energy Storage Systems for Salt Hydrates Based on Material Classification: A Review. *Sol. Energy* **2021**, *214*, 149–178. [CrossRef]
42. Clark, R.J.; Mehrabadi, A.; Farid, M. State of the Art on Salt Hydrate Thermochemical Energy Storage Systems for Use in Building Applications. *J. Energy Storage* **2020**, *27*, 101145. [CrossRef]
43. Marín, P.E.; Milian, Y.; Ushak, S.; Cabeza, L.F.; Grágeda, M.; Shire, G.S.F. Lithium Compounds for Thermochemical Energy Storage: A State-of-the-Art Review and Future Trends. *Renew. Sustain. Energy Rev.* **2021**, *149*, 111381. [CrossRef]
44. Molenda, M.; Stengler, J.; Linder, M.; Wörner, A. Reversible Hydration Behavior of $CaCl_2$ at High H_2O Partial Pressures for Thermochemical Energy Storage. *Thermochim. Acta* **2013**, *560*, 76–81. [CrossRef]
45. Rammelberg, H.U.; Schmidt, T.; Ruck, W. Hydration and Dehydration of Salt Hydrates and Hydroxides for Thermal Energy Storage—Kinetics and Energy Release. *Energy Procedia* **2012**, *30*, 362–369. [CrossRef]
46. Van Essen, V.M.; Cot Gores, J.; Bleijendaal, L.P.J.; Zondag, H.A.; Schuitema, R.; Bakker, M.; Van Helden, W.G.J. Characterization of Salt Hydrates for Compact Seasonal Thermochemical Storage. In Proceedings of the ASME 3rd International Conference on Energy Sustainability, San Francisco, CA, USA, 19–23 July 2009; Volume 2, pp. 825–830.
47. Posern, K.; Kaps, C. Calorimetric Studies of Thermochemical Heat Storage Materials Based on Mixtures of $MgSO_4$ and $MgCl_2$. *Thermochim Acta* **2010**, *502*, 73–76. [CrossRef]
48. Fopah Lele, A.; N'Tsoukpoe, K.E.; Osterland, T.; Kuznik, F.; Ruck, W.K.L. Thermal Conductivity Measurement of Thermochemical Storage Materials. *Appl. Therm. Eng.* **2015**, *89*, 916–926. [CrossRef]
49. Michel, B.; Neveu, P.; Mazet, N. Comparison of Closed and Open Thermochemical Processes, for Long-Term Thermal Energy Storage Applications. *Energy* **2014**, *72*, 702–716. [CrossRef]
50. N'Tsoukpoe, K.E.; Le Pierrès, N.; Luo, L. Numerical Dynamic Simulation and Analysis of a Lithium Bromide/Water Long-Term Solar Heat Storage System. *Energy* **2012**, *37*, 346–358. [CrossRef]

51. Donkers, P.A.J.; Beckert, S.; Pel, L.; Stallmach, F.; Steiger, M.; Adan, O.C.G. Water Transport in $MgSO_4 \cdot 7H_2O$ during Dehydration in View of Thermal Storage. *J. Phys. Chem. C* **2015**, *119*, 28711–28720. [CrossRef]
52. N'Tsoukpoe, K.E.; Schmidt, T.; Rammelberg, H.U.; Watts, B.A.; Ruck, W.K.L. A Systematic Multi-Step Screening of Numerous Salt Hydrates for Low Temperature Thermochemical Energy Storage. *Appl. Energy* **2014**, *124*, 1–16. [CrossRef]
53. Donkers, P.A.J.; Pel, L.; Adan, O.C.G. Experimental Studies for the Cyclability of Salt Hydrates for Thermochemical Heat Storage. *J. Energy Storage* **2016**, *5*, 25–32. [CrossRef]
54. De Jong, A.J.; Trausel, F.; Finck, C.; Van Vliet, L.; Cuypers, R. Thermochemical Heat Storage—System Design Issues. *Energy Procedia* **2014**, *48*, 309–319. [CrossRef]
55. Gaeini, M.; Shaik, S.A.; Rindt, C.C.M. Characterization of Potassium Carbonate Salt Hydrate for Thermochemical Energy Storage in Buildings. *Energy Build.* **2019**, *196*, 178–193. [CrossRef]
56. Donkers, P.A.J.; Sögütoglu, L.C.; Huinink, H.P.; Fischer, H.R.; Adan, O.C.G. A Review of Salt Hydrates for Seasonal Heat Storage in Domestic Applications. *Appl. Energy* **2017**, *199*, 45–68. [CrossRef]
57. Solé, A.; Martorell, I.; Cabeza, L.F. State of the Art on Gas-Solid Thermochemical Energy Storage Systems and Reactors for Building Applications. *Renew. Sustain. Energy Rev.* **2015**, *47*, 386–398. [CrossRef]
58. Zhao, Y.J.; Wang, R.Z.; Zhang, Y.N.; Yu, N. Development of SrBr2 Composite Sorbents for a Sorption Thermal Energy Storage System to Store Low-Temperature Heat. *Energy* **2016**, *115*, 129–139. [CrossRef]
59. Zhao, Q.; Lin, J.; Huang, H.; Wu, Q.; Shen, Y.; Xiao, Y. Optimization of Thermochemical Energy Storage Systems Based on Hydrated Salts: A Review. *Energy Build.* **2021**, *244*, 111035. [CrossRef]
60. Ait Ousaleh, H.; Sair, S.; Zaki, A.; Faik, A.; Mirena Igartua, J.; El Bouari, A. Double Hydrates Salt as Sustainable Thermochemical Energy Storage Materials: Evaluation of Dehydration Behavior and Structural Phase Transition Reversibility. *Sol. Energy* **2020**, *201*, 846–856. [CrossRef]
61. D'Entremont, A.; Corgnale, C.; Hardy, B.; Zidan, R. Simulation of High Temperature Thermal Energy Storage System Based on Coupled Metal Hydrides for Solar Driven Steam Power Plants. *Int. J. Hydrogen Energy* **2018**, *43*, 817–830. [CrossRef]
62. Harries, D.N.; Paskevicius, M.; Sheppard, D.A.; Price, T.E.C.; Buckley, C.E. Concentrating Solar Thermal Heat Storage Using Metal Hydrides. *Proc. IEEE* **2012**, *100*, 539–549. [CrossRef]
63. Felderhoff, M.; Bogdanović, B. High Temperature Metal Hydrides as Heat Storage Materials for Solar and Related Applications. *Int. J. Mol. Sci.* **2009**, *10*, 325–344. [CrossRef]
64. Zhang, J.; Sun, L.Q.; Zhou, Y.C.; Peng, P. Dehydrogenation Thermodynamics of Magnesium Hydride Doped with Transition Metals: Experimental and Theoretical Studies. *Comput. Mater. Sci.* **2015**, *98*, 211–219. [CrossRef]
65. Zhou, C.; Zhang, J.; Bowman, R.C.; Fang, Z.Z. Roles of Ti-Based Catalysts on Magnesium Hydride and Its Hydrogen Storage Properties. *Inorganics* **2021**, *9*, 36. [CrossRef]
66. Khan, D.; Zou, J.; Zeng, X.; Ding, W. Hydrogen Storage Properties of Nanocrystalline Mg_2Ni Prepared from Compressed $2MgH_2$–Ni Powder. *Int. J. Hydrogen Energy* **2018**, *43*, 22391–22400. [CrossRef]
67. Banrejee, S.; Kumar, A.; Ruz, P.; Sudarsan, V. Improvement of Hydrogen Storage Characteristics of Catalyst Free Magnesium Nanoparticles Prepared by Wet Milling. *Int. J. Energy Res.* **2021**, *45*, 17597–17608. [CrossRef]
68. Chen, X.; Zou, J.; Zeng, X.; Ding, W. Hydrogen Storage Properties of a Mg-La-Fe-H Nano-Composite Prepared through Reactive Ball Milling. *J. Alloys Compd.* **2017**, *701*, 208–214. [CrossRef]
69. Silva, R.A.; Leal Neto, R.M.; Leiva, D.R.; Ishikawa, T.T.; Kiminami, C.S.; Jorge, A.M.; Botta, W.J. Room Temperature Hydrogen Absorption by Mg and Mg–TiFe Nanocomposites Processed by High-Energy Ball Milling. *Int. J. Hydrogen Energy* **2018**, *43*, 12251–12259. [CrossRef]
70. Kim, K.C. A Review on Design Strategies for Metal Hydrides with Enhanced Reaction Thermodynamics for Hydrogen Storage Applications. *Int. J. Energy Res.* **2018**, *42*, 1455–1468. [CrossRef]
71. Bogdanovi, B.; Reiser, A.; Schlichte, K.; Spliethoff, B.; Tesche, B. Thermodynamics and Dynamics of the Mg-Fe-H System and Its Potential for Thermochemical Thermal Energy Storage. *J. Alloys Compd.* **2002**, *345*, 77–89. [CrossRef]
72. Sulaiman, N.N.; Juahir, N.; Mustafa, N.S.; Halim Yap, F.A.; Ismail, M. Improved Hydrogen Storage Properties of MgH_2 Catalyzed with K_2NiF_6. *J. Energy Chem.* **2016**, *25*, 832–839. [CrossRef]
73. Abdul Majid, N.A.; Maeda, N.; Notomi, M. Improved Hydrogen Desorption Properties of Magnesium Hydride with $TiFe_{0.8}Mn_{0.2}$, Graphite and Iron Addition. *Int. J. Hydrogen Energy* **2019**, *44*, 29189–29195. [CrossRef]
74. Kumar, E.M.; Rajkamal, A.; Thapa, R. Screening Based Approach and Dehydrogenation Kinetics for MgH2: Guide to Find Suitable Dopant Using First-Principles Approach. *Sci. Rep.* **2017**, *7*, 15550. [CrossRef]
75. Jain, P.; Dixit, V.; Jain, A.; Srivastava, O.N.; Huot, J. Effect of Magnesium Fluoride on Hydrogenation Properties of Magnesium Hydride. *Energies* **2015**, *8*, 12546–12556. [CrossRef]
76. Zhu, M.; Lu, Y.; Ouyang, L.; Wang, H. Thermodynamic Tuning of Mg-Based Hydrogen Storage Alloys: A Review. *Materials* **2013**, *6*, 4654–4674. [CrossRef]
77. Darkwa, K. Thermochemical Energy Storage in Inorganic Oxides: An Experimental Evaluation. *Appl. Therm. Eng.* **1998**, *18*, 387–400. [CrossRef]
78. Schaube, F.; Koch, L.; Wörner, A.; Müller-Steinhagen, H. A Thermodynamic and Kinetic Study of the De- and Rehydration of $Ca(OH)_2$ at High H_2O Partial Pressures for Thermo-Chemical Heat Storage. *Thermochim. Acta* **2012**, *538*, 9–20. [CrossRef]

79. Schaube, F.; Kohzer, A.; Schütz, J.; Wörner, A.; Müller-Steinhagen, H. De- and Rehydration of $Ca(OH)_2$ in a Reactor with Direct Heat Transfer for Thermo-Chemical Heat Storage. Part A: Experimental Results. *Chem. Eng. Res. Des.* **2013**, *91*, 856–864. [CrossRef]
80. Criado, Y.A.; Alonso, M.; Abanades, J.C. Kinetics of the $CaO/Ca(OH)_2$ Hydration/Dehydration Reaction for Thermochemical Energy Storage Applications. *Ind. Eng. Chem. Res.* **2014**, *53*, 12594–12601. [CrossRef]
81. Dai, L.; Long, X.F.; Lou, B.; Wu, J. Thermal Cycling Stability of Thermochemical Energy Storage System $Ca(OH)_2/CaO$. *Appl. Therm. Eng.* **2018**, *133*, 261–268. [CrossRef]
82. Criado, Y.A.; Alonso, M.; Abanades, J.C. Enhancement of a $CaO/Ca(OH)_2$ Based Material for Thermochemical Energy Storage. *Sol. Energy* **2016**, *135*, 800–809. [CrossRef]
83. Funayama, S.; Takasu, H.; Zamengo, M.; Kariya, J.; Kim, S.T.; Kato, Y. Composite Material for High-temperature Thermochemical Energy Storage Using Calcium Hydroxide and Ceramic Foam. *Energy Storage* **2019**, *1*, e53. [CrossRef]
84. Funayama, S.; Takasu, H.; Kim, S.T.; Kato, Y. Thermochemical Storage Performance of a Packed Bed of Calcium Hydroxide Composite with a Silicon-Based Ceramic Honeycomb Support. *Energy* **2020**, *201*, 117673. [CrossRef]
85. Kariya, J.; Ryu, J.; Kato, Y. Reaction Performance of Calcium Hydroxide and Expanded Graphite Composites for Chemical Heat Storage Applications. *ISIJ Int.* **2015**, *55*, 457–463. [CrossRef]
86. Huang, C.; Xu, M.; Huai, X. Experimental Investigation on Thermodynamic and Kinetic of Calcium Hydroxide Dehydration with Hexagonal Boron Nitride Doping for Thermochemical Energy Storage. *Chem. Eng. Sci.* **2019**, *206*, 518–526. [CrossRef]
87. Shkatulov, A.; Aristov, Y. Calcium Hydroxide Doped by KNO_3 as a Promising Candidate for Thermochemical Storage of Solar Heat. *RSC Adv.* **2017**, *7*, 42929–42939. [CrossRef]
88. Wang, T.; Zhao, C.Y.; Yan, J. Investigation on the $Ca(OH)_2/CaO$ Thermochemical Energy Storage System with Potassium Nitrate Addition. *Sol. Energy Mater. Sol. Cells* **2020**, *215*, 110646. [CrossRef]
89. Gollsch, M.; Afflerbach, S.; Angadi, B.; Linder, M. Investigation of Calcium Hydroxide Powder for Thermochemical Storage Modified with Nanostructured Flow Agents. *Sol. Energy* **2020**, *201*, 810–818. [CrossRef]
90. Cosquillo Mejia, A.; Afflerbach, S.; Linder, M.; Schmidt, M. Experimental Analysis of Encapsulated $CaO/Ca(OH)_2$ Granules as Thermochemical Storage in a Novel Moving Bed Reactor. *Appl. Therm. Eng.* **2020**, *169*, 114961. [CrossRef]
91. Afflerbach, S.; Afflerbach, K.; Trettin, R.; Krumm, W. Improvement of a Semipermeable Shell for Encapsulation of Calcium Hydroxide for Thermochemical Heat Storage Solutions: Material Design and Evaluation in Laboratory and Reactor Scale. *Sol. Energy* **2021**, *217*, 208–222. [CrossRef]
92. André, L.; Abanades, S. Investigation of Metal Oxides, Mixed Oxides, Perovskites and Alkaline Earth Carbonates/Hydroxides as Suitable Candidate Materials for High-Temperature Thermochemical Energy Storage Using Reversible Solid-Gas Reactions. *Mater Today Energy* **2018**, *10*, 48–61. [CrossRef]
93. André, L.; Abanades, S. Evaluation and Performances Comparison of Calcium, Strontium and Barium Carbonates during Calcination/Carbonation Reactions for Solar Thermochemical Energy Storage. *J. Energy Storage* **2017**, *13*, 193–205. [CrossRef]
94. Lu, S.; Wu, S. Calcination-Carbonation Durability of Nano $CaCO_3$ Doped with Li_2SO_4. *Chem. Eng. J.* **2016**, *294*, 22–29. [CrossRef]
95. Daud, F.D.M.; Vignesh, K.; Sreekantan, S.; Mohamed, A.R. Improved CO_2 Adsorption Capacity and Cyclic Stability of CaO Sorbents Incorporated with MgO. *New J. Chem.* **2016**, *40*, 231–237. [CrossRef]
96. Wang, K.; Gu, F.; Clough, P.T.; Zhao, P.; Anthony, E.J. Porous MgO-Stabilized CaO-Based Powders/Pellets via a Citric Acid-Based Carbon Template for Thermochemical Energy Storage in Concentrated Solar Power Plants. *Chem. Eng. J.* **2020**, *390*, 124163. [CrossRef]
97. Benitez-Guerrero, M.; Valverde, J.M.; Perejon, A.; Sanchez-Jimenez, P.E.; Perez-Maqueda, L.A. Low-Cost Ca-Based Composites Synthesized by Biotemplate Method for Thermochemical Energy Storage of Concentrated Solar Power. *Appl. Energy* **2018**, *210*, 108–116. [CrossRef]
98. Chen, X.; Jin, X.; Liu, Z.; Ling, X.; Wang, Y. Experimental Investigation on the $CaO/CaCO_3$ Thermochemical Energy Storage with SiO_2 Doping. *Energy* **2018**, *155*, 128–138. [CrossRef]
99. Da, Y.; Xuan, Y.; Teng, L.; Zhang, K.; Liu, X.; Ding, Y. Calcium-Based Composites for Direct Solar-Thermal Conversion and Thermochemical Energy Storage. *Chem. Eng. J.* **2020**, *382*, 122815. [CrossRef]
100. Sun, H.; Li, Y.; Yan, X.; Zhao, J.; Wang, Z. Thermochemical Energy Storage Performance of Al_2O_3/CeO_2 Co-Doped CaO-Based Material under High Carbonation Pressure. *Appl. Energy* **2020**, *263*, 114650. [CrossRef]
101. Raganati, F.; Chirone, R.; Ammendola, P. Calcium-Looping for Thermochemical Energy Storage in Concentrating Solar Power Applications: Evaluation of the Effect of Acoustic Perturbation on the Fluidized Bed Carbonation. *Chem. Eng. J.* **2020**, *392*, 123658. [CrossRef]
102. Møller, K.T.; Ibrahim, A.; Buckley, C.E.; Paskevicius, M. Inexpensive Thermochemical Energy Storage Utilising Additive Enhanced Limestone. *J. Mater. Chem. A Mater.* **2020**, *8*, 9646–9653. [CrossRef]
103. Block, T.; Schmücker, M. Metal Oxides for Thermochemical Energy Storage: A Comparison of Several Metal Oxide Systems. *Sol. Energy* **2016**, *126*, 195–207. [CrossRef]
104. Silakhori, M.; Jafarian, M.; Arjomandi, M.; Nathan, G.J. Thermogravimetric Analysis of Cu, Mn, Co, and Pb Oxides for Thermochemical Energy Storage. *J. Energy Storage* **2019**, *23*, 138–147. [CrossRef]
105. Deutsch, M.; Horvath, F.; Knoll, C.; Lager, D.; Gierl-Mayer, C.; Weinberger, P.; Winter, F. High-Temperature Energy Storage: Kinetic Investigations of the CuO/Cu_2O Reaction Cycle. *Energy Fuels* **2017**, *31*, 2324–2334. [CrossRef]

106. Alonso, E.; Pérez-Rábago, C.; Licurgo, J.; Fuentealba, E.; Estrada, C.A. First Experimental Studies of Solar Redox Reactions of Copper Oxides for Thermochemical Energy Storage. *Sol. Energy* **2015**, *115*, 297–305. [CrossRef]
107. Gigantino, M.; Brunser, S.S.; Steinfeld, A. High-Temperature Thermochemical Heat Storage via the CuO/Cu_2O Redox Cycle: From Material Synthesis to Packed-Bed Reactor Engineering and Cyclic Operation. *Energy Fuels* **2020**, *34*, 16772–16782. [CrossRef]
108. André, L.; Abanades, S.; Cassayre, L. Mixed Co, Cu and Mn-Based Metal Oxides for Thermochemical Energy Storage Application. In Proceedings of the AIP Conference Proceedings, Maharashtra, India, 8 November 2018; American Institute of Physics Inc.: College Park, MD, USA, 2018; Volume 2033.
109. Bielsa, D.; Zaki, A.; Faik, A.; Arias, P.L. Efficiency Improvement of Mn_2O_3/Mn_3O_4 Redox Reaction by Means of Different Operation Strategies. In Proceedings of the AIP Conference Proceedings, Bogor, Indonesia, 25 July 2019; American Institute of Physics Inc.: College Park, MD, USA, 2019; Volume 2126.
110. André, L.; Abanades, S.; Cassayre, L. Experimental and Thermodynamic Study of Co-Fe and Mn-Fe Based Mixed Metal Oxides for Thermochemical Energy Storage Application. In Proceedings of the AIP Conference Proceedings, Bydgoszcz, Poland, 27 June 2017; American Institute of Physics Inc.: College Park, MD, USA, 2017; Volume 1850.
111. Neises, M.; Tescari, S.; De Oliveira, L.; Roeb, M.; Sattler, C.; Wong, B. Solar-Heated Rotary Kiln for Thermochemical Energy Storage. *Sol. Energy* **2012**, *86*, 3040–3048. [CrossRef]
112. Jahromy, S.S.; Birkelbach, F.; Jordan, C.; Huber, C.; Harasek, M.; Werner, A.; Winter, F. Impact of Partial Pressure, Conversion, and Temperature on the Oxidation Reaction Kinetics of Cu_2O to CuO in Thermochemical Energy Storage. *Energies* **2019**, *12*, 508. [CrossRef]
113. Liu, H.; Wang, W.; Zhang, Y. Performance Gap between Thermochemical Energy Storage Systems Based on Salt Hydrates and Materials. *J. Clean. Prod.* **2021**, *313*, 127908. [CrossRef]
114. Ferchaud, C.J.; Zondag, H.A.; Veldhuis, J.B.J.; De Boer, R. Study of the Reversible Water Vapour Sorption Process of $MgSO_4.7H_2O$ and $MgCl_2.6H_2O$ under the Conditions of Seasonal Solar Heat Storage. *J. Phys.* **2012**, *395*, 12069.
115. Opel, O.; Rammelberg, H.U.; Gerard, M.; Ruck, W. Thermalchemical Storage Materials Research—TGA/DSC-Hydration Studies. In Proceedings of the International Conference on Sustainable Energy Storage, Belfast, Ireland, 21–24 February 2011.
116. Zondag, H.; Kikkert, B.; Smeding, S.; De Boer, R.; Bakker, M. Prototype Thermochemical Heat Storage with Open Reactor System. *Appl. Energy* **2013**, *109*, 360–365. [CrossRef]
117. Verga, M.; Armanasco, F.; Guardamagna, C.; Valli, C.; Bianchin, A.; Agresti, F.; Lo Russo, S.; Maddalena, A.; Principi, G. Scaling up Effects of Mg Hydride in a Temperature and Pressure-Controlled Hydrogen Storage Device. *Int. J. Hydrogen Energy* **2009**, *34*, 4602–4610. [CrossRef]
118. Paskevicius, M.; Sheppard, D.A.; Williamson, K.; Buckley, C.E. Metal Hydride Thermal Heat Storage Prototype for Concentrating Solar Thermal Power. *Energy* **2015**, *88*, 469–477. [CrossRef]
119. Xia, G.; Tan, Y.; Chen, X.; Sun, D.; Guo, Z.; Liu, H.; Ouyang, L.; Zhu, M.; Yu, X. Monodisperse Magnesium Hydride Nanoparticles Uniformly Self-Assembled on Graphene. *Adv. Mater.* **2015**, *27*, 5981–5988. [CrossRef]
120. Yan, J.; Zhao, C.Y. Experimental Study of $CaO/Ca(OH)_2$ in a Fixed-Bed Reactor for Thermochemical Heat Storage. *Appl. Energy* **2016**, *175*, 277–284. [CrossRef]
121. Yang, Y.; Li, Y.; Yan, X.; Zhao, J.; Zhang, C. Development of Thermochemical Heat Storage Based on $CaO/CaCO_3$ Cycles: A review. *Energies* **2021**, *14*, 6847. [CrossRef]
122. Hutchings, K.N.; Wilson, M.; Larsen, P.A.; Cutler, R.A. Kinetic and Thermodynamic Considerations for Oxygen Absorption/Desorption Using Cobalt Oxide. *Solid State Ion* **2006**, *177*, 45–51. [CrossRef]
123. Agrafiotis, C.; Roeb, M.; Schmücker, M.; Sattler, C. Exploitation of Thermochemical Cycles Based on Solid Oxide Redox Systems for Thermochemical Storage of Solar Heat. Part 1: Testing of Cobalt Oxide-Based Powders. *Sol. Energy* **2014**, *102*, 189–211. [CrossRef]
124. Mette, B.; Kerskes, H.; Drück, H. Experimental and Numerical Investigations of Different Reactor Concepts for Thermochemical Energy Storage. *Energy Procedia* **2014**, *57*, 2380–2389. [CrossRef]
125. Foutch, G.L.; Johannes, A.H. Reactors in Process Engineering. In *Encyclopedia of Physical Science and Technology*; Academic Press Inc.: Cambridge, MA, USA, 2003.
126. Schmidt, M.; Szczukowski, C.; Roßkopf, C.; Linder, M.; Wörner, A. Experimental Results of a 10 KW High Temperature Thermochemical Storage Reactor Based on Calcium Hydroxide. *Appl. Therm. Eng.* **2014**, *62*, 553–559. [CrossRef]
127. Ranjha, Q.A.; Vahedi, N.; Oztekin, A. Numerical Study of Thermochemical Storage Using $Ca(OH)_2/CaO$—High Temperature Applications. In Proceedings of the ASME International Mechanical Engineering Congress and Exposition, Proceedings (IMECE), Phoenix, AZ, USA, 11–17 November 2016; Volume 6B-2016.
128. Funayama, S.; Takasu, H.; Zamengo, M.; Kariya, J.; Kim, S.T.; Kato, Y. Performance of Thermochemical Energy Storage of a Packed Bed of Calcium Hydroxide Pellets. *Energy Storage* **2019**, *1*, e40. [CrossRef]
129. Schmidt, M.; Linder, M. A Novel Thermochemical Long Term Storage Concept: Balance of Renewable Electricity and Heat Demand in Buildings. *Front. Energy Res.* **2020**, *8*, 00137. [CrossRef]
130. Schmidt, M.; Gutierrez, A.; Linder, M. Thermochemical Energy Storage with $CaO/Ca(OH)_2$– Experimental Investigation of the Thermal Capability at Low Vapor Pressures in a Lab Scale Reactor. *Appl. Energy* **2017**, *188*, 672–681. [CrossRef]
131. Peng, X.; Yao, M.; Root, T.W.; Maravelias, C.T. Design and Analysis of Concentrating Solar Power Plants with Fixed-Bed Reactors for Thermochemical Energy Storage. *Appl. Energy* **2020**, *262*, 114543. [CrossRef]

132. Schaube, F.; Utz, I.; Wörner, A.; Müller-Steinhagen, H. De- and Rehydration of $Ca(OH)_2$ in a Reactor with Direct Heat Transfer for Thermo-Chemical Heat Storage. Part B: Validation of Model. *Chem. Eng. Res. Des.* **2013**, *91*, 865–873. [CrossRef]
133. Schmidt, M.; Gollsch, M.; Giger, F.; Grün, M.; Linder, M. Development of a Moving Bed Pilot Plant for Thermochemical Energy Storage with CaO/Ca(OH)2. *AIP Conf. Proc.* **2016**, *1734*, 050041.
134. Shirzad, M.; Karimi, M.; Silva, J.A.C.; Rodrigues, A.E. Moving Bed Reactors: Challenges and Progress of Experimental and Theoretical Studies in a Century of Research. *Ind. Eng. Chem. Res.* **2019**, *58*, 9179–9198. [CrossRef]
135. Angerer, M.; Becker, M.; Härzschel, S.; Kröper, K.; Gleis, S.; Vandersickel, A.; Spliethoff, H. Design of a MW-Scale Thermo-Chemical Energy Storage Reactor. *Energy Rep.* **2018**, *4*, 507–519. [CrossRef]
136. Preisner, N.C.; Bürger, I.; Wokon, M.; Linder, M. Numerical Investigations of a Counter-Current Moving Bed Reactor for Thermochemical Energy Storage at High Temperatures. *Energies* **2020**, *13*, 772. [CrossRef]
137. Preisner, N.C.; Linder, M. A Moving Bed Reactor for Thermochemical Energy Storage Based on Metal Oxides. *Energies* **2020**, *13*, 1232. [CrossRef]
138. Huang, W.; Korba, D.; Randhir, K.; Petrasch, J.; Klausner, J.; AuYeung, N.; Li, L. Thermochemical Reduction Modeling in a High-Temperature Moving-Bed Reactor for Energy Storage: 1D Model. *Appl. Energy* **2022**, *306*, 118009. [CrossRef]
139. Di Renzo, A.; Scala, F.; Heinrich, S. Recent Advances in Fluidized Bed Hydrodynamics and Transport Phenomena—Progress and Understanding. *Processes* **2021**, *9*, 639. [CrossRef]
140. Zondag, H.A.; Van Essen, V.M.; Bakker, M. Application of $MgCl_2{\cdot}6H_2O$ for Thermochemical Seasonal Solar Heat Storage. In Proceedings of the 5th International Renewable Energy Storage Conference IRES 2010, Berlin, Germany, 22–24 November 2010.
141. Delhomme, B.; De Rango, P.; Marty, P.; Bacia, M.; Zawilski, B.; Raufast, C.; Miraglia, S.; Fruchart, D. Large Scale Magnesium Hydride Tank Coupled with an External Heat Source. *Int. J. Hydrogen Energy* **2012**, *37*, 9103–9111. [CrossRef]
142. Meier, A.; Bonaldi, E.; Cella, G.M.; Lipinski, W.; Wuillemin, D.; Palumbo, R. Design and Experimental Investigation of a Horizontal Rotary Reactor for the Solar Thermal Production of Lime. *Energy* **2004**, *29*, 811–821. [CrossRef]
143. Meroueh, L.; Yenduru, K.; Dasgupta, A.; Jiang, D.; AuYeung, N. Energy Storage Based on $SrCO_3$ and Sorbents—A Probabilistic Analysis towards Realizing Solar Thermochemical Power Plants. *Renew. Energy* **2019**, *133*, 770–786. [CrossRef]
144. Flegkas, S.; Birkelbach, F.; Winter, F.; Groenewold, H.; Werner, A. Profitability Analysis and Capital Cost Estimation of a Thermochemical Energy Storage System Utilizing Fluidized Bed Reactors and the Reaction System $MgO/Mg(OH)_2$. *Energies* **2019**, *12*, 4788. [CrossRef]
145. Flegkas, S.; Birkelbach, F.; Winter, F.; Freiberger, N.; Werner, A. Fluidized Bed Reactors for Solid-Gas Thermochemical Energy Storage Concepts—Modelling and Process Limitations. *Energy* **2018**, *143*, 615–623. [CrossRef]
146. Rougé, S.; Criado, Y.A.; Soriano, O.; Abanades, J.C. Continuous $CaO/Ca(OH)_2$ Fluidized Bed Reactor for Energy Storage: First Experimental Results and Reactor Model Validation. *Ind. Eng. Chem. Res.* **2017**, *56*, 844–852. [CrossRef]
147. Mu, L.; Buist, K.A.; Kuipers, J.A.M.; Deen, N.G. Hydrodynamic and Heat Transfer Study of a Fluidized Bed by Discrete Particle Simulations. *Process.* **2020**, *8*, 463. [CrossRef]
148. Darkwa, K.; Ianakiev, A.; O'Callaghan, P.W. Modelling and Simulation of Adsorption Process in a Fluidised Bed Thermochemical Energy Reactor. *Appl. Therm. Eng.* **2006**, *26*, 838–845. [CrossRef]
149. Kim, J.; Han, G.Y. Effect of Agitation on Fluidization Characteristics of Fine Particles in a Fluidized Bed. *Powder Technol.* **2006**, *166*, 113–122. [CrossRef]
150. Wang, J.J.; Han, Y.; Gu, X.P.; Feng, L.F.; Hu, G.H. Effect of Agitation on the Fluidization Behavior of a Gas-Solid Fluidized Bed with a Frame Impeller. *AIChE J.* **2013**, *59*, 1066–1074. [CrossRef]
151. Lv, B.; Deng, X.; Shi, C.; Fang, C. Effect of Agitation on Hydrodynamics and Separation Performance of Gas–Solid Separation Fluidized Bed. *Powder Technol.* **2021**, *388*, 129–138. [CrossRef]
152. Dong, K.; Zhou, Y.; Huang, Z.; Wang, J.; Yang, Y. Gas Bubble Behaviors in a Gas-Solid Fluidized Bed with an Arch Agitator. *Powder Technol.* **2014**, *266*, 38–44. [CrossRef]
153. Shi, D.P.; Luo, Z.H.; Guo, A.Y. Numerical Simulation of the Gas-Solid Flow in Fluidized-Bed Polymerization Reactors. *Ind. Eng. Chem. Res.* **2010**, *49*, 4070–4079. [CrossRef]
154. Tafrishi, H.; Sadeghzadeh, S.; Ahmadi, R. Molecular Dynamics Simulations of Phase Change Materials for Thermal Energy Storage: A Review. *RSC Adv.* **2022**, *12*, 14776–14807. [CrossRef] [PubMed]
155. Shkatulov, A.; Becit, B.; Zahn, D. Molecular Dynamics Simulations of Nitrate/MgO Interfaces and Understanding Metastability of Thermochemical Materials. *ACS Omega* **2022**, *7*, 16371–16379. [CrossRef] [PubMed]
156. Xu, M.; Huai, X.; Cai, J. Agglomeration Behavior of Calcium Hydroxide/Calcium Oxide as Thermochemical Heat Storage Material: A Reactive Molecular Dynamics Study. *J. Phys. Chem. C* **2017**, *121*, 3025–3033. [CrossRef]
157. Alvares, C.M.S.; Deffrennes, G.; Pisch, A.; Jakse, N. Thermodynamics and Structural Properties of CaO: A Molecular Dynamics Simulation Study. *J. Chem. Phys.* **2020**, *152*, 084503. [CrossRef]
158. Wang, S.; Shen, Y. Coarse-Grained CFD-DEM Modelling of Dense Gas-Solid Reacting Flow. *Int. J. Heat Mass Transf.* **2022**, *184*, 122302. [CrossRef]
159. Shen, J.; Liu, X.; Ho, W.H. CFD Simulation of a Fluidized Bed as a Solar Receiver. In Proceedings of the AIP Conference Proceedings, College Park, MD, USA, 25 July 2019; American Institute of Physics Inc.: College Park, MD, USA, 2019; Volume 2126.
160. Bowen, Y.; Jun, Y.; Changying, Z. Investigating the Performance of a Fluidized Bed Reactor for a Magnesium Hydroxide Thermochemical Energy Storage System. *Energy Storage Sci. Technol.* **2021**, *10*, 1735–1744. [CrossRef]

161. Li, Z.; Xu, M.; Huai, X.; Huang, C.; Wang, K. Simulation and Analysis of Thermochemical Seasonal Solar Energy Storage for District Heating Applications in China. *Int. J. Energy Res.* **2021**, *45*, 7093–7107. [CrossRef]
162. Ghosh, S.; Fennell, P.S. Design and Techno-Economic Analysis of a Fluidized Bed-Based $CaO/Ca(OH)_2$ Thermochemical Energy Combined Storage/Discharge Plant with Concentrated Solar Power. In Proceedings of the AIP Conference Proceedings, Daegu, Republic of Korea, 1–4 October 2019; American Institute of Physics Inc.: College Park, MD, USA, 2020; Volume 2303.
163. Alobaid, F.; Almohammed, N.; Massoudi Farid, M.; May, J.; Rößger, P.; Richter, A.; Epple, B. Progress in CFD Simulations of Fluidized Beds for Chemical and Energy Process Engineering. *Prog. Energy Combust. Sci.* **2021**, *19*, 100930.

energies

MDPI

Review

Improved Performance of Latent Heat Energy Storage Systems in Response to Utilization of High Thermal Conductivity Fins

Wenwen Ye, Dourna Jamshideasli and Jay M. Khodadadi *

Department of Mechanical Engineering, Auburn University, 1418 Wiggins Hall, Auburn, AL 36849-5341, USA
* Correspondence: khodajm@auburn.edu; Tel.: +1-334-844-3333; Fax: +1-334-844-3307

Abstract: Analytical, computational and experimental investigations directed at improving the performance of latent heat thermal energy storage systems that utilize high thermal conductivity fins in direct contact with phase change materials are reviewed. Researchers have focused on waste heat recovery, thermal management of buildings/computing platforms/photovoltaics/satellites and energy storage for solar thermal applications. Aluminum (including various alloys), brass, bronze, copper, PVC, stainless steel and steel were the adopted fin materials. Capric-palmitic acid, chloride mixtures, dodecanoic acid, erythritol, fluorides, lauric acid, naphthalene, nitrite and nitrate mixtures, paraffins, potassium nitrate, salt hydrates, sodium hydrate, stearic acid, sulfur, water and xylitol have been the adopted fusible materials (melting or fusion temperature T_m range of -129.6 to 767 °C). Melting and solidification processes subject to different heat exchange operating conditions were investigated. Studies of thawing have highlighted the marked role of natural convection, exhibiting that realizing thermally unstable fluid layers promote mixing and expedited melting. Performance of the storage system in terms of the hastened charge/discharge time was strongly affected by the number of fins (or fin-pitch) and fin length, in comparison to fin thickness and fin orientation. Strength of natural convection, which is well-known to play an important role on thawing, is diminished by introduction of fins. Consequently, a designer must consider suppression of buoyancy and the extent of sacrificed PCM in selecting the optimum positions and orientation of the fins. Complex fin shapes featuring branching arrangements, crosses, Y-shapes, etc. are widely replacing simple planar fins, satisfying the challenge of forming short-distance conducting pathways linking the temperature extremes of the storage system.

Keywords: extended surfaces; fins; fusible materials; melting; phase change materials; phase transformation; solidification; thermal conductivity enhancers

Citation: Ye, W.; Jamshideasli, D.; Khodadadi, J.M. Improved Performance of Latent Heat Energy Storage Systems in Response to Utilization of High Thermal Conductivity Fins. *Energies* **2023**, *16*, 1277. https://doi.org/10.3390/en16031277

Academic Editors: Alon Kuperman and Alessandro Lampasi

Received: 30 September 2022
Revised: 9 January 2023
Accepted: 11 January 2023
Published: 25 January 2023

1. Introduction

Uninterrupted supply of dispatchable energy to residential/commercial/industrial sectors that is generally provided by the nuclear/fossil fuel power plants has required great technological advances in recent decades when compared to prior practices [1]. Concurrently, greater concerns with the safety of nuclear reactors and the instability of pricing/supply of fossil fuels, in addition to more focus on the environment, brought about re-evaluation of alternative sources of energy, and adoption of renewable sources of energy, i.e., solar, wave, wind, etc., has emerged. A number of conferences in response to the energy crisis of early 1970's were convened [2–10], focusing on alternative and renewable sources of energy. Intermittency of the output of renewable energy systems demands design and utilization of effective, reliable and robust storage units. The development of energy storage systems has long constituted the major bottleneck for deeper penetration of renewable sources of energy into the market. Among various forms of energy, thermal energy is a by-product of a diverse group of man-made machinery/processes and energy conversion systems. Despite its ubiquity, thermal energy is characterized as a low-grade form of energy, leading to the usage of terms such as waste heat. Based on the above arguments, storage of thermal energy (or waste-heat recovery) can be viewed as a capacitor/accumulator system (thermal battery or thermal buffering) capable of providing thermal comfort in buildings/transportation systems, conserving energy in various sectors of the economy, adding to the operational life of electronics and improving the efficiency of industrial processes.

Thermal Energy Storage in Fusible Materials

Broad classifications of approaches adopted for thermal energy storage (TES) have not changed over decades [11,12]. TES techniques involving manipulation of sensible or latent energy through heating or cooling a bulk of material are identified as *thermophysical* approaches. Phase change materials (PCM), also known as fusible materials, are widely used for TES at a constant temperature by taking advantage of their latent heat (heat of fusion) during phase transition. Paraffins, fatty acids, sugar alcohols, salt hydrates, etc., are examples of PCM, with their T_m varying over a wide range. In effect, a variety of materials are convenient for adoption in low-, medium- and high-temperature TES applications. However, a great percentage of PCM exhibit relatively low thermal conductivity (TC) that, in effect, degrades the rates of energy discharge/charge. Inserting materials with high thermal conductivity leading to a composite of PCM/additives is a logical approach. Introducing metallic fins/foams/wools into PCM has long been practiced. Interactions of heat transport mechanisms with the adopted configurations of these inserts remain as challenging issues to be investigated.

Different classes of PCM, their thermophysical/transport properties, encapsulation, improvement of heat transfer, effects of container shape and system-related topics were discussed in previous review papers [13–15]. Analysis and performance evaluation of PCM-based TES during the thawing/freezing cycles are inherently time-dependent. Elucidating the relevant heat transfer mechanisms (i.e., conduction and convection) encountered in these phase transition systems and their competing/cooperating roles discussed in these reviews are limited (except [15]). Prompted by this shortcoming, and due to the greater importance of TES in recent years, Fan and Khodadadi [16] reviewed literature focused on improving the effective TC of PCM through the introduction of *fixed, non-moving* high-conductivity inserts. Dhaidan and Khodadadi [17] published a review of 63 studies (dating back to 1966) devoted exclusively to the role of high TC fins in modifying the performance of latent heat thermal energy storage systems (LHTESS). The present review serves as the addendum to [17]. Moreover, a number of ignored/hard-to-access papers on the topic, including 26 articles in Japanese and Korean, are also discussed. In comparison to Abdulateef et al. [18], the present work covers far greater number of articles published on this topic. Carbon- and metal-based foam structures are not treated in this review. Discussion of micron- and nano-scale particles introduced into PCM systems is not attempted, and the interested reader is referred to Khodadadi et al. [19].

2. Depth of Coverage/Scope and Organization of the Current Review

Providing reviews of research work reporting improvement to performance of LHTESS through introduction of *fixed, non-moving* high TC inserts/fins/structures not covered in [17] is the objective of this paper. In effect, an exhaustive review of this topic will be available for researchers by combining [17] and the current article.

Organization of the Review

Readers of this review should be aware that: (a) Reviews will be discussed in a *chronological* order and not *thematic*, as this is the approach taken by majority of researchers in getting acquainted with a new topic. The authors believe that students and others being initiated to this topic can consult this document as a systematic and complete account of the established knowledge; (b) Each article cited below was fully read by at least one of the authors; (c) Each article was then discussed among all authors in order to arrive at a consensus related to its contributions and novelty related to prior work, any innovations/duplications, anomalies, shortcomings, etc.; and (d) Artwork in this paper was taken from the reviewed papers, and the authors do not assume responsibility for their possible lack of quality.

Summaries of the reviewed articles in relation to the classification of the LHTES units, specific PCM and its properties, classification of the type of fins, dominant heat transfer mechanisms, details of the experiments and/or computational/analytical methodologies, etc., are listed in appropriate tables near the end of the article. Some readers might choose to consult that information at the outset and proceed to become acquainted with a certain group of papers that share similar attributes/themes.

3. Research on Improving Performance of LHTESS through Introducing Inserts/Fins/Structures with High Thermal Conductivity

Improvement of performance of LHTESS, through the addition of extended surfaces or fins with high TC and various geometric configurations, is widely practiced. In accommodating direct contact of the PCM with high TC fins and the active heat transfer surfaces, the main goal is to remove the shortcomings related to the weak TC of PCM. In addition, diffusive and convective heat transport within the PCM that are realized without these fins are extremely important. Through modifying heat transport in logically re-designed TES systems, charge/discharge characteristics of such units should be improved at the expense of a smaller amount of PCM and generally heavier weight of the system.

According to [17], introducing fins into PCM was practiced as early as 1966. Numerous studies were performed since then to investigate the consequences of using fins on melting and solidification. Many of these studies are discussed in this review, serving as an update to [17]. After providing a chronological review of the targeted papers, their highlights will be provided in a tabular form. Classification of these papers will then follow in a tabular form. Combining reviews of relevant papers in this document and those of [16,17], which provided reviews/classifications of 75 fin-assisted latent heat thermal energy storage systems, researchers will have access to comprehensive reviews of 206 papers reported since 1966. The cumulative frequency of the pre-2016 publications (Figure 1) suggests uninterrupted ongoing interest in the topic of this review article.

Ismail et al. [20] investigated solidification and melting of a PCM (paraffin and sulfur) filled in a tube-in-tank setup experimentally. A transparent cuboid tank was located at the center of a bigger insulated cuboid with a hot working fluid maintained at the T_m of the PCM circulating through the spacing between these two tanks. Vertical tubes with circular and square cross sections that were fitted with four ES plate longitudinal fins (LF) attached to the outer surface of the tubes were inserted at the middle of the smaller tank. The heat transfer fluid (HTF), silicone, was introduced through an inner circular tube at the top of the vertical tubes and was then extracted at an outlet port on the periphery of the vertical tubes. The PCM was filled into the space between the smaller cuboid tank and the finned tube. Circulation of cold or hot HTFs within the vertical tubes triggered freezing or melting of the PCM. A lifting mechanism was installed on the top of the setup which was used for raising the finned tube during the experiment in order to measure the instantaneous thickness of the frozen layer. The development of the solidified layer was also measured using a camera system, and also a technique employing recorded temperatures. The varied frozen layers with respect to time for both circular and square tube cases were presented. Thickness of the fin did not have major effects on the phase transition process, and further suggested that a greater number of fins tend to promote the heat transfer rate and solidification rate. Without showing results, it was mentioned that the improvement of heat transfer was less marked with the number of fins increasing beyond six. A greater number of fins led to suppression of convection currents within the molten PCM and decreased storage capacity. Higher ratio of the inner hydraulic diameter over the outer hydraulic diameter of the tanks and greater fin height were observed to lead to acceleration of solidification, whereas the higher ratio of hydraulic diameters resulted in suppression of the convective currents.

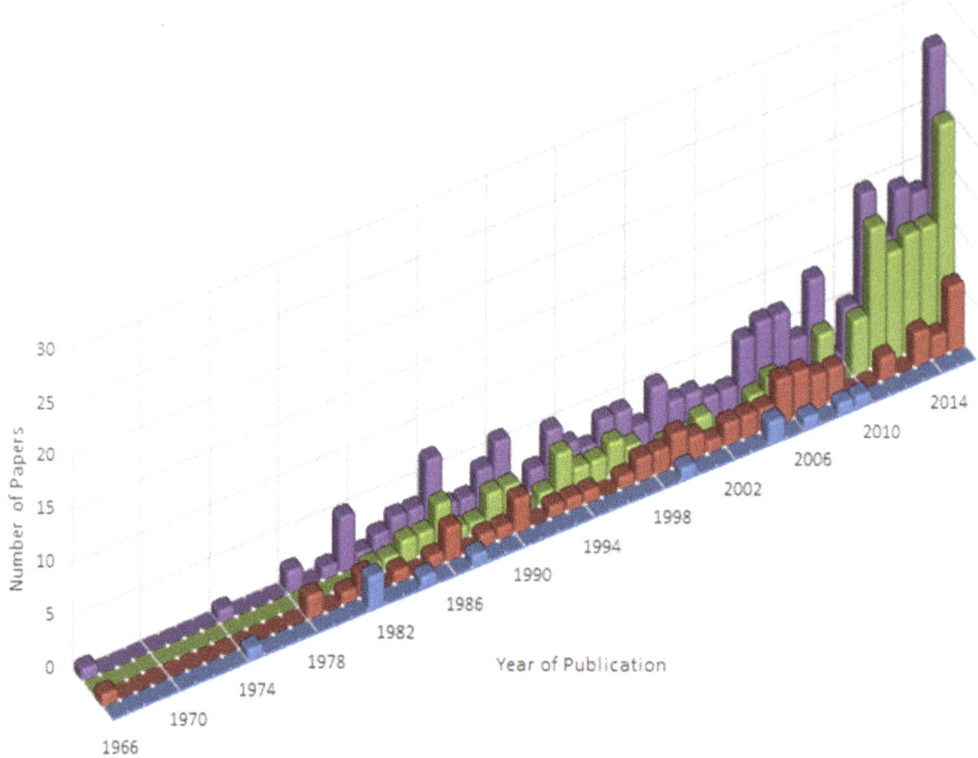

Figure 1. Pre-2016 published papers vs. year of publication reviewed by Fan and Khodadadi [16] (blue row), Dhaidan and Khodadadi [17] (red row), current study (green row) and total per year (purple row).

Ismail and Alves [21] performed both numerical and experimental analyses of solidification of eicosane around one of a staggered array of LF tubes within an LHTES unit, with the HTF passing inside the tubes. Increased fin height, greater value of the number of fins, lowered compactness ratio and higher degree of superheating (the difference between T_m and wall temperature) were observed to lead to reduced freezing time, and the effects of the fin thickness on the freezing time was negligible. Positions of the experimental and numerical liquid–solid interface (LSI) agreed well. Differences between the two approaches in the case of long fins can be explained by the inappropriate use of a linear approximation of the temperature gradient due to the big inclination of the LSI near the fins. Differences between the two approaches at later instants were due to ignorance of the increased heat transfer area resulting from the motion of the dendrites near the LSI, which increased the solidification rate. The emergence of the dendrites accounted for the domination of conduction during solidification, and the adoption of fins were observed to suppress convection and speed up the heat transfer rate.

Ho and Viskanta [22] experimentally investigated the thermal performance of n-Octadecane in a cavity (copper bottom plate, two aluminum vertical walls and two vertical Plexiglas® windows) during melting and freezing realized by circulating HTF through channels in the copper plate. Degassed liquid PCM was syphoned into the chamber such that there was an air gap above the PCM. Profiles of the LSI were recorded photographically. Conduction was the dominant heat transfer mechanism at the early stage of melting, whereas density-induced melt motion forced the liquid to overflow above the unmelted part. Wavy-shaped LSI appeared below the solid core due to unsteady vortex circulation.

A thicker melt layer was observed at the top, where melting was faster due to natural convection. Empirical correlation of the molten fraction with respect to dimensionless time was obtained, showing that subcooling of PCM could delay melting because part of the energy was transferred to sensible heat to increase the temperature of the solid core, and specimens with a higher aspect ratio (*AR*) were observed to induce slower melting rate. A Mach–Zehnder interferometer was introduced to determine 2-D temperature distribution and convective recirculation patterns. Convection was observed to intensify gradually based on the observed fringe lines deflection. 'Cellular' flow patterns appeared near the surface of heat source, but were then suppressed by intensified recirculation with the progression of melting (Figure 2a,b). A shadowgraph technique was applied to investigate the local heat transfer coefficient. It was demonstrated that $Nu_y/Ra_y{}^{1/4}$ (*Nu* and *Ra*, being the Nusselt and Rayleigh numbers, are measures of convective to conductive heat transfer across a boundary, and the importance of natural convection, respectively) decreased as melting continued, indicating transition from conduction-dominated melting to convection mode, whereas an opposite trend was observed with the raising of the *AR*. A periodic variation of the local *Nu* number along the base indicated the presence of multiple recirculating patterns. These fluctuations weakened as the melt zone became larger, resulting in decreased vortex cells. The energy balance method based on the molten volume fraction (VF) was adopted to obtain the average heat transfer data from the heated surface, due to the poor performance of the shadowgraph method at the corner of the test cell. An empirical correlation of the average *Nu* as a function of the *Ra* and *Ste* (the Stefan number is a measure of sensible heat compared to the latent heat) numbers, in addition to the *AR* and initial subcooling parameters, was presented. As for the freezing experiments, the solidified layer was initiated at the bottom, then propagated along the conducting walls. For a smaller *AR*, a more uniform solidified layer with shorter walls was observed. Superheating was observed to only influence the early stage of the cooling process. Small *AR* was observed to enhance the solidification rate.

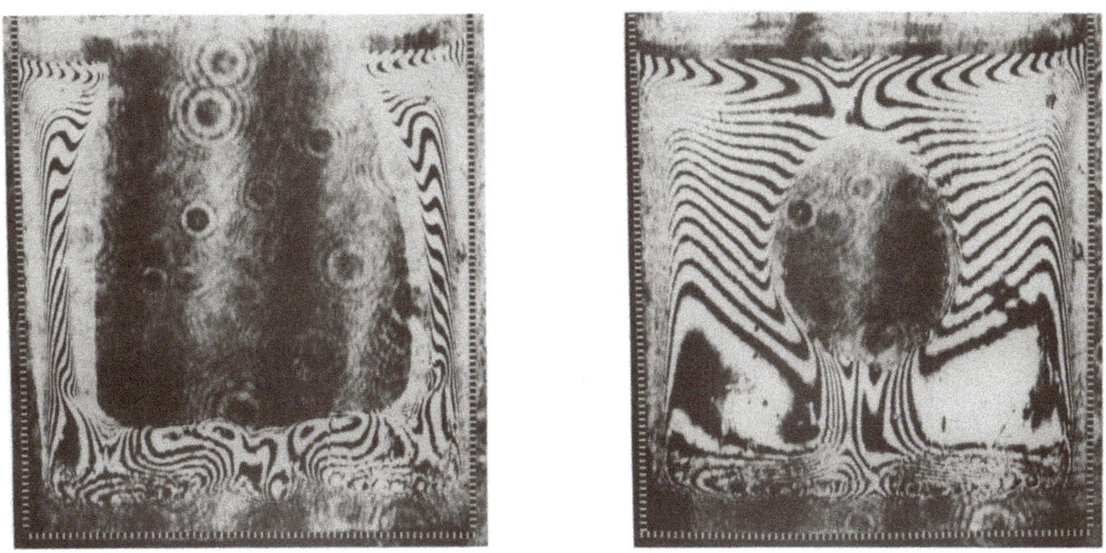

(**a**) (**b**)

Figure 2. Photographs of the interference fringe images in the liquid n-Octadecane during melting in a rectangular cavity with conducting bottom and side walls, T_w = 30.4 °C, *AR* = 1: (**a**) t = 240 min; and (**b**) t = 428 min. Reprinted/adapted with permission from Ho and Viskanta [22]. Copyright 1984, Elsevier.

Details of [21] were reported by [23]. For the experimental component, the HTF passed through a frame-supported vertical single tube which was placed inside the PCM contained within an insulated shell. At certain time instants, the test model was removed to measure the radial and circumferential distributions of the solidified layer. The fraction of solidified PCM, while considering the effects of fin height, number of fins, compactness ratio and degree of superheating was presented.

Okada [24] investigated melting due to a vertical copper cylinder placed concentrically in a horizontal disk-like solid n-Octadecane initially at its T_m that was then subjected to a step change in temperature. Thawing of the PCM, including the effect of natural convection, was studied by a finite-difference method. Positions of LSI and the temperatures along the radial direction on the mid-plane of the PCM were measured and agreed well to computational data. Variations of the *Nu* on the heated cylinder and dependence of volume of the molten liquid with the *Ra* and time were discussed.

Saito et al. [25] reported on the performance of a TES unit incorporating 106 parallel vertical plate brass fins saturated with naphthalene. A single active wall of the unit was controlled by the flow of a heating oil (90–100 °C) through an oil jet box, in which the oil was directed into jets impinging on a heat transfer surface. A 2-D vorticity-stream function-based finite-difference model utilizing apparent TC and specific heat was employed for simulating the flow and thermal fields in a single PCM cell. The optimum *AR* of the PCM cell that gave rise to the maximum dimensionless average heat flux was found to exist for a given combination of the Grashof number (measure of the relative importance of the buoyancy and viscous forces), Prandtl number (measure of a fluid's diffusivities of momentum and heat) and dimensionless fin pitch.

Freezing around a circular hollow tube (inside of which an HTF was flowing) with annular fins (AF) placed on the outer surface was studied by Imura and Yoshida [26] using a 2-D model of a single PCM cell. Among the fin's radius, length, thickness and the *Ste* parameters, the fin length exhibited the greatest variation, whereas the *Ste* was observed to have little influence. In addition, a numerical approach utilizing a 1-D quasi-steady state diffusion analysis was also presented, exhibiting close comparison with the findings of the 2-D analysis. Imura and Yoshida [27] reported on phase change within an LHTES unit consisting of a hollow horizontal circular brass tube with brass AF. Water was circulated through the hollow tube, and n-Octadecane served as the PCM. The experimental data were then compared to [26]. During freezing, experimentally determined heat fluxes were found to be 20 percent greater than the numerical results, due to the occurrence of dendrites. The effect of the *St* was found to be small, as shown earlier [26]. For the thawing experiments, convection became more dominant in relation to conduction as the experiments progressed. Heat fluxes obtained from the experiments were far greater than the values predicted by a computational approach that ignored flow.

Melting of n-Octadecane around a horizontal rod that included four axial holes drilled into it as HTF passageways with and without LF was investigated by Betzel and Beer [28]. The PCM was held within a thermally insulated unit and the LSI was visualized through Plexiglas® plates. Fin materials of PVC and copper were adopted, and three LF in two patterns (Y and reversed Y) were studied. Based on instantaneous melting contours and flow patterns, convection was observed to dominate after a short time, prevailing over conduction. The heat transfer rate was observed to be marked at the top section of the PCM-filled annulus space between the fins, whereas melting progressed slowly within the space below the tube and next to lower fins. Symmetric convection cells were found along the vertical axis. Compared to the bare rode, the PCM melting rate was decreased on the top annulus sector; however, it was increased within the lower sector for PVC-fins with the Y-pattern. For the case PVC-fins with the reversed Y-pattern, the melting rate was even slower than the case with the Y-pattern. A high melting rate and a nearly isothermal liquid surface was realized with copper fins. The presence of six convection cells contributed to expedited melting rate for the Y-pattern, whereas for the reversed Y-pattern, only two convection cells were generated, resulting in a slower melting rate. Correlating the dimensionless melting rate to time, copper fins resulted in highest melting rates, whereas the PVC fins contributed to a slightly higher melting rate compared to bare rod. Y-pattern of fins led to a faster melting rate than the units with the reversed Y-pattern. Spanwise Gortler vortices (Figure 3), detected at the concave LSI, enhanced the heat transfer rate. Correlations of the mean *Nu* as a function of the *Ste* and *Ra* were obtained.

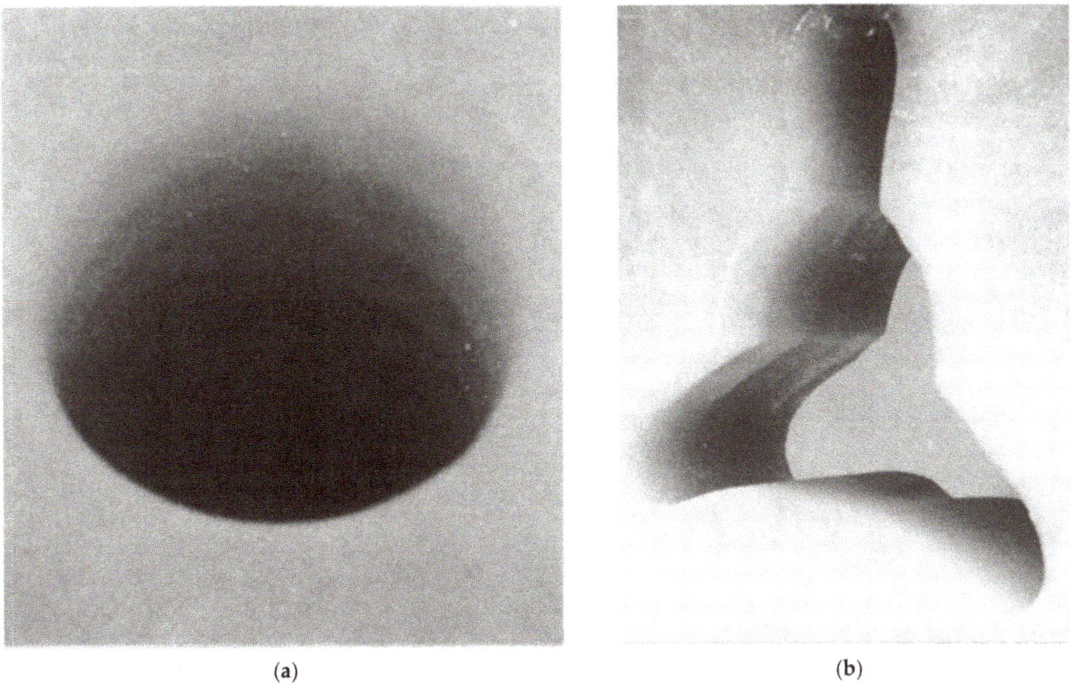

Figure 3. Isometric views of spanwise Gortler vortices corresponding to (**a**) bare copper rod and (**b**) copper rod with reverse Y-arrangement. Reprinted/adapted with permission from Betzel and Beer [28]. Copyright 1986, Elsevier.

Okada [29] extended [24] by performing experimental and computational analyses of thawing around a vertical copper cylinder with or without subcooling. Measured positions of the LSI agreed well with results of the simulations. By lowering the initial temperature of the PCM below its T_m (dimensionless subcooling factors −0.5, −1, −2 and −3), the melting rate was suppressed markedly. The subcooling factor lowered the rate of the total thermal energy storage. Dependence of the average *Nu* on the surface of the heated cylinder and the average thickness of the molten liquid region were independent of the subcooling factor. Volume of the molten PCM for different cylinder diameters, for the case of no subcooling, were correlated to a dimensionless time that included the *AR* of the cylinder.

The English translation of Saito et al. [30], i.e., ref. [31], reported results of experimental and computational studies on melting in a rectangular unit were very similar to [25]. The unit was mounted on a platform that could be inclined. Comparing the unit to the one used by [25], the main distinction was that the plate fins in the present study were inclined to the horizontal, whereas only vertical plate fins were studied by [25]. It was observed that the shape of the moving solid phase varied as the PCM received heat from the surrounding walls at the contact points. A numerical method that assumed a parabolic velocity distribution in the liquid phase was developed. Results of the experiments using n-Octadecane agreed well with computational findings. The average melting rate was found to be independent of the inclination angle. However, the contacting mode and the heat flux variation on the walls were dependent on the inclination angle.

An experimental study in combination with an approximate solution to the phase transition phenomenon in an LHTES unit consisting of a horizontal circular tube through which the HTF flowed was investigated by Ito et al. [32]. The specific PCM was n-Octadecane that surrounded the tube, which was fitted with plate AF. A dimensionless heat extraction rate, or an apparent *Bi* (the Biot number is a measure of relative importance of heat transfer resistances within and at the surface of an object), as a function of a dimensionless time (obtained by analysis or experiment for the finned tube at a constant wall temperature) was utilized. Apparent performance of the thermal energy storage unit with the finned tube was analyzed, exhibiting good agreement with the experimental results.

Sasaguchi and Sakamoto [33] focused on studying thawing in an elemental annular cylindrical region bounded by two LF and conducting inner and outer tube walls. Melting that was initiated on the heated inner tube wall depended on the orientation of the test region ($\varphi = -90°$ to $+90°$), the angle between the two LF ($\alpha = 20°$ to $90°$) and the temperature of the tube wall. The effects of the orientation of the elemental test region, the angle between the two fins and the temperature of the heated tube wall on the position of the evolving LSI, including the influence of convection, were discussed. Addressing limited computational findings in relation to LHTES systems having tubes with AF (radial), Sasaguchi [34] reported on the relevant results of a new method. Moreover, the predicted performance of the units was compared with those of LHTES systems having a tube with LF and AF (radial) in addition to an unfinned tube. Melting of eicosane in an elemental annular cylindrical region bounded by two LF and conducting inner and outer tube walls was studied experimentally by Sasaguchi [35]. Melting was controlled by an array of impinging radial jets of a hot fluid supplied from a constant-temperature bath. Thawing experiments were conducted to determine the effects of the orientation of the elemental test region ($\varphi = -90°$ to $+90°$), the angle between the two fins ($\alpha = 20°$ to $90°$) and the temperature of the heated tube wall. Expedited thawing next to the fins and the role of convection in promoting melting are clearly observed. Measurements of temperature of the PCM were also compared to [33].

The appropriateness of a similarity curve for evaluating the performance of LHTES units was investigated by Kaino [36]. A system of concentric circular tubes with LF in the annular space was analyzed the freezing process. A similarity rule was shown to be valid, as well as to an unfinned unit, while its appropriateness deteriorated with the increasing number of fins. Differences among similarity curves for various number of fins were so marked that the configuration effect can also be accounted for. Heat exchanger effectiveness was compared to existing experimental data, exhibiting good agreement. Computational studies of Kaino [37] showed that, in a storage unit with LF undergoing freezing, the influence of the *Ste* varied with the number of plate fins. Contributions of the sensible heat released within the PCM, the heat transfer tube and the fins were evaluated separately. Relation between the solidified fraction and sensible heat for the *Bi* range 0.1–1000 was presented. The interplay of the latent and sensible heat was examined for various *Ste* numbers, exhibiting an inverse relation. Given the existence of a similarity function varying with the frozen fraction and independent of the *Bi* to be linked to uniformity of heat flux on a heat transfer surface, Kaino [38] focused on the previously studied LHTES units featuring LF, in which such uniformity was perturbed. For a heat transfer tube made of a high TC material, the existence of such a similarity rule can be assured for a wide range of tube thicknesses. However, given a heat transfer tube with a low TC material, the similarity rule was valid only with considerably thick tube walls. Kaino [39] extended [38] to include the effects of thickness, TC and the number of fins. It was shown that upon reducing the thermal conductance of the fin, thickness and/or TC of the fins, the similarity rule becomes more applicable.

Phase transition within porous media saturated with a PCM contained between two parallel planar fins was studied by Sasaguchi and Takeo [40] (English translation [41]). Heat conduction dominated the freezing process. The effect of the orientation of the fins with the vertical direction was also investigated. Velocity vectors and isotherms at three dimensionless time instants during thawing for a case with the fins positioned in the vertical direction were discussed, and the strengthening effect of natural convection at later time instants were clearly observed. Similar behavior was observed at the other extreme with the fins being in the horizontal position.

Al-Jandal and Sayigh [42] studied the performance of a proposed solar tube collector (STC). Experiments were performed to simulate a storage system using two vertical cylindrical concentric tubes with stearic acid filling the annular space. This instrumented set-up shared many features of earlier experimental set-ups going back to 1981 [17], and discussed above [20,21,23] and later in this paper, since the HTF was introduced within a vertical end-closed tube. In this work, the researchers also incorporated a path for another HTF to circulate on the outer shell of the unit. Experimental results corresponding to the utilization of 13 AF and 3 LF provided quantitative information concerning heat transfer and transient evolution of the LSI, pointing to the role of buoyancy-driven convection.

Choi and Kim [43] performed a study for unfinned, circular stainless steel 5-finned and 10-finned tube systems. The experimental apparatus was composed of a vertical cylindrical Pyrex glass TES vessel and a stainless-steel double-tube, with 5 or 10 AF welded on the its outer surface, with 20 or 10 mm axial pitch. Temperatures of the five-finned-tube system were always higher than those of the unfinned-tube system. Thermal performance during melting in all three tube systems were more strongly affected by the inlet temperature than by the flow rate of the HTF. The volume of melted PCM in the 5- and 10-finned-tube systems were nearly 25% greater than the unfinned-tube, since the annular finned-tube system expedited melting and inhibited convective motion. Heat storage in the 5- and the 10-finned-tube systems were 37% and 48% greater than the unfinned-tube system, respectively. For the unfinned-tube at low molten VF, the outside heat transfer coefficient (h_o) decreased by increasing the melted volume in the region affected by conduction, while beyond this region, the melted volume increased by increasing the melted liquid VF. For the systems with fins, within the finned section, h_o decreased sharply since the presence of fins inhibited convection. On the other hand, beyond this section, h_o did not increase as much as the unfinned-tube system since the fins partially inhibited convection. In the unfinned-tube system, the heat transfer coefficient was much greater than the value calculated for steady conduction except near the tube wall. Melting from an outside wall of the convectively heated unfinned-tube was modeled. The measured melting-front velocity enhanced by increasing the *Ste* and *Bi* exhibited good agreement with the analytical solution. Therefore, the effect of convection on melting was negligible, and it only contributed to increasing the sensible heat of the melted liquid PCM. Choi et al. [44] investigated the melting thermal performance of a low-temperature vertical cylindrical shell-and-tube LHTES. The HTF-carrying tube consisted of a concentric double tube, with the HTF (water) from the top inlet of the inner tube flowing downward and then re-directed in the opposite direction, flowing out through the top outlet placed at the periphery of the outer tube. A helical type silicon wire was inserted as a turbulence promoter to reduce the thermal resistance at the HTF side. Twelve equally spaced AF were attached to the outer surface of the HTF tube, and the tube's fins and shell were all made of stainless steel. The PCM was kept at 5 °C lower than its T_m. The rate of heat storage decreased sharply at the beginning during thawing for three types of geometries (thin-finned-tube, thick-finned-tube and unfinned-tube system) and then tended to slow down gradually. Higher inlet temperature of the HTF was observed to lead to enhanced rates of heat storage due to improved TD. The effect of the flow rate was negligible due to existence of the turbulence promoter, which made the HTF flow turbulent even at low flow rates. The thick-finned-tube exhibited 70% greater heat storage compared to the unfinned-tube system, whereas the thin-finned-tube did not promote the heat storage rate noticeably. No transition point was observed in the monotonically increasing trend during melting due to supercooling. The heat transfer coefficient between the PCM and the heat-transfer tube surface decreased with time, and the thick fins were observed to enhance the heat transfer coefficient by two times greater than the system without fins. The *Bi* had no effects on the melting front velocity, and faster melting front velocity was observed to be closer to the heat transfer surface due to higher heat transfer coefficient at the initial stage of charging. Melting front velocity increased with the *Ste*. Improved consistency was observed between the experimental results and predictions obtained from an unsteady-state approximation compared to the predictions from a quasi-stationary approximation. Correlations of the amount of heat storage in terms of the *Fo* (dimensionless time), *Ste* and *Re* (the Reynolds number that is a measure of the relative importance of the inertia and viscous forces) were proposed for unfinned-tube and finned-tube system, respectively.

Freezing of paraffin surrounding an aluminum heat pipe with LF was studied by Horbaniuc et al. [45]. Considering only diffusive transport, radial and angular solidification LSIs between two adjacent fins that were independently contributed by heat transfer from the heat pipe and fins, respectively, were discussed. Thicknesses of the solidified layer corresponding to two different directions were independently calculated. Moreover, for determining the angular position of the freezing layer, since the temperature distribution is unknown, two different temperature variations (exponential and polynomial) were assumed. Freezing time with 6 fins was more than 180 min, whereas for 12 fins, the elapsed times were 120 min (exponential) and 150 min (polynomial).

The freezing process was investigated by Choi et al. [46] using the same experimental apparatus [44]. Supercooling was found to be greater, while, closer to the heat-transfer surface, and enhanced cooling rate led to a higher degree of supercooling, which is why the degree of supercooling was observed to be higher for the thick finned-tube system compared to the unfinned-tube. Variations in temperature within the PCM was unnoticeable along the axis of the tube. The bottom portion of the PCM with lower temperature exhibited a greater degree of supercooling. For both the thick finned-tube and unfinned-tube systems, the rates of heat recovery decreased sharply during the initial sensible heat recovery phase, and then increased rapidly when the latent heat recovery initiated, followed by a decreased rate after reaching a maximum value. Lower inlet temperature of the HTF resulting in greater TD led to a higher rate and amount of heat recovery, whereas the effects of the HTF flow rate were negligible due to presence of an inserted turbulent promoter. The cumulative amount of heat recovery highly depended on the HTF inlet temperature, but not on the flow rate. Transition points existed at the beginning of discharging, and the presence of fins, which prevent crystal growth, delayed the occurrence of transition points. Thin finned-tube systems did not exhibit significantly greater amounts of heat recovery, since a higher degree of supercooling offsets the enhanced heat transfer rate due to the presence of fins. Fluidity was prevented by the fins, the presence of voids due to shrinkage and addition of the thickening agent, in addition to decreasing the heat transfer rate. On the other hand, the thick finned-tube system with the same extended area as the thin finned-tube behaved favorably, indicating the important effects of the fin's thickness. Similar variation trends of the heat transfer coefficient were observed for the heat recovery rate. Discharging was divided into three continuous stages, i.e., initial sensible heat transfer, latent heat recovery and the unsteady-state conduction phase after the emergence of the solid layer on the heat transfer surface. Whereas the measured PCM heat transfer coefficient was smaller than the calculated one from the steady-state conduction equation at the beginning due to the supercooling, the measured and calculated values agreed well once the latent heat transfer started. A faster cooling rate was observed for the finned-tube systems, which led to more severe supercooling, thus lowering the PCM-side heat transfer coefficient. The unsteady-state approximation provided more accurate results in terms of the freezing front velocity than a quasi-stationary approximation. The LSI velocity did not depend on the *Bi* number for the range of HTF flow rate considered, while this velocity increased with the *Ste*. The amount of heat recovery as a function of the *Fo*, *Ste* and *Re* numbers for both unfinned-tube and thick finned-tube system were obtained for discharging.

Han and Han [47] examined heat transfer characteristics of low-temperature LHTESS with annular finned and unfinned tubes. The heat storage vessel was 530 mm high (inner diameter of 74 mm), whereas the end-capped inner HTF (water)-handling tube was 480 mm high, with an outer diameter of 13.5 mm, similar to [42–44,46]. The heat recovery rate was affected by the HTF's flow rates and inlet temperature. Heat transfer improvement provided by fins, compared to the unfinned tube system, was found to be negligible in the case thin-finned systems; however, the heat transfer coefficient of the thick-finned system was about 60% higher than the unfinned system. The experimentally determined heat transfer coefficients for the unfinned tube and thick-finned tube systems were 150–260 W/m^2K and 230–530 W/m^2K, respectively. The fin efficiency based on the heat transfer coefficient and increase in area provided by fins was found to be 0.05 and 0.26 for the thin- and the thick-finned systems, respectively.

Freezing and thawing of water within a rectangular box with a vertical heat transfer plate consisting of pin fins with square cross-sections were investigated by Hirasawa et al. [48]. The influence of the pin pitch and number of pin fins placed in a square pattern were studied. Distributions of temperature, ice/water VFs and variations of heat flux were measured, and the flow patterns in the water were observed. During solidification, the rate of phase change increased monotonically as the number of fins was increased. For melting, position of the LSI within the unit with 25 pin fins after 65 or 66 min of start of melting still exhibited the non-uniformity of distance from the heated wall for the case of no fins. The relationship between the modified *Nu* and the *Ra* summarized the role of natural convection on melting.

Chen et al. [49] conducted a 3-D analysis of freezing and melting of water around a vertical heat transfer plate consisting of pin fins with square cross-sections. Shape of the LSI, temperature and velocity fields were evaluated for different numbers of fins and comparisons against the experimental results [48] were made. Heat conduction was dominant when the number of fins was high during solidification. For melting, convection next to the heat transfer plate between the fins was confirmed.

Wirtz et al. [50] studied temperature stabilization of electronic modules by hybrid coolers using a dry waxy granulate solid–solid organic compound. A modification of a commercial aluminum plate-fin heat sink, where the lower portion of the space between the fins was filled with PCM, was employed as the prototype hybrid cooler. An elastomer coating was used to encapsulate the PCM, and the heat source was bonded to the underside of the base plate. A mathematical model of a "half-fin" segment of the hybrid cooler was formulated using a thermal circuit composed of seven-element heat transfer components. Different heating and cooling strategies were evaluated, and a figure-of-merit characteristic of the cooler/PCM system was introduced. The simulations suggested that the transient response of a given hybrid cooler can be characterized by temperature stabilization time (employed figure-of-merit) that can be measured using a single experiment. Numerical simulations exhibited that an efficient hybrid cooler should have a high PCM conductance and small transition temperature interval in order to provide tight thermal control.

Given high heat fluxes observed during the close contact-melting (CCM), Hong and Kim [51] studied the utilization of split fins in order to enhance the melting speed in an LHTES unit. The split fin system was an array of fins separated from each other by a gap distance of 1 mm (total horizontal depth of 53 mm) placed on a heating copper plate, which was activated by impinging jets of a brine mixture HTF. Melting of ice for both split and non-split (without 1 mm gaps) fins was studied, showing that CCM by split fins increases the melting rate compared to non-split ones.

Inaba et al. [52] investigated convection within an inclined rectangular LHTES unit having one copper heating wall with plate fins. Emergence of the molten phase and its position in relation to the effects of plate fin length, inclination angle and heated wall temperature were identified. Flow patterns for various inclination angles exhibiting the marked role of natural convection were elucidated. Relevant dimensionless correlations were presented for this storage unit.

Using an LHTES system to the liquefied natural gas (LNG) vaporization process can level the fluctuations in the cold energy generation rate, due to daily and seasonal variations in NG consumption. Yamashita et al. [53] performed liquefaction tests by melting n-Pentane. Using a rectangular LHTES unit, which had 24 horizontal tubes, each having 12 equally spaced LF, performance of the unit and the heat transfer characteristics around the finned tubes were investigated. Liquefaction time of the boil-off gas in the unit varied inversely with the duty of the unit, but this quantity became shorter with the increase in the duty due to effect of the PCM solid that remained near the outlet of the storage unit.

Thermal performance of PCM incorporated in a photovoltaic (PV) system, to regulate the temperature rise and provide building heating, was studied by Huang et al. [54]. A single flat aluminum plate system, two PV/PCM systems without internal fins (system I with a height 40 mm and depths of 20, 30 and 50 mm; system II with a height 132 mm and a depth of 20) and a PV/PCM system with internal fins (system III) were selected. The front plate was exposed to the insolation, and both the front and rear plates were exposed to convection. For experiment and simulations, and RT25 and for simulations, paraffin wax was selected. An experimental set-up (0.3 m $\times$ 0.132 m $\times$ 0.0045 m) aluminum plate, covered by selective solar absorbing film, was utilized. Two aluminum plate fins (0.0045 mm thick) fixed to the front wall were employed to validate another part of the simulation. The resultant melt front and isotherms from experiments were in good agreement with the simulation results (Figure 4). For system I, flow was upward, adjacent to the front heated plate of the PV/PCM system, and downward at the LSI, where heat was transferred to the solid and the fluid cooled. The PCM in the top portion of the system melted faster due to convection compared to its base. Vertical velocity components were greater adjacent to the front plate of PV/PCM system and at the melt front. The largest vertical velocity component in the PCM, either for the rising flow or falling flow, was at the mid-height within PV/PCM system I. Cooled by the high TC of the rear plate, the density of the molten PCM adjacent to the rear plate of the PCM container increased, leading to improved flow. By increasing the ambient temperature, the temperature of the front surface did not change significantly; however, since less heat was lost from the system, the time required for melting was reduced. Moreover, with enhancing the incident insolation intensity, the temperature at the front surface increased and the t_m was reduced. For system III, after 45 min, convective flow of hot molten PCM passed through the gap between the fin and rear plate at the end of the fins into upper sections, whereas cooler PCM moved downward through the gap. The flow pattern was maintained until the PCM was fully molten. Temperatures at the front surface of PV/PCM system I were maintained at a low value for a longer period than that for the base case plate. For system I, increasing the depth of PCM beyond 30 mm had an insignificant impact on its performance. Comparing systems I and II, the average cell temperature was higher, due to the greater amount of the liquid PCM circulation. Electrical efficiency for system I with 30 mm PCM depth and an aluminum plate for real operating conditions (3 days starting 21st of June, in SE England) was obtained. The regulated temperature increased the efficiency of the system. For those three days, the efficiency was the same for the aluminum plate. However, the effect of PCM on the performance of the PV/PCM system was slightly different for the first day compared to the other two days.

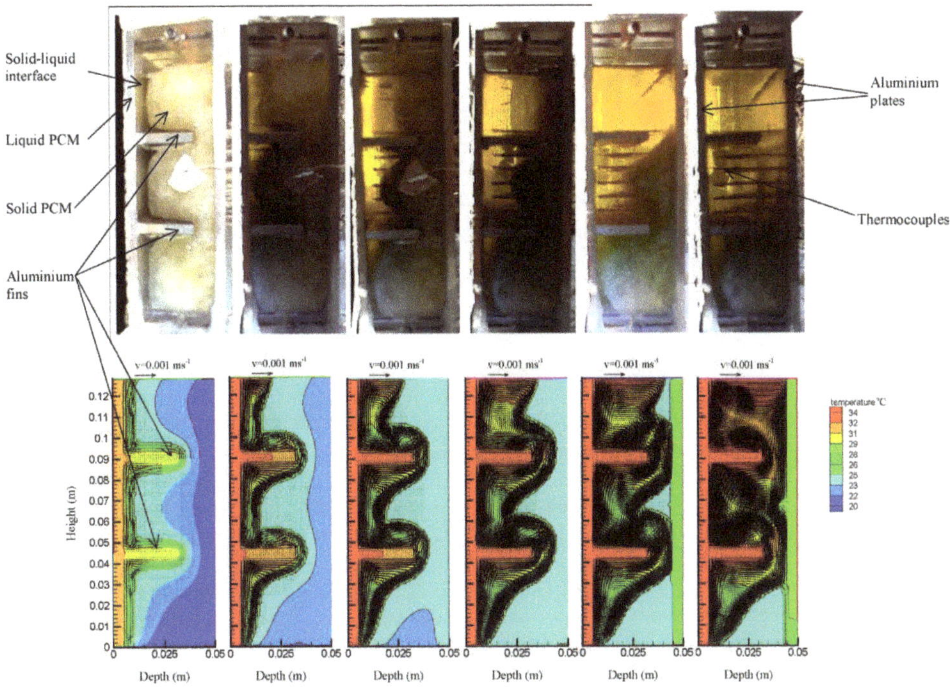

Figure 4. The **top** figure is photograph of an experiment and they are generally of this caliber. As for the **bottom** figure, the extent of "darkness" of the black vectors carries the message even though it might appear as sign of being illegible. Also, we are borrowing these from the source, and have not no way of improving resolutions. Instantaneous photographs of the melting PCM showing the progression of the LSI (above row, left to right) and predicted isotherms/velocities (bottom row, left to right) for a PV/PCM system with fins ("darkness" of the zones due to the black velocity vectors suggests greater dominance of natural convective motions). Reprinted/adapted with permission from Huang et al. [54]. Copyright 2004, Elsevier.

Yamashita et al. [55] extended earlier work [53] by reporting experimental results of their n-Pentane-based TES tests. Behavior of the PCM freezing was elucidated by studying the thermal performance of the storage unit and the heat transfer characteristics around the finned tubes. It was observed that the ratio of stored cold energy to storable energy was effective for correlating the experimental results during freezing as well as thawing. In addition, the thermal conductance of the finned tubes, which was lower than that in melting, was represented by a simple cylindrical model. Yamashita et al. [56] performed computational analysis of phase change processes in the LHTES system they studied experimentally [55]. Results of numerical calculations agreed well with the findings of the n-Pentane-based pilot-plant tests. Moreover, results of the analysis exhibited that the liquefaction time and amount of discharged cold energy for thawing decreased markedly when the length of the liquefaction zone within the finned tubes exceeded the total length of the store.

The thermal performance of stearic acid filled within the shell side of a vertical shell-and-tube energy storage unit (Figure 5) during solidification was studied by Liu et al. [57]. The set-up consisted of an electrically heated rod surrounded with PCM, which was sealed with a concentric stainless steel inner tube. The HTF flowed through the annulus between the inner PCM container tube and the outer coaxial stainless steel pipe, and the whole unit was insulated. The PCM was initially charged by hot water (at higher temperature than T_m) to a complete molten state with ES temperature field, followed by freezing initiated by circulating water at a lower temperature. A drastic temperature drop was observed during the initial stage of solidification due to a great TD between the PCM and HTF, domination of convection in the liquid PCM and quick release of sensible heat. This was followed by a slowed decreasing rate of temperature after initiation of freezing, resulting from small TC of the solidified PCM outer shell, shrinkage of area and slower absorption of latent heat. The temperature declined faster for the solid phase sensible heat diffusion stage after completion of freezing, though with smaller TD. It was noted that the temperature distribution was even at the beginning of freezing, influenced by the initial uniform temperature. Thereafter, the slope of the temperature became steeper as solidification was in progress, and it tended to become flatter near the end of freezing. The duration of time needed for complete freezing was drastically shortened with decreased inlet temperature of the HTF, due to the induced greater TD. The temperature of the PCM was found to decrease with lowering of the inlet temperature, and the influence became marked as time progressed. Alteration of the *Re* (200–500, laminar flow) was observed to not influence temperature measurements noticeably. Thermal resistance induced by the low convection coefficient was negligible compared to the resistance due to low TC of the PCM. An enhanced heat transfer rate was realized with the addition of a spiral twisted split copper AF (Figure 5c) attached on the outer surface of the inner tube and spanning the whole annular gap. Improvement of the freezing rate was more pronounced at the initial stage, and convection was suppressed with the progression of solidification. Fins with thinner widths were reported to be more effective in enhancing the freezing rate, compared to the thicker fins with same total volume of fin, due to the availability of a more effective heat transfer surface.

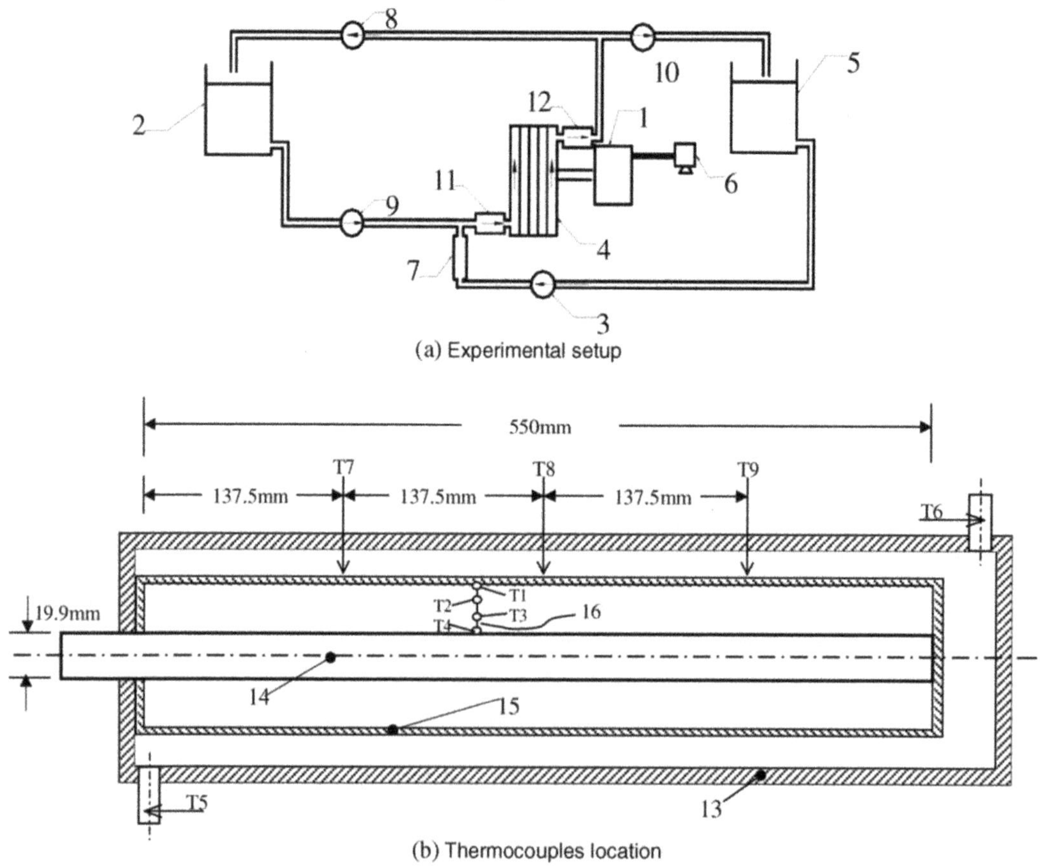

Figure 5. *Cont.*

(c)

Figure 5. Schematic diagrams of the (**a**) experimental set-up, (**b**) locations of 9 thermocouples (T1–T9) in the storage unit (sub-components 1–16 are identified in the original publication) and (**c**) spiral twisted tape copper fins [private communication]. Reprinted/adapted with permission from Liu et al. [57]. Copyright 2005, Elsevier.

Kayansayan and Acar [58] studied solidification of distilled water filled in the annular space between the finned HTF-carrying tube (*Re* in the range 500–7000) and the rectangular cell container in a shell-and-tube heat exchanger. Two opposite walls of the container were made of Plexiglas® for ease of recording the images of freezing, whereas all other surfaces of the container were not Plexiglas® (Figure 6a). One-piece finned tubes (49.2 cm long with inner and outer radii of 20 and 30 mm, respectively) were manufactured from solid bronze to eliminate any contact resistance between the tube and the fins. The HTF-carrying tube accommodated AF (thickness of 3 mm; 54 and 64 mm in diameter). Predictions of the HTF exit temperature and molten VF of this numerical study agreed well with previous literature, and a small discrepancy due to negligible wall heat capacitance and wall temperature was observed. The experimentally recorded diameter of ice (Figure 6b) was compared to the results of the numerical predictions. Symmetry of the wavy experimental profiles of the LSIs identified the accuracy of considering conduction solely in the PCM. The solidification rate was observed to increase with enhanced fin density, the *Ste* (by lowering the inlet temperature of the HTF) and the Peclet number (*Re Pr*), whereas the maximum deviation was observed for high *Re*. Overprediction of the thickness of the frozen layer compared to the observed slower advance of the LSI at high *Re* of the HTF was attributed to inevitable heat gains from the ambient temperatures. The outer wall temperature of the tube as influenced by the fin diameter was apparent at higher *Fo*, and the solidification rate was enhanced with greater fin diameter. The fusion rate was higher with a greater number of fins, whereas this relation tended to be similar for different numbers of fins whose effects were better felt at high *Re*. The greatest discrepancies between the experimental results of the stored energy and predictions were noted at higher *Fo* and low *Re*. More uniform distribution in frozen layer diameter along the flow direction was obtained with greater *Re* of HTF. Based on a parametric study of the thermal performance, improvement in energy storage with enhanced fin diameter ratio was more distinguished at high *Re*. In effect, higher fin density was observed to attain higher energy storage capacity, and its effects were more apparent at high *Re*.

Huang et al. [59] extended an earlier study [54] of a PCM-assisted PV unit by incorporating a 3-D model. The rectangular cuboid system (Figure 7a) was composed of an RT25 paraffin wax compartment with one of its sides exposed to insolation. Results of three cases ($3D_1$, $3D_2$, $3D_f$) with different boundary conditions were compared with the previous 2-D numerical results. For the $3D_1$ case, the side walls were adiabatic, and for the $3D_2$ case, the side walls had a heat transfer coefficient of 5 $Wm^{-2}K^{-1}$. For the $3D_f$ case, five evenly spaced aluminum square cross-section pin fins were placed on the front active face, and the remaining walls were adiabatic. For the $3D_1$ model, fluid movement occurred in all directions, which subsequently led to the more visible phase change on the rear surface of the 3-D system compared to the 2-D simulations. Moreover, the average temperatures on the front surfaces of the systems during melting were slightly lower and more stable for the $3D_1$ model prediction compared to the 2-D results. Because of heat loss from the side faces, the maximum vertical component of velocity predicted at the center of the unit for the $3D_2$ case was greater during melting in comparison with the 2-D simulations. Based on the same reasoning, the rate of temperature increase predicted on the front surface by the $3D_2$ model was lower than the 2-D model; however, the predicted movement of the LSI agreed well (Figure 7b,c). Moreover, higher velocities were predicted adjacent to both the front surface and the LSI with the $3D_2$ simulations compared to 2-D results. After the PCM was fully molten, the fluid adjacent to the sides slowed down due to the non-slip boundary condition, and fluid recirculating zones were formed at corners of the cuboid. For the $3D_f$ model, the presence of high TC pin fins improved heat transfer into the bulk of the PCM and encouraged thermal homogeneity in the system, but these fins simultaneously acted as barriers to natural convection.

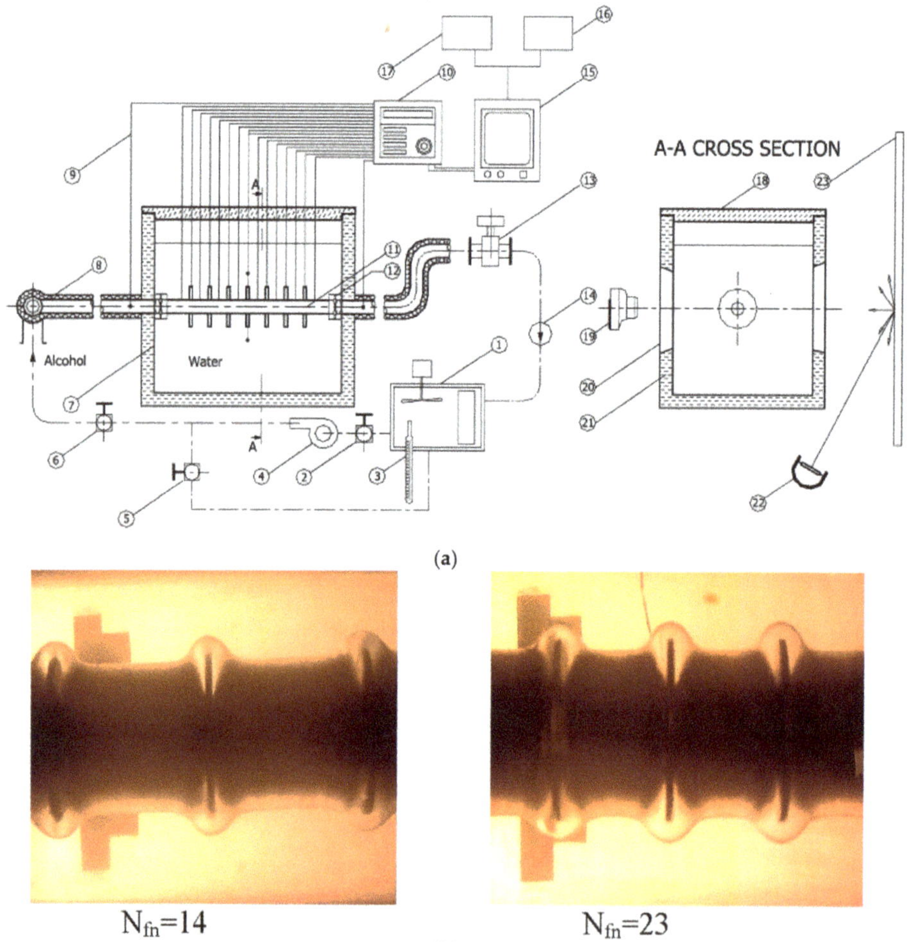

Figure 6. Schematic diagram of the (**a**) experimental set-up (sub-components 1–23 are identified in the original publication) and (**b**) photographs of the ice fronts on finned tubes (arrays of 14 and 23 fins). Reprinted/adapted with permission from Kayansayan and Acar [58]. Copyright 2006, Elsevier.

Wang et al. [60] performed a 2-D study of the effect of orientation of a hybrid aluminum heat sink, with five aluminum vertical plate fins, with a paraffin wax and air layer occupying the spaces between the fins. With the heat source at the top identified as $\theta = 0°$, the unit was suddenly placed in four different orientations of the heat source ($\theta = 45°, 90°, 135°$ and $180°$). During melting, due to the low density of air and compression of the expanding PCM, the air phase must move upward. The results illustrated that the effect of orientation on the thermal performance was limited.

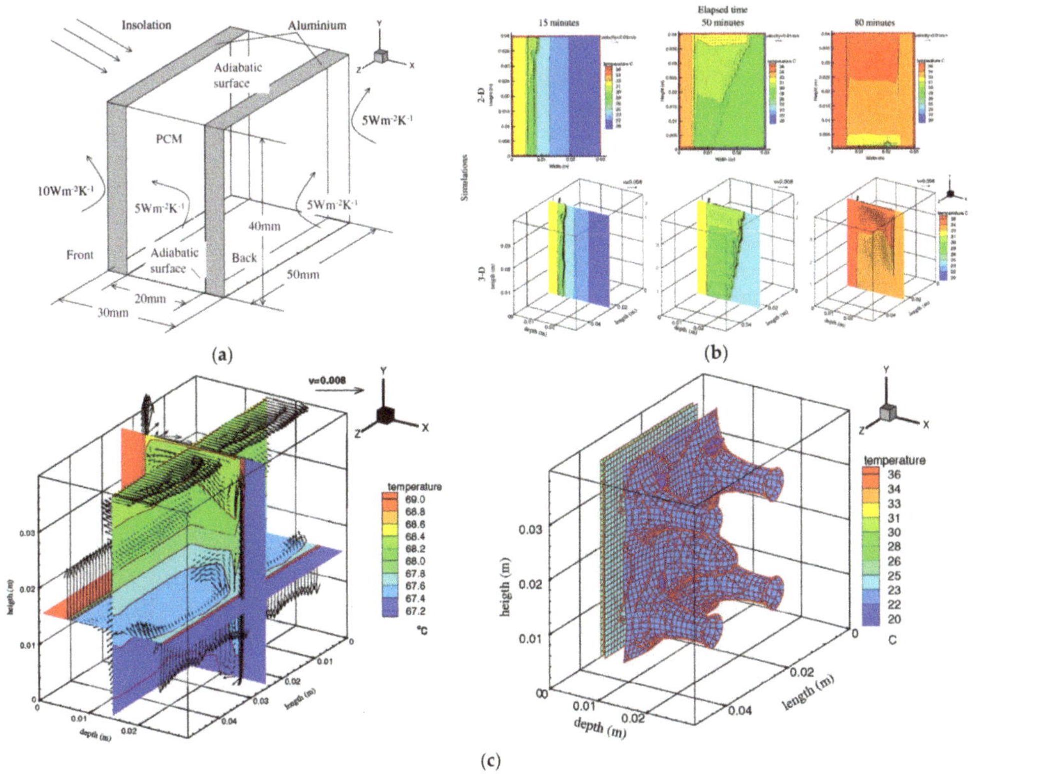

Figure 7. Schematic diagram of the (**a**) model PV/PCM, (**b**) 2D vs. 3D predictions of temperature and velocity fields and (**c**) combined temperature contours/velocity vectors on selected planes along with isothermal surface for the case of five pin fins. Reprinted/adapted with permission from Huang et al. [59]. Copyright 2006, Elsevier.

An LHTES unit composed of an annular finned HTF-carrying horizontal tube and five distinct PCM cells, with application to solar power generation, was studied by Seeniraj and Narasimhan [61]. Individual sealed thermal storage cells surrounding the HTF-pipe were arranged in the flow direction based on a descending order of their T_ms. Two different eutectic mixtures of ($LiF\text{-}CaF_2$ and $LiF\text{-}MgaF_2$) were utilized as the first and second PCM. The third to fifth model PCM were assumed to possess declining T_m, but had the same thermo-physical properties as the second PCM. The HTF was liquid sodium (laminar fully developed). Performance of the multiple PCM unit was compared with that of a single unit having PCM-2. The predicted LSI for both systems exhibited expedited melting on the upstream fin of each compartment in comparison to the downstream fin, with the multiple-PCM unit showing faster melting at later stages. Whereas the sensible and total energy for a single PCM system were greater than the multiple PCM unit, the latent heat for a multiple PCM system was greater than the corresponding value for a single PCM unit. Utilizing multiple PCM outweighs implementation of a single PCM in terms of more uniform exit temperatures of the HTF and expedited thawing.

Saha et al. [62] investigated the contribution of a TES unit (two arrangements of aluminum pin and plate fins) for thermal management of electronic devices. Considering the duration of melting and lower operating temperatures, the optimal VF of TC enhancer in the TES unit (not including the base) was 8%. Maintaining the optimal VF, the influence of fin geometries and their multiplicity on the performance of TES unit were discussed. It was found that a greater number (36) of small cross-sectional area pin fins were the desirable design. Three notable time spans corresponding to the temporal temperature variations of the heater and PCM were noted, i.e., (i) temperature increase due to the heat gain, (ii) fairly uniform temperature and (iii) temperature rise due to sensible heat of molten PCM.

Wang et al. [63] conducted a study of thermal management of an electronic device (similar to [60]) to investigate the performance of extruded heat sinks filled with PCM. The expansion of PCM upon melting, convection in the fluid phase and motion of the solid phase within the liquid were considered. Conduction through the planar fins and the end walls, and convection within the air/liquid hybrid system, were also simulated. Increasing the initial PCM VF led to shorter t_m that varied with a more gradual slope than a linear proportionality with the initial VF, since the increased height of the PCM could promote free convection. Moreover, the temperature rise in the middle of the PCM was delayed. A higher TD between the base temperature and the melting point caused shorter t_m and a steeper heat transfer coefficient decline with time. Imposition of a square wave temperature variation led to a faster local temperature rise and a greater maximum stored energy compared to a sinusoidal waveform. As for the effect of the *AR* for a given base heat flux, t_m was greater for higher values of the *AR*. For a given *AR*, the LSI moves away parallel to the vertical fin for low values of melt fraction, indicating uniform melting along the fin surface for *AR* of 0.5. For a small *AR*, a wedge-shaped liquid phase was present at the bottom of the cavities, showing that melting more likely took place close to the bottom.

Kandasamy et al. [64] investigated improving thermal performance of a paraffin-based heat sink electronic device. Molten PCM was filled in the heat sink case modules with fin arrangements identified as HS1, HS2 and HS3, which were maintained at 95 °C using a hot plate. The case surface and chip temperatures rose with time due to the heat input at first, followed by steady-state trends, while a 4–5 °C TD was maintained. The transient die junction temperature with the inclusion of heat sinks was observed to decrease compared to that of original package, and the HS2 and HS3 modules attained greater temperature reductions compared to the HS1 configuration. It was observed that the embedded PCM did not make an apparent difference to the die junction temperature response of the package for the low level of input power, which was unable to activate the melting of the PCM. When the input power was raised from 2 to 4 W, the presence of the PCM was clearly observed, in effect extending the time required to reach the steady-state. Furthermore, the thermal resistance of the package with a greater power input (4–6 W) was higher. A 3-D study with the HS1 geometry was conducted. Considering the expansion of the PCM, a small fraction of the cavity was occupied by air. The simulated temperature at a specific location was compared to the experimental data, and a good agreement was observed with only a small discrepancy, which may be due to the difference between the actual and assumed boundary conditions. The evolution of the contours of the liquid fraction of the PCM exhibited that the entrapped air promoted the heat transfer rate to PCM in contact with it, i.e., the presence of the air assisted in melting the PCM from the upper portion in contact with it when its temperature was above the melting point. Moreover, a curved interface between the molten PCM and air was observed.

Agyenim et al. [65] studied melting followed by freezing in four shell-and-tube configurations. The base case consisted of a PCM-filled shell space embedded in a horizontal HTF-carrying copper concentric tube. Three modified configurations were eight AF or eight LF welded onto the surfaces of the heat transfer tubes and a multi-tube system consisting of four cylindrical heat transfer tubes (four planes of symmetry). The presence of the LF improved conduction heat transfer significantly during charging, due to increased heat transfer surface area. Moreover, a small region at the bottom of the shell was not melted completely due to the buoyancy effect, which transferred heat effectively through the formed liquid channels. The same fin-system led to reduction of the level of supercooling during discharging because the fins acted as nucleating sites as solidification was initiated, leading the authors to recommend it for discharge of erythritol in a shell and tube system. The radially finned system exhibited no significant improvement in t_m, and there was no supercooling since the monitored temperatures at the end of charging were not above the phase transition temperature. Complete melting times for the configurations with radial and LFs were longer compared to the multi-tube system due to suppression of convection. The multi-tube system had the shortest t_m due to presence of convection, but demonstrated a rapid temperature drop to a supercooling temperature of 102.4 °C during discharging.

The experimental study of Lee and Chun [66] sought to optimize heat rejection from an in-situ solar panel. Six 12-Watt panels consisting of different heat rejection schemes, i.e., PCM (melting point of 44 °C) with no fins, PCM with arrays of profiled aluminum fins (placed either inward or outward from the panel) and honeycomb at the back of the panel, were tested. Aluminum honeycomb was imbedded in the back container to improve the TC of PCM. The solar panel consisting of honeycomb and outward fins with PCM exhibited best performance in terms of controlling panel temperature and its efficiency.

Agyenim and Hewitt [67] explored the thermal characteristics of a copper-based horizontally oriented longitudinally finned shell-and-tube LHTES with HTF flowing through the inner tube. Various isotherm diagrams exhibiting progress with time at the midway cross-section are shown in Figure 8. The trend of the average temperature suggested three phases during discharging, i.e., solid sensible heating, phase change and liquid sensible heating. Increased inlet HTF temperature was observed to lead to enhanced heat transfer rate, more unequal temperature distribution and greater cumulative amount of energy charged. At the beginning of the heat recovery phase, the supercooling effect was not observed. Faster temperature changing rates were noticed at the start for both charging and discharging. More molten PCM was observed at the upper part of the unit compared to the lower part due to the existence of convection. PCM at the outer periphery of the store exhibited a higher temperature, resulting from the end effect from the high-TC copper container. Considering the condenser efficiency, with an air source heat pump, an inlet HTF temperature of 62.9 °C was chosen, though 76.7 °C achieved higher average PCM temperature. The higher inlet HTF temperature was also observed to reduce the average percentage energy lost to the ambient temperatures. Though a value of 1.19 for the ratio of energy charged to the theoretical maximum amount of energy available was achieved after 24 h of charging, not all of the PCM were in the molten state. This observation suggested lower HTF temperature and effective heat transfer mechanism were favorable for improved melting. The value of U increased faster in terms of the enhanced HTF inlet temperature during charging compared to discharging. Integration of a PCM-based storage unit to an air source heat pump to meet 100% residential heating energy load for buildings in the UK exhibited a 30% reduction of the store size compared to the case using an oil medium. Radiator surface temperature peaked at the beginning of discharging and then dropped gradually, and higher values were observed with the increased HTF temperature during charging.

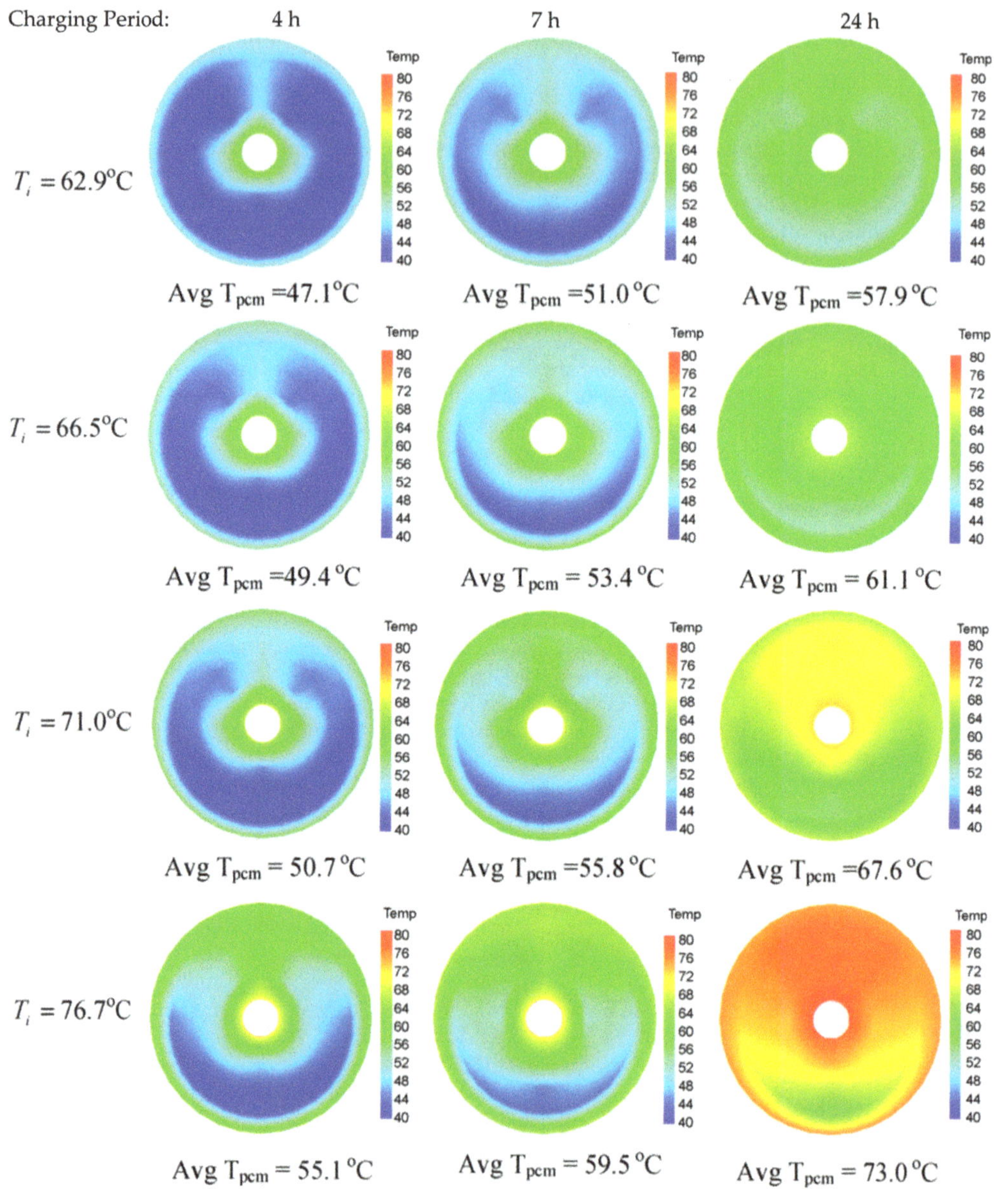

Figure 8. Instantaneous isotherms obtained from thermocouple measurements on the vertical mid-section of the storage unit. Reprinted/adapted with permission from Agyenim and Hewitt [67]. Copyright 2010, Elsevier.

Fok et al. [68] reported experimental results on the cooling performance of different PCM-based heat sinks for portable hand-held electronic devices. Each heat sink was attached to a plate heater providing input power of 3–5 W. The system was then enclosed in a plastic casing made of 2 mm thick polycarbonate. The heater was insulated from the casing, whereas the top surface was not insulated. One set of experiments was conducted at constant power levels (i.e., 3, 4 and 5 W) lasting 150 min. Comparing the TD at same positions, the temperatures of the heat sinks with fins were generally lower than that of the heat sink without fins. Surface temperatures increased drastically during the experiments and quickly exceeded the human's bearable limit. The observed temperatures rose more slowly in the heat sinks with fins, making them desirable as it extends the usage time of the portable device. In the PCM-based heat sink, a greater number of internal fins can help to lower the device temperature because more fins can distribute excessive heat to the PCM. The temperature rose slowest in the heat sink with the greatest number of internal fins. Comparing temperature rise for the configurations with different orientations of the heat sink, it was shown that the device orientations did not affect the phase change process markedly. Transient thermal performance of the heat sinks in the frequent, heavy and light usage modes was also studied. For the frequent and heavy usage modes during the charging stage, temperature rose more rapidly for heat sinks without fins, whereas during the discharging stage, the heat sink with a PCM displayed a slower cooling rate. While there was little TD for the light usage mode, in this case, PCM did not play an important role on the cooling rate. Thermal performance in the heavy usage mode was examined for PCM-based heat sinks with various numbers of fins. During charging, heat sinks with more fins reached a lower peak temperature. This indicated that the increasing surface area of fins will promote the heat transfer rate. It was shown that the fins had negligible effect on the cooling of the mobile devices during discharging where heat dissipation depends on convection.

Saha and Dutta [69] conducted a numerical study to explore the effects of the geometric *AR*s and heat flux on the melting of an n-eicosane-based aluminum heat sink with aluminum plate type fins. A single relation for the *Nu* was not suitable for all *AR*s when the melt convection was taken into account. Three different correlations of the *Nu* that involve the *Ra*, *Ste* and *Fo* were derived, corresponding to three various ranges of the *AR*s.

Wei et al. [70] investigated charging and discharging characteristics of an LHTES unit with a staggered cluster of parallel HTF-carrying annularly finned tubes submerged in a PCM filled in an insulated rectangular shell. HTF flowed inside these tubes with a fixed inlet temperature (55 °C for charging and 40 °C for discharging). HTF with higher flow rate was observed to lead to higher HTF outlet temperature, higher temperature of the PCM and subsequent reduced melting time. During discharging, a higher flow rate of the HTF led to a lower HTF outlet temperature. Stored and released heat energy was marked at early phase of charging and discharging, respectively, due to the large TD difference between the HTF and PCM. Later on, these quantities tended to constant values. Addition of fins was observed to contribute to the uniformity of the temperature of the PCM during both melting and solidification, and reduced local overheating.

Sugawara et al. [71] conducted a study of solidification and 2-D numerical analysis of melting of water around a cooled copper tube with surrounding copper AF with two porosities (0.025 and 0.05). The experimental set-up was composed of a Styrofoam-insulated cavity placed in a low-temperature cell maintained at near 0 °C, and disk-like copper foils were surrounded with water. After confirming an initial temperatures (0, 4 and 8 °C) in the water, freezing started by circulating the coolant maintained at about −18 °C in a tank. It was difficult to measure ice formation in the cavity including copper foil disks. To overcome this, the freezing mass was measured by the volume dilatation using a manometer placed on the side of the freezing cell. Thermal resistance within the clearance of foil disks exhibited by numerical results was comparatively large at the beginning of freezing; however, it decreased at later times. Indeed, the thermal resistance in the clearance was estimated at about 1% compared with the total resistance in the copper foil region including ice. The results illustrated that the experimentally obtained surface temperature was not noticeably affected by the initial water temperature (i.e., superheating), but changed greatly with the porosity. Freeze-out and melt-out times were shortened by using copper foils. Superheating/sub-cooling affected the freezing/melting; however, the effect on the freeze-/melt-out time was not considerable. The copper foils contribute more to melting enhancement than to promoting freezing. Moreover, mere heat conduction due to melting of the ice layer which remained at the underside of the heating tube decreased the melting rate.

Agyenim et al. [72] studied the enhancement of thermal performance of medium-T_m erythritol with inclusion of LF to an HTF-carrying horizontal concentric tube system. Eight ES fins were welded onto outer wall of the inner tube with 3 mm gaps between the tips of the fins and the inside wall of the shell. This system was used to power a LiBr/H_2O absorption cooling unit. Melting and freezing experiments were conducted by means of two fluid loops in which hot (hot silicone oil, charging) and cold (cold water, discharging) HTF were circulated. The optimum inlet temperature for the charging process was selected at 140 °C to avoid incomplete melting of the PCM, due to the low inlet temperature (130 − 135 °C) of the HTF, overheating at upper section of the shell due to natural convection and unequal heat distribution caused by high inlet temperature of the HTF. Isotherms exhibited that greater mass flow rate of the HTF promoted the rate of melting. This was linked to a prolonged entrance length to achieve the fully developed flow that, in turn, led to greater velocity fluctuations in the molten flow. Optimum mass flow rate was 30 kg/min, which led to the shortest melting time and near-complete melting. Variation of the average temperature of the PCM with respect to time demonstrated different stages of melting, and no supercooling was found at the start of discharging. Discharge time was observed to be less than that for charging due to the higher TD between the PCM and HTF. The average temperature of the PCM dropped rapidly at the start of heat recovery, and then levelled off. The calculated recovered heat energy was 70.9% of the maximum heat storage, whereas 29.1% of heat was due to supercooling heat and heat losses. Temperature readings were found to be consistently lower along the radial and axial directions during charging. A large TD between the upper and lower parts was observed because of convection currents existing between neighboring fins. Temperature variation and gradient with respect to time along the radial direction showed that the heat transfer in other two directions was weak.

Both 1-D analytical and 2-D numerical solutions based on the enthalpy method were conducted by Talati et al. [73] to investigate freezing of PCM filled in a rectangular enclosure divided by horizontal aluminum fins. The PCM was at its solidification temperature initially and a constant heat flux was applied on the vertical end-wall. The investigated single cell was divided into two regions. One zone, comparatively far from the fin (region 1), was only exposed to constant heat flux at the end-wall, and thus the heat transfer was only in the horizontal direction. The other region (region 2) was where heat transfer in the vertical direction prevailed, due to incorporation of conducting fin effects. Three cases with different values of area ratio (*AR*) were investigated. Predicted position of the LSI demonstrated that the 1-D analytical method was in a good agreement with the numerical analysis in region 1, while the mismatch happened in region 2 due to neglected horizontal heat transfer. A sharp corner was observed in the 2-D analysis results, corresponding to high temperature at the corners. The 1-D model predicted effectively for smaller *AR*, indicating greater depth of the PCM. TD between the end-wall and the symmetry plane was observed to increase with the raising of the length of fins. Differences between the 1-D and 2-D analyses were small, while the comparatively largest error was present for higher *AR* and greater duration of conduction. The observed rising rate of solidification of the PCM was faster for the smaller *AR* due to the dominant heat transfer through the wall. The solidification fraction increased steeply at the beginning and then slowed down.

Huang et al. [74] employed PCM to enhance the solar-to-electrical conversion efficiency of a building-integrated photovoltaic device (BIPV) by lowering the operating temperatures. The thermal performance of different internal metal fin arrangements (Figure 9a) were presented. An experimental evaluation of the presence of impurities that cause heterogeneous nucleation during solidification was conducted. Only 85% of the total volume of the test system was filled with PCM. Systems without fins and with fins (thickness of 0.5 mm) were investigated. PCM RT27 and RT35, having the same liquid density as PCM Waksol A, but with solid densities higher than Waksol A, were used. It was noted that the addition of internal fins improved the temperature control of the PV in a

PV/PCM system. This was enhanced by employing demountable metal fins extending into the PCM from the front wall. The PCM first solidified in the area adjacent to the cooled wall. Volume contraction upon freezing led to formation of voids in the center of the PCM (Figure 9b). A relation for the variation of the ratio of fin spacing to depth, with the time period of temperature control, to the establishment of a stable temperature was developed. When the fin spacing was more than 33 mm, convection in the molten PCM led to packing of the temperature variations. Due to convection, temperature of the melting layer next to the active wall increased sharply towards the value of the front surface temperature, while the temperature in the solid phase maintained their slow conduction-dominated rising trends.

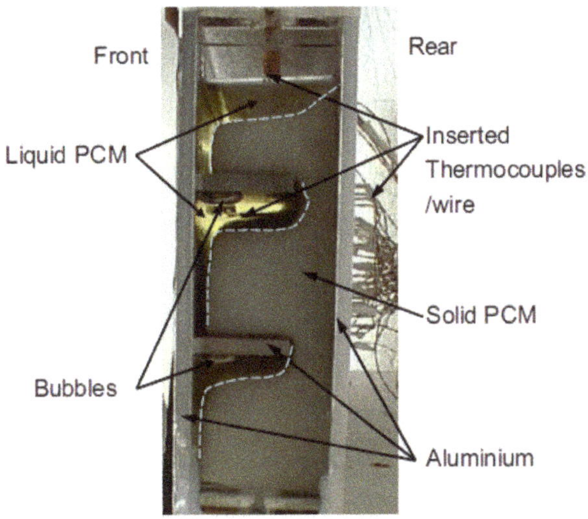

(a)

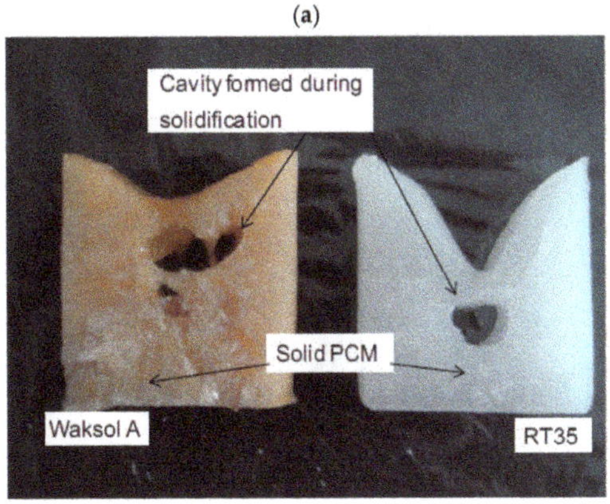

(b)

Figure 9. Schematic diagram of the (**a**) model PV/PCM with horizontal fins and (**b**) void formed within Waksol A and RT35 PCM. Reprinted/adapted with permission from Huang et al. [74]. Copyright 2011, Elsevier.

Robak et al. [75] studied charging and discharging performance of various n-Octadecane-based LHTES units that employed heat pipes or fins. The units utilized distilled water as the HTF that was circulated within the base heat exchanger. A heat exchanger with a plane top surface for benchmark experiments and a heat exchanger with a modified top plate to accommodate heat pipes or fins were utilized. Five heat pipes and five steel rod fins were placed in threaded holes. One hole was centered in the cylindrical test cell, while four holes were placed in a square pattern. For the benchmark configuration, heat diffused upward within the PCM, leading to a planar LSI (Figure 10) with the heat pipe-assisted unit exhibiting similar characteristics in the lower regions of the test cell. Secondary LSIs were established around the periphery of the warm heat pipes that provided pathways for molten PCM. Fin-assisted thawing was similar to those of the benchmark, except for the waviness in the LSI along the wall of the test cell in the vicinity of the fins (Figure 10). This structure had a slower melting rate compared to the heat pipe-assisted unit, since liquid PCM was not provided with a clear pathway to the top of the solid PCM and no tertiary melting ensued. Overall melting rates for the heat pipe-assisted cases were on average 70% greater than the benchmark and 50% greater than the fin-assisted scenario. As for freezing experiments, the benchmark experiments exhibited a planar LSI with slight waviness adjacent to the top plate of the heat exchanger (Figure 11). For the heat pipe-assisted experiments, multiple LSIs formed both along the top of the heat exchanger and around the peripheries of heat pipes. Relative to the solidification rate of the benchmark case, the heat pipes doubled the rate of freezing, whereas presence of fins led to little augmentation of the overall rate.

(a)

(b)

(c)

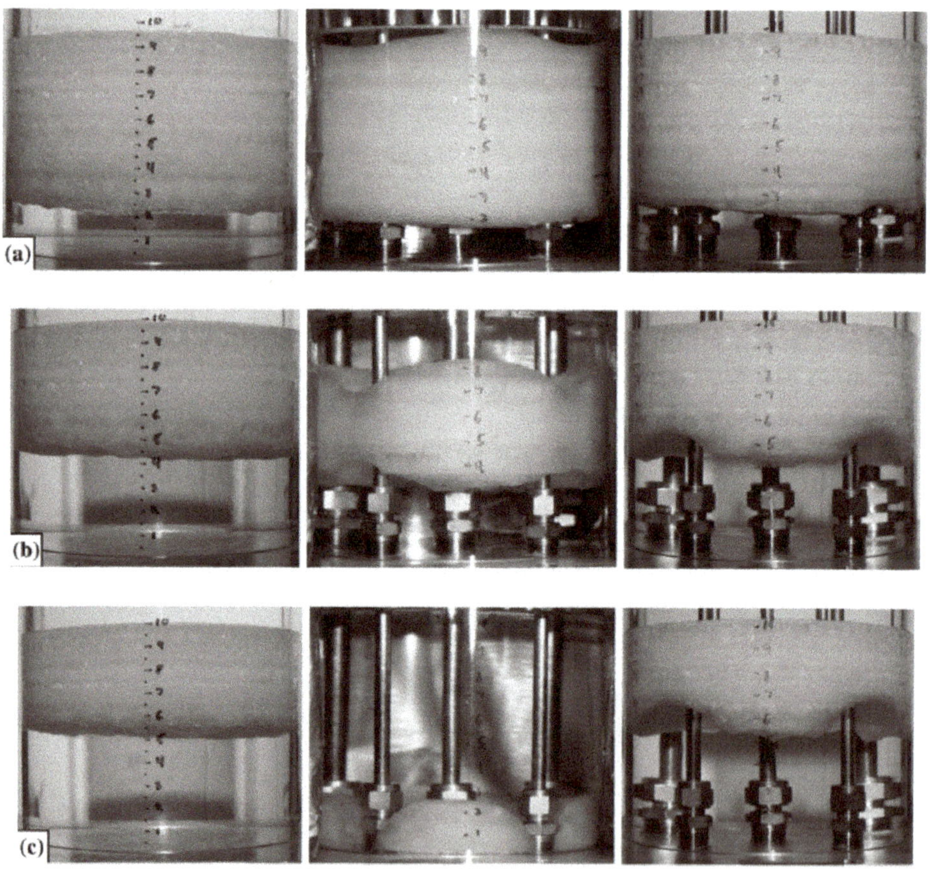

Figure 10. Transient evolution of thawing for the benchmark (**left**), heat pipe-assisted (**middle**) and fin-assisted (**right**) LHTES units corresponding to time instants (**a**) t = 60 min, (**b**) t = 120 min, (**c**) t = 150 min, ($T_{HTF,in}$ = 45 °C, $\dot{m}_{HTF}$ = 0.0026 kg/s, with the vertical scale given in cm). Reprinted/adapted with permission from Robak et al. [75]. Copyright 2011, Elsevier.

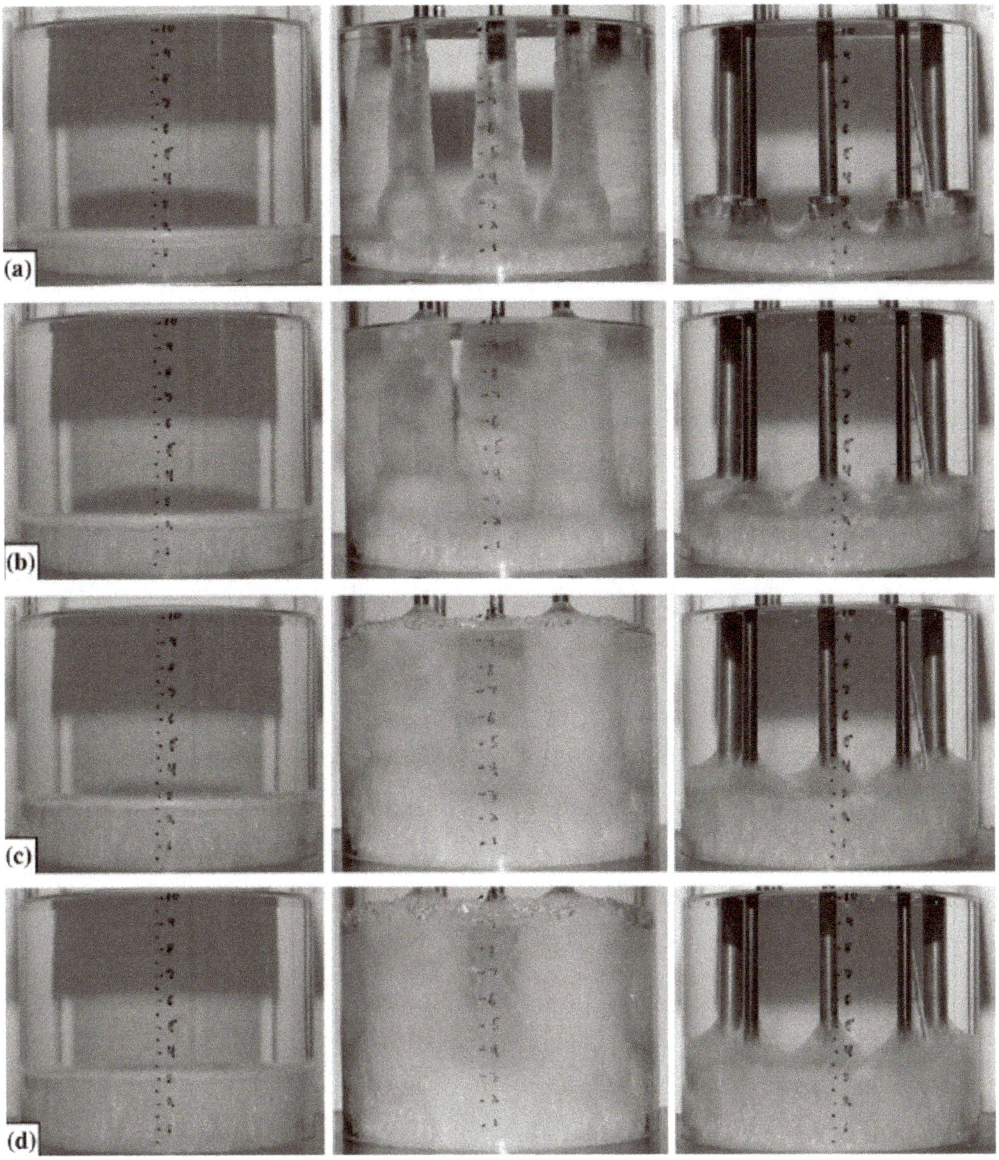

Figure 11. Transient evolution of freezing for the benchmark (**left**), heat pipe-assisted (**middle**) and fin-assisted (**right**) LHTES units corresponding to time instants (**a**) t = 60 min, (**b**) t = 120 min, (**c**) t = 180 min and (**d**) t = 240 min ($T_{HTF,in}$ = 10 °C, $\dot{m}_{HTF}$ = 0.0022 kg/s, with the vertical scale given in cm). Reprinted/adapted with permission from Robak et al. [75]. Copyright 2011, Elsevier.

Bauer [76] studied the solidification time associated with an aluminum finned plane isothermal wall and a single tube with AF with n-Octadecane as the PCM. A quasi-stationary approximation was adopted, where sensible heat was neglected compared to the influence of latent heat. Small *Ste* numbers ($0.01 < St < 0.1$), long fins (half-width of the PCM to the fin length > 0.5), ideal contact of fin and wall and no free convection in the melt were assumed. Both approximate analytical solutions and computational results were based on adopting effective properties for density, specific heat and latent heat. Numerical solutions were obtained using a commercial software package (Fluent, ANSYS Inc., Canonsburg, PA, USA, Version 6.2.16) and applying the enthalpy–porosity method. In the simulation, 1-D heat conduction in the fins and 2-D conduction in the PCM were modeled. A fin factor was defined to evaluate heat flow in the fin, which increased proportionally with thickness (or VF) and the TC of the fin in relation to heat flow within the PCM without phase change. For the plane wall case, the solution was confirmed by experimental results.

To investigate the effects of AF on solidification, Ismail and Lino [77] experimentally studied the case of a horizontal finned tube submersed in a water tank with ethanol HTF circulating through the tube. Measurements of temperature and flow visualization focused on the third fin region of five fins. Tendency of growth of the LSI position varied from fast to slow, and the LSI velocity slowed down because of the thermal resistance between the HTF and the PCM. Relations of both LSI position and velocity with respect to different parameters were fitted for specific conditions. Lower HTF temperature was observed to lead to more solidified PCM and an increase in the LSI velocity due to high TD between the HTF and PCM. High mass flow rate of HTF (greater *Re* and heat transfer coefficient) led to an increase in solidified mass and the LSI velocity. Using a coiled wire turbulence promoter can increase the pressure drop within the HTF tube and the heat transfer coefficient and, hence, more frozen PCM, but it was not as efficient as the AF.

Hosseinizadeh et al. [78] compared the effects of various parameters such as power levels, number of planar fins, fin height, fin thickness and utilization of Rubitherm RT80 on the performance of aluminum heat sinks with 0–7 fins. Eighty five percent of the heat sink height contained the PCM and the remaining 15% encompassed the atmospheric air region needed for expansion of PCM. Based on experimental findings, it was observed that for thicker fins, the LSI moved away from the fin surfaces uniformly, whereas, for thin fins, the movement of the LSI was seen to be non-uniform. The computational results for the same system did not exhibit the observed trends. Moreover, it was noticed that during earlier periods, the lower regions of the PCM for the case of thin fins thaw faster due to the heating from the base of heat sink. For both fin thicknesses, at later periods of melting, the upper regions of the PCM next to the air layer melted faster, indicating internal fluid convection, and increasing the number of and height of fins led to an appreciable increase in overall thermal performance. However, increasing the fin thickness led to a slight improvement. There was an optimum fin thickness, above which the heat sink performance showed no further improvement. As for increasing the power level input, the melting rate of the PCM was expedited. In all cases, heat conduction was the primary mode of heat transfer at the initial stage of melting. At later stages, free convection played a more crucial role in enhancing the melting of the PCM.

A numerical study was conducted by Ye et al. [79] using Fluent software package on thermal performance of a paraffin-based TES unit composed of aluminum inner plates, outer plates and plate fins, separating the system into uniform cavities. Assuming similar performance of each cell, only half of a cavity was used as the computational domain, with PCM filling 85% of the cavity. Uniform temperature was applied on the bottom surface by circulating water through the inner plates, whereas with the top plate was insulated. The volume-of-fluid (VOF) model was employed to resolve the PCM-air system, and the enthalpy–porosity approach was used for modeling phase change. During melting, a greater difference between the T_m and heating wall temperature resulted in a rapid growth of the liquid fraction and a higher wall heat flux initially. Wall heat flux then decreased due to increasing thermal resistance of the growing liquid layer. The rate of decay then slowed down, followed by a period of no variation due to buoyancy-driven flow and eventual negligible heat transfer. During freezing, physically unrealistic formation of liquid pockets was observed in the solidified PCM.

Three shell-and-tube LHTES units with different spacing between neighboring aluminum AF were numerically analyzed by Long [80]. HTF (water) flowed within the inner tube and a composite of paraffin and a nano-structure (aluminum) was filled in the annulus region. Fins originated from the inner tube extending all the way to the outer shell. During charging, a higher number of fins caused steeper outlet temperature drop of the HTF, and more heat release capacity resulted from a lower thermal resistance between the PCM and HTF, with decreasing fin distances inducing improved heat transfer. Only the 12 fin/inch arrangement could satisfy the heat requirement (more than 1800 Wh in 15 min) of a household shower. During discharging, the average temperature of the PCM dropped faster, and the LSI moved rapidly with more fins. A lower inlet temperature of the HTF led to lower outlet temperatures of HTF and shorter heat release time, leading to a shorter needed time for phase change and faster moving of the LSI.

A 2-D computational analysis was conducted to assess the melting performance of paraffin wax filled in rectangular casings with different internal aluminum fin shapes (planar, T-shape, Y-shape, cross-shape, keeping fin volume constant) by Tan et al. [81]. Constant heat flux was applied at the bottom and the side wall of the casing. The numerical investigation was validated by observing agreement with in-house experimental temperature data for straight fins. Similar melting patterns for different fin shapes were observed at early stages (1000 s) of melting, where the LSI formed near the fins and active walls. As more PCM melted, convection became the dominant heat transfer mechanism. Cross and T-shape fins promoted expedited melting near the bottom and middle sections, respectively, while the Y-shape fins exhibited a different promoted melting performance. Comparison of variance of the bulk melt fraction with time did not demonstrate apparent differences among different fins. At a later time instance of 4000 s, three non-straight shapes of fins exhibited greater radius of recirculating vortices compared to the straight fin, resulting in thicker LSIs. In other words, using T- and Y-shaped fins, side vortices extended into the core of PCM, leading to lower TD, thus boosting convection. Conversely, recirculating vortices formed under the horizontal parts of the cross-shaped fins were obstructed from growth due to smaller gap spacing from the bottom wall. Due to similar temperature distribution as straight fin melting, the cross-shape fin configuration was selected for comparison. Whereas finned enclosures exhibited faster liquid formation compared to fin-free cases, a smaller number of long straight fins improved melting, which was achieved in a comparative study with a higher number of shorter fins and cross-shape fins.

The thermal characteristics of a lauric acid-based vertical shell-and-tube LHTES unit, coupled with a solar domestic hot water unit, was investigated by Murray et al. [82] numerically and experimentally. Uniformly spaced copper AF were attached to the outer surface of the inner tube. The outer shell was made of acrylic plastic and kept un-insulated. PCM was kept at room temperature in solid state initially, and hot water from a constant-temperature water bath was pumped through the inner copper tube. Once the temperature of the PCM reached the steady-state, cold water was introduced to solidify the PCM. The COMSOL Multiphysics package (version 4.0a) was used to model the 2-D computational domain without considering convection. A fast temperature increase at the upper corner of the shell was monitored after a time interval upon initiation of melting, due to the onset of convection. Numerical predictions agreed well with the observed results, but the melting time was slightly longer compared to the experimental findings. Higher effective heat transfer rates were observed initially, due to the assumption of ideal contact in numerical study. However, recorded temperature discrepancies at same height and spaced 180° apart indicated the asymmetry of the fin layout.

The influence of convection on the performance of heat sinks with PCM was investigated by Saha and Dutta [83]. The heat sink consisted of aluminum plate fins embedded in PCM, and it was subjected to heat flux supplied from the bottom. A single-domain enthalpy-based computational fluid dynamics (CFD) model was coupled with a genetic algorithm for performing optimization. Two cases, one without melt convection and the other with convection, were considered. Geometrical optimizations of heat sinks were different for the two cases, indicating the importance of convection. In the case of conduction analysis, the optimum width of the half fin was a constant that was in good agreement with results reported in the literature. On the other hand, once convection was considered, the optimum half fin width depended on the effective thermal diffusivity due to conduction and convection. With melt convection, the optimized design led to a marked improvement of operational time.

A PCM-based TES design-to-validation procedure was proposed and verified in an experimental prototype by Chiu and Martin [84]. The PCM was filled in the annulus of a vertical shell-and-tube configuration with the HTF flowing downward inside the inner tube, bearing AF at uniform intervals. A fixed-grid enthalpy-method based on an explicit finite-difference approach was adopted for handling conduction heat transfer through fins and PCM, thus ignoring buoyancy-driven convection. To assure the reliability of the numerical model, the T-history method was employed to examine the thermal-physical properties of two PCM. Three alternative schemes were identified in integrating the enthalpy method. The fixed phase change temperature scheme was observed to lead to over-estimated enthalpy values and, hence, over-prediction of the phase transition time. Direct use of the measured specific heat made it difficult to model supercooling, while the adapted Dirac delta function curve provided a good approximation of the PCM properties. Thawing and solidification were conducted with HTF at 10 °C below and above T_m, respectively. Frozen PCM was found to build up uniformly on all fins, and the LSI moved from the inner tube toward the outer shell (Figure 12). With the constant inner tube temperature assumption, numerical and experimental data were comparable to within small discrepancies, and supercooling was not observed during the freezing at monitored locations. The numerical model was observed to attain a shorter phase transition time than experimental data, which may have resulted from the uninsulated TES tank.

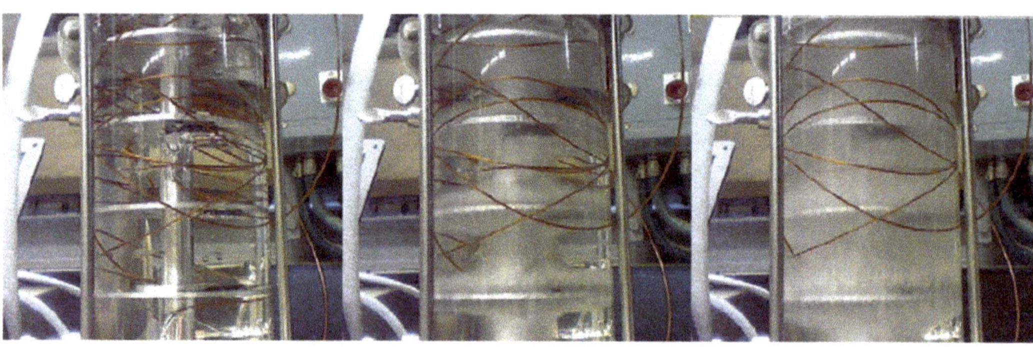

(a) (b) (c)

Figure 12. Instantaneous photographs during freezing of paraffin at (**a**) $t = 0$, (**b**) $t = 1$ h and (**c**) $t = 3$ h. Reprinted/adapted with permission from Chiu and Martin [84]. Copyright 2012, Elsevier.

Effects of the number and distribution of AF on thermal performance of a shell-and-tube LHTESS were investigated numerically by Ogoh and Groulx [85]. Uniformly distributed copper AF were mounted on the inner copper pipe, which extended all the way to the outer pipe and divided the unit into several smaller cells, leading to neglecting convection. Hot water HTF flowed through the inner tube, and paraffin wax was filled in the annulus region. The amount of melted PCM increased with number of fins (up to 27), while the average HTF velocity was observed not to affect the thermal characteristics with smaller number of fins (1 and 5). On the other hand, given a higher initial number of fins, the addition of fins and higher HTF velocity imposed a stronger effect on enhancing heat transfer. The storage unit with 15 fins was identified as the optimal design to balance the total amount of stored energy, HTF velocity and fin materials.

A PCM-based heat sink in the thermal management of portable electronic devices was investigated experimentally by Baby and Balaji [86]. The PCM was n-eicosane that was embedded in a rectangular heat sink (aluminum). All the sides were insulated with cork except the top transparent Perspex® sheet (Figure 13). Effectiveness of pin (square cross-section) and plate fins with same 9% VF were investigated. Addition of the PCM to the heat sink stretched the time duration of the average wall temperature to reach a fixed value. Different tendencies of average wall temperatures indicated the shifting of stages of melting. Higher input power shortened the time required for complete melting, and the pin fins led to the maximum latent heating time due to more heat transfer area for convection. Temperature uniformity due to the presence of pin fins was exhibited by the consistence of evolution of temperature readings. The heat sink with fins greatly enhanced the time taken to reach a fixed temperature of the base at a higher power level. Longer time to reach a set temperature corresponded to increased *Ste*. The duration of the latent heating phase varied almost linearly with the *Ste*, and the latent heat played the most important role in the prolonged operating time of the heat sink for low *Ste* numbers.

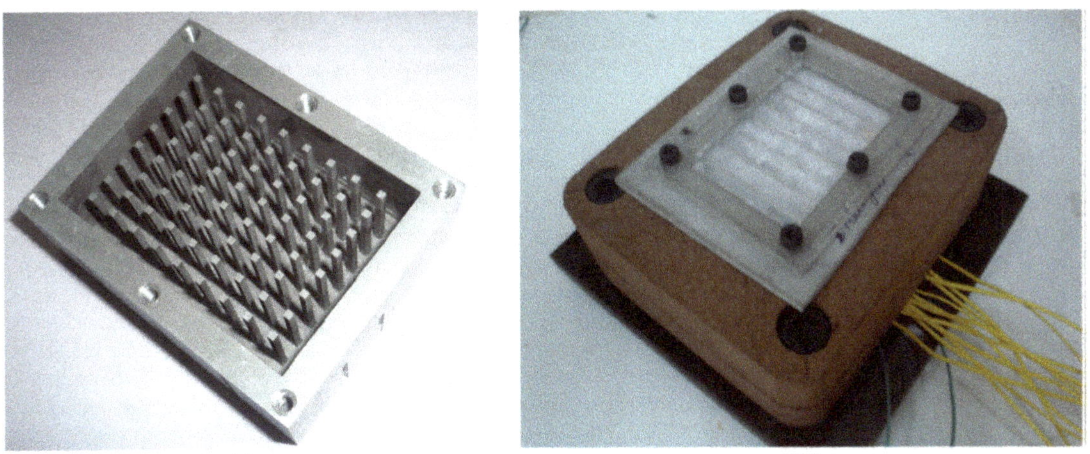

(a) (b)

Figure 13. Heat sink module (**a**) with 72 aluminum square cross-section pin fins (**b**) enclosed by insulation. Reprinted/adapted with permission from Baby and Balaji [86]. Copyright 2012, Elsevier.

Mosaffa et al. [87] provided an approximate 1-D analytical solution and a 2-D numerical model based on the enthalpy method for the solidification of paraffin in a rectangular container with internal horizontal aluminum fins exposed to the HTF's convective cooling boundaries. For small values of the cell *AR*, the store PCM solidified quickly, and heat transfer was mainly through the walls of the cell. Reducing the temperature of the HTF had a greater influence on increasing the PCM solid fraction than enhancing the convective heat transfer coefficient of the HTF. The accuracy of the proposed analytical solution was validated by its comparison with computational results of the temperature distribution within the fin and the location of the LSI for three different values of the cell *AR*. For the *AR* greater than unity, improved comparison between the analytical and computational results was achieved.

The effect of the position and length of a horizontal extended surface mounted on the single active wall of a square cavity filled with a solid PCM on the melting phenomenon with natural convection was studied by Jourabian et al. [88]. The enthalpy-based lattice Boltzmann method was employed, along with the bounce-back boundary scheme. The D2Q9 and D2Q5 models were employed for the velocity and temperature fields, respectively. At dimensionless time of 0.2 ($Fo \times Ste$), adding a fin with any length gave rise to the increase in conduction heat transfer adjacent to fin surfaces, while it had no significant effect on heat transfer in other regions due to weak natural convection. For a dimensionless time of 0.5 and 0.8, although the existence of the fin with any length intensified conduction at the bottom and adjacent areas of the fin surfaces, it diminished natural convection at the top section of the cavity in comparison with a cavity without the fin. Placing the fin below the mid-plane (for dimensionless time of 0.2) had an insignificant effect on the melting rate and the LSI because conduction heat transfer was not dominating. At the same time, with the TC of the fin being greater than that of the fluid, the surfaces of the fin acted identically like a hot wall, whereas enhancement of the melting time was observed when the fin was mounted near the top of the cavity (dimensionless vertical position of fin equal to 0.75), due to extreme diminution of natural convection for this position of the fin. As time progressed, the observed streamlines exhibited more considerable effects of convection. Accordingly, adding the fin led to the formation of two vortices (recirculating in opposite directions) above and below the fin, in effect reducing convection within the top section of the cavity. Furthermore, the recirculating vortex formed below the fin led to improving heat transfer. The aforementioned effects became more noticeable when the length of the fin was increased. Improving the TC of the fin enhanced heat conduction and subsequently the melting rate. Moreover, the extent of the liquid fraction was observed to improve in comparison with a cavity without a fin.

Mosaffa et al. [89] developed an approximate analytical model to study the 2-D solidification of an LHTES unit composed of a shell and tube geometry with AF. Calcium chloride hexahydrate was the PCM, and aluminum fins were employed on the inner HTF-carrying tube extending all the way to the shell. Each sub-zone of the TES unit was divided into two regions. In the inner region, away from the two fins of the sub-zone, the heat sink was the HTF, and the fins did not influence solidification. Moreover, heat was transferred from the wall in the radial direction, and the conduction equation for the solid PCM was applicable. In region 2 next to the fins, heat was released by the extended surfaces (fins). The analytical solution was compared to that obtained via a 2-D numerical method based on an enthalpy formulation for prediction of the location of the LSI. The solidification rate of the PCM exceeded that in the finned rectangular store, having the same volume and heat transfer surface area reported in the literature. For a cell *AR* of 0.5, it was noticed that heat extraction decreased and the temperature of air in the flow direction was reduced as time elapsed. During this period, the thickness of the solid PCM increased, leading to higher thermal resistance. Radial temperature distributions within the PCM suggested insensitivity to the *Re*. For high values of the cell *AR*, heat flows mainly through the wall from the LSI to the HTF. When the cell *AR* was small, the fin had a major role on heat transfer. For each configuration, the enclosure with a lower value of the cell *AR* exhibited a higher solidification rate. For a cell *AR* of 0.5, the effect of a higher air velocity on thermal storage performance was not considerable, whereas it increased solidification rate of the PCM. The solid fraction increased with increasing the air flow rate, since increasing the flow rate resulted in a higher *Re* and greater heat transfer from the PCM to air. Increasing the air flow rate lowered the difference between the inlet and outlet air temperatures. The effect of increasing air mass flow rate was more significant than increasing the heat extracted from the storage. As the inlet air temperature decreased, PCM solidification rate improved. It was found that the effect of inlet air temperature was more significant than that of air velocity on the outlet temperature.

Thermal performance of paraffin wax (RT80) filled in a plate-fin TES unit was studied during melting and solidification by Xu et al. [90]. The PCM was filled in a periodically stacked passageway formed with staggered/serrated aluminum fins above a horizontal clapboard, and the HTF (water) passed through similar fins with same configuration below the clapboard. Employing a staggered arrangement of the serrated fins was meant to disturb the growth of both hydrodynamic and thermal boundary layers. The temperature and *Re* of the HTF were regulated to study their effects on the characteristics of the PCM. CFD code Fluent was employed to model a portion of the unit considering its symmetry. Natural convection was neglected during simulations, and only the laminar regime of the HTF was considered. Higher temperatures of the molten PCM were observed to be near the region adjacent to the conductive fins, indicating the heat transfer enhancement potential of the fins. Both solidification and t_m were shortened with the higher *Re* of the HTF. However, the effects of the HTF's *Re* were not obvious while it increased to the range of 0.15–0.25 m/s during charging. The melting time was observed to be reduced markedly with the inlet temperature increasing from 356 to 363 K, whereas the decrease in t_m was negligible when the inlet temperature exceeded 363 K. The rate of change of the freezing time in terms of the inlet temperature of the HTF was noticed to be faster for lower *Re*.

Shokouhmand and Kamkari [91] investigated enhanced melting performance of paraffin wax stored in a horizontal shell-and-tube unit with HTF flowing in the inner tube. Results of numerical simulations with aluminum LF on the inner tube were compared to that of the bare inner tube. Good agreement between the experimental findings from literature and numerical simulations of the LSI was demonstrated. Tubes with fins were observed to lead to faster melting rates. It was observed that the PCM melted faster in the upper section of the unit due to improved convection. The melting rate was found to be faster at the beginning of thawing for all the three cases, due to a higher TD between the hot wall and the PCM, and the melt fraction for 2 and 4 fins were 2 and 2.9 times of that of the bare tube, respectively.

Hamdani and Mahlia [92] experimentally explored thermal performance of a vertical shell-and-tube LHTES unit with two copper LF and AF attached to the outer surface of the inner downward-flowing HTF-carrying tube. Pure paraffin wax was filled in the space between the inner tube and the stainless steel outer shell (insulated with a porous polythene insulator) and water was the HTF. Due to the dominating heat transfer mode changing from conduction to natural convection, PCM temperatures rose fast initially and tended to become constant gradually, and temperatures at lower positions were observed to be greater and to increase faster than those at upper locations. Axial temperature gradient of the PCM tended to be uniform as melting progressed. Compared to AF, LF exhibited shorter time durations required to attain T_m due to its greater ability of spreading heat.

Tan et al. [93] reported findings of a numerical investigation focusing on the melting characteristics of PCM in a TES unit with vertical planar fins including aluminum spiral fillers. Melting behaviors assisted by natural convection currents were simulated using Fluent 6.3 software. Adding spiral fillers improved performance to some extent when compared to the fin-only case.

Levin et al. [94] conducted a numerical optimization study for the design of an LHTES system used for cooling an electronic device. The rectangular case contained sodium hydrate-based PCM and ES internal plate fins originating from the base of the active wall, whereas other surfaces were insulated. Volume change was ignored, and the presence of convection was neglected (*Ra* well below the range of 10^6–10^7). A uniform power density was applied on the active wall for 30 min and then the PCM was re-solidified. The sole aim of optimization was to attain the minimized height of the unit needed to fulfill the requirement for the critical time, i.e., the time duration required for the interface temperature to reach 60 °C, while preserving the capability of absorbing the heat without exceeding the maximum allowable temperature (60 °C). Both the number and thickness of the fins, and thus the PCM volumetric percentage, were varied systematically for specific heights of unit. Inclusion of fins was observed to lead to higher critical times, whereas there was no apparent improvement in the optimal critical times while the number of fins was greater than 0.2 fins/mm.

Tay et al. [95] investigated improvement of the effectiveness of shell-and-tube LHTES units using radially pointing pins of circular cross-section (Figure 14a) and AF attached to the inner tube. The coolant HTF (a non-combustible, aqueous-based fluid with dissolved ionic solids) flowed inside the inner copper tube surrounded by water (PCM). The effect of buoyancy was neglected, in that conduction was the chief heat transfer mechanism during solidification. Two parameters were introduced, namely the average effectiveness (ε) being the ratio of the differences in the average inlet and outlet temperatures to the differences in the average inlet temperature, and T_m indicating the amount of useful PCM and the compactness factor (*CF*), defined as ratio of the volume of PCM over the total volume suggesting the effective energy storage density. For both pins and AF designs, 12 configurations were examined for a similar range of *CF*. For instance, the mass fraction of ice for a system with 16 equally spaced pins at three time instants are shown in Figure 14b. The finned tube configurations were observed to attain an improved ε and greater heat transfer surface area, which induced a shorter phase change duration. The finned tube exhibited an additional 20–40% rise of $\varepsilon \bullet CF$ and 25% reduced time in terms of the phase change duration compared to the pinned tube configuration. Moreover, the phase change duration decreased with greater pin/fin volume, whereas the variation became negligible as the ratio of the tube volume to the total volume exceeded 0.03.

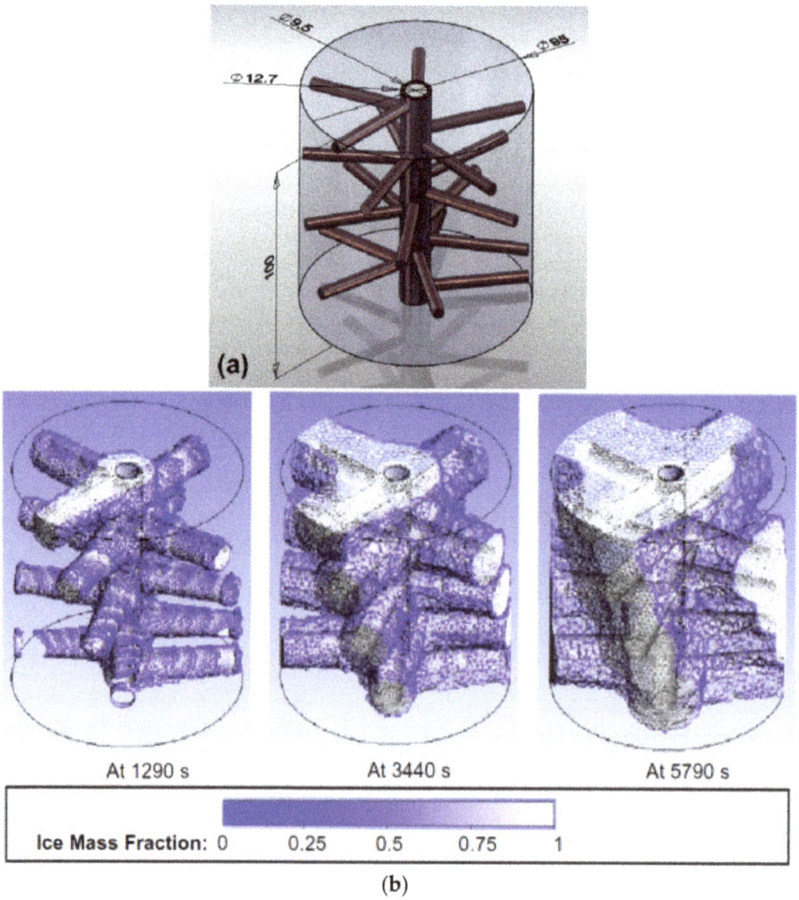

Figure 14. Shell-and-tube storage unit featuring (**a**) radially pointing pins of circular cross-section and (**b**) mass fraction of ice for a system with 16 equally spaced pins at three time instants during freezing signifying build-up of ice around the pins. Reprinted/adapted with permission from Tay et al. [95]. Copyright 2013, Elsevier.

Configurations of paraffin-based LHTES units, i.e., U-tube, U-tube with in-line fins, U-tube with staggered fins and a novel festoon design were studied by Kurina et al. [96]. Conjugate heat transfer between the HTF and PCM undergoing a cyclic melting and freezing was solved using the CFD approach utilizing the enthalpy–porosity formulation. Comparing these designs, the novel festoon channel unit yielded improved heat transfer rate for both charging (Figure 15) and discharging. In order to improve heat transfer performance, units with no fins with combinations of the base PCM and PCM with identical thermophysical properties but with T_m 10 °C above and below that of the base PCM placed in horizontal and vertical arrangements were investigated.

An optimization analysis on the heat transfer characteristics of paraffin wax and n-eicosane-based pin fin heat sinks was conducted by Baby and Balaji [97] experimentally. The effects of the number of pin fins with square cross-sections, varying power levels at the base, VFs of the PCM and inserted TCE were analyzed using an earlier set-up [86]. Vertical variation of temperature exhibited a uniform distribution due to natural convection, except for a slightly higher temperature near the base. A dimensionless study was performed, and the 72-pin fin unit exhibited the maximum time needed to reach a set temperature (t_{max}). The time to reach the set temperature was noticed to increase with increasing VF of the PCM. Both the volume of PCM and the TCE had strong influences on the unit performance. By assigning a higher set point temperature, it was observed that the difference in time required to attain it was narrowed as the VF of the PCM was decreased. An optimized configuration of the heat sink was obtained using a hybrid optimization technique completely based on the experimental results, which combines ANN (artificial neural network) and GA (Genetic Algorithm) for maximizing the time consumed by the heat sink to reach a set point temperature. Experimental data was used to develop the ANN, and the trained network became a replacement for the experiments with high accuracy, and was used to drive the GA-based optimization to maximize the operating time. The predicted optima were validated with additional measurements.

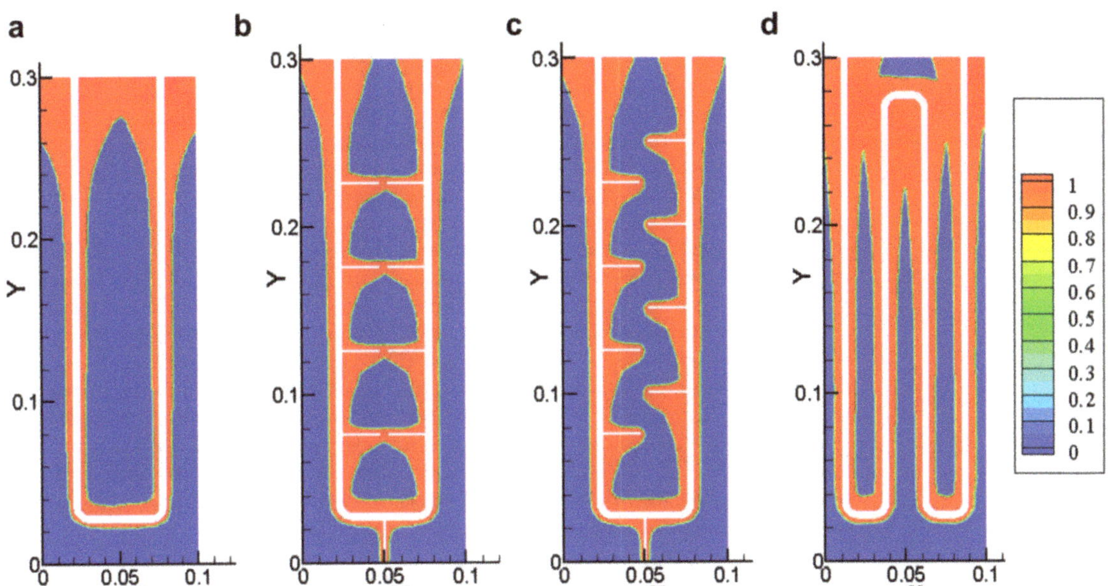

Figure 15. *Cont.*

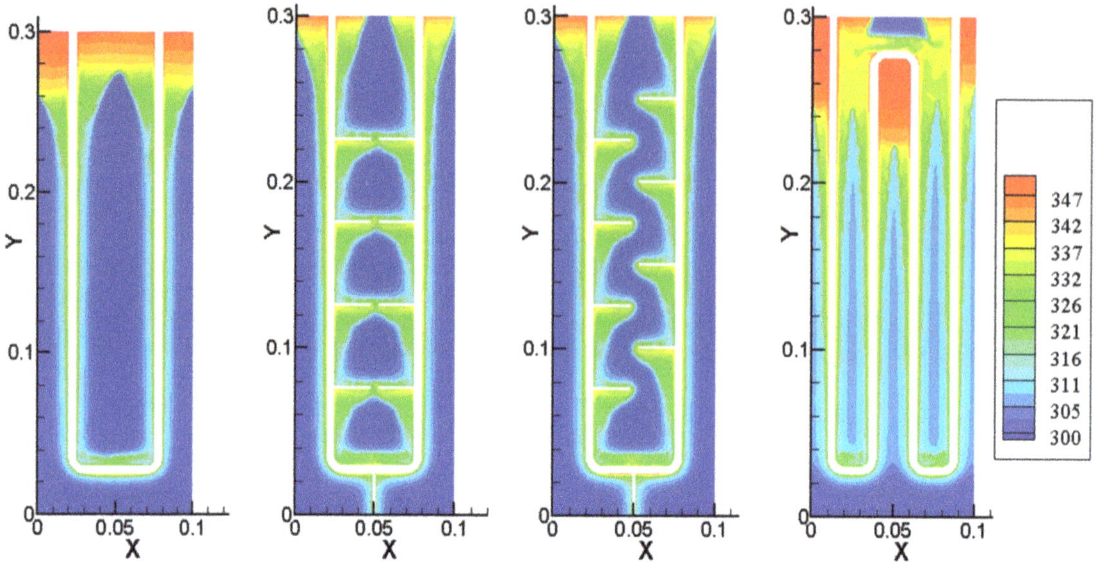

Figure 15. Phase (**top row**) and temperature (**bottom row**) fields during charging at t = 300 s corresponding to U-tube with (**a**) no fins, (**b**) in-line fins and (**c**) staggered fins, in addition to the (**d**) festoon design. Variables X and Y are dimensionless axes, whereas the colorized scales for the phase and temperatures fields are dimensionless and K, respectively. Reprinted/adapted with permission from Kurnia et al. [96]. Copyright 2013, Elsevier.

Mahmoud et al. [98] investigated the effects of heat sink geometry and PCM type on the thermal performance of a heat rejection unit experimentally. An aluminum heat sink was embedded in an insulation material, whereas a plate heater supplying 3, 4, and 5 W was attached to the base of the heat sink. Investigations on six similar heat sinks were conducted to study the effects of various fin configurations: a baseline single cavity, two heat sinks with different numbers of vertical plate fins, two heat sinks with different number of vertical crossed fins and a heat sink inserted with vertical hexagonal honeycomb foil (Figure 16). The addition of PCM into the cavities of the heat sinks was found to reduce the heating rate and the peak temperature of the heat sinks, while the heat sinks cooled down slower. Performance improvements exhibited by the 6-cavity parallel-fin unit in terms of lower temperature of heat sinks was comparable to the 36-cavity crossed-fin configuration. These two units were found to be superior to other structures, exhibiting that a greater number of fins could promote heat transfer (Figure 17). Between these two units, the six-cavity unit was advantageous in terms of the cost of materials and greater convection. Whereas these two units showed improved performance during charging, a reverse trend was observed with respect to the rate of cooling down after the removal of the heat source. The honeycomb-based heat sink exhibited acceptable performance. Five types of PCM with higher or lower T_ms compared to paraffin wax (RT42) were adopted in the heat sinks. PCM with lower melting points were observed to give rise to lower peak temperatures and later starts of solidification during a thermal cycle.

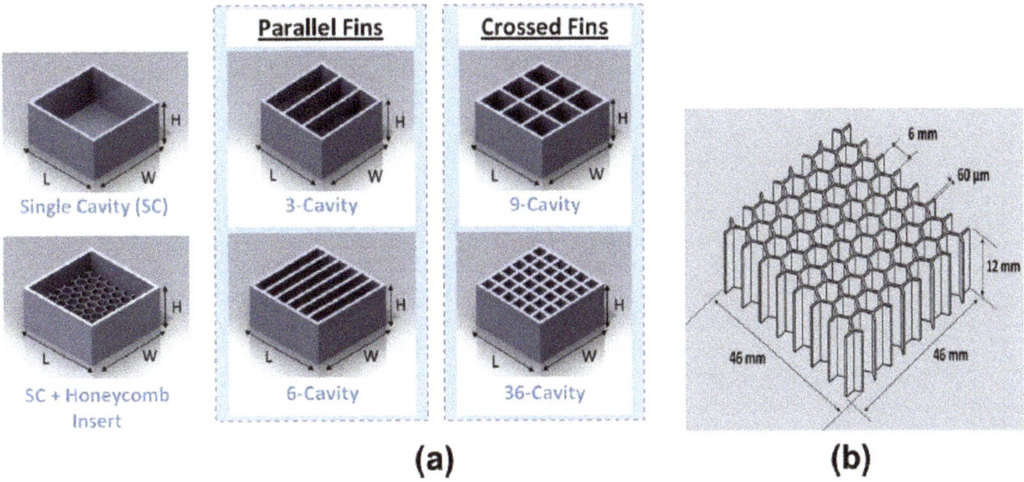

Figure 16. Baseline single cavity heat sink modified to (**a**) 3-, 6-, 9- and 36-cavity modules using parallel and crossed fins, in addition to (**b**) honeycomb insert. Reprinted/adapted with permission from Mahmoud et al. [98]. Copyright 2013, Elsevier.

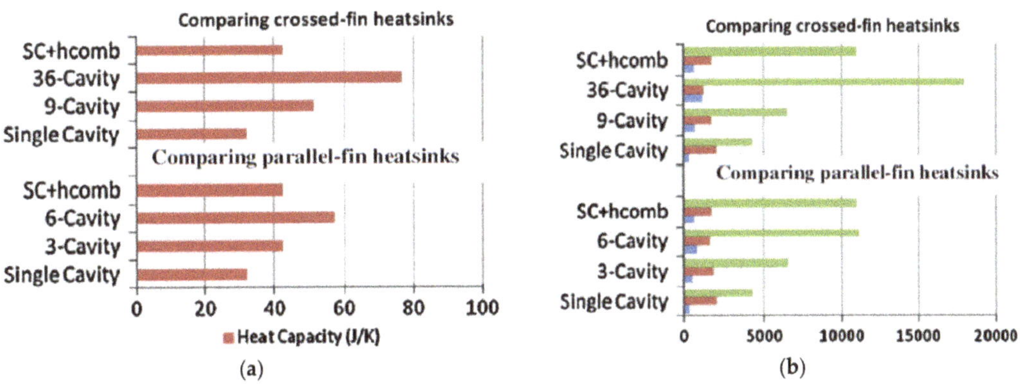

Figure 17. Comparison among different heat sink designs in relation to (**a**) heat capacity and (**b**) associated geometrical parameters (units of the surface areas are mm^2 with green, red and blue bars corresponding to contact area with PCM, PCM area exposed to air and top heat sink area, respectively.). Reprinted/adapted with permission from Mahmoud et al. [98]. Copyright 2013, Elsevier.

Mat et al. [99] studied melting in a triple-tube heat exchanger (TTHX) with RT82 PCM in the annular tube, sandwiched with water HTF streams in the outer and inner tubes, using the FLUENT software. Horizontal copper pipes were employed due to their high TC, and three heating approaches without fins were practiced. The inside heating case involved an active inner tube and an insulated outer tube, whereas the reverse constituted the outside heating case. With the third case, i.e., heating both sides, the inner and outer tubes were both active. Melting times of the unit without fins for the heating of both sides method, outside heating and inside heating were 110, 230 and >300 min, respectively. The effects of the radial length of LF placed on the inner/outer PCM-wetted walls of the annular tube, and their staggered combination on heat transfer performance, were compared. The highest value of enhanced energy charge rate (43.4%) in comparison with the TTHX without fins was the case with the combination fins (42 mm fin length). Coalescence of convection cells in the PCM expedited melting. A larger *Ste* number led to a higher melting fraction and a lower melting time. The numerical results of both heated sides of the TTHX were validated with measurements obtained from an experimental set-up.

Design improvements of a shell-and-tube LHTES unit using an approach based on the analysis of entropy generation was reported by Guelpa et al. [100] utilizing CFD. Contributions to the local entropy generation rate were evaluated for both un-finned and AF systems. Arrangement of the fins was modified to improve the efficiency of the system. The improved system accommodated reduced PCM solidification time and increased second-law efficiency.

Three mathematical models for simulation of a compact finned-plate LHTES system were proposed by Campos-Celador et al. [101], namely, numerical model (N), simplified analytical model (SA) and simplified numerical model (SN). For a rectangular unit finned plate configuration, the RT60 paraffin was filled in the space between the neighboring parallel fins. The thermal storage units were then placed in parallel, whereas HTF water flowed between the units normal to the fins. HTF channels were considered to be 1-D in the direction of the flow for all the models, and only laminar regime was considered. In the other extreme of complexity, the storage unit was discretized as a single isothermal node in the SA model. As for the SN model, the whole system was only discretized in the direction of flow of the HTF, thus constituting as an intermediate solution compared to the previous two models. The results drawn from the three approaches were validated through observing good agreement with experimental data. Small discrepancies of stored/released energy for the models were observed at the beginning of both processes, due to the effect of the heat stored in the inlet and outlet manifolds. Results of the SA model were found to be sensitive to the length of the unit. On the other hand, the simplified approaches required less CPU time, and the SN model agreed fairly well with the results of the N model, suggesting that the intermediate SN model was the most suitable model for optimization design.

Solomon and Velraj [102] investigated enhancement of the thermal behavior of the RT21 paraffin filled in the annulus region of a double-pipe heat exchanger with eight LF fitted on the outer surface of the inner copper tube. The container was sealed with acrylic bottom and top plates. Cool air generated in a climatic simulator was used as the HTF. The PCM was heated up to 30.3 ± 0.05 °C initially, and then cooled with air (temperature of 12 and 14 °C, and inlet velocities of 3–6 m/s). Solidification time was observed to decrease with greater fin radii, and the reduction of freezing time was more prominent with higher heat flux (HHF, due to a lower inlet temperature and higher *Re* of the HTF). On the other hand, for the case with lower heat flux (LHF), the rate of freezing was slower. The reduction of freezing time with the fin radii increasing from 20 to 26 mm was negligible, due to suppression of convection during initial sensible cooling phase. Similarly, a higher temperature of the PCM was obtained with a finned configuration, compared to non-fin unit with higher cooling rate for the case of LHF. The effects of the addition of fins on the temperature field were not marked within the upper part of the PCM, where suppression of free convection was trivial in comparison to the lower segment. The difference was lowered for the HHF case because of the offset of enhanced heat conduction effects on suppressed convection. The HHF case combined with the non-fin configuration was observed to lead to a lower onset of solidification temperature compared to the case of LHF with fin configuration. This was due to the induced suppression of the convection resulting from presence of the fin that decelerated sensible cooling, thus delaying nucleation and reducing the supercooling effect. The enhanced *Re* of the air was noted to accelerate the temperature drop during the sensible cooling process and reduce its duration. The effects of variation of the *Re* were more noticeable within the lower part of the PCM due to a greater heat transfer coefficient near the inlet of the HTF and higher driving TD. It was observed that lower inlet HTF temperature led to declined solidification time, and the effect was more observable with enhanced HTF's *Re*. Mosaffa et al. [103] investigated freezing of a salt hydrate filled in a shell-and-tube TES storage with AF using a conduction-only analytical model. In one case, the inner wall was kept at a constant temperature, whereas in the second case, a fixed heat flux was applied on the inner wall, with the outer wall kept insulated in both cases. A two-zone approach similar to [89] was utilized, and the results were compared with a 2-D model. The total solidification time was also investigated for two configurations (finned cylindrical and rectangular shell). Configurations with different fin lengths and spacings were investigated, and longer fins were observed to contribute more to conduction, with greater discrepancies found in region 2. For the case of constant temperature at the inner wall, solidification time of the PCM encapsulated in a cylindrical shell was less than that in a rectangular shell on the basis of equal volume of the PCM, surface area of heat transfer and height of the TES. This was due to the greater rate of the thermal resistance for the rectangular unit. On the other hand, for the case of a constant heat flux applied at the inner wall, the cylindrical configuration exhibited higher total freezing time because of the higher release of heat from the store in region 1 with greater spacing between the fins. In addition, the shortest time was obtained for the case of shortest fin length and longest fin spacing. A unit capable of absorbing heat by eicosane, held within rectangular cells and dissipating it to the ambient air, was studied experimentally by Kozak et al. [104] in both room- and elevated-temperature environments. Experimentally and numerically determined base temperatures and predicted melt fractions exhibited fairly good agreement. In addition, findings obtained using a more complex model accounting for convection in the melt were discussed. Dimensional analysis relating the melt fraction and *Nu* number in terms of the *Fo* and modified *Ste* numbers highlighted the effects of the latent, sensible accumulation and heat transfer to the air. Dimensionless curves for latent heat-based accumulation rates are rather similar in various cases. However, it was found that the share of sensible-heat-based accumulation rates tended to increase when the heat input increased. PCM-based thermal control hardware for the electro-optical payload of low earth-orbit satellites, as a substitute for existing thermal buffer units, was studied by Kim et al. [105]. Utilizing planar fins, it was shown n-Hexadecane and n-Pentadecane-based units, with only 12% of mass of existing thermal buffer units, were adequate to control the payload module temperature.

Chen et al. [106] investigated the thermal characteristics of the PCM cold storage integrated in an ejector cooling system experimentally. A hollow vertical tube with AF on the outer surface was inserted in the PCM-containing insulated cylindrical tank. HTF flowed vertically upward and the PCM was S15 (EPS, UK). The PCM temperature dropped quickly initially, during the liquid sensible cooling stage, followed by a gradually reducing temperature range during the latent heat storage period. A higher *Re* was observed to lead to shorter solidification duration and lower PCM temperature at the end of freezing. The cold storage rate was noted to rise quickly initially, and then dropped slowly during the latent heat storage process, followed by a notable decrease at the end of freezing. A greater average PCM cold storage rate was obtained with higher HTF *Re* during the latent storage phase. The cold storage capacity increased markedly during the liquid sensible storage stage, then rose gradually during the latent storage process, followed by an increase with a smaller slope at the end of freezing. During discharging mode, with the HTF inlet temperature at 25 °C, the temperature of the PCM was observed to grow quickly during the initial solid sensible heat storage period, and then increased slowly during the sensible stage, followed by a subsequent faster increase. The cold storage rate varied in a reversed way. A higher *Re* was observed to shorten the melting time and led to a greater average PCM cold storage rate. Coefficient of performance of the ejector cooling system remained almost constant with the integration of the PCM cold storage unit. Effectiveness of both charging and discharging were observed to decline with greater *Re*, which led to a lower temperature between the outlet and inlet of the HTF. Both charging and discharging were conducted with the same TD between the HTF and PCM for natural convection, whereas the effectiveness of charging was noticed to be higher than that of discharging due to higher TC of the frozen PCM. Two correlations were obtained for the effectiveness in terms of the ratio of *Re* for both charging and discharging.

Shon et al. [107] investigated thermal performance improvement due to combining an LHTES unit enclosing a conventional automotive heat core fin-tube heat exchanger with louver fins, in order to store a vehicle's waste heat from the coolant. Xylitol ($CH_2OH(CHOH)_3CH_2OH$) was the PCM because of its nontoxicity and desired T_m. A PCM-filled heat exchanger with fins distributed in a U-shape within it was connected in series with the heat core line and 100 copper tubes passed through the fins. During the time period that heat was transferred from the coolant to the PCM, the heat transfer rate was observed to increase with higher values of the coolant *Re*. On the other hand, a weak reverse trend was found for the change of the coolant temperature. Heat transfer through natural convection in the thin layer of the liquid-phase PCM could not be neglected during melting, because the heat resistance of the molten PCM plays a very important role in lowering the heat transfer efficiency. As observed in Figure 18, upon initiation of melting, fins led to broadening of the heat transfer surface area, and natural convection resulted in decreased heat resistance. The melting rate of the PCM was observed to be affected only by the *Re* of the coolant. Discrepancies between the analysis and experiments were found, due to neglecting of other heat losses. The improved heat exchanger attained 33.7% compared to heat storage system without PCM, with respect to its warm-up performance.

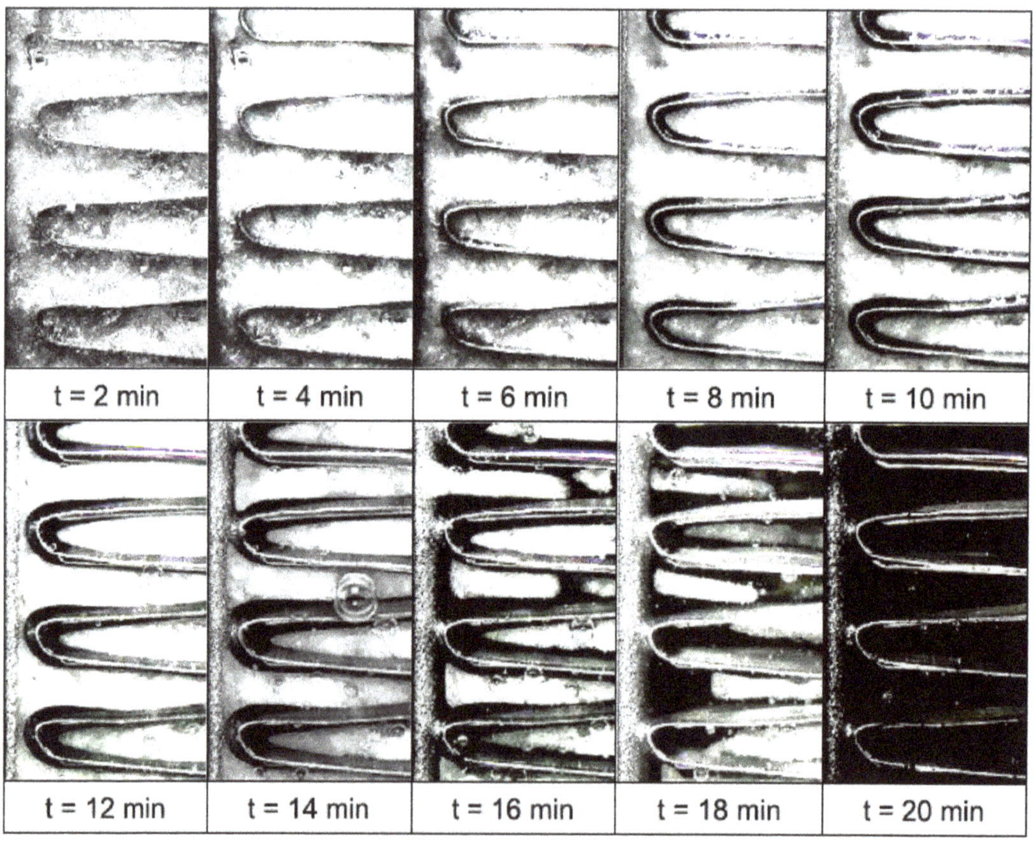

Figure 18. Time-lapse photographs of thawing of PCM next to U-shape fins. Reprinted/adapted with permission from Shon et al. [107]. Copyright 2014, Elsevier.

Experiments were conducted by Murray and Groulx [108] on a cylindrical LHTES which could be charged and discharged simultaneously by utilizing independent HTF water streams, complementing earlier work on the consecutive mode of operation detailed in [17]. Dodecanoic acid ($CH_3(CH_2)_{10}COOH$, Alfa Aesar (Ward Hill, MA, USA)) was filled in the cylindrical acrylic plastic container insulated with fiberglass. Two copper tubes with AF were oriented vertically through the container, with the hot and cold HTF entering from the upper inlet stations. Regardless of the initial state of the PCM, the *Re* was observed not to strongly affect the various temperature versus time variations recorded. On the other hand, starting with the PCM in solid phase, the temperature of the upper part of the PCM was observed to be higher than the lower part due to the strengthening of natural convection (Figure 19a). For a given HTF *Re*, the PCM within the lower part of the side of the cold HTF stream was observed to be frozen at the end of the 24 h experiment (Figure 19b), whereas the PCM remained completely molten on the side of the hot HTF. The melting rate was faster, and a higher temperature was achieved, with higher *Re* at the upper location of the cold HTF stream. In another set of experiments, an LHTES unit was charged and discharged simultaneously for a fixed time duration, followed by only charging for a fixed interval in a cycle. For a fixed hot HTF *Re*, as the cold HTF *Re* was varied in the thermal cycles, a greater temperature drop was observed on the lower part of the PCM due to prominence of natural convection next to the downward-moving hot HTF stream. PCM on the side of cold HTF stream solidified and melted within each cycle, whereas it remained in the solid state after three cycles with the higher *Re* of cold HTF.

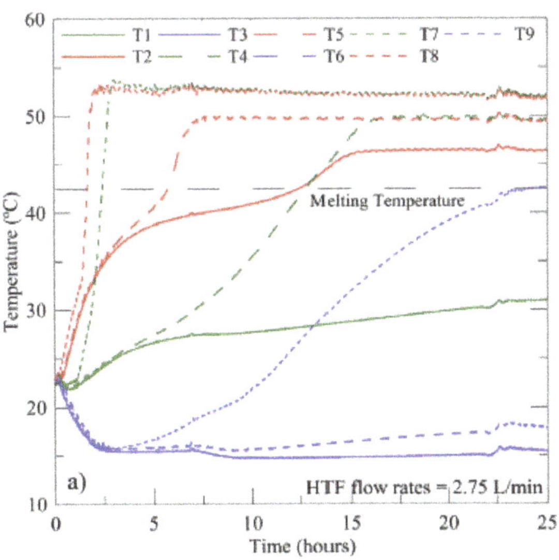

(a)

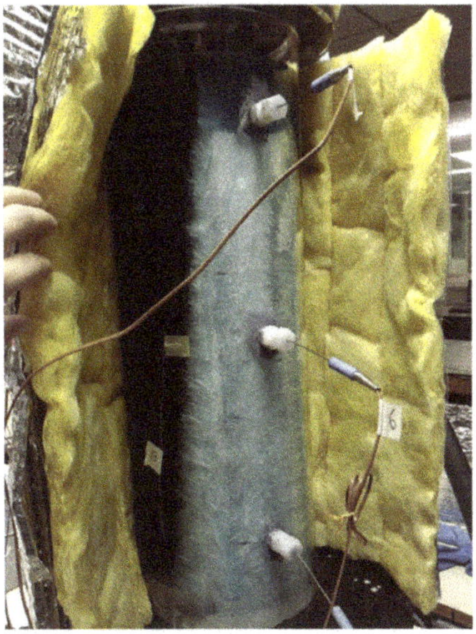

(b)

Figure 19. Results for an initially solid PCM subjected to a simultaneous charging/discharging experiment for a time duration of 24 h exhibiting (**a**) temperature vs. time history of 9 thermocouples (T1–T9) and (**b**) simultaneous coexistence of two molten and solid halves after 24 h. Reprinted/adapted with permission from Murray and Groulx [108]. Copyright 2014, Elsevier.

Al-Abidi et al. [109] studied melting and solidification of a PCM in a both-side heating TTHX horizontally aligned tube heat exchanger with internal-external LF, using an experimental set-up [99]. The effects of the *Re* and inlet temperature of the two HTF streams on melting, with steady-state/non-steady HTF inlet temperature variation, were studied. For the steady-state case, the total melting times of the PCM were reduced to 86% by increasing the steady HTF inlet temperature from 85 to 100 °C. For the unsteady case, the melting rate was affected noticeably when the *Re* was higher. For both cases, the effect of the *Re* was insignificant relative to that of the charging temperature. The solidification rate of the PCM was improved as the *Re* increased. Closer to the inner and outer tubes, the measured temperatures during thawing were lower, since conduction was the dominant heat transfer mechanism between the HTF and PCM. The average temperature near the surface of the inner HTF tube was high due to the early effects of natural convection. Within the lower part of the annulus, melting was achieved early on compared with those in the other angular positions, due to higher thermal diffusion and natural convection. During solidification, it was observed that the greater the radial distance from the inner HTF tube, the lower the corresponding PCM temperature was. The cooling rate was reduced as time progressed because of the increase in the thermal resistance, owing to the increase in the solid layer. Moreover, the upper part of the PCM possessed a lower temperature, whereas the cooling rate within the upper part was higher because of the entrapped air in the upper zone.

Liu and Groulx [110] conducted an experimental study of heat transfer characteristics of a horizontal shell-and-tube LHTESS with water circulating through the inner LF (four fins with two orientations, i.e., + and ×) copper tube to charge or discharge heat from dodecanoic acid. Before the experiments, 50 °C liquid PCM was completely filled in the unit, whereas a 6% vacancy at the top of the container was observed upon complete freezing. PCM was cooled to 10 °C at the start of charging. A conductive heat transfer mechanism was observed to be dominant at the start of melting, as evidenced from the linearly increasing tendency of the temperature profiles. Natural convection started making a difference once more liquid PCM emerged near the active surface. The melting rate was found to be faster within the upper two quadrants of the +-fin configuration, in comparison to the lower quadrants, due to convection enhancement originating from both vertical and horizontal fins (compared to only one contributing vertical fin in the lower quadrants), and the upward-distributing heat resulting from convection. Increased inlet temperature not only enhanced heat transfer at the start of charging due to induced larger TD, but also due to the faster onset of natural convection. Increasing of *Re* did not affect the heat transfer rate at the initial stage of thawing, and the whole period of discharging, due to the trivial contribution of the enhancing forced convection coefficient inside the copper pipe compared to the controlling factor of the thermal resistance of solid PCM. Upon conclusion of the initial melting stage, the heat transfer rate was promoted slightly when natural convection emerged. Discharging started instantly once the charging ended. During discharging, similar temperature profiles were observed for both the upper and lower quadrants, since conduction was the dominant heat transfer mechanism. A constant temperature plateau occurred soon after the beginning of discharging, almost at the same time for different probe positions, due to the thermocouples being perfect nucleation sites. For the ×-fin configuration, the most intense convection was noticed at the upper quadrant, followed by side and lower quadrants. Similar behavior for the ×-fin configuration were presented in response to the effects of the HTF inlet temperature and *Re* to the +-fin system during charging. Temperature profiles were similar for positions displaced at the same radial coordinates within all quadrants during freezing, until the temperature of 30 °C was reached, and temperature took a longer time to decrease at the side quadrant where more energy was extracted. Phase change behavior of the upper and bottom quadrants of the +-fin unit were similar to the upper and bottom quadrants of the ×-fin unit, respectively.

Tay et al. [111] studied a class of industrial-level tube-in-tank LHTES units with AF. Ignoring convection, a simulation model of phase change of two PCM (water and salt hydrate), with 500 fins and 8 copper tubes, was developed using the CFX-PRE (Ansys, Inc, Canonsburg, PA, USA) package version 12.1. A 2-D CFD model was also studied, focusing on a copper HTF-carrying unit including convection. Experimental validation of the CFD models was made using a finned tube inside a vertical cylindrical heat exchanger filled with the RT35 PCM and water as the HTF. Fourteen equally spaced fins were attached on the copper tube. CFD-based heat flow rate predictions were in good agreement with the experimental results. Average effectiveness (ε) and the thermal resistance ratio derived from the CFD results compared well with the calculation of the 2D ε-NTU (effectiveness–number of transfer units) method applied to a finned tube configuration with the use of an appropriately adjusted P-factor (a parameter that identifies the proportion of heat flow which was parallel or isothermal). It was observed that the average effectiveness decreased by enhancing the *Re*. Moreover, the thermal resistance ratio of the PCM dramatically reduced with a finned tube for both freezing and melting cases, such that the dominant resistance became the resistance of the HTF. A correlation of the P-factor as a function of the TC of the PCM for three numbers of fins per unit length (9.2, 18.8 and 36.8) was derived, which held for the freezing case; however, it led to an underestimation for the melting cases, where natural convection was significant.

Hasan et al. [112] investigated the performance and incurred costs/benefits of a PV-PCM system including a fixed internal aluminum heat sink consisting of vertical plate fins. One PV panel was used as the reference, and two other units were manufactured as PV-PCM systems. Both eutectic mixtures of capric-palmitic acid and $CaCl_2{\cdot}6H_20$, initially kept heated at 70 °C and stirred for 12 h, were used. At the top of the container, free space was allowed for volume expansion. The PV-PCM systems were kept at 16 °C for 48 h until the PCM was completely solidified before laying them outdoors to absorb excessive heat of the PV panel. Two cities in Ireland and Pakistan with different latitudes were chosen to conduct these experiments. Open-circuit voltage and short-circuit currents of the units were recorded to calculate the electrical energy output. Solar radiation intensity, ambient temperature and wind speed were also measured for the subsequent calculation of the total solar energy falling onto the PV panels and heat losses. Inclusion of the PCM lowered the surface temperature of the PV panel. Thermal energy stored in the PCM was converted to electrical energy with 30% conversion efficiency. However, considering the total cost and benefits of mass production of the PV-PCM systems, the system was not economically effective in Dublin, but proved to be viable in Pakistan.

Khalifa et al. [113] investigated the enhanced performance during solidification of a high-temperature LHTES system using heat pipes with LF. This served as a building block for a system consisting of stainless steel HTF tubes carrying Therminol® within a rectangular PCM-filled shell, with each tube penetrated evenly with four heat pipes, circumferentially, and repeated with a fixed distance along the tube. In the numerical model, thermal resistances related to the HTF tube or the heat pipes were incorporated. An experiment employing RT82 as the PCM instead of KNO_3, for safety reasons, validated the numerical predictions of freezing solely around the heat pipes with good agreement. In the experimental set-up, paraffin was filled in a cylindrical stainless-steel vessel with a copper-water-charged heat pipe cooled by a water jacket, and tests for a bare heat pipe and a heat pipe with LF were conducted. The fins were attached to the heat pipes, exploiting thermal epoxy to minimize interface thermal resistance. The recorded temperature drop was observed to be faster at positions near the heat pipe, whereas the numerical and experimental data were closer at locations away from the heat pipe. This was possibly due to the occurrence of phase change outside the specified T_m range, with the position near heat pipe not being able to accommodate the excess heat caused by the expedited temperature drop. Numerical simulation for a KNO_3-based LHTES unit was performed, with stainless steel-mercury-charged bare and finned heat pipes. The LF on the heat pipe were found to lead to accelerated freezing of PCM and thicken the solidified layer compared to a bare heat pipe. The increasing rate of the solidified layer with stored energy was faster at the beginning of freezing and then slowed down. In addition, more energy was extracted by using of the finned heat pipes compared to the bare heat pipes, while the thermal resistance between the PCM and the HTF was diminished. Thermal enhancement was estimated by the heat pipe effectiveness, ε_{hp} (ratio between extracted energy of the test case and that of no heat pipes case) and the fin effectiveness ε_{fin} (ratio of extracted energy in finned heat pipes case with that in bare heat pipe). Combining fins with heat pipes was found to lead to improved ε_{hp}, which increased with a greater number of fins. On the other hand, the number of fins should be raised below a limit to avoid losses of the potential of storage energy, because the excess fins would occupy the space for PCM.

Rahimi et al. [114] studied experimentally the effects of changing the inlet temperature of HTF (water) and its flow rates on the melting and solidification of RT35 (Rubitherm) as the PCM, filling the shell side for finned-tube and bare heat exchangers. The inlet temperatures of HTF for charging tests were 50, 60 and 70 °C, whereas for discharging tests it was 10 °C. The set-up for the finned-tube heat exchangers was composed of horizontal planar fins and six vertical copper tubes, which were connected using U-tubes. The storage unit was placed inside a Plexiglas® shell, which was insulated by glass wool. For the fin-less heat exchanger system, the inlet temperature's rise from 50 to 60 °C and then to 70 °C led to a t_m reduction from 429 to 250 and then to 177 min, respectively. On the other hand, the presence of fins led to faster reduction of t_m, i.e., 211 to 121 and then 92 min. When the flow rate was set to 1.6 L/min, flow became turbulent, and, subsequently, t_m was lowered more than 52% for the bare heat exchanger unit. However, for the finned-tube system, the dependence of the melting time on the *Re* exhibited similar trends to its dependence on HTF temperature. An increase in the *Re* lowered the average temperature of the PCM during discharging of the finned-tube heat exchanger system more intensely. For the bare tube heat exchanger unit, the effect was more substantial. Rahimi et al. [115] extended [114] to study the influence of spacing of horizontal fins.

The effects of CCM in a vertical shell-and-tube storage unit with AF on the inner tube were studied by Kozak et al. [116], exhibiting that CCM significantly improves the heat transfer rate and lowers the melting durations by almost 2.5 times. CCM can be initiated by supplying heat to the outer shell of an LHTES unit. A single-cell enclosure filled with a PCM was modeled computationally and validated experimentally. The solid bulk of the PCM was able to sink, thus enabling CCM on the non-isothermal fin surface, in effect elevating the fins to be more than just extended surfaces. Total melting time and instant melting patterns of the numerical predictions and experimental findings were in good agreement. Results of the numerical model were compared with findings of a simplified analytical model, which accounts for the CCM only, revealing effects not predicted by common CCM modeling approaches in the literature. Theoretical expressions for the dimensionless time-dependent melt fraction, heat transfer rate and molten layer thickness were obtained by the analytical model. The melt fraction depended on $FoSte^{3/4}$, whereas the *Nu* and the normalized layer thickness depended also on the additional group $Ste^{1/4}$. A CCM-assisted melting of eicosane (Roper Thermals, Clinton, CT, USA) filled in the annular space of a horizontal double-pipe concentric LHTES unit with three evenly spaced LF (Y-configuration) attached to the inner tube was investigated by Rozenfeld et al. [117]. With the unit exposed to the ambient air (below T_m), melting was observed to only occur next to the HTF tube and the fins. The asymmetric shape of the solid PCM at a given instant was due to the rising of liquid, which enhanced the melting rate (Figure 20a). With the unit placed in a static heated water bath (about 5 °C above T_m) without HTF, once melting next to the shell surface was observed and the solid was free to move, the HTF stream was turned on. In the upper V-shaped part of the unit, the thin molten layer that formed between the shell and the fins did not grow perceptibly during melting, and the solid PCM moved vertically, melting on the inclined fins. On the other hand, within the lower part of the unit, the motion of the solid PCM was noticed to be rotating toward the vertical fin, while the radial thickness of the solid block sliding on the shell remained constant (Figure 20b). The rate of CCM was observed to be more than 2.5 times quicker than that of "regular" melting. Focusing on the lower part, a numerical model was developed for melting on the vertical surface with "rotation" of the solid approaching it, considering non-uniform temperature distribution of the fin, secondary melting from the fin and frictional resistance at the shell. Further ignoring the resistance of the fluid envelope and the temperature distribution in the fin led to an analytical solution that exhibited good agreement with predictions of the numerical model in terms of the variation of the melt fraction. It was shown that the melt fraction was a function of $FoSte^{3/4}$. As for the case with a fin angle of $\Phi = \pi/3$ (six LF "star" * configurations), similar results were obtained. However, overprediction of the melt fraction was greater with the case of the six-fin configuration compared to the Y-system. In addition, similar trends of the melt fractions for different fin angles were observed, and a mathematical dependence on $FoSte^{3/4}\ \Phi^{\frac{1}{2}}$ was derived. The local Nu increased with greater radial distance along the fin and lowered with the elapsed time. Time-dependent, spatially averaged Nu_{avg} varied with $FoSte^{3/4}$ for various fin angles. A linear dependence of $Nu_{avg}\ \Phi^{-7/45}$ on $FoSte^{3/4}\ \Phi^{-3/5}$ was also approximated. The Nu declined with time for CCM more moderately than for the case of regular melting.

Rathod and Banerjee [118] investigated thermal enhancement due to adding 3 LF to a stearic acid-based shell-and-tube heat energy storage with an air-gap space at the top. Hot and cold water (HTF) were allowed to flow through the inner tube, and the outer shell was insulated with cerawool. Temperature of the PCM rose faster during melting at the top sections, due to convection of the molten PCM and circulation of heated air in the air-gap for cases with and without fins. The temperature drop rate of the PCM was noticeable initially during solidification due to a great system TD, and temperature variations on different axial stations were almost uniform for both cases. The unit with fins was observed to attain a faster temperature rise during melting compared to the case without fins, while the temperature variation trend was similar. During freezing, the effects of the addition of fins to the unit was negligible before the PCM was totally solidified. On the other hand, the solidification rate was noted to accelerate due to the application of fins, and thus the freezing time was reduced. Influence of the inlet Re on temperature variation within the lower section of the unit was also investigated. The reduction of t_m due to the addition of fins was noticeable, with greater Re; however, the influence of the increase in mass flow rate (2, 3, 4, and 5 kg/min, both laminar and turbulent regimes) was negligible for both charging and discharging. Enhanced inlet temperature was noted to lead to a shorter t_m due to higher system TD.

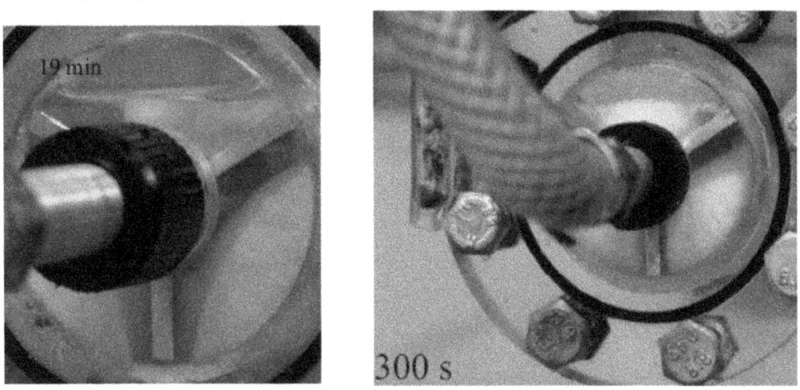

(a) (b)

Figure 20. Isometric views of the melting patterns within the Y-shaped shell-and-tube LHTES unit after (**a**) 19 min when the PCM sticks to the outer shell (no CCM) and (**b**) 300 s = 5 min subject to CCM. Reprinted/adapted with permission from Rozenfeld et al. [117]. Copyright 2015, Elsevier.

An experimental study of the melting and freezing of paraffin (Sigma-Aldrich) in an LHTES unit was performed by Paria et al. [119]. A shell-and-tube configuration with the heated/cooled HTF circulating through the radially finned inner copper tube originating from a constant temperature bath was studied. A numerical study was also conducted using the CFD code FLUENT. As for charging, a shorter time to reach the T_m was obtained with greater values of the *Re*. Melting of the PCM expanded peripherally away from the inner tube initially, however, as time progressed, less time was taken for the PCM at the upper part of the annular space to achieve T_m, due to occurrence of natural convection. As for discharging, the PCM near the HTF inlet froze more rapidly. The temperature decrease rate tended to be flat after some time, due to the dominant mechanism transforming from natural convection to thermal diffusion. The solidification rate of the PCM was noted to enhance with the *Re*.

The charging performance of a PCM-based finned HTF-carrying tube cold storage unit for high-temperature cooling application in buildings was investigated experimentally by Zhai et al. [120]. An inner copper pipe for vertical flow of the HTF was placed concentrically within the shell of the storage unit. The annulus region was divided evenly into five parts using four AF (thickness of 1 mm) and then four LF divided each segment into four similar segments (Figure 21). An in-house PCM was used. The storage unit, originally maintained at the ambient temperature of 20 °C, was cooled by an unspecified HTF with an inlet temperature of 12 °C and flow rate of 300 Lh^{-1}. Measured temperatures of the farthest points were regarded as the PCM temperature of each segment. At a given time instant, the average PCM temperatures were observed to increase along the flow direction, whereas the phase transition of the PCM started earlier at positions that were closer to the symmetry axis. It was found that the phase change rate tended to slow down with increased circumferential angle away from the conductive fins. One segment was studied numerically due to symmetry of the test unit. In addition, three other units with different fin configurations (the unit without fins, AF only, LF only) were studied. The simulated results agreed well with experimental data for the transient temperature. In comparison to the unit with AF, the LF storage unit attained a much shorter freezing time than that of the unit with no fins, due to greater contact area with the PCM and shorter distance between fins. A parametric study was conducted for optimization of the structural parameters. The solidification time was noted to rise with an increase in the AF pitch, and the rate change was the greater when the fin pitch was between 40 and 100 mm. Increasing the number of LF led to smaller included angles, thus improving heat transfer along the radial and circumferential directions, whereas the slope of phase change time variation tended to be flat after the number of LF exceeded nine. It was also observed that the lower the fin's height, the faster the phase transition of the PCM was realized, whereas to ensure enough volume of PCM for heat capacity, the length of the storage unit should be increased. The freezing time was observed to decrease with thicker fins, but the influence of the thickness was not remarkable. The optimized parameter for AF pitch, the number of rectangular fins and fin's thickness were 40 mm, 9, and 1 mm, respectively, whereas the optimized scheme is faster than that of the experimental unit by 26.3%.

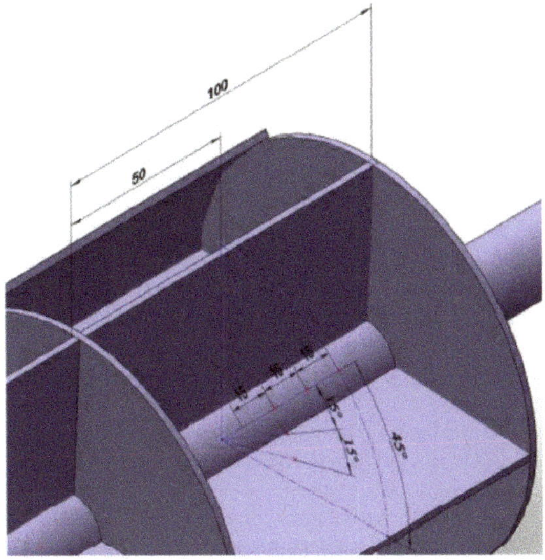

Figure 21. Isometric view of the shell-and-tube LHTES unit featuring both annular and longitudinal plate fins (units are in mm). Reprinted/adapted with permission from Zhai et al. [120]. Copyright 2015, Elsevier.

Khalifa et al. [121] studied a new thermal enhancement design with horizontal heat pipes in a suspended arrangement adjacent to horizontal HTF channels. Rectangular vertical plate fins were utilized, whereas the heat pipes were fixed to HTF channels. A thermal resistance network was developed for mathematical modeling of solidification of the LiCl-KCl and RT60 (experiment). Since the fins were thin and long, a model considering the fin-PCM combination with effective thermo-physical properties was employed. The experimental results exhibited the same trend as the numerical findings, with a difference of about 4%. The numerical results indicated that longer time durations during solidification, and deviation from the experiments, were related to the moving mesh technique adopted in the numerical model. A numerical simulation of a large-scale LHTES unit for a concentrated solar power application was attempted. In the proposed unit, stainless steel naphthalene-charged heat pipes were adopted for their appropriate temperature range and insensitivity to corrosion, whereas stainless steel was chosen for the HTF channel material. Anodized aluminum and Therminol® were selected for fins and HTF, respectively. At an early stage, effectiveness of the heat pipes for cases with 6, 8, 10 and 12 fins were less than one, indicating reduction in both heat transfer rate and storage volume. The values of effectiveness grew steadily at an approximately constant rate proportional to the number of fins. Variations of the effectiveness of the fins as a function of the product of the fin factor and *Ste*, which took into account the fin VF, the sensible heat and the latent heat, was presented. Since the fin effectiveness was recommended to be greater than 2, the use of fins was justified when the product of the fin factor and the *Ste* was greater than 20. A correlation was derived for the prediction of the time required when the effectiveness of the heat pipe reached unity for 90 geometries, 3 different fin VFs and 4 values of fin's TC. The TC of PCM and the *Ste* were fixed at 0.5 W/mK and 0.4, respectively. The size of an LHTES unit, based on the proposed finned heat pipes corresponding to 9 h of operation for a 50 MW electric power output of a concentrated solar power plant, was estimated.

The constructal theory was adopted by Kalbasi and Salimpour [122] to design the optimal structure for a PCM-based cooling system for electronic devices with vertical plate fins attached to the heated bottom of a rectangular case (insulated top and side walls), which was divided into several enclosures. The effects of the contact surface area between the fins and RT27 (Rubitherm), as well as convection of the molten PCM on thawing, were investigated numerically. A 2-D model of an enclosure initially filled with 80% subcooled PCM (20% air) in the space between two neighboring fins was analyzed with the liquid density varying with temperature. This numerical approach was validated with experimental results [78]. Two cases identified as "wide" (fin spacing was much greater than the height of fin) and "narrow" (the reverse case) were considered. The main part of heat transfer was through the base surface for the "wide" case, and wider fin spacing led to less TD between the base and the PCM and, thus, longer safe operation time. The primary route for heat transfer occurred through the fins for the "narrow" case, and greatest TD was observed for the case of $AR \cong 1$. By altering the geometrical parameters, the time needed to reach the maximum temperature (t_{max}) could be maximized. The effect of the ratio of fin and bottom plate thickness on t_{max} in terms of the AR was shown to be negligible. Shortest t_{max} was obtained for a square shape ($AR \cong 1$) compared to the "wide" (large AR) and "narrow" (very small AR) cases. For the same value of contact surface area between the fins and PCM, greater t_{max} was attained for the wider case due to the presence of convection. For small AR, t_{max} decreased with the growing of AR due to declined contact surface area, while it increased with greater AR due to the enhanced contact surface area and convection of the molten PCM. It was shown that greater VF of the PCM led to longer t_{max}. While the investigation was extended from one enclosure to an assembly of enclosures, the results were observed to be similar to the case of a single enclosure. With AR rising from 1 to 3, while the minimum contact surface area occurred at $AR = 2$, t_{max} increased because the effect of improved convection reduced the influence of diminished contact surface. For a small value of AR, convection was negligible and t_{max} mostly depended on the contact surface area. The rate of increase in contact surface area, hence the rate of growth of t_{max}, decreased with greater number of enclosures (N_e). At higher AR values, raising N_e from a small value reduced convection, hence t_{max}, while with further increase in N_e, the contact surface area and t_{max} both increased. For lower N_e, a high value of AR was recommended due to the marked effects of convection, whereas, for higher N_e, a small AR ratio was appropriate, since thawing depended mostly on the contact surface area.

Tao and He [123] performed a 3-D numerical study of a He/Xe mixture as the HTF inside a horizontal concentric tube surrounded by LiF/CaF_2 mixture PCM with/without LF placed on the lower half of HTF tube. A shortcoming of the analysis was the imposition of the vertical symmetry plane that inhibited formation of non-symmetric vortices. Effects of the number, height, thickness of fins and natural convection on the position of the LSI, heat storage rate, heat storage capacity and temperature distribution were analyzed. By increasing the geometric parameters of the fins, heat storage capacity decreased due to the reduced PCM mass. For smaller height, thickness and any number of fins, the heat storage rate increased, since the effect of TC was more significant than that related to the reduced natural convection effect. However, for greater values of height and thickness of fins, the heat storage rate was lowered. For smaller height and thickness of fins, the effective TC was enhanced in the bottom half region and the melting fraction was enhanced, whereas the melting rate was decreased within the top half zone. However, for greater values of height, thickness of fins and any number of fins, the melting rate in the lower part was enhanced and the uniformity of temperature was weakened.

A study was conducted by Liu and Li [124] on the impact of seven design and operational parameters on the performance of an RT42-based solar chimney. The setup consisted of a vertical glass cover, a PCM container, an air channel between the glass cover and the absorber of the PCM container and two openings (along the top façade of the LHTES unit, i.e., air outlet and the bottom air inlet). Horizontal plate fins were inserted into the container to accelerate heat transfer. The front wall opposite to the glass cover was painted black to serve as a high thermal absorber, whereas the other five walls were well insulated. The vents of the chimney were closed during charging to maximize storage of solar energy, and the vents were kept open during discharging to allow the heated air to pass through the air channel. A solar simulator composed of 400 W tungsten halogen lamps was designed to generate the solar radiation flux, and the heat flux was varied by changing the distance between the system and heat absorber. As for the effects of the latent heat of fusion, a greater value of latent heat of fusion was observed to lead to longer melting and freezing times and higher enhancement of the absorber surface temperature during melting compared to freezing. The absorber surface temperature dropped faster for the PCM with lower latent heat. In addition, similar variation trends of the airflow rate and the TD between the inlet and outlet were noted, with these values being high initially and then dropping abruptly followed by a constant interval and a subsequent great drop. These values were higher at the initial instant of freezing for lower latent heat. Greater latent heat led to a longer duration of constant airflow rate and TD. Imposition of higher heat flux was found to shorten t_m, prolong the freezing time and attain greater absorber surface temperature during melting in comparison to the freezing period. A slight increase in the airflow rate and air TD were observed with higher heat flux for a certain time instant. Lowering of the inlet air temperature was noted to result in shorter freezing time, lower absorber surface temperature during freezing compared to the melting stage and higher air flow rate and TD. It was observed that extremely enhanced absorber TC will only shorten the melting period and absorber surface temperature during melting slightly. Higher insulation TC contributed to longer melting and shorter freezing times, as well as less airflow rate and TD at initial and later periods of freezing. Increased transmissivity of the glass cover was observed to shorten t_m, as well as increase the air flow rate and outlet temperature. Improved absorptivity of the absorber surface also proved to lead to increased t_m, higher surface temperature, prolonged freezing time and raised air flow rate and outlet air temperature.

The implicit lattice Boltzmann method (LBM) was employed to analyze freezing of pure paraffin within a rectangular finned container by Talati and Taghilou [125], utilizing the D2Q9 lattice structure. For the first problem, freezing in a container with constant wall temperature was studied. With the container consisting of repeated PCM cells separated by metal fins and due to symmetry, only a 2-D analysis of half of one cell was performed. Solutions were obtained by splitting the 2-D problem into two 1-D problems. Predictions of temperature at four points for different values of the square root of thermal diffusivity ratio of solid and liquid phases were in agreement with analytical solutions. The LBM results of temperature distribution at different time instants agreed well with results using ANSYS FLUENT 14. The observed discrepancies between the LBM and analytical results of the distribution of dimensionless temperature along the center of the fin were attributed to 1-D heat transfer assumption along the fin in the analytical method. Noticeable running time was saved by using the LBM method compared to FVM method. While keeping constant volume of the PCM, freezing time was found to decrease with greater value of γ (ratio of half height of PCM cell and fin length, ranging from 0.1 to 10) when $\gamma \geq 0.5$ due to smaller conduction resistance, but was reduced for cases of $\gamma < 0.5$ resulting from high TC of the fin. The analytical solution over-estimated the time for complete solidification in all cases in comparison to the LBM results, and the largest error of 25% was found when $\gamma = 0.5$. The effects of fin material on solidification were negligible. For a second problem, thermal performance of a composite plane wall was explored with PCM containers and bricks aligned in tandem series between an external marble wall and an internal plaster wall. The walls were exposed to air of different temperatures and heat transfer coefficients. A 2-D cross-section of the composite wall with symmetric boundary conditions of the container was simulated. The LBM and FVM results of temperature distribution within the PCM were in good agreement. Adoption of PCM container was perceived to eliminate the temperature drop in the internal wall, leading to consequent reduced convective heat loss from the internal wall. Assumption of PCM container with aluminum wall led to a much lower freezing time, in contrast to the assumption of the negligible effect of the PCM container wall.

Jmal and Baccar [126] numerically investigated solidification in a paraffin-air TTHX unit (Figure 22). By capping the outer cylinder at top, the unit was different than other TTHX systems [99,109], with the distinction of just using one fluid for the two HTF (inner and outer) passageways. Staggered arrangements of AF on the inner and outer PCM-wetted walls of the annular tube were studied with the cold air entering vertically upward from the inner HTF tube inlet and, finally, flowing in the reverse direction within the outer annular second HTF passageway. Despite of asymmetry of the configuration near the capped end, a 2-D numerical model only within the PCM segment subject to assignment of convective heat transfer boundary conditions on both sides of the PCM container were studied. The addition of fins contributed to enhanced heat transfer from the PCM to the HTF, in comparison to heat storage without fins. The temperature fields after 1 h, shown in the top row of Figure 23, were observed to be more uniform as the number of fins was raised, leading to formation of nearly repeating cells. Raising the number of fins weakened the convection, which was then confined to the narrow vertical space between the fins (bottom row of Figure 23). In effect, temperature rise of outlet air corresponding to nine fins was slightly higher than that of five fins. During solidification with nine fins, frozen PCM appeared in the vicinity of the tube walls and the fins, initially, and eddies formed due to natural convection contributed to uniformity of temperature and heat extraction from the PCM. The heat transfer rate was observed to decline with time, as indicated from the decreasing rate of the outlet air temperature. Heat extraction was inhibited due to increasing thermal resistance with thicker solidified PCM, and the dominant heat transfer was replaced by conduction. Consequently, this insulating layer restrained the inner core of the liquid PCM from freezing.

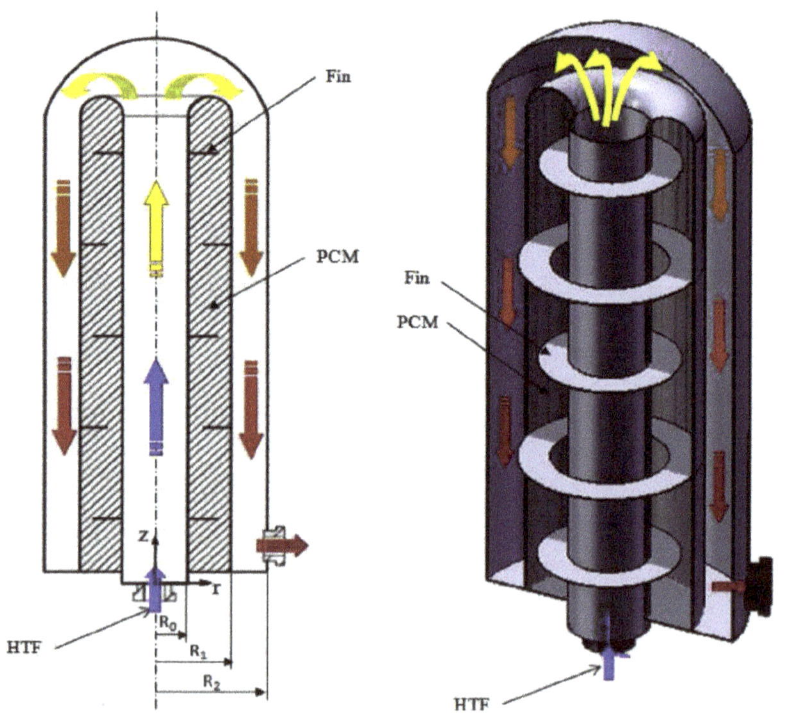

Figure 22. Views of the PCM-air TTHX unit featuring staggered arrangement of annular fins on the inner and outer PCM-wetted walls with capping the outer cylinder making it different than other systems [99,109], allowing air to serve both the inner (cold blue arrow) and outer (hot orange/red arrows) HTF passageways. Reprinted/adapted with permission from Jmal and Baccar [126]. Copyright 2015, Elsevier.

Hosseini et al. [127] investigated the effects of AF's length and the *Ste* on the behavior of the RT50-based horizontal shell-and-tube heat exchanger. The analysis concentrated on 2.5 h of melting followed by 3 h of freezing. During charging, the temperature of the upper part of the PCM was generally observed to be higher than that of the bottom part, due to the effects of the evolving buoyant hot PCM front. Before the emergence of the molten layer, the TD between the upper and middle segments were more noticeable than that between the middle and bottom sections. Longer fins were found to contribute to heat penetration into the PCM, thus leading to an increased melting rate and a reduced melting time, whereas melting was also promoted with greater *Ste*. During discharging, longer fins was noted to lead to faster temperature reduction rates and lower temperatures at the end of discharging. It took a longer time to observe the solid layer covering the bottom part of the fins for the long-fin system. The higher temperature at the end of melting with longer fins resulted in a greater *Ste* during discharging, which then delayed initiation of freezing. On the other hand, given an elapsed time period, the freezing rate of the heat exchanger with a doubled fin length exceeded that of a unit with short fins. Change of the *Ste* did not contribute to the variance of the total solidification time.

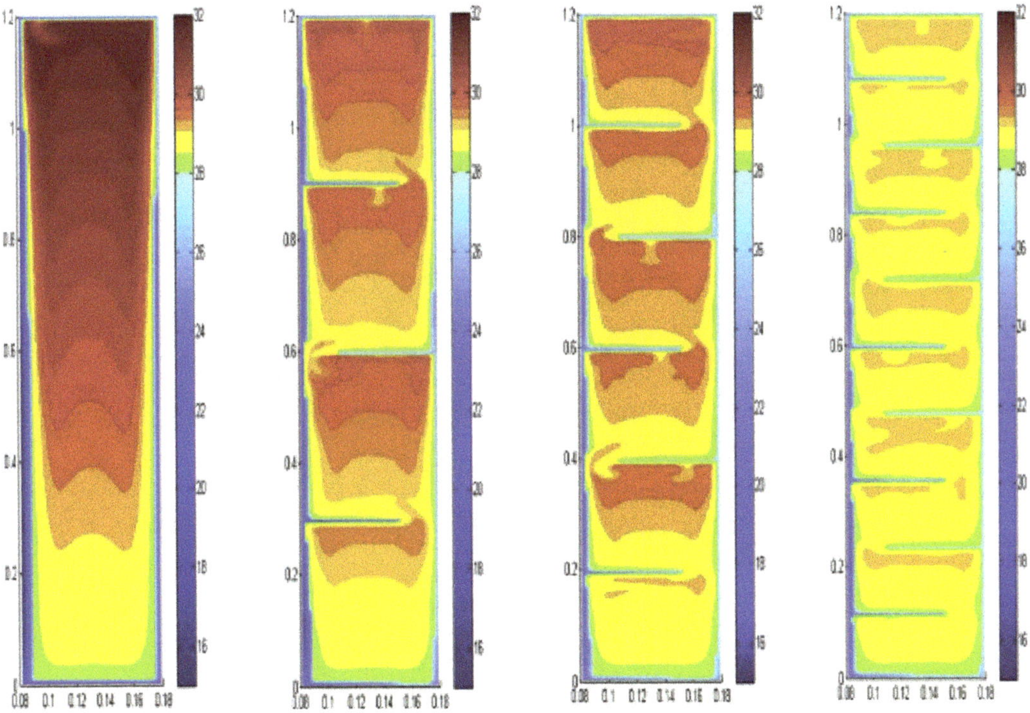

Figure 23. *Cont.*

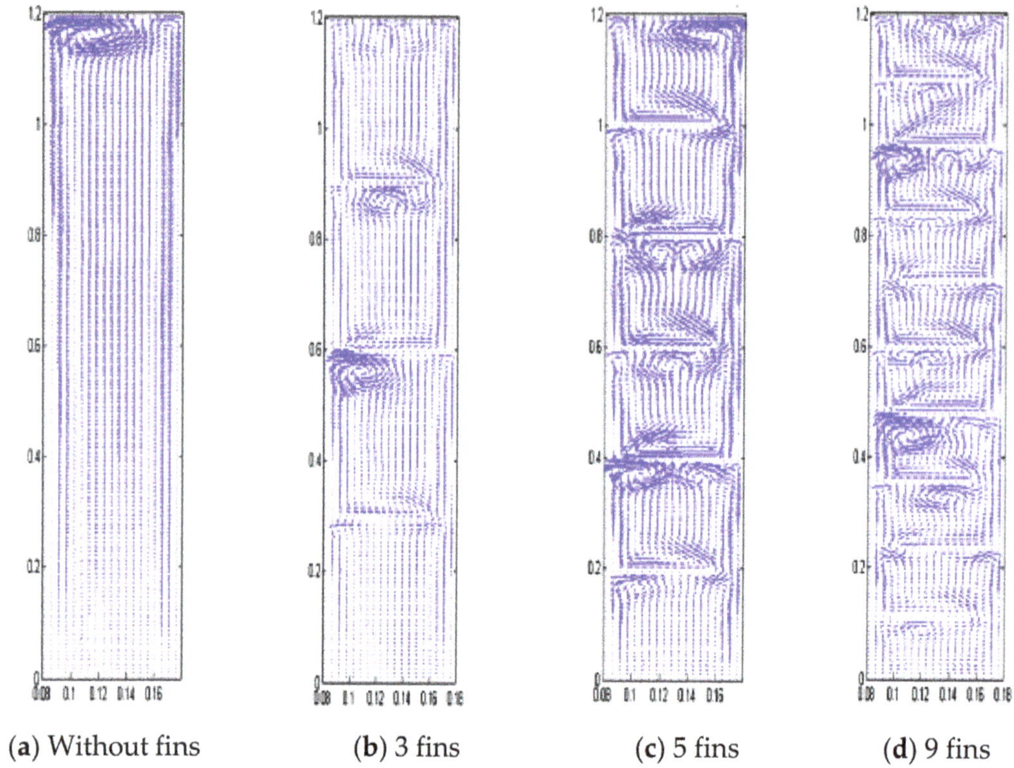

Figure 23. Instantaneous temperature contours (**top row**) and velocity vectors (**bottom rows**) after 1 h corresponding to (**a**) 0, (**b**) 3, (**c**) 5 and (**d**) 9 staggered annular fins plate fins; dimensionless axes are used in the r and z directions and the contour levels for temperature are in °C; overcrowded velocity vectors signify marked intensity of convection. Reprinted/adapted with permission from Jmal and Baaccar [126]. Copyright 2015, Elsevier.

Experiments were conducted by Gharbi et al. [128] on the thermal characteristics of utilizing PCM in heat sinks used as the cooling system of electronic devices for four configurations (pure PCM, PCM in silicon matrix, PCM in a graphite matrix and pure PCM unit with fins) tested in two positions. Plastic paraffin with a T_m below the electronic threshold critical temperature (60 °C) was the PCM of choice. The test system was composed of a container with one wall made of a copper slab contacting an electric heater (4 W) and the rest of the walls were made of Plexiglas® sheets to allow for photographic observation. As for the behavior of the LSI of cases of PCM with silicon matrix and pure PCM, at an early time instant of 60 min, the LSI was nearly parallel to the heated wall for both tested positions, indicating a dominant heat conduction mechanism. At a later time instant of 120 min, the LSI still remained somehow parallel to the wall for the PCM-silicon matrix unit, as the movement of the LSI due to convective cell was prohibited by the matrix. On the other hand, for the case of the vertical heated wall, a deformed front with more molten PCM at the top was present for the case of pure PCM. For both vertical and horizontal modes of operation, fast temperature rise on the surface of the copper slab was observed for the benchmark case for the system with only air convection, compared to cases of PCM combined with two matrices. Considering the performance of the graphite-based PCM heat sink against that of silicon matrix, two decreasing temperature slopes were observed at two instants, signifying the beginning and the end of melting.

The effects of LF, PCM ($NaNO_3$ and $NaNO_3$/expanded graphite) and HTF (synthetic oil and molten salt) flowing within a horizontal inner tube were studied by Li and Wu [129] through employing a 3-D numerical model of a shell-and-tube LHTES unit. During melting of $NaNO_3$ with fins, molten salt as the HTF performed slightly more effectively than synthetic oil in terms of the thawing time. For the same case, t_m was shortened compared to the case of $NaNO_3$ without fins by 14% at low HTF velocity, while this reduction was lowered to 20% as the HTF velocity was increased. Melting times for $NaNO_3$ with fins and its composite without fins were 40% and 25% less than the melting time of $NaNO_3$ without fins. During melting, the reduced temperature gradient due to the liquid layer of PCM around the outer surface of the HTF tube and more significant contribution of natural convection caused the heat flux, for all cases, to decrease progressively. The effective Nu number was lower for molten salts compared to synthetic oil. Inspecting the corresponding velocity fields for cases without and with fins, one and three pairs of vortices on the left and right sides were generated, respectively. For the case of no fins, molten PCM gradually occupied the bottom portion of the shell. The greater combined size of the three pairs of vortices led the $NaNO_3$ PCM to melt faster for the case of finned system. During solidification, molten salt performed slightly more effectively compared to that in charging. With growth of a solid layer formed on the outer surface of the HTF tube, the temperature in this region became closer to that of the tube wall, which reduced the heat flux.

Feng et al. [130] numerically investigated the melting of Docosane impregnated in a vertically finned high-porosity metal foam heat sink with two approaches, i.e., pore-scale and volume-averaged simulations. Plate fins were attached to the bottom surface of an enclosure, which was kept at 347 K, with other surfaces assumed adiabatic. The PCM and the finned metal foam were initially kept at 300 K. Foam geometry was modeled with sphere-centered tetrakaidekahedron cell packing into a network structure in the pore-scale simulation, and the computations were conducted with ANSYS FLUNET 14.0. Only half of a fin-foam channel (3-D) was considered, due to periodicity and symmetry of the computational domain, and volume-averaged simulations relied on the one-temperature model based on the local thermal equilibrium assumption. The PCM-infiltrated foam was regarded as a homogeneous medium and a 2-D model of half of a fin-foam channel was considered. Results of the two approaches predicted qualitatively similar LSIs (Figure 24a,b), temperature distributions (Figure 24c,d) and flow fields due to convection, whereas the simulations could only capture the overall U-shaped LSI separating two distinct phases. The higher temperature gradient was observed to be in the liquid phase of the PCM, especially at the LSI, indicating a greater portion of heat transition into latent heat. Predictions of melt fraction and bottom surface heat flux using the two methods agreed well quantitatively (with and without natural convection). It was suggested that the local thermal equilibrium between the foam and the PCM was observed with natural convection in the liquid PCM. The total melting time was reduced by 28% and the surface heat flux was increased by 10–25% compared to the case without natural convection. The presence of natural convection slightly led to increased heat transfer from the bottom surface to the PCM and foam, but decreased heat transfer to the fins was observed. Streamlines within the liquid pool predicted by the pore-level and volume-averaged approaches are shown in Figure 25a,b, respectively. Compared to conventional plate-fin and metal foam without fin insertions heat sinks, the proposed finned foam unit exhibited the highest melting rate and heat transfer coefficient. This reinforces the knowledge that, whereas plate fins are very effective in removing heat normally away from an active surface [16], the presence of the porous media will in turn contribute to greater mixing of the liquid PCM in between neighboring fins.

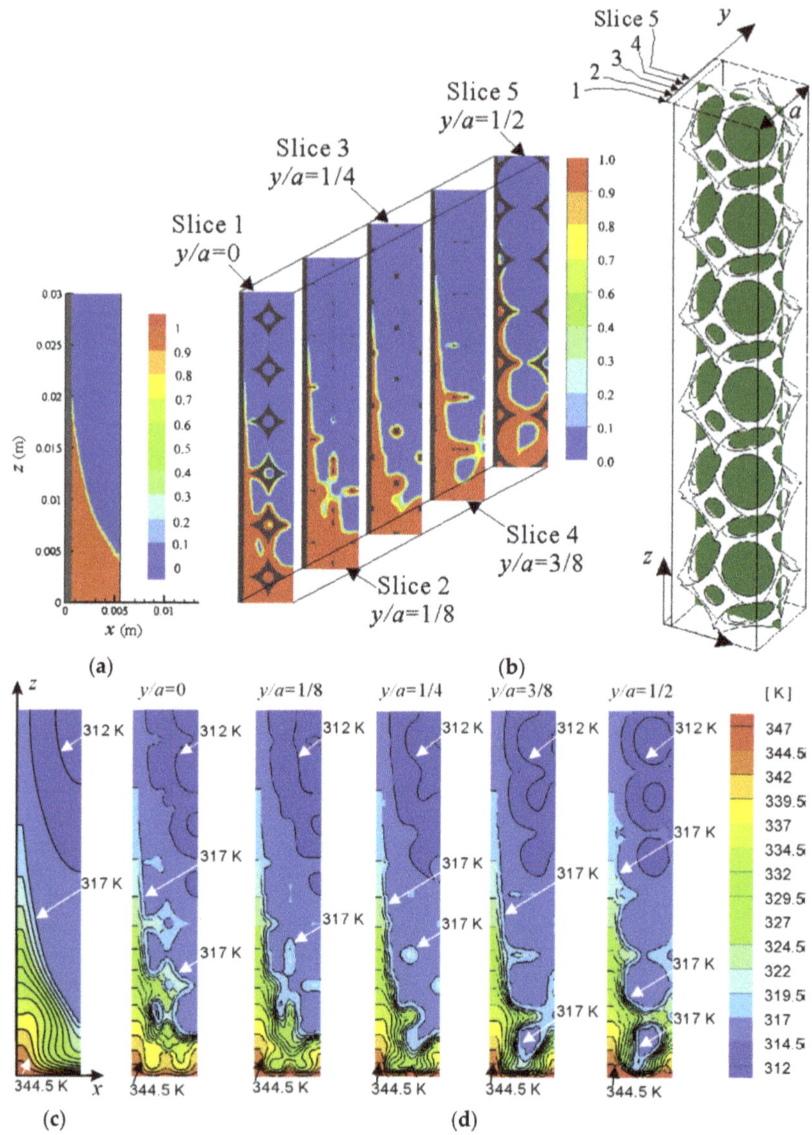

Figure 24. Instantaneous dimensionless solid–liquid phase distributions in the top row predicted by (**a**) volume-averaged (left column) and (**b**) pore-scale simulation (5 slices) methods, along with the corresponding temperature fields in the bottom row predicted by (**c**) volume-averaged (left column) and (**d**) pore-scale simulation (5 slices) methods, shown at the time instant of 30 s (melt fraction = 0.25 for the volume-averaged case). Reprinted/adapted with permission from Feng et al. [130]. Copyright 2015, Elsevier.

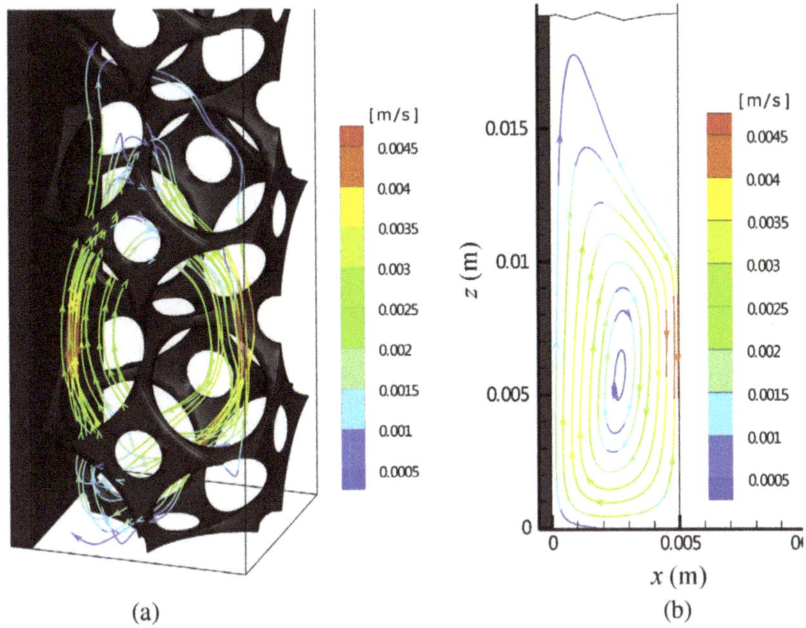

Figure 25. Instantaneous liquid streamlines predicted by (**a**) 3-D pore-scale simulation and (**b**) 2-D volume-averaged methods, shown at the time instant of 60 s (melt fraction = 0.5 for the volume-averaged case). Reprinted/adapted with permission from Feng et al. [130]. Copyright 2015, Elsevier.

Pakrouh et al. [131] performed a numerical study to optimize the geometry of an RT44HC-based heat sink, including an air gap at the top with 25–100 vertical aluminum pin fins by means of monitoring the critical temperature (50, 60, 70 and 80 °C), with the time required for the base to reach these values designated as critical times. The Taguchi algorithm for three-level parameters was employed to determine the configuration that provided the maximum operating time. An orthogonal array of 27 cases with combinations of the number, thickness, height of fins and base thickness were utilized in order to obtain their effects on the parameter, called the "contribution ratio," which indicated the effect of each quantity on the objective function. Variations of base temperature with time during heating were discussed. The deliberation of base temperature indicated that the optimum state always corresponded to the greatest fin height. The enhancement ratio, defined as the ratio of the time duration taken by the fin-based heat sink to reach the critical temperature to the time duration taken by a finless unit, was obtained for four critical temperatures. The statistical measure for performance, known as analysis of variance (ANOVA) tables, including sums of squares, variances, Fisher ratios and contribution ratios for four different critical temperatures were also employed to find the optimum conditions. Optimum PCM percentages were determined to be 60.61%, corresponding to a case of the 100 pin fins heat sink with 4 mm thick fins for critical temperature of 50 °C, and 82.65%, corresponding to the 100 pin fins heat sink with 2 mm thick fins for other critical temperatures.

Tiari and Qiu [132] developed a 3-D model (ANSYS FLUENT14.0) to investigate charging of a eutectic mixture of $NaNO_3$ and KNO_3 (T_m of 220 °C) LHTESS with three configurations (Figure 26a) of embedded heat pipes with 1 mm thick AF attached to the condenser section of each heat pipe. The number of heat pipes for Cases 1, 2 and 3 were 5, 9 and 9, with the shared evaporator of the heat pipes covering the bottom surface of the container. For the first case, including the effects of natural convection led to a higher melting rate as well as a lower container base wall temperature. The effect of natural convection on PCM temperature for Case 1 at two different elevations (4 and 8 cm) for different time instants were obtained. At the early stage of melting, when conduction was the dominant mechanism, similar temperature distributions were seen at both elevations, with and without natural convection. As charging progressed and the liquid layer became thicker, thermal resistance between the wall and molten salt was reduced. A more uniform temperature distribution was realized due to diffusion of thermal energy. It was seen that convection effects were stronger at lower elevations close to the heated base, and a relatively more uniform temperature field was observed in comparison to that at higher elevations. This was due to the presence of a higher amount of molten PCM at lower elevations early on during melting. As heating continued and the melt zone became bigger, the role of natural convection became even stronger. Progress of the LSI at the same elevations for Case 1, with and without the effects of natural convection, was obtained. All the PCM turned to liquid over the cross-sectional area at 4 cm for the case with convection, while the PCM was still partially in the solid state if buoyancy was ignored. At the elevation of 8 cm, the PCM was observed to be not fully molten regardless of the presence of buoyancy. At the early stage of melting, a layer of molten PCM was formed close to the base wall in Case 1, whereas the two other cases still contained mushy zones at lower elevations (Figure 26b). As the charging continued, part of the PCM located between the heat pipes and close to the container wall in Cases 1 and 2 remained solid because the heat pipes were far apart. However, in Case 3, although the number of heat pipes was the same as in Case 2, the maximum distance between heat pipes was minimized. Heat was spread out more uniformly, and less solid PCM remained in Case 3, and the container base wall temperature was reduced.

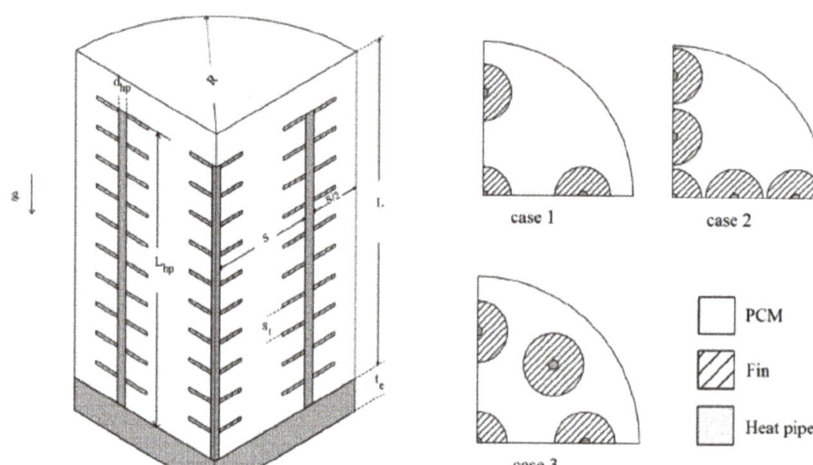

(a)

Figure 26. *Cont.*

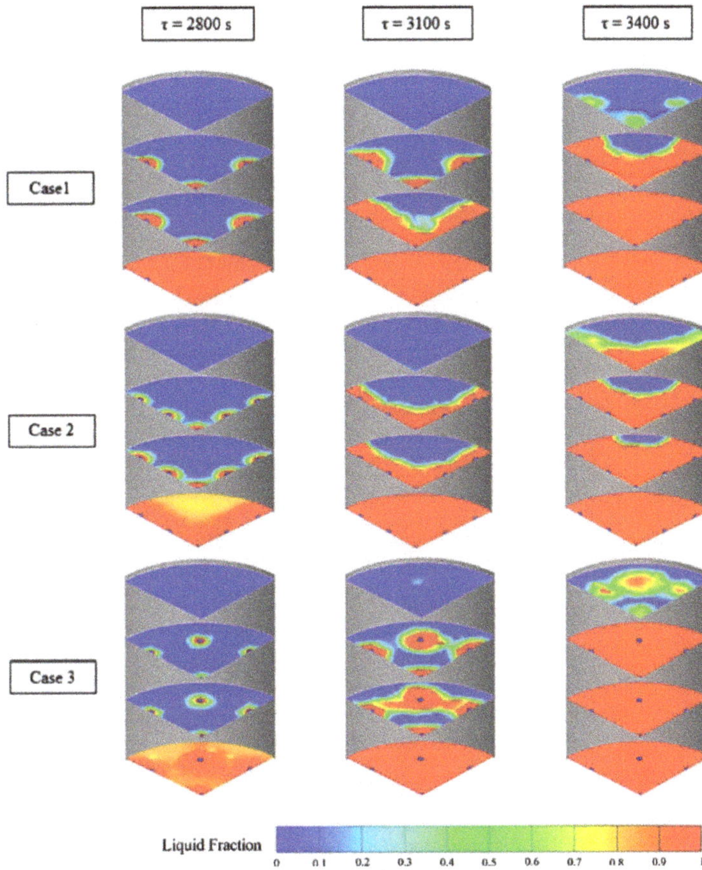

(**b**)

Figure 26. Schematic diagrams of (**a**) three cutaway heat pipe/PCM/fin configurations of the thermal energy storage units studied and the corresponding (**b**) contours of dimensionless liquid fraction field for different cases on four vertical axial planes at 2800, 3100 and 3400 s time instants. Reprinted/adapted with permission from Tiari and Qiu [132]. Copyright 2015, Elsevier.

Kalbasi and Salimpour [133] conducted a numerical analysis to study the effectiveness of adding horizontal fins (second construct level) to vertical fins (first construct order, Kalbasi and Salimpour [122]) in a 2-D rectangular heat sink. Effects of different degrees of freedom (N_e, number of horizontal fins, *AR* and complexity of the heat sink) on t_{max} were investigated. Vertical fins were attached to a horizontal active base, thus dividing the heat sinks into several enclosures, whereas only half of one enclosure was considered. Horizontal fins were fitted on the vertical fins with uniform spacing. Only 80% of the heat sink was filled with RT-27, and the top was open to the ambient air with other surfaces, except the base, which was insulated. In order to keep a constant PCM VF, increasing N_e led to shorter length. For small *AR*s and small N_e, reduction of uniformity of temperature distribution occurred with a greater number of fins and resulted in a shorter t_{max}. For greater value of *AR*s, an optimum value of the number of fins existed to obtain the longest t_{max}. Increasing the number of fins led to greater t_{max}, when the number was below the optimum number. This was due to greater continuous connectivity of the partitioned PCM, leading to higher chance of convection, but a further addition of fins lowered t_{max} due to reduced uniformity of the temperature. The quantity t_{max} was also observed to grow with increased N_e for a certain number of *AR*s. Accordingly, the variation of t_{max} as a function of the fin length ratio was also deduced. For greater N_e, increased *AR* value reduced the contact surface between the PCM and fins and, hence, lowered t_{max}. On the other hand, for lower N_e, greater *AR* was observed to increase t_{max}, due to promotion of convection with the wide range of enclosures. The quantity t_{max} was again lowered beyond the optimum value of the N_e due to suppressed convection by horizontal fins. Parameter t_{max} was observed to rise with the increase in geometrical complexities for a low value of the *AR*, due to the improvement of temperature uniformity brought about by highly conductive horizontal fins. For larger *AR*, convection was noticed to be dominant, and the increasing complexity to the first design order and further to the second order led to a decrease in convection, and, thus, decreased t_{max}. Only for greater N_e was the second order of design observed to be more effective than the first order design, whereas it was still less effective than the elementary order (without fins).

Samanta et al. [134] numerically studied the solidification characteristics of $CaCl_2$, $6H_2O$ infused in the shell side of a shell-and-tube unit with the cold HTF flowing inside an inner tube hosting AF. The PCM was initially assumed to be at the ambient temperature above its T_m. The boundary of the domain touching the HTF inner tube was maintained at a constant temperature below the T_m, while the other three boundaries were adiabatic. Rayleigh–Bernard convection developed next to the LSI zone during solidification, indicating a sudden fall of temperature of the PCM. For the experiment, a transparent borosilicate glass tube was utilized as the outer shell. Measured temperature variations within the PCM agreed well with the numerical data. The rate of decrease in temperature was observed to be very fast, due to sub-cooling of the system.

Numerical simulations of the thermal response of an LHTES unit for a high-temperature solar tower application was validated by Malan et al. [135] with experiments using paraffin wax. The simulation model was then used for a high-temperature LHTES unit using salts, incorporating a solar tower as the heat input. Heat pipes and two types of fins were employed to increase the effective heat transfer surface area in order to enhance the heat transfer rate. This experimental storage unit consisted of a container with a charging multichannel rectangular micro heat pipe (Furukawa Electric, Japan) connected to the heating side, and a discharging identical heat pipe connected to the heat removing side. During charging, a water-carrying kettle inserted with the charging heat pipe was heated electrically, whereas during discharging, hot water was extracted, and subsequent cold water was allowed to flow through the top heat exchanger connected to the discharging heat pipe. Horizontal plate fins were embedded in the PCM container, with zigzag-shaped fins placed between those channels forming small compartments which reduced the heat transfer distance. A 2-D conduction model was simulated with the C++ package, since the symmetry of the fins formed repetitive channels, and convection was neglected in the simulation. Temperature variations and absorbed/removed heat at three points at the mid-height of the container were recorded, and agreed well with the numerical results. During discharging, the numerical model exhibited faster cooling than the experimental unit, which was suggested to be due to different flow rates used. Adoption of fins were observed to reduce the melting time sufficiently, and led to the extraction of more energy. The validated numerical model was then used for a high-temperature PCM proposed for adoption in a concentrating solar thermal application. The response of the simulation exhibited a cyclic behavior in response to realistic solar radiation conditions. The favorable response suggested the potential of this model's physical utilization, whereas the great heat loss was observed due to excessive high temperature of the PCM.

Khatra et al. [136] numerically investigated the thermal and flow performance of an initially superheated n-Octadecane within an internally finned enclosure during solidification. This LHTES unit was used for a cooling application, in which the PCM stored outdoor coolness during the night to supply indoor cooling during the day. A 2-D unit was analyzed with three parallel horizontal equidistant plate fins originating from a vertical wall set at 255.16 K (same as fins), with other walls adiabatic. A comparative study of finned and unfinned enclosures demonstrated that introducing fins weakened natural convection, improved the dimensionless heat transfer and led to significant reduced solidification time. Effects of the *AR*s of the enclosure (3 to 8) and fins (2.69 to 13.89) on heat transfer enhancement were investigated, while keeping the mass of the PCM and fins mass constant. Greater value of the *AR* was observed to result in enhanced heat flux from the cold wall and consequent expedited freezing, due to depressed free convection, and the expanded heat transfer surface. Increased *AR* of fins promoted freezing, since longer fins attained greater heat transfer surface and weakened the convection within the liquid phase. For the enclosure *AR* of 4 and explored range of fin *AR*, a 17.74% reduction of the nondimensional solidification time was realized. It was also observed that, for the fin *AR* of 5, the nondimensional freezing time was reduced by 49.48% for the studied enclosure *AR*.

In terms of the melt front speed and phase change performance of n-eicosane within a concentric annulus, Darzi et al. [137] sought an improvement method of adding different numbers (4, 10, 15 and 20) of metal LF fins, compared to approaches of changing the inner tube shape or adding nanoparticles to the PCM. Taking the circular inner tube as a benchmark, horizontally oriented elliptical (major axis lying on the horizontal plane) HTF tubes attained a faster melting rate in the initial 20 min, but slower speed afterwards, and longer full melting time. Utilization of vertically oriented elliptical tubes (major axis lying on the vertical plane) promoted low-speed stable melting beneath the inner tube and decreased the full melting time, in comparison with the circular inner tube and the overall performance improved with greater *AR*s of the ellipse. However, the above scheme was not beneficial to the solidification process. Increasing the Cu-nanoparticle VF enhanced both the melting and solidification rates. Nevertheless, the heat conduction dominating the stable heat transfer at the bottom section was not ameliorated. In summary, adding fins on the hot or cold inner tube outperformed the other methods chosen to expedite melting and solidification, respectively. Increasing the fin number in the melting process is less efficient than that in the solidification, due to the intensified suppression of the natural convection effect with greater number of fins.

Optimized nature-inspired snowflake-shaped fins with four main branches facilitating thermal penetration depth into the corners of a square-shaped LHTES unit were proposed by Sheikholeslami et al. [138] to enhance freezing. Considering only the diffusive transport and keeping the fin surface area constant, it was concluded that the angle of the smaller branch (β_2) and its position (X_2) identified in Figure 27 did not have considerable effect on solidification rate of the PCM, and the optimized values were $5\pi/12$ and 0.66 L, respectively. Nevertheless, opposite trend for the angle of the bigger branch (β_1) and its position (X_1) was observed, and the best values were $3.69\pi/12$ and 0.16 L, respectively. Regarding expediting of solidification or maximum energy storage capacity, the optimized snowflake-shaped fin configuration has the best performance, compared to the unit with Cu nanoparticles dispersed in PCM and the unit with a simple LF configuration, which indicated the effectiveness of the snowflake-shaped fin configuration, especially in the acceleration of discharging process in the corner.

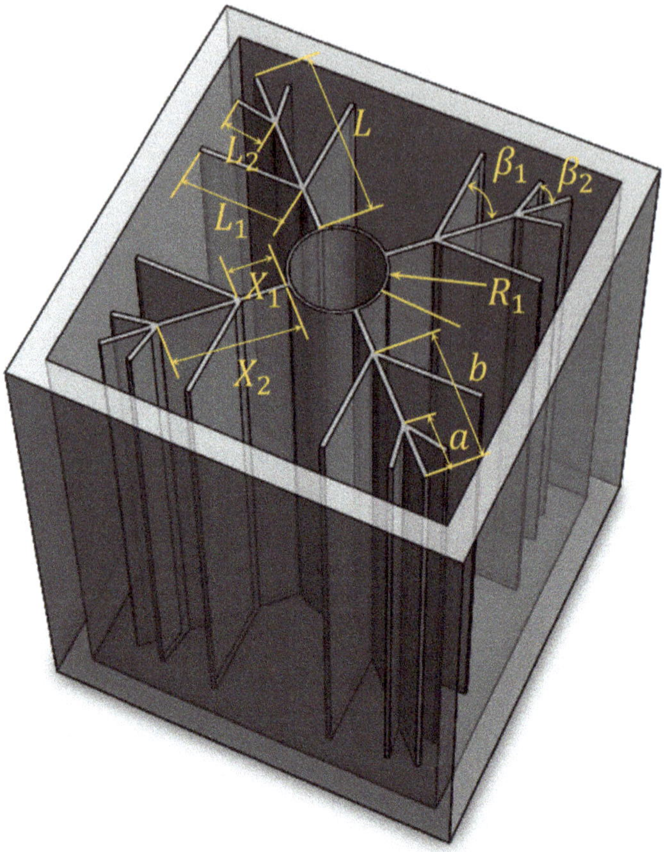

Figure 27. Isometric view of the proposed snowflake fins (*L*s are fin lengths, *a* & *b* are the small and big fin branch distances measured from the fin's end, respectively, βs are branch angles, X_1 and X_2 are fin branch distances measured from the cold surface inner pipe of outer radius R_1). Reprinted/adapted with permission from Sheikholeslami et al. [138]. Copyright 2016, Elsevier.

A shell-and-tube LHTES configuration with helical fins attached to an inner tube was proposed by Rozenfeld et al. [139]. It was demonstrated experimentally that a slightly heated shell could make CCM possible and shorten t_m by a factor of three in comparison to the units exposed to ambient air. A hybrid analytical-numerical model was developed to describe the phenomenon of CCM on the helical surface. An analytical model, without incorporating sensible heat, was initially built under the assumption that the fin was isothermal, which over-predicted the melting rate. Melting on helical fin was stated to be more expedited than that on radial fins. Expanding to a numerical model that considered the time-dependent fin temperature distribution and the sensible heat of the PCM, good agreement was achieved between the numerical predictions and experimental results. Governing dimensionless parameters were identified analytically and applied to numerical results. Dimensionless parameters, including the *St*, *Fo*, Archimedes and *Pr* numbers, and the group representing the geometry of configuration, were combined with dimensional analysis to obtain a complete generalization of the results. The charging performance was presented in terms of certain factors, including the HTF temperature, unit pitch, fin radii, fin thickness and relative fin and PCM volumes. The structure of the helical fins was advantageous, since it avoided local increase in pressure related to volume expansion during phase transition and solidification voids. It also enabled a continuous volume of the PCM, promoting natural convection in the annulus.

Performance improvement during the solidification process was investigated by Kuboth et al. [140] by varying distributions of copper AF within shell-and-tube LHTES units. Storage units of different fin allocations contained identical volume and number of fins (100) and RT42. Thirty different fin arrangements were examined, including the linearly and exponentially increased fin density towards the outlet of the HTF (water) pipe, section-by-section change of the fin density (2 and 3 segments) with denser fins close to the outlet of the units and equidistant fin distribution. Whereas an FVM-based 2-D numerical model was adopted within the PCM store, 1-D axial convection within the HTF was assumed. In-house code was validated with a 2-D model in ANSYS Fluent, showing good agreement except for a small discrepancy due to the differences in discretization schemes. It was observed that storage performance was affected by varying the fin arrangements. Most of non-equidistant distributions of fins can achieve a higher average output power than equidistant ones, except for certain cases with fin density increasing exponentially with the growth rate factor greater than 1.023. In comparison to uniform fin arrangements, the average output power at total discharge was improved by three percent using the most efficient fin allocation, with a linear growth rate of 10. Exponentially increased fin densities towards the storage unit outlet also resulted in enhanced discharge performance with proper growth rate factor. The studied intermediate discharge process demonstrated that distributions of fins could induce a more uniform discharge.

To enhance the overall melting of PCM (RT42) filled in a rectangular cavity, Ji et al. [141] investigated computationally the effects of two parallel fins with inclination angles (0°, +15°, +30°, −15° and −30°) attached to the vertical side wall contacting the heater plate (Figure 28). Overheating of the PCM at the top side of the cavity without fins was firstly observed during the melting process due to the natural convection effect, which resulted in reduced overall melting rate. Considering the melting rate and temperature distribution for the case with 0° fins (Figure 28a), the fins with upward angles of +15° and +30° (Figure 28b,c) exhibited non-uniformity of those quantities since more heat was transferred to top part of the cavity. It was noticed that fins with −15° (downward, Figure 28d) exhibited a faster melting rate and more uniform temperature distribution compared to the case with 0° fins. These were attributed to the heat being transferred to the bottom side of the enclosure along the downward-pointing 15° inclined fins, and the buoyancy effect was mitigated. From the enhancement ratio data, the angled fins primarily take effect at the second half of the melting process. The effects of the fin length and heat flux were further explored on the PCM melting enhancement, with the fin angle setting from 0° to 15°. It was shown that the increased melting rate became more apparent with increasing the dimensionless fin length from 0.25 to 0.75 when the fin angle changed from 0° to 15°. The greatest improvement and more uniform temperature distribution was achieved by the dimensionless fin length of 0.75 and −15° fins, for which heat was spread to the top and bottom of the container simultaneously. The −15° fins slightly enhanced the melting rate, with a lower heat flux input compared to fins of 0°, but the improvement diminished as heat flux increased.

Dendritic fins proposed by Luo and Liao [142] were attached to the inner tube of the shell-and-tube LHTES unit to enhance melting performance. The improved melting rate with the dendritic structure was compared with the LF fin unit. The results indicated that the dendritic fin greatly enhanced the melting rate and led to uniformity of the temperature distribution. This was attributed to formation of multiple independent PCM zones that dispersed heat rapidly in metallic fin bifurcations.

Duan et al. [143] investigated constrained melting rates of PCM within honeycomb cores of non-hexagonal (triangular, trapezoidal, rectangular and circular) cells, in comparison to hexagonal cells, for three *Ra* numbers. The melting time-saving ratios of PCM in triangular and quadrilateral cells compared to the hexagonal cell decreased for a low aspect ratio. It was observed that the PCM in cells with a smaller geometrical factor ($\sqrt{Area}/perimeter$) melted faster, as conduction dominates the heat transfer at low *Ra* number, so the melting rate of PCM in triangular, trapezoidal and rectangular cells are greater than hexagonal cells. As for higher *Ra* numbers, the melting time of the PCM is affected by both the geometrical factor and the orientation of the cores, due to enhanced natural convection.

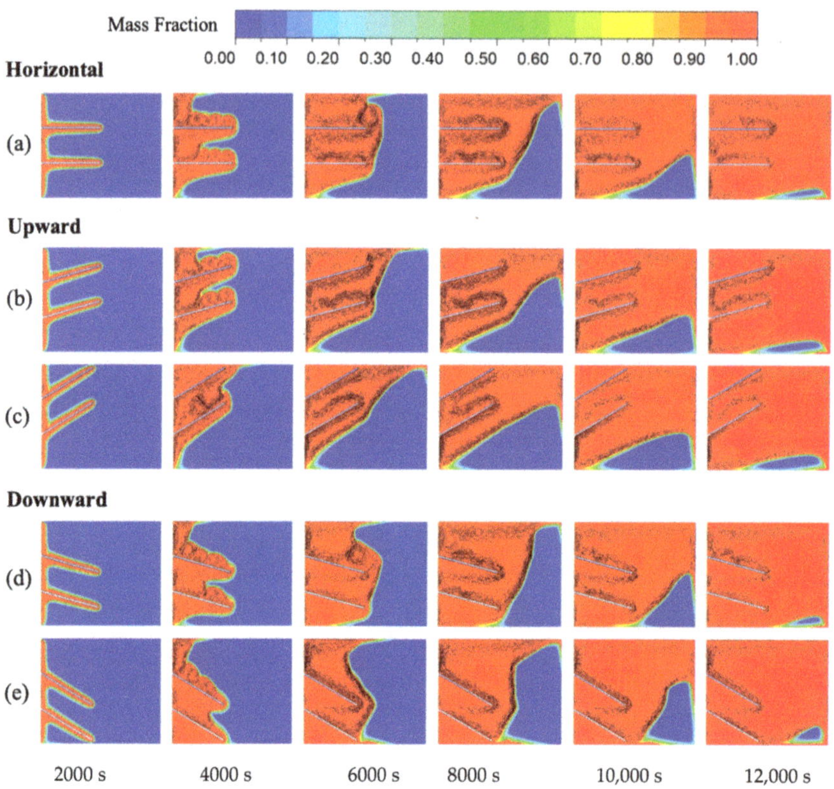

Figure 28. Instantaneous contours of the dimensionless melt fraction with superimposed fluid velocity vectors for the fin-PCM cases: (**a**) $\theta = 0°$, (**b**) $\theta = +15°$, (**c**) $\theta = +30°$, (**d**) $\theta = -15°$ and (**e**) $\theta = -30°$, with the fixed dimensionless fin length equal to 0.50 and heat flux input of 2500 W/m^2 on the left wall. Reprinted/adapted with permission from Ji et al. [141]. Copyright 2018, Elsevier.

Deng et al. [144] explored the effect of arrangements of plate LF on the melting within a shell-and-tube LHTES unit. Melting improvement of different arrangements were explored for six ES straight, angled, lower-half and upper-half fins. The shortest t_m was obtained for the case with lower-half fins, followed by angled, straight and upper-half fins. Effects of the number of fins, dimensionless fin length, HTF temperature and outer pipe material on melting performance of the four arrangements were then investigated. Melting enhancement by increasing the number of fins was most effective in the case of angled fins, followed by straight, lower-half and upper-half fins. Based on complete t_m and heat storage capacity data, the best type of arrangements to increase the performance of LHTES were: (a) with six fins or less, the recommended arrangement was lower-half fins and (b) given greater than six fins, it was the angled fins. For 6 fins and dimensionless fin lengths of 0.5 and 0.95, the optimum arrangement was the angled case. Raising the HTF temperature was observed to be more effective for the case with angled fins. Shorter t_m and comparably higher heat storage capacity made aluminum to be advantageous as the outer pipe material, due to its high conductivity and relative small sensible specific heat. It is also observed that heat storage capacity was the lowest compared to other three configurations.

Shahsavar et al. [145] investigated the effects of fin configurations on the performance of the shell-and-tube LHTES units, in both melting and solidification processes, considering the locations, thickness and diameter of the five AF placed on the outer wall of the HTF inner tube. RT-35 was filled within the annulus region while water HTF flowed upward vertically through the inner tube, and the total volume of the fins was kept constant. It was observed that the addition of fins was more advantageous for the melting process rather than the freezing process. Charging and discharging time periods reduced by 41.4 and 9.7%, respectively, for the ES finned-case compared with the non-finned case. Shortening the distance between the first fin and the bottom of the unit from 40 to 10 mm led to shorter thawing time and enhanced heat storage rate. Compared with the ES case, the melting time reduced by 23.9% with the best fin array. Moreover, increasing the diameter of the fins from 5 to 6 mm led to faster melting, but further increasing to 7 mm brought about adverse effects, due to suppression of the natural convection. For the freezing process, the optimal case is the ES arrangement, while the heat recovery rate increased by 11.4% compared with the non-finned case.

Melting enhancement of RT-82 within the annulus of a shell-and-tube LHTES unit by adding LF or metal foams was investigated by Zhao et al. [146]. Concentric inner and outer copper pipes were subjected to a constant temperature of 363.15 K. Carbon fiber (XN-100) fins penetrated into the PCM from the walls of both pipes. The constraint of constant fin volume and thickness remained unchanged during the investigation. After exploring the effects of the fin number density, a function for t_m in terms of the number of fins was fitted. It was observed that t_m was shortened while the fin number increased to 16, but further reduction was not realized with greater number of fins. The optimum fin number was obtained at 16, with over 60% of t_m reduction compared to the no-finned case. Additional 8 and 4% time reductions were attained by utilizing longer and denser bottom fins, respectively. However, the addition of tree-like fins resulted in a longer t_m, which indicated that it was not a good option for small PCM domains. It was also noticed that the trunk length effects on t_m were more pronounced than the bifurcation morphology. Finally, the melting performances of the units with LF were compared with those using three metal foams (nickel, aluminum and copper). It was observed that t_m of the optimized strategy with longer bottom fins was rather less than those of Cu and Al foams, and greatly shorter than that of Ni foams. These results indicated that well-arranged fins could be as efficient as metal foams.

Charging and discharging processes inside a shell-and-tube type LHTES unit (RT 25 paraffin as PCM and water as the HTF) featuring LF fins was studied by Kirinic et al. [147]. In-house experimental temperature measurements at three axial stations and three radial positions were then compared to the numerical predictions, exhibiting satisfactory agreement between the two approaches. Compared to the plain tube unit, a 52% reduction in the total melting time and a 43% reduction in the total freezing time was realized through adoption of LF. Relative accumulated/released energy, defined as the ratio of accumulated/discharged energy and maximum storage capacity for the plain tube configuration, for the charging/discharging scenarios were 0.949 and 0.948, respectively, whereas the same performance was realized for the no-fin systems after 3 and 12 h, respectively.

The influence of adding twisted fins in a triple-tube LHTES unit compared with using straight fins and no fins was investigated by Ghalambaz et al. [148]. PCM is placed between the inner and outer annuli, through which vertical water HTF streams flow in opposite directions, whereas the copper fins could be attached to the two vertical walls of the PCM container neighboring the HTF streams. Keeping the PCM mass constant, utilization of four twisted fins reduced the melting time by 18% compared with the same number of straight fins, and 25% compared with the no-fins unit. Increasing the number of fins from two to four and six, the heat storage rate rose 14.2% and 25.4%.

Melting of the lauric acid in a rectangular thermal storage unit featuring vertically and horizontally oriented HTF-heated sides with three, five, seven and nine anchored fins (total fin volume fixed) was numerically studied by Safari et al. [149]. With a fixed number of fins, increasing the fin length improves the melting rate due to a higher surface area and boosting of the thermal penetration depth. Examining the instantaneous average velocities (Figure 29) indicates that the horizontal enclosures accommodate development of convection currents until near the end of the melting process, whereas in vertical units, the strength of the convection currents is diminished earlier due to the melt interface shrinkage. The observed decremental trend of the surface-averaged *Nu* number with increasing the number of fins led to the conclusion that heat transfer rate is controlled by the trade-off between the increase in the heat transfer area and the hampering effect of the fins on the *Nu*. Horizontal enclosures store about the same amount of thermal energy at a relatively lower temperature and shorter melting time than vertical enclosures, making them more desired in thermal management applications.

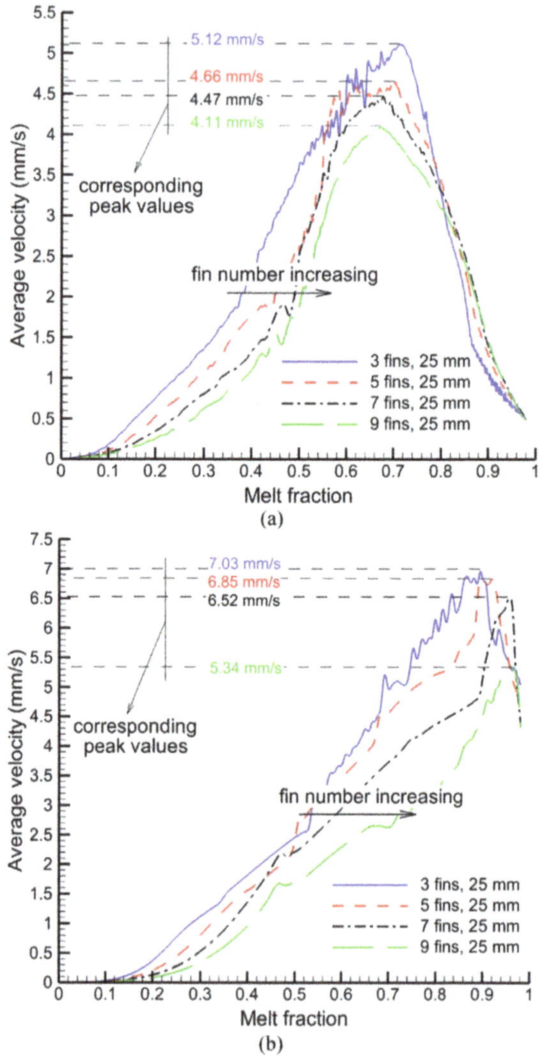

Figure 29. Instantaneous average melt velocities for (**a**) vertically and (**b**) horizontally oriented heated wall exhibiting the persistence of natural convective current late into the melting process for the horizontally oriented unit. Reprinted/adapted with permission from Safari et al. [149]. Copyright 2022, Elsevier.

A 2-D study was performed by Ye and Khodadadi [150] to analyze the enhancement of the melting by adding arrow-shaped fins in a horizontal shell-and-tube LHTES system (Figure 30a). With γ signifying the dimensionless gap distance between the bottom of the heated tube surface and the fin branch, it was observed that increasing the fin angle while using a fixed fin length ratio of $\gamma = 0$ improved the melting rate. Despite varying γ, among the six cases studied, the one with $\theta = 60°$ and $\gamma = 0$ exhibited the highest heat transfer enhancement, compared to the other arrangements, due to the greatest amount of PCM being subjected to convection. In order to highlight the role of unsteady thermal plumes in the top half of the unit and above the branched fin, close-up views of the velocity field colorized by temperature in the vicinity of hot surfaces where the upwelling thermal plumes were initiated for the case of $\theta = 60°$, $\gamma = 0$ are presented in Figure 30b at time instants of 1, 4 and 7 min. At the time instant of 1 min, one pair of vortices that rotate simultaneously in opposite directions were observed near the top surface of the inner heated tube. In effect, hot fluid rose upward, featuring an upwelling thermal plume that impinged on the receding solid PCM. Another pair of vortices was found at the vicinity of the upper tip of the branched fin. At the 4 min time instant, expanded vortices were still rotating in the same pattern as captured at the earlier instant, causing a greater amount of the solid PCM to undergo melting. By the 7 min time instant, the upper pair of vortices have undergone further expansion, whereas the lower pair of vortices shifted closer toward the main vertical fin.

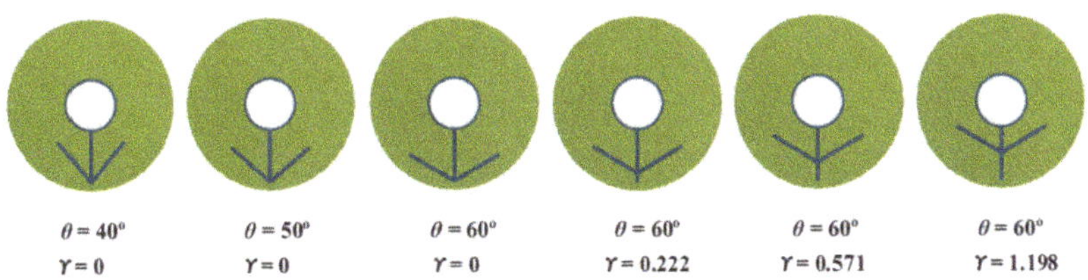

(a)

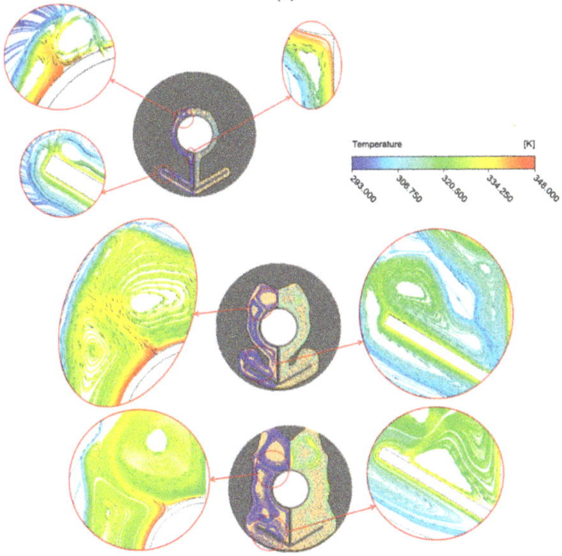

(b)

Figure 30. (**a**) End views of 6 horizontal LHTES units featuring arrow-shape fins and (**b**) velocity field of the molten PCM at time instants of 1, 4 and 7 min for the case of $\theta = 60°$, $\gamma = 0$ colorized by temperature. Graph (**a**) was Reprinted/adapted with permission from Ye and Khodadadi [150]. Copyright 2022, Elsevier.

4. Highlights of Reviewed Work

A summary of the 131 reviewed studies targeted at improving the performance of LHTESS, realized through inserting fins with high values of TC to improve heat transfer, are presented in Table 1. Specifically, the container shape, the imposed boundary conditions, phase-change mode (melting (charging) and/or solidification (discharging)), PCM, fin material, geometry and orientation and the adopted methodologies/techniques are summarized. The importance of geometrical parameters (similar to [15]) and operational factors on the characteristics of phase change conversion in melting and solidification modes are noted. Introducing fins is viewed as a significant geometrical modification to enhance the effective TC of PCM. Adding fins will enhance the thawing and freezing rates, shorten the charging and discharging times, realize uniform and stable operating temperature and assist safe operation of the heat sink. Moreover, design parameters of the fins (number, length, thickness and orientation) influence the performance of LHTESS to different degrees. It was found that the number of fins (or fin-pitch) and fin length have stronger effects on the system performance compared to that caused by fin thickness and fin orientation. On the other hand, insertion of fins will restrain natural convection, which is well-known to play an important role on thawing. Therefore, interacting but opposing influences of enhancement of the effective TC and simultaneous suppression of buoyancy should be decided by the designer through selecting the optimum location and orientation of planned fins.

Table 1. Summary of studies on fin-assisted latent heat thermal energy storage systems in a chronological order (conference papers are listed according to the year of presentation that might be different from the year of publication); Listing of the abbreviations used are summarized at the bottom of the table. Other references involving utilization of fins in LHTESS reported between 1966 and 2015 can be found in [16,17].

Year Authors	Container Shape and BC (Mode of Phase Transition)	PCM (T_m); Thermal Conductivity and Latent Heat	Fin Specifications; Fin Properties	Type of Study
1982				
Ismail et al. [20]	Insulated PCM-filled RU (0.4 × 0.4 × 0.3 m^3); Vertical double tube with silicone HTF (M and mainly S)	Paraffin k = 0.21 W/mK, L = 147 kJ/kg; Sulfur k = 0.17 W/mK, L = 38.5 kJ/kg	Four equally spaced longitudinal copper plate fins fitted at the outer surface of the outer	E (10 ThCs placed on symmetry plane of 2 neighboring fins and HTF stream)
1983				
Ismail and Alves [21]	Staggered array of finned tubes within a latent heat storage unit with the HTF passing inside the tubes (S)	Paraffin (n-eicosane) (55.8–57.2 °C) k = 0.09 W/mK, L = 162 kJ/kg	Four evenly spaced longitudinal fins	E and N (2-D; FVM; enthalpy approach; CON only)
1984				
Ho and Viskanta [22]	RU with air gap above the PCM; hot and cold working fluid circulating through channels in the bottom copper plate (M and S)	N-Octadecane (purity of 99%)	Aluminum vertical walls	E
Ismail and Alves [23]	A staggered array of finned tubes (S)	Paraffin (n-eicosane) (55.8–57.2 °C) k = 0.09 W/mK, L = 162 kJ/kg	Four evenly spaced longitudinal fins	E and N (2-D; FVM; enthalpy approach; CON only)
1985				
Okada [24]	A vertical copper cylinder placed concentrically in a horizontal disk-like PCM (M)	N-Octadecane 99% pure (28.3 °C) k_l = 0.155 W/mK, L = 241 J/g	Copper vertical cylinder placed concentrically in a horizontal disk-like solid PCM	E and N (2-D; FDM)
1986				
Saito et al. [25]	RU heated on the Cu-constantan ThC-installed side with oil jets (M)	Naphthalene	A total of 106 brass vertical plate fins 0.3 mm thick spaced 1.5 mm apart	E and N (2-D; FDM using apparent k and c_p)
Imura and Yoshida [26]	CU with a horizontal cooled tube (S)			N (1-D and 2-D CON)
Imura and Yoshida [27]	CU with a horizontal water-cooled brass tube (M and S)	N-Octadecane (28 °C) k_l = 0.149 W/mK, k_s = 0.42 W/mK L = 241 J/g	Annular brass fins k = 98.9 W/mK	E (ThCs)
Betzel and Beer [28]	Insulated RU (25 × 30 × 4 cm^3) initially held at 0.5 °C below T_m holding a horizontal copper rod carrying HTF (M)	N-Octadecane 98% pure (28.2 °C) Pr = 50	PVC (glued on copper rod) and copper fins (soldered on copper rod); Y-shaped and reversed Y-shaped of three longitudinal fins	E
1987				
Okada [29]	Cylindrical disk initially at fusion temperature (M)	N-Octadecane 99% pure (28.3 °C) k_l = 0.155 W/mK, L = 241 J/g	Copper vertical cylinder placed concentrically in a horizontal disk-like solid PCM	E and N (2-D; FDM)
1989				
Saito et al. [30,31]	RU heated on one side with water (M)	N-*Octadecane* (28.1 °C)	A total of 6 copper plate fins 2 mm thick spaced 10 mm apart with variable angle of inclination	E and N (2-D; FDM); a parabolic velocity was used

Table 1. *Cont.*

Year Authors	Container Shape and BC (Mode of Phase Transition)	*PCM* (T_m); Thermal Conductivity and Latent Heat	Fin Specifications; Fin Properties	Type of Study
Ito et al. [32]	RU with a horizontal fluid-carrying tube (N/A)	*N-Octadecane* (28 °C) k_l = 0.15 W/mK, k_s = 0.42 W/mK L = 241 J/g	Annular fins	E and A (approximate solution)
1990				
Sasaguchi and Sakamoto [33]	Elemental annular cylindrical zone (M)		Two longitudinal fins	N (2-D; FDM)
Sasaguchi [34]	CU with IT for HTF (M and S)		Annular and longitudinal fins	N (2-D)
Sasaguchi [35]	Annular cylindrical zone heated by impinging jets on the tube wall; varied orientation and angle between the two fins (M)	N-eicosane (36 °C) k_l = 0.15 W/mK, k_s = 0.3 W/mK L = 245 J/g	Two longitudinal fins	E (ThCs)
1992–1993				
Kaino [36–39]	CUs with IT for HTF (S)		Longitudinal fins	N (2-D; FDM; CON only)
1994				
Sasaguchi and Takeo [40,41]	RU with one wall held at CWT (M and S)		Two planar parallel fins	N (2-D; FDM; volume-averaged media)
Al-Jandal and Sayigh [42]	Vertical CU with a vertical water- cooled end-closed tube and HTF jacket outer shell (M)	Stearic acid ($C_{18}H_{38}O_2$) (70 °C)	Annular (13) and longitudinal (3) fins	E
1995				
Choi and Kim [43]	Vertical CU with a vertical water- cooled end-closed tube (M)	Magnesium chloride hexahydrate ($MgCl_2 \cdot H_2O \equiv$ MCHH) (116.7 °C)	Annular fins; stainless steel	E (12 ThCs within PCM)
Choi et al. [44]	Vertical CU with a vertical water-heated/cooled end-closed tube (M)	Sodium acetate trihydrate ($CH_3COONa{\cdot}3H_2O$) (60 °C) k_s = 0.6 W/mK, L = 226 kJ/kg	Twelve equally spaced annular fins made of stainless steel	E (T-type ThCs on HTF tube)
1996				
Horbaniuc et al. [45]	Sectional test cells split by longitudinal fins along the perimeter of hp; constant hp wall and fin base temperature (S)	Paraffin (60 °C) k_f = 0.25 W/mK, L = 180 kJ/kg	Longitudinal plate fins (6 and 12) on the outer surface of hp	A (exponential and polynomial approximation; CON only)
Choi et al. [46]	Vertical CU with a vertical water-cooled end-closed tube (S)	Sodium acetate trihydrate ($CH_3COONa{\cdot}3H_2O$) (60 °C) k_s = 0.6 W/mK, L = 226 kJ/kg	Twelve equally spaced annular fins made of stainless steel	E (ThCs on HTF tube) and A (unsteady-state approximation and quasi-stationary solution)
Han and Han [47]	Vertical CU with a vertical water-cooled end-closed tube (S)	$Na_2SO_4/10H_2O$ (32.4 °C) k_s = 0.54 W/mK, L = 241 kJ/kg	Twelve equally spaced annular fins made of stainless steel	E
1997				
Hirasawa et al. [48]	RU with a vertical HT plate consisting of pin fins (M and S)	Water	Fins with 25, 49, 81 and 121 pins, with square CSs, placed horizontally in a square pattern	E
Chen et al. [49]	RU with a vertical HT plate consisting of pin fins (M and S)	Water	Fins with 25, 49, 81 and 121 pins, with square CSs, placed horizontally in a square pattern	N (3-D; FDM under quasi steady-state condition)

Table 1. *Cont.*

Year Authors	Container Shape and BC (Mode of Phase Transition)	*PCM* (T_m); Thermal Conductivity and Latent Heat	Fin Specifications; Fin Properties	Type of Study
1999				
Wirtz et al. [50]	RU with vertical fins and heat source at base (M)	Waxy granulate of 0.1–1 mm (81 °C) L = 212 J/cc	Vertical aluminum plate-fins of a commercial heat sink	E and N (thermal circuits)
2000				
Hong and Kim [51]	RU with a vertical heating copper plate activated by impinging jets of brine HTF (M)	Ice	Split fin system of ten 44×4.4×2 mm^3 sheets separated 1 mm apart placed horizontally on a heating plate	E
2001				
Inaba et al. [52]	RU with one copper heating wall (M)	Paraffin wax (42 °C) k = 0.25 W/mK, L = 128.4 J/g	Plate fins	E (ThCs)
2004				
Yamashita et al. [53]	RU of 24 horizontal tubes with natural gas flowing through (M)	N-*Pentane* (−129.6 °C) L = 116.8 J/g	Twelve equally spaced longitudinal fins (2, 2.5 and 3.5 mm in thickness)	E
Huang et al. [54]	RU with the aluminum front side exposed to insolation (M and S)	RT25 (E) (26.6 °C) k_l = 0.18 W/mK, k_s = 0.2 W/mK L = 232 kJ/kg Paraffin wax (N) (32 °C) k_l = 0.22 W/mK, k_s = 0.5 W/mK L = 251 kJ/kg	Aluminum horizontal plate fins	E (51 ThCs within PCM) and N (2-D; FVM)
2005				
Yamashita et al. [55]	RU of 24 horizontal tubes with natural gas flowing through (S)	N-*Pentane* (−129.6 °C) L = 116.8 J/g	Twelve equally spaced longitudinal fins (2, 2.5 and 3.5 mm in thickness)	E
Yamashita et al. [56]	RU of 24 horizontal tubes with natural gas flowing through (M and S)	N-Pentane (−129.6 °C) L = 116.8 J/g	Twelve equally spaced longitudinal fins (2, 2.5 and 3.5 mm in thickness)	N (2-D; FDM); equivalent thick-wall cylinder model
Liu et al. [57]	Vertical CU with a PCM-filled concentric annulus surrounding an electrically heated rod and HTF jacket outer shell (S)	Stearic acid of analytical purity (67–70 °C) L = 224.3 kJ/kg	Spiral twisted split annular copper fins spanning the whole annular gap (widths 0.25 and 0.5 cm)	E (9 ThCs placed in PCM, PCM container and HTF stream); DSC
2006				
Kayansayan and Acar [58]	RU (50×570×42 cm^3) filled with PCM; horizontal 99% pure ethyl-alcohol-carrying 1-piece finned-tube (S)	Distilled water	Annular fins made of solid bronze (87.2 Cu, 6.57 Sn, 4.13 Zn and 1.97% Pb) with axial density of 14, 23, and 31 fins/m)	E (T-type ThCs at fin tips, base and HTF stream); N (2-D; FVM; CON only)
Huang et al. [59]	RU cell with the aluminum front side exposed to insolation (M)	RT25 (26.6 °C) k_l = 0.18 W/mK, k_s = 0.2 W/mK L = 232 kJ/kg	Aluminum square CS pin fins evenly spaced (five)	N (2-D and semi-implicit 3-D finite volume model)
2007				
Wang et al. [60]	RU heated from one surface under CWT or CHF conditions (M)	Paraffin wax (46–48 °C) k_l = 0.12 W/mK, k_s = 0.2 W/mK L = 173 kJ/kg	Aluminum plate fins (five) k = 202.4 W/mK	N (2-D; VOF; EPM)

Table 1. *Cont.*

Year Authors	Container Shape and BC (Mode of Phase Transition)	*PCM* (T_m); Thermal Conductivity and Latent Heat	Fin Specifications; Fin Properties	Type of Study
2008				
Seeniraj and Narasimhan [61]	CU with concentric horizontal sodium salt HTF-carrying tube (M)	PCM-1 (1040 K) LiF-CaF_2 (80–20 wt%) k_s = 0.06W/mK, k_l = 0.45 W/mK PCM-2 (1008 K) LiF–MgF_2 (eutectic mixture) k_s = 0.06W/mK, k_l = 1 W/mK PCM-3 to 5 (973, 923, 873 K) LiF–MgF_2 (eutectic mixture)	Four annular fins of uniform thickness (0.25 × tube diameter) with various sealed PCM in between fins k =0.58 W/mK	N (2-D; EPM)
Saha et al. [62]	RUI with four insulated walls and bottom wall attached to a constant power heater; NC cooling from the top surface (M)	Eicosane (35 °C) k = 0.23 W/mK, L = 241 kJ/kg	Fins with 9, 36, 81, and 121 pins (2 × 2 × 25 mm^3); 3 plate fins (1.14 × 42 × 25 mm^3) and 9 pin fins (4 × 4 × 25 mm^3); aluminum k = 180 W/mK	E and N (EPM; FVM)
Wang et al. [63]	RU heated from one surface under CWT or CHF conditions; top to unit exposed to ambient air (M)	Paraffin wax (46–48 °C) k = 0.21/0.12 W/mK, L = 173 kJ/kg Eicosane (36 °C) k = 0.15 W/mK, L = 247 kJ/kg Heneicosane (40.4 °C) k = 0.15 W/mK, L = 213 kJ/kg Suntech P116 (47 °C) k = 0.24 W/mK, L = 266 kJ/kg	Aluminum planar fins with PCM in between k = 202.4 W/mK	N (2-D; EPM; FVM; VOF)
Kandasamy et al. [64]	RU heated from bottom CWT surface (M)	Paraffin wax (53–57 °C) k_l = 0.12 W/mK, k_s = 0.2 W/mK L = 173 kJ/kg.	Aluminum fins with three arrangements (6 small plate fins, 10 large plate fins, 3 × 11 row of elliptical fins) k = 202.4 W/mK	E and N (3-D; VOF)
Agyenim et al. [65]	Horizontal CU with HTF-carrying copper tube (M and S)	Erythritol (117.7 °C)	Eight annular, eight longitudinal copper fins; four multi-tube configurations	E
2010				
Lee and Chun [66]	RUs placed on solar panels (M and S)	N/A (44 °C)	Profiled aluminum fins (inward or outward); aluminum honeycomb on back of panel	E
Agyenim and Hewitt [67]	Horizontal CU with HTF-carrying copper tube (M and S)	Paraffin (RT58)	Longitudinal copper fins (eight)	E
Fok et al. [68]	RU with plate heater attached below (M and S)	N-eicosane (36 °C) k = 0.15 W/mK, L = 2460 J/kg	Plate fins were fabricated with aluminum k = 202.4 W/mK	E (6 K-type ThCs on perimeter of the heat sink)
Saha and Dutta [69]	RU of planar vertical fins surrounded by PCM with CHF (M)	N-eicosane (35 °C) k = 0.23 W/mK, L = 241 kJ/kg	Aluminum planar vertical fins; k = 180 W/m K	N (2-D; EPM; FEM)
Wei et al. [70]	Insulated RU PCM-filled shell with a staggered cluster of parallel HTF-carrying annularly finned tubes (M and S)	Paraffin (42–47 °C) k = 0.27 W/mK, L = 141 kJ/kg	Annular fins	E (17 ThCs in PCM and HTF stream)

Table 1. *Cont.*

Year Authors	Container Shape and BC (Mode of Phase Transition)	*PCM* (T_m); Thermal Conductivity and Latent Heat	Fin Specifications; Fin Properties	Type of Study
2011				
Sugawara et al. [71]	RU with circular copper foil disks around a 50 wt% propylene-glycol aqueous solution-carrying tube (M and S)	Water (0 °C)	166 annular copper foil disks of 94 mm outer diameter, 19.1 mm inner diameter and 0.03 mm thickness. k = 180 W/m K	E and N (2-D; Darcy law for flow resistance; effective TC)
Agyenim et al. [72]	Horizontal CU with HTF-carrying (hot silicone oil/cold water) copper tube (M and S)	Erythritol (117.7 °C) k_l = 0.33 W/mK, k_s = 0.7 W/mK L = 340 kJ/kg (observed to vary upon thermal cycling)	Eight evenly spaced longitudinal fins, 1 mm thick	E (41 ThCs in PCM)
Talati et al. [73]	PCM-filled RU divided uniformly with plate metal fins; CHF on the vertical wall (S)	Salt hydrate ClimSel C28 (28 °C) k = 0. 6 W/mK, L = 162 kJ/kg.	Parallel plate aluminum horizontal fins k = 177 W/mK (0.01, 0.015, 0.02 m long)	A (1-D analytical) and N (2-D; EPM)
Huang et al. [74]	RU (132.4×40×300 mm^3) with the aluminum front side exposed to insolation (M and S)	RT27 (25–28 °C) k = 0.2 W/mK, L = 184 kJ/kg RT35 (35 °C) k = 0.2 W/mK, L = 157 kJ/kg *Waksol A* (32–36 °C) k_l = 0.33 W/mK, k_s = 0.31 W/mK L = 162 kJ/kg	Steel plate fins (4, 8, 12 and 24 mm spacings, fin thickness of 0.5 mm)	E (7 T-type ThCs); solar insolation measured using a Kipp and Zonen class 2 pyranometer
Robak et al. [75]	CU with the bottom base accommodating HTF stream of distilled water; HTF flow rate set using an Omega FLV-4605A 0–2 LPM calibrated flow controller (M and S)	Paraffin, n-Octadecane 99% purity ($C_{18}H_{38}$) (27.5 °C) k_l = 0.15 W/mK, k_s = 0.36 W/mK L = 244 kJ/kg	Five 316 stainless steel vertical rods as fins were placed in five 13-mm diameter threaded holes; Five vertical 175-mm long and 6-mm outer diameter copper-water hps	E (4 Teflon-coated, 254 µm diameter chromel–alumel (K-type) ThCs at the inlet and outlet of the heat exchanger); Temperature regulated by a RM5 Lauda bath to within an accuracy of ±0.1 °C
Bauer [76]	RU CWT wall; CWT tube with annular fins (S)	N-Octadecane k = 0.5 W/mK, L = 100 kJ/kg	Plate fins in rectangular cells; Annular fins on tubes k = 150 W/m K	A and N (1-D in the fin and 2-D in PCM; EPM)
Ismail and Lino [77]	RU (700×500×500 mm^3) with a horizontal HTF-carrying (ethanol) copper tube (S)	Water	Annular fins (5) of 1 mm thickness (outer diameters 40, 60, 120 and 180 mm)	E
Hosseinizadeh et al. [78]	RU with CHF base and top adiabatic (M)	RT80 (81 °C) k = 0.2 W/mK, L = 175 kJ/kg	Vertical aluminum planar fins (3, 5, 7), thickness (2, 4 and 6 mm), height (10, 20, 40 mm) k = 180 W/mK	E (16 ThCs) and N (2-D/3-D; EPM)
Ye et al. [79]	RU with CWT on the bottom surface by circulating water through the inner plates (M and S)	Paraffin (33–35 °C) k_l = 0.15 W/mK, k_s = 0.2 W/mK L = 176 kJ/kg	Aluminum plate fins separating the system into uniform cavities k = 202 W/mK	N (EPM; VOF)
Long [80]	Horizontal CU with HTF (water)-carrying tube (M and S)	Composite of paraffin and a nano-structure (aluminum) (326–330 K) k_l = 0.3 W/mK, k_s = 0.35 W/mK L = 165.1 kJ/kg	Plane annular aluminum fins, 0.15 mm thick (6, 9 and 12 fin/inch)	N (2-D; CON only)
Tan et al. [81]	RU 300 × 300 × 25 mm^3) with CHF bottom and sides (M)	*Paraffin wax* (43–49 °C) k = 0.2 W/mK, L = 140 kJ/kg	Aluminum planar fins with straight, T-, Y- and cross shapes k = 202.4 W/mK	N (2-D)

Table 1. *Cont.*

Year Authors	Container Shape and BC (Mode of Phase Transition)	*PCM* (T_m); Thermal Conductivity and Latent Heat	Fin Specifications; Fin Properties	Type of Study
Murray et al. [82]	CU with HTF (water)-carrying copper tube (M and S)	Lauric acid (42 °C) k_l = 0.148 W/mK, k_s = 0.15 W/mK L = 182 kJ/kg	Five uniformly spaced annular copper fins	E (8 T-type ThCs in PCM and HTF stream); N (2-D; CON only)
Saha and Dutta [83]	RU with CHF base (M)	Eicosane (35 °C) k = 0.23 W/mK, L = 241 kJ/kg	Plane vertical fins (3 and 4), 0.5 and 2.9 mm thick k = 180 W/mK	N (2-D, EPM) coupled with a genetic algorithm
2012				
Chiu and Martin [84]	Vertical CU with HTF flowing downward through the IT (M and S)	Paraffin A Gelled salt-hydrate (12–14 °C) k = 0.6 W/mK, L = 140 kJ/kg	Annular fins; Aluminum Alloy 6082 (AL 4212)	E and N (2-D; CON only; fixed-grid EPM, explicit FDM)
Ogoh and Groulx [85]	CU with hot water-carrying copper pipe (M)	Paraffin wax	Uniformly distributed annular copper fins (maximum of 27)	N (2-D FEM; CON only; effective heat capacity method)
Baby and Balaji [86]	Rectangular heat sink (80 × 62 × 25 (height) mm^3); plate heater (ranging from 2–7 W) at bottom (M)	N-eicosane (36.5 °C) k = 0.15 W/mK, L = 238 kJ/kg	72 aluminum pin fins (2 × 2 × 20 mm^3) 3 aluminum plate fins (48 × 20 mm^2, 2 mm thick) k = 202.4 W/mK	E (Modulated DSC)
Mosaffa et al. [87]	RU subject to convective boundaries (S)	Paraffin (32 °C) k_l = 0.22 W/mK, k_s = 0.5 W/mK L = 251 Jk/kg	Plate aluminum fins; k = 177 W/m K	A (1-D) and N (2-D, EPM), CON only.
Jourabian et al. [88]	Square cavity filled initially at uniform T_m; one active wall kept at a higher temperature (M)	N/A	Plate horizontal fin on the active wall	N (2-D; Boussinesq approximation; lattice Boltzmann method, EPM)
Mosaffa et al. [89]	CU with IT HTF (air) (S)	$CaCl_2 \cdot 6H_2O$ (29.7 °C) k = 1.1 W/mK, L= 187.5 kJ/kg	Annular aluminum fins; k= 177 W/mK	A (2-D; EPM, CON only).
Xu et al. [90]	RU (1500 × 800 × 29 mm^3); Two layers of square plate-fin unit, with upper part filled with PCM and the bottom part with HTF (water) (M and S)	Paraffin wax RT80 (353 K) k = 0.2 W/mK, L = 175 kJ/kg	Staggered/serrated aluminum plate fins stacked in periodically formed passageway k = 202.4 W/mK	E and N (3-D; CON only.
Shokouhmand and Kamkari [91]	CU with HTF-carrying horizontal pipe (M)	Paraffin wax (46–48 °C) k_l = 0.12 W/mK, k_s = 0.2 W/mK L = 173 kJ/kg	Two and four evenly spaced longitudinal aluminum fins	N (2-D) EPM and Boussinesq approximation
Hamdani and Mahlia [92]	PCM-filled CU with HTF (hot water) flowing downward in the IT (M)	Pure paraffin wax (56–60.1 °C)	Longitudinal and annular copper fins	E
Tan et al. [93]	Rectangular heat sink heated from bottom (M)	Paraffin wax (47 °C) k = 0.2 W/mK, L = 140 kJ/kg	Aluminum vertical plate fins and spiral fillers	N (2-D) EPM
2013				
Levin et al. [94]	RU; uniform power density inactive wall (M and S)	Sodium hydrate (45–52.5 °C) k = 0.6 W/mK, L = 113 kJ/kg	Plate aluminum fins	N (2-D; FEM; CON only)
Tay et al. [95]	Vertical CU with a concentric HTF-carrying copper tube (S)	Water	Sixteen copper radially pointing pins Five uniformly spaced copper annular fins	N (3-D; CON only)

Table 1. *Cont.*

Year Authors	Container Shape and BC (Mode of Phase Transition)	*PCM* (T_m); Thermal Conductivity and Latent Heat	Fin Specifications; Fin Properties	Type of Study
Kurina et al. [96]	RU with HTF-carrying U-tube, U-tube with in-line fins, U-tube with staggered fins and a novel festoon design (M and S)	Paraffin wax (36–38 °C) k_l = 0.12 W/mK, k_s = 0.2 W/mK L = 173.4 kJ/kg Two similar PCM with T_m higher and lower (26–28 °C) and (46–48 °C)	Horizontal fins (in-line and staggered) attached to vertical legs of the U-tube design HTF-carrying channel	N (2-D) EPM
Baby and Balaji [97]	Rectangular heat sink (80 × 62 × 25 mm^3); plate heater (ranging from 2–7 W) at bottom (M and S)	N-eicosane (36.5 °C) k_l = 0.16 W/mK, k_s = 0.4 W/mK L = 237.4 kJ/kg Paraffin wax (53–57 °C) k_l = 0.12 W/mK, k_s = 0.2 W/mK L = 173.4 kJ/kg	Aluminum pin fins (33, 72, 120) attached to the bottom have 2 × 2 mm^2 CS and 20 mm length k = 202.4 W/mK	E (Artificial Neutral Network-Genetic Algorithm hybrid optimization technique)
Mahmoud et al. [98]	Rectangular heat sink (50 × 50 × 25 mm^3); plate heater (ranging from 2–7 W) at bottom (M and S)	HS29P (29 °C) k_l = 0.54 W/mK, k_s = 1.1 W/mK L = 190 kJ/kg; HS34P (34 °C) k_l = 0.47 W/mK, k_s = 0.5–0.6 W/mK L = 150 kJ/kg; *OM37P* (37 °C), L = 218 kJ/kg; OM46P (46 °C), L = 245 kJ/kg; HS58P (58 °C), L = 250 kJ/kg; *RT42* (42 °C), k = 0.2 W/mK, L = 174 kJ/kg	Uniformly spaced parallel fins (three, six cavities) Cross fins (9, 36 cavities), 2 mm thick Hexagonal honeycomb made of extra-hard 3003 alloy aluminum foil of 60 µm thickness	E
Mat et al. [99]	Horizontal TTHX with PCM in annular tube sandwiched with water streams in outer and inner tubes (M)	RT82 (350–358 K) k = 0.2 W/mK, L = 176 kJ/kg	Copper longitudinal fins attached staggeredly on the inner and outer PCM-wetted walls of the annular tube k = 387.6 W/mK	E, N (2-D; FVM; EPM; Boussinesq approximation)
Guelpa et al. [100]	Vertical CU with a concentric HTF-carrying copper tube (S)	RT55 (51–56 °C) k = 0.2 W/mK, L = 249 kJ/kg	Annular fins (12, 27) distributed uniformly and non-uniformly	N (2-D), EPM
Campos-Celador et al. [101]	12 finned flat rectangular PCM-filled plates; 6 arranged in parallel and 2 in series; water flowing through the channel between the plates (M and S)	RT60 paraffin (53–61 °C) k = 0.2 W/mK, L = 123.5 kJ/kg	Uniformly spaced plate aluminum fins	A, E, N (NC treated using an effective TC in 2-D)
Solomon and Velraj [102]	Vertical CU with cold air flowing upward through the copper IT (S)	RT21 Paraffin (18–23 °C) k = 0.2 W/mK, L = 134 kJ/kg	Eight evenly spaced longitudinal copper fins having thickness of 3 mm with radii of 10,15, 20 and 26 mm	E
Mosaffa et al. [103]	Vertical CU with CWT or CHF IT HTF (S)	Salt hydrate ClimSel *C23* (23 °C) k = 0.6 W/mK, L = 148 kJ/kg	Aluminum annular fins of 1 mm thickness	A (1-D analytical; 2-D; EPM; CON only)
Kozak et al. [104]	RUs filled fully with PCM and exposed to a CHF base at bottom (M and S)	Eicosane (36.7 °C) k = 0.15 W/mK, L = 248 kJ/kg	Vertical aluminum 6061 plate fins	E and N (2-D) EPM
2014				
Kim et al. [105]	RU with varying temperature active wall (M and S)	N-Pentadecane (9.7 °C) k = 0.513 W/mK, L = 230 kJ/kg N-Hexadecane (18 °C) k = 0.15 W/mK, L = 237 kJ/kg	Horizontal aluminum 6010 plate fins	E and N (2-D)

Table 1. *Cont.*

Year Authors	Container Shape and BC (Mode of Phase Transition)	PCM (T_m); Thermal Conductivity and Latent Heat	Fin Specifications; Fin Properties	Type of Study
Chen et al. [106]	Vertical CU with HTF (water) flowing upward inside the concentric copper IT (M and S)	*S15* (EPS, www.epsltd.co.uk (accessed on 29 September 2022)) (15 °C) $k = 0.43$ W/mK, $L = 142$ kJ/kg	Copper annular plate fins (nine) equally spaced along the axis	E
Shon et al. [107]	Insulated PCM-filled RU enclosing an automotive heat core fin-tube HX with louver fins (M and S)	Xylitol (93–94.5 °C) $L = 265$ kJ/kg	Corrugated fin–tube-type heat core made of AL1100 aluminum alloy; U-shape copper fins within HX	E
Murray and Groulx [108]	Vertical CU with hot and cold HTF copper ITs flowing axially downward simultaneously (M and S)	Dodecanoic acid, 98% pure (42.5 °C) $k_l = 0.148$ W/mK, $k_s = 0.150$ W/mK $L = 182$ kJ/kg	Evenly spaced four longitudinal copper fins attached to each of the HTF tubes	E
Al-Abidi et al. [109]	Horizontal TTHX with PCM in annular tube sandwiched with water streams (HTF) in the outer and inner tubes (M and S)	RT82 (82 °C)	Copper longitudinal fins attached staggeredly on the inner and outer PCM-wetted walls of the annular tube $k = 387.6$ W/mK	E
Liu and Groulx [110]	Horizontal cylindrical LHTES unit with HTF (water) circulating through the IT (M and S)	Dodecanoic acid (lauric acid) (44 °C) $k_l = 0.148$ W/mK, $k_s = 0.150$ W/mK $L = 182$ kJ/kg	Four evenly spaced longitudinal copper fins in + and × arrangements	E
Tay et al. [111]	Vertical cylindrical tube-in-tank industrial LHTES unit; Model shell-and-tube system for experiments (M and S)	Water $k_l = 0.6$ W/mK, $k_s = 2.2$ W/mK Salt hydrate (−11 °C) $k_l = 0.56$ W/mK, $k_s = 2.2$ W/mK RT35 (E) $k = 0.2$ W/mK, $L = 157$ kJ/kg	Evenly spaced annular fins (500 with 92 mm diameter for N) (14 with 53 mm diameter and 1.5 mm thick for E)	E and N (3-D; CON only; 2-D EPM)
Hasan et al. [112]	Aluminum RU PCM-filled heat sink integrated to the back of the PV panels (M)	Capric-palmitic acid (22.5 °C) $k = 0.14$ W/mK, $L = 173$ kJ/kg $CaCl_2{\cdot}6H_20$ (Salt hydrate) (29.8 °C) $k_l = 0.56$ W/mK, $k_s = 1.08$ W/mK $L = 191$ kJ/kg	Vertical aluminum plate fins fitted internally on the heat sink	E (T-type ThCs)
Khalifa et al. [113]	RU shell with HTF-carrying (water for E, therminol® for N) horizontal tubes; HTF tubes were penetrated with four hps repeated axially (S)	RT82 (82–83 °C) $k = 0.2$ W/mK, $L = 176$ kJ/kg (experiment); Potassium nitrate (335 °C) $k_l = 0.425$ W/mK, $k_s = 0.5$ W/mK $L = 95$ kJ/kg	Four evenly spaced longitudinal fins (A1100 aluminum) attached to the hps $k = 220$ W/mK	E (T-type ThCs) and N (CON only; FVM based on the effective heat capacity method; the thermal resistances)
Rahimi et al. [114]	RU shell with HTF-carrying (water) vertical copper tubes connected by U-tubes (M and S)	RT35 (302–308 K) $k = 0.2$ W/mK, $L = 170$ kJ/kg	Horizontal planar aluminum copper fins separated 5 mm apart	E (flow rates of HTF were 0.2, 0.4, 0.6 and 1.6 L/min)
Rahimi et al. [115]	RU shell with HTF (water)-carrying vertical copper tubes connected by U-tubes (M and S)	RT35 (302–308 K) $k = 0.2$ W/mK, $L = 170$ kJ/kg	Horizontal planar aluminum copper fins separated 5, 10 and 15 mm apart	E
Kozak et al. [116]	Vertical CU with HTF-carrying IT (M)	Eicosane (36.7 °C) $k_l = 0.15$ W/mK, $k_s = 0.4$ W/mK $L = 248$ kJ/kg	Annular aluminum fins	A, E and N (2-D; EPM)

Table 1. *Cont.*

Year Authors	Container Shape and BC (Mode of Phase Transition)	*PCM* (T_m); Thermal Conductivity and Latent Heat	Fin Specifications; Fin Properties	Type of Study
2015				
Rozenfeld et al. [117]	CU (immersed in air or heated water) with a horizontal water-heated one-piece finned aluminum tube (M)	Commercial-grade Eicosane (96% $C_{20}H_{42}$) (36.7 °C) k_l = 0.15 W/mK, k_s = 0.4 W/mK L = 248 kJ/kg	Y-shape (three) and *-shape (six) planar longitudinal fins made of aluminum 7075	A, E and N
Rathod and Banerjee [118]	CU with a vertical water-heated/cooled brass tube (M and S)	Stearic acid (fatty acid) (57.5 °C)	Three equally spaced brass longitudinal fins (0.003 m thick)	E
Paria et al. [119]	CU with a horizontal water (heated and cooled) copper tube at constant temperature (M and S)	Paraffin ASTM D 87 (53–57 °C) k_l = 0.35 W/mK, k_s = 0.17 W/mK L = 160 kJ/kg	Uniform-spaced annular copper fins k = 52 W/mK	E and N (FVM)
Zhai et al. [120]	CU with a vertical water-cooled copper tube initially cooled to 12 °C (M)	In-house C–L acid (capric and lauric acids) with oleic acid as the additive (15–16.9 °C) k_l = 0.42 W/mK, k_s = 0.57 W/mK L = 115.1 kJ/kg	Uniformly spaced four annular and four longitudinal copper fins divided the annular space into 20 sealed segments	E (five temperature sensors at four angular portions at different radial distances) and N (3-D; CON only)
Khalifa et al. [121]	Rectangular shell with HTF-carrying (water for E, therminol® for N) horizontal channels; hs were fixed horizontally to HTF channels (S)	LiCl-KCl (348 °C) k_l = 0.42 W/mK, k_s = 0.48 W/mK L = 170 kJ/kg RT60 (E) (60 °C) k = 0.2 W/mK, L = 140 kJ/kg	Rectangular vertical anodized aluminum plate fins k = 50 W/mK	A (thermal resistance with effective thermophysical properties for PCM-fin) and E
Kalbasi and Salimpour [122]	RUs (CHF bottom wall, insulated on other sides) (M)	RT27 (28–30 °C) k_l = 0.15 W/mK, k_s = 0.24 W/mK L = 175 kJ/kg	Vertical aluminum alloy 6061 fins attached to active surface k = 180 W/mK	A (constructal theory) and N (2-D; VOF; EPM)
Tao and He [123]	Horizontal cylinder storage unit with He/Xe-carrying HTF tube (M)	LiF (80.5%) and CaF_2 (19.5%) (1040 K) k = 3.8 W/mK, L = 816 kJ/kg	Longitudinal evenly spaced fins (three, five and seven) of varying radial length and thickness	N (3-D; FVM)
Liu and Li [124]	RUs heated by solar radiation flux on one side adjacent to vertical air channels and other sides insulated (M and S)	RT42 (38–43 °C) k = 0.2 W/mK, L = 174 kJ/kg	Uniformly spaced stainless steel horizontal plate fins (30) along the air channels (1 mm thick and 50 mm spacing)	E and N (1-D; effective heat capacity formulation; approximate thermos-physical property method for PCM/fin)
Talati and Taghilou [125]	RU with CWT or CHF active wall; Composite wall with PCM filled container (S)	Pure paraffin (25 °C) k = 0.185 W/mK, L = 124 kJ/kg	Aluminum and copper plate fins separating PCM into different cells k_{Al} = 177 W/mK; k_{Cu} = 401 W/mK	N (1-D, 2-D; Implicit lattice Boltzmann, analytical, and FVM)
Jmal and Baccar [126]	CU of coaxial vertical TTHX with top-end of the outer tube closed and one air stream forming the two HTF passageways (S)	Paraffin C_{18} (28–28.5 °C) k_l = 0.148 W/mK, k_s = 0.15 W/mK L = 244 kJ/kg	Annular aluminum plate fins (3, 5, 9) of 3 mm thickness staggered on the inner and outer PCM-wetted walls of the annular tube; k = 204 W/mK	N (2-D; FVM; enthalpy method)
Hosseini et al. [127]	CU with a horizontal laminar water-heated/cooled concentric IT (M and S)	RT50 (318–324 °C) k = 0.2 W/mk, L = 168 kJ/kg	8 evenly spaced annular fins (lengths 13 and 26 mm)	E and N (2-D; EPM)

Table 1. *Cont.*

Year Authors	Container Shape and BC (Mode of Phase Transition)	PCM (T_m); Thermal Conductivity and Latent Heat	Fin Specifications; Fin Properties	Type of Study
Gharbi et al. [128]	RUs with CHF copper slabs; unit tested in horizontal and vertical positions (M and S)	Plastic paraffin (47.84–51.75 °C) k_l = 0.16 W/mK, k_s = 0.23 W/mK L = 138 kJ/kg; PCM/silicon matrix (37 °C) k = 1 W/mK, L = 88 kJ/kg; PCM/graphite matrix (37 °C) k = 1 W/mK, L = 188 kJ/kg	Copper plate fins placed uniformly along the copper slabs (14 1 mm thick and 9 mm long fins; 6 1 mm thick and 21 mm long fins) k = 401 W/mK	E
Li and Wu [129]	CU with a horizontal concentric HTF tube (synthetic oil and molten salt) (M and S)	$NaNO_3$ k = 0.5 W/mK, L = 173 kJ/kg $NaNO_3$/expanded graphite (90/10 wt%) k = 10 W/mK, L= 156 kJ/kg	Longitudinal plate fins	N (3-D; EPM)
Feng et al. [130]	RU with CWT base (M)	Paraffin wax (Docosane) (317 K) k= 0.4 W/mK, L= 260 kJ/kg	Equally spaced aluminum plate fins attached to the base; high-porosity open-cell aluminum foam in between fins k = 202.4 W/mK	N (2-D volume-averaged simulations and 3-D pore-scale simulation assuming thermal equilibrium)
Pakrouh et al. [131]	RU with a CHF base consisting of vertical pin fins (M)	RT44HC (41–45 °C) k = 0.2 W/mK, L = 255 kJ/kg;	Aluminum pin fins of square CS (25, 49 and 100) k =202.4 W/mK	N (3-D; FVM, EPM, VOF to handle the air-PCM system; Taguchi algorithm for 3-level parameters)
Tiari and Qiu [132]	CU with CHF base on which evaporator-shared vertical hps (five, nine and nine identified as Cases 1–3) are placed (M)	Eutectic mixture of $NaNO_3$ and KNO_3 (60:40% molar ratio) (493 K) k = 0.8 W/mK, L= 109 kJ/kg	Annular nickel fins (10) k = 92 W/mK	N (3-D; FVM, Boussinesq approximation EPM)
Kalbasi and Salimpour [133]	RUs with CHF base initially at 27 °C (M)	RT27 (28–30 °C) k_l = 0.15 W/mK, k_s = 0.24 W/mK L = 175 kJ/kg	Vertical plate aluminum alloy 6061 fins attached to the base of enclosures; horizontal aluminum plate fins (second construct level) attached to vertical fins k = 180 W/mK	N (2-D; constructal theory; VOF to handle the air-PCM system; EPM)
Samanta et al. [134]	CU with a vertical water- cooled copper tube (S)	$CaCl_2{\cdot}6H_2O$ (29.7–29.9 °C) k_l = 0.54 W/mK, k_s = 1.1 W/mK L = 187.4 kJ/kg	Annular copper fins k = 386 W/mK	E and N (2-D; EPM; FVM)
Malan et al. [135]	RU with a multi-channel micro hp on heating side and identical hp on heat removing side (M and S)	Paraffin wax (E) KCl/KF, 45/55% on a molar basis (N)	Equally spaced vertical aluminum plate fins with zigzag-shaped fins between those plates	E and N (2-D; CON only)
Khatra et al. [136]	RU with CWT surface and three walls insulated (S)	N-Octadecane (301.16 K) k = 0.378 W/mK, L = 243 kJ/kg	Aluminum plate fins (three) originating from the active surface	N (2-D; FVM; EPM)
2016–2022				
Darzi et al. [137]	Horizontal shell-and-tube CU; inner tube CWT and outer adiabatic wall (M and S)	N-eicosane (36 (35–37) °C) k_l = 0.15 W/mK, k_s = 0.24 W/mK L = 247.6 kJ/kg	Uniformly spaced plate longitudinal fins (4–20), 1.5 cm long and 0.1 cm thick	N (2-D; FVM; EPM)
Sheikholeslami et al. [138]	RU shell-and-tube; CWT at the inner tube and insulated shell (S)	Water (273 K) k_l = 0.60 W/mK, k_s = 2.24 W/mK L = 335 kJ/kg	Snowflake-shaped longitudinal aluminum fins placed on HTF tube	N (2-D; CON only; Standard Galerkin FEM)

Table 1. *Cont.*

Year Authors	Container Shape and BC (Mode of Phase Transition)	PCM (T_m); Thermal Conductivity and Latent Heat	Fin Specifications; Fin Properties	Type of Study
Rozenfeld et al. (2017) [139]	CU shell-and-tube unit; CWT on the inner HTF tube and insulated shell (M)	Eicosane (36.7 °C) c_p = 2 kJ/kg K, L = 176 kJ/kg	Helical annular fins (pitch = 40 mm and length of 4.5 threads); 1-piece aluminum 7075 fin tube	A and N
Kuboth et al. [140]	CU shell-and-tube (S)	RT42 (40–44 °C) c_p = 2.0 kJ/kg K, L = 176 kJ/kg	100 annular copper fins of 30 different arrangements (thickness of fins was 1 mm)	N (2-D; FVM; EPM)
Ji et al. [141]	RU heated on one vertical wall (M)	RT42 (42 °C) k = 0.2 W/m K, L = 165 kJ/kg	Two parallel copper fins attached to the active vertical side; inclined angle of 0°, +15°, +30°, −15° and −30°	N (2-D; FVM; EPM)
Luo and Liao [142]	CU shell-and-tube; CWT at the inner tube and insulated shell (M)	Lauric acid (315–317 K) k = 0.15 W/mK, L = 178 kJ/kg	Dendritic longitudinal aluminum fins placed on HTF tube	N (2-D; CON only)
Duan et al. [143]	Honeycombs of different cells with CWT boundaries (M)	Pure paraffin (323 K) k = 0.2 W/mK, L = 224 kJ/kg	Aluminum plate fins of varying orientations and ARs	N (2-D; FVM; EPM)
Deng et al. [144]	CU shell-and-tube; CWT at the inner tube and outer shell was insulated (M)	Lauric acid (317.37 K) k= 0.147 W/mK, L = 173.8 kJ/kg	Longitudinal copper plate fins placed equally spaced in four arrangements	N (2-D; FVM; EPM)
Shahsavar et al. [145]	CU shell-and-tube unit; CWT at the inner tube and outer shell was insulated (M and S)	RT-35 (302–309 K) K = 0.2 W/mK, L = 170 kJ/kg	Annular copper fins (5); thickness (1, 2 and 3 mm) and outer radii (5, 6 and 7 mm)	N (2-D; FVM; EPM)
Zhao et al. [146]	CU shell-and-tube unit; CWT on both sides of the annular PCM (M)	RT-82 (350.15–358.15 K) K = 0.2 W/mK, L = 176 kJ/kg	Longitudinal carbon fiber (XN-100) plate fins; copper, aluminum and nickel metal foams	N (2-D; FVM; EPM)
Kirincic et al. [147]	CU shell-and-tube unit (M and S)	RT-25 (18–25 °C) K = 0.2 W/mK, L = 170 kJ/kg	Longitudinal aluminum plate fins	E and N (3-D; FVM; EPM)
Ghalambaz et al. [148]	CU TTHX with two HTF passageways (M)	RT-35 (35 °C) k= 0.2 W/mK, L = 170 kJ/kg	Longitudinal twisted and planar straight copper fins (two, four and six)	N (3-D; FVM; EPM)
Safari et al. [149]	RU with an HTF-heated side wall oriented vertically or horizontally (M)	Lauric acid (43.5–48,2 °C) k_l = 0.14 W/mK, k_l= 0.16 W/mK L = 187.2 kJ/kg	Planar aluminum fins (three, five, seven and nine)	N (2-D; FVM; EPM)
Ye and Khodadadi [150]	CU shell-and-tube unit; CWT on the inner HTF tube and insulated shell (M)	N-octadecane (301 K) k = 0.15 W/mK, L = 242.5 kJ/kg	Arrow-shaped longitudinal copper fins (length ratios γ = 0, 0.222, 0.571 and 1.198) and branch angles (θ = 40°, 50°, and 60°)	N (2-D; FVM; EPM)

BC = boundary condition; Bi = Biot number; CHF = constant heat flux; CON = conduction heat transfer; CS = cross-section; CU = cylindrical unit or test cell; CWT = constant wall temperature; DSC = differential scanning calorimetry; E = experimental; EPM = enthalpy-porosity method; FDM = finite difference method; FEM = finite element method; FR = flow rate; hp= heat pipe; HT = heat transfer; IT = inner tube; M = melting; NC = natural convection; N/A = not applicable or not available; NF = number of fins; RU = Rectangular unit or test cell; S = solidification; SH = sensible heat; T_m = melting temperature; T_w = wall temperature; ThC = thermocouple.

5. Classification of Similar Work (Themes), Performance Indicators and Challenges

Similar to [17], which discussed 75 fin-assisted LHTES systems dating back to 1966, the reviewed studies here [20–150] were classified and summarized in Table 2. Themes of "*Rectangular cuboid thermal storage units* and *shell and tube heat exchangers*" are the broadest groupings, whereas a few outliers are listed separately. In studies with the theme of "Rectangular cuboid storage systems with horizontal/vertical/other types of fins in contact with PCM", phase transition was activated on a boundary subjected to a constant heat flux, constant wall temperature, heat transfer fluid stream(s) or jet-cooling, whereas HTF stream(s) initiate phase transition in the shell and tube heat exchangers, for which AF/LF are in direct contact with PCM. Given the variety of configurations, fin/PCM materials, lack of widely accepted thermophysical properties, etc., the widely sought-after correlation:

$$\text{Efficiency} = f(\text{PCM properties, shape, boundary conditions, fin type/material, etc.})$$

does not exist at this time.

Table 2. Classification of the reviewed studies [20–150] on fin-assisted LHTESS based on similarity of work (theme); listing of the abbreviations other than those used in Table 1 are summarized at the bottom of the table.

Rectangular Cuboid Storage Units [Other Similar Configurations are Summarized at the Bottom]			
Thermal Conditions of the Active Wall on which Fins are Anchored	**Fin Orientation/Type**		
	Horizontal	**Vertical**	**Other**
CHF	[73], [124] simulated insolation. [125,128]. [135] attached on vertical hps and filled with zigzag-shaped fins in between.	[50,62,63,69,78,81,83,86,93,94,98,104,122,128]	[52,60,68] planar fins VIA, [62,86,97,131] square CS pin fins, [81] T-, Y- and cross shapes, [93] spiral fillers. [98] crossed and honeycomb fins. [132] annular fins on vertical hps. [133] vertical/horizontal fins. [141] slanted fins.
CWT	[88,125,136].	[25] JA, [63,64,76,79] [130] with high-porosity metal foam between fins.	[40,41] inclined plate fins. [60] planar fins VIA. [143] honeycomb cells.
HTF	[51] split fins. [87,149].	[22,101,149]	[30,31] VIA, [48,49] horizontal square CS pin fins, [90] Staggered/serrated aluminum plate fins.
Other	[54,74] insolation. [105] varying active wall temperature.	[66,112] insolation.	[59] insolation/horizontal square CS pin fins.

Shell and Tube Heat Exchanger (STHE) Units (1 HTF and 2 HTF Streams) [Other Similar STHE Configurations are Summarized at the Bottom]				
HTF Stream(s)	**Shell Geometry**	**(Original) Direction of Flow of the HTF Stream(s)**	**Fin Orientation/Type**	
			Annular	**Longitudinal**
Vertical	Cylindrical	With Gravity	[42] HPH, inner HTF stream end-capped. [43,44,46,47] HP, HTF stream end-capped. [82,84,92,95,111,114,115] HP. [116] HP-CCM. [139] HP and helical fins. [140] HP.	[21,23] HP, HTF stream end-capped. [42] HPH, inner HTF stream end-capped. [92,147] HP, [108] HHP. [148] TTHX, HPH.
		Against Gravity	[100,106,120,134,145] HP. [126] TTHX, HPH.	[102,118,120] HP. [148] TTHX, HPH.
	Rectangular Cuboid	With Gravity	N/A	[20] HP
		Against Gravity	N/A	N/A
Horizontal	Cylindrical		[26,27,32,34,65,71,76,85,89,103,119,127] HP. [61] 5 PCM, HP.	[33,34,36–39,65,67,72,91,110,123,129] HP. [35] HP, HTF stream end-capped feeding JA. [71] hpPH; [99,109] TTHX, HPH; [117] HP, CCM. [137] HP and elliptical HTF pipe. [142] HP and dendritic fins. [144] HP. [146] HP and metal foams. [150] HP and arrow-fins.
	Rectangular Cuboid		[58,70,77,103] HP. [121] HP including hps.	[28,53,55,56] HP. [113] HP including hps. [138] HP and snowflake-fins.
[24,29] Disk-like insulated PCM with a copper concentric vertical cylinder subjected to a step change in temperature. [45] Cylindrical heat pipes with longitudinal fins. [57] STHE (vertical) with spiral twister split annular fins on electrically heated rod, HPH. [75] Cylindrical container (HTF at base) with five vertical rods (also studied five heat pipes). [95] Cylindrical shell-and-tube storage unit with radially pointing pins of circular cross-section, HP. [96] Rectangular "shell-and-tube" storage unit with HTF-carrying U-tubes (no fin, in-line horizontal fins and staggered horizontal fins) and a festoon design, HP. [107] Curved copper plate U-shape fins, HP. [120] Combination of both annular and longitudinal fins forming sealed segments, HP.				

CS = cross-section; H = HTF; JA = jet arrays, P = PCM, VIA = varying inclination angles. Terms such as HPH refer to the order of the constituents encountered moving away from within the unit to the outside.

Researchers have sought improved performance of LHTES units through shortening charge/discharge time periods, in connection with the sacrificed PCM, due to introducing fins. Adoption of simple planar fins has diminished over the years, while more complicated shapes, such as branching arrangements, crosses and Y-shapes, etc., are being reported, at times with the aid of the constructal theory. However, the fundamental challenge of utilizing high TC fins remains the promotion of conducting pathways with minimum distance that connect the high and low temperatures of a heat storage system.

6. Concluding Remarks

Analytical, computational and experimental investigations focused on improving the performance of LHTES systems that utilize generally metal-based high TC fins/extended surfaces were reviewed. A variety of PCM, including capric-palmitic acid, chloride mixtures, dodecanoic acid, erythritol, fluorides, lauric acid, naphthalene, nitrite and nitrate mixtures, paraffins, potassium nitrate, salt hydrates, sodium hydrate, stearic acid, sulfur, water and xylitol, covering T_m in the range of −129.6 to 767 °C, have been reported. Freezing and thawing within various TES vessel geometries and heat exchange operating conditions were studied. The unifying findings/observations of these studies are:

(a) Length and number of fins (or fin-pitch) markedly affected the performance of the TES units (reduced charge/discharge times) in comparison to the fin's thickness and orientation.
(b) Presence of the fins generally diminished the role of buoyancy-driven convection, which plays a significant role during melting and practically no role in freezing scenarios.
(c) Conflicting trends between enhancing the effective TC of the PCM and weaker natural convection should be considered by the designer through selecting the optimum positions and orientations of the fins.
(d) The extent of sacrificed PCM that is replaced by fins, thus lowering the storage capacity of the TES unit and contributing to the sensible stored heat, is another concern.

Whereas simple planar fins are still being studied, more complicated shapes (e.g., branching arrangements, crosses, Y-shapes, slanted, dendritic, snowflake-shapes, arrow-shapes, helical, varied honeycomb cells, etc.) are being explored, at times with the wider adoption of the constructal theory. Promoting short conducting pathways linking the high and low temperatures of the storage system through innovative approaches still remains the ultimate challenge.

Author Contributions: Conceptualization, J.M.K.; Methodology, J.M.K.; Formal Analysis, J.M.K.; Investigation, W.Y., D.J. and J.M.K.; Writing—Original Draft Preparation, W.Y., D.J. and J.M.K.; Writing—Review & Editing, W.Y., D.J. and J.M.K.; Visualization, W.Y.; Supervision, J.M.K.; Project Administration, J.M.K. All authors have read and agreed to the published version of the manuscript.

Funding: This research received no external funding.

Data Availability Statement: Presented data in the form of graphs/tables were taken with permission from original publications.

Conflicts of Interest: The authors declare no conflict of interest.

List of Symbols

AR	Aspect ratio
N_e	Number of enclosures
t_m	Melting time, s
t_{max}	Maximum time needed to reach a set temperature, s

Abbreviations

AF	Annular fins
ES	Equally or uniformly spaced
HTF	Heat transfer fluid
LF	Longitudinal fins
LSI	Liquid-solid interface
PCM	Phase change materials
TC	Thermal conductivity
TD	Temperature difference
TTHX	Triple-tube or triplex-tube heat exchanger
VF	Volume fraction

References

1. Considine, D.M. *Energy Technology Handbook*; McGraw-Hill: New York, NY, USA, 1977; 1702p.
2. Hartnett, J.P. Alternative energy sources. In Proceedings of the Conference Sponsored by the International Centre of Heat and Mass Transfer, Dubrovnik, Yugoslavia, 25–30 August 1975; Academic Press: Cambridge, MA, USA, 1976; p. 328.
3. Veziroglu, T.N. Alternative energy sources: An international compendium. In Proceedings of the First Miami International Conference, Miami Beach, FL, USA, 5–7 December 1977; Hemisphere Publishing Corporation Washington: Washington, DC, USA, 1978; Volume 11, p. 5170.
4. Veziroglu, T.N. Alternative energy sources II. In Proceedings of the Second Miami International Conference, Miami Beach, FL, USA, 10–13 December 1979; Hemisphere Publishing Corporation Washington: Washington, DC, USA, 1981; Volume 9, p. 4171.
5. Veziroglu, T.N. Alternative energy sources III. In Proceedings of the Third Miami International Conference, Miami Beach, FL, USA, 15–17 December 1980; Hemisphere Publishing Corporation Washington: Washington, DC, USA, 1983; Volume 9.
6. Veziroglu, T.N. Alternative energy sources IV. In Proceedings of the Fourth Miami International Conference, Miami Beach, FL, USA, 14–16 December 1981; Ann Arbor Science Publishers: Ann Arbor, Michigan, 1982; Volume 8.
7. Veziroglu, T.N. Alternative energy sources V. In Proceedings of the Fifth Miami International Conference, Miami Beach, FL, USA, 13–15 December 1982; Elsevier: Amsterdam, The Netherlands, 1983; Volume 6.
8. Veziroglu, T.N. Alternative energy sources VI. In Proceedings of the Sixth Miami International Conference, Miami Beach, FL, USA, 12–14 December 1983; Hemisphere Publishing Corporation Washington: Washington, DC, USA, 1985; Volume 4.
9. Veziroglu, T.N. Alternative energy sources VII. In Proceedings of the Seventh Miami International Conference, Miami Beach, FL, USA, 9–11 December 1985; Hemisphere Publishing Corporation Washington: Washington, DC, USA, 1987; Volume 6.
10. Veziroglu, T.N. Alternative energy sources VIII. In Proceedings of the Eighth Miami International Conference, Miami Beach, FL, USA, 14–16 December 1987; Hemisphere Publishing Corporation Washington: Washington, DC, USA, 1989; Volume 2.
11. Bahadori, M.N. Thermal energy storage. *Iran. J. Sci. Technol.* **1976**, *5*, 159–171.
12. Gur, I.; Sawyer, K.; Prasher, R. Searching for a better thermal battery. *Science* **2012**, *335*, 1451–1452. [CrossRef] [PubMed]
13. Abhat, A. Low temperature latent heat thermal energy storage-heat storage materials. *Solar Energy* **1983**, *30*, 313–332. [CrossRef]
14. Zalba, B.; Marin, J.M.; Cabeza Mehling, H. Review on thermal energy storage with phase change: Materials, heat transfer analysis and applications. *Appl. Therm. Eng.* **2003**, *23*, 251–283. [CrossRef]
15. Dhaidan, N.S.; Khodadadi, J.M. Melting and convection of phase change materials in different shape containers: A review. *Renew. Sustain. Energy Rev.* **2015**, *43*, 449–477. [CrossRef]
16. Fan, L.; Khodadadi, J.M. Thermal conductivity enhancement of phase change materials for thermal energy storage: A review. *Renew. Sustain. Energy Rev.* **2011**, *15*, 24–46. [CrossRef]
17. Dhaidan, N.S.; Khodadadi, J.M. Improved performance of latent heat energy storage systems utilizing high thermal conductivity fins: A review. *J. Renew. Sustain. Energy* **2017**, *9*, 034103. [CrossRef]
18. Abdulateef, A.M.; Mat, S.; Abdulateef, J.; Sopian, K.; Al-Abidi, A.A. Geometric and design parameters of fins employed for enhancing thermal energy storage systems: A review. *Renew. Sustain. Energy Rev.* **2018**, *82*, 1620–1635. [CrossRef]
19. Khodadadi, J.M.; Fan, L.; Babaei, H. Thermal conductivity enhancement of nanostructure-based colloidal suspensions utilized as phase change materials for thermal energy storage: A review. *Renew. Sustain. Energy Rev.* **2013**, *24*, 418–444. [CrossRef]
20. Ismail, K.A.R.; Liu, C.Y.; Correa, G.E. Experimental investigation of solidification and fusion around cylindrically finned tubes, Alternative Energy Sources V. Part A: Solar Radiation/Collection/Storage. In Proceedings of the Fifth Miami International Conference, Miami Beach, FL, USA, 13–15 December 1982; Elsevier: Amsterdam, The Netherlands, 1983; pp. 271–276.
21. Ismail, K.A.R.; Alves, C.L. Heat Transfer with Phase Change around Finned Cylindrical Tubes: Theory and Experiments, Alternative Energy Sources VI. Volume 1: Solar Energy and Applications. In Proceedings of the Sixth Miami International Conference, Miami Beach, FL, USA, 12–14 December 1983; Hemisphere Publishing Corporation: Washington, DC, USA, 1985; pp. 289–304.
22. Ho, C.-J.; Viskanta, R. Inward solid-liquid phase-change heat transfer in a rectangular cavity with conducting vertical walls. *Int. J. Heat Mass Transf.* **1984**, *27*, 1055–1065. [CrossRef]

23. Ismail, K.A.R.; Alves, C.L. PCM storage using externally finned circular tubes. In Proceedings of the 22nd National Heat Transfer Conference & Exhibition, Niagara Falls, NY, USA, 5–8 August 1984; ASME: New York, NY, USA, 1984; p. 8.
24. Okada, M. Heat Transfer During Melting from a Vertical Cylinder (1st Report, Analysis and Experiments of the Melting without Subcooling). *Trans. JSME Ser. B* **1985**, *51*, 382–387. (In Japanese) [CrossRef]
25. Saito, A.; Nagakubo, S.I.; Utaka, Y.; Katayama, K. A Study on The Heat Transfer of Latent Heat Thermal Energy Storage (3rd Report, Natural Convection Effects on the Performance of a Thermal Energy Storage Apparatus with Dense Plate-fins). *Trans. JSME Ser. B* **1986**, *51*, 2148–2157. reprinted in *Bull. JSME* **1986**, *29*, 845–853(In Japanese) [CrossRef]
26. Imura, H.; Yoshida, M. Latent heat storage in a unit with a circular-finned tube (1st report, numerical analysis in the solidification process). *Trans. JSME Ser. B* **1986**, *52*, 1843–1849. (In Japanese) [CrossRef]
27. Imura, H.; Yoshida, M. Latent heat storage in a unit with a circular-finned tube (2nd report, experimental study). *Trans. JSME Ser. B* **1986**, *52*, 1850–1855. (In Japanese) [CrossRef]
28. Betzel, T.; Beer, H. Experimental investigation of heat transfer during melting around a horizontal tube with and without axial fins. Int. Comm. *Heat Mass Transf.* **1986**, *13*, 639–649. [CrossRef]
29. Okada, M. Heat Transfer During Melting from a Vertical Cylinder (2nd report, Effects of Aspect Ratio of the Cylinder and Subcooling). *Trans. JSME Ser. B* **1987**, *53*, 1055–1060. (In Japanese) [CrossRef]
30. Saito, A.; Imamura, T.; Utaka, Y.; Saito, A. On The Contact Heat Transfer of with Melting (4th Report, Direct-Contact Melting Process within an Inclined Rectangular Cross Section). *Trans. JSME Ser. B* **1988**, *54*, 1123–1130. (In Japanese) [CrossRef]
31. Saito, A.; Imamura, T.; Utaka, Y.; Saito, A. On the contact heat transfer with melting (direct contact melting process within an inclined rectangular cross-section). *JSME Int. J. Ser. 2* **1989**, *32*, 411–419. [CrossRef]
32. Ito, S.; Miura, N.; Yano, Y. Approximate Solution and Experimental Investigation of the Thermal Performance of a Latent Heat Thermal Energy Storage Unit with a Finned Tube. *Trans. JSME Ser. B* **1989**, *55*, 782–790. (In Japanese) [CrossRef]
33. Sasaguchi, K.; Sakamoto, Y. Effects of Natural Convection on Melting of a Phase Change Material around A Finned Tube. *Trans. JSME Ser. B* **1989**, *55*, 1418–1425. reprinted in *Heat Transf. Jpn. Res.* **1990**, *19*, 474–491(In Japanese) [CrossRef]
34. Sasaguchi, K. Heat Transfer Enhancement in a Latent Heat Thermal Energy Storage Unit Using a Tube with Radial Fins. *Trans. JSME Ser. B* **1990**, *56*, 2461–2468. (In Japanese) [CrossRef]
35. Sasaguchi, K. An Experimental Study on Melting of a Phase-Change Material around Finned Tubes. *Trans. JSME Ser. B* **1990**, *56*, 2785–2792. (In Japanese) [CrossRef]
36. Kaino, K. Usefulness of a Similarity Curve as a Chart for Thermal Characteristics of Latent Heat Energy Storage Unit. *Trans. JSME Ser. B* **1992**, *58*, 229–235. (In Japanese) [CrossRef]
37. Kaino, K. Similarity curve in the solidification process of latent heat energy storage unit with straight fins (1st report, effect of Stefan number on the formation of the similarity rule). *Trans. JSME Ser. B* **1993**, *59*, 236–242. (In Japanese) [CrossRef]
38. Kaino, K. Similarity curve in the solidification process of latent heat energy storage unit with straight fins (2nd report, effect of heat transfer tube on the formation of the similarity rule). *Trans. JSME Ser. B* **1993**, *59*, 571–577. (In Japanese) [CrossRef]
39. Kaino, K. Similarity Curve in the Solidification Process of Latent Heat Energy Storage Unit with Straight Fins: 3rd Report, Effect of Fin on the Formation of the Similarity Rule). *Trans. JSME Ser. B* **1993**, *59*, 1228–1235. (In Japanese) [CrossRef]
40. Sasaguchi, K.; Takeo, H. Solid/Liquid Phase Change Heat Transfer in Porous Media (Effect of the Orientation of a Hot Wall with Fins on the Melting Process). *Trans. JSME Ser. B* **1993**, *59*, 1678–1684. (In Japanese) [CrossRef]
41. Sasaguchi, K.; Takeo, H. Effect of the orientation of a finned surface on the melting of frozen porous media. *Int. J. Heat Mass Trans.* **1994**, *37*, 13–26. [CrossRef]
42. Al-Jandal, S.S.; Sayigh, A.A.M. Thermal performance characteristics of STC system with phase change storage. *Renew. Energy* **1994**, *5*, 390–399. [CrossRef]
43. Choi, J.C.; Kim, S.D. Heat transfer in a latent heat-storage system using $MgCl_2{\cdot}6H_2O$ at the melting point. *Energy* **1995**, *20*, 13–25. [CrossRef]
44. Choi, J.C.; Kim, S.D.; Han, G.Y. Heat Transfer Characteristics in Low-Temperature Latent Heat Storage Systems Using Salt Hydrates. *Korean J. Chem. Eng.* **1995**, *12*, 258–263. [CrossRef]
45. Horbaniuc, B.; Popescu, A.; Dumitraşcu, G. The correlation between the number of fins and the discharge time for a finned heat pipe latent heat storage system. *Renew. Energy* **1996**, *9*, 605–608. [CrossRef]
46. Choi, J.C.; Kim, S.D.; Han, G.Y. Heat transfer characteristics in low-temperature latent heat storage systems using salt-hydrates at heat recovery stage. *Sol. Energy Mater Sol. Cells* **1996**, *40*, 71–87. [CrossRef]
47. Han, S.K.; Han, G.Y. Heat transfer enhancement by fins in a latent heat storage system using phase change material. *J. Energy Eng.* **1996**, *5*, 115–122. (In Korean)
48. Hirasawa, Y.; Chen, D.; Watanabe, K.; Takegoshi, E. An Experimental Study on the Solidification and Melting of Water around a Vertical Heat Transfer Plate with Pin Fins. *Trans. JSME Ser. B* **1997**, *63*, 3715–3721. (In Japanese) [CrossRef]
49. Chen, D.; Hirasawa, Y.; Takegoshi, E. A Numerical Study on the Solidification and Melting of Water Around a Vertical Heat Transfer Plate with Pin Fins. *Trans. JSME Ser. B* **1997**, *63*, 3722–3728. (In Japanese) [CrossRef]
50. Wirtz, R.A.; Zheng, N.; Chandra, D. Thermal management using dry phase change materials. In Proceedings of the 15th IEEE Semiconductor Thermal Measurement and Management Symposium, San Diego, CA, USA, 9–11 March 1999; pp. 74–82.
51. Hong, H.; Kim, M.G. Melting of Ice on the Heating Plate with Split Fins. *Korean J. Air-Cond. Refrig. Eng.* **2000**, *12*, 67–74. reprinted in *Int. J. Air-Cond. Refrig.* **2001**, *9*, 1–7(In Korean)

52. Inaba, H.; Matsuo, K.; Horibe, A. Natural Convection Behavior in an Inclined Rectangular Latent Heat Storage Vessel Having One Heating Wall with Plate Fins. *Trans. JSME Ser. B* **2001**, *67*, 2113–2120. (In Japanese) [CrossRef]
53. Yamashita, Y.; Hirata, Y.; Iwata, Y.; Yamazaki, K.; Ito, Y. Performance and heat transfer characteristics of a latent heat storage unit with finned tubes: Experimental study on liquefaction of LNG Boil-off Gas by melting n-pentane as a phase-change material. *Kagaku Kogaku Ronbunshu* **2004**, *30*, 399–406. (In Japanese) [CrossRef]
54. Huang, M.J.; Eames, P.C.; Norton, B. Thermal regulation of building-integrated photovoltaics using phase change materials. *Int. J. Heat Mass Transf.* **2004**, *47*, 2715–2733. [CrossRef]
55. Yamashita, Y.; Hirata, Y.; Iwata, Y.; Yamazaki, K.; Ito, Y. Performance and heat transfer characteristics of a latent heat storage unit with finned tubes: Experimental study on storage of LNG cold energy by freezing n-pentane as a phase-change material. *Kagaku Kogaku Ronbunshu* **2005**, *31*, 144–150. (In Japanese) [CrossRef]
56. Yamashita, Y.; Hirata, Y.; Iwata, Y.; Yamazaki, K.; Ito, Y. Performance and heat transfer characteristics of a latent heat storage unit with finned tubes: Numerical analysis on freezing and melting processes of n-pentane as a phase-change material. *Kagaku Kogaku Ronbunshu* **2005**, *31*, 151–158. (In Japanese) [CrossRef]
57. Liu, Z.; Sun, X.; Ma, C. Experimental study of the characteristics of solidification of stearic acid in an annulus and its thermal conductivity enhancement. *Energy Convers. Manag.* **2005**, *46*, 971–984. [CrossRef]
58. Kayansayan, N.; Ali Acar, M. Ice formation around a finned-tube heat exchanger for cold thermal energy storage. *Int. J. Therm. Sci.* **2006**, *45*, 405–418. [CrossRef]
59. Huang, M.J.; Eames, P.C.; Norton, B. Comparison of a small-scale 3D PCM thermal control model with a validated 2D PCM thermal control model. *Sol. Energy Mater. Sol. Cells* **2006**, *90*, 1961–1972. [CrossRef]
60. Wang, X.Q.; Mujumdar, A.S.; Yap, C. Effect of orientation for phase change material (PCM)-based heat sinks for transient thermal management of electric components. *Int. Commun. Heat Mass Trans.* **2007**, *34*, 801–808. [CrossRef]
61. Seeniraj, R.V.; Narasimhan, N.L. Performance enhancement of a solar dynamic LHTS module having both fins and multiple PCMs. *Sol. Energy* **2008**, *82*, 535–542. [CrossRef]
62. Saha, S.K.; Srinivasan, K.; Dutta, P. Studies on optimum distribution of fins in heat sinks filled with phase change materials. *J. Heat Transf.* **2008**, *130*, 034505. [CrossRef]
63. Wang, X.-Q.; Yap, C.; Mujumdar, A.S. A parametric study of phase change material (PCM)-based heat sinks. *Int. J. Therm. Sci.* **2008**, *47*, 1055–1068. [CrossRef]
64. Ravi, K.; Wang, X.-Q.; Mujumdar, A.S. Transient cooling of electronics using phase change material (PCM)-based heat sinks. *Appl. Therm. Eng.* **2008**, *28*, 1047–1057.
65. Agyenim, F.; Eames, P.; Smyth, M. A comparison of heat transfer enhancement in medium temperature thermal energy storage heat exchanger using fins and multitubes. In Proceedings of the ISES World Congress 2007, Beijing, China, 8–21 September 2007; Springer: Berlin/Heidelberg, Germany, 2008; pp. 2726–2730.
66. Lee, H.J.; Chun, J.H. Temperature Control for PV Panel Absorbing Heat by Phase Change Material and its Estimation. *J. Korean Sol. Energy Soc.* **2010**, *30*, 10–15. (In Korean)
67. Agyenim, F.; Hewitt, N. The development of a finned phase change material (PCM) storage system to take advantage of off-peak electricity tariff for improvement in cost of heat pump operation. *Energy Build* **2010**, *42*, 1552–1560. [CrossRef]
68. Fok, S.C.; Shen, W.; Tan, F.L. Cooling of portable hand-held electronic devices using phase change materials in finned heat sinks. *Int. J. Therm. Sci.* **2010**, *49*, 109–117. [CrossRef]
69. Saha, S.K.; Dutta, P. Heat transfer correlations for PCM-based heat sinks with plate fins. *Appl. Therm. Eng.* **2010**, *30*, 2485–2491. [CrossRef]
70. Li, W.; Li, X.; Zhao, J. Experimental study of a finned-tubes phase change heat storage system. In Proceedings of the Asia Pacific Power and Energy Engineering Conference (APPEEC), Chengdu, China, 28–31 March 2010; pp. 1–4.
71. Sugawara, M.; Komatsu, Y.; Takahashi, Y.; Beer, H. Freezing enhancement around a horizontal tube using copper foil disks. *Heat Mass Transf.* **2011**, *47*, 1691–1698. [CrossRef]
72. Francis, A.; Eames, P.; Smyth, M. Experimental study on the melting and solidification behaviour of a medium temperature phase change storage material (Erythritol) system augmented with fins to power a LiBr/H_2O absorption cooling system. *Renew. Energy* **2011**, *36*, 108–117.
73. Talati, F.; Mosaffa, A.H.; Rosen, M.A. Analytical approximation for solidification processes in PCM storage with internal fins: Imposed heat flux. *Heat Mass Transf.* **2011**, *47*, 369–376. [CrossRef]
74. Huang, M.J.; Eames, P.C.; Norton, B.; Hewitt, N.J. Natural convection in an internally finned phase change material heat sink for the thermal management of photovoltaics. *Sol. Energy Mater. Sol. Cells* **2011**, *95*, 1598–1603. [CrossRef]
75. Robak, C.W.; Bergman, T.L.; Faghri, A. Enhancement of latent heat energy storage using embedded heat pipes. *Int. J. Heat Mass Transf.* **2011**, *54*, 3476–3484. [CrossRef]
76. Bauer, T. Approximate analytical solutions for the solidification of PCMs in fin geometries using effective thermophysical properties. *Int. J. Heat Mass Transf.* **2011**, *54*, 4923–4930. [CrossRef]
77. Ismail, K.A.R.; Lino, F.A.M. Fins and turbulence promoters for heat transfer enhancement in latent heat storage systems. *Exp. Therm. Fluid Sci.* **2011**, *35*, 1010–1018. [CrossRef]
78. Hosseinizadeh, S.F.; Tan, F.L.; Moosania, S.M. Experimental and numerical studies on performance of PCM-based heat sink with different configurations of internal fins. *Appl. Therm. Eng.* **2011**, *31*, 3827–3838. [CrossRef]

79. Ye, W.-B.; Zhu, D.-S.; Wang, N. Numerical simulation on phase-change thermal storage/release in a plate-fin unit. *Appl. Therm. Eng.* **2011**, *31*, 3871–3884. [CrossRef]
80. Long, J. Simulation Investigation for Heat Transfer in Fin-Tube Thermal Storage Unit with Phase Change Material. *Adv. Mater. Res.* **2011**, *168*, 895–899. [CrossRef]
81. Tan, L.; Kwok, Y. Date A and Akbarzadeh A, Numerical analysis of natural convection effects in latent heat storage using different fin shapes. *Int. J. Energy Sci.* **2011**, *1*, 162–168.
82. Murray, R.E.; Desgrosseilliers, L.; Stewart, J.; Osbourne, N.; Marin, G.; Safatli, A.; Groulx, D.; White, M.A. Design of a latent heat energy storage system coupled with a domestic hot water solar thermal system. In Proceedings of the World Renewable Energy Congress 2011 (WREC 2011), Linköping, Sweden, 8–13 May 2011.
83. Saha, S.K.; Dutta, P. Effect of melt convection on the optimum thermal design of heat sinks with phase change material. *J. Enhanc. Heat Transf.* **2011**, *18*, 249–259. [CrossRef]
84. Chiu, J.N.W.; Viktoria, M. Submerged finned heat exchanger latent heat storage design and its experimental verification. *Appl. Energy* **2012**, *93*, 507–516. [CrossRef]
85. Wilson, O.; Groulx, D. Effects of the number and distribution of fins on the storage characteristics of a cylindrical latent heat energy storage system: A numerical study. *Heat Mass Transf.* **2012**, *48*, 1825–1835.
86. Rajesh, B.; Balaji, C. Experimental investigations on phase change material based finned heat sinks for electronic equipment cooling. *Int. J. Heat Mass Transf.* **2012**, *55*, 1642–1649.
87. Mosaffa, A.H.; Talati, F.; Rosen, M.A.; Tabrizi, H.B. Approximate analytical model for PCM solidification in a rectangular finned container with convective cooling boundaries. *Int. Commun. Heat Mass Transf.* **2012**, *39*, 318–324. [CrossRef]
88. Jourabian, M.; Farhadi, M.; Sedighi, K.; Darzi, A.R.; Vazifeshenas, Y. Simulation of natural convection melting in a cavity with fin using lattice Boltzmann method. *Int. J. Numer. Methods Fluids* **2012**, *70*, 313–325. [CrossRef]
89. Mosaffa, A.H.; Talati, F.; Tabrizi, H.B.; Rosen, M.A. Analytical modeling of PCM solidification in a shell and tube finned thermal storage for air conditioning systems. *Energy Build.* **2012**, *49*, 356–361. [CrossRef]
90. Xu, S.; Ling, X.; Peng, H. Simulation on the Plate-fin Thermal Storage in a New Desalination System. *J. Converg. Inf. Technol.* **2012**, *7*, 387–395.
91. Shokouhmand, H.; Kamkari, B. Numerical simulation of phase change thermal storage in finned double-pipe heat exchanger. *Appl. Mech. Mater.* **2012**, *232*, 742–746. [CrossRef]
92. Hamdani, I.; Mahlia, T.M.I. Investigation of melting heat transfer characteristics of latent heat thermal storage unit with finned tube. *Procedia Eng.* **2012**, *50*, 122–128.
93. Tan, L.; Kwok, Y.; Date, A.; Akbarzadeh, A. Numerical study of natural convection effects in latent heat storage using aluminum fins and spiral fillers. *Int. J. Mech. Mechatron. Eng.* **2012**, *6*, 1438–1445.
94. Levin, P.P.; Avraham, S.; Gad, H. Numerical optimization of a PCM-based heat sink with internal fins. *Int. J. Heat Mass Transf.* **2013**, *61*, 638–645. [CrossRef]
95. Tay, N.H.S.; Bruno, F.; Belusko, M. Comparison of pinned and finned tubes in a phase change thermal energy storage system using CFD. *Appl. Energy* **2013**, *104*, 79–86. [CrossRef]
96. Kurnia, J.C.; Sasmito, A.P.; Jangam, S.V.; Mujumdar, A.S. Improved design for heat transfer performance of a novel phase change material (PCM) thermal energy storage (TES). *Appl. Therm. Eng.* **2013**, *50*, 896–907. [CrossRef]
97. Rajesh, B.; Balaji, C. Thermal optimization of PCM based pin fin heat sinks: An experimental study. *Appl. Therm. Eng.* **2013**, *54*, 65–77.
98. Mahmoud, S.; Tang, A.; Toh, C.; Raya, A.D.; Soo, S.L. Experimental investigation of inserts configurations and PCM type on the thermal performance of PCM based heat sinks. *Appl. Energy* **2013**, *112*, 1349–1356. [CrossRef]
99. Mat, S.; Al-Abidi, A.A.; Sopian, K.; Sulaiman, M.Y.; Mohammad, A.T. Enhance heat transfer for PCM melting in triplex tube with internal–external fins. *Energy Convers. Manag.* **2013**, *74*, 223–236. [CrossRef]
100. Guelpa, E.; Sciacovelli, A.; Verda, V. Entropy generation analysis for the design improvement of a latent heat storage system. *Energy* **2013**, *53*, 128–138. [CrossRef]
101. Campos-Celador, A.; Diarce, G.; González-Pino, I.; Sala, J.M. Development and comparative analysis of the modeling of an innovative finned-plate latent heat thermal energy storage system. *Energy* **2013**, *58*, 438–447. [CrossRef]
102. Ravikumar, S.G.; Velraj, R. Analysis of the heat transfer mechanisms during energy storage in a Phase Change Material filled vertical finned cylindrical unit for free cooling application. *Energy Convers. Manag.* **2013**, *75*, 466–473.
103. Mosaffa, A.; Talati, F.; Rosen, M.A.; Tabrizi, H.B. Phase change material solidification in a finned cylindrical shell thermal energy storage: An approximate analytical approach. *Therm. Sci.* **2013**, *17*, 407–418. [CrossRef]
104. Kozak, Y.; Abramzon, B.; Ziskind, G. Experimental and numerical investigation of a hybrid PCM–air heat sink. *Appl. Therm. Eng.* **2013**, *59*, 142–152. [CrossRef]
105. Kim, T.Y.; Seo, J.G.; Hyun, B.S.; Cheon, H.Y.; Lee, J.J. Study on the Thermal Buffer Mass and Phase Change Material for Thermal Control of the Periodically Working Satellite Component. *J. Korean Soc. Aeronaut. Space Sci.* **2014**, *42*, 1013–1019. (In Korean)
106. Chen, X.; Worall, M.; Omer, S.; Su, Y.; Riffat, S. Experimental investigation on PCM cold storage integrated with ejector cooling system. *Appl. Therm. Eng.* **2014**, *63*, 419–427. [CrossRef]
107. Shon, J.; Kim, H.; Lee, K. Improved heat storage rate for an automobile coolant waste heat recovery system using phase-change material in a fin–tube heat exchanger. *Appl. Energy* **2014**, *113*, 680–689. [CrossRef]

108. Murray, R.E.; Groulx, D. Experimental study of the phase change and energy characteristics inside a cylindrical latent heat energy storage system: Part 2 simultaneous charging and discharging. *Renew. Energy* **2014**, *63*, 724–734. [CrossRef]
109. Al-Abidi, A.A.; Mat, S.; Sopian, K.; Sulaiman, M.Y.; Mohammad, A.T. Experimental study of melting and solidification of PCM in a triplex tube heat exchanger with fins. *Energy Build.* **2014**, *68*, 33–41. [CrossRef]
110. Liu, C.; Groulx, D. Experimental study of the phase change heat transfer inside a horizontal cylindrical latent heat storage system. *Int. J. Therm. Sci.* **2014**, *82*, 100–110. [CrossRef]
111. Tay, N.H.S.; Belusko, M.; Castell, A.; Cabeza, L.F.; Bruno, F. An effectiveness-NTU technique for characterising a finned tubes PCM system using a CFD model. *Appl. Energy* **2014**, *131*, 377–385. [CrossRef]
112. Hasan, A.; McCormack, S.J.; Huang, M.J.; Norton, B. Energy and Cost Saving of a Photovoltaic-Phase Change Materials (PV-PCM) System through Temperature Regulation and Performance Enhancement of Photovoltaics. *Energies* **2014**, *7*, 1318–1331. [CrossRef]
113. Khalifa, A.; Tan, L.; Date, A.; Akbarzadeh, A. A numerical and experimental study of solidification around axially finned heat pipes for high temperature latent heat thermal energy storage units. *Appl. Therm. Eng.* **2014**, *70*, 609–619. [CrossRef]
114. Rahimi, M.; Ranjbar, A.A.; Ganji, D.D.; Sedighi, K.; Hosseini, M.J. Experimental investigation of phase change inside a finned-tube heat exchanger. *J. Eng.* **2014**, *2014*, 641954. [CrossRef]
115. Rahimi, M.; Ranjbar, A.A.; Ganji, D.D.; Sedighi, K.; Hosseini, M.J.; Bahrampoury, R. Analysis of geometrical and operational parameters of PCM in a fin and tube heat exchanger. *Int. Commun. Heat Mass Transf.* **2014**, *53*, 109–115. [CrossRef]
116. Kozak, Y.; Rozenfeld, T.; Ziskind, G. Close-contact melting in vertical annular enclosures with a non-isothermal base: Theoretical modeling and application to thermal storage. *Int. J. Heat Mass Transf.* **2014**, *72*, 114–127. [CrossRef]
117. Rozenfeld, T.; Kozak, Y.; Ziskind, G. Close-contact melting in a horizontal cylindrical enclosure with longitudinal plate fins: Demonstration, modeling and application to thermal storage. *Int. J. Heat Mass Transf.* **2015**, *86*, 465–477. [CrossRef]
118. Rathod, M.K.; Banerjee, J. Thermal performance enhancement of shell and tube Latent Heat Storage Unit using longitudinal fins. *Appl. Therm. Eng.* **2015**, *75*, 1084–1092. [CrossRef]
119. Paria, S.; Sarhan, A.A.D.; Goodarzi, M.S.; Baradaran, S.; Rahmanian, B.; Yarmand, H.; Alavi, M.A.; Kazi, S.N.; Metselaar, H.S.C. Indoor solar thermal energy saving time with phase change material in a horizontal shell and finned-tube heat exchanger. *Sci. World J.* **2015**, *2015*, 291657. [CrossRef] [PubMed]
120. Zhai, X.Q.; Cheng, X.W.; Wang, C.; Wang, R.Z. Experimental investigation and performance analysis of a fin tube phase change cold storage unit for high temperature cooling application. *Energy Build.* **2015**, *89*, 9–17. [CrossRef]
121. Khalifa, A.; Tan, L.; Date, A.; Akbarzadeh, A. Performance of suspended finned heat pipes in high-temperature latent heat thermal energy storage. *Appl. Therm. Eng.* **2015**, *81*, 242–252. [CrossRef]
122. Rasool, K.; Salimpour, M.R. Constructal design of phase change material enclosures used for cooling electronic devices. *Appl. Therm. Eng.* **2015**, *84*, 339–349.
123. Tao, Y.B.; He, Y.L. Effects of natural convection on latent heat storage performance of salt in a horizontal concentric tube. *Appl. Energy* **2015**, *143*, 38–46. [CrossRef]
124. Liu, S.; Li, Y. Heating performance of a solar chimney combined PCM: A numerical case study. *Energy Build.* **2015**, *99*, 117–130. [CrossRef]
125. Talati, F.; Taghilou, M. Lattice Boltzmann application on the PCM solidification within a rectangular finned container. *Appl. Therm. Eng.* **2015**, *83*, 108–120. [CrossRef]
126. Jmal, I.; Baccar, M. Numerical study of PCM solidification in a finned tube thermal storage including natural convection. *Appl. Therm. Eng.* **2015**, *84*, 320–330. [CrossRef]
127. Hosseini, M.J.; Rahimi, M.; Bahrampoury, R. Thermal analysis of PCM containing heat exchanger enhanced with normal annular fines. *Mech. Sci.* **2015**, *6*, 221–234. [CrossRef]
128. Salma, G.; Harmand, S.; Jabrallah, S.B. Experimental comparison between different configurations of PCM based heat sinks for cooling electronic components. *Appl. Therm. Eng.* **2015**, *87*, 454–462.
129. Li, Z.; Wu, Z.G. Analysis of HTFs, PCMs and fins effects on the thermal performance of shell–tube thermal energy storage units. *Sol. Energy* **2015**, *122*, 382–395. [CrossRef]
130. Feng, S.; Shi, M.; Li, Y.; Lu, T.J. Pore-scale and volume-averaged numerical simulations of melting phase change heat transfer in finned metal foam. *Int. J. Heat Mass Transf.* **2015**, *90*, 838–847. [CrossRef]
131. Pakrouh, R.; Hosseini, M.J.; Ranjbar, A.A.; Bahrampoury, R. A numerical method for PCM-based pin fin heat sinks optimization. *Energy Convers. Manag.* **2015**, *103*, 542–552. [CrossRef]
132. Tiari, S.; Qiu, S. Three-dimensional simulation of high temperature latent heat thermal energy storage system assisted by finned heat pipes. *Energy Convers. Manag.* **2015**, *105*, 260–271. [CrossRef]
133. Kalbasi, R.; Salimpour, M.R. Constructal design of horizontal fins to improve the performance of phase change material rectangular enclosures. *Appl. Therm. Eng.* **2015**, *91*, 234–244. [CrossRef]
134. Samanta, H.; Roy, P.C.; Barman, N. Modeling of Solidification of CCHH ($CaCl_2$, $6H_2O$) in a Shell-and-Tube PCM based Heat Storage Unit. *Procedia Eng.* **2015**, *127*, 816–823. [CrossRef]
135. Malan, D.J.; Dobson, R.T.; Dinter, F. Solar thermal energy storage in power generation using phase change material with heat pipes and fins to enhance heat transfer. *Energy Procedia* **2015**, *69*, 925–936. [CrossRef]
136. Khatra, L.; El Qarnia, H.; El Ganaoui, M.; Lakhal, E.K. Numerical investigation of heat transfer during solidification in a rectangular enclosure with internally horizontal partial fins. *Comput. Therm. Sci.* **2015**, *7*, 293–312. [CrossRef]

137. Darzi, A.A.R.; Jourabian, M.; Farhadi, M. Melting and solidification of PCM enhanced by radial conductive fins and nanoparticles in cylindrical annulus. *Energy Convers. Manag.* **2016**, *118*, 253–263. [CrossRef]
138. Sheikholeslami, M.; Lohrasbi, S.; Ganji, D.D. Response surface method optimization of innovative fin structure for expediting discharging process in latent heat thermal energy storage system containing nano-enhanced phase change material. *J. Taiwan Inst. Chem. Eng.* **2016**, *67*, 115–125. [CrossRef]
139. Rozenfeld, A.; Kozak, Y.; Rozenfeld, T.; Ziskind, G. Experimental demonstration, modeling and analysis of a novel latent-heat thermal energy storage unit with a helical fin. *Int. J. Heat Mass Transf.* **2017**, *110*, 692–709. [CrossRef]
140. Kuboth, S.; König-Haagen, A.; Brüggemann, D. Numerical analysis of shell-and-tube type latent thermal energy storage performance with different arrangements of circular fins. *Energies* **2017**, *10*, 274. [CrossRef]
141. Ji, C.; Qin, Z.; Low, Z.; Dubey, S.; Choo, F.H.; Duan, F. Non-uniform heat transfer suppression to enhance PCM melting by angled fins. *Appl. Therm. Eng.* **2018**, *129*, 269–279. [CrossRef]
142. Luo, X.; Liao, S. Numerical study on melting heat transfer in dendritic heat exchangers. *Energies* **2018**, *11*, 2504. [CrossRef]
143. Duan, J.; Xiong, Y.; Yang, D. Melting behavior of phase change material in honeycomb structures with different geometrical cores. *Energies* **2019**, *12*, 2920. [CrossRef]
144. Deng, S.; Nie, C.; Jiang, H.; Ye, W.B. Evaluation and optimization of thermal performance for a finned double tube latent heat thermal energy storage. *Int. J. Heat Mass Transf.* **2019**, *130*, 532–544. [CrossRef]
145. Shahsavar, A.; Goodarzi, A.; Mohammed, H.I.; Shirneshan, A.; Talebizadehsardari, P. Thermal performance evaluation of non-uniform fin array in a finned double-pipe latent heat storage system. *Energy* **2020**, *193*, 116800. [CrossRef]
146. Zhao, C.; Opolot, M.; Liu, M.; Bruno, F.; Mancin, S.; Hooman, K. Numerical study of melting performance enhancement for PCM in an annular enclosure with internal-external fins and metal foams. *Int. J. Heat Mass Transf.* **2020**, *150*, 19348. [CrossRef]
147. Kirincic, M.; Trp, A.; Lenic, K. Numerical evaluation of the latent heat thermal energy storage performance enhancement by installing longitudinal fins. *J. Energy Storage* **2021**, *42*, 103085. [CrossRef]
148. Ghalambaz, M.; Mahdi, J.M.; Shafaghat, A.; Eisapour, A.H.; Younis, O.; Talebizadeh Sardari, P.; Yaïci, W. Effect of twisted fin array in a triple-tube latent heat storage system during the charging mode. *Sustainability* **2021**, *13*, 2685. [CrossRef]
149. Safari, V.; Kamkari, B.; Hooman, K.; Khodadadi, J.M. Sensitivity analysis of design parameters for melting process of lauric acid in the vertically and horizontally oriented rectangular thermal storage units. *Energy* **2022**, *255*, 124521. [CrossRef]
150. Ye, W.; Khodadadi, J.M. Effects of arrow-shape fins on the melting performance of a horizontal shell-and-tube latent heat thermal energy storage unit. *J. Energy Storage* **2022**, *54*, 105201. [CrossRef]

MDPI
St. Alban-Anlage 66
4052 Basel
Switzerland
www.mdpi.com

MDPI Books Editorial Office
E-mail: books@mdpi.com
www.mdpi.com/books

www.ingramcontent.com/pod-product-compliance
Lightning Source LLC
Chambersburg PA
CBHW041015270926
51850CB00024B/46
9783036593227